# WORLD
# WEAPON
# DATABASE

# WORLD WEAPON DATABASE ADVISORY BOARD

**Hayward Alker, Jr.,** Professor of International Relations
*Department of Political Science, Massachusetts Institute of Technology*

**Ashton Carter,** Associate Professor of Public Policy and Associate Director
*Center for Science and International Affairs*
*Kennedy School of Government, Harvard University*

**Bonita Dombey,** Strategic Programs Analyst
*National Security Division*
*US Congressional Budget Office*

**Matthew Evangelista,** Visiting Fellow
*Brookings Institution*

**Stephen van Evera,** Managing Editor
*International Security*
*Center for Science and International Affairs*
*Kennedy School of Government, Harvard University*

**David Holloway,** Senior Research Fellow
*Center for International Security and Arms Control, Stanford University*

**Frank von Hippel,** Professor of Public and International Affairs
*Woodrow Wilson School, Princeton University*

**Alan Kay,** President
*Institute for Defense and Disarmament Studies*

**Steven Miller,** Assistant Professor of Political Science
*Defense Studies Program*
*Center for International Studies, Massachusetts Institute of Technology*

**Philip Morrison,** Institute Professor
*Massachusetts Institute of Technology*

**Christopher Paine,** Staff for Nuclear Nonproliferation Policy
*Subcommittee on Energy Conservation and Power*
*US House Committee on Energy and Commerce*

**William Potter,** Executive Director
*Center for International and Strategic Affairs*
*University of California at Los Angeles*

**George Rathjens,** Professor of Political Science
*Defense Studies Program*
*Center for International Studies, Massachusetts Institute of Technology*

**Judith Reppy,** Associate Director
*Peace Studies Program, Cornell University*

**Jane Sharp,** Research Fellow
*Center for European Studies, Harvard University*

**Ronald Siegel,** Visiting Fellow
*International Institute for Strategic Studies*

**Howard Stoertz,** Defense Consultant
*(former National Intelligence Officer for Strategic Programs)*

# WORLD WEAPON DATABASE

Randall Forsberg
*Series Editor*

## VOLUME I
## SOVIET MISSILES

*By* Barton Wright
*Assisted by* John Murphy

INSTITUTE FOR DEFENSE
& DISARMAMENT STUDIES
*Brookline, Massachusetts*

LEXINGTON BOOKS
*D.C. Heath and Company/Lexington, Massachusetts/Toronto*

*Library of Congress Cataloging-in-Publication Data*
Main entry under title:

World weapon database.

   Includes indexes.
   Contents: v. 1. Soviet missiles / by Barton Wright.
   1. Weapons systems.   I. Forsberg, Randall.
II. Institute for Defense and Disarmament Studies (U.S.)
UF500.W67   1986      623.4      85-45506
ISBN 0-669-11798-6 (v. 1 : alk. paper)

Copyright © 1986 by D.C. Heath and Company

All rights reserved. No part of this publication may be reproduced
or transmitted in any form or by any means, electronic or mechanical,
including photocopy, recording, or any information storage or retrieval
system, without permission in writing from the publisher.

Published simultaneously in Canada
Printed in the United States of America
Casebound International Standard Book Number: 0-669-11798-6
Library of Congress Catalog Card Number: 85-45506

The paper used in this publication meets the minimum requirements of
American National Standard for Information Sciences—Permanence of
Paper for Printed Library Materials, ANSI Z39.48-1984.

The last numbers on the right below indicate the number and date of printing.

10 9 8 7 6 5 4 3 2 1

95 94 93 92 91 90 89 88 87 86

# CONTENTS

World Weapon Database   Randall Forsberg   viii
Acknowledgments   x
User's Guide   xi
Military Acronyms   xv
NATO Codenames and US Designations   xvii

## I. SOURCES AND METHODS OF THE DATABASE   1

Introduction to Database Sources   3
Acronyms and Complete References for Full Sources   7
Acronyms and Complete References for Partial Sources   11
Data-Entry Conventions   15
Generic Missiles and Modified Versions   19
Definitions of Data Categories   21

## II. A SHORT HISTORY OF SOVIET MISSILES   27

### FIRST BALLISTIC AND CRUISE MISSILES   29

### SOVIET BALLISTIC MISSILES   31
Intercontinental Ballistic Missiles   31
Submarine-Launched Ballistic Missiles   33
Intermediate-Range Ballistic Missiles   34
Short-Range Ballistic Missiles   35

### SOVIET CRUISE MISSILES   37
Cruise Missile Propulsion and Guidance   37
Land-Attack Cruise Missiles   38
Antiship and Antisubmarine Missiles   39
Antiballistic Missiles   40
Antiaircraft Missiles   41

## III. BEST ESTIMATES OF SOVIET NUCLEAR MISSILES   45

### INTRODUCTION TO THE BEST ESTIMATES   47
Table 1. Best Estimates of the Characteristics of Soviet Nuclear Ballistic Missiles   50
Table 2. Best Estimates of the Characteristics of Soviet Nuclear Cruise Missiles   52
Table 3. Best Estimates of the Numbers of Soviet Land-Based Nuclear Missiles, 1955–1985   54

### DERIVATION OF BALLISTIC MISSILE ESTIMATES   57
Intercontinental Ballistic Missiles   57
Submarine-Launched Ballistic Missiles   71
Intermediate-Range Ballistic Missiles   79
Short-Range Ballistic Missiles   83

### DERIVATION OF CRUISE MISSILE ESTIMATES   89
Strategic Cruise Missiles   89
Antiship and Antisubmarine Missiles   91
   Sea-Launched Antiship Missiles   91
   Air-Launched Antiship Missiles   96
   Sea-Launched Antisubmarine Missiles   99
Antiballistic Missiles   101
Antiaircraft Missiles   103
   Strategic Surface-to-Air Missiles   103

## IV. SOVIET MISSILE DATABASE   105

### BALLISTIC MISSILES   107

Intercontinental Ballistic Missiles   107

| | | | |
|---|---|---|---|
| SS-6 | 107 | SS-7 Mod 3 | 120 |
| SS-7 | 112 | SS-8 | 121 |
| SS-7 Mod 1 | 118 | SS-9 | 127 |
| SS-7 Mod 2 | 119 | SS-9 Mod 1 | 135 |

| | | | | |
|---|---|---|---|---|
| SS-9 Mod 2 | 137 | SS-17 Mod 3 | 194 | |
| SS-9 Mod 3 | 139 | SS-18 | 196 | |
| SS-9 Mod 4 | 142 | SS-18 Mod 1 | 203 | |
| SS-9 Mod 5 | 145 | SS-18 Mod 2 | 206 | |
| SS-10 | 146 | SS-18 Mod 3 | 209 | |
| SS-11 | 149 | SS-18 Mod 4 | 212 | |
| SS-11 Mod 1 | 158 | SS-18 Mod 5 | 215 | |
| SS-11 Mod 2 | 161 | SS-19 | 216 | |
| SS-11 Mod 3 | 163 | SS-19 Mod 1 | 224 | |
| SS-11 Mod 4 | 167 | SS-19 Mod 2 | 226 | |
| SS-13 | 168 | SS-19 Mod 3 | 228 | |
| SS-13 Mod 1 | 175 | SS-19 VarRange | 231 | |
| SS-13 Mod 2 | 176 | SS-24 | 232 | |
| SS-16 | 177 | SS-25 | 234 | |
| SS-17 | 183 | SS-26 | 237 | |
| SS-17 Mod 1 | 190 | SS-27 | 238 | |
| SS-17 Mod 2 | 192 | | | |

Submarine-Launched Ballistic Missiles 239

| | | | |
|---|---|---|---|
| SS-N-4 | 239 | SS-N-8 Mod 3 | 282 |
| SS-N-5 | 246 | SS-N-17 | 283 |
| SS-N-6 | 254 | SS-N-18 | 287 |
| SS-N-6 Mod 1 | 262 | SS-N-18 Mod 1 | 292 |
| SS-N-6 Mod 2 | 265 | SS-N-18 Mod 2 | 294 |
| SS-N-6 Mod 3 | 268 | SS-N-18 Mod 3 | 296 |
| SS-N-8 | 271 | SS-N-20 | 297 |
| SS-N-8 Mod 1 | 280 | SS-N-23 | 301 |
| SS-N-8 Mod 2 | 281 | | |

Intermediate-Range Ballistic Missiles 303

| | | | |
|---|---|---|---|
| SS-1 | 303 | SS-15 | 331 |
| SS-2 | 305 | SS-20 | 334 |
| SS-3 | 307 | SS-20 Mod 1 | 343 |
| SS-4 | 311 | SS-20 Mod 2 | 344 |
| SS-5 | 319 | SS-20 Mod 3 | 345 |
| SS-14 | 326 | | |

Short-Range Ballistic Missiles 346

| | | | |
|---|---|---|---|
| FROG | 346 | SS-1 Scud | 368 |
| FROG 1 | 349 | SS-1b Scud A | 371 |
| FROG 2 | 352 | SS-1c Scud B | 376 |
| FROG 3 | 355 | SS-1c Scud C | 381 |
| FROG 4 | 358 | SS-12 | 382 |
| FROG 5 | 361 | SS-21 | 387 |
| FROG 6 | 363 | SS-22 | 391 |
| FROG 7 | 364 | SS-23 | 394 |

**CRUISE MISSILES** 397

Ground-Launched Surface-to-Surface Missiles 397

| | | | |
|---|---|---|---|
| SSC-1 | 397 | SSC-4 | 405 |
| SSC-2a | 401 | GLCM | 407 |
| SSC-2b | 402 | | |

Antitank Missiles 408

| | | | |
|---|---|---|---|
| AT-1 | 408 | AT-3 | 418 |
| AT-2 | 412 | AT-4 | 423 |
| AT-2 Swatter A | 416 | AT-5 | 425 |
| AT-2 Swatter B | 417 | AT-6 | 427 |

Air-to-Surface Missiles 429

| | | | |
|---|---|---|---|
| AS-1 | 429 | AS-8 | 461 |
| AS-2 | 432 | AS-9 | 462 |
| AS-3 | 436 | AS-10 | 464 |
| AS-4 | 441 | AS-11 | 466 |
| AS-5 | 447 | AS-12 | 467 |
| AS-6 | 452 | AS-15 | 468 |
| AS-7 | 458 | TASM | 470 |

Sea-Launched Surface/Sub-to-Surface/Sub Missiles 471

| | | | |
|---|---|---|---|
| SS-N-1 | 471 | SS-N-10 | 508 |
| SS-N-2 | 475 | SS-N-12 | 510 |
| SS-N-2a | 481 | SS-N-13 | 514 |
| SS-N-2b | 483 | SS-N-14 | 516 |
| SS-N-2c | 485 | SS-N-15 | 520 |
| SS-N-3 | 488 | SS-N-16 | 523 |
| SS-N-3a | 495 | SS-N-19 | 525 |
| SS-N-3b | 497 | SS-N-21 | 528 |
| SS-N-3c | 498 | SS-N-22 | 530 |
| SS-N-7 | 499 | SS-N-24 | 532 |
| SS-N-9 | 503 | | |

Antiballistic Missiles 533

| | | | |
|---|---|---|---|
| ABM-1 | 533 | SH-4 | 541 |
| ABM-3 | 540 | SH-8 | 542 |

Ground-Launched Surface-to-Air Missiles 543

| | | | |
|---|---|---|---|
| SA-1 | 543 | SA-5 | 569 |
| SA-2 | 547 | SA-6 | 576 |
| SA-3 | 556 | SA-7 | 583 |
| SA-4 | 563 | SA-7b | 588 |

|  |  |  |  |
|---|---|---|---|
| SA-8 | 589 | SA-11 | 605 |
| SA-8b | 595 | SA-12 | 608 |
| SA-9 | 596 | SA-13 | 610 |
| SA-10 | 601 | SA-14 | 613 |

Sea-Launched Surface-to-Air Missiles 614

|  |  |  |  |
|---|---|---|---|
| SA-N-1 | 614 | SA-N-6 | 631 |
| SA-N-2 | 618 | SA-N-7 | 634 |
| SA-N-3 | 621 | SA-N-8 | 636 |
| SA-N-4 | 625 | SA-N-9 | 637 |
| SA-N-5 | 629 | | |

Air-to-Air Missiles 638

|  |  |  |  |
|---|---|---|---|
| A-1 | 638 | AA-3IR | 650 |
| AA-2 | 641 | AA-3RADAR | 651 |
| AA-2-2 | 645 | AA-4 | 652 |
| AA-3 | 647 | AA-5 | 653 |
| AA-5IR | 656 | AA-8 | 668 |
| AA-5RADAR | 657 | AA-8IR | 671 |
| AA-6 | 658 | AA-8RADAR | 672 |
| AA-6IR | 661 | AA-9 | 673 |
| AA-6RADAR | 662 | AA-10 | 675 |
| AA-7 | 663 | AA-XP-1 | 676 |
| AA-7IR | 666 | AA-XP-2 | 677 |
| AA-7RADAR | 667 | AA-X | 678 |

**Data Index**   **679**

**Missile Index**   **681**

**About the Author and Editor**   **699**

**About the Institute**   **700**

**IDDS Publications**   **701**

# WORLD WEAPON DATABASE

*By* RANDALL FORSBERG

The World Weapon Database is a large research project of the Institute for Defense and Disarmament Studies. This book is the first volume of the Database to be published. The World Weapon Database will fill a dozen volumes, which together will cover all major weapons produced worldwide since 1945: guided missiles, military aircraft, naval ships, and armed or tracked military vehicles.

## Origin of the World Weapon Database

The concept of the World Weapon Database grew out of my frustration with the open literature on world weapons and military forces. Between 1969 and 1982, I undertook several comparative international studies and long-term surveys of military forces. At the Stockholm International Peace Research Institute (SIPRI), I was responsible for a decade for the estimates of US and Soviet strategic nuclear weapons published in the annual *SIPRI Yearbook*. For a research project, *Resources Devoted to Military Research and Development*, I analyzed all types of new weapons produced worldwide in the 1960s. For other projects, I surveyed worldwide antisubmarine warfare forces; compiled a history of US strategic and intermediate-range nuclear weapons since 1945; traced the evolution of US bomber aircraft since 1910; surveyed trends since 1945 in major US force components (ships, tactical aircraft squadrons, and army divisions); and studied the problems of assessing the East-West military balance in Europe.

For each study, I scanned scores of sources, sometimes several hundred. I found that on every conceivable aspect of weaponry—roles, numbers, dates, and characteristics—authoritative sources differed from one another, yet none attempted to explain or comment on the differences. Often sources published annually would change their own estimates from one year to the next without explanation or comment. Over and over again, assembling reliable information on a given aspect of armed forces required painstaking comparison of numbers published in somewhat different formats and with slightly different definitions in different editions of various annual sources: government documents, reference books, and trade journal surveys.

Over the course of a decade, I found myself scanning some sources a second, third, or even fourth time to gather slightly different information or to refresh my memory on some point of controversy. I learned that other researchers around the world, attempting to offer an independent view of some aspect of world armaments, were engaged in the same tedious process. All of us were wasting time compiling, comparing, and reevaluating the same basic statistics in slightly different contexts for projects with somewhat different emphases. Furthermore, although comprehensive, unclassified information existed on all types of weaponry, it was not available in a readily accessible, regularly updated form, with matters in dispute clearly identified. The reason was simple: no institution outside government intelligence agencies had ever invested the research time needed to compile and maintain a comprehensive database on modern weapons and armed forces.

Thus, in 1980, I founded the Institute for Defense and Disarmament Studies (IDDS), in part with the specific purpose of assembling and maintaining the World Weapon Database: a computerized, regularly updated compilation of source data on the numbers and characteristics of all major weapons produced worldwide since 1945.

## Structure, Sources, and Methods

The World Weapon Database is organized by area of technology and producing country. Production of major weapons is heavily concentrated in the United States and the Soviet Union; together the two superpowers probably account for two-thirds of world weapon production. Thus the working model for the Database involves the following twelve volumes:

1. Soviet Missiles
2. Soviet Military Aircraft
3. Soviet Naval Vessels
4. Soviet Armed and Tracked Vehicles

5. US Missiles
6. US Military Aircraft
7. US Naval Vessels
8. US Armed and Tracked Vehicles
9. Missiles of Europe, Asia, Latin America, and Africa
10. Military Aircraft of Europe, Asia, Latin America, and Africa
11. Naval Vessels of Europe, Asia, Latin America, and Africa
12. Armed and Tracked Vehicles of Europe, Asia, Latin America, and Africa

For each area of technology (missiles, aircraft, ships, and tanks), the IDDS researchers have established a set of categories in which they compile data. The categories cover weapon design (stages, dimensions, propulsion, guidance); performance (speed, range, payload, and so on); history of development, production, and deployment; numbers produced and areas of deployment; importing countries; and design, use, and strategy. For missiles, information is compiled in about sixty categories. For aircraft, there are about one hundred categories. For each class of ship, plus its weapons and equipment, we assemble data on about two hundred features.

For each weapon system, the research group identifies a wide range of sources with substantive information. The information in each source is then allocated to the categories in the format. Using first paper forms and then a computer, researchers record the information and then generate listings under each category of what all sources have said about that category. This permits researchers at IDDS and elsewhere to compare the estimates in various sources. In addition, it makes readily available far more source material on any topic than most analysts have time to assemble for specific articles.

## Timing, Updates, and Computer Subscription Service

The historical Database, covering the period from 1945 to 1985, is expected to take about twenty person-years to complete: three years each for US and Soviet missiles and military aircraft and one year each for the other eight volumes.

Since 1982, the Institute for Defense and Disarmament Studies has had several researchers working on various segments of the Database. The manuscript for the second volume, *Soviet Military Aircraft*, should be finished by mid-1986; those for US missiles and military aircraft should be completed a year later. We hope to complete the remaining eight volumes by the end of 1988.

When the historical World Weapon Database is completed, or perhaps sooner, the Institute will publish updates of all its parts. Gradually we hope to expand source coverage to include more topical articles and to publish updated editions of each volume every year or two.

In addition to publishing the Database in book form, IDDS plans to make it available through a computer subscription service, with updates twice a year. Readers who would like to subscribe should let the Institute know of their interest. Since the Database is planned as a permanent and constantly expanding system, the Institute would also be grateful for comments from users on ways to improve its design.

## Purposes of the World Weapon Database

Any vast database has myriad uses. The World Weapon Database will increase public understanding of the military forces and policies of all nations, help correct widespread misconceptions about certain aspects of military forces, and clarify long-term trends in military technology and force structure. Perhaps most important, it will provide a solid foundation for projecting likely future world military forces—and for developing safer alternatives.

The Database makes clear which aspects of weapon systems are uncontroversial and which are controversial, and for the latter, it shows the range of opinion. It permits research on systematic bias in certain sources compared with others. It is useful not only for those concerned about the truth but also for those concerned about what people believe is the truth and how these perceptions influence their views on policy issues.

We hope that eventually the Database will help build a broad consensus on the basic facts about world armaments, for if informed debate moves beyond petty disputes about the forces of various nations, we can focus on the truly important issue: how to create a safer, more productive, and more democratic world.

# ACKNOWLEDGMENTS

I am indebted to Randall Forsberg for her vision of the World Weapon Database and for her valuable advice and expertise on military affairs. I am also grateful to her for her review and extensive editing of the final manuscript.

John Murphy excelled at the unenviable tasks of updating the Database and shepherding the manuscript out the door after I left the Institute. Over a period of several months, he added the 1985 US Department of Defense data, entered additional data from sources used in preparing best estimates, helped update the best estimates, and printed out the Database reproduced in this book. I am especially grateful for his contribution, which made him virtually a second author.

It was with Peter Steven that I tackled the problem of making small computers perform large tasks, and to him I owe a great debt on that account.

I thank all my fellow staff members at the Institute for Defense and Disarmament Studies—whether in research, public education, or support—for creating an uncommonly friendly and supportive atmosphere in which to work. My thanks go especially to Neta Crawford and Laura Reed, who worked on the World Weapon Database with me, studying airplanes and ships, during the early years when the Database was taking shape. Sharing the joys, challenges, and frustrations of gathering data was a great help.

This volume has been made possible by grants from the CS Fund, the Levinson Foundation, the HKH Fund, the Rockefeller Family Fund, and several individual donors who wish to remain anonymous. I gratefully acknowledge their generous support.

# USER'S GUIDE

The User's Guide is intended for readers experienced in compiling data on weapons who want to start using the book quickly. The sources and methods of the Soviet Missile Database are described more fully in part I, which should be read carefully by less experienced users.

This book brings together in one place data on Soviet missiles published in many other authoritative or widely used sources. The Soviet Missile Database (part IV) is the printed version of the data. The Institute for Defense and Disarmament Studies (IDDS) plans to update the printed Database every few years and to provide more frequent updates in a computer subscription service. Please contact IDDS (2001 Beacon St., Brookline, Massachusetts 02146) for information on when the Soviet Missile Database will be available in online computer form and when other parts of the World Weapon Database will be published.

The Soviet Missile Database covers all Soviet guided missiles ever produced, along with unguided FROGs (free-rockets-over-ground). It does not cover bombs, torpedoes, artillery, or manned airplanes.

## Parts of the Book

Part I, Sources and Methods of the Database, describes the sources included in the Database and the conventions used in entering data from them. All sources are identified by four-letter acronyms. For quick use, a short reference for each source acronym is provided on the inside of the back cover. Full Sources, from which *all* data were recorded, have fully alphabetic acronyms. Partial Sources, from which only unusual data were recorded, have acronyms made up of two letters followed by two numbers. Complete references for Full and Partial Sources are listed in part I. In the Database, sources appear in chronological order. The year of publication is shown next to the source acronym; for example, the acronym JAWA 84 represents *Jane's All the World's Aircraft 1984–85*, published in 1984.

Data for each missile or modified version of a missile start on a new page. If modified versions of a missile exist, data are presented first for "generic" entries, which apply to all models or are not specified by model; then data are given which explicitly or implicitly refer to a particular model, such as Mod 1 or Mod 2. For missiles with modified versions, there is no physical object corresponding to what we call the generic missile; all actual missiles represent one specific model or another. The generic description simply lists features common to all (or most) modified versions. Where data are not identified as pertaining to a particular model, but are judged unquestionably to do so, they are listed with that model rather than with the generic entry, with a note in the comment column on the right of the entry indicating that this editorial judgment has been made.

The source data on each missile are presented in sixty data categories, which are explained in part I. Sources use many formats in publishing information on Soviet missiles. An important part of preparing the Database was to devise a standard format for diverse data and then adapt the source data to the format without losing important information or distorting the original meaning. The data categories are grouped in several clusters: designation; aspects of performance, warhead, dimensions, propulsion system, and guidance system; production history; numbers produced and deployed; importing countries; launch platform; and strategy. For each missile and missile model, the categories always appear in the same order. If none of the sources reviewed contained information on a particular characteristic of a missile, that category is omitted.

Quantitative data are reproduced in the unit of measure employed in the original source. Then they are converted by computer to a standard metric unit, with the conversion results in a column to the right.

Soviet missiles are listed in the Database by the designations used by the US government: these represent the only complete set of designations available. The list of NATO codenames and US designations following this section shows the US designations that correspond to NATO codenames, such as Sapwood or Grail. Although the official US practice is to add an "X" to designations for Soviet missiles under development, we have omitted this notation. To find data on the SS-NX-17, for example, see SS-N-17; for the SS-X-24, see SS-24.

Part II, A Short History of Soviet Missiles, gives some perspective on the mass of numbers that make up most of this book. Like the Database, the history is arranged by type of missile. Within each type, the history stresses capabilities that have changed over the years as successive new generations of missiles have been introduced. Readers unfamiliar with missiles in general or with certain types of Soviet missile may want to read this part first.

Part III, Best Estimates of Soviet Nuclear Missiles, was prepared after the Database was assembled, edited, and printed out. These estimates do not represent the final word on these matters since they will be revised and updated in the future. What is unique about the estimates is the scope of data on which they are based, the availability of the source data for readers to scan at first hand, and the detailed explanation of how each number was chosen. Unlike most estimates, which give readers a simple choice of agreeing or disagreeing, believing or disbelieving, these estimates provide an opportunity for informed discussion of the evidence for and against given figures. In addition, this format facilitates clarification of corrections or updates, when new information warrants a change.

Part IV, Soviet Missile Database, is the compilation of source data on Soviet missiles. It is arranged by general type of missile (ICBMs, SLBMs, and so on); within each type, missiles appear in chronological order.

Data and missile indexes appear at the end of the book. To locate the page with data about a certain characteristic of a missile, readers should consult the Data Index, which lists categories down the left and missiles across the top. A dash means no data on that characteristic of that missile were found in any source in the Database. The Missile Index lists the pages in the various sections of the book where there is information on a given missile.

## How to Read the Database

On the facing page is an annotated page from a Database entry. The IDDS designation (generally the official US designation) is followed by the acronyms of all sources with data on this missile. Next comes the first data category—in this example, *US Designation*. Each line below the name of the category gives data from one source. If a source is not listed under a particular category, it did not contain information about that category (subject to minor qualifications, described in part I).

The left-most columns contain the source acronym and publication date. For brief references to source acronyms, readers should consult the inside of the back cover. For complete references, see part I.

Since sources are listed in chronological order, readers can see the evolution of understanding, judgment, or information about a particular missile or, alternatively, the signs of development of modified or improved versions that differ from the original version.

The column (or columns) to the right of the source acronym gives a "short value" for the category: quantitative data or, if the category is not quantitative, a word or phrase. The column farthest to the right on each line, the comment column, explains the short value or describes it more fully. If the data for a category are such that they cannot be entered in a narrow column, the comment column takes up the entire line. This is the case, for example, with the category *Strategy* (not shown on the sample page). For quantitative categories, such as *Range* or *Nuclear Yield*, the short-value columns are labeled with the units of measure used in the sources. The next column to the right is a computer-generated conversion to a standard metric value, underlined and labeled "conv" for *conversion*. For example, if certain sources give a *Range* estimate in miles or nautical miles, while others give estimates in kilometers, all original data in miles or nautical miles will appear as such; then all estimates will appear converted to kilometers under the heading *km conv*. For sources that give both metric and nonmetric values, the converted nonmetric value appears in the conversion column.

The comment column starts wherever the short-value columns end, runs to the right margin, and may continue over several lines.

Occasionally there are two or more different entries from the same source for the same category. For example, a given source may show one value in a text section and a conflicting value in a table. In such cases, the data appear in the Database on successive lines, but the source is not repeated. Instead the comment lines explain why more than one value is listed.

## US Government Sources

Unlike Western weapons, which are described by official sources, Soviet missiles are rarely mentioned by Soviet government sources. As a result, most of the unclassified information on Soviet missiles comes originally from Western military or intelligence agencies. Trade journals, news articles, and reference books may contain some information on the smaller Soviet missiles gleaned from exported or captured

USER'S GUIDE xiii

# Annotated Data Page

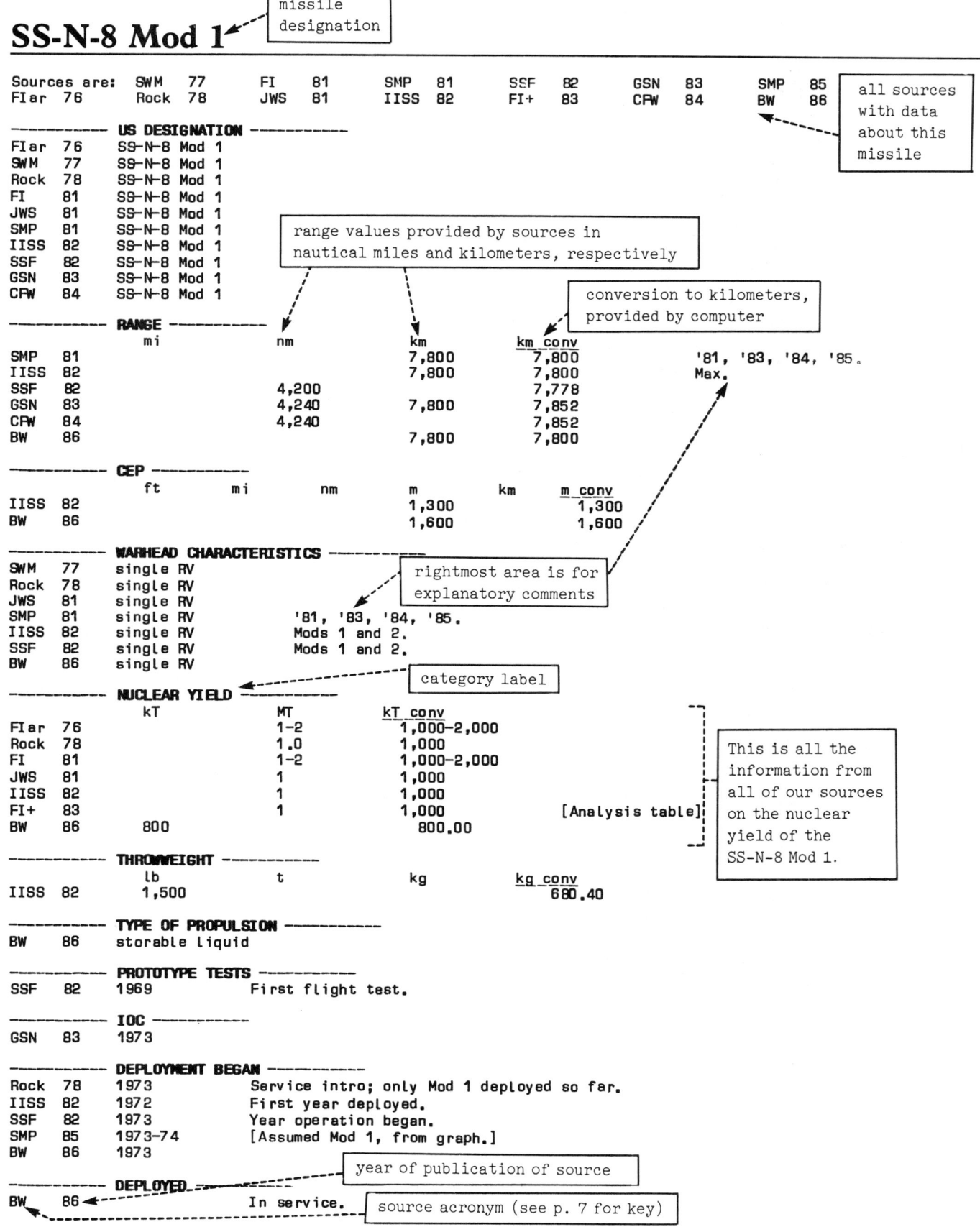

weapons, from exhibitions at trade fairs, or even from Soviet agencies seeking to promote foreign sales. Such information is rare, however.

Because there are so few official Soviet data, because physical evidence about missiles is harder to obtain than that for ships or aircraft, and because the subject is politically loaded, there is probably more confusion and dispute about Soviet missiles than about any other major type of weaponry in the world. The need to survey and compare many sources is particularly great.

The most authoritative (though still not always accurate) sources of public information on Soviet missiles are three annual publications of the US Department of Defense: the *Annual Report of the Secretary of Defense;* the annual statement on US military posture by the Joint Chiefs of Staff; and the annual booklet *Soviet Military Power,* published by the US Department of Defense. The acronyms used for these three sources are SecD, JCS, and SMP, respectively.

As one of the first steps in this project, the author prepared an index to the three US annual reports, entitled *Index of Soviet and Chinese Military Affairs in Annual US Defense Department Reports, 1969–1983.* Updates of the *Index* for 1984 and 1985 were used to update the Database. In addition, the author prepared an unpublished index of formerly classified annual reports of the secretary of defense for fiscal years (FY) 1963 through 1977, declassified in response to an IDDS Freedom of Information Act request. (In the Database, the *formerly classified* annual reports are referred to as SecC.) *Because of the carefully prepared indexes to these documents, we can state with some confidence that all data on specific Soviet missiles contained in SecC or SecD for 1963 through 1985, in JCS for 1969 through 1985, and in SMP for 1981 through 1985 are reproduced in this book.*

## Questionable Sources and Cutoff Date

We have included in the Database a few books and articles that have little original material and do not seem reliable. They are included for two reasons. First, when we started the project, we did not consider any source worthless, and we do not want to deny to others the opportunity to draw their own conclusions. Second, the World Weapon Database documents not only what is true about military forces but also what people believe about them.

Some readers will undoubtedly know of sources we should have included but did not. The Institute would be grateful for references, all of which would be reviewed for future editions. The cutoff date for sources was August 1984, except for the 1985 editions of the three annual US Department of Defense documents.

## Computer Hardware and Software

For this first segment of the World Weapon Database, data have been entered, stored, and processed in the Institute's CMC International Superbrain microcomputers. Floppy disks with 338K storage were used initially, supplemented recently by a 15 megabyte hard disk. Because the Institute uses microcomputers, sophisticated data handling programs were not practicable. We employed a text editor (Mince) for data entry and editing in a database system devised by the author. To sort, select, and format the data for printing, we worked with a battery of special-purpose programs, written by the author and Peter Steven in C.

# MILITARY ACRONYMS

| | |
|---|---|
| AA | Antiaircraft |
| AAM | Air-to-air missile |
| ABM | Antiballistic missile |
| ALCM | Air-launched cruise missile |
| APC | Armored personnel carrier |
| ASAT | Antisatellite |
| ASM | Air-to-surface missile |
| ASW | Antisubmarine warfare |
| ATM | Antitank missile |
| AWACS | Airborne Warning and Control System |
| BMEWS | Ballistic Missile Early Warning System |
| C3I or CCCI | Command, control, communication, and intelligence |
| CEP | Circular error probable |
| CIA | Central Intelligence Agency (US) |
| CLOS | Command to line of sight |
| CNO | Chief of Naval Operations (US) |
| CONUS | Continental United States |
| CRS | Congressional Research Service (US) |
| CV | Aircraft carrier |
| CW | Chemical warfare |
| DIA | Defense Intelligence Agency (US) |
| DoD | Department of Defense (US) |
| ECCM | Electronic counter-countermeasures |
| ECM | Electronic countermeasures |
| ELINT | Electronic intelligence |
| EMP | Electromagnetic pulse |
| END | European Nuclear Disarmament |
| EW | Electronic warfare |
| FOBS | Fractional Orbit Bombardment System |
| FRG | Federal Republic of Germany (West Germany) |
| FROG | Free rocket over ground |
| FY | Fiscal year |
| GLCM | Ground-launched cruise missile |
| GNP | Gross national product |
| HE | High explosive |
| HEAP | High-explosive armor piercing |
| HEAT | High-explosive antitank |
| ICBM | Intercontinental ballistic missile |
| IFF | Identification, friend or foe |
| INF | Intermediate-range nuclear forces |
| IOC | Initial operational capability |
| IR | Infrared |
| IRBM | Intermediate-range ballistic missile |
| JATO | Jet-assisted take-off |
| LOX | Liquid oxygen |
| LRINF | Longer-range intermediate (-range) nuclear forces |
| LRTNF | Longer-range theater nuclear forces |
| M(B)FR | Mutual (and Balanced) Force Reduction Talks |
| MIRV | Multiple independently targetable reentry vehicle |
| MIT | Massachusetts Institute of Technology |
| MoD | Ministry of Defense (UK) |
| Mod | Modification |
| MRBM | Medium-range ballistic missile |
| MRV | Multiple reentry vehicle |
| NATO | North Atlantic Treaty Organization |
| NBC | Nuclear, biological, chemical |
| NSWP | Non-Soviet Warsaw Pact |
| OTH | Over the horizon |
| OTH-B | Over-the-horizon backscatter |
| PBV | Post-boost vehicle |
| POMCUS | Prepositioned overseas materiel, configured to unit sets |
| PRC | People's Republic of China |
| R&D | Research and development |
| RV | Reentry vehicle |
| SAC | Strategic Air Command (US) |
| SACLOS | Semiautomatic command to line of sight |
| SALT | Strategic Arms Limitation Treaty |
| SAM | Surface-to-air missile |
| SAR | Semiactive radar |
| SLBM | Submarine-launched ballistic missile |
| SLCM | Sea-launched cruise missile |
| SP | Self-propelled [artillery] |
| SRAM | Short-Range Attack Missile |
| SSB | Submarine, ballistic missile equipped |
| SSBN | Submarine, ballistic missile equipped, nuclear powered |
| SSGN | Submarine, guided (cruise) missile equipped, nuclear powered |
| SSM | Surface-to-surface missile |
| SSN | Submarine, nuclear powered |
| START | Strategic Arms Reduction Talks |

| | | | |
|---|---|---|---|
| **TNF** | Theater nuclear forces | **VRBM** | Variable-range ballistic missile |
| **TOW** | Tube launched, optically tracked, wire-guided (antitank missile) | **VSTOL** | Vertical and short take-off and landing |
| **UK** | United Kingdom | **VTOL** | Vertical take-off and landing |
| **US** | United States | **WTO** | Warsaw Treaty Organization (Warsaw Pact) |
| **USSR** | Union of Soviet Socialist Republics | | |

# NATO CODENAMES AND US DESIGNATIONS

| NATO Codename | US Designation | NATO Codename | US Designation |
|---|---|---|---|
| Acrid | AA-6 | Samlet | SSC-2b |
| Alkali | AA-1 | Sandal | SS-4 |
| Anab | AA-3 | Sandbox | SS-N-12 |
| Apex | AA-7 | Sapwood | SS-6 |
| Aphid | AA-8 | Sark | SS-N-4 |
| Ash | AA-5 | Sasin | SS-8 |
| Atoll | AA-2 | Savage | SS-13 |
| Awl | AA-4 | Sawfly | SS-12 |
| Gainful | SA-6 | Scaleboard | SS-12 |
| Galosh | ABM-1 | Scamp | SS-14 |
| Gammon | SA-5 | Scapegoat | SS-14 |
| Ganef | SA-4 | Scarp | SS-9 |
| Gaskin | SA-9 | Scrag | SS-10 |
| Gecko | SA-8 | Scrooge | SS-15 |
| Goa | SA-3, SA-N-1 | Scrubber | SS-N-1 |
| Goblet | SA-N-3 | Scud | SS-1b, SS-1c |
| Grail | SA-7 | Scunner | SS-1 |
| Guideline, | SA-2, SA-N-2 | Sego | SS-11 |
| Guild | SA-1 | Sepal | SSC-1 |
|  |  | Serb | SS-N-5 |
| Kangaroo | AS-3 | Shaddock | SS-N-3 |
| Kelt | AS-5 | Shyster | SS-3 |
| Kennel | AS-1 | Sibling | SS-2 |
| Kerry | AS-7 | Silex | SS-N-14 |
| Kingfish | AS-6 | Skean | SS-5 |
| Kipper | AS-2 | Snapper | AT-1 |
| Kitchen | AS-4 | Spandrel | AT-5 |
| Saddler | SS-7 | Spigot | AT-4 |
| Sagger | AT-3 | Spiral | AT-6 |
| Salish | SS-2a | Styx | SS-N-2 |
|  |  | Swatter | AT-2 |

# PART I

## SOURCES AND METHODS OF THE DATABASE

# INTRODUCTION TO DATABASE SOURCES

The Soviet Missile Database covers all major, independent sources that we could find: that is, all sources that give reasonably detailed information on many types of Soviet missiles. The Database also covers many minor sources, which contain less information and cover only one or a few missiles. The sources include government reports, journal articles, and books. Some are one-time publications; others are part of a series, usually annual. Many journal articles are excluded; most of them use data derived from sources in the Database. We use only public sources, which we can document. Classified sources are excluded.

## Cutoff Dates

The source cutoff date is August 1984, except for three annual US Department of Defense documents described on page xiv. Data from the 1985 editions of these three documents were entered by John Murphy, using an index to the documents prepared at the Institute for Defense and Disarmament Studies by Stephen Messinger. (For detail on these three sources and the IDDS index to them, see the User's Guide.) The estimates of Soviet nuclear missiles given in part III were updated by Randall Forsberg, using the data taken from the 1985 US Department of Defense documents.

## Source Acronyms

Because the Database covers so many sources, a system of short references is useful. In both the Database and the text, we use acronyms to identify the sources. Each acronym has four or fewer letters and is derived mnemonically from the author or the title. Many acronyms, such as JCS (Joint Chiefs of Staff), CBO (Congressional Budget Office), and AWST (*Aviation Week and Space Technology*), are already in widespread use, and we expect that many others will soon become familiar to frequent users of the Database. To help readers distinguish between Full Sources and Partial Sources, the acronyms for Partial Sources always consist of two letters followed by two numbers (MR03, US18, CH01), whereas acronyms for Full Sources never contain numerals.

## Full and Partial Sources

For some sources, which we call Full Sources, we recorded all information given on Soviet missiles. For other sources, which we call Partial Sources, we recorded only information not already found in the Full Sources. Complete citations for both types of sources, listed by acronym, follow this section.

In the citations for Full Sources, the acronym is followed by what the acronym refers to (author, title, or working reference) and then the rest of the information in a standard bibliographic citation. Annotated below the citation is information on the parts of the source covered in the Database. In some cases, the source mentions Soviet missiles irregularly throughout the text but has condensed information in tables on just a few pages. We may have taken information only from those tables, without scanning the text for occasional nuggets of hard data. In other cases, we may have started with information given in a table but then modified or added to it based on the text. When information from the text is consistent with information from tables, it is simply added to the Database. Where two parts of the same source are contradictory, however, both are shown, with the notations "[text]" and "[table]" to indicate the relevant part of the source. Users who want to refer back to the original source should check the complete citations for the parts used in the Database.

The Partial Sources are mainly journal articles on nonstrategic weapons found by scanning the *Air University Index* and *Applied Science and Technology Index* for 1945–1984. In addition, a few books were treated as Partial Sources. Unlike Full Sources, from which all information about Soviet missiles is recorded automatically (even information given in every other source), information provided by Partial Sources was recorded only if it differed from data taken from six Full Sources compiled by the summer of 1982: the annual surveys in *Aviation Week* (AWST 82), *Air Force Magazine* (AFM 82), and *Flight International* (FI 81), plus three books: *Missiles of the World* (MOW 76), *Soviet War Machine* (SWM 77), and *Jane's Weapon*

*Systems* (JWS 80). Sometimes data from a Partial Source were recorded if they had been reported in only one of these Full Sources.

## Baseline Year for Periodical Full Sources

Between December 1981 when data collection started and August 1984 when it stopped, new editions of periodical Full Sources that had been entered in the Database were published. In most cases, the original entries were updated. For instance, *Aviation Week and Space Technology* (AWST) publishes a comprehensive table on Soviet missiles in March each year. The 1982 AWST survey was originally entered into the Database; and new data appearing in the 1983 and 1984 AWST surveys were added later. For all periodical Full Sources except the annual US Defense Department documents, data from later editions adding to or changing information from the baseline year are identified in the Database by a plus sign (+) next to the year of publication. This should be interpreted as follows: "Update of data given in the baseline year of a periodical source."

## Year of Publication

Next to the source acronym in the Database is the year of publication in two digits, such as 85. For a few annual sources—the *Military Balance* or volumes of the various *Jane's* series—this may be a bit confusing, since their titles include two years (1982–1983). In all such cases, the year of publication, which is given in the Database, is the earlier of the two years in the title.

## Dates and Updates for US Defense Department Documents

The annual US Department of Defense documents SecD (*Annual Report of the Secretary of Defense*) and JCS (posture statement of the Joint Chiefs of Staff) include in their titles a reference to the coming fiscal year (FY). Both documents are published early in the preceding calendar year. (The FY 1982 documents, for example, were published in January or February 1981). For both documents, we follow the standard practice of showing the year of publication (*not* the year in the title) in the date column next to the source acronym. But when information is repeated in several successive annual reports, we list the information with the first source and then note in the comment column the initial and later reports giving these data, identified by title year. For instance, data beside the source SecD 81 might include the following note in the comment column: FY82–84. This means that the information appeared not only in the FY 82 report (that is, the SecD 81 first source) but also in the FY 83 and FY 84 reports, published in 1982 and 1983, respectively.

Because the Database lists sources in order of publication, it is often possible to determine which source reported a piece of information first. One exception, however, is nongovernmental annual sources. These have been scanned starting in a recent baseline year and not checked for earlier entries of the same data.

## Obsolete Missiles (★)

In some cases, the acronym for an entry is followed by an asterisk, thus: AWST★ 68. This symbol alerts readers that the information does not come from the baseline year for the source but from an earlier year. When a baseline source did not list an older missile, we scanned earlier years to determine the most recent or most complete entry on the obsolete missile. The year with the starred acronym indicates when the source last gave its most complete information on that missile.

## Source Updates (+)

The Database uses two methods of updating baseline sources. For the US Department of Defense sources (SecD, SecC, JCS, and SMP) and for data on numbers deployed taken from the annual *Military Balance*, all information in each annual issue, including old information, is indicated in the Database. New information is recorded under the current year entry; repeated old information is noted by an indication in the comment column (such as FY 83–85) alongside the reference for the first year the information appeared. Information that was previously reported but has been dropped is indicated implicitly; FY 81–83, for example, indicates that the information was not repeated in FY 84. (The + sign is not used for these sources, since they are updated comprehensively.)

For other repeating sources, Database updates are not equally sensitive to the repetition of information from one year to the next. For the baseline year, we record all information. For later years, we record new information, but we do not mark the repetition of old information. Where a baseline update appears, a + sign is added to the acronym. When information from a previous year is no longer reported in the more recent year, the comment "[final report]" is usually added to the old data.

# ACRONYMS AND COMPLETE REFERENCES FOR FULL SOURCES

Full Sources are those from which *all* information on Soviet missiles was recorded in the Database. Acronyms for Full Sources consist of letters only (no numerals). The acronyms are intended to help readers recall the sources. They may refer to either the author or the title. If they are true acronyms (the first letters of words in the title) or if they refer to multiple authors, they are all capital letters. If they represent an abbreviated (shortened) version of a name, title, or short reference, they appear in corresponding upper- and lower-case letters.

The references listed here begin with the part of the citation the acronym refers to, and then they give the rest of the bibliographic information.

Acronyms that end with two numerals, referring to Partial Sources, are listed in the next section.

**ABFS** W. Arkin, A. Burrows, R. Fieldhouse, and J. Sands. *Nuclearization of the Oceans: Symposium on the Denuclearization of the Oceans.* Stockholm, Sweden: Myrdal Foundation, May 1984.

**AFM** *Air Force Magazine.* "Gallery of Soviet Aerospace Weapons." March 1982, pp. 107–110.
*Update:* "Gallery of Soviet Aerospace Weapons." March 1984, pp. 123–127.
*Obsolete Missiles:* "Gallery of Soviet Aerospace Weapons." March issues, 1978, 1979, 1980.

**ArkS** W. Arkin and J. Sands. "The Soviet Nuclear Stockpile." *Arms Control Today* 14:5, June 1984, pp. 1, 4–7.

**AWar** *Aviation Week* article: "Soviets' Nuclear Arsenal Continues to Proliferate." *Aviation Week and Space Technology*, 16 June 1980, pp. 67ff.

**AWST** *Aviation Week and Space Technology.* "Soviet Missiles," 8 March 1982, pp. 146–147.
*Update:* "Soviet Missiles," 12 March 1984, pp. 164–165.
*Obsolete Missiles:* "Soviet Missiles," Aerospace Forecast and Inventory Issue (always a March issue), 1971, 1973, 1975, 1978.

**Bras** *Brassey's Artillery.* Brian Blunt and Talley Taylor. Shelford Bidwell, ed. New York: Bonanza, 1977.
Every entry for a Soviet missile was used. The list of nations using various weapons in the back of the book was used only to check what was in service in the USSR. Information on the nations that have imported Soviet missiles was derived from the missile description section, not from the list of weapons in service for various nations.

**BW** Barton Wright, Institute for Defense and Disarmament Studies. Best estimates, assembled in part III with comments on their derivation from the other data in the Database.

**CBO** Congressional Budget Office. *Modernizing US Strategic Offensive Forces: The Administration's Programs and Alternatives.* May 1983.
Data from pp. 90–91 only.

**CFW** *Combat Fleets of the World 1982/83.* J. L. Couhat, ed. Annapolis, MD: US Naval Institute Press, 1982. (Original French edition published in 1981).
Only weapon subentry under USSR entered, pp. 583–585, 591.
*Update: 1984/85* edition; data from pp. 673–675, 681.
In the data, unmarked years in parentheses assumed to be when deployment began.

**Clns** John M. Collins. *U.S.–Soviet Military Balance: Concepts and Capabilities 1960–1980.* Congressional Research Service. New York: McGraw-Hill, 1980.
Data recorded on specifications, numbers (1970–1979 series), and carriers (for air- and sea-launched missiles). All data recorded came from pp. 443–469, 474, 485–487, 497–498, 517–523. A map on p. 133 showing ICBM and INF deployment areas was not used.
*Update:* John M. Collins and T. P. Glakas. *US/Soviet Military Balance: Statistical Trends 1970–1982.* Congressional Research Service update, 1 January 1983.
All data taken from pp. 14–36, 47, 57–59, 67, 81–82, 88–90, 95.

**COTS** *The Challenge of the Sputniks.* R. Witkin, ed. Garden City, NY: Doubleday, 1958.
Only data from pp. 68-69 entered.

**FI** *Flight International.* "World Missile Directory," 30 May 1981, pp. 1608ff.
Data from both tables and text entered; discrepancies noted.
*Update:* "World Missile Directory," 5 February 1983, pp. 315-357.
*Obsolete Missiles:* "World Missile Directory," 14 March 1974 (pp. 338ff.) and 2 June 1979 (pp. 1820 ff.).

**FIar** *Flight International* article: D. Richardson, "Soviet Strategic Nuclear Rockets Guide," 11 December 1976, pp. 1729-1733.

**GSN** *Guide to the Soviet Navy,* 3rd ed. N. Polmar. Annapolis, MD: US Naval Institute Press, 1983.
Data recorded from Missiles section, pp. 356-367, and one coastal missile entry, p. 36.

**HRST** *History of Rocketry and Space Travel.* W. von Braun and F. I. Ordway. New York: Crowell, 1975.
Data entered only from tables on pp. 141 and 149. A note below these tables says: "All figures estimated."

**IISS** International Institute for Strategic Studies. *The Military Balance 1982-83.* London: IISS, 1982.
Data on nuclear delivery vehicles taken from pp. 113, 115, and 117. Data on serviceability, survivability, reliability, and discrepancies with nuclear delivery vehicle tables from p. 136. The data on p. 140 were not used because the same information is in the main tables.
*Update:* *The Military Balance 1983-84,* published 1983, pp. 119, 121, 123.
*Historical Series:* Year-by-year numbers of individual missiles were entered using most editions of *The Military Balance.* All issues from 1959 (the first year of publication) to 1983 were consulted except 1961, 1964, 1966, 1971, and 1972. The baseline was the earliest deployment year for each system. Each number is recorded in the first year it was given, with later years shown in a note.

**IMSG** *International Missile and Spacecraft Guide.* F. I. Ordway and R. C. Wakeford. New York: McGraw-Hill, 1960.
Only data on pp. 1-8 of USSR section entered. Information on missiles in general and air-to-surface missiles was not recorded. Source says that data are from "nonofficial sources" and should be treated with "considerable caution."

**JAWA** *Jane's All the World's Aircraft 1982-83.* J. W. R. Taylor, ed. London: Jane's Publishing Company, 1982.
Data entered from missile listings, pp. 700-703, and associated picture captions.
*Update:* 1983-84 edition, published 1983, pp. 742-745.
*Obsolete Missiles:* editions published in 1962 (p. 415), 1964 (p. 445), and 1968 (p. 531).

**JCS** Joint Chiefs of Staff. *United States Military Posture for FY 19__.* Organization [earlier, Chairman] of the Joint Chiefs of Staff. Washington, D.C.: US Department of Defense, annual.
Reviewed for FY 72-86. When years are listed in the comment column, such as FY 74-76, this means that the data appeared in the FY 1974, FY 1975, and FY 1976 reports, not before and not after.
The year without "FY" before it refers to the calendar year of publication, which is always one year earlier than the fiscal year in the title. "JCS 81" in the Database, for example, refers to the JCS posture statement for FY 1982.

**JFS** *Jane's Fighting Ships 1982-83.* John Moore, ed. London: Jane's Publishing Company, 1982.
All data from table on pp. 162 and 169-171 were entered (page numbers with square brackets). Information from the ship entries was not included.
*Update:* 1983-84 edition, published 1983, pp. 174, 181-183.

**JWS** *Jane's Weapon Systems.* R. Pretty, ed. London: Jane's Publishing Company, annual.
*1981-82* edition, published 1981, used for strategic weapons. Data from text entries only.
*1982-83* edition, published 1982, used for tactical weapons.
Entries were made first for the analysis tables in the back of the book and then expanded based on text. If text elaborates in a way that is consistent, data are simply added or modified; otherwise discrepancies

are noted. Differences in unit of measure, such as centimeters versus meters, are not noted if the values are identical. Some information on pp. 228–229 on numbers of strategic SAMs was used, though not within a weapon entry.

*Update: 1983–84* edition, published 1983, for strategic and tactical weapons. Strategic weapons updated comprehensively, tactical weapons selectively (characteristics and major changes noted). Table data in the back entered as part of update for strategic data, although they were not in the *1981–82* original entry. (Estimates reproduced from *Soviet Military Power* (SMP) were not considered JWS data.)

**Obsolete Missiles:** Editions published in 1975, 1977, 1979, 1980.

**Ley** Willy Ley. *Rockets, Missiles, and Men in Space.* New York: Viking Press, 1968.

Data entered only from tables on pp. 474–476.

**Meyr** Stephen M. Meyer. *Soviet Theatre Nuclear Forces: Part II.* Adelphi Paper 188. London: International Institute for Strategic Studies, 1984.

Only data from p. 54 entered.

**MOW** *Missiles of the World.* M. J. H. Taylor and J. W. R. Taylor. New York: Scribners, 1976.

**Nitz** Paul H. Nitze. Testimony before Senate Foreign Relations Committee. Hearings on the SALT II Treaty. 96th Congress, 1st session, 1979, Part 1, pp. 439–482.

Data only from tables pp. 459–461, 463–465. Projections of numbers for 1982 and 1985 not entered. Projected future single-warhead ICBM not entered.

**NWG** [Nuclear Weapons in Germany]: "Factors Affecting the Withdrawal of Nuclear Weapons from East and West Germany." William Arkin. Paper delivered at a meeting of peace researchers at the European Nuclear Disarmament (END) Convention, May 1983.

**Rock** *The Rocket: The History and Development of Rocket and Missile Technology.* D. Baker. New York: Crown, 1978.

Data first entered from tables, pp. 270–273, and then expanded on the basis of text descriptions, pp. 225–230.

**SecC** Secretary of Defense (originally classified, later declassified under a Freedom of Information Act request). *Report of Secretary of Defense _____ to the Congress on the FY 19__ Budget, FY__ Authorization Request and FY 19__–FY 19__ Defense Programs.*

Reviewed for FY 63–77. All data recorded for FY 63–69; data from FY 70–77 recorded only when they did not appear in the unclassified version of the report (*see* SecD. For FY 68–77 the only sections available were on strategic forces. Because the document was declassified, "[deleted]" appears in some Database entries, referring to material explicitly deleted in the declassification process. Other data-entry conventions are identical to those used for JCS (*see* JCS).

**SecD** Secretary of Defense (unclassified version). *Report of Secretary of Defense _____ to the Congress on the FY 19__ Budget, FY 19__ Authorization Request and FY 19__–19__ Defense Programs.* Washington, D.C.: US Government Printing Office.

Reviewed for FY 70–86. Entry conventions identical to those used for JCS.

**SMP** *Soviet Military Power.* US Department of Defense. Washington, D.C.: US Government Printing Office. Editions published in 1981, 1983, 1984, 1985.

The notation '83 in comment column refers to the edition published in 1983.

**SSF** *Soviet Strategic Forces: Requirements and Responses.* Robert P. Berman and John C. Baker. Washington, D.C.: Brookings Institution, 1982.

Only the following data entered (including footnotes): tables B–1, B–2, B–3, B–4 (pp. 102–108); table B–5, "tubes" and "in operation in 1980" only; tables C–4, C–7 (pp. 136, 138). Aggregate numbers on p. 42 not recorded.

**SWM** *The Soviet War Machine: An Encyclopedia of Russian Military Equipment and Strategy.* R. Bonds, ed. New York: Hamlyn, 1977.

Data entered first from table entries; later expanded with text entries.

**USND** *Understanding Soviet Naval Developments,* 4th ed. Washington, D.C.: US Department of Defense, Office of CNO [Chief of Naval Operations], January 1981.

Data entered only from Missile Guide, pp. 131–132; nothing entered from text.

**vanD**  V. H. van Diepen. "Strategic Force Survivability and the Soviet Union." Master's thesis, Massachusetts Institute of Technology, February 1983.

Data only from tables on pp. 186–187, 192–194, and 209.

**Zaeh**  A. J. Zaehringer. *Soviet Space Technology.* New York: Harper & Brothers, 1961.

Data entered only from tables on pp. 75, 88–89.

# ACRONYMS AND COMPLETE REFERENCES FOR PARTIAL SOURCES

Partial Sources are those from which only information not found in basic reference sources was recorded in the Database. Acronyms for Partial Sources end with two numerals. Some Partial Sources are books; others are journal articles. For journal articles, the first two letters of the acronym refer exclusively to a particular journal. For example, AW always refers to *Aviation Week and Space Technology*, and US always refers to *US Naval Institute Proceedings*.

All sources whose acronyms do not end with numerals are Full Sources, listed in the preceding section.

**A140** Colin S. Gray. *The Future of Land-Based Missile Forces.* Adelphi Paper 140. London: International Institute for Strategic Studies, 1977.

**A168** Gregory Treverton. *Nuclear Weapons in Europe.* Adelphi Paper 168. London: International Institute for Strategic Studies, 1981, p. 6.

**AF01** D. K. Malone. "Air Defense of Soviet Ground Forces." *Air Force Magazine*, March 1978, pp. 78–83.

**AJ01** "Thurmond Details Soviet Naval Threat." *Armed Forces Journal International*, 23 May 1970, p. 7.

**AJ02** R. W. Forsyth and J. P. Forsyth. "Yom Kippur War Signal . . . Is Anybody Peeking at the Other Guy's Hand?" *Armed Forces Journal International*, August 1974, pp. 22–23.

**AJ03** B. F. Schemmer. "Army Delays 'Launch & Leave' Hellfire as DoD Reveals USSR Version Is Operational." *Armed Forces Journal International*, April 1981, p. 20.

**AW01** "Modernization Marks Soviet Missiles." *Aviation Week and Space Technology* 107, 12 December 1977, pp. 82–83.

**AW02** "Eastern Bloc Augments Attack Force." *Aviation Week and Space Technology* 108, 6 February 1978, pp. 57–61.

**AW03** C. A. Robinson. "Soviets Push Advanced Naval Weapons." *Aviation Week and Space Technology* 111, 24 September 1979, pp. 139ff.

**AW04** "Soviets Stage Integrated Test of Weapons." *Aviation Week and Space Technology*, 28 June 1983, pp. 20–21.

**AW05** "Soviet, Polish Aircraft over Baltic Sea." *Aviation Week and Space Technology* 110, 5 February 1979, pp. 16–17.

**BA01** J. E. Anderson. "First Strike: Myth or Reality." *Bulletin of the Atomic Scientists*, November 1981, pp. 6–11.

**BA02** Les Aspin. "Judge Not by Numbers Alone." *Bulletin of the Atomic Scientists*, June 1980, pp. 28–33.

**BA03** Frank Barnaby. "War-Fighting Weapons for Europe." *Bulletin of the Atomic Scientists*, March 1980, pp. 8–10.

**BA04** Christopher Paine. "Pershing II: The Army's Strategic Weapon." *Bulletin of the Atomic Scientists*, October 1980, pp. 25–31.

**Bl01** Desmond Ball. *Politics and Force Levels: Strategic Missile Programs of the Kennedy Administration.* Los Angeles: University of California Press, 1980.

**CB01** Congressional Budget Office. US Congress. *Counterforce Issues for the US Strategic Nuclear Forces.* January 1978.

**Ch01** C. Chant, ed. *The World's Air Forces.* London: David and Charles, 1979, pp. 167–175.

**GA01** K. Gatland. *Missiles and Rockets.* New York: Macmillan, 1975, p. 150.

**Gu01** B. Gunston. *The Encyclopedia of World Air Power.* New York: Crescent Books, 1980.

**Hi01** R. Higham and J. Kipp. *Soviet Aviation and Air Power: A Historical View.* Boulder, Colo: Westview Press, 1977.

**IA01** "Soviet Missile Inventory: I—Missiles for the Land and Marine Forces." *Interavia* 20, March 1965, pp. 377–381.

**IA02** Photograph caption. *Interavia* 20, April 1965, p. 528.

**IA03** "Snapper and Swatter." *Interavia* 20, April 1965, p. 529.

**IA04** "Soviet Missile Inventory—II." *Interavia* 21, February 1966, pp. 188–189.

**IA05** "Modern Weapon Systems of the Soviet Army." *Interavia* 21, May 1966, pp. 590–593.

**IA06** H. T. Simmons. "Problems and Prospects for a SALT II Arms Control Agreement. *Interavia* 30, August 1975, p. 856.

| | | | |
|---|---|---|---|
| **ID01** | "A Threat to the Nuclear Missile Balance? New Weapons for the Soviet Armed Forces." *International Defense Review* 1, January 1968, pp. 50-56. | **MR03** | "Air Defense Missile Development." *Military Review* 46, March 1966, pp. 104-105. |
| **ID02** | L. M. Rivers. "The Soviet Threat." *International Defense Review* 4, February 1971, pp. 39-40. | **MR04** | "Soviet Missile-Armed Naval Forces, Part I." *Military Review* 46, May 1966, pp. 102-106. |
| **ID03** | "SA-6—Arab Ace in the 20-day War." *International Defense Review* 6, December 1973, pp. 779-781. | **MR05** | "Soviet Missile-Armed Naval Forces, Part II." *Military Review* 46, June 1966, pp. 105-108. |
| **ID04** | "Soviet Notebook." *International Defense Review* 7, March 1974, pp. 285-286. | **NA01** | E. van Veen. "The Soviet Tactical Air Force and Tactical Nuclear Weapons." *NATO's Fifteen Nations* 17, August-September 1972, pp. 42-44. |
| **ID05** | "New Details on the SA-6." *International Defense Review* 7, April 1974, p. 529. | **NA02** | R. W. Forsyth and J. P. Forsyth. "The Lesson of the Cheap Shot." *NATO's Fifteen Nations* 19, August-September 1974, pp. 82-84. |
| **ID06** | "SA-8: The Latest Soviet Mobile SAM System." *International Defense Review* 8, December 1975, pp. 805-806. | **NA03** | "Military Production in the USSR." *NATO's Fifteen Nations* 21, June-July 1976, pp. 46-47. |
| **ID07** | SA-9 picture and caption. *International Defense Review* 8, June 1975, p. 804. | **OR01** | J. C. Snyder. "European Security, East-West Policy, and the INF Debate." *Orbis*, Winter 1984, pp. 913-970. |
| **ID08** | "Three New Soviet Air-to-Air Missiles in Service." *International Defense Review* 9, June 1976, p. 400. | **Pr01** | John Prados. *The Soviet Estimate: US Intelligence Analysis and Russian Military Strength.* New York: Dial Press, 1982. |
| **ID09** | "Already in Service: The 2nd Generation of Soviet Anti-tank Missiles." *International Defense Review* 11, January 1978, pp. 15-17. | **SA01** | B. M. Blechman and M. R. Moore. "A Nuclear-Weapon-Free Zone in Europe." *Scientific American*, April 1983, pp. 37-43. |
| **ID10** | A. Malzeyev. "Soviet Air-Launched Cruise Missiles." *International Defense Review* 11, January 1978, pp. 41-45. | **SU01** | R. Garthoff. "The Soviet SS-20 Decision." *Survival* 25:3, May/June 1983, pp. 110-119. |
| **ID11** | G. Rogers. "Anti-Ship Warfare." *International Defense Review* 11, March 1978, pp. 347-350. | **Ts01** | K. Tsipis and M. Bunn. *Ballistic Missile Guidance and Technical Uncertainties of Countersilo Attacks.* MIT Program in Science and Technology for International Security, Report 9, August 1983. |
| **ID12** | J. Hansen. "The Development of Soviet Tactical Air Defense." *International Defense Review* 14, May 1981, pp. 531-535. | | |
| **ID13** | D. Richardson. "AA-7 Apex and AA-8 Aphid." *International Defense Review* 15, January 1982, p. 14. | **US01** | S. Terzibaschitsch. "From 'Kildin' to 'Kresta.'" *US Naval Institute Proceedings*, June 1970, pp. 127-128. |
| **JP01** | S. Yusti. "The SS-20 Missiles Are Aimed at the Gulf from Kabul." Joint Publications Research Service 2518. Foreign Broadcast Information Service, 5 April 1982. | **US02** | C. J. Eliot. "Ship-to-Ship Missiles." *US Naval Institute Proceedings*, November 1972, pp. 108-114. |
| **Lu01** | Edward Luttwak. *The Grand Strategy of the Soviet Union.* New York: St. Martin's Press, 1983. | **US03** | M. Hewish. "Weapon Systems: Soviet Navy Surface-to-Surface Missiles." *US Naval Institute Proceedings*, January 1977, p. 115. |
| **MR01** | "USSR Air-to-Surface Missile." *Military Review* 36, November 1956, p. 80. | **US04** | N. Polmar. "Soviet Navy Surface-to-Surface Missiles." *US Naval Institute Proceedings*, July 1977, pp. 90-91. |
| **MR02** | "Soviet Missiles." *Military Review* 41, May 1961, pp. 100-107. | | |

**US05** N. Polmar and D. A. Paolucci. "Sea-Based 'Strategic' Weapons for the 1980s and Beyond." *US Naval Institute Proceedings,* May 1978, pp. 98–113.

**US06** R. T. Ackley. "The Wartime Role of Soviet SSBNs." *US Naval Institute Proceedings,* June 1978, pp. 34–42.

**US07** G. Charbonneau. "The Soviet Navy and Forward Deployment." *US Naval Institute Proceedings,* March 1979, pp. 35–40.

**US08** J. L. George. "SALT and the Navy." *US Naval Institute Proceedings,* June 1979, pp. 28–37.

**US09** B. Hahn. "PRC Submarine-Launched Ballistic Missile Development." *US Naval Institute Proceedings,* October 1979, pp. 132–135.

**US10** S. Gorton. "SS–N–14." *US Naval Institute Proceedings,* July 1979, p. 24.

**US11** L. F. Brooks. "Tactical Nuclear Weapons: The Forgotten Facet of Naval Warfare." *US Naval Institute Proceedings,* January 1980, pp. 28–33.

**US12** C. H. Clawson. "The Wartime Role of Soviet SSBNS—Round Two." *US Naval Institute Proceedings,* March 1980, pp. 64–71.

**US13** M. N. Vego. "Tactical Employment of Soviet FPBs: Part 1" *US Naval Institute Proceedings,* June 1980, pp. 95–99.

**US14** J. W. Keho and K. S. Brower. "Their New Cruiser." *US Naval Institute Proceedings,* December 1980, pp. 121–126.

**US15** G. Johannsohn. "Needed: A Defensive Capability for Diego Garcia." *US Naval Institute Proceedings,* January 1981, pp. 17–18.

**US16** H. D. Connell. "NATO Tacair Has a Tough Nut to Crack." *US Naval Institute Proceedings,* February 1981, pp. 33–39.

**US17** A. W. Hull. "Action–Reaction." *US Naval Institute Proceedings,* February 1981, pp. 31–39.

**US18** N. Polmar. "Soviet Nuclear Submarines." *US Naval Institute Proceedings,* July 1981, pp. 31–39.

**US19** J. W. Kehoe and K. S. Brower. "Another One of Their New Destroyers: Udaloy." *US Naval Institute Proceedings,* February 1982, pp. 115–119.

**US20** M. Vego. "Their SSGs/SSGNs." *US Naval Institute Proceedings,* October 1982, pp. 60–68.

# DATA-ENTRY CONVENTIONS

## Comment Column

As far as possible, we have tried to present the source data so that the essence of the story is told in simple, tabular form, in a number, a word, or a short phrase (a "short value" for each category). Where this might result in the loss of important subtleties, however, we add comments in the column farthest to the right, which we call the comment column. The comments are generally one of the following types:

▲ **Further Information:** The data stand alone, but there is some elaboration in the source, which appears as a comment. For example, if the carrier of a missile is reported to be the Backfire bomber, we list this in the Database in tabular form, but if the source adds, "New weapons for Backfire that fit in weapons bay are expected," the additional information about possible replacements for the missile is shown in the comment column.

▲ **Qualification:** A condition applies to the data. For instance, the range of a cruise missile may be 300 miles but only with a "high flight profile." The qualification is given in the comment column.

▲ **Context:** If it is not certain what the source meant by the short value, the longer passage in which it appears is quoted in the comment column to allow readers to observe the ambiguity.

▲ **Basis of Categorization:** If there is doubt whether the source data belong in a particular category, the comment column gives the basis for putting the information there. For example, if a source says, "this missile has an accuracy of 0.5 nm," it is probably referring to CEP. But since one cannot be certain, we add "accuracy" in the comment column while putting 0.5 in the CEP category.

▲ **Years:** For the two main US Defense Department sources, SecD and JCS, there may be data for each year from 1969 through 1985, and for the declassified version of SecD, which we call SecC, for each year from 1963 through 1977. Rather than clutter the tables with repetition, we list information only for the first year in which it appears and then put a note in the comment column showing the fiscal year reports in which it is repeated. "FY73–80" means the information appeared originally in the FY 1973 report (published in 1972) and was reported in the fiscal year 1974–1980 reports, but not thereafter.

## Notational Conventions

The conventions used throughout the Database for relative size (greater than, about) and for degrees of uncertainty are as follows:

| | |
|---|---|
| $> x$ | greater than x, more than x, above x, exceeds x |
| $\geq x$ | x or more, at least x, not less than x (think of $\geq$) |
| $<$ | less than x, not as high as x, under x |
| $\leq x$ | up to x, x or less, not more than x (think of $\leq$) |
| c. x | circa x, approximately x, around x (imprecision) |
| ? x | perhaps x, maybe x (uncertainty) |
| [x] | short value not given in source but inferred by the author |
| [x, ed] | comment not in source but by the author; a few common author's comments are not so labeled: [table], [text], [unmarked], [from graph] |

For sources that present data in both tabular and text format, the Database does not note the specific part unless they conflict; in that case, data are marked "[table]" or "[text]". "[Unmarked]" appears when a source provides one value with a qualification and another with no qualification where one would expect a different one; "[unmarked]" means that the value is not marked with a constraint or definition. "[From graph]" means data were read from a graph in the source. Generally in such cases a great deal of interpolation was required, and the data should be treated with special caution.

## Quotation Marks

Text that is not in quotation marks or in square brackets may be either paraphrased or quoted directly from the source. Quotation marks guarantee precise transcription from the source when the author felt the information was particularly interesting or controversial, when the meaning was unclear (saving the reader a trip to the source only to discover that the ambiguity is in the original), or when the passage quoted is long. Our quotations do not respect the more restrictive conventions; we may change capitalization of the first word and omit words that precede or follow in the same sentence without using ellipses. Words are never removed from the middle of a quotation without ellipses, however. Single quotation marks indicate either single or double quotation marks that appear in the source.

## Dates

To indicate dates, we use the day, the first few letters of the month, and the year in two digits. Thus 5 September 1972 appears as 5 Sept 72; partial dates such as 5 Sept and Sept 72 are also used.

## Units of Measure

Abbreviations for units of measure used in the Database are as follows:

*Length*
- mm — millimeters
- cm — centimeters
- m — meters
- km — kilometers
- in — inches
- ft — feet
- yd — yards
- mi — statute miles
- nm — nautical miles

*Mass or weight*
- kg — kilograms
- lb — pounds
- t — metric tons (1000 kg)

*Speed*
- mps — meters per second
- kmph — kilometers per hour
- fps — feet per second
- mph — miles per hour
- Mach — multiples of the speed of sound

*Explosive force*
- kT — kilotons
- MT — megatons

A persistent, irritating difficulty in comparing estimates from different sources is that they are given in different units of measure. Moreover, precise conversions of round-number estimates, whether rounded or not, can introduce a distortion of fact or interpretation. Thus we give data in the original units of measure and then, by computer, convert all source estimates to a standard metric measure. This is the column on the data sheet to the right labeled "conv" (for "conversion") and underlined, thus: km conv. Numbers in the metric conversion column have two digits after the decimal point if they are less than 1,000 and no digits after the decimal if they are greater than 1,000.

The columns to the left of the conversion column contain sources' original values. Where sources provide estimates in several units of measure, we reproduce all of them. In such cases, the metric conversion is from the source's English unit of measure. Retaining original units keeps perspective on the precision intended by the source; 453.60 kg, for example, implies a precise estimate of great accuracy; if the source's original figure is 1,000 lb, the information is more likely to be approximate. In addition, it is likely that a common cause of minor disparities among various sources is repeated conversions back and forth between English and metric units. In some cases, the Database shows that conversion and rounding account for all of the differences among the sources on a particular feature of a given missile.

The Database short-value columns accommodate all widely used units of measure for each parameter. In a few cases, however, sources use uncommon units, usually differing in scale from those more widely used. In such cases, we enter the appropriate value in the Database in the column employing the same system of measure, and we show the source's original unit of measure in the comment column. For the *Diameter* category, for example, the Database provides short-value columns for source estimates given in centimeters, meters, inches, or feet. If a source listed "500 mm" for the diameter of a missile, we would enter "50" in the centimeter column and put the note "[500 mm in source]" in the comment column.

There are two ambiguous units of measure that sources commonly do not distinguish. The term *miles* can refer to either statute miles or nautical miles; we have recorded such ambiguous cases as statute miles, although we have made an effort for major sources to resolve the ambiguity. The term *tons* can refer to either metric tons (1,000 kg, which equals 2,260 lb) or English tons, which can be either short (2,000 lb) or long (2,200 lb). A small *t* used as an abbreviation, as in "4,500 t," we have assumed to indicate metric tons and have recorded as such without comment. The terms *tons* and *tonnes* we have also assumed to mean metric tons, pending further research, but where they are used, we have preserved the exact word in the comment column.

## Editing and Editorial Judgments

We seek to provide users with what each source has said about each missile, intruding as little as possible our own judgment of what the source meant. We also seek to present the data from various sources in a standardized, concise format to allow easy comparison. Reconciling these two aims requires some compromises. We have freely used well-known acronyms in place of full words (such as ICBM for intercontinental ballistic missile and SAR homing for semiactive radar homing).

Another area where judgment is required is missile designation. For early reports of testing or development, often no designation or what ultimately turns out to be the wrong designation is applied to a missile. We have exercised our judgment in an effort to present all data about the same missile in one place, regardless of what the source may have called it. We do, however, have categories for recording designations where we preserve the designation the source used. Thus, readers who disagree with our classification have the information they need to change it.

Sources often give a comparative measurement—for example, "The SS-17's payload exceeds that of the SS-11." We always record the relation in the comment column. In addition, when the same source gives a value for the number compared to, we usually combine the value and the relation to construct a short value, which is then entered in square brackets. Continuing the example used above, suppose the same source shows 3000 lb for the SS-11 payload. The *Payload* entry for the SS-17 would then read "[> 3,000]" in the "lb" column and "Exceeds SS-11 payload" in the comment column to the right. For annual sources (including SecD, SecC, and JCS), only the same year of the same source counts as being the same document. If SecD has a comparative statement in 1977 referring to a value reported only in 1976, we do not insert an inferred value.

# GENERIC MISSILES AND MODIFIED VERSIONS

Many Soviet missiles have been modified several times. The successive modifications are called Mod 1, Mod 2, and so on. SS-N-6, for example, referred originally to a single missile design—initially the only design for this missile. Later two variants were introduced, both with an improved propulsion system. In addition, the Mod 2 had a larger single warhead, and the Mod 3 had two MRV (multiple reentry vehicle) warheads. When the first variant (Mod 2) was introduced, the original missile became known as Mod 1.

Because differences among modified versions are often of military significance, it is preferable to gather information about a missile under the specific version. Sometimes this is not possible, however. When the source identifies information as pertaining to a specific version, we list it with that version. When the source identifies information as pertaining to all versions, we repeat it under each one. When the source does not identify which version some information refers to but we can determine this with confidence, we categorize it accordingly, with the associated comment: "[Mod inferred, ed]." Otherwise the information is retained in an entry, labeled only with the basic missile designation, which we refer to as the generic missile.

For annual US government sources (JCS, SecC, SecD, and SMP), the *Editor's Notes* category for the earliest year in which a missile is mentioned shows the later years when the missile appears in that source. For missiles with modified versions, this information pertains to the generic missile and its versions; it is not given separately under the modified versions.

# DEFINITIONS OF DATA CATEGORIES

For all parts of the World Weapon Database, we gather a variety of hard data on weapons. Most of this information is relevant to gauging military or political significance. For Soviet missiles, virtually all that is reported in the sources is recorded somewhere in the Database. This section lists the categories in which we gather information and provides a detailed statement of what each category covers. The categories are grouped into several clusters (titles should not be taken too literally): Designations, Performance, Warhead, Dimensions, Propulsion, Guidance, History, Numbers, Platform, and Design, Use, and Strategy.

## Designations

▲ **US Designation:** The US system of referring to Soviet missiles is an alphanumeric code, with the first letter generally referring to the launch environment (*S* for surface or subsurface, *A* for air); the second letter referring to the target, using the same letters; and a number referring to the order of this missile in the sequence of all similar missiles ever produced. Examples are SS-9 (surface-to-surface), AA-6 (air-to-air), and AS-6 (air-to-surface). Naval missiles, launched from ships or submarines, have an *N* in the designation, such as SA-N-4 (ship-based antiaircraft missile) or SS-N-19 (ship-based antiship missile). The FROGs (free rockets over ground) do not have standard US missile designations because they are unguided rockets. Nonetheless, we treat their designations (FROG 1, and so on) as US designations. Sources rarely say that they are using US codes, but when we recognize them, we put the information in this category without comment.

▲ **NATO Codename:** The NATO system for referring to Soviet missiles is by single words, such as Scaleboard (surface-to-surface), Gecko (surface-to-air), Atoll (air-to-air), Styx (ship-to-ship), and Sego (surface-to-surface). As with US designations, sources rarely acknowledge they are using NATO codenames, but we record them as such without comment.

▲ **Soviet Designation:** The Soviet designation for a missile (such as RS-20), recorded on the rare occasions when one is reported.

▲ **Other Designation:** Ways of referring to missiles not obviously part of the US, NATO, or Soviet designation system. This includes designations in the original Western system used in the late 1950s and early 1960s (such as T-3, T-5A, T-8, J-1, M-100A, and M-2).

▲ **Description:** Any brief description of what kind of missile is being described, such as ICBM, ABM, naval cruise missile.

▲ **Editor's Notes:** This category contains comments by the author about the entirety of what a source has said about the missile and especially about how the missile is presented in the source. For SecC, SecD, SMP, and JCS, this category shows which years contain any mention of the missile at all.

## Performance

▲ **Minimum Altitude:** The minimum altitude at which the missile can intercept a target. This information is important mainly for surface-to-air missiles.

▲ **Maximum Altitude:** The maximum altitude at which the missile can intercept a target, also important mainly for surface-to-air missiles.

▲ **Range:** The effective range of the missile. Sources usually refer to the maximum range of a missile. When the range is specified as minimum or maximum, this information is noted in the comment. When sources provide a range of values, such as "1,000–2,000 miles," it is unclear whether this represents uncertainty about the maximum range or values for minimum and maximum range. We convey the uncertainty by repeating the estimates as they appear in the source.

▲ **Speed:** Maximum or typical speed of the missile. Source conditions are reported. For modern missiles, speed is usually given only for nonballistic missiles since the maximum speed of modern ballistic missiles, close to 20,000 mph (or Mach 20), is so great that it defies conventional defenses, regardless of small variations. For cruise missiles, maximum speed will vary with the flight path. Speeds in Mach numbers (multiples of the speed of sound) are not converted to kilometers per hour because, like the speed of sound, they vary with altitude. At sea level, Mach 1 (equal to the speed of sound) is around 714 mph.

▲ **CEP (Circular Error Probable):** The radius within which half of the warheads fired at a target are expected to fall. (According to a more precise definition

CEP is the radius of a circle—not necessarily centered on the target—within which half of the warheads are expected to fall. This would allow for systematic bias in the trajectory between the launcher and the target; but in real life conditions, it is impossible to measure such bias.) We preserved references such as "accuracy" in the comment column; they are probably references to CEP, but we cannot be certain. Specific information on hard target capability is also listed here because it depends primarily on CEP.

## Warhead

▲ **Warhead Type:** This category shows information on whether the missile is capable of carrying a nuclear, conventional, or chemical warhead. The data shown here most often are "yes" (it is capable of carrying this warhead type) and "maybe" (it may be capable of carrying the warhead type). Rarely does any source explicitly deny a certain capability. We enter data in this category only when the source is explicit on the matter. We do not infer from a yield in the range of kilotons or megatons a "yes" under nuclear. This category also contains any specification of the type of conventional warhead: HE (high explosive), HEAT (high-explosive antitank), HEAP (high-explosive, armor-piercing), shaped charge, or fragmentation.

▲ **Penetration** (for antitank missiles only): The amount of armor plate that the missile can penetrate to destroy the tank.

▲ **Warhead Characteristics:** Contains information about the warhead not covered by other categories. For ballistic missiles, it contains information on the number of warheads, independent targeting, and the bus system.

▲ **Nuclear Yield:** The explosive power of nuclear warheads. For multiwarhead missiles, *total* means the total for all warheads carried on the missile added together, and *each* means the yield for each warhead carried.

▲ **Warhead Weight:** Includes the weight of high explosives. It may also include other parts of the warhead, such as fuses or casing. Usually it applies to conventional warheads. Any specification of meaning given in the source is shown as a note.

▲ **Throwweight:** This category is used only when specifically mentioned by the source, since we have been unable to discern a consistent usage for the term.

According to military dictionaries, *payload* is the warhead (US Department of Defense usage) plus container and activating devices (NATO usage), while *throwweight* is the mass of the payload.

▲ **Payload:** Used only when specifically named by the source. See the dictionary definition under throwweight.

## Dimensions

▲ **Weight:** The weight of the entire missile, usually before launch. Any information on the weight of fuel or the empty weight is also recorded, with explanation.

▲ **Length:** Length of the missile.

▲ **Diameter:** Diameter of the missile. If different diameters are given for different stages, they are noted in the comment.

▲ **Wingspan:** Wingspan or finspan of the missile. If different spans are given for different fin sets, they are noted in the comment.

## Propulsion

▲ **Number of Stages:** The number of stages of propulsion.

▲ **Type of Propulsion:** Types of propulsion include liquid- and solid-fueled rockets, ramjets, and turbojets. Other information about propulsion, such as thrust or engine designation, also appears here.

## Guidance

▲ **Midcourse (or Main) Guidance:** *Missiles* in the modern sense are guided missiles, with a means for the weapon to correct its flight path after launch in order to bring it to a moving target or fixed aim point. All weapons in the Database except the FROG (free rocket over ground) series have some form of guidance. Midcourse guidance keeps the missile headed toward its aim point; terminal guidance involves more precise direction at the target. The types of midcourse guidance are inertial and radio command.

▲ **Terminal Guidance:** A means of guiding a missile directly to its target, which may allow for changes of target position. Terminal guidance systems include active, semiactive, or passive radar homing; infrared homing; and television homing.

▲ **Transmission Method** (for antitank missiles only): A means of transmitting guidance instructions from a vehicle or installation to a missile: either radio or fine wires that unwind from the missile in flight.

▲ **Control System:** A means on the missile of correcting the flight path, such as aerodynamic surfaces or vanes near the rocket efflux.

▲ **Radar:** Any radar associated with the operation of the missile, usually not located on the missile itself but on an associated launch vehicle, ship, or ground facility.

# History

The short value for every category in this section is a date (a year or some other time period).

▲ **First Observed:** The first time the missile was noted by Western observers.

▲ **Observed:** Any other significant observation of the missile by the West.

▲ **Observed in Parade:** The Soviet Union routinely displays weapon systems at national parades held in May and November each year. This category is used for weapons observed on display during one of these parades.

▲ **Development Began:** When development of the missile began. The category may also have information on when design began, noted in the comment column.

▲ **Development Underway:** When development was in progress. Development in this sense overlaps with testing.

▲ **Prototype Tests:** When testing began or a period during which tests were conducted. The comment gives any specific information about the testing program, including whether the test was the first one noted.

▲ **Production Began:** When production of the missile began.

▲ **Production Underway:** A time period when the missile was in production; this includes information such as "still being produced."

▲ **Production Ended:** When production of the missile stopped.

▲ **IOC (initial operational capability):** This term can refer either to an early date when a new weapon has been successfully tested and is ready to begin production or to a date one or two years later when the first operational units have been deployed and entered active service with troops. Often US Defense Department documents use the term differently for the USA and USSR: in the early sense for Soviet weapons and the late sense for US weapons. Lacking precise information, we use the category whenever the source indicates something related, such as "year operational" or "first became operational." The source's phrase is retained as a comment unless it is simply "IOC" or "initial operational capability."

▲ **Deployment Began:** When the missile was first deployed operationally. This includes dates described as "service introduction," "service entry," and "first deployed." Since these terms may not describe the same event, the source's phrasing is usually retained as well. This category also includes information denying that deployment has begun.

▲ **Deployed:** A time period when the missile is deployed operationally. This includes such data as "in service" or "still in service."

▲ **Deployment Complete:** When new deployments stopped being made.

▲ **Retired:** Used for any dates relevant to the weapon's being retired: when this began, when it was underway, or when it was completed, with no units of the weapon any longer in service. The comment specifies which of these the date applies to, if the source specifies.

▲ **Deployment History:** Used for information about deployment more general than that covered in previous categories. It includes references to changes in deployment status, such as the replacement of one missile by another or preparation for additional or modified deployments. Numbers, however, appear below under *Numbers in Service*.

## Numbers

▲ **Numbers Produced:** The number of units of the missile produced.

▲ **Numbers in Service:** The number of units of the missile in service each year and changes in the number in service, such as "60 were added last year." The numbers are entered by year, in chronological order, with a subheading for the year. Numbers are entered under the heading "undated" when the source does not give a specific year, as in "There were at one time 288." When the source indicates that a number pertains to the present, we enter the number under the year of publication of the source, though the source may intend a somewhat earlier time.

## Platform

▲ **Launcher/Silo:** Information about any fixed basing mode for a missile, such as a fixed launcher or an underground silo.

▲ **Carrier:** Any mobile platform that carries the missile. For AAMs and ASMs, carriers are airplanes; for SAMs and SSMs, they are mobile vehicles, tracked launchers, ships, or submarines. Data about several different carriers are grouped under sequential *Carrier* headings. Comments such as "N = 15" mean there are (or were) 15 of that particular ship, either in service or produced. Also associated with the carrier is information on numbers of tubes and reloads.

▲ **Tubes:** The number of spaces for missiles that can be mounted for firing without reloading. Ships and submarines sometimes have several distinct launchers mounted, each with a certain number of tubes. By convention, we record them in the form "number of mounts × number of tubes per mount." For instance, if a ship has three double mounts, "3 × 2" is entered. The plus notation, as in "1 + 1," means that two qualitatively different weapons are carried; the specific weapons are indicated in the comment column, where they are again separated by a plus (as in "AS-5 + AS-6"). For air-to-air missiles, where it is common for planes to carry a mix of radar-homing and IR-homing versions, "2 + 2" means 2 radar plus 2 IR unless otherwise noted.

▲ **Reloads:** The number of reload rounds carried or "yes" to indicate that reloads are possible. In many cases, the total number of missiles per vehicle may be given, including missiles ready to fire on the launcher(s). This number is entered under reloads but with the comment "total."

## Design, Use, and Strategy

▲ **Designer:** The group responsible for the design of the missile, usually a design bureau.

▲ **Design and Engineering:** Qualitative information about the development, design, or production of the missile, such as another missile that it was based on or conjecture about design requirements.

▲ **Use and Configuration:** Configuration (including ancillary equipment), procedures for firing, and flight path. Includes how the missile is operated, how long it takes to set up, what auxiliary vehicles are required, whether a submarine must surface to fire the missile, and how many launchers are deployed in a battery or an army or along a front.

▲ **Reloads in General:** Information about reloads for fixed missiles. For mobile missiles, reload information appears under *Carrier*. This category also includes whether ICBMs are hot launched (motors firing in the silo) or cold launched (missile ejected before the motor fires), which determines the feasibility of rapid reloading.

▲ **Target Type:** The broad class(es) of target the missile is intended to destroy, such as ships, submarines, land-based targets, aircraft, or tanks. Data indicating what would be expected on the basis of the US designation code are not repeated here. For SAMs, for example, we do not specify "aircraft" as a potential target.

▲ **Strategy:** The weapon's role in a broad military context and how it might be used with other weapons in a campaign.

▲ **Combat Reports and Effectiveness:** Reports on what happened when the missile was used in combat and how effective it was; also specific data on how effective it is expected to be, based on tests.

▲ **Geographical Deployments:** Information on where the weapon is deployed within the USSR, at sea, or in other nations, with the number deployed in specific regions (unless it is the entire force).

▲ **Exported to:** Nations the missile has been exported to and the missile's status in those nations.

▲ **Relation to Other Missiles:** Any comparison to another missile that is not specific enough to belong under another category, such as "resembles the US Sidewinder." If a comparison is drawn for a specific feature, such as length, throwweight, or accuracy, the information appears under the relevant category.

▲ **Other Information:** Unusual information that does not belong under any other category. One example is the space-launch rockets derived from a particular missile.

# PART II

## A SHORT HISTORY OF SOVIET MISSILES

# FIRST BALLISTIC AND CRUISE MISSILES

Missiles are unmanned weapons that fly through the air or space with their own propulsion and have some guidance mechanism to alter their flight path after launch. The term includes weapons that range from small antitank weapons carried in suitcases by soldiers to gigantic ICBMs that can hurl nuclear warheads halfway around the earth. The relation of missiles to military strategy and tactics is different for each kind of missile. Thus the history is organized by type of missile; strategic concerns are addressed as they arise.

The first guided missiles used in warfare—that is, not merely short-range, unguided rockets—were the German V-1 and V-2 missiles used in World War II. The V-1 buzz bomb was a small, pilotless airplane with an autopilot that allowed it to keep on a course set before launch. When the V-1 reached the target area it was set for, often London, the engine shut off, and the plane dove to the ground and exploded. The V-2 operated on a completely different principle. It took off almost vertically, with a rocket engine that delivered all its thrust very quickly, propelling the missile out into space at great speed. From there the earth's gravity took over, pulling the weapon back down to earth on a parabolic curve covering a long distance—for example, from the Netherlands to London. The V-2 operated on the ballistic principle like an artillery shell or even a ball thrown by a person, but on a much bigger scale.

The V-1, which flew through the air like an airplane, was the first cruise missile, and the V-2, which rose up into outer space on powerful, fast-burning rocket motors, was the first ballistic missile—representing the two basic technological types in the world of missiles. At the close of World War II, some of the German scientists who had worked on these weapons were taken to the United States; others were taken to the Soviet Union. On both sides, the early missiles were derived from the V-1 and the V-2, which were more advanced than missiles then under development in any other country.

Ballistic missiles and cruise missiles have different strengths and weaknesses. Ballistic missiles can travel long distances at great speeds since their rocket motors impart a great initial velocity in the few minutes before they burn out and fall off, and almost all of the nose cone's flight is in outer space, where there is no atmospheric drag. When the missile nose cone (or reentry vehicle) reenters the atmosphere, it approaches the target at such a high speed (4,000–20,000 mph) that intercepting and destroying it is very difficult. Ballistic missiles are well suited to inertial guidance, the most reliable form of guidance. In this system, the accelerations acting on the missile are measured with precision and fed into a computer, which plots course corrections. A major source of error is *drift*, which builds up as time passes. With a ballistic missile, all the corrections to its flight path are completed shortly after it leaves the atmosphere, a few minutes after launch. In space, the warhead encounters no unpredictable force (except the very slight uneven pull of the earth's magnetic field), and it comes back down through the atmosphere so quickly that little divergence from course arises there.

In contrast, a cruise missile must travel for several hours through the atmosphere, where the drift of the inertial guidance is much greater. Cruise missiles, however, are much better suited to terminal guidance, in which the missile senses the target and corrects its course while approaching it. Moving at speeds comparable to aircraft, cruise missiles are still under power and maneuverable when they approach their targets. As a result, ballistic missiles are generally used for traveling great distances and attacking fixed targets whose exact position is known, while cruise missiles are used at shorter range to attack moving targets, such as airplanes, ships, or tanks.

# SOVIET BALLISTIC MISSILES

## INTERCONTINENTAL BALLISTIC MISSILES

At the close of World War II, the USSR faced a hostile United States armed with nuclear weapons and with bombers that could reach Soviet territory. The USSR had neither nuclear weapons nor a military means of reaching US territory. Along with developing nuclear weapons, the Soviet Union had to develop a means of delivering them over a long distance. Although the Soviets did build some bombers for this task, they have clearly preferred missiles with their inherent advantages of speed and invulnerability to defenses. And although the United States had bases near the Soviet Union from which shorter-range planes could attack, the Soviet Union had no such bases near the United States.

The V-2, which the Soviets inherited from the Germans, had a range of a few hundred kilometers—far short of the 10,000 km needed to reach US territory. Thus a number of prototypes and missiles of intermediate range were developed before the first intercontinental ballistic missile (ICBM). The earliest rockets are not well known. Popular books of the 1950s and 1960s often described in detail whole series of Soviet missiles that never existed. Around 1947 there appeared the SS-1 Scunner with a range of 250 km (a V-2 clone), followed soon after by the 500-km SS-2 Sibling (which may have been an atmospheric sounding rocket not in the line of ICBM development). Neither of these rockets is believed to have been deployed in a military role. In the mid-1950s, the SS-3 Shyster, with a range of about 1,200 km, is thought to have entered service. This missile could reach West Germany from the western USSR. Even the most basic facts about this missile, such as the number deployed, are not known with any certainty. Operational by 1959 was the SS-4 Sandal, a missile that remains in service today. This missile has a range of 2,000 km, enough to cover most of France and Britain. With its deployment, the Soviets, still unable to reach US territory with nuclear weapons, could threaten the most powerful US allies in Europe.

The Soviets alarmed the West in 1957 with the launch of Sputnik, the first artificial satellite. Its booster was constructed out of the first ICBM, the SS-6 Sapwood, which the Soviets tested over intercontinental ranges the same year. Since the United States had not yet tested an ICBM, this development raised the specter of a Soviet ability to destroy the United States with nuclear weapons, and it fed the fear that the Soviets were ahead in military technology and might keep surprising the West with threatening military developments. A careful look at the SS-6, however, suggests that it was a virtually unusable weapon and was not built in quantity. It was so large and heavy that it could be deployed only on rail lines. It used an unstorable liquid fuel that took 12 hours to load. Deployed on a launch pad above ground, it was highly vulnerable to damage by a nuclear missile landing nearby. In addition, it was probably highly unreliable and inaccurate—possibly so inaccurate as not to offer certainty of hitting even a city over a course of 10,000 km (6,000 miles). The Soviets deployed only a few SS-6s, probably just four, at test centers, and these did not become operational until 1961.

In late 1962, the SS-7 Saddler entered service. Several advances combined to make this the USSR's first real ICBM, with 197 deployed. The SS-7 had storable liquid fuel, making it much less vulnerable to a preemptive attack. About a third of the SS-7s were deployed horizontally in concrete coffins, providing them with some protection against nuclear explosions nearby. (It seems likely, however, that this third did not have their fuel loaded until after they were raised to the vertical position.) Finally, advances in nuclear warhead technology probably made possible a lighter, smaller warhead. This design would allow greater range, permitting the missile to cover targets throughout much of the continental United States.

Shortly after the SS-7, the SS-8 entered service, a similar rocket but with the older, unstorable fuel. Only 23 were deployed. This pattern is fairly common in the history of Soviet strategic weapons: two similar missiles are developed simultaneously, one with a more advanced technological feature (in this case, the storable fuel of the SS-7) and the other conservatively retaining older technology. If the more advanced weapon works well, it dominates the production series; if not, the less advanced weapon is emphasized. In either case, a few of the less successful type may be deployed.

With the deployment of the SS-7 and SS-8, the Soviets achieved their early strategic goal of maintaining a nuclear force that could reach the United States. In the second generation of Soviet ICBMs, advances in fuel efficiency and warhead weight could be put to purposes other than simply increasing range: carrying larger warheads to the target (SS-9), producing a cheaper missile that could be deployed in great numbers (SS-11), and experimenting with solid fuel (SS-13). Like the US Minuteman I ICBMs deployed starting in 1960, these second-generation Soviet ICBMs were placed in hardened underground silos, made of steel-reinforced concrete, to survive the effects of nearby nuclear blasts. The SS-9 and SS-11 were deployed beginning in 1966, the SS-13 in 1969.

The SS-9, which carried a very large warhead (perhaps 18–20 MT), may have been the first Soviet ICBM built specifically to target US nuclear weapons. Former US Secretary of Defense Harold Brown has said the SS-9s were almost certainly targeted on the US Minuteman ICBM Launch Control Centers.

The SS-10 was apparently a missile design of the same general size and mission as the SS-9 but with nonstorable liquid fuel. It was never produced or deployed.

According to Berman and Baker, authors of the Brookings study *Soviet Strategic Forces,* the SS-11 was improvised based on a missile originally intended for naval use. By the early 1960s it was apparent that the US would deploy a large number of Minuteman ICBMs (ultimately 1,000). The Soviet Union could not build as many large, expensive SS-9s and needed a smaller, cheaper missile to match the United States in numbers. They built 1,030 SS-11s, about half of which remain in service today.

The SS-11 was not a perfect counter, however, since the Minuteman was a solid-fueled missile. Solid fuel has more energy per unit weight, is more reliable, and requires less maintenance than storable liquid fuel, which degrades over time. The USSR's first attempt at a solid-fueled ICBM was the SS-13, but the SS-13 had such a short range and was so inaccurate that it was in essence abandoned, with only 60 deployed.

The first units of the third generation of Soviet ICBMs became operational around 1975. They introduced MIRVs (multiple, independently targetable reentry vehicles), a technology first deployed by the United States in 1970. In MIRVed missiles, a device called a bus (or postboost vehicle) maneuvers in space with small rocket motors, releasing nuclear warheads in slightly different directions, very precisely, so that they will hit separate targets when they reach the earth. In theory, MIRVs can allow a single missile on one side to destroy several missiles on the other side. Without MIRVs and with comparable numbers of ICBMs, a potential attack would destroy only one opposing warhead for each warhead used, and, thus, preemptive strikes would offer no military gain.

MIRVs gave an already existing tendency new energy. Each side constantly makes its missiles more accurate, threatening the silos on the other side. Then each side shields its silos with more layers of concrete and steel to protect against ever-closer nuclear explosions. Overall the theoretical attackers on each side are winning the battle of accurate, MIRVed missiles versus fixed silos but losing the war of creating a reasonable expectation that they could carry out such immense, complicated, never-rehearsed attacks, subject to enormous uncertainties, to destroy most of the opponent's ICBM silos—and to do so without the opposing side's launching its ICBMs on warning of the incoming attack.

Along with MIRVs, third-generation Soviet ICBMs—the SS-17, SS-18, and SS-19—introduced harder silos and improved accuracy. Further improvements in accuracy were made in the later models of the third-generation missiles. This generation has not changed the number of ICBMs deployed. Limited by the SALT agreements, new ICBMs have replaced older ICBMs on a one-for-one basis. The destructive power of the new missiles is much greater, however, since their MIRVs have raised the number of separately deliverable ICBM warheads from about 1,400 to about 6,400.

The SS-16 was the second attempt at a solid-fueled ICBM, following the unsuccessful SS-13. But the SS-16, too, may have been a failure in terms of range and accuracy since the Soviets agreed in SALT II not to deploy it. All of the deployed third-generation missiles are liquid fueled.

The SS-18 is the "heavy ICBM," in arms control terminology, that replaced the SS-9. A total of 308 were deployed, most believed to carry eight warheads at first and now 10 warheads.

The SS-17 with four MIRVs and the SS-19 with six MIRVs replaced about half of the SS-11s. The SS-17 apparently has a serious flaw that limits its accuracy. More SS-19s were deployed initially (360 to the 150 SS-17s), and continued improvements have been made in the SS-19 but not in the SS-17.

Initially some of all three new ICBMs were believed to be equipped with single high-yield warheads. The 1984 and 1985 editions of *Soviet Military Power* imply, however, that no one-warhead versions remain in service.

A decade after the SS-17, -18, -19 series was introduced, the Soviets are preparing to deploy two new solid-fueled ICBMs: the SS-24, a ten-warhead missile that will replace the SS-17s and SS-19s, and the SS-25, a one-warhead mobile missile that will replace the remaining old, one-warhead SS-11s and SS-13s. (The SALT II treaty permits only one new ICBM, and the USSR claims that the SS-25 is not a new missile but an improved SS-13.) In addition, the Soviets are preparing to test two improved missiles, follow-ons to the SS-24 and the SS-18. These have been unofficially referred to as the SS-26 and SS-27, respectively; but whether the ultimate US designations will treat them as new missiles or as modified versions of the SS-24 and SS-18 remains to be seen.

# SUBMARINE-LAUNCHED BALLISTIC MISSILES

The United States and the Soviet Union both began testing submarine-launched ballistic missiles (SLBMs) in the late 1950s. The great attraction of these missiles is that detecting the submarines that carry them is difficult or impossible. Another initial attraction of SLBMs to the Soviet Union might have been that they offered a way to reach the United States with shorter-range missiles before the USSR had ICBMs.

In the 1950s the Soviets experimented with missiles fired from chambers towed by submarines. For firing, the chamber would be righted to vertical and the missile launched from the container. Little is known about the experimental missiles fired in this manner, called Golem.

The first operational Soviet SLBM was the SS-N-4 Sark, introduced in 1958 or 1959. It was deployed in three types of submarine, which carried three missiles each. The SS-N-4 had a range of only 650 km, and the submarine had to surface to fire it, making it a vulnerable strategic system. As with the SS-6 ICBM, the Soviets seem to have rushed to deploy a system of limited military value—a kind of weapon the United States could have built but chose not to.

The follow-on SS-N-5 Serb became operational in 1963, three years after the US Polaris A-1. Like the SS-N-4, the SS-N-5 was carried three missiles to a submarine, but it could be fired while the submarine remained submerged and had a range of 1,300 km. (The Polaris A-1 had a range of 2,200 km and was carried 16 to a submarine.) Eventually SS-N-5s replaced all SS-N-4s, and some remain operational today, aimed at targets in Europe and Asia.

Not until 1968 did the Soviets deploy a missile comparable to the Polaris A-1: The SS-N-6, with a range of 2,400 km, deployed 16 each on nuclear-powered submarines. In the 1970s, improved versions of the SS-N-6 were introduced with ranges of 3,000 km. The SS-N-6 was widely deployed and still accounts for nearly half of the Soviet SLBMs but is being retired in favor of new missiles with much longer ranges.

Although similar to the Polaris A-1, the SS-N-6 did not give the Soviet Union a system of equal invulnerability because of differences between the two superpowers in geography, submarine technology, and antisubmarine warfare (ASW) technology. The United States has very quiet submarines with easy access to the open oceans. In contrast, the noisy Soviet Yankee-class submarines carrying SS-N-6s had to move through chokepoints bordering Western nations, guarded by sensitive Western ASW systems. Then they had to patrol the deep ocean areas off the US coasts, where the deep sound channel would transmit their noise to US ASW systems. As a result SS-N-6s did not offer a secure strategic retaliatory force.

In 1973 the Soviet Union deployed a new SLBM, the SS-N-8, with such a long range (7,800 km initially) that it could strike US targets from the Barents Sea and the Sea of Okhotsk, close to Soviet territory. Soviet strategic submarines equipped with the SS-N-8 and later SLBMs need no longer pass through heavily patrolled chokepoints or remain on station near the deep sound channel. The SS-N-8, based on Delta-class submarines, now constitutes about a third of the Soviet SLBM force.

In 1978 the USSR introduced a MIRVed SLBM, the SS-N-18. (The US Poseidon, with 10 MIRVs, was deployed beginning in 1970.) The SS-N-18 was deployed in three versions: Mod 1 with three MIRVs, Mod 2 with a single warhead, and Mod 3 with seven MIRVs. Although the MIRVed versions have a maximum range less than that of the SS-N-8 (6,500 km), they can still strike US targets from the northern seas. Currently about a quarter of the Soviet SLBMs are SS-N-18s.

Unlike US SLBMs, which used solid fuel from the start in 1960, all Soviet SLBMs through the SS-N-18 were propelled by storable liquid fuel. Around 1975 the Soviets tested the solid-fuel SS-NX-17, but the tests were evidently a failure: the weapon was deployed only on the one submarine used for tests. The SS-NX-17 had a range of only 3,900 km even when limited to one warhead, and it would have perpetuated the chokepoint detection problems of earlier Soviet SLBMs.

The first solid-fueled Soviet SLBM produced in quantity is the SS-N-20, deployed starting in 1983 on Typhoon-class submarines. Two of these submarines are now operational and more are being built, but there are reports that the missile has not done well in tests and that a follow-on missile or modified version is under development. A 1985 US Defense Department report says that the missile carries six to eight MIRVs, down from the six to nine reported earlier. There are no US government comments on accuracy. Our estimate (in part III) of 1,000 m for CEP assumes a modest improvement over the reported 1,400 m of the SS-N-18. It is possible, however, that the accuracy is no better than 1,400 m and that the weapon is deployed exclusively as a second-strike deterrent, not aimed at military targets.

The only new SLBM known to be under development is the SS-NX-23, a successor to the SS-N-18, with longer range, permitting Soviet strategic submarines to operate even closer to their northern coasts, in shallow, protected waters where it is difficult, if not impossible, for US ASW forces to operate.

# INTERMEDIATE-RANGE BALLISTIC MISSILES

The first two Soviet medium- and intermediate-range ballistic missiles (MRBMs and IRBMs), the SS-3 and SS-4, were described earlier as stepping-stones to ICBMs. The SS-3 Shyster, deployed around 1955, had an estimated range of 1,200 km; the SS-4 Sandal, deployed in 1959, had a 2,000 km range, covering most of France and the United Kingdom. The longer range was evidently important to the Soviets: they deployed 600 SS-4s. Although Soviet leaders were no doubt concerned with these potential adversaries and with US nuclear arms in Western Europe, SS-4s may have been intended mainly as an anti-US deterrent; they were deployed between 1959 and 1962, when the USSR still lacked a reliable way to attack US territory with nuclear weapons. The SS-4s and medium-range bombers were probably intended to hold Western Europe hostage against a US nuclear attack, absent long-range Soviet nuclear missiles.

The SS-5 doubled the range of the SS-4, reaching 4,100 km. Relatively few SS-5s were deployed, however—only 97—probably because there were only a few places in the UK beyond range of the SS-4 that were deemed to warrant nuclear targeting. Early SS-4s had unstorable liquid fuel. Later SS-4s and SS-5s had storable liquid fuel. The SS-5 was fired from fixed sites. There is some speculation that the SS-4 was mobile but in a cumbersome configuration requiring a long time to set the weapon up for firing.

An interest in truly mobile weapons of this range was probably the impetus behind the next two developmental IRBMs: the SS-14 Scamp/Scapegoat, tested beginning in 1965, and the SS-15 Scrooge, tested beginning in 1968. The SS-14 comprised the top two stages of the solid-fueled SS-13 ICBM, and the SS-15 may have been derived from the SS-13. Both the SS-14 and SS-15 had solid fuel and were based on tracked vehicles, and both were technical failures that never went into production.

In the view of several analysts, the failure of the SS-14 and SS-15, in addition to the shortcomings of the SS-4s and SS-5s, led to the variable-range deployment of SS-11s. In 1970, the last 120 of the 970 SS-11 ICBMs deployed initially (another 60 were deployed later) were placed in SS-4 fields. In the West, this was taken to indicate that, like the SS-4s, they were aimed at targets in Europe or Asia, not the United States.

The most recent Soviet IRBM is the SS-20, a missile caught up in the political turmoil in Europe since 1980. Like the SS-14, the SS-20 comprises two stages of a solid-fuel ICBM, in this case, the SS-X-16. But although the SS-X-16 was never deployed, the SS-20 has been treated as a highly successful missile. A total of 378 mobile launchers had been deployed by 1984, and construction of new bases started in 1984 suggests that 450 launchers may be deployed in 1987.

The SS-20 is the first IRBM with MIRVs (three each). The Soviet Union no doubt wanted the SS-20 to replace the SS-4s, SS-5s, and variable-range SS-11s because the SS-20 has solid fuel, which offers greater reliability, mobility, and invulnerability, and because MIRVs allow the SS-20 to cover the same targets much more cheaply. In addition, since the variable-range SS-11s are technically ICBMs, they are counted against SALT limits on strategic arms, and to exploit the full measure of the ceilings on strategic weapons, the USSR would have to fill the IRBM role with a nonstrategic weapon, the SS-20.

The SS-20 is reported to be much more accurate than the SS-4s, SS-5s, and SS-11s, but it is not clear there are enough hardened targets in Europe (or Asia) to translate its greater hard-target kill potential into some meaningful military advantage. Except for a few nuclear storage bunkers and command and control centers, there is little that SS-20s can destroy that the SS-4s and SS-5s could not.

# SHORT-RANGE BALLISTIC MISSILES

The USSR has deployed several short-range, mobile, land-based missiles for use against targets on or near a battlefield. All of the longer-range weapons discussed previously are fitted with nuclear warheads; the short-range ballistic missiles (SRBMs) are believed to be capable of carrying nuclear, conventional (high explosive), or chemical warheads. The maximum ranges of different SRBMs vary between 25 km and 900 km. Based on range, the missiles can be divided into three groups.

The short-range group comprises mainly unguided FROGs (free rockets over ground or free range over ground). The first five, FROG 1 through FROG 5, which entered service between 1957 and 1964, had estimated nuclear yields of 10–20 kt and ranges of 25–50 km, and all were based on tracked vehicles like tanks. In fact, whether five different versions were actually deployed is doubtful. The principal evidence for their existence was their appearing in the annual military parades in Moscow. In 1965, FROG 7 was introduced. It had a range of 70 km and was based on a wheeled vehicle, providing road mobility. At the peak of deployment, an estimated 480 FROG 7s were in service. Since 1983, the SS-21, a more accurate, nuclear-capable missile with a 120 km range, has been augmenting and replacing the FROG 7s. An estimated 170 SS-21s had been deployed by early 1985.

Middle-range SRBMs started out with the SS-1b Scud A, deployed in 1957. The designation suggests it was derived from an early V-2 copy, the SS-1a Scunner, with a 150 km range. The SS-1c Scud B with a range of 300 km, introduced beginning in 1965, had completely replaced the Scud A by 1977. Both missiles use storable liquid fuel. Scud As and a few early Scud Bs were based on tracked vehicles, but most Scud Bs are on wheeled vehicles. The SS-23, a guided missile with a range of 500 km, was introduced in 1983.

Long-range SRBMs began with the 900 km SS-12 Scaleboard, first deployed in 1969. This had storable liquid fuel and was based on a wheeled vehicle. Its successor, the SS-22, which entered service around 1984, is believed to be a more accurate, solid-fueled missile.

# SOVIET CRUISE MISSILES

## CRUISE MISSILE PROPULSION AND GUIDANCE

Unlike ballistic missiles, whose rocket motors fall off after sending them into outer space, cruise missiles fly through the air in powered flight for the duration of their trip. The term *cruise missile*, which is ill defined, is sometimes used quite restrictively. *Jane's Aerospace Dictionary* limits cruise missiles to weapons that are "wing supported"; other sources use the term for long-range weapons with air-breathing engines.

Two main kinds of propulsion are possible for cruise missiles. One is a rocket motor that burns fuel (either solid or liquid) and oxidizer, both contained in the missile. The other is an air-breathing motor in which the fuel is carried in the missile, and the oxidizer is oxygen from the air. Because the oxidizer need not be carried, air-breathing engines generally offer greater range. For short-range missions, rocket motors provide more thrust and greater acceleration. A common configuration combines the two: a rocket booster that provides initial acceleration, followed by an air-breathing motor to sustain the missile for a long flight.

Because cruise missiles can use terminal guidance to home in on the target, they are preferred for use against moving targets and targets whose position is not known precisely at the time of launch. Often the critical performance question for a cruise missile is how well its terminal guidance system works. This can involve questions of basic electronic capability, electronic countermeasures (ECM, such as radar jamming), and counter-counter measures (ECCM). Little can be said about the competition in electronic technology, which is highly secret and constantly changing without visible changes in missiles.

A few generalizations about guidance systems can be offered, however. Because ships are large, metal, heat-emitting objects, set against a relatively uniform surface background, they offer easy targets. Similarly airplanes flying high and viewed from below are easy to pick out physically against empty sky. For these reasons, as Soviet guidance systems have improved, Soviet cruise missiles have posed a growing threat to Western combat aircraft and ships. Airplanes flying low, viewed from above, are hard to identify because they must be distinguished from the ground. The USSR may be on the verge of solving this problem, which the United States solved earlier. Picking out smaller objects on the ground—tanks, buildings, or bridges—is still more difficult. The only reliable way to do so is for a human being to identify the target visually and direct the missile to it using wire guidance, laser target illuminators for homing devices, or a television camera in the nose linked with radio command. Completely automated guidance—for "fire and forget" weapons—is still at the forefront of US technology and not yet in sight for the USSR.

Two basic types of terminal guidance are used against ships and airplanes: infrared (IR) and radar. In IR guidance, a heat-sensitive device in the front of the missile detects heat emitted by the engines of the target and homes in on the heat source. Radar guidance can use radar signals to localize the target in several different ways. In active radar, the missile sends out radar signals and processes the echoes, independent of the missile launch platform. In semiactive radar, radar signals are sent out by a friendly source (usually the ship, plane, or ground installation that launched the missile), and the missile homes in on the echoes. In the oldest configuration, the launch platform tracks both the missile and the target and controls the missile by radio command. The proliferation of radars on ships and aircraft has resulted in yet another form of radar guidance: passive radar, in which the missile homes on signals from enemy radars.

IR guidance is not susceptible to electronic countermeasures, though it can be fooled by alternate heat sources such as flares. One main limitation is that it works over comparatively short ranges; another is that it can perform only a tail-chase interception, a problem when attacking aircraft. Radar guidance works over much longer ranges, and it permits interception courses where the missile aims for where the target will be when it arrives, not where the target is now. With radar guidance, however, active radar is necessary for the missile to travel beyond the radar range of the launch platform, as in seeking out distant ships; and active radar can be employed only on large missiles since it exacts a size and weight penalty. Finally, radar and radio command are vulnerable to interference and jamming.

# LAND-ATTACK CRUISE MISSILES

## Strategic Cruise Missiles

The AS-3 Kangaroo is the largest Soviet air-to-surface missile. Resembling a small jet fighter, it is carried under the belly of the large strategic Bear bombers. The AS-3, which entered service in 1961, is a strategic standoff weapon with a range of 650 km. It lacks terminal guidance and is intended for attacking large area targets, such as cities. The AS-3s are being phased out and replaced by AS-4s. It is not yet clear whether the AS-4 antiship missiles are being adapted to the strategic mission or the Bear bombers are being directed to the antiship role.

The AS-X-15, SS-NX-21, and SSC-X-4 are modern, long-range cruise missiles, corresponding to the US ALCM, SLCM (Tomahawk), and GLCM, launched from air, sea, and ground platforms, respectively. All have ranges of about 3,000 km. (Although classified here as strategic, like the US GLCMs deployed in Europe, the SSC-X-4 is not based in an area that would permit it to reach US territory.) The main innovation that resulted in modern US cruise missiles was the TERCOM (terrain comparison or terrain contour matching) guidance system. TERCOM allows the missile to update its inertial guidance by taking a rapid series of altitude readings at intervals, comparing the readings to a bank of stored altitude profiles, and identifying its precise position. The Soviet Union has supposedly developed a similar guidance system; but given the difficulty the United States had with this technology, it is questionable how effective Soviet TERCOM is. US reconnaissance satellites can observe signs of operational capability, flight test range, and other features; assessing a cruise missile guidance system remotely is much harder.

## Tactical Air-to-Surface Missiles

There is a constant stream of reports about new Soviet tactical ASMs, sometimes describing the same weapon with different designations. The designations AS-8 and AS-10, for example, are now both believed to have been confused reports about the AT-6 air-launched antitank missile. This situation and others raise a question whether the Soviets have any effective, widely deployed tactical air-to-surface weapons. The only two weapons on which there is a modicum of agreement are the AS-7 Kerry and the AS-9.

First deployed in the 1970s, the AS-7, with a range of 10 km, is launched from a tactical aircraft that guides the missile using a joystick and radio command. The AS-9, deployed around 1975, is a passive radar missile with a range of some 80 km. Its other characteristics are hotly disputed.

## Antitank Missiles

Historically the advantage that the machinegun gave to the defense in ground combat was lost to the offense posed by the tank and the tank tactics developed by the Germans in World War II. Today the advantage in ground combat is thought to be shifting back toward the defense, and the most important reason is the antitank missile. With ATMs, an infantry team can destroy a tank with high probability, using a weapon that costs a small fraction of what a tank costs. Although the West probably leads in the technology of antitank missiles, the most notable combat use of these weapons was the use of Soviet missiles by Egyptian troops in the 1973 Arab-Israeli war. The Soviet AT-3 Sagger missiles inflicted heavy losses on Israeli tanks.

Recent Soviet ATMs have improvements in range and armor penetration capability as tanks acquire more armor. The most crucial area of technology is guidance, however, and the Soviets have made only modest advances in this area. The most primitive system, used in the first three Soviet ATMs, is called command to line of sight (CLOS). The operator steers the missile using a joystick and must keep track of both the missile and the target and steer skillfully. The next step up is called semiautomatic CLOS (SACLOS). In this system the operator keeps the cross-hairs of a sight on the target, and an automated system steers the missile. This is the system used in the three later Soviet ATMs and in all modern Western systems. Both guidance techniques require the user to stay exposed for the entire flight of the missile. Neither side has been able to develop a system that would allow the user to launch the missile and then take cover.

Most ATMs are wireguided; command signals are transmitted through fine wires that unravel from the rear of the missile as it flies. This avoids the vulnerability to jamming that can block radio transmission of signals. But the wires are fragile, so this guidance system is not suited to launch from helicopters. One countermeasure that tanks have adopted on detection of an ATM is to fire a round in the direction of the missile to disrupt the command wires or distract the operator.

Remarkably little information is available on the service entry dates of Soviet ATMs. The AT-1 Snapper may have appeared around 1960, the AT-2 Swatter and AT-3 Sagger during the 1960s, the AT-4 Spigot and AT-5 Spandrel in the mid-1970s, and the AT-6 Spiral since 1980. Armor penetration has advanced from 35 cm for the AT-1 up to 50 cm for the more modern weapons, and range has increased from around 2,300 m to perhaps 4,000 m. The AT-2 Swatter has radio transmission of command signals. The AT-6 Spiral may use a laser target illuminator; the missile detects a laser reflection from the target and homes on it. Otherwise all Soviet ATMs use wire guidance. The AT-2 and AT-6 are the main helicopter-launched ATMs, though AT-3s have also been fitted on helicopters for export. The AT-4 is fired exclusively by infantry teams and the AT-5 exclusively from vehicles; the AT-1, AT-2, and AT-3 seem to be adaptable to either configuration.

# ANTISHIP AND ANTISUBMARINE MISSILES

## Sea-Launched Antiship Missiles

The first antiship cruise missile, the SS-N-1 Scrubber, appeared around 1958 on two classes of destroyer. It was essentially an unmanned airplane, lanched from a large, unwieldy apparatus consisting of a hangar and a 17 m launch rail. Based on the German V-1, it had many operational limitations and was never widely deployed. Both IR and active radar have been suggested as terminal guidance systems, though it is doubtful how well they worked. Basic facts, such as range and nuclear capability, are in dispute.

The SS-N-2 Styx became operational around 1960. A small antiship weapon with a 40 km range, it is intended for launch from patrol boats and is one of the few Soviet antiship weapons that cannot carry a nuclear warhead. It is also the first antiship weapon ever used in combat; in 1967 an Egyptian patrol boat sank an Israeli destroyer 20 km away using several SS-N-2s. Three versions of the SS-N-2 have appeared (one initially called the SS-N-11), and it is still widely deployed. Both active radar and IR guidance are used. In the Soviet Union, where the SS-N-2 is fitted only on short-range patrol boats, the missile is clearly a defensive weapon.

The 400 km SS-N-3 Shaddock, the first in a family of long-range antiship cruise missiles, also became operational around 1960. It is unclear whether this missile would be able to get close enough to hostile ships for its active radar guidance to be effective. For the missile to find a target at ranges much over 50 km, there must be a cooperating aircraft that can correct the missile flight path en route. The SS-N-3 has been widely deployed on Soviet submarines and surface ships and on land (with the designation SSC-1 Sepal) for coastal defense. It is capable of carrying a nuclear warhead.

When the SS-N-3 was first deployed, US aircraft carriers represented as potent a nuclear threat to the USSR as land-based bombers and a more potent threat than missiles. Thus the SS-N-3 was probably aimed mainly against US aircraft carriers. Although US carriers are no longer considered a component of strategic nuclear forces, they are still equipped for nuclear attacks on the USSR. Thus the Soviet Union has modernized its long-range nuclear antiship capability: the SS-N-12 Sandbox entered service in 1973 and the SS-N-19 in 1980 on newer classes of ships and submarines. These missiles have greater speed and possibly greater range, but their main improvements are probably in guidance and reliability.

The 50 km SS-N-7 and 110 km SS-N-9, which entered service in 1968 and 1969, respectively, seem intended to fill a gap between the short-range SS-N-2 Styx and the long-range SS-N-3/12/19 family. The SS-N-9 is deployed mainly on corvettes, intermediate in size between patrol boats and destroyers. The SS-N-7 is deployed on submarines, and the SS-N-9 may be deployed on submarines as well. With SS-N-3s submarines had to surface to fire; with SS-N-7s they can fire while submerged, an important consideration because the latter's shorter range makes it more likely that hostile forces will be nearby. The SS-N-7 and the SS-N-9 are believed to be nuclear capable. The SS-N-22, which entered service in 1982, seems to be an updated version of the SS-N-9.

Around 1970 the Soviets tested the SS-NX-13, a submarine-launched ballistic antiship missile with a range of perhaps 700 km. The system did not work and was abandoned.

## Sea-Launched Antisubmarine Missiles

The USSR has sought long and in vain for a way to find and destroy US submarines, particularly SSBNs (the nuclear-powered, strategic-missile-equipped subs). With the advent of long-range Soviet SLBMs that operate from home waters, the USSR has defined a more manageable antisubmarine warfare (ASW) requirement: to find and destroy the US attack submarines (SSNs) that seek Soviet SSBNs.

In 1968 the Soviets introduced the SS-N-14 Silex, which was eventually fitted on most large Soviet surface ships. With a range of 55 km, the missile carries either an acoustic homing torpedo with a conventional warhead or a nuclear depth bomb, which it drops in the vicinity of the target submarine.

In the early 1970s the USSR deployed the SS-N-15 and SS-N-16, which are similar to the SS-N-14 but based on submarines rather than ships. With a range of 45 km, the SS-N-15 carries a nuclear depth bomb, and the SS-N-16 carries an acoustic homing torpedo. Both are launched from standard submarine torpedo tubes and may be widely deployed. Unlike regular torpedoes, missiles that carry torpedoes or depth charges through the air can cover great distances and do so quickly; however, many sources doubt that these long-range Soviet ASW weapons are effective.

## Air-Launched Antiship Missiles

At least half of Soviet air-to-surface missiles (ASMs) are antiship weapons. They use either IR or active radar homing, and some can use passive radar. The early ASMs used turbojet propulsion; more recent missiles use rocket propulsion for much greater speed. The ranges of these weapons vary radically with flight path: if they fly high and slowly, they can go a very long way, but then they are vulnerable to ships' defenses. The ranges given here are for low, fast flights.

The first two, the AS-1 Kennel and AS-2 Kipper, which entered service around 1960, had ranges of about 90 km. The AS-4 Kitchen and the AS-5 Kelt, which entered service around 1966, and the AS-6 Kingfish, introduced around 1977, are characterized by higher speed and longer ranges—up to 300 km for the AS-4 Kitchen. The AS-11 may be another weapon in this series. Two obscure early coastal defense missiles may have been based on the AS-1 Kennel: the SSC-2a Salish and the SSC-2b Samlet.

# ANTIBALLISTIC MISSILES

The first antiballistic missile, the ABM-1 Galosh, became operational at a site near Moscow in 1968, and within a year all 64 Galosh missiles were operational. (Earlier twice that many had evidently been planned.) In 1979, 32 of these missiles were dismantled. The ABM-1 is a relatively large missile that carries a nuclear warhead out of the atmosphere to destroy incoming warheads in outer space. The Joint Chiefs of Staff FY 1979 posture statement says this system could provide adequate protection against small attacks by unsophisticated missiles without penetration aids. This has been the US government's unwavering estimation of the system's capabilities from the beginning; it means that the system is worthless against a US attack.

Once again, the evidence points to the Soviets' deploying a system the US could have built earlier but elected not to. In 1969 US officials said that in some important respects, the Galosh resembles the Nike-Zeus strategic SAMS, which the United States abandoned years earlier. As in the case of bomber defense, however, the Soviet strategic position is very different from that of the United States: France, the United Kingdom, and China each has some 100 ballistic missiles aimed mainly or exclusively at the USSR. The ABM-1 might have provided significant protection for Moscow against attacks by these countries (especially China), and it undoubtedly complicates their nuclear planning.

The Soviets have been continuing research on new ABM systems, and now the Moscow system is being revamped. Although the permanent US designations of the new missiles are not available yet, a longer-range replacement for the ABM-1 is now called the SH-4, and a short-range missile for use within the atmosphere is the SH-8. It is virtually certain that this newer system will also be unable to protect Moscow from a large-scale nuclear attack by advanced missiles. As long as the 1972 ABM Treaty is in effect, the United States and the USSR are prohibited from building more than 100 ABM launchers in a single location (reduced from the original treaty limits of 200 missiles in two locations).

# ANTIAIRCRAFT MISSILES

## Strategic Surface-to-Air Missiles

In no other area are US and Soviet missile deployments more strikingly different than strategic surface-to-air missiles (SAMs): generally large, fixed missiles, based inside a country to defend against bomber and attack aircraft. In the late 1950s and early 1960s, the United States had about 1,000 strategic SAMs. All but a handful were dismantled in the 1970s, however. In contrast, the USSR built up a force of about 10,000 strategic SAM launchers and has maintained this force with no sign of a decline. The difference between the two countries is not surprising since Soviet territory was and is far more threatened by bomber and attack aircraft. In 1960, the United States had over 2,000 nuclear-armed bombers within range of the USSR, while the Soviet Union had just 150 bombers that could reach the USA. Although most US strategic bombers were later phased out, Soviet territory is also threatened by US tactical land- and carrier-based attack aircraft and by British, French, and Chinese bombers and attack aircraft. Even if Soviet air defenses were never able to thwart a concerted US bomber attack, they may be very effective against nations with fewer and less advanced aircraft.

The first Soviet SAM bases, for the SA-1 Guild, were built starting in 1954 in two concentric rings around Moscow, the outer one 45 nm from the heart of the city. By 1962, if not sooner, 3,200 SA-1 launchers were deployed, all in the vicinity of Moscow. Little else is reported about the SA-1 except that like the earliest US strategic SAMs, it was probably not very effective.

The SA-2 Guideline has seen widespread use ever since it entered service in 1958. It was deployed throughout the USSR, exported to many other countries, and used in warfare in third world nations. The SA-2 was probably responsible for downing Gary Powers's U-2 flight in 1960. (The SA-1 was incapable of operating effectively at the very high altitudes of U-2 flights.) About 4,500 SA-2 launchers were deployed at the peak, although today the number is below 3,000. The possibility that some SA-2s might be armed with nuclear warheads arose in connection with a model displayed in 1967 with an enlarged warhead. Most of the SA-2s, however, are not nuclear armed.

The wartime use of SA-2s reflects the invisible battle between ECM and ECCM. As described by *Jane's Weapon Systems 1982–83*, Israeli ECM devices were at first adequate to counter the SA-2s deployed by Arab nations. Then improved SA-2s were introduced with a terminal guidance system that used a wider range of radar frequencies than Israeli ECM could handle. As a result, Israel lost many planes before it obtained US ECM pods to jam the new SA-2 guidance system. Although it is likely that the United States has an overall technological edge in the ECM-ECCM competition, the relative effectiveness of given weapons and countermeasures in any future war would vary and could easily be unknown before the fact.

In the 1960s, faced with a stiff array of Soviet SAMs and improved Soviet interceptor aircraft, US bombers shifted from high-altitude to low-altitude flight. Although the US bombers could not travel as fast at low altitude, they could escape detection by radar and avoid Soviet SAMs, whose guidance systems could not pick them out at low altitude. The minimum altitude of the SA-2 was 3,000 ft initially, 1,500 ft in a later version. Thus one main improvement sought by the Soviets in their next strategic SAM was an ability to operate more effectively at low altitudes. It is unlikely whether even today the Soviets can effectively attack US bombers flying just a few hundred feet off the ground, equipped with advanced ECM and with nuclear-armed missiles to attack defenses. Nonetheless, the USSR has been moving to improve its technology in this area.

The SA-3 Goa was introduced around 1961. Since SA-2s operated at high altitudes, the SA-3 (a smaller and cheaper missile) had a much lower maximum altitude. There are no firm figures on the minimum altitude of the SA-3, which has decreased with time, but sources agree that it is lower than the minimum altitude for the SA-2. Some 1,400 SA-3 launchers were deployed by the late 1970s. Initially there were two launch rails per launcher. A four-rail launcher introduced in the late 1970s has resulted in over 4,000 SA-3s deployed on launchers by 1985.

Construction began on the SA-5 SAM in 1963, with the first sites operational in 1967. During those four years and for some years after, this missile was a source of great controversy in the United States. Many US analysts thought it foreshadowed a giant Soviet ABM network. It was eventually shown, however, to be a SAM aimed at high-altitude aircraft. One reason for the ABM speculation is that it was unclear why the Soviets needed another high-altitude SAM; this is still a perplexing question. It is possible that the SA-5 was designed to counter the proposed US B-70 bomber of the 1960s, which never was developed, and then the SA-5 was deployed anyway out of bureaucratic inertia. Alternatively, there may have been flaws in the SA-2 the Soviets wanted to correct. In any event, at least 1,200 SA-5s were ultimately deployed throughout the USSR. The weapon has a maximum altitude of about 30,000 m and a range of 300 km. A few may carry nuclear warheads.

The SA-10 is a new strategic SAM, deployed in 1980 after a protracted period of development. Hailing the SA-10 as a major advance in air defense, the US government claims that it can operate at any altitude. It is a high acceleration missile, with a guidance system capable of distinguishing among multiple targets. US reports say that is was initially designed to attack aircraft and then adapted to attack cruise missiles, although that capability is in question. *Soviet Military Power 1984* suggests that it "may have the potential to intercept some types of US strategic ballistic missiles" and that it could play a supporting part in a future ABM system. The qualifications of this statement taken in conjunction with the reasons the Defense Department has to play up such a threat also make the ABM capability suspect, however. A mobile SA-10 launcher is under development, a departure for strategic SAMs. This is of concern to the United States because bomber penetration tactics call for the use of long-range missiles to destroy fixed air defense bases, thus cutting corridors for the bombers to fly safely through. If the SA-10s are mobile, bomber strategy will be made more difficult, although it is likely that the radars and command centers critical to the operation of strategic air defenses will remain fixed and vulnerable.

It is likely that the SA-10 will be deployed in large numbers, replacing the older SA-2, SA-3, and SA-5 strategic SAMs.

## Tactical Surface-to-Air Missiles

Unlike strategic SAMs, which protect a nation's cities and people, tactical SAMs protect ground troops on the field of battle or naval forces at sea. Because ground forces are mobile, tactical SAMs must be mobile as well, and improved mobility is one of the main improvements in tactical SAMs over the years. Also, because ground battles may take place near enemy air bases, the radars of tactical SAMs must deal with larger numbers of small planes. Moreover, since defense against aircraft carrying nuclear arms is so difficult, tactical air defense concentrates on conventionally armed aircraft, which must come close to their targets to attack. Thus, short-range weapons are stressed in tactical air defense. Soviet tactical SAMs do not carry nuclear warheads.

NATO has maintained powerful and constantly improving tactical air forces since 1945, and the Soviets have steadily improved their tactical air defense. Where the balance stands is not clear. The 1982 Israeli invasion of Lebanon pitted US countermeasures against Soviet air defense technology, with a victory for US technology. Western commentators often portray the increasing mobility of Soviet air defense as part of an offensive strategy. Mobile air defense could defend rapidly advancing Soviet troops from the NATO air strikes designed to stop them. But as the Soviets learned so painfully in the early stages of World War II, mobility can be critical to a successful defense. In fact, it improves both offensive capability and defensive capability.

The first Soviet tactical SAM was designed to attack high-altitude aircraft; later systems have concentrated on low altitudes. The SA-4 Ganef, deployed in 1967, has a maximum altitude of 25,000 m and a range of 70 km. It is carried in pairs on tracked launching vehicles. The SA-X-12, a planned replacement, has raised concern in the Pentagon because of its supposed capability against tactical ballistic missiles and SLBMs.

In 1968 the USSR introduced the SA-7 Grail, at the other end of the spectrum from the SA-4 in size and range. With a maximum altitude of about 1,500 m and a range of 3 km, it is carried by soldiers and fired from the shoulder. It has proved effective against helicopters and has been used many times in warfare in third world nations. The SA-14, its replacement, is expected to enter service soon, but the improvements it incorporates are not known.

Also introduced in 1968 was the SA-9 Gaskin,

with a range of 8 km and a maximum altitude of 4,000 m, mounted on a light, jeep-like vehicle. The SA-13, which entered service in the late 1970s, is an SA-9 follow-on, with improvements that include a better guidance system.

The SA-6 Gainful, which entered service in the early 1970s, was very successful in the 1973 Arab-Israeli war. With a range of 30 km and a maximum altitude around 15,000 m, the SA-6 is carried on a tracked vehicle with three missiles each. It uses semiactive radar homing. A recent follow-on, the SA-11, has unknown improvements.

The SA-8 Gecko was introduced around 1975. With half the range and minimum altitude of the SA-6, it represents a further step toward shorter-range, lower-altitude systems. Its main innovation was mobility. Previous systems were mobile, but they needed auxiliary vehicles such as radar vans to accompany them. Before they could fire, the vehicles would have to stop and string cables among the vans. The SA-8 has all the equipment it needs to fire, including radars, on one vehicle. Although it cannot actually fire while moving, the time needed to prepare for firing is much reduced. Four missiles were carried on each vehicle in the original version. A more recent version has six per vehicle.

## Naval Surface-to-Air Missiles

Recognizing the potent threat of NATO airpower, the Soviets have equipped most of their naval ships with SAMs. Every naval SAM except the SA-N-8 is believed to have been derived from one of the land-based SAMs already discussed and to have similar range and altitude characteristics. No naval SAM is nuclear capable, although William Arkin claims that the SA-N-6 will soon be nuclear capable.

In 1961 the SA-N-1 Goa, based on the SA-3, was introduced. It is deployed on several classes of cruisers and destroyers, with two launch tubes on smaller ships and four on the larger ships.

The cruiser *Dzerzhinski* was converted to the SA-N-2, based on the SA-2, in 1962. But since only one ship was converted, this system was probably a failure.

The SA-N-3 Goblet entered service around 1967, with four tubes per ship on the Moskva- and Kiev-class helicopter carriers and on some cruisers. The SA-N-3 fills the same role as the SA-N-1, which it succeeded. It may be based on the land-based SA-6.

Around 1970 the SA-N-4 became operational. Based on the SA-8, it has been deployed on virtually all new Soviet combatants of appreciable size, providing short-range defense.

The SA-N-5 appeared in the mid-1970s. It is based on the SA-7, but where the latter is hand carried by ground troops, the SA-N-5 is swivel mounted on patrol boats and other small vessels.

The SA-N-6, based on the SA-10, began deployment around 1980 on Kirov- and Slava-class ships. It is probably the most capable Soviet naval SAM.

The SA-N-7, based on the SA-11, entered service in 1981. It should be a replacement for the SA-N-3.

The SA-N-8, which has appeared on the *Udaloy*, is not yet associated with any land-based predecessor. It has been suggested that the SA-N-8 is a replacement for the SA-N-4.

## Air-to-Air Missiles

The Soviet Union has produced air-to-air missiles (AAMs) in three range groupings: short (about 6 km), medium (16–25 km), and long (30–37 km). Most AAMs exist in both an infrared-homing (IR-homing) and a semiactive radar-homing (SAR-homing) version; often both are carried on one airplane for use in a single engagement. The SAR-homing versions usually have a longer range than the IR homing. These weapons are usually fitted on interceptor aircraft.

The first generation of Soviet AAMs came into service around 1961: the AA-1 Alkali and the AA-2 Atoll in the short-range category; the AA-3 Anab of medium range; and the AA-5 Ash in the long-range category. The AA-2, which has been used in conflicts in the third world, has performed very poorly, with poor guidance being the problem.

Second-generation Soviet AAMs came into service around 1975: the long-range AA-6 Acrid, medium-range AA-7 Apex, and short-range AA-8 Aphid. Presumably the guidance systems of these missiles are more capable, although they have not been tested in combat.

Both the first- and the second-generation AAMs lacked the ability to intercept aircraft flying below them at low altitude; their guidance systems could not separate the target out from the ground clutter. This deficiency is purportedly being corrected in third-generation AAMs, just coming into service on the most advanced interceptors. The AA-9, a long-range weapon now in service, is reported to have this "look-down/shoot-down" capability. The AA-X-10 is a new medium-range weapon, and there are reports of a new short-range weapon as well.

# PART III

## BEST ESTIMATES OF SOVIET NUCLEAR MISSILES

# INTRODUCTION TO THE BEST ESTIMATES

Tables 1 and 2, showing best estimates of selected characteristics of all Soviet nuclear missiles ever made, and table 3, showing the number of each Soviet land-based nuclear missile, year by year since 1955, are more complete than the data published in most other sources: they cover more characteristics for more missiles and give longer time series than virtually any other source. In addition, a special feature of the three tables is that in the next two sections of the book, on the derivation of the estimates, there is for each number series and characteristic of each missile an explanatory paragraph setting out the origin of the estimate. These explanations refer to and analyze large parts of the Database that comprises part IV of the book. Thus, instead of a single value for a characteristic or number, readers have available the reasoning that led to our estimate, as well as the figures published in other major sources.

## Weapons Covered

The Database includes data on *all* Soviet missiles, but the estimates in this section cover only *nuclear-armed* missiles. They cover all nuclear missiles ever deployed operationally or expected to be deployed by the USSR. They exclude missiles tested but never deployed (the SS-9 Mod 3), missiles that exist but are not operational (the SS-N-17), and missiles that never actually existed (the SS-N-10). Some missiles whose nuclear capability is uncertain are included (the SS-N-9).

We include in table 1 expected future systems, such as the SS-24 and SS-25. *All Soviet air-to-air missiles and antitank missiles, most surface-to-air missiles, and various other missiles are omitted because they are not nuclear armed.*

## Introduction to Estimates of Characteristics

Tables 1 and 2 present data on characteristics that give a sense of the varying military capabilities of the missiles. Parentheses indicate estimates with a greater degree of uncertainty than most.

The columns under *Service, Entered* and *Retired*, give the years when the first units of the weapon entered active service and when the last units were retired. We are conservative in judging the year when the weapon entered service: the completion of testing or the digging of silos is not enough; operational missiles must be deployed and ready to fire.

The column *Warhead Number* indicates how many warheads are carried on each missile. If more than one, *Warhead Type* indicates whether they can be aimed independently at separate targets (MIRVs) or are fired at a single target (MRVs). *Yield (kT)* gives the yield of *each* warhead on the missile in kilotons. Megaton yields appear as thousands of kilotons.

The column *CEP (m)* refers to accuracy (circular error probable). A warhead fired at a target is expected to fall this close or closer half the time. *Maximum Range (km)* is the farthest the missile can travel and still have a reasonable chance of destroying the target. *Maximum Speed (Mach)* gives the maximum speed of the weapon in Mach numbers (multiples of the speed of sound, which varies with altitude and other conditions and is typically 714 mph at sea level).

The *Propulsion* column gives the principal form of propulsion, using the following abbreviations:

| | |
|---|---|
| S | solid-fueled rocket |
| SL | storable liquid-fueled rocket |
| UL | unstorable liquid-fueled rocket |
| TJ | turbojet |
| RJ | ramjet |

The five columns under *Launch Platform* have information about the launch platform for the missile. The first, *Number*, indicates the number of the platform deployed (for ships and submarines). *Type* gives the type of platform. Unusual entries here are "fixed" for a fixed site on land; "track" for a tracked vehicle such as a tank; "wheel" for a wheeled vehicle; and "6-wh" for a 6-wheeled vehicle. *Name or Class* specifies the name of the launching ship, submarine, aircraft, or vehicle. Under *Launchers* is the number of this kind of missile carried by each launch platform (exclusive of reloads). In the *Reloads* column, "yes" or "no" indicates whether reloading is possible. Occasional numbers here are the number of reloads, when reloading is possible. "Some" indicates that only some of the missiles are capable of reloading.

The final columns of the table show *Peak Number Deployed*, the largest number of this kind of missile ever deployed at one time, and *Comments*, which has other relevant information about the missile.

## Introduction to Estimates of Numbers

All Soviet land-based nuclear missiles ever made are shown in table 3. The entries under each year are estimates of the numbers of weapons on launchers operationally deployed *as of 1 January* that year. Empty cells indicate our estimate that none were in service on that date. To preclude uncertainty, a zero is shown in the year preceding the first year when the missile was in service on 1 January; and a zero is shown in the year following the last year of service.

## Making Military Estimates with Public Data

In making each estimate in the tables, the author worked from a printed copy of the Database. He considered arguments for different values, hypotheses about the differences among sources, and so forth. This reasoning about the data in the Database, set out in the next section, is the actual origin of the estimates, not a *post hoc* reconstruction.

Because of its origin and purpose, the section explaining the best estimates uses terse, informal prose. We expect that readers not interested in detail will skip the section, and that those who use it will understand the style.

Each explanation in this section is designed to be read while referring to the relevant part of the Database. To be concise, we use the source acronyms used in the Database. They may seem strange at first, particularly abbreviations such as Meyr 84 (Meyer), Rock 78 (*The Rocket*), Nitz 79 (Nitze) or vanD 83 (van Diepen). But IDDS staff members have found that the acronyms soon become familiar references that are easy to read and interpret. When range and CEP are discussed, all comparisons are made in metric values. If a source gives an English-system value, we give first that value and then in parentheses the metric conversion provided in the Database.

There are no set procedures for making estimates of military capabilities based on open literature; if there were, the Database would not be needed. There are, however, a few general guidelines worth mentioning.

Our estimates keep close to the Database. No value lies outside the range reflected in our sources.

The most important factors in choosing the estimate in one source over another—apart from plausibility of the figure—are the degree of impartiality of the sources and their closeness to the original source data. Most information on Soviet missiles derives ultimately from the intelligence activities of Western nations, especially the United States. US satellite coverage and electronic listening posts are the original source of most raw data. These data, processed by the US Central Intelligence Agency (CIA), National Security Agency (NSA), and Defense Intelligence Agency (DIA), find their way into the public realm in several ways. Some are published in the annual reports of the Joint Chiefs of Staff (JCS) and the secretary of defense (SecD). Other information is provided to Congress by the secretary of defense in classified form (SecC), but with the passage of time, it is declassified and released to the public under the Freedom of Information Act. Government officials also testify before Congress, and their testimony in hearings is available to the public. And the large number of officials who have access to and use the information often end up revealing it, either inadvertently or intentionally in deliberate leaks.

Once in the public domain, information will be reproduced and spread by military and trade journal magazines, books, and reference works. Other scholars may pick up their information and analyze and republish it. There is room in this process for individuals to expand on published tidbits, with both informed and uninformed analysis and speculation. Frequently people simply guess. Throughout the entire process, there is a marked tendency for people not to quote specific sources. This is understandable when the source is a leak, but this understandable exception seems to have provided many authors an excuse for never citing any sources.

Given this situation, published government sources are by far the most reliable ones and the ones we prefer in cases of conflict over information: specifically the sources SecC, SecD, JCS, and SMP. AWar 80 is an unofficial source, but data released later in official sources have revealed that it contains a high-quality leak of intelligence information. Other sources vary considerably in quality, but we are reluctant to share our general impressions without having made a more systematic study of the matter. It will often be apparent to readers of our explanations how we regard different sources. Also, sources that deal with a wide variety of missiles often vary considerably in reliability depending on what they are discussing.

The other principles used in deriving our estimates are not based on any special knowledge. For instance, if 5 sources give a range as 300 km, another 5 give it as 700 km, and 1 source explains that the range is 300 km at low altitude and 700 km at high altitude, we take advantage of the sense this source has made out of what appeared to be a conflict and judge accordingly. If little information is available about a new missile, but it is known to be an advanced version of an older one, we will evaluate the little information there is relative to information on the older missile. When such criteria as these are not applicable, we fall back on simpler measures such as reporting the view of the majority of sources. Or if sources' values cover a range, we pick one in the middle of the range.

We provide quantitative estimates with a reasonable degree of precision. We rarely provide more than two significant digits for continuous data (range, CEP, yield). Sometimes even less precision is appropriate. If sources vary between 1 kiloton and 1 megaton for the yield of a weapon, for instance, we are more likely to pick an order-of-magnitude value such as 10 kt than a precise value such as 25 kt. This problem arises in a different way when converting from a round value in the English system to a metric value. (All our best estimates are metric.) We choose a metric conversion that implies roughly the same degree of precision as the original estimate; for example, to avoid giving an unwarranted impression of precision, a source estimate of 350 nautical miles is translated as a best estimate of 650 kilometers, although a precise conversion would be 648.20 kilometers.

## Table 1
### Best Estimates of the Characteristics of Soviet Nuclear Ballistic Missiles

|  | Service | | Warhead | | Yield (kT) | CEP (m) | Maximum Range (km) | Maximum Speed (Mach) |
|---|---|---|---|---|---|---|---|---|
|  | Entered | Retired | Number | Type | | | | |
| **ICBMs** | | | | | | | | |
| SS-6 Sapwood | 1961 | 1968 | 1 | | 5,000 | 3,700 | 8,000 | |
| SS-7 Saddler Mods 1, 2 | 1962 | 1978 | 1 | | 3,000 | 2,800 | 11,000 | |
| SS-7 Saddler Mod 3 | 1963 | 1979 | 1 | | 6,000 | 2,800 | 11,000 | |
| SS-8 Sasin | 1964 | 1977 | 1 | | 3,000 | 1,900 | 11,000 | |
| SS-9 Scarp Mods 1, 2 | 1966 | 1979 | 1 | | 18,000 | 900 | 12,000 | |
| SS-11 Sego Mod 1 | 1966 | 1979 | 1 | | 950 | 1,400 | 11,000 | |
| SS-11 Sego Mod 2 | (1974) | | 1 | | 1,100 | 1,100 | 13,000 | |
| SS-11 Sego Mod 3 | 1973 | | 3 | MRV | 350 | 1,100 | 10,600 | |
| SS-13 Savage Mod 1 | 1969 | (1975) | 1 | | 600 | 1,900 | 9,400 | |
| SS-13 Savage Mod 2 | 1973 | | 1 | | 600 | 1,500 | 9.400 | |
| SS-17 Mod 1 | 1975 | (1982) | 4 | MIRV | 750 | 440 | 10,000 | |
| SS-17 Mod 2 | 1978 | | 1 | | 3,600 | 430 | 11,000 | |
| SS-17 Mod 3 | 1981 | | 4 | MIRV | 750 | 370 | 10,000 | |
| SS-18 Mod 1 | 1974 | (1981) | 1 | | 24,000 | 430 | 12,000 | |
| SS-18 Mod 2 | 1977 | (1983) | (8) | MIRV | 900 | 430 | 11,000 | |
| SS-18 Mod 3 | 1977 | (1983) | 1 | | 20,000 | 350 | 16,000 | |
| SS-18 Mod 4 | 1979 | | 10 | MIRV | 500 | 260 | 11,000 | |
| SS-19 Mod 1 | 1975 | (1983) | 6 | MIRV | 550 | 350 | 9,600 | |
| SS-19 Mod 2 | (1978) | | 1 | | 4,300 | 300 | 10,100 | |
| SS-19 Mod 3 | (1981) | | 6 | MIRV | 550 | 250 | 10,000 | |
| SS-24 | 1986 | | 10 | MIRV | | (250) | 10,000 | |
| SS-25 | 1985 | | 1 | | | | 10,500 | |
| **SLBMs** | | | | | | | | |
| SS-N-4 Sark | (1958) | (1979) | 1 | | 3,000 | 4,000 | 650 | |
| SS-N-5 Serb | 1963 | | 1 | | 1,000 | 2,800 | 1,300 | |
| SS-N-6 Mod 1 | 1968 | | 1 | | 700 | 1,900 | 2,400 | |
| SS-N-6 Mod 2 | 1973 | | 1 | | 650 | 1,900 | 3,000 | |
| SS-N-6 Mod 3 | 1974 | | 2 | MRV | 350 | 1,900 | 3,000 | |
| SS-N-8 Mod 1 | 1973 | | 1 | | 800 | 1,600 | 7,800 | |
| SS-N-8 Mod 2 | 1977 | | 1 | | 800 | 1,600 | 9,100 | |
| SS-N-18 Mod 1 | 1978 | | 3 | MIRV | 200 | 1,400 | 6,500 | |
| SS-N-18 Mod 2 | 1978 | | 1 | | 450 | 1,400 | 8,000 | |
| SS-N-18 Mod 3 | (1978) | | 7 | MIRV | (100) | (1,400) | 6,500 | |
| SS-N-20 | 1983 | | 6–8 | MIRV | (100) | (1,000) | 8,300 | |
| SS-NX-23 | 1987 | | | MIRV | (100) | (1,000) | 9,300 | |
| **IRBMs** | | | | | | | | |
| SS-3 Shyster | 1955 | (1967) | 1 | | (300) | (3,700) | (1,200) | |
| SS-4 Sandal | 1959 | | 1 | | 1,000 | 2,300 | 2,000 | |
| SS-5 Skean | 1961 | 1984 | 1 | | 1,000 | 1,900 | 4,100 | |
| SS-20 | 1977 | | 3 | MIRV | 150 | 430 | 5,000 | |
| **SRBMs** | | | | | | | | |
| FROG 1 | 1957 | (1970) | 1 | | 25 | 800 | 32 | |
| FROG 2 | 1957 | (1970) | 1 | | (10) | 800 | 25 | |
| FROG 3 | 1960 | | 1 | | (20) | 650 | 40 | |
| FROG 4 | (1959) | (1970) | 1 | | (20) | 650 | 50 | |
| FROG 5 | (1964) | | 1 | | (20) | 550 | 50 | |
| FROG 7 | 1965 | | 1 | | (200) | (400) | 70 | |
| SS-21 | (1983) | | 1 | | (100) | 300 | 120 | |
| SS-1b Scud A | 1957 | (1973) | | | (40) | 930 | 150 | |
| SS-1c Scud B | 1965 | | | | (10) | 930 | 300 | |
| SS-23 | (1986) | | | | (100) | (370) | 500 | |
| SS-12 Scaleboard | 1969 | | | | (500) | 740 | 900 | |
| SS-22 | (1984) | | | | 500 | 370 | 900 | |

| | | Launch Platform | | | | Peak | | |
|---|---|---|---|---|---|---|---|---|
| Propulsion | Number | Type | Name or Class | Launchers | Reloads | Number Deployed | Comments | |
| UL | | fixed | | | yes | 4 | | **ICBMs**<br>SS-6 Sapwood |
| SL | | fixed | | | some | 197 { | | SS-7 Saddler Mods 1, 2 |
| SL | | fixed | | | some | | | SS-7 Saddler Mod 3 |
| UL | | fixed | | | some | 23 | | SS-8 Sasin |
| SL | | fixed | | | no | 288 | | SS-9 Scarp Mods 1, 2 |
| SL | | fixed | | | no | 970 | | SS-11 Sego Mod 1 |
| SL | | fixed | | | no | (360) | | SS-11 Sego Mod 2 |
| SL | | fixed | | | no | (60) | | SS-11 Sego Mod 3 |
| S | | fixed | | | no | 60 | | SS-13 Savage Mod 1 |
| S | | fixed | | | no | 60 | | SS-13 Savage Mod 2 |
| SL | | fixed | | | | (130) | | SS-17 Mod 1 |
| SL | | fixed | | | | (20) | | SS-17 Mod 2 |
| SL | | fixed | | | | (130) | | SS-17 Mod 3 |
| SL | | fixed | | | | (56) | | SS-18 Mod 1 |
| SL | | fixed | | | | (132) | | SS-18 Mod 2 |
| SL | | fixed | | | | (24) | | SS-18 Mod 3 |
| SL | | fixed | | | | (308) | | SS-18 Mod 4 |
| SL | | fixed | | | | (210) | | SS-19 Mod 1 |
| SL | | fixed | | | | (60) | | SS-19 Mod 2 |
| SL | | fixed | | | | (360) | | SS-19 Mod 3 |
| S | | fixed | | | | | | SS-24 |
| S | | wheel | | 1 | yes | | | SS-25 |
| SL | 7 | SSB | Zulu V | 2 | no | | Subs surface to fire. | **SLBMs**<br>SS-N-4 Sark |
| | | SSB | Golf I | 3 | no | | | |
| | | SSBN | Hotel I | 3 | no | | | |
| SL | | SSB | Golf II | 3 | no | | Subs fire submerged. | SS-N-5 Serb |
| | | SSBN | Hotel II | 3 | no | | | |
| SL | 1 | SSB | Golf IV | 4 | no | | | { SS-N-6 Mod 1 |
| SL | 33 | SSB | Yankee I | 16 | no | | | SS-N-6 Mod 2 |
| SL | | | | | | | | SS-N-6 Mod 3 |
| SL | 1 | SSB | Golf III | 6 | no | | | { SS-N-8 Mod 1 |
| | 1 | SSBN | Hotel III | 6 | no | | | |
| | 18 | SSBN | Delta I | 12 | no | | | |
| SL | 4 | SSBN | Delta II | 16 | no | | | SS-N-8 Mod 2 |
| SL | | | | | | | | { SS-N-18 Mod 1 |
| SL | 14 | SSBN | Delta III | 16 | no | | | SS-N-18 Mod 2 |
| SL | | | | | | | | SS-N-18 Mod 3 |
| S | 3 | SSBN | Typhoon | 20 | no | | | SS-N-20 |
| SL | 2 | SSBN | Delta IV | 16 | no | | May replace SS-N-18. | SS-NX-23 |
| UL | | | | | yes | (48) | | **IRBMs**<br>SS-3 Shyster |
| SL | | fixed | | | some | (600) | | SS-4 Sandal |
| SL | | fixed | | | some | (97) | | SS-5 Skean |
| S | | wheel | | | yes | (400) | | SS-20 |
| S | | track | JS-3 | 1 | yes | | | **SRBMs**<br>FROG 1 |
| S | | track | PT-76 | 1 | yes | | | FROG 2 |
| S | | track | PT-76 | 1 | yes | | | FROG 3 |
| S | | track | PT-76 | 1 | yes | | | FROG 4 |
| S | | track | PT-76 | 1 | yes | | | FROG 5 |
| S | | wheel | ZIL-135 | 1 | yes | (480) | | FROG 7 |
| S | | 6-wh | TEL | 1 | yes | (130) | | SS-21 |
| SL | | track | JS-3 | 1 | yes | (100) | | SS-1b Scud A |
| SL | | wheel | MAZ-543 | 1 | yes | (575) | | SS-1c Scud B |
| S | | wheel | | 1 | yes | | | SS-23 |
| SL | | wheel | MAZ-543 | 1 | yes | (120) | | SS-12 Scaleboard |
| S | | wheel | (MAZ-543) | 1 | yes | (20) | | SS-22 |

## Table 2
### Best Estimates of the Characteristics of Soviet Nuclear Cruise Missiles

|  | Service Entered | Service Retired | Warhead Number | Warhead Type | Yield (kT) | CEP (m) | Maximum Range (km) | Maximum Speed (Mach) |
|---|---|---|---|---|---|---|---|---|
| **Strategic Cruise Missiles** | | | | | | | | |
| SSC-4 | (1985) | | 1 | | | | 3,000 | |
| SS-NX-21 | (1986) | | 1 | | (200) | | 3,000 | (0.7) |
| AS-3 Kangaroo | 1960 | | 1 | | 2,000 | (1,000) | 650 | 0.8 |
| AS-15 | 1984 | | 1 | | | | 3,000 | |
| **Naval Cruise Missiles** | | | | | | | | |
| *Sea-launched antiship* | | | | | | | | |
| SS-N-1 Scrubber | 1959 | | 1 | | (10) | | 85 | 0.9 |
| SS-N-3a Shaddock | 1958 | | 1 | | 350 | | 410 | 1.0 |
| SS-N-3b Shaddock | 1962 | | 1 | | 350 | | 280 | 1.0 |
| SS-N-7 | 1968 | | 1 | | 200 | | 50 | 0.9 |
| SS-N-9 | 1969 | | 1 | | 200 | | 110 | 1.4 |
| SS-N-12 Sandbox | 1973 | | 1 | | 350 | | 550 | 2.5 |
| SS-N-19 | 1980 | | 1 | | (350) | | 550 | 2.5 |
| SS-N-22 | 1982 | | 1 | | (200) | | 110 | 2.5 |
| SSC-1b Sepal | 1962 | | 1 | | (100) | (650) | 450 | 1.0 |
| *Air-launched antiship* | | | | | | | | |
| AS-2 Kipper | 1961 | | 1 | | (400) | (1,000) | 100 | 1.2 |
| AS-4 Kitchen | (1967) | | 1 | | 350 | (1,000) | 300 | (2.5) |
| AS-5 Kelt | 1966 | | 1 | | (500) | | 160 | 0.9 |
| AS-6 Kingfish | (1977) | | 1 | | (200) | (350) | 280 | (1.2) |
| *Sea-launched antisub* | | | | | | | | |
| SS-N-14 Silex | 1968 | | 1 | | (10) | | 55 | 0.9 |
| SS-N-15 | 1972 | | 1 | | (10) | | 45 | |
| **ABMs** | | | | | | | | |
| ABM-1 Galosh | 1968 | | 1 | | (3,000) | | (300) | |
| **SAMs** | | | | | | | | |
| *Strategic Surface-to-Air* | | | | | | | | |
| SA-2 Guideline | 1958 | | 1 | | | | (50) | 3.5 |
| SA-5 Gammon | 1967 | | 1 | | | | 300 | 3.5 |

Note: Most numerical entries in the table are estimates with some degree of uncertainty. Parentheses indicate a greater degree of uncertainty. For an explanation of the sources and reliability of each estimate, see the following section.

| Propulsion | No. | Type | Launch Platform Name or Class | Launchers | Reloads | Peak Number Deployed | Comments | |
|---|---|---|---|---|---|---|---|---|
| | | | | | | | | **Strategic Cruise Missiles** |
| TJ | | | several | | | | | SSC-4 |
| TJ | | sub | | | yes | | Victor III, Mike, Sierra, Akula, and/or Yankee | |
| TJ | | bomber | Bear B, C | 1 | no | | | AS-3 Kangaroo |
| TJ | | bomber | Bear H | 4 | | | Future carrier: Blackjack. | AS-15 |
| | | | | | | | | **Naval Cruise Missiles** |
| | | | | | | | | *Sea-launched antiship* |
| TJ | (8) | ship | Krupnyy | 2 | yes | | | SS-N-1 Scrubber |
| | 4 | ship | Kildin | 1 | yes | | | |
| TJ | 5 | sub | W Twin Cyl | 2 | no | | 50 km range from lone subs. | SS-N-3a Shaddock |
| | 7 | sub | W Long Bin | 4 | no | | | |
| | 16 | sub | Juliett | 4 | no | | | |
| | 5 | sub | Echo I | 6 | no | | | |
| | 27 | sub | Echo II | 8 | no | | | |
| TJ | 4 | sub | Kynda | 8 | 8 | | | SS-N-3b Shaddock |
| | 4 | ship | Kresta I | 4 | no | | | |
| S | 12 | Charlie I | | 8 | no | | | SS-N-7 |
| S | | ship | Nanuchka 1, 3 | 6 | no | | 50 km range from lone subs. | SS-N-9 |
| | | ship | Sarancha | (4) | no | | | |
| | | sub | Charlie II | 8 | no | | | |
| | | sub | Papa | (10) | no | | | |
| TJ | | sub | Echo II | 8 | no | | | SS-N-12 Sandbox |
| | 1 | ship | Kiev | 8 | yes | | | |
| | | ship | Slava | | yes | | Total of 16 missiles carried. | |
| TJ | | sub | Oscar | 24 | no | | | SS-N-19 |
| | | ship | Kirov | 20 | | | | |
| S | | ship | Sovremennyy | | | | Improved SS-N-9; 8 carried. | SS-N-22 |
| | | ship | Tarantul II | (4) | | | | |
| TJ | | 8-wh | | 1 | yes | (100) | Coastal defense. | SSC-1b Sepal |
| | | | | | | | | *Air-launched antiship* |
| TJ | | bomber | Badger C, G | 1 | no | | 210 km at high altitude. | AS-2 Kipper |
| SL | | bomber | Blinder B | 1 | no | | 720 km at high altitude. | AS-4 Kitchen |
| | | bomber | Backfire B | 1–2 | no | | | |
| | | bomber | Bear G | 2 | no | | | |
| SL | | bomber | Badger G, (C) | 2 | no | | 320 km, Mach 1.2 at high altitude. | AS-5 Kelt |
| (SL) | | bomber | Badger C, G | 2 | no | | 700 km, Mach 3.0 at high altitude. | AS-6 Kingfish |
| | | | | | | | | *Sea-launched antisub* |
| S | | ship | Kara | 8 | no | | | SS-N-14 Silex |
| | | ship | Krivak I, II | 4 | no | | | |
| | | ship | Kresta II | 8 | no | | | |
| | 1 | ship | Kirov | 2 | yes | | Only on first of class. | |
| | | ship | Udaloy | 8 | no | | | |
| S | 38 | sub | Victor 1, 2, 3 | | yes | | | SS-N-15 |
| | | sub | Alfa, Mike | | yes | | | |
| | | | | | | | | **ABMs** |
| (S) | | fixed | | | | 64 | Reloading takes 30 min. | ABM-1 Galosh |
| | | | | | | | | **SAMs** |
| | | | | | | | | *Strategic Surface-to-Air* |
| SL | | fixed | | | | | | SA-2 Guideline |
| S | | fixed | | | | | | SS-5 Gammon |

**Table 3**
**Best Estimates of the Numbers of Soviet Land-Based Nuclear Missiles, 1955–1985**

| | 1955 | 1956 | 1957 | 1958 | 1959 | 1960 | 1961 | 1962 | 1963 | 1964 | 1965 | 1966 | 1967 | 1968 | 1969 | 1970 |
|---|---|---|---|---|---|---|---|---|---|---|---|---|---|---|---|---|
| **ICBMs** | | | | | | | | | | | | | | | | |
| SS-6 Sapwood | | | | | | | 0 | 4 | 4 | 4 | 4 | 4 | 4 | 2 | 0 | |
| SS-7 Saddler, soft | | | | | | | | 0 | 78 | 128 | 128 | 128 | 128 | 128 | 128 | 128 |
| SS-7 Saddler, hard | | | | | | | | | 0 | 12 | 57 | 69 | 69 | 69 | 69 | 69 |
| SS-8 Sasin, soft | | | | | | | | | | 0 | 14 | 14 | 14 | 14 | 14 | 14 |
| SS-8 Sasin, hard | | | | | | | | | | 0 | 6 | 9 | 9 | 9 | 9 | 9 |
| SS-9 Scarp Mods 1,2 | | | | | | | | | | | 0 | 54 | 120 | 156 | 198 | |
| SS-11 Sego Mod 1 | | | | | | | | | | | | 0 | 160 | 390 | 570 | 750 |
| SS-11 Sego Mod 2 | | | | | | | | | | | | | | | | |
| SS-11 Sego Mod 3 | | | | | | | | | | | | | | | | |
| SS-13 Savage Mod 1 | | | | | | | | | | | | | | | 0 | 20 |
| SS-13 Savage Mod 2 | | | | | | | | | | | | | | | | |
| SS-17 Mod 1 | | | | | | | | | | | | | | | | |
| SS-17 Mod 2 | | | | | | | | | | | | | | | | |
| SS-17 Mod 3 | | | | | | | | | | | | | | | | |
| SS-18 Mod 1 | | | | | | | | | | | | | | | | |
| SS-18 Mod 2 | | | | | | | | | | | | | | | | |
| SS-18 Mod 3 | | | | | | | | | | | | | | | | |
| SS-18 Mod 4 | | | | | | | | | | | | | | | | |
| SS-19 Mod 1 | | | | | | | | | | | | | | | | |
| SS-19 Mod 2 | | | | | | | | | | | | | | | | |
| SS-19 Mod 3 | | | | | | | | | | | | | | | | |
| **IRBMs** | | | | | | | | | | | | | | | | |
| SS-3 Shyster | 0 | 24 | 48 | 48 | 48 | 48 | 48 | 48 | 48 | 38 | 28 | 18 | 8 | 0 | | |
| SS-4 Sandal | | | | | 0 | 150 | 300 | 450 | 600 | 600 | 600 | 600 | 600 | 600 | 570 | 540 |
| SS-5 Skean | | | | | | | 0 | 25 | 50 | 75 | 97 | 97 | 97 | 97 | 97 | 93 |
| SS-20 | | | | | | | | | | | | | | | | |
| **SRBMs** | | | | | | | | | | | | | | | | |
| FROGs 1–5 | | | | 0 | 20 | 40 | 80 | 120 | 160 | 200 | 240 | 280 | 310 | 310 | 310 | 310 |
| FROG 7 | | | | | | | | | | | | 0 | 10 | 50 | 90 | 130 | 170 |
| SS-21 | | | | | | | | | | | | | | | | |
| SS-1b Scud A | | | | 0 | 20 | 40 | 80 | 100 | 100 | 100 | 100 | 100 | 90 | 80 | 70 | 60 | 50 |
| SS-1c Scud B | | | | | | | | | | | | 0 | 50 | 100 | 150 | 200 | 250 |
| SS-23 | | | | | | | | | | | | | | | | |
| SS-12 Scaleboard | | | | | | | | | | | | | | | 0 | 27 |
| SS-22 | | | | | | | | | | | | | | | | |
| **Antiship Missile** | | | | | | | | | | | | | | | | |
| SS-1b Sepal | | | | | | | 0 | 20 | 40 | 60 | 80 | 100 | 100 | 100 | 100 | |
| **ABM** | | | | | | | | | | | | | | | | |
| ABM-1 Galosh | | | | | | | | | | | | | | 0 | 48 | 64 |

Note: Most numerical entries in the table are estimates with some degree of uncertainty. For an explanation of the sources and reliability of each estimate, see the following section.

| 1971 | 1972 | 1973 | 1974 | 1975 | 1976 | 1977 | 1978 | 1979 | 1980 | 1981 | 1982 | 1983 | 1984 | 1985 | |
|---|---|---|---|---|---|---|---|---|---|---|---|---|---|---|---|
| | | | | | | | | | | | | | | | **ICBMs** |
| 124 | 124 | 124 | 124 | 124 | 100 | 80 | 7 | 0 | | | | | | | SS-6 Sapwood |
| 66 | 66 | 66 | 66 | 66 | 66 | 66 | 48 | 0 | | | | | | | SS-7 Saddler, soft |
| 10 | 10 | 10 | 10 | 10 | 4 | 0 | | | | | | | | | SS-7 Saddler, hard |
| 9 | 9 | 9 | 9 | 9 | 9 | 6 | 0 | | | | | | | | SS-8 Sasin, soft |
| | | | | | | | | | | | | | | | SS-8 Sasin, hard |
| 258 | 288 | 288 | 288 | 278 | 252 | 216 | 120 | 60 | 0 | | | | | | SS-9 Scarp Mods 1, 2 |
| 940 | 970 | 970 | 970 | 880 | 740 | 620 | 510 | 380 | 310 | 270 | 230 | 160 | 130 | 100 | SS-11 Sego Mod 1 |
| | | | 0 | 20 | 50 | 90 | 130 | 170 | 210 | 250 | 290 | 330 | 360 | 360 | SS-11 Sego Mod 2 |
| | | 0 | 30 | 60 | 60 | 60 | 60 | 60 | 60 | 60 | 60 | 60 | 60 | 60 | SS-11 Sego Mod 3 |
| 40 | 60 | 60 | 40 | 20 | 0 | | | | | | | | | | SS-13 Savage Mod 1 |
| | | 0 | 20 | 40 | 60 | 60 | 60 | 60 | 60 | 60 | 60 | 60 | 60 | 60 | SS-13 Savage Mod 2 |
| | | | | | 0 | 20 | 40 | 60 | 80 | 130 | 130 | 65 | 0 | | SS-17 Mod 1 |
| | | | | | | | 0 | 20 | 20 | 20 | 20 | 20 | 20 | 0 | SS-17 Mod 2 |
| | | | | | | | | | | 0 | 65 | 130 | 130 | 150 | SS-17 Mod 3 |
| | | | 0 | 10 | 30 | 56 | 32 | 32 | 32 | 32 | 0 | | | | SS-18 Mod 1 |
| | | | | | | 0 | 36 | 132 | 132 | 132 | 104 | 44 | 0 | | SS-18 Mod 2 |
| | | | | | | 0 | 24 | 24 | 24 | 24 | 24 | 24 | 0 | | SS-18 Mod 3 |
| | | | | | | | | 0 | 60 | 120 | 180 | 240 | 308 | 308 | SS-18 Mod 4 |
| | | | | 0 | 70 | 140 | 200 | 210 | 210 | 210 | 130 | 80 | 0 | | SS-19 Mod 1 |
| | | | | | | | 0 | 30 | 60 | 60 | 60 | 60 | 60 | 0 | SS-19 Mod 2 |
| | | | | | | | | | 0 | 30 | 110 | 190 | 270 | 360 | SS-19 Mod 3 |
| | | | | | | | | | | | | | | | **IRBMs** |
| | | | | | | | | | | | | | | | SS-3 Shyster |
| 508 | 508 | 508 | 508 | 508 | 508 | 508 | 488 | 428 | 388 | 364 | 300 | 232 | 224 | 120 | SS-4 Sandal |
| 90 | 90 | 90 | 90 | 90 | 90 | 90 | 85 | 75 | 60 | 35 | 35 | 16 | 13 | 0 | SS-5 Skean |
| | | | | | | 0 | 9 | 72 | 135 | 180 | 270 | 333 | 378 | 400 | SS-20 |
| | | | | | | | | | | | | | | | **SRBMs** |
| 310 | 310 | 310 | 270 | 230 | 190 | 150 | 140 | 140 | 140 | 140 | 140 | 140 | 140 | 140 | FROGs 1–5 |
| 210 | 250 | 290 | 330 | 370 | 410 | 450 | 480 | 480 | 480 | 480 | 480 | 480 | 480 | 480 | FROG 7 |
| | | | | | | | | | | | 0 | 30 | 80 | 130 | SS-21 |
| 40 | 20 | 10 | 0 | | | | | | | | | | | | SS-1b Scud A |
| 260 | 280 | 340 | 400 | 400 | 450 | 500 | 530 | 530 | 550 | 575 | 575 | 575 | 575 | 575 | SS-1c Scud B |
| | | | | | | | | | | | | | | 0 | SS-23 |
| 54 | 81 | 100 | 120 | 120 | 120 | 120 | 120 | 120 | 120 | 120 | 120 | 120 | 110 | 100 | SS-12 Scaleboard |
| | | | | | | | | | | | | 0 | 10 | 20 | SS-22 |
| | | | | | | | | | | | | | | | **Antiship Missile** |
| 100 | 100 | 100 | 100 | 100 | 100 | 100 | 100 | 100 | 100 | 100 | 100 | 100 | 100 | 100 | SSC-1b Sepal |
| | | | | | | | | | | | | | | | **ABM** |
| 64 | 64 | 64 | 64 | 64 | 64 | 64 | 64 | 64 | 32 | 32 | 32 | 32 | 32 | 32 | ABM-1 Galosh |

# DERIVATION OF BALLISTIC MISSILE ESTIMATES

## INTERCONTINENTAL BALLISTIC MISSILES

### SS-6 Sapwood

▲ **Numbers and Service Dates:** Generally, our numbers estimates do not take account of launchers at test sites; however, we make an exception in principle in cases where a weapon is very new and the possibility of a few operational ones at a test site as opposed to none could be strategically significant. But for the SS-6 such problems may not arise because the range values cited for it are around 8,000 km, and this is not enough to reach CONUS from the test base at Tyuratam (near the Aral Sea). (In fairness, though, all sources giving range values are rather unreliable.) So even if some were operational there, they might not count as ICBMs. Only those based at Plesetsk should be considered operational.

This strict definition of operational helps decide on the initial deployment date as well. In the range between 1957 and 1962, later dates are much more likely. Although this subject is of great interest, there is still controversy, and I have low confidence in any number I might pick in this range. On pp. 117–118 of PR01 is cited a CIA report prepared for the Berlin crisis dated 6 September 1961: "We are confident that neither [the SS-6 nor the SS-7] will be operational as ICBM weapon systems during the coming autumn and winter." This is the exact criterion for service entry we are looking for and, if true, implies a 1962 initial deployment, but other passages in the same work muddy this picture. And Bl01 gives spring 1961 as the first deployment at Plesetsk (pp. 46, 55), giving as sources for p. 55 interviews with Roger Hilsman, Lt. Gen. Daniel Graham, and Herbert Scoville. Bl01's quotation is, "Only four of the SS-6 ICBMs were actually ever deployed, and these would not have been operational before the spring of 1961." This leaves open the possibility of their being operational later, but nonetheless we settle on spring 1961 as the operational date.

As for the number of systems, Bl01's 4 is our best guess, bolstered by a fair amount of evidence. Regarding the other estimates, FIar 76 gives about 10 and Rock 78 gives about 12, and vanD gives 2 but cites as his source Bl01. On p. 46, which he cites, the only number given is 4, though in a footnote on p. 55, Bl01 quotes Scoville as saying there were "only one or two missiles, certainly not more than four" at Plesetsk. The higher numbers given by FIar 76 and Rock 78 are discredited because both sources indicate the weapon was retired in the early 1960s, while SecC 67 says, "We believe the Soviets will begin to phase out the SS-8s and the soft SS-6s and SS-7s" soon. SecC 67 also says, "Only a handful were deployed." 1967 is the last year SecC gives any explicit mention of SS-6. But SecC 68 reports the number of soft ICBMs (in general) declining from 146 to 144, and SecC 69 reports it declining from 144 to 142. SecC 70 indicates the same 142 soft launchers and gives their breakdown as 128 SS-7 and 14 SS-8, allowing no room for SS-6s. SMP 83 backs up the conclusion that all SS-6s were gone by 1970 by saying that the first generation of ICBMs was phased out in the 1960s. If SecC 67 is read as indicating the real beginning of soft ICBM phaseout (and not the beginning of some second phase of it), there must have been no more than 4 SS-6s, since only 4 soft ICBM launchers were phased out. There could have been fewer than 4, with the successive pairs of missiles phased out being a mix of SS-6s, SS-7s and SS-8s, but this seems less parsimonious than assuming 4 SS-6s, especially since soft ICBM numbers were steady for a year after the dismantling in question. It is also consistent with other data (though we should consider the possibility that other sources reached the same conclusion by going through the same exercise with this SecC data).

The resulting pattern is very clear. In beginning-of-year data, we have 0 for 1961, 4 for 1962 through 1967, 2 for 1968, and 0 for 1969. The 4 for 1962 is uncertain.

▲ **Yield:** 5 MT is given by SecC in 65, and there is little controversy.

▲ **CEP:** 3.7 km is converted from 2 nm of SecC 65. SSF and vanD follow this, two others estimating 8 km and 16 km.

▲ **Range:** 4 unofficial sources give 5,000 mi or 8,000 km as the range. A few give slightly higher ranges (8,800 km or 8,500 km). HRST gives 4,800 km and SSF gives 3,200 km. These lower figures are hard to believe because they imply that even when based as far north as Plesetsk, SS-6 could not reach CONUS. Even the 8,000 km range is proof that SS-6s at Tyuratam could not have been operational since they would lack the range to get to the United States.

▲ **Propulsion:** SecC 65–67 has nonstorable liquid. No other source disagrees.

▲ **Silo Hardness:** Bl01 gives the hardness of the first soft Atlas Ds "emplaced vertically on exposed launching pads" as 2 psi; the SS-6, similarly configured, would probably have a similar hardness. Everyone agrees it was very soft.

▲ **Reloads:** Our estimate is based on common sense: above-ground ICBM pads can be reloaded. No source says anything on this matter.

## SS-7 Saddler

▲ **Mods of the SS-7:** The official basis of SS-7 mods is from JCS 75, which identifies the 3 mods, tells us that Mods 1 and 2 entered service in 1962 and Mod 3 in 1963, and that all have single warheads. SecC 65 states that SS-7s entering the force "this year" may carry a 6 MT warhead, some others may be retrofitted with it. 1963 is two years earlier than 1965, but unofficial sources seem to have identified the 6 MT warhead with Mod 3. In the characteristics chart, we have recognized Mods 1 and 2 as one entry and Mod 3 as another, yield being the only difference.

For numbers, we can get a breakdown by hard and soft launchers. I have seen no evidence linking these launcher differences to the mods, though perhaps the otherwise undistinguished Mods 1 and 2 correspond to launcher types. The best we can do is give a breakdown by hard and soft launchers and live with the fact that this breakdown does not correspond to the characteristics breakdown.

▲ **Numbers and Service Dates:** ICBM numbers is the best documented part of vanD's appendix, and the methodology looks good. The author is working from many sources I do not have available at the time of this writing, and where comparable, his numbers are consistent with the best of my sources. Thus I follow his numbers, with one exception: I believe there were 4 SS-6s instead of 2, and as a result I believe the soft SS-7s did not decline from 130 to 128 in 1969 but were never above 128 before that.

It is also worth noting that there are no estimates grossly at variance with these; SS-7 and SS-8 numbers are relatively uncontroversial. The disagreements are about the exact buildup and the exact phase-out patterns, and since both were quite rapid for these missiles, they do not have too much significance.

▲ **Yield:** SecC 65 mentions the 6 MT version that may enter service; SecC 66 talks of 3 MT, and SecC 70–71 talk of 3–6 MT. I follow SSF in inferring that 3 MT is for Mods 1 and 2 and 6 MT for Mod 3. The other dominant estimate is the vague 5 MT.

▲ **CEP:** SecC 65–66 has 1–2 nm, and SecC 67 has 1–1.5 nm. It is hard to know whether there was an accuracy improvement or a retroactive estimate change. Also note that 2 nm is what SecC gives for SS-6. 1.5 nm is a safe guess by being midpoint of the first range, top of the second and below SS-6. SSF makes the same estimate but gives 1 nm for Mod 3. That is reasonable, but I think the evidence does not warrant it on balance.

▲ **Range:** Unofficial sources have estimates ranging from 10,500–11,000 km, and Clns with 12,000 km. 11,000 is in the middle and rounder than 10,500. If SSF wants to give Mod 3 a better CEP due to greater RV weight associated with higher yield, the max range should also go down.

▲ **Propulsion:** Our estimate follows SecC. There is no controversy among sources.

▲ **Silo Hardness:** SecC has them deployed at both hard and soft sites, with about two-thirds at soft sites. The hard ones are described as triple silo and are contrasted to the SS-9, in a new type of single silo launch site. Pr01 also has the hard SS-7s in the coffin-type arrangement. This presumably leaves the soft ones on exposed pads above ground, much like SS-6. SecC 65 confuses things by referring to the hard ones as silos and giving 300 psi as the design hardness. But coffins can be referred to as silos, and the 200–400 psi SecC gives is a likely design hardness based on US practice, which is a rather tenuous connection, leaving open the possibility of failure to achieve that hardness. Once again I go to Bl01, which has hardness values of 5 psi for Atlas D in above-ground coffins and 25 psi for Atlas E in coffins flush to ground surface. I assume the SS-7 coffins are the harder kind, to be somewhat more in line with the SecC hardness figure.

▲ **Reloads:** SecC 65 has soft sites with refire missiles and hard sites without (within what has the tone of a counting exercise).

## SS-8 Sasin

▲ **Numbers and Service Dates:** As for SS-7, we follow the vanD numbers series.

▲ **Yield:** SecC has about 3 MT in 1966 and 3–5 MT in 1970–1971. Other sources tend to give 5 MT. SSF has 3 MT, and Clns curiously gives 2–4 MT.

▲ **CEP:** SecC 65 gives 1 nm. Others use this value. We might worry that this early value is just assuming that the SS-8 is more advanced than the SS-7; however, it is a reasonable estimate given that unstorable liquid has a more reliable burn than storable liquid.

▲ **Range:** This is difficult to specify. There are no official sources. The available sources clump at 10,000 km, 10,500 km, and 11,000 km. I compared sources that give a range for both SS-7 and SS-8, widely discussed together. Some gave the same range, but of the rest, as many gave each the longer range. Introducing a difference with SS-7 seems unwarranted. 10,500 also seems too precise.

▲ **Propulsion:** SecC 65–67 give unstorable liquid, and SSF follows these sources. Every other open source gives storable liquid. We follow the official view but consider a propellant change after 1967 possible.

▲ **Silo Hardness:** *See* SS-7.

▲ **Reloads:** *See* SS-7.

## SS-9 Scarp

▲ **Mods 1 and 2:** Mods 1 and 2 were single RV mods deployed around 1966. Mod 2 was the bulk of the SS-9 force according to JCS 74, which also says Mod 1 had a slightly lower yield than Mod 2. SSF follows JCS 75 in the counterintuitive claim that Mod 1 was first deployed in 1967 and Mod 2 in 1966. Since no source describes any other differences between the two mods, we list them as one entry.

▲ **Mod 4:** There is considerable controversy over whether this system was ever deployed. FIar 76, AFM* 78, and JWS* 79 suggest it was never deployed. SSF 82, Clns 80, and Rock 78 say it was deployed, Rock going so far as to say 38 were deployed. The IOC is uncontested as 1971. SecC 71 says, "The extent of MRV deployment, if any, is not known at this time." In my view, all indications point to a system never deployed. First, official sources never say it was deployed. Second, it was tested not only in 1969–1970 but again in January 1973 when the imminence of the SS-18 would have made production of an operational Mod 4 pointless. Indeed since the MIRV program was clearly in the works, it seems most likely that Mod 4 all along was a development in support of the SS-18. Finally, SecC has the number of SS-9s as 276 by March 1971. With an IOC of 1971 (and likely not in the first two months), there were at most 12 silos left to take the Mod 4. Any program of replacing earlier mods with Mod 4 would have been wasteful with the SS-18 so close and would surely have engendered some comment by US intelligence.

Although the testing of Mod 4 was watched very closely and raised gret fears, the tendency to think it must have been deployed should be resisted.

▲ **Numbers and Service Dates:** SS-9 has 3 distinct phases to its numbers: the buildup, the peak level, and the decline and phase-out.

*Buildup:* As with SS-7 and SS-8, we rely heavily on the figures of vanD but not without scrutiny. His estimates are consistent with all the SecC data we have available; he also relies on some data we do not have available. The service he has performed is to provide a set of numbers fixed at the beginning of the year rather than scattered around, as the SecC numbers are.

Both Clns 80 and SecC 77 give 1967 as the IOC, which is inconsistent with vanD's figure of 54 operational at the beginning of 1967. While SecC 77 seems a very reliable source, JCS 75 has 1966 for the IOC of Mod 2, which if anything is more reliable because more detailed.

*Peak:* JCS 75, IISS, and vanD indicate 288 as the peak, supported by JCS 73, which gives 313, with 25 under construction. Clns 80 looks odd in this regard, giving 308 for the years 1972 through 1974.

*Decline:* The decline of a system like SS-9, which was being replaced one for one by the SS-18, is not directly remarked on very much. One gets the impression that sources simply take the number of SS-18s, subtract from the number of silos, and give that as the remaining number of SS-9s. This is quite clear in the data of Clns and vanD, where the sum of SS-9s and SS-18s is a constant 308 from 1975 onward. (To vanD's credit, he notes this as a methodological assumption.) But it is apparent that missile replacement is not instantaneous, and many sources recognize this. For instance, while Clns gives 68 SS-9s for 1979, it notes that none are operational, as all are being converted to SS18s. IISS notes that the 288 SS-9s in 1975 are being replaced. And SMP's contention that SS-18s went into much harder silos than SS-9 implies a conversion period.

There is a strong orderly pressure to assume that replacements are instantaneous; this assumption in some sense captures what is happening, but is not relevant to the real question of how many weapons were operational at each date To do that, our best method seems to be to estimate the time required for a conversion and assume that for each SS-18 becoming operational, an SS-9 went out of service that much before. What is this time lag? For the original construction of SS-9 silos, the figure was apparently 18 months (see Randall Forsberg). JCS 77 gives some evidence, saying the conversion of the first group began in 1973 and that the first operational launchers appeared in 1975. This indicates a 2-year delay, though the first conversions would be expected to go more slowly than later ones. If we add the further commonsense idea that converting a silo would take less time than building it originally, we settle on one year as our guess for the conversion interval. With this assumption in mind, the SS-9 decline will be a function of the SS-18 buildup. 20 additional silos were built for SS-18 that never contained SS-9s, and we assume that these were filled first.

▲ **Yield:** 18 MT is given by vanD. We take this estimate, and for the same reason: it is in the middle of a string of values given by SecC from 1965 through 1971. A number of sources give 25 MT for Mod 2 and 20 MT for Mod 1, but these estimates are from before the declassification of SecC.

▲ **CEP:** SecC 67 gives 0.5–1.0 nm, and SecC 70 hints strongly at accuracy improvements over the next year. Nitz also gives 0.5 nm, and Clns's 0.4 nm is close. SSF uses 0.5 nm too.

▲ **Range:** More than 7 recent sources agree on this range, the exception being Clns with 13,000 km. No official values are given.

▲ **Propulsion:** No dissent among the sources.

▲ **Silo hardness:** Extrapolation from the SecC 65 figure of 200–400 psi will be followed for SS-9, -11, and -13. But note that CB78 p. 37, assumes 550 psi for SS-11s.

# SS-11 Sego

▲ **Numbers:** Like the SS-9, the SS-11 has a buildup, a peak, and a decline, though unlike the SS-9, the SS-11 has not declined to 0. There are two additional complications for SS-11s. A number were deployed in MR-IRBM fields, apparently to fulfill an intermediate-range role. In some tallies these are treated separately; in others they are included. A168 (Treverton) says the SS-11s were deployed in this way when the SS-14 (designed for the theater role) failed—also implying that SS-4s and SS-5s were inadequate to the task. But with the 1972 SALT I accords, the Soviets had an incentive to build SS-20s to take over the theater role, freeing theater SS-11s for strategic targets again. Given this ambiguity about role, we treat the MR-IRBM SS-11s as strategic, lumping them with other SS-11s. Also, 60 silos built for SS-19s had SS-11s put back into them (AW04 82, among other sources). These, too, we count as SS-11s, until they finally received SS-19s in the last few years.

*Buildup:* As for the SS-9, we rely mostly on vanD numbers, which are consistent with our best sources. Clns differs in extending the number 970 back to 1970, although the original buildup was not yet complete then. SSF has 290 VRBMs and 720 other SS-11s, for a total of 1,010 for 1970. This is inconsistent with SecC's 90 VRBMs and 840 others for 30 December 1970.

*Mods:* 970 were deployed before Mods 2 and 3 became available. JCS 77 says that a mix of Mods 2 and 3 was believed deployed in modified silos (the 60 SS-19 silos). Suddenly SMP 84 says there are 100 Mod 1 and 420 Mods 2 and 3, and there are no intervening data. Evidently a great deal of modification went on. We are left doing a great deal of guessing. Since Mod 3 involves 3 MRVs and more fissile material, we assume fewer of those were converted than of Mod 2; ArkS says as many as 60 were deployed, and we assume 60. We also assume the 60 Mod 3s were deployed first. This estimate is consistent with a number of sources suggesting that the Mod 3 was deployed in the modified silos. It also makes sense operationally for the Mod 3s to have been deployed early, for the main advantage of a MRV system is its ability to penetrate ABMs, and the US ABM was built and dismantled around 1975. But if JCS 74 is right that 20 Mod 3s were installed in modified silos by mid-1973, then either these were dismantled or moved to other SS-11 silos in the past few years as SS-19s have been placed in these silos. This level of complexity we cannot resolve. We simply do not know the exact details of all SS-11 Mods.

The assumption of 60 Mod 3s leaves 360 Mod 2. We assume the Mod 2 conversion has been going on linearly since 1975. Actually it is more likely to have happened rapidly during a few years, but since we have no idea what years those were, the best guess seems the linear one.

*Peak:* The SS-11 reached a plateau of 970 before replacement began, but the 60 additional, unfilled silos complicate matters. The exact replacement schedule will determine whether a higher peak was reached.

*Replacement:* As we did for the SS-9, we will assume a 1-year time for conversion of the SS-11s to SS-17s and SS-19s, and the decline of SS-11s will be computed based on that. For the 60 SS-19 silos that had SS-11s put in them, we would expect that the time to convert to SS-19 missiles would be less than for the ordinary SS-11 silo, but we do not know how much and use the simplifying assumption of 1 year for all conversions.

## SS-11 Sego Mod 1

▲ **Yield:** A number of unofficial sources use 1–2 MT, but official and semiofficial sources never go as high as 2 MT. SecC 68 has 1 MT, and AWar 80 has 950 kt, as have SSF, CBO, and Clns. We use 950 kt because of its specificity.

▲ **CEP:** The earliest estimate is SecC 67–68, with 1–1.5 nm. Later sources seem to agree on figures nearly the same or lower: 1.5 km from a few earlier unofficial sources; 0.76 nm (1,407 m) from AWar, followed by IISS, SSF, CBO, and Clns. Nitz uses 1 nm. One muses whether Nitz portrayed the SS-11 as worse to make the SS-17, -18, and -19 look better. It is also likely that SS-11 accuracy improved since SecC 67, so we use 1,400 m. (*See also* discussion under Mod 2.)

▲ **Range:** Estimates cluster tightly, with some older, unofficial sources saying 10,000 km, a fair number of unofficial sources saying 10,500 km, and the best sources saying 11,000 km, including SecC 71, AWar, Clns, and SMP 84. SSF p. 99 says the first mod may have had an initial range of only 3,000 nm (5,556 km), however. Perhaps this is a mod before Mod 1.

▲ **Propulsion:** Storable liquid has no dissent.

▲ **Silo Hardness:** The SecC 65 figure.

## SS-11 Sego Mod 2

▲ **Nuclear Load:** Mods 1 and 2 carry a single RV, but Mod 2 was tested with penetration aids, unlike Mod 1, according to JCS 75–76. FIar, JWS 81, and AFM 82 have taken this to be a certainty rather than just a test.

▲ **Yield:** Data are scanty, with 1.1 MT from AWar 80, followed by SSF.

▲ **CEP:** Data are scanty, with AWar giving 0.59 nm (1,092 m), followed by SSF. Interestingly, AFM+ 84 gives 1.4 km. vanD gives 1.25 nm for Mod 1 (midpoint of the early SecC figures) and then 1.4 km for some improvement (we assume it is Mod 2). All sources seem to agree that 1.4 km is the CEP for some SS-11 Mod. The question is whether it is Mod 2, with Mod 1 worse, or Mod 1, with Mod 2 better. AWar is quite explicit on the matter, so we stick with them. It is also possible that the Mod 1 had an accuracy improvement but remained Mod 1.

▲ **Range:** AWar 80 has 12,000 km, followed by SSF; AFM 84 has 10,500 km, but SMP 84 has 13,000 km. We have to believe SMP, but with some suspicion.

▲ **Propulsion:** No separate data, but Mod 2 has the same booster as Mod 1.

## SS-11 Mod 3

▲ **Service Dates:** All sources agree on 1973 for the IOC and year first deployed. Unlike Mod 2, JCS never indicates they have not yet been deployed. No source suggests it is retired.

▲ **Nuclear Load:** 3 MRVs is unanimous.

▲ **Yield:** 200, 300, and 500 kt are common values cited; IISS has 100 to 300, and Clns has 500 to 1,000. AWar has 350, followed by SSF. We follow AWar because it is the most authoritative source and it gives a mid-range number.

▲ **CEP:** A few earlier estimates suggest less than 1 km, but AWar 80 has 0.59 m (1,092 m), and 4 other sources follow around 1,100 m.

▲ **Range:** Pairs of sources give 10,000 and 8,800 km; Clns gives 10,200; and AWar, SSF, and SMP 84 give 10,600. The fact that SMP 84 gives the same number as AWar did 4 years earlier is evidence suggesting AWar had a good leak.

▲ **Propulsion:** Storable liquid: no controversy.

▲ **Silo hardness:** We repeat our 300 psi guess. A140 (Colin Gray) says that Mod 3 requires silo modification from Mod 1. He does not suggest this is a hardness change, but it is interesting in evaluating the data on the mod mix, if true.

## SS-13 Savage Mod 1

▲ **Service Dates:** Work on digging the silos began before missile testing was complete, as SecC 69 says. This explains how SMP 83 can say deployment was underway by 1966, but SecD 77 and JCS 75–78 give the IOC as 1969. FIar, SWM, IISS, and Rock have 1968 as the date, and SSF has 1967–69, but official sources overrule them.

SMP 84 says there are 60 Mod 2s, and there are only 60 in all, so it is apparent they consider Mod 1 to be retired. No source gives any indication of when it was phased out of service. Our method is to consider when Mod 2 came into service and assume a reasonable linear schedule of replacement, perhaps using the same 3-year period it took to deploy the missiles initially. Of course, that involved silo construction rather than just nose cone replacement, but the SS-13 was also by then an unimportant ICBM, and probably vast resources were not allocated to its upgrading. Warhead-section replacement is easy, however, so our estimate is two years after the first Mod 2 came into service, or 1975. We end up following vanD's numbers exactly for the overall deployment.

▲ **Yield:** Many unofficial sources have yield at 1 MT, and a few set it at 2 or 3 MT. But SecC 69 says the yield is most likely smaller than that of the SS-11, and AWar 80 has 600 kt, followed by SSF, Clns, and CBO. Nitz and IISS have 750 kt. We stick with AWar.

▲ **CEP:** Except for Rock (with an optimistic 1,300 m), all sources have either 2 km or 1 nm (1,852 m). AWar gives 1 nm, as do SSF and CBO and Nitz, so we follow with 1,900 m.

▲ **Range:** AWar 80 and SMP 84 have 9,400 km for Mod 2, but no semiofficial source has anything on the range of Mod 1. There is agreement among unofficial sources at about 8,000 km. The difference is great, and no mention is made of a range improvement being a property of Mod 2. IISS, the only source to give a range explicitly for Mod 1 (rather than the generic), gives 10,000 km. We put our determination at 9,400 km for both.

▲ **Propulsion:** Unanimous, including official sources.

▲ **Silo Hardness:** *See* SS-11.

## SS-13 Savage Mod 2

Specific data on Mod 2 are spotty, but the sources that do give data are the ones we respect most.

▲ **Numbers and Service Dates:** A modified version is discussed by JCS 73 and other sources; SMP 84 indicates it was first deployed in 1973 and that all 60 SS-13s now deployed are Mod 2. Our assumption is that the Mod 2s replaced the Mod 1s at the same rate of 20 per year as the Mod 1s were initially deployed.

▲ **Warheads:** No one says Mod 2 has anything other than 1 RV.

▲ **Yield:** SSF gives 600 kt for Mod 2; no source suggests a different value.

▲ **CEP:** AWar has 0.82 nm (1,518 m), followed by SSF. JCS 73 says the Mod 2 will be more accurate than Mod 1, and no source offers a different number.

▲ **Range:** 9,400 km is given by AWar and SMP 84; no source dissents.

▲ **Propulsion:** No data but same booster as Mod 1.

▲ **Silo Hardness:** *See* SS-11.

## SS-17

▲ **Overall Numbers:** We base our numbers on vanD, with a few corrections. Although JCS 78 and SecC 78 give greater than 60 for early 1978, vanD's 60 is appropriate since it applies to the beginning of the year. But when JCS 79 says nearly 100 for 1979, vanD must have a figure less than 100; we use 90. The plateau of 150, reached by early 1980, is completely unanimous.

▲ **Numbers Breakdown by Mods:** We follow the general shape of Clns 83 data. ArkS report the DIA saying that some SS-17s and SS-19s are still configured to carry single warheads, so we assume that 20 Mod 2s came into service by the beginning of 1979 and that they remained in service through 1984. Since SMP 84 clearly implies that Mod 3 is the dominant mod, we assume that all Mod 1s were replaced in equal portions in 1981 and 1982 (approximating the Clns figure). Although it seems likely that some single-warhead versions remain in service, we follow SMP 85 in showing the 20 Mod 2s as converted to Mod 3 during 1984 so that all SS-17, -18, and -19 missiles are the latest MIRV version by 1985.

This leaves a question of why the IOC of Mod 3 is given as 1979 when deployment did not begin until 1981. The evidence on Mod 3 service entry is scanty. SSF gives 1979, as does Clns 83. But Clns 83 also gives numbers and lists none until 1982. IISS gives 1982 as the first year deployed. SMP 84 says that deployment of the most recent round of ICBM Mods began only five years ago, but that might refer to the SS-19 Mod 3 or SS-18 Mod 4. IISS, however, gives 1982 as the service entry date for all 3 missiles, which makes its claims inconsistent with SMP 84 and therefore suspect. Perhaps the Soviets have been making incremental accuracy improvements in two stages, and IISS is confused about which constitutes the next generation of mods. We go with 1979.

▲ **Silo Hardness:** No one suggests that the SS-17, -18, and -19 have different hardnesses from each other, so we discuss them together. The government record starts with JCS 76 saying that every new missile goes into "new modernized, hard silos," and the associated launch systems are in silos, not bunkers. SecD 77 casts some doubt on the consensus that seems to emerge later by saying that some modified silos are designed to resist high overpressures, implying that perhaps not all are. SecD 80 reports that all "converted silos may be capable of withstanding very high static overpressures." And SMP 83 says the missiles are "housed in the world's hardest silos." As for specific numbers, AWar 80 gives 3,500 to 4,500 psi for the SS-17, -18, and -19. This figure was repeated by AWST on 12 October 1981 and followed with the 4,000 psi midpoint in CBO 83 (p. 18). But in January 1978 CBO was using 2,000 psi (p. 37 in Counterforce Issues). The June 1980 *Bulletin of the Atomic Scientists* reported 2,500 psi. The change from 2,000 to 4,000 psi has not accompanied any news of a further Soviet silo upgrade and is thus presumably an estimate change rather than a silo change. The official number is 4,000 psi, but it is highly suspect.

## SS-17 Mod 1

▲ **Nuclear Load:** 4 MIRVs is unanimous for Mod 1.

▲ **Yield:** A number of unofficial estimates range from 800 kt to 1 MT, but with no particular pattern and suggesting they are guesses. IA06 75, Rock 78, and JWS give 200 kt. This guess is also consistent with SecD 77, which says the yield exceeds Minuteman III per warhead (170 kt). However, a great variety of the best sources—AWar 80, Clns, IISS, SSF, AFM, and CBO—use 750 kt. We choose it but note that the 200 kt figure is very different, a discrepancy worth more research.

▲ **CEP:** Almost every CEP that is specifically attributed to Mod 1 seems to flow from the AWar 80 value of 0.24 nm (444.48 m). This appears to have been rounded by IISS and AFM 84 to 450 and by AFM 82 to 500. FIar 76 and Rock 78 give 500 m and 600 m, respectively, but they are not talking specifically about Mod 1. Our 440 m is a careful rounding of 0.24 nm.

▲ **Range:** AWar 80 and SMP 81 and 83 use 10,000 km. It is closely followed by almost everyone who talks of Mod 1 and is the middle of a tight range of estimates for SS-17 with no mod specified.

▲ **Propulsion:** It is unanimously storable liquid fuel, with cold launch.

## SS-17 Mod 2

▲ **Yield:** We use 3.6 MT, given by AWar and SSF. JWS+ 83 gives 1–2 MT. But the contending estimate is 6 MT by Clns 80, IISS, Lu01, and AFM 84, with AFM 82 giving 5 MT. 3.6 MT is an odd value for yield, and it seems possible that AWar mixed it up with throwweight, which IISS gives as 3,600 lb. AWar, however, gives a similarly odd value for the SS-19 Mod 2, which is not the throwweight.

▲ **CEP:** AWar gives 0.23 nm (425.96 m), as do Clns and SSF. IISS and Lu01 give 450 m, probably a rounded derivative. We round to the nearest 10 m, or 430 m.

▲ **Range:** AWar 80 and SMP 81 cite 11,000 km, as do the majority of other sources. A few values dip as low as 10,500 km.

## SS-17 Mod 3

▲ **Nuclear Load:** 4 MIRVs is unanimous, including among government sources.

▲ **Yield:** Data are scanty on Mod 3 in particular. There is little reason to expect a change from Mod 1, but the estimates for Mod 3 are 750 kt by SSF and vanD (who also give 750 kt for Mod 1), 500 kt by Clns (who gives 750 kt for Mod 1), and 2 kt (an obvious typo, perhaps for 200 kt) by IISS (which gives 750 kt for Mod 1). The changes by Clns and IISS leave open the possibility that after the initial 750 kt estimate, the yield estimate for the SS-17 went down, affecting both Mods 1 and 3. SMP and most other sources show no range change from Mod 1 to 3, so there is no reason to expect a lighter RV on Mod 3.

▲ **CEP:** The only CEP estimates attributed specifically to Mod 3 are 0.20 nm (370.40 m) by Clns 83 and 0.24 nm (440 m) by vanD. CBO 83 gives a CEP of 0.17 nm (314.84 m) for the SS-17 in general for 1983, which might be assumed to be the Mod 3. vanD quotes as his source AWar 80, but AWar 80 recognizes only one MIRVed mod for the SS-17, presumably Mod 1. vanD would seem to have to agree, since he says Mod 3 came into service only in 1982. CBO cites as its sources AWar, Nitz, and Clns 83 for CEPs in general. The 0.17 nm value must be from Nitz, the value he gives for the SS-17 on 18 June 1979. But Nitz must then have been referring to Mod 1, and for Mod 1 his estimate is highly divergent from everyone who has written since. Nitz's data are also suspect since (for each of 4 different dates) he gives identical accuracy figures for the SS-17, -18, and -19; in fact the missiles differ in accuracy, with the SS-17 widely recognized as less accurate. Clns 83 gives no source, but his estimate seems the most consistent with modest incremental improvements in accuracy.

▲ **Range:** 10,000 km is from SMP 83 and 84. IISS concurs, and Clns 83 gives 6,000 nm (11,112 km).

▲ **Propulsion:** *See* Mod 1.

## SS-18

▲ **Overall Numbers:** The first job is to settle on the total for all mods, because this is the figure most often discussed and also best known. Once that is known, we proceed to the speculative business of breaking down this figure by mods.

There is unanimity that the plateau for numbers of SS-18s is 308. The latter part of the buildup is less controversial because official numbers are available. There we go with the vanD figures, which are consistent with the official numbers. But the early history is uncertain. vanD's 26 for the beginning of 1975 seems high compared to a range of other sources, 3 of which cite 10, and at least 1 citing it as of midyear. One other adjustment in the numbers is required to meet a condition set forth by JCS 79: that for the fifth straight year, the number of conversions increased. Although open to various interpretations, this condition is more likely to hold if the 1976 figure is fewer than the 38 of vanD or the 36 of IISS and Clns. We use 30. The complete series from 1974 to 1981 is then: 0, 10, 30, 56, 92, 188, 248, 308. This makes the net additions from one year to the next as follows: 10, 20, 26, 36, 96, 60, 60.

▲ **Breakdown by Mods:** The only source that gives explicit numbers breakdown by mods is Clns 83. Since his numbers are midyear (or later) they are not comparable with our beginning-of-year figures, but we try to stay roughly in line with Clns figures. Otherwise our mod breakdown is guided by these principles: The first 56 must have been Mod 1, since no other mod was ready for service before the start of 1977. We assume that all the new SS-18s deployed during 1977 and 1978 were Mod 2, whereas Mod 3 replaced 24 of the Mod 1s during 1977 (a relatively minor change). Then we assume that the remaining 120 deployed during 1979 and 1980 were Mod 4. This leaves us at the beginning of 1981 with 32 Mod 1, 132 Mod 2, 24 Mod 3, and 120 Mod 4. By the beginning of 1984 we need to get to 308 Mod 4s (ArkS got DIA to confirm that this implication of SMP 84 was correct). We assume the increase was roughly linear, with 60 Mod 4s added in 1981, 60 in 1982, and 68 in 1983, and we assume the mods were replaced in this order: 1, 2, then 3. The reason for assuming that some single RV Mod 3s were retained while Mod 2s were replaced is that for some missions, a single large warhead is useful.

The main difference between these numbers and Clns numbers is that he assumes that MIRVed mods replaced single RV mods soon (in FY 78 and 80), while I assume the replacement came only after all silos were filled. His assumption makes sense if the silo conversion process was the limiting factor on new deployments; my assumption makes sense if missile or warhead production was the limiting factor. (Why would the Soviets choose to replace an SS-18 Mod 1 with a Mod 2 if they could replace an SS-9 with a Mod 2?)

▲ **Silo Hardness:** *See* SS-17.

## SS-18 Mod 1

▲ **Nuclear Load:** Single RV is unanimously agreed upon, including by JCS and SMP.

▲ **Yield:** Estimates vary from 18 to 50 MT. IISS and AFM 84 suggest 20 MT. AWar 80, Clns, FI, and SSF give 24 MT, as we do.

▲ **CEP:** After Rock estimates 550 m in 1978, all estimates stay very close to AWar's 0.23 nm (425.96 m), varying from 400 to 457 m.

▲ **Range:** JCS 75–76 have more than 5,500 nm (10,186 km), but by 1980, SecD was giving 12,000 km, and almost everyone else was following, including AWar, SMP, IISS, and AFM.

▲ **Propulsion:** The booster is unanimously agreed to be storable liquid.

## SS-18 Mod 2

▲ **Service Entry:** JCS 75–76 give the IOC as 1975, but JCS 77–78 have it as 1976. Clns, FI, and SSF identify 1976 as the date of service entry. Rock and IISS use 1977. But more important, JCS 77 said, "Initial deployment may already have begun," and Clns numbers charts show the first nonzero in 1977. I suspect that the missile had had enough testing to be ready to deploy in 1976, and the silos were started, which many sources took to mean deployment began. But only in 1977 did any become operational, which is our definition.

▲ **Retired:** SMP 84 says there are 308 SS-18 Mod 4s in service, and ArkS say that "according to DIA... the SS-18 is now definitely fully MIRVed to 3,080 operational warheads." Recent government sources have consistently described the SS-18 Mod 2 as having 8 to 10 warheads, so the ArkS quote is not inconsistent with there still being some Mod 2s. But given the quotation in a draft by Chris Paine, it seems unlikely the Mod 2 is still deployed: "The Mod-2 had serious problems. The guy who designed the post-boost vehicle is probably in Siberia because everything you could do wrong in the design of a post boost vehicle he did. He really goofed it." (Jim Miller of DIA, testimony to HASC, FY 1980, pt. 3, pp. 132, 129.) Once again the specific date is uncertain and is extrapolated from Clns 83's numbers estimates.

▲ **Nuclear Load:** Until 1978, all accounts, including SecD, said the Mod 2 carried up to 8 RVs. Since 1978, starting with a JCS report of 8–10 RVs, virtually every source has indicated 8–10 RVs. No source has indicated just 10 RVs, but 3 have indicated just 8: vanD, IISS, and Clns. Perhaps Mod 1s were deployed some with 8 and some with 10 (and perhaps some with 9), but given the problems with the PBV indicated under "Retired," it seems somewhat more likely they settled for 8.

▲ **Yield:** Among those who do not specify a mod but are talking of a MIRVed version and those before 1980 and a few after, yield estimates tend to be 2 MT. Sources include Lu01 83, IA06 75, Rock 78, and US08 79. Nitz 79 gives 750 kt, a number corroborated by no one. AWar in 1980 came out with 0.55–0.90 MT, with yield varying depending on the number in the bus. JWS 81 and AFM 82 still use 2 MT but revise their numbers downward by JWS+ 83 and AFM+ 84, indicating strongly that 2 MT is an old estimate. FI 81 and SSF take the 0.55–0.90 MT range as it stands. Of the 3 sources that take the 0.90 MT top, 2 (Clns and IISS) are those who settled on 8 RVs instead of 10. We follow the same line, assuming 8 900 MT RVs. JWS+ 83 has 500 kt in its analysis table, though it gives less than 1 MT in its text.

▲ **CEP:** AWar gives 0.23 nm (425.96 m), which is followed by Clns and SSF, and near the middle of a range of estimates from 400–450 m. vanD has a CEP of 0.30 nm (556 m), but he cites his sources, and this value is higher than those in any of them.

▲ **Range:** SMP 81 and 83 give 11,000 km, which is also the majority view. FI 81 and Rock 78 have 12,000 km, and SWM 77 and JWS 81 have 9,250 km.

▲ **Propulsion:** *See* SS-18 Mod 1.

## SS-18 Mod 3

▲ **Nuclear Load:** Single RV is unanimous, including by JCS and SMP (except for an apparent AWST typo giving 14 MIRVs).

▲ **Yield:** Yield estimates are nearly unanimous at 20 MT, starting with AWar 80. Rock 78 has 25 MT.

▲ **CEP:** AWar 80 gives 0.19 nm (351.88 m). This estimate is followed by Clns, FI, IISS, SSF, and AFM+ 84. An earlier AFM 82 had suggested that 180 m was achieved in trials. Lu01 is the only divergent estimate, at 200 m.

▲ **Range:** AWar 80, SecD 80, and SMP 81 and 83 give 16,000 km. Clns has 8,000 nm (14,816 km), and FI curiously goes from 16,000 km in 81 to 9,250 km in 83.

▲ **Propulsion:** *See* SS-18 Mod 1.

## SS-18 Mod 4

▲ **Nuclear Load:** 10 MIRVs is completely unanimous among the sources including SMP 83, 84.

▲ **Yield:** 500 kt is the majority favorite for yield for this weapon, including AWar 80, Clns, FI, IISS, JWS, AFM, and Lu01. SSF and vanD use 550 kt, but vanD is not recognizing specific mods, so he is probably copying the 550 kt from Mod 1. Nitz uses 750 kt, but again this is not for an identified mod, and he is using this same value indiscriminately for all MIRVed mods of SS-17, -18, and -19. CB01 and BA01 use 1.5 MT. These are presented as more in the line of assumptions than serious assertions as to the truth.

SMP 84 also has a polemical assertion about SS-18 yield, saying that each warhead has over 20 times the destructive power of nuclear devices developed during World War II. Using the figure of 12.5 kt they are asserting that SS-18 yield exceeds 250 kt (if 15 kt is assumed, it exceeds 300 kt). Of course, 500 kt meets this criterion, but then why not speak of over 30 times the destructive power or even 40? Possibly they are using a different figure for the Hiroshima bomb or referring to bombs that were developed but not built. A slight corroborating suggestion that the SS-18 yield estimate has gone down is from range data. The 500 kt yield figure is from AWar 80, which gives a range of 10,000 km. AWar range figures have generally been born out exactly by SMP figures released later. But SMP 83 and 84 give 11,000 km for Mod 4 range. Perhaps this range increase results from a lower weight (and thus yield) for SS-18 RVs than was estimated in 1980. This line of reasoning is too speculative to influence our own estimate.

▲ **CEP:** AWar 80 gives 0.14 nm (259.28 m), as do Clns 80, FI 81, and SSF. Numbers very close to 300 m are given by IISS, AFM, and Lu01. vanD gives the 0.14 nm figure for 1979 but uses 0.10 nm (185m) for 1982. This 0.10 nm figure, however, is not contained in any of the sources he cites, except possibly Clns 78 (and if so, the year of publication makes figures for 1982 necessarily a guess). Other estimates that do not refer specifically to the Mod 4 but probably do so by referring to MIRVed SS-18s are from CB01 78, 1,200–1,500 ft (366–457 m); BA01 81, 600–1,000 ft (183–305 m); and a paper by LANAC (Lawyers' Alliance for Nuclear Arms Control, "The Case against the MX Missile," 1983) giving 800 ft (244 m). Nitz gives 0.17 nm (314.84 m) for 1979 and 0.15 nm (277.80 m) for 1982. The CB01 numbers are higher than the others, but it is an early source. The BA01 range is centered precisely at the figure given by LANAC, which is also the official value for the SS-19 Mod 3 (in fact LANAC is referring to the SS-18 and -19 jointly). In summary, the main values are 300 m, 259 m, and 245 m, the last of which is associated primarily with the SS-19 Mod 3. We use 259 m (rounded to 260).

These numbers, all of which originate around 1980–1981 (and are assumed to refer to the state of deployed systems at the time, perhaps not a sound assumption), do not allow for the basic point suggested by Nitz and vanD: minor guidance changes may have decreased CEP retroactively since Mod 4 was fielded. But it seems that the approach Nitz and vanD take is to guess, since vanD's number does not come from his cited sources, and Nitz is predicting the future. It seems safer to keep the old numbers rather than arbitrarily decreasing them, when it is possible that Soviet CEP has not decreased, at least not substantially.

▲ **Range:** AWar 80 gives 10,000 km, as do FI, and SSF; Clns, who gives 5,500 nm (10,186 km), is very close. But SMP 83 and 84 give 11,000 km and are followed by JWS, AFM, and IISS+83, the last giving up its earlier 9,000 km estimate. We follow the SMP estimate.

▲ **Propulsion:** *See* SS-18 Mod 1.

## SS-19

▲ **Overall Numbers:** Only 2 years are suggested for initial deployment, 1974 and 1975, but the evidence fits better with 1975. First consider the evidence for 1974. JCS 76–77 give 1974 as IOC for the SS-19 in general, but JCS 78 gives 1975 as IOC for Mod 1. Clns 80 gives 1974 as the year it was first deployed, but in Clns+ 83 his first nonzero entry is for 1975. Standing evidence is SMP 81, JWS 81, and IISS 82. On the other side are Rock 78, Meyr 84, and SSF 82, along with the 2 switches from 1974. JCS 75 says that it may be starting operational deployment, and SecD 75 says it has started or soon will start, operational deployment. These suggest that by early 1975 it had not yet become operational.

As before, we use the vanD series, with modifications. One modification is required for his value of 20 at the start of 1975; we are fairly confident none was operational during 1974, so we give 0 for 1975. The numbers for 1979 and 1980 are also controversial. vanD gives 240. IISS gives 300. JCS 79 gives "nearly 300" but includes the SS-19/11 silos. SecC 79 gives about 300 for the SS-17 and -19 combined. Clns gives

240. Clns on the whole is lagging, even behind official sources in this part of the buildup. We stick with vanD's 240. For 1980 vanD gives 270, while AWar, SSF, and IISS give 300, although these three estimates may be later than vanD's January number. Inexplicably Clns still gives 240. With some doubt, we stick with the vanD 270. The last phase of the ICBM buildup is the relatively recent conversion of the remaining 60 SS-19/11 silos to SS-19s. Consistent with the data, we give 300 for 1982, 330 for 1983 and 1984, with the 360 figure reached only in the first few months of 1984. (JCS 84 gives 330 as of 1 January 1984, and SMP 84 gives 360.)

▲ **Breakdown by Mods:** JCS 77 says that SS-19s deployed to date carried 6 MIRVs, meaning no Mod 2s were deployed before 1977. Clns has the first 20 deployed in 1977, though he says elsewhere that deployment began in 1978. SSF also says deployment began in 1978, and IISS has deployment beginning in 1979. Testing data are scanty, but JCS 77–78 have that the Soviets were then testing the single RV mod. Assuming that testing began in 1977, operational deployment the same year seems unlikely, though not impossible. We give an uncertain 1978 as our estimate for Mod 2 deployment. Clns+ 83 and JWS 83 agree on a figure of 60 Mod 2s. Clns then goes on to show a decline from the 60 Mod 2s down to 10 in 1982; we do not agree. With the first Mod 2s becoming operational during 1978, we assume 30 were deployed during 1978 and 30 during 1979. After that, the 90 new deployments were Mod 3 (we have a first operational date of 1980). We also have the problem of Mod 1 to Mod 3 conversion. SMP 84 indicates that all 360 are Mod 3, while Arkin and Sands said the DIA admitted that some single RV are still deployed. We assume that all 60 single RV remain, while all Mod 1s have been converted to Mod 3. We assume conversion was complete by the beginning of 1984 and that after starting out more slowly (30 during 1980), it has been roughly linear since. This implies 240 conversions between early 1981 and early 1984. (There are a total of 330 in January 1984, of which 270 were Mod 3; there were 30 Mod 3 in early 1981.) This divides well into 80 per year. During 1984, the conversion of the final 30 SS-11s remaining in modified silos to SS-19 Mod 3s is assumed to have taken place, along with conversion of the 60 Mod 2s to Mod 3. Although some Mod 2s may in fact remain, we follow SMP 84 and SMP 85 in showing all SS-17, -18, and -19 as finally converted to MIRVs.

These estimates are roughly consistent with Clns but not in detail. And they are clearly highly uncertain in detail as well.

▲ **Silo Hardness:** *See* SS-17.

## SS-19 Mod 1

▲ **Nuclear Load:** JCS 74 started with an estimate of 4–6 MIRVs, but at the same time SecD 74 and 75 were saying it had been tested only with a MIRV payload of 6 RVs. A few sources picked up the 4–6 MIRVs, but the only one to do so recently is FI, which switched from 6 MIRVs in 1981 to 4 or 6 MIRVs in 1983. All other sources, including JCS by 1976 and SMP 81 and 83, give 6 MIRVs only.

▲ **Yield:** AWar 80 gives 0.55 MT, as do Clns, AFM, IISS, SSF, and FI, initially. FI+ 83 gives 200 kt. JWS 81 gives an uncertain 200 kt. It will be of great interest to find out on what basis FI 83 has revised their yield estimate down to 200 kt. SecD 77 says the yield exceeds that of Minuteman III per warhead. Among sources that do not recognize the Mod 1, yield estimates vary from 200 kt to 1 MT. At least until we hear from FI, we must stick with the 550 kt figure.

▲ **CEP:** There is more discrepancy for the SS-19 Mod 1 than the SS-17 and -18 mods. AWar 80 gives 0.19 nm (351.88 m), as does SSF. Clns gives 0.21 nm (259.28 m), and IISS gives 500 m. For the missile in general, Rock 78 gives 450 m. AWST gives 0.14 nm, but they are probably referring to Mod 3. Meyr 84 gives 100–300 m, but he is concerned with the variable range variant, which would be expected to have better accuracy over a shorter range. Lu01 gives 300–450 m, and vanD gives 556 m. Although it is the lowest plausible estimate, we follow AWar with 350 m.

▲ **Range:** Of the official estimates, JCS 74 starts out with < 5,500 nm (10,186 km), but JCS 75–76 has 5,000 nm (9,260 km), explicitly saying it can deliver 6 MIRVs at that range. SecD 80–81 and SMP 81 give 9,600 km as the maximum range. SMP 81 also says it is capable of delivering six RVs to a range of about 9,000 km. Unofficial estimates fall within the range of these estimates, with the exception of IISS, which gives 11,000 km for Mod 1. We use the 9,600 km figure.

▲ **Propulsion:** Storable liquid is the unanimous judgment.

## SS-19 Mod 2

▲ **Nuclear Load:** Single RV is unanimous, including by SecD, JCS, and SMP.

▲ **Yield:** 4.3 MT is given by AWar 80 and also by SSF and Lu01. 5 MT is given by AFM and IISS, 5–10 MT by JWS– 83, and 10 MT by Clns. We use the 4.3 MT because of the authority of AWar and because it is a precise number (perhaps too precise).

▲ **CEP:** AWar gives 0.21 nm (388.92 m), as does SSF. Clns gives 0.14 nm (259.28 m), and IISS and AFM+ 84 give 300 m. Lu01 gives 200 m. It is hard to imagine why a high-yield single RV missile would have a worse CEP than the MIRVed Mod 1 that was previously deployed, but that it what AWar implies. We use the middlish 300 m.

▲ **Range:** AWar 80 gives 10,100 km, as do SecD 80–81 and SSF, and Clns 80 is close. SMP 81 and 83 give 10,000 km, as do IISS and AFM+ 84. Although 10,000 km is the most recent estimate, we use 10,100 km because it is more precise, and probably the 10,000 km figure is just rounding.

▲ **Propulsion:** *See* SS-19 Mod 1.

## SS-19 Mod 3

▲ **Nuclear Load:** 6 MIRVs is unanimous, including by JCS and SMP.

▲ **Yield:** The consensus is 550 kt, held to by IISS, SSF, Clns, and Lu01. JWS+ 83 has 500 kt, probably a rounded figure. Nitz has an undifferentiated 750 kt. *See* Mod 1 on why this value might be less. AWar had nothing to say about the yield of Mod 3, and all other sources have yields the same as Mod 1. This is circumstantial evidence that all sources are copying AWar (or receiving independent leaks of the same information).

▲ **CEP:** AWar gives 0.14 nm (259.28 m), as do SSF and vanD. IISS, AFM 84, and Lu01 give numbers close to 300 m. Inexplicably Clns gives 0.15 nm (277.80 m). James Miller of the DIA, testifying before the Senate Appropriations Committee on 16 June 1981 (cited by Ts01), gives 245 m, a number available to the public because of an error in declassification. The LANAC paper uses 800 ft (244 m), which is probably the same number as Miller's. This offers a rare opportunity to compare open sources with the inside truth on CEP. AWar's 0.14 nm is not just the same value rounded because 0.13 nm would be closer to 245 m. Possibly AWar, published a year earlier, reflects an older and slightly higher CEP estimate. It is also possible that the leaker changed his data slightly. The discrepancy is, in any event, small enough to maintain the credibility of AWar. Nonetheless, we use the official figure of 245 m, rounded up to 250 m because that is nearer other sources.

For a broader discussion of possible improvements, as suggested in Nitz and vanD data, *see* SS-18 Mod 4.

▲ **Range:** SMP 83 and 84 give 10,000 km. In fact, SMP 83 in a table contrasts the 9,600 km range of Mod 1 with the 10,000 km of Mod 3. SSF 82 gives 5,200 nm (9,630 km), before SMP 83 appeared. Clns 83 gives 5,500 nm (10,186 km). We go with SMP's 10,000.

▲ **Propulsion:** *See* SS-19 Mod 1.

## SS-24

▲ **Service Entry:** AFM 84 says that 4 or 5 test launches were apparently unsuccessful and at least two were successes. SMP 84 says testing is proceeding and also that it is in preseries production and that IOC is expected in 1985. Because IOC for ICBMs may precede operational deployment and because the USSR has yet to deploy a successful solid-fueled ICBM, we move this date back to 1986. Our estimate is supported by SMP 85, which projects deployment in fixed sites for 1986, although that implies that true operational capability may not be achieved by 1 January.

▲ **Nuclear Load:** SMP 85 says "up to 10 MIRVs": we take this number as the most recent. (AFM 84 says it carried 8 RVs during its second successful test flight, 22 November 1982.)

▲ **Yield:** No sources give data, and because SS-24 is solid fuel and thus unlike the SS-17–19s, it seems best not to guess.

▲ **CEP:** FI 81 suggests perhaps less than 260 m; SMP 84 says it is likely to be even more accurate than SS-18 Mod 4 and SS-19 Mod 3. Given the new solid-fuel technology, we are skeptical of any real improvement and give the same 250 m that we did for SS-19 Mod 3, which is appropriately vague.

▲ **Range:** SMP 85 shows a range of 10,000 km, the only hard figure available and one typical of recent Soviet MIRVed ICBMs.

▲ **Propulsion:** Solid is given by FI 81, AFM 84, and SMP 84. No source disagrees.

▲ **Silo Hardness:** We expect the same value as for the other recent ICBMs because no sources give any data. FI 81 does say it is expected to be deployed in superhardened silos.

▲ **Launch Platform:** FI 81, FI+83, and AFM 84 indicate silo basing. SMP 84 indicates silo basing at first, though "mobile deployment could follow several years after" IOC in 1985. SMP 84 also says, "Available evidence suggests mobile as well as silo deployments for . . . SS-24." The correct answer for the question of mobile basing appears to be "probably later."

## SS-25

▲ **Service Entry:** SMP 85 says that two bases formerly used for SS-20 deployments are nearing completion of conversion to road-mobile deployment of the single-warhead SS-25, which should be operational by the end of 1985. We accept this.

▲ **Nuclear Load:** SMP 84 says single RV; no source disagrees.

▲ **Yield:** No source gives any data, and it seems insufficiently like any other missile (SS-13 is much older) to warrant a guess.

▲ **CEP:** FI 81 indicates a CEP like SS-18 and SS-19, but no other source has any comment. We omit any CEP reference because the failure of SMP 84 to mention CEP may indicate it is worse than previously thought.

▲ **Range:** SMP gives 10,500 km, the only reported hard figure.

▲ **Propulsion:** FI 81, AFM 84, AWST+ 84, and SMP 84 give solid, with no one dissenting.

▲ **Silo Hardness:** Along with mobile deployment recognized by AFM 84 and FI 81, SMP 84 also says, "Available evidence suggests mobile as well as silo deployments for . . . SS-25." This silo deployment seems tenuous enough not to warrant guessing on silo hardness.

▲ **Launch Platform:** AFM 84, FI 81, and SMP 84 indicate a mobile deployment. SMP 84 says it is likely designed for mobile deployment, "with a home base with launcher garages equipped with sliding roofs; massive off-road, wheeled (TELs]; and necessary mobile support equipment for refires from the launcher." We accept this as far as the wheels of the launcher and the reload capability are concerned. We also assume that only one missile will be carried by the launcher.

# SUBMARINE-LAUNCHED BALLISTIC MISSILES

## SS-N-4 Sark

▲ **Service Entry:** Except for the 1955 and 1956 figures given by Rock, sources vary between 1958 and 1961 for service entry. Five sources, including JFS and SSF, give 1958. No more than two sources give any of the other years. I suspect there is unanimity that the first Zulu subs were armed with SS-N-4s in 1958 but considerable doubt as to when they became operational after that. Meyr 84 gives 1959, GSN 83 gives 1959–1960, US12 gives 1960, and SWM and Clns give 1961. JWS* 79 says that some believe it never was operational. We pick the year 1958 but with considerable uncertainty.

▲ **Retired:** Clns 80 and Clns+ 83 provide the only numbers series, and they show SS-N-4 dropping to 0 after 1979; 1979 is the last year JWS lists the weapon, but no other source says anything definite on the subject.

▲ **Nuclear Load:** Singe RV is unanimous.

▲ **Yield:** SecC 65 gave 2–3.5 MT, a figure echoed by SSF 82 and Meyr 84. All other sources give a vague 1 MT, often with indications of uncertainty. We use 3 MT as the closest value to a midpoint but still appropriately vague.

▲ **CEP:** Lu01 uses 1.8 km for the SS-N-4, -5, and -6, an estimate that seems insensitive to advancing technology. Aside from US12's 7–10 nm (13–18.5 km), the other 3 CEP estimates group fairly closely: 1.5 nm (2,778 m) by Clns 80 and vanD, 2.0 nm (3,704 m) by SSF 82, and 3–5 km by Meyr 84. 4 km is the midpoint of Meyr and the rounding of SSF 82, so we use that. 3 km would be not much worse than SS-N-5, which we would expect to be a substantial improvement over SS-N-4.

▲ **Range:** SecC 64–68 give 350 nm (648.20 km). Except for a few much higher estimates in the early years, other sources stay within the range of 555–650 km. Meyr 84 gives 500 km, with the note, "Values pertain to initial service period." Perhaps the range before SecC's first report in 1964 was less, but this line of evidence seems speculative. We use 650 km.

▲ **Propulsion:** This is a rare missile, and there is controversy over its type of fuel. Sources before 1980 say solid (except Rock 78), and sources after say liquid (except JFS 82). The most compelling evidence comes from data on the SS-N-17, which JCS 77–78 declare to be the first Soviet SLBM using a solid propellant.

▲ **Zulu Carrier:** 7 sources give Zulu as a carrier, 5 say it has 2 tubes, and FIar 76 and Rock 78 say it was installed on 7 boats.

▲ **Golf and Hotel Carriers:** These are very widely agreed as carriers with 3 tubes each.

## SS-N-5 Serb

▲ **Service Entry:** 1963 is given as IOC by JCS 75–78 and as when SS-N-5 first entered force by JCS 75–77, and this is followed by almost every source, with three exceptions giving 1964: SWM 77, IISS 82, and Meyr 84. We stick with 1963.

▲ **Retired:** No one suggests it is retired.

▲ **Nuclear Load:** Single warhead is unanimous.

▲ **Yield:** 8 sources give 1 MT, most indicating uncertainty. Clns 80 and Meyr 84 give 1–2 MT, SSF 82 gives 4 MT, and CFW 82, JFS 82, and GSN 83 give around 800 kt. With no good reason to choose, we keep to the median and vague 1 MT value.

▲ **CEP:** Except for the 7–10 nm given by US12 80 (13,000–18,500 m), we have 1.5 nm (2,778 m) by Clns 80 and SSF 82, 2,800 m by IISS 82, and 2.7–4.0 km by Meyr 84. Lu01 uses 1.8 km for SS-N-4, -5, and -6, which seems insensitive to advancing technology. 2,800 m seems to be the appropriately rounded consensus estimate, and that is what we use.

▲ **Range:** Official sources are abundant on this point. From SecC 65 through JCS 78, they give 700 nm (1,296 km). There is a 1-year anomaly where the estimate goes to 650 nm (1,203 km), found in SecC 70 and JCS 71 (and not a typo because the decrease is recognized in the text). Clns 80 gives 900 nm (1,666 km), and SMP 81 and 83 give 1,400 km. Almost all unofficial sources follow one of these values. Meyr 84 gives 1,100 km, specifying that the values apply to the initial service period. The choice is between the earlier 1,300 km and SMP's 1,400 km. There is a suggestion in GSN 83 that the range is believed to have increased (rather than the estimate's just changing), and so it seems more representative to use the 1,300 km that applied for most of the missile's life.

▲ **Propulsion:** The situation is much like that for SS-N-4, with FIar 76 and Rock 78 indicating that it was earlier thought to be solid fueled and JFS+ 83 switching to liquid fuel. JCS asserts that the SS-N-17 was the first Soviet solid-fueled SLBM.

▲ **Golf II and Hotel II:** These as carriers are unanimous, as is the 3-tube load. (JWS 81 suggests 2 or 3 tubes.)

## SS-N-6

▲ **Designation:** It is unclear whether *Sawfly* applies to the SS-N-6. SecC 70 says that in the past year occurred "the testing of a new, probably naval-oriented, ballistic missile. This could possibly be the Sawfly missile that was noted in a Soviet parade in 1967, and which at that time was described as a new naval missile." But a 1969 testing start is not suitable for the SS-N-6, which became operational in 1968. GSN 83 follows this with, "The SLBM given the NATO code name Sawfly was apparently a competitive prototype to the SS-N-6 and not the same missile." Although many sources use *Sawfly* as a designation, government source never do, and it is probably a spurious designation.

▲ **Propulsion:** Like the SS-N-4 and -5, there is some tendency to suggest solid fuel, but it is less pronounced for the SS-N-6. SecC 67 says solid or liquid fuel, but SecC 68-70 say storable liquid. JCS 75-76 say the liquid fuel is for all mods.

▲ **Yankee I:** The 16-tube Yankee as carrier is unanimous. SecD 75 confirm that production ended with 34 built.

▲ **Golf IV:** This is widely recognized as a test platform for Yankee. Clns 80 gives 4 tubes, and SSF 82 gives 5. We arbitrarily go with the 4 tubes.

## SS-N-6 Mod 1

▲ **Service Entry:** Although JCS 77 and 78 state that the Yankee sub was introduced in late 1967, JCS 75-78 and SMP 81 say the SS-N-6 entered service in 1968. Only vanD and US12 give another date, 1969. We use 1968.

▲ **Retired:** Mods 2 and 3 may have replaced some Mod 1s. JCS 75-76 say that it is expected Mods 1 and 2 will be phased out. FIar 76, FI 81, and FI+ 83 say that all Yankees have SS-N-6 Mod 3, implying that Mods 1 and 2 are retired. We need more evidence to conclude that it has been completely retired.

▲ **Nuclear Load:** Single RV is unanimous.

▲ **Yield:** The vague 1 MT is given by a number of sources. Clns 80 and Meyr 84 give 500 kt–1 MT. FIar 76 and FI 81 give 1–2 MT. But CBO gives 750 kt, and AWar 80 gives 0.70 MT, as does SSF 82. We use 0.70 MT because of AWar's reputation, its specificity and its consistency with the other estimates.

▲ **CEP:** Rock 78 gives 2.8 km, and Clns 80 gives 0.7 nm (1,296 m). Nitz 79 and CBO 83 give 0.5 nm (926 m), IISS 82 gives 900 m, and Meyr gives the wide range 0.9–2.0 km. But AWar 80, SSF, and CFW+ 84 give 1 nm (1,852 m). Lu01 uses 1.8 km for SS-N-4, -5, and -6, and this seems reasonable for SS-N-6. The 1,829 m of CB01 also seems close. We use the AWar value, rounded to 1,900 m.

▲ **Range:** After some earlier speculations of the same general magnitude, SecC 70 gave 1,300 nm (2,407 km). Thereafter this value is used by almost every source for the Mod 1. AWar 80 has an obvious typo, giving 2,400 nm rather than 2,400 km as the range.

## SS-N-6 Mod 2

▲ **Service Entry:** JCS 75 gives 1974 as IOC, and in JCS 76 both 1974 and 1973 are given, in different places. But JCS 77–78 cite 1973. FIar 76 and Rock 78 give 1974, probably misled by the first JCS report; JWS 81, IISS 82, and SSF 82 give 1973, which we use.

▲ **Retired:** FIar 76, FI 81, and FI+ 83 say that all Yankees have SS-N-6 Mod 3, implying that Mods 1 and 2 are retired. We need more evidence to conclude in our estimate that it has been completely retired.

▲ **Nuclear Load:** Single RV is unanimous.

▲ **Yield:** Aside from some 1–2 MT estimates (FIar 76 and FI 81) and a few vague 1 MTs, we have AWar 80 with 0.65 MT, followed by SSF 82. Not only is AWar 80's number precise, but it is less than the Mod 1, as one might expect for a missile with longer range. But it is so slightly less that it raises worries too. We use it nonetheless.

▲ **CEP:** The data are virtually identical to Mod 1, as is our judgment.

▲ **Range:** SecC 75 and JCS 75–76 give 1,600 nm (2,963 km). All other sources follow this value.

## SS-N-6 Mod 3

▲ **Service Entry:** For IOC, JCS 75 gives 1974, JCS 76 gives both 1974 and 1975 in different places, and JCS 77–78 give 1973. For actual service entry, most sources give 1974, with one estimate of 1973 and a few of 1975. We use 1974, based on the latest JCS's 1973 for IOC and the likelihood that operational deployment follows a bit later.

▲ **Retired:** No one suggests it is retired.

▲ **Nuclear Load:** SecC 74–75, JCS 77, and SecD 77 say 2 or 3 MRVs. JCS 78 says 2 MRVs, as do AWar 80 and SMP 83 and 84. SMP 81 says 2 MIRVs, the MIRV part surely a typo, though IISS 82 also gives MIRVs. All other sources give either 2 or 3 MRVs. The right answer is clearly 2 MRVs.

This may be an interesting case to test the biases of sources (or of the leaks that feed them), where no official source distinguishes between 2 and 3 RVs until JCS 78 and where even SecC 75 gives a bias for 2 by saying, "Two or possibly three MIRVs." The sources that noncommittally report 2–3 MRVs are MOW 76, FI 81, and SSF 82 (the last two after JCS 78 came out). The sources that report 3 MRVs are FIar 76, SWM 77, Rock 78, Clns 80, JWS 81, and JFS 82. Not one source in our sample gave 2 MRVs before JCS 78 did, and even then none caught on until after the SMP series came out.

▲ **Yield:** Estimates range from 200 kt by IISS 82 to 0.9 MT by Rock 78 and 500 kt–1 MT by Clns 80. Nitz 79 gives 500 kt. In the middle is AWar 80 with 0.35 MT, followed by SSF 82. We go with AWar 80.

▲ **CEP:** Estimates range from 0.7 nm (1,296 m) by Clns up to "nearly 2 km" for Rock 78. Nitz gives 0.5 nm (1,389 m). IISS has a higher 1,400 m for Mod 3 (than its 900 m for the other two). AWar has a rather vague 1 nm, followed by SSF. We stick with AWar's estimate, though we might expect the MRV weapon to be less accurate than the single RV, because we have no other higher plausible estimate.

▲ **Range:** SecC 74–75 and JCS 75–76 start with 1,600 nm (2,963 km), which all other sources follow. JFS+ 83 abandons the 1,600 nm it gave in 1982, citing 1,800 nm instead. There is no apparent reason for this shift, and possibly it is a typo. We stick with 3,000 km.

## SS-N-8

▲ **Mods:** SS-N-8 mods are a confusing muddle. Mod 3 is recognized only by unofficial sources and is universally said to carry 3 MIRVs. None of the sources implies it is deployed, and 1 (Rock 78) explicitly says it is not yet deployed.

The Mod 2 is recognized by 4 unofficial sources (SWM 77, Rock 78, FI 81, and JWS 81) as carrying 3 MRVs. But SMP 81, 83, 84, followed by IISS 82 and SSF 82, have it as carrying a single RV, distinguished from Mod 1 by a range increased to 9,100 km. These are obviously 2 different mods. Since the official SMP is using Mod 2 to refer to a single RV vehicle, it seems likely the MRVed and MIRVed mods of the SS-N-8 are either garbled reports of the SS-N-18 or were test programs designed to support the SS-N-18. But none of these sources believes Mod 2 is deployed. The only indirect hint is from SMP 81, which says the maximum operational range of the SS-N-8 (in general) is about 9,000 km. "Operational" here probably means that the weapon is operational but could be contrastive to a yet-longer range (of this still experimental vehicle) attainable without an operational payload. And although SMP still recognizes the range of Mod 1 as being 7,800 km, all official sources after 1978 have the range above 8,000 km. Possibly DoD saw some longer-range tests of SS-N-8 and, not having any idea whether they were a new real weapon, is assuming the worst. But on balance, we are forced to assume that the single RV Mod 2 is a deployed mod.

▲ **Propulsion:** Storable liquid is unanimous.

▲ **Delta I:** There is no controversy, with the number of submarines coming from AWar 80, FI 81, SSF 82, and JWS+ 83.

▲ **Delta II:** There is no controversy, with the number of submarines coming from AWar 80, FI 81, SSF 82, and JWS+ 83.

▲ **Golf III:** Recognized by 7 sources, including SecD 81. 6 tubes is reported by Clns 80 and SSF 82. 1 submarine is estimated by SSF 82 and GSN 83 and implied by CFW+ 84.

▲ **Hotel III:** Recognized by 6 sources, including SecD 81, and SMP 81, 83, 84. 6 tubes is by SMP 81, 83, 84 and SSF 82 and Clns+ 83. 1 sub is by SSF 82 and JWS+ 83.

## SS-N-8 Mod 1

▲ **Service Entry:** 1973 is given for IOC by JCS 75–78. CFW+ 84 gives the mysterious "1973/77," the 1977 probably referring to Mod 2, which is recognized for the first time by CFW in 84. The only discrepant estimate is 1972 from JFS 82.

▲ **Retired:** No one suggests it has left service, but with the Mod 2 coming in, this would be possible.

▲ **Nuclear Load:** As for the deployed system, no source suggests anything other than a single RV. SecC 74 says a MRV or MIRV will probably appear on SS-N-8 in the next year or so, but this is not followed up. FIar 76 lists 3 MRVs as an alternate payload, and FI 81 says it has been tested with 3 MIRVs. (*See* SS-N-8, "Mods.")

▲ **Yield:** JCS 77 and 78 say it carries a "relatively large warhead" but relative to what is not certain. Some sources have a vague 1 MT, some have 1–2 MT, and GSN 83 has 0.8–1.5 MT, the upper part of which is followed in CFW+ 84. The lower 0.8 MT part is from AWar 80, followed by AWST 82 and JFS+ 83. We go with AWar's 800 kt.

▲ **CEP:** FIar 76 gives a CEP of 400 m, with the comment that this is similar to Minuteman III and makes the SS-N-8 a counterforce weapon. But JCS 73 says it has not demonstrated hard target capability, and SecD 77 says that current SLBMs do not have significant hard target capability. This and SecD 73's statement that the accuracy is somewhat better than SS-N-6, presumably ruling out a dramatic improvement, make the 400 m figure not plausible, though it is followed by SWM 77, FI 81, and JWS 81. Rock 78 also gives 500 m. Aside from this low end of the scale, all 5 estimates vary between IISS's 1,300 m and AWar's 0.84 nm (1,555 m). Clns gives 0.8 nm (1,481 m), Rock 78 gives 1.4 km (in a different place from the 500 m figure), and CFW+84 gives 1,500 m. We base our result on AWar's figure, the only question being whether to adhere to strict rounding rules of the 1,555 m up to 1,600 or to bend them and round down to 1,500 m in deference to the other not-very-reliable sources. A peril to using 1,500 m is that it suggests far more vagueness than AWar intended, so we stick with 1,600 m.

▲ **Range:** The long range of this missile has been mentioned as constituting a surprise for the West. SecC 70–71 had the range at about 3,000 nm (5,556 km). JCS 73 has it at about 4,000 nm (7,408 km). JCS 74–77 and SecD 75–76 have it at 4,200 nm (7,778 km). All sources that recognize the Mod 1 give essentially this last value, including SMP 81, 83, 84. For sources writing about the generic SS-N-8, estimates tend upward, with SecD 78 giving 7,800 km or more, SecD 79 giving 8,000 km or more, and SecD 80 giving about 8,000 km. SecD seems to have taken the position that by 1980, at least, the Mod 2 was not deployed or did not exist. AWar 80 gives 9,000 km, and then there are the SMP 9,200 km figures. We ignore the later upward quavers as incorrect or else referring to Mod 2 and stick with the 7,800 km figure.

## SS-N-8 Mod 2

▲ **Service Entry:** Assuming the Mod 2 is deployed at all, there are a few indications that the year is 1977. GSN 83 gives 1977 as IOC, and CFW+ 84 gives "1973/77" for when deployment began for the generic SS-N-8, with the first matching the Mod 1 deployment date and 1977 left as the Mod 2 date. CFW+ 84 does (elsewhere) recognize the two mods. And SSF 82 gives 1976 for first flight test.

▲ **Retired:** If it is deployed, it is almost certainly not retired; no one suggests it is retired.

▲ **Nuclear Load:** *See* SS-N-8, "Mods," for a discussion of why this is single RV.

▲ **Yield:** Both sources that recognize Mod 2 as a single RV (IISS 82 and SSF 82) give a yield of 800 kt.

▲ **CEP:** The only source that gives an explicit CEP for Mod 1 is IISS 82, with 1,300 m. For Mod 2, IISS gives 900 m. The only other CEP value for Mod 2 is SSF 82's 0.84 nm (1,555 m), which is the AWar 80 figure for the generic. SSF 82 does not give a CEP for Mod 1. AWar 80 does have its CEP value in the same sentence with the 9,000 km range (and under the Delta II sub, though this is less important), which would support the Mod 2 interpretation. Reviewing Mod 1 CEP values, there is no higher CEP given by any source to select plausibly as the Mod 1 CEP, so we must conclude that the two mods have the same CEP.

▲ **Range:** The defining feature of Mod 2 is its range, and all sources follow (within rounding errors) SMP 81, 83, 84's 9,100 km.

## SS-N-17—Deployed?

Whether SS-N-17 even in late 1985 has operational status is a matter in question. It is known that it was installed in one modified Yankee (a test bed) in 1977 and that it has remained on that one boat and no others. As such, it counts against the SALT limits on SLBM launchers. All government publications have at least implied that it has operational capability by removing the X from the designation around 1982. Perhaps it does have operational capability, but this seems militarily unlikely because it has not been deployed on any other submarines despite ample time to do so. JWS+ 83 says it is believed to have had periodic sea trials since 1977, suggesting it is probably not outfitted with nuclear warheads. Two motivations come to mind for why DoD might consider this system operational even if it is not. First, it may be that because it is counted against SALT limits, it would be embarrassing to admit that it is not operational and thus penalizing the Soviets. Second, consistent with a propaganda approach that compares numbers of different US and Soviet systems, this chocks up the Soviet tally.

## SS-N-18

▲ **Propulsion:** Liquid fuel is nearly unanimous; by this point only two sources are specifying that the liquid is storable, probably because unstorable liquid has disappeared as an option. The exceptions are solid specified for Mod 1 by IISS 82 and for the generic by JFS+ 83, changing its earlier liquid estimate from JFS 82.

▲ **Delta III:** Delta III is unanimous, as is its 16 tubes. 15 for the number of submarines is simply the 14 that SMP 83 says are launched, plus one. A few sources suggest they may be retrofitted on Delta IIs, but none suggests this has actually been done.

## SS-N-18 Mod 1

▲ **Service Entry:** Dates given by sources vary from 1977 through 1979. ArkS also report that DoD revised its estimate of initial deployment of Delta III from 1978 to 1976. US06 78 and Clns 80 cite 1977. Winter 1977–1978 is given by FI 81 (for "became operational"). 1978 is given by GSN 83 (IOC), IISS 82, vanD, and CFW+ 84. And 1979 is given by JWS 81 and US17 81. The only official estimates (aside from graphs, which are in the same general range) are the blank in the IOC chart in JCS 77 and 78 and the SecD 79 statement that the Soviets have begun deployment, which would seem to allow either 1978 or 1979. In the midst of this confusion, we take the median of 1978 but with no great confidence.

▲ **Retired:** No one suggests that Mod 1 is retired, except DoD in assumptions for a warhead count, which ArkS counter with "DoD admits that all three versions of the SS-N-18 are fielded."

▲ **Nuclear Load:** 3 MIRVs is universally recognized for Mod 1, including by AWar 80 and SMP 81, 83, 84.

▲ **Yield:** AWar 80 gives 200 kt, as do 5 other sources. Clns+ 83, Nitz 79, and CBO 83 give 500 kt. We follow AWar 80.

▲ **CEP:** AWar 80 gives 0.76 nm (1,407 m), as do IISS and SSF. For the generic SS-N-18, Clns 80, AWST 82, GSN 83, and JFS+ 83 follow this number. Clns+ 83 updates his number to 0.5 nm (926 m), now pertaining to the Mod 1. CFW 82 gives the generic CEP as 800 m but then updates this to 1,100 m in CFW+ 84. Nitz 79 gives 0.3 nm (555.60 m), which is also given by CBO 83 for the generic. Lu01 gives 1,350 m; CB01 gives 914 m for the generic. In sum, aside from the basic consensus estimate of 1,400 m, there are some lower varying estimates (and also vanD with 2,590 m), none attracting more than 2 votes and involving 2 updates, 1 down by Clns and 1 up by CFW. We stick with AWar's 1,400 m.

▲ **Range:** AWar 80 gives 16,600 km, which is surely a typo, probably for 6,600 km. SMP 81, 83, 84 give 6,500 km. All other sources agree except for IISS 82, which cites 7,400 km, but IISS+ 83 adopted 6,500 km.

## SS-N-18 Mod 2

▲ **Service Entry:** Clns 83 gives 1977, and IISS+ 83 gives 1978, the same years they give for Mod 1. There is no other indication of when deployment began. We use 1978.

▲ **Retired:** No one suggests directly that it is retired, though SMP 83 and 84 say each Delta III carries 16 MIRVed SLBMs. But ArkS say DoD admits that all 3 mods are deployed.

▲ **Nuclear Load:** Single RV is unanimous, including among government sources.

▲ **Yield:** AWar 80 gives 450 kt, which is followed by 5 other sources. The only exception is Clns, citing a vague 500 kt–1 MT.

▲ **CEP:** AWar 80 and SSF 82 give 0.76 nm (1,407 m), Lu01 gives 1,350 m, and Clns 83 gives 0.5 nm (926 m). All of these sources give the same value for Mods 1 and 2. Only IISS 82 goes from 1,400 m to 600

m. One would expect the single RV version to have better accuracy than the MIRV, but because Mod 2 has achieved greater range by offloading MIRVs, there are some factors working against this tendency: greater range decreases accuracy, and the Mod 2 warhead, which is only twice the yield of Mod 1, may not be much heavier. We use AWar's 1,400 m.

▲ **Range:** AWar 80 reports 8,000 km, as do SMP 81, 83, 84, and almost all other sources. FI 81 says 16,000 km has been reported for the single RV version (with some skepticism in its own report). And IISS 82 gives 8,300 km, though it corrects to 8,000 km in IISS+ 83. We use 8,000 km.

## SS-N-18 Mod 3

▲ **Service Entry:** The only estimate for deployment beginning is Clns 83, who gives 1978, one year later than his 1977 figure for Mods 1 and 2. We go with this figure but with great uncertainty.

▲ **Retired:** No one suggests it is retired.

▲ **Nuclear Load:** 7 MIRVs is unanimous, including by SMP 81, 83, 84.

▲ **Yield:** The only yield values are 200 kt by IISS 82 and Nitz 79 and 500 kt by Clns 83. These are the same values these sources give for Mod 1. Since neither gives a shorter range for Mod 3 (IISS gives the same range and Clns gives a longer one) and both mods were deployed at nearly the same time, something is awry in the estimates. How Mod 3 can lift 7 RVs the same distance Mod 1 can lift 3 of equal yield (and weight) is a mystery. Range here is a critical variable, since variations between 6,500 and 8,000 km crucially affect the USSR's ability to hit US targets from within Soviet SSBN sanctuary zones. There is no way to resolve this discrepancy while remaining true to the source data. But US intelligence should be excellent on missile range and nuclear load. Yield is far less reliably known. Since there are only two (discrepant) values for Mod 3 yield, this seems the logical place to apply an adjustment. Yield and RV weight are not directly proportional, so our guess would be around 50–75 kt for Mod 3, but to keep appropriate vagueness in the answer, we choose 100 kt.

▲ **CEP:** The only 3 CEP values given for Mod 3 are Clns 83 with 0.5 nm (926 m), IISS with 600 m, and Nitz 79 with 0.3 nm (555.60 m). For each of these sources, the Mod 3 value is the same as the Mod 1 value. In that judgment, I would agree; there is no reason why a 7 RV version of a missile deployed at the same time as a 3 RV version should have better accuracy (especially if it has a lighter RV). (Lu01 goes from 1,350 m for Mods 1 and 2 to 600 m for Mod 3.) But I have seen no reason to accept any of these sources' numbers for other mods and thus do not accept them now. Higher CEP figures are given for the generic SS-N-18, some of which may recognize the 7 RV mod. We keep the same 1,400 m value we used for the other mods. If there were evidence that the 7 RV mod was deployed significantly more recently and thus could incorporate guidance improvements, a lower CEP would be worth considering.

▲ **Range:** SMP 81, 83, 84 give 6,500 km as the range, as does every other source except Clns 83, who gives 4,500 nm (8,334 km).

## SS-N-20

▲ **Service Entry:** After all the speculation in sources about then-future events is dispensed with, the data are that SMP 83 says the first Typhoon will be fully operational by the end of 1983; IISS+ 83 speculates it is now operational, and SMP 84 says it is now fully operational. CFW+ 84 also gives 1983. Although SMP data are consistent with 1984, 1983 is more likely, and that is what we use.

▲ **Retired:** No one suggests it is retired.

▲ **Nuclear Load:** We use 6–8 MIRVs because this estimate was first reported in SMP 85 and is lower than earlier official and unofficial estimates. (We assume the tendency of SMP is to report the highest plausible number.) Early sources, including SMP 81, gave 12 RVs, and Nitz 79 assumed 14 RVs. These figures may have been simple projections from the 10–14 MIRV capacity of the US Poseidon SLBM. SMP 83, 84, and JCS 84, and most other sources, use 6–9 MIRVs, presumably based on observed tests with some uncertainty. JFS+ 83 and JWS+ 83 seem slow to catch up, only in 1983 updating to 12 MIRVs. SMP 84 suggests that each Typhoon adds "some 200" RVs to the force, implying 10 RVs per SS-N-20 missile. It seems likely that the interpretation of test data led to the downward revision from 6–9 to 6–8 and that this is the latest and most accurate information.

▲ **Yield:** Nitz 79 gives 200 kt, and CBO 83 gives 100 kt. There are no other data. Since Nitz is obviously speculating, we use CBO's 100 kt.

▲ **CEP:** Data are very scanty, with Nitz 79 giving an

obviously conjectural 0.25 nm (463 m), CBO 83 giving 0.3 nm (555.60 m), and CFW+ 84 giving 600 m. An accuracy improvement over SS-N-18 would make sense because SS-N-20 is considerably newer; however, the missile is solid fueled, and the USSR has not to date constructed a successful solid-fueled ICBM or SLBM. CFW 82 and GSN 83 speak of significant development problems. And although official sources do not give CEPs, they often suggest that accuracy of the latest missile will be better than that of the previous ones. For the SS-N-20, SMP 81 says that it is assumed Typhoon's missile will be more capable than SS-N-18, "possibly having greater range, better accuracy, higher payload, and more warheads." But SMP 83 and 84 are silent on the question of accuracy of the SS-N-20, though SMP 84 does suggest that the experimental SS-NX-23 will be more accurate than the SS-N-18. With strong evidence that the sources are just guessing or conjecturing, there is no harm in guessing and conjecturing ourselves. 1,000 m is appropriately vague, better than SS-N-18 but not much better. (Recall that SLBM accuracy is worse than ICBM accuracy to the degree that submarines cannot know their own position accurately.)

▲ **Range:** Estimates vary from "> 4,000 nm" (7,408 km) to 5,000 nm (9,260 km). JCS 85 says "approximately 5,000 nm," but three main unofficial sources (FI+ 83, JWS+ 83, and CFW+ 84) use 8,000 km. All 4 editions of SMP, however, give 8,300 km, as does JCS 84 ("booster range"). Since no other number is as widely used, we accept 8,300.

▲ **Propulsion:** Solid fuel is unanimous.

▲ **Typhoon:** Typhoon carrying 20 tubes is unanimous; 3 in service is from SMP 85.

## SS-N-23

▲ **Service Entry:** We follow SMP 85, which says that 2 Delta IV submarines designed to carry the SS-N-23 have been launched, and JCS 85, which says that the first Delta IV "should become operational in 1986." This leads to January 1987 as the date in service.

▲ **Nuclear Load:** SMP 84 indicates it will carry more MIRVs than the SS-N-18, which can carry 1, 3, or 7 warheads. No new information on warheads was reported in 1985, though flight tests have been reported since 1984. We report only that it will carry MIRVs.

▲ **Yield:** No source speculates. We use the same vague 100 kt we used for the SS-N-18 Mod 3 and SS-N-20.

▲ **CEP:** SMP 84 and 85 say it is more accurate than the SS-N-18, and SMP 85 says it is the "most accurate" Soviet SLBM. We conservatively give the same round 1,000 m estimate that we give for the SS-N-20, however, roughly a 30 percent accuracy increase over SS-N-18.

▲ **Range:** As for MIRVed warheads' yield and CEP, published estimates of range are vague. SMP 84 says it is "long range," and SMP 85 calls it the "most capable long-range" MIRVed SLBM. We use the 5,000 nm (9,300 km) reported in JCS 85 for the SS-N-20 as a likely round figure for gradually increasing SLBM range.

▲ **Propulsion:** One of the solid facts about this system is its liquid propulsion, which is given explicitly by SMP 84 and also implied by the statement that it is a follow-on to the liquid-fueled SS-N-18.

▲ **Delta IV:** JCS 85 and SMP 85 report the launching of a new class of strategic submarine, the Delta IV, designed to carry the SS-N-23. SMP 85 specifies that it will carry 16 SS-N-23s and that 2 have been launched.

# INTERMEDIATE-RANGE BALLISTIC MISSILES

## SS-3 Shyster

▲ **Numbers and Service Dates:** 1955 is given for the year of service entry by the only 3 sources who give a year: FIar 76, Rock 78, and Meyr 84. It does fit the constraints of being after the USSR had nuclear weapons and before the SS-4, but these are weak constraints.

No source gives a year for retirement, though no source suggests it is still in service. The SSF 82 numbers chart shows 28 in 1965 and none in 1970, indicating a late 1960s retirement. We choose 1968 as the middle of that range, though this is highly uncertain. It is also near the completion of SS-4 deployments.

Numbers data are virtually nonexistent. SSF gives 24 in 1955, 48 in 1960, and 28 in 1965. SMP 81 says there were over 700 fixed SS-3, -4, and -5 launchers at their peak in the mid-1960s, and this is consistent (from what is known of SS-4 and SS-5 numbers) with the 28 of SSF. With our assumption of 1968 as the last year in service, we construct a very tentative numbers series. If 1955 is the year of service entry, then the beginning-of-year number must have been 0. We assume increments of 24 over the next 2 years and a stable 48 until the decline began during 1963. At that point we reduce 10 a year until all are gone. This estimate is in line with the SSF data but appropriately does not give any impression of precision.

Peter Almquist of MIT's Political Science Department has written an unpublished paper, "Data on RVSN M/IRBM Forces," revised 26 September 1983. Based on early declassified official reports and on *USSR Facts and Figures 1980* (John Scherer), he has constructed tables for SS-3s, -4s, and -5s through 1977. Although there are many differences in detail, the main difference is that SS-3s build up to a peak of 125 that is maintained from 1960 to 1963 rather than SSF's 48. Apparently *USSR Facts and Figures 1980* has a graph of SS-4 and SS-5 deployments that is consistent with certain known official information, precisely drawn, and itself from an official source. Almquist's SS-3 numbers are derived from this graph's numbers, subtracted from the early declassified official numbers that aggregate SS-3s, -4s, and -5s. This methodology is reasonable, but not at all certain since these sources may disagree on their notions of how many SS-4s and -5s there were. Perhaps the main point is that available data are consistent with a wide variety of hypotheses about such basic facts as how many SS-3s were ever deployed.

▲ **Nuclear Load:** Single RV is given by 2 sources and is undoubtedly left unsaid by others for such an early missile.

▲ **Yield:** The only source estimating yield is Meyr 84, giving 100–500 kt. We use the midpoint of this range, with great uncertainty.

▲ **CEP:** Meyr 84 gives the only CEP estimate, 2.5–5.0 km. This range seems reasonable given that the CEP value for the follow-on SS-4 is around 2.5 km and the SS-6 ICBM that came out 2 years later had a 3.7 km CEP (over a much longer range). 3.7 km is both the center of the Meyr 84 range and the same as the SS-6, so we use it.

▲ **Range:** Range estimates vary widely, a situation also recognized by JWS* 75 and FIar 76, both of which suggest that variations may be due to a change from LOX-alcohol to LOX-kerosene fuel. But having recognized this, JWS* 75 gives 800–1,200 km, and FIar 76 gives 900 km, FIar clearly indicating that it is giving the longer-range, later value. Thus the fuel change explanation does not remove all discrepancies. Perhaps the values it does explain are those below 800 km, such as the 400 mi (643.60 km) of COTS 58 and the 250 mi (402.25 km) of Zaeh 61. AWST* 71 and HRST give values around 1,200 km, as does Meyr 84. Rock 78 gives 900 km, and US09 uses the 800–1,200 km figure.

In sum, there are 3 values used by sources: 900 km, 1,000 km (midpoint of 800–1,200), and 1,200 km. None of these sources has great authority, so we use 1,200 km since this figure is recognized by the most sources.

▲ **Propulsion:** Propulsion is universally thought to be liquid oxygen and either alcohol or kerosene. Both of these combinations are unstorable.

▲ **Reloads:** No source explicitly mentions reloads, but the above-ground firing table that has been suggested as its basing mode implies a reload capability.

## SS-4 Sandal

▲ **Numbers and Service Dates:** The year operational is given by JCS in 1982 as 1960, in 1983 as 1959, and in 1984 as 1958. This progressive reevalution 20 years after the fact is puzzling and does not inspire confidence in the most recent number. 8 sources give 1959 as the year for service entry; 2 (JWS 81 and SSF 82) give 1958. We follow the majority and use 1959 (though none of the 8 sources is particularly authoritative).

The history of the early deployments of SS-4s is unclear. SSF gives 0 in 1955, 200 in 1960, 608 in 1965, and 508 in 1970. Having no better idea what to do, we assume a deployment of 150 each year, starting with service entry in 1959, which is consistent with SSF's 200 figure for 1960 (later in the year). Thus the plateau of 600 is reached by the beginning of 1963 and stays constant thereafter.

SSF's 508 for 1970 is anchored quite well by SecC 71, which gives 424 in soft silos and 84 in hardened silos for the end of 1970. The decline from 608 also makes sense given the story in A168 that about a quarter of SS-4s and -5s were transferred east, mostly in 1968, and that all of those were dismantled during 1969 and 1970, replaced with SS-11s. SSF's precise 608 in 1965 sounds like the precise 508 with an imprecise 100 added to it. Clns has 550 for 1970 and 500 for 1971 and succeeding years. In line with the "shift east and dismantle" story, we assume the number stayed at 600 to the beginning of 1968 and then dropped roughly 30 a year to the 508 value of 1971.

From this point on the quality of the data is better. Clns and IISS have annual series, SMP 83 has a graph of deployments, and there are official numbers for the latest years. Unofficial estimates are generally 500 through 1976; SSF has 508 for 1975, and we assume that remains the correct answer until the decline begins. Clns has the decline beginning in 1977 and proceeding faster than the others, while IISS has it beginning in 1980 and then declining rapidly. (The IISS estimates, appearing each year for that year, are less likely to result in a consistent series than a retrospective study.) The SMP 83 graph generally lies between these 2 poles. We use numbers taken from this graph (where official numbers are not actually available), bolstered by a certain amount of calibration. JCS 82 gives 300 for December 1981, which is just what my reading of the graph shows for 1982 (I interpret the graph as giving January figures). SMP 83 gives 232 as of 1 January 1983; I read the graph as 245. SMP 84 gives 224 as of 1 January 1984; I read 220. SMP 81 gives 320, while my graph reading was 365 for 1981, but SMP 81 appeared in September and probably reflected a midyear state of the decline. One final operation is performed on the graph numbers. I had earlier estimated only to the nearest 5; here I round to the nearest 4, reflecting the launcher groupings of 4. The figure for 1985, 120, is taken from SMP 85.

▲ **Nuclear Load:** Single RV is unanimous, including official sources.

▲ **Yield:** SecC 65 says MR-IRBM yields are in "the kiloton to the six MT range," which is quite broad. 9 sources indicate the vague 1 MT. SSF 82 indicates 2 MT, and Meyr 84 gives 1–3 MT. These higher values have the flavor of SecC values, but at least in our declassification of SecC, there are no yield values for SS-4. We stick with the vague 1 MT value.

▲ **CEP:** Estimates by Clns 80 of 1.25 nm (2,315 m), by IISS 82 of 2,300 m, and by JWS+ 83 of 2,400 m are quite closely clustered, and they are very near the midpoint of Meyr 84's 1.5–3.0 km. A higher 1.5 nm (2,778 m) is provided by SSF 82, and that is near the midpoint of BA04's value of 6,000–12,000 ft (1,828.8–3,647.6 m). We go with the Clns 80 value, rounded.

▲ **Range:** Except for Zaeh 61 (which may have misidentified the missile) giving a much shorter range, the range of estimates varies from 900 mi (1,448 km) to 1,400 nm (2,592 km). The official estimates start with 1,020 nm (1,889 km) by SecC 65–70. Next is 1,050 nm (1,944 km) from SecC 71 and JCS 77–78. Finally there is 2,000 km by JCS 82–84 and SMP 81, 83, 84. These numbers do not differ much, and we go with the most recent (and most rounded) 2,000 km. The bulk of nonofficial sources have values between 1,800 and 2,000 km.

▲ **Propulsion:** Everyone agrees on liquid fuel. FIar 76 recognizes a LOX-kerosene version (which is an unstorable liquid) originally, followed by a nitric acid–kerosene version (which is storable) more recently. All other sources mention only the later version. Assuming the earlier version was deployed, both fuels are correct, but in picking one short answer, storable fuel gives a better picture of the missile.

▲ **Launch Platform:** Rock 78 and FIar 76 are the only sources to state explicitly that the system can be mobile, though others (IA01 65, MOW 76, JWS 81, and AFM 82) follow the gist of the rest of FIar 76: "In the mobile role, the complete weapon system consists of a convoy of about a dozen vehicles carrying ground-support equipment and propellants. A 20-man firing crew is required to erect and fuel the missile." On the other hand, all the official and semiofficial sources, including SecC 65 and 71, SMP 81, and JCS 82–84, speak only of fixed sites. It is unlikely that the more recent official sources are omitting mobile SS-4s since they are contrasting the SS-4 to the SS-20 in mobility. The earliest story of extensive auxiliary vehicles with a mobile system is IA01 from 1965, and my conclusion is that this was an early speculation that turned out to be false but is still permeating the literature.

▲ **Time to Launch:** The oft-repeated quotation about ancillary equipment is associated with the mobile deployment mode and makes little sense for fixed site deployments. But the Special National Intelligence Estimate of 19 November 1962 gave the firing time as 4–6 hours for SS-4, according to Pr01. That source also says the Soviets could launch SS-4s within 8 hours of a decision to do so. Gr01 says launching would take several hours. And Treverton, in A168, says a day or more is needed to prepare for firing.

▲ **Reloads:** 4 sources, including SecC 65 and JCS 81, indicate that soft SS-4 sites have a refire capability, though hard ones do not.

## SS-5 Skean

▲ **Numbers and Service Dates:** 1961 is given as the year operational by JCS 82–84 and by 9 other sources. Only SWM 77 suggests with uncertainty that it was first deployed in 1964, while recognizing 1961 as the date of service entry, a story that seems internally inconsistent.

Working backward, we have a firm figure of 90 by SecC in 1971. SSF also gives 90 in 1970 and 97 in 1965. Given the "move east and dismantle" story (*see* SS-4), 97 is reasonable for 1965. We assume a linear increment from the 0 of 1961 to the 97 of 1965 (in increments of 25 to avoid any unwarranted impression of precision). Of course, the buildup could have occurred much more quickly. As for SS-4s, we assume that the first decline took place during 1969 and 1970.

SS-4 and SS-5 are so often discussed together that data available for one are nearly always available for the other; in particular the kinds of data available on the decline are just as they are for SS-4. But Clns, IISS, and graph data from SMP for the SS-5 are more consistent than for the SS-4. The graph data calibrate as well as they did before with official numbers, and we use them in the same way, with one exception. SS-5s occur in groups of both 3 and 4, so there is no reason to round further than the nearest 5.

▲ **Retired:** 13 are listed by JCS 84 for 1 January 1984. SMP 84 says "no longer operational," and "By remaining missiles were not operational on 1 January, but we give 1984 because SMP 84 does not say that. Complete retirement not later than 1984 is confirmed by SMP 85.

▲ **Nuclear Load:** Single RV is unanimous, including by government sources.

▲ **Yield:** SecC 65 has "the kiloton to the six MT range" for MR-IRBMs generally. 7 sources give the vague 1 MT, and Meyr 84 gives 500 kt–2 MT. Rock 78 gives 1–5 MT, and SSF 82 gives 4–6 MT, as does OR01. Not knowing where the higher estimates come from, we stick with the majority's 1 MT.

▲ **CEP:** Clns 80 gives 0.6 nm (1,111 m), as does IISS 82 with 1,100 m. SSF 82 gives 1 nm (1,852 m) as does BA04, with 6,000 ft (1,828.8 m). Meyr 84 encompasses this range with his 0.9–1.9 km. We go with SSF 82's 1 nm, rounded to 1,900 m, but for no particularly good reason.

▲ **Range:** Official estimates are completely consistent over time, with SecC 65–71 giving 2,200 nm (4,074 km), as do JCS 77–78. SecD 80 gives 1,900–4,100 km for SS-4 and -5, with SS-5 having the longer range, and SMP 81, 83, 84 and JCS 82–84 give 4,100 km. Other estimates vary from 2,413 km by HRST 75 to 4,907 km by Clns+ 83, with a tendency to be lower than the official estimate, though this may be due to a confusion of statute with nautical miles.

▲ **Propulsion:** All sources agree on liquid fuel. FIar 76, SWM 77, and AWST 82 say storable, while Rock 78 implies unstorable. We go with the majority.

▲ **Launch Platform:** Sources mention soft and hard sites, but none suggests it is mobile, and Rock 78 explicitly denies mobility.

▲ **Time to Launch:** All estimates for SS-5 are the same as for SS-4, except that the 19 November 1962 SNIE gives 6–8 hours for SS-5 instead of 4–6. We go with the 7 hour midpoint. (The 8 hour figure applies only to SS-4 as well.)

▲ **Reloads:** 4 sources, including SecC 65 and JCS 81, indicate that soft sites have a refire capability, implying (though none stating) that hard sites do not.

## SS-20

▲ **SS-20 Mods:** IISS 82 lists Mods 1, 2, and 3, and the distinctions, if not the designations, are shared by FIar 76 and JWS 83. Mod 1 carries a single large warhead, Mod 2 carries 3 MIRVs, and Mod 3 carries a single small warhead over a longer range. Since every source, including official ones, gives the warhead count as 3 MIRVs, we infer that only this mod is deployed. There seems no need to use a mod designation, however, when the other mods are not deployed or expected to be deployed.

▲ **Numbers and Service Dates:** Every source since 1980 gives 1977 as the IOC or year operation began. A few earlier sources gave 1976, but this is not consistent with official sources. JCS 77 says initial deployment is expected "in the near future," and JCS 78 says initial deployment has begun, and SecD 78 says it is assessed to be operational. 1977 is the year we choose, with the only uncertainty being whether deployed systems were operational before 1978 began.

As a much discussed weapon, the numbers data for SS-20 are some of the best available for any Soviet weapon during its buildup. Although the SS-20 has replaced the SS-4 and SS-5 in some sense, it is not a 1-for-1 replacement, and so (unlike ICBMs) the rate of SS-4 and -5 decline is not linked to the rate of SS-20 buildup. All SS-20 numbers are in nines because that is the operational unit for the missile.

With 1977 as the year it entered service, we start with 0 for 1977. The 1978 figure we choose is 9, which is close to the 10 of OR01 and the rounded version of the 12 I got from the SMP 83 graph. For 1979 I use 72, which is rounded from SMP 83's 75 and close to OR01's 70. The 1980 figure is the most uncertain. The SMP 83 graph seems to indicate 130, while SecD 84 and OR01 say there were 140. My solution is to adopt 135, which is a multiple of 9 between the two. My 1981 value of 180 is from SecD 81, though there is some concern because that may be later in 1981 than the beginning of the year. For 1982 we have JCS giving 260+ for December 1981 and OR01 giving 270 for December 1981. Although 261 is a possibility, we go with the 270. The 1983 and 1984 numbers are from official sources, labeled for 1 January. During 1984, SMP 85 reports, the Soviets started construction on many new SS-20 sites, more than offsetting the dismantling of some SS-20 launchers at 2 bases being prepared for deployment of SS-25 mobile ICBMs. Following SMP 85, we give 400 peak deployed (as of 1 January 1985), up from 378 in 1983 and 1984, but this number is expected to rise with the completion of additional SS-20 launchers under construction.

▲ **Nuclear Load:** All official sources, including SecD 77–81, JCS 79, 81–84, and SMP 81, 83, 84, have recently suggested 3 MIRVs. Some sources suggest single RV mods, but absent any official comment on them, we assume they were never deployed.

▲ **Yield:** Clns 80, AFM 82, IISS 82, JWS+ 83, and a few other sources give 150 kt. SSF 82 gives 150–500 kt, as do BA03 and OR01. Meyr 84 gives 100–300 kt.

We have no explanation for these estimates, but because they are ranges, they imply uncertainty. We take the majority 150 kt.

▲ **CEP:** Estimates vary widely, the lowest being 300 ft (91.44 m) by US15 81. But this source's accuracy is in doubt since it gives a throwweight for SS-20 of 12,000 lb, nearly double that of other sources for the SS-19 ICBM. Other low estimates are 0.1–0.4 km by Meyr 84 and 150 m by JP01 82, 440 ft (134 m) by Treverton, in A168. High estimates are 750 m by FI 81 if launched from presurveyed sites and the 2,500 ft (762 m) by AFM 82 with the same condition. The other 6 estimates vary between 0.16 nm (296.32 m) (by SSF 82) and 457 m. One interesting change is Clns 80, giving 0.2 nm (370.40 m), and Clns+ 83, giving 0.23 nm (425.96 m). Clns is the most authoritative source speaking on the subject, official sources noting only marked improvements over previous systems (a condition that 762 m would satisfy). We go with Clns's latest value, rounded to 430 m. This is the same as the early SS-17s and -18s, which seems reasonable considering Soviet trouble with solid fuel, countered by the longer range of the ICBMs.

▲ **Range:** Official estimates of range generally tend upward over time. SecD 77 says it has demonstrated a range of 2,000 nm (3,704 km) or more; SecD 78 says at least 3,000 km; SecD 79 says well over 3,000 km; SecD 80–81 and JCS 81 say 4,400 km or more; and SMP 81, 83, 84 and JCS 82–84 give 5,000 km. JWS+ 83 reports the Soviet figure for its range as 4,000–4,500 km. Clns 80 gives 2,700 nm (5,000 km, sic), and Clns+ 83 gives 3,100 miles. If these are nautical miles, this is a higher estimate (5,741 km), but if they are statute miles, it is 4,999 km. Other unofficial estimates stay closely within the bounds of official estimates. We go with the official 5,000 km, though the Soviet 4,500 km is worth considering.

▲ **Propulsion:** 12 sources, including official ones, give solid propulsion. No one differs.

▲ **Launch Platform:** All sources, including official ones, indicate a mobile system, with no dissent.

▲ **Time to Launch:** Although all sources agree that the responsiveness of SS-20 is much better than that of SS-4 or SS-5, the only hard estimate is by SMP 81: "Reaction time (max.)" of "1 hour+." What maximum reaction time contrasts with is unkown and perhaps cancels the "+" on the 1 hour. Perhaps 1 hour is the best guess.

▲ **Reloads:** All sources, including official ones, indicate a refire capability. No one disagrees. 1 reload per launcher is given by SMP 81, 83, 84 and IISS 82.

# SHORT-RANGE BALLISTIC MISSILES

## FROGs

▲ **FROG Models:** 6 different models of this unguided rocket are recognized, bearing the designations FROG 1 through FROG 5 and FROG 7. (FROG 6 is recognized to have been a nonoperational training model.) FROG 1 and FROG 7 are clearly distinct rockets with distinct characteristics. It is also clear that FROGs 3, 4, and 5 are closely related. It is my impression that these rockets have been identified by their appearances in Moscow parades, with each different model receiving a different designation. How these parade models correspond to deployed models is not clear. I think that from an unofficial source's point of view, covering all the bases requires listing separately each model seen in Moscow because the data are concrete. But only some of the models may have been deployed, and recognizing all the minor variants is making much out of small differences. Probably bigger differences exist in other weapons that are unremarked because less is known concretely about them.

Sources show a tendency to give more attention to FROGs 3, 5, and 7, particularly the 7. FROGs 3, 5, and 7 are the only models ever referred to by JCS, and JCS 81 says that non-Soviet Warsaw Pact (NSWP) forces have older FROG 3s and 5s, being gradually replaced in NSWP forces by FROG 7s. And IISS+83 refers to FROGs 3 and 5 in its NSWP data. (IISS 82 referred only to FROG 3 in this context.) If we were assigned to pare down the number of FROG models, the best candidates would be FROGs 2 and 4. Possibly some of the various models were for export only.

▲ **Numbers:** There are different ways of counting FROGs. One is to count those facing Europe versus those facing Asia. Another is to count totals versus those in Soviet service versus those in NSWP service. The number we are seeking is a national total of those in Soviet service. The only official numbers we have are for WTO totals facing Europe, where SMP 83 gives 650 for the end of 1981, and SMP 84 gives 700 for the end of 1983. These figures also include SS-21s.

IISS gives figures consistent with these official ones. For 1983 they give 198 NSWP and 620 Soviet, of which 440 are oriented to Europe (leaving 180 for Asia). The 198 and the 440 combine to yield 638, which leaves room for 62 SS-21s in the SMP 700 figure. This is exactly the figure IISS gives for SS-21s (though it may be a bit high).

Clns gives a historical series going the furthest back (to 1970), but he does not describe what his numbers represent. In a book on the US-Soviet military balance, they should be national totals, and as such they are on the same order as IISS's. His 1982 figures are 80 higher than the IISS figures for 1983, and for 1973–1976 his figures are 50 higher than the corresponding IISS figures.

With no data before the Clns series begins in 1970 and IISS going back only to 1973, we are left with an interpolation problem. If we assume a linear increase from the 0 of 1957 to the 600 of 1973, we come up with 37.5 each year; this is not far from the Clns growth rate from 1970 to 1973 of 50 per year. Our ultimate solution is to assume 40 per year, with growth of only 20 in each of the first 2 years of 1957 and 1958.

There are probably 4 FROG launchers for every Soviet motorized rifle and armored division. If this figure has been used as the basis for estimating the number of Soviet FROG and SS-21 launchers (with 180 divisions, the resultant 720 is close to SMP's number), then the entire exercise is in doubt, for it is unlikely that many Soviet cadre-strength divisions have such equipment.

While NSWP FROG numbers are not directly relevant here, it is worth noting that the IISS numbers for these between 1975 and 1983 vary between 198 and 208.

▲ **Numbers for Models:** Ideally we would give numbers histories for each FROG model, but available data allow only a tenuous distinction between all earlier FROGs (1–5) and FROG 7s. IISS gives 600 FROGs between 1973 and 1976. In 1977 they give about 450 FROG 7s. Between 1978 and 1980 they give no totals for FROGs alone but in 1981 and 1982 estimate 482 for FROG 7. In 1983 they give 620 for FROGs in general. Although it is possible that these numbers measure the same thing (all Soviet FROGs) but differ on the authors' views of which FROGs are in service, it is striking that the numbers are substantially less when and only when FROG 7 alone is mentioned. We interpret these to be FROG 7 numbers that exclude FROGs 1–5 and use them to construct a FROG 7 series and a residual FROG 1–5 series. In the 1976–1977 period, 150 FROG 1–5s seem to be left over; in the 1982–1983 period, 138 seem to be left over. We also know the FROG 7 did not enter service until 1965. To reach 450 by 1977, we need nearly 40 per year; we correct this by assuming 10 for the first year. In the absence of other information, we assume that in the next year, 1978, FROG 7s reach their peak of 480, which is their current level. This is also near the time when the first of the follow-on SS-21s appeared.

The resulting pattern is a buildup of FROG 1–5s from 1958 to 1966 and a buildup of FROG 7s until 1973, which continued until 1978 but with a drawdown in FROG 1–5s. Since 1978 FROG levels have been stable, with the addition of SS-21s.

A number of sources believe that the SS-21 is replacing FROG 7. If so, we would expect FROG 7 numbers to decline. But it is likely that FROG 7s are shipped elsewhere to replace older FROGs. And the overall FROG totals do not show a decline yet.

▲ **Propulsion:** For all FROGs, sources are unanimous on solid fuel.

## FROG 1

▲ **Service Entry:** 4 sources give 1957; only Meyr 84 gives 1958. We use 1957.

▲ **Retired:** It is assumed retired, but no source gives a date. 1970 is a highly provisional guess.

▲ **Yield:** Rock 78 gives up to 25 kt, AWST 82 gives 25 kt, and Meyr 84 gives 20–100 kt. We go with 25 kt.

▲ **CEP:** Meyr gives 0.6–1.0 km, the only source even to speculate. We give the 0.8 km midpoint but with uncertainty.

▲ **Range:** Range estimates generally are in 3 groups: 3 sources with 15 mi (24 km), 5 sources with 20 mi (32 km), and 3 sources at 40 mi (65 km). We go with 32 km.

▲ **JS-3:** Tracked JS-3 is unanimous; 2 sources even say it carries only 1.

## FROG 2

▲ **Service Entry:** 2 sources give 1957 as the service entry date, and one notes that it was paraded in Moscow in November 1957. We use 1957.

▲ **Retired:** Sources imply it is no longer in service. We use 1970 as a highly provisional guess for year of retirement.

▲ **Yield:** No source gives any value for a yield. SWM 77 and Rock 78 give values around 550 kg for warhead weight, and IMSG 60 gives a payload in this same range. There is some question whether the Soviets had a nuclear warhead this light in 1957; the SS-6 of the same year had a payload of 7,000–9,000 lb (3,175–4,082 kg) for a 5 MT warhead. But to make a guess, we choose the vague 10 kt, since it is less than half of FROG 1's 25 kt, having about half the throw-weight.

▲ **CEP:** No source gives values for CEP. We use the value of Mod 1, deployed in the same year.

▲ **Range:** Except for HRST 75's 30 mi (48.27 km) and Zaeh 61's 5 mi (8.05 km), the other 8 or so sources vary between 16 and 32 km. The 15 mi (24.14 km) value is used by 4 sources, and rounded to 25 km it is used by another, so that is our guess, and it is also central in the range of estimates. (NWG 83 gives 14–70 km as the range for FROGs in general, and since FROG 2 has the shortest range, that could be interpreted as a 14 km estimate here.)

▲ **PT-76:** There is no disagreement with the PT-76 tracked vehicle carrying 1.

## FROG 3

▲ **Service Entry:** 3 sources give 1960, and 1 gives 1959. We go with 1960.

▲ **Retired:** It is presumably still in service in NSWP states, and since warheads are presumably under exclusive Soviet control, these could be construed as Soviet weapons. It also may be still in service in the Soviet Union.

▲ **Yield:** The only yield value is 20–100 kt by Meyr 84. The warhead weight given by 2 of 3 sources is about 450 kg, about the same as the FROG 2. But it is 3 years later than the FROG 2, better yield-to-weight ratios may have been obtained, and we go with the lower end of Meyr 84's range, 20 kt.

▲ **CEP:** Only Meyr 84 speculates on CEP, giving 0.5–0.8 km. We use the middle of this range but with reservations.

▲ **Range:** HRST 75 gives 50 mi (80.45 km), and AWST 82 gives 45–55 mi (72.41–88.50 km). Rock 78 also gives 80 km in one place but 45 km in another. 8 other sources cluster strongly around 40 km, 7 either giving that value or a range of values centered on it. We go with 40 km.

▲ **PT-76:** 4 sources give PT-76, with no disagreement.

## FROG 4

▲ **Service Entry:** The only estimate is 1959 from Rock 78. We go with this estimate, though it is uncertain.

▲ **Retired:** No source gives retirement dates, but we assume it is retired and give the wild guess of 1970.

▲ **Yield:** No source gives a value, so we use the yield of FROG 3, a similar missile of the same vintage.

▲ **CEP:** No source gives a value, so we use FROG 3's CEP, a similar missile of the same vintage.

▲ **Range:** HRST 75 gives 70 mi (112.63 km), as does AWST 82. Rock 78 gives 100 km in one place but 50 km in another. The other 7 sources give values between 45 and 50 km. We use the 50 km because it fits those 7 sources slightly better.

▲ **PT-76:** PT-76 is suggested by 4 sources.

## FROG 5

▲ **Service Entry:** Only 2 years are given: 1961 by Rock 78 and 1964 by Meyr 84. There is little reason to choose between them, but since we are using Meyr's numbers for yield and CEP, we might as well be internally consistent and use his service entry date.

▲ **Retired:** *See* FROG 3.

▲ **Yield:** Meyr 84 provides 20–100 kt, the only yield estimate. We use an uncertain 20 kt, providing consistency with other FROGS.

▲ **CEP:** The only estimate is Meyr 84 with 0.4–0.7 km. We use the 0.55 km midpoint.

▲ **Range:** 6 sources give range values from 35 km to 60 km, no two the same. But there is a central tendency around 50 km, which we use. This is also the value we had for FROG 4.

▲ **PT-76:** PT-76 is given by 2 sources, with no dissent.

## FROG 7

▲ **Numbers:** *See* FROGs.

▲ **Service Entry:** 1965 is given by 5 sources, without dissent, with MOW 76 having it first observed on 7 November 1965. Bras 77 and JWS 82 believe it was first shown to the public in 1967, but this does not affect our estimate of 1965.

▲ **Retired:** No one suggests it is retired, though SMP says it is being replaced by the SS-21.

▲ **Yield:** Bras 77 suggests 25 kt, Meyr 84 gives 50–300 kt, and IISS 82 gives 200 kt. NWG 83 says, "The FROG-7 can deliver 3 separate nuclear warheads, in the one, 100 and 200 kiloton range." We go with the 200 kt, though with some suspicion and a desire to know what range it attains with that warhead.

▲ **CEP:** IISS gives 400 m for USSR forces and 380 m for NSWP forces, including in the latter category FROG 3s. Meyr 84 gives 0.4–0.7 km. We go with 400 m. Although FROG 7 appeared about the same time as FROG 5, it has been the favored model and may have had improvements over its service life.

▲ **Range:** All estimates are between 55 and 72 km, with 9 sources giving very close to 60 km. But SMP 83 indirectly and SMP 84 give the range as 70 km, so that is what we use.

▲ **ZIL-135:** Three sources give ZIL-135; 4 say the carrier is wheeled, and 2 say 1 missile is carried. MOW 76 indicates that 2 reloads are visible on the latest picture.

## SS-21

▲ **Numbers and Service Dates:** An IOC of 1976 is given by JCS 81, SMP 81, and JWS. Clns 80, IISS 82, and Meyr 84 suggest service entry in 1978. JWS changes its description from "ready for operation" in the 1982 edition to "potentially ready for operation" in the 1983 edition. Reports of numbers deployed suggest a more recent service entry date. We use 1983.

JCS and SMP reported that only a few were deployed. Clns gives 10 for 1978 through 1980, 20 for 1981, and 25 for 1982. IISS gives an uncertain 10 for 1982 and an uncertain 62 for 1983. Recent press reports mention the deployment of SS-21s in Eastern Europe, and the *New York Times* of 15 September 1983 quotes a NATO source as saying SS-21s are being deployed in Eastern Europe at a rate of 4 a month. This means about 50 a year. SMP 84 also says that new production capability is available to support an annual production of SS-21s and SS-23s at over 200 missiles. This is an outside limit on future production and, given that production figures are always greater than deployment figures and that it applies to SS-23s as well, is consistent with a 50 per year deployment rate.

Starting with 30 in 1983, we increase in 1984 and 1985 by 50 each. SMP 85 specifies a mix of FROGs and SS-21s for short-range missiles facing Europe but FROGs only for weapons in this category facing China and south Asia.

▲ **Yield:** The only values for yield are Clns 80 with 1 kt, updated to 100 kt in Clns+ 83, 10–100 kt by NWG 83, and 20–100 kt by Meyr 84. In the light of the NWG 83 statement of 3 warheads for the FROG 7 (1, 100, and 200 kt), perhaps there are three for the SS-21 as well. We cite 100 kt but with suspicion.

▲ **CEP:** We have 300 m from IISS 82, 0.15 nm (277.80 m) from Clns+ 83, and 0.1–0.3 km by Meyr 84. AWST 82 also cites a "35% accuracy increase" over FROG; 300 m approximates this for FROG 7's 400 m, so that is what we go with.

▲ **Range:** SMP 83, 84 give 120 km, as do many other sources. A few early sources (Clns 80 and FI 81) have as low as 100 km, and JWS 82 has 60 km but corrects to 120 km or below in JWS+ 83. We use 120 km. *Aviation Week* on 9 April 1984 (p. 46) has 70 nm (129.64 km).

▲ **Carrier:** 3 sources show a wheeled vehicle; none gives a designation, though 1 says it is the same vehicle first used with SA-8 Gecko.

## SS-1 Scud

▲ **Overall Numbers:** The first step is to establish an overall profile for Scud numbers where data are available and then later consider mod breakdowns. Clns provides the longest series, showing a constant 300 Scuds from 1970 through 1972 and then increasing at around 50 a year to 500 in 1977 and then more slowly to 550 in 1982. JCS 76 says Scuds are still being added to the forces, supporting qualitatively the buildup Clns describes. IISS aggregate Scuds and SS-12s at 300 for 1973–1976 (then 750 for 1977), and since there are about 75 SS-12s in 1973–1976, this leaves about 225 Scuds. But after a few years of no Scud estimates at all, IISS gives 540 for 1981 and 570 for 1983. These are very close to the Clns estimates for these years.

Because IISS numbers are so spotty and erratic and because they end up agreeing with Clns numbers, we base our estimates on the Clns series. Other scraps of data on recent numbers are a British Foreign Office report quoted in OR01 that 250 Scuds are deployed in East Europe. NWG has 54 deployed in East Germany.

Neither of these is obviously inconsistent with Clns data. Reinforcement for Clns series is offered by the detailed 1985 estimate given in SMP 85: a total of 575, with 400 opposite NATO Europe, 100 in the Far East, and 75 opposite south Asia. (In addition, SMP 85 estimates that a brigade of 12 to 18 missiles is held in strategic reserve.) Assuming continuous growth, we project that a peak of 575 was reached in 1981 (a year after 550 was reached) and has been maintained since then.

▲ **Breakdown by Mods:** No one has any data between 1957, when Scud A first became operational, and 1970, when Clns first gives his numbers. Our reconstruction is based on this reasoning: As a rather primitive weapon deployed when the USSR was not emphasizing tactical nuclear weapons, we assume (a very round) 100 were built (20 a year over 5 years). When the doubled-range Scud B became available after 1965, that weapon was produced faster and in greater quantities, accompanied by a leisurely (10 per year) drawdown of the Scud As (this leads well into the Clns figures).

While not counted in the current estimates, it is worth noting that IISS has some data on NSWP Scuds, citing an approximate 100 in 1975 and 1977 and between 130 and 142 since (with the one exception of 163 in 1980, when the figure also included the purported KY-3 Scud C).

## SS-1b Scud A

▲ **Service Entry:** 1957 is provided by 5 sources, including Clns+ 83. No source dissents. 1957 is also when the missile was first observed.

▲ **Retired:** Rock 78 and Clns 80 say it has been retired, while JWS 82 says it is believed withdrawn from first-line service. Rock 78 gives 1976 as the date of retirement, while Clns 80 gives 1978. But Clns+ 1983's numbers show the last nonzero entry in 1973. We choose an uncertain 1976 as the middle of the Clns values and Rock 78's value.

▲ **Yield:** Blechman and Moore, in SA01 (April 1983), give the yield as 1 kt. Clns+ 83 gives it as 40 kt, Bras 77 gives it as about 100 kt, and Meyr 84 gives 20–100 kt. We go with Clns+ 83's 40 kt, with uncertainty.

▲ **CEP:** The only CEP values are Clns+ 83 with 0.5 nm (926 m) and Meyr 84 with 0.8–1.5 km. We go with the Clns+ 83 value, being within Meyr's range but rounded to 930 m.

▲ **Range:** 6 sources provide their estimate as a range between 2 numbers, the lower end being 50 mi (80.45 km) or 80 km and the upper end somewhere between

150 km and 177 km. 1 of these (Ley 68) specifically says that 50 mi is the minimum range and 102 mi (164.12 km) is the maximum. The only sources to give a solitary 80 km figure since 1965 are Clns 80 and Clns+ 83. All other solitary estimates are between 130 km and 180 km. We tentatively conclude that 80 km is a minimum range and choose 150 km for the maximum because it is a middle figure of the remaining numbers, because it is appropriately vague, and because it is given by 3 sources (Bras 77, IISS 82, and Meyr 84).

▲ **Propulsion:** All sources agree on storable liquid.

▲ **Time to Launch:** 2 sources give a time to set up, MOW 76 saying more than 1 hour after reaching launch site and Bras 77 saying that reaction time is claimed as 1 hour. This is the only time to launch in our data for SRBMs, and it might profitably be extrapolated to other SRBMs.

▲ **Carrier:** 5 sources give JS-3, and 4 others say a tracked vehicle.

## SS-1c Scud B

▲ **Service Entry:** 5 sources agree on 1965 as the service entry date, with no dissent. It was also first shown in November 1965. NWG 83 says it was introduced in 1968, but that probably refers to its introduction into East Germany.

▲ **Retired:** No one suggests it has been retired, though some suggest it is being replaced by SS-23.

▲ **Yield:** Estimates are few and varied. Clns 80 has 1 kt, IISS 82 has less than 1 MT, NWG 83 has 1–10 kt, and Meyr 84 has 100–500 kt. We go with a highly uncertain 10 kt.

▲ **CEP:** Clns+ 83 gives 0.5 nm (926 m), and Meyr 84 gives 0.5–1.0 km. There are no other estimates, and we go with Clns+ 83, rounded to 930 m.

▲ **Range:** As for Scud A, 4 sources give their estimates as a range, the lower being either 165 or 160 km and the upper between 270 and 300 km (except for IISS's 450 km value for NSWP forces, which is puzzling). Of sources since 1970 that give single estimates, all are between 280 and 300 km. JCS 82 is the sole official source, giving 300 km. We conclude that 160 km is the minimum range and 300 km the maximum.

▲ **Propulsion:** Except for the report of solid of IA05 66, all other sources (8 in all) are unanimous about storable liquid.

▲ **Carriers:** MAZ-543 is given by 7 sources, 4 giving JS-3, with 2 indicating it was first seen on the JS-3 but that MAZ-543 is the standard. Nothing explicit is said about reloads, but this type of weapon can be assumed to have reloads.

## SS-23

▲ **Numbers and Service Dates:** IISS 82 gives 1979–1980. NWG 83 says it was first declared operational in 1980. SMP 81 says it has been introduced, and Clns 83 gives 1981. Meyr 84 gives 1982. On the basis of these data alone, we would pick 1981 as our estimate; however, as with the SS-21, the numbers of deployed systems for these early years are very small. In addition, official sources are slow in removing the "X" (meaning still in the experimental stage) from the designation. Clns 83 gives 48 in service in 1982, but IISS+ 83 gives only 10 for July 1983. JCS removes the "X" in 1983, as does SecD, but SecD reinserts the "X" in 1984. SMP 81 and 83 use the "X," first removed from this source in 1984. SMP 85 still says, however, "Initial deployment is anticipated opposite NATO and China." On this basis, we project deployments during 1985, leading to an initial in-service date of January 1986.

▲ **Yield:** The only 2 estimates are Clns 83 with 100 kt and Meyr 84 with 100–500 kt. We go with 100 kt.

▲ **CEP:** Aside from numerous statements that the accuracy is better than for the Scud and AWST 82's claim of a 50 percent accuracy increase over Scud, there are only two estimates: 0.2 nm (370.40 m) by Clns 83 and 0.2–0.6 km by Meyr 84. These estimates are not disparate and roughly meet the AWST 82 constraint, so we go with Clns 83, rounded to 370 m.

▲ **Range:** Meyr 84 gives 350 km, and Clns 83 gives 300 mi (482.70 km), but 5 other sources (including JCS 82 and SMP 83, 84) give 500 km. IISS switches from 350 km in 1982 to 500 km in 1983.

▲ **Propulsion:** Clns 83 and JWS+ 83 give solid, and there is no dissent.

▲ **Carrier:** Clns 83 is alone in describing the carrier at all, saying it is a wheeled vehicle.

## SS-12 Scaleboard

▲ **Service Entry:** Many sources agree that this system was first shown to the public in November 1967. JWS 82 and Clns+ 83 (in one place) give 1965 as the date first deployed, while Clns+ 83 (in another place) and 4 other sources give 1969. Rock 78 cites

1968. SecC 69 indicates that a new missile is being deployed near the Chinese border, and although SecC 69 does not say it is the SS-12, it is identified by a similar passage in SecC 70 noting deployment in the past year. This reinforces the majority view of 1969, on the one assumption that it was not yet operational at the time of SecC 69.

▲ **Numbers:** By the beginning of 1971, SecC reported 54 in service. What has happened since is less clear. Clns gives a series from 1970 to the present, with 60 for 1970 and 80 for 1971 (higher than the SecC figure), growing to a plateau of 120 by 1975. This has remained constant until SS-22s began to be introduced in 1984. IISS 81 and 83 report totals of 120. SSF, on the other hand, follows the SecC 71 estimate of 54 for 1970 and then gives 72 for 1975 and 1980. There are no other data. Given a conflict pitting Clns and IISS against SSF, we go with the majority, though it is a close call. We revise the early years in Clns series to take account of the SecC data point. We assume that in the first 3 years, equal portions of 27 per year were installed, followed by 2 years of 20 each, bringing the force to the Clns value of 120 by 1974.

If JWS 82 is right that the SS-22 replaces the SS-12 "on Scaleboard launchers," then the replacement is 1 for 1, and the decline of the SS-12 is linked to the rise of SS-22. We assume this and assume no time lost for conversion between systems.

▲ **Yield:** 5 sources indicate a yield of 1 MT, and AWST gives 500 kt, IISS 82 gives 200 kt, and Meyr 84 gives 200 kt–1 MT. We use 500 kt, largely because it is consistent with its SS-22 replacement.

▲ **CEP:** IISS 82 gives 900 m, Clns+ 83 gives 0.4 nm (740.80 m), and Meyr 84 gives 0.4–1.0 km. We go with the Clns+ 83 value, rounded to 740 m, because it is in the middle.

▲ **Range:** Official sources are quite consistent, SecC 70 and SecD 71 giving 500 nm (926 km) and SMP 81, 83, 84 and JCS 82 giving 900 km. Unofficial estimates range from 700–1,000 km, with the exception of 490 km, provided by IISS 82 as the bottom end of a range. It appears that the 800 km values might be due to misunderstanding of the original SecC value as statute rather than nautical miles (500 statute miles is 804.50 km). We go with the official 900 km.

▲ **Propulsion:** This is a rare case of genuine controversy over propellant type. SecC 71 and 4 other sources indicate it is liquid, 3 further specifying it as storable liquid. But 4 other sources indicate solid fuel. 2 of these (FIar 76 and FI* 79) indicate uncertainty, and the other two are Rock 78 and Clns 80. We go with the official liquid answer but are somewhat disturbed that the government has not spoken on the subject since 1971.

▲ **MAZ-543:** 7 sources give MAZ-543, and SecC 70 confirms it is mobile.

## SS-22

▲ **Service Entry:** Clns 83 indicates deployment in 1977; IISS 82, inconsistently, gives both 1978 (deployment in Europe) and 1979 (deployment generally); and Meyr 84 says 1980. JWS 82 also says the US DoD reported deployment in 1978, and official US sources have no "X" in the designation from the start (JCS 79), which makes 1978 plausible. SMP 81, however, says the SS-12 is being replaced by the SS-22; SMP 83 and 84 editions say the SS-12 is expected to be replaced; and then SMP 85 says the SS-12 has been replaced by SS-22. We interpret these statements to mean that despite the expectation in 1981 of large-scale replacement, it did not take place then. As a result we show a first-in-service date of 1984.

▲ **Numbers:** Since we believe the first deployments, expected much earlier, did not occur until 1984, we ignore numbers given by Clns and IISS. Although these are the only sources giving hard data, both show a large number in service by 1982. With service entry in 1984, we assume the first 10 in service that year, and 20 in 1985.

▲ **Yield:** IISS 82 and Clns 83 give 500 kt, with Meyr 84 giving 100 kt–1 MT. We use 500 kt.

▲ **CEP:** Clns 83 gives 0.2 nm (370.40 m), while IISS+ 83 gives 300 m, and Meyr 84 gives 0.2–0.4 km. We go with Clns 83's figure, rounded to 370 m.

▲ **Range:** Official sources (SMP 81, 83, 84, JCS 82) are consistent in their estimates of 900 km. Unofficial sources vary between 700 and 1,000 km. We use 900 km.

▲ **Propulsion:** FI 81 and Clns 83 give solid, and it would indeed seem strange to introduce a new liquid-fueled tactical missile unless SS-22 is just a slight upgrade of SS-12. We presume solid fuel.

▲ **Carrier:** JWS 82 says it is to replace SS-12 on Scaleboard launchers, and Clns 83 is consistent to the extent of saying it is a wheeled vehicle. We use an uncertain copy of the SS-12 carrier.

# DERIVATION OF CRUISE MISSILE ESTIMATES

## STRATEGIC CRUISE MISSILES

### SSC-X-4

▲ **Service Dates:** SMP 84 says "nearly deployed" and that it may not be ready for operational deployment until about 1985. FI+ 83's statement that it is unlikely to appear before the late 1980s holds less weight, and we estimate an uncertain 1985.

▲ **CEP:** SMP 84 says "capable of threatening hardened targets," but in the absence of yield information, this is too scanty a basis for even estimating a CEP.

▲ **Range:** SMP 84 gives 3,000 km, the only estimate.

▲ **Propulsion:** No source gives any explicit information, but we can safely assume turbojet, since it is the propulsion of the similar SS-N-21 and AS-15 and because of FI+ 83's saying it is similar in concept to the US GLCM.

### SS-NX-21

▲ **Service Entry:** This missile is listed as SS-NX-21 by the 1984 versions of SMP, AWST, and CFW, implying it is not yet operational. SMP 85 and JCS 85 still retain the "X," and SMP 85 repeats the expectation of SMP 84 that it is expected to become operational during the year. We put an uncertain 1986 as the initial January in-service date.

▲ **Yield:** ABFS 84 estimate 200 kt, which we accept with uncertainty.

▲ **Range:** 3,000 km is given by GSN 83, JWS 83, and SMP 83, 84. The only other estimate is 900–1,200 nm (1,666–2,222 km) by CFW 84. We use 3,000 km.

▲ **Speed:** CFW 84 gives Mach 0.7. We use this, with uncertainty, in the absence of any other information.

▲ **Propulsion:** Although no source gives information explicitly, it is generally understood that these modern cruise missiles use some sort of air-breathing engine, either turbojet or fanjet.

▲ **Carrier:** SMP 84 speculates about Victor III, Mike, Sierra, or new Yankee subs as carriers, and SMP 85 adds Akula. Since it is a torpedo-tube-launched weapon, reloads are implicitly possible.

### AS-3 Kangaroo

▲ **Service Entry:** 3 sources agree it was observed on Aviation Day in 1961. JFS 82 gives 1961 for the date operational but corrects to 1960 in JFS+ 83. Clns 80 gives 1960 under the fighter-attack aircraft section and 1961 in the strategic section. Ch01 79 gives 1963, IISS gives 1961, and Meyr 84 gives 1960. vanD gives 1961. One can wonder whether any Western source knows when this weapon became operational. We have little to guide us in our choice. The majority (barely) says 1961, but JFS made a recent switch, and Meyr 84 is also a recent source. We go with 1960, supposing that some new information has become available.

▲ **Retired:** No one suggests it is retired, though many people think it may be headed that way. SMP 84 implies it may soon be retired because AS-4 is replacing it on Bears.

▲ **Yield:** SecC 70–71 give 1–3 MT, as does Meyr 84. Clns 80 gives 1 MT, and IISS gives greater than 1 MT. AFM 82 and JFS 82 give 500 kt, while JAWA 82, JWS 82, and AFM+ 84 give 800 kt. We go with the official 2 MT.

▲ **CEP:** Meyr 84 gives 0.5–1.5 km. We go with his midpoint of 1 km. vanD gives 1 nm (1.852 km), which seems equally vague and not particularly preferable.

▲ **Range:** Estimates vary widely. Except for the lower values of estimate ranges (which are likely to be minimum ranges), all estimates are between 370 and 650 km. For official sources, SecC 70 and 71 give 275 nm (510 km). JCS 77–78 and SMP 81 give 350 nm (648.20 km) and 650 km, respectively. The semiofficial USND 81 gives 200–300 nm (370.40–555.60 km). Allowing for possible improvements over the years, we go with the most recent official 650 km.

▲ **Speed:** 9 unofficial sources give estimates between Mach 1.6 and Mach 2. SMP 84 says it is subsonic. We assume the SMP 84 statement is correct and not available to the earlier unofficial sources.

▲ **Propulsion:** 8 sources agree on turbojet, without dissent.

▲ **Carriers:** 9 sources indicate Bear B; 7 indicate Bear C.

## AS-15

▲ **Service Entry:** JCS 85 says "now operational," and a SecD 85 graph appears to show 1984 as the date of introduction.

▲ **Yield:** No source speculates, and neither do we.

▲ **CEP:** SMP 84 says it is capable of threatening hardened targets. AFM 84 says 150 ft is possible with TERCOM-type guidance system. This is clearly labeled as mirror-image speculation (on the basis of US performance), and we do not use it. Having no data, we make no estimate.

▲ **Range:** 3 estimates are reported: 1,200 km, 2,400 km, and 3,000 km. JAWA 82 says 1,200 km is the range of tests, while the maximum range could be as high as 2,400 km. SMP 84 and AFM 84 give 3,000 km. This being the most recent and authoritative estimate, we use it. ABFS 84 gives greater than 1,500 nm (2,778 km), which is close to the SMP 84 estimate, but the difference is intriguing; 3,000 km could be a rounding of the very vague 1,500 nm (from the nearest 500 nm to the nearest 1,000 km).

▲ **Speed:** No data are given; we do not guess.

▲ **Propulsion:** AWST 82 and JWS 82 say turbojet; FI 81 says air breathing. We assume turbojet is correct.

▲ **Bear and Blackjack:** SMP 84 says Bear H is the initial carrier, confirmed by SecD 85 and JCS 85; and an artist's drawing in SMP 85 shows a Bear H carrying 4 AS-15s under the wings. SMP 85 and JCS 85 believe that the Blackjack bomber, with a possible deployment date of 1988, may also carry this missile, but that is clearly speculative.

# ANTISHIP AND ANTISUBMARINE MISSILES

## ■ SEA-LAUNCHED ANTISHIP MISSILES

### SS-N-1 Scrubber

▲ **Nuclear Capability:** MR05 66, Clns 80, and CFW 82 say it is nuclear capable. FI* 74, MOW 76, JWS* 77, SWM 77, and Rock 78 do not say so, though none denies it. ABFS 84, in providing a yield value, imply it is nuclear. We would not be surprised to find that it really was not nuclear capable but for now assume it was.

▲ **Service Entry:** MR05 says it was operational in 1957–1958. All other sources give either 1958 or 1959, with MOW 76 and SWM 77 giving both. Rock 78 and Clns 80 give 1958, and GSN 83 gives 1959, also indicating that it was introduced on the Kildin in 1959. We choose 1959 because of the explicitness of GSN 83 on this point and because we are taking a comparatively strict definition of operational status.

▲ **Retired:** GSN 83 says that by the late 1960s, ships armed with SS-N-1 were being discarded or converted. FI* 74 is the last entry from FI, and Jane's stops reporting it with JWS* 77. But CFW 82 says it may still be fitted in the remaining Kildin-class *Neulovimyy*. It is tempting to figure that this 1 remaining instance of the older of the 2 classes of ships it was fitted on is not operational and being kept for experimental or historical purposes. In this case we might choose an uncertain date of 1975 for retirement, based on JWS* 77's statement that by 1975 only a single Krupny remained and was presumed withdrawn by now (1977). But we cannot pass over the 1 remaining ship, so we call it still operational.

▲ **Yield:** Clns 80 provides a rather unhelpful "KT." ABFS 84 give a yield of low kt. We translate this into an uncertain 10 kt.

▲ **Range:** Except for AWST 82's 13.8 mi (22.20 km) figure, which is almost certainly a minimum, range estimates vary between 45 and 241 km. The 2 sources with figures around 45 km contrast this value with a higher one over 200 km, AJ01 70 saying it is the normal range, compared to one with forward-observing aircraft, and CFW 82 saying it is the range for surface targets as opposed to land targets. The next value in increasing order is 45 nm (83.34 km) from SecC 71, the only official source. Most sources give maximum values between 180 and 240 km. JWS* 77, which gives 100 nm (185 km), says that the practical operating range is probably much less. The picture that emerges is of a weapon limited not by how far it can physically fly but by the range at which it can find a target. We go with the SecC figure rounded to 85 km, it being broadly consistent with the limitations other sources suggest.

▲ **Speed:** MR05 66 suggests Mach 1.2, but other sources are consistent, with FI* 74 giving an uncertain Mach 0.95, SWM 77 and Rock 78 giving Mach 0.9, and JWS* 77 and CFW 82 giving subsonic. We use Mach 0.9.

▲ **Propulsion:** All 6 sources agree on some sort of air-breathing engine, 4 indicating turbojet and 2 ramjet. 5 also indicate a solid booster. We go with turbojet, partly because the 2 most recent sources do not indicate uncertainty and give turbojet.

▲ **Krupny and Kildin:** 8 sources indicate each carrier; 5 indicate the number carried. MR04 66 suggest 6 reloads for each launcher, but this sounds too speculative, yet descriptions of the system's configuration suggest that reloads are possible. GSN 83 is the source of the numbers of ship I use; this is confirmed by 2 other sources for Kildin but not for Krupny (where FI* 74 actually gives the conflicting number 4).

### SS-N-3 Shaddock

▲ **Mods:** 3 mods are recognized by a number of sources. A is carried by submarines, B by surface ships, and C is said to be a strategic variant carried by submarines. Since there is no controversy that the missile is carried by both submarines and surface ships, the basic distinction of versions A and B is clear. Version C is more questionable.

▲ **Version C:** There is no suggestion that particular classes of submarines carried C rather than A. The missile has never been revealed to the public (MOW 76), so the notion of a strategic variant would have to come from intelligence estimates or else be hypothesized to fit the strategic mission. But although SecC 66–68 list the SS-N-3 in a way implying that it is part of a strategic threat, SecC 69 says the Soviets do not appear to consider cruise missile submarines as a strategic attack system. This in itself does not rule out the strategic variant, for the Soviets may have designed a missile for that role but never deployed it or deployed it in the past but retired it.

What are the functional differences between mods A and C? GSN 83 and CFW 84 credit C with a range of 400 nm instead of 250 nm for A, but US20, while giving the same range for A, shows a decreased range of 170 nm for C. And JWS 82 and JFS 82 credit the strategic variant with a yield of 800 kt instead of 350 kt for the others. CFW 84 credits A with active radar homing, while C has only inertial guidance.

Assuming the same propulsion system, one can imagine trading off the weight of active radar guidance for either greater yield or greater range. But it appears that JWS and JFS have opted for a higher yield (saying nothing about range), and GSN 83 and CFW 84 have opted for longer range (saying nothing about yield). One can speculate that US20 82 took the increased yield line (it gives a higher warhead weight for C) and as a result decreased its range value, perhaps not figuring on any savings from the removal of terminal guidance. SecC 68–69 indicate a maximum range of 450 nm for SS-N-3 in general but a normal operating range against ships of 250 nm. This could be implying that 450 nm is against land targets and be the basis for the C version speculation. But SecC 68–69 say nothing about mods, this perhaps being an operational option for the same missile.

On the whole, the C version seems too tentative and too uncertain in its most basic characteristics to be worth listing separately.

▲ **Versions A and B:** Aside from the different carriers, are there any functional differences between A and B? The only possibilities are in range, and even there the only differences seem to be a function of the radio-radar links available. SecC 71 gives 220 nm (407.44 km) for submarines and 150 nm (277.80 km) for surface combatants. No other source makes a distinction of this kind; other sources suggesting a difference propose a dramatically lower figure (40–60 km) for submarines without cooperating aircraft. Whether we end up accepting some distinction of this kind, it does not harm in the present table layout to list the 2 mods, even if they have no distinguishing functional characteristics.

Data following apply to all mods unless otherwise specified.

▲ **Designation:** Shaddock is given as a NATO name by 7 sources and Sepal by 2, with 2 others listing both Sepal and Shaddock. We believe that Sepal is associated with SSC-1, and so we use simply Shaddock.

▲ **Service Entry:** Estimates vary between 1958 and 1962. Apparently it first appeared on surface ships in 1962. SMP 81 says, "The Soviets began their submarine cruise missile programs in the 1950s converting existing submarines to fire the long-range SS-N-3 missile." It seems agreed that the first installation was on the Whisky Twin Cylinder in 1958. But FI 81 says it is doubtful this configuration was ever operational. No other source makes this allegation, and FI 81 itself lists 5 Whisky Twin Cylinder submarines; it seems unlikely it would have been installed on 5 submarines if it did not work. We end up listing 1958 for the A mod and 1963 for the B.

▲ **Yield:** 4 sources give 350 kt, though 2 of these (JFS 82 and JWS 82) contrast this basic yield to an 800 kt strategic yield. (*See* "Version C" for why we do not list the strategic variant.) Clns+ 83 gives 10 kt, and Meyr 84 gives 50–200 kt. We use 350 kt.

▲ **Range:** Ranges given without qualification vary between 240 and 840 km. Lower ranges have some indication that they are limited by guidance. Official figures change over time: SecC 66 gives 300 nm (555.60 km), SecC 67 gives 350–450 nm (648.20–833.40 km); SecC 68–69 have 450 nm (833.40 km) maximum and 250 nm (463 km) as the normal operating range against ships. SecC 71 has 220 nm (407.44 km) arming subs and 150 nm (277.80) arming surface combatants. The SecC 68–69 figures seem to have been picked up most widely by other sources, though the later SecC 71 figures would seem more accurate. It is notable that Clns 80, who may be using DIA information, gives 150–250 mi (241.35–402.25 km). For maximum range, we give SecC 71's figures, rounded to 410 km and 280 km. The first we give to the submarine-based A version and the second to the ship-based B version.

JWS 82 gives the practical range as 180 km for cruisers and less for submarines. US02 72 says 25 mi (40.23 km) has been mentioned for submarines. US20 gives 25–30 nm (46.30–55.56 km) as the effective autonomous range. Along with the maximum range, we give a vague 50 km as the range for a lone submarine, not daring to speculate about the range for a cruiser without cooperating aircraft.

▲ **Speed:** MOW 76 gives Mach 0.95, and JWS 82 gives transonic. FI 81 and JFS+ 83 give Mach 1.4, JFS moving downward slightly from Mach 1.5 of JFS 82. MR05 66 gives M2 and Rock 78 gives Mach 2.5. GSN 83, in discussing guidance, also gives an implicit speed value, saying that the submarine must remain on the surface for 25 minutes when firing at targets

250 nm distant. This implies 600 knots, or close to Mach 1. We choose the lower Mach 1 because it is imprecise and because the system is so old.

▲ **Propulsion:** Along with 2 solid boosters, recognized by 7 sources, the sustainer is said to be turbojet or ramjet by another 2 sources. We choose turbojet.

▲ **Carriers:** Every carrier is indicated by at least 4 sources, except the specific Whisky mods, which are given by 3. All numbers of platforms are from FI 81, though these are confirmed by MOW 76 for Kynda and Kresta I. Reloads are assumed impossible from submarines, and the reload data for Kynda are from MR04 66 and MOW 76; for Kresta I data are from MOW 76. Information on tubes is always from at least 2 sources. For Kynda and Kresta I, CFW 82 seems to give half the number of other sources, but this may be because I have interpreted a statement that it is launched from "a quad launcher" to mean there is only 1 quad launcher, while they may mean that whatever launchers there are are quad.

## SS-N-7

▲ **Designation:** There is disagreement over whether the name *Siren* pertains to the SS-N-7 or SS-N-9. IISS 82, GSN 83, and CFW+ 84 put it with the SS-N-7; FI 81, USND 81, AWST 82, JFS 82, and JWS 82 put it with the SS-N-9. With this confusion so apparent, we decline to associate the name with either system.

▲ **Service Entry:** Estimates vary from 1967 through 1971. JFS 82 gives 1969–1970 for the date operational but corrects to 1968 in JFS+ 83. Clns 80 and IISS 82 also give 1968; FI 81 gives 1967–1968. 1969 is given by Rock 78, 1969–1970 by SWM 77, 1970 by CFW+ 84, and 1971 by GSN 83. 1968 seems to be the majority estimate, so we go with it.

▲ **Yield:** Clns 80 gives "KT" and Clns+ 83 gives 10 kt. 5 other sources give 200 kt, which is what we use. Operationally, however, a lower yield is not out of the question for such a short-range weapon.

▲ **Range:** All estimates fall between 45 and 65 km. Of the 4 sources that give 35 nm (64.82 km), 1 (US20 82) contrasts that in-theory value to a 25–30 nm (46.30–55.56 km) value in practice because of the sonar acquisition range of the submarines. We are interested in the actual militarily useful range and thus choose 50 km, the midpoint of a more or less even distribution of estimates between 45 and 55 km.

▲ **Speed:** SWM 77, Rock 78, and FI 81 give Mach 1.5, but JFS 82 and GSN 83 give Mach 0.9, while JWS 82 and US20 82 indicate subsonic speed. We follow the 4 more recent sources and estimate Mach 0.9.

▲ **Propulsion:** 6 sources give solid, among them FI 81, which also says that some sources report turbofan, though this is not repeated in FI 83. We go with solid.

▲ **Charlie I:** 11 sources indicate some version of the C or Charlie submarines, 5 specifying Charlie I. 8 tubes is given by 7 sources, with FI+ 83 saying 10 tubes. 12 submarines is given by FI 81 and FI+ 83.

▲ **Rejected Carriers Victor, Papa, and Charlie II:** Victor is suggested only by ID04 74, but we assume this is a mistake since no one confirms it. Papa is given by Clns 80 and FI 81 and FI+ 81 and FI+ 83, this last update saying that perhaps Papa carries SS-N-9 instead. Indeed 8 sources have the Papa carrying the SS-N-9, so we assume it does not carry the SS-N-7. Charlie II is suggested only by FI 81 and FI+ 83, the update saying it may carry SS-N-9 instead; indeed 7 sources (including SMP 84) have Charlie II carrying SS-N-9.

## SS-N-9

▲ **Designation:** *See* SS-N-7 regarding the name Siren.

▲ **Service Entry:** Clns 80 gives 1971, but all other sources give 1968 or 1969. The vote is 1 source for 1968, 4 for 1969, and 4 for 1968–1969. We give 1969 on the basis of the vote and our bias toward later years because of our strict definition of operational status.

▲ **Yield:** Clns 80 says "KT," and 3 other sources say 200 kt, which is also our estimate.

▲ **Range:** There are two common classes of higher figures and a range of lower ones. The highest class varies between 240 and 280 km. JWS 82 says this figure is the original estimate, with external midcourse guidance by aircraft, subsequently revised downward.

The second and overall most common figure is 60 nm (111.12 km) or its rounded conversion of 110 km. JWS 82 says this figure was quoted by US and British sources as the maximum range. US20 82 says this is the range in theory. CFW 82 says it "can reach 60 miles with an aerial relay (aircraft fitted with Video Data Link system)." JFS+ 83 provides a slightly higher 70 nm (129.64 km), saying it requires third party to reach maximum range. The maximum range seems to be 110 km but only with cooperating aircraft.

JWS 82 gives 40 nm (74.08 km) as the normal operating range in an entry oriented to surface ship fittings. US20 82 gives 25–30 nm (46.30–55.56 km) as the range for submarines in practice due to acquisition range limitations. How should we present this complex picture? Since the ships on which the SS-N-7 is fitted are not open water vessels, help from cooperating aircraft seems reasonable, whereas for submarines it does not. We give the range as 110 km, with a note that from lone submarines the range is 50 km (close to the midpoint of the US20 figure).

▲ **Speed:** FI 81 gives Mach 0.8, and GSN 83 gives Mach 0.9, though earlier in its text it describes the missile as supersonic. JFS 82 gives the speed as Mach 0.9 but changes this to Mach 1.4 in JFS+ 83, joining SWM 77 and AWST 82 in that value. We choose Mach 1.4.

▲ **Propulsion:** Solid is given by 6 sources. FI 81 suggests liquid and perhaps air breathing. We use solid.

▲ **Nanuchka:** 12 sources recognize Nanuchka, 2 in addition distinguishing between the Nanuchka I and Nanuchka III. 5 sources give the figure of 6 tubes. No source gives reliable numbers. Pictures make clear the impossibility of reloading.

▲ **Sarancha:** 6 sources give Sarancha, Clns+ 83 also giving 4 tubes, which we report with uncertainty. We assume there are no reloads on such a small boat. No source gives numbers of the ship classes.

▲ **Charlie II:** 8 sources give Charlie II, 3 (including SMP 81) giving 8 tubes. We assume reloads are impossible. FI+ 83 gives 6 ships, but this we consider too uncertain to report. CFW 82 gives Charlie I, and GSN 83 gives Charlie III, but since these are lone sources, we do not accept them.

▲ **Papa:** 8 sources suggest Papa, FI+ 83 giving 10 tubes, which we accept with uncertainty. FI+ 83 also says there is 1 submarine, which we again consider too unreliable to report. We assume reloads are impossible.

▲ **Rejected Sovremennyy:** ABFS 84 gives Sovremennyy, but since no other source does, we do not take it. Sovremennyy is a carrier of the follow-on SS-N-22, which makes the ABFS mistake understandable.

# SS-N-12 Sandbox

▲ **Service Entry:** 1973 is given by JFS 82, JWS 82, GSN 83 and CFW+ 84. 1975 is given by Meyr 84 and 1976 by Clns 80 and IISS 82. On this matter we go with 1973, not only because it is the majority but because the naval-oriented source, who may keep more careful track of ship deployment dates, are unanimous in support.

▲ **Yield:** IISS 82, JFS 82, and US20 82 give 350 kt. Clns 80 gives "KT," while Clns+ 83 gives 10 kt. Meyr 84 gives 100–200 kt. We go with the 350 kt figure partly because it is the majority but also because this is the value chosen for the SS-N-3 forerunner.

▲ **Range:** SMP 81 gives 550 km, and sources after that are nearly unanimous in following it (or the 300 nm it is probably derived from). This 300 nm (555.60 km) figure first arises with US03 77, which specifies this is cruising Mach 2.5 at 35,000 ft. This source also indicates up to 2,000 nm (3,704 km) range if flying higher at transonic speed. We use SMP 81's 550 km.

▲ **Speed:** 4 sources give Mach 2.5, with none dissenting. GSN 83 says the SS-N-12 is approximately twice as fast as SS-N-3, a condition that Mach 2.5 meets relative to the Mach 1 value that we chose for SS-N-3.

▲ **Propulsion:** 3 sources indicate turbojet, while CFW 82 indicates liquid. We use turbojet.

▲ **Echo II and Kiev:** 9 sources give Echo II, 8 give Kiev. 8 tubes for Echo II is not indicated explicitly for SS-N-12, but it is given by 4 sources for the SS-N-3 it is replacing. Kiev's 8 tubes are confirmed by SMP 81, 83, and 84, as well as FI 81. We provide no numbers of ship for Echo II because it is a replacement program, with no data on how far advanced it is. No source provides data on numbers for Kiev. We assume no reloads for Echo II. SMP 81 indicates reloads are carried by Kiev; no one denies it.

▲ **Rejected Juliett and Slava:** Clns 80 gives Juliett loaded with "SS-N-3/12," but since no other source corroborates SS-N-12, we assume the weapon is the SS-N-3. Slava, the first of a new class of cruiser, is suggested by CFW+ 84 and confirmed by SMP 85 as an SS-N-12 form. SMP 85 gives 16 missiles, which, we assume, included reloads as well as missiles in tubes.

## SS-N-19

▲ **Service Entry:** 1971 is given by GSN 83 and CFW+ 84, but this seems impossible because its predecessor SS-N-12 became operational at least 2 years later and because it is so far advanced in the numeric series. We assume these 2 1971 figures are the result of 1 typographic error that was copied into 2 without thinking. If so, it is most likely a typo for 1981. IISS 82 gives 1980, while Clns 83 gives 1980 in a discussion of submarines and 1981 in a discussion of ships. SMP 81 says the Oscar class was introduced in 1980, and although that does not guarantee it was operational, we assume it was and use 1980 ourselves.

▲ **Yield:** Clns 83 provides the only yield estimate, 500 kt. We prefer to use the same yield we used for the predecessor SS-N-3 and SS-N-12 systems, 350 kt, though with uncertainty.

▲ **Range:** Except for the early 300–400 km of FI 81, all estimates fall between 400 and 556 km. SMP 81 says it is estimated to exceed 450 km, SMP 83 gives 500 km, and JCS 84 gives 300 nm (555.60 km). Other sources do not cluster around any other number. We use SMP 85's 550 km. US20 82 indicates 25–30 nm (46.30–55.56 km) for submarines because of acquisition range limitations (as for all weapons of this type). Since this is uncorroborated, we omit it.

▲ **Speed:** JWS 82 and ABFS 84 give Mach 2.5, FI+ 83 gives up to Mach 2.5, and GSN 83 gives Mach 1+. We go with the majority Mach 2.5, which is also the speed we have for the predecessor SS-N-12.

▲ **Propulsion:** CFW gives liquid, but CFW is the source that previously went against the majority in suggesting liquid fuel for the SS-N-12. GSN 83 gives turbojet, which we go with. FI+ 83 says that the US DoD drawing suggests a winged missile "with under-fuselage powerplant and twin rocket boosters, but such a configuration is incompatible with the size and shape of the hatches covering the Kirov's launch tubes, or with the launch tubes of the Oscar class." This does not rule out turbojet propulsion but might suggest a different configuration of the booster(s).

▲ **Oscar and Kirov:** Both Oscar and Kirov are given by at least 9 sources. 20 tubes for the Kirov is given by 3 sources, and 24 tubes for Oscar is given by 5, though JWS 82 dissents with 20. We assume no reloads for Oscar. There are no data for Kirov.

▲ **Rejected Bal-Com 1 and Yankee:** FI+ 83 gives Bal-Com 1, but presumably it is speculation because the ship is still under construction. ABFS 84 gives Y-class submarines, but since this is uncorroborated, we do not include it.

## SS-N-22

▲ **Service Entry:** GSN 83 has the IOC as 1981. IISS 83 has its first year deployed as an uncertain 1982. SMP 84 says it is in production. For now we go with IISS's 1982.

▲ **Yield:** No source gives data, so we use the 200 kt of SS-N-9, with uncertainty.

▲ **Range:** GSN 83 and CFW 84 give figures close to their estimates for SS-N-9, GSN with 60 nm and 111 km; CFW with 55–68 nm (101.86–125.94 km). JFS+ 83 gives 120 nm (222.24 km), considerably higher than its 70 nm (129.64 km) figure for SS-N-9. Because there is no suggestion of why it should have a longer range, we use the same figure we did for SS-N-9: 110 km.

▲ **Speed:** CFW 84 gives Mach 2.5, and GSN 83 says it is reported to be a much higher speed version of SS-N-9. Mach 2.5 meets that criterion relative to SS-N-9's Mach 1.4, so we use it.

▲ **Propulsion:** GSN 83 gives the only estimate: solid. We accept this, especially since it is the same as SS-N-9.

▲ **Sovremennyy:** 5 sources list it, including SMP 85, which estimates that 8 are carried.

▲ **Tarantul II:** 3 sources list Tarantul, 2 specifying Tarantul II and the other referring to later Tarantul classes. FI 83 gives 4 tubes, which we take with uncertainty. We have no data on reloads.

▲ **Rejected Krasina:** ABFS 84 list Krasina, more recently known as Slava. Since no other source lists this, we do not either.

## SSC-1b Sepal

▲ **Designation:** SSC as one letter grouping is given by 6 sources; only 2 give SS-C (and all sources for SSC-2a, -2b, and -4 use the letters without separation). 3 sources give SSC-1, while 5 give SSC-1b or SSC-1B, and these tend to be the later sources. The name *Shaddock* is used by 4 earlier sources, including SecD 71-73; but *Sepal* is used by 4 more recent sources. The name *Shaddock* is not surprising since SSC-1b is thought to be a land-based variant of the SS-N-3 Shaddock that was more widely deployed on ships.

▲ **Numbers and Service Dates:** Little is known about numbers. SecD 72 says it is deployed in "small" numbers. IISS gives an uncertain 100 for the past 11 years (1973-1983). And GSN says there were (in the early 1970s) 19 battalions of this weapon, each having 15 to 18 missiles. JWS 82 and FI 81 cite the same figure and say it is inclusive of reloads. The GSN figures imply 285 to 354 missiles in the early 1970s. They also cite a manpower drop in the coastal missile force from 18,000 to 10,000. If missiles were proportional to manpower, there would be roughly 160 to 190 missiles now; if the savings are in men but not weapons, the number would remain unchanged. If there were roughly 2 reloads for each launcher, the early 1970s figure would roughly match the IISS figure. What limits, if any, SecD 72 intended to set with "small" numbers is unknown. We simply follow IISS, indicating just as they do that we do not know the answer. No one suggests it is retired; JWS 82 says it is deployed. The missile does not receive much attention, though, and its being retired would not be out of the question.

3 sources specify 1962 as the year of service introduction; no one dissents. As for a plausible buildup curve, we assume that numbers increased from 0 during the 1962 IOC year by 20s until it reached its peak of 100. This has the advantage of at least putting the buildup during a period when US carrier-based strategic strikes were a salient Soviet fear. We have no confidence in such a history.

▲ **Yield:** Estimates vary widely. SWM 77 suggests 1 kt, FI 81 says it is in the kiloton range, IISS 82 says less than 1 MT, and Meyr 84 says 50 to 200 kt. Our uncertain guess is Meyr's midpoint of 100 kt, which meets the other constraints (except SWM 77) as well.

▲ **CEP:** Meyr 84 gives 0.5-0.8 km, the only estimate. We take his midpoint of 650 m with great uncertainty. Obviously CEP means something different when applied to a cruise missile than to a ballistic missile, but it undoubtedly still is a measure of how close the missile is expected to be to its target when it explodes.

▲ **Range:** 5 sources give values at or very close to 450 km. FI 81 says the maximum range could be up to 850 km but that 200 km is optimum. AWST gives 200 mi (321.80 km), and Meyr 84 gives 300 km. The estimate here should correspond to that for the SS-N-3, but with a land-based missile the chance of cooperating aircraft seems much greater. We use the majority estimate of 450 km.

▲ **Speed:** This should presumably be the same as for SS-N-3. For SSC-1b, SWM 77 gives Mach 1.5, Rock 78 gives Mach 2.5, and FI 81 gives Mach 1.4. We withhold judgment pending SS-N-3's results.

▲ **Propulsion:** 5 sources are unanimous that the rocket is initially boosted by 2 JATO solid rockets but is sustained by some air-breathing engine, either turbojet or ramjet. It is from examination of SS-N-3 that we choose turbojet.

▲ **Carrier:** 6 sources agree on a wheeled vehicle that is transporter and erector for the missile. It is obvious that only 1 missile is carried, though no source says so. No source explicitly mentions reloads, though JWS refers to a number as "inclusive of reloads," implying a reload capability.

## ■ AIR-LAUNCHED ANTISHIP MISSILES

## AS-2 Kipper

▲ **Service Entry:** 4 sources agree that the missile was observed in 1961. GSN 83, IISS 82, and CFW+ 84 give 1961 as the service entry date; JFS 82 and Meyr 84 give 1960; and ID10 78 and Ch01 79 give 1965. We use 1961, because the sources that give that date seem more reputable than the others.

▲ **Retired:** No source suggests it is retired.

▲ **Yield:** The only data are from IISS, with less than 1 MT, and Meyr 84, with 200-600 kt. We go with 400 kt, which is Meyr's midpoint and satisfies the IISS constraint. ID10 78 says the nuclear warhead is presumably the standard 1,000 kg nuclear bomb of the

Tactical Air Force, which, if true, would offer a shortcut to a more reliable answer.

▲ **CEP:** Meyr 84 gives 0.5–1.5 km. No other source speaks, so we give a highly uncertain 1 km.

▲ **Range:** Except for AWST 82 with 31 mi (49.88 km), all estimates are between 160 and 212 km. ID10 78 in addition has 100 km for a low-level or very low-level flight, while giving 210 km for high altitude. While 200 km as a median and vague value is possible, 210 km is listed 7 times, is not the rounding of some other number, and is our estimate for high-altitude flight. But we do take ID10's point about low-level flight and feel this is the militarily more significant number, since a high-flying missile is more detectable and vulnerable than a low-flying one.

▲ **Speed:** Four sources give Mach 1.2, JFS 82 gives Mach 1.4. We go with the Mach 1.2 figure.

▲ **Propulsion:** 10 sources give turbojet, including CFW+ 84, correcting from its earlier solid.

▲ **Carriers:** 12 sources give Badger C. 3 give Badger G (including USND 81 and GSN 83). A load of 1 is given by 3 sources.

## AS-4 Kitchen

▲ **Service Entry:** 3 sources say it was displayed in 1961, and 2 of these also say it was displayed on the Tu-22 in 1967. When the missile became operational is controversial. JFS 82 gives 1965, and Rock 78 gives 1966. IISS 82 and Meyr 84 give 1962. GSN 83, Ch01 79, Clns 80, and CFW+ 84 state 1967. Partly because it is the majority view and partly because the 2 stray estimates are closer to it than to 1962, we choose 1967, but with uncertainty.

▲ **Retired:** No source suggests it is retired.

▲ **Yield:** 3 sources give 350 kt, 2 give 200 kt, and Meyr 84 gives 200–600 kt (which is close to 350 kt). We use 350 kt.

▲ **CEP:** Meyr 84 gives 0.5–1.5 km. Because it is the only source, we use 1 km.

▲ **Range:** Estimates vary between 277 and 800 km. SecC 68 gives 300 nm (555.60 km). The best sense of the situation again comes from ID10 78, suggesting that 720 km is the maximum range at high altitude and 300 km is the maximum for low-level flight. We use 300 km as our estimate since more sources tend toward that end of the range (and thus implicitly assume a low flight profile). We note 720 km as the maximum range at high altitude.

▲ **Speed:** SMP 84 says it is supersonic, as does JCS 83. 5 sources suggest values around Mach 2.5: ID10 78, Ch01 79, Gu01 80, FI 81, and AFM 82. 3 sources suggest values around Mach 3.5: Clns+ 83, with Mach 3.3, correcting downward from the Mach 4 of Clns 80; CFW 82; and JFS 82. For no strong reason we choose Mach 2.5; it is the majority choice.

▲ **Propulsion:** 8 sources suggest liquid fuel, 5 specifying a liquid rocket. GSN 83 says turbojet. We go with (storable) liquid.

▲ **Blinder:** 15 sources report it carried by the Blinder, 8 suggesting the Blinder B. 1 of these also reports the Blinder C in one place, but we assume this to be a typo. 6 sources have it carrying a load of 1. ABFS 84 says Blinder carries only bombs, but it is referring to SNA Blinders.

▲ **Backfire:** 18 sources, including official ones, recognize the Backfire. It can carry 1 or 2 missiles, according to JCS 81 and SMP 84 (this range is not uncertainty but a range of options).

▲ **Bear:** 6 sources recognize Bear, including SMP 84. While GSN 83 and IISS+ 83 mention the Bear B and C, SMP 84 suggests that those conversions from AS-3 that have been completed are called Bear G. Because it is common to change the designation of a plane receiving new armament, we believe that this is correct. They were B and C before getting AS-4; now they are G.

## AS-5 Kelt

▲ **Service Entry:** 2 sources give 1965, 1 gives 1965–1966, 3 give 1966, 1 gives 1966–1967, 1 gives 1968, and 1 gives 1969. With no official sources, we go with the median value of 1966.

▲ **Retired:** No source suggests it is retired.

▲ **Nuclear Capability:** There is some question whether this missile has a nuclear capability. Only these sources suggest it does: IA06 75, Clns 80 and 83, CFW 82, and GSN 83. On balance we agree with them.

▲ **Yield:** Only Clns 83 gives a value: 500 kt. We use

it, though we consider the possibility of choosing one uniform value for all nonstrategic ASMs.

▲ **CEP:** No source speculates. Neither do we.

▲ **Range:** AFM+ 84 gives 200 mi (321.80 km) for "at height" and 100 mi (160.90 km) for low altitude. 2 earlier sources make roughly this same distinction, and all source values fall within this range. The only official estimate is SecC 69, with 120 nm (222.24 km). This value has no particular weight since we know nothing of SecC 69's flight profile assumptions. As with AS-4, it appears that most sources give values closer to the low-altitude number, suggesting that is the expected Soviet attack plan. We give the low-altitude value (160 km), with a note about the high-altitude value, rounded to 320 km.

▲ **Speed:** All estimates vary between Mach 0.9 and Mach 1.3. AFM 82 gives Mach 0.9 for low-level flight and Mach 1.2 for 30,000 ft altitude. Since we are giving the low-altitude range, we give the low altitude speed as well but note in the comments the high-altitude speed of Mach 1.2.

▲ **Propulsion:** 9 sources give liquid, which we assume to be storable liquid. CFW 82 gives solid but does not repeat this in CFW 84.

▲ **Carrier:** 9 sources give Badger G, 3 of them indicating it carries 2 missiles. 3 sources give Badger C but in a rather tenuous fashion. JFS 82 gives Badger C but does not repeat this in JFS+ 83; CFW 82 does not give Badger C, but CFW+ 84 does. GSN 83 gives Badger C. We give Badger C without much certainty.

## AS-6 Kingfish

▲ **Service Entry:** There is a remarkable controversy on this system, with a bimodal distribution: 6 sources give 1970, 1 gives 1970–1971, and 5 give figures ranging from 1975 to 1977. No source gives any value from 1972 to 1974. The supporting evidence is not helpful: 2 sources report it first seen in 1975, while a third says it was first seen by a Japanese pilot in 1977. The 1 source that gives both an early date and a late date is Gu01 80, saying it was reported operational as early as 1970 but the "definitive version" entered service in 1975–1976. Of the 7 sources for the earlier period, 4 say "operational" rather than deployed. The best hypothesis I can construct is that DoD observed the missile in testing around 1970, but it had a prolonged testing phase and developmental difficulties and was not deployed operationally until the later period. Of the later dates, IISS 82 and Meyr 84 give 1977. We go with their figure (with uncertainty) because they are comparatively recent and respectable sources.

▲ **Retired:** No one suggests it is retired.

▲ **Yield:** 5 sources give 200 kt. Clns 80 gives 1 MT, JWS 82 gives 350 kt, and Meyr 84 gives 100–500 kt. We use the majority 200 kt.

▲ **CEP:** Meyr gives 0.2–0.5 km, the only source to make an estimate. We use the midpoint of his range with uncertainty.

▲ **Range:** 3 early sources—MOW 76, ID10 78, and Ch01 79—indicate ranges for both high and low altitude. Their low estimates range from 217 to 250 km and their high estimates from 700 to 800 km. Later sources tend to give figures in the 200 km range, suggesting once again that a low flight profile is assumed for actual use. USND 81 gives a figure of 150–250 nm (277.80–463.00 km), which is picked up by GSN 83 and CFW+ 84, while JFS+ 83 picks up the lower end. JWS 82 gives 135+ mi (217.22 km), attributing this to UK MoD Mason, in early 1976. Our estimate for low-altitude flight is the lower USND figure, rounded to 280 km. It is unclear whether we should take the older 700 km figure for high-altitude flight or USND 81's 250 nm (463 km). USND is more recent but does not say on what basis it gives either of its figures. We stick with 700 km, which is the low end of the range given by early sources for high-altitude flight.

▲ **Speed:** 8 sources give either Mach 3 or a range of which Mach 3 is the midpoint. 3 sources indicate Mach 2.5 for cruise. Gu01 80 gives Mach 1.2 for a low-level mission. We take Mach 3 for high-altitude flight and an uncertain Mach 1.2 for low-altitude flight.

▲ **Propulsion:** There is marked controversy, with SWM 77, ID10 78, Gu01 80, and JFS 82 giving solid, and MOW 76, Rock 78, AFM 82, JAWA 82, and JWS 82 giving liquid. GSN 83 gives turbojet. We give an uncertain (storable) liquid because the sources seem slightly more credible for liquid fuel.

▲ **Badger:** 6 sources recognize Badger C; 7 recognize Badger G. JFS+ 83 mentions the Badger D, but we leave this out since no other source suggests it. 6 sources have versions of Badger carrying 2 AS-6s. One source gives a load of 1 (ID10 78). We go with 2.

▲ **Rejected Backfire:** 4 sources list the Backfire without comment, but 4 others indicate this as an expectation that has not yet been confirmed. AFM+

84 says SMP refers to Backfire carrying 3,300 km, Mach 3 antishipping ASMs; it also says AS-6 must have this range capability if not flying a low profile for the entire mission. The passage is in SMP 83, p. 60, in a discussion of naval aviation. But earlier in the same SMP volume and again in SMP 84, Backfire is said to carry AS-4. Although it never indicates explicitly that 3 can be carried, SMP 84 says Backfire is designed to carry AS-4 mounted partially in its fueslage and that the Backfire can also carry 2 wing-mounted AS-4s on wing pylons. I would guess that all 3 can be carried at once but at a prohibitive cost in range. There is insufficient evidence to conclude Backfire carries AS-6.

# ■ SEA-LAUNCHED ANTISUBMARINE MISSILES

## SS-N-14 Silex

▲ **Service Entry:** Sources differ widely, with 3 indicating it was operational in 1968 and 5 others indicating some date around 1974. No explanation is offered, but we can construct one based on our knowledge of the SS-N-10 designation. According to JWS* 77 and GSN 83, the SS-N-10 designation was originally applied to what turned out to be the SS-N-14. Both JWS* 77 and SWM 77 have the SS-N-10 as first operational in 1968. We can speculate with some confidence that 1974 is when some other instantiation of SS-N-14 was recognized and that it actually came into service in 1968.

▲ **Yield:** Clns 80 and IISS 82 indicate a value in the kt range, and JWS 82 further specifies the low kt range. We guess 10 kt.

▲ **Range:** Except for 3 minimum range estimates of 4 nm (7.4 km), all estimates fall between 30 and 56 km. 7 estimates are of 30 nm (55.56 km) or its 55 km conversion. 4 more are in the 46–48 km range. And there are 3 lower estimates. We go with the majority 55 km, especially since some of the lower estimates may be the result of statute versus nautical mile confusion.

▲ **Speed:** JFS 82 gives Mach 0.9, JWS 82 gives Mach 0.95, and FI 81 gives subsonic. We go with Mach 0.9 because it is more clearly subsonic and a rounder number.

▲ **Propulsion:** 5 sources give solid, including JFS 82, which is updated in JFS+ 83 to cruise missile assisted in the entry under ASW weapons. Despite this unexplained change, we keep to solid.

▲ **Carriers:** Every ship class we list is given by at least 8 sources, except the newer Udaloy, which is given by US19 82, GSN 83, and JWS+ 83. SMP 85 says that only the lead ship of the Kirov class carries the missile. The number of tubes is from at least 2 sources, often Clns 80 and FI 81. Reload information is primarily from GSN 83, though it is sometimes confirmed by 1 other source. As for models, Krivak II is specified by 3 sources and Krivak I by 2.

▲ **Rejected Moskva, Kiev, and Krasina:** 3 older sources list the Moskva and 2 of these also list the Kiev, FI 81 being one and specifying that it is launched from the multipurpose SUW-N-1 launcher. But according to JWS 82, it is FRAS-1, an unguided ASW rocket, that is launched from SUW-N-1. The 2 other sources are even older (SWM 77 and US03 77). We do not list these ships, primarily because no other source has picked them up.

## SS-N-15

▲ **Service Entry:** All estimates fall between 1972 and 1975. Rock 78 indicates an uncertain 1975; JFS 82 indicates 1974 in its entry for missiles and 1972 in its entry for ASW weapons. Clns+ 83 indicates 1973, and GSN 83 and CFW+ 84 indicate 1972. Since the 1974 and 1975 estimates have problems, we go with the majority 1972.

▲ **Yield:** The only estimate is 10 kt by Clns+ 83. This is reasonable for as ASW weapon, and we use it, though with uncertainty.

▲ **Range:** All estimates fall between 32 and 46 km, except for JWS+ 83 with 45–50 km, calling these numbers official US figures that supersede the previous 35 km. Although we do not know what these official US figures are, we accept them, taking into account the lower estimates to the extent of choosing the lower edge of the JWS+ 83 figures, 45 km.

▲ **Speed:** No value is given for speed, and we do not guess in this case.

▲ **Propulsion:** Rock 78 and GSN 83 give solid. No source differs.

▲ **Victor:** FI 81, ABFS 84, and GSN 83 list just Victor, while JWS+ 83 and CFW 82 specify Victor I, II, and III. JFS 82 lists Victor II. ABFS 84 give 38 as the number of submarines, which we accept.

▲ **Alfa:** 5 sources, including SMP 84, list Alfa. ABFS 84 give 6, which we accept.

▲ **Mike:** This is found only in SMP 85.

▲ **Reloads:** Since the missile is launched from the submarine's torpedo tubes, a reload capability is inherent, though not specified with each submarine in the raw data.

▲ **Charlie II, Tango, Papa:** JFS 82 and JWS+ 83 list Charlie II, JWS+ 83 with uncertainty; ABFS 84 list Charlie. JFS 82 and JWS 82 list Tango, and JWS and ABFS 84 list Papa. ABFS 84 alone list Echo. While we decline to list these less-well-confirmed instances, we do put "others?" with Alfa to suggest the possibility. Since it is launched from standard torpedo tubes, it may be hard to rule out any new submarine class from carrying it.

# ANTIBALLISTIC MISSILES

## ABM-1 Galosh

▲ **Designation:** The current system is called ABM-1B by many sources, including all government sources; however, the system is called ABM-1 by earlier SecCs and SecDs. Since we have nothing reliable to differentiate models of the ABM-1, we use the less specific designation.

▲ **Numbers and Service Dates:** The Soviet ABM was closely watched and commented on, especially during it construction, so numbers here are quite certain. The entire story can be told by just a few passages (with which other data are generally consistent, though SMP 81 gives the IOC as 1968, Rock 78 gives 1970 for service introduction, and Clns 80 gives 1964 for first deployed). SecC 70 said that during 1969, the first 3 complexes were brought to operational status; all sources agree there are 16 launchers per complex. SecC 70 also reported that the construction of the 4 complexes appears complete, but they may not have received all the missiles. SecC 71 reported 64 were operational by 1 October 1970. All sources agree on 64 until around 1979, of which JCS 82 reports that in late 1979, 32 launchers were dismantled. AFM 82 and AFM+ 84 say the launchers were deactivated during 1980, but we believe JCS. And while SMP 84 reports that the system is being upgraded to the 100 launchers permitted by the ABM Treaty, they do not indicate that these are operational yet. We assume that the 32 are still in place, not out of commission for dismantling.

▲ **Yield:** Only 3 sources speculate on the yield, Rock 78 saying MT size, Ch01 79 giving 2–3 MT, and JWS 82 saying multimegaton. We use an uncertain 3 MT, that fitting Ch01 and JWS 82.

▲ **Range:** JWS* 75 gives 200–400 nm and 350–700 km (but the nm figure actually converts as 370.40–740.80 km). But JWS 82 gives 300 km or more. And while Clns 80 gives 200 mi (321.80 km), Clns+ 83 gives 350 mi (563.15 km). The other 5 sources give values between 300 and 322 km, though three of these indicate that the value could be more. Of our sources, Clns+ 83 is fairly authoritative, though the possibility that he meant 350 km instead of 350 mi is disquieting. We go with a vague and uncertain 300 km, which fits most sources, because although we expect that the true value is greater, we have no reliable clue as to what it is. Inventing a number like 400 km seems a worse option.

▲ **Propulsion:** ID01 68 and Ch01 79 give solid, but no other sources give a value. We go with solid fuel but with uncertainty.

▲ **Reloads:** SecC 67, 69, 70, and 71 agree that reloading is possible. SecC 67 questions whether it would have any value in a real engagement, taking 10 to 30 minutes after arrival of the missile at the launcher. SecC 69 estimates 30 minutes to reload, and SecC 70 estimates 15–30 minutes. Our estimate of 30 minutes for reloading appears as a comment, since this sets a clear upper limit on the performance of the system.

# ANTIAIRCRAFT MISSILES

## ■ STRATEGIC SURFACE-TO-AIR MISSILES

### SA-2 Guideline

▲ **Service Entry:** Except for Rock 78, who gives 1957, the other 6 independent sources give 1958 or 1959. 1959 is given by SMP 81 and AFM 82, while 1958 is given by SecC 71, JCS 75–76, Clns 80, and JWS+ 83. We use 1958.

▲ **Retired:** No source suggests it is retired.

▲ **Numbers:** Although good data are available on the numbers of SA-2s and SA-5s deployed, they are not included in the summary chart because the great bulk of these systems are not nuclear armed by anyone's estimation. In our judgment, including them in a chart of Soviet nuclear systems would obscure more than it reveals. If any numbers were available on the number of nuclear-armed SA-2s and SA-5s, we would have used these subtotals.

▲ **Nuclear Capability:** A group of 5 sources allows for nuclear warheads, though an additional 5 sources list HE warheads with no mention of nuclear. Jeff Sands, of the Arkin group in Washington, D.C., indicates that in a government hearing, they found confirmation that the SA-2 does have a nuclear version.

▲ **Yield:** No source speculates on the yield. Neither do we.

▲ **Range:** All of the numerous estimates for range of this system fall between 29 km and 50 km, with more recent estimates tending to be higher. And JCS 75 says that SA-2 has been modernized over its life, with new versions increasing in range. We go with the SMP 81, 83, 84 estimate of 50 km. The 5 sources from the 1960s recognize a value near 30 km and one near 40 km. MR03 66 identifies the shorter one with the 1957 version and the longer one with the 1960 version. The first mention of a number near 50 km is HRST's 30 mi (48.27 km) in 1975, so this 50 km version has been around for a fair part of the SA-2's service life. But there is an argument for using a shorter range as more representative of the missile when it was at its peak military significance.

▲ **Speed:** 5 sources give Mach 3.5, while MR03 66 gives Mach 2.5 for the 1957 version and Mach 3 for the 1960 version. We go with Mach 3.5, though there is an argument for using a smaller value.

▲ **Propulsion:** 7 of 8 sources since 1974 say it has solid boosters and liquid sustainers. Rock 78 says it has a ramjet sustainer, while 2 earlier sources give solid. We invoke a general rule that the sustainer is the more critical propulsion system and use that in our chart.

▲ **Carrier:** 3 sources give the Zil 157 as its carrier, though sources also note that various auxiliary equipment is needed and that the missile is not very mobile. But presumably in its strategic defense role, it is fixed, and one would be surprised to find nuclear warheads with the tactical defense SA-2s. So we consider it effectively not mobile.

### SA-5 Gammon

▲ **Designation:** This was originally known as Griffon. This certainly sounds like a NATO code name. Pr01 says the name was changed because of the controversy over its ABM capabilities. It was also originally known as the Tallinn system, after the city where it was (first?) noted.

▲ **Service Dates:** 4 sources agree that it was displayed in a parade in 1963. But most sources have the year first operational as 1967, including SecC 71, JCS 75, and JWS 82. Clns 80 has 1963, though that is corrected to 1967 in Clns+ 83. The only outstanding dissent is SMP 81, which also gives 1963. We speculate that both Clns 80 and SMP 81 got their number from the same place, and Clns has now found it to be wrong, whereas no recent government source has bothered to discuss the SA-2 in this respect. SecC 68's "A few may now be operational" (the best evidence of all) if anything would suggest 1968 (not 1963). We go with 1967.

▲ **Retired:** No source suggests it is retired.

▲ **Numbers:** *See* SA-2, "Numbers," for why none are given.

▲ **Nuclear Capability:** For the SA-5, all 5 sources that say anything about the warhead type allow for the nuclear possibility.

▲ **Yield:** No source speculates. Neither do we. A fairly small yield, enough to destroy aircraft nearby but not so big as to cause damage to the territory to be protected, would make sense.

▲ **Range:** Except for figures that are best interpreted as minimum ranges, all estimates vary between 160 and 300 km. The only official pronouncements are 100 nm by SecC 69 (185.20 km), 50–100 nm by SecC 70, and 300 km by SMP 81, 83, 84. SecC's phrasing in 1969 makes it clear that it has very low confidence in its number. No other number gets widespread support, so we use SMP's 300 km. Even if the early SecC estimates are right, it is possible that performance improvements have occurred over the missile's life.

▲ **Speed:** MR03 66 gives Mach 3–5, and FI 81 and AFM 82 give Mach 3.5 (FI gives Mach 3.5+). We use Mach 3.5.

▲ **Propulsion:** 7 sources give solid, the only dissent being the early MR03 66.

▲ **Carrier:** AW02 78 and Clns 80 say it operates from fixed sites, but Clns+ 83 says it is mobile. Other sources suggest it is fixed by implication. We assume that even if Clns+ 83 is right, it is not the nuclear models that are mobile.

## Possibly Nuclear: SA-1, SA-3, SA-10, and SA-N-6

ABFS have said, "It is known that there are nuclear-capable surface-to-air naval missiles either deployed or under development and near deployment. The SA-N-6 is thought to be adapted from the dual-capable SA-10 missile, though there is no indication that it itself is dual-capable. If the SA-N-6 is not nuclear-capable, then a nuclear-capable surface-to-air missile is near deployment stage." No source I have except Clns 83 claims SA-10 is nuclear capable, but of the rejected systems, SA-10 is the strongest candidate for nuclear capability.

As for the SA-1, -3, and -10, Clns is the only source (except for ABFS on the SA-10) that claims a nuclear capability for them, though he recognizes a conventional capability as well. The care with which he has done his analysis for SAMs is put in question by his statement that it is the SA-3 that had an improved, possibly nuclear version first displayed in 1967. No other source indicates this, though it is just right if pertaining to the SA-2. This may be a typo, but it may not.

# PART IV

## SOVIET MISSILE DATABASE

# SS-6

| Sources are: | MRO2 | 61 | SecC | 65 | Ley | 68 | Flar | 76 | BL01 | 80 | SSF | 82 | SecD | 84 |
|---|---|---|---|---|---|---|---|---|---|---|---|---|---|---|
| COTS | 58 | | Zaeh | 61 | SecC | 66 | JCS | 72 | SWM | 77 | AWST | 82 | SMP | 83 | SMP | 85 |
| IMSG | 60 | | JAWA* | 64 | SecC | 67 | HRST | 75 | Rock | 78 | Pr01 | 82 | vanD | 83 | BW | 86 |

---------- **US DESIGNATION** ----------

| SecC | 65 | SS-6 |
| JCS | 72 | SS-6 |
| HRST | 75 | SS-6 |
| Flar | 76 | SS-6 |
| SWM | 77 | SS-6 |
| Rock | 78 | SS-6 |
| AWST | 82 | SS-6 |
| SSF | 82 | SS-6 |
| SMP | 83 | SS-6 |
| vanD | 83 | SS-6 |
| SecD | 84 | SS6 |

---------- **NATO CODENAME** ----------

| HRST | 75 | Sapwood |
| Flar | 76 | Sapwood |
| SWM | 77 | Sapwood |
| Rock | 78 | Sapwood |
| AWST | 82 | Sapwood |
| SSF | 82 | Sapwood |

---------- **OTHER DESIGNATION** ----------

| COTS | 58 | T-3 |
| IMSG | 60 | T-3 |
| MRO2 | 61 | T-3 |
| Zaeh | 61 | T-3 |
| JAWA* | 64 | T-3 |
| Ley | 68 | T-3 |
|  |  | M-104 |

---------- **DESCRIPTION** ----------

| IMSG | 60 | ICBM |
| JCS | 72 | [Listed as] first generation Soviet strategic missile. |
| HRST | 75 | "First-generation ICBM". |
| Rock | 78 | ICBM |
| SSF | 82 | First generation. |
| SMP | 83 | First generation ICBM. |

---------- **EDITOR'S NOTES** ----------

| SecC | 65 | Listed FY66-68. |
| JCS | 72 | Listed FY73. |
| SMP | 83 | Listed '83, '85. |
| SecD | 84 | Listed FY85. |

---------- **RANGE** ----------

|  |  | mi | nm | km | km conv |  |
|---|---|---|---|---|---|---|
| IMSG | 60 | 5,000 | | | 8,045 | |
| MRO2 | 61 | ≤ 8,000 | | | 12,872 | |
| Zaeh | 61 | 5,500 | | | 8,849 | |
| JAWA* | 64 | 5,000 | | 8,000 | 8,045 | |
| Ley | 68 | 5,000 | | | 8,045 | |
| HRST | 75 | 3,000 | | | 4,827 | |
| Flar | 76 | | | 8,000+ | 8,000 | |
| SWM | 77 | | | | | ICBM range |
| Rock | 78 | | | 8,500 | 8,500 | [Table] |
|  |  | | | > 8,000 | 8,000 | [Text] |
| AWST | 82 | | | | | ICBM range |
| SSF | 82 | | 3,200 | | 5,926 | |
| BW | 86 | | | 8,000 | 8,000 | |

---------- **SPEED** ----------

|  |  | mph | kmph | Mach | kmph conv |  |
|---|---|---|---|---|---|---|
| IMSG | 60 | 15,500 | | | 24,939 | |
| JAWA* | 64 | | | 25 | | At burnout. |

## 108  SOVIET MISSILES  IDDS

```
---------- CEP ----------
              ft         mi        nm        m        km      m conv
IMSG   60                10                                   16,090      "Guided to within 10 miles
                                                                          of its target" by inertial
                                                                          guidance.
SecC   65                          2                          3,704       FY66-67.
FIar   76                                          ? 8        8,000
SSF    82                          2.0                        3,704
vanD   83                          2.0                        3,704
BW     86                                    3,700            3,700

---------- WARHEAD TYPE ----------
              nuclear    conven    chem
IMSG   60     yes                            Thermonuclear.
MR02   61     yes
JAWA*  64     yes

---------- WARHEAD CHARACTERISTICS ----------
Rock   78     single RV
SSF    82     single RV
vanD   83     single RV
BW     86     single RV

---------- NUCLEAR YIELD ----------
              kT                   MT                 kT conv
SecC   65                          5                  5,000
FIar   76                          ? 5                5,000
Rock   78                          5.0-10.0           5,000-10,000
SSF    82                          5.0                5,000
vanD   83                          5.0                5,000
BW     86     5,000                                   5,000

---------- THROWWEIGHT ----------
              lb                   t         kg       kg conv
SSF    82     7,000-9,000                             3,175-4,082

---------- PAYLOAD ----------
              lb                   t         kg       kg conv
IMSG   60     2,200                                   997.92             "Payload weight".
SecC   66     7,000-9,000                             3,175-4,082        "Lifted" in 1957-1958 satellite
Rock   78                                    2,000    2,000              launches. Sustainer went into
                                                                         orbit too, giving total orbit
                                                                         mass of about 8,000 kg.

---------- WEIGHT ----------
              lb                   t         kg       kg conv
COTS   58                          150                150,000            "150-ton".
IMSG   60     c. 175,000                              79,380
MR02   61                          c. 85              85,000
Zaeh   61     180,000                                 81,648
JAWA*  64     176,000                        80,000   79,833
SecC   65     500,000                                 226,800            Gross lift-off weight.
Ley    68                          85                 85,000             Tons.

---------- LENGTH ----------
              in         ft        cm        m        m conv
COTS   58                c. 150                       45.72
IMSG   60                c. 90                        27.43
MR02   61                110                          33.53
Zaeh   61                100-125                      30.48-38.10
JAWA*  64                88'0"               27       26.82
Ley    68                90-120                       27.43-36.58
FIar   76                                    30       30.00
SWM    77                100                 30.5     30.48
Rock   78                                    29       29.00
AWST   82                100                          30.48
```

BALLISTIC MISSILES  ICBMs  SS-6  109

```
------------ DIAMETER ------------
                in       ft         cm        m      cm conv
IMSG   60       190                                   482.60
Zaeh   61                16                           487.68
JAWA*  64                11'6"               3.5      350.52     First stage.
Ley    68                16                           487.68
FIar   76                                    8.5      850.00
SWM    77                9.7                 2.95     295.66     Sustainer only. 4 strap-on boosters,
                                                                 19m x 3m max diameter.
Rock   78                                                        [Table]
                                             3.0      300.00     [Text:] base diameter "across the
                                            10.3    1,030        fins".
AWST   82                26                           792.48
```

```
------------ NUMBER OF STAGES ------------
COTS   58    2 or 3
IMSG   60    3
MR02   61    3
Zaeh   61    3
JAWA*  64    3
Ley    68    3
SWM    77    1 1/2
AWST   82    1 1/2
```

```
------------ TYPE OF PROPULSION ------------
IMSG   60    liquid         Liquid oxygen and kerosene (1st stages); liquid oxygen and alcohol, (2nd and
                            3rd stages). Thrust: 480,000 lb, 1st stage; 268,000 lb, 2nd stage; 78,100 lb,
                            3rd stage. Firing time [perhaps first stage, ed] 315 sec.
MR02   61    liquid
Zaeh   61                   Thrust: 78,000 lb, "#3"; 268,000 lb, "#2"; 440,000 lb, "#1".
JAWA*  64                   Booster is 2R-14, liquid oxygen and kerosene, has thrust of 200,000 kg
                            (440,000 lb). operational about as long as US counterparts, probably in small
                            numbers. Adaptations of the missile used for satellite launch.
SecC   65    nonstorable liquid   FY66-68.
Ley    68    liquid
HRST   75    liquid
FIar   76    Lox/kerosene   Central core vehicle contained 4-chamber RD-107 engine with 96,000 kg thrust.
                            Surrounded by 4 boosters, each powered by 4-chamber RD-108 engine of 102,000
                            kg thrust.
SWM    77    liquid         Liquid oxygen/kersene. Had central core and 4 strap-on boosters. No fewer than
                            20 main thrust chambers and 12 swivel-mounted verniers (small motors for fine
                            control of speed and direction) fire at liftoff.
Rock   78    liquid         [Unstorable]; stages 0 and 1. Thrust, tonnes: 408, stage 0 [sic]; 96, stage 1.
                            Central liquid sustainer, 4 liquid boosters attached; all use same liquid
                            oxygen/kerosene propellant. Booster and sustainer motors developed
                            specifically for SS-6 requirements. In its time, most powerful rocket in
                            development in world. Sustainer supports RD-108 motor with "four combustion
                            chambers, a single-shaft turbine and four steerable vernier motors". Each
                            booster has RD-107 motor with 4 combustion chambers and 2 steerable verniers,
                            102 tonnes per booster. All 5 motors ignite at launch, 20 main and 12 vernier
                            combustion chambers firing at once. [Source gives more detail on propulsion
                            operation, ed].
AWST   82    cryogenic
SSF    82    nonstorable liquid
BW     86    unstorable liquid
```

```
------------ MIDCOURSE (OR MAIN) GUIDANCE ------------
IMSG   60    inertial        Guided by "inertial guidance system contained in the second stage".
JAWA*  64    radio-inertial
SSF    82    radio command
```

```
------------ DEVELOPMENT BEGAN ------------
FIar   76    early 1950s
SSF    82    1949-50         Year design began.
SMP    85    1949            Technological development began. [From graph.]
```

```
------------ DEVELOPMENT UNDERWAY ------------
Rock   78                    Developed early to mid-1950s.
```

```
----------- PROTOTYPE TESTS -----------
COTS   58                "On Aug. 26, 1957, the Russians announced they have successfully fired an ICBM",
                         their T-3. A special T-3 rocket launched Sputnik, the earth's first artificial
                         satellite, on Oct. 4. More sophisticated Sputnik with dog aboard was launched in
                         November.
IMSG   60    1957        Fired successfully over its full range in 1957. Developmental vehicle with same
                         designation is undergoing tests.
FIar   76    3 Aug 1957  First test flight.
             27 Aug 1957 First full-range test; 6 weeks later used to launch Sputnik 1.
SWM    77    Aug 1957    First Soviet ICBM test at long range.
Rock   78    3 Aug 1957  First flown.
             27 Aug 1957 Full range test.
                         Launched Sputnik 1 on 4 Oct 1957.
SSF    82    1957        First flight test.
SMP    85    1955-56     Engineering and testing began. [From graph.]

----------- PRODUCTION UNDERWAY -----------
IMSG   60                Military T-3 is in large-scale production.

----------- IOC -----------
vanD   83    1962        [P209]
             1961        Spring, [p186]

----------- DEPLOYMENT BEGAN -----------
COTS   58                "Despite public statements, some Washington experts privately believe the T-3 is
                         operational".
IMSG   60                Military T-3 is operational.
MR02   61    1959        One version operational since 1959.
Rock   78    1957        Service intro.
BL01   80    1961        First Soviet ICBMs became operational with "deployment of four SS-6 Sapwood
                         missiles at Plesetsk in the spring of 1961".
SSF    82    1959-61     Year operation began.
SMP    83                "Deployment of the Soviets' first (SS-6) and second (SS-7 and SS-8) generation
                         ICBMs began in the late 1950s and early 1960s".
SecD   84    1960        Introduction of strategic systems [graph].
SMP    85    1960-61     [From graph.]
BW     86    1961

----------- RETIRED -----------
FIar   76    early 1960s Retired as soon as possible.
Rock   78    1962        [Table]
                         [Text:] deployed "only up to the early 1960's".
SMP    83    1960s       "The first generation [of ICBMs] was phased out in the 1960s".
SMP    85    1969        [From graph.]
BW     86    1968

----------- DEPLOYMENT HISTORY -----------
SecC   67                Believed Soviets will begin to phase out soft SS-6s.

----------- NUMBERS IN SERVICE -----------

----------- NUMBER IN 1960 -----------
SSF    82    4           In 1960 and 1965.

----------- NUMBER IN 1961 -----------
BL01   80    4           Only SS-6 missiles actually ever deployed; may be an upper figure. According to former
                         senior US intelligence officer [Herbert Scoville, ed.] "'only one or two missiles,
                         certainly not more than four'" deployed at Plesetsk.
Pr01   82    0           "We are confident that neither [the SS-6 nor the SS-7] will be operational as ICBM
                         weapon systems during the coming autumn and winter." From CIA report of 6 Sept 1961.

----------- NUMBER IN 1962 -----------
vanD   83    2           1962-67, 1 Jan.

----------- NUMBER IN 1964 -----------
JAWA*  64                Numbers probably small.

----------- NUMBER IN 1965 -----------
SecC   65                "A very few", FY66-67.

----------- NUMBER IN 1967 -----------
SecC   67                "Only a handful were deployed".

----------- NUMBER IN 1968 -----------
vanD   83    none        1 Jan.

----------- NUMBER IN 1970 -----------
SSF    82    none        In 1970, 1975, 1980.
```

------------- UNDATED NUMBER -------------
| FIar | 76 | ? 10 | Only small number deployed as nominal strategic force. |
| Rock | 78 | c. 12 | "Only a dozen or so" were ever deployed. |
| BW | 86 | 4 | Peak. |

------------- LAUNCHER/SILO -------------
FIar 76  Size of missile dictated 'soft' launch sites.
SSF  82  Basing mode: Fixed site.

------------- DESIGNER -------------
SSF  82  Korolev       Design Bureau.

------------- DESIGN AND ENGINEERING -------------
IMSG  60  Utilizes equipment designed for T-2 and T-1; 2nd stage thought powered by single T-2 motor; 3rd stage thought powered by T-1 propulsion unit.
JAWA* 64  Consists of T-2 plus additional booster.

------------- USE AND CONFIGURATION -------------
FIar 76  Difficulties with electronic systems resulted in low reliability and poor accuracy. Use of lox/kerosene in such large quantities resulted in long reaction time.
SWM  77  Required "immense support facilities".
vanD 83  Reliability: .20; Penetration 1.0. 2 deployed per site.

------------- RELOADS IN GENERAL -------------
BW  86  Yes.

------------- STRATEGY -------------
JAWA* 64  Russia's original ICBM.
FIar  76  When US first thought of ICBM in 1951, vehicle weighing more than 200 tons would have been required due to size and weight of existing warheads. US decided to wait until lighter warhead available, but Russians, faced with similar problems, opted to continue development to field ICBM force at the earliest possible date as counter to US manned bombers.
Rock  78  First operational Soviet ICBM. Strategic answer to US manned bomber force. Huge size made it strategically obsolete "before very long".

------------- OTHER INFORMATION -------------
Zaeh 61  "Probably modified for Sputnik and Lunik".
FIar 76  With one or more upper stages added, Sapwood has been used as launch vehicle for Sputnik, Luna, Vostok, Voskhod, Soyuz, Prognoz, Zond, Mars, and Venera. US refers to these as A-series launchers.
SWM  77  "A virtually unmodified SS-6" launched first Sputnik; modified ones launched Vostok I, Soyuz, and various unmanned flights.
Rock 78  As space launch vehicle, called A series; launched from Tyuratam and Plesetsk. Variants A-1, A-2-e, A-2 (main type), A-m, A-1-m, A-2-m. Has launched all Soviet manned space flights.
AWST 82  "Developed as core for Vostok booster".

## SS-7

```
Sources are:   IISS  70      IISS  73      IISS  75      JCS   76      AFM*  78      USO8  79      JCS   83
SecC  65       SecC  70      JCS   73      JCS   75      MOW   76      JCS   78      Clns  80      SMP   83
SecC  66       SecC  70      FI*   74      SecC  75      IISS  77      Rock  78      SecD  80      vanD  83
SecC  67       SecC  71      JCS   74      SecD  75      JCS   77      JWS*  79      AWST  82      SecD  84
SecC  68       JCS   72      SecC  74      FIar  76      SecD  77      Nitz  79      PrO1  82      SMP   85
SecC  69       SecC  72      SecD  74      IISS  76      SWM   77      SecD  79      SSF   82      BW    86
```

### ———— US DESIGNATION ————

| | | |
|---|---|---|
| SecC | 65 | SS-7 |
| SecD | 70 | SS-7         FY71,75-76,78-81,85. |
| JCS  | 72 | SS-7         FY73-79,84. |
| FI*  | 74 | SS-7 |
| FIar | 76 | SS-7 |
| MOW  | 76 | SS-7 |
| SWM  | 77 | SS-7 |
| AFM* | 78 | SS-7 |
| Rock | 78 | SS-7 |
| JWS* | 79 | SS-7 |
| Nitz | 79 | SS-7/8 |
| Clns | 80 | SS-7 |
| AWST | 82 | SS-7 |
| SSF  | 82 | SS-7 |
| SMP  | 83 | SS-7         '84, '85. |
| vanD | 83 | SS-7 |

### ———— NATO CODENAME ————

| | | |
|---|---|---|
| FI*  | 74 | Saddler |
| FIar | 76 | Saddler |
| MOW  | 76 | Saddler |
| SWM  | 77 | Saddler |
| AFM* | 78 | Saddler |
| Rock | 78 | Saddler |
| JWS* | 79 | Saddler |
| AWST | 82 | Saddler |
| SSF  | 82 | Saddler |

### ———— DESCRIPTION ————

| | | |
|---|---|---|
| JCS | 72 | Listed as first generation Soviet strategic missile. |
| JCS | 78 | Old ICBM |
| Rock | 78 | ICBM |
| SSF | 82 | Second generation. |
| SMP | 83 | Second generation ICBM. |

### ———— EDITOR'S NOTES ————

| | | |
|---|---|---|
| SecC | 65 | Listed FY66-73,75-76. |
| SecD | 70 | Listed FY71,75-76,78-81,85. |
| JCS  | 72 | Listed FY73-79,84. |
| SMP  | 83 | Listed '83, '85. |

### ———— RANGE ————

| | | mi | nm | km | km conv | |
|---|---|---|---|---|---|---|
| FI*  | 74 | | | | | medium |
| FIar | 76 | | | 11,000 | 11,000 | |
| SWM  | 77 | 6,500 | | 10,460 | 10,458 | |
| AFM* | 78 | 6,800 | | | 10,941 | |
| Rock | 78 | | | 10,500 | 10,500 | |
| JWS* | 79 | | | c. 11,000 | 11,000 | |
| Clns | 80 | | 6,500 | | 12,038 | |
| AWST | 82 | | | | | ICBM range |
| SSF  | 82 | | 5,900 | | 10,926 | |

### ———— CEP ————

| | | ft | mi | nm | m | km | m conv | |
|---|---|---|---|---|---|---|---|---|
| SecC | 65 | | | 1-2 | | | 1,852-3,704 | FY66-67. |
| SecC | 67 | | | 1.0-1.5 | | | 1,852-2,778 | |
| JCS  | 72 | | | | | | | Insufficient for "good hard-target capability". |
| FIar | 76 | | | | | < 2 | 2,000 | |
| AFM* | 78 | | < 1.25 | | | < 2 | 2,011 | |
| Rock | 78 | | | | | 1.4 | 1,400 | "The warhead is accurate to only 1.4 km"; [may not be CEP, ed]. |
| Nitz | 79 | | | 1.0 | | | 1,852 | |
| Clns | 80 | | | 1.0 | | | 1,852 | |
| vanD | 83 | | | 1.5 | | | 2,778 | |

BALLISTIC MISSILES  ICBMs  SS-7  113

```
----------  WARHEAD TYPE ----------
           nuclear  conven  chem
MOW   76   yes
JWS*  79   yes
AWST  82   yes

----------  WARHEAD CHARACTERISTICS ----------
JCS   72                      Large.
SecD  77   single RV          Number of independently targetable RVs: 1.
SWM   77   single RV          "Warhead 1 x 20/25 MT or 1961" [source misprint, ed].
Rock  78   single RV
JWS*  79                      Mods 1,2,3 exist.
Nitz  79   1 RV
USD8  79   single RV
Clns  80   single RV
SSF   82   single RV
vanD  83   single RV

----------  NUCLEAR YIELD ----------
           kT       MT         kT_conv
SecC  65            c. 6       6,000          SS-7s entering force "this year" may carry 6 MT
                                              warhead; some others may be retrofitted with it.
SecC  66            c. 3       3,000
SecC  70            3-6        3,000-6,000    FY71-72.
Flar  76            5          5,000
MOW   76            5          5,000
SWM   77                                      "Warhead 1 x 20/25 MT or 1961" [source misprint,
                                              ed].
AFM*  78            5          5,000
Rock  78            5.0        5,000          [Unmarked in table and text]
                    25         25,000         Capable of carrying 25 MT charge [in development
                                              flights].
JWS*  79            c. 5       5,000
Nitz  79            3          3,000
USD8  79            5          5,000
Clns  80            3-5        3,000-5,000
AWST  82            5-10       5,000-10,000
vanD  83            3.0        3,000

----------  WARHEAD WEIGHT ----------
           lb       t          kg          kg_conv
Flar  76                       c. 1,200    1,200        Reported, far less than
                                                        previously available warheads.
Rock  78                       1,200       1,200        Weight of nuclear warhead.

----------  THROWWEIGHT ----------
           lb       t          kg          kg_conv
Flar  76                       1,200       1,200
Rock  78            1.2                    1,200        Tonnes [from SS-9 entry, ed].
Nitz  79   4,000                           1,814
USD8  79   8,000                           3,628
Clns  80   4,000                           1,814
SSF   82   3,000-4,000                     1,360-1,814

----------  PAYLOAD ----------
           lb            t     kg          kg_conv
SecC  66   3,000-4,000                     1,360-1,814   FY67-68.

----------  WEIGHT ----------
           lb       t          kg          kg_conv
SecC  65   280,000                         127,008       Lift-off weight.

----------  LENGTH ----------
           in       ft         cm     m        m_conv
SecC  70            100                        30.48     [From graph, unit inferred].
Flar  76                              32.5     32.50
SWM   77            104.5             31.8     31.85
AFM*  78            107                        32.61
Rock  78                              32.5     32.50
JWS*  79                              c. 35    35.00
AWST  82            100                        30.48
```

114  SOVIET MISSILES  IDDS

---------- DIAMETER ----------
|       |    | in  | ft  | cm  | m    | cm conv |
|-------|----|-----|-----|-----|------|---------|
| FIar  | 76 |     |     |     | 3.1  | 310.00  |
| SWM   | 77 |     | 9.0 |     | 2.74 | 274.32  |
| Rock  | 78 |     |     |     | 3.1  | 310.00  |
| JWS*  | 79 |     |     | c. 3|      | 300.00  | Max, upper stage significantly less.
| AWST  | 82 | 10  |     |     |      | 304.80  |

---------- NUMBER OF STAGES ----------
FIar   76    2
SWM    77    2
AFM*   78    2
Rock   78    2
JWS*   79    2
AWST   82    2

---------- TYPE OF PROPULSION ----------
SecC   65    storable liquid    FY66-63.
FIar   76    storable liquid
SWM    77    storable liquid
AFM*   78    liquid
Rock   78    storable liquid    Stages 1 and 2.
JWS*   79    liquid
Clns   80    liquid
AWST   82    storable
SSF    82    liquid

---------- MIDCOURSE (OR MAIN) GUIDANCE ----------
SWM    77    inertial
Rock   78    inertial           Despite poor accuracy.
JWS*   79    radio command      Probably initially.
             inertial           ? Updated to inertial.
SSF    82    radio command

---------- OBSERVED ----------
FIar   76           Never shown in Red Square parade.
AFM*   78           Never revealed in public.
Rock   78           Much secrecy surrounded missile; has never been publicly displayed.
JWS*   79           Never shown in a Moscow parade, little known.

---------- DEVELOPMENT BEGAN ----------
SSF    82    1954           Year design began.
SMP    85    1950           Technological development began. [From graph.]

---------- PROTOTYPE TESTS ----------
SSF    82    1961           First flight test.
SMP    85    1956-57        Engineering and testing began. [From graph.]

---------- IOC ----------
SecD   77    1962
vanD   83    1963           [P209], silos [p186]
             1962           Coffins [p136]

---------- DEPLOYMENT BEGAN ----------
JCS    73    < 1964         Deployed prior to 1964, FY74-78.
MOW    76    1961           ICBM in service since 1961.
JCS    77                   "SS-7 and SS-8 are old ICBMs that were designed and largely deployed before the
                            mid-1960s", FY78-79.
SWM    77    1961           Entered service.
Rock   78    1961           Service intro, [table].
             by 1961        "In service by 1961", [text].
Clns   80    1962           First deployed.
JCS    83    c. 1961        Year of introduction [from graph].
SMP    83                   "Deployment of the Soviets' first (SS-6) and second (SS-7 and SS-8) generation
                            ICBMs began in the late 1950s and early 1960s".
SecD   84    1962           Introduction of strategic systems [graph].
SMP    85    1961-62        [From graph.]

---------- DEPLOYED ----------
FI*    74           Operated by WTO.
JWS*   79           Still believed operational.

---------- DEPLOYMENT COMPLETE ----------
SecC   66           "With the cessation of the SS-6/7/8 deployment programs...".
SecC   67           Last year said Soviets "appeared to be completing" SS-7 and SS-8 deployments [not
                    contradicted, ed].

---------- **RETIRED** ----------

| | | | |
|---|---|---|---|
| FIar | 76 | | In May US State Dept confirmed that dismantling had not taken place within mandatory 4 months of service entry of replacement SLBMs. Dismantling began in June, however. |
| JCS | 76 | | In September, USSR [informed and/or began] dismantling [SS-7 or SS-8]. |
| JCS | 77 | | In March, 1975 SCC Session, "Soviets notified us that dismantling or destruction" had begun. |
| AFM* | 78 | | Dismantling began in 1976. |
| SecD | 80 | | 209 deactivated under SALT I [SS-7 and SS-8]. |
| SMP | 85 | 1978 | [From graph.] |

---------- **DEPLOYMENT HISTORY** ----------

| | | |
|---|---|---|
| SecC | 67 | No significant changes expected in soft launchers [SS-7 and 8]; believed Soviets will begin to phase out SS-7s, perhaps "a few years later" the hard SS-7s too. |
| SecC | 68 | Believe all soft missiles will be phased out by mid-1972; may possibly have started to phase out hard SS-7s and 8s by mid-1972. |
| SecC | 70 | By mid-72, anticipated "all or most of the older, soft SS-7s and SS-8s will have been phased out, leaving only the hard SS-7s in the force". |
| JCS | 73 | If SLBM replacement option taken, expect SS-7 and SS-8 phase-out to begin by mid-1970s, when SLBM ceiling will be reached. |
| SecC | 74 | "A few soft SS-7 sites have been partially dismantled and a number of others appear inactive, but none has as yet been made completely inoperable". |
| SecD | 74 | Under SALT, can be replaced by SLBMs. |
| JCS | 75 | Believed all 209 SS-7s and SS-8s will eventually be phased out and replaced by SLBMs; No notification of this to SCC yet, or of destruction. |
| SecC | 75 | Although not yet notified by Soviets, evidence suggests Soviets intend to phase out all the old SS-7 and SS-8 launchers. |
| SecD | 75 | Probably most (if not all) of SS-7s and SS-8s will be phased out for SLBMs. |
| FIar | 76 | Now being phased out and replaced by SLBMs. In May US State Dept confirmed that dismantling had not taken place within mandatory 4 months of service entry of replacement SLBMs. Dismantling began in June, however. |
| JCS | 76 | In 1975, USSR announced in SCC that dismantling or destruction had begun [SS-7 or SS-8]. In September, USSR [informed and/or began] dismantling [SS-7 or SS-8]. [SS-7] "is being replaced" by SLBMs. |
| MOW | 76 | SS-7 and SS-8 being replaced with sub-launched missiles under SALT. |
| JCS | 77 | [Soviets] are currently reducing SS-7s and SS-8s (launchers), FY78-79. In March, 1975 SCC Session, "Soviets notified us that dismantling or destruction" had begun. |
| SecD | 77 | They have been deactivating SS-7s and SS-8s as SLBMs expand, FY78-79. |
| SWM | 77 | Being replaced by SLBMs under SALT I. |
| AFM* | 78 | To be replaced by SLBMs. Dismantling began in 1976. |
| JCS | 78 | As of 1 Oct 1977, "a number" of SS-7s and SS-8s dismantled to allow SLBM increases. |
| Rock | 78 | Being replaced by SLBMs. All will be replaced by early 1980s. |
| JWS* | 79 | Being dismantled to substitute SLBMs under SALT [SS-7 and SS-8]. |
| SecD | 79 | Soviets "have deactivated" a large number of SS-7s and SS-8s to build SLBMs. |

---------- **NUMBERS PRODUCED** ----------

| | | | |
|---|---|---|---|
| SSF | 82 | | Reload capability [SS-7,8]: 142 in 1965 and 1970, 131 in 1975. |

---------- **NUMBERS IN SERVICE** ----------

---------- **NUMBER IN 1960** ----------

| | | |
|---|---|---|
| SSF | 82 | none |

---------- **NUMBER IN 1961** ----------

| | | | |
|---|---|---|---|
| PrO1 | 82 | 0 | "We are confident that neither [the SS-6 nor the SS-7] will be operational as ICBM weapon systems during the coming autumn and winter." From CIA report of 6 Sept 1961. |

---------- **NUMBER IN 1963** ----------

| | | | |
|---|---|---|---|
| vanD | 83 | 78 | 1 Jan, coffins [soft] |

---------- **NUMBER IN 1964** ----------

| | | | |
|---|---|---|---|
| SecC | 70 | | 195 in mid-64, operational launchers [read from graph, SS-7,8]. |
| SecC | 71 | | 220, mid-64 through mid-72, operational launchers [read from graph, SS-7,8]. |
| vanD | 83 | 130 | 1964-68, 1 Jan, coffins [soft] |
| | | 12 | 1 Jan, silos [hard] |

---------- **NUMBER IN 1965** ----------

| | | | |
|---|---|---|---|
| SecC | 65 | | ICBM force is primarily SS-7, FY66-67. |
| SecC | 70 | | 210 in mid-65, operational launchers [read from graph, SS-7,8]. |
| SecC | 72 | | 209 operational launchers in 1965, 1970, and 1972 [SS-7,8]. |
| SSF | 82 | 197 | In 1965 and 1970. |
| vanD | 83 | 57 | 1 Jan, silos [hard] |

---------- **NUMBER IN 1966** ----------

| | | | |
|---|---|---|---|
| SecC | 70 | | 220, mid-66 through mid-70, operational launchers [read from graph, SS-7,8]. |
| vanD | 83 | 69 | 1966-70, 1 Jan, silos [hard] |

116  SOVIET MISSILES  IDDS

---------- **NUMBER IN 1967** ----------
SecC  68           78 in hard silos (1 Oct 1967, unchanged from last year) [SS-7,8].
                   144 soft ICBMs (1 Oct 1967, down 2 from 146 of last year) [SS-7,8,?6].
FIar  76           200 by 1967, in hard and soft sites.
Rock  78    < 200  "Nearly" 200 by 1967.

---------- **NUMBER IN 1968** ----------
SecC  69           78 in hard silos (1 Sept 1968, same as reported last year) [SS-7,8].
                   142 soft ICBMs (1 Sept 1968, down 2 from 144 of last year) [SS-7,8,?6].

---------- **NUMBER IN 1969** ----------
SecC  70           78 in hard silos, 1 Sept 1969 [SS-7,8].
                   142 in soft configuration, 1 Sept 1969 [SS-7,8].
SecD  70           Estimated number as of 1 Sept 1969 same as 1 Sept 1968.
vanD  83    128    1969-70, 1 Jan, coffins [soft]

---------- **NUMBER IN 1970** ----------
IISS  70    220    July, total deployed, SS-7,8.
SecC  70    69     Hard, for Jan 1970, mid-70, and mid-71.
            128    Soft, for Jan 1970, mid-70, and mid-71.
SecC  71    66     Hard, for 30 Dec 1970, 1 Mar 1971, mid-71, and mid-72.
            124    Soft, for 30 Dec 1970, 1 Mar 1971, mid-71.
Clns  80    190    1970-1975.

---------- **NUMBER IN 1971** ----------
vanD  83    124    1971-75, 1 Jan, coffins [soft]
            66     1971-77, 1 Jan, silos [hard]

---------- **NUMBER IN 1972** ----------
SecC  71           "124-60" soft, for mid-72.
JCS   72           ICBM force includes over 200 SS-7s and SS-8s on launchers.

---------- **NUMBER IN 1973** ----------
IISS  73    209    1973-74, July, total deployed, SS-7,8.
JCS   73           209 SS-7s and SS-8s operational and under construction or conversion.
SWM   77    98     "One 1973 report quotes 98" in service.

---------- **NUMBER IN 1974** ----------
JCS   74    209    [probably SS-7s and SS-8s, ed].
SecD  74           209 SS-7s and SS-8s, FY75-76.

---------- **NUMBER IN 1975** ----------
IISS  75    190    July, total deployed.
JCS   75    190    190 SS-7s.
SSF   82    190

---------- **NUMBER IN 1976** ----------
FIar  76    c. 90  Now.
IISS  76    140    July, total deployed.
MOW   76           C. 100, of 209 SS-7s + SS-8s at the time of SALT.
Clns  80    140
vanD  83    100    1 Jan, coffins [soft]

---------- **NUMBER IN 1977** ----------
IISS  77    109    July, total deployed, SS-7,8.
Nitz  79    139    Launchers, SS-7.
Clns  80    80
vanD  83    80     1 Jan, coffins [soft]

---------- **NUMBER IN 1978** ----------
AFM*  78           109 or less; according to Salt I, 209 total SS-7s and SS-8s. To be replaced by SLBMs.
                   Dismantling began in 1976, and no more than 109 SS-7s and SS-8s remain.
Rock  78           Currently only half the peak number.
Clns  80    none   1978-1979.
vanD  83    7      1 Jan, coffins [soft]
            48     1 Jan, silos [hard]

---------- **NUMBER IN 1979** ----------
JWS*  79           209, being dismantled to substitute SLBMs under SALT [SS-7 and SS-8].
vanD  83    none   1 Jan, coffins [soft]
            none   1 Jan, silos [hard]

---------- **NUMBER IN 1980** ----------
SSF   82    none

## BALLISTIC MISSILES  ICBMs  SS-7

### ———— UNDATED NUMBER ————
| | | | |
|---|---|---|---|
| FIar | 76 | | First Russian ICBM deployed in large numbers. |
| SWM | 77 | 190 | As many as 190 in service over the years. Widely deployed in early years. |
| JCS | 78 | | Originally 209 launchers at operational complexes (SS-7 and SS-8). |
| SecD | 80 | | 209 deactivated under SALT I (SS-7 and SS-8). |
| SSF | 82 | 197 | "Number of missiles deployed". |
| BW | 86 | 197 | Peak. |

### ———— LAUNCHER/SILO ————
| | | |
|---|---|---|
| SecC | 65 | Deployed in both a soft and a hard configuration. Silo design hardness would fall in 200-400 psi range. |
| SecC | 69 | "Hard (triple silo)". |
| JCS | 72 | Most SS-7s and SS-8s are in soft configuration. |
| JCS | 75 | [SS-7] deployed at both hard and soft sites, FY76-77. |
| FIar | 76 | 200 by 1967, in hard and soft sites. |
| SWM | 77 | Deployed in both hard and soft sites with SS-8. |
| Rock | 78 | Deployed "at various locations and in a variety of modes"; some at silos, others are at surface launch sites and thus obsolete because vulnerable to attack. |
| JWS* | 79 | Hard and soft facilities. |
| SSF | 82 | Basing mode: Fixed site. |

### ———— DESIGNER ————
| | | | |
|---|---|---|---|
| SSF | 82 | Yangel | Design Bureau. |

### ———— DESIGN AND ENGINEERING ————
| | | |
|---|---|---|
| Rock | 78 | Developed as direct result of technology in mid-1950s reducing weight of nuclear charge. |

### ———— USE AND CONFIGURATION ————
| | | |
|---|---|---|
| SecC | 65 | 2 launchers per soft site plus probably one refire missile; 3 silos per hard site and probably no refire missiles. |
| SecC | 67 | 2 launchers per soft site; 3 launchers per hard site. |
| PrO1 | 82 | Hardened sites for SS-7s... "were clusters of several ballistic missiles in 'bin' type shelters. The rocket has first to be raised to an upright position and then fueled, a process that could take hours." |
| vanD | 83 | Reliability: .72; Penetration 1.0. 2 deployed per site, "SS-7/-8 coffins". 3 deployed per site, "SS-7/-8 silos". |

### ———— RELOADS IN GENERAL ————
| | | |
|---|---|---|
| SecC | 65 | 2 launchers per soft site plus probably one refire missile; 3 silos per hard site and probably no refire missiles. |
| SecC | 71 | Total of all ICBMs that could be made available "includes 140 refire missiles for SS-7/8 soft sites" [probably based on 1 per launcher, ed]. |
| Clns | 80 | Cold launch: no. |
| SSF | 82 | Reload capability (SS-7,8): 142 in 1965 and 1970, 131 in 1975. |

### ———— STRATEGY ————
| | | |
|---|---|---|
| FI* | 74 | Targeted mainly against Europe and China/Japan. |
| SWM | 77 | Now obsolescent. |
| Rock | 78 | First successful Soviet ICBM. |
| JWS* | 79 | Oldest Soviet ICBM that is still believed operational. |

### ———— GEOGRAPHICAL DEPLOYMENTS ————
| | | |
|---|---|---|
| Rock | 78 | Deployed "at various locations and in a variety of modes". |

### ———— EXPORTED TO ————
| | | |
|---|---|---|
| FI* | 74 | Operated by WTO. |

### ———— OTHER INFORMATION ————
| | | |
|---|---|---|
| JWS* | 79 | US Posture (FY74 and FY75) contain a drawing. |

# SS-7 Mod 1

Sources are:  JCS 75    SSF 82    BW 86

### RANGE

|       |    | mi | nm | km | km conv |
|-------|----|----|----|-----|---------|
| BW    | 86 |    |    | 11,000 | 11,000 |

### CEP

|       |    | ft | mi | nm | m | km | m conv | |
|-------|----|----|----|-----|---|----|--------|---|
| SSF   | 82 |    |    | 1.5 |   |    | 2,778  | Mods 1 and 2. |
| BW    | 86 |    |    |     | 2,800 |  | 2,800 | |

### WARHEAD CHARACTERISTICS

| JCS | 75 | single warhead | All 3 mods, FY76-79. |
|-----|----|----|----|
| BW  | 86 | single RV | |

### NUCLEAR YIELD

|       |    | kT | MT | kT conv | |
|-------|----|----|-----|---------|---|
| SSF   | 82 |    | 3.0 | 3,000   | Mods 1 and 2. |
| BW    | 86 | 3,000 |  | 3,000 | |

### TYPE OF PROPULSION

| BW | 86 | storable liquid |
|----|----|----|

### IOC

| JCS | 75 | 1962 | Mods 1 and 2, FY76-79. |
|-----|----|------|----|

### DEPLOYMENT BEGAN

| SSF | 82 | 1962 | Mods 1 and 2, Year operation began. |
|-----|----|------|----|
| BW  | 86 | 1962 | |

### RETIRED

| BW | 86 | 1978 |
|----|----|------|

### RELOADS IN GENERAL

| BW | 86 | Some. |
|----|----|-------|

# SS-7 Mod 2

Sources are:  JCS  75      SSF  82

---------- **CEP** ----------
```
                  ft          mi         nm         m         km       m conv
SSF   82                                 1.5                          ‾2,778‾      Mods 1 and 2.
```

---------- **WARHEAD CHARACTERISTICS** ----------
JCS   75    single warhead     All 3 mods, FY76-79.

---------- **NUCLEAR YIELD** ----------
```
                  kT                     MT                  kT conv
SSF   82                                 3.0                 ‾3,000‾      Mods 1 and 2.
```

---------- **IOC** ----------
JCS   75    1962          Mods 1 and 2, FY76-79.

---------- **DEPLOYMENT BEGAN** ----------
SSF   82    1962          Mods 1 and 2, Year operation began.

# SS-7 Mod 3

Sources are:  JCS 75    FIar 76    Rock 78    SSF 82    BW 86

**———————— RANGE ————————**
|        | mi | nm | km | km conv |
|--------|----|----|----|---------|
| BW 86  |    |    | 11,000 | 11,000 |

**———————— CEP ————————**
|        | ft | mi | nm | m | km | m conv |
|--------|----|----|----|---|----|--------|
| SSF 82 |    |    | 1.0 |   |    | 1,852 |
| BW 86  |    |    |    | 2,800 |    | 2,800 |

**———————— WARHEAD CHARACTERISTICS ————————**
JCS 75   single warhead    All 3 mods, FY76-79.
BW  86   single RV

**———————— NUCLEAR YIELD ————————**
|        | kT | MT | kT conv |
|--------|----|----|---------|
| SSF 82 |    | 6.0 | 6,000 |
| BW  86 | 6,000 |  | 6,000 |

**———————— TYPE OF PROPULSION ————————**
BW  86   storable liquid

**———————— IOC ————————**
JCS 75   1963    FY76-79.

**———————— DEPLOYMENT BEGAN ————————**
FIar 76   1963    Entered service.
Rock 78   1963    Entered service.
SSF  82   1963    Year operation began.
BW   86   1963

**———————— RETIRED ————————**
BW  86   1979

**———————— RELOADS IN GENERAL ————————**
BW  86   Some.

# SS-8

```
Sources are:  SecC 68    JCS  72    SecD 74    Flar 76    SWM  77    SecD 79    JCS  83
IAD1 65       SecC 69    SecC 72    HRST 75    JCS  76    AFM* 78    USO8 79    SMP  83
SecC 65       IISS 70    IISS 73    IISS 75    MOW  76    JCS  78    Clns 80    vanD 83
SecC 66       SecC 70    JCS  73    JCS  75    IISS 77    Rock 78    SecD 80    SecD 84
SecC 67       SecD 70    FI*  74    SecC 75    JCS  77    JWS* 79    AWST 82    SMP  85
ID01 68       SecC 71    JCS  74    SecD 75    SecD 77    Nitz 79    SSF  82    BW   86
```

---------- **US DESIGNATION** ----------

| | | |
|---|---|---|
| SecC 65 | SS-8 | |
| SecD 70 | SS-8 | FY71,75-76,78-81,85. |
| JCS 72 | SS-8 | FY73-79,84. |
| FI* 74 | SS-8 | |
| HRST 75 | SS-8 | |
| Flar 76 | SS-8 | |
| MOW 76 | SS-8 | |
| SWM 77 | SS-8 | |
| AFM* 78 | SS-8 | |
| Rock 78 | SS-8 | |
| JWS* 79 | SS-8 | |
| Nitz 79 | SS-7/8 | |
| Clns 80 | SS-8 | |
| AWST 82 | SS-8 | |
| SSF 82 | SS-8 | |
| SMP 83 | SS-8 | '83, '85 |
| vanD 83 | SS-8 | |

---------- **NATO CODENAME** ----------

| | |
|---|---|
| FI* 74 | Sasin |
| HRST 75 | Sasin |
| Flar 76 | Sasin |
| MOW 76 | Sasin |
| SWM 77 | Sasin |
| AFM* 78 | Sasin |
| Rock 78 | Sasin |
| JWS* 79 | Sasin |
| AWST 82 | Sasin |
| SSF 82 | Sasin |

---------- **DESCRIPTION** ----------

| | |
|---|---|
| JCS 72 | Listed as first generation Soviet strategic missile. |
| HRST 75 | ICBM |
| MOW 76 | ICBM |
| SWM 77 | In ICBM class. |
| AFM* 78 | ICBM |
| JCS 78 | Old ICBM |
| Rock 78 | ICBM |
| JWS* 79 | ICBM |
| SSF 82 | Second generation. |
| SMP 83 | Second generation ICBM. |

---------- **EDITOR'S NOTES** ----------

| | |
|---|---|
| SecC 65 | Listed FY66-73,75-76. |
| SecD 70 | Listed FY71,75-76,78-81,85. |
| JCS 72 | FY73-79,84. |
| SMP 83 | Listed '83, '85. |

---------- **RANGE** ----------

| | | mi | nm | km | km conv | |
|---|---|---|---|---|---|---|
| IAD1 | 65 | 6,000+ | | | 9,654 | |
| ID01 | 68 | 6,000 | | | 9,654 | |
| FI* | 74 | | | c. 10,000 | 10,000 | |
| HRST | 75 | 3,500 | | | 5,631 | |
| Flar | 76 | | | 10,500 | 10,500 | |
| MOW | 76 | 6,500 | | 10,500 | 10,458 | Max. |
| SWM | 77 | 6,500 | | 10,460 | 10,458 | Range [chart] |
| | | 6,800 | | 11,000 | 10,941 | Max range [text] |
| AFM* | 78 | [c. 6,800] | | | 10,941 | Warhead, range, accuracy same order as SS-7. |
| Rock | 78 | | | 11,000 | 11,000 | |
| JWS* | 79 | | | c. 10,000 | 10,000 | |
| Clns | 80 | | 6,900 | | 12,778 | |
| AWST | 82 | | | | | ICBM range |
| SSF | 82 | | 5,400 | | 10,000 | |
| BW | 86 | | | 11,000 | 11,000 | |

122  SOVIET MISSILES  IDDS

### ---------- CEP ----------

|  |  | ft | mi | nm | m | km | m conv |  |
|---|---|---|---|---|---|---|---|---|
| SecC | 65 |  |  | c. 1 |  |  | 1,852 | Insufficient for "good hard-target capability". |
| JCS | 72 |  |  |  |  |  |  |  |
| Flar | 76 |  |  |  |  | 2 | 2,000 |  |
| AFM* | 78 |  | [< 1.25] |  |  | [< 2] | 2,011 | Warhead, range, accuracy same order as SS-7. |
| Rock | 78 |  |  |  |  | < 2 | 2,000 | "Error radius on impact probably less than 2 km". |
| Nitz | 79 |  |  | 1.0 |  |  | 1,852 |  |
| Clns | 80 |  |  | 1.0 |  |  | 1,852 |  |
| SSF | 82 |  |  | 1.0 |  |  | 1,852 |  |
| vanD | 83 |  |  | 1.0 |  |  | 1,852 |  |
| BW | 86 |  |  |  | 1,900 |  | 1,900 |  |

### ---------- WARHEAD TYPE ----------

|  |  | nuclear | conven | chem |
|---|---|---|---|---|
| FI* | 74 | yes |  |  |
| MOW | 76 | yes |  |  |
| Rock | 78 | yes |  |  |
| JWS* | 79 | yes |  |  |

### ---------- WARHEAD CHARACTERISTICS ----------

| JCS | 72 | Large. |
|---|---|---|
| JCS | 75 | single RV |
| SecD | 77 | single RV — Number of independently targetable RVs: 1. |
| Rock | 78 | single RV — Warhead has same spherical cap as SS-5; probably nuclear charge and RV same as, or later derivative of, SS-5's. |
| Nitz | 79 | 1 RV |
| USDB | 79 | single RV |
| Clns | 80 | single RV |
| SSF | 82 | single RV |
| vanD | 83 | single RV |
| BW | 86 | single RV |

### ---------- NUCLEAR YIELD ----------

|  |  | kT | MT | kT conv |  |
|---|---|---|---|---|---|
| SecC | 66 |  | c. 3 | 3,000 |  |
| SecC | 70 |  | 3-5 | 3,000-5,000 | FY71-72. |
| FI* | 74 |  | 5-10 | 5,000-10,000 |  |
| Flar | 76 |  | 5 | 5,000 |  |
| MOW | 76 |  | c. 5 | 5,000 |  |
| SWM | 77 |  | 5 | 5,000 |  |
| AFM* | 78 |  | [c 5] | 5,000 | Warhead, range, accuracy same order as SS-7. |
| Rock | 78 |  | 5.0 | 5,000 |  |
| JWS* | 79 |  | c. 5 | 5,000 |  |
| Nitz | 79 |  | 3 | 3,000 |  |
| USDB | 79 |  | 5 | 5,000 |  |
| Clns | 80 |  | 2-4 | 2,000-4,000 |  |
| AWST | 82 |  | 5-10 | 5,000-10,000 |  |
| SSF | 82 |  | 3.0 | 3,000 |  |
| vanD | 83 |  | 3.0 | 3,000 |  |
| BW | 86 | 3,000 |  | 3,000 |  |

### ---------- WARHEAD WEIGHT ----------

|  |  | lb | t | kg | kg conv |
|---|---|---|---|---|---|
| Flar | 76 |  |  | 1,200 | 1,200 |

### ---------- THROWWEIGHT ----------

|  |  | lb | t | kg | kg conv |  |
|---|---|---|---|---|---|---|
| Flar | 76 |  |  | 1,200 | 1,200 |  |
| Rock | 78 |  | 1.2 |  | 1,200 | Tonnes [from SS-9 entry, ed]. |
| Nitz | 79 | 4,000 |  |  | 1,814 |  |
| USDB | 79 | 8,000 |  |  | 3,628 |  |
| Clns | 80 | 3,500 |  |  | 1,587 |  |
| SSF | 82 | 2,500-4,000 |  |  | 1,134-1,814 |  |

### ---------- PAYLOAD ----------

|  |  | lb | t | kg | kg conv |  |
|---|---|---|---|---|---|---|
| SecC | 65 |  |  |  |  | Last year believed it had very large payload; now believe its payload is similar to SS-7. |
| SecC | 66 | 2,500-4,000 |  |  | 1,134-1,814 |  |
| SecC | 67 | 3,000-4,000 |  |  | 1,360-1,814 |  |

## BALLISTIC MISSILES  ICBMs  SS-8

### WEIGHT

|  |  | lb | t | kg | kg conv |  |
|---|---|---|---|---|---|---|
| IA01 | 65 |  | c. 130 |  | 130,000 |  |
| SecC | 65 | 180,000 |  |  | 81,648 | Lift-off weight. |
| ID01 | 68 |  | c. 120 |  | 120,000 |  |

### LENGTH

|  |  | in | ft | cm | m | m conv |  |
|---|---|---|---|---|---|---|---|
| IA01 | 65 |  | 82 |  |  | 24.99 |  |
| SecC | 70 |  | 78 |  |  | 23.77 | [From graph, unit inferred]. |
| FI* | 74 |  |  | c. 25 |  | 25.00 |  |
| HRST | 75 |  | 85 |  |  | 25.91 |  |
| FIar | 76 |  |  |  | 25.5 | 25.50 |  |
| MOW | 76 |  | 80'0" |  | 24.40 | 24.38 |  |
| SWM | 77 |  | 80+ |  | 24.4+ | 24.38 |  |
| AFM* | 78 |  | 83 |  |  | 25.30 |  |
| Rock | 78 |  |  |  | 25.5 | 25.50 |  |
| JWS* | 79 |  |  | c. 25 |  | 25.00 |  |
| AWST | 82 |  | 85 |  |  | 25.91 |  |

### DIAMETER

|  |  | in | ft | cm | m | cm conv |  |
|---|---|---|---|---|---|---|---|
| IA01 | 65 |  | 10.3 |  |  | 313.94 | First stage. |
| FI* | 74 |  |  | c. 275 |  | 275.00 |  |
| HRST | 75 | 108 |  |  |  | 274.32 |  |
| FIar | 76 |  |  |  | 2.9 | 290.00 |  |
| MOW | 76 |  | 9'0" |  | 2.75 | 274.32 |  |
| SWM | 77 |  | 9.0 |  | 2.74 | 274.32 |  |
| Rock | 78 |  |  |  | 2.9 | 290.00 | First stage; second is smaller. |
| JWS* | 79 |  |  | c. 2.75 |  | 275.00 | Max. |
| AWST | 82 |  | 9 |  |  | 274.32 |  |

### NUMBER OF STAGES

| FI* | 74 | 2 |
| HRST | 75 | 2 |
| FIar | 76 | 2 |
| MOW | 76 | 2 |
| SWM | 77 | 2 |
| AFM* | 78 | 2 |
| Rock | 78 | 2 |
| JWS* | 79 | 2 |
| AWST | 82 | 2 |

### TYPE OF PROPULSION

| SecC | 65 | nonstorable liquid | FY66-68. |
| FI* | 74 | liquid |  |
| HRST | 75 | liquid |  |
| FIar | 76 | storable liquid |  |
| MOW | 76 | storable liquid |  |
| SWM | 77 | storable liquid |  |
| AFM* | 78 | liquid |  |
| Rock | 78 | storable liquid | Stages 1 and 2. 1st stage probably has 4 motors or single motor with 4 combustion chambers. "Both stages feature separate systems tunnels". |
| JWS* | 79 | storable liquid |  |
| Clns | 80 | liquid |  |
| AWST | 82 | storable |  |
| SSF | 82 | nonstorable liquid |  |
| BW | 86 | unstorable liquid |  |

### MIDCOURSE (OR MAIN) GUIDANCE

| FI* | 74 | inertial |
| MOW | 76 | probably inertial |
| SWM | 77 | inertial |
| JWS* | 79 | probably inertial |
| SSF | 82 | radio command |

### CONTROL SYSTEM

| SWM | 77 | ? Steerable exhaust vanes. |
| Rock | 78 | Vanes in efflux, no stabilizing fins because of silo launch. |

### OBSERVED IN PARADE

| MOW | 76 | Nov 1964 |  |
| SWM | 77 | Nov 1964 |  |
| AFM* | 78 | Nov 1964 | Display suggests it may have been regarded as a backup to SS-7. |
| Rock | 78 | Nov 1964 | First displayed. Rear enclosed by cover; on articulated conveyance. |
| JWS* | 79 | 1964 | Not seen in recent parades. |

SOVIET MISSILES   IDDS

---------- **DEVELOPMENT BEGAN** ----------
SSF   82   1954         "Year design began".
SMP   85   1951-52      Technological development began. [From graph.]

---------- **PROTOTYPE TESTS** ----------
SSF   82   1961         "First flight test".
SMP   85   1957-58      Engineering and testing began. [From graph.]

---------- **IOC** ----------
JCS   75   1963         FY76-78.
MOW   76   c. 1963      Operational since c. 1963.
SecD  77   1963
vanD  83   1963

---------- **DEPLOYMENT BEGAN** ----------
FIar  76   1963
JCS   77                "SS-7 and SS-8 are old ICBMs that were designed and largely deployed before the mid-1960s", FY78-79.
SWM   77   1963         Service entry.
Rock  78   1963         Service intro.
JWS*  79   1963
Clns  80   1963         First deployed.
SSF   82   1963         Year operation began.
JCS   83   c. 1962      Year of introduction [from graph].
SMP   83                "Deployment of the Soviets' first [SS-6] and second [SS-7 and SS-8] generation ICBMs began in the late 1950s and early 1960s".
SecD  84   1963         Introduction of strategic systems [graph].
SMP   85   1962-63      [From graph.]
BW    86   1964

---------- **DEPLOYED** ----------
FI*   74                Operated by WTO.
MOW   76                In service.

---------- **DEPLOYMENT COMPLETE** ----------
SecC  65                Deployment now appears to have been curtailed.
SecC  66                "With the cessation of the SS-6/7/8 deployment programs...".
SecC  67                Last year said Soviets "appeared to be completing" SS-7 and SS-8 deployments [not contradicted, ed].

---------- **RETIRED** ----------
JCS   76                In September, USSR [informed and/or began] dismantling [SS-7 or SS-8].
JCS   77                In March, 1975 SCC Session, "Soviets notified us that dismantling or destruction" had begun.
AFM*  78                Dismantling began in 1976.
SecD  80                209 deactivated under SALT I [SS-7 and SS-8].
SMP   85   1977         [From graph.]
BW    86   1977

---------- **DEPLOYMENT HISTORY** ----------
SecC  67   No significant changes expected in soft launchers [SS-7 and 8]; believed Soviets will begin to phase out SS-8s.
SecC  68   Believe all soft missiles will be phased out by mid-1972; may possibly have started to phase out hard SS-7s and 8s by mid-1972.
SecC  70   By mid-72, anticipated "all or most of the older, soft SS-7s and SS-8s will have been phased out, leaving only the hard SS-7s in the force".
JCS   73   If SLBM replacement option taken, expect SS-7 and SS-8 phase-out to begin by mid-1970s, when SLBM ceiling will be reached.
SecD  74   Under SALT, can be replaced by SLBMs.
JCS   75   Believed all 209 SS-7s and SS-8s will eventually be phased out and replaced by SLBMs; No notification of this to SCC yet, or of destruction.
SecC  75   Although not yet notified by Soviets, evidence suggests Soviets intend to phase out all the old SS-7 and SS-8 launchers.
SecD  75   Probably most (if not all) of SS-7s and SS-8s will be phased out for SLBMs.
FIar  76   Being replaced by the new SLBMs.
JCS   76   In 1975, USSR announced in SCC that dismantling or destruction had begun [SS-7 or SS-8]. In September, USSR [informed and/or began] dismantling [SS-7 or SS-8]. [SS-8] "expected to be replaced" by SLBMs.
MOW   76   Due for replacement.
JCS   77   [Soviets] are currently reducing SS-7s and SS-8s [launchers], FY78-79. In March, 1975 SCC Session, "Soviets notified us that dismantling or destruction" had begun.
SecD  77   They have been deactivating SS-7s and SS-8s as SLBMs expand, FY78-79.
SWM   77   Being replaced by SLBMs under SALT I.
AFM*  78   To be replaced by SLBMs. Dismantling began in 1976.
JCS   78   As of 1 Oct 1977, "a number" of SS-7s and SS-8s dismantled to allow SLBM increases.
JWS*  79   Being phased out; being dismantled to substitute SLBMs under SALT [SS-7 and SS-8].
SecD  79   Soviets "have deactivated" a large number of SS-7s and SS-8s to build SLBMs.

---------- NUMBERS PRODUCED ----------
SSF    82           Reload capability [SS-7,8]: 142 in 1965 and 1970, 131 in 1975.

---------- NUMBERS IN SERVICE ----------

---------- NUMBER IN 1960 ----------
SSF    82           none

---------- NUMBER IN 1964 ----------
SecC   70           195 in mid-64, operational launchers [read from graph, SS-7,8].
SecC   71           220, mid-64 through mid-72, operational launchers [read from graph, SS-7,8].

---------- NUMBER IN 1965 ----------
SecC   65           A small number, FY66-68.
SecC   70           210 in mid-65, operational launchers [read from graph, SS-7,8].
SecC   72           209 operational launchers in 1965, 1970, and 1972 [SS-7,8].
SSF    82    23     In 1965 and 1970.
vanD   83    14     1965-70, 1 Jan, coffins [soft]
              6     1 Jan, silos [hard]

---------- NUMBER IN 1966 ----------
SecC   70           220, mid-66 through mid-70, operational launchers [read from graph, SS-7,8].
vanD   83    9      1966-76, 1 Jan, silos [hard]

---------- NUMBER IN 1967 ----------
SecC   68           78 in hard silos (1 Oct 1967, unchanged from last year) [SS-7,8].
                    144 soft ICBMs (1 Oct 1967, down 2 from 146 of last year) [SS-7,8,?6].

---------- NUMBER IN 1968 ----------
SecC   69           78 in hard silos (1 Sept 1968, same as reported last year) [SS-7,8].
                    142 soft ICBMs (1 Sept 1968, down 2 from 144 of last year) [SS-7,8.?6].

---------- NUMBER IN 1969 ----------
SecC   70           78 in hard silos, 1 Sept 1969 [SS-7,8].
                    142 in soft configuration, 1 Sept 1969 [SS-7,8].
SecD   70           Estimated number as of 1 Sept 1969 same as 1 Sept 1968.

---------- NUMBER IN 1970 ----------
IISS   70    220    July, total deployed, SS-7,8.
SecC   70    9      Hard, for Jan 1970 and mid-70.
              14    Soft, for Jan 1970 and mid-70.
SecC   71    9      Hard, for 30 Dec 1970, 1 Mar 1971, mid-71, and mid-72.
              10    Soft, for 30 Dec 1970, 1 Mar 1971, mid-71.
Clns   80    19     1970-1976.

---------- NUMBER IN 1971 ----------
SecC   70           "9-0" hard, for mid-71.
                    "14-0" soft, for mid-71.
vanD   83    10     1971-75, 1 Jan, coffins [soft]

---------- NUMBER IN 1972 ----------
SecC   71           "10-0" soft, for mid-72.
JCS    72           ICBM force includes over 200 SS-7s and SS-8s on launchers.
MOW    76           209 SS-7s and SS-8s as of SALT agreement.
AFM*   78           109 or less; according to Salt I, 209 total SS-7s and SS-8s. To be replaced by SLBMs.

---------- NUMBER IN 1973 ----------
IISS   73    209    1973-74, July, total deployed, SS-7,8.
JCS    73           209 SS-7s and SS-8s operational and under construction or conversion.

---------- NUMBER IN 1974 ----------
JCS    74           209 [probably SS-7s and SS-8s].
SecD   74           209 SS-7s and SS-8s, FY75-76.

---------- NUMBER IN 1975 ----------
IISS   75    19     1975-76, July, total deployed.
JCS    75    19
SWM    77    19     In service mid-1975.
SSF    82    19

---------- NUMBER IN 1976 ----------
vanD   83    4      1 Jan, coffins [soft]

## NUMBER IN 1977
| | | | |
|---|---|---|---|
| IISS | 77 | 109 | July, total deployed, SS-7,8. |
| Nitz | 79 | 19 | Launchers, SS-8. |
| Clns | 80 | 9 | |
| vanD | 83 | none | 1 Jan, coffins [soft] |
| | | 6 | 1 Jan, silos [hard] |

## NUMBER IN 1978
| | | | |
|---|---|---|---|
| AFM* | 78 | | Dismantling began in 1976, and no more than 109 SS-7s and SS-8s remain. |
| Clns | 80 | none | 1978-79. |

## NUMBER IN 1979
| | | | |
|---|---|---|---|
| JWS* | 79 | | 209, being dismantled to substitute SLBMs under SALT [SS-7 and SS-8]. |

## NUMBER IN 1980
| | | | |
|---|---|---|---|
| SSF | 82 | none | |

## UNDATED NUMBER
| | | | |
|---|---|---|---|
| FIar | 76 | 200 | Deployed [in all] |
| JCS | 78 | | Originally 209 launchers at operational complexes [SS-7 and SS-8]. |
| Rock | 78 | 200 | Only 200 "ever in service at one time". |
| SecD | 80 | | 209 deactivated under SALT I [SS-7 and SS-8]. |
| SSF | 82 | 23 | "Number of missiles deployed". |
| BW | 86 | 23 | Peak. |

## LAUNCHER/SILO
| | | |
|---|---|---|
| SecC | 65 | Deployed in both a soft and a hard configuration. Silo design hardness would fall in 200-400 psi range. |
| SecC | 70 | "Hard [triple silo]", FY71-72. |
| JCS | 72 | Most SS-7s and SS-8s are in soft configuration. |
| JCS | 75 | Deployed at both hard and soft sites, FY76-77. |
| Rock | 78 | Silo launch. |
| SSF | 82 | Basing mode: Fixed site. |

## DESIGNER
| | | | |
|---|---|---|---|
| SSF | 82 | Korolev | Design Bureau. |

## DESIGN AND ENGINEERING
| | | |
|---|---|---|
| FIar | 76 | Similar in appearance to SS-5; may use same engine installation. |
| SWM | 77 | Technology closely resembles that of SS-5 Skean. |
| Rock | 78 | Superficially similar to SS-5. |

## USE AND CONFIGURATION
| | | |
|---|---|---|
| SecC | 65 | 2 launchers per soft site plus probably one refire missile; 3 silos per hard site and probably no refire missiles. |
| SecC | 67 | 2 launchers per soft site; 3 launchers per hard site. |
| vanD | 83 | Reliability: .70; Penetration 1.0. 2 deployed per site, "SS-7/-8 coffins". 3 deployed per site, "SS-7/-8 silos". |

## RELOADS IN GENERAL
| | | |
|---|---|---|
| SecC | 65 | 2 launchers per soft site plus probably one refire missile; 3 silos per hard site and probably no refire missiles. |
| SecC | 71 | Total of all ICBMs that could be made available "includes 140 refire missiles for SS-7/8 soft sites" [probably based on 1 per launcher, ed]. |
| Clns | 80 | Cold Launch: no. |
| SSF | 82 | Reload capability [SS-7,8]: 142 in 1965 and 1970, 131 in 1975. |
| BW | 86 | Some. |

## STRATEGY
| | | |
|---|---|---|
| FI* | 74 | Targeted mainly against Europe and China/Japan. |
| FIar | 76 | Probably intended as back-up to SS-7. |
| Rock | 78 | Played complementary operational role to SS-7 as SS-5 did to SS-4. |
| JWS* | 79 | Obsolescent. |

## EXPORTED TO
| | | | |
|---|---|---|---|
| FI* | 74 | | Operated by WTO. |

BALLISTIC MISSILES   ICBMs   SS-9   127

# SS-9

```
Sources are:   IISS    70   SecC    72   GAO1    75   IISS    77   FI*     79   AWar    80   Clns+   83
SecC    65     SecC    70   SecD    72   HRST    75   JCS     77   IISS    79   Clns    80   JCS     83
SecC    66     SecD    70   IISS    73   IAO6    75   SecD    77   JCS     79   IISS    80   SMP     83
SecC    67     IDO2    71   JCS     73   JCS     75   SWM     77   JWS*    79   SecD    80   vanD    83
IDO1    68     JCS     71   SecC    73   FIar    76   IISS    78   Nitz    79   JCS     81   SecD    84
SecC    68     SecC    71   SecD    73   IISS    76   JCS     78   SecD    79   SMP     81   SMP     84
IISS    69     SecD    71   JCS     74   MOW     76   Rock    78   USD8    79   AWST    82   SMP     85
SecC    69     JCS     72   SecD    74   SecD    76   SecD    78   AFM*    80   SSF     82
```

---------- US DESIGNATION ----------

| | | |
|---|---|---|
| SecC | 65 | SS-9 |
| SecD | 70 | SS-9    FY71-81,85. |
| JCS  | 71 | SS-9    FY72-82,83. |
| HRST | 75 | SS-9 |
| FIar | 76 | SS-9 |
| MOW  | 76 | SS-9 |
| SWM  | 77 | SS-9 |
| Rock | 78 | SS-9 |
| FI*  | 79 | SS-9 |
| JWS* | 79 | SS-9 |
| Nitz | 79 | SS-9 |
| AFM* | 80 | SS-9 |
| AWar | 80 | SS-9 |
| Clns | 80 | SS-9 |
| SMP  | 81 | SS-9    '81, '83, '84, '85. |
| AWST | 82 | SS-9 |
| SSF  | 82 | SS-9 |
| vanD | 83 | SS-9 |

---------- NATO CODENAME ----------

| | | |
|---|---|---|
| HRST | 75 | Scarp |
| FIar | 76 | Scarp |
| MOW  | 76 | Scarp |
| JCS  | 77 | Scarp |
| SWM  | 77 | Scarp |
| Rock | 78 | Scarp |
| FI*  | 79 | Scarp |
| JWS* | 79 | Scarp |
| AFM* | 80 | Scarp |
| AWST | 82 | Scarp |
| SSF  | 82 | Scarp |

---------- DESCRIPTION ----------

| | | | |
|---|---|---|---|
| JCS  | 72 |      | "SS-9 Mod 1,2, & 3" [listed as 2nd generation Soviet strategic missile forces]. |
| MOW  | 76 | ICBM | And "fractional orbital bombardment system". |
| Rock | 78 | ICBM | First operational second-generation ICBM, and typical of the type. |
| FI*  | 79 |      | First Soviet operational 2nd generation ICBM. |
| JWS* | 79 | ICBM | |
| SSF  | 82 |      | Third generation. |

---------- EDITOR'S NOTES ----------

| | | |
|---|---|---|
| SecC | 65 | Listed FY66-77. |
| SecD | 70 | Listed FY71-81,85. |
| JCS  | 71 | Listed FY72-82,83. |
| SMP  | 81 | Listed '81, '83, '84, '85. |

---------- RANGE ----------

| | | mi | nm | km | km conv | |
|---|---|---|---|---|---|---|
| IDO1 | 68 | 12,000+ | | | 19,308 | [Sic, miles] |
| HRST | 75 | 4,000 | | | 6,436 | |
| MOW  | 76 | 7,500 | | 12,070 | 12,067 | Max. |
| SWM  | 77 | 7,500+ | | 12,000+ | 12,067 | |
| Rock | 78 | | | 12,000 | 12,000 | |
| FI*  | 79 | | | c. 12,000 | 12,000 | |
| JWS* | 79 | | | c. 12,000 | 12,000 | |
| Clns | 80 | | 7,000 | | 12,964 | |
| AWST | 82 | 7,500 | | | 12,067 | |
| SSF  | 82 | | 6,500 | | 12,038 | |

128  SOVIET MISSILES  IDDS

## CEP

| Source | Year | ft | mi | nm | m | km | m conv | Notes |
|---|---|---|---|---|---|---|---|---|
| SecC | 67 | | | 0.5-1.0 | | | 926.00-1,852 | Hard target capability for MIRVs would require "much greater accuracies" than Soviets ICBMs so far, FY68-69. |
| SecC | 70 | | | | | | | CEP is currently [deleted]; could be reduced to [deleted] in the next year or so. To reduce CEP below that figure, Soviets would probably have to develop and deploy new guidance and RV package for SS-9, and probably would not be available before 1972 at earliest. |
| SecD | 70 | | | | | | | With single large warhead, considered sufficient to destroy Minuteman in silo. |
| SecD | 71 | | | | | | | Accuracy "could be substantially improved by 1975/76", and projected force could be substantial threat to Minuteman. |
| JCS | 75 | | | | | | | SS-9 [no version given, ed] has hard-target capability, FY76-77. |
| FIar | 76 | | | | | 0.8-1.5 | 800.00-1,500 | "Better than 1.5km and could be as low as 0.8km". |
| Rock | 78 | | | | | 0.5 | 500.00 | "Impact error radius". |
| Nitz | 79 | | | .5 | | | 926.00 | |
| Clns | 80 | | | 0.4 | | | 740.80 | |
| vanD | 83 | | | .5 | | | 926.00 | |

## WARHEAD TYPE

| Source | Year | nuclear | conven | chem |
|---|---|---|---|---|
| MOW | 76 | yes | | |
| JWS* | 79 | yes | | |

## WARHEAD CHARACTERISTICS

| Source | Year | | Notes |
|---|---|---|---|
| SecC | 67 | | "No direct evidence of such an effort", but Soviets might put MIRVs on SS-9s. |
| SecC | 68 | | "We still have no evidence of such an effort" [MRV or MIRV]. |
| SecC | 69 | | SS-9 MIRV hard target capability would require "much greater accuracy" than thus far credited to Soviet ICBMs. Still no evidence of Soviet flight test program for very high accuracy RVs; nor any firm evidence of their testing missile penetration aids. |
| JCS | 73 | | Reasonable to assume "SS-9 follow-on" would be MIRVed. |
| MOW | 76 | | 4 warheads have been tested: single 20-25 MT charge, an unguided MRV with 3 4-5 MT RVs, FOBS space-bomb, and (under development) a MIRV warhead. FOBS payloads 6'6" long and 4' in diameter. FOBS thought not deployed at any regular SS-9 sites. |
| SecD | 77 | single RV | Number independently targetable RVs: 1. |
| SWM | 77 | single RV | |
| Rock | 78 | single RV | |
| Nitz | 79 | 1 RV | |
| USO8 | 79 | single RV | |
| Clns | 80 | single RV | |
| vanD | 83 | single RV | |

## NUCLEAR YIELD

| Source | Year | kT | MT | kT conv | Notes |
|---|---|---|---|---|---|
| SecC | 65 | | 12-25 | 12,000-25,000 | Might have yield this high. |
| SecC | 66 | | 18 | 18,000 | |
| SecC | 67 | | 18-25 | 18,000-25,000 | |
| ID01 | 68 | | 20-50 | 20,000-50,000 | |
| SecC | 68 | | 12-25 | 12,000-25,000 | FY69-72. |
| MOW | 76 | | | | 4 warheads have been tested: single 20-25 MT charge, an unguided MRV with 3 4-5 MT RVs... |
| SWM | 77 | | 20-25 | 20,000-25,000 | "20/25" MT. |
| Rock | 78 | | 18.0-25.0 | 18,000-25,000 | |
| Nitz | 79 | | 20 | 20,000 | |
| USO8 | 79 | | 18-25 | 18,000-25,000 | |
| Clns | 80 | | 20 | 20,000 | |
| AWST | 82 | | 20 | 20,000 | |
| vanD | 83 | | 18.0 | 18,000 | |

BALLISTIC MISSILES   ICBMs   SS-9   129

```
------------ WARHEAD WEIGHT ------------
              lb              t           kg           kg conv
JWS* 79                                   5000-7000    5,000-7,000

------------ THROWWEIGHT ------------
              lb              t           kg           kg conv
Flar 76                                   5,000        5,000
Rock 78                       5.5                      5,500          Tonnes.
Nitz 79       13,500                                   6,123
USO8 79       10,000                                   4,536
Clns 80       11,000                                   4,989
SSF  82       9000-11000                               4,082-4,989

------------ PAYLOAD ------------
              lb              t           kg           kg conv
SecC 66       9000-11000                               4,082-4,989
AWST 82       13,500                                   6,123

------------ WEIGHT ------------
              lb              t           kg           kg conv
ID01 68                       c. 200                   200,000
FI*  79                                   c. 200,000   200,000

------------ LENGTH ------------
              in       ft       cm       m           m conv
ID01 68                111.5                         33.99
SecC 70                115                           35.05        [From graph, unit inferred].
GA01 75                113.7                         34.66
HRST 75                120                           36.58
Flar 76                                   37         37.00
MOW  76                113'6"             34.5       34.59
SWM  77                113.5              34.6       34.59        [Unmarked, contrasts with Mod 4, ed]
Rock 78                                   34         34.00
FI*  79                                   37         37.00
JWS* 79                          c. 35               35.00
AWST 82                120                           36.58

------------ DIAMETER ------------
              in       ft       cm       m           cm conv
Flar 76                                   3.4        340.00
MOW  76                10'0"              3.05       304.80
SWM  77                10.0               3.05       304.80
Rock 78                                   3.4        340.00       First stage.
                                          1.15       115.00       Terminal stage and warhead assembly.
FI*  79                          3.4 [sic]                3.40
JWS* 79                                   c. 3      300.00
AWST 82                10                            304.80

------------ NUMBER OF STAGES ------------
Flar 76       2 or 3   "Has been described by different sources as two or three-staged". First stage powered by
                       cluster of 6 engines, plus 4 vernier motors faired into surrounding skirt.
MOW  76       3
SWM  77       2
Rock 78       2        Certain amount of speculation over precise number of stages.
FI*  79       3
JWS* 79       3        First stage has 6 nozzles and 4 vernier nozzles.
AWST 82       3

------------ TYPE OF PROPULSION ------------
SecC 66       storable liquid
JCS  72       liquid                      FY73,76-77.
HRST 75       liquid
Flar 76       storable liquid
MOW  76       liquid                      6 first-stage nozzles surrounded by 4 vernier nozzles.
SWM  77       storable liquid             6 first-stage main thrust chambers.
Rock 78       storable liquid             Stages 1 and 2. First stage has 6 engines within the base, plus 4 verniers.
FI*  79       liquid
JWS* 79       liquid
Clns 80       liquid
AWST 82       liquid
SSF  82       liquid
```

130   SOVIET MISSILES   IDDS

```
----------- MIDCOURSE (OR MAIN) GUIDANCE -----------
FIar   76                 Guidance system better than other 2nd generation ICBMs.
MOW    76    inertial
SWM    77    inertial
Rock   78    inertial
FI*    79    inertial
JWS*   79    inertial
SSF    82                 Fly-the-wire, inertial.

----------- CONTROL SYSTEM -----------
SWM    77    4 swivel-mounted verniers.
Rock   78    Verniers.
FI*    79    Stage 1: vernier nozzles, other stages unknown.

----------- OBSERVED -----------
SMP    84    [Picture described in caption as:] Soviet missile crew loads SS-9 into silo in late
              1960s.

----------- OBSERVED IN PARADE -----------
MOW    76    7 Nov 1967
SWM    77    Nov 1967
Rock   78    7 Nov 1967    First displayed publicly, though in service earlier.
JWS*   79    7 Nov 1967

----------- DEVELOPMENT BEGAN -----------
SSF    82    1957          Year design began.
SMP    85    1953          Technological development began. [From graph.]

----------- PROTOTYPE TESTS -----------
JCS    71                  [Is or was testing pause, ed:] "While some question may remain as to the nature and
                            duration of the SS-9 pause, ...".
JCS    72                  Very extensive and vigorous flight-testing program of improved versions underway
                            "for some time".
JCS    73                  "SS-9 follow-on" being actively tested.
JCS    74                  Competitive design and flight test program believed used prior to SS-9 deployment
                            decision [like expected SS-X-17, SS-X-19 competition].
SMP    85    1959-60       Engineering and testing began. [From graph.]

----------- IOC -----------
SecC   65    1965          Expected to become operational.
SecC   66                  Expected to become operational this year.
SecD   77    1967
JCS    83    c. 1964-65    Year of introduction [from graph].
vanD   83    1966          Latter 1966.

----------- DEPLOYMENT BEGAN -----------
JCS    73    after 1964    Deployed after 1964, FY74,76-78.
FIar   76                  First of 2nd generation ICBMs to be deployed.
             1965          Entered service.
MOW    76    1965
SWM    77    1965          Service entry.
Rock   78    1965          Service intro.
JWS*   79    1965          Installation began.
Clns   80    1967          First deployed.
SMP    83                  Deployment underway by 1966.
SecD   84    1966          Introduction of strategic systems [graph].
SMP    85    1964-65       [From graph.]

----------- DEPLOYED -----------
MOW    76                  In service. FOBS thought not deployed at any regular SS-9 sites.

----------- DEPLOYMENT COMPLETE -----------
SecD   72                  This phase of missile deployment may have been completed.
SecD   73                  Deployment appears to be completed. New large silos are probably for SS-9
                            follow-on.

----------- RETIRED -----------
FIar   76    2nd half 1975 Some SS-9s withdrawn from service.
JWS*   79                  These silos converting to SS-18, beginning 1973.
JCS    81                  "During the year, the Soviets converted the remaining deployed SS-9 launchers, some
                            300 missiles, to the SS-18" [sic].
SMP    85    1979-80       [From graph.]
```

# BALLISTIC MISSILES ICBMs SS-9

## DEPLOYMENT HISTORY

| | | |
|---|---|---|
| SecC | 67 | Program moving at faster rate than expected last year. |
| SecC | 68 | New SS-9 construction appears to be tapering off. |
| SecC | 70 | Deployment rate at 1969 has increased over 1968, but is less than "peak development" year of 1965. |
| SecC | 71 | Work on at least 3 groups may have been suspended, and work has slowed on the sites of 2 other groups. Several groups of SS-9s and SS-11s have become operational during the past few months. |
| SecD | 71 | Deployment continued, though deployment rate decreased [during 1970]. Work suspended on some sites, slowed on several others. Suspension may be a) end of planned deployment, b) to introduce newer versions, c) preparing for new missiles. |
| SecC | 72 | No construction starts on standard SS-9 sites during past year. New construction includes 25 large silos for missile "similar to, or slightly larger than" SS-9. Most of these silos could be completed during 1972. |
| SecD | 72 | Very little construction activity during the past year. |
| SecC | 73 | New silos expected completed "a year later" than "middle of this year" [mid-74, ed]. |
| JCS | 75 | Some already converted to launchers for "more advanced missiles"; others undergoing conversion, FY76-77. Work underway at SS-9 complexes to permit SS-18 deployment, FY76-77. |
| MOW | 76 | Being replaced by SS18. |
| SecD | 76 | Being replaced by SS-18. |
| JCS | 77 | Conversion of first group began in 1973. SS-9 launch group conversion to SS-18 continues. |
| JCS | 78 | For 4th straight year, number of SS-9 launch groups "being converted" to SS-18 increased. Anticipated all SS-9s will be converted to SS-18. |
| Rock | 78 | Being replaced by SS-18. |
| FI* | 79 | 5 existing versions expected to be replaced by SS-18. |
| JCS | 79 | For 4th straight year, "number of conversion has increased". Conversion continues at a rapid pace, FY80-81. |
| JWS* | 79 | These silos converting to SS-18, beginning 1973. |

## NUMBERS IN SERVICE

### NUMBER IN 1960

| | | | |
|---|---|---|---|
| SSF | 82 | none | In 1960 and 1965. |

### NUMBER IN 1965

| | | | |
|---|---|---|---|
| IDO2 | 71 | > 200 | |

### NUMBER IN 1966

| | | | |
|---|---|---|---|
| SecC | 70 | none | In mid-66, operational launchers [from graph]. |
| SecC | 71 | none | In mid-66, operational launchers [from graph]. |

### NUMBER IN 1967

| | | | |
|---|---|---|---|
| SecC | 66 | | 60-70 launchers expected deployed by mid-1967. |
| SecC | 68 | 114 | As of 1 Oct. |
| SecC | 70 | 60 | In mid-67, operational launchers [from graph]. |
| SecC | 71 | 70 | In mid-67, operational launchers [from graph]. |
| vanD | 83 | 54 | 1 Jan. |

### NUMBER IN 1968

| | | | |
|---|---|---|---|
| SecC | 66 | | 100-110 expected for mid-1968 [reported in FY68]. |
| SecC | 67 | | 130-140 expected for mid-1968. |
| SecC | 69 | 156 | As of 1 Sept. |
| SecC | 70 | 125 | In mid-68, operational launchers [from graph]. |
| SecC | 71 | 130 | In mid-68, operational launchers [from graph]. |
| JWS* | 79 | 225 | Deployed by spring 1968. After 9 mos. of apparent inactivity, installation recommenced. These silos converting to SS-18, beginning 1973. 200 converted by early 1979. |
| vanD | 83 | 120 | 1 Jan. |

### NUMBER IN 1969

| | | | |
|---|---|---|---|
| SecC | 68 | | 180-222 estimated by mid-1969. |
| IISS | 69 | c. 200 | July, total deployed. |
| SecC | 70 | 174-198 | Operational on 1 Sept. |
| | | 270 | On 1 Sept, including those estimated to be under construction. |
| | | 185 | In mid-69, operational launchers [from graph]. |
| SecD | 70 | 230 | Was estimate less than one year ago [1969]. |
| SecC | 71 | 175 | In mid-69, operational launchers [from graph]. |
| vanD | 83 | 156 | 1 Jan. |

```
---------- NUMBER IN 1970 ----------
SecC  69              228-240 estimated by mid-1970.
IISS  70    240       July, total deployed.
SecC  70              222 expected operational by mid-70.
            198       Operational in Jan 1970.
                      250 in mid-70, operational launchers [from graph].
SecD  70    > 275     Over 275 estimated to be deployed or under construction. Over 275 [of 1,100 operational
                      launchers] for SS-9.
                      Increase of 160 ICBMs in one year--almost all accounted for by new SS-9s and SS-11s.
SecC  71    258       Operational 30 Dec.
            215       In mid-70, operational launchers [from graph].
SecC  72    228       Operational in 1970.
FIar  76    288       By 1970.
Clns  80    228
SSF   82    240
vanD  83    198       1 Jan.

---------- NUMBER IN 1971 ----------
SecC  70              270-282 expected operational by mid-71.
SecC  71    276       Operational 1 Mar.
                      276-282 operational mid-71.
                      245 in mid-71, operational launchers [from graph].
Clns  80    270
vanD  83    258       1 Jan.

---------- NUMBER IN 1972 ----------
SecC  71              288-306 operational mid-72.
                      265 in mid-72, operational launchers [from graph].
SecC  72    288       Operational in 1972.
AFM*  80    288       In May 72, suitable for Mods 1,2, or 5, plus 25 silos under construction.
Clns  80    308       In 1972-1974.
vanD  83    288       1972-74, 1 Jan.

---------- NUMBER IN 1973 ----------
IISS  73    288       1973-75, July, total deployed.
JCS   73    313       Plus new silos; launchers "operational and under construction or conversion".

---------- NUMBER IN 1974 ----------
JCS   74    313       Of which 25 under construction when Interim Agreement signed.
                      Mod 2 is bulk of SS-9 force; relatively small number deployed.

---------- NUMBER IN 1975 ----------
IAD6  75    313       "313 silos for modern heavy SS-9".
JCS   75    288       Originally deployed; some already converted to launchers for "more advanced missiles";
                      others undergoing conversion, FY76-77.
SWM   77    288       In service mid-1975.
Rock  78              "In all, some 250 SS-9 Mod 1 and Mod 2 missiles were deployed with 38 Mod 4 variants
                      between 1970 and 1975".
USO8  79    "288-313" Early 1975.
Clns  80    298
SSF   82    278
vanD  83    282       1 Jan.

---------- NUMBER IN 1976 ----------
FIar  76    210       Remaining at present.
IISS  76    252       July, total deployed.
Clns  80    272
vanD  83    270       1 Jan.

---------- NUMBER IN 1977 ----------
IISS  77    238       July, total deployed.
SWM   77              Widely deployed in silos.
Rock  78    only 210  In service by 1977.
Nitz  79    164       Launchers.
Clns  80    208
vanD  83    252       1 Jan.

---------- NUMBER IN 1978 ----------
IISS  78    190       July, total deployed.
SecD  78              Over 100 SS-18s now exist, converted from SS-9.
Clns  80    312       [Sic]
vanD  83    216       1 Jan.
```

BALLISTIC MISSILES  ICBMs  SS-9  133

```
                  ---------- NUMBER IN 1979 ----------
IISS   79   100         July, total deployed.
Nitz   79   120         18 June, launchers.
SecD   79               Nearly 200 SS-18s now exist, converted from SS-9.
AFM*   80   100
Clns   80    68         All are in process of conversion to SS-18, none operational.
vanD   83   120         1 Jan.

                  ---------- NUMBER IN 1980 ----------
AWar   80   only 18    Expected still deployed by end of 1980.
                       164 "silos" with single RVs and 100 "Launchers" with multiple RVs until recently.
IISS   80   none       July, total deployed.
SecD   80              More than 200 SS-18s now exist, converted from SS-9.
SSF    82   none
Clns+  83   none       In 1980-1982.
vanD   83   60         1 Jan.

                  ---------- NUMBER IN 1981 ----------
JCS    81              "During the year, the Soviets converted the remaining deployed SS-9 launchers, some 300
                       missiles, to the SS-18" [sic].
vanD   83   none       1 Jan.

                  ---------- UNDATED NUMBER ----------
IDO2   71   300        Expected ultimately.
MOW    76   288        Peak.
JWS*   79   288        "Originally", US source.
SSF    82   288        "Number of missiles deployed".

                  ---------- LAUNCHER/SILO ----------
SecC   65   Deployment expected in one silo per site hard configuration.
SecC   66   Soviets constructing new type of single silo launch site, believed for SS-9.
SecC   70   Hard single silos.
SecC   72   New construction includes 25 large silos for missile "similar to, or slightly larger than" SS-9.
SecC   73   New silos being constructed in SS-9 missile complexes.
SecD   73   New large silos are probably for SS-9 follow-on.
SecD   74   SS-X-18s can be deployed in SS-9 silos with some modification.
Flar   76   Silo-based.
SWM    77   Widely deployed in silos.
Rock   78   Silo launched.
SSF    82   Basing mode: Hardened silo.
SMP    84   "Starting in 1970, SS-9 silos were rebuilt to increase their survivability and accommodate the
            SS-18".

                  ---------- DESIGNER ----------
SSF    82   Yangel         Design Bureau.

                  ---------- USE AND CONFIGURATION ----------
vanD   83   Reliability: .80; Penetration 1.0. 6 missiles in a battalion.

                  ---------- RELOADS IN GENERAL ----------
MOW    76   Hot launch.
SecD   76   Hot launch.
Clns   80   Cold launch: no.
SSF    82   Launching mode: Hot.
SMP    84   SS-9 is not transported and loaded into silo in a canister.

                  ---------- STRATEGY ----------
SecC   67   "More expensive and [for hard targets] a much more effective missile" than SS-11. Hard target
            capability for MIRVs [on SS-9 and generally] would require "much greater accuracies" than Soviets
            ICBMs so far; probably could achieve "operational capability" by 1971-72; US would probably detect
            testing perhaps 2 years earlier.
SecC   68   Suitable for use against hardened missile silos. Hard target capability with MIRVs would require
            "much greater accuracy" than Soviet ICBMs so far; would take 4-5 years from start of development
            to operational capability; US probably would detect testing of it "at least two years" before that
            happened.
JCS    73   Deployment of 300 "SS-9 follow-on" ICBMs would greatly enhance hard-target capabilities.
HRST   75   "Most advanced ICBM". May place payload into fractional orbit trajectory.
Rock   78   Dramatic improvement over SS-7 and SS-8.
JWS*   79   Obsolescent.

                  ---------- GEOGRAPHICAL DEPLOYMENTS ----------
SMP    83              [Shown on map at] Tyuratam.

                  ---------- RELATION TO OTHER MISSILES ----------
SecC   65   Probably larger than SS-7/SS-8.
SecC   68   About the size of Titan II.
SWM    77   Largest Soviet missile before SS-18.
```

---------- **OTHER INFORMATION** ----------
JWS* 79   Best known of Soviet ICBMs.

# SS-9 Mod 1

```
Sources are:  JCS   73    JCS   75    SWM   77    FI*   79    USO9  79    AWar  80    BW    86
JCS   72      JCS   74    FIar  76    Rock  78    JWS*  79    AFM*  80    SSF   82
```

### US DESIGNATION

|  | |  |
|---|---|---|
| JCS | 72 | SS-9 Mod 1 |
| FIar | 76 | SS-9 Mod 1 |
| SWM | 77 | SS-9 Mod 1 |
| Rock | 78 | SS-9 Mod 1 |
| FI* | 79 | SS-9 Mod 1 |
| JWS* | 79 | SS-9 Mod 1 |
| AFM* | 80 | SS-9 Mod 1 |
| AWar | 80 | SS-9 mod |
| SSF | 82 | SS-9 Mod 1 |

### EDITOR'S NOTES

|  | |  |
|---|---|---|
| JCS | 72 | Mod 1 mentioned FY73-79. |

### RANGE

| Source | | mi | nm | km | km conv | Notes |
|---|---|---|---|---|---|---|
| FIar | 76 | | | 12,000 | 12,000 | Mods 1-3. |
| USO9 | 79 | | | c. 12,000 | 12,000 | |
| BW | 86 | | | 12,000 | 12,000 | |

### CEP

| Source | | ft | mi | nm | m | km | m conv | Notes |
|---|---|---|---|---|---|---|---|---|
| JCS | 72 | | | | | | | Only "Mods 1 and 2 appear to have a hard-target kill capability". |
| JCS | 73 | | | | | | | Mods 1 and 2 have hard-target capability; not enough deployed to threaten Minuteman force, FY74-75. |
| JCS | 75 | | | | | | | SS-9 [no version given] has hard-target capability, FY76-77. |
| AFM* | 80 | | | | | | | Before current generation of ICBMs, these [Mod 1?, ed] were only operational Soviet ICBMs with hard-target capability. |
| AWar | 80 | | | 0.5 | | | 926.00 | |
| SSF | 82 | | | 0.5 | | | 926.00 | |
| BW | 86 | | | | 900 | | 900.00 | |

### WARHEAD TYPE

| Source | | nuclear | conven | chem |
|---|---|---|---|---|
| Rock | 78 | yes | | |

### WARHEAD CHARACTERISTICS

| Source | | Type | Notes |
|---|---|---|---|
| JCS | 72 | single RV | Large warhead, FY73-74. |
| JCS | 75 | single RV | FY76-79. |
| FIar | 76 | single RV | |
| SWM | 77 | single RV | |
| Rock | 78 | single RV | |
| FI* | 79 | single RV | Mods 1 and 2. |
| JWS* | 79 | single RV | Mods 1 and 2. |
| AFM* | 80 | single RV | |
| AWar | 80 | single RV | |
| SSF | 82 | single RV | |
| BW | 86 | single RV | |

### NUCLEAR YIELD

| Source | | kT | MT | kT conv | Notes |
|---|---|---|---|---|---|
| JCS | 74 | | | | Slightly smaller yield than Mod 2. |
| FIar | 76 | | 20 | 20,000 | |
| SWM | 77 | | 20 | 20,000 | |
| Rock | 78 | | 20 | 20,000 | |
| JWS* | 79 | | 20-25 | 20,000-25,000 | |
| USO9 | 79 | | 20 | 20,000 | |
| AFM* | 80 | | [< 25] | 25,000 | Yield slightly less than Mod 2. |
| AWar | 80 | | 20 | 20,000 | |
| SSF | 82 | | 20.0 | 20,000 | |
| BW | 86 | 18,000 | | 18,000 | |

SOVIET MISSILES  IDDS

```
─────────── WEIGHT ───────────
             lb            t              kg          kg conv
US09  79                              c. 200,000     200,000
─────────── LENGTH ───────────
             in     ft     cm          m            m conv
US09  79                              37            37.00
─────────── DIAMETER ───────────
             in     ft     cm          m           cm conv
US09  79                              3.4           340.00
─────────── NUMBER OF STAGES ───────────
US09  79     3
─────────── TYPE OF PROPULSION ───────────
US09  79     liquid
BW    86     storable liquid
─────────── MIDCOURSE (OR MAIN) GUIDANCE ───────────
US09  79     inertial
─────────── OBSERVED IN PARADE ───────────
AFM*  80     7 Nov 1967
─────────── PROTOTYPE TESTS ───────────
SSF   82     1964              First flight test.
─────────── IOC ───────────
JCS   75     1967              FY76-79.
US09  79     1966
─────────── DEPLOYMENT BEGAN ───────────
FIar  76     1966              In service since 1966.
SWM   77     1965              Entered service.
AFM*  80     ? 1965            Deployment started.
SSF   82     1967              Year operation began.
BW    86     1966
─────────── RETIRED ───────────
BW    86     1979
─────────── NUMBERS IN SERVICE ───────────

─────────── NUMBER IN 1975 ───────────
Rock  78           "In all, some 250 SS-9 Mod 1 and Mod 2 missiles were deployed with 38 Mod 4 variants
                   between 1970 and 1975".
─────────── NUMBER IN 1976 ───────────
FIar  76           Few remain operational.
─────────── NUMBER IN 1980 ───────────
AFM*  80           "Relatively small number still emplaced".
─────────── UNDATED NUMBER ───────────
BW    86     288   Peak.
─────────── RELOADS IN GENERAL ───────────
BW    86     No.
─────────── STRATEGY ───────────
SWM   77     Principal objectives are Titan and Minuteman silos.
```

# SS-9 Mod 2

```
Sources are:  JCS  73    JCS  75    A140 77    Rock 78    JWS* 79    SSF  82
JCS  72       JCS  74    Flar 76    SWM  77    FI*  79    AFM* 80
```

## US DESIGNATION

|  | |  |
|---|---|---|
| JCS | 72 | SS-9 Mod 2 |
| Flar | 76 | SS-9 Mod 2 |
| SWM | 77 | SS-9 Mod 2 |
| Rock | 78 | SS-9 Mod 2 |
| FI* | 79 | SS-9 Mod 2 |
| JWS* | 79 | SS-9 Mod 2 |
| AFM* | 80 | SS-9 Mod 2 |
| SSF | 82 | SS-9 Mod 2 |

## EDITOR'S NOTES

JCS 72 — Mod 2 mentioned FY73-79.

## RANGE

| Source | | mi | nm | km | km conv | |
|---|---|---|---|---|---|---|
| Flar | 76 | | | 12,000 | 12,000 | Mods 1-3. |
| AFM* | 80 | 7,500 | | | 12,067 | |

## CEP

| Source | | ft | mi | nm | m | km | m conv | Notes |
|---|---|---|---|---|---|---|---|---|
| JCS | 72 | | | | | | | Only "Mods 1 and 2 appear to have a hard-target kill capability". |
| JCS | 73 | | | | | | | Mods 1 and 2 have hard-target capability; not enough deployed to threaten Minuteman force, FY74-75. |
| JCS | 75 | | | | | | | SS-9 [no version given] has hard-target capability, FY76-77. |
| AFM* | 80 | | | < 0.9 | | < 1.5 | 1,448 | |
| SSF | 82 | | | | 0.5 | | 926.00 | |

## WARHEAD TYPE

| Source | | nuclear | conven | chem |
|---|---|---|---|---|
| AFM* | 80 | yes | | |

## WARHEAD CHARACTERISTICS

|  | | | |
|---|---|---|---|
| JCS | 72 | single RV | Large warhead, FY73-74. |
| JCS | 75 | single warhead | FY76-79. |
| Flar | 76 | single RV | |
| SWM | 77 | single RV | |
| FI* | 79 | single RV | Mods 1 and 2. |
| JWS* | 79 | single RV | Mods 1 and 2. |
| AFM* | 80 | single RV | |
| SSF | 82 | single RV | |

## NUCLEAR YIELD

| Source | | kT | MT | kT conv | Notes |
|---|---|---|---|---|---|
| JCS | 74 | | | | Largest yield of any known ICBM. Mod 1 has slightly smaller yield than Mod 2. |
| Flar | 76 | | 25 | 25,000 | |
| SWM | 77 | | 25 | 25,000 | |
| Rock | 78 | | 25 | 25,000 | |
| FI* | 79 | | 25 | 25,000 | |
| JWS* | 79 | | < 20-25 | 20,000-25,000 | Mod 2 is less than 20-25 MT of Mod 1. |
| AFM* | 80 | | 25 | 25,000 | Largest yield of any known ICBM prior to SS-18. |
| SSF | 82 | | 20.0 | 20,000 | |

## LENGTH

| Source | | in | ft | cm | m | m conv |
|---|---|---|---|---|---|---|
| AFM* | 80 | | 121'5" | | | 37.01 |

## DIAMETER

| Source | | in | ft | cm | m | cm conv |
|---|---|---|---|---|---|---|
| AFM* | 80 | | 11'2" | | | 340.36 |

## NUMBER OF STAGES

AFM* 80 — 3

## TYPE OF PROPULSION

AFM* 80 — liquid

---------- **MIDCOURSE (OR MAIN) GUIDANCE** ----------
AFM*  80     inertial

---------- **PROTOTYPE TESTS** ----------
SSF   82     1964-65     First flight test.

---------- **IOC** ----------
JCS   75     1966        FY76-79.

---------- **DEPLOYMENT BEGAN** ----------
SSF   82     1966        Year operation began.

---------- **NUMBERS IN SERVICE** ----------

---------- **NUMBER IN 1974** ----------
JCS   74                 Mod 2 is bulk of SS-9 force; relatively small number deployed [of ICBMs are SS-9s?, ed].

---------- **NUMBER IN 1975** ----------
Rock  78                 "In all, some 250 SS-9 Mod 1 and Mod 2 missiles were deployed with 38 Mod 4 variants between 1970 and 1975".

---------- **NUMBER IN 1976** ----------
FIar  76                 Most widely deployed variant.

---------- **NUMBER IN 1977** ----------
A140  77                 Composes the "bulk of the diminishing total of SS-9 deployed".
SWM   77                 "The standard weapon" [standard SS-9 Mod].

---------- **UNDATED NUMBER** ----------
AFM*  80                 Mod 2 was the bulk of the SS-9 force.

---------- **RELATION TO OTHER MISSILES** ----------
Rock  78     Similar performance to Mod 1.

# SS-9 Mod 3

| Sources are: | SecC | 70 | SecD | 71 | JCS | 75 | AFM* | 78 | JWS* | 79 | SSF | 82 |
|---|---|---|---|---|---|---|---|---|---|---|---|---|
| SecC | 68 | | SecD | 70 | JCS | 72 | FIar | 76 | Rock | 78 | AFM* | 80 | |
| SecC | 69 | | SecC | 71 | JCS | 74 | SWM | 77 | FI* | 79 | AWST | 82 | |

### US DESIGNATION

| | | |
|---|---|---|
| SecC | 70 | SS-9 Mod 3 |
| SecD | 70 | SS-9 Mod 3 |
| JCS | 72 | SS-9 Mod 3 |
| FIar | 76 | SS-9 Mod 3 |
| SWM | 77 | SS-9 Mod 3 |
| AFM* | 78 | SS-9 Mod 3 |
| Rock | 78 | SS-9 Mod 3 |
| FI* | 79 | SS-9 Mod 3 |
| JWS* | 79 | SS-9 Mod 3 |
| AWST | 82 | SS-9 Mod 3 |
| SSF | 82 | SS-9 Mod 3 |

### NATO CODENAME

| | | |
|---|---|---|
| AWST | 82 | Scarp |
| SSF | 82 | Scarp |

### OTHER DESIGNATION

| | | | |
|---|---|---|---|
| SecC | 70 | SS-X-6 | Former name. |
| SecD | 70 | | "Retrofire weapon". |
| Rock | 78 | F-l-r | Redesignated this to indicate orbital capability. |

### DESCRIPTION

| | | |
|---|---|---|
| SSF | 82 | "Global-range missile" [and in other listing] "Intercontinental-range missile". |

### EDITOR'S NOTES

| | | |
|---|---|---|
| SecC | 68 | Mod 3 mentioned FY69-72. |
| JCS | 72 | Mod 3 mentioned FY73-79. |
| SSF | 82 | SS-9 Mod 3 listed in table separately under ICBMs and global missiles. Data is nevr inconsistent (including range); records are thus collapsed. |

### RANGE

| | | mi | nm | km | km conv | |
|---|---|---|---|---|---|---|
| FIar | 76 | | | 12,000 | 12,000 | Mods 1-3. |

### CEP

| | | ft | mi | nm | m | km | m conv | |
|---|---|---|---|---|---|---|---|---|
| SecC | 68 | | | 1-2 | | | 1,852-3,704 | Across northern approaches. |
| | | | | 1.5-3 | | | 2,778-5,556 | From the south. |
| JCS | 72 | | | | | | | Mod 3 with large CEP considered soft-target, FY73-74. |
| SSF | 82 | | | 1.0-2.0 | | | 1,852-3,704 | Depressed trajectory. |
| | | | | 1.5-3.0 | | | 2,778-5,556 | FOBS. |

### WARHEAD CHARACTERISTICS

| | | | |
|---|---|---|---|
| JCS | 72 | | FOBS or depressed trajectory version, FY73-74. |
| JCS | 75 | single RV | FY76-79. |
| FI* | 79 | FOBS | |
| AWST | 82 | FOBS | |
| SSF | 82 | single RV | |

### NUCLEAR YIELD

| | | kT | MT | kT conv |
|---|---|---|---|---|
| SSF | 82 | | 20.0 | 20,000 |

### THROWWEIGHT

| | | lb | t | kg | kg conv |
|---|---|---|---|---|---|
| SSF | 82 | 9000-11000 | | | 4,082-4,989 |

### PAYLOAD

| | | lb | t | kg | kg conv | |
|---|---|---|---|---|---|---|
| Rock | 78 | | | 4,500 | 4,500 | Can be lifted to low earth orbit. |

### TYPE OF PROPULSION

| | | |
|---|---|---|
| SSF | 82 | liquid |

### MIDCOURSE (OR MAIN) GUIDANCE

| | | |
|---|---|---|
| SSF | 82 | Fly-the-wire, inertial. |

140  SOVIET MISSILES  IDDS

---------- **DEVELOPMENT UNDERWAY** ----------
SecD 70                     Development of retrofire weapon continues.

---------- **PROTOTYPE TESTS** ----------
SecC 68                     "I announced last November" that Soviets intensively testing FOBS [not contradicted, ed].
SecC 69                     Test program discussed last year apparently ended in Oct 1967. Another test in Apr 1968 was identical to earlier flights. This test was followed by 2 additional flights of 'depressed trajectory' missile, with RVs deboosted prior to completion of ballistic course. In October, system tested again in fractional orbit mode. "However, in none of these tests have we noted the modifications which the intelligence community thinks are necessary if the FOBS is to be launched over either the North or South Pole to hit targets in the U.S. on the first orbit".
SecC 70  1965              Testing of FOBS: 1 failure in 1965; 2 successes and 2 failures in 1966; 8 successes and 2 failures in 1967; 4 successes in 1968; 1 success in 1969. 20 tests in all.
SecD 71  1965+             Extensive testing since 1965, FY72-73. Believed to be for FOBS or retrofired depressed trajectory ICBM, FY72-73.
JCS  74                    Last test of Mod 3 was over 2 years ago; still uncertain capabilities and mission.
JCS  75                    No known tests of Mod 3 since Aug 1971, mission and capability uncertain, FY76-77.
FIar 76  1966              First test; no trials detected since Aug 1971.
AFM* 78                    Under test until 1971, both in depressed trajectory mode and as FOBS.
Rock 78  1967-1971         14 tests in these years, orbits between 130 and 300 km. "Used exclusively, between 1967 and 1971, to demonstrate a FOBS...role". No FOBS tests since 1971. All tests launched from Tyuratam.
                           Have been more than 50 "F-1" launches [Mod 3 and Mod 5]; used only for military payloads.
JWS* 79                    Tested both in depressed trajectory mode and as FOBS. No test known since Aug 1971.
SSF  82  1965              First flight test.

---------- **IOC** ----------
JCS  75  1969              FY76-79.

---------- **DEPLOYMENT BEGAN** ----------
SecC 70                    One firing was in Sept 1969, followed by 11-month hiatus, and it may have been crew training launch. If so, system may be operational or nearly so.
SecD 70                    Small number may be deployed already.
SecC 71                    1 firing in Sept 1969 may have been crew training launch; if so, system may be operational. 2 firings, apparently both for crew training, were conducted in 1970. Total testing: 22.
JCS  74                    Mod 3 believed not deployed at any regular SS-9 complex, FY75-77.
FIar 76                    US DoD believes not operational.
AFM* 78                    "There is no evidence that this version became operational".
Rock 78                    "Appeared in 1967". Not deployed operationally.
JWS* 79                    US believes it is not deployed in any operational SS-9 complexes, but obviously hard to tell.
SSF  82  1969              Year operation began.

---------- **NUMBERS IN SERVICE** ----------

---------- **NUMBER IN 1969** ----------
SecC 68                    Soviets might have 10-25 FOBS by mid-1969.

---------- **NUMBER IN 1972** ----------
SecC 68                    Soviets might have 20-75 FOBS by mid-1972.

---------- **NUMBER IN 1982** ----------
SSF  82  18                "Number of missiles deployed".

---------- **LAUNCHER/SILO** ----------
SSF  82                    Basing mode: Hardened silo.

---------- **DESIGNER** ----------
SSF  82                    Yangel    Design Bureau.

---------- **DESIGN AND ENGINEERING** ----------
SecD 70                    Uses SS-9 booster.

BALLISTIC MISSILES   ICBMs   SS-9 Mod 3   **141**

---------- **STRATEGY** ----------

SecC 68   Could be low trajectory across northern approaches, or around the southern approaches. Poor accuracy, and would pay "heavy penalty" in payload; thus would be soft target weapon. US rejected such system years ago, but Soviets may think it useful for "surprise nuclear strike against our bomber bases or as a penetration tactic against ABM systems".

SecC 70   "Limitations of the present system would restrict its use as a FOBS to targets on the US East Coast".

SecD 70   Could perform as depressed trajectory ICBM, FOBS, or dual system.

Flar 76   Tested as depressed-trajectory and FOBS. In depressed-trajectory, lower-angle trajectory lowers apogee and allows warhead to cover greater distance before rising above radar horizon of the target. In FOBS mode, warhead is in low orbit, decelerated as it approaches target, and placed on suitable trajectory to impact. Defending radar's warning time is greatly reduced and the attack can be mounted from any direction, but accuracy tends to suffer unless terminal guidance is used.

SWM 77   Employed in tests of FOBS and depressed-trajectory. FOBS intended to enter US from South; previously not covered by radar, but is as of 1968.

Rock 78   1967-1971 tests: Payload in structure (8 m long, 2 m diameter) placed in low orbit, returned to earth before completion of first earth revolution. This displays capability for depressed trajectory or low orbit insertion, keeping flight path below US radar screens, enabling surprise attack or reduced warning time. If used operationally, FOBS would place warhead in low orbit, retro-fire some distance uprage of target. Test orbits of 130-300 km are considerably below 1,500 km of conventional trajectory. Probably reason not deployed is in part defence alert radars on southern flank of US, negating South Pole surprise route.

AFM* 80   FOBS banned under SALT II.

# SS-9 Mod 4

```
Sources are:   JCS   71      SecD  72     JCS   75     AFM*  78     Nitz  79     SSF   82
SecC  69       SecC  71      JCS   73     JCS   75     Rock  78     AWar  80
SecC  70       SecD  71      JCS   74     FIar  76     FI*   79     Clns  80
SecD  70       JCS   72      IAD6  75     SWM   77     JWS*  79     AWST  82
```

---------- **US DESIGNATION** ----------

| | | |
|---|---|---|
| SecD | 70 | SS-9 MRV variant. |
| JCS  | 71 | SS-9 Mod 4 |
| FIar | 76 | SS-9 Mod 4 |
| SWM  | 77 | SS-9 Mod 4 |
| AFM* | 78 | SS-9 Mod 4 |
| Rock | 78 | SS-9 Mod 4 |
| FI*  | 79 | SS-9 Mod 4 |
| JWS* | 79 | SS-9 Mod 4 |
| Nitz | 79 | SS-9 |
| AWar | 80 | SS-9 mod |
| Clns | 80 | SS-9 Mod 4 |
| AWST | 82 | SS-9 Mod 4 |
| SSF  | 82 | SS-9 Mod 4 |

---------- **NATO CODENAME** ----------

| | | |
|---|---|---|
| Rock | 78 | Scarp |
| AWST | 82 | Scarp |

---------- **DESCRIPTION** ----------

| | | |
|---|---|---|
| Rock | 78 | ICBM |
| JWS* | 79 | ICBM |

---------- **EDITOR'S NOTES** ----------

| | | |
|---|---|---|
| SecC | 69 | Mod 4 mentioned FY70-72. |
| JCS  | 72 | Mod 4 mentioned FY73-79. |

---------- **RANGE** ----------

| | | mi | nm | km | km conv |
|---|---|---|---|---|---|
| Rock | 78 | | | 12,000 | 12,000 |
| Clns | 80 | | 5,500 | | 10,186 |

---------- **CEP** ----------

| | | ft | mi | nm | m | km | m conv | |
|---|---|---|---|---|---|---|---|---|
| JCS  | 72 | | | | | | | "Mod 4 is still an enigma" [in discussion of hard-target kill capability, ed]. |
| Nitz | 79 | | | 1 | | | 1,852 | |
| AWar | 80 | | | 1 | | | 1,852 | |
| Clns | 80 | | | 1.0 | | | 1,852 | |
| SSF  | 82 | | | 1.0 | | | 1,852 | |

---------- **WARHEAD CHARACTERISTICS** ----------

| | | | |
|---|---|---|---|
| SecD | 70 | 3 RVs | Unclear whether 3 RVs MRV or MIRV [terminology problem]. If SS-9 not aimed at MIRV capability, follow on system could be MIRVed as early as 1972. |
| JCS  | 71 | | One version [of SS-9] has 3 RVs. |
| SecC | 71 | | "In two recent tests, the system appeared to have demonstrated a capability to target the RVs independently although the pattern in which the RVs land is considerably smaller [33 chars deleted]". With the development of MIRVs for the SS-9, 2600-3900 warheads could be deliverable by 1976 [this is 478-1182 more than the delivery systems shown in the companion chart, ed]. |
| JCS  | 72 | 3 RVs | FY73-74. Failed to demonstrate MIRV capability; could be deployed as MRV, FY73-74. |
| JCS  | 73 | | Reasonable to assume "SS-9 follow-on" would be MIRVed. |
| IAD6 | 75 | 4 MIRVs | "Possible Mod 4 with 4 MIRVs". |
| JCS  | 75 | 3 warheads | FY76-79. |
| FIar | 76 | 3 MRVs | |
| SWM  | 77 | 3 RVs | |
| Rock | 78 | 3 x MRV | MIRVs never deployed. |
| FI*  | 79 | 3 MRVs | |
| Nitz | 79 | 3 RVs | |
| AWar | 80 | 3 MRVs | |
| Clns | 80 | 3 MRVs | |
| AWST | 82 | 3 MRVs | |
| SSF  | 82 | 3 warheads | |

## NUCLEAR YIELD

|  |  | kT | MT | kT_conv |  |
|---|---|---|---|---|---|
| SecC | 71 |  | 5 | 5,000 | Triple warhead. |
| SWM | 77 |  | 5 | 5,000 |  |
| Rock | 78 |  | 12.0-15.0 | 12,000-15,000 | [Table; total yield for missile?, ed] |
|  |  |  | 5 | 5,000 | [Text] |
| JWS* | 79 |  | c. 4 | 4,000 |  |
| Nitz | 79 |  | 3.5 | 3,500 |  |
| AWar | 80 |  | 3.5 | 3,500 | Per RV. |
| Clns | 80 |  | 3.5 | 3,500 | Each. |
| SSF | 82 |  | 3.5 | 3,500 | Per warhead. |

## THROWWEIGHT

|  |  | lb | t | kg | kg_conv |
|---|---|---|---|---|---|
| Nitz | 79 | 13,500 |  |  | 6,123 |
| Clns | 80 | 12,500 |  |  | 5,670 |

## LENGTH

|  |  | in | ft | cm | m | m_conv |  |
|---|---|---|---|---|---|---|---|
| SWM | 77 |  | 115 |  | 35.0 | 35.05 | MRV version [Mod 4, ed] |
| Rock | 78 |  |  |  | 37 | 37.00 |  |

## DIAMETER

|  |  | in | ft | cm | m | cm_conv |
|---|---|---|---|---|---|---|
| Rock | 78 |  |  |  | 3.4 | 340.00 |

## TYPE OF PROPULSION

| Rock | 78 | storable liquid | Stages 1 and 2. |
|---|---|---|---|
| Clns | 80 | liquid |  |

## PROTOTYPE TESTS

| SecC | 69 | [1968] | Now we have evidence Soviets working on some form of advanced reentry system. In August, September, October and December of last year, 4 SS-9s launched with 3 RVs each, all of which survived to impact. Indicate MRV testing, too early to assess ultimate operational configuration. At moment appear similar to US Polaris A-3 with its 3 RVs. Not incompatible with MIRV development. |
|---|---|---|---|
| SecC | 70 | 1968 | Testing of MRVs: none in 1965, 1966 or 1967; 4 in 1968; 7 in 1969; and 2 in 1970 [so far, both in January]. All successful, except 1 unresolved in 1969. Total of 13 tests since program began flight testing in Aug 1968. |
| SecD | 70 |  | MRV testing continues (3 RVs). Unclear whether 3 RVs MRV or MIRV [terminology problem]. |
| JCS | 71 | Aug 1968 | Flight tests began in Aug 1968, continued in 1969 and 1970 (3 RV Mod). |
| SecC | 71 |  | MRVs tested over 20 times on SS-9 since Aug 1968. Total testing: 22. "In two recent tests, the system appeared to have demonstrated a capability to target the RVs independently although the pattern in which the RVs land is considerably smaller [33 chars deleted]". |
| SecD | 71 |  | MRVs tested many times since Aug 1968, FY72-73. |
| JCS | 72 |  | Mod 4 flight tests Aug 1968 to Nov 1970. |
| SecD | 72 |  | Last MRV tests were late in 1970. Probably USSR has not tested MIRV so far. |
| JCS | 73 | Jan 1973 | First flight test since Nov 1970 occurred in January of this year [1973]. |
| JCS | 74 | Jan 1973 | Jan 1973 test had RVs of much different design; parachutes ensured soft landing and recovery; More tests in 1973; some improvement in targeting flexibility noted; In light of SS-X-18, questionable if it will be deployed as MIRV. |
| JCS | 75 |  | Mod 4 tests resumed in 1973, but none detected in 1974. Possible reasons for 1973 test bunch and 1974 halt: achieved R&D objective or testing supported an independently-cancelled program, FY76-77. |
| JCS | 75 |  | Mod 4 tests resumed in 1973, but none detected in 1974 or 1975. |
| FIar | 76 | 1969-70 | First tested; further trials in 1973, perhaps as part of MIRV development program; no tests since. 'Footprint' of three impact points is similar to the dispersal of a typical cluster of Minuteman silos. |
| SWM | 77 |  | Employed for tests of MRVs and MIRVs. |
|  |  | Jan 1973 | Early tests ended Nov 1970; new tests began Jan 1973. |
| AFM* | 78 |  | Test vehicle for Soviet MIRVs. |
| Rock | 78 | 1969 | First tested. |
|  |  | [1973] | Further tests 2 years after deployment in 1971, as part of Soviet MIRV developments, but MIRVs never deployed on SS-9. |
| JWS* | 79 |  | Important element of progress in Soviet MIRV development. |
|  |  | Jan 1969 | Tests began at least as long ago as Jan 1969 (3 MRVs over 8,000 km). No tests after Nov 1970 for a very long time. Footprint of RVs corresponded to Minuteman layout, but idea that that was the Mod 4 function dropped in favor of coincidence as an explanation. No tests in 1971 or 1972 implied delay or failure in MIRV development. Tests in 1973 had parachutes for soft landing, and later 1973 tests showed improvement in targeting flexibility. No tests 1974 or 1975. |
| SSF | 82 | 1968 | First flight test. |

---------- IOC ----------
SecC 69        IOC of SS-9 with 3 MRVs might be achieved by late CY 69. MIRV system suitable for soft target attack probably achievable by 1970. System capable of attacking hard targets like Minuteman silos could probably not attain IOC until 1972.
SecC 70        System hasn't yet demonstrated capability of independent targeting of RVs. If program goal is simple MRV, could be operational in next few months. MIRV would require at least another year of testing.
JCS  75  1971  FY76-79.

---------- DEPLOYMENT BEGAN ----------
SecD 70        MIRV is achievable for SS-9. MRV (simple) possible by late 1969. More advanced MRV possible by late 1970. If SS-9 not aimed at MIRV capability, follow on system could be MIRVed as early as 1972.
SecC 71        The extent of MRV deployment, if any, is not known at this time.
SecD 71        Probably no operational MIRV at present.
JCS  72        If USSR still developing MIRV for Mod 4, "could be available for deployment relatively soon".
FIar 76        Not thought to be operational.
AFM* 78        Despite an improvement in targeting flexibility, never deployed.
Rock 78  1971  Service intro.
JWS* 79        Not known to be deployed. Probably just for development.
Clns 80  1971  First deployed.
SSF  82  1971  Year operation began.

---------- NUMBERS IN SERVICE ----------

---------- NUMBER IN 1975 ----------
Rock 78  38    "In all, some 250 SS-9 Mod 1 and Mod 2 missiles were deployed with 38 Mod 4 variants between 1970 and 1975".

---------- NUMBER IN 1977 ----------
Nitz 79  100   Launchers.

---------- NUMBER IN 1980 ----------
AWar 80        164 "silos" with single RVs and 100 "launchers" with multiple RVs until recently.

---------- RELOADS IN GENERAL ----------
Clns 80        Cold launch: no.

---------- STRATEGY ----------
JCS  71        SS-9 with 3 RVs might be threat to Minuteman if deployed [in text as a presupposition, ed].
AFM* 78        Despite an improvement in targeting flexibility, never deployed.

# SS-9 Mod 5

Sources are:    FIar 76        SWM    77        Rock    78        AFM*    80

------------- US DESIGNATION -------------
FIar 76     SS-9 Mod 5
Rock 78     SS-9 Mod 5
AFM* 80     SS-9 Mod 5

------------- OTHER DESIGNATION -------------
Rock 78     F-1-m        In satellite-destroying role, indicating orbital capability.

------------- NUMBER OF STAGES -------------
Rock 78     4        2 stages of Scarp plus 3rd orbit-insertion stage, and "a fourth manoeuvering vehicle with the payload" 3rd stage probably similar in size to F-1-r [SS-9 Mod 3]; maneuvering vehicle probably 5 m long, 2 m diameter.

------------- PROTOTYPE TESTS -------------
FIar 76              SS-9 boosters carrying satellite-killing payloads have been launched from Tyuratam; interceptions against satellites launched from Plesetsk.
SWM 77               Fifth model identified with 'anti-satellite' tests.
Rock 78    1968      "Demonstrated in 1968"; "used to send a satellite to inspect the target vehicle, move away and explode". Tests continued [from 1968] until 1971, ceased, then began again in 1976. Last flights in 1971 and series since 1976 employed C-1 (SS-5 based) to launch target. All tests launched from Tyuratam.
                     Have been more than 50 "F-1" launches [Mod 3 and Mod 5]; used only for military payloads.

------------- DESIGN AND ENGINEERING -------------
AFM* 80     Effect of SS-9 deactivation on this program not known.

------------- USE AND CONFIGURATION -------------
FIar 76     According to AWST, SS-9 missiles wheeled from hangars, erected, fuelled and launched on ASAT; done on schedule "which on actual missions would result in a launch in less than 90 min from the start of booster transport to the pad".
AFM* 80     Tests suggest operational launch of Mod 5 could be made within 90 minutes of order to intercept.

------------- STRATEGY -------------
AFM* 80     Launch vehicle for satellite-killing payloads.

------------- OTHER INFORMATION -------------
Rock 78     F-1-m also used for launch of ocean surveillance satellites.

# SS-10

```
Sources are:  IAD4  66     Ley   68     HRST  75     SWM   77     JWS*  79     SSF   82
              SecC  65     ID01  68     FI*   74     FIar  76     Rock  78     AWST  82     SMP   85
```

---------- US DESIGNATION ----------
```
SecC  65    SS-10
FI*   74    SS-10
FIar  76    SS-10
SWM   77    SS-10
Rock  78    SS-10
JWS*  79    SS-10
AWST  82    SS-10
SSF   82    SS-X-10
SMP   85    SS-X-10
```

---------- NATO CODENAME ----------
```
Ley   68    Scrag
FI*   74    Scrag
HRST  75    Scrag
FIar  76    Scrag
SWM   77    Scrag
Rock  78    Scrag
JWS*  79    Scrag
AWST  82    Scrag
SSF   82    Scrag
```

---------- DESCRIPTION ----------
```
HRST  75    ICBM
Rock  78    ICBM              Long range ICBM.
JWS*  79    ICBM
SSF   82                      "Global-range missile".
SMP   85    ICBM
```

---------- EDITOR'S NOTES ----------
```
SecC  65    Listed FY66-67.
SMP   85    Listed '85.
```

---------- RANGE ----------

|  | mi | nm | km | km conv | |
|---|---|---|---|---|---|
| IAD4 66 | 6,000+ | | | 9,654 | |
| Ley 68 | 6,000+ | | | 9,654 | |
| HRST 75 | 5,000 | | | 8,045 | |
| FIar 76 | | | 12,000 | 12,000 | |
| SWM 77 | 7,500+ | | 12,000+ | 12,067 | |
| Rock 78 | | | 12,000 | 12,000 | |
| JWS* 79 | | | 8,000 | 8,000 | Seems conservative, due to size. 'Global range' was claimed in parade. |
| AWST 82 | | | | | ICBM range |

---------- CEP ----------

|  | ft | mi | nm | m | km | m conv | |
|---|---|---|---|---|---|---|---|
| SSF 82 | | | 1.0-2.0 | | | 1,852-3,704 | Depressed trajectory. |
|  | | | 1.5-3.0 | | | 2,778-5,556 | FOBS. |

---------- WARHEAD CHARACTERISTICS ----------
```
SSF   82    single RV
```

---------- NUCLEAR YIELD ----------

|  | kT | MT | kT conv |
|---|---|---|---|
| ID01 68 | | 20-50 | 20,000-50,000 |
| SSF 82 | | 20.0 | 20,000 |

---------- THROWWEIGHT ----------

|  | lb | t | kg | kg conv |
|---|---|---|---|---|
| SSF 82 | 9000-11000 | | | 4,082-4,989 |

---------- WEIGHT ----------

|  | lb | t | kg | kg conv | |
|---|---|---|---|---|---|
| IAD4 66 | | c. 200 | | 200,000 | |
| ID01 68 | | c. 185 | | 185,000 | |
| Ley 68 | | 200 | | 200,000 | Tons. |

## LENGTH

|       |    | in   | ft    | cm | m     | m conv |         |
|-------|----|------|-------|----|-------|--------|---------|
| IA04  | 66 |      | > 100 |    |       | 30.48  |         |
| ID01  | 68 |      | 115   |    |       | 35.05  |         |
| Ley   | 68 |      | 100+  |    |       | 30.48  |         |
| HRST  | 75 |      | 120   |    |       | 36.58  |         |
| FIar  | 76 |      |       |    | 38.5  | 38.50  |         |
| SWM   | 77 |      | 124   |    | 37.8  | 37.80  |         |
| Rock  | 78 |      |       |    | 38.5  | 38.50  | [Table] |
|       |    |      |       |    | 38    | 38.00  | [Text]; tallest ICBM to date. 1st stage, 18.2 m; 2nd, 7.7 m; 3rd, 6.4 m. |
| JWS*  | 79 |      |       |    | c. 37 | 37.00  | [Text]  |
|       |    |      |       |    | c. 39 | 39.00  | [Specs] |
| AWST  | 82 |      | 124   |    |       | 37.80  |         |

## DIAMETER

|       |    | in   | ft    | cm  | m     | cm conv |         |
|-------|----|------|-------|-----|-------|---------|---------|
| IA04  | 66 |      | 8.4   |     |       | 256.03  | Minus skirt. |
| ID01  | 68 |      | 9.5   |     |       | 289.56  | Without tail cone; first stage. |
| Ley   | 68 |      | 8 1/2 |     |       | 259.08  |         |
| HRST  | 75 | 108  |       |     |       | 274.32  |         |
| FIar  | 76 |      |       |     | 3.4   | 340.00  |         |
| SWM   | 77 |      | 9.0   |     | 2.74  | 274.32  |         |
| Rock  | 78 |      |       |     | 3.4   | 340.00  | 1st stage; 2nd 1.8 m [sic]; 3rd 2.4 m [sic]. |
| JWS*  | 79 |      |       |     | < 3   | 300.00  | "Nearly" 3 m, base of 1st stage, [text] |
|       |    |      |       |     | c. 2.75 | 275.00 | [Specs] |
| AWST  | 82 |      | 9     |     |       | 274.32  |         |

## NUMBER OF STAGES

| HRST | 75 | 3 |
| FIar | 76 | 3 |
| SWM  | 77 | 3 |
| Rock | 78 | 3 |
| JWS* | 79 | 3 |
| AWST | 82 | 3 |

## TYPE OF PROPULSION

| HRST | 75 | liquid | |
| FIar | 76 | Lox/kerosene | First stage powered by 4 large rocket engines, probably the first Russian example of gimballed motors; 2nd and 3rd stages each had single motor. Believed to have been last Russian missile to use cryogenic propellants; less advanced than SS-9. |
| SWM  | 77 | nonstorable liquid | Used nonstorable cryogenic propellants. First stage engine may have been related to RD-111 which employed LOX/kerosene (vacuum thrust 166 tonnes, chamber pressure 80 atmospheres, specific impulse 317 sec). |
| Rock | 78 | nonstorable liquid | Stages 1, 2, and 3. Liquid oxygen and kerosene. 4 large rocket engines in first stage. SS-9 and all subsequent [to SS-10] liquid propellant ICBMs used storable propellants. |
| JWS* | 79 | liquid | 1st stage has 4 gimballed nozzles, 2nd and 3rd have single fixed nozzles, 2nd very large and 3rd quite small. |
| AWST | 82 | storable | |
| SSF  | 82 | nonstorable liquid | |

## MIDCOURSE (OR MAIN) GUIDANCE

| SWM  | 77 | inertial | |
| JWS* | 79 | inertial | Presumed. |
| SSF  | 82 |          | Fly-the-wire inertial. |

## CONTROL SYSTEM

| SWM  | 77 | First-stage control by swivelling the 4 thrust chambers. |
| JWS* | 79 | 1st stage has 4 gimballed nozzles, 2nd and 3rd have single fixed nozzles, 2nd very large and 3rd quite small. |

## OBSERVED

| IA04 | 66 | US photo reconnaissance detected underground launch facilities years ago. |

## OBSERVED IN PARADE

| ID01 | 68 | May 1965   | First shown. |
| Rock | 78 | 9 May 1965 | First displayed publicly. |
| JWS* | 79 | 1965       | In 1965 parade was called sister vehicle of launch vehicle for Vostok and Voshkod spacecraft. |

## DEVELOPMENT BEGAN

| SSF | 82 | 1957-58 | Year design began. |
| SMP | 85 | 1957    | Technological development began. [From graph.] |

## PROTOTYPE TESTS

| | | | |
|---|---|---|---|
| SecC | 65 | | Currently undergoing tests; new system, DoD has little information about it, FY66-67. |
| SSF | 82 | 1964-65 | First flight test. |
| SMP | 85 | 1963-64 | Engineering and testing began. [From graph.] |

## IOC

| | | | |
|---|---|---|---|
| SecC | 65 | late 1965 | Could become operational. |
| IAD4 | 66 | early 60s | "Missile itself has already been operational since the beginning of the sixties". |

## DEPLOYMENT BEGAN

| | | |
|---|---|---|
| FI* | 74 | "Thought not to be operational". |
| FIar | 76 | Not thought to be operational. |
| SWM | 77 | Not put into service. |
| Rock | 78 | Not deployed operationally. Deleted in favor of SS-9, possibly because of unstorable propellants. |
| JWS* | 79 | Probably now abandoned; believed not operational. |

## DEPLOYMENT HISTORY

| | | |
|---|---|---|
| SMP | 85 | Cancelled without deployment in 1968. [From graph.] |

## LAUNCHER/SILO

| | | |
|---|---|---|
| IAD4 | 66 | US photo reconnaissance detected underground launch facilities years ago. |

## DESIGNER

| | | | |
|---|---|---|---|
| SSF | 82 | Korolev | Design Bureau. |

## DESIGN AND ENGINEERING

| | | |
|---|---|---|
| FIar | 76 | Developed in parallel with SS-9. |
| SWM | 77 | Developed in parallel with SS-9. |
| Rock | 78 | Developed in parallel with SS-9. |
| JWS* | 79 | Maybe was intermediate stage towards the SS-18. |

## STRATEGY

| | | |
|---|---|---|
| IAD4 | 66 | In capabilities and reaction time, giant missile can't compare with the most modern designs. |
| Rock | 78 | May have had similar design objective to SS-9; commentator at display parade said that "missiles of the SS-10 type could 'find their targets literally from any direction'". |

## RELATION TO OTHER MISSILES

| | | |
|---|---|---|
| SWM | 77 | Less advanced than SS-9. |

# SS-11

```
Sources are:   SecC  71    IISS  74    MOW   76    JCS   78    Clns  80    AFM   82    SuO1  83
SecC  66       SecD  71    JCS   74    NAO3  76    Rock  78    IISS  80    AWO4  82    vanD  83
SecC  67       JCS   72    SecC  74    SecD  76    SecD  78    JCS   80    AWST  82    AFM+  84
SecC  68       SecD  72    GAO1  75    IISS  77    FI*   79    SecD  80    JCS   82    JCS   84
SecC  69       IISS  73    IAO6  75    JCS   77    IISS  79    A168  81    SSF   82    Meyr  84
IISS  70       JCS   73    IISS  75    SecD  77    JCS   79    JCS   81    CBO   83    SecD  84
SecC  70       SecD  73    JCS   75    SWM   77    Nitz  79    JWS   81    Clns+ 83    SMP   84
SecD  70                   FIar  76    CBO1  78    USO8  79    SecD  81    JWS+  83    JCS   85
JCS   71                   IISS  76    IISS  78    AWar  80    SMP   81    SMP   83    SMP   85
```

```
---------- US DESIGNATION ----------
SecC  66    SS-11
SecD  70    SS-11           FY71-82,85.
JCS   71    SS-11           FY72-86.
FIar  76    SS-11
MOW   76    SS-11
SWM   77    SS-11
Rock  78    SS-11
FI*   79    SS-11
Nitz  79    SS-11
AWar  80    SS-11
Clns  80    SS-11
JWS   81    SS-11
SMP   81    SS-11           '81, '83, '84, '85.
AFM   82    SS-11
AWST  82    SS-11
SSF   82    SS-11
CBO   83    SS-11
vanD  83    SS-11
Meyr  84    SS-11
```

```
---------- NATO CODENAME ----------
FIar  76    Sego
MOW   76    Sego
JCS   77    Sego
SWM   77    Sego
Rock  78    Sego
FI*   79    Sego
JWS   81    Sego
AFM   82    Sego
AWST  82    Sego
SSF   82    Sego
```

```
---------- DESCRIPTION ----------
JCS   71    VRBM            Variable range ballistic missiles.
JCS   72                    2nd generation strategic missile; "SS-11 improved" is 3rd generation.
JCS   73    Light ICBM      FY74-75.
MOW   76    ICBM
Rock  78    ICBM
JCS   81                    Listed as significant USSR initiative (strategic offensive), ICBM, FY82,84.
JWS   81    ICBM
AFM   82    "Light" ICBM
SSF   82                    Third generation.
JWS+  83                    Third generation ICBM.
```

```
---------- EDITOR'S NOTES ----------
SecC  66    Listed FY67-77.
SecD  70    Listed FY71-82,85.
JCS   71    Listed FY72-86.
SMP   81    Listed '81, '83, '84, '85.
Meyr  84    Note to chart: "Values pertain to initial service period".
```

150  SOVIET MISSILES  IDDS

```
             ---------- RANGE ----------
                     mi           nm           km            km_conv
SecC   71                      600-5900                    1,111-10,926   [Those deployed in MR/IRBM
                                                                           complexes].
GA01   75      6,000                                       9,654
FIar   76                                   10,000         10,000
MOW    76      6,525                        10,500         10,498         Max.
NA03   76      6,500                                       10,458
Rock   78                                   10,000         10,000
FI*    79                                   10,000         10,000
Clns   80                      6,000                       11,112
JWS    81                                   10000-11000    10,000-11,000
AWST   82                      5,500                       10,186
JWS+   83                                   10,500         10,500         [Analysis table]

             ---------- CEP ----------
                     ft      mi       nm        m      km     m_conv
SecC   67                            1.0-1.5                  1,852-2,778  FY68-69.
                                                                            Because of "relatively poor
                                                                            CEP and small payload"
                                                                            little value against hard
                                                                            targets such as Minuteman
                                                                            silos; essentially for use
                                                                            against cities, FY68-70.
                                                                            Improvements in accuracy and
                                                                            penetration capability could
                                                                            be made in SS-11s.
JCS    73                                                                   New version probably more
                                                                            accurate than original, but
                                                                            believed to lack hard target
                                                                            capability even so.
JCS    74                                                                   Considerably less accurate
                                                                            than Minuteman.
JCS    75                                                                   No version has hard target
                                                                            capability, FY76-77.
FIar   76                                                                   No version of SS-11 is
                                                                            effective against hard
                                                                            targets such as Minuteman
                                                                            silos.
CB01   78      3000                                           914.40        Estimated, 1985.
FI*    79                                                                   Insufficient for hard silos.
Nitz   79                              1                      1,852
Clns   80                              0.76                   1,407
CB0    83                              0.76                   1,407         In 1983, 1990, 1996.
Meyr   84                                              0.4-0.7  400.00-700.00

             ---------- WARHEAD TYPE ----------
                     nuclear    conven    chem
MOW    76    yes
JWS    81    yes
Meyr   84    yes

             ---------- WARHEAD CHARACTERISTICS ----------
SecD   71                  Warhead variants are being developed, MRVs and penetration aids are possible.
MOW    76    single RV
SecD   77                  Number of independently targetable warheads: 1.
SWM    77                  "Warhead 1 x 20/25 MT or of 500 kT", [source misprint, ed].
Rock   78    single RV
Nitz   79    1 RV
USO8   79    single RV
Clns   80    single RV
SecD   80    single RV
AWST   82    single RV
CB0    83    single RV
Meyr   84    single RV
```

## BALLISTIC MISSILES  ICBMs  SS-11

### NUCLEAR YIELD

| Source | Year | kT | MT | kT conv | Notes |
|---|---|---|---|---|---|
| SecC | 68 | | c. 1.0 | 1,000 | |
| JCS | 74 | | | | Slightly higher than Minuteman. |
| MOW | 76 | | 1-2 | 1,000-2,000 | Single warhead version. |
| SWM | 77 | | | | "Warhead 1 x 20/25 MT or of 500 kT", [source misprint, ed]. |
| CBO1 | 78 | | 1.5 | 1,500 | Estimated, 1985. |
| Rock | 78 | | 1.0-2.0 | 1,000-2,000 | |
| Nitz | 79 | | 1.0 | 1,000 | |
| USO8 | 79 | | 1 | 1,000 | |
| Clns | 80 | 950 | | 950.00 | |
| AWST | 82 | | 1 | 1,000 | Single warhead version. |
| CBO | 83 | 950 | | 950.00 | |
| JWS+ | 83 | | 1 | 1,000 | [Analysis table] |
| Meyr | 84 | 500-2,000 | | 500.00-2,000 | 500 kT-2 MT. |

### THROWWEIGHT

| Source | Year | lb | t | kg | kg conv |
|---|---|---|---|---|---|
| Flar | 76 | | | 700 | 700.00 |
| Nitz | 79 | 2,000 | | | 907.20 |
| USO8 | 79 | 2,000 | | | 907.20 |
| Clns | 80 | 2,200 | | | 997.92 |
| SSF | 82 | 1,000-2,000 | | | 453.60-907.20 |
| CBO | 83 | 2,200 | | | 997.92 |

### PAYLOAD

| Source | Year | lb | t | kg | kg conv |
|---|---|---|---|---|---|
| SecC | 66 | 1,000-2,000 | | | 453.60-907.20 |

### WEIGHT

| Source | Year | lb | t | kg | kg conv |
|---|---|---|---|---|---|
| FI* | 79 | | | c. 50,000 | 50,000 |

### LENGTH

| Source | Year | in | ft | cm | m | m conv | Notes |
|---|---|---|---|---|---|---|---|
| SecC | 70 | | 60 | | | 18.29 | [From graph, unit inferred]. |
| Flar | 76 | | | | 19 | 19.00 | |
| | | | | | 20 | 20.00 | Container. |
| SWM | 77 | | 65.6 | | 20.0 | 19.99 | |
| Rock | 78 | | | | 19 | 19.00 | |
| | | | | | 20 | 20.00 | Container. |
| FI* | 79 | | | | c. 19 | 19.00 | |
| JWS | 81 | | | | c. 20 | 20.00 | |
| AFM | 82 | | [63] | | | 19.20 | 3' shorter than SS-13. |
| AWST | 82 | | 64 | | | 19.51 | |

### DIAMETER

| Source | Year | in | ft | cm | m | cm conv | Notes |
|---|---|---|---|---|---|---|---|
| Flar | 76 | | | | 2.4 | 240.00 | |
| | | | | | 3 | 300.00 | Container. |
| SWM | 77 | | 8.0 | | 2.4 | 243.84 | |
| Rock | 78 | | | | 2.4 | 240.00 | |
| | | | | | 3 | 300.00 | Container. |
| FI* | 79 | | | c. 200 | | 200.00 | |
| JWS | 81 | | | | c. 2.5 | 250.00 | Max. |
| AWST | 82 | | 6 | | | 182.88 | |

### NUMBER OF STAGES

| Source | Year | Stages | Notes |
|---|---|---|---|
| JCS | 75 | 2 | FY76-77. |
| Flar | 76 | 2 | |
| SWM | 77 | 2 | |
| Rock | 78 | 2 | |
| FI* | 79 | 2 | |
| JWS | 81 | 2 | Suggested that RV has propulsion, making 3 stages; can't tell though. |
| AWST | 82 | 3 | |
| JWS+ | 83 | 3 | [Analysis table] |

## SOVIET MISSILES

### ———— TYPE OF PROPULSION ————
| | | | |
|---|---|---|---|
| SecC | 66 | storable liquid | |
| JCS | 72 | liquid | FY73,76-77. |
| FIar | 76 | storable liquid | 4 first-stage thrust chambers. |
| MOW | 76 | storable liquid | |
| SWM | 77 | storable liquid | |
| Rock | 78 | storable liquid | Stages 1 and 2. 4 rocket motors in first stage. |
| FI* | 79 | liquid | |
| Clns | 80 | liquid | |
| JWS | 81 | storable liquid | |
| AWST | 82 | storable | |
| SSF | 82 | liquid | |

### ———— MIDCOURSE (OR MAIN) GUIDANCE ————
| | | | |
|---|---|---|---|
| MOW | 76 | radio | Radio instructions from ground-based computer. |
| SWM | 77 | inertial | |
| FI* | 79 | inertial | |
| JWS | 81 | inertial | |
| SSF | 82 | | Fly-the-wire, inertial. |

### ———— OBSERVED ————
| | | |
|---|---|---|
| MOW | 76 | Little is known of the missile. |
| AFM | 82 | No photo of SS-11 has ever been identified [final report]. |

### ———— OBSERVED IN PARADE ————
| | | | |
|---|---|---|---|
| FIar | 76 | | Displayed only within cylindrical container. |
| SWM | 77 | ? Nov 1973 | Not shown in military parades with other rockets of its day. Enclosed missile in Nov 1973 parade may have been SS-11. |
| Rock | 78 | | Developed in "utmost secrecy", never publicly displayed; has been seen "shrouded by its cylindrical container". |
| SMP | 83 | | [Picture shown of SS-11 in parade]. |

### ———— DEVELOPMENT BEGAN ————
| | | | |
|---|---|---|---|
| SSF | 82 | 1955-58 | Year design began. |
| SMP | 85 | 1955 | Technological development began. [From graph.] |

### ———— PROTOTYPE TESTS ————
| | | | |
|---|---|---|---|
| SecC | 70 | | Modifications underway for SS-11. Possible it is MRV capability or more accurate single RV, but "more likely development would be penetration aids". |
| | | 1969-1970 | Testing of modifications commenced in 1969. 7 tests in 1969; 2 in 1970 [so far]; all 9 were successes. Tests involve modifications to missile and a new RV; multiple objects detected. |
| SecD | 70 | | USSR continues to test modified SS-11 payloads. |
| JCS | 71 | | Active test program [for improvements] continues, FY72-73. |
| SecC | 71 | 1969 | Testing of modifications commenced in 1969. Total of 28 tests to date. Tests involve modifications to missile and a new RV; multiple objects detected, and suggest purpose is penetration aids or MRVs. |
| SecD | 71 | 1969 | Testing of modified SS-11s continues; began in 1969. |
| JCS | 72 | 1969 | Improved version testing began in 2nd half of 1969. |
| SecD | 72 | 1969 | Flight testing of SS-11 mods began in 1969. |
| JCS | 73 | | Active testing of new SS-11 version continued in 1972, including 2 flights into Pacific Ocean. 'SS-11 follow-on' being tested actively. |
| SecC | 73 | 1969 | Flight testing of SS-11 modifications began in 1969; have progressed to improved SS-11 version employing 3 MRVs. Soviet MRVs could be deployed now [might not be SS-11 MRVs, ed]. |
| SecD | 73 | | SS-11 follow-on being developed and tested [not SS-11]. |
| MOW | 76 | July 1970 | US monitoring service reported 2 with "warheads equivalent to a one-megaton charge" launched 5,715 mi (9,200 km) from Tyuratam to Pacific, July 27 and 28, 1970. |
| | | 21 Aug 1970 | 2 SS-11s identified at end of 6,525 mi (10,500 km) flights with MRV warheads. |
| | | Feb 1973 | 9 flights from near Lake Baikal to Kamchatka Peninsula. |
| SuO1 | 83 | 1968 | SS-11 VRBM, began to be tested at IRBM ranges. |
| SMP | 85 | 1961-62 | Engineering and testing began. [From graph.] |

### ———— IOC ————
| | | | |
|---|---|---|---|
| SecD | 77 | 1966 | |
| vanD | 83 | 1968 | Fall, for SS-11 VRBM. |
| | | 1973 | For SS-11/-19 silos. |

BALLISTIC MISSILES  ICBMs  SS-11   153

```
----------- DEPLOYMENT BEGAN -----------
SecC   66              Soviets constructing new type of single silo launch site, believed for SS-11.
                       Estimated to become operational this year.
MOW    76   1966       Operational in camouflaged launchers since 1966.
SWM    77   1966       Service entry.
Rock   78   1966       Service intro.
Clns   80   1966       First deployed.
JWS    81   1966       Entered service, single RV.
SMP    83   by 1966    Deployment underway by 1966.
Su01   83   1968       For SS-11 VRBM, "against potential targets both in China and in Europe".
Mayr   84   1966       Service.
SecD   84   1966-67    Introduction of strategic systems [graph].
SMP    84   1966       Initial deployment, '84, '85.
SMP    85   1966-67    [From graph.]

----------- DEPLOYED -----------
JCS    74              3 versions tested, but only Mod 1 and Mod 3 deployed.
FI *   79              Mod 1 and Mod 3 deployed.
AW04   82              Deployed in 3 versions.

----------- DEPLOYMENT COMPLETE -----------
SecC   71              Deployment of launchers appears completed at present time.
                       Several groups of SS-11s became operational during the past few months.
JCS    72              Deployment probably complete.
SecD   73              Deployment [of one Mod] completed. Deployment of new SS-11 version expected
                       (initially) in new silos.

----------- DEPLOYMENT HISTORY -----------
SecC   67              Significant change in US projections of Soviet forces in past year is "faster-than-anticipated
                       rate of construction of hard ICBM silos, particularly for the new small SS-11".
SecC   70              Appears Soviets deploying some SS-11s in IRBM role at existing MR/IRBM sites. Rate of SS-11
                       deployment in IRBM complexes greater in 1969 than 1968; rate of SS-11 deployment in ICBM complexes
                       less in 1969 than 1968.
SecD   70              "Modifications are underway" on SS-11, objective unclear. Increased significantly in past year.
                       Expected to increase substantially by mid-1970.
JCS    71              Deployment appears to have leveled off at present.
SecC   71              All sites now believed operational [at MRBM/IRBM complexes]; SS-4 deactivation may mean SS-11s [at
                       MR/IRBM complexes] being deployed to replace SS-4s and -5s.
SecD   71              "Soviets are deploying" SS-11s at MRBM/IRBM complexes, began in 1968.
SecC   72              No construction starts on standard SS-11 sites during past year; new construction includes 60
                       smaller silos compatible "with an SS-11 size missile"; most of these silos could be completed
                       during 1972.
SecD   72              Very little construction on standard SS-11 missile sites during the past year.
JCS    73              New version estimated ready for deployment.
SecC   73              Believed that SS-11 Mod 3 will initially be deployed in new small silos.
SecC   74              "It would take at least [11 chars deleted] years to replace all of the SS-11s [deleted] with the
                       follow-on missiles".
JCS    75              "SS-11 silos have been or are in the process of being prepared for the SS-17 and the SS-19",
                       FY76-77.
MOW    76              Being replaced by SS-17 and SS-19.
SecD   76              Being replaced by [SS-17, SS-19].
JCS    78              SS-11 silos being converted to SS-17, FY79-80. SS-11 silos being converted to SS-19.
SecD   78              SS-17 and SS-19 are in old SS-11 silos, FY79-81.
FI *   79              Being replaced by SS-17 and SS-19.
JCS    79              Some SS-19 silos contain SS-11; SS-19 probably will replace.
JCS    80              Conversion to SS-17 silos complete, FY81-82. Program to replace SS-11s in SS-19 silos with SS-19
                       has begun.
A168   81              Some 120 SS-11 variable range missiles deployed in M/IRBM fields as interim measure between
                       failure of SS-14 and deployment of SS-20.
AFM    82              Replacement of some SS-11s by SS-17 completed. SS-19s expected to replace other SS-11s.
Su01   83              Some 180 SS-11 VRBM deployed at two former SS-4,5 fields in European USSR. 60 silos begun each
                       year in 1968, 1970 and 1971. These "initially supplemented, and then replaced, SS-4 and SS-5
                       deployments in those fields." Last 60 converted to SS-19s. First "100 or so" SS-4,5 deactivated
                       from 1968 to 1972, as replaced by SS-11 VRBM. In 1969-70 all 70 SS-4,5 missiles in Far East
                       deactivated, replaced in early 1970s by 120 SS-11s.
JCS    84              [SS-11,13,17] expected to be replaced with new ICBMs like SS-X-24,25. FY84-85.
SMP    85              Will be deployed at least through 1990. [From graph.]

----------- NUMBERS IN SERVICE -----------

----------- NUMBER IN 1960 -----------
SSF    82   none       In 1960 and 1965, "includes only those believed to be primarily dedicated to covering
                       intercontinental-range targets".

----------- NUMBER IN 1964 -----------
SecC   70   none       In mid-1964, -65, and -66; operational launchers [from graph].
SecC   71   none       In mid-1964, -65, and -66; operational launchers [from graph, VRBM tally is separate].
            none       From mid-64 to mid-69, operational launchers [from graph, VRBM tally].
```

## 154  SOVIET MISSILES  IDDS

```
----------- NUMBER IN 1967 -----------
SecC 66            140-180 estimated operational by mid-1967.
SecC 68   330-380  As of 1 Oct.
SecC 70   225      In mid-67; operational launchers [from graph].
SecC 71   260      In mid-67, operational launchers [from graph, VRBM tally is separate].
vanD 83   160      1 Jan, includes VRBMs.

----------- NUMBER IN 1968 -----------
SecC 66            200-250 expected operational by mid-1968 [reported in FY68, ed].
SecC 67            320-500 expected operational by mid-1968.
SecC 68            New "SS-11 starts are now down to an annual rate of about 150".
SecC 69   520      Operational, 1 Sept.
SecC 70   505      In mid-68; operational launchers [from graph].
SecC 71   470      In mid-68, operational launchers [from graph, VRBM tally is separate].
vanD 83   390      1 Jan, includes VRBMs.

----------- NUMBER IN 1969 -----------
SecC 68            560-610 expected by mid-1969.
SecC 70   640-650  Operational 1 Sept.
                   820 counting those estimated to be under construction, 1 Sept.
          620      In mid-69; operational launchers [from graph].
SecD 70            160-ICBM increase in 1 year up to 1 Sept 1969, almost all accounted for by SS-9 and
                   SS-11.
SecC 71   590      In mid-69, operational launchers [from graph, VRBM tally is separate].
vanD 83   570      1 Jan, includes VRBMs.

----------- NUMBER IN 1970 -----------
SecC 69            680-720 expected by mid-70.
IISS 70   800      July, total deployed.
SecC 70   720      Operational, Jan.
                   780-830 expected operational by mid-70.
                   705 in mid-70; operational launchers [from graph].
                   30-40 expected operational at MR/IRBM sites by mid-70.
SecC 71   840      Operational 30 Dec.
          635      In mid-70, operational launchers [from graph, VRBM tally is separate].
          90       30 Dec, operational at MR/IRBM complexes.
          60       In mid-70, operational launchers [from graph, VRBM tally].
SecC 72   830      Operational.
Clns 80   970      1970-1973.
SSF  82   720      In 1970, "includes only those believed to be primarily dedicated to covering
                   intercontinental-range targets".
vanD 83   750      1 Jan, includes VRBMs.

----------- NUMBER IN 1971 -----------
SecC 70            820-900 expected operational by mid-71.
                   80-120 expected operational at MR/IRBM sites by mid-71.
SecC 71   850      Operational in ICBM fields, 1 Mar, expected mid-71.
                   705 in mid-71, operational launchers [from graph, VRBM tally is separate].
          120      Operational at MR/IRBM complexes, 1 Mar, also expected for mid-71 and mid-72.
                   120-170 in mid-71, operational at MR/IRBM complexes.
                   110 in mid-71, operational launchers [from graph, VRBM tally].
SecD 71   > 900    For the moment, launchers leveled off at over 900.
vanD 83   940      1 Jan, includes VRBMs.

----------- NUMBER IN 1972 -----------
SecC 68            By mid-1972 they could have 250-300 [SS-13s] in addition to some 560-610 SS-11s.
SecC 71            850-860 expected operational mid-72.
                   755 in mid-72, operational launchers [from graph, VRBM tally is separate].
                   180-240 in mid-72, operational at MR/IRBM complexes.
                   150 in mid-72, operational launchers [from graph, VRBM tally].
SecC 72   970      operational.
JCS  73            1096 [SS-11s and SS-13s] launchers operational, under construction or conversion, "plus
                   new silos" (small silos); mid-1972.
IA06 75   970      As of SALT I.
vanD 83   970      1972-74, 1 Jan, includes VRBMs. Of 970 peak, 120 were VRBMs.

----------- NUMBER IN 1973 -----------
IISS 73   970      July, total deployed, includes SS-11 MR/IRBMs.

----------- NUMBER IN 1974 -----------
IISS 74   1018     July, total deployed, includes 100 SS-11 MR/IRBMs.
JCS  74            1096 launchers associated with SAL; 66 estimated under construction on date Interim
                   Agreement signed [SS-11 and SS-13 together]; 1030 operational.
Clns 80   1030
vanD 83   30       1 Jan, "-11/19s".
```

BALLISTIC MISSILES  ICBMs  SS-11  155

```
------------- NUMBER IN 1975 -------------
GA01   75        > 900       Operational.
IA06   75                    1,096 for SS-11 and SS-13 together.
IISS   75          991       July, total deployed, includes 100 SS-11 MR/IRBMs.
JCS    75                    Remains most extensively deployed ICBM, FY76-77.
SWM    77          990       In service mid-1975.
USD8   79        1,012       In early 1975.
Clns   80          960
SSF    82          650       In 1975, "includes only those believed to be primarily dedicated to covering
                             intercontinental-range targets".
vanD   83          950       1 Jan, includes VRBMs.
                    60       1975-81, 1 Jan, "-11/19s".

------------- NUMBER IN 1976 -------------
FIar   76                    Most extensively deployed Soviet missile.
IISS   76          900       July, total deployed, includes 100 SS-11 MR/IRBMs.
MOW    76        1,018       Of 1,607 ICBMs.
Clns   80          910
vanD   83          880       1 Jan, includes VRBMs.

------------- NUMBER IN 1977 -------------
IISS   77          840       July, total deployed.
Nitz   79          455       Launchers.
Clns   80          850
vanD   83          790       1 Jan, includes VRBMs.

------------- NUMBER IN 1978 -------------
IISS   78          780       July, total deployed.
Clns   80          750
vanD   83          710       1 Jan, includes VRBMs.

------------- NUMBER IN 1979 -------------
FI*    79        > 1,000     Over 85% are Mod 1.
                             Most extensively deployed Russian ICBM.
IISS   79          638       July, total deployed.
Nitz   79          215       18 June, launchers.
Clns   80          650       "70 SS-11s are being converted to SS-19s and are non-operational".
vanD   83          630       1 Jan, includes VRBMs.

------------- NUMBER IN 1980 -------------
AWar   80          520       In hardened silos.
IISS   80          580       1980-81, July, total deployed.
SSF    82          320       In 1980, "includes only those believed to be primarily dedicated to covering
                             intercontinental-range targets".
Clns+  83          640
vanD   83          550       1 Jan, includes VRBMs.

------------- NUMBER IN 1981 -------------
JWS    81        c. 500      In early 1981. Most widely deployed Soviet ICBM (at one time). Remains very common.
                             Widely deployed after c. 1970.
                             At least 500 silos converted to SS-17 and SS-19.
SecD   81        > 500       In current force.
SMP    81          580
Clns+  83          580
vanD   83          520       1 Jan, includes VRBMs.

------------- NUMBER IN 1982 -------------
AFM    82          580       Remaining in silos.
AW04   82          520       Deployed in 3 versions. 60 silos built for SS-19 still contain SS-11.
AWST   82        c. 1,000
JCS    82          580       Inventory.
SSF    82        1,030       "Number of missiles deployed".
Clns+  83          550
JWS+   83                    At least 120 deployed at SS-4 and SS-5 sites at Derazhnya and Pervomaysk, as of 1982-83.

------------- NUMBER IN 1983 -------------
CBO    83          550
JWS+   83                    SIPRI gives 260, but also recognizes US estimates of 550 and 580; US DoD expects 520 in
                             mid-1980s.
SMP    83          550

------------- NUMBER IN 1984 -------------
AFM+   84        c. 550      Remaining in silos. 30 more expected converted to SS-19 by mid-1980s.
JCS    84          550       As of 1 Jan.

------------- NUMBER IN 1985 -------------
SMP    83          520       When current ICBM modernization complete, by mid-1980s.
JCS    85          520       As of 1 Jan.
```

156  SOVIET MISSILES  IDDS

---------- **NUMBER IN 1990** ----------
CBO     83          520 in 1990 and 1996.

---------- **LAUNCHER/SILO** ----------
SecC    66   Soviets constructing new type of single silo launch site, believed for SS-11.
SecC    70   Hard single silos.
SecC    72   New construction includes 60 smaller silos compatible "with an SS-11 size missile"; most of these silos could be completed during 1972.
SecC    73   Smaller silos expected completed by middle of this year.
SecD    73   Deployment of new SS-11 version expected [initially] in new silos.
MOW     76   Operational in camouflaged launchers since 1966.
JCS     77   Substantial number of SS-11 silos have been modernized.
CBO1    78   Assumed silo hardness of 550 psi, as opposed to 2000 psi for "Soviet modernized silos".
SecD    78   SS-17 and SS-19 are in old SS-11 silos, FY79-81.
JCS     79   Some SS-19 silos contain SS-11; SS-19 probably will replace.
JWS     81   Silo-launch.
AW04    82   60 silos built for SS-19 still contain SS-11.
SSF     82   Basing mode: Hardened silo.
SMP     84   Housed in less survivable silos [than SS-17,18,19], '84, '85.

---------- **DESIGNER** ----------
SSF     82   Chelomei       Design Bureau.

---------- **USE AND CONFIGURATION** ----------
JCS     75   Extensive troop testing has included 100 launches from operational sites, FY76-77.
CBO     83   System availability: 0.85.
vanD    83   10 missiles in a battalion.
Meyr    84   Penetration: 1.0; reliability: 0.6-0.8, "includes system reliability multiplied by operational readiness of deployed system".

---------- **RELOADS IN GENERAL** ----------
IA06    75   May be cold-launchable.
FIar    76   "Most sources claim that Sego is cold-launched from the silo, but the recent installation of SS-17 and -19 missiles in SS-11 silos suggests that hot-launch techniques may be used". [In SS-17 entry, lengthy silo-depth vs cold/hot launch question [for SS-11]; conclusion, ed:] SS-11 is hot launched from 30 m deep silo, or designer allowed extra room for future developments.
SecD    76   Hot launch.
SWM     77   SS-11 sites suitable for adaptation to cold launch [source has cold launch details, ed].
Rock    78   Thought to adopt cold-launch technique; but retrofitting of SS-17s and SS-19s indicates enough room was left to vent [hot launch] exhaust; on other hand perhaps SS-11 cold-launched from container in silo, it is only non-mobile ICBM displayed in container [this last from SS-17 entry, ed].
Clns    80   Cold launch: no.
SSF     82   Launching mode: Hot.
JWS+    83   US report says all Soviet liquid-fueled ICBMs contained in launch canister offering possibility of reloading. Might take several days, but thought worthwhile advantage; probably there are provisions for reload missiles and warheads.
SMP     83   "Contained in a launch canister within the silo. This and the silo design minimize damage to the launcher during" firing, giving reload capability.

---------- **STRATEGY** ----------
SecC    70   SS-11s deployed at MR/IRBM sites not included in ICBM totals, but could reach many targets in US.
JCS     71   Some deployed in MR/IRBM complexes, FY72-73.
SecC    71   Believed [those deployed in MR/IRBM complexes] aimed primarily at Europe, but could cover many targets in the US, "and could be retargeted in [about 15 chars deleted]".
SecD    71   Some launchers associated with MR/IRBM fields.
JCS     74   Reduced-range tests support PRC and Europe as possible targets.
FI*     79   May have IRBM role; tested at reduced ranges.
Meyr    84   "Initially developed for intercontinental strikes, but likely to be assigned to missions in the European theatre of military operations". [Ed: chart is labeled] Systems "assigned primary operational and strategic missions within the European theatre(s) of military operations".

---------- **GEOGRAPHICAL DEPLOYMENTS** ----------
SWM     77          Deployed in silos "West and East of the Urals".
JWS     81          Most of 500 remaining in early 1981 deployed opposite borders with China, Europe, Scandinavia, and Middle East. Two fields east of Carpathians house mixture of SS-11 and SS-19, proportions unknown.
SMP     83   [Shown on map at] Derazhnya, Kozelsk, Teykovo, Kostroma, Perm, Gladkaya, Drovyanaya, Olovyannaya, Svobodnyy. [Same, but Derazhnya omitted, in '84, '85 reports].

---------- **RELATION TO OTHER MISSILES** ----------
SecC    67   About the size of US Minuteman, FY68-70. Relatively simple and cheap missile [vs SS-9].
JCS     72   Larger than Minuteman.
JCS     74   Probably regarded by USSR as Minuteman counterpart.

---------- **OTHER INFORMATION** ----------
JCS 78   [Space in chart for Mod 4; no entry or label].
JWS 81   4 distinct versions.

# SS-11 Mod 1

```
Sources are:  FIar 76    FI*  79    AFM  82    IISS+ 83   SMP 85
JCS 74        SWM  77    AFM* 80    IISS 82    vanD  83   BW  86
JCS 75        Rock 78    AWar 80    SSF  82    SMP   84
```

---------- **US DESIGNATION** ----------

| | | |
|---|---|---|
| JCS  | 74 | SS-11 Mod 1 |
| FIar | 76 | SS-11 Mod 1 |
| SWM  | 77 | SS-11 Mod 1 |
| Rock | 78 | SS-11 Mod 1 |
| FI*  | 79 | SS-11 Mod 1 |
| AFM* | 80 | SS-11 Mod1 |
| AWar | 80 | SS-11 mod |
| AFM  | 82 | SS-11 Mod 1 |
| IISS | 82 | SS-11 Mod 1 |
| SSF  | 82 | SS-11 Mod 1 |
| vanD | 83 | SS-11 |
| SMP  | 84 | SS-11 Mod 1     '84, '85. |

---------- **NATO CODENAME** ----------

IISS 82    Sego

---------- **DESCRIPTION** ----------

SSF 82     Second generation, variable range.

---------- **EDITOR'S NOTES** ----------

SSF 82     SS-11 Mod 1 is listed twice, once under "Medium-range missiles" (VRBM entry) and again under "Intercontinental-range missiles" (ICBM entry). Data is from VRBM entry except where indicated.

vanD 83    Source contrasts two unnamed Mods.

---------- **RANGE** ----------

| | | mi | nm | km | km conv | |
|---|---|---|---|---|---|---|
| SWM  | 77 | 6,525 |       | 10,500 | 10,498 | |
| Rock | 78 |       |       | 10,500 | 10,500 | |
| AWar | 80 |       |       | 11,000 | 11,000 | |
| IISS | 82 |       |       | 10,500 | 10,500 | Max. |
| SSF  | 82 |       | 5,900 |        | 10,926 | [Both entries] |
| SMP  | 84 |       |       | 11,000 | 11,000 | Max '84, '85. |
| BW   | 86 |       |       | 11,000 | 11,000 | |

---------- **CEP** ----------

| | | ft | mi | nm | m | km | m conv | |
|---|---|---|---|---|---|---|---|---|
| JCS  | 74 | | | | | | | Mods 1 and 3 both lack hard target capability. |
| FIar | 76 | | | | | c. 1.5 | 1,500 | Would be improved if used with shorter range (IRBM). |
| Rock | 78 | | | | | 1.5 | 1,500 | "Error radius at the target". |
| AFM* | 80 | | | | | | | Slightly higher yield than US Minuteman, but considerably less accurate. |
| AWar | 80 | | | 0.76 | | | 1,407 | |
| IISS | 82 | | | | 1,400 | | 1,400 | |
| SSF  | 82 | | | 0.76 | | | 1,407 | [Both entries] |
| vanD | 83 | | | 1.25 | | | 2,315 | |
| BW   | 86 | | | | 1,400 | | 1,400 | |

---------- **WARHEAD CHARACTERISTICS** ----------

| | | | |
|---|---|---|---|
| JCS  | 75 | single RV | FY76-79. |
|      |    |           | Mod 2 tested with penetration aids (unlike Mod 1), FY76-77. |
| FIar | 76 | single RV | |
| Rock | 78 | single RV | |
| FI*  | 79 | single RV | |
| AFM* | 80 | single RV | |
| AWar | 80 | single RV | |
| SSF  | 82 | single RV | [Both entries] |
| vanD | 83 | single RV | |
| SMP  | 84 | single RV | '84, '85. |
| BW   | 86 | single RV | |

BALLISTIC MISSILES   ICBMs   SS-11 Mod 1   159

### NUCLEAR YIELD

| | | kT | MT | kT conv | |
|---|---|---|---|---|---|
| FIar | 76 | | 1-2 | 1,000-2,000 | |
| Rock | 78 | | 1-2 | 1,000-2,000 | |
| FI* | 79 | | 1-2 | 1,000-2,000 | |
| AFM* | 80 | | | | Slightly higher yield than US Minuteman, but considerably less accurate. |
| AWar | 80 | | 0.95 | 950.00 | |
| IISS | 82 | | 1 | 1,000 | |
| SSF | 82 | | 0.95 | 950.00 | [Both entries] |
| vanD | 83 | | 1.0 | 1,000 | |
| BW | 86 | 950 | | 950.00 | |

### THROWWEIGHT

| | | lb | t | kg | kg conv | |
|---|---|---|---|---|---|---|
| Rock | 78 | | < 0.7 | | 700.00 | "Nearly" 0.7 tonnes, "all variants". |
| IISS | 82 | 2,000 | | | 907.20 | |

### LENGTH

| | | in | ft | cm | m | m conv | |
|---|---|---|---|---|---|---|---|
| SMP | 84 | | | | 20 | 20.00 | [From graph], '84, '85 |

### TYPE OF PROPULSION

| SSF | 82 | liquid |
| BW | 86 | storable liquid |

### MIDCOURSE (OR MAIN) GUIDANCE

| Rock | 78 | inertial |
| SSF | 82 | Fly-the-wire, inertial. |

### DEVELOPMENT BEGAN

| SSF | 82 | 1955-58 | Year design began. |

### PROTOTYPE TESTS

| JCS | 74 | | Tested at reduced range as well as ICBM (Mod 1). |
| SWM | 77 | | Mod 1 tested over wide range of distances with single RV. |
| SSF | 82 | 1965 | First flight test, [both entries]. |

### IOC

| JCS | 74 | 1966 | FY75-79. |
| AFM* | 80 | 1966 | Operational. |
| vanD | 83 | 1966 | Latter 1966. |

### DEPLOYMENT BEGAN

| FIar | 76 | 1966 | First deployed. |
| Rock | 78 | 1966 | Introduced. |
| FI* | 79 | 1966 | First deployed. |
| IISS | 82 | 1966 | First year deployed. |
| SSF | 82 | 1966 | Year operation began [ICBM entry]. |
| | | 1970 | "Year operation began" [VRBM entry] |
| SMP | 85 | 1966-67 | [Presumed Mod 1, from graph, ed.] |
| BW | 86 | 1966 | |

### DEPLOYED

| JCS | 74 | 3 versions tested, but only Mod 1 and Mod 3 deployed. |
| FI* | 79 | Mod 1 and Mod 3 deployed. |

### RETIRED

| AFM | 82 | | Mod 1 is now retired. |
| BW | 86 | 1979 | |

### DEPLOYMENT HISTORY

| Rock | 78 | By 1976 about 50 Mod 1s replaced by the improved Mod 3, and total force was converting to SS-17 and SS-19. |
| AFM* | 80 | More than 60 replaced by SS-11 Mod 3; others have been superceded by SS-17s and SS-19s. |

### NUMBERS IN SERVICE

### NUMBER IN 1955

| SSF | 82 | none | In 1955, 1960, 1965, includes only those missiles facing Western Europe and China. |

### NUMBER IN 1970

| SSF | 82 | 290 | Includes only those missiles facing Western Europe and China. |

## 160   SOVIET MISSILES   IDDS

---------- **NUMBER IN 1971** ----------
FIar 76   970   Between 1971 and 1975 a total of 970 rounds were in service, but this year [1976] about 50 retired and replaced by Mod 3. Others are being replaced by SS-17 and -19.
Rock 78   970   970 put into service by 1971.

---------- **NUMBER IN 1975** ----------
SSF 82   320   Includes only those missiles facing Western Europe and China.

---------- **NUMBER IN 1976** ----------
FIar 76   [> 850]   Most (more than 85%) of the existing force (of more than 1,000) is this variant.
Rock 78            By 1976 about 50 Mod 1s replaced by the improved Mod 3.

---------- **NUMBER IN 1979** ----------
FI* 79   [> 850]   Over 85% of over 1,000 SS-11s are Mod 1.

---------- **NUMBER IN 1980** ----------
AFM* 80            More than 60 of 970 originally deployed replaced by SS-11 Mod 3.
SSF 82   260   Includes only those missiles facing Western Europe and China.

---------- **NUMBER IN 1982** ----------
IISS 82   $\leq$ 570   July, total deployed.
SSF 82   320   "Number of missiles deployed".

---------- **NUMBER IN 1983** ----------
IISS+ 83   $\leq$ 550   July, total deployed.

---------- **NUMBER IN 1984** ----------
SMP 84   100   '84, '85.

---------- **UNDATED NUMBER** ----------
AFM* 80   970   Originally deployed.
BW 86   970   Peak.

---------- **LAUNCHER/SILO** ----------
SSF 82   Basing mode: Hardened silo.

---------- **DESIGNER** ----------
SSF 82   Chelomei   Design Bureau.

---------- **USE AND CONFIGURATION** ----------
vanD 83   Reliability: .85; Penetration 1.0.

---------- **RELOADS IN GENERAL** ----------
SMP 84   Hot launch, '84, '85.
BW 86   No.

---------- **STRATEGY** ----------
FIar 76   According to some reports, part of force used in IRBM role, perhaps with heavier warhead.
Rock 78   "Several Mod 1 variants have been set up among ICBM forces in the western USSR and this had [sic] led to speculation that the SS-11 has found application, itself, as an intermediate range weapon". If so, would carry heavier warhead, and have improved accuracy due to range.

# SS-11 Mod 2

```
Sources are:   FIar  76      SWM   77      AWar  80      SSF   82      SMP   84
JCS   74       JCS   76      Rock  78      JWS   81      vanD  83      SMP   85
JCS   75       JCS   77      FI*   79      AFM   82      AFM+  84      BW    86
```

## US DESIGNATION

|  | |  |
|---|---|---|
| JCS  | 74 | SS-11 Mod 2 |
| FIar | 76 | SS-11 Mod 2 |
| SWM  | 77 | SS-11 Mod 2 |
| Rock | 78 | SS-11 Mod 2 |
| FI*  | 79 | SS-11 Mod 2 |
| AWar | 80 | SS-11 Mod 2 |
| JWS  | 81 | SS-11 Mod2 |
| AFM  | 82 | SS-11 Mod 2 |
| SSF  | 82 | SS-11 Mod 2 |
| vanD | 83 | SS-11 |
| SMP  | 84 | SS-11 Mod 2    '84, '85. |

## EDITOR'S NOTES

| vanD | 83 | Source contrasts two unnamed Mods. |

## RANGE

| Source | | mi | nm | km | km conv | |
|---|---|---|---|---|---|---|
| AWar | 80 | | | 12,000 | 12,000 | |
| SSF  | 82 | | 6,500 | | 12,038 | |
| AFM+ | 84 | c. 6,500 | | | 10,458 | |
| SMP  | 84 | | | 13,000 | 13,000 | Max, '84, '85. |
| BW   | 86 | | | 13,000 | 13,000 | |

## CEP

| Source | | ft | mi | nm | m | km | m conv | |
|---|---|---|---|---|---|---|---|---|
| AWar | 80 | | | 0.59 | | | 1,092 | |
| AFM  | 82 | | | | | | | "Considerably less accurate" than US Minuteman. |
| SSF  | 82 | | | 0.59 | | | 1,092 | |
| vanD | 83 | | | .76 | | | 1,407 | |
| AFM+ | 84 | | 0.87 | | | 1.4 | 1,399 | |
| BW   | 86 | | | | 1,100 | | 1,100 | |

## WARHEAD CHARACTERISTICS

| JCS  | 75 | single RV | FY76-79. Mod 2 tested with penetration aids (unlike Mod 1), FY76-77. |
| FIar | 76 | | Reported to have more accurate RV incorporating countermeasures to aid penetration of ABM system. |
| JWS  | 81 | single RV | Plus penaids. |
| AFM  | 82 | single RV | Mod 2 differs from Mod 1 in having penetration aids. |
| SSF  | 82 | single RV | |
| vanD | 83 | single RV | |
| SMP  | 84 | single RV | '84, '85. |
| BW   | 86 | single RV | |

## NUCLEAR YIELD

| Source | | kT | MT | kT conv | |
|---|---|---|---|---|---|
| FIar | 76 | | 1-2 | 1,000-2,000 | |
| AWar | 80 | | 1.1 | 1,100 | |
| AFM  | 82 | | | | Yield greater than US Minuteman. |
| SSF  | 82 | | 1.10 | 1,100 | |
| vanD | 83 | | 1.0 | 1,000 | |
| BW   | 86 | 1,100 | | 1,100 | |

## THROWWEIGHT

| Source | | lb | t | kg | kg conv | |
|---|---|---|---|---|---|---|
| Rock | 78 | | < 0.7 | | 700.00 | "Nearly" 0.7 tonnes, "all variants". |

## LENGTH

| Source | | in | ft | cm | m | m conv | |
|---|---|---|---|---|---|---|---|
| SMP | 84 | | | | 20 | 20.00 | [From graph '84, '85.] |

## TYPE OF PROPULSION

| BW | 86 | storable liquid |

## PROTOTYPE TESTS

| FIar | 76 | | No test flights recently, suggesting program terminated. |
| SSF  | 82 | 1969 | First flight test. |

162  SOVIET MISSILES  IDDS

---------- **IOC** ----------
JCS    75    1973         FY76-79.
JWS    81    ? 1973       Became operational.
SSF    82    1973         Year operation began.
vanD   83    1974

---------- **DEPLOYMENT BEGAN** ----------
JCS    74                 Program probably terminated. 3 versions tested, but only Mod 1 and Mod 3 deployed.
JCS    75                 "Lack of identifiable operational testing" led to belief Mod 2 program terminated; recent firings suggest crew training, FY76-77. Deployment "possibly may have been initiated".
FIar   76                 No test flights recently, suggesting program terminated.
JCS    76                 Belief that Mods 2 and 3 "already have been deployed" in silos under construction at time of Interim Agreement.
                          Deployment "may have been initiated".
JCS    77                 A mix of Mods 2 and 3 believed deployed in modified silos.
SWM    77                 Mod 2 is a non-operational test vehicle.
Rock   78                 Was a test development and not operationally adapted to service requirements.
FI*    79                 Mod 1 and Mod 3 deployed.
SMP    85    c. 1976      [From graph for Mods 2 and 3 together.]
BW     86    ? 1974

---------- **DEPLOYED** ----------
BW     86                 In service.

---------- **DEPLOYMENT HISTORY** ----------
JCS    77    Mod 2 and 3 force probably will decline in next 5-10 years; [will be replaced].
SMP    85    Will be deployed at least through 1990. [From graph for Mods 2 and 3 together.]

---------- **NUMBERS IN SERVICE** ----------

---------- **NUMBER IN 1984** ----------
SMP    84              420, Mods 2 and 3 together.

---------- **NUMBER IN 1985** ----------
SMP    85              420, Mods 2 and 3 together.

---------- **UNDATED NUMBER** ----------
BW     86    ? 360    Peak.

---------- **LAUNCHER/SILO** ----------
JCS    76    Belief that Mods 2 and 3 "already have been deployed" in silos under construction at time of Interim Agreement.
JCS    77    A mix of Mods 2 and 3 believed deployed in modified silos.

---------- **USE AND CONFIGURATION** ----------
vanD   83    Reliability: .85; Penetration 1.0.

---------- **RELOADS IN GENERAL** ----------
SMP    84    Hot launch, '84, '85.
BW     86    No.

# SS-11 Mod 3

```
Sources are:  JCS    74    JCS    76    SWM   77    AWar  80    AWST  82    SMP   84
JCS    72    IA06   75    MOW    76    Rock  78    Clns  80    IISS  82    SMP   85
JCS    73    JCS    75    A140   77    FI *  79    JWS   81    SSF   82    BW    86
SecC   73    FIar   76    JCS    77    Nitz  79    AFM   82    AFM+  84
```

---------- **US DESIGNATION** ----------

|  | |  |
|---|---|---|
| JCS | 72 | SS-11 Mod 3 |
| SecC | 73 | SS-11 Mod 3 |
| FIar | 76 | SS-11 Mod 3 |
| SWM | 77 | SS-11 Mod 3 |
| Rock | 78 | SS-11 Mod 3 |
| FI * | 79 | SS-11 Mod 3 |
| Nitz | 79 | SS-11 (MRV) |
| AWar | 80 | SS-11 Mod 3 |
| Clns | 80 | SS-11 Mod 3 |
| JWS | 81 | SS-11 Mod3 |
| AFM | 82 | SS-11 Mod 3 |
| IISS | 82 | SS-11 Mod 3 |
| SSF | 82 | SS-11 Mod 3 |
| SMP | 84 | SS-11 Mod 3  '84, '85. |

---------- **NATO CODENAME** ----------

| Source | | Codename |
|---|---|---|
| Rock | 78 | Sego |
| IISS | 82 | Sego |

---------- **OTHER DESIGNATION** ----------

|  | | |
|---|---|---|
| AWST | 82 | MRV variant. |

---------- **DESCRIPTION** ----------

|  | | |
|---|---|---|
| Rock | 78 | ICBM |

---------- **RANGE** ----------

| Source | | mi | nm | km | km conv | |
|---|---|---|---|---|---|---|
| Rock | 78 | | | 10,000 | 10,000 | |
| AWar | 80 | | | 10,600 | 10,600 | |
| Clns | 80 | | 5,500 | | 10,186 | |
| AFM | 82 | c. 6,200 | | | 9,975 | |
| IISS | 82 | | | 8,800 | 8,800 | Max. |
| SSF | 82 | | 5,700 | | 10,556 | |
| AFM+ | 84 | c. 5,450 | | | 8,769 | |
| SMP | 84 | | | 10,600 | 10,600 | Max, '84, '85. |
| BW | 86 | | | 10,600 | 10,600 | |

---------- **CEP** ----------

| Source | | ft | mi | nm | m | km | m conv | |
|---|---|---|---|---|---|---|---|---|
| JCS | 74 | | | | | | | Mods 1 and 3 both lack hard target capability. |
| FIar | 76 | | | | ? < 1 | | 1,000 | This version may have more accurate guidance system, giving better CEP. |
| A140 | 77 | | | | | | | "Known to be more accurate" than SS-11 Mod 1. |
| Rock | 78 | | | | < 1 | | 1,000 | "Target error radius". |
| AWar | 80 | | | 0.59 | | | 1,092 | |
| Clns | 80 | | | 0.6 | | | 1,111 | |
| IISS | 82 | | | | 1,100 | | 1,100 | |
| SSF | 82 | | | 0.59 | | | 1,092 | |
| AFM+ | 84 | | 0.7 | | | 1.1 | 1,126 | |
| BW | 86 | | | | 1,100 | | 1,100 | |

164  SOVIET MISSILES  IDDS

## WARHEAD CHARACTERISTICS

| | | | |
|---|---|---|---|
| JCS | 73 | MRV | New version has MRV payload. |
| IA06 | 75 | 3 MRVs | |
| JCS | 75 | 3 RVs | FY76-79. |
| | | 3 RVs | Max. |
| Flar | 76 | 3 MRVs | First operational Russian missile to carry MRVs. |
| MOW | 76 | | MRV with 3 separate charges and decoys. |
| SWM | 77 | 3 RVs | |
| Rock | 78 | 3 x MRV | First Soviet ICBM to carry MRVs in operational mode. |
| FI * | 79 | 3 MRVs | First Soviet MRV. |
| Nitz | 79 | [3 RVs] | [Derived by warheads/launchers, ed] |
| Clns | 80 | 3 MRVs | |
| JWS | 81 | 3 MRVs | |
| AFM | 82 | 3 MRVs | First Soviet MRV. |
| AWST | 82 | 3 MRVs | |
| IISS | 82 | 3 MRVs | |
| SSF | 82 | 3 RVs | |
| SMP | 84 | 3 MRVs | '84, 85. |
| BW | 86 | 3 MRVs | |

## NUCLEAR YIELD

| | | kT | MT | kT conv | |
|---|---|---|---|---|---|
| IA06 | 75 | 200 | | 200.00 | Each. |
| Flar | 76 | 300 | | 300.00 | |
| SWM | 77 | 500 | | 500.00 | Each. |
| Rock | 78 | | 1.5 | 1,500 | [Table] |
| | | 500 | | 500.00 | Each of 3 warheads, [text]. |
| FI * | 79 | 300 | | 300.00 | Each, [text] |
| | | 500 | | 500.00 | Each, [table] |
| AWar | 80 | | 0.35 | 350.00 | |
| Clns | 80 | 500-1,000 | | 500.00-1,000 | "500 KT-1 MT". |
| JWS | 81 | ? 200+ | | 200.00 | [Text] |
| | | c. 200 | | 200.00 | [Specs] |
| AFM | 82 | 300 | | 300.00 | |
| AWST | 82 | | | | c. 250 KT |
| IISS | 82 | 100-300 | | 100.00-300.00 | |
| SSF | 82 | | 0.35 | 350.00 | |
| BW | 86 | 350 | | 350.00 | |

## THROWEIGHT

| | | lb | t | kg | kg conv | |
|---|---|---|---|---|---|---|
| Rock | 78 | | < 0.7 | | 700.00 | "Nearly" 0.7 tonnes, "all variants". |
| Clns | 80 | 2,500 | | | 1,134 | |
| IISS | 82 | 2,500 | | | 1,134 | |

## LENGTH

| | | in | ft | cm | m | m conv | |
|---|---|---|---|---|---|---|---|
| Rock | 78 | | | | 19 | 19.00 | |
| SMP | 84 | | | | 20 | 20.00 | [From graph], '84, '85. |

## DIAMETER

| | | in | ft | cm | m | cm conv |
|---|---|---|---|---|---|---|
| Rock | 78 | | | | 2.4 | 240.00 |

## TYPE OF PROPULSION

| | | | |
|---|---|---|---|
| Rock | 78 | storable liquid | Stages 1 and 2. |
| Clns | 80 | liquid | |
| BW | 86 | storable liquid | |

## FIRST OBSERVED

| | | |
|---|---|---|
| JWS | 81 | Detected soon after Mod 2 [1973]. |

## PROTOTYPE TESTS

| | | | |
|---|---|---|---|
| JCS | 74 | | Extensive testing since 1969; very successful. |
| A140 | 77 | 1969-73 | Tested. |
| AFM | 82 | 1969 | Tests began 1969. |
| AWST | 82 | | Tested with MRV. |
| SSF | 82 | 1969 | First flight test. |

## IOC

| | | | |
|---|---|---|---|
| JCS | 75 | 1973 | FY76-79. |

BALLISTIC MISSILES  ICBMs  SS-11 Mod 3   165

```
---------- DEPLOYMENT BEGAN ----------
JCS   75              Belief that Mod 3s "already have been deployed" in silos under construction at time
                        of Interim Agreement.
JCS   76              Belief that Mods 2 and 3 "already have been deployed" in silos under construction
                        at time of Interim Agreement.
Rock  78    1973      Service intro.
Clns  80    1973      First deployed.
IISS  82    1973      First year deployed.
SSF   82    1973      Year operation began.
SMP   85    c. 1976   [From graph for Mods 2 and 3 together.]
BW    86    1973

---------- DEPLOYED ----------
JCS   74              3 versions tested, but only Mod 1 and Mod 3 deployed.
JCS   77              A mix of Mods 2 and 3 believed deployed in modified silos.
SWM   77              Mod 3 is the operational model.
FI*   79              Mod 1 and Mod 3 deployed.
BW    86              In service.

---------- DEPLOYMENT HISTORY ----------
SecC  73    Believed that SS-11 Mod 3 will initially be deployed in new small silos.
JCS   74    MRV SS-11 version "being deployed rapidly" in some of new "Light" ICBM silos.
JCS   77    Mod 2 and 3 force probably will decline in next 5-10 years; [will be replaced].
IISS  82    Mod 3 replaced some Mod 1.
SMP   85    Will be deployed at least through 1990. [From graph for Mods 2 and 3 together.]

---------- NUMBERS IN SERVICE ----------

---------- NUMBER IN 1969 ----------
AFM   82              Tests began 1969, over 60 deployed fast.

---------- NUMBER IN 1973 ----------
JCS   74              20 Mod 3s, as of mid-1973, operational in new small silos that were not operational at
                        1972 Interim Agreement time.

---------- NUMBER IN 1976 ----------
FIar  76    > 60      Now in service.

---------- NUMBER IN 1977 ----------
Nitz  79    455       Launchers.

---------- NUMBER IN 1978 ----------
Rock  78    c. 66     Some 66 silos prepared, and "effectively uprates the warhead complement to 198" if all
                        installed with Mod 3.

---------- NUMBER IN 1979 ----------
Nitz  79    455       18 June, launchers.

---------- NUMBER IN 1982 ----------
IISS  82    "some"    July, total deployed.

---------- NUMBER IN 1984 ----------
SMP   84              420, Mods 2 and 3 together.

---------- NUMBER IN 1985 ----------
SMP   85              420, Mods 2 and 3 together.

---------- UNDATED NUMBER ----------
BW    86    ? 60      Peak.

---------- LAUNCHER/SILO ----------
SecC  73    Smaller silos expected completed by middle of this year.
JCS   74    20 Mod 3s, as of mid-1973, operational in new small silos that were not operational at 1972
              Interim Agreement time.
JCS   75    Belief that Mod 3s "already have been deployed" in silos under construction at time of Interim
              Agreement.
JCS   76    Belief that Mods 2 and 3 "already have been deployed" in silos under construction at time of
              Interim Agreement.
A140  77    Requires slight modification to silos built for SS-11 Mod 1.
JCS   77    A mix of Mods 2 and 3 believed deployed in modified silos.
Rock  78    Some 66 silos prepared, and "effectively uprates the warhead complement to 198" if all installed
              with Mod 3.

---------- RELOADS IN GENERAL ----------
Clns  80    Cold launch: no.
SMP   84    Hot launch, '84, '85.
BW    86    No.
```

――――――― **STRATEGY** ―――――――
JCS 72    Improvements [testing] believed aimed at enhancing penetration capabilities against ABM-defended soft targets, FY73-74.
JCS 74    Mod 3 believed "initially was developed" for MRV effect on penetrating ABMs. Mod 3's rapid deployment despite ABM Treaty implies USSR sees advantages in it against undefended targets: probably increased accuracy and greater targeting flexibility.

# SS-11 Mod 4

Sources are:  JWS  81      SSF  82

---------- **US DESIGNATION** ----------
| | | |
|---|---|---|
| JWS | 81 | SS-11 Mod4 |
| SSF | 82 | SS-11 Mod 4 |

---------- **WARHEAD CHARACTERISTICS** ----------
| | | |
|---|---|---|
| JWS | 81 | 3 or 6 small MRVs   Mysterious payload. |
| SSF | 82 | 3-6 RVs |

---------- **FIRST OBSERVED** ----------
| | | |
|---|---|---|
| JWS | 81 | Appeared late 1970s. |

---------- **PROTOTYPE TESTS** ----------
| | | | |
|---|---|---|---|
| SSF | 82 | 1974 | First flight test. |

# SS-13

| Sources are: | | IISS | 70 | SecD | 72 | IA06 | 75 | SecD | 77 | AWar | 80 | JCS | 82 | AFM+ | 84 |
|---|---|---|---|---|---|---|---|---|---|---|---|---|---|---|---|
| IA01 | 65 | SecC | 70 | IISS | 73 | JCS | 75 | SWM | 77 | Clns | 80 | SSF | 82 | JCS | 84 |
| SecC | 66 | SecD | 70 | JCS | 73 | SecC | 75 | JCS | 78 | JCS | 81 | CBO | 83 | SecD | 84 |
| SecC | 67 | JCS | 71 | SecC | 73 | FIar | 76 | Rock | 78 | JWS | 81 | Clns+ | 83 | JCS | 85 |
| ID01 | 68 | SecC | 71 | SecD | 73 | JCS | 76 | FI * | 79 | SecD | 81 | JWS+ | 83 | SMP | 85 |
| Ley | 68 | SecD | 71 | JCS | 74 | MOW | 76 | JCS | 79 | SMP | 81 | SMP | 83 | | |
| SecC | 68 | JCS | 72 | GA01 | 75 | NA03 | 76 | Nitz | 79 | AFM | 82 | Su01 | 83 | | |
| SecC | 69 | SecC | 72 | HRST | 75 | JCS | 77 | US08 | 79 | AWST | 82 | vanD | 83 | | |

---------- **US DESIGNATION** ----------

| SecC | 69 | SS-13 | FY70-76. |
|---|---|---|---|
| SecD | 70 | SS-13 | FY71-76,78,82,85. |
| JCS | 71 | SS-13 | FY72-80,82-86. |
| FIar | 76 | SS-13 | |
| MOW | 76 | SS-13 | |
| SWM | 77 | SS-13 | |
| Rock | 78 | SS-13 | |
| FI * | 79 | SS-13 | |
| Nitz | 79 | SS-13 | |
| AWar | 80 | SS-13 | |
| Clns | 80 | SS-13 | |
| JWS | 81 | SS-13 | |
| SMP | 81 | SS-13 | '81, '83, '84, '85. |
| AFM | 82 | SS-13 | |
| AWST | 82 | SS-13 | |
| SSF | 82 | SS-13 | |
| CBO | 83 | SS-13 | |
| vanD | 83 | SS-13 | |

---------- **NATO CODENAME** ----------

| Ley | 68 | Savage |
|---|---|---|
| HRST | 75 | Savage |
| FIar | 76 | Savage |
| MOW | 76 | Savage |
| JCS | 77 | Savage |
| SWM | 77 | Savage |
| Rock | 78 | Savage |
| FI * | 79 | Savage |
| JWS | 81 | Savage |
| AFM | 82 | Savage |
| AWST | 82 | Savage |
| SSF | 82 | Savage |

---------- **DESCRIPTION** ----------

| JCS | 72 | | 2nd generation strategic missile; "SS-13 improved" is 3rd generation. |
|---|---|---|---|
| JCS | 73 | | "Light" ICBM, FY74-75. |
| HRST | 75 | ICBM | |
| MOW | 76 | ICBM | |
| Rock | 78 | ICBM | |
| JCS | 81 | | [Listed as "recent generation ICBMs" with SS-17, 18, 19, ed]. |
| JWS | 81 | ICBM | |
| JCS | 82 | | [Listed as "older generation" with SS-11, ed]. |
| SSF | 82 | | Third generation. |

---------- **EDITOR'S NOTES** ----------

| SecC | 66 | Listed FY67-76, not by name FY67-69. |
|---|---|---|
| SecD | 70 | Listed FY71-76,78,82,85. |
| JCS | 71 | Listed FY72-80,82-86. |
| JCS | 78 | Source has space in a chart for Mod 2, but no entries or label, FY79. Source also has heading "SS-13/SS-16", but no reference to SS-13 under it, only to SS-16, FY79. |
| SMP | 81 | Listed '81, '83, '84, '85. |

BALLISTIC MISSILES  ICBMs  SS-13  169

## RANGE

|  |  | mi | nm | km | km conv | |
|---|---|---|---|---|---|---|
| Ley | 68 | 5,000+ | | | 8,045 | |
| HRST | 75 | 5,000 | | | 8,045 | |
| IAD6 | 75 | | | 8000-10500 | 8,000-10,500 | "8-10,500" km. |
| FIar | 76 | | | 8,000 | 8,000 | |
| MOW | 76 | 5,000 | | 8,000 | 8,045 | Max. |
| NAO3 | 76 | 5,000 | | | 8,045 | |
| SWM | 77 | 5,000+ | | 8,000+ | 8,045 | |
| Rock | 78 | | | 8,000 | 8,000 | |
| FI* | 79 | | | 8,000 | 8,000 | |
| Clns | 80 | | 5,000 | | 9,260 | |
| JWS | 81 | | | 8000-10000 | 8,000-10,000 | |
| AFM | 82 | 6,200 | | | 9,975 | |
| AWST | 82 | | 5,000 | | 9,260 | |
| SSF | 82 | | 5,075 | | 9,398 | |

## CEP

|  |  | ft | mi | nm | m | km | m conv | |
|---|---|---|---|---|---|---|---|---|
| SecC | 68 | | | | | | | Expected about the same as SS-11. |
| SecC | 69 | | | | | | | [Deleted], about same as SS-11. Would not have much value against hard targets. |
| SecC | 75 | | | | | | | "The use of a bus on the SS-X-16, therefore, may simply represent the Soviet solution to the accuracy problems encountered with the SS-13, as well as provide the additional energy required for range flexibility". |
| FIar | 76 | | | | 2 | | 2,000 | Because of poor CEPs, reports that the force has been fitted with MRVs are unlikely to be correct. |
| Rock | 78 | | | | | 1.3 | 1,300 | "Target error radius". |
| FI* | 79 | | | | 2 | | 2,000 | |
| Nitz | 79 | | | 1 | | | 1,852 | |
| Clns | 80 | | | 1.0 | | | 1,852 | |
| AFM | 82 | | 1.25 | | 2 | | 2,011 | |
| CBO | 83 | | | 1.0 | | | 1,852 | In 1983, 1990, 1996. |
| vanD | 83 | | | 1.0 | | | 1,852 | |

## WARHEAD TYPE

|  |  | nuclear | conven | chem |
|---|---|---|---|---|
| MOW | 76 | yes | | |
| JWS | 81 | yes | | |
| AFM | 82 | yes | | |

## WARHEAD CHARACTERISTICS

| SecC | 71 | | Warhead variants being developed for SS-13 [and SS-9 and SS-11]. |
|---|---|---|---|
| SecD | 71 | | Warhead variants (MRV or decoys) being developed. |
| JCS | 75 | single RV | FY76-79. |
| FIar | 76 | | Because of poor CEPs, reports that the force has been fitted with MRVs are unlikely to be correct. |
| SecD | 77 | single RV | Number of independently targetable warheads: 1. |
| SWM | 77 | | "Warhead 1-2 MT or 3 x MRV". |
| Rock | 78 | single RV | Normally. |
|  |  | MRVs | "Conflicting reports indicate that for a time some of the 60 SS-13s carried [MRVs]" similar to SS-11 Mod 3 design. |
| FI* | 79 | 1 RV or 3 MRVs | |
| Nitz | 79 | 1 RV | |
| USD8 | 79 | single RV | |
| AWar | 80 | single RV | |
| Clns | 80 | single RV | |
| JWS | 81 | single RV | ?Never equipped with other than single warhead. |
| SSF | 82 | single RV | |
| CBO | 83 | single RV | |
| vanD | 83 | single RV | |

## 170  SOVIET MISSILES  IDDS

### ────── NUCLEAR YIELD ──────

|  |  | kT | MT | kT conv |  |
|---|---|---|---|---|---|
| ID01 | 68 |  | 2-3 | 2,000-3,000 | Most likely smaller than SS-11. |
| SecC | 69 |  |  |  |  |
| Flar | 76 |  | 1 | 1,000 |  |
| MOW | 76 |  | 1 | 1,000 |  |
| SWM | 77 |  | 1-2 | 1,000-2,000 | "Warhead 1-2 MT or 3 x MRV". |
| Rock | 78 |  | 1.0 | 1,000 | [Table] |
|  |  |  | >_ 1 | 1,000 | [Text] |
|  |  | 300 |  | 300.00 | Each warhead of possible MRV configuration. For single RV variant. |
| FI* | 79 |  | 1 | 1,000 |  |
| Nitz | 79 |  | .75 | 750.00 |  |
| USO8 | 79 |  | 1 | 1,000 |  |
| AWar | 80 |  | 0.60 | 600.00 |  |
| Clns | 80 | 600 |  | 600.00 |  |
| JWS | 81 |  | c. 1 | 1,000 |  |
| AFM | 82 |  | 1 | 1,000 |  |
| CBO | 83 | 600 |  | 600.00 |  |
| vanD | 83 |  | .6 | 600.00 |  |
| AFM+ | 84 | 750 |  | 750.00 |  |

### ────── THROWWEIGHT ──────

|  |  | lb | t | kg | kg conv |  |
|---|---|---|---|---|---|---|
| Flar | 76 |  |  | 500 | 500.00 |  |
| Rock | 78 |  | < 0.5 |  | 500.00 | "Nearly" 0.5 tonnes. |
| Nitz | 79 | 1,000 |  |  | 453.60 |  |
| USO8 | 79 | 2,000 |  |  | 907.20 |  |
| Clns | 80 | 1,500 |  |  | 680.40 |  |
| SSF | 82 | 1,000 |  |  | 453.60 |  |
| CBO | 83 | 1,500 |  |  | 680.40 |  |

### ────── WEIGHT ──────

|  |  | lb | t | kg | kg conv |  |
|---|---|---|---|---|---|---|
| IA01 | 65 |  | c. 40 |  | 40,000 |  |
| ID01 | 68 |  | 36-38 |  | 36,000-38,000 |  |
| Ley | 68 |  | 40 |  | 40,000 | Tons. |
| FI* | 79 |  |  | 35,000 | 35,000 |  |

### ────── LENGTH ──────

|  |  | in | ft | cm | m | m conv |  |
|---|---|---|---|---|---|---|---|
| ID01 | 68 |  | 64.8 |  |  | 19.75 |  |
| Ley | 68 |  | 64 |  |  | 19.51 |  |
| SecC | 70 |  | 64 |  |  | 19.51 | [Read from graph, unit inferred]. |
| GA01 | 75 |  | 64 |  |  | 19.51 |  |
| HRST | 75 |  | 65 |  |  | 19.81 |  |
| Flar | 76 |  |  |  | 20 | 20.00 |  |
| MOW | 76 |  | 66'0" |  | 20.0 | 20.12 |  |
| SWM | 77 |  | 65.6 |  | 20.0 | 19.99 |  |
| Rock | 78 |  |  |  | 20 | 20.00 |  |
| FI* | 79 |  |  |  | 20 | 20.00 |  |
| JWS | 81 |  |  |  | 20 | 20.00 | Stages 1-3: 8.7m, 4m, 3.5m. |
| AFM | 82 |  | 66'0" |  |  | 20.12 |  |
| JWS+ | 83 |  |  |  |  |  | Nose cone and RV, 2.3 m. |

### ────── DIAMETER ──────

|  |  | in | ft | cm | m | cm conv |  |
|---|---|---|---|---|---|---|---|
| IA01 | 65 |  | 5.75 |  |  | 175.26 | First stage, minus skirt. |
| Ley | 68 |  | 5 3/4 |  |  | 175.26 |  |
| GA01 | 75 |  | 6 |  |  | 182.88 |  |
| HRST | 75 | 84 |  |  |  | 213.36 |  |
| Flar | 76 |  |  |  | 2.0 | 200.00 |  |
| MOW | 76 |  | 5'6" |  | 1.68 | 167.64 | Stages 1-3: 5'6", 4'7.5", 3'2.5"; stages 1-3 (metric): 1.68 m, 1.40 m, 0.98 m. |
| SWM | 77 |  | 5.5 |  | 1.7 | 167.64 |  |
| Rock | 78 |  |  |  | 2.0 | 200.00 | [Table] |
|  |  |  |  |  | 1.7 | 170.00 | [Text] |
| FI* | 79 |  |  | 170 |  | 170.00 |  |
| JWS | 81 |  |  |  |  |  | Body diameters, stages 1-3: 1.7m, 1.4m, 1m, (nosecone and RV 1m). Missile base diameters, stages 1-3: 2m, 1.9m, 1.4m. |
| AFM | 82 |  | 6'6" |  |  | 198.12 | Max, first-stage skirt. |

## BALLISTIC MISSILES ICBMs SS-13

```
------------ NUMBER OF STAGES ------------
HRST   75    3
MOW    76    3
SWM    77    3
Rock   78    3
FI*    79    3
JWS    81    3
AFM    82    3
AWST   82    3

------------ TYPE OF PROPULSION ------------
SecC   68    solid       FY69-72.
SecD   70    solid
JCS    72    solid       FY73,75-77.
JCS    74    solid       Only solid fuel missile in inventory.
HRST   75    solid
FIar   76    solid       Only solid-propellant Russian ICBM.
MOW    76    solid       4 nozzles each.
SWM    77    solid
Rock   78    solid       Stages 1, 2, and 3. "First solid propellant ICBM to enter operational
                         service".
FI*    79    solid
Clns   80    solid
JWS    81    solid       All 3 stages have 4 nozzles, upper stages believed identical with SS-14.
AFM    82    solid
AWST   82    solid
SSF    82    solid

------------ MIDCOURSE (OR MAIN) GUIDANCE ------------
MOW    76    inertial    Probably.
SWM    77    inertial
FI*    79    inertial
JWS    81    inertial    Presumed.
AFM    82    inertial
SSF    82                Fly-the-wire, inertial.

------------ CONTROL SYSTEM ------------
FI*    79    Movable nozzles.

------------ OBSERVED IN PARADE ------------
MOW    76    9 May 1965
SWM    77    May 1965
Rock   78    May 1965    First displayed.
JWS    81    1965
SMP    81                [Picture shows missile on display in Moscow].

------------ DEVELOPMENT BEGAN ------------
SSF    82    1958-62     Year design began.
SMP    85    1958        Technological development began. [From graph.]

------------ PROTOTYPE TESTS ------------
SecC   66                "We have had no indication up to now that a full scale flight test program of [a
                         solid fuel ICBM] has been initiated".
SecC   68                We now have good evidence Soviets are indeed working on small solid-fuel ICBM,
                         anticipated for many years.
SecC   69                Still in development flight test phase.
JCS    71                Active test program for improvements continues, FY72-73.
SecD   71                Tests of modification of SS-13 (and SS-11 and SS-9) have continued; warhead
                         variants (MRV or decoys) being developed.
JCS    72                Improvements believed aimed at improving accuracy, FY73-74.
SMP    85    1964-65     Engineering and testing began. [From graph.]

------------ IOC ------------
JCS    75    1969        [From graph in FY84], FY76-79,84.
SecD   77    1969
vanD   83    1970        [P209]
              1969       [P186]
```

## DEPLOYMENT BEGAN

| | | | |
|---|---|---|---|
| SecC | 67 | | Reasonable to expect Soviets could have solid fuel ICBM by 1971, either in fixed mode as follow-on to SS-11 or in mobile mode. |
| SecC | 69 | | Start of program comes about a year earlier than previously estimated—while missile is still in development flight test phase. |
| FIar | 76 | 1968 | Entered service. |
| | | 1971 | Stationed in silos at Plesetsk since 1971. |
| SWM | 77 | 1968 | Service entry. |
| Rock | 78 | 1968 | Service intro. |
| Clns | 80 | 1969 | First deployed. |
| SMP | 83 | by 1966 | Deployment [of SS-9,11,13] underway. |
| SecD | 84 | 1969 | Introduction of strategic systems [graph]. |
| SMP | 85 | 1969-70 | [From graph.] |

## DEPLOYED

| | | |
|---|---|---|
| JCS | 72 | Still deployed. |
| IA06 | 75 | Deployed as of 1972. |
| JWS | 81 | In service for some time. |

## DEPLOYMENT COMPLETE

| | | |
|---|---|---|
| JCS | 72 | Deployment complete (probably). |
| SecC | 72 | No construction starts on standard sites during the past year. |
| SecC | 73 | Deployment program appears completed. |
| SecD | 73 | Deployment appears completed. |

## DEPLOYMENT HISTORY

| | | |
|---|---|---|
| SecC | 69 | Still being in flight testing might explain why silo construction relatively slow and only at one complex. Perhaps Soviets plan only limited deployment at hard sites, giving preference to later mobile deployment. |
| SecC | 70 | Deployment rate for 1969 was the same as past two years, much slower than SS-9 and SS-11. |
| SecD | 70 | Small increase predicted by mid-1970. |
| SecC | 71 | Deployment rate [about 20 per year] continues as it has for past 4 years. As many as 20 additional launchers expected completed by mid-year. |
| SecD | 71 | Deployment rate continues as it has the past 4 years; some indication it may be slowing. Deployment continues, but at a reduced pace. |
| SecD | 72 | Very little construction in past year on standard SS-13 sites. |
| JCS | 74 | SS-X-16 is likely successor, FY75-76. |
| JCS | 76 | Estimated that by 1977, all SS-13s will be replaced by SS-X-16. |
| MOW | 76 | SS-13s are expected to be replaced by 1977. |
| SWM | 77 | Likely to be superceded by SS-16. Compares with Minuteman I, but never widely deployed. |
| Rock | 78 | Will eventually be replaced by "SS-16 models". |
| FI* | 79 | To be replaced by SS-16. |
| JCS | 79 | Little doubt SS-16 developed in part as SS-13 replacement. |
| JCS | 81 | Little change in program in past year, FY82-83. |
| JWS | 81 | Never deployed statically on a large scale. |
| AWST | 82 | Limited deployment SS-11 replacement. |
| SMP | 83 | Numbers will remain unchanged as current ICBM modernization is completed. |
| JCS | 84 | [SS-11,13,17] expected to be replaced with new ICBMs like SS-X-24,25. FY84-85. |
| SMP | 85 | Will be deployed through 1990. [From graph.] |

## NUMBERS IN SERVICE

### NUMBER IN 1960

| | | | |
|---|---|---|---|
| SSF | 82 | none | In 1960 and 1965. |

### NUMBER IN 1969

| | | | |
|---|---|---|---|
| SecC | 70 | c. 10-20 | 1 Sept. |

### NUMBER IN 1970

| | | | |
|---|---|---|---|
| SecC | 69 | | 30-50 expected installed in hard silos by mid-70. |
| IISS | 70 | 40 | July, total deployed. |
| SecC | 70 | 20 | Operational in hard single silos in Jan 1970. 40 expected operational by mid-70. |
| SecD | 70 | | Some deployed. 160 new ICBMs in year before 1 Sept 1969; most were SS-9s and SS-11s; remainder were SS-13s. |
| SecC | 71 | 40 | Operational in hard single silos, 30 Dec 1970 and 1 Mar 1971. |
| SecC | 72 | 20 | Operational in 1970. |
| Clns | 80 | 20 | |
| SSF | 82 | 40 | |
| vanD | 83 | 20 | 1 Jan. |

### NUMBER IN 1971

| | | | |
|---|---|---|---|
| SecC | 70 | | 50-60 expected operational by mid-71. |
| SecC | 71 | | 60 expected operational in mid-71. |
| Clns | 80 | 40 | |
| vanD | 83 | 40 | 1 Jan. |

BALLISTIC MISSILES  ICBMs  SS-13  173

```
---------- NUMBER IN 1972 ----------
SecC  68           250-300 by mid-1972 (possibility), including as many as 100 in mobile mode, in addition
                       to 560-610 SS-11s.
SecC  71           60-80 expected operational in mid-72.
SecC  72           60 operational in 1972.
JCS   73           1096 (SS-11s and SS-13s), launchers operational, under construction or conversion, "plus
                       new silos" (small silos); mid-1972.
Clns  80    60     1972-1979.
vanD  83    60     1972-81, 1 Jan.

---------- NUMBER IN 1973 ----------
IISS  73    60     1973-81, July, total deployed.

---------- NUMBER IN 1974 ----------
JCS   74    60     Only 60 launchers "have been deployed". 1096 (SS-11s and SS-13s) launchers associated
                       with SAL; 66 estimated under construction on date Interim Agreement signed, 1030
                       launchers operational.

---------- NUMBER IN 1975 ----------
JCS   75    60     Currently deployed, FY76-77.
SWM   77    60     In service mid-1975, based Plesetsk.
SSF   82    60     In 1975 and 1980.

---------- NUMBER IN 1976 ----------
FIar  76    60     In silos at Plesetsk.
MOW   76    60     In service.

---------- NUMBER IN 1977 ----------
Nitz  79    60     Launchers.

---------- NUMBER IN 1978 ----------
Rock  78    60     Only 60 "silo launched models" deployed to date.

---------- NUMBER IN 1979 ----------
FI*   79    60     All at Plesetsk.

---------- NUMBER IN 1980 ----------
AWar  80    60     Deployed.
Clns+ 83    60     In 1980-1982.

---------- NUMBER IN 1981 ----------
JWS   81    50     Deployed, early 1981.
SecD  81    50     In current ICBM force.
SMP   81    60     '81, '83, '84, '85.

---------- NUMBER IN 1982 ----------
AFM   82    only 60
JCS   82    60     Inventory.
JWS+  83    60     Early 1982.

---------- NUMBER IN 1983 ----------
CBO   83    60     1983, 1990, 1996.

---------- NUMBER IN 1984 ----------
JCS   84    60     As of 1 Jan.

---------- NUMBER IN 1985 ----------
JCS   85    60     As of 1 Jan.

---------- UNDATED NUMBER ----------
SSF   82    60     "Number of missiles deployed".

---------- LAUNCHER/SILO ----------
SecC  68     "Might be deployed in both fixed (hard) and mobile modes".
FIar  76     60 in silos at Plesetsk.
Rock  78     All 60 deployed in silos are around Plesetsk.
JCS   79     Silo-based.
SSF   82     Basing mode: Hardened silo.

---------- CARRIERS ----------

---------- CARRIER ---------- TUBES ---- RELOADS ----------
SecC  68                                      "Might be deployed in both fixed (hard) and mobile modes".
SecC  69                                      Might be deployed in mobile mode, perhaps as early as
                                                  mid-71.

---------- DESIGNER ----------
SSF   82     Nadiradize   Design Bureau.
```

------- **DESIGN AND ENGINEERING** -------
| | | |
|---|---|---|
| SecD | 73 | SS-13 follow-on being developed and tested [not SS-13]. |
| FIar | 76 | Developed in parallel with SS-11. Project not very successful, "several reports telling of technical difficulties with the guidance and propulsion systems". |
| SWM | 77 | Developed in parallel with better-performance SS-11. |
| FI* | 79 | Developed in parallel with SS-11. Seems to have suffered technical problems with guidance and propulsion. |

------- **USE AND CONFIGURATION** -------
| | | |
|---|---|---|
| CBO | 83 | System availability: 0.85. |
| vanD | 83 | Reliability: .80; Penetration 1.0. 10 missiles in a battalion. |

------- **RELOADS IN GENERAL** -------
| | | |
|---|---|---|
| Clns | 80 | Cold Launch: no. |
| SSF | 82 | Launching mode: Hot. |

------- **STRATEGY** -------
| | | |
|---|---|---|
| SecC | 69 | SS-13 in fixed mode does not appear to offer "any particular performance advantages over the SS-11". |

------- **COMBAT REPORTS/EFFECTIVENESS** -------
| | | |
|---|---|---|
| Rock | 78 | As "first solid propellant ICBM to enter operational service", achieved "certain historical distinction, but very little else about its operational deployment is complimentary". |

------- **GEOGRAPHICAL DEPLOYMENTS** -------
| | | | |
|---|---|---|---|
| FIar | 76 | | Stationed in silos at Plesetsk since 1971. |
| | | Plesetsk | 60 in silos. |
| SWM | 77 | Plesetsk | |
| Rock | 78 | | All 60 deployed in silos are around Plesetsk. |
| FI* | 79 | Plesetsk | All 60. |
| JWS | 81 | | Most of 50 believed based in 2 areas: region around Ivanovo, northeast of Moscow; and near Yoshkar Ola, between Gorki and Kazan. |
| SMP | 83 | | [Shown on map at] Yoshkar Ola, '83, '84, '85. |

------- **RELATION TO OTHER MISSILES** -------
| | | |
|---|---|---|
| SecC | 68 | About the size of SS-11. |
| SecC | 69 | Appears no better than earliest Minuteman missiles. Also [in addition to SS-14] possible Soviets will use 1st and 3rd stages of SS-13 to achieve new IRBM with 3,000 nm range. No flight tests yet, but some evidence will start soon. Could achieve IOC in 1970-71 and could be deployed in both fixed and mobile modes. |
| JCS | 72 | Larger than Minuteman. |
| MOW | 76 | Russian counterpart of Minuteman. |
| SWM | 77 | Compares with Minuteman I. |
| JWS | 81 | Said to be comparable to US Minuteman. SS-16 resembled SS-13 closely. |
| AFM | 82 | In Minuteman category. |

------- **OTHER INFORMATION** -------
| | | |
|---|---|---|
| SuO1 | 83 | "Not considered satisfactory", [with SS-14 and SS-15]. |

# SS-13 Mod 1

Sources are:   AWar 80     IISS 82     SSF 82     SMP 85     BW 86

---------- US DESIGNATION ----------
| | | |
|---|---|---|
| IISS | 82 | SS-13 Mod 1 |
| SSF | 82 | SS-13 Mod 1 |

---------- NATO CODENAME ----------
| | | |
|---|---|---|
| IISS | 82 | Savage |

---------- RANGE ----------

| | | mi | nm | km | km conv | |
|---|---|---|---|---|---|---|
| IISS | 82 | | | 10,000 | 10,000 | Max. |
| BW | 86 | | | 9,400 | 9,400 | |

---------- CEP ----------

| | | ft | mi | nm | m | km | m conv | |
|---|---|---|---|---|---|---|---|---|
| AWar | 80 | | | 1 | | | 1,852 | [Basic version] |
| IISS | 82 | | | | 2,000 | | 2,000 | |
| SSF | 82 | | | 1.0 | | | 1,852 | |
| BW | 86 | | | | 1,900 | | 1,900 | |

---------- WARHEAD CHARACTERISTICS ----------
| | | |
|---|---|---|
| IISS | 82 | single RV |
| BW | 86 | single RV |

---------- NUCLEAR YIELD ----------

| | | kT | MT | kT conv | |
|---|---|---|---|---|---|
| IISS | 82 | 750 | | 750.00 | |
| SSF | 82 | | 0.6 | 600.00 | Mods 1 and 2, per warhead. |
| BW | 86 | 600 | | 600.00 | |

---------- THROWWEIGHT ----------

| | | lb | t | kg | kg conv |
|---|---|---|---|---|---|
| IISS | 82 | 1,000 | | | 453.60 |

---------- TYPE OF PROPULSION ----------
| | | |
|---|---|---|
| BW | 86 | solid |

---------- PROTOTYPE TESTS ----------
| | | | |
|---|---|---|---|
| SSF | 82 | 1965-69 | First flight test. |

---------- DEPLOYMENT BEGAN ----------
| | | | |
|---|---|---|---|
| IISS | 82 | 1968 | First year deployed. |
| SSF | 82 | 1967-69 | Year operation began. |
| SMP | 85 | 1979 | [Presumed Mod 1, from graph.] |
| BW | 86 | 1969 | |

---------- RETIRED ----------
| | | |
|---|---|---|
| BW | 86 | ? 1975 |

---------- NUMBERS IN SERVICE ----------

---------- NUMBER IN 1982 ----------
| | | | |
|---|---|---|---|
| IISS | 82 | 60 | 1982-83, July, total deployed. |

---------- UNDATED NUMBER ----------
| | | | |
|---|---|---|---|
| BW | 86 | 60 | Peak. |

---------- RELOADS IN GENERAL ----------
| | | |
|---|---|---|
| BW | 86 | No. |

# SS-13 Mod 2

Sources are: JCS 73  AWar 80  SSF 82  SMP 84  SMP 85  BW 86

## US DESIGNATION
SSF 82  SS-13 Mod 2

## RANGE

|  | mi | nm | km | km conv |  |
|---|---|---|---|---|---|
| AWar 80 |  |  | 9,400 | 9,400 | Improved version. |
| SMP 84 |  |  | 9,400 | 9,400 | Max, '84, '85. |
| BW 86 |  |  | 9,400 | 9,400 |  |

## CEP

|  | ft | mi | nm | m | km | m conv |  |
|---|---|---|---|---|---|---|---|
| JCS 73 |  |  |  |  |  |  | New version is more accurate than Mod 1, but strictly a soft-target weapon due to yield. |
| AWar 80 |  |  | 0.82 |  |  | 1,518 | Improved version. |
| SSF 82 |  |  | 0.82 |  |  | 1,518 |  |
| BW 86 |  |  |  | 1,500 |  | 1,500 |  |

## WARHEAD CHARACTERISTICS
SMP 84  single RV  '84, '85
BW 86  single RV

## NUCLEAR YIELD

|  | kT | MT | kT conv |  |
|---|---|---|---|---|
| JCS 73 |  |  |  | New version has relatively small warhead [generally, not relative to Mod 1, ed]. |
| SSF 82 |  | 0.6 | 600.00 | Mods 1 and 2, per warhead. |
| BW 86 | 600 |  | 600.00 |  |

## LENGTH

|  | in | ft | cm | m | m conv |  |
|---|---|---|---|---|---|---|
| SMP 84 |  |  |  | 20 | 20.00 | [From graph, '84, '85.] |

## TYPE OF PROPULSION
BW 86  solid

## PROTOTYPE TESTS
JCS 73         Flight testing continued in 1972 on a very modest scale.
SSF 82  1970   First flight test.

## DEPLOYMENT BEGAN
SSF 82  1972  Year operation began.
SMP 84  1973  Initial deployment, '84, '85.
SMP 85  1979  [From graph.]
BW 86   1973

## DEPLOYED
BW 86         In service.

## DEPLOYMENT HISTORY
JCS 73  New version may be ready for deployment.
SMP 85  Will be deployed through 1990. [Presumed Mod 2, from graph.]

## NUMBERS IN SERVICE

## NUMBER IN 1984
SMP 84  60

## NUMBER IN 1985
SMP 85  60

## UNDATED NUMBER
BW 86   60  Peak.

## LAUNCHER/SILO
SMP 84  Housed in less survivable silos [than SS-17,18,19], '84, '85.

## RELOADS IN GENERAL
SMP 84  Hot launch, '84, '85.
BW 86   No.

BALLISTIC MISSILES  ICBMs  SS-16  177

# SS-16

```
Sources are:    GAO1  75    JCS   76    SWM   77    SecD  79    FI    81    AWST  82
JCS   73        IAO6  75    MOW   76    CBO1  78    USO8  79    JCS   81    SSF   82
SecD  73        JCS   75    SecC  76    JCS   78    AWar  80    JWS   81    SecD  84
JCS   74        SecC  75    SecD  76    Rock  78    Clns  80    SecD  81    SMP   84
SecC  74        SecD  75    JCS   77    SecD  78    JCS   80    SMP   81    SMP   85
SecD  74        FIar  76    SecD  77    JCS   79    SecD  80    AFM   82
```

---------- **US DESIGNATION** ----------

| | | | |
|---|---|---|---|
| SecD | 73 | SS-X-16 | FY74-78. |
| SecC | 74 | SS-16 | |
| JCS  | 75 | SS-X-16 | FY76-78. |
| FIar | 76 | SS-X-16 | [Text] |
|      |    | SS-16   | [Table] |
| MOW  | 76 | SS-X-16 | |
| SecD | 77 | SS-16   | FY78-82,85. |
| SWM  | 77 | SS-16   | |
| JCS  | 78 | SS-16   | FY79-82. |
| Rock | 78 | SS-16   | |
| AWar | 80 | SS-16   | |
| Clns | 80 | SS-16   | |
| FI   | 81 | SS-16   | |
| JWS  | 81 | SS-16   | |
| SMP  | 81 | SS-16   | '81, '84, '85. |
| AFM  | 82 | SS-X-16 | |
| AWST | 82 | SS-16   | |
| SSF  | 82 | SS-X-16 | |

---------- **SOVIET DESIGNATION** ----------

FI    81    RS-14

---------- **DESCRIPTION** ----------

| | | | |
|---|---|---|---|
| JCS  | 74 |      | Listed as significant initiative, strategic offensive systems. |
| FIar | 76 |      | 3rd generation. |
| MOW  | 76 | ICBM | |
| Rock | 78 |      | First in series of 3rd generation ICBMs. |
| AWar | 80 |      | Mobile ICBM. |
| JWS  | 81 | ICBM | 2nd generation, 'light' ICBM. |
| SMP  | 81 |      | Small ICBM. |
| SSF  | 82 |      | Fourth generation. |

---------- **EDITOR'S NOTES** ----------

| | | |
|---|---|---|
| JCS  | 73 | Listed FY74-82, but not by name in FY74. |
| SecD | 73 | Listed FY74-82,85. |
| SecC | 74 | Listed FY75-77. |
| JWS  | 81 | Final report of SS-16; not reported in 1983. |
| SMP  | 81 | Listed '81, '84, '85. |
| AFM  | 82 | Final report. |

---------- **RANGE** ----------

| | | mi | nm | km | km conv | |
|---|---|---|---|---|---|---|
| JCS  | 74 |        | > 5000 |        | 9,260  | [Also] exceeds SS-13 range. |
| SecD | 75 |        |        |        |        | About the same range as SS-13. |
| FIar | 76 |        |        | 9,000  | 9,000  | |
| SWM  | 77 | 5,600  |        | 9,000  | 9,010  | |
| Rock | 78 |        |        | 9,500  | 9,500  | [Table] |
|      |    |        |        | 9,600  | 9,600  | [Text] |
| AWar | 80 |        |        | 9,150  | 9,150  | |
| Clns | 80 |        | 5,450  |        | 10,093 | |
| SecD | 80 |        |        | 9,200  | 9,200  | Exclusive of range imparted by PBV. |
| JWS  | 81 |        |        | > 8,000| 8,000  | Said to be greater than SS-13's range. |
| AFM  | 82 | > 5,000|        |        | 8,045  | |
| AWST | 82 |        | 5,000  |        | 9,260  | |
| SSF  | 82 |        | 4,970  |        | 9,204  | |
| SMP  | 84 |        |        | 9,000  | 9,000  | Max, '84, '85. |

178  SOVIET MISSILES  IDDS

## CEP

|  |  | ft | mi | nm | m | km | m conv |  |
|---|---|---|---|---|---|---|---|---|
| JCS | 75 |  |  |  |  |  |  | Accuracy appreciably [or much] greater than SS-13, FY76-77. |
| SecC | 75 |  |  |  |  |  |  | Would not have significant hard target capability. |
| Flar | 76 |  |  |  |  |  |  | Better than Savage [SS-13, ed]. |
| SecC | 76 |  |  |  |  |  |  | System accuracy considerably better than presented in last year's estimates; apparently because older guidance system components used in earlier tests replaced with much improved components in all the more recent tests. |
| CBO1 | 78 | 1200-1500 |  |  |  |  | 365.76-457.20 | Estimated for new generation of ICBMs in 1985 [SS-16, SS-17, SS-18, SS-19, ed]. |
| AWar | 80 |  |  | 0.26 |  |  | 481.52 |  |
| Clns | 80 |  |  | 0.26 |  |  | 481.52 |  |
| SSF | 82 |  |  | 0.26 |  |  | 481.52 |  |

## WARHEAD TYPE

|  |  | nuclear | conven | chem |
|---|---|---|---|---|
| JWS | 81 | yes |  |  |

## WARHEAD CHARACTERISTICS

| JCS | 74 |  | MIRV probable, but tested so far only with single RV. Indications that MIRV is planned. Has digital computer. |
|---|---|---|---|
| SecD | 74 |  | Employs a PBV, FY75-77, 79-81. Has been tested thus far with only one RV, but PBV strongly suggests future MIRV. |
| JCS | 75 | single RV | FY76-79. |
|  |  |  | Has PBV associated with MIRVs, but tested to date only with single RV, FY76-79. |
| SecC | 75 |  | The use of a bus may simply represent the Soviet solution to the accuracy problems encountered with the SS-13, as well as provide the additional energy required for range flexibility. |
| SecD | 75 |  | "Cannot preclude" possibility of eventual MIRV mode deployment. So far tested only with single RV. |
| Flar | 76 | single RV<br>3 MRVs | Final stage carries PBV with on-board digital computer, enabling MIRVs. So far only MRVs flown on the missile. |
| MOW | 76 |  | Uses a computer-operated PBV. Tested so far only with single RV. |
| SecD | 76 | single RV | FY77-81. |
|  |  |  | Presently no evidence it has been tested with MIRV. |
| SecD | 77 |  | Probably will have single RV. |
| SWM | 77 | single RV or MIRVs | MIRVs from computer-controlled PBV. |
| Rock | 78 | 3 x MRV | MRV used in tests; but deployed version may have MIRV. Uses post-boost maneuvering stage and onboard computer, like all other 3rd generation missiles. |
| SecD | 78 | single RV | Currently carries a single warhead, FY79-81. |
| USO8 | 79 | single RV |  |
| AWar | 80 | single RV |  |
| Clns | 80 | single RV |  |
| FI | 81 | single RV |  |
| JWS | 81 |  | PBV suggests MIRVs [official US statement], but testing to date suggests far off if at all. |
| AFM | 82 |  | Fitted with PBV, like US bus, but tested only with single RV. |
| AWST | 82 |  | MIRV capability. |
| SSF | 82 | single RV |  |
| SMP | 84 | single RV | '84, '85. |

## NUCLEAR YIELD

|  |  | kT | MT | kT conv |  |
|---|---|---|---|---|---|
| Flar | 76 |  | 1 | 1,000 | Single RV version, [table] |
|  |  |  | ≥ 1 | 1,000 | Single RV version, [text] |
| SWM | 77 |  | 1+ | 1,000 | Single RV. |
| CBO1 | 78 |  | 1.0 | 1,000 | Estimated, 1985. |
| Rock | 78 |  | ? 1.0 | 1,000 |  |
| USO8 | 79 |  | 1 | 1,000 |  |
| AWar | 80 |  | 0.65 | 650.00 |  |
| Clns | 80 | 650 |  | 650.00 |  |
| JWS | 81 |  | > 1 | 1,000 | Probably. |
| SSF | 82 |  | 0.65 | 650.00 |  |

BALLISTIC MISSILES  ICBMs  SS-16  179

|  |  | THROWWEIGHT |  |  |  |  |  |
|---|---|---|---|---|---|---|---|
|  |  | lb | t | kg | kg conv |  |  |
| JCS | 75 |  |  |  |  | Appreciably [or much] greater than SS-13, FY76-77. |  |
| SecD | 75 |  |  |  |  | About twice that of SS-13. |  |
| Flar | 76 |  |  | 900 | 900.00 | [Also] twice that of Savage. |  |
| Rock | 78 |  | < 1 |  | 1,000 | "Nearly" 1 tonne; [also] twice that of SS-13. |  |
| USD8 | 79 | 2,000 |  |  | 907.20 |  |  |
| Clns | 80 | 2,000 |  |  | 907.20 |  |  |
| JWS | 81 |  |  |  |  | Much greater than SS-13. |  |
| SSF | 82 | 2,000 |  |  | 907.20 |  |  |

|  |  | PAYLOAD |  |  |  |  |
|---|---|---|---|---|---|---|
|  |  | lb | t | kg | kg conv |  |
| JCS | 74 |  |  |  |  | Exceeds SS-13. |

|  |  | WEIGHT |  |  |  |
|---|---|---|---|---|---|
|  |  | lb | t | kg | kg conv |
| FI | 81 |  |  | c. 36,000 | 36,000 |

|  |  | LENGTH |  |  |  |  |  |
|---|---|---|---|---|---|---|---|
|  |  | in | ft | cm | m | m conv |  |
| Flar | 76 |  |  |  | 20 | 20.00 |  |
| SWM | 77 |  | 65.6 |  |  | 19.99 |  |
| Rock | 78 |  |  |  | 20 | 20.00 |  |
| FI | 81 |  |  |  | c. 20 | 20.00 |  |
| JWS | 81 |  |  |  | c. 20 | 20.00 |  |
| AWST | 82 |  | 65 |  |  | 19.81 |  |
| SMP | 84 |  |  |  | 18 | 18.00 | [From graph], '84, '85. |

|  |  | DIAMETER |  |  |  |  |
|---|---|---|---|---|---|---|
|  |  | in | ft | cm | m | cm conv |
| Flar | 76 |  |  |  | 2.1 | 210.00 |
| SWM | 77 |  | 6.5 |  | 2.0 | 198.12 |
| Rock | 78 |  |  |  | 2.1 | 210.00 |
| FI | 81 |  |  | ?c 170 |  | 170.00 |
| AWST | 82 |  | 6.5 |  |  | 198.12 |

|  |  | NUMBER OF STAGES |  |
|---|---|---|---|
| JCS | 77 | 3 | FY78-80. |
| SWM | 77 | 3 |  |
| Rock | 78 | 3 |  |
| SecD | 78 | 3 | FY79-81. |
| FI | 81 | 3 |  |
| JWS | 81 | 3 |  |
| AWST | 82 | 3 |  |
| SMP | 84 | 3 | '84, '85. |

|  |  | TYPE OF PROPULSION |  |
|---|---|---|---|
| JCS | 74 | solid | FY75-77. |
| SecC | 74 |  | PBV is powered by small solid fuel motors. |
| SecD | 74 | solid | FY75-76,79-81. |
| Flar | 76 | solid |  |
| MOW | 76 | solid |  |
| SWM | 77 | solid | Pentagon sources describe SS-16 as slightly smaller than SS-13, but with twice the throw weight per given range; the advance implied by this may show up in other missiles too. |
| Rock | 78 | solid | Stages 1, 2, and 3. |
| Clns | 80 | solid |  |
| FI | 81 | solid |  |
| JWS | 81 | solid |  |
| SMP | 81 | solid | '81, '84, '85. |
| AFM | 82 | solid | Only solid-propellant missile in USSR's new generation. |
| AWST | 82 | solid |  |
| SSF | 82 | solid |  |

|  |  | MIDCOURSE (OR MAIN) GUIDANCE |  |
|---|---|---|---|
| JCS | 75 |  | "Advanced navigation guidance system", FY76-77. |
| FI | 81 | inertial |  |
| JWS | 81 | inertial | "Advanced navigation guidance system" [official US statement]. |
| AWST | 82 |  | Advanced navigation. |
| SSF | 82 |  | Fly-the-wire, onboard digital computer. |

|  |  | CONTROL SYSTEM |
|---|---|---|
| FI | 81 | ? Fluid injection. |

---------- **FIRST OBSERVED** ----------
| | | | |
|---|---|---|---|
| FIar | 76 | 1972-73 | First detected by US reconnaissance. |
| SWM | 77 | 1972-73 | Detected by US space reconnaissance. |
| Rock | 78 | 1972 | First appeared. |

---------- **DEVELOPMENT BEGAN** ----------
| | | | |
|---|---|---|---|
| SSF | 82 | 1965 | Year design began. |
| SMP | 85 | 1964 | Technological development began. [From graph.] |

---------- **PROTOTYPE TESTS** ----------
| | | | |
|---|---|---|---|
| JCS | 73 | | 'SS-13 follow-on' being actively tested. |
| SecD | 73 | | USSR is developing and testing it. |
| SecD | 74 | | Tested during the past year. |
| JCS | 75 | Dec 19[74] | Two 4200 nm launches to Pacific were made. |
| FIar | 76 | | So far only MRVs flown on the missile. |
| MOW | 76 | | Tested so far only with single RV. |
| SecD | 76 | | SS-X-16 and its SS-X-20 derivative continue testing, but "emphasis" recently on SS-X-20. |
| | | 1974-75 | In 1974, extensive flight testing; continued in 1975. |
| SecD | 79 | | Tested only once since 1975. |
| SecD | 80 | | Tested only once since 1975, and unsuccessfully then. |
| JWS | 81 | 1972-74 | US reports only one (unsuccessful) test flight since 1975. |
| SSF | 82 | 1972 | First flight test. |
| SMP | 84 | 1972 | First test; last known test in 1976, '84, '85. |
| SMP | 85 | 1971-72 | Engineering and testing began. [From graph.] |

---------- **IOC** ----------
| | | | |
|---|---|---|---|
| JCS | 74 | 1975 | FY75-76,78. |
| JCS | 76 | | [Blank or unknown], FY77,79. |
| SecD | 77 | beyond 1975 | Unknown, but beyond 1975. |

---------- **DEPLOYMENT BEGAN** ----------
| | | | |
|---|---|---|---|
| JCS | 74 | | Could be ready to deploy by 1975, but MIRV version would need high priority testing program soon to meet that goal. |
| SecD | 74 | | Land-mobile version may be under development; might eventually be deployed, FY75-76. |
| IA06 | 75 | 1975 | Deployment. |
| JCS | 75 | | MIRV version may be deployed, but 1 year or more after testing begins. "Single warhead version probably may be operational this year [sic]". |
| SecC | 75 | | Can be deployed in SS-13 silos without any major external modifications. Absence of any evidence of [SS-13] silo modification would not preclude deployment. |
| SecD | 75 | | "We have evidence that all four of these new ICBMs have started, or soon will start, operational deployment". |
| FIar | 76 | | Not yet in service. |
| JCS | 76 | | MIRV version would require a considerable amount of flight testing. |
| SecC | 76 | | Lack of deployment may reflect Soviet decision to hold further development of SS-X-16 pending outcome of mobile ICBM issue in SALT. |
| SecD | 76 | | "Probably capable of being deployed at any time". |
| JCS | 77 | | Last year's assessment that deployment was imminent—has not occurred [sic]. |
| Rock | 78 | 1977 | Service intro. |
| | | | Developed, but not yet deployed. |
| | | | "Not as yet deployed" [early in text]; "SS-16 has already been deployed in small numbers in the mobile configuration" [later in text]. |
| JCS | 79 | | "Considerable uncertainty exists" regarding status of "the silo-based version of the SS-16". |
| AWar | 80 | | None deployed in mobile mode (according to US analysts). |
| Clns | 80 | 1978 | First deployed. |
| SecD | 80 | | None deployed. |
| FI | 81 | | US DoD claims never deployed. |
| JWS | 81 | | Not deployed by mid-1981 (banned under SALT II). |
| | | | Deployment was expected 1975. |
| SecD | 81 | | Developed but not deployed. |
| SMP | 81 | | "The system was never deployed". |
| SecD | 84 | 1976-77 | Introduction of strategic systems [graph]. "Experimental system, with operational capability". Soviets "probably are deploying SS-16 missiles in violation of SALT II". |
| SMP | 84 | | Soviets claim it has not been deployed, '84, '85. ["Number deployed" given as "operationally capable"]. "Available information does not allow a conclusive judgment on whether the Soviets deployed the SS-16, but does indicate probable deployment". |
| SMP | 85 | | "While the evidence is somewhat ambiguous, it indicates that the SS-16 activities at Plesetsk are a probable violation of SALT II, which banned SS-16 deployment." "The 1985 report [second US government report on Soviet noncompliance, ed.] also reaffirmed that the Soviet Union has probably violated the SS-16 deployment prohibition of SALT II." ["Number deployed" given as "undetermined"]. |
| | | 1976-77 | [From graph.] |

## DEPLOYMENT HISTORY

| | | |
|---|---|---|
| SMP | 85 | Will be deployed at least through 1990. [From graph.] |

## NUMBERS PRODUCED

| | | | |
|---|---|---|---|
| FIar | 76 | | "Some US defence officials believe that it has been manufactured in quantity over the last year and stockpiled pending a deployment decision". |
| Clns | 80 | c. 60 | "The Soviets have produced perhaps 60 SS-16s, but none are counted as deployed". |

## LAUNCHER/SILO

| | | |
|---|---|---|
| SecD | 75 | Would be deployed first in silos, only later in land-mobile mode. |
| FIar | 76 | Mobile or hot-launched from silos. |
| SecD | 76 | Could be either "silo-based or mobile. |
| SWM | 77 | Could be deployed in fixed or mobile launchers. |
| Rock | 78 | Could be silo launched, or mobile. |
| SecD | 78 | Intended as land-mobile; could be deployed in silos. |
| JCS | 79 | "Considerable uncertainty exists" regarding status of "the silo-based version of the SS-16". |
| JWS | 81 | Can't tell form of deployment, or if to be employed at all. Official US opinion says either mobile or silo-based, opinion "hardening to" mobile mode. |
| SSF | 82 | Basing mode: Mobile and hardened silo. |
| SMP | 84 | Probably intended originally for both silo and mobile deployment, '84, '85. |

## CARRIERS

| | | CARRIER | TUBES | RELOADS | |
|---|---|---|---|---|---|
| JCS | 74 | | | | No direct evidence of mobile mode of deployment. "Indications suggest" development includes land-mobile deployment option. |
| SecD | 74 | | | | Land-mobile version may be under development; might eventually be deployed, FY75-76. |
| JCS | 75 | | | | "Circumstantial evidence" indicates "being developed as a land-based mobile system", FY76-77. |
| SecD | 75 | | | | Would be deployed first in silos, only later in land-mobile mode. |
| FIar | 76 | | | | Can be placed on mobile launchers or in silos. |
| MOW | 76 | | | | ? Being developed in land-mobile form. |
| SecD | 76 | | | | Could be either "silo-based or mobile. |
| JCS | 77 | | | | No evidence in past year against idea of mobile version of SS-16. Was developed with mobile deployment capabilities, FY78-80. |
| SecD | 77 | | | | Soviets could deploy SS-X-16 in land-mobile mode to replace or add to SS-13. |
| SWM | 77 | | | | Could be deployed in fixed or mobile launchers. |
| JCS | 78 | | | | "Apparently decided not to deploy it in its mobile mode at this time" [direct quote only for FY79], FY79-80. |
| Rock | 78 | | | | Could be silo launched, but SS-16 "can be deployed in a land mobile system which would undoubtedly be called upon to carry heavier payloads over a shorter distance". |
| SecD | 78 | | | | Intended as land-mobile; could be deployed in silos. |
| SecD | 79 | | | | Designed as a mobile system, but not deployed that way, FY80-81. |
| AWar | 80 | | | | Mobile ICBM. |
| JCS | 80 | | | | Soviets "have tested a land-mobile" SS-16. |
| FI | 81 | | | | Can be used in a mobile role. |
| JWS | 81 | | | | Can't tell form of deployment, or if to be employed at all. Official US opinion says either mobile or silo-based, opinion "hardening to" mobile mode. |
| SMP | 81 | | | | "Developed by the Soviets in the early 1970s for mobile deployment. The system was never deployed". |
| AWST | 82 | | | | Mobile system. |
| SSF | 82 | | | | Basing mode: Mobile and hardened silo. |
| SMP | 84 | | | | Probably intended originally for both silo and mobile deployment, using equipment and basing arrangement comparable to that used with SS-20, '84, '85. |

## DESIGNER

| | | | |
|---|---|---|---|
| SSF | 82 | Nadiradize | Design Bureau. |

## DESIGN AND ENGINEERING

| | | |
|---|---|---|
| JCS | 75 | Apparently some difficulty with SS-X-16; believed now overcome. |
| JCS | 76 | Development phase essentially completed. |
| MOW | 76 | Under development. |
| JCS | 77 | Development phase essentially completed by the end of 1975. |
| JCS | 78 | Development was essentially complete by the end of 1974. Status of SS-16 program "is unclear". |
| SecD | 78 | Development essentially completed. |
| JCS | 81 | Little or no change in program status over last year. |
| JWS | 81 | In advanced development stage. |
| SMP | 84 | Development of SS-16 founded on technologies represented by SS-X-14 and SS-X-15. |

------- **RELOADS IN GENERAL** -------
GAO1 75  'Cold Launch' technique.
SecD 75  Hot-Launched, FY76-77,81.
FIar 76  Hot-launched.
SWM  77  Hot-launched from silo.
Rock 78  Hot-launch, in silo-launched mode.
Clns 80  Cold Launch: no.
JWS  81  Hot Launch.
SSF  82  Launching mode: Hot.
SMP  84  Cold Launch, '84, '85.

------- **STRATEGY** -------
JCS  75  [Pp 14-15 of FY76 give extensive information on relation of SS-16 to arms control, ed].
         Expected to replace SS-13.
FIar 76  Replacement for SS-13.
MOW  76  Intended to replace the 60 SS-13s.
SecD 76  Because smaller and single RV, less threat to Minuteman than SS-17,18,19.
JCS  77  Upgrading of SS-X-20 to SS-X-16 could be done "relatively quickly"; could "significantly increase"
         ICBMs against US; would degrade peripheral capabilities, FY78-79.
SecD 77  Soviets could deploy SS-X-16 in land-mobile mode to replace or add to SS-13.
Rock 78  Designed to replace SS-13. "Can be deployed in a land mobile system which would undoubtedly be
         called upon to carry heavier payloads over a shorter distance". Specialists believe mobile
         configuration may be most lethal application.
JCS  79  As part of SALT II, Soviets agreed "not to produce, test, or deploy the SS-16 missile or" its
         "third stage, post-boost vehicle, or reentry vehicle" [direct quote only for FY81], FY80-82. "By
         upgrading deployed SS-20s to the SS-16, the Soviets could achieve a mobile ICBM capability",
         FY80-81. Little doubt SS-16 developed in part as a replacement for silo-based SS-13s.
Clns 80  Moscow agreed not to test, produce, or deploy if SALT II approved.
SecD 80  Production, deployment, and testing expressly banned by SALT II, FY81-82.
FI   81  Developed as replacement for silo-based SS-13s.
JCS  81  Upgrade of SS-20 to SS-16 "rather rapidly"; but would degrade theater arsenal.
JWS  81  Likely successor to SS-13.
SMP  81  "Developed by the Soviets in the early 1970s for mobile deployment".
AFM  82  "It promised particular problems for the US at one time", since storing third stages would allow
         rapid conversion of SS-20s to SS-16 ICBMs.
SMP  84  USSR agreed in SALT II not to produce, test, or deploy SS-16s or key components, '84, '85.

------- **RELATION TO OTHER MISSILES** -------
SecD 73  Follow on to SS-13.
JCS  74  Follow-on to SS-13. Size about the same as SS-13.
SecD 74  Light SS-13-class missile, FY75-76.
SecD 75  May be slightly smaller than SS-13.
JCS  77  SS-20 is first 2 stages of SS-16, FY78-82.
Rock 78  "Believed by some to be slightly smaller than the SS-13".
SecD 78  SS-20 consists of first two SS-16 stages, FY79-80.
AWar 80  First 2 stages used for SS-20.
SecD 80  SS-20 is a derivative of SS-16.
JWS  81  Looks similar to SS-13. First 2 stages are basis of SS-20.
SMP  81  About the size of Minuteman.
AFM  82  About same size as Minuteman.
SMP  84  "The SS-20 LRINF missile is closely related to the SS-16", '84, '85.

------- **OTHER INFORMATION** -------
FIar 76  "The US objected during the Salt 1 talks to any deployment of land-based mobile ICBMs by the USSR,
         but has since unilaterally declared that such systems are acceptable provided no attempt is made
         to conceal the number in service. These objections are difficult to understand in view of the
         existence of the SS-15 mobile system".
JWS  81  US intelligence data "subject to more rather than less security constraints" in published
         information.

# SS-17

| Sources are: | JCS 75 | SecC 76 | IISS 78 | USO8 79 | JCS 81 | JCS 82 | SMP 83 |
|---|---|---|---|---|---|---|---|
| SecD 73 | SecD 75 | SecD 76 | SecD 78 | AWar 80 | JWS 81 | SSF 82 | vanD 83 |
| JCS 74 | FIar 76 | IISS 77 | Rock 78 | Clns 80 | SecD 81 | CBO 83 | AFM+ 84 |
| SecC 74 | IISS 76 | JCS 77 | SecD 78 | IISS 80 | SMP 81 | FI+ 83 | JCS 84 |
| SecD 74 | JCS 76 | SecD 77 | IISS 79 | JCS 80 | AFM 82 | IISS+ 83 | SecD 84 |
| IA06 75 | MOW 76 | SWM 77 | JCS 79 | SecD 80 | AWST 82 | JCS 83 | JCS 85 |
| IISS 75 | NA03 76 | CBO1 78 | Nitz 79 | FI 81 | IISS 82 | JWS+ 83 | SMP 85 |

---------- **US DESIGNATION** ----------

| | | |
|---|---|---|
| SecD 73 | SS-X-17 | FY74-75. |
| JCS 74 | SS-X-17 | |
| | SS-17 | FY76-86. |
| SecC 74 | SS-17 | |
| SecD 75 | SS-17 | FY76-82,85. |
| FIar 76 | SS-17 | |
| MOW 76 | SS-17 | |
| SWM 77 | SS-17 | |
| Rock 78 | SS-17 | |
| Nitz 79 | SS-19/SS-17 | |
| AWar 80 | SS-17 | |
| Clns 80 | SS-17 | |
| FI 81 | SS-17 | |
| JWS 81 | SS-17 | |
| SMP 81 | SS-17 | '81, '83, '84, '85. |
| AFM 82 | SS-17 | |
| AWST 82 | SS-17 | |
| IISS 82 | SS-17 | |
| SSF 82 | SS-17 | |
| CBO 83 | SS-17 | |
| vanD 83 | SS-17 | |

---------- **SOVIET DESIGNATION** ----------

| | |
|---|---|
| FI 81 | RS-16 |
| JWS 81 | ? RS-16 |
| AFM 82 | RS-16 |
| IISS 82 | RS-16 |

---------- **DESCRIPTION** ----------

| | | |
|---|---|---|
| JCS 74 | | Listed as significant initiative, strategic offensive systems. |
| SecD 74 | | Medium missile, FY75-76. |
| FIar 76 | | 3rd generation. |
| MOW 76 | ICBM | |
| Rock 78 | ICBM | |
| JWS 81 | ICBM | |
| AFM 82 | "Light" ICBM | |
| SSF 82 | | Fourth generation. |
| JWS+ 83 | | Fourth generation ICBM. |

---------- **EDITOR'S NOTES** ----------

| | |
|---|---|
| SecD 73 | Listed FY74-82,85. |
| JCS 74 | Listed FY75-86. |
| SecC 74 | Listed FY75-77. |
| SMP 81 | Listed '81, '83, '84, '85. |

---------- **RANGE** ----------

| | mi | nm | km | km conv | |
|---|---|---|---|---|---|
| JCS 74 | | > 5500 | | 10,186 | |
| IA06 75 | | | > 9,000 | 9,000 | |
| FIar 76 | | | 10,000 | 10,000 | |
| NA03 76 | 6,500 | | | 10,458 | |
| SWM 77 | 6,214+ | | 10,000+ | 9,998 | |
| Rock 78 | | | 10,500 | 10,500 | [Qualified in text as:] potential range. |
| FI 81 | | | 10,000+ | 10,000 | |
| AWST 82 | | 5,000 | | 9,260 | |

## SOVIET MISSILES

| | | CEP | | | | | | |
|---|---|---|---|---|---|---|---|---|
| | | ft | mi | nm | m | km | m_conv | |
| SecD | 74 | | | | | | | Clearly designed for higher accuracy; may achieve hard-target capability in "early part of next decade". |
| JCS | 75 | | | | | | | More accurate than SS-11, FY76-77. |
| SecD | 75 | | | | | | | Accuracy worse than SS-19. Certain features more advanced than SS-19, but high accuracy is not a prime objective at present. |
| Flar | 76 | | | | | | | Capable of attacking hard targets. |
| MOW | 76 | | | | | 500 | | 500.00 | "As small as 500" m. Single RV accuracy gives hard-target capability but not MIRV. Employs high-speed RVs, and "guidance refinement may confer a counter-force capability". |
| SecC | 76 | | | | | | | System accuracy considerably better than presented in last year's estimates; apparently because older guidance system components used in earlier tests replaced with much improved components in all the more recent tests. |
| SWM | 77 | | | | | | | "Described as suitable for use against point targets". |
| CBO1 | 78 | 1200-1500 | | | | | 365.76-457.20 | Estimated for new generation of ICBMs in 1985 [SS-16, SS-17, SS-18, SS-19, ed]. |
| Rock | 78 | | | | 600 | | 600.00 | Target accuracy. |
| SecD | 78 | | | | | | | SS-17,18,19 all have "potential, with feasible accuracy improvements, to attain high single-shot kill probabilities against US silos". |
| Nitz | 79 | | | .2 | | | 370.40 | |
| SMP | 81 | | | | | | | Much more accurate than SS-11, but not as accurate as SS-18 and SS-19. |
| CBO | 83 | | | 0.17 | | | 314.84 | In 1983. |
| | | | | 0.14 | | | 259.28 | In 1990. |
| | | | | 0.10 | | | 185.20 | In 1996. |

| | | WARHEAD TYPE | | |
|---|---|---|---|---|
| | | nuclear | conven | chem |
| JWS | 81 | yes | | |

| | | WARHEAD CHARACTERISTICS | |
|---|---|---|---|
| JCS | 74 | MIRV | FY75-77. |
| | | 4 RVs | Estimated. |
| SecD | 74 | | Has PBV, FY75-76. Flown with 4 MIRVs and single RV, FY75-76. |
| IA06 | 75 | 4 MIRVs | |
| JCS | 75 | 4 RVs | Max, FY76-78. MIRVs from PBV, FY76-77. |
| SecD | 75 | 4 RVs | Believed it will be deployed with 4 RVs, FY76-78. |
| Flar | 76 | 4 MIRVs | |
| MOW | 76 | | 1 large RV or 4 MIRVs. Employs high-speed RVs. |
| JCS | 77 | | "If the single RV variant is deployed, the determination of the mix [of single RV and MIRVed missiles] will be difficult". |
| SWM | 77 | 4 MIRVs | "Post-boost 'bus' dispensing 4 x kT MIRVs". |
| JCS | 78 | | All SS-17s will be counted as MIRVed under US SALT II proposals, because SS-17 tested as MIRV. |
| Rock | 78 | 4 x MIRV | Tested with various warhead combinations, including "triple payload multiple re-entry vehicle" and MIRVs. |
| SecD | 78 | | Can carry MIRVs, or single high-yield RV, FY79-80. |
| Nitz | 79 | 1 RV | |
| USO8 | 79 | 4 MIRVs | |
| JWS | 81 | 4-6 MIRVs | On-board computer reported to control RVs. |
| AWST | 82 | 4 MIRVs | |
| CBO | 83 | 4 RVs | |

BALLISTIC MISSILES  ICBMs  SS-17

## NUCLEAR YIELD

| Source | Year | kT | MT | kT conv | Notes |
|---|---|---|---|---|---|
| IAD6 | 75 | 200 | | 200.00 | Each. |
| FIar | 76 | | 1 | 1,000 | Exceeds Minuteman III [per warhead] [170 kT, ed]. |
| SecD | 77 | | | | KT range. |
| SWM | 77 | | | | Estimated, 1985. |
| OBO1 | 78 | | 0.6 | 600.00 | Total [table]. |
| Rock | 78 | | 0.8 | 800.00 | |
| | | 200 | | 200.00 | Each of 4 MIRVs [table; also text, referring to those 20 already deployed]. |
| | | | 1 | 1,000 | Each of 4 MIRVs [text, sic; SS-17 "can deliver" this payload; upper limit?, ed]. |
| Nitz | 79 | | 15 | 15,000 | |
| USO8 | 79 | | .9 | 900.00 | |
| OBO | 83 | 750 | | 750.00 | Each. |

## THROWWEIGHT

| Source | Year | lb | t | kg | kg conv | Notes |
|---|---|---|---|---|---|---|
| SecC | 74 | | | | | 2-3 times SS-11. |
| SecD | 74 | | | | | 3-5 times early model SS-11s. |
| JCS | 75 | | | | | Increased over SS-11, FY76-77. |
| FIar | 76 | | | 2,000 | 2,000 | |
| MOW | 76 | | | | | 3-5 times throwweight of SS-11. |
| SecD | 77 | | | | | Greater than Minuteman III. |
| Rock | 78 | | 2 | | 2,000 | Tonnes. |
| Nitz | 79 | 7,000 | | | 3,175 | |
| USO8 | 79 | 6,000 | | | 2,721 | |
| FI | 81 | | | | | twice SS-11s |
| SSF | 82 | 8,000 | | | 3,628 | |
| OBO | 83 | 6,025 | | | 2,732 | |

## PAYLOAD

| Source | Year | lb | t | kg | kg conv | Notes |
|---|---|---|---|---|---|---|
| SecD | 75 | | | | | 4 times SS-11 Mod 1. |
| AWar | 80 | c. 8,000 | | | 3,628 | |

## WEIGHT

| Source | Year | lb | t | kg | kg conv |
|---|---|---|---|---|---|
| FI | 81 | | | c. 65,000 | 65,000 |

## LENGTH

| Source | Year | in | ft | cm | m | m conv | Notes |
|---|---|---|---|---|---|---|---|
| FIar | 76 | | | | 23 | 23.00 | [Table] |
| | | | | | 25 | 25.00 | [Text] Reported to be 25% longer than SS-11. |
| SWM | 77 | | 80 | | 24.0+ | 24.38 | [+ Sign for m, not ft, ed] |
| Rock | 78 | | | | 23 | 23.00 | [Table] |
| | | | | | c. 24 | 24.00 | [Text] |
| FI | 81 | | | | ? c. 24 | 24.00 | |
| JWS | 81 | | | | ? c. 24 | 24.00 | |
| SMP | 81 | | | | 20 | 20.00 | [From chart], '81, '84, '85. |
| AFM | 82 | | 75'0" | | | 22.86 | |
| AWST | 82 | | 80 | | | 24.38 | |
| AFM+ | 84 | | 66'0" | | | 20.12 | |

## DIAMETER

| Source | Year | in | ft | cm | m | cm conv | Notes |
|---|---|---|---|---|---|---|---|
| FIar | 76 | | | | 2.6 | 260.00 | |
| SWM | 77 | | 8.2 | | 2.5 | 249.94 | |
| Rock | 78 | | | | 2.6 | 260.00 | [Table] |
| | | | | | 2.5 | 250.00 | [Text] |
| FI | 81 | | | ?c 250 | | 250.00 | |
| JWS | 81 | | | | ? c. 2.5 | 250.00 | Base. |
| AFM | 82 | | 8'6" | | | 259.08 | Max. |
| AWST | 82 | | 8.5 | | | 259.08 | |

## NUMBER OF STAGES

| Source | Year | Stages | Notes |
|---|---|---|---|
| FIar | 76 | 2 | |
| SWM | 77 | 2 | |
| Rock | 78 | 2 | Excluding bus. |
| FI | 81 | 2 | |
| JWS | 81 | 2 | |
| AFM | 82 | 2 | |
| AWST | 82 | 2 | |

186  SOVIET MISSILES  IDDS

```
----------  TYPE OF PROPULSION  ----------
SecC   74                          PBV powered by liquid fuel engines.
SecD   74   liquid                 FY75-76,81-82.
JCS    75   liquid                 FY76-77.
FIar   76   storable liquid
MOW    76   liquid
SWM    77   liquid storable
Rock   78   storable liquid        Stages 1 and 2.
FI     81   liquid
JWS    81   storable liquid
AFM    82   liquid
AWST   82   liquid-storable
SSF    82   liquid

----------  MIDCOURSE (OR MAIN) GUIDANCE  ----------
JCS    74                          Has digital computer.
FI     81   inertial
JWS    81   inertial
SSF    82                          Fly-the-wire, onboard digital computer.

----------  CONTROL SYSTEM  ----------
FI     81   ?Gimballed nozzles.

----------  DEVELOPMENT BEGAN  ----------
SSF    82   1965                   Year design began.
SMP    85   1964                   Technological development began. [From graph.]

----------  DEVELOPMENT UNDERWAY  ----------
Rock   78                          Developed in early 1970s.

----------  PROTOTYPE TESTS  ----------
SecD   73                          Being tested and developed.
SecD   74   [1973]                 Tested in past year.
FIar   76   2nd half 1974          First flight.
SecD   76   1974-75                In 1974, extensive flight testing, continuing in 1975.
SWM    77                          17 test flights by April 1974; early trials were with MRV of 3 RVs.
Rock   78   1972                   First flight.
JWS    81   1972-74
SMP    85   1970-71                Engineering and testing began. [From graph.]

----------  PRODUCTION UNDERWAY  ----------
MOW    76                          In production.

----------  IOC  ----------
JCS    74   1975                   FY75,77-78.
JCS    75   1975                   Estimated.
SecD   76                          Has now achieved operational status.
            1975
SMP    81   1975                   First became operational.
JCS    83   c. 1976                [Inferred from graph].

----------  DEPLOYMENT BEGAN  ----------
JCS    74                          Could be deployed in 1975.
IA06   75   1975                   "Deployment 1975".
JCS    75                          Last year was reported SS-17 and SS-19 would not be operational before 1975, wrong,
                                   FY76-77.
SecD   75                          Started, or soon will start, operational deployment.
FIar   76                          Has now entered service.
MOW    76   1975                   MIRV became operational.
JCS    77                          MIRVed variant has been deployed, FY78-79.
Rock   78   1975                   Service intro.
JWS    81   1975
JCS    84   1975-76                Major strategic system deployments [graph], FY85-86.
SecD   84   1974-75                Introduction of strategic systems [graph].
SMP    85   1975-75                [From graph.]

----------  DEPLOYMENT COMPLETE  ----------
JCS    80                          The SS-17 conversion program has now been completed, FY81-82.
FI     81                          "Fielding of this type has been completed".
JWS    81   1980
SecD   81                          [Implies deployment has stopped].
AFM    82                          Deployment believed complete.
```

BALLISTIC MISSILES  ICBMs  SS-17  187

---------- DEPLOYMENT HISTORY ----------
| | | |
|---|---|---|
| SecD | 73 | Follow-on to SS-11. |
| JCS | 75 | "SS-11 silos have been or are in the process of being prepared for the SS-17 and the SS-19", FY76-77. |
| FIar | 76 | Replacement for the SS-11 Sego. |
| JCS | 76 | Being deployed at the expected moderate pace. |
| MOW | 76 | Successor to SS-11. |
| JCS | 77 | Conversions of SS-11 to SS-17 will probably be limited at each complex to avoid too many launchers being off-line, FY78-79. Pace of SS-11 to SS-17 conversion proceeding not quite as rapidly as expected last year. |
| SecD | 77 | Deployment continues in modified and upgraded silos at a rapid rate. |
| SWM | 77 | Replacement for SS-11. |
| JCS | 78 | Pace of SS-11 to SS-17 conversion proceeding slower than expected last year. Conversion of some SS-11 silos is underway. |
| Rock | 78 | Began to replace SS-11 in SS-11 silos in 1975. Is [still] replacing Sego. |
| JCS | 79 | SS-11 to SS-17 conversion continues "at a slow pace". |
| FI | 81 | Both SS-17 and SS-19 have replaced SS-11. |
| JWS | 81 | SS-17 (with SS-19) to replace SS-11, in SS-11 silos. Official US view is that SS-17 and SS-19 were deployed in competition. |
| JCS | 82 | Little or no change in deployment [past year]. |
| SMP | 83 | Continuing modernization of SS-17,18,19 force. |
| JCS | 84 | [SS-11,13,17] expected to be replaced with new ICBMs like SS-X-24,25, FY84-85. |
| SMP | 85 | Will be deployed at least through 1990. [From graph.] |

---------- NUMBERS PRODUCED ----------
| | | |
|---|---|---|
| SecD | 80 | No evidence that SS-17,18,19 production "is significantly greater" than the number of their launchers. |

---------- NUMBERS IN SERVICE ----------

---------- NUMBER IN 1960 ----------
| | | | |
|---|---|---|---|
| SSF | 82 | none | In 1960, 1965, 1970. |

---------- NUMBER IN 1970 ----------
| | | | |
|---|---|---|---|
| Clns | 80 | none | In 1970-1974. |

---------- NUMBER IN 1975 ----------
| | | | |
|---|---|---|---|
| IISS | 75 | 10 | July, total deployed. |
| SWM | 77 | 10 | In service mid-1975. |
| Clns | 80 | 10 | |
| SSF | 82 | 10 | |

---------- NUMBER IN 1976 ----------
| | | | |
|---|---|---|---|
| FIar | 76 | 30 | Being installed in modified SS-11 silos. |
| IISS | 76 | 20 | July, total deployed. |
| Clns | 80 | 20 | |
| vanD | 83 | 20 | 1 Jan. |

---------- NUMBER IN 1977 ----------
| | | | |
|---|---|---|---|
| IISS | 77 | 40 | July, total deployed. |
| SecD | 77 | c. 40 | |
| Rock | 78 | 20 | SS-17s emplaced in SS-11 silos by 1977, each with 4 200 kT warheads. |
| Nitz | 79 | 10 | Launchers. |
| Clns | 80 | 50 | |
| vanD | 83 | 40 | 1 Jan. |

---------- NUMBER IN 1978 ----------
| | | | |
|---|---|---|---|
| IISS | 78 | 60 | July, total deployed. |
| JCS | 78 | > 60 | Now operational in converted SS-11 silos. |
| SecD | 78 | | Deployment of SS-17,18,19 continues at c. 124 per year, FY79-81. |
| | | > 60 | More than 60 launchers (converted from SS-11). |
| | | | About 300 SS-17,19s (together) in converted SS-11 silos. |
| Clns | 80 | 100 | |
| vanD | 83 | 60 | 1 Jan. |

---------- NUMBER IN 1979 ----------
| | | | |
|---|---|---|---|
| IISS | 79 | 100 | July, total deployed. |
| JCS | 79 | | Believed "most of the SS-17s" have MIRV capability, FY80-81. |
| | | < 100 | "Nearly" 100 operational in converted SS-11 silos. |
| Clns | 80 | 140 | |
| vanD | 83 | 100 | 1 Jan. |

188  SOVIET MISSILES  IDDS

```
----------- NUMBER IN 1980 -----------
AWar    80    150        Deployed.
IISS    80    150        1980-81, July, total deployed.
JCS     80    c. 150     "A total of about 150 silos" converted "from the old SS-11 configuration".
SecD    80    c. 150     Deployed in converted SS-11 silos.
JWS     81    150        By 1980.
SSF     82    150
vanD    83    150        1980-81, 1 Jan.

----------- NUMBER IN 1981 -----------
FI      81    c. 150     In SS-11 silos.
                         "Most operational rounds are Mod 2" [sic, final report].
JCS     81    150        "150 silos converted" from SS-11 configuration.
SecD    81    c. 150
                         SS-17,18,19s mostly equipped with MIRVs.
SMP     81    150        '81, '83, '84, '85.

----------- NUMBER IN 1982 -----------
AFM     82    150        Since 1975, 150 SS-11 silos modified to accept SS-17s.
AWST    82    150        Operational in 150 silos.
JCS     82    150        Inventory.
JWS+    83    150        Early 1982, apparently still valid. Converted SS-11 silos now thought to contain
                         majority with MIRV warheads; 30 single RV and 120 MIRVed has been suggested.

----------- NUMBER IN 1983 -----------
CBO     83    150        1983, 1990, 1996.
FI+     83             Most operational rounds are Mod 1.
SMP     83             Will remain unchanged as current ICBM modernization is completed.
                       Great majority of SS-17,18,19 equipped with MIRVs.

----------- NUMBER IN 1984 -----------
JCS     84    150        As of 1 Jan.

----------- NUMBER IN 1985 -----------
JCS     85    150        As of 1 Jan.

----------- UNDATED NUMBER -----------
SWM     77             SS-17 and SS-19 restricted to 1,036 under SALT I.
SSF     82    150      "Number of missiles deployed".

----------- LAUNCHER/SILO -----------
SecD    74    Can be deployed in new-type silos, or SS-11 silos.
JCS     75    More survivable than SS-11, FY76-77.
JCS     76    Every SS-17 deployed goes into "new modernized, hard silos"; associated launch systems are in
              silos, not bunkers.
SecD    77    Deployment continues in modified and upgraded silos at a rapid rate. Some modified silos designed
              to resist high overpressures.
SWM     77    Cold launched from modified SS-11 silos.
Rock    78    Began to replace SS-11 in SS-11 silos in 1975.
AWar    80    3,500-4,500 psi hardness [SS-17,18,19].
SecD    80    All "converted silos may be capable of withstanding very high static overpressures".
JWS     81    SS-17 (with SS-19) to replace SS-11, in SS-11 silos.
SMP     81    Deployed in 150 converted SS-11 silos. More than half of 1,398 ICBM launchers rebuilt to house
              SS-17,18,19 in "vastly more survivable, hardened silos".
AFM     82    DoD says some silos for SS-17 and other ICBMs "hardened to resist very high over-pressure".
SSF     82    Basing mode: Hardened silo.
SMP     83    SS-17,18,19 "housed in the world's hardest silos", '83, '84, '85.

----------- DESIGNER -----------
SSF     82    Yangel        Design Bureau.

----------- DESIGN AND ENGINEERING -----------
JCS     74    SS-17 and SS-19 may be competing designs to replace SS-11, but possible both will be deployed.
SecD    74    SS-X-19 and SS-X-17 apparently competitive developments to replace SS-11.
JCS     75    It was reported last year that SS-17 was the most advanced vehicle; now appears to be wrong.
Flar    76    Believed related in technology to SS-11.
JCS     76    It was reported at one time that SS-17 was the most advanced vehicle; now appears to be wrong.
Rock    78    Conceived (with SS-19) as successor to SS-11.

----------- USE AND CONFIGURATION -----------
CBO     83    System availability: 0.95.
vanD    83    10 missiles in a battalion.
```

BALLISTIC MISSILES ICBMs SS-17

------- **RELOADS IN GENERAL** -------

| | | |
|---|---|---|
| SecC | 74 | Cold launched: boosted out of silos on piston (called sabot) powered by gas produced by burning of small solid fuel charge before main booster ignited. |
| JCS | 75 | Sabot "cold" launch, FY76-77. |
| SecD | 75 | Cold-launched, FY76-77,79-82. |
| FIar | 76 | Designed to be cold-launched. |
| MOW | 76 | Cold Launched. |
| SWM | 77 | Cold launched from modified SS-11 silos. |
| JCS | 78 | Cold Launch. |
| Rock | 78 | Cold Launch. |
| SecD | 80 | No evidence USSR has any plan or capability to use excess missiles as reserves or refires; quite confident they have not tested or trained in these ways. |
| FI | 81 | Cold launch. |
| JCS | 81 | Cold Launch technique could allow reload and refire in protracted nuclear conflict, FY82-83. USSR probably cannot reload silos "in less than several days". |
| JWS | 81 | Cold Launched. |
| SMP | 81 | Cold Launch [all mods], '81, '83. Cold launch is "consistent with the notion of building in the capability to reload and refire missiles during a protracted nuclear conflict". "The Soviets probably cannot refurbish and reload silo launchers in a period less than several days". |
| AFM | 82 | Cold Launch. Silo could be reloaded, but "this would be a slow process". |
| AWST | 82 | Cold Launched. |
| SSF | 82 | Launching mode: Cold. |
| IISS+ | 83 | Cold launch. |
| JWS+ | 83 | US report says all Soviet liquid-fueled ICBMs contained in launch canister offering possibility of reloading. |
| SMP | 83 | Cold Launch [of SS-17,18] lends itself to a reload and refire capability. "The Soviets probably cannot refurbish and reload silo launchers in a period less than a few days". "Contained in a launch canister within the silo. This and the silo design minimize damage to the launcher during" firing, giving reload capability. |

------- **STRATEGY** -------

| | | |
|---|---|---|
| Rock | 78 | Developed as long-range ICBM able to destroy hardened silos. |
| SMP | 81 | "Improved reliability, range, payload accuracy and survivability". |

------- **GEOGRAPHICAL DEPLOYMENTS** -------

| | | |
|---|---|---|
| SMP | 83 | [Shown on map at] Yedrovo, Kostroma, '83, '84, '85. |

------- **RELATION TO OTHER MISSILES** -------

| | | |
|---|---|---|
| SecD | 74 | Technically more advanced than SS-19. |
| SecD | 75 | Not much larger in volume than SS-11. |
| FIar | 76 | Believed more advanced in technology than rival SS-19. |

# SS-17 Mod 1

| Sources are: | JCS | 78 | Clns | 80 | FI | 81 | AFM | 82 | SSF | 82 | JWS+ | 83 | AFM+ | 84 |
|---|---|---|---|---|---|---|---|---|---|---|---|---|---|---|
| SecD | 74 | | JCS | 79 | SecD | 80 | JWS | 81 | IISS | 82 | Clns+ | 83 | SMP | 83 | SMP | 85 |
| JCS | 77 | | AWar | 80 | SecD | 80 | SMP | 81 | JCS | 82 | FI+ | 83 | vanD | 83 | BW | 86 |

---------- **US DESIGNATION** ----------

| | | |
|---|---|---|
| SecD | 74 | SS-17 Mod 1 |
| AWar | 80 | SS-17 mod |
| Clns | 80 | SS-17 Mod 1 |
| FI | 81 | SS-17 Mod 1 |
| JWS | 81 | SS-17 Mod 1 |
| SMP | 81 | SS-17 Mod 1 |
| AFM | 82 | SS-17 Mod 1 |
| IISS | 82 | SS-17 Mod 1 |
| SSF | 82 | SS-17 Mod 1 |
| vanD | 83 | SS-17 |

---------- **EDITOR'S NOTES** ----------

| | | |
|---|---|---|
| vanD | 83 | Source contrasts two unnamed Mods. |

---------- **RANGE** ----------

| | | mi | nm | km | km conv | |
|---|---|---|---|---|---|---|
| AWar | 80 | | | 10,000 | 10,000 | |
| Clns | 80 | | 5,500 | | 10,186 | |
| SecD | 80 | | | 10,000 | 10,000 | Max, FY81-82. [Qualification:] "exclusive of range imparted by" PBV. |
| JWS | 81 | | | 10,000 | 10,000 | |
| SMP | 81 | | | 10,000 | 10,000 | Max, '81, '83. |
| AFM | 82 | 6,200 | | | 9,975 | |
| IISS | 82 | | | 10,000 | 10,000 | Max. |
| SSF | 82 | | 5,400 | | 10,000 | |
| BW | 86 | | | 10,000 | 10,000 | |

---------- **CEP** ----------

| | | ft | mi | nm | m | km | m conv | |
|---|---|---|---|---|---|---|---|---|
| SecD | 74 | | | | | | | MIRVed version would be soft-target weapon. |
| AWar | 80 | | | 0.24 | | | 444.48 | |
| Clns | 80 | | | 0.24 | 500 | | 444.48 | |
| AFM | 82 | | .3 | | 500 | | 482.70 | |
| IISS | 82 | | | | 450 | | 450.00 | |
| SSF | 82 | | | 0.24 | | | 444.48 | |
| vanD | 83 | | | .32 | | | 592.64 | |
| AFM+ | 84 | | 0.3 | | c. 450 | | 482.70 | |
| BW | 86 | | | | 440 | | 440.00 | |

---------- **WARHEAD CHARACTERISTICS** ----------

| | | | |
|---|---|---|---|
| JCS | 78 | 4 warheads | Mod 1. |
| AWar | 80 | 4 RVs | |
| Clns | 80 | 4 MIRVs | |
| SecD | 80 | 4 warheads | |
| FI | 81 | 4 MIRVs | [Text] |
| | | 4 MRVs | [Table] |
| JWS | 81 | 4 MIRVs | |
| SMP | 81 | 4 MIRVs | '81, '83. |
| AFM | 82 | 4 MIRVs | Mod 1 shaped for high speed reentry for greater accuracy, Mod 2 RV is for hard target capability. |
| IISS | 82 | 4 MIRVs | |
| SSF | 82 | 4 RVs | |
| vanD | 83 | 4 RVs | |
| BW | 86 | 4 MIRVs | |

---------- **NUCLEAR YIELD** ----------

| | | kT | MT | kT conv | |
|---|---|---|---|---|---|
| AWar | 80 | | 0.75 | 750.00 | |
| Clns | 80 | 750 | | 750.00 | Each. |
| FI | 81 | | 1 | 1,000 | |
| JWS | 81 | > 200 | | 200.00 | |
| AFM | 82 | 900 | | 900.00 | |
| IISS | 82 | 750 | | 750.00 | Per warhead. |
| SSF | 82 | | 0.75 | 750.00 | Per warhead. |
| JWS+ | 83 | 200 | | 200.00 | [Analysis table] |
| vanD | 83 | | .75 | 750.00 | |
| AFM+ | 84 | 750 | | 750.00 | |
| BW | 86 | 750 | | 750.00 | |

BALLISTIC MISSILES   ICBMs   SS-17 Mod 1   191

```
----------- THROWWEIGHT -----------
              lb          t          kg         kg conv
Clns  80     6,025                              2,732
IISS  82     6,000                              2,721

----------- TYPE OF PROPULSION -----------
Clns  80     liquid
BW    86     storable liquid

----------- PROTOTYPE TESTS -----------
SSF   82     1972           First flight test.

----------- IOC -----------
JCS   78     1975
vanD  83     1976           [P209]
             1975           [P186]

----------- DEPLOYMENT BEGAN -----------
JCS   77                    MIRVed variant has been deployed, FY78-79.
Clns  80     1975           First deployed.
FI    81     1975           Entered service.
IISS  82     1975           First year deployed.
SSF   82     1975           Year operation began.
SMP   85     1975-75        [Presumed Mod 1, from graph, ed.]
BW    86     1975

----------- RETIRED -----------
BW    86     ? 1982

----------- NUMBERS IN SERVICE -----------

----------- NUMBER IN 1970 -----------
Clns+ 83     none     In 1970-1974.

----------- NUMBER IN 1975 -----------
Clns+ 83     10

----------- NUMBER IN 1976 -----------
Clns+ 83     20

----------- NUMBER IN 1977 -----------
Clns+ 83     50

----------- NUMBER IN 1978 -----------
Clns+ 83     80

----------- NUMBER IN 1979 -----------
JCS   79                  Believed "most of the SS-17s" have MIRV capability, FY80-81.
Clns  80     120          120 of 140 total in 1979 are Mod 1. "All SS-17s and SS-19s were MIRVed until 1978"; 20
Clns+ 83     120          single-shot SS-17s since installed.

----------- NUMBER IN 1980 -----------
SecD  80                  SS-17,18,19s mostly equipped with MIRVs.
Clns+ 83     130          In 1980-1981.

----------- NUMBER IN 1982 -----------
IISS  82     <_ 150       1982-83, July, total deployed.
JCS   82                  Majority of 150 SS-17s contain the 4-RV MIRVs.
Clns+ 83     30

----------- NUMBER IN 1983 -----------
FI+   83                  Most operational rounds are Mod 1.
JWS+  83     ? 120        Of 150.
SMP   83                  Great majority of SS-17,18,19 equipped with MIRVs.

----------- UNDATED NUMBER -----------
BW    86     ? 130    Peak.

----------- LAUNCHER/SILO -----------
IISS  82     In mod SS-11 silos.

----------- USE AND CONFIGURATION -----------
vanD  83     Reliability: .80; Penetration 1.0.

----------- RELOADS IN GENERAL -----------
Clns  80     Cold launch: yes.
```

# SS-17 Mod 2

```
Sources are:   SecD  77      SecD  80      SMP   81      SSF   82      Lu01  83      SMP   85
SecD  74       JCS   77      AWar  80      FI    81      AFM   82      Clns+ 83      AFM+  84      BW    86
JCS   77                     Clns  80      JWS   81      IISS  82      JWS+  83      ArkS  84
```

## US DESIGNATION

|  |  |
|---|---|
| SecD 74 | SS-17 Mod 2 |
| JCS 77 | SS-17 Mod 2 |
| AWar 80 | SS-17 mod |
| Clns 80 | SS-17 Mod 2 |
| FI 81 | SS-17 Mod 2 |
| JWS 81 | SS-17 Mod 2 |
| SMP 81 | SS-17 Mod 2 |
| AFM 82 | SS-17 Mod 2 |
| IISS 82 | SS-17 Mod 2 |
| SSF 82 | SS-17 Mod 2 |

## RANGE

| Source | mi | nm | km | km conv | Notes |
|---|---|---|---|---|---|
| AWar 80 | | | 11,000 | 11,000 | |
| Clns 80 | | 5,700 | | 10,556 | |
| SecD 80 | | | 11,000 | 11,000 | Max, FY81-82. [Qualification:] "exclusive of range imparted by" PBV. |
| JWS 81 | | | 11,000 | 11,000 | |
| SMP 81 | | | 11,000 | 11,000 | Max. |
| AFM 82 | 6,800 | | | 10,941 | |
| IISS 82 | | | 11,000 | 11,000 | Max. |
| SSF 82 | | 5,900 | | 10,926 | |
| BW 86 | | | 11,000 | 11,000 | |

## CEP

| Source | ft | mi | nm | m | km | m conv | Notes |
|---|---|---|---|---|---|---|---|
| SecD 74 | | | | | | | Single RV version could carry large warhead, and probably has sufficient accuracy for very effective hard target capability. |
| AWar 80 | | | 0.23 | | | 425.96 | |
| Clns 80 | | | 0.23 | | | 425.96 | |
| IISS 82 | | | | 450 | | 450.00 | |
| SSF 82 | | | 0.23 | | | 425.96 | |
| Lu01 83 | | | | 450 | | 450.00 | |
| BW 86 | | | | 430 | | 430.00 | |

## WARHEAD CHARACTERISTICS

|  | |  |
|---|---|---|
| JCS 77 | single RV | FY78-79. |
| SecD 77 | single RV | Single RV high-yield warhead also developed. |
| AWar 80 | single RV | |
| Clns 80 | single RV | |
| SecD 80 | 1 warhead | FY81-82. |
| FI 81 | single RV | |
| JWS 81 | single large RV | |
| SMP 81 | single RV | '81, '83. |
| AFM 82 | single RV | Mod 1 shaped for high speed reentry for greater accuracy, Mod 2 RV is for hard target capability. |
| IISS 82 | single RV | |
| SSF 82 | single RV | |
| BW 86 | single RV | |

## NUCLEAR YIELD

| Source | kT | MT | kT conv | Notes |
|---|---|---|---|---|
| AWar 80 | | 3.6 | 3,600 | |
| Clns 80 | | 6 | 6,000 | |
| JWS 81 | | | | High kT. |
| AFM 82 | | 5 | 5,000 | |
| IISS 82 | | 6 | 6,000 | |
| SSF 82 | | 3.6 | 3,600 | |
| JWS+ 83 | | 1-2 | 1,000-2,000 | [Analysis table] |
| Lu01 83 | | 6 | 6,000 | |
| AFM+ 84 | | 6 | 6,000 | |
| BW 86 | 3,600 | | 3,600 | |

## THROWWEIGHT

|      |    | lb    | t | kg | kg conv |
|------|----|-------|---|----|---------|
| Clns | 80 | 6,000 |   |    | 2,721   |
| IISS | 82 | 3,600 |   |    | 1,632   |

## TYPE OF PROPULSION
Clns 80    liquid  
BW    86    storable liquid

## PROTOTYPE TESTS
JCS    77    Feb 1976    Testing of single RV mod began, FY78-79.  
SSF    82    1976    First flight test.

## DEPLOYMENT BEGAN
Clns    80    1977    First deployed.  
                      "All SS-17s and SS-19s were MIRVed until 1978"; 20 single-shot SS-17s since installed.  
IISS    82    1977    First year deployed.  
SSF    82    1977    Year operation began.  
SMP    85    c. 1979    [From graph.]  
BW    86    1978

## DEPLOYED
BW    86    In service.

## NUMBERS IN SERVICE

## NUMBER IN 1970
Clns+    83    none    In 1970-1977.

## NUMBER IN 1978
Clns+    83    20    In 1978-81.

## NUMBER IN 1979
Clns    80    20

## NUMBER IN 1981
FI    81    "Most operational rounds are Mod 2" [sic, final report].  
SMP    81    Single RV versions developed, but "few if any" deployed.

## NUMBER IN 1982
IISS    82    "few"    July, total deployed.  
Clns+    83    10

## NUMBER IN 1983
JWS+    83    ? 30    Of 150.

## NUMBER IN 1984
AFM+    84    Few only.  
ArkS    84    Some SS-17s are still configured to carry single warheads, according to DIA officials; omitted from SMP 84 because of 'space considerations' and because number of single warhead SS-17s and SS-19s is 'too few to matter'.

## UNDATED NUMBER
BW    86    ? 20    Peak.

## LAUNCHER/SILO
IISS    82    In mod SS-11 silos [final report].

## RELOADS IN GENERAL
Clns    80    Cold launch: yes.

# SS-17 Mod 3

```
Sources are:   SSF    82       IISS   83      vanD   83      SMP    84      BW    86
Nitz  79       Clns   83       SMP    83      AFM    84      SMP    85
```

---------- **US DESIGNATION** ----------

|  |  |
|---|---|
| Nitz 79 | SS-17 |
| SSF 82 | SS-17 Mod 3 |
| Clns 83 | SS-17 Mod 3 |
| IISS 83 | SS-17 Mod 3 |
| SMP 83 | SS-17 Mod 3 |
| vanD 83 | SS-17 |
| AFM 84 | SS-17 Mod 3 |
| SMP 84 | SS-17 Mod 3 |
| SMP 85 | SS-17 Mod 3 |

---------- **EDITOR'S NOTES** ----------

Nitz 79 — This missile appears in source on 4 lines with dates in the designation, but differing only in CEP. It is presumed Mod 3 because the earliest date is 1977.

vanD 83 — Source contrasts two unnamed Mods.

---------- **RANGE** ----------

| Source | mi | nm | km | km conv | |
|---|---|---|---|---|---|
| Clns 83 | | 6,000 | | 11,112 | |
| IISS 83 | | | 10,000 | 10,000 | |
| SMP 83 | | | 10,000 | 10,000 | Max, '83, '84, '85. |
| AFM 84 | 6,200 | | | 9,975 | |
| BW 86 | | | 10,000 | 10,000 | |

---------- **CEP** ----------

| Source | ft | mi | nm | m | km | m conv | |
|---|---|---|---|---|---|---|---|
| Nitz 79 | | | .2 | | | 370.40 | 1977. |
| | | | .17 | | | 314.84 | 18 Jun 1979. |
| | | | .15 | | | 277.80 | 1982. |
| | | | .12 | | | 222.24 | 1985 [literally 1982, ed] |
| Clns 83 | | | 0.20 | | | 370.40 | |
| vanD 83 | | | .24 | | | 444.48 | |
| BW 86 | | | | 370 | | 370.00 | |

---------- **WARHEAD CHARACTERISTICS** ----------

|  | | |
|---|---|---|
| Nitz 79 | 4 RVs | |
| SSF 82 | 4 RVs | |
| Clns 83 | 4 MIRVs | |
| IISS 83 | 4 MIRVs | |
| SMP 83 | 4 MIRVs | '83, '84, '85. |
| vanD 83 | 4 RVs | |
| AFM 84 | 4 MIRVs | |
| BW 86 | 4 MIRVs | |

---------- **NUCLEAR YIELD** ----------

| Source | kT | MT | kT conv | |
|---|---|---|---|---|
| Nitz 79 | | .75 | 750.00 | |
| SSF 82 | | 0.75 | 750.00 | Per warhead. |
| Clns 83 | 500 | | 500.00 | Each. |
| IISS 83 | 2 | | 2.00 | [Sic] |
| vanD 83 | | .75 | 750.00 | |
| BW 86 | 750 | | 750.00 | |

---------- **THROWEIGHT** ----------

| Source | lb | t | kg | kg conv |
|---|---|---|---|---|
| Nitz 79 | 7,000 | | | 3,175 |
| Clns 83 | 6,000 | | | 2,721 |

---------- **LENGTH** ----------

| Source | in | ft | cm | m | m conv | |
|---|---|---|---|---|---|---|
| SMP 84 | | | | 20 | 20.00 | [From graph], '84, '85. |

---------- **TYPE OF PROPULSION** ----------

|  | |
|---|---|
| Clns 83 | liquid |
| BW 86 | storable liquid |

---------- **IOC** ----------

|  | |
|---|---|
| vanD 83 | 1982 |

########## DEPLOYMENT BEGAN ##########

| | | | |
|---|---|---|---|
| SSF | 82 | 1979 | Year operation began. |
| Clns | 83 | 1979 | First deployed. |
| IISS | 83 | 1982 | First year deployed. |
| SMP | 84 | | Deployment [of SS-17 Mod 3, SS-18 Mod 4, SS-19 Mod 3] began only 5 years ago, [6 years in '85 report]. |
| SMP | 85 | c. 1982 | [From graph.] |
| BW | 86 | 1981 | |

########## DEPLOYED ##########

BW   86            In service.

########## DEPLOYMENT HISTORY ##########

SMP  85   Will be deployed at least through 1990. [Presumed Mod 3, from graph.]

########## NUMBERS IN SERVICE ##########

########## NUMBER IN 1970 ##########
Clns 83   none     In 1970-1981.

########## NUMBER IN 1977 ##########
Nitz 79   10       Launchers.

########## NUMBER IN 1979 ##########
Nitz 79   140      18 June, launchers.

########## NUMBER IN 1982 ##########
Clns 83   110

########## NUMBER IN 1983 ##########
IISS 83   "few"    July, total deployed.

########## NUMBER IN 1984 ##########
SMP  84   150

########## NUMBER IN 1985 ##########
SMP  85   150

########## UNDATED NUMBER ##########
BW   86   ? 130    Peak.

########## USE AND CONFIGURATION ##########
vanD 83   Reliability: .80; Penetration 1.0.

########## RELOADS IN GENERAL ##########
Clns 83   Cold launch: yes.
SMP  84   Cold launch, '84, '85.

########## RELATION TO OTHER MISSILES ##########
AFM  84   Probably Mod 1 upgraded.
SMP  84   Somewhat less capable than SS-19, but has similar targeting flexibility, '84, '85.

# SS-18

| Sources are: | IISS | 75 | SecD | 76 | Rock | 78 | Clns | 80 | SecD | 81 | JCS | 83 | JCS | 85 |
|---|---|---|---|---|---|---|---|---|---|---|---|---|---|---|
| SecC | 73 | JCS | 75 | IISS | 77 | SecD | 78 | SMP | 81 | JWS+ | 83 | SMP | 85 | | |
| SecD | 73 | SecD | 75 | JCS | 77 | IISS | 79 | AFM | 82 | SMP | 83 | | | | |
| JCS | 74 | FIar | 76 | SecD | 77 | JCS | 79 | AWST | 82 | vanD | 83 | | | | |
| SecC | 74 | IISS | 76 | SWM | 77 | Nitz | 79 | IISS | 82 | AFM+ | 84 | | | | |
| SecD | 74 | JCS | 76 | CB01 | 78 | SecD | 79 | JCS | 82 | JCS | 84 | | | | |
| GA01 | 75 | MOW | 76 | IISS | 78 | US08 | 79 | SSF | 82 | SecD | 84 | | | | |
| IA06 | 75 | NA03 | 76 | JCS | 78 | AWar | 80 | FI+ | 83 | SMP | 84 | | | | |

---------- **US DESIGNATION** ----------

| SecC | 73 | SS-18 | |
| SecD | 73 | SS-X-18 | FY74-75. |
| JCS | 74 | SS-X-18 | |
| GA01 | 75 | SSX-18 | |
| JCS | 75 | SS-18 | FY76-86. |
| SecD | 75 | SS-18 | FY76-82,85. |
| FIar | 76 | SS-18 | |
| MOW | 76 | SS-18 | |
| SWM | 77 | SS-18 | |
| Rock | 78 | SS-18 | |
| Nitz | 79 | SS-18 Single | |
| AWar | 80 | SS-18 | |
| Clns | 80 | SS-18 | |
| FI | 81 | SS-18 | |
| JWS | 81 | SS-18 | |
| SMP | 81 | SS-18 | '81, '83, '84, '85. |
| AFM | 82 | SS-18 | |
| AWST | 82 | SS-18 | |
| IISS | 82 | SS-18 | |
| SSF | 82 | SS-18 | |
| vanD | 83 | SS-18 | |

---------- **SOVIET DESIGNATION** ----------

| FI | 81 | RS-20 |
| JWS | 81 | ? RS-20 |
| AFM | 82 | RS-20 |
| IISS | 82 | RS-20 |

---------- **DESCRIPTION** ----------

| JCS | 74 | | Listed as significant initiative, strategic offensive systems. Improved heavy ICBM. |
| MOW | 76 | ICBM | |
| JCS | 78 | MLBM | Modern large ballistic missile, FY79-81. |
| Rock | 78 | | Versatile heavyweight ICBM. |
| JWS | 81 | ICBM | |
| AFM | 82 | | "Heavy" missile. |
| SSF | 82 | | Fourth generation. |
| JWS+ | 83 | | Fourth generation ICBM. |
| SecD | 84 | | World's largest ICBMs. |

---------- **EDITOR'S NOTES** ----------

| SecC | 73 | Listed FY74-77. |
| SecD | 73 | Listed FY74-82,85. |
| JCS | 74 | Listed FY75-86. |
| SMP | 81 | Listed '81, '83, '84, '85. |

---------- **RANGE** ----------

| | | mi | nm | km | km conv | |
|---|---|---|---|---|---|---|
| JCS | 74 | | > 5500 | | 10,186 | |
| GA01 | 75 | 6,000+ | | | 9,654 | |
| IA06 | 75 | | | >_ 10,500 | 10,500 | |
| JCS | 75 | | 5500 | | 10,186 | MIRVed version believed capable of deploying as many as 8 warheads to 5500 nm. |
| FIar | 76 | | | 12,000 | 12,000 | |
| MOW | 76 | > 6,300 | | > 10,000 | 10,136 | |
| NA03 | 76 | 7,500 | | | 12,067 | |
| Rock | 78 | | | 12,000 | 12,000 | |
| AWST | 82 | | 7,000 | | 12,964 | |

BALLISTIC MISSILES  ICBMs  SS-18  197

### CEP

| | | ft | mi | nm | m | km | m_conv | |
|---|---|---|---|---|---|---|---|---|
| JCS | 74 | | | | | | | Increase in CEP is "a definite goal". |
| SecD | 74 | | | | | | | Even with MIRVs, will have very respectable hard target capability. |
| JCS | 75 | | | | | | | More accurate than SS-9. |
| SecD | 75 | | | | | | | Accuracy could be improved with guidance refinements. Designed with accuracy as "prime objective". |
| Flar | 76 | | | | [c 500+] | | 500.00 | Comparable to SS-17. |
| CB01 | 78 | 1200-1500 | | | | | 365.76-457.20 | Estimated for new generation of ICBMs in 1985 [SS-16, SS-17, SS-18, SS-19, ed]. |
| Nitz | 79 | | | .2 | | | 370.40 | |
| SMP | 81 | | | | | | | "Each warhead of the ten RV variants has a better than 50 percent chance of destroying a MINUTEMAN silo"; even more destructive if used in pairs. Single RV versions of SS-18 "capable of destroying any known fixed target with high probability". |
| AFM | 82 | | | | | | | Greater accuracy and flexibility than SS-9. |
| AWST | 82 | 600 | | | | | 182.88 | |
| JCS | 82 | | | | | | | Hard target capability. Certain versions of SS-18 and SS-19 have "significantly improved accuracies". |
| SMP | 83 | | | | | | | "The most accurate versions of the SS-18 and SS-19 are capable of destroying hard targets. Together, these systems have the capability to destroy most of the 1,000 US MINUTEMAN ICBMs, using only a portion of the warheads available". |
| SMP | 84 | | | | | | | "At least as accurate and possibly more accurate" than Minuteman III. |

### WARHEAD TYPE

| | | nuclear | conven | chem |
|---|---|---|---|---|
| JWS | 81 | yes | | |

### WARHEAD CHARACTERISTICS

| | | | |
|---|---|---|---|
| JCS | 74 | 5-8 MIRVs | [Mod not specified, ed]. PBV; bus-type MIRV system, similar to Minuteman III and Poseidon. |
| SecD | 74 | | Has PBV, FY75-76. |
| GA01 | 75 | ? 4-6 MIRVs | Possibly. |
| IA06 | 75 | 5-8 MIRVs | |
| JCS | 75 | "6/8" warheads | Max. MIRVed version has dispensed at least 6 RVs; believed capable of deploying as many as 8 warheads to 5500 nm. |
| JCS | 76 | 8 RVs | Varying number of RVs tested, but it appears 8 RVs will be basic payload. 8 warheads max. Has PBV. |
| JCS | 77 | 8 RVs | 8 warheads. |
| SecD | 77 | single RV | High yield, single RV version also developed. |
| Rock | 78 | single RV | |
| SecD | 78 | | Can deploy either high-yield single RV, or MIRVs, FY79-80. |
| JCS | 79 | <_ 10 | "Can carry up to 10 high-yield warheads and deliver them with great accuracy", [Mod not specified, ed]. |
| Nitz | 79 | 1 RV | |
| USD8 | 79 | 8 MIRVs | |
| JWS | 81 | | Computer-controlled RVs. |
| AWST | 82 | 8-10 MIRVs | |
| JWS+ | 83 | | In one model may carry PBV. |
| JCS | 84 | | SS-18,19 carry multiple warheads. |

# SOVIET MISSILES  IDDS

## NUCLEAR YIELD

| Source | Year | kT | MT | kT conv | Notes |
|---|---|---|---|---|---|
| IA06 | 75 | | 2 | 2,000 | Each. |
| FIar | 76 | | 1-2 | 1,000-2,000 | |
| SecD | 77 | | 50 | 50,000 | "50 MT or 8 MIRVs", table. Exceeds Minuteman III (per warhead) [170 kT, ed]. |
| OB01 | 78 | | 1.5 | 1,500 | Estimated, 1985. |
| Rock | 78 | | 18.0-25.0 | 18,000-25,000 | |
| Nitz | 79 | | 20 | 20,000 | |
| US08 | 79 | | 2 | 2,000 | |
| BA01 | 81 | | 1.5 | 1,500 | "Reported" to yield 1.5 MT, largest Soviet ICBM warheads. |

## THROWWEIGHT

| Source | Year | lb | t | kg | kg conv | Notes |
|---|---|---|---|---|---|---|
| SecD | 74 | | | | | 30% greater than SS-9. |
| FIar | 76 | | | 7,000 | 7,000 | |
| MOW | 76 | | | | | 30% greater than SS-9. |
| SecD | 77 | | | | | Exceeds Minuteman III. |
| SWM | 77 | | | | | Possibly 30% greater than SS-9 Scarp (Pentagon sources). |
| Rock | 78 | | 7 | | 7,000 | Tonnes. |
| | | | [> 7.15] | | 7,150 | Tonnes, > 30% more than Scarp. |
| Nitz | 79 | 16,000 | | | 7,257 | |
| US08 | 79 | 15000-18000 | | | 6,804-8,164 | |
| SecD | 80 | | | | | Without SALT II, SS-18 could be equipped with 20 or even 30 MIRVs. |
| FI | 81 | | | | | 30% greater than SS-9. |
| AFM | 82 | | | | | Greater than any US or Soviet ICBM. |
| SSF | 82 | 16,000 | | | 7,257 | |
| JCS | 84 | | | | | The 308 SS-18s have "more throw-weight potential than the combined force of all US ICBMs and SLBMs"; [expected to continue through 1993, graph], FY85-86. |

## PAYLOAD

| Source | Year | lb | t | kg | kg conv | Notes |
|---|---|---|---|---|---|---|
| AWar | 80 | c. 16,000 | | | 7,257 | "Useful payload". |

## WEIGHT

| Source | Year | lb | t | kg | kg conv | Notes |
|---|---|---|---|---|---|---|
| JCS | 75 | | | | | Heavier than SS-9. |
| Rock | 78 | | | | | Heaviest and largest ICBM built. |
| FI | 81 | | | c. 220,000 | 220,000 | |

## LENGTH

| Source | Year | in | ft | cm | m | m conv | Notes |
|---|---|---|---|---|---|---|---|
| FIar | 76 | | | | 36 | 36.00 | |
| SWM | 77 | | 121.0 | | 37.0 | 36.88 | |
| Rock | 78 | | | | 36 | 36.00 | |
| FI | 81 | | | | c. 36 | 36.00 | |
| JWS | 81 | | | | c. 35 | 35.00 | |
| SMP | 81 | | | | 31 | 31.00 | [From graph], '81, '83. |
| AFM | 82 | | 118'0" | | | 35.97 | |
| AWST | 82 | | 120 | | | 36.58 | |
| FI+ | 83 | | | | c. 35 | 35.00 | |
| AFM+ | 84 | | 104'0" | | | 31.70 | |

## DIAMETER

| Source | Year | in | ft | cm | m | cm conv | Notes |
|---|---|---|---|---|---|---|---|
| FIar | 76 | | | | 3.4 | 340.00 | |
| SWM | 77 | | 11.0 | | 3.3 | 335.28 | |
| Rock | 78 | | | | 3.4 | 340.00 | [Table] |
| | | | | | > 3 | 300.00 | [Text] |
| FI | 81 | | | c. 300 | | 300.00 | |
| JWS | 81 | | | | c. 3 | 300.00 | Base. |
| AFM | 82 | | 10'0" | | | 304.80 | Max. |
| AWST | 82 | | 11 | | | 335.28 | |

---------- **NUMBER OF STAGES** ----------
| | | |
|---|---|---|
| JCS | 74 | 2 |
| MOW | 76 | 2 |
| SWM | 77 | 2 |
| Rock | 78 | 2 |
| FI | 81 | 2 |
| JWS | 81 | 2 |
| AFM | 82 | 2 |
| AWST | 82 | 2 |

---------- **TYPE OF PROPULSION** ----------
| | | | |
|---|---|---|---|
| JCS | 74 | liquid | FY75-77. |
| SecC | 74 | | PBV powered by liquid fuel engines. |
| SecD | 74 | liquid | FY75-76,81-82. |
| FIar | 76 | storable liquid | Largest liquid-propellant missile in the world. |
| MOW | 76 | liquid | |
| SWM | 77 | liquid storable | |
| Rock | 78 | storable liquid | Stages 1 and 2. |
| FI | 81 | liquid | |
| JWS | 81 | liquid | |
| AFM | 82 | liquid | |
| AWST | 82 | liquid-storable | |
| SSF | 82 | liquid | |

---------- **MIDCOURSE (OR MAIN) GUIDANCE** ----------
| | | | |
|---|---|---|---|
| JCS | 74 | | Has digital computer. |
| FI | 81 | inertial | |
| AWST | 82 | | "Advanced guidance system". |
| SSF | 82 | | Fly-the-wire, onboard digital computer. |

---------- **CONTROL SYSTEM** ----------
| | | |
|---|---|---|
| Rock | 78 | Gimballed nozzles. |
| FI | 81 | ? Gimballed nozzles. |

---------- **DEVELOPMENT BEGAN** ----------
| | | | |
|---|---|---|---|
| SSF | 82 | 1965 | Year design began. |
| SMP | 85 | 1964 | Technological development began. [From graph.] |

---------- **DEVELOPMENT UNDERWAY** ----------
| | | | |
|---|---|---|---|
| SecD | 73 | | Being developed and tested. |
| IAD6 | 75 | | [Currently] "in advanced development". |
| Rock | 78 | 1972-1974 | Developed. |

---------- **PROTOTYPE TESTS** ----------
| | | | |
|---|---|---|---|
| JCS | 74 | | Now being tested at Tyuratam. Recent tests have used single RV; indicates continuing interest in large warhead with greater accuracy. |
| SecD | 74 | | Has been tested with single large RV and 5 relatively large MIRVs. Tested during past year. |
| GA01 | 75 | Feb 1974 | Test, Tyuratam-Pacific. |
| IAD6 | 75 | end, 1972 | Tests observed. |
| SecD | 75 | | Flight tested in both a MIRV and single RV mode, FY76-77. |
| SecD | 76 | 1974-75 | In 1974 extensive flight testing; in 1975 testing continued. |
| JWS | 81 | | Recent tests with single RV. |
| SMP | 85 | 1969-70 | Engineering and testing began. [From graph.] |

---------- **PRODUCTION UNDERWAY** ----------
| | | |
|---|---|---|
| MOW | 76 | In production. |

---------- **IOC** ----------
| | | | |
|---|---|---|---|
| SecD | 77 | [1974-75] | [Midway between 74 and 75 lines on graph, ed]. |
| JCS | 83 | c. 1974 | Year of introduction [from graph, Mod not specified, ed]. |

---------- **DEPLOYMENT BEGAN** ----------
| | | | |
|---|---|---|---|
| JCS | 74 | | By mid-1975, new large silos estimated to become operational with SS-X-18. |
| SecD | 75 | | Started, or soon will start, operational deployment. |
| JCS | 76 | | Single RV version now being deployed. |
| JCS | 77 | 1973 | Began converting first SS-9 launch group to SS-18. |
| JCS | 78 | | Probably all 3 mods will be deployed. |
| Rock | 78 | 1976 | Service intro. |
| SecD | 80 | | Soviets have "only just begun to deploy a version" of SS-18 with 10 MIRVs. |
| JWS | 81 | 1974 | Deployed operationally. |
| JCS | 84 | 1974-75 | Major strategic system deployments [graph], FY85-86. |
| SecD | 84 | 1973-74 | Introduction of strategic systems [graph]. |
| SMP | 85 | 1974-75 | [From graph.] |

## 200 SOVIET MISSILES IDDS

---------- **DEPLOYMENT COMPLETE** ----------

| | | | |
|---|---|---|---|
| JCS | 81 | | During the year, USSR "converted the remaining deployed SS-9 launchers, some 300 missiles, to the SS-18". |
| JWS | 81 | | Virtually all SS-9 silos now converted to SS-18. |
| SecD | 81 | | Deployment expected to end in early 1980s with deployment of remaining planned SS-18s. |
| AFM | 82 | | Replacement of SS-9 with SS-18 complete. |
| JCS | 82 | | SS-9 to SS-18 conversion has been completed. |

---------- **DEPLOYMENT HISTORY** ----------

| | | |
|---|---|---|
| SecD | 73 | Follow-on to SS-9. |
| JCS | 74 | Follow-on to SS-9. |
| JCS | 75 | Work underway at SS-9 complexes to permit SS-18 deployment, FY76-77. |
| FIar | 76 | Now starting to replace hot-launched SS-9. |
| JCS | 77 | SS-9 launch group conversion to SS-18 continues. |
| SecD | 77 | "Continue to be deployed in modified and upgraded silos at a rapid rate". |
| JCS | 78 | For the 4th straight year, "number of SS-9 launch groups being converted to the SS-18 increased". Anticipated that all SS-9 silos will be converted to SS-18. |
| JCS | 79 | For the 5th straight year, "the number of conversions has increased" (SS-9 to SS-18); conversion program continues at a rapid pace. |
| JCS | 80 | Conversion of SS-9 to SS-18 "is continuing at a rapid pace". |
| JWS | 81 | Functional sucessor of SS-9. |
| SMP | 81 | Replaced SS-9. |
| JWS+ | 83 | Conversion from SS-9 completed by 1982. |
| SMP | 83 | Continuing modernization of SS-17,18,19 force. |
| SMP | 84 | "Modernization...nears an end". "At least one additional version of both the SS-18 and SS-19, however, is likely to be produced and deployed in existing silos in the future." |
| SMP | 85 | Will be deployed at least through 1990. [From graph.] "Modified versions of the SS-18...are likely to be produced and deployed in existing silos in the future." |

---------- **NUMBERS PRODUCED** ----------

| | | |
|---|---|---|
| SecD | 80 | No evidence that SS-17,18,19 production is significantly greater than the number of their launchers. |

---------- **NUMBERS IN SERVICE** ----------

---------- **NUMBER IN 1960** ----------

| | | | |
|---|---|---|---|
| SSF | 82 | none | In 1960, 1965, 1970. |

---------- **NUMBER IN 1970** ----------

| | | | |
|---|---|---|---|
| Clns | 80 | none | In 1970-1974. |

---------- **NUMBER IN 1975** ----------

| | | | |
|---|---|---|---|
| IISS | 75 | 10 | July, total deployed. |
| FIar | 76 | 40 | Deployed by September of last year [1975], according to Rumsfeld, US SecDef. |
| SWM | 77 | 10 | All Mod 1, deployed mid-75. Restricted to 310 under SALT I. |
| Clns | 80 | 10 | |
| SSF | 82 | 10 | "All SS-18s counted one warhead in 1975-76". |
| vanD | 83 | 26 | 1 Jan. |

---------- **NUMBER IN 1976** ----------

| | | | |
|---|---|---|---|
| IISS | 76 | 36 | July, total deployed. |
| Clns | 80 | 36 | |
| vanD | 83 | 38 | 1 Jan. |

---------- **NUMBER IN 1977** ----------

| | | | |
|---|---|---|---|
| IISS | 77 | 50 | July, total deployed. |
| SecD | 77 | > 50 | Deployed. |
| Rock | 78 | 40 | Declared by Russians operational in SS-9 silos by 1977. |
| Nitz | 79 | 22 | Launchers. |
| Clns | 80 | 100 | "About 50 out of 100 [SS-18s] were MIRVed in 1977". |
| vanD | 83 | 56 | 1 Jan. |

---------- **NUMBER IN 1978** ----------

| | | | |
|---|---|---|---|
| IISS | 78 | 110 | July, total deployed. |
| JCS | 78 | c. 100 | Currently, silos operational. |
| SecD | 78 | > 100 | Launchers (converted from SS-9s). Deployment of SS-17,18,19 continues at c. 125 per year, FY79-81. |
| Clns | 80 | 176 | |
| vanD | 83 | 92 | 1 Jan. |

## NUMBER IN 1979

| | | | |
|---|---|---|---|
| IISS | 79 | 200 | July, total deployed. |
| JCS | 79 | < 200 | Currently, "nearly" 200 operational silos. |
| SecD | 79 | < 200 | "Nearly" 200 launchers (converted from SS-9s). |
| Clns | 80 | 240 | Also, 68 SS-9s are being converted to SS-18. "All but 26 of 240 [SS-18s] were MIRVed in 1979". |
| vanD | 83 | 188 | 1 Jan. |

## NUMBER IN 1980

| | | | |
|---|---|---|---|
| AWar | 80 | 248 | Deployed; 60 being prepared. |
| IISS | 80 | 308 | 1980-83, July, total deployed. |
| JCS | 80 | | When complete, force will be "300 SS-18s of various models". |
| | | 308 | "Boosters of the SS-18 class". |
| SecD | 80 | > 200 | Launchers (converted from SS-9s). Vast majority of deployed SS-18s are of 8 or 10 warhead variety. |
| SSF | 82 | 308 | |
| vanD | 83 | 248 | 1 Jan. |

## NUMBER IN 1981

| | | | |
|---|---|---|---|
| SecD | 75 | | Sizeable force projected; [ICBM-threatening force] could be operational by early 80s. |
| FI | 81 | 308 | Deployed. |
| JWS | 81 | c. 300 | |
| SecD | 81 | > 300 | SS-17,18,19s mostly equipped with MIRVs. |
| SMP | 81 | 308 | '81, '83, '84, '85. |
| vanD | 83 | 308 | 1 Jan. |

## NUMBER IN 1982

| | | | |
|---|---|---|---|
| AFM | 82 | 308 | |
| AWST | 82 | 300 | Deployed. |
| JCS | 82 | 308 | Inventory. |

## NUMBER IN 1983

| | | | |
|---|---|---|---|
| JWS+ | 83 | 308 | |
| SMP | 83 | | Will remain unchanged as current ICBM modernization is completed. Great majority of SS-17,18,19 equipped with MIRVs. |

## NUMBER IN 1984

| | | | |
|---|---|---|---|
| JCS | 84 | 308 | As of 1 Jan. |

## NUMBER IN 1985

| | | | |
|---|---|---|---|
| JCS | 85 | 308 | As of 1 Jan. |

## UNDATED NUMBER

| | | | |
|---|---|---|---|
| SSF | 82 | 308 | "Number of missiles deployed". |

## LAUNCHER/SILO

| | | |
|---|---|---|
| JCS | 74 | By mid-1975, new large silos estimated to become operational with SS-X-18. |
| SecD | 74 | Can be deployed in the "new type large silos". Can be deployed in SS-9 silos (with some modification). |
| FIar | 76 | Since cold-launched SS-18 takes up less room than SS-9, room for more steel and concrete. SS-18 could be deployed in Russian IIIZ C3I silos, but Pentagon now says only 30 of these, not 150 as reported earlier. |
| JCS | 76 | Every SS-18 deployed goes into new modernized, hard silos; the associated launch control facilities are in silos, not bunkers. |
| SecD | 77 | Some modified silos hardened to resist very high overpressure. |
| AWar | 80 | Silo hardness [SS-17,18,19] 3,500-4,500 psi. |
| SecD | 80 | All "converted silos may be capable of withstanding very high static overpressures". |
| SMP | 81 | More than half of 1,398 ICBM launchers rebuilt to house SS-17,18,19 in "vastly more survivable, hardened silos". |
| SSF | 82 | Basing mode: Hardened silo. |
| SMP | 83 | SS-17,18,19 "housed in the world's hardest silos", '83, '84, '85. |

## DESIGNER

| | | |
|---|---|---|
| SSF | 82 | Yangel Design Bureau. |

## DESIGN AND ENGINEERING

| | | |
|---|---|---|
| FIar | 76 | Intended to replace the existing SS-9 force. |
| MOW | 76 | Designed to replace SS-9. |

## USE AND CONFIGURATION

| | | |
|---|---|---|
| vanD | 83 | Missiles in a battalion: "6 or 10(2 groups)". |
| SMP | 84 | Transported and loaded into silo in a canister. Special silo-loading equipment is used with SS-18 [shown in drawing]. |

## RELOADS IN GENERAL

| | | |
|---|---|---|
| SecC | 74 | Cold launched: boosted out of silos on piston (called sabot) powered by gas produced by burning of small solid fuel charge before main booster ignited. |
| JCS | 75 | "Cold" launch, FY76-79,82. |
| SecD | 75 | Cold-launched, FY76-77,79-82. |
| MOW | 76 | Cold launched. |
| SWM | 77 | Cold launched. |
| JCS | 78 | Cold launch. |
| Rock | 78 | Cold launch. |
| SecD | 80 | No evidence USSR has any plan or capability to use excess missiles as reserves or refires; quite confident they have not tested or trained in these ways. |
| FI | 81 | Cold-launched. |
| JCS | 81 | Cold launch technique could allow reload and refire in protracted nuclear conflict, FY82-83. USSR probably cannot reload silos "in less than several days". |
| JWS | 81 | Cold-launch. |
| SMP | 81 | Cold launch (all mods), '81, '83. "The Soviets probably cannot refurbish and reload silo launchers in a period less than several days". |
| AFM | 82 | Cold-launched. |
| AWST | 82 | Cold-launch missile. |
| IISS | 82 | Cold launch. |
| SSF | 82 | Launching mode: cold. |
| JWS+ | 83 | US report says all Soviet liquid-fueled ICBMs contained in launch canister offering possibility of reloading. |
| SMP | 83 | Cold launch [of SS-17,18] lends itself to a reload and refire capability. "The Soviets probably cannot refurbish and reload silo launchers in a period less than a few days". "Contained in a launch canister within the silo. This and the silo design minimize damage to the launcher during" firing, giving reload capability. |

## STRATEGY

| | | |
|---|---|---|
| SecD | 73 | A force of 300 could pose a serious threat to US ICBMs in silos. |
| JCS | 75 | "More lethal" than SS-9. |
| SecD | 75 | Sizeable force projected; [ICBM-threatening force] could be operational by early 80s. |
| FIar | 76 | Warhead/accuracy combinations make it potent weapon against US Minuteman force. According to Air Force magazine, SS-18 has suffered persistent technical problems not yet solved, despite weapon's service entry; report not confirmed by any other source. |
| SWM | 77 | "Every indication of being a first strike weapon". |
| JCS | 78 | Cold launch allows refires and greater throw-weight [per silo]. |
| Rock | 78 | Replacement for SS-9; with requirement to provide technology for a first strike capability. Can carry very high yield charge or multitude of MIRVs for "concussion attack, penetration to generate a seismic disturbance (earthquake) or multiple attack on industrial areas as required". |
| SecD | 78 | SS-17,18,19 all have "potential, with feasible accuracy improvements, to attain high single-shot kill probabilities against US silos". Uncertain how fast threat to Minuteman might become serious, presently function of SS-18,19. |
| SecD | 79 | Major progress [in past year] in understanding threat to ICBMs; intelligence data on new SS-18,19 versions indicates substantial threat to Minuteman by early 1980s. |
| SecD | 80 | "The numbers of high quality warheads on new versions of the SS-18 and SS-19 seriously threaten our MINUTEMAN force in the early 1980s...". |
| SMP | 81 | "Improved reliability, range, payload accuracy and survivability". |
| JCS | 84 | Today, most accurate versions of SS-18,19 capable of destroying time-urgent and hardened targets in an initial attack, FY85-86, ["most" such targets in FY86 report, ed]. Improvements in deployed SS-18,19s projected for 1983-93. If Soviets increase number of warheads on SS-18s, EMT and HTKP [hard target kill capability] "could approach the capabilities possessed by the entire US ICBM and SLBM forces", [also shown on graph 1983-93, ed.], FY85-86. |
| SecD | 84 | SS-18,19s carry 4,000-5,000 "highly accurate warheads designed specifically to attack our missile silos"; with only a portion of warheads, can attack Minuteman force. |

## GEOGRAPHICAL DEPLOYMENTS

| | | |
|---|---|---|
| JWS+ | 83 | Deployed centrally in Dombarovskiy, Imeni Gastello, Aleysk, Zhangiz Tobe, and Uzhur missile fields. |
| SMP | 83 | [Shown on map at] Dombarovskiy, Kartaly, Imeni Gastello, Aleysk, Zhangiz Tobe, Uzhur, '83, '84, '85. |

## RELATION TO OTHER MISSILES

| | | |
|---|---|---|
| JCS | 75 | Comparable in volume to SS-9, FY76-77. |
| SecD | 75 | Volume comparable to SS-9. |
| JWS | 81 | US drawings indicate same size as SS-9. |
| SecD | 81 | MX will have "military capability" "equivalent of the much larger" SS-18. |
| SMP | 81 | Largest of current Soviet ICBMs, similar in dimensions to SS-9, about twice the size of MX. |

# SS-18 Mod 1

| Sources are: | | SecD | 75 | JCS | 77 | AWar | 80 | JWS | 81 | SSF | 82 | AFM+ | 84 |
|---|---|---|---|---|---|---|---|---|---|---|---|---|---|
| JCS | 74 | FIar | 76 | SWM | 77 | Clns | 80 | SMP | 81 | CBO | 83 | ArkS | 84 |
| JCS | 75 | MOW | 76 | JCS | 78 | SecD | 80 | AFM | 82 | Clns+ | 83 | SMP | 85 |
| SecC | 75 | SecD | 76 | Rock | 78 | FI | 81 | IISS | 82 | JWS+ | 83 | BW | 86 |

---------- **US DESIGNATION** ----------

| JCS | 74 | SS-18 Mod 1 |
|---|---|---|
| SecC | 75 | SS-18 Mod 1 |
| SecD | 75 | SS-18 Mod 1 |
| FIar | 76 | SS-18 Mod 1 |
| MOW | 76 | SS-18 Mod 1 |
| SWM | 77 | SS-18 Mod 1 |
| Rock | 78 | SS-18 Mod 1 |
| AWar | 80 | SS-18 mod |
| Clns | 80 | SS-18 Mod 1 |
| FI | 81 | SS-18 Mod1 |
| JWS | 81 | SS-18 Mod1 |
| SMP | 81 | SS-18 Mod 1 |
| AFM | 82 | SS-18 Mod1 |
| IISS | 82 | SS-18 Mod 1 |
| SSF | 82 | SS-18 Mod 1 |
| CBO | 83 | SS-18 MOD 1 |

---------- **RANGE** ----------

| | | mi | nm | km | km conv | |
|---|---|---|---|---|---|---|
| JCS | 75 | | > 5500 | | 10,186 | FY76-77. |
| SWM | 77 | 6,525 | | 10,500+ | 10,498 | [Sic, + for km, not for mi, ed] |
| AWar | 80 | | | 12,000 | 12,000 | |
| Clns | 80 | | 6,000 | | 11,112 | |
| SecD | 80 | | | 12,000 | 12,000 | FY81-82. [Qualification:] "exclusive of range imparted by" PBV. |
| FI | 81 | | | 12,000 | 12,000 | |
| JWS | 81 | | | 10,500 | 10,500 | |
| SMP | 81 | | | 12,000 | 12,000 | Max, '81, '83. |
| AFM | 82 | 7,450 | | | 11,987 | |
| IISS | 82 | | | 12,000 | 12,000 | Max. |
| SSF | 82 | | 6,500 | | 12,038 | |
| JWS+ | 83 | | | 12,000 | 12,000 | Max. |
| BW | 86 | | | 12,000 | 12,000 | |

---------- **CEP** ----------

| | | ft | mi | nm | m | km | m conv | |
|---|---|---|---|---|---|---|---|---|
| JCS | 75 | | | | | | | Improved. Capable of destroying any known fixed target, FY76-77. |
| SecD | 75 | | | | | | | Believed to have better CEP than SS-9; has good hard target capability. |
| Rock | 78 | | | | 550 | | 550.00 | Target accuracy. |
| AWar | 80 | | | 0.23 | | | 425.96 | |
| Clns | 80 | | | 0.23 | | | 425.96 | |
| FI | 81 | | | | c. 400 | | 400.00 | |
| IISS | 82 | | | | 450 | | 450.00 | |
| SSF | 82 | | | 0.23 | | | 425.96 | |
| CBO | 83 | | | 0.15 | | | 277.80 | In 1983. |
| | | | | 0.12 | | | 222.24 | In 1990. |
| | | | | 0.08 | | | 148.16 | In 1996. |
| AFM+ | 84 | 1,500 | | | | | 457.20 | |
| BW | 86 | | | | 430 | | 430.00 | |

```
---------- WARHEAD CHARACTERISTICS ----------
JCS   75    single RV       FY76-79.
SecD  75    single RV
FIar  76    single RV       Mods 1 and 3; Mod 3 RV is lighter and more accurate than Mod 1.
MOW   76    single RV
SecD  76    single RV       FY77,81-82.
SWM   77    single RV
Rock  78    single RV
AWar  80    single RV
Clns  80    single RV
FI    81    single RV
SMP   81    single RV       '81, '83.
AFM   82    single RV
IISS  82    single RV
SSF   82    single RV
CBO   83    single RV
JWS+  83    single RV
BW    86    single RV

---------- NUCLEAR YIELD ----------
              kT        MT        kT conv
JCS   75                                      "Much greater destructive power than Titan",
                                              FY76-77.
FIar  76                25        25,000
SWM   77                40-50     40,000-50,000
Rock  78                <_ 50     50,000      Maximum yield charge.
                        18-25     18,000-25,000   "18 or 25 MT" is its most flexible application.
AWar  80                24        24,000
Clns  80                24        24,000
FI    81                24        24,000
JWS   81                ?> 25     25,000
                        50        50,000      As high as 50 MT has been suggested.
AFM   82                18-25     18,000-25,000
IISS  82                20        20,000
SSF   82                24.0      24,000
CBO   83    2,500                 2,500
JWS+  83                20-25     20,000-25,000   [Analysis table]
AFM+  84                20        20,000
BW    86    24,000                24,000

---------- THROWWEIGHT ----------
              lb        t         kg         kg conv
                                             7,484
Clns  80    16,500                           7,484
IISS  82    16,500                           7,484
CBO   83    16,500                           7,484

---------- TYPE OF PROPULSION ----------
Clns  80    liquid
BW    86    storable liquid

---------- MIDCOURSE (OR MAIN) GUIDANCE ----------
JCS   75    Employs sophisticated version of "fly-the-wire", with on-board digital
            computer, FY76-77.
SecD  75    Has computer aboard.
Rock  78    Guidance system with 18 or 25 MT warhead improves accuracy and increases
            penetration velocity [over 50 MT charge].
JWS   81    Report of sophisticated variant of Soviet 'fly-the-wire' control system, with
            onboard computer.

---------- PROTOTYPE TESTS ----------
SecD  75           Mod 1 testing more advanced [than Mod 2].
SSF   82    1972   First flight test.

---------- IOC ----------
JCS   74    1975
JCS   75    1974   FY76-79.
FI    81    1974   Operational since 1974.

---------- DEPLOYMENT BEGAN ----------
JCS   75           Now operational, FY76-77.
SecD  75           Believed operational.
FIar  76           Mod 1 is now in service, Mods 2 and 3 aren't yet.
SecD  76           Now operational.
Clns  80    1974   First deployed.
IISS  82    1975   First year deployed.
SSF   82    1974   Year operation began.
SMP   85    1974-75   [Presumed Mod 1, from graph.]
BW    86    1974
```

## BALLISTIC MISSILES ICBMs SS-18 Mod 1

|  |  | **RETIRED** |
|---|---|---|
| ArkS | 84 | "According to DIA...the SS-18 is now definitely fully MIRVed to 3,080 operational warheads". |
| BW | 86 | ? 1981 |

|  |  | **DEPLOYMENT HISTORY** |
|---|---|---|
| SecC | 75 | Mod 1 being deployed first, to be followed, and at least partially replaced by Mod 2. |
| MOW | 76 | Being deployed in new silos as well as SS-9 silos. |
| JCS | 78 | Soviets "may plan to deploy a mix of the single-RV variants"; although Mod 3 may eventually replace Mod 1. |

### NUMBERS IN SERVICE

**NUMBER IN 1970**
| Clns+ | 83 | None in 1970-1974, Mods 1 and 3. |
|---|---|---|

**NUMBER IN 1975**
| SWM | 77 | 10 | Deployed mid-75. |
|---|---|---|---|
| Clns+ | 83 | | 10, Mods 1 and 3. |

**NUMBER IN 1976**
| Clns+ | 83 | 36, Mods 1 and 3. |
|---|---|---|

**NUMBER IN 1977**
| JCS | 77 | Since Mod 1 was only version where testing and crew training were complete by time of first SS-18 silo completion, believed Mod 1 is installed in these first launchers. Soviets apparently deploy mix of single-RV SS-18 mods; Mod 3 advantages suggest Mod 1 deployment will be limited; Mod 3 may eventually replace Mod 1. |
|---|---|---|
| Clns+ | 83 | 60, Mods 1 and 3. |

**NUMBER IN 1978**
| Clns+ | 83 | 36 in 1978-1979, Mods 1 and 3. |
|---|---|---|

**NUMBER IN 1980**
| Clns | 80 | 214 are "Mod 2, 4"; 26 are "Mod 1, 3" [of 240 deployed]. |
|---|---|---|
| Clns+ | 83 | 26 in 1980-1981, Mods 1 and 3. |

**NUMBER IN 1982**
| AFM | 82 | | Some are operational. |
|---|---|---|---|
| Clns+ | 83 | | 16, Mods 1 and 3. |

**NUMBER IN 1983**
| CBO | 83 | 24 | 1983, 1990, 1996. "Estimate based on top-line Soviet RV count provided in Department of State, 'Fact Sheet on START', May 1982, and the number of RVs deployed on other systems". |
|---|---|---|---|

**UNDATED NUMBER**
| BW | 86 | ? 56 | Peak. |
|---|---|---|---|

|  |  | **LAUNCHER/SILO** |
|---|---|---|
| MOW | 76 | Being deployed in new silos as well as SS-9 silos. |

|  |  | **USE AND CONFIGURATION** |
|---|---|---|
| CBO | 83 | System availability: 0.95. |

|  |  | **RELOADS IN GENERAL** |
|---|---|---|
| Clns | 80 | Cold launch: yes. |

|  |  | **STRATEGY** |
|---|---|---|
| AFM | 82 | For use against deep underground shelters. |

# SS-18 Mod 2

```
Sources are:    FIar 76      JCS  77      AWar 80      JWS  81      SSF  82      LuO1 83      SMP  85
JCS  75         JCS  76      SWM  77      Clns 80      SMP  81      CBO  83      vanD 83      BW   86
SecC 75         MOW  76      JCS  78      SecD 80      AFM  82      Clns+ 83     AFM+ 84
SecD 75         SecD 76      Rock 78      FI   81      IISS 82      JWS+ 83      ArkS 84
```

---------- **US DESIGNATION** ----------

|  |  |
|---|---|
| JCS 75 | SS-18 Mod 2 |
| SecC 75 | SS-18 Mod 2 |
| SecD 75 | SS-18 Mod 2 |
| FIar 76 | SS-18 Mod 2 |
| MOW 76 | SS-18 Mod 2 |
| SWM 77 | SS-18 Mod 2 |
| Rock 78 | SS-18 Mod 2 |
| AWar 80 | SS-18 mod |
| Clns 80 | SS-18 Mod 2 |
| FI 81 | SS-18 Mod2 |
| JWS 81 | SS-18 Mod2 |
| SMP 81 | SS-18 Mod 2 |
| AFM 82 | SS-18 Mod2 |
| IISS 82 | SS-18 Mod 2 |
| SSF 82 | SS-18 Mod 2 |
| CBO 83 | SS-18 MOD 2/Follow-on. |
| vanD 83 | SS-18 |

---------- **DESCRIPTION** ----------

Rock 78  ICBM

---------- **EDITOR'S NOTES** ----------

vanD 83  Source has 4 separate SS-18 lines, for 1974, 1977, 1979, and 1982. We have assumed the first two to be Mod 2, the second two to be Mod 4. Where the two lines for each Mod differ, they are identified by year.

---------- **RANGE** ----------

| Source | mi | nm | km | km conv | Notes |
|---|---|---|---|---|---|
| SWM 77 | 5,750 | | 9,250+ | 9,251 | [Sic; + for km, not for mi, ed] |
| Rock 78 | | | 12,000 | 12,000 | |
| AWar 80 | | | 11,000 | 11,000 | |
| Clns 80 | | 5,500 | | 10,186 | |
| SecD 80 | | | 11,000 | 11,000 | FY81-82. [Qualification:] "exclusive of range imparted by" PBV. |
| FI 81 | | | 12,000 | 12,000 | |
| JWS 81 | | | 9,250 | 9,250 | |
| SMP 81 | | | 11,000 | 11,000 | Max, '81, '83. |
| AFM 82 | 6,800 | | | 10,941 | |
| IISS 82 | | | 11,000 | 11,000 | Max. |
| SSF 82 | | 5,900 | | 10,926 | |
| JWS+ 83 | | | 11,000 | 11,000 | Max. |
| BW 86 | | | 11,000 | 11,000 | |

---------- **CEP** ----------

| Source | ft | mi | nm | m | km | m conv | Notes |
|---|---|---|---|---|---|---|---|
| MOW 76 | | | | | | | Likely has hard target capability. |
| Rock 78 | | | | 450 | | 450.00 | Target accuracy. |
| AWar 80 | | | 0.23 | | | 425.96 | |
| Clns 80 | | | 0.23 | | | 425.96 | |
| FI 81 | | | | [c 400] | | 400.00 | Similar to Mod 1. |
| IISS 82 | | | | 450 | | 450.00 | |
| SSF 82 | | | 0.23 | | | 425.96 | |
| CBO 83 | | | 0.15 | | | 277.80 | In 1983. |
| | | | 0.12 | | | 222.24 | In 1990. |
| | | | 0.08 | | | 148.16 | In 1996. |
| vanD 83 | | | .30 | | | 555.60 | [1974 entry] |
| | | | .23 | | | 425.96 | [1977 entry] |
| BW 86 | | | | 430 | | 430.00 | |

BALLISTIC MISSILES   ICBMs   SS-18 Mod 2   207

### WARHEAD CHARACTERISTICS

| Source | Yr | Value | Notes |
|---|---|---|---|
| SecD | 75 | ≤ 8 MIRVs | |
| Flar | 76 | 8 MIRVs | |
| MOW | 76 | 8 MIRVs | Has 8 "fairly large MIRVs". |
| SecD | 76 | 8 RVs | |
| JCS | 77 | | Mix of payloads probably will be deployed on Mod 2. |
| SWM | 77 | ≤ 8 MIRVs | From computerised post-boost 'bus'. |
| JCS | 78 | 8-10 RVs | [Also] "8/10" warheads. |
| Rock | 78 | 8-10 x MIRV | |
| AWar | 80 | 8-10 RVs | |
| Clns | 80 | 8 MIRVs | |
| SecD | 80 | "8/10" warheads | FY81-82. |
| FI | 81 | 8-10 MIRVs | |
| JWS | 81 | 8-10 MIRVs | Bus-type, with onboard computer. PBV is spin-stabilized, and thought similar to Minuteman III and Poseidon. |
| SMP | 81 | "8/10" MIRVs | '81, '83. |
| AFM | 82 | 8-10 RVs | On PBV similar to that for US Minuteman III and Poseidon. |
| IISS | 82 | 8 MIRVs | |
| SSF | 82 | 8-10 RVs | |
| CBO | 83 | 10 | |
| JWS+ | 83 | "8/10" RVs | |
| vanD | 83 | 8 RVs | [1974 entry] |
| | | 10 RVs | [1977 entry] |
| BW | 86 | ? 8 MIRVs | |

### NUCLEAR YIELD

| Source | Yr | kT | MT | kT conv | Notes |
|---|---|---|---|---|---|
| SWM | 77 | | 1+ | 1,000 | Each. |
| Rock | 78 | | 16.0-20.0 | 16,000-20,000 | Total. |
| | | 2,000 | | 2,000 | Each of 8-10 MIRVs [currently available]. |
| | | | c. 5 | 5,000 | Warhead under development, 8 of which could go on Mod 2, and would constitute greatest SS-18 threat. |
| AWar | 80 | | 0.55-0.90 | 550.00-900.00 | Varies depending on number in bus. |
| Clns | 80 | 900 | | 900.00 | Each. |
| FI | 81 | 550-900 | | 550.00-900.00 | |
| JWS | 81 | | ? 2 | 2,000 | |
| AFM | 82 | | 2 | 2,000 | |
| IISS | 82 | 900 | | 900.00 | Per warhead. |
| SSF | 82 | | 0.9-0.55 | 900.00-550.00 | Per warhead. |
| CBO | 83 | 500 | | 500.00 | Each. |
| JWS+ | 83 | | < 1 | 1,000 | [Text] |
| | | 500 | | 500.00 | [Analysis table] |
| Lu01 | 83 | | 2 | 2,000 | |
| vanD | 83 | | .90 | 900.00 | [1974 entry] |
| | | | .55 | 550.00 | [1977 entry] |
| AFM+ | 84 | 900 | | 900.00 | |
| BW | 86 | 900 | | 900.00 | |

### THROWWEIGHT

| Source | Yr | lb | t | kg | kg conv |
|---|---|---|---|---|---|
| Clns | 80 | 16,700 | | | 7,575 |
| IISS | 82 | 16,700 | | | 7,575 |
| CBO | 83 | 16,700 | | | 7,575 |

### LENGTH

| Source | Yr | in | ft | cm | m | m conv |
|---|---|---|---|---|---|---|
| Rock | 78 | | | | 36 | 36.00 |

### DIAMETER

| Source | Yr | in | ft | cm | m | cm conv |
|---|---|---|---|---|---|---|
| Rock | 78 | | | | 3.4 | 340.00 |

### TYPE OF PROPULSION

| Source | Yr | | Notes |
|---|---|---|---|
| Rock | 78 | storable liquid | Stages 1 and 2. |
| Clns | 80 | liquid | |
| BW | 86 | storable liquid | |

### PROTOTYPE TESTS

| Source | Yr | Notes |
|---|---|---|
| JCS | 75 | Mod 2 test program proceeding at somewhat slower pace [than Mod 1?, ed]. |
| JCS | 76 | Pace of test program has increased in past year. |
| JCS | 77 | Testing continues on alternative payloads for Mod 2. |

### IOC

| Source | Yr | Value | Notes |
|---|---|---|---|
| JCS | 75 | 1975 | FY76-77. Could be operational in 1975. |
| JCS | 77 | 1976 | FY78-79. |
| vanD | 83 | 1974 | [1974 entry] |
| | | 1977 | [1977 entry] |

```
------------ DEPLOYMENT BEGAN ------------
SecD   75                   Mod 1 believed operational; mod 2 to follow; because Mod 1 testing more advanced.
FIar   76                   Mod 1 is now in service, Mods 2 and 3 aren't yet.
JCS    77                   "Initial deployment may already have begun".
Rock   78     1977          Service intro.
Clns   80     1976          First deployed.
FI     81     1976          Entered service.
IISS   82     1977          First year deployed.
SSF    82     1976          Year operation began.
SMP    85     c. 1976       [Mods 2 and 3 together, from graph.]
BW     86     1977

------------ RETIRED ------------
ArkS   84                   "According to DIA...the SS-18 is now definitely fully MIRVed to 3,080 operational
                            warheads".
BW     86     ? 1983

------------ DEPLOYMENT HISTORY ------------
SecC   75     Mod 1 being deployed first, to be followed, and at least partially replaced by Mod 2.

------------ NUMBERS IN SERVICE ------------

------------ NUMBER IN 1970 ------------
Clns+  83     none          In 1970-1976.

------------ NUMBER IN 1976 ------------
MOW    76                   Mod 2 expected to have widest deployment.
SecD   76                   Deployment of Mod 2 (MIRVed) will be substantially more than single RV mods.

------------ NUMBER IN 1977 ------------
JCS    77                   Mod 2 probably will be most widely deployed.
Clns+  83     40

------------ NUMBER IN 1978 ------------
JCS    78                   Mod 2 expected to be most widely deployed, largely because of MIRV capability.
Clns+  83     140

------------ NUMBER IN 1979 ------------
Clns+  83     154

------------ NUMBER IN 1980 ------------
Clns   80                   214 are "Mod 2, 4"; 26 are "Mod 1, 3" [of 240 deployed].
Clns+  83     162           In 1980-1981.

------------ NUMBER IN 1982 ------------
AFM    82                   Major current operational version.
Clns+  83     92

------------ NUMBER IN 1983 ------------
CBO    83     285           1983, 1990, 1996.

------------ UNDATED NUMBER ------------
BW     86     ? 132         Peak.

------------ USE AND CONFIGURATION ------------
CBO    83     System availability: 0.95.
vanD   83     Reliability: .85; Penetration 1.0.

------------ RELOADS IN GENERAL ------------
Clns   80     Cold launch: yes.
```

# SS-18 Mod 3

| Sources are: | SecD | 76 | AWar | 80 | JWS | 81 | SSF | 82 | Lu01 | 83 | SMP | 85 |
|---|---|---|---|---|---|---|---|---|---|---|---|---|
| FIar | 76 | JCS | 77 | Clns | 80 | SMP | 81 | Clns+ | 83 | AFM+ | 84 | BW | 86 |
| JCS | 76 | JCS | 78 | SecD | 80 | AFM | 82 | FI+ | 83 | ArkS | 84 | | |
| MOW | 76 | Rock | 78 | FI | 81 | IISS | 82 | JWS+ | 83 | AWST | 84 | | |

---------- **US DESIGNATION** ----------

| FIar | 76 | SS-18 Mod 3 |
|---|---|---|
| JCS | 76 | SS-18 Mod 3 |
| MOW | 76 | SS-18 Mod 3 |
| SecD | 76 | SS-18 Mod 3 |
| Rock | 78 | SS-18 Mod 3 |
| AWar | 80 | SS-18 mod |
| Clns | 80 | SS-18 Mod 3 |
| FI | 81 | SS-18 Mod3 |
| JWS | 81 | SS-18 Mod3 |
| SMP | 81 | SS-18 Mod 3 |
| AFM | 82 | SS-18 Mod3 |
| IISS | 82 | SS-18 Mod 3 |
| SSF | 82 | SS-18 Mod 3 |

---------- **DESCRIPTION** ----------

| JWS | 81 | ICBM |
|---|---|---|
| SSF | 82 | "Global-range missile". |

---------- **EDITOR'S NOTES** ----------

SSF 82   Source has 2 separate listings for SS-18 Mod 3 under Intercontinental-range missiles and Global-range missiles. Data is from the "Global" entry unless otherwise indicated.

---------- **RANGE** ----------

| | | mi | nm | km | km conv | |
|---|---|---|---|---|---|---|
| JCS | 76 | | | | | Mod 3 has longer range than others, FY77-79. |
| AWar | 80 | | | 16,000 | 16,000 | |
| Clns | 80 | | 8,000 | | 14,816 | |
| SecD | 80 | | | 16,000 | 16,000 | FY81-82. [Qualification:] "exclusive of range imparted by" PBV. |
| FI | 81 | | | 16,000 | 16,000 | |
| JWS | 81 | | | 10,500 | 10,500 | |
| SMP | 81 | | | 16,000 | 16,000 | Max, '81, '83. |
| AFM | 82 | 9,940 | | | 15,993 | |
| IISS | 82 | | | 10,500 | 10,500 | Max. |
| SSF | 82 | | 8,640 | | 16,001 | [Intercontinental entry] |
| | | | 8,600 | | 15,927 | [Global entry] |
| FI+ | 83 | | | 9,250 | 9,250 | |
| JWS+ | 83 | | | 16,000 | 16,000 | Max. |
| AFM+ | 84 | 9,950 | | | 16,009 | |
| BW | 86 | | | 16,000 | 16,000 | |

---------- **CEP** ----------

| | | ft | mi | nm | m | km | m conv | |
|---|---|---|---|---|---|---|---|---|
| JCS | 76 | | | | | | | Mod 3 RV somewhat lighter and more accurate than Mod 1, FY77-79. |
| JCS | 78 | | | | | | | Mod 3 more accurate than Mod 1. |
| Rock | 78 | | | | | | | Accuracy to be improved over Mod 1, to put 25 MT warhead "virtually on top of any specific missile silo" in US. |
| AWar | 80 | | | 0.19 | | | 351.88 | |
| Clns | 80 | | | 0.19 | | | 351.88 | |
| FI | 81 | | | | ? 350 | | 350.00 | [Also] more accurate than Mod 1. |
| AFM | 82 | < 590 | | | | | 179.83 | Achieved in trials. |
| IISS | 82 | | | | 350 | | 350.00 | |
| SSF | 82 | | | 0.19 | | | 351.88 | [Both entries] |
| Lu01 | 83 | | | | 200 | | 200.00 | |
| AFM+ | 84 | 1,150 | | | | | 350.52 | |
| BW | 86 | | | | 350 | | 350.00 | |

210 SOVIET MISSILES IDDS

---------- **WARHEAD CHARACTERISTICS** ----------
| | | | |
|---|---|---|---|
| FIar | 76 | single RV | Mods 1 and 3; Mod 3 RV is lighter and more accurate than Mod 1. |
| JCS | 76 | single RV | FY77-79. |
| MOW | 76 | single RV | |
| SecD | 76 | single RV | FY77,81-82. |
| Rock | 78 | single RV | |
| AWar | 80 | single RV | |
| Clns | 80 | single RV | |
| FI | 81 | single RV | |
| JWS | 81 | | RV is high-yield, but lighter and more accurate than Mod 1. |
| SMP | 81 | single RV | '81, '83. |
| AFM | 82 | single RV | Lighter and more accurate than for Mod 1. |
| IISS | 82 | single RV | |
| SSF | 82 | single RV | [Both entries] |
| JWS+ | 83 | single RV | |
| AWST | 84 | > 14 MIRVs | [Sic] |
| BW | 86 | single RV | |

---------- **NUCLEAR YIELD** ----------
| | | kT | MT | kT conv | |
|---|---|---|---|---|---|
| Rock | 78 | | 25 | 25,000 | |
| AWar | 80 | | 20 | 20,000 | |
| Clns | 80 | | 20 | 20,000 | |
| FI | 81 | | 20 | 20,000 | |
| IISS | 82 | | 20 | 20,000 | |
| SSF | 82 | | 20.0 | 20,000 | [Both entries] |
| JWS+ | 83 | | 20 | 20,000 | |
| AFM+ | 84 | | 20 | 20,000 | |
| BW | 86 | 20,000 | | 20,000 | |

---------- **THROWWEIGHT** ----------
| | | lb | t | kg | kg conv |
|---|---|---|---|---|---|
| Clns | 80 | 16,500 | | | 7,484 |
| IISS | 82 | 16,000 | | | 7,257 |
| SSF | 82 | 16,000 | | | 7,257 |

---------- **TYPE OF PROPULSION** ----------
| | | |
|---|---|---|
| Clns | 80 | liquid |
| SSF | 82 | liquid |
| BW | 86 | storable liquid |

---------- **MIDCOURSE (OR MAIN) GUIDANCE** ----------
| | | | |
|---|---|---|---|
| JWS | 81 | inertial | |
| SSF | 82 | | Fly-the-wire, onboard digital computer. |

---------- **PROTOTYPE TESTS** ----------
| | | | |
|---|---|---|---|
| JWS | 81 | | Recent tests of single RV suggest USSR has continuing interest in very powerful warhead. |
| SSF | 82 | 1975 | First flight test. |

---------- **IOC** ----------
| | | | |
|---|---|---|---|
| JCS | 76 | 1975 | |
| JCS | 77 | 1976 | FY78-79. |

---------- **DEPLOYMENT BEGAN** ----------
| | | | |
|---|---|---|---|
| FIar | 76 | | Mod 1 is now in service, Mods 2 and 3 aren't yet. |
| JCS | 77 | | May now be ready for deployment. |
| Rock | 78 | | Not yet sufficiently developed for operational service. |
| Clns | 80 | 1977 | First deployed. |
| FI | 81 | 1976 | Deployed since 1976. |
| IISS | 82 | 1979 | First year deployed. |
| SSF | 82 | 1976 | Year operation began, [both entries] |
| SMP | 85 | c. 1976 | [Mods 2 and 3 together, from graph.] |
| BW | 86 | 1977 | |

---------- **RETIRED** ----------
| | | | |
|---|---|---|---|
| ArkS | 84 | | "According to DIA...the SS-18 is now definitely fully MIRVed to 3,080 operational warheads". |
| BW | 86 | ? 1983 | |

---------- **DEPLOYMENT HISTORY** ----------
| | | |
|---|---|---|
| JCS | 77 | Soviets apparently deploy mix of single-RV SS-18 mods; Mod 3 advantages suggest Mod 1 deployment will be limited; Mod 3 may eventually replace Mod 1. |
| JCS | 78 | Soviets "may plan to deploy a mix of the single-RV variants"; although Mod 3 may eventually replace Mod 1. |
| AFM | 82 | Mod 3 may replace Mod 1. |

---------- **NUMBERS IN SERVICE** ----------

------------ **NUMBER IN 1970** ------------
Clns+ 83           None in 1970-1974, Mods 1 and 3.

------------ **NUMBER IN 1975** ------------
Clns+ 83           10, Mods 1 and 3.

------------ **NUMBER IN 1976** ------------
Clns+ 83           36, Mods 1 and 3.

------------ **NUMBER IN 1977** ------------
Clns+ 83           60, Mods 1 and 3.

------------ **NUMBER IN 1978** ------------
Clns+ 83           36 in 1978-1979, Mods 1 and 3.

------------ **NUMBER IN 1980** ------------
Clns  80           214 are "Mod 2, 4"; 26 are "Mod 1, 3" [of 240 deployed].
Clns+ 83           26 in 1980-1981, Mods 1 and 3.

------------ **NUMBER IN 1982** ------------
Clns+ 83           16, Mods 1 and 3.

------------ **UNDATED NUMBER** ------------
BW    86    ? 24   Peak.

------------ **LAUNCHER/SILO** ------------
SSF   82           Basing mode: Hardened silo.

------------ **DESIGNER** ------------
SSF   82           Yangel          Design Bureau.

------------ **USE AND CONFIGURATION** ------------
JCS   77           First crew training launch occurred Feb 1976.
AFM   82           Crew training launches began Feb 1976.

------------ **RELOADS IN GENERAL** ------------
Clns  80           Cold launch: yes.

# SS-18 Mod 4

```
Sources are:   Clns  80      SSF     82      JWS    83      vanD   83      SMP    84
Nitz  79       FI    81      Clns+   83      Lu01   83      AFM    84      SMP    85
AWar  80       IISS  82      IISS+   83      SMP    83      ArkS   84      BW     86
```

---------- **US DESIGNATION** ----------

|  | |  |
|---|---|---|
| Nitz | 79 | SS-18 |
| AWar | 80 | SS-18 mod |
| Clns | 80 | SS-18 Mod 4 |
| FI | 81 | SS-18 Mod4 |
| IISS | 82 | SS-18 Mod 4 |
| SSF | 82 | SS-18 Mod 4 |
| SMP | 83 | SS-18 Mod 4 |
| vanD | 83 | SS-18 |

---------- **EDITOR'S NOTES** ----------

Nitz 79 — This missile presumed Mod 4 because of 10 RVs; it appears in source on 4 lines with dates in the designation, but differing only in CEP.

vanD 83 — Source has 4 separate SS-18 lines, for 1974, 1977, 1979, and 1982. We have assumed the first two to be Mod 2, the second two to be Mod 4. Where the two lines for each Mod differ, they are identified by year.

---------- **RANGE** ----------

| Source | | mi | nm | km | km conv | |
|---|---|---|---|---|---|---|
| AWar | 80 | | | 10,000 | 10,000 | |
| Clns | 80 | | 5,500 | | 10,186 | |
| FI | 81 | | | 10,000 | 10,000 | |
| IISS | 82 | | | 9,000 | 9,000 | Max. |
| SSF | 82 | | 5,400 | | 10,000 | |
| IISS+ | 83 | | | 11,000 | 11,000 | |
| JWS | 83 | | | 11,000 | 11,000 | Max. |
| SMP | 83 | | | 11,000 | 11,000 | Max, '83, '84, '85. |
| AFM | 84 | 6,800 | | | 10,941 | |
| BW | 86 | | | 11,000 | 11,000 | |

---------- **CEP** ----------

| Source | | ft | mi | nm | m | km | m conv | |
|---|---|---|---|---|---|---|---|---|
| Nitz | 79 | | | .2 | | | 370.40 | 1977. |
| | | | | .17 | | | 314.84 | 18 Jun 1979. |
| | | | | .15 | | | 277.80 | 1982. |
| | | | | .12 | | | 222.24 | 1985. |
| AWar | 80 | | | 0.14 | | | 259.28 | |
| Clns | 80 | | | 0.14 | | | 259.28 | |
| FI | 81 | | | | 260 | | 260.00 | |
| IISS | 82 | | | | 300 | | 300.00 | |
| SSF | 82 | | | 0.14 | | | 259.28 | |
| Lu01 | 83 | | | | 300 | | 300.00 | |
| vanD | 83 | | | .14 | | | 259.28 | [1979 entry] |
| | | | | .10 | | | 185.20 | [1982 entry] |
| AFM | 84 | < 1,000 | | | | | 304.80 | |
| BW | 86 | | | | 260 | | 260.00 | |

---------- **WARHEAD CHARACTERISTICS** ----------

|  | | | |
|---|---|---|---|
| Nitz | 79 | 10 RVs | |
| AWar | 80 | 10 MIRVs | |
| Clns | 80 | 10 MIRVs | |
| FI | 81 | 10 MIRVs | 14 payloads, 10 warheads (MIRVs) plus decoys or other penetration aids. |
| IISS | 82 | 10 MIRVs | |
| SSF | 82 | 10 RVs | |
| JWS | 83 | 10 RVs | |
| SMP | 83 | 10 MIRVs | '83, '84, '85. |
| vanD | 83 | 10 RVs | |
| AFM | 84 | 10 MIRVs | |
| BW | 86 | 10 MIRVs | |

BALLISTIC MISSILES  ICBMs  SS-18 Mod 4

### NUCLEAR YIELD

|  |  | kT | MT | kT conv |  |
|---|---|---|---|---|---|
| Nitz | 79 |  | .75 | 750.00 |  |
| AWar | 80 |  | 0.50 | 500.00 |  |
| Clns | 80 | 500 |  | 500.00 | Each. |
| FI | 81 | 500 |  | 500.00 |  |
| IISS | 82 | 500 |  | 500.00 | Per warhead. |
| SSF | 82 |  | 0.55 | 550.00 | Per warhead. |
| JWS | 83 |  | 0.5 | 500.00 |  |
| Lu01 | 83 | 500 |  | 500.00 |  |
| vanD | 83 |  | .55 | 550.00 |  |
| AFM | 84 | 500 |  | 500.00 |  |
| SMP | 84 |  |  |  | Each warhead > 20 times destructive power of nuclear devices developed during WW II, '84, '85. |
| BW | 86 | 500 |  | 500.00 |  |

### THROWWEIGHT

|  |  | lb | t | kg | kg conv |
|---|---|---|---|---|---|
| Nitz | 79 | 16,000 |  |  | 7,257 |
| Clns | 80 | 16,700 |  |  | 7,575 |
| IISS | 82 | 16,700 |  |  | 7,575 |

### LENGTH

|  |  | in | ft | cm | m | m conv |  |
|---|---|---|---|---|---|---|---|
| SMP | 84 |  |  |  | 32 | 32.00 | [From graph], '84, '85. |

### TYPE OF PROPULSION

| Clns | 80 | liquid |
| BW | 86 | storable liquid |

### DEVELOPMENT BEGAN

| SMP | 85 | 1964 |

### IOC

| vanD | 83 | 1979 | [1979 entry] |
|  |  | 1982 | [1982 entry] |

### DEPLOYMENT BEGAN

| Clns | 80 | 1979 | First deployed. |
| IISS | 82 | 1982 | First year deployed. |
| SMP | 84 |  | Deployment [of SS-17 Mod 3, SS-18 Mod 4, SS-19 Mod 3] began only 5 years ago, [6 years in '85 report]. |
| SMP | 85 | c. 1981 | [From graph.] |
| BW | 86 | 1979 |  |

### DEPLOYED

| BW | 86 |  | In service. |

### DEPLOYMENT HISTORY

| SMP | 83 | [Since Sept-81], replacement of older missiles with Mod 4. Deployment of Mod 4 continues in superhardened silos. |
| AFM | 84 | Mod 2 vehicles may be converted to Mod 4. |
| SMP | 85 | Will be deployed at least through 1990. [Presumed Mod 4, from graph.] |

### NUMBERS IN SERVICE

### NUMBER IN 1970

| Clns+ | 83 | none | In 1970-1978. |

### NUMBER IN 1977

| Nitz | 79 | 22 | Launchers. |

### NUMBER IN 1979

| Nitz | 79 | 188 | 18 June, launchers. |
| Clns+ | 83 | 50 |  |

### NUMBER IN 1980

| Clns | 80 |  | 214 are "Mod 2, 4"; 26 are "Mod 1, 3" [of 240 deployed]. |
| Clns+ | 83 | 120 | In 1980-1981. |

### NUMBER IN 1982

| Clns+ | 83 | 200 |

### NUMBER IN 1984

| ArkS | 84 | [308] | "According to DIA...the SS-18 is now definitely fully MIRVed to 3,080 operational warheads". |
| SMP | 84 | 308 |  |

---------- **NUMBER IN 1985** ----------
SMP    85    308

---------- **UNDATED NUMBER** ----------
BW     86    ? 308    Peak.

---------- **USE AND CONFIGURATION** ----------
vanD   83    Reliability: .85; Penetration 1.0.

---------- **RELOADS IN GENERAL** ----------
Clns   80    Cold launch: yes.
SMP    84    Cold launch, '84, '85.

---------- **STRATEGY** ----------
SMP    84    "Specifically designed to attack and destroy ICBM silos and other hardened targets in the United States", '84, '85. "The force of SS-18 Mod 4s currently deployed has the capability to destroy more than 80 percent of the US ICBM silo launchers using two nuclear warheads against each US silo", '84, '85.

---------- **RELATION TO OTHER MISSILES** ----------
AFM    84    Similar to Mod 2.

# SS-18 Mod 5

Sources are:   IISS 82      IISS+ 83     SMP 85

---------- **US DESIGNATION** ----------
IISS 82      SS-18 (Mod 5)

---------- **EDITOR'S NOTES** ----------
IISS 82      All data appears in parentheses, indicating uncertainty, even as to the existence of Mod 5.

---------- **RANGE** ----------
|        | mi | nm | km | km conv |      |
|--------|----|----|------|---------|------|
| IISS 82 |    |    | ? 9,000 | 9,000 | Max. |

---------- **CEP** ----------
|        | ft | mi | nm | m | km | m conv |
|--------|----|----|----|------|----|--------|
| IISS 82 |   |   |   | ? 250 |   | 250.00 |

---------- **WARHEAD CHARACTERISTICS** ----------
IISS 82      ? 10 MIRVs
IISS+ 83     ? 10 MIRVs

---------- **NUCLEAR YIELD** ----------
|        | kT | MT | kT conv |              |
|--------|-----|----|---------|--------------|
| IISS 82 | ? 750 |   | 750.00 | Per warhead. |

---------- **THROWWEIGHT** ----------
|        | lb | t | kg | kg conv |
|--------|------|---|----|---------|
| IISS 82 | ? 16,000 |   |   | 7,257 |

---------- **DEPLOYMENT BEGAN** ----------
IISS 82      ? 1985          First year deployed.
SMP 85                       "Modified versions of the SS-18...are likely to be produced and deployed in existing silos in the future."

# SS-19

```
Sources are:    SecC 75    IISS 77    SecD 78    JCS  80    AWST 82    SMP   83    SMP 85
JCS  74         SecD 75    JCS  77    IISS 79    SecD 80    IISS 82    vanD  83
SecC 74         FIar 76    SecD 77    JCS  79    FI   81    JCS  82    AFM+  84
SecD 74         IISS 76    SWM  77    Nitz 79    JCS  81    SSF  82    JCS   84
GA01 75         JCS  76    CB01 78    SecD 79    JWS  81    CB 0 83    Meyr  84
IA06 75         MOW  76    IISS 78    US08 79    SecD 81    IISS+ 83   SecD  84
IISS 75         NA03 76    JCS  78    AWar 80    SMP  81    JCS  83    SMP   84
JCS  75         SecD 76    Rock 78    Clns 80    AFM  82    JWS+ 83    JCS   85
```

---------- **US DESIGNATION** ----------

```
JCS  74    SS-X-19
SecC 74    SS-19
SecD 74    SS-X-19
GA01 75    SSX-19
JCS  75    SS-19         FY76-86.
SecD 75    SS-19         FY76-82,85.
FIar 76    SS-19
MOW  76    SS-19
SWM  77    SS-19
Rock 78    SS-19
Nitz 79    SS-19/SS-17
AWar 80    SS-19
Clns 80    SS-19
FI   81    SS-19
JWS  81    SS-19
SMP  81    SS-19         '81, '83, '84, '85.
AFM  82    SS-19
AWST 82    SS-19
IISS 82    SS-19
SSF  82    SS-19
CB 0 83                  SS-19/Follow-on.
vanD 83    SS-19
Meyr 84    SS-19
```

---------- **SOVIET DESIGNATION** ----------

```
FI   81    RS-18
JWS  81    ? RS-18
AFM  82    RS-18
IISS 82    RS-18
```

---------- **DESCRIPTION** ----------

```
JCS  74                  Listed as significant initiative, strategic offensive systems.
MOW  76    ICBM
Rock 78    ICBM          Long-range ICBM.
JWS  81    ICBM
AFM  82    "Light" ICBM
SSF  82                  Fourth generation.
```

---------- **EDITOR'S NOTES** ----------

```
JCS  74    Listed FY75-86.
SecC 74    Listed FY75-77.
SecD 74    Listed FY75-82,85.
SMP  81    Listed '81, '83, '84, '85.
Meyr 84    Note to chart: "Values pertain to initial service period".
```

---------- **RANGE** ----------

|  |  | mi | nm | km | km conv |
|---|---|---|---|---|---|
| JCS  | 74 |  | > 5500 |  | 10,186 |
| GA01 | 75 | 6,000+ |  |  | 9,654 |
| IA06 | 75 |  |  | 9,000+ | 9,000 |
| FIar | 76 |  |  | 10,000 | 10,000 |
| MOW  | 76 | 5,750 |  | 9,250 | 9,251 |
| NA03 | 76 | 6,500 |  |  | 10,458 |
| SWM  | 77 | 6,214 |  | 10,000 | 9,998 |
| Rock | 78 |  |  | 10,000 | 10,000 |
| FI   | 81 |  |  | 10,000+ | 10,000 |
| JWS  | 81 |  |  | ?> 9,000 | 9,000 |
| AFM  | 82 | 5,950-6,200 |  |  | 9,573-9,975 |
| AWST | 82 |  | 5,500 |  | 10,186 |

BALLISTIC MISSILES  ICBMs  SS-19  217

## CEP

| Source | Year | ft | mi | nm | m | km | m conv | Notes |
|---|---|---|---|---|---|---|---|---|
| SecD | 74 | | | | | | | Clearly designed for higher accuracy, FY75-76. Not yet sufficiently accurate to be hard-target weapon; likely to do so early part of next decade. |
| JCS | 75 | | | | | | | More accurate than SS-11, FY76-77. |
| FIar | 76 | | | | | | | Accuracy probably similar to SS-17. |
| MOW | 76 | | | | | | | "May eventually have a hard-target capability". High speed RVs are employed; uprated guidance could be used. |
| CBO1 | 78 | 1200-1500 | | | | | 365.76-457.20 | Estimated for new generation of ICBMs in 1985 [SS-16, SS-17, SS-18, SS-19, ed]. |
| Rock | 78 | | | | 450 | | 450.00 | "Target error". |
| Nitz | 79 | | | .2 | | | 370.40 | |
| AWST | 82 | | | 0.14 | | | 259.28 | |
| JCS | 82 | | | | | | | Certain versions of SS-19 have "significantly improved accuracies". |
| CBO | 83 | | | 0.15 | | | 277.80 | In 1983. |
|  |  | | | 0.12 | | | 222.24 | In 1990. |
|  |  | | | 0.08 | | | 148.16 | In 1996. |
| SMP | 83 | | | | | | | "The most accurate versions of the SS-18 and SS-19 are capable of destroying hard targets. Together, these systems have the capability to destroy most of the 1,000 US MINUTEMAN ICBMs, using only a portion of the warheads available". |
| Meyr | 84 | | | | | 0.1-0.3 | 100.00-300.00 | |
| SMP | 84 | | | | | | | "At least as accurate and possibly more accurate" than Minuteman III. |

## WARHEAD TYPE

| Source | Year | nuclear | conven | chem |
|---|---|---|---|---|
| JWS | 81 | yes | | |
| Meyr | 84 | yes | | |

## WARHEAD CHARACTERISTICS

| Source | Year | | Notes |
|---|---|---|---|
| JCS | 74 | MIRV | FY75-77. |
|  |  | Est. 4-6 RVs | |
| SecD | 74 | MIRV, 6 RVs | FY75, 77-78. Has PBV, FY75-76. Tested only with MIRV payload of 6 RVs, FY75-76. RVs designed for high-speed re-entry. |
| IA06 | 75 | 4-6 MIRVs | |
| JCS | 75 | 6 RVs | 6 RVs max, FY76-78. |
| SecD | 75 | 6 RVs | Probably will be deployed with 6 RVs. |
| FIar | 76 | 6 MIRVs | |
| JCS | 76 | 6 MIRVs | Has PBV. |
| MOW | 76 | 6 MIRVs | |
| SWM | 77 | 6 MIRVs | RVs dispensed by post-boost 'bus'. |
| Rock | 78 | 6 MIRVs | |
| SecD | 78 | | Can carry either high-yield single RV, or MIRVs, FY79-80. |
| Nitz | 79 | 1 RV | |
| USD8 | 79 | 6 MIRVs | |
| AWST | 82 | 6 MIRVs | |
| CBO | 83 | 6 RVs | |
| JCS | 84 | | SS-18, 19 carry multiple warheads. |
| Meyr | 84 | "1/6MIRV" | |

218  SOVIET MISSILES  IDDS

```
------------ NUCLEAR YIELD ------------
              kT          MT      kT_conv
IAD6  75   ? 200                  200.00          Each.
FIar  76               1          1,000           Each.
SecD  77                                          Exceeds Minuteman III per warhead.
SWM   77   400-500                400.00-500.00
CBO1  78               0.8        800.00          Estimated, 1985.
Rock  78               1.2        1,200
           200                    200.00          Each of 6 MIRVs in preferred configuration.
           400                    400.00          Missile can carry 6 of these warheads,
                                                  [non-preferred configuration].
Nitz  79               15         15,000
USO8  79               .5-1       500.00-1,000
CBO   83   550                    550.00          Each.
Meyr  84   500-5,000              500.00-5,000    500 kT-5 MT.

------------ THROWWEIGHT ------------
              lb          t       kg     kg conv
SecC  74                                          2-3 times SS-11.
SecD  74                                          3-5 times early-model SS-11s.
JCS   75                                          3-4 times SS-11, FY76-77.
SecC  75                                          Throwweight about 3-5 times
                                                  greater than SS-11.
SecD  75                                          3-4 times greater than the SS-11.
FIar  76                          3,000  3,000
SecD  77                                          Exceeds Minuteman III.
Rock  78                                          Nearly 4 times greater than
                                                  SS-11.
Nitz  79   7,000                         3,175
SecD  79                                          Without SALT II, SS-19 "could
                                                  carry more than the 6 RVs it has
                                                  now".
USO8  79   7,000                         3,175
JWS   81                                          3-4 times that of SS-11.
SSF   82   8,000                         3,628
CBO   83   7,525                         3,413

------------ PAYLOAD ------------
              lb          t       kg     kg conv
AWar  80   c. 8,000                      3,628
SMP   81                                          3 to 4 times SS-11.

------------ WEIGHT ------------
              lb          t       kg     kg conv
Clns  80                                          SALT II recognizes Soviet SS-19s
                                                  as "Light", although much heavier
                                                  than Minutemen.
FI    81                          78,000  78,000

------------ LENGTH ------------
              in    ft    cm      m       m_conv
SecD  74                                          Longer than SS-11 and SS-X-17.
FIar  76                          27      27.00
SWM   77          80              24.4    24.38
Rock  78                          27      27.00
FI    81                          c. 27   27.00
JWS   81                          c. 25   25.00
SMP   81                          24      24.00   [From graph], '81, '83.
AWST  82          80              24.38
AFM+  84          75'0"                   22.86

------------ DIAMETER ------------
              in    ft    cm      m       cm_conv
FIar  76                          2.8     280.00
SWM   77          8.2             2.5     249.94
Rock  78                          2.8     280.00  [Table]
                                  2.5     250.00  [Text]
FI    81                c. 250            250.00
JWS   81                          c. 2.75 275.00
AWST  82          8.5                     259.08
```

##### NUMBER OF STAGES

| | | |
|---|---|---|
| MOW | 76 | 2 |
| SWM | 77 | 2 |
| Rock | 78 | 2 |
| FI | 81 | 2 |
| JWS | 81 | 2 |
| AFM | 82 | 2 |
| AWST | 82 | 2 |

##### TYPE OF PROPULSION

| | | | |
|---|---|---|---|
| SecC | 74 | | PBV powered by liquid fuel engines. |
| SecD | 74 | liquid | FY75-76, 81-82. |
| JCS | 75 | liquid | FY76-77. |
| FIar | 76 | storable liquid | |
| MOW | 76 | liquid | |
| SWM | 77 | liquid storable | ? Propellants unsymmetrical dimethyl hydrazine (UDMH) and nitrogen tetroxide (N2O4). |
| Rock | 78 | storable liquid | Stages 1 and 2. |
| FI | 81 | liquid | |
| JWS | 81 | liquid | |
| AFM | 82 | liquid | |
| AWST | 82 | liquid-storable | |
| SSF | 82 | liquid | |

##### MIDCOURSE (OR MAIN) GUIDANCE

| | | | |
|---|---|---|---|
| JCS | 74 | | Digital computer. |
| JCS | 75 | | Guidance combines navigation with refined Soviet 'fly-the-wire' technique, FY76-77. On-board computer determines deviation from course; directs correction or plots new course, FY76-77. |
| Rock | 78 | inertial | |
| FI | 81 | inertial | |
| JWS | 81 | inertial | Combination of inertial and refined Soviet 'fly-the-wire' technique. Computer determines deviation from planned course and directs correction or plots new course. |
| SSF | 82 | | Fly-the-wire, onboard digital computer. |

##### CONTROL SYSTEM

| | | |
|---|---|---|
| FI | 81 | ? Gimballed nozzles. |

##### DEVELOPMENT BEGAN

| | | | |
|---|---|---|---|
| SSF | 82 | 1966 | Year design began. |
| SMP | 85 | 1964 | Technological development began. [From graph.] |

##### DEVELOPMENT UNDERWAY

| | | |
|---|---|---|
| IAO6 | 75 | In advanced development. |

##### PROTOTYPE TESTS

| | | | |
|---|---|---|---|
| SecD | 74 | [1973] | Tested during past year. |
| GAO1 | 75 | Jan 1974 | Pacific test firing. |
| IAO6 | 75 | Spring, 1973 | Tests [sic] observed. |
| JCS | 75 | Apr 1973 | First tested, FY76-77. |
| FIar | 76 | 1974 | First tested. |
| SecD | 76 | 1974-75 | In 1974 extensive flight testing; in 1975 testing continued. |
| JWS | 81 | Apr 1973 | First test. |
| SMP | 85 | 1970-71 | Engineering and testing began. [From graph.] |

##### IOC

| | | | |
|---|---|---|---|
| JCS | 74 | | IOC expected to be 1975. |
| JCS | 75 | | IOC ? |
| JCS | 76 | 1974 | FY77-78. |
| MOW | 76 | 1975 | Became operational. |
| SecD | 76 | | MIRVed SS-19 now operational. |
| SecD | 77 | [1973-1974] | [Midway between 73 and 74 lines on graph]. |
| SMP | 81 | 1974 | Became operational. |

SOVIET MISSILES

---------- **DEPLOYMENT BEGAN** ----------
| | | | |
|---|---|---|---|
| JCS | 74 | | Could be deployed in 1975. |
| JCS | 75 | | Last year was reported SS-17 and SS-19 would not be operational before 1975, wrong, FY76-77. |
| | | | May be starting (now) operational deployment. |
| SecD | 75 | | Started, or soon will start, operational deployment. Probably has started deployment. |
| FIar | 76 | 1975 | Entered service a year after first test. |
| Rock | 78 | 1975 | Service intro. |
| JWS | 81 | 1974 | Deployed operationally. |
| JCS | 83 | c. 1975 | Year of introduction [from graph]. |
| JCS | 84 | 1976-77 | Major strategic system deployments [graph], FY85-86. |
| Meyr | 84 | 1975 | Service. |
| SecD | 84 | 1975 | Introduction of strategic systems [graph]. |
| SMP | 85 | 1975-76 | [From graph.] |

---------- **DEPLOYED** ----------
| | | | |
|---|---|---|---|
| JCS | 76 | | "Now is believed to be deployed" in operational silos. |

---------- **DEPLOYMENT COMPLETE** ----------
| | | | |
|---|---|---|---|
| FI | 81 | 1980 | Deployment ended. |
| JCS | 81 | 1980 | Appears deployment of SS-19 was completed in 1980. |
| JWS | 81 | 1980 | |
| SecD | 81 | | [Implies deployment has stopped, ed]. |

---------- **DEPLOYMENT HISTORY** ----------
| | | |
|---|---|---|
| SecD | 74 | Successor to SS-11. |
| JCS | 75 | "SS-11 silos have been or are in the process of being prepared for the SS-17 and the SS-19", FY76-77. |
| SecD | 75 | Will be deployed in new-type silos. |
| JCS | 76 | Being deployed at the expected moderate pace. |
| JCS | 77 | SS-19 silo conversion program is further along than SS-17 and SS-18. |
| SecD | 77 | "Continue to be deployed in modified and upgraded silos at a rapid rate". |
| SWM | 77 | [Like SS-17] an SS-11 replacement. |
| Rock | 78 | Currently replacing SS-11 on a one-to-one basis. |
| JCS | 79 | SS-19 silos with SS-11s "probably will be retrofitted to accommodate the SS-19". |
| AWar | 80 | Believed 60 silos built for SS-19 have SS-11s; located at Derazhnya and Pervomaysk. Soviets agreed in SALT II to dismantle 12 of them. |
| JCS | 80 | Soviets have initiated program "to retrofit SS-19 silos which previously contained" SS-11s. |
| FI | 81 | "3 versions have been reported". "All are expected to be retrofitted to the Mod 3 standard". |
| JWS | 81 | Successor to SS-11 (with SS-17). |
| AFM | 82 | Is replacing older SS-11s. |
| SMP | 83 | Continuing modernization of SS-17,18,19 force. |
| SMP | 84 | "Modernization...nears an end". "At least one additional version of both the SS-18 and SS-19, however, is likely to be produced and deployed in existing silos in the future." |
| SMP | 85 | Will be deployed at least through 1990. [From graph.] |

---------- **NUMBERS PRODUCED** ----------
| | | |
|---|---|---|
| SecD | 80 | No evidence that SS-17,18,19 production is significantly greater than the number of their launchers. |

---------- **NUMBERS IN SERVICE** ----------

---------- **NUMBER IN 1960** ----------
| | | | |
|---|---|---|---|
| SSF | 82 | none | In 1960, 1965, 1970; "includes only those believed to be primarily dedicated to covering intercontinental-range targets". |

---------- **NUMBER IN 1970** ----------
| | | | |
|---|---|---|---|
| Clns | 80 | none | 1970-1974. |

---------- **NUMBER IN 1975** ----------
| | | | |
|---|---|---|---|
| IISS | 75 | 50 | July, total deployed. |
| SWM | 77 | 50 | In service mid-1975. |
| Clns | 80 | 60 | |
| SSF | 82 | 50 | "Includes only those believed to be primarily dedicated to covering intercontinental-range targets". |
| vanD | 83 | 20 | 1 Jan. |

---------- **NUMBER IN 1976** ----------
| | | | |
|---|---|---|---|
| FIar | 76 | > 100 | Deployed so far. |
| IISS | 76 | 100 | July, total deployed. |
| Clns | 80 | 100 | |
| vanD | 83 | 70 | 1 Jan. |

---------- **NUMBER IN 1977** ----------
| | | | |
|---|---|---|---|
| IISS | 77 | 140 | July, total deployed. |
| JCS | 77 | | SS-19s deployed to date carried 6 MIRVs. |
| SecD | 77 | c. 140 | Around 140 deployed. |
| Nitz | 79 | 50 | Launchers. |
| Clns | 80 | 120 | |
| vanD | 83 | 140 | 1 Jan. |

---------- **NUMBER IN 1978** ----------
| | | | |
|---|---|---|---|
| IISS | 78 | 200 | July, total deployed. |
| JCS | 78 | > 200 | Operational silos for SS-19. |
| Rock | 78 | > 100 | Currently installed. |
| SecD | 78 | | Deployment of SS-17,18,19 continues at c. 125 per year, FY79-81. |
| | | > 200 | Over 200 launchers (converted from SS-11). |
| Clns | 80 | 180 | |
| vanD | 83 | 200 | 1 Jan. |

---------- **NUMBER IN 1979** ----------
| | | | |
|---|---|---|---|
| IISS | 79 | 300 | 1979-81, July, total deployed. |
| JCS | 79 | | Nearly 300 operational SS-19 silos contain "a mix of SS-19 and SS-11 ICBMs", FY80-81. |
| SecD | 79 | | About 300 SS-17,19 (combined) launchers, converted from SS-11. |
| Clns | 80 | 240 | (210 of 240 are Mod 1; 30 are Mod 2; and "ALL SS-17s and SS-19s were MIRVed until 1978"; 30 single-shot SS-19s since installed). |
| vanD | 83 | 240 | 1 Jan. |

---------- **NUMBER IN 1980** ----------
| | | | |
|---|---|---|---|
| AWar | 80 | 300 | Operational; 60 in preparation. |
| | | | Believed 60 silos built for SS-19 have SS-11s; located at Derazhnya and Pervomaysk. Soviets agreed in SALT II to dismantle 12 of them. |
| JCS | 80 | | Soviets have initiated program "to retrofit SS-19 silos which previously contained" SS-11s; when complete, expected to have "well over 300" SS-19s. |
| SecD | 80 | > 200 | More than 200 now deployed (in converted SS-11 silos). |
| SSF | 82 | 180 | "Includes only those believed to be primarily dedicated to covering intercontinental-range targets". |
| vanD | 83 | 270 | 1 Jan. |

---------- **NUMBER IN 1981** ----------
| | | | |
|---|---|---|---|
| SecD | 75 | | Sizable force projected; [ICBM-threatening force] could be operational by early 80s. |
| FI | 81 | 300 | |
| JCS | 81 | 360 | "There are 360 SS-19-type silos containing both SS-19s and SS-11s". |
| JWS | 81 | 360 | SS-19 type silos as of 1980, containing both SS-19s and SS-11s. |
| | | | Most deployed SS-19s are Mod 3. |
| SecD | 81 | c. 300 | |
| | | | SS-17,18,19 mostly equipped with MIRVs. |
| SMP | 81 | 300 | |
| | | | MIRVed version "makes up most of the SS-19 force". |
| vanD | 83 | 300 | 1 Jan. |

---------- **NUMBER IN 1982** ----------
| | | | |
|---|---|---|---|
| AFM | 82 | > 300 | |
| JCS | 82 | 300 | Inventory. |
| SSF | 82 | 300 | "Number of missiles deployed". Another 60 in preparation. |
| JWS+ | 83 | | 60 deployed at SS-4 and SS-5 sites at Derazhnya and Pervomaysk, in 1982-83 period. |

---------- **NUMBER IN 1983** ----------
| | | | |
|---|---|---|---|
| CBO | 83 | 330 | |
| JWS+ | 83 | c. 300 | [Analysis table] |
| SMP | 83 | 330 | |
| | | | Great majority of SS-17,18,19 equipped with MIRVs. |

---------- **NUMBER IN 1984** ----------
| | | | |
|---|---|---|---|
| AFM+ | 84 | c. 330 | Now; 360 expected by mid-1980s. |
| JCS | 84 | 330 | As of 1 Jan. |

---------- **NUMBER IN 1985** ----------
| | | | |
|---|---|---|---|
| SMP | 83 | 360 | When current ICBM modernization complete, by mid-1980s. |
| JCS | 85 | 360 | As of 1 Jan. |

---------- **NUMBER IN 1990** ----------
| | | | |
|---|---|---|---|
| CBO | 83 | | 360 in 1990, 1996. |

---------- **UNDATED NUMBER** ----------
| | | | |
|---|---|---|---|
| SWM | 77 | | SS-17 and SS-19 restricted to 1,036 under SALT I. |

## LAUNCHER/SILO

| | | |
|---|---|---|
| SecD | 74 | Can be deployed in new-type silos. Can be deployed in SS-11 silos (with modification). |
| JCS | 75 | More survivable than SS-11, FY76-77. |
| SecD | 75 | Will be deployed in new-type silos. |
| FIar | 76 | "Some reports claim that the weapon is being installed in modified SS-11 silos, but—as the -19 is hot-launched and would require a silo at least 40m deep—this seems unlikely. Either new silos are to be provided, or the US Department of Defence is mistaken in classing the SS-19 as hot-launched." |
| JCS | 76 | Every SS-19 deployed goes into "new modernized, hard silos"; associated launch systems are in silos, not launchers. |
| MOW | 76 | "SS-19 could fit into SS-11 silos although it is a longer missile". |
| SecD | 77 | Some silos hardened to resist very high overpressure. |
| SWM | 77 | Modified SS-11 silos. |
| JCS | 79 | SS-19 silos with SS-11s "probably will be retrofitted to accommodate the SS-19". |
| AWar | 80 | Silo hardness [SS-17,18,19] 3,500-4,500 psi. Believed 60 silos built for SS-19 have SS-11s; located at Derazhnya and Pervomaysk. Soviets agreed in SALT II to dismantle 12 of them. |
| SecD | 80 | All "converted silos may be capable of withstanding very high static overpressures". |
| JWS | 81 | 360 SS-19 type silos as of 1980, containing both SS-19s and SS-11s. |
| SMP | 81 | More than half of 1,398 ICBM launchers rebuilt to house SS-17,18,19 in "vastly more survivable, hardened silos". |
| AFM | 82 | Since longer than SS-11, more extensive modification of silos required. |
| AWST | 82 | Uses converted SS-11 silos. |
| SSF | 82 | Basing mode: Hardened silo. |
| SMP | 83 | "Contained in a launch canister within the silo. This and the silo design minimize damage to the launcher during" firing, giving reload capability. SS-17,18,19 "housed in the world's hardest silos", '83, '84, '85. |

## CARRIERS

### CARRIER — TUBES — RELOADS

| | | |
|---|---|---|
| IA06 | 75 | "Could be mobile". |

### DESIGNER

| | | |
|---|---|---|
| SSF | 82 | Chelomei Design Bureau. |

### DESIGN AND ENGINEERING

| | | |
|---|---|---|
| JCS | 74 | SS-17 and SS-19 may be competing designs to replace SS-11, but possible both will be deployed. |
| SecD | 74 | Less advanced than SS-17 [technically]. Clearly most successful of new ICBMs. |
| FIar | 76 | Back-up to SS-17, believed to embody less advanced technology than SS-17, in spite of similar configuration. Nevertheless, has achieved high success rate during trials. |
| SWM | 77 | Less advanced technically than SS-17. |
| Rock | 78 | Companion development with SS-17. "Has enjoyed singular success in all its design trials and production activities". "More favoured than the SS-17". Utilizes less advanced technology than SS-17. |

### USE AND CONFIGURATION

| | | |
|---|---|---|
| CBO | 83 | System availability: 0.95. |
| vanD | 83 | 10 missiles in a battalion. |
| Meyr | 84 | Penetration: 1.0; reliability: 0.7-0.9, "includes system reliability multiplied by operational readiness of deployed system". |

### RELOADS IN GENERAL

| | | |
|---|---|---|
| GA01 | 75 | 'Cold Launch' technique. |
| JCS | 75 | Hot launch, FY76-77. |
| SecD | 75 | Hot launch, FY76-77,79-82. |
| MOW | 76 | Hot launch. |
| SWM | 77 | Cold launch. |
| Rock | 78 | Cold launch technique similar to SS-17 and SS-18. |
| JWS | 81 | Hot launch. |
| SMP | 81 | Hot launch, '81, '83. Uses hot-launch, "engine ignition occurring while the missile is in its silo". "The Soviets probably cannot refurbish and reload silo launchers in a period less than several days". |
| AFM | 82 | Hot-launch. |
| AWST | 82 | Hot launch. |
| IISS | 82 | Cold launch [sic]. |
| SSF | 82 | Launching mode: Hot. |
| IISS+ | 83 | Hot launch. |
| JWS+ | 83 | US report says all Soviet liquid-fueled ICBMs contained in launch canister offering possibility of reloading. |
| SMP | 83 | Has reload capability. |

BALLISTIC MISSILES  ICBMs  SS-19  223

---------- **STRATEGY** ----------
| | | |
|---|---|---|
| SecD | 75 | Sizable force projected; [ICBM-threatening force] could be operational by early 80s. |
| Rock | 78 | Utilizes less advanced technology than SS-17, but "could probably be used against hardened targets" if needed. |
| SecD | 78 | "We believe" SS-19 to be most capable of SS-17,18,19 due to combination of accuracy and yield; even though fewer RVs than SS-18. SS-17,18,19 all have "potential, with feasible accuracy improvements, to attain high single-shot kill probabilities against U.S. silos". Uncertain how fast threat to Minuteman might become serious, presently function of SS-18,19. |
| SecD | 79 | Major progress [in past year] in understanding threat to ICBMs; intelligence data on new SS-18,19 versions indicates substantial threat to Minuteman by early 1980s. |
| SecD | 80 | "The numbers of high quality warheads on new versions of the SS-18 and SS-19 seriously threaten our MINUTEMAN force in the early 1980s...". |
| FI | 81 | USSR's most effective ICBM. |
| JCS | 81 | Soviets may have plans to reconstitute hot-launched as well as cold-launched silos [not restricted to SS-19]. |
| SMP | 81 | "Improved reliability, range, payload accuracy and survivability". |
| AFM | 82 | Large-scale deployment despite silo difficulty supports DoD belief that SS-19 is most capable ICBM. |
| JCS | 84 | Today, most accurate versions of SS-18,19 capable of destroying time-urgent and hardened targets in an initial attack, FY85-86, ["most" such targets in FY86 report, ed]. Improvements in deployed SS-18,19s projected for 1983-93. |
| Meyr | 84 | "Initially developed for intercontinental strikes, but likely to be assigned to missions in the European theatre of military operations". [Ed: chart is labeled] Systems "assigned primary operational and strategic missions within the European theatre(s) of military operations". |
| SecD | 84 | SS-18,19s carry 4,000-5,000 "highly accurate warheads designed specifically to attack our missile silos"; with only a portion of warheads, can attack Minuteman force. |

---------- **GEOGRAPHICAL DEPLOYMENTS** ----------
| | | |
|---|---|---|
| AWar | 80 | Believed 60 silos built for SS-19 have SS-11s; located at Derazhnya and Pervomaysk. Soviets agreed in SALT II to dismantle 12 of them. |
| JWS+ | 83 | Mostly in major Derazhnya, Kozelsk, Pervomaysk and Tatishchevo missile fields. |
| SMP | 83 | [Shown on map at] Derazhnya, Kozelsk, Pervomaysk, Tatishchevo, '83, '84, '85. |
| SMP | 84 | [Shown on map at Teykovo; in addition to places shown in '83]. |
| SMP | 85 | [Teykovo, added to map of deployments for '84 report, removed in '85.] |

---------- **RELATION TO OTHER MISSILES** ----------
| | | |
|---|---|---|
| JCS | 75 | Exceeds SS-11 volume, "much larger", FY76-77. |
| SMP | 81 | Volume much larger than SS-11. Comparable in size to MX. |

# SS-19 Mod 1

| Sources are: | AWar 80 | FI 81 | AFM 82 | Clns+ 83 | vanD 83 | BW 86 |
|---|---|---|---|---|---|---|
| JCS 75 | Clns 80 | JWS 81 | IISS 82 | FI+ 83 | AFM+ 84 | |
| JCS 78 | SecD 80 | SMP 81 | SSF 82 | Lu01 83 | SMP 85 | |

---------- **US DESIGNATION** ----------

| AWar | 80 | SS-19 mod |
| Clns | 80 | SS-19 Mod 1 |
| SecD | 80 | SS-19 Mod 1 |
| FI | 81 | SS-19 Mod 1 |
| JWS | 81 | SS-19 Mod 1 |
| SMP | 81 | SS-19 Mod 1 |
| AFM | 82 | SS-19 Mod 1 |
| IISS | 82 | SS-19 Mod 1 |
| SSF | 82 | SS-19 Mod 1 |
| vanD | 83 | SS-19 |

---------- **EDITOR'S NOTES** ----------

vanD 83 — Source has 4 separate SS-19 lines, for 1974, 1977, 1979, and 1982. We have assumed the first two to be Mod 1, the second two to be Mod 3. Where the two lines for each Mod differ, they are identified by year.

---------- **RANGE** ----------

| | mi | nm | km | km conv | |
|---|---|---|---|---|---|
| JCS 75 | | 5000 | | 9,260 | Believed can deliver 6 MIRVs at 5000 nm [Mod 1, ed], FY76-77. |
| AWar 80 | | | 9,550 | 9,550 | |
| Clns 80 | | 5,000 | | 9,260 | |
| SecD 80 | | | 9,600 | 9,600 | Max, FY81-82. [Qualification:] "exclusive of range imparted by" FBV. |
| SMP 81 | | | 9,600 | 9,600 | Max, '81, '83. |
| | | | 9,000 | 9,000 | MIRVed version [Mod 1, ed] "believed capable of delivering six RVs to a range of about 9,000" km. |
| IISS 82 | | | 11,000 | 11,000 | Max. |
| SSF 82 | | 5,200 | | 9,630 | |
| AFM+ 84 | 5,950 | | | 9,573 | |
| BW 86 | | | 9,600 | 9,600 | |

---------- **CEP** ----------

| | ft | mi | nm | m | km | m conv | |
|---|---|---|---|---|---|---|---|
| AWar 80 | | | 0.19 | | | 351.88 | |
| Clns 80 | | | 0.21 | | | 388.92 | |
| IISS 82 | | | | 500 | | 500.00 | |
| SSF 82 | | | 0.19 | | | 351.88 | |
| Lu01 83 | | | | 300-450 | | 300.00-450.00 | |
| vanD 83 | | | .30 | | | 555.60 | [1974 entry] |
| | | | .23 | | | 425.96 | [1977 entry] |
| BW 86 | | | | 350 | | 350.00 | |

---------- **WARHEAD CHARACTERISTICS** ----------

| JCS | 78 | 6 MIRVs | |
| AWar | 80 | 6 MIRVs | |
| Clns | 80 | 6 MIRVs | |
| SecD | 80 | 6 warheads | FY81-82. |
| FI | 81 | 6 MIRVs | |
| JWS | 81 | 4 or 6 MIRVs | |
| SMP | 81 | 6 MIRVs | '81, '83. |
| AFM | 82 | MIRV with 6 RVs | |
| IISS | 82 | 6 MIRVs | |
| SSF | 82 | 6 RVs | |
| FI+ | 83 | 4 or 6 MIRVs | |
| vanD | 83 | 6 RVs | |
| BW | 86 | 6 MIRVs | |

BALLISTIC MISSILES   ICBMs   SS-19 Mod 1   225

```
            ---------- NUCLEAR YIELD ----------
                      kT         MT        kT conv
  AWar  80                       0.55       550.00
  Clns  80            550                   550.00     Each.
  FI    81            550                   550.00
  JWS   81          ? 200                   200.00
  AFM   82            550                   550.00
  IISS  82            550                   550.00     Per warhead.
  SSF   82                       0.55       550.00     Per warhead.
  FI+   83            200                   200.00     Each.
  Lu01  83            500                   500.00
  vanD  83                       .55        550.00
  BW    86            550                   550.00

            ---------- THROWEIGHT ----------
                      lb         t          kg         kg conv
  Clns  80            7,525                            3,413
  IISS  82            8,000                            3,628

            ---------- TYPE OF PROPULSION ----------
  Clns  80    liquid
  BW    86    storable liquid

            ---------- PROTOTYPE TESTS ----------
  AFM   82    1974        Testing began 1974, Mod 1 rapidly deployed shortly after.
  SSF   82    1973        First flight test.

            ---------- IOC ----------
  JCS   78    1975
  vanD  83    1974        [1974 entry], Dec 1974 [p186]
              1977        [1977 entry]

            ---------- DEPLOYMENT BEGAN ----------
  Clns  80    1974        First deployed.
  AFM   82                Testing began 1974, Mod 1 rapidly deployed shortly after.
  IISS  82    1974        First year deployed.
  SSF   82    1975        Year operation began.
  SMP   85    1975-76     [Presumed Mod 1, from graph.]
  BW    86    1975

            ---------- RETIRED ----------
  IISS  82                Out of service.
  BW    86    ? 1983

            ---------- DEPLOYMENT HISTORY ----------
  AFM+  84    Has been replaced by Mod 3.

            ---------- NUMBERS IN SERVICE ----------

            ---------- NUMBER IN 1970 ----------
  Clns+ 83    none        In 1970-1974.

            ---------- NUMBER IN 1975 ----------
  Clns+ 83    60

            ---------- NUMBER IN 1976 ----------
  Clns+ 83    100         In 1976-1977.

            ---------- NUMBER IN 1978 ----------
  Clns+ 83    120

            ---------- NUMBER IN 1979 ----------
  Clns+ 83    180         In 1979-1981.

            ---------- NUMBER IN 1982 ----------
  IISS  82    none        July, total deployed.
  Clns+ 83    80

            ---------- UNDATED NUMBER ----------
  BW    86    ? 210       Peak.

            ---------- USE AND CONFIGURATION ----------
  vanD  83    Reliability: .85; Penetration 1.0.

            ---------- RELOADS IN GENERAL ----------
  Clns  80    Cold launch: no.
```

# SS-19 Mod 2

```
Sources are:  AWar  80      FI    81      AFM   82      Clns+ 83      AFM+  84      BW    86
JCS   77      Clns  80      JWS   81      IISS  82      JWS+  83      ArkS  84
SecD  77      SecD  80      SMP   81      SSF   82      Lu01  83      SMP   85
```

## US DESIGNATION

|  | |  |
|---|---|---|
| JCS | 77 | SS-19 Mod 2 |
| AWar | 80 | SS-19 mod |
| Clns | 80 | SS-19 Mod 2 |
| SecD | 80 | SS-19 Mod 2 |
| FI | 81 | SS-19 Mod 2 |
| JWS | 81 | SS-19 Mod 2 |
| SMP | 81 | SS-19 Mod 2 |
| AFM | 82 | SS-19 Mod 2 |
| IISS | 82 | SS-19 Mod 2 |
| SSF | 82 | SS-19 Mod 2 |

## RANGE

| Source | | mi | nm | km | km conv | Notes |
|---|---|---|---|---|---|---|
| AWar | 80 | | | 10,100 | 10,100 | |
| Clns | 80 | | 5,500 | | 10,186 | Max, FY81-82. [Qualification:] "exclusive of range imparted by" PBV. |
| SecD | 80 | | | 10,100 | 10,100 | Max, '81, '83. |
| SMP | 81 | | | 10,000 | 10,000 | Max. |
| IISS | 82 | | | 10,000 | 10,000 | |
| SSF | 82 | | 5,450 | | 10,093 | |
| AFM+ | 84 | 6,200 | | | 9,975 | |
| BW | 86 | | | 10,100 | 10,100 | |

## CEP

| Source | | ft | mi | nm | m | km | m conv | Notes |
|---|---|---|---|---|---|---|---|---|
| AWar | 80 | | | 0.21 | | | 388.92 | |
| Clns | 80 | | | 0.14 | | | 259.28 | |
| JWS | 81 | | | | | | | Perhaps high-accuracy, high-yield. |
| IISS | 82 | | | | 300 | | 300.00 | |
| SSF | 82 | | | 0.21 | | | 388.92 | |
| Lu01 | 83 | | | | 200 | | 200.00 | |
| AFM+ | 84 | < 1,000 | | | | | 304.80 | |
| BW | 86 | | | | 300 | | 300.00 | |

## WARHEAD CHARACTERISTICS

|  | | |  |
|---|---|---|---|
| JCS | 77 | single RV | FY78-79. |
| SecD | 77 | single RV | |
| AWar | 80 | single RV | |
| Clns | 80 | single RV | |
| SecD | 80 | single RV | FY81-82. |
| FI | 81 | 1 RV | |
| JWS | 81 | single RV | |
| SMP | 81 | single RV | '81, '83. |
| AFM | 82 | single RV | |
| IISS | 82 | single RV | |
| SSF | 82 | single RV | |
| BW | 86 | single RV | |

## NUCLEAR YIELD

| Source | | kT | MT | kT conv | Notes |
|---|---|---|---|---|---|
| SecD | 77 | | | | High-yield. |
| AWar | 80 | | 4.3 | 4,300 | |
| Clns | 80 | | 10 | 10,000 | |
| JWS | 81 | | | | Perhaps high-accuracy, high-yield. |
| AFM | 82 | | 5 | 5,000 | |
| IISS | 82 | | 5 | 5,000 | |
| SSF | 82 | | 4.3 | 4,300 | Per warhead. |
| JWS+ | 83 | | 5-10 | 5,000-10,000 | [Analysis table] |
| Lu01 | 83 | | 4.3 | 4,300 | |
| BW | 86 | 4,300 | | 4,300 | |

## THROWWEIGHT

| Source | | lb | t | kg | kg conv |
|---|---|---|---|---|---|
| Clns | 80 | 7,000 | | | 3,175 |
| IISS | 82 | 7,500 | | | 3,402 |

## BALLISTIC MISSILES  ICBMs  SS-19 Mod 2

---------- **TYPE OF PROPULSION** ----------
Clns 80    liquid
BW   86    storable liquid

---------- **OBSERVED** ----------
JCS  77              If single RV mod deployed, "there will be difficulty in distinguishing it from the MIRVed variant".

---------- **PROTOTYPE TESTS** ----------
JCS  77              Soviets are now testing single RV mod, FY78-79.
AFM  82              Has been tested.

---------- **DEPLOYMENT BEGAN** ----------
Clns 80    1978      First deployed.
IISS 82    1979      First year deployed.
SSF  82    1978      Year operation began.
SMP  85    c. 1979   [From graph.]
BW   86    ? 1978

---------- **DEPLOYED** ----------
BW   86              In service.

---------- **NUMBERS IN SERVICE** ----------

---------- **NUMBER IN 1970** ----------
Clns+ 83   none      In 1970-1976.

---------- **NUMBER IN 1977** ----------
Clns+ 83   20

---------- **NUMBER IN 1978** ----------
Clns+ 83   60        In 1978-1979.

---------- **NUMBER IN 1980** ----------
Clns+ 83   40        In 1980-1981.

---------- **NUMBER IN 1982** ----------
IISS  82   "few"     July, total deployed.
Clns+ 83   10

---------- **NUMBER IN 1983** ----------
JWS+  83   c. 60

---------- **NUMBER IN 1984** ----------
AFM+  84             Few deployed.
ArkS  84             Some SS-19s are still configured to carry single warheads, according to DIA officials; omitted from SMP 84 because of 'space considerations' and because number of single warhead SS-17s and SS-19s is 'too few to matter'.

---------- **UNDATED NUMBER** ----------
BW    86   ? 60      Peak.

---------- **LAUNCHER/SILO** ----------
IISS  82             In modified SS-11 silos.

---------- **RELOADS IN GENERAL** ----------
Clns  80             Cold launch: no.

# SS-19 Mod 3

| Sources are: | FI   | 81 | IISS  | 82 | Clns  | 83 | Lu01 | 83 | vanD | 83 | SMP | 85 |
|---|---|---|---|---|---|---|---|---|---|---|---|---|
| Nitz | 79 | JCS | 81 | JCS | 82 | IISS+ | 83 | SMP | 83 | AFM | 84 | BW | 86 |
| AWar | 80 | JWS | 81 | SSF | 82 | JWS+ | 83 | Ts01 | 83 | SMP | 84 | | |

---------- **US DESIGNATION** ----------

| Nitz | 79 | SS-19 | |
| AWar | 80 | SS-19 mod | Improved 6 RV mod. |
| FI | 81 | SS-19 Mod 3 | |
| JCS | 81 | SS-19 Mod 3 | |
| JWS | 81 | SS-19 Mod 3 | |
| IISS | 82 | SS-19 Mod 3 | |
| SSF | 82 | SS-19 Mod 3 | |
| Clns | 83 | SS-19 Mod 3 | |
| SMP | 83 | SS-19 Mod 3 | |
| vanD | 83 | SS-19 | |
| AFM | 84 | SS-19 Mod 3 | |

---------- **EDITOR'S NOTES** ----------

Nitz 79 — This missile appears in source on 4 lines with dates in the designation, but differing only in CEP. It is presumed Mod 3 because the earliest date is 1977.

vanD 83 — Source has 4 separate SS-19 lines, for 1974, 1977, 1979, and 1982. We have assumed the first two to be Mod 1, the second two to be Mod 3. Where the two lines for each Mod differ, they are identified by year.

---------- **RANGE** ----------

| | | mi | nm | km | km conv | |
|---|---|---|---|---|---|---|
| IISS | 82 | | | ? 10,000 | 10,000 | Max. |
| SSF | 82 | | 5,200 | | 9,630 | |
| Clns | 83 | | 5,500 | | 10,186 | |
| SMP | 83 | | | 10,000 | 10,000 | Max, '83, '84, '85. |
| AFM | 84 | 6,200 | | | 9,975 | |
| BW | 86 | | | 10,000 | 10,000 | |

---------- **CEP** ----------

| | | ft | mi | nm | m | km | m conv | |
|---|---|---|---|---|---|---|---|---|
| Nitz | 79 | | | .2 | | | 370.40 | 1977. |
| | | | | .17 | | | 314.84 | 18 Jun 1979. |
| | | | | .15 | | | 277.80 | 1982. |
| | | | | .12 | | | 222.24 | 1985. |
| AWar | 80 | | | 0.14 | | | 259.28 | Highly accurate. |
| JWS | 81 | | | | | | | |
| IISS | 82 | | | | 300 | | 300.00 | |
| SSF | 82 | | | 0.14 | | | 259.28 | |
| Clns | 83 | | | 0.15 | | | 277.80 | |
| Lu01 | 83 | | | | 300 | | 300.00 | |
| Ts01 | 83 | | | | 245 | | 245.00 | DIA estimate from Senate Appropriations Committee testimony. |
| vanD | 83 | | | .14 | | | 259.28 | [1979 entry] |
| | | | | .10 | | | 185.20 | [1982 entry] |
| AFM | 84 | < 1,000 | | | | | 304.80 | |
| BW | 86 | | | | 250 | | 250.00 | |

---------- **WARHEAD CHARACTERISTICS** ----------

| Nitz | 79 | 6 RVs | |
| FI | 81 | MIRV payload | Number and yield of warheads unknown. |
| JCS | 81 | MIRV | |
| JWS | 81 | MIRV | |
| IISS | 82 | 6 MIRVs | |
| SSF | 82 | 6 RVs | |
| Clns | 83 | 6 MIRVs | |
| JWS+ | 83 | 6 warheads | |
| SMP | 83 | 6 MIRVs | '83, '84, '85. |
| vanD | 83 | 6 RVs | |
| AFM | 84 | | Similar to Mod 1. |
| BW | 86 | 6 MIRVs | |

BALLISTIC MISSILES  ICBMs  SS-19 Mod 3  229

## ─── NUCLEAR YIELD ───

|  |  | kT | MT | kT conv |  |
|---|---|---|---|---|---|
| Nitz | 79 |  | .75 | 750.00 |  |
| IISS | 82 | 550 |  | 550.00 | Per warhead. |
| SSF | 82 |  | 0.55 | 550.00 | Per warhead. |
| Clns | 83 | 550 |  | 550.00 | Each. |
| JWS+ | 83 |  | 0.5 | 500.00 |  |
| Lu01 | 83 | 550 |  | 550.00 |  |
| vanD | 83 |  | .55 | 550.00 |  |
| BW | 86 | 550 |  | 550.00 |  |

## ─── THROWWEIGHT ───

|  |  | lb | t | kg | kg conv |
|---|---|---|---|---|---|
| Nitz | 79 | 8,000 |  |  | 3,628 |
| IISS | 82 | 8,000 |  |  | 3,628 |
| Clns | 83 | 7,500 |  |  | 3,402 |

## ─── LENGTH ───

|  |  | in | ft | cm | m | m conv |  |
|---|---|---|---|---|---|---|---|
| SMP | 84 |  |  |  | 24 | 24.00 | [From graph], '84, '85. |

## ─── TYPE OF PROPULSION ───

| Clns | 83 | liquid |
| BW | 86 | storable liquid |

## ─── IOC ───

| vanD | 83 | 1979 | [1979 entry] |
|  |  | 1982 | [1982 entry] |

## ─── DEPLOYMENT BEGAN ───

| AWar | 80 |  | Improved version [over Mod 1, ed] now being deployed. |
| JWS | 81 | 1980 | Entered service. |
| IISS | 82 | 1982 | First year deployed. |
| SSF | 82 | 1979 | Year operation began. |
| Clns | 83 | 1980 | First deployed. |
| SMP | 84 |  | Deployment [of SS-17 Mod 3, SS-18 Mod 4, SS-19 Mod 3] began only 5 years ago, [6 years in '85 report]. |
| SMP | 85 | c. 1982 | [From graph.] |
| BW | 86 | ? 1981 |  |

## ─── DEPLOYED ───

| JCS | 81 | Deployed. |
| BW | 86 | In service. |

## ─── DEPLOYMENT HISTORY ───

| FI | 81 | "All [SS-19s] are expected to be retrofitted to the Mod 3 standard". |
| JCS | 81 | "The remaining SS-19s probably will be retrofitted with this version [MIRVed Mod 3] over the next several years". |
| JCS | 82 | "The retrofit of SALT II-accountable MIRVed launcher silos associated with the SS-19 is underway. The program to convert silos from SS-11 to SS-19 Mod 3 configuration will probably require several years to complete". |
| SMP | 83 | [Since last edition in Sept-81], replacement of older missiles with Mod 3. Deployment of Mod 3 continues in superhardened silos. |
| AFM | 84 | Has replaced Mod 1. |
| SMP | 85 | Will be deployed at least through 1990. [Presumed Mod 3, from graph.] |

## ─── NUMBERS IN SERVICE ───

## ─── NUMBER IN 1970 ───

| Clns | 83 | none | In 1970-1979. |

## ─── NUMBER IN 1977 ───

| Nitz | 79 | 50 | Launchers. |

## ─── NUMBER IN 1979 ───

| Nitz | 79 | 280 | 18 June, launchers. |

## ─── NUMBER IN 1980 ───

| Clns | 83 | 20 |

## ─── NUMBER IN 1981 ───

| JCS | 81 |  | As with SS-18, "many of the SS-19s are now of the new, more accurate, MIRVed Mod 3". |
| Clns | 83 | 80 |  |

## ─── NUMBER IN 1982 ───

| IISS | 82 | ≤ 310 | July, total deployed. |
| Clns | 83 | 240 |  |

---------- **NUMBER IN 1983** ----------
IISS+ 83    ≤ 330    July, total deployed.
JWS+  83    c. 300

---------- **NUMBER IN 1984** ----------
SMP   84    360

---------- **NUMBER IN 1985** ----------
SMP   85    360

---------- **UNDATED NUMBER** ----------
BW    86    ? 360    Peak.

---------- **LAUNCHER/SILO** ----------
IISS  82    In mod SS-11 silos.
SMP   83    Deployment of Mod 3 continues in superhardened silos.

---------- **USE AND CONFIGURATION** ----------
vanD  83    Reliability: .85; Penetration 1.0.

---------- **RELOADS IN GENERAL** ----------
Clns  83    Cold launch: no.
SMP   84    Hot launch, '84, '85.

---------- **STRATEGY** ----------
SMP   84    "Has nearly identical capabilities" to SS-18 Mod 4, [re:] "The force of SS-18 Mod 4s currently deployed has the capability to destroy more than 80 percent of the US ICBM silo launchers using two nuclear warheads against each US silo". SS-19 Mod 3 "could be used against targets in Eurasia", '84, '85.
SMP   85    "Could be assigned similar missions" to SS-18 Mod 4, [re:] "The force of SS-18 Mod 4s currently deployed has the capability to destroy more than 80 percent of the US ICBM silo launchers using two nuclear warheads against each US silo".

# SS-19 VarRange

Sources are:   SSF   82

---------- **US DESIGNATION** ----------
SSF   82   SS-19           (Variable range).

---------- **DESCRIPTION** ----------
SSF   82                   Third generation. Variable range.

---------- **EDITOR'S NOTES** ----------
SSF   82   Unlike other SS-19s, listed under "Medium-range missiles".

---------- **RANGE** ----------
|       | mi | nm | km | km conv |
|-------|----|----|----|---------|
| SSF 82 |    | 5,200-5,450 |    | 9,630-10,093 |

---------- **CEP** ----------
|       | ft | mi | nm | m | km | m conv |
|-------|----|----|----|---|----|--------|
| SSF 82 |    |    | 0.14-0.19 |   |    | 259.28-351.88 |

---------- **WARHEAD CHARACTERISTICS** ----------
SSF   82   single or 6 RVs

---------- **NUCLEAR YIELD** ----------
|       | kT | MT | kT conv |       |
|-------|----|----|---------|-------|
| SSF 82 |    | 0.55 | 550.00 | Per warhead. |

---------- **TYPE OF PROPULSION** ----------
SSF   82   liquid

---------- **MIDCOURSE (OR MAIN) GUIDANCE** ----------
SSF   82   fly-the-wire    Onboard digital compuuter.

---------- **DEVELOPMENT BEGAN** ----------
SSF   82   1966            Year design began.

---------- **PROTOTYPE TESTS** ----------
SSF   82   1973            First flight test.

---------- **DEPLOYMENT BEGAN** ----------
SSF   82   1975            Year operation began.

---------- **NUMBERS IN SERVICE** ----------

---------- **NUMBER IN 1955** ----------
SSF   82   none            In 1955, 1960, 1965, 1970, 1975, Includes only those missiles facing Western Europe and China.

---------- **NUMBER IN 1980** ----------
SSF   82   120             Includes only those missiles facing Western Europe and China. "Number of missiles deployed".

---------- **LAUNCHER/SILO** ----------
SSF   82   Basing mode: Hardened silo.

---------- **DESIGNER** ----------
SSF   82   Chelomei        Design Bureau.

232  SOVIET MISSILES  IDDS

# SS-24

```
Sources are:   Rock  78      AWST  82      FI+   83      AWST+ 84      SMP   84      BW    86
FIer  76       FI    81      CBO   83      AFM   84      JCS   84      SMP   85
```

```
---------- US DESIGNATION ------------
Rock   78     SS-X-
FI     81     SS-X-?
AWST   82     SSX-?
CBO    83                    New Solid 1 - Silo.
AFM    84     SS-X-24
AWST+  84     SSX-24
JCS    84     SS-X-24        FY85-86.
SMP    84     SS-X-24        '84, '85.

---------- OTHER DESIGNATION ------------
AWST+  84     PL-04          "Formally known as".

---------- DESCRIPTION ------------
FI     81     ICBM
AWST+  84                    New cold-launched small ICBM.
SMP    84                    Fifth-generation ICBM, '84, '85.

---------- EDITOR'S NOTES ------------
SMP    84     Listed '84, '85.

---------- RANGE ------------
              mi         nm           km          km conv
SMP    85                             10,000      10,000
BW     86                             10,000      10,000

---------- CEP ------------
              ft    mi         nm          m           km    m conv
FI     81                                  ?< 260            260.00
CBO    83                      0.10                          185.20      In 1990.
                               0.08                          148.16      In 1996.
SMP    84                                                                Likely even more accurate
                                                                         than SS-18 Mod 4 and SS-19
                                                                         Mod 3.
BW     86                                  ? 250             250.00

---------- WARHEAD CHARACTERISTICS ------------
CBO    83     10 RVs
AFM    84     8 RVs          During 2nd successful test flight, 22 Nov 1982.
SMP    85     Up to 10 MIRVs
BW     86     10 MIRVs

---------- NUCLEAR YIELD ------------
              kT         MT          kT conv
CBO    83     500                    500.00       Each.

---------- THROWWEIGHT ------------
              lb         t           kg          kg conv
CBO    83     8,000                              3,628

---------- LENGTH ------------
              in         ft          cm          m           m conv
SMP    84                                        22          22.00       [From graph], '84, '85.

---------- TYPE OF PROPULSION ------------
FI     81     solid
AFM    84     solid
SMP    84     solid                  '84, '85.
BW     86     solid

---------- DEVELOPMENT BEGAN ------------
SMP    85     1974           Technological development began. [From graph.]

---------- DEVELOPMENT UNDERWAY ------------
SMP    84                    In development, '84, '85. Development program underway for many years.
```

## PROTOTYPE TESTS

| | | | |
|---|---|---|---|
| FI | 81 | | Beginning trials. Ground testing reported under way late 1980. |
| AWST | 82 | 1980 | "New series of four ICBMs. Flight tests started 1980". |
| FI+ | 83 | 26 Oct 1982 | First test, a first-stage failure. |
| AFM | 84 | | Jan 1984 AFM has details of test firings. 4 or 5 test launches were apparently unsuccessful; [at least 2] successes. |
| SMP | 84 | | Testing is proceeding, from Plesetsk. |
| SMP | 85 | | Test firings continue. "Well along" in flight test program, from Plesetsk. |
| | | 1980-81 | Engineering and testing began. [From graph.] |

## PRODUCTION UNDERWAY

SMP 84   "Now in pre-series production".

## IOC

SMP 84   Expected in 1985.
SMP 85   Expected in 1986.

## DEPLOYMENT BEGAN

SMP 85   1985-86   [Possible deployment, from graph.]
BW 86   1986

## DEPLOYMENT HISTORY

JCS 84   Soviets expected to "replace older SS-11, SS-13, and SS-17 missiles with greater numbers of a new generation of ICBMs like the SS-X-24 and SS-X-25", FY85-86.
SMP 85   Possible deployment may last through 1990. [From graph.]

## NUMBERS IN SERVICE

### NUMBER IN 1983

CBO 83   none

### NUMBER IN 1990

CBO 83   100 in 1990.

### NUMBER IN 1996

CBO 83   300 in 1996.

## LAUNCHER/SILO

FI 81   Expected to be deployed in super-hardened silos.
FI+ 83   Silo-based.
AFM 84   Intended for silo deployment.
SMP 84   Probably will be silo-deployed at first, '84, '85. "Mobile deployment could follow several years after" IOC in 1985.
SMP 85   "Will probably be silo-deployed at first." "Rail-mobile deployment could follow [IOC in 1986, ed.] by one to two years."

## CARRIERS

### CARRIER —— TUBES —— RELOADS

SMP 84   "Available evidence suggests mobile as well as silo deployments for both systems [SS-24 and SS-25]".
SMP 85   Illustration depicts SS-24 launcher being erected from railcar.

## DESIGN AND ENGINEERING

FIar 76   "Three or more fourth-generation ICBMs are under development for service in the mid-1980s. At least one is reported to use solid propellant. Flight trials of some of these missiles should start in the near future".
Rock 78   At least 3 fourth generation missiles currently under development. Trials of one expected in 1978.

## USE AND CONFIGURATION

CBO 83   System availability: 0.95.

## RELOADS IN GENERAL

AWST+ 84   Cold-launched.
SMP 85   Cold launch.

## RELATION TO OTHER MISSILES

FI 81   Similar in size to SS-17.
AFM 84   About same size as US Peacekeeper.

# SS-25

```
Sources are:   Rock  78      AWST  82      FI+   83      AWST+ 84      SecD  84      SMP   85
               FIer  76      FI    81      CBO   83      AFM   84      JCS   84      SMP   84      BW    86
```

---------- US DESIGNATION ----------
|  | |  |
|---|---|---|
| Rock 78 | SS-X- | |
| FI 81 | SS-X-? | |
| AWST 82 | SSX-? | |
| CBO 83 | | New Solid 2 - Mobile. |
| AFM 84 | SS-X-25 | |
| AWST+ 84 | SSX-25 | |
| JCS 84 | SS-X-25 | FY85-86. |
| SecD 84 | SS-X-25 | |
| SMP 84 | SS-X-25 | '84, '85. |

---------- OTHER DESIGNATION ----------
AWST+ 84    PL-05       "Formally known as".

---------- DESCRIPTION ----------
AWST+ 84    New medium solid ICBM.
SMP   84    Fifth-generation ICBM, '84, '85.

---------- EDITOR'S NOTES ----------
SecD 84     Listed FY85.
SMP  84     Listed '84, '85.

---------- RANGE ----------
| | mi | nm | km | km conv |
|---|---|---|---|---|
| SMP 85 | | | 10,500 | 10,500 |
| BW 86 | | | 10,500 | 10,500 |

---------- CEP ----------
| | ft | mi | nm | m | km | m conv | |
|---|---|---|---|---|---|---|---|
| FI 81 | | | | [< 260] | | 260.00 | "When fired from pre-surveyed launch sites, it will have a CEP similar to that of the missile mentioned above". |
| CBO 83 | | | 0.10 | | | 185.20 | In 1990. |
| | | | 0.08 | | | 148.16 | In 1996. |

---------- WARHEAD CHARACTERISTICS ----------
CBO 83    4 RVs          According to Defense Daily, 22 Feb 1983, p. 276. Newsweek (14 Mar 1983, p. 15) suggests single RV.
SMP 84    single RV      '84, '85.
BW  86    single RV

---------- NUCLEAR YIELD ----------
| | kT | MT | kT conv | |
|---|---|---|---|---|
| CBO 83 | 500 | | 500.00 | Each. |

---------- THROWWEIGHT ----------
| | lb | t | kg | kg conv |
|---|---|---|---|---|
| CBO 83 | 3,000 | | | 1,360 |

---------- LENGTH ----------
| | in | ft | cm | m | m conv | |
|---|---|---|---|---|---|---|
| SMP 84 | | | | 18 | 18.00 | [From graph], '84, '85. |

---------- TYPE OF PROPULSION ----------
FI    81    solid
AFM   84    solid
AWST+ 84    solid
SMP   84    solid       '84, '85.
BW    86    solid

---------- DEVELOPMENT BEGAN ----------
SMP 85    1974    Technological development began. [From graph.]

---------- DEVELOPMENT UNDERWAY ----------
SecD 84    Currently under development.
SMP  84    In development, '84, '85. Development program underway for many years.

## BALLISTIC MISSILES ICBMs SS-25

### ---------- PROTOTYPE TESTS ----------
| | | | |
|---|---|---|---|
| FI | 81 | | Being prepared for flight test. |
| AWST | 82 | 1980 | "New series of four ICBMs. Flight tests started 1980". |
| FI+ | 83 | | None yet. |
| AFM | 84 | | Currently under test. |
| SMP | 84 | | Testing is proceeding, from Plesetsk. |
| SMP | 85 | | Test firings continue. "Well along" in flight test program, from Plesetsk. |

### ---------- PRODUCTION UNDERWAY ----------
SMP 84      "Now in pre-series production".

### ---------- IOC ----------
SMP 85      Two bases, probably for SS-25, nearing operational capability. [See USE AND CONFIGURATION, ed.]

### ---------- DEPLOYMENT BEGAN ----------
| | | | |
|---|---|---|---|
| SMP | 85 | 1985-86 | [Possible deployment, from graph.] |
| BW | 86 | 1985 | |

### ---------- DEPLOYMENT HISTORY ----------
JCS 84      Soviets expected to "replace older SS-11, SS-13, and SS-17 missiles with greater numbers of a new generation of ICBMs like the SS-X-24 and SS-X-25", FY85-86.
SMP 85      Possible deployment may last through 1990. [From graph.]

### ---------- NUMBERS IN SERVICE ----------

### ---------- NUMBER IN 1983 ----------
CBO 83      none

### ---------- NUMBER IN 1990 ----------
CBO 83      100 in 1990.

### ---------- NUMBER IN 1996 ----------
CBO 83      300 in 1996.

### ---------- LAUNCHER/SILO ----------
FI+ 83      US satellites have detected what could be multiple-shelter basing scheme at Plesetsk. A network of railway lines running from a central track terminate in what could be protective shelters.
SMP 84      "Available evidence suggests mobile as well as silo deployments for both systems [SS-24 and SS-25]".

### ---------- CARRIERS ----------

### ---------- CARRIER ---------- TUBES ---- RELOADS ----------
| | | CARRIER | TUBES | RELOADS | |
|---|---|---|---|---|---|
| FI | 81 | | | | Mobile missile. |
| AFM | 84 | | | | Expected deployed on mobile launcher. |
| | | | | | reloads: Inherent capability. |
| SMP | 84 | | | | Likely designed for mobile deployment, '84, '85. |
| BW | 86 | | 1 | yes | Mobile missile. |

### ---------- DESIGN AND ENGINEERING ----------
FIer 76      "Three or more fourth-generation ICBMs are under development for service in the mid-1980s. At least one is reported to use solid propellant. Flight trials of some of these missiles should start in the near future".
Rock 78      At least 3 fourth generation missiles currently under development. Trials of one expected in 1978.
SMP 84      Development of [mobile] SS-X-25 based on technologies represented by SS-X-14 and SS-X-15.
SMP 85      1980-81 Engineering and testing began. [From graph.]

### ---------- USE AND CONFIGURATION ----------
CBO 83      System availability: 0.90.
AFM 84      Deployment on mobile launcher will increase survivability.
SMP 84      Likely designed for mobile deployment, "with a home base with launcher garages equipped with sliding roofs; massive, off-road, wheeled [TELs]; and necessary mobile support equipment for refires from the launcher".
SMP 85      "Two bases, probably for the SS-X-25, are nearing operational capability. They consist of launcher garages equipped with sliding roofs and several support buildings to house the necessary mobile support equipment." [Illustration provided, ed.]

### ---------- RELOADS IN GENERAL ----------
SMP 85      Cold launch.

## SOVIET MISSILES

------- **STRATEGY** -------

| | | |
|---|---|---|
| FI+ | 83 | Development of a 2nd ICBM would contravene SALT II. |
| SecD | 84 | "Probably is a second new ICBM type, prohibited by the SALT II agreement". "Even if it is not a new type, it still violates SALT II provisions regarding the permitted ratio between the weight of an ICBM reentry vehicle and the missile's total throw-weight". |
| SMP | 84 | "Probably a second new ICBM type, prohibited by the SALT II agreement; if it is not, it violates the SALT II provisions regarding the permitted ratio between the weight of an ICBM reentry vehicle and the missile's total throwweight". |
| SMP | 85 | "The SS-X-25 violates Soviet obliigations under SALT II." "Some shifting of the SS-20 force has recently been observed as the Soviets prepare for deployment of the SS-X-25 ICBM." The Soviet Union has violated SALT II by "testing an ICBM [SS-25, ed]. with a single reentry vehicle whose weight is less than 50 percent of the ICBM throwweight, if we were to accept the Soviet argument that the SS-X-25 is not a prohibited second new type." |

------- **RELATION TO OTHER MISSILES** -------

| | | |
|---|---|---|
| FI | 81 | Similar in size to US MX. |
| FI+ | 83 | Similar in size to US Minuteman. |
| AFM | 84 | Smaller than SS-X-24. |
| SMP | 84 | Approximately same size as US Minuteman, '84, '85. |

# SS-26

Sources are:  FIar 76    Rock 78    FI 81    AWST 82    SMP 85

---------- **US DESIGNATION** ----------
Rock 78   SS-X-
FI   81   SS-X-?
AWST 82   SSX-?

---------- **CEP** ----------
             ft        mi        nm        m         km       m conv
SMP  85

"Both of these missiles [SS-26, SS-27, ed.] are likely to have better accuracy and greater throwweights than their predecessors."

---------- **THROWWEIGHT** ----------
             lb                  t                   kg       kg conv
SMP  85

"Both of these missiles [SS-26, SS-27, ed.] are likely to have better accuracy and greater throwweights than their predecessors."

---------- **TYPE OF PROPULSION** ----------
FI   81   liquid
SMP  85   solid

---------- **DEVELOPMENT UNDERWAY** ----------
SMP  85   "Recent activity at the Soviet ICBM test ranges indicates that two additional new ICBMs are under development."

---------- **PROTOTYPE TESTS** ----------
AWST 82   1980         "New series of four ICBMs. Flight tests started 1980".
SMP  85                Will begin flight testing in the next few years.

---------- **DESIGN AND ENGINEERING** ----------
FIar 76   "Three or more fourth-generation ICBMs are under development for service in the mid-1980s. At least one is reported to use solid propellant. Flight trials of some of these missiles should start in the near future".
Rock 78   At least 3 fourth generation missiles currently under development. Trials of one expected in 1978.
FI   81   "The final member of the next generation of Soviet... ICBMs may be a backup to the silo-launched solid-propellant design".

---------- **RELATION TO OTHER MISSILES** ----------
SMP  85   May be larger than the SS-24.

238 SOVIET MISSILES IDDS

# SS-27

Sources are:   FIar 76     Rock 78     FI 81     AWST 82     SMP 85

---------- **US DESIGNATION** ----------
Rock 78   SS-X-
FI 81     SS-X-?
AWST 82   SSX-?

---------- **DESCRIPTION** ----------
FI 81     Large ICBM

---------- **CEP** ----------
          ft        mi        nm        m         km        m conv
SMP 85                                                              "Both of these missiles [SS-26, SS-27, ed.] are likely to have better accuracy and greater throwweights than their predecessors."

---------- **THROWWEIGHT** ----------
          lb        t                   kg                  kg conv
SMP 85                                                              "Both of these missiles [SS-26, SS-27, ed.] are likely to have better accuracy and greater throwweights than their predecessors."

---------- **TYPE OF PROPULSION** ----------
FI 81     Liquid

---------- **DEVELOPMENT UNDERWAY** ----------
SMP 85    "Recent activity at the Soviet ICBM test ranges indicates that two additional new ICBMs are under development."

---------- **PROTOTYPE TESTS** ----------
AWST 82   1980        "New series of four ICBMs. Flight tests started 1980".

---------- **DEPLOYMENT HISTORY** ----------
SMP 85    Nearing the flight test stage of development.

---------- **DESIGN AND ENGINEERING** ----------
FIar 76   "Three or more fourth-generation ICBMs are under development for service in the mid-1980s. At least one is reported to use solid propellant. Flight trials of some of these missiles should start in the near future".
Rock 78   At least 3 fourth generation missiles currently under development. Trials of one expected in 1978.
FI 81     "Presumably intended as an eventual replacement of the SS-18". "In the early stages of devevlopment".

---------- **STRATEGY** ----------
SMP 85    To replace the SS-18.

# SS-N-4

| Sources are: | SecC | 67 | FI* | 74 | MOW | 76 | JWS* | 79 | US17 | 81 | Clns+ | 83 | BW | 86 |
|---|---|---|---|---|---|---|---|---|---|---|---|---|---|---|
| SecC | 64 | Ley | 68 | IISS | 74 | SecD | 76 | US08 | 79 | USND | 81 | GSN | 83 | | |
| IA01 | 65 | SecC | 68 | HRST | 75 | JCS | 77 | Clns | 80 | AWST | 82 | Lu01 | 83 | | |
| SecC | 65 | JCS | 72 | IA06 | 75 | SWM | 77 | IISS | 80 | CFW | 82 | vanD | 83 | | |
| MR05 | 66 | IISS | 73 | JCS | 75 | Rock | 78 | US12 | 80 | JFS | 82 | Meyr | 84 | | |
| SecC | 66 | JCS | 73 | FIar | 76 | IISS | 79 | JCS | 81 | SSF | 82 | SMP | 85 | | |

---------- US DESIGNATION ----------

| SecC | 64 | SS-N-4 | |
|---|---|---|---|
| JCS | 72 | SS-N-4 | FY73-74,76-79,82. |
| FI* | 74 | SS-N-4 | |
| FIar | 76 | SS-N-4 | |
| MOW | 76 | SS-N-4 | |
| SecD | 76 | SS-N-4 | |
| SWM | 77 | SS-N-4 | |
| Rock | 78 | SS-N-4 | |
| JWS* | 79 | SSN-4 | |
| Clns | 80 | SS-N-4 | |
| USND | 81 | SS-N-4 | |
| AWST | 82 | SS-N-4 | |
| CFW | 82 | SS-N-4 | |
| JFS | 82 | SS-N-4 | |
| SSF | 82 | SS-N-4 | |
| GSN | 83 | SS-N-4 | |
| vanD | 83 | SS-N-4 | |
| Meyr | 84 | SS-N-4 | |
| SMP | 85 | SS-N-4 | |

---------- NATO CODENAME ----------

| Ley | 68 | Snark |
|---|---|---|
| FI* | 74 | Sark |
| HRST | 75 | Sark |
| FIar | 76 | Sark |
| MOW | 76 | Sark |
| SWM | 77 | Sark |
| Rock | 78 | Sark |
| JWS* | 79 | Sark |
| USND | 81 | Sark |
| AWST | 82 | Sark |
| CFW | 82 | Sark |
| SSF | 82 | Sark |
| GSN | 83 | Sark |

---------- DESCRIPTION ----------

| JCS | 72 | | Listed as 1st generation strategic missile. |
|---|---|---|---|
| FI* | 74 | | Prototype Russian SLBM. |
| SecD | 76 | | Listed under "land-based and sea-based" MRBMs and IRBMs. |
| Rock | 78 | SLBM | |
| JWS* | 79 | | The prototype Soviet SLBM. |
| JCS | 81 | SLBM | |
| USND | 81 | SLBM | |
| SSF | 82 | | First generation. |

---------- EDITOR'S NOTES ----------

| SecC | 64 | Listed FY65-69,76-77, not by name FY65-66. |
|---|---|---|
| JCS | 72 | Listed FY73-74,76-79,82. |
| MOW | 76 | Listed with sub-launched ballistic missiles. |
| SecD | 76 | Listed FY77. |
| AWST | 82 | Listed with sub-launched ballistic missiles. |
| CFW | 82 | Final report of SS-N-4. |
| JFS | 82 | Final report. |
| Meyr | 84 | Note to chart: "Values pertain to initial service period". |
| SMP | 85 | Listed '85. |

## RANGE

| | | mi | nm | km | km conv | |
|---|---|---|---|---|---|---|
| SecC | 64 | | 350 | | 648.20 | FY65-69. |
| IA01 | 65 | c. 1,250 | | | 2,011 | |
| Ley | 68 | 1,200 | | | 1,930 | |
| FI* | 74 | | | 1,500 | 1,500 | |
| IA06 | 75 | | | ? 650 | 650.00 | |
| FIar | 76 | | | 650 | 650.00 | |
| MOW | 76 | c. 350 | | c. 565 | 563.15 | |
| SecD | 76 | | [350] | | 648.20 | "SS-N-4/5 (350-700 NM)"; [since SS-N-5 range is 700 nm, SS-N-4 range is 350 nm?, ed]. |
| JCS | 77 | | [350] | | 648.20 | Range of "SS-N-5/4" is "350-700 NM"; [since SS-N-5 range is 700 nm, SS-N-4 range is 350 nm?, ed]. |
| SWM | 77 | 360 | | | 579.24 | [Table] |
| | | 350 | | 580 | 563.15 | [Text] |
| Rock | 78 | | | 600 | 600.00 | "Moderately effective range". |
| JWS* | 79 | | | c. 650 | 650.00 | |
| | | | | 600 | 600.00 | According to "one well-known observer" of USSR Navy. |
| US08 | 79 | | 300 | | 555.60 | |
| Clns | 80 | | 300 | | 555.60 | |
| US17 | 81 | | 370 | | 685.24 | |
| USND | 81 | | c. 350 | | 648.20 | |
| AWST | 82 | | 350 | | 648.20 | |
| CFW | 82 | | 300 | | 555.60 | |
| JFS | 82 | | 370 | | 685.24 | Max. |
| SSF | 82 | | 350 | | 648.20 | |
| GSN | 83 | | c. 350 | c. 650 | 648.20 | |
| Meyr | 84 | | | 500 | 500.00 | |
| BW | 86 | | | 650 | 650.00 | |

## CEP

| | | ft | mi | nm | m | km | m conv | |
|---|---|---|---|---|---|---|---|---|
| Clns | 80 | | | 1.5 | | | 2,778 | |
| US12 | 80 | | | 7-10 | | | 12,964-18,520 | "Should be possible" [at 300 nm]. |
| SSF | 82 | | | 2.0 | | | 3,704 | |
| Lu01 | 83 | | | | | 1.8 | 1,800 | |
| vanD | 83 | | | 1.5 | | | 2,778 | |
| Meyr | 84 | | | | | 3.0-5.0 | 3,000-5,000 | |
| BW | 86 | | | | 4,000 | | 4,000 | |

## WARHEAD TYPE

| | | nuclear | conven | chem |
|---|---|---|---|---|
| FI* | 74 | yes | | |
| MOW | 76 | yes | | |
| SWM | 77 | yes | | |
| US08 | 79 | yes | | |
| CFW | 82 | yes | | |
| JFS | 82 | yes | | |
| GSN | 83 | yes | | |
| Meyr | 84 | yes | | |

## WARHEAD CHARACTERISTICS

| | | |
|---|---|---|
| Rock | 78 | single RV |
| US08 | 79 | single RV |
| Clns | 80 | single RV |
| SSF | 82 | single RV |
| GSN | 83 | single RV |
| vanD | 83 | single RV |
| Meyr | 84 | single RV |
| BW | 86 | single RV |

## NUCLEAR YIELD

|  |  | kT | MT | kT conv |  |
|---|---|---|---|---|---|
| SecC | 65 |  | 2-3.5 | 2,000-3,500 |  |
| FI* | 74 |  | ? 1 | 1,000 |  |
| FIar | 76 |  | ? 1 | 1,000 |  |
| Rock | 78 |  | 1.0 | 1,000 |  |
| USO8 | 79 |  | 1+ | 1,000 |  |
| Clns | 80 |  | 1 | 1,000 |  |
| JFS | 82 |  |  |  | "Megaton". |
| SSF | 82 |  | 2.0-3.5 | 2,000-3,500 |  |
| GSN | 83 |  | c. 1 | 1,000 |  |
| venD | 83 |  | 2.75 | 2,750 |  |
| Meyr | 84 |  | 2-3.5 | 2,000-3,500 |  |
| BW | 86 | 3,000 |  | 3,000 |  |

## WEIGHT

|  |  | lb | t | kg | kg conv |  |
|---|---|---|---|---|---|---|
| IA01 | 65 |  | c. 19.5 |  | 19,500 |  |
| Ley | 68 |  | 19.5 |  | 19,500 | Tons. |
| FI* | 74 |  |  | 18000-20000 | 18,000-20,000 |  |
| IA06 | 75 |  |  | 18000-20000 | 18,000-20,000 |  |
| SWM | 77 | 42,000 |  | 19,050 | 19,051 |  |
| Rock | 78 |  | 19 |  | 19,000 | Tonnes, launch weight. |
| JWS* | 79 |  | 18-20 |  | 18,000-20,000 | Tons. |
| JFS | 82 | c. 41,000 |  |  | 18,597 | Launch weight. |
| GSN | 83 | c. 44,000 |  | c. 19,800 | 19,958 |  |

## LENGTH

|  |  | in | ft | cm | m | m conv |
|---|---|---|---|---|---|---|
| IA01 | 65 |  | 46 |  |  | 14.02 |
| Ley | 68 |  | 46 |  |  | 14.02 |
| FI* | 74 |  |  |  | c. 15 | 15.00 |
| HRST | 75 |  | 45 |  |  | 13.72 |
| FIar | 76 |  |  |  | 15 | 15.00 |
| MOW | 76 |  | 48'0" |  | 14.50 | 14.63 |
| SWM | 77 |  | 49 |  | 15.0 | 14.94 |
| Rock | 78 |  |  |  | 15 | 15.00 |
| JWS* | 79 |  |  |  | 15 | 15.00 |
| JFS | 82 |  | 42.5 |  |  | 12.95 |
| GSN | 83 |  | c. 46 |  | c. 14 | 14.02 |

## DIAMETER

|  |  | in | ft | cm | m | cm conv |  |
|---|---|---|---|---|---|---|---|
| IA01 | 65 |  | 5.4 |  |  | 164.59 |  |
| Ley | 68 |  | 5.4 |  |  | 164.59 |  |
| FI* | 74 |  |  | 180 |  | 180.00 |  |
| HRST | 75 | 72 |  |  |  | 182.88 |  |
| FIar | 76 |  |  |  | 1.9 | 190.00 |  |
| MOW | 76 |  | 5'9" |  | 1.75 | 175.26 |  |
| SWM | 77 |  | 5.9 |  | 1.8 | 179.83 |  |
| Rock | 78 |  |  |  | 1.8 | 180.00 |  |
| JWS* | 79 |  |  | 180 |  | 180.00 |  |
| GSN | 83 | 71 |  | 180 |  | 180.34 | [1,800 mm in source] |

## NUMBER OF STAGES

| FI* | 74 | 2 |  |
|---|---|---|---|
| MOW | 76 | 2 | Probably. |
| SWM | 77 | 2 |  |
| Rock | 78 | 1 |  |
| JWS* | 79 | 2 |  |
| US12 | 80 | 1 |  |
| AWST | 82 | 1 |  |
| JFS | 82 | 2 |  |

```
                ------ TYPE OF PROPULSION ------
FI*     74      solid
HRST    75      solid
Flar    76      ? solid
MOW     76      solid              7 first-stage nozzles.
SWM     77      solid
Rock    78      storable liquid
JWS*    79      solid
Clns    80      solid
US12    80      liquid
AWST    82      liquid storable
CFW     82      liquid
JFS     82      solid
SSF     82      liquid
GSN     83      liquid
BW      86      storable liquid

                ------ MIDCOURSE (OR MAIN) GUIDANCE ------
FI*     74      inertial
SWM     77      inertial
Rock    78      inertial           First generation inertial guidance.
JWS*    79      inertial
JFS     82      inertial
SSF     82      inertial
GSN     83      inertial

                ------ OBSERVED IN PARADE ------
MOW     76      7 Nov 1962
SWM     77      Nov 1962
Rock    78      7 Nov 1962         First public display.

                ------ DEVELOPMENT BEGAN ------
SSF     82      1949-50            Year design began.
SMP     85      1947               Technological development began. [From graph.]

                ------ DEVELOPMENT UNDERWAY ------
Flar    76      mid-1950s          Developed.

                ------ PROTOTYPE TESTS ------
JWS*    79      1955               Test firing.
JFS     82      1955
SMP     85      1953-54            Engineering and testing began. [From graph.]

                ------ IOC ------
SWM     77      1958               Operational.
US12    80      1960
JFS     82      1958               Operational.
GSN     83      1959-1960
vanD    83      1956

                ------ DEPLOYMENT BEGAN ------
Flar    76                         First Russian "submarine-launched missile" [apparently distinct from SLBM, ed].
                1958               Installed from 1958 onwards.
SWM     77                         First Soviet SLBM.
                1961               First deployed.
Rock    78      1955               Service intro [table].
                1956               "Considered operational in 1956" [text].
                                   First SLBM developed for Soviet Navy. World's first SLBM.
JWS*    79                         Some say it never was operational.
Clns    80      1961               First deployed.
US17    81      1958               Year of introduction.
SSF     82      1958               Year operation began.
GSN     83                         Soviet Navy's first operational SLBM.
Meyr    84      1959               Service.
SMP     85      1958-59            [From graph.]
BW      86      ? 1958

                ------ DEPLOYED ------
MOW     76                         In service.

                ------ DEPLOYMENT COMPLETE ------
Rock    78                         "Appearing and equipping [subs] between 1955 and 1960".

                ------ RETIRED ------
JWS*    79                         Almost certainly not operational now.
US12    80                         Status: operational.
SMP     85      1975               [From graph.]
BW      86      ? 1979
```

## BALLISTIC MISSILES  SLBMs  SS-N-4

---------- **DEPLOYMENT HISTORY** ----------

| | | |
|---|---|---|
| FI* | 74 | Considered by most sources obsolescent or obsolete. |
| IA08 | 75 | Almost certainly obsolete. |
| FIar | 76 | The system "must now be considered completely obsolete". Only a small number of missiles remain in service. |
| Rock | 78 | Very few still in use. |
| CRW | 82 | Obsolete. |
| JFS | 82 | Obsolete. |

---------- **NUMBERS IN SERVICE** ----------

---------- **NUMBER IN 1955** ----------

| | | | |
|---|---|---|---|
| SSF | 82 | | None in 1955, SS-N-4,5; not including forward-deployed units. |

---------- **NUMBER IN 1960** ----------

| | | | |
|---|---|---|---|
| SSF | 82 | | None in 1960, 1975, 1980, SS-N-4,5; forward-deployed units. |
| | | | 36 in 1960, SS-N-4,5; not including forward-deployed units. |

---------- **NUMBER IN 1965** ----------

| | | | |
|---|---|---|---|
| SSF | 82 | | 15 in 1965, SS-N-4,5; forward deployed units. |
| | | | 105 in 1965, SS-N-4,5; not including forward-deployed units. |

---------- **NUMBER IN 1970** ----------

| | | | |
|---|---|---|---|
| Clns | 80 | 27 | In 1970-1972. |
| SSF | 82 | | 9 in 1970, SS-N-4,5; forward deployed units. |
| | | | 365 in 1970, SS-N-4,5; not including forward-deployed units. |
| Clns+ | 83 | 21 | In 1970-1972. |

---------- **NUMBER IN 1973** ----------

| | | | |
|---|---|---|---|
| IISS | 73 | 36 | July, total deployed. |
| JCS | 73 | | Mid-1973 560 SLBM total excludes "some 60 SS-N-4 and -5 launchers" on diesel subs. |
| Clns | 80 | 21 | In 1973-1975. |

---------- **NUMBER IN 1974** ----------

| | | | |
|---|---|---|---|
| IISS | 74 | 27 | 1974-78, July, total deployed. |

---------- **NUMBER IN 1975** ----------

| | | | |
|---|---|---|---|
| USD8 | 79 | 27 | Aggregate warheads, early 1975. |
| SSF | 82 | | 569 in 1975, SS-N-4,5; not including forward-deployed units. |

---------- **NUMBER IN 1976** ----------

| | | |
|---|---|---|
| Clns | 80 | 18 |
| Clns+ | 83 | 15 |

---------- **NUMBER IN 1977** ----------

| | | |
|---|---|---|
| Clns | 80 | 12 |

---------- **NUMBER IN 1978** ----------

| | | |
|---|---|---|
| Clns | 80 | 9 |

---------- **NUMBER IN 1979** ----------

| | | | |
|---|---|---|---|
| IISS | 79 | 18 | July, total deployed. |
| Clns | 80 | 3 | |

---------- **NUMBER IN 1980** ----------

| | | | |
|---|---|---|---|
| IISS | 80 | c. 9 | July, total deployed; withdrawn but not reported as scrapped. |
| SSF | 82 | | 445 in 1980, SS-N-4,5; not including forward-deployed units. |
| Clns+ | 83 | none | In 1980-1982. |

---------- **CARRIERS** ----------

---------- **CARRIER** ---------- **TUBES** --- **RELOADS** ----------

| | | | | | |
|---|---|---|---|---|---|
| SecC | 64 | | 3 | | |
| | | | | | tubes: Most current Soviet subs estimated to carry 3 missiles apiece. |
| SecC | 65 | | | | "Almost all" Soviet ballistic missile subs equipped with this missile. |
| | | G and H | | | Some evidence that construction of G- and H-class subs has stopped. |
| SecC | 66 | [G and H] | | | Except for 2-3 nuclear powered with SS-N-5, all other ballistic missile subs (of 43-48 total, 8-10 SSBNs) carry SS-N-4. |
| SecC | 67 | | | | All operational ballistic missile subs not carrying SS-N-5 carry SS-N-4. |
| SecC | 68 | | | | The remaining ballistic missile subs [other than the many being converted to SS-N-5] will probably continue to carry SS-N-4. |

|  |  | CARRIER | TUBES | RELOADS |  |
|---|---|---|---|---|---|
| MR05 | 66 | Z | 2 | | About 10 subs. Modified 1958-59 for ballistic missiles, Z subs primarily training ships. |
| FIar | 76 | Z V | 2 | | Installed on 7. |
| SWM | 77 | Zulu V | 2 or 3 | | In extended bridge fin. |
| Rock | 78 | Zulu | 2 | | N=7; only 1 remains. |
| JWS* | 79 | Z-V | 2 | | Followed missile similar or identical to Scud A in this sub. |
| USND | 81 | Zulu V | | | |
| GSN | 83 | Zulu-V | 2 | | All have been discarded. |
| BW | 86 | Zulu V | 2 | no | N=7. |

|  |  | CARRIER | TUBES | RELOADS |  |
|---|---|---|---|---|---|
| MR05 | 66 | G | 3 | | N=20-30. |
| JCS | 75 | G | 3 | | G-class carries 3 "SS-N-4/5", FY76-77. tubes: FY76-78. |
| FIar | 76 | Golf 1 | 3 | | Installed on 22. |
| MOW | 76 | G-1 | 3 | | N=9. |
| SecD | 76 | G | | | |
| JCS | 77 | Golf | 3 | | Golf class carries 3 "SS-N-5/4". |
| SWM | 77 | Golf | 2 or 3 | | In extended bridge fin. |
| Rock | 78 | Golf | 3 | | N=22; service phase-out underway by 1976. |
| JWS* | 79 | G-1 | 3 | | Followed missile similar or identical to Scud A in this sub. |
| Clns | 80 | Golf I | 3 | | |
| US12 | 80 | Golf I | | | |
| USND | 81 | Golf | | | |
| AWST | 82 | Golf | | | "Deployed on some Golf-class submarines". |
| CFW | 82 | Golf I | | | |
| JFS | 82 | Golf I | | | |
| SSF | 82 | Golf I | 3 | | None in operation in 1980. |
| GSN | 83 | Golf I | | | 13 converted to SS-N-5. |
| BW | 86 | Golf I | 3 | no | |

|  |  | CARRIER | TUBES | RELOADS |  |
|---|---|---|---|---|---|
| MR05 | 66 | H | 3 | | N=8-13. |
| FIar | 76 | Hotel 1 | 3 | | Installed on 15. |
| JCS | 77 | Hotel | | | "The first of eight HOTEL" SSBNs was launched in 1959 and carried SS-N-4s, FY78-79. |
| SWM | 77 | Hotel | 2 or 3 | | In extended bridge fin. |
| Rock | 78 | Hotel | 3 | | N=15. |
| JWS* | 79 | H-1 | 3 | | |
| US12 | 80 | Hotel I | | | |
| USND | 81 | Hotel | | | |
| SSF | 82 | Hotel I | 3 | | None in operation in 1980. |
| GSN | 83 | Hotel | | | 7 converted to SS-N-5. |
| BW | 86 | Hotel I | 3 | no | |

---------- **DESIGNER** ----------
SSF  82  Yangel    Design Bureau.

---------- **USE AND CONFIGURATION** ----------
SecC 64  Surface-launched, FY65-69. Missile must be fired from the surface.
SecC 66  Only recently have Soviet ballistic missile subs "regularly carried out ocean patrols, but these appear to be to staging areas rather than to strike stations".
HRST 75  "Submarine-launched".
FIar 76  Missiles housed in vertical launch tubes which passed through sail to accommodate extreme length of missile. Subs had to surface to fire missiles.
MOW  76  [Soviet] commentator said submerged or surface firing possible.
SWM  77  Surface launched.
Rock 78  Surface launch only, for all 3 types of subs.
JWS* 79  Surface launch only, in all 3 sub classes.
CFW  82  Launched only from surface.
JFS  82  Surface launched.
GSN  83  Surface launched.
vanD 83  Reliability: .70; Penetration 1.0.
Meyr 84  Penetration: 1.0; reliability: 0.2-0.5, "includes system reliability multiplied by operational readiness of deployed system".

---------- **STRATEGY** ----------
SecC 68  "The intelligence community now estimates that the diesel-powered ballistic missile submarines are primarily intended for use against Eurasian targets".
Rock 78  Plays secondary, reserve role.
Meyr 84  [Ed: chart is labeled] Systems "assigned primary operational and strategic missions within the European theatre(s) of military operations".

---------- **GEOGRAPHICAL DEPLOYMENTS** ----------
SSF  82  Forward deployed units (SS-N-4,5): none in 1960, 1975, 1980; 15 in 1965; 9 in 1970.

---------- **EXPORTED TO** ----------
MOW   76   China         A single Chinese G-class sub (Soviet-built) may carry Sarks.

---------- **RELATION TO OTHER MISSILES** ----------
MOW   76   "First-generation Soviet counterpart of Polaris".
JWS*  79   Similar in appearance to Polaris A-2, but longer.

246  SOVIET MISSILES  IDDS

# SS-N-5

```
Sources are:   SecC  70    IAD6  75    JCS   77    IISS  81    IISS  82    LuO1  83    SMP  85
SecC  64       JCS   71    JCS   75    SWM   77    JCS   81    JFS   82    vanD  83    BW   86
IAD1  65       SecC  71    SecC  75    JCS   78    JWS   81    SSF   82    ABFS  84
SecC  65       JCS   72    FIar  76    Rock  78    SecD  81    Clns+ 83    CFW+  84
SecC  66       IISS  73    IISS  76    IISS  79    SMP   81    GSN   83    JCS   84
SecC  67       JCS   73    JCS   76    USO8  79    US17  81    IISS+ 83    Meyr  84
IDO1  68       FI*   74    MOW   76    Clns  80    USND  81    JCS   83    SecD  84
SecC  68       IISS  74    SecD  76    US12  80    AWST  82    JFS+  83    SMP   84
SecC  69       HRST  75    A140  77    FI    81    CFW   82    JWS+  83    JCS   85
```

```
----------  US DESIGNATION ----------
SecC  64     SS-N-5
JCS   71     SS-N-5       FY73-79,82,85-86.
FI*   74     SS-N-5
FIar  76     SS-N-5
MOW   76     SS-N-5
SecD  76     SS-N-5       FY77,82,85.
SWM   77     SS-N-5
Rock  78     SS-N-5
Clns  80     SS-N-5
FI    81     SS-N-5
JWS   81     SSN-5
SMP   81     SS-N-5       '81, '83, '84, '85.
USND  81     SS-N-5
AWST  82     SS-N-5       [SLBM entry only]
CFW   82     SS-N-5
IISS  82     SS-N-5
JFS   82     SS-N-5
SSF   82     SS-N-5
GSN   83     SS-N-5
vanD  83     SS-N-5
Meyr  84     SS-N-5
```

```
----------  NATO CODENAME ----------
FI*   74     Serb
HRST  75     Serb
FIar  76     Serb
MOW   76     Serb
SWM   77     Serb
Rock  78     Serb
FI    81     Serb
JWS   81     Serb
USND  81     Serb
AWST  82     Serb
CFW   82     Serb
IISS  82     Serb
SSF   82     Serb
GSN   83     Serb
JFS+  83     Serb
```

```
----------  DESCRIPTION ----------
JCS   72                       Listed as 1st generation strategic missile.
SecD  76                       Listed under "Land-based and sea-based" MRBMs and IRBMs.
JCS   77     IRBM
SWM   77                       Second generation missile.
Rock  78     SLBM
JWS   81                       Second-generation SLBM.
USND  81     SLBM
AWST  82     IRBM range        "Second generation submarine IRBM", [submarine IRBM entry only]
SSF   82                       Second generation.
GSN   83                       Improved SLBM.
JCS   83     SLBM
```

```
----------  EDITOR'S NOTES ----------
SecC  64     Listed FY65-72,76-77, not by name FY65-66.
JCS   71     Listed FY72-79,82,85-86; referred to not by name in FY72: USSR has G class and H class subs
             carrying "submerged launched 650 n.mi. range SLBMs".
SecD  76     Listed FY77,82,85.
FI    81     Final report.
SMP   81     Listed '81, '83, '84, '85.
AWST  82     Source has 2 separate table entries for Serb, one among the SLBMs (and referred to below as "SLBM
             entry"); the other among the ICBMs (and referred to as "submarine IRBM entry"). Data is common to
             the two entries unless marked.
Meyr  84     Note to chart: "Values pertain to initial service period".
```

BALLISTIC MISSILES  SLBMs  SS-N-5  247

## RANGE

| Source | Year | mi | nm | km | km conv | Notes |
|---|---|---|---|---|---|---|
| IA01 | 65 | c. 1,250 | | | 2,011 | |
| SecC | 65 | | 700 | | 1,296 | FY66-70,72. |
| ID01 | 68 | ? < 2,000 | | | 3,218 | Probably under 2,000 mi. |
| SecC | 70 | | 650 | | 1,203 | |
| JCS | 71 | | 650 | | 1,203 | |
| FI* | 74 | | | 1,300 | 1,300 | |
| IA06 | 75 | | | 1,200-2,400 | 1,200-2,400 | |
| JCS | 75 | | 700 | | 1,296 | FY76-78. |
| FIer | 76 | | | 1,300 | 1,300 | Only half that of Polaris 1, despite similarity in configuration and size. |
| MOW | 76 | 750 | | 1,200 | 1,206 | Max. |
| SecD | 76 | | 700 | | 1,296 | |
| JCS | 77 | | | | | Relatively short range, FY78-79. |
| SWM | 77 | 750 | | 1,200 | 1,206 | [Table] |
|  |  |  |  | 1,210 | 1,210 | [Text] |
| JCS | 78 | | | 1,300 | 1,300 | |
| Rock | 78 | | | 1,250 | 1,250 | [Table] |
|  |  |  |  | 1,200 | 1,200 | [Text] |
| USD8 | 79 | | 650 | | 1,203 | |
| Clns | 80 | | 900 | | 1,666 | |
| FI | 81 | | | 1,300 | 1,300 | |
| JWS | 81 | | | 1,200-2,400 | 1,200-2,400 | Range of estimates, lower figure more likely. |
| SMP | 81 | | | 1,400 | 1,400 | '81, '83. |
| USND | 81 | | c. 700 | | 1,296 | |
| AWST | 82 | | 700 | | 1,296 | [SLBM entry only] |
| CPW | 82 | | 900 | | 1,666 | |
|  |  |  | 700 | | 1,296 | Originally . |
| IISS | 82 | | | 1,400 | 1,400 | Max. |
| JFS | 82 | | 900 | | 1,666 | Max. |
| SSF | 82 | | 750 | | 1,389 | |
| GSN | 83 | | 900 | 1,650 | 1,666 | Range believed increased to this during service life. |
|  |  |  | c. 700 | c. 1,300 | 1,296 | Originally. |
| JFS+ | 83 | | 850 | | 1,574 | Max. |
| JWS+ | 83 | | | 1,300 | 1,300 | [Analysis table] |
| CPW+ | 84 | | 900 | | 1,666 | |
| Meyr | 84 | | | 1,100 | 1,100 | |
| BW | 86 | | | 1,300 | 1,300 | |

## CEP

| Source | Year | ft | mi | nm | m | km | m conv | Notes |
|---|---|---|---|---|---|---|---|---|
| SecC | 70 | | | | | | | Range estimate revised downward this year, but yield and CEP estimates remain the same, [deleted]. |
| JCS | 77 | | | | | | | [Relatively] large CEP, FY78-79. |
| Clns | 80 | | | 1.5 | | | 2,778 | |
| US12 | 80 | | | 7-10 | | | 12,964-18,520 | "Should be possible" [at 700 nm] |
| IISS | 82 | | | | 2,800 | | 2,800 | |
| SSF | 82 | | | 1.5 | | | 2,778 | |
| Lu01 | 83 | | | | | 1.8 | 1,800 | |
| vanD | 83 | | | 1.5 | | | 2,778 | |
| Meyr | 84 | | | | | 2.7-4.0 | 2,700-4,000 | |
| BW | 86 | | | | 2,800 | | 2,800 | |

## WARHEAD TYPE

| Source | Year | nuclear | conven | chem | Notes |
|---|---|---|---|---|---|
| MOW | 76 | yes | | | |
| SWM | 77 | yes | | | |
| JCS | 81 | yes | | | Nuclear delivery system. |
| CPW | 82 | yes | | | |
| JFS | 82 | yes | | | |
| GSN | 83 | yes | | | |
| JWS+ | 83 | yes | | | |
| Meyr | 84 | yes | | | |

248  SOVIET MISSILES  IDDS

```
─────── WARHEAD CHARACTERISTICS ───────
JCS    75    single RV           FY76-79.
                                 Effective against soft targets "because of the warhead size", FY76,78-79.
MOW    76    single RV
Rock   78    single RV
Clns   80    single RV
JWS    81    single RV
SMP    81    single RV           '81, '83.
IISS   82    single RV
SSF    82    single RV
GSN    83    single RV
vanD   83    single RV
Meyr   84    single RV
BW     86    single RV

─────── NUCLEAR YIELD ───────
             kT          MT           kT conv
SecC   70                                          Range estimate revised downward this year, but
                                                   yield and CEP estimates remain the same,
                                                   [deleted].
FI*    74                ? 1          1,000
FIar   76                ? 1          1,000
MOW    76                             MT range.
SWM    77                ? 1          1,000
Rock   78                1.0          1,000
USD8   79                1+           1,000
Clns   80                1-2          1,000-2,000
FI     81                ? 1          1,000
JWS    81                             MT range
CFW    82    c. 800                   800.00
IISS   82                c. 1         1,000
JFS    82    800                      800.00
SSF    82                4.0          4,000
GSN    83    c. 800                   800.00
JWS+   83                ? 1          1,000        [Analysis table]
vanD   83                1.5          1,500
Meyr   84                1-2          1,000-2,000
BW     86    1,000                    1,000

─────── WEIGHT ───────
             lb          t            kg           kg conv
IAD1   65                c. 15                     15,000       Tons.
FI*    74                             18,000       18,000
SWM    77    40,000                   18,144       18,144
Rock   78                18                        18,000       Tonnes; slightly lighter (and
                                                                smaller) than Sark.
FI     81                             18,000       18,000
JWS    81                c. 18                     18,000       Tons.
JFS    82    c. 41,000                             18,597       Launch weight.
GSN    83    c. 37,000                c. 16,650    16,783
CFW+   84                             16,500       16,500

─────── LENGTH ───────
             in          ft      cm   m            m conv
IAD1   65                35                        10.67
ID01   68                35                        10.67        Includes casing over nozzles.
FI*    74                             10           10.00
HRST   75                31                        9.45
FIar   76                             10           10.00
MOW    76                35'1"        10.7         10.69
SWM    77                35           10.7         10.67        [Table]
                         33           10           10.06        [Text]
Rock   78                             10.7         10.70
FI     81                             10           10.00
JWS    81                             c. 12.9      12.90
SMP    81                             13           13.00        [From graph], '81, '83.
AWST   82                31                        9.45         [Submarine IRBM entry only]
JFS    82                42.5                      12.95
GSN    83                c. 43        c. 13        13.11
CFW+   84                             13           13.00
```

## DIAMETER

|      |    | in    | ft    | cm     | m    | cm conv |                            |
|------|----|-------|-------|--------|------|---------|----------------------------|
| IA01 | 65 |       | 5.4   |        |      | 164.59  |                            |
| ID01 | 68 |       | 5.25  |        |      | 160.02  |                            |
| FI*  | 74 |       |       | c. 150 |      | 150.00  |                            |
| HRST | 75 | 54    |       |        |      | 137.16  |                            |
| FIar | 76 |       |       |        | 1.7  | 170.00  |                            |
| MOW  | 76 |       | 5'0"  |        | 1.50 | 152.40  | Max.                       |
| SWM  | 77 |       | 4.9   |        | 1.5  | 149.35  |                            |
| Rock | 78 |       |       |        | 1.5  | 150.00  |                            |
| FI   | 81 |       |       | c. 150 |      | 150.00  |                            |
| JWS  | 81 |       |       |        | 1.42 | 142.00  | Max.                       |
| AWST | 82 |       | 4.5   |        |      | 137.16  | [Submarine IRBM entry only]|
|      |    |       | 5     |        |      | 152.40  | [SLBM entry only]          |
| GSN  | 83 | c. 48 |       | c. 121.7|     | 121.92  | [1,217 mm in source]       |
| CFW+ | 84 |       |       |        | 1.2  | 120.00  |                            |

## NUMBER OF STAGES

| FI*  | 74 | 2 |                              |
|------|----|---|------------------------------|
| FIar | 76 | 2 |                              |
| MOW  | 76 | 2 |                              |
| SWM  | 77 | 2 |                              |
| Rock | 78 | 1 |                              |
| US12 | 80 | 1 |                              |
| FI   | 81 | 2 |                              |
| JWS  | 81 | 2 |                              |
| AWST | 82 | 2 | [Submarine IRBM entry only] |
|      |    | 1 | [SLBM entry only]            |
| JFS  | 82 | 2 |                              |

## TYPE OF PROPULSION

| FI*  | 74 | liquid          |                                                                 |
|------|----|-----------------|-----------------------------------------------------------------|
| HRST | 75 | solid           |                                                                 |
| FIar | 76 | storable liquid | Contrary to early reports, both stages use liquid fuel.         |
| MOW  | 76 | solid           |                                                                 |
| SWM  | 77 | solid           | Recent info indicates "increasingly probable" it is storable liquid. |
| Rock | 78 | storable liquid | Thought when first displayed to be solid fuel.                  |
| Clns | 80 | solid           |                                                                 |
| US12 | 80 | liquid          |                                                                 |
| FI   | 81 | liquid          |                                                                 |
| JWS  | 81 | solid           |                                                                 |
| AWST | 82 | solid           | [Submarine IRBM entry only]                                     |
|      |    | liquid storable | [SLBM entry only]                                               |
| CFW  | 82 | liquid          |                                                                 |
| JFS  | 82 | solid           |                                                                 |
| SSF  | 82 | liquid          |                                                                 |
| GSN  | 83 | liquid          |                                                                 |
| JFS+ | 83 | liquid          |                                                                 |
| BW   | 86 | storable liquid |                                                                 |

## MIDCOURSE (OR MAIN) GUIDANCE

| FI*  | 74 | inertial |
| MOW  | 76 | inertial |
| SWM  | 77 | inertial |
| FI   | 81 | inertial |
| JWS  | 81 | inertial |
| JFS  | 82 | inertial |
| SSF  | 82 | inertial |
| GSN  | 83 | inertial |
| CFW+ | 84 | inertial |

## OBSERVED IN PARADE

| MOW  | 76 | Nov 1964 |                      |
| SWM  | 77 | Nov 1964 |                      |
| Rock | 78 | Nov 1964 | First public display.|
| JWS  | 81 | Nov 1967 |                      |

## DEVELOPMENT BEGAN

| SSF | 82 | 1954-55 | Year design began.                              |
| SMP | 85 | 1952    | Technological development began. [From graph.] |

## DEVELOPMENT UNDERWAY

| SecC | 64 | | There is evidence a longer-range "submerged launch" missile is now under development. |

## PROTOTYPE TESTS

| JWS | 81 | ? Mar 1962 | First submarine launch.                     |
| SMP | 85 | 1958-59    | Engineering and testing began. [From graph.]|

250 SOVIET MISSILES IDDS

```
————————  IOC ————————
JCS    75    1963         FY76-79.
SWM    77    1963         Operational.
US12   80    1963
JWS    81    1963         Operational.
US17   81    1963         Year of introduction.
JFS    82    1963         Operational.
GSN    83    1963
vanD   83    1966

————————  DEPLOYMENT BEGAN ————————
JCS    75    1963         First SS-N-5 entered force, FY76-78.
FIar   76    1963         Entered service.
                          First Russian SLBM.
SWM    77    1964         Entered service.
Rock   78    1963         Service intro.
Clns   80    1963         First deployed.
IISS   82    1964         First year deployed.
SSF    82    1963         Year operation began.
JCS    83    c. 1963      Year of introduction [from graph].
CFW+   84    1963
Meyr   84    1964         Service.
SecD   84    1963-64      Introduction of strategic systems [graph].
SMP    85    1963-64      [From graph.]
BW     86    1963

————————  DEPLOYED ————————
MOW    76                 In service.
US12   80                 Status: operational.
BW     86                 In service.

————————  RETIRED ————————
SMP    85    1987-88      [From graph.]

————————  DEPLOYMENT HISTORY ————————
FI*    74    Considered obsolescent. Removal of launch tubes from Hotel subs began 1978.
JCS    75    Expect 24 SAL launchers to be deleted from inventory.
FIar   76    Golf boats expected to be withdrawn soon, but not Hotel boats.
JCS    76    Expect some of 24 SAL launchers to be deleted from inventory.
JWS    81    Successor to Sark.
CFW    82    Obsolescent.
JWS+   83    Regarded as obsolescent.

————————  NUMBERS IN SERVICE ————————

————————  NUMBER IN 1955 ————————
SSF    82                 None, SS-N-4,5; not including forward-deployed units.

————————  NUMBER IN 1960 ————————
SSF    82                 None in 1960, 1975, 1980, SS-N-4,5; forward-deployed units.
                          36, SS-N-4,5; not including forward-deployed units.

————————  NUMBER IN 1965 ————————
SSF    82                 15, SS-N-4,5; forward-deployed units.
                          105, SS-N-4,5; not including forward-deployed units.

————————  NUMBER IN 1970 ————————
Clns   80    54           1970-1972.
SSF    82                 9, SS-N-4,5; forward-deployed units.
                          365, SS-N-4,5; not including forward-deployed units.
Clns+  83    60           In 1970-1972.

————————  NUMBER IN 1973 ————————
IISS   73    60           July, total deployed (30 on diesel subs, 30 on nuclear).
JCS    73                 Mid-1973 560 SLBM total excludes "some 60 SS-N-4 and -5 launchers".
Clns   80    60           1973-1977.

————————  NUMBER IN 1974 ————————
IISS   74    57           1974-75, July, total deployed (33 on diesel subs, 24 on nuclear).

————————  NUMBER IN 1975 ————————
JCS    75    24           Currently, SAL-accountable launchers.
SWM    77    24-33        In service mid-1975.
USD8   79    54           Early 1975.
SSF    82                 569, SS-N-4,5; not including forward-deployed units.
```

BALLISTIC MISSILES  SLBMs  SS-N-5  251

```
------------ NUMBER IN 1976 ------------
IISS   76   54          1976-78, July, total deployed (33 on diesel subs, 21 on nuclear).
JCS    76   21          Currently, SAL-accountable launchers.
JWS    81   21          Launchers remaining in H-class subs, expected to decline, DoD 1976.

------------ NUMBER IN 1978 ------------
Clns   80   57          1978-1979.
Clns+  83   60

------------ NUMBER IN 1979 ------------
IISS   79   60          1979-80, July, total deployed (39 on diesel subs, 21 on nuclear).

------------ NUMBER IN 1980 ------------
SSF    82              445, SS-N-4,5; not including forward-deployed units.
Clns+  83   57          In 1980-1982.

------------ NUMBER IN 1981 ------------
IISS   81   57          1981-82, July, total deployed (39 on diesel subs, 18 on nuclear).

------------ NUMBER IN 1983 ------------
IISS+  83   48          July, total deployed (39 on diesel subs, ? 9 on nuclear).
JWS+   83   18          At beginning of 1983, according to SIPRI.

------------ NUMBER IN 1984 ------------
JCS    84   45          As of 1 Jan.
SMP    84   45

------------ NUMBER IN 1985 ------------
JCS    85   45          As of 1 Jan.
SMP    85   42

------------ CARRIERS ------------

------------ CARRIER ------------ TUBES --- RELOADS ------------
SecC   65   G and H                              Soviets will probably retrofit "all of their present force
                                                 of H-class submarines and at least some G-class
                                                 submarines" with [SS-N-5].

------------ CARRIER ------------ TUBES --- RELOADS ------------
SecC   65   G                                    One G-class sub recently converted to serve as test
                                                 vehicle for [SS-N-5].
SecC   67   [G]                                  Believed at least one of diesel subs retrofitted with
                                                 SS-N-5, and "possibly a few more" may be in future.
SecC   68   [G]                                  Most SSBs [Golfs] being retrofitted to SS-N-5.
SecC   69   [G]                                  We now estimate "that the diesel-powered ballistic missile
                                                 submarines are also being converted to the SS-N-5".
SecC   70   [G]                                  Some SSBs also apparently being converted to SS-N-5.
JCS    71   G
SecC   71   G II              3
JCS    72   G                                    "Presently deployed in the older G and H-class" subs.
JCS    75   G                 3                  G carries 3 "SS-N-4/5", FY76-77.
                                                 tubes: FY76-79.
FIar   76   Golf II           3                  Remains on 9.
MOW    76   G II              3
SecD   76   G
AI40   77   Golf II                              10 in operation in 1977.
JCS    77   Golf                                 Carries "3 SS-N-5/4" [sic].
            Golf II                              Currently deployed in Hotel II and Golf II.
SWM    77   Golf II           3                  "Golf III" [in text, assumed misprint, ed].
JCS    78   Golf II                              Deployed on Hotel II and Golf II.
Rock   78   Golf                                 N=11; converted from SS-N-4.
Clns   80   Golf II           3
JWS    81   G              2 or 3
USND   81   Golf II
AWST   82   Golf                                 Deployed, [SLBM entry only]
CRW    82   Golf II                              6 based in Baltic.
IISS   82   G-II                                 N=13.
JFS    82   Golf II
SSF    82   Golf II           3                  13 in operation in 1980.
GSN    83   Golf II           3                  Replaced SS-N-4 in 13 Golfs.
ABFS   84   Golf II                              N=6, assigned to Baltic Fleet.
SMP    84   Golf II           3                  N=13 still maintained and operated; 6 based in Baltic,
                                                 remaining 7 patrol Sea of Japan, '84, '85.
                                                 tubes: '84, '85.
BW     86   Golf II           3        no
```

## 252 SOVIET MISSILES IDDS

### CARRIER — TUBES — RELOADS

| Source | Yr | Carrier | Tubes | Reloads | Notes |
|---|---|---|---|---|---|
| SecC | 66 | [H] | | | Carried only by: 2-3 nuclear-powered ballistic missile subs. |
| SecC | 67 | [H] | | | All nuclear-powered subs apparently being retrofitted to SS-N-5 "on a relatively slow schedule". |
| SecC | 68 | [H] | | | All older SSBNs being retrofitted to SS-N-5. |
| SecC | 70 | [H] | | | Retrofit of all 8 older SSBNs to SS-N-5 could be complete by mid-70. |
| JCS | 71 | H | | | |
| SecC | 71 | H | 3 | | tubes: One H class has 6 tubes instead of 3. |
| JCS | 72 | H | | | "Presently deployed in the older G and H-class" subs. |
| FI* | 74 | Hotel II | | | N=6 or 7. |
| JCS | 75 | H | 3 | | FY76-77. tubes: FY76-79. |
| SecC | 75 | H-II | | | Expected all H-II subs will be phased out in next few years to stay within 950 modern SLBM launcher ceiling. |
| FIar | 76 | Hotel II | 3 | | Remains on 9. |
| MOW | 76 | H II | 3 | | |
| SecD | 76 | H | | | |
| A140 | 77 | Hotel | | | 8 in operation in 1977. |
| JCS | 77 | Hotel II | 3 | | Currently deployed in Hotel II and Golf II. |
| SWM | 77 | Hotel II | | | |
| JCS | 78 | Hotel II | | | Deployed on Hotel II and Golf II. |
| Rock | 78 | Hotel | | | N=9; converted from SS-N-4. |
| Clns | 80 | Hotel | 3 | | |
| JWS | 81 | H | 2 or 3 | | |
| SecD | 81 | Hotel | | | Deployed. |
| SMP | 81 | Hotel II | 3 | | '81, '83, '84. tubes: '81, '83, '84. |
| USND | 81 | Hotel II | | | |
| AWST | 82 | Hotel | | | Deployed, [SLBM entry only] |
| CFW | 82 | Hotel II | | | |
| IISS | 82 | H-II | | | N=6. |
| JFS | 82 | Hotel II | | | |
| SSF | 82 | Hotel II | 3 | | 7 in operation in 1980. |
| GSN | 83 | Hotel II | 3 | | Replaced SS-N-4 in 7 Hotels. |
| JWS+ | 83 | Hotel II | | | Number declining. |
| ABFS | 84 | Hotel II | | | N=3, Northern Fleet. |
| SMP | 84 | Hotel II | | | N=2 still maintained and operated. |
| BW | 86 | Hotel II | 3 | no | |

### DESIGNER
SSF 82  Yangel  Design Bureau.

### DESIGN AND ENGINEERING
JWS 81  Originally intended for E-class subs.

### USE AND CONFIGURATION
SecC 64  Submerged Launch, FY65-71.
SecC 66  Only recently have Soviet ballistic missile subs "regularly carried out ocean patrols, but these appear to be to staging areas rather than to strike stations".
JCS 71  Submerged Launch.
HRST 75  "Submarine-launched".
FIar 76  Tubes mounted in sail. 18 cold-gas generators on base of missile eject it from launch tube, dropped when first stage ignites.
MOW 76  "Eighteen small gas jets at base for ejection from launch-tube"; electrically-detonated cold-gas system. Suitable for underwater launch.
SWM 77  18 electrically-operated nozzles of a gas generator visible in parade.
Rock 78  Launched by 18 cold-gas generators on base of first stage; when maximum height reached, generators jettisoned and 1st stage ignites.
JWS 81  Ejection by 18 electrically-fired cold-gas nozzles.
CFW 82  Can be launched while submerged.
JFS 82  Dived launch.
GSN 83  Underwater launch.
vanD 83  Reliability: .70; Penetration 1.0.
ABFS 84  "Unreliable".
Meyr 84  Penetration: 1.0; reliability: 0.2-0.5, "includes system reliability multiplied by operational readiness of deployed system".

### STRATEGY
SecC 68  "The intelligence community now estimates that the diesel-powered ballistic missile submarines are primarily intended for use against Eurasian targets".
ABFS 84  Too short a range to hit US targets.
Meyr 84  [Ed: chart is labeled] Systems "assigned primary operational and strategic missions within the European theatre[s] of military operations".

---------- **COMBAT REPORTS/EFFECTIVENESS** ----------
Rock  78   Smaller size than Sark "decidedly advantageous" to its role.
IISS  82   Serviceability: 0.45; Survivability: 0.8; Reliability: 0.6.

---------- **GEOGRAPHICAL DEPLOYMENTS** ----------
Rock  78   6 SS-N-5 equipped Golf subs were seen to move into Baltic in Nov 1976, "thereby becoming the first SLBM vessels to enter those waters".
CFW   82   6 Golf II based in Baltic.
SSF   82   Forward deployed units (SS-N-4,5): none in 1960, 1975, 1980; 15 in 1965; 9 in 1970.

---------- **RELATION TO OTHER MISSILES** ----------
MOW   76   Polaris-type weapon.
JWS   81   Looks like Polaris A-2.

254 SOVIET MISSILES IDDS

# SS-N-6

| Sources are: | JCS 71 | HRST 75 | JCS 77 | CLns 80 | USND 81 | IISS+ 83 | Meyr 84 |
|---|---|---|---|---|---|---|---|
| SecC 65 | SecC 71 | IISS 75 | SWM 77 | IISS 80 | AWST 82 | JCS 83 | SecD 84 |
| SecC 66 | JCS 72 | JCS 75 | CBO1 78 | SecD 80 | CFW 82 | JFS+ 83 | SMP 84 |
| SecC 67 | IISS 73 | SecC 75 | IISS 78 | US12 80 | IISS 82 | JWS+ 83 | JCS 85 |
| ID01 68 | JCS 73 | SecD 75 | JCS 78 | FI 81 | JFS 82 | Lu01 83 | SMP 85 |
| Ley 68 | SecD 73 | FIer 76 | Rock 78 | IISS 81 | SSF 82 | SMP 83 | BW 86 |
| SecC 68 | ID04 74 | JCS 76 | SecD 78 | JCS 81 | CBO 83 | vanD 83 | |
| SecC 69 | IISS 74 | MOW 76 | Nitz 79 | JWS 81 | Clns+ 83 | ABFS 84 | |
| SecC 70 | JCS 74 | NA03 76 | USD8 79 | SecD 81 | FI+ 83 | CFW+ 84 | |
| SecD 70 | SecD 74 | SecD 76 | AWar 80 | SMP 81 | GSN 83 | JCS 84 | |

────────── **US DESIGNATION** ──────────

| SecC | 65 | SS-N-6 | |
| SecD | 70 | SS-N-6 | FY74-78,80-82,85. |
| JCS | 71 | SS-N-6 | FY73-79,82,85-86. |
| FIer | 76 | SS-N-6 | |
| MOW | 76 | SS-N-6 | |
| SWM | 77 | SS-N-6 | |
| Rock | 78 | SS-N-6 | |
| Nitz | 79 | SS-N-6 | |
| AWar | 80 | SS-N-6 | |
| Clns | 80 | SS-N-6 | |
| FI | 81 | SS-N-6 | |
| JWS | 81 | SSN-6 | |
| SMP | 81 | SS-N-6 | '81, '83, '84, '85. |
| USND | 81 | SS-N-6 | |
| AWST | 82 | SS-N-6 | [SLBM entry only] |
| CFW | 82 | SS-N-6 | |
| IISS | 82 | SS-N-6 | |
| JFS | 82 | SS-N-6 | |
| SSF | 82 | SS-N-6 | |
| CBO | 83 | SS-N-6 | |
| GSN | 83 | SS-N-6 | |
| vanD | 83 | SS-N-6 | |
| Meyr | 84 | SS-N-6 | |

────────── **NATO CODENAME** ──────────

| SecD | 70 | Sawfly | [May not be SS-N-6, see Prototype Test information, ed]. |
| HRST | 75 | Sawfly | |
| FIer | 76 | Sawfly | |
| MOW | 76 | Sawfly | |
| SWM | 77 | Sawfly | |
| Rock | 78 | Sawfly | |
| FI | 81 | Sawfly | |
| JWS | 81 | Sawfly | |
| USND | 81 | Sawfly | |
| AWST | 82 | Sawfly | [Submarine IRBM entry only] |
| IISS | 82 | Sawfly | |
| SSF | 82 | Sawfly | |
| GSN | 83 | [not Sawfly] | "The SLBM given the NATO code name Sawfly was apparently a competitive prototype to the SS-N-6 and not the same missile". |

────────── **OTHER DESIGNATION** ──────────

| Ley | 68 | "New type" |

────────── **DESCRIPTION** ──────────

| JCS | 72 | | Listed as 2nd generation strategic missile. |
| JCS | 74 | | Listed as significant initiative in strategic offensive systems: "SS-N-6 improvements". |
| SWM | 77 | | Third-generation missile, "much improved SLBM". |
| USND | 81 | SLBM | |
| AWST | 82 | IRBM range | [Submarine IRBM entry only] |
| | | | "Third generation submarine IRBM", [submarine IRBM entry only] |
| SSF | 82 | | Third generation. |
| JCS | 83 | SLBM | |

BALLISTIC MISSILES  SLBMs  SS-N-6  255

---------- EDITOR'S NOTES ----------
| | | |
|---|---|---|
| SecC | 65 | Listed FY66-72,74-77, not by name FY66-72. |
| Ley | 68 | Evidence that source is referring to SS-N-6 is tenuous. |
| SecD | 70 | Listed FY71,74-82,85. Referred to not by name, FY71,79. |
| JCS | 71 | Listed FY72-79,82,85-86. Referred to not by name, FY72: Y class submerged launch missiles. |
| Rock | 78 | In the table: 3 separate entries for unmarked, Mod 2, and Mod 3. |
| SMP | 81 | Listed '81, '83, '84, '85. |
| AWST | 82 | The source has 2 separate table entries, one with the SLBMs (and referred to below as the "SLBM entry"); and one with the ICBMS (and referred to below as the "submarine IRBM entry". Data is common to both unless marked. |
| Meyr | 84 | Note to chart: "Values pertain to initial service period". |

---------- RANGE ----------

| | | mi | nm | km | km conv | |
|---|---|---|---|---|---|---|
| SecC | 67 | | 1,000 | | 1,852 | [Range of hypothetical missile to go in new sub, ed]. |
| | | | 1,000-2,000 | | 1,852-3,704 | [Range of hypothetical missile to go in new sub, ed]. |
| Ley | 68 | 1,200 | | | 1,930 | |
| SecC | 68 | | c. 1,500 | | 2,778 | New missile has been tested out to about "1,500 miles". |
| SecC | 70 | | 1,300 | | 2,407 | Y-class missile has "nominal range of about 1,300 vice 1,500 n.mi.". |
| SecD | 70 | | > 1,200 | | 2,222 | [Y class missile]. |
| JCS | 71 | | 1,300 | | 2,407 | [Mod not specified], FY72,75. |
| IDO4 | 74 | | 1,300 | | 2,407 | |
| JCS | 74 | | | | | New MRV variant expected to have slightly longer range. |
| HRST | 75 | 1,500 | | | 2,413 | |
| MOW | 76 | | | | | Larger than US Navy Poseidon, but shorter range. |
| NAO3 | 76 | 1,750 | | | 2,815 | |
| SecD | 76 | | 1300-1600 | | 2,407-2,963 | |
| JCS | 77 | | 1,300 | | 2,407 | Range depends on Mod. |
| | | | 1,600 | | 2,963 | Range depends on Mod. |
| SWM | 77 | 1,500 | | 2,400 | 2,413 | [Unmarked]. 3 versions: "Mod 1 with limited range"; Mod 2 with greater range, single RV; Mod 3 with still longer range and MRVs. |
| JCS | 78 | | | 2,400-3,000 | 2,400-3,000 | |
| USO8 | 79 | | 1,500 | | 2,778 | |
| Clns | 80 | | 1,600 | | 2,963 | Both [unmarked] and Mod 3. |
| SecD | 80 | | | 3,000 | 3,000 | |
| US12 | 80 | | 100-200 | | 185.20-370.40 | This minimum range should be possible with liquid fuel. |
| USND | 81 | | 1,600 | | 2,963 | |
| AWST | 82 | | 1,300-1,600 | | 2,407-2,963 | [SLBM entry only] |
| CFW | 82 | | 1,300 | | 2,407 | Initially. |
| | | | 1,600 | | 2,963 | Current version [final report] |
| Meyr | 84 | | | 2,500 | 2,500 | |

---------- CEP ----------

| | | ft | mi | nm | m | km | m conv | |
|---|---|---|---|---|---|---|---|---|
| SecC | 70 | | | | | | | [Though range estimate revised,] estimated yield and CEP remains about the same, [deleted]. |
| CBO1 | 78 | 6000 | | | | | 1,828 | Estimated, with SS-N-8, 1985. |
| Rock | 78 | | | | | 2.8 | 2,800 | With single 1 MT warhead; but yield more than compensates and makes this more effective loading [than 3 RV]. |
| Nitz | 79 | | | .5 | | | 926.00 | |
| Clns | 80 | | | 0.7 | | | 1,296 | Both [unmarked] and Mod 3. |
| JWS | 81 | | | | | | | No version has hard target capability, according to US DoD. |
| CBO | 83 | | | 0.5 | | | 926.00 | In 1983; lack of data on future trends. |
| LuO1 | 83 | | | | | 1.8 | 1,800 | |
| vanD | 83 | | | 1.0 | | | 1,852 | |
| CFW+ | 84 | | | | 1,850 | | 1,850 | |
| Meyr | 84 | | | | | 0.9-2.0 | 900.00-2,000 | |

## SOVIET MISSILES

### WARHEAD TYPE

|     |    | nuclear | conven | chem |                          |
|-----|----|---------|--------|------|--------------------------|
| MOW | 76 | yes     |        |      |                          |
| SWM | 77 | yes     |        |      |                          |
| JCS | 81 | yes     |        |      | Nuclear delivery system. |
| SMP | 81 | yes     |        |      |                          |
| CFW | 82 | yes     |        |      |                          |
| JFS | 82 | yes     |        |      |                          |
| GSN | 83 | yes     |        |      |                          |
| JWS+| 83 | yes     |        |      |                          |
| Meyr| 84 | yes     |        |      |                          |

### WARHEAD CHARACTERISTICS

| JCS  | 75 | "2-3 (MRV)" | FY76-78. |
| SecC | 75 |             | Possible Soviets plan to MIRV SS-N-6, but SS-N-8 appears more likely candidate. |
| Nitz | 79 | 1 RV        |          |
| Clns | 80 | single RV   | [Unmarked] |
| CFW  | 82 | MRV         | Current version [final report] |
| CBO  | 83 | single RV   |          |
| SMP  | 83 | [3 RVs]     | Dismantling one Yankee results in loss of 48 RVs [implies 3 RVs per missile, ed]. |
| vanD | 83 | single RV   |          |
| Meyr | 84 | "1/3MRV"    |          |

### NUCLEAR YIELD

|      |    | kT        | MT   | kT conv        |                                                                             |
|------|----|-----------|------|----------------|-----------------------------------------------------------------------------|
| SecC | 70 |           |      |                | [Though range estimate revised,] estimated yield and CEP remains about the same, [deleted]. |
| MOW  | 76 |           |      |                | MT range.                                                                   |
| SWM  | 77 |           | 1    | 1,000          |                                                                             |
| CBO1 | 78 |           | 1.0  | 1,000          | Estimated, with SS-N-8, 1985.                                               |
| Nitz | 79 |           | 1.0  | 1,000          |                                                                             |
| USO8 | 79 |           | 1+   | 1,000          |                                                                             |
| Clns | 80 | 500-1000  |      | 500.00-1,000   | "500 KT - 1 MT" [both unmarked and Mod 3]                                   |
| CFW  | 82 |           | c. 1 | 1,000          |                                                                             |
| JFS  | 82 |           |      |                | "Megaton".                                                                  |
| CBO  | 83 | 750       |      | 750.00         |                                                                             |
| GSN  | 83 |           | c. 1 | 1,000          |                                                                             |
| JFS+ | 83 |           | 1    | 1,000          |                                                                             |
| vanD | 83 |           | .70  | 700.00         |                                                                             |
| Meyr | 84 | 500-1,000 |      | 500.00-1,000   | 500 kT-1 MT.                                                                |

### THROWWEIGHT

|      |    | lb    | t | kg | kg conv |
|------|----|-------|---|----|---------|
| Nitz | 79 | 1,600 |   |    | 725.76  |
| CBO  | 83 | 1,600 |   |    | 725.76  |

### WEIGHT

|     |    | lb        | t    | kg         | kg conv |                |
|-----|----|-----------|------|------------|---------|----------------|
| Ley | 68 |           | 15.0 |            | 15,000  | Tons.          |
| SWM | 77 | 42,000    |      | 19,050     | 19,051  |                |
| FI  | 81 |           |      | c. 19,000  | 19,000  |                |
| JFS | 82 | 44,000    |      |            | 19,958  | Launch weight. |
| GSN | 83 | c. 42,000 |      | c. 18,900  | 19,051  |                |
| CFW+| 84 |           |      | 18,900     | 18,900  |                |

### LENGTH

|      |    | in | ft    | cm | m      | m conv |                                  |
|------|----|----|-------|----|--------|--------|----------------------------------|
| ID01 | 68 |    | c. 43 |    |        | 13.11  |                                  |
| Ley  | 68 |    | 35    |    |        | 10.67  |                                  |
| HRST | 75 |    | 34    |    |        | 10.36  |                                  |
| FIar | 76 |    |       |    | 13     | 13.00  |                                  |
| MOW  | 76 |    | 42'0" |    | 12.8   | 12.80  |                                  |
| SWM  | 77 |    | 42    |    | 12.8   | 12.80  |                                  |
| FI   | 81 |    |       |    | 13     | 13.00  |                                  |
| JWS  | 81 |    |       |    | c. 9.65| 9.65   | [From graph], '81, '83, '84, '85.|
| SMP  | 81 |    |       |    | 9      | 9.00   | [Submarine IRBM entry only]      |
| AWST | 82 |    | 39    |    |        | 11.89  | [SLBM entry only]                |
|      |    |    | 30    |    |        | 9.14   |                                  |
| JFS  | 82 |    | 31.6  |    |        | 9.63   |                                  |
| FI+  | 83 |    |       |    | 9.65   | 9.65   |                                  |
| GSN  | 83 |    | c. 33 |    | c. 10  | 10.06  |                                  |
| JFS+ | 83 |    | 30    |    |        | 9.14   |                                  |
| CFW+ | 84 |    |       |    | 10     | 10.00  |                                  |

## DIAMETER

|  |  | in | ft | cm | m | cm conv |  |
|---|---|---|---|---|---|---|---|
| Ley | 68 |  | 5.4 |  |  | 164.59 |  |
| HRST | 75 | 72 |  |  |  | 182.88 |  |
| FIar | 76 |  |  |  | 1.8 | 180.00 |  |
| MOW | 76 |  | 5'9" |  | 1.75 | 175.26 |  |
| SWM | 77 |  | 5.9 |  | 1.8 | 179.83 |  |
| FI | 81 |  |  | c. 180 |  | 180.00 |  |
| JWS | 81 |  |  |  | c. 1.65 | 165.00 |  |
| AWST | 82 |  | 6 |  |  | 182.88 | [Submarine IRBM entry only] |
|  |  |  | 5 |  |  | 152.40 | [SLBM entry only] |
| FI+ | 83 |  |  | c. 165 |  | 165.00 |  |
| GSN | 83 | c. 71 |  | c. 180 |  | 180.34 | [1,800 mm in source] |
| CFW+ | 84 |  |  |  | 1.8 | 180.00 |  |

## NUMBER OF STAGES

| MOW | 76 | 2 | Probably. |
|---|---|---|---|
| SWM | 77 | 2 |  |
| Rock | 78 | 2 | First 2-stage SLBM introduced by Soviets. |
| FI | 81 | 2 |  |
| AWST | 82 | 2 |  |
| JFS | 82 | 2 |  |
| GSN | 83 | 1 |  |

## TYPE OF PROPULSION

| SecC | 67 | solid or liquid |  |
|---|---|---|---|
| SecC | 68 | storable liquid | FY69-70. |
| JCS | 72 | liquid | [Implied, ed]. |
| HRST | 75 | solid |  |
| FIar | 76 | storable liquid |  |
| MOW | 76 | liquid | 4 first-stage nozzles. |
| SWM | 77 | solid | Recent info indicates "increasingly probable" it is storable liquid. |
| Clns | 80 | liquid | Both [unmarked] and Mod 3. |
| SecD | 80 | liquid |  |
| US12 | 80 | liquid |  |
| FI | 81 | liquid |  |
| JWS | 81 | liquid |  |
| SMP | 81 | liquid |  |
| AWST | 82 | solid | [Submarine IRBM entry only] |
|  |  | liquid storable | [SLBM entry only] |
| CFW | 82 | liquid |  |
| JFS | 82 | liquid |  |
| SSF | 82 | liquid |  |
| GSN | 83 | liquid |  |

## MIDCOURSE (OR MAIN) GUIDANCE

| MOW | 76 | inertial |
|---|---|---|
| SWM | 77 | inertial |
| FI | 81 | inertial |
| JFS | 82 | inertial |
| SSF | 82 | inertial |
| GSN | 83 | inertial |
| CFW+ | 84 | inertial |

## OBSERVED IN PARADE

| MOW | 76 |  | None of the 3 sub-launched ballistic missiles in Moscow parades since 1962 is believed to be truly representative of an operational weapon. So, the nose cone of Sawfly is "almost certainly an inert test package". |
|---|---|---|---|
| SWM | 77 | Nov 1967 |  |

## DEVELOPMENT BEGAN

| SSF | 82 | 1960 | Year design began. |
|---|---|---|---|
| SMP | 85 | 1957 | Technological development began. [From graph.] |

## PROTOTYPE TESTS

| SecD | 70 |  | Development of past year has been "the testing of a new, probably naval-oriented, ballistic missile. This could possibly be the Sawfly missile that was noted in a Soviet parade in 1967, and which at that time was described as a new naval missile". |
|---|---|---|---|
| SecD | 74 |  | No evidence of depressed-trajectory testing. |
| SMP | 85 | 1963-64 | Engineering and testing began. [From graph.] |

## PRODUCTION UNDERWAY

| SMP | 84 | "The Soviets continue to produce four SLBMs—the SS-N-6, SS-N-8, SS-N-17 and SS-N-18—at the rate of 200 a year. One new system, the SS-N-20/Typhoon, is in series production". |
|---|---|---|

258  SOVIET MISSILES  IDDS

```
---------------- IOC ----------------
SWM    77    1967           Operational.
SMP    81    1968           "SS-N-6/Yankee I weapon system became operational".
vanD   83    1969

---------------- DEPLOYMENT BEGAN ----------------
JCS    74    1968           Yankees first deployed.
JCS    75    1968           SS-N-6 first entered naval inventory, FY76-79.
JCS    77    late 1967      "YANKEE class SSBN weapon system" introduced.
JCS    78    late 1967      Yankee SSBN introduced.
Clns   80    1968           [Unmarked Mod], First deployed.
JWS    81    1968           Entered service.
JCS    83    c. 1968        Year of introduction [from graph].
CFW+   84    1968
Meyr   84    1968           Service.
SecD   84    1968-69        Introduction of strategic systems [graph].
SMP    85    1968-69        [From graph.]

---------------- DEPLOYED ----------------
MOW    76                   In service.
SecD   81                   Part of Soviet SSBN force.

---------------- DEPLOYMENT HISTORY ----------------
JCS    75    Most widely deployed SLBM [in USSR], FY76-79.
SecD   80    Several Yankees have had missile tubes removed; may be converted to SSNs.
FI     81    USSR is withdrawing SS-N-6 armed Yankee subs as Delta subs increase.
SMP    85    Will be deployed at least through 1990. [From graph.]

---------------- NUMBERS IN SERVICE ----------------

---------------- NUMBER IN 1955 ----------------
SSF    82    none           In 1955, 1960, 1965; not including forward-deployed units.

---------------- NUMBER IN 1960 ----------------
SSF    82    none           In 1960, 1965; forward deployed units.

---------------- NUMBER IN 1968 ----------------
SecC   68                   1 sub expected operational by end of this fiscal year.
SecC   69                   46 total SLBM launchers in nuclear sub force in Sept 1968, FY70-71.

---------------- NUMBER IN 1969 ----------------
SecC   68                   3-4 subs could be operational by mid-69.
SecC   70                   104-120 SLBM launchers [not just SS-N-6] on 1 Sept, all of growth is Y-class.
SecD   70                   Increase of 65 SLBM launchers on deployable submarines from 1 Sept 1968 to 1 Sept 1969
                            is due to Y-class subs.
SecC   71    112            Oct 1969, operational.

---------------- NUMBER IN 1970 ----------------
SecC   69                   About 158-238 SLBM launchers [not just SS-N-6] by mid-70, based on 4-8 subs produced per
                            year.
SecC   70                   184-248 SLBM launchers [not just SS-N-6] expected by mid-70, all of growth is Y-class.
SecC   71    224            1 Oct, "Polaris Type" launch tubes.
Clns   80    208
SSF    82    32             Forward deployed units.
             272            Not including forward-deployed units.

---------------- NUMBER IN 1971 ----------------
SecC   70                   296-376 SLBM launchers [not just SS-N-6] expected by mid-71, all of growth is Y-class.
SecC   71    272            1 Jan, operational.
                            336-352, mid-71, "Polaris Type" launch tubes.
Clns   80    320

---------------- NUMBER IN 1972 ----------------
SecC   68                   15-18 subs could be operational by mid-72.
SecC   69                   About 286-494 SLBM launchers [not just SS-N-6] by mid-72, based on 4-8 subs produced per
                            year.
SecC   71                   448-480, mid-72, operational.
JCS    72                   Still being deployed at 9-10 subs per year (16 missiles each).
Clns   80    416

---------------- NUMBER IN 1973 ----------------
IISS   73    496            July, total deployed.
Rock   78    544            Total operational Sawfly inventory, constant since 1973.
Clns   80    496
Clns+  83    480
```

## BALLISTIC MISSILES SLBMs SS-N-6

------------ **NUMBER IN 1974** ------------
| | | | |
|---|---|---|---|
| SecC | 70 | | Early 1969 projection was for 35-50 Y class, 560-800 SLBM launchers, in 1975-1977; now projected to reach this 'end strength' in 1974-1975. |
| IISS | 74 | 528 | July, total deployed. |
| Clns | 80 | 528 | In 1974-1976. |
| Clns+ | 83 | 512 | |

------------ **NUMBER IN 1975** ------------
| | | | |
|---|---|---|---|
| IISS | 75 | 544 | 1975-77, July, total deployed. |
| JCS | 75 | 544 | Current level of 544 launchers, some MRVs, expected to remain until "a new SS-N-6 sized missile" appears. |
| USOB | 79 | 544 | Early 1975. |
| SSF | 82 | 64 | In 1975 and 1980; forward deployed units. |
| | | 480 | Not including forward-deployed units. |

------------ **NUMBER IN 1976** ------------
| | | | |
|---|---|---|---|
| JCS | 76 | 540 | Launch tubes on 34 units. |
| Clns+ | 83 | 548 | |

------------ **NUMBER IN 1977** ------------
| | | | |
|---|---|---|---|
| Nitz | 79 | 272 | Launchers. |
| Clns | 80 | 532 | |

------------ **NUMBER IN 1978** ------------
| | | | |
|---|---|---|---|
| IISS | 78 | 528 | 1978-79, July, total deployed. |
| SecD | 78 | 540 | Tubes on 34 Yankees, FY79-80. |
| Clns | 80 | 500 | |

------------ **NUMBER IN 1979** ------------
| | | | |
|---|---|---|---|
| Nitz | 79 | 254 | 18 June, launchers. |
| Clns | 80 | 468 | |
| Clns+ | 83 | 484 | |

------------ **NUMBER IN 1980** ------------
| | | | |
|---|---|---|---|
| IISS | 80 | 469 | July, total deployed. |
| SSF | 82 | 388 | Not including forward-deployed units. |
| Clns+ | 83 | 468 | |

------------ **NUMBER IN 1981** ------------
| | | | |
|---|---|---|---|
| IISS | 81 | [453] | July, total deployed. |
| Clns+ | 83 | 448 | |

------------ **NUMBER IN 1982** ------------
| | | | |
|---|---|---|---|
| IISS | 82 | 400 | July, total deployed. |
| Clns+ | 83 | 384 | |

------------ **NUMBER IN 1983** ------------
| | | | |
|---|---|---|---|
| CBO | 83 | 448 | 1983, 1990. |
| IISS+ | 83 | 384 | July, total deployed. |

------------ **NUMBER IN 1984** ------------
| | | | |
|---|---|---|---|
| JCS | 84 | 368 | As of 1 Jan. |
| SMP | 84 | 368 | |

------------ **NUMBER IN 1985** ------------
| | | | |
|---|---|---|---|
| JCS | 85 | 336 | As of 1 Jan. |
| SMP | 85 | 336 | |

------------ **NUMBER IN 1996** ------------
| | | | |
|---|---|---|---|
| CBO | 83 | | 160 in 1996. |

------------ **LAUNCHER/SILO** ------------
| | | |
|---|---|---|
| SecC | 68 | Land-based version could be deployed in MR/IRBM mode. |

------------ **CARRIERS** ------------

------------ **CARRIER** ------------ **TUBES** ---- **RELOADS** ------------
| | | | | |
|---|---|---|---|---|
| SecC | 65 | | ? 4-8 | Soviets have under construction a sub estimated as first of new SSBN class, probably will enter service during 1965. tubes: May carry more missiles than 3 of G or H; possibly 4-8. |
| SecC | 66 | | 6-12 | By 1968 Soviets could have SSBN [for this missile]; by mid-1970 could have as many as 7 boats. |
| SecC | 67 | | >_ 8 | Could have as many as 10 boats operational by 1971. |

| | | CARRIER | TUBES | RELOADS | |
|---|---|---|---|---|---|
| Clns | 80 | Golf IV | 4 | | |
| SecD | 81 | Golf | | | Also deployed on a Golf sub. |
| USND | 81 | Golf IV | | | Test platform. |
| CFW | 82 | Golf IV | | | Experimental sub [final report] |
| SSF | 82 | Golf IV | 5 | | 1 in operation in 1980, for testing. |
| GSN | 83 | Golf IV | | | N=1. |
| BW | 86 | Golf IV | 4 | | N=1. |

| | | CARRIER | TUBES | RELOADS | |
|---|---|---|---|---|---|
| SecC | 68 | | 16 | | First sub expected operational by end of this fiscal year, 3-4 could be operational by mid-1969, 15-18 by mid-1972. tubes: FY69-72. |
| SecC | 69 | | | | We believe Soviets could produce 4-8 of these subs per year. |
| SecC | 70 | Y | | | Estimate 5-6 Y subs operational 1 Sept 1969; 11 as of 1 Feb 1970; 4 in Jan 1969. |
| JCS | 71 | Y | 16 | | FY72-74,76-79. tubes: FY72-74,76-79. |
| SecC | 71 | Y | | | 17-18 Y subs operational; another 15-16 in various stages of assembly and fitting out. 7 operational Oct 1969, 17 operational 1 Jan 1971, 16 under construction, 28-30 estimated operational, mid-72; [launch tubes given for above are exactly 16 times numbers of subs]. |
| JCS | 73 | Yankee | | | C. 29 launched by mid-1972. |
| SecD | 73 | Yankee | | | [Either Yankee or Y class for FY74-78,81-82, ed] |
| JCS | 74 | Yankee | | | |
| SecD | 74 | Yankee | 16 | | |
| | | Y | | | tubes: FY75-76. |
| JCS | 75 | Yankee | | | N=33 completed. |
| SecD | 75 | Y | | | Reason to believe production has stopped. |
| FIar | 76 | Yankee | 16 | | Production ended with 34. |
| JCS | 76 | Yankee | | | 34 in service. |
| | | Yankee | | | Production terminated in 1974. |
| MOW | 76 | Y | 16 | | N=34. |
| JCS | 77 | Yankee | | | N=33 as of mid-1975. |
| SWM | 77 | Yankee | 16 | | Between 1966 and 1974, 34 produced, FY78-79. |
| | | | | | 15 Ys were under construction according to 1969 Pentagon report. |
| JCS | 78 | Yankee | | | One now carries SS-NX-17, 33 left with SS-N-6. |
| Rock | 78 | Yankee | 16 | | N=34 maximum in 1973. First subs with missiles in hull aft of sail. |
| SecD | 78 | Yankee | | | Construction stopped at 34 units and 540 tubes, FY79-80. |
| AWar | 80 | Yankee 1 | 16 | | N=28. |
| Clns | 80 | Yankee I | 16 | | |
| SecD | 80 | Yankee | | | Construction stopped "five years ago" at 34. |
| FI | 81 | Yankee | 16 | | N=29. |
| JWS | 81 | Y | | | N=28 in 1981; there were 34 when production ended in 1975. |
| SMP | 81 | Yankee I | 16 | | '81, '83, '84, '85. |
| | | | | | tubes: '81, '83, '84, '85. |
| | | Yankee | | | 34 in service by end of 1974. |
| USND | 81 | Yankee I | | | |
| AWST | 82 | Yankee | | | [SLBM entry only] |
| CFW | 82 | Yankee I | | | |
| JFS | 82 | Yankee | | | |
| SSF | 82 | Yankee I | 16 | | 28 in operation in 1980. |
| FI+ | 83 | Yankee | | | Eight withdrawn, converted to SSNs; 26 remaining have entirely Mod 3s. |
| GSN | 83 | Yankee I | | | 1 converted to SS-N-17, others have had missile section removed. |
| JWS+ | 83 | Y | | | Of 34 built, 24 carry SS-N-6, 1 has SS-N-17, 9 had tubes removed, may be SSNs [text]. |
| | | Yankee I and II | | | [Sic], 22 carry SS-N-6 Mods 1 and 3, [figure caption]. |
| BW | 86 | Yankee I | 16 | no | N=34. |

| | | DESIGNER | | | |
|---|---|---|---|---|---|
| SSF | 82 | Chelomei | Design Bureau. | | |

## USE AND CONFIGURATION

| | | |
|---|---|---|
| SecC | 66 | Salvo-fired. |
| HRST | 75 | "Submarine-launched". |
| JCS | 75 | 3 mods are "relatively interchangeable", FY76-77. |
| SWM | 77 | Missiles are in pressure hull, as with US Polaris. |
| US12 | 80 | Submerged-launch. |
| AWST | 82 | "Variable range mode", [submarine IRBM entry only] |
| CRW | 82 | Can be launched while submerged. |
| CBO | 83 | System availability: 0.40. |
| GSN | 83 | Underwater launch. |
| vanD | 83 | Reliability: .75; Penetration 1.0. |
| Meyr | 84 | Penetration: 1.0; reliability: 0.4-0.8, "includes system reliability multiplied by operational readiness of deployed system". |

## STRATEGY

| | | |
|---|---|---|
| SecC | 68 | Land-based version could be deployed in MR/IRBM mode. |
| Rock | 78 | First Soviet SLBM system of truly IRBM range. |
| Meyr | 84 | "Initially developed for intercontinental strikes, but likely to be assigned to missions in the European theatre of military operations". [Ed: chart is labeled] Systems "assigned primary operational and strategic missions within the European theatre(s) of military operations". |

## GEOGRAPHICAL DEPLOYMENTS

| | | |
|---|---|---|
| SSF | 82 | Forward deployed units: none in 1960, 1965; 32 in 1970; 64 in 1975 and 1980. |
| ABFS | 84 | Yankee submarines began patrols off the eastern and western seaboards of the US in 1968 and 1970 respectively. |

## RELATION TO OTHER MISSILES

| | | |
|---|---|---|
| FIer | 76 | Matches Polaris A1 and A2 in performance, though operational same year Polaris production line closed. |
| JCS | 77 | Follow-on to SS-N-6 "projected after 1981"; expected to be comparable in size [appears not to be SS-NX-17,18, ed]. |
| JCS | 78 | During early 1980s, a follow-on could be deployed. |
| Rock | 78 | Polaris-equivalent strike force. |

# SS-N-6 Mod 1

```
Sources are:  FIar  76    USD6  78    FI    81    SMP   81    JFS   82    JWS+  83    BW    86
SecC  74      MOW   76    AWar  80    IISS  81    US17  81    SSF   82    CFW   84
JCS   75      Rock  78    US12  80    JWS   81    IISS  82    GSN   83    SMP   85
```

## US DESIGNATION

|  | |  |
|---|---|---|
| SecC | 74 | SS-N-6 Mod 1 |
| JCS | 75 | SS-N-6 Mod 1 |
| FIar | 76 | SS-N-6 Mod 1 |
| MOW | 76 | SS-N-6 Mod 1 |
| USD6 | 78 | SS-N-6 Mod 1 |
| US12 | 80 | SS-N-6 Mod 1 |
| FI | 81 | SS-N-6 Mod 1 |
| JWS | 81 | SS-N-6 Mod 1 |
| SMP | 81 | SS-N-6 Mod 1 |
| US17 | 81 | SS-N-6 Mod 1 |
| IISS | 82 | SS-N-6 Mod 1 |
| SSF | 82 | SS-N-6 Mod 1 |
| GSN | 83 | SS-N-6 Mod 1 |
| CFW | 84 | SS-N-6 Mod 1 |

## DESCRIPTION

| | | |
|---|---|---|
| Rock | 78 | SLBM |

## EDITOR'S NOTES

| | | |
|---|---|---|
| Rock | 78 | This mod unmarked in table. |
| AWar | 80 | Source provides no designations for any mods. |
| JFS | 82 | Unmarked case, no designation. |

## RANGE

| Source | | mi | nm | km | km conv | |
|---|---|---|---|---|---|---|
| SecC | 74 | | 1,300 | | 2,407 | FY75-76. |
| JCS | 75 | | 1,300 | | 2,407 | FY76-77. |
| FIar | 76 | | | 2,400 | 2,400 | |
| MOW | 76 | 1,500 | | 2,451 | 2,413 | |
| Rock | 78 | | | 2,400 | 2,400 | |
| USD6 | 78 | | 1,300 | | 2,407 | |
| AWar | 80 | | 2,400 | | 4,444 | [Sic; Mod inferred, ed] |
| FI | 81 | | | 2,400 | 2,400 | |
| JWS | 81 | | | 2400 | 2,400 | |
| SMP | 81 | | | 2,400 | 2,400 | '81, '83, '84, '85. |
| IISS | 82 | | | 2,400 | 2,400 | Max. |
| JFS | 82 | | 1,300 | | 2,407 | Max. |
| SSF | 82 | | 1,300 | | 2,407 | |
| GSN | 83 | | 1,300 | 2,400 | 2,407 | |
| CFW | 84 | | 1,300 | | 2,407 | |
| BW | 86 | | | 2,400 | 2,400 | |

## CEP

| Source | | ft | mi | nm | m | km | m conv | |
|---|---|---|---|---|---|---|---|---|
| JCS | 75 | | | | | | | Mods 1,2,3 all soft-target weapons, FY76-78. |
| AWar | 80 | | | 1 | | | 1,852 | Mod 1; Mod 2 same as Mod 1; Mod 3 same as Mod 2 [sic]. |
| IISS | 82 | | | | 900 | | 900.00 | Mods 1 and 2. |
| SSF | 82 | | | 1.0 | | | 1,852 | All 3 mods. |
| BW | 86 | | | | 1,900 | | 1,900 | |

## WARHEAD CHARACTERISTICS

| | | | |
|---|---|---|---|
| JCS | 75 | single RV | FY76-79. |
| FIar | 76 | single RV | Mods 1 and 2. |
| MOW | 76 | single RV | Mods 1 and 2. |
| Rock | 78 | single RV | Mods 1 and 2. |
| AWar | 80 | single RV | |
| JWS | 81 | single RV | |
| SMP | 81 | single RV | '81, '83, '84, '85. |
| IISS | 82 | single RV | Mods 1 and 2. |
| SSF | 82 | single RV | Mods 1 and 2. |
| GSN | 83 | single RV | Mods 1 and 2. |
| BW | 86 | single RV | |

BALLISTIC MISSILES  SLBMs  SS-N-6 Mod 1  263

---------- NUCLEAR YIELD ----------
|        |    | kT  | MT    | kT conv     |               |
|--------|----|-----|-------|-------------|---------------|
| FIar   | 76 |     | 1-2   | 1,000-2,000 | Mods 1 and 2. |
| Rock   | 78 |     | 1.0   | 1,000       | Mods 1 and 2. |
| AWar   | 80 |     | 0.70  | 700.00      |               |
| FI     | 81 |     | 1-2   | 1,000-2,000 | Mods 1 and 2. |
| IISS   | 82 |     | 1     | 1,000       | Mods 1 and 2. |
| SSF    | 82 |     | 0.7   | 700.00      |               |
| JWS+   | 83 |     | ? 1   | 1,000       | [Analysis table] |
| CFW    | 84 |     | c. 1  | 1,000       |               |
| BW     | 86 | 700 |       | 700.00      |               |

---------- THROWWEIGHT ----------
|       |    | lb    | t | kg | kg conv |               |
|-------|----|-------|---|----|---------|---------------|
| IISS  | 82 | 1,500 |   |    | 680.40  | Mods 1 and 3. |

---------- WEIGHT ----------
|       |    | lb | t  | kg | kg conv |         |
|-------|----|----|----|----|---------|---------|
| Rock  | 78 |    | 19 |    | 19,000  | Tonnes. |

---------- LENGTH ----------
|       |    | in | ft | cm | m  | m conv |
|-------|----|----|----|----|----|--------|
| Rock  | 78 |    |    |    | 13 | 13.00  |

---------- DIAMETER ----------
|       |    | in | ft | cm | m   | cm conv |
|-------|----|----|----|----|-----|---------|
| Rock  | 78 |    |    |    | 1.8 | 180.00  |

---------- NUMBER OF STAGES ----------
| US12 | 80 | 1 | All 3 mods. |

---------- TYPE OF PROPULSION ----------
| JCS  | 75 | liquid          | In all Mods, FY76-77.     |
| Rock | 78 | storable liquid | Stages 1 and 2, all 3 mods. |
| IISS | 82 | liquid          | All 3 mods.               |
| BW   | 86 | storable liquid |                           |

---------- MIDCOURSE (OR MAIN) GUIDANCE ----------
| US12 | 80 | inertial | All 3 mods. |

---------- PROTOTYPE TESTS ----------
| SSF | 82 | 1967 | First flight test. |

---------- IOC ----------
| JCS  | 75 | 1968    | FY76-79.                          |
| MOW  | 76 | 1968    | Became operational.               |
| US12 | 80 | 1968    |                                   |
|      |    | c. 1969 | Yankee/SS-N-6 system became operational. |
| JFS  | 82 | 1967    | Operational.                      |
| GSN  | 83 | 1968    |                                   |

---------- DEPLOYMENT BEGAN ----------
| FIar | 76 | 1968    | Became operational with Russian Navy. |
| Rock | 78 | 1968    | Service intro.                  |
| US17 | 81 | 1967    | Year of introduction.           |
| IISS | 82 | 1968    | First year deployed.            |
| SSF  | 82 | 1968    | Year operation began.           |
| SMP  | 85 | 1968-69 | [Presumed Mod 1, from graph.]   |
| BW   | 86 | 1968    |                                 |

---------- DEPLOYED ----------
| JWS | 81 | Current Yankees carry "either Mod 1 or Mod 3". |
| BW  | 86 | In service. |

---------- RETIRED ----------
| FI | 81 | All Yankees have SS-N-6 Mod 3. |

---------- DEPLOYMENT HISTORY ----------
| JCS  | 75 | Expected Mod 1 and Mod 2 will be phased out, FY76-77. |
| FIar | 76 | Mod 3 is expected to replace all Mod 1 and 2 rounds. |
| MOW  | 76 | Mod 2 started replacing Mod 1 in 1974. |
| Rock | 78 | Mod 2 and Mod 3 variants are being installed on Mod 1 equipped subs, but it is thought Soviet Navy is using single 1 MT warhead of Mod 2 rather than "low yield triple warhead" MRV of Mod 3; [unclear whether Mod 3 being deployed or not, ed]. |
| US12 | 80 | "Obsolete (1976?)". |
| JWS  | 81 | Mods 2 and 3 likely are replacing Mod 1. |

---------- NUMBERS IN SERVICE ----------

---------- **NUMBER IN 1981** ----------
IISS 81    165      July, total deployed, Mods 1 and 2.

# SS-N-6 Mod 2

```
Sources are:   FIar  76      SWM   77     US12  80     SMP   81     SSF   82     BW    86
JCS   75       JCS   76      Rock  78     FI    81     US17  81     GSN   83
SecC  75       MOW   76      US06  78     IISS  81     IISS  82     CFW   84
SecD  75       JCS   77      AWar  80     JWS   81     JFS   82     SMP   85
```

---------- US DESIGNATION ----------

|  | |  |
|---|---|---|
| JCS | 75 | SS-N-6 Mod 2 |
| SecC | 75 | SS-N-6 Mod 2 |
| SecD | 75 | SS-N-6 Mod 2 |
| FIar | 76 | SS-N-6 Mod 2 |
| MOW | 76 | SS-N-6 Mod 2 |
| SWM | 77 | SS-N-6 Mod 2 |
| Rock | 78 | SS-N-6 Mod 2 |
| US06 | 78 | SS-N-6 Mod 2 |
| US12 | 80 | SS-N-6 Mod 2 |
| FI | 81 | SS-N-6 Mod 2 |
| JWS | 81 | SS-N-6 Mod 2 |
| SMP | 81 | SS-N-6 Mod 2 |
| US17 | 81 | SS-N-6 Mod 2 |
| IISS | 82 | SS-N-6 Mod 2 |
| JFS | 82 | SS-N-6 Mk II |
| SSF | 82 | SS-N-6 Mod 2 |
| GSN | 83 | SS-N-6 Mod 2 |
| CFW | 84 | SS-N-6 Mod 2 |

---------- DESCRIPTION ----------

Rock 78 — SLBM

---------- EDITOR'S NOTES ----------

SecC 75 — Information recorded for FY76. Note that the "Mod 2" of the FY75 report is actually a mistaken report of Mod 3.

AWar 80 — Source provides no designations for any mods.

---------- RANGE ----------

| Source | | mi | nm | km | km conv | |
|---|---|---|---|---|---|---|
| JCS | 75 | | 1,600 | | 2,963 | FY76-77. |
| SecC | 75 | | 1,600 | | 2,963 | |
| SecD | 75 | | | | | Mods 2 and 3 have slightly longer range than Mod 1. |
| FIar | 76 | | | 3,000 | 3,000 | Mods 2 and 3. |
| MOW | 76 | 1,840 | | 2,960 | 2,960 | |
| SWM | 77 | 1,860 | | 3,000 | 2,992 | Mods 2 and 3. |
| Rock | 78 | | | 3,000 | 3,000 | Mods 2 and 3. |
| US06 | 78 | | 1,600 | | 2,963 | |
| AWar | 80 | | | 3,000 | 3,000 | [Mods 2 and 3, inferred, ed]. |
| FI | 81 | | | 3,000 | 3,000 | Mods 2 and 3. |
| JWS | 81 | | | 3000 | 3,000 | |
| SMP | 81 | | | 3,000 | 3,000 | '81, '83, '84, '85. |
| IISS | 82 | | | 3,000 | 3,000 | Max, Mods 2 and 3. |
| JFS | 82 | | 1,600 | | 2,963 | Max, Mks II and III. |
| SSF | 82 | | 1,600 | | 2,963 | |
| GSN | 83 | | 1,600 | 2,950 | 2,963 | Mods 2 and 3. |
| CFW | 84 | | 1,600 | | 2,963 | Mods 2 and 3. |
| BW | 86 | | | 3,000 | 3,000 | |

---------- CEP ----------

| Source | | ft | mi | nm | m | km | m conv | |
|---|---|---|---|---|---|---|---|---|
| JCS | 75 | | | | | | | Mods 1,2,3 all soft-target weapons, FY76-78. |
| AWar | 80 | | | [1] | | | 1,852 | 1 nm for Mod 1; Mod 2 same as Mod 1; Mod 3 same as Mod 2 [sic]. |
| IISS | 82 | | | | 900 | | 900.00 | Mods 1 and 2. |
| SSF | 82 | | | 1.0 | | | 1,852 | All 3 mods. |
| BW | 86 | | | | 1,900 | | 1,900 | |

```
---------- WARHEAD CHARACTERISTICS ----------
JCS    75    single RV         FY76-79.
SecC   75    single RV
SecD   75    single RV
FIar   76    single RV         Mods 1 and 2.
MOW    76    single RV         Mods 1 and 2.
SWM    77    single RV
Rock   78    single RV         Mods 1 and 2.
AWar   80    single RV         [Mods 1 and 2].
JWS    81    single RV
SMP    81    single RV         '81, '83, '84, '85.
IISS   82    single RV         Mods 1 and 2.
SSF    82    single RV         Mods 1 and 2.
GSN    83    single RV         Mods 1 and 2.
BW     86    single RV

---------- NUCLEAR YIELD ----------
             kT         MT         kT conv
FIar   76               1-2        1,000-2,000     Mods 1 and 2.
Rock   78               1.0        1,000           Mods 1 and 2.
AWar   80               0.65       650.00
FI     81               1-2        1,000-2,000     Mods 1 and 2.
IISS   82               1          1,000           Mods 1 and 2.
SSF    82               0.65       650.00
CFW    84               c. 1       1,000
BW     86    650                   650.00

---------- WEIGHT ----------
             lb        t          kg         kg conv
Rock   78              19                    19,000          Tonnes.

---------- LENGTH ----------
             in        ft         cm         m         m conv
Rock   78                                    13        13.00

---------- DIAMETER ----------
             in        ft         cm         m         cm conv
Rock   78                                    1.8       180.00

---------- NUMBER OF STAGES ----------
US12   80    1         All 3 mods.

---------- TYPE OF PROPULSION ----------
JCS    75                        Mod 2 advance over Mod 1 was primarily propulsion, FY76-77.
             liquid              In all Mods, FY76-77.
Rock   78    storable liquid     Stages 1 and 2, all 3 mods.
JWS    81                        Mod 2 improved propulsion over Mod 1.
IISS   82    liquid              All 3 mods.
BW     86    storable liquid

---------- MIDCOURSE (OR MAIN) GUIDANCE ----------
US12   80    inertial            All 3 mods.

---------- PROTOTYPE TESTS ----------
JCS    75    Oct 1972            Testing began, FY76-77.
US12   80    Oct 1972            Began testing.
JWS    81    Oct 1972            Test flights began.
SSF    82    1972                First flight test.

---------- IOC ----------
JCS    75    1974                [Sic, FY77 report gives both 1973 and 1974 as IOC, ed], FY76-77.
JCS    76    1973                FY77-79.
US12   80    1976                Mods 2 and 3.
JFS    82    1974-75             Mks II and III.
GSN    83    1972-1973           Mods 2 and 3.

---------- DEPLOYMENT BEGAN ----------
FIar   76    1974                Entered service, Mods 2 and 3.
MOW    76                        Mod 2 started replacing Mod 1 in 1974.
Rock   78    1974                Service intro, Mods 2 and 3.
US06   78    by 1975             Came into service.
US12   80    1973                Operational on Yankee.
JWS    81    1973                Began operational service 1973.
US17   81    1972                Year of introduction.
IISS   82    1973                First year deployed.
SSF    82    1973                Year operation began, Mods 2 and 3.
SMP    85    1977                [From graph.]
BW     86    1973
```

## BALLISTIC MISSILES  SLBMs  SS-N-6 Mod 2

---------- **DEPLOYED** ----------
| | | |
|---|---|---|
| US12 | 80 | Mods 2 and 3 are operational. |
| BW | 86 | In service. |

---------- **RETIRED** ----------
| | | |
|---|---|---|
| FI | 81 | All Yankees have SS-N-6 Mod 3. |
| JWS | 81 | Current Yankees carry "either Mod 1 or Mod 3". |

---------- **DEPLOYMENT HISTORY** ----------
| | | |
|---|---|---|
| JCS | 75 | Expected Mod 1 and Mod 2 will be phased out, FY76-77. |
| FIar | 76 | Mod 3 is expected to replace all Mod 1 and 2 rounds. |
| Rock | 78 | Mod 2 and Mod 3 variants are being installed on Mod 1 equipped subs, but it is thought Soviet Navy is using single 1 MT warhead of Mod 2 rather than "Low yield triple warhead" MRV of Mod 3; [unclear whether Mod 3 being deployed or not, ed]. |
| JWS | 81 | Mods 2 and 3 likely are replacing Mod 1. |
| SMP | 85 | Will be deployed at least through 1990. [From graph.] |

---------- **NUMBERS IN SERVICE** ----------

---------- **NUMBER IN 1981** ----------
| | | | |
|---|---|---|---|
| IISS | 81 | 165 | July, total deployed, Mods 1 and 2. |

---------- **STRATEGY** ----------
| | | |
|---|---|---|
| JCS | 75 | Believed capable of hitting any US target from 100 fathom curve. |
| SecD | 75 | SS-N-6 Mod 2 can bypass US FSS-7 radars. |
| FIar | 76 | Mod 2 variant gave ability to hit US targets from 100-fathom depth-curve. |
| JCS | 76 | Mod 2 believed capable of hitting targets from 100 fathom curve. |

# SS-N-6 Mod 3

```
Sources are:  SecD  75    SecD  77    Nitz  79    IISS  81    JFS   82    JWS+  83
SecC  74      FIar  76    SWM   77    AWar  80    JWS   81    SSF   82    SMP   83
SecD  74      JCS   76    JCS   78    Clns  80    SMP   81    FI+   83    CFW   84
JCS   75      MOW   76    Rock  78    US12  80    US17  81    GSN   83    SMP   85
SecC  75      JCS   77    US06  78    FI    81    IISS  82    JFS+  83    BW    86
```

---------- **US DESIGNATION** ----------

|  | |  |
|---|---|---|
| SecC | 74 | SS-N-6 Mod 3 |
| SecD | 74 | SS-N-6 Mod 3 |
| JCS  | 75 | SS-N-6 Mod 3 |
| FIar | 76 | SS-N-6 Mod 3 |
| MOW  | 76 | SS-N-6 Mod 3 |
| SWM  | 77 | SS-N-6 Mod 3 |
| Rock | 78 | SS-N-6 Mod 3 |
| US06 | 78 | SS-N-6 Mod 3 |
| Nitz | 79 | SS-N-6 |
| Clns | 80 | SS-N-6 Mod 3 |
| US12 | 80 | SS-N-6 Mod 3 |
| FI   | 81 | SS-N-6 Mod 3 |
| JWS  | 81 | SS-N-6 Mod 3 |
| SMP  | 81 | SS-N-6 Mod 3 |
| US17 | 81 | SS-N-6 Mod 3 |
| IISS | 82 | SS-N-6 Mod 3 |
| JFS  | 82 | SS-N-6 Mk III |
| SSF  | 82 | SS-N-6 Mod 3 |
| GSN  | 83 | SS-N-6 Mod 3 |
| CFW  | 84 | SS-N-6 Mod 3 |

---------- **DESCRIPTION** ----------

| | | |
|---|---|---|
| Rock | 78 | SLBM |

---------- **EDITOR'S NOTES** ----------

| | | |
|---|---|---|
| SecC | 74 | Information recorded for FY75-76. Note that the "Mod 2" of the FY75 report is actually Mod 3. |
| AWar | 80 | Source provides no designations for any mods. |

---------- **RANGE** ----------

| Source | | mi | nm | km | km conv | Notes |
|---|---|---|---|---|---|---|
| SecC | 74 | | 1,600 | | 2,963 | FY75-76. |
| JCS  | 75 | | 1,600 | | 2,963 | FY76-77. |
| SecD | 75 | | | | | Mods 2 and 3 have slightly longer range than Mod 1. |
| FIar | 76 | | | 3,000 | 3,000 | Mods 2 and 3. |
| SWM  | 77 | 1,860 | | 3,000 | 2,992 | Mods 2 and 3. |
| Rock | 78 | | | 3,000 | 3,000 | Mods 2 and 3. |
|      |    | | 1,600 | | 2,963 | With MRV. |
| US06 | 78 | | 1,600 | | 2,963 | Mods 2 and 3. |
| AWar | 80 | | | 3,000 | 3,000 | [Mods 2 and 3, Mods inferred, ed]. |
| Clns | 80 | | 1,600 | | 2,963 | Both [unmarked] and Mod 3. |
| FI   | 81 | | | 3,000 | 3,000 | Mods 2 and 3. |
| JWS  | 81 | | | 3000 | 3,000 | |
| SMP  | 81 | | | 3,000 | 3,000 | '81, '83, '84, '85. |
|      |    | | | c. 3,000 | 3,000 | Max operational range. |
| IISS | 82 | | | 3,000 | 3,000 | Max, Mods 2 and 3. |
| JFS  | 82 | | 1,600 | | 2,963 | Max, Mks II and III. |
| SSF  | 82 | | 1,600 | | 2,963 | Mods 2 and 3. |
| GSN  | 83 | | 1,600 | 2,950 | 2,963 | Mods 2 and 3. |
| JFS+ | 83 | | 1,800 | | 3,333 | Max. |
| CFW  | 84 | | 1,600 | | 2,963 | |
| BW   | 86 | | | 3,000 | 3,000 | |

---------- **CEP** ----------

| Source | | ft | mi | nm | m | km | m conv | Notes |
|---|---|---|---|---|---|---|---|---|
| JCS  | 75 | | | | | | | Mods 1,2,3 all soft-target weapons, FY76-78. |
| Rock | 78 | | | | | < 2 | 2,000 | "Nearly" 2 km, with 3 MRVs. |
| Nitz | 79 | | | .75 | | | 1,389 | |
| AWar | 80 | | | [1] | | | 1,852 | 1 nm for Mod 1; Mod 2 same as Mod 1; Mod 3 same as Mod 2 [sic]. |
| Clns | 80 | | | 0.7 | | | 1,296 | Both [unmarked] and Mod 3. |
| IISS | 82 | | | | 1,400 | | 1,400 | |
| SSF  | 82 | | | 1.0 | | | 1,852 | All 3 mods. |
| BW   | 86 | | | | 1,900 | | 1,900 | |

## WARHEAD CHARACTERISTICS

| Source | Year | Value | Notes |
|---|---|---|---|
| SecC | 74 | 2 or 3 MRVs | |
| SecC | 75 | | "Two or possibly three MRVs". |
| SecD | 75 | MRVs | |
| FIar | 76 | 3 MRVs | |
| JCS | 76 | not MIRV | FY77-79. |
| MOW | 76 | 2 or 3 RVs | |
| JCS | 77 | 2-3 RVs | "At least two and possibly three RVs". |
| SecD | 77 | 2 or 3 MRVs | |
| SWM | 77 | 3 MRVs | |
| JCS | 78 | "2 (MRV)" | |
| Rock | 78 | 3 MRVs | |
| Nitz | 79 | 3 RVs | |
| AWar | 80 | 2 RVs | |
| Clns | 80 | 3 MRVs | |
| FI | 81 | 2 (or 3) MRVs | [Final report] |
| JWS | 81 | 3 MRVs | |
| SMP | 81 | 2 MIRVs | [Sic]; "first Soviet SLBM to carry multiple RVs". |
| IISS | 82 | 2 MIRVs | [Sic] |
| JFS | 82 | 3 MRV | |
| SSF | 82 | 2-3 RVs | |
| GSN | 83 | 2 MRVs | |
| JWS+ | 83 | 2 MRVs | |
| SMP | 83 | 2 MRVs | '83, '84, '85. |
| CFW | 84 | 2 RVs | |
| BW | 86 | 2 MRVs | |

## NUCLEAR YIELD

| Source | Year | kT | MT | kT conv | Notes |
|---|---|---|---|---|---|
| Rock | 78 | | ? 0.9 | 900.00 | |
| Nitz | 79 | | .5 | 500.00 | |
| AWar | 80 | | 0.35 | 350.00 | |
| Clns | 80 | 500-1000 | | 500.00-1,000 | "500 KT - 1 MT" [both unmarked and Mod 3; "each" for Mod 3] |
| IISS | 82 | 200 | | 200.00 | Per warhead. |
| SSF | 82 | | 0.35 | 350.00 | Per warhead. |
| BW | 86 | 350 | | 350.00 | |

## THROWEIGHT

| Source | Year | lb | t | kg | kg conv | Notes |
|---|---|---|---|---|---|---|
| Nitz | 79 | 1,600 | | | 725.76 | |
| IISS | 82 | 1,500 | | | 680.40 | Mods 1 and 3. |

## WEIGHT

| Source | Year | lb | t | kg | kg conv | Notes |
|---|---|---|---|---|---|---|
| Rock | 78 | | 19 | | 19,000 | Tonnes. |

## LENGTH

| Source | Year | in | ft | cm | m | m conv |
|---|---|---|---|---|---|---|
| Rock | 78 | | | | 13 | 13.00 |

## DIAMETER

| Source | Year | in | ft | cm | m | cm conv |
|---|---|---|---|---|---|---|
| Rock | 78 | | | | 1.8 | 180.00 |

## NUMBER OF STAGES

| Source | Year | Value | Notes |
|---|---|---|---|
| US12 | 80 | 1 | All 3 mods. |

## TYPE OF PROPULSION

| Source | Year | Value | Notes |
|---|---|---|---|
| JCS | 75 | liquid | In all Mods, FY76-77. |
| Rock | 78 | storable liquid | Stages 1 and 2, all 3 mods. |
| Clns | 80 | liquid | Both [unmarked] and Mod 3. |
| IISS | 82 | liquid | All 3 mods. |
| BW | 86 | storable liquid | |

## MIDCOURSE (OR MAIN) GUIDANCE

| Source | Year | Value | Notes |
|---|---|---|---|
| US12 | 80 | inertial | All 3 mods. |

## PROTOTYPE TESTS

| Source | Year | Value | Notes |
|---|---|---|---|
| SecD | 74 | [1973] | In past year, flight test of a new "multiple RV" version of SS-N-6 with MRVs. |
| SSF | 82 | 1973 | First flight test. |

## IOC

| | | | |
|---|---|---|---|
| JCS | 75 | 1974 | [Sic; FY77 report gives both 1974 and 1975 as IOC, ed], FY76-77. |
| JCS | 76 | 1975 | May have reached IOC. |
| MOW | 76 | 1975 | Thought to have reached operational capability. |
| JCS | 77 | 1973 | FY78-79. |
| US12 | 80 | 1976 | Mods 2 and 3. |
| JFS | 82 | 1974-75 | Mks II and III. |
| GSN | 83 | 1972-1973 | Mods 2 and 3. |

## DEPLOYMENT BEGAN

| | | | |
|---|---|---|---|
| SecD | 74 | | Improvements for SS-N-6 [means MRV, ed] "is nearing operational status". New MRV variant "may be deployed" in Yankees. |
| FIar | 76 | 1974 | Entered service, Mods 2 and 3. |
| Rock | 78 | 1974 | Service intro, Mods 2 and 3. |
| Clns | 80 | 1974 | First deployed. |
| US12 | 80 | 1975 | Operational on Yankee. |
| JWS | 81 | | Mod 2 began operational service 1973, closely followed by Mod 3. |
| US17 | 81 | 1975 | Year of introduction. |
| IISS | 82 | 1974 | First year deployed. |
| SSF | 82 | 1973 | Year operation began, Mods 2 and 3. |
| SMP | 85 | 1977 | [From graph.] |
| BW | 86 | 1974 | |

## DEPLOYED

| | | |
|---|---|---|
| US12 | 80 | Mods 2 and 3 are operational. |
| JWS | 81 | Current Yankees carry "either Mod 1 or Mod 3". |
| BW | 86 | In service. |

## DEPLOYMENT HISTORY

| | | |
|---|---|---|
| FIar | 76 | Mod 3 is expected to replace all Mod 1 and 2 rounds. |
| Rock | 78 | Mod 2 and Mod 3 variants are being installed on Mod 1 equipped subs, but it is thought Soviet Navy is using single 1 MT warhead of Mod 2 rather than "low yield triple warhead" MRV of Mod 3; [unclear whether Mod 3 being deployed or not, ed]. |
| JWS | 81 | Mods 2 and 3 likely are replacing Mod 1. |
| SMP | 85 | Will be deployed at least through 1990. [From graph.] |

## NUMBERS IN SERVICE

### NUMBER IN 1977

| | | | |
|---|---|---|---|
| Nitz | 79 | 272 | Launchers. |

### NUMBER IN 1979

| | | | |
|---|---|---|---|
| Nitz | 79 | 272 | 18 June, launchers. |

### NUMBER IN 1981

| | | | |
|---|---|---|---|
| IISS | 81 | 288 | July, total deployed. |

## CARRIERS

### CARRIER — TUBES — RELOADS

| | | | |
|---|---|---|---|
| FIar | 76 | Yankee | All Yankees have SS-N-6 Mod 3. |
| FI | 81 | Yankee | All Yankees have SS-N-6 Mod 3. |
| FI+ | 83 | Yankee | N=26, all with Mod 3 missiles. |

# SS-N-8

| Sources are: | ID04 | 74 | FIar | 76 | CBO1 | 78 | USO7 | 79 | SecD | 81 | Clns+ | 83 | SMP | 84 |
|---|---|---|---|---|---|---|---|---|---|---|---|---|---|---|
| SecC | 70 | IISS | 74 | IISS | 76 | IISS | 78 | USO8 | 79 | SMP | 81 | FI+ | 83 | JCS | 85 |
| JCS | 71 | JCS | 74 | JCS | 76 | JCS | 78 | AWar | 80 | USND | 81 | GSN | 83 | SMP | 85 |
| SecC | 71 | SecC | 74 | MOW | 76 | Rock | 78 | Clns | 80 | AWST | 82 | JCS | 83 | BW | 86 |
| JCS | 72 | SecD | 74 | NA03 | 76 | SecD | 78 | IISS | 80 | CFW | 82 | JFS+ | 83 | | |
| SecD | 72 | IA06 | 75 | SecD | 76 | USO6 | 78 | SecD | 80 | IISS | 82 | JWS+ | 83 | | |
| IISS | 73 | IISS | 75 | IISS | 77 | AWO3 | 79 | US12 | 80 | JCS | 82 | vanD | 83 | | |
| JCS | 73 | JCS | 75 | JCS | 77 | IISS | 79 | FI | 81 | JFS | 82 | CFW+ | 84 | | |
| SecC | 73 | SecC | 75 | SecD | 77 | Nitz | 79 | IISS | 81 | SSF | 82 | JCS | 84 | | |
| SecD | 73 | SecD | 75 | SWM | 77 | SecD | 79 | JWS | 81 | CBO | 83 | SecD | 84 | | |

------------ US DESIGNATION ------------

| JCS | 72 | SS-NX-8 | |
| SecD | 72 | SS-NX-8 | FY73-74. |
| JCS | 73 | SS-N-8 | FY74-79,82-86. |
| SecC | 73 | SS-N-8 | FY74-77. |
| SecD | 73 | SS-N-8 | FY74-82,85. |
| FIar | 76 | SS-N-8 | |
| MOW | 76 | SS-N-8 | |
| SWM | 77 | SS-N-8 | |
| Rock | 78 | SS-N-8 | |
| Nitz | 79 | SS-N-8 | |
| AWar | 80 | SS-N-8 | |
| Clns | 80 | SS-N-8 | |
| FI | 81 | SS-N-8 | |
| JWS | 81 | SSN-8 | |
| SMP | 81 | SS-N-8 | '81, '83, '84, '85. |
| USND | 81 | SS-N-8 | |
| AWST | 82 | SS-N-8 | |
| CFW | 82 | SS-N-8 | |
| IISS | 82 | SS-N-8 | |
| JFS | 82 | SS-N-8 | |
| SSF | 82 | SS-N-8 | |
| CBO | 83 | SS-N-8 | |
| GSN | 83 | SS-N-8 | |
| vanD | 83 | SS-N-8 | |

------------ NATO CODENAME ------------

| ID04 | 74 | Sawfly | [Sic] |
| CFW | 82 | Sawfly | |

------------ DESCRIPTION ------------

| JCS | 72 | | Listed as 3rd generation strategic missile. |
| JCS | 74 | | Listed as significant initiative, strategic offensive systems. |
| MOW | 76 | SLBM | |
| SWM | 77 | | Advanced missile system. |
| Rock | 78 | SLBM | First of 3rd generation SLBMs. |
| JWS | 81 | SLBM | |
| USND | 81 | SLBM | |
| SSF | 82 | | Fourth generation. |
| JCS | 83 | SLBM | |

------------ EDITOR'S NOTES ------------

| SecC | 70 | Listed FY71-72,74-77, not by name FY71-72. |
| JCS | 71 | Listed FY72-79,82-86. Mentioned not by name in FY72: Soviets "have also been actively testing a new SLBM, with a much longer range than the Y class missile". |
| SecD | 72 | Listed FY73-82,85. |
| MOW | 76 | [Info on D class subs] |
| Rock | 78 | 3 separate entries for Mod 1, Mod 2, and Mod 3. Info identical for 3 mods is shown unmarked. |
| SMP | 81 | Listed '81, '83, '84, '85. |

| | | RANGE | | | | |
|---|---|---|---|---|---|---|
| | | mi | nm | km | km conv | |
| SecC | 70 | | c. 3,000 | | 5,556 | FY71-72. |
| JCS | 73 | | c. 4,000 | | 7,408 | |
| SecD | 73 | | | | | More than 3 times SS-N-6 range. |
| ID04 | 74 | | 4,200 | | 7,778 | |
| JCS | 74 | | 4,200 | | 7,778 | FY75-78. |
| SecD | 75 | | 4,200 | | 7,778 | FY76-77. |
| Flar | 76 | | | 7,800 | 7,800 | |
| MOW | 76 | 4,800 | | 7,725 | 7,723 | |
| NA03 | 76 | 4,800 | | | 7,723 | |
| SecD | 77 | | >_ 4,200 | | 7,778 | Comparable to full-payload Trident I range. |
| SWM | 77 | 4,847 | | 7,800 | 7,798 | Range [table]. |
| | | 4,900 | | 7,890 | 7,884 | 2 test launches Oct 1974 [text]. |
| JCS | 78 | | | 7,800 | 7,800 | |
| Rock | 78 | | | 7,800 | 7,800 | [Table] |
| | | | | > 7,800 | 7,800 | [Text:] Had demonstrated capability of sending a useful payload more than 7,800 km. Exceeds current US SLBMs. |
| | | | | 9,200 | 9,200 | Achieved in tests; uncertain whether fully loaded. |
| SecD | 78 | | | >_ 7,800 | 7,800 | Range exceeds Trident I, FY79-80. |
| USD6 | 78 | | 5,600 | | 10,371 | By Nov 1976, flight tested to this range. |
| | | | 4,000 | | 7,408 | [Range] |
| AW03 | 79 | | > 4,000 | | 7,408 | About equal to Trident I. |
| SecD | 79 | | | > 8,000 | 8,000 | Now. |
| USD8 | 79 | | 4,300 | | 7,963 | Expected later. |
| | | | 4,800 | | 8,889 | |
| AWar | 80 | | | 9,000 | 9,000 | |
| Clns | 80 | | 4,800 | | 8,889 | |
| SecD | 80 | | | c. 8,000 | 8,000 | |
| FI | 81 | | | 8,000 | 8,000 | Range better than all US SLBMs, including Trident. |
| | | | | <_ 9,200 | 9,200 | Range for trial round. |
| JWS | 81 | | | < 8,000 | 8,000 | "Almost" 8,000 km, longest of any US or USSR SLBM. |
| SMP | 81 | | | c. 9,000 | 9,000 | Max operational range. |
| USND | 81 | | 4,000+ | | 7,408 | |
| AWST | 82 | | 5,400 | | 10,000 | |
| CFW | 82 | | 4,200 | | 7,778 | [Says "miles", ed] |
| JFS | 82 | | 4,200 | | 7,778 | Max. |
| JWS+ | 83 | | | c. 7,800 | 7,800 | [Analysis table] |

BALLISTIC MISSILES  SLBMs  SS-N-8

## CEP

| Source | Year | ft | mi | nm | m | km | m_conv | Notes |
|---|---|---|---|---|---|---|---|---|
| JCS | 73 | | | | | | | Has not demonstrated hard-target capability [said indirectly, ed]. |
| SecD | 73 | | | | | | | Somewhat better than SS-N-6. |
| FIar | 76 | | | | 400 | | 400.00 | Similar to Minuteman III; makes SS-N-8 a counterforce weapon. |
| JCS | 77 | | | | | | | "Accuracy potential" together with long range make it greatest SLBM threat to US, FY78-79. |
| SecD | 77 | | | | | | | Current SLBMs do not have significant hard-target capability but accurate enough against bomber bases. |
| SWM | 77 | | 0.25 | | | 0.4 | 402.25 | |
| CBO1 | 78 | 6000 | | | | | 1,828 | Estimated, with SS-N-6, 1985. |
| Rock | 78 | | | | 500 | | 500.00 | |
| | | | | | | 1.4 | 1,400 | "Target error of probably no more than 1.4 km". |
| Nitz | 79 | | | .5 | | | 926.00 | |
| AWar | 80 | | | 0.84 | | | 1,555 | |
| Clns | 80 | | | 0.8 | | | 1,481 | |
| FI | 81 | | | | ? 400 | | 400.00 | |
| JWS | 81 | | | | | 0.4 | 400.00 | |
| CBO | 83 | | | 0.5 | | | 926.00 | In 1983; lack of data on future trends. |
| vanD | 83 | | | .84 | | | 1,555 | |
| CPW+ | 84 | | | | 1,500 | | 1,500 | [Unit inferred, ed] |

## WARHEAD TYPE

| Source | Year | nuclear | conven | chem |
|---|---|---|---|---|
| SWM | 77 | yes | | |
| AWST | 82 | yes | | |
| CPW | 82 | yes | | |
| JFS | 82 | yes | | |
| GSN | 83 | yes | | |
| JWS+ | 83 | yes | | |

## WARHEAD CHARACTERISTICS

| Source | Year | | Notes |
|---|---|---|---|
| JCS | 72 | single RV | FY73-74,76-79. |
| SecC | 74 | | SS-N-8 MRV or MIRV will probably appear in the next year or so. |
| SecD | 74 | | No evidence of MRV or MIRV (as yet). |
| IAD6 | 75 | single RV | |
| SecC | 75 | | Possibly USSR to MIRV both SS-N-6 and SS-N-8, but SS-N-8 seems more likely candidate; "it is much the larger of the two missiles in volume, and some range could be sacrificed for more payload". |
| FIar | 76 | single RV / 3 MRVs | |
| SecD | 76 | single RV | FY77-81. |
| Nitz | 79 | 1.0 RVs | [Sic] |
| AWar | 80 | single RV | |
| Clns | 80 | single RV | |
| FI | 81 | | Has been tested with 3 MIRVs. |
| AWST | 82 | single RV | |
| CBO | 83 | single RV | |
| GSN | 83 | single RV | |
| vanD | 83 | single RV | |
| CPW+ | 84 | single RV | |

## SOVIET MISSILES

### NUCLEAR YIELD

| | | kT | MT | kT conv | | |
|---|---|---|---|---|---|---|
| IA06 | 75 | | | | | "Single small warhead". |
| JCS | 77 | | | | | Delivers "relatively large warhead" to 4,200 nm max range. |
| SWM | 77 | | 1-2 | 1,000-2,000 | | |
| CB01 | 78 | | 1.0 | 1,000 | | Estimated, with SS-N-6, 1985. |
| JCS | 78 | | | | | Relatively large warhead. |
| Nitz | 79 | | 1.0 | 1,000 | | |
| USDB | 79 | | 1+ | 1,000 | | |
| AWar | 80 | | 0.80 | 800.00 | | |
| Clns | 80 | 500-1000 | | 500.00-1,000 | | "500 KT-1 MT". |
| AWST | 82 | | 0.8 | 800.00 | | |
| CRW | 82 | | c. 1 | 1,000 | | |
| JFS | 82 | | | | | "Megaton". |
| CBO | 83 | 750 | | 750.00 | | Each. |
| GSN | 83 | | c. 0.8-1.5 | 800.00-1,500 | | |
| JFS+ | 83 | | 0.8 | 800.00 | | |
| vanD | 83 | | .80 | 800.00 | | |
| CRW+ | 84 | | c. 1.5 | 1,500 | | |

### THROWWEIGHT

| | | lb | t | kg | kg conv |
|---|---|---|---|---|---|
| Nitz | 79 | 1,800 | | | 816.48 |
| CBO | 83 | 1,800 | | | 816.48 |

### WEIGHT

| | | lb | t | kg | kg conv | |
|---|---|---|---|---|---|---|
| SWM | 77 | 45,000 | | 20,410 | 20,412 | Literally "4-5,000" [misprint, ed] |
| Rock | 78 | | 20+ | | 20,000 | Tonnes. |
| FI | 81 | | | c. 20,000 | 20,000 | |
| JFS | 82 | 45,000 | | | 20,412 | Launch weight. |
| GSN | 83 | c. 45,000 | | c. 20,450 | 20,412 | |
| CRW+ | 84 | | | 20,400 | 20,400 | |

### LENGTH

| | | in | ft | cm | m | m conv | |
|---|---|---|---|---|---|---|---|
| Flar | 76 | | | | ? 17 | 17.00 | |
| SWM | 77 | | 42.6-45.9 | | 13-14 | 12.98-13.99 | |
| Rock | 78 | | | | 17 | 17.00 | |
| FI | 81 | | | | c. 17 | 17.00 | |
| JWS | 81 | | | | c. 12.95 | 12.95 | |
| SMP | 81 | | | | 13 | 13.00 | [From graph], '81, '83, '84, '85. |
| AWST | 82 | | 40 | | | 12.19 | |
| JFS | 82 | | 42.5 | | | 12.95 | |
| FI+ | 83 | | | | 12.95 | 12.95 | |
| GSN | 83 | | c. 43 | | c. 13 | 13.11 | |
| CRW+ | 84 | | | | 13 | 13.00 | |

### DIAMETER

| | | in | ft | cm | m | cm conv |
|---|---|---|---|---|---|---|
| Flar | 76 | | | | ? 2.0 | 200.00 |
| Rock | 78 | | | | 1.8 | 180.00 |
| FI | 81 | | | c. 200 | | 200.00 |
| JWS | 81 | | | | c. 1.65 | 165.00 |
| AWST | 82 | | 6 | | | 182.88 |
| FI+ | 83 | | | c. 165 | | 165.00 |

### NUMBER OF STAGES

| | | | |
|---|---|---|---|
| SWM | 77 | 2 | |
| JCS | 78 | 2 | |
| Rock | 78 | 2 | |
| FI | 81 | 2 | |
| SMP | 81 | 2 | First Soviet 2-stage SLBM. |
| AWST | 82 | 2 | |
| JFS | 82 | 2 | |
| GSN | 83 | 2 | First 2-stage submarine missile. |

BALLISTIC MISSILES  SLBMs  SS-N-8  275

```
————————— TYPE OF PROPULSION —————————
JCS    75    liquid              FY76-77.
FIar   76    storable liquid
MOW    76    liquid
SWM    77    storable liquid
Rock   78    storable liquid     Stages 1 and 2.
SecD   79    liquid              FY80-81.
Clns   80    liquid
FI     81    liquid
JWS    81    liquid
SMP    81    liquid
AWST   82    liquid storable
CFW    82    liquid
JFS    82    liquid
SSF    82    liquid
GSN    83    liquid

————————— MIDCOURSE (OR MAIN) GUIDANCE —————————
FIar   76    steller-inertial
SWM    77    "steller-inertial"
Rock   78    steller-inertial
FI     81    steller-inertial
JWS    81    steller inertial
JFS    82    steller-inertial
SSF    82    steller inertial
GSN    83    inertial
CFW+   84    inertial

————————— DEVELOPMENT BEGAN —————————
SSF    82    1962        Year design began.
SMP    85    1962        Technological development began. [From graph.]

————————— DEVELOPMENT UNDERWAY —————————
SecD   72                Development continues.

————————— PROTOTYPE TESTS —————————
JCS    72                Soviet "very extensive and vigorous flight-test program" going on "for some time"
                         includes SS-NX-8. Soviets "have been actively testing" new, longer-range SS-NX-8.
                         If test program has been successfully completed, may soon be ready for operational
                         deployment. Almost certain to be part of future SLBM force.
JCS    73                "Intensively flight tested in the last year"; including 3 flights into the Pacific.
                         Probably ready for deployment.
SecD   73                Qualitative upgrade "appears on the horizon with the testing" of SS-N-8.
SecC   75                This year "rather substantial evidence" that USSR preparing to test new version of
                         SS-N-8 or a follow on missile, and "some evidence" that this new missile may carry
                         MIRVs.
FI     81                Has been tested with 3 MIRVs.
SMP    85    1968-69     Engineering and testing began. [From graph.]

————————— PRODUCTION UNDERWAY —————————
SMP    84                "The Soviets continue to produce four SLBMs—the SS-N-6, SS-N-8, SS-N-17 and
                         SS-N-18—at the rate of 200 a year. One new system, the SS-N-20/Typhoon, is in
                         series production".

————————— IOC —————————
JCS    75    1973        FY76-79.
US12   80    1973
JWS    81    1973        Operational.
JFS    82    1972        Operational.
vanD   83    1974

————————— DEPLOYMENT BEGAN —————————
FIar   76    1973
MOW    76    1973
USD7   79    since 1972  Delta entering service.
Clns   80    1973        First deployed.
SMP    81    early 1970s Introduced.
JCS    83    c. 1973     Year of introduction [from graph].
CFW+   84    "1973/77"
JCS    84    1973-74     Major strategic system deployments [graph], FY85-86.
SecD   84    1973        Introduction of strategic systems [graph].
SMP    85    1973-74     [From graph.]

————————— DEPLOYED —————————
MOW    76    in service
```

276  SOVIET MISSILES  IDDS

```
----------- DEPLOYMENT HISTORY -----------
JCS    73   Believe it is "being installed" in new Delta boat.
SecD   73   USSR "introducing into its sea-based forces" SS-NX-8. First Delta should soon become operational.
SecD   74   Being deployed.
SMP    85   Will be deployed at least through 1990. [From graph.]

----------- NUMBERS PRODUCED -----------
MOW    76              "Virtually all SS-N-8 missiles built by 1976 had been installed" in the 2 kinds of D
                       subs.

----------- NUMBERS IN SERVICE -----------

----------- NUMBER IN 1960 -----------
SSF    82   none       In 1960, 1965, 1970.

----------- NUMBER IN 1970 -----------
Clns   80   none       In 1970-1972.

----------- NUMBER IN 1973 -----------
IISS   73   36         July, total deployed.
Clns   80   12
Clns+  83   28

----------- NUMBER IN 1974 -----------
IISS   74   108        July, total deployed.
Clns   80   60
Clns+  83   80

----------- NUMBER IN 1975 -----------
IISS   75   156        July, total deployed.
USO8   79   220        Early 1975.
Clns   80   132
SSF    82   132        Including 6 missiles in testbed launchers.
Clns+  83   146

----------- NUMBER IN 1976 -----------
IISS   76   220        July, total deployed.
Clns   80   220

----------- NUMBER IN 1977 -----------
IISS   77   284        July, total deployed, might include one Delta III sub.
Nitz   79   220        Launchers.
Clns   80   274
Clns+  83   286        In 1977-1978.

----------- NUMBER IN 1978 -----------
IISS   78   370        July, total deployed.
Clns   80   292

----------- NUMBER IN 1979 -----------
IISS   79   266        July, total deployed.
Nitz   79   268        18 June, launchers.
Clns   80   289
Clns+  83   292        In 1979-1982.

----------- NUMBER IN 1980 -----------
IISS   80   302        July, total deployed.
SSF    82   282        Including 6 missiles in testbed launchers.

----------- NUMBER IN 1981 -----------
IISS   81   291        July, total deployed.

----------- NUMBER IN 1982 -----------
IISS   82   292        1982-83, July, total deployed.

----------- NUMBER IN 1983 -----------
CBO    83   280        1983, 1990.

----------- NUMBER IN 1984 -----------
JCS    84   292        As of 1 Jan.
SMP    84   292

----------- NUMBER IN 1985 -----------
JCS    85   292        As of 1 Jan.
SMP    85   292

----------- NUMBER IN 1996 -----------
CBO    83              268 in 1996.
```

BALLISTIC MISSILES  SLBMs  SS-N-8  277

|  |  | CARRIERS |  |  |  |
| --- | --- | --- | --- | --- | --- |
|  |  | CARRIER | TUBES | RELOADS |  |
| SecC | 70 |  |  |  | Extensive modifications needed to fit into Y-class sub, or else new sub class required. |
| SecC | 71 |  |  |  | "Missile is probably too large (based on estimated size) for the Y-class submarine". |
| SWM | 77 |  |  |  | "Still larger boat carrying more missiles is under construction". |
| JCS | 78 |  |  |  | Carried predominantly by Delta I and Delta II. |
| Rock | 78 |  |  |  | Shipyard production promises 10 boats per year in near future. |
|  |  | CARRIER | TUBES | RELOADS |  |
| Clns | 80 | Golf III | 6 |  |  |
| SecD | 81 | Golf |  |  | Also deployed on Golf and Hotel classes. |
| USND | 81 | Golf III |  |  |  |
| CFW | 82 | Golf II |  |  | Experimental sub [final report] |
| SSF | 82 | Golf III | 6 |  | 1 in operation in 1980. |
| GSN | 83 | Golf III |  |  | 1 unit, test bed. |
| CFW+ | 84 | Golf III |  |  | Experimental sub. |
| BW | 86 | Golf III | 6 | no | N=1. |
|  |  | CARRIER | TUBES | RELOADS |  |
| JCS | 73 | Delta | 12 |  | FY74,76-77. One launched by mid-1972. tubes: FY74,76-77. |
| SecC | 73 | Delta |  |  | FY74,76. 2 Deltas launched thus far; first unit should soon become operational. |
| SecD | 73 | Delta | 12 |  | FY74-78. tubes: FY74-78. |
| JCS | 74 | Delta |  |  | Deployed aboard at least 3. |
| IA06 | 75 | Delta | 12 |  | N=8. |
| SecC | 75 |  |  |  | 11 of 12-tube Deltas have been launched. |
| SecD | 75 | Delta |  |  | 8 of original Delta launched. |
| Flar | 76 | Delta I | 12 |  | 8 in service. |
| MOW | 76 | D subs | 12 |  |  |
| JCS | 77 | Delta I |  |  | FY78-79. |
| SWM | 77 | Delta 1 | 12 |  | Operational 1972. |
| JCS | 78 | Delta I |  |  | Construction believed ended. |
| Rock | 78 | Delta I | 12 |  | N=10, by 1977. |
| SecD | 78 | Delta I |  |  | FY79-82. |
| AWar | 80 | Delta 1 | 12 |  | N=18. |
| Clns | 80 | Delta I | 12 |  |  |
| SecD | 80 | Delta I | 12 |  | Total of 32 operational DELTA-class subs. |
| US12 | 80 | Delta I |  |  |  |
| FI | 81 | Delta I | 12 |  | N=18. |
| JWS | 81 | Delta I | 12 |  |  |
| SMP | 81 | Delta I | 12 |  | '81, '83, '84, '85. First Delta completed during 1973. tubes: '81, '83, '84, '85. |
| USND | 81 | Delta I |  |  |  |
| AWST | 82 | Delta 1 subs |  |  |  |
| CFW | 82 | Delta I |  |  |  |
| JCS | 82 | Delta I |  |  |  |
| JFS | 82 | Delta I |  |  |  |
| SSF | 82 | Delta I | 12 |  | 18 in operation in 1980. |
| GSN | 83 | Delta I |  |  |  |
| JWS+ | 83 | Delta I |  |  | N=18, mid-1983. |
| JCS | 85 | Delta I |  |  |  |
| BW | 86 | Delta I | 12 | no | N=18. |

| | | CARRIER | TUBES | RELOADS | |
|---|---|---|---|---|---|
| JCS | 74 | | | | Tenuous evidence of longer Deltas, might take more SLBMs. |
| JCS | 75 | | | | "Larger launch vehicle" for SS-N-8 required to allow USSR to build to SALT limits as they have announced; longer Delta soon to be launched. |
| SecD | 75 | New Delta | | | FY76-77. Under construction. |
| Flar | 76 | Delta II | | | 4 or 5 in service; 6 or more under construction. |
| JCS | 76 | Delta [II] | 16 | | Larger 16-tube Delta probably will carry SS-N-8. |
| MOW | 76 | mod D subs | 16 | | |
| SecD | 76 | New Delta | 16 | | Several units have been launched. |
| JCS | 77 | Delta II | | | FY78-79. |
| SWM | 77 | Delta 2 | 16 | | Put to sea in 1976. |
| JCS | 78 | Delta II | | | |
| Rock | 78 | Delta II | | | N=5, by 1977. |
| SecD | 78 | Delta II | | | FY79-82. |
| AWer | 80 | Delta 2 | 16 | | N=4. |
| Clns | 80 | Delta II | 16 | | |
| SecD | 80 | Delta II | 16 | | Total of 32 operational DELTA-class subs. |
| US12 | 80 | Delta II | | | |
| FI | 81 | Delta II | 16 | | N=4. |
| JWS | 81 | Delta II | 16 | | |
| SMP | 81 | Delta II | 16 | | '81, '83, '84, '85. |
| | | | | | tubes: '81, '83, '84, '85. |
| USND | 81 | Delta II | | | |
| AWST | 82 | Delta 2 subs | | | |
| CPW | 82 | Delta II | | | |
| JCS | 82 | Delta II | | | |
| JFS | 82 | Delta II | | | |
| SSF | 82 | Delta II | 16 | | 4 in operation in 1980. |
| GSN | 83 | Delta II | | | |
| JWS+ | 83 | Delta II | | | N=4, mid-1983. |
| JCS | 85 | Delta II | | | |
| BW | 86 | Delta II | 16 | no | N=4. |

| | | CARRIER | TUBES | RELOADS | |
|---|---|---|---|---|---|
| SecD | 79 | DELTA | | | Around 29 operational. |

| | | CARRIER | TUBES | RELOADS | |
|---|---|---|---|---|---|
| SecC | 70 | | | | Possible that H-class sub currently undergoing modification may be employed as test platform. |
| US12 | 80 | Hotel III | | | |
| SecD | 81 | Hotel | | | Also deployed on Golf and Hotel classes. |
| SMP | 81 | Hotel III | 6 | | '81, '83, '84. |
| | | | | | tubes: '81, '83, '84. |
| SSF | 82 | Hotel III | 6 | | 1 in operation in 1980. |
| Clns+ | 83 | Hotel III | 6 | | |
| JWS+ | 83 | Hotel III | | | N=1, trials. |
| BW | 86 | Hotel III | 6 | no | N=1. |

| | | DESIGNER | |
|---|---|---|---|
| SSF | 82 | | Design Bureau: Chelomei derivative. |

| | | DESIGN AND ENGINEERING | |
|---|---|---|---|
| Flar | 76 | SS-N-8 meant elimination of Russian technological lag in SLBMs. | |
| GSN | 83 | Significantly larger than SS-N-6, indicating it was developed from the SS-N-4/5 series. | |

| | | USE AND CONFIGURATION | |
|---|---|---|---|
| MOW | 76 | Underwater launch. | |
| CBO | 83 | System availability: 0.40. | |
| venD | 83 | Reliability: .75; Penetration 1.0. | |

## STRATEGY

| | | |
|---|---|---|
| JCS | 74 | SS-N-8 range increases ocean areas for SLBM attack on US; some new areas not covered by existing US SLBM detectors. Gives USSR a "temporary qualitative advantage" [in strategic forces]. |
| SecD | 74 | No evidence of depressed-trajectory testing. US 474N radars and satellites have limitations against SS-N-8. |
| JCS | 75 | From Yankee patrol areas, can target entire CONUS; from Barents, can cover any eastern seaboard target now covered by Yankee. |
| SecD | 75 | SS-N-8 can bypass US FSS-7 radars. |
| FIar | 76 | Can cover whole of US from Barents Sea. |
| JCS | 76 | From "existing patrol areas" in "Barents and Greenland Seas", targets "essentially all" of CONUS. In North Atlantic, coverage of Central and northern South America as well as CONUS. |
| JCS | 77 | Can hit "most US targets" from home ports, FY78-79. |
| SecD | 77 | Can cover major US targets from Barents Sea and North Pacific. |
| JCS | 78 | Delta operating capabilities contribute to threat too. |
| SecD | 78 | Permits Soviets to cover US targets from Barents Sea and waters of the North Pacific (as does SS-NX-18). |
| SecD | 79 | Permits Soviets to cover US targets from Barents Sea and waters of the western Pacific (as does SS-NX-18). |
| SecD | 80 | Permits Soviets to cover targets in CONUS from Barents Sea and Sea of Okhotsk. |
| SMP | 81 | Can hit targets in the US from their home ports. |
| JCS | 82 | Can hit US targets from adjacent Soviet waters or even home ports. |

## GEOGRAPHICAL DEPLOYMENTS

| | | |
|---|---|---|
| JCS | 76 | "Existing patrol areas" in "Barents and Greenland Seas". |

## RELATION TO OTHER MISSILES

| | | |
|---|---|---|
| JCS | 75 | Much greater volume than SS-N-6, FY76-77. |
| JWS | 81 | SSN-8 used to be thought a development of SSN-6, but US Defence now says it is "sufficiently different" to have its own name. |

## OTHER INFORMATION

| | | |
|---|---|---|
| JCS | 77 | Follow-on to SS-N-8 expected, of same volume [apparently not SS-NX-17,18, ed]. |
| JCS | 78 | During early 1980s, a follow-on could be deployed. |

# SS-N-8 Mod 1

```
Sources are:   SWM   77      FI    81      SMP   81      SSF   82      GSN   83      SMP   85
FIar  76       Rock  78      JWS   81      IISS  82      FI+   83      CFW   84      BW    86
```

## US DESIGNATION

|  | |  |
|---|---|---|
| FIar | 76 | SS-N-8 Mod 1 |
| SWM | 77 | SS-N-8 Mod 1 |
| Rock | 78 | SS-N-8 Mod 1 |
| FI | 81 | SS-N-8 Mod 1 |
| JWS | 81 | SS-N-8 Mod 1 |
| SMP | 81 | SS-N-8 Mod 1 |
| IISS | 82 | SS-N-8 Mod 1 |
| SSF | 82 | SS-N-8 Mod 1 |
| GSN | 83 | SS-N-8 Mod 1 |
| CFW | 84 | SS-N-8 Mod 1 |

## RANGE

| Source | | mi | nm | km | km conv | Notes |
|---|---|---|---|---|---|---|
| SMP | 81 | | | 7,800 | 7,800 | '81, '83, '84, '85. |
| IISS | 82 | | | 7,800 | 7,800 | Max. |
| SSF | 82 | | 4,200 | | 7,778 | |
| GSN | 83 | | 4,240 | 7,800 | 7,852 | |
| CFW | 84 | | 4,240 | | 7,852 | |
| BW | 86 | | | 7,800 | 7,800 | |

## CEP

| Source | | ft | mi | nm | m | km | m conv |
|---|---|---|---|---|---|---|---|
| IISS | 82 | | | | 1,300 | | 1,300 |
| BW | 86 | | | | 1,600 | | 1,600 |

## WARHEAD CHARACTERISTICS

|  | | |  |
|---|---|---|---|
| SWM | 77 | single RV | |
| Rock | 78 | single RV | |
| JWS | 81 | single RV | |
| SMP | 81 | single RV | '81, '83, '84, '85. |
| IISS | 82 | single RV | Mods 1 and 2. |
| SSF | 82 | single RV | Mods 1 and 2. |
| BW | 86 | single RV | |

## NUCLEAR YIELD

| Source | | kT | MT | kT conv | Notes |
|---|---|---|---|---|---|
| FIar | 76 | | 1-2 | 1,000-2,000 | |
| Rock | 78 | | 1.0 | 1,000 | |
| FI | 81 | | 1-2 | 1,000-2,000 | |
| JWS | 81 | | 1 | 1,000 | |
| IISS | 82 | | 1 | 1,000 | |
| FI+ | 83 | | 1 | 1,000 | [Analysis table] |
| BW | 86 | 800 | | 800.00 | |

## THROWEIGHT

| Source | | lb | t | kg | kg conv |
|---|---|---|---|---|---|
| IISS | 82 | 1,500 | | | 680.40 |

## TYPE OF PROPULSION

|  | | |
|---|---|---|
| BW | 86 | storable liquid |

## PROTOTYPE TESTS

|  | | | |
|---|---|---|---|
| SSF | 82 | 1969 | First flight test. |

## IOC

|  | | |
|---|---|---|
| GSN | 83 | 1973 |

## DEPLOYMENT BEGAN

|  | | |  |
|---|---|---|---|
| Rock | 78 | 1973 | Service intro; only Mod 1 deployed so far. |
| IISS | 82 | 1972 | First year deployed. |
| SSF | 82 | 1973 | Year operation began. |
| SMP | 85 | 1973-74 | [Assumed Mod 1, from graph.] |
| BW | 86 | 1973 | |

## DEPLOYED

|  | | |
|---|---|---|
| BW | 86 | In service. |

# SS-N-8 Mod 2

```
Sources are:  SWM   77      FI    81      SMP   81      SSF   82      JFS+  83      SMP   85
FIar   76     Rock  78      JWS   81      IISS  82      GSN   83      CRW   84      BW    86
```

---------- **US DESIGNATION** ----------

|  | |  |
|---|---|---|
| FIar | 76 | SS-N-8 Mod 2 |
| SWM | 77 | SS-N-8 Mod 2 |
| Rock | 78 | SS-N-8 Mod 2 |
| FI | 81 | SS-N-8 Mod 2 |
| JWS | 81 | SS-N-8 Mod 2 |
| SMP | 81 | SS-N-8 Mod 2 |
| IISS | 82 | SS-N-8 Mod 2 |
| SSF | 82 | SS-N-8 Mod 2 |
| GSN | 83 | SS-N-8 Mod 2 |
| CRW | 84 | SS-N-8 Mod 2 |

---------- **RANGE** ----------

| Source | | mi | nm | km | km conv | |
|---|---|---|---|---|---|---|
| SMP | 81 | | | 9,100 | 9,100 | '81, '83, '84, '85. |
| IISS | 82 | | | 9,100 | 9,100 | Max. |
| SSF | 82 | | 4,900 | | 9,074 | |
| GSN | 83 | | 4,950 | 9,100 | 9,167 | |
| JFS+ | 83 | | 4,900 | | 9,074 | Max. |
| CRW | 84 | | 4,950 | | 9,167 | |
| BW | 86 | | | 9,100 | 9,100 | |

---------- **CEP** ----------

| Source | | ft | mi | nm | m | km | m conv |
|---|---|---|---|---|---|---|---|
| IISS | 82 | | | | 900 | | 900.00 |
| SSF | 82 | | | 0.84 | | | 1,555 |
| BW | 86 | | | | 1,600 | | 1,600 |

---------- **WARHEAD CHARACTERISTICS** ----------

|  | | | |
|---|---|---|---|
| SWM | 77 | 3 MRVs or MIRVs | |
| Rock | 78 | 3 x MRV | |
| FI | 81 | 3 MRVs | |
| JWS | 81 | 3 MRVs | |
| SMP | 81 | single RV | '81, '83, '84, '85. |
| IISS | 82 | single RV | Mods 1 and 2. |
| SSF | 82 | single RV | Mods 1 and 2. |
| BW | 86 | single RV | |

---------- **NUCLEAR YIELD** ----------

| Source | | kT | MT | kT conv | |
|---|---|---|---|---|---|
| Rock | 78 | | ? 0.9 | 900.00 | [Table]; RVs in "kiloton range" [text]. |
| IISS | 82 | 800 | | 800.00 | |
| SSF | 82 | | 0.8 | 800.00 | Per warhead. |
| BW | 86 | 800 | | 800.00 | |

---------- **THROWWEIGHT** ----------

| Source | | lb | t | kg | kg conv |
|---|---|---|---|---|---|
| IISS | 82 | 8,000 | | | 3,628 |

---------- **TYPE OF PROPULSION** ----------

| BW | 86 | storable liquid |
|---|---|---|

---------- **PROTOTYPE TESTS** ----------

| SSF | 82 | 1976 | First flight test. |
|---|---|---|---|

---------- **IOC** ----------

| GSN | 83 | 1977 |
|---|---|---|

---------- **DEPLOYMENT BEGAN** ----------

| FIar | 76 | | Mod 2 not yet in service. |
|---|---|---|---|
| Rock | 78 | | Only Mod 1 deployed so far. |
| SMP | 85 | 1981 | [From graph.] |
| BW | 86 | 1977 | |

---------- **DEPLOYED** ----------

| BW | 86 | In service. |
|---|---|---|

---------- **DEPLOYMENT HISTORY** ----------

| SMP | 85 | Will be deployed at least through 1990. [Assumed Mod 2, from graph.] |
|---|---|---|

# SS-N-8 Mod 3

```
Sources are:   FIar  76      Rock  78      FI   81      JWS   81      IISS  82      FI+   83      JWS+  83
```

```
------------- US DESIGNATION -------------
FIar  76      SS-N-8 Mod 3
Rock  78      SS-N-8 Mod 3
FI    81      SS-N-8 Mod 3
JWS   81      SS-N-8 Mod 3
IISS  82      SS-N-8 Mod 3

------------- EDITOR'S NOTES -------------
IISS  82      Final report.

------------- CEP -------------
              ft        mi        nm        m         km        m conv
IISS  82                                    450                 450.00

------------- WARHEAD CHARACTERISTICS -------------
FIar  76      3 MIRVs              "Mod 3?" has been tested with 3 MIRVs.
Rock  78      "? x MIRV"
FI    81      3 MIRVs              ? Mod 3.
JWS   81      3 MIRVs
IISS  82      3 MIRVs
FI+   83      3 MIRVs
JWS+  83      3 MIRVs              "Was believed to have" 3 MIRVs, but no confirmation of this version.

------------- NUCLEAR YIELD -------------
              kT        MT        kT conv
Rock  78                ? 0.8     800.00
IISS  82      200                 200.00      Per warhead.

------------- DEPLOYMENT BEGAN -------------
Rock  78                Only Mod 1 deployed so far.
```

# SS-N-17

| Sources are: | CB01 | 78 | Nitz | 79 | SecD | 81 | IISS | 82 | IISS+ | 83 | AWST+ | 84 | SMP | 85 |
|---|---|---|---|---|---|---|---|---|---|---|---|---|---|---|
| SecC | 75 | IISS | 78 | SecD | 79 | SMP | 81 | JFS | 82 | JCS | 83 | CFW+ | 84 | | |
| FIer | 76 | JCS | 78 | Clns | 80 | US17 | 81 | SSF | 82 | JFS+ | 83 | JCS | 84 | | |
| SecC | 76 | Rock | 78 | SecD | 80 | USND | 81 | Clns+ | 83 | JWS+ | 83 | SecD | 84 | | |
| JCS | 77 | SecD | 78 | FI | 81 | AWST | 82 | FI+ | 83 | SMP | 83 | SMP | 84 | | |
| SecD | 77 | USD6 | 78 | JWS | 81 | CFW | 82 | GSN | 83 | vanD | 83 | JCS | 85 | | |

---------- **US DESIGNATION** ----------

| FIer | 76 | SS-XN-17 | [Sic] |
| JCS | 77 | SS-NX-17 | FY78-79. |
| | | SS-N-X-17 | FY78-79. |
| SecD | 77 | SS-NX-17 | FY78-82. |
| Rock | 78 | SS-NX-17 | |
| Nitz | 79 | SS-NX-17 | |
| Clns | 80 | SS-N-17 | |
| FI | 81 | SS-N-17 | |
| JWS | 81 | SSN-X-17 | |
| SMP | 81 | SS-NX-17 | |
| USND | 81 | SS-N-17 | |
| AWST | 82 | SS-NX-17 | |
| CFW | 82 | SS-N-17 | |
| IISS | 82 | SS-NX-17 | |
| JFS | 82 | SS-N-17 | |
| SSF | 82 | SS-NX-17 | |
| GSN | 83 | SS-N-17 | |
| IISS+ | 83 | SS-N-17 | |
| JCS | 83 | SS-N-17 | FY84-85. |
| JWS+ | 83 | SS-NX-17 | Official US sources have dropped X, but Jane's sees no justification for it. |
| SMP | 83 | SS-N-17 | '83, '84, '85. |
| vanD | 83 | SS-N-17 | |
| AWST+ | 84 | SS-NX-7 | [Sic] |
| SecD | 84 | SSN17 | |

---------- **DESCRIPTION** ----------

| Rock | 78 | | 4th generation SLBM. |
| JWS | 81 | SLBM | |
| USND | 81 | SLBM | |
| AWST | 82 | SLBM | |
| SSF | 82 | | Fifth generation. |
| JCS | 83 | SLBM | |

---------- **EDITOR'S NOTES** ----------

| SecC | 75 | Referred to, not by name, FY76-77. |
| JCS | 77 | Listed FY78-79, 84-85. |
| SecD | 77 | Listed FY78-82, 85. |
| SMP | 81 | Listed '81, '83, '84, '85. |
| AWST+ | 84 | Mislabeled chart entry appears in 1984, under ballistic missiles. |

---------- **RANGE** ----------

| | | mi | nm | km | km conv | |
|---|---|---|---|---|---|---|
| Rock | 78 | | | 3,200 | 3,200 | |
| SecD | 78 | | | | | Greater than SS-N-6. |
| USD6 | 78 | | 2,400 | | 4,444 | |
| Clns | 80 | | 2,000 | | 3,704 | |
| SecD | 80 | | | 3000-4000 | 3,000-4,000 | |
| FI | 81 | | | ? 5,000+ | 5,000 | |
| SMP | 81 | | | 3,900 | 3,900 | '81, '83, '84, '85. |
| US17 | 81 | | 2,400 | | 4,444 | |
| USND | 81 | | 1,600+ | | 2,963 | |
| AWST | 82 | | 1,700 | | 3,148 | |
| CFW | 82 | | 2,000 | | 3,704 | |
| IISS | 82 | | | 3,900 | 3,900 | Max. |
| JFS | 82 | | 2,400 | | 4,444 | Max. |
| SSF | 82 | | 2,100 | | 3,889 | |
| GSN | 83 | | 2,000+ | 3,700+ | 3,704 | |
| JCS | 84 | | | | | All new SLBMs except SS-N-17 capable of striking targets throughout most of US from Soviet home waters, '84, '85. |

## CEP

|  |  | ft | mi | nm | m | km | m conv | |
|---|---|---|---|---|---|---|---|---|
| CB01 | 78 | 3000 |  |  |  |  | 914.40 | Estimated, with SS-N-18, 1985. |
| SecD | 78 |  |  |  |  |  |  | Accuracy greater than SS-N-6, FY79-81. |
| Nitz | 79 |  |  | .25 |  |  | 463.00 | |
| Clns | 80 |  |  | 0.75 |  |  | 1,389 | |
| IISS | 82 |  |  |  | 1,500 |  | 1,500 | |
| GSN | 83 |  |  |  |  |  |  | Greater accuracy than previous Soviet SLBMs. |
| vanD | 83 |  |  | .25 |  |  | 463.00 | |

## WARHEAD TYPE

|  |  | nuclear | conven | chem |
|---|---|---|---|---|
| CRW | 82 | yes |  |  |
| JFS | 82 | yes |  |  |
| GSN | 83 | yes |  |  |
| JWS+ | 83 | yes |  |  |
| CRW+ | 84 | yes |  |  |

## WARHEAD CHARACTERISTICS

| FIar | 76 |  | Has a PBV, permitting MIRVs. |
|---|---|---|---|
| SecC | 76 |  | "It is too early to determine characteristics of the small missiles with any confidence, but the large missile has been MIRVed". |
| JCS | 77 | single RV | May have MIRV capability; first SLBM (of USSR) with PBV for RVs, FY78-79. |
| SecD | 77 | single warhead | Large PBV. |
| Rock | 78 | "? x MIRV" | Has PBV capable of carrying MIRVs. |
| SecD | 78 |  | PBV, FY79-81. |
| Nitz | 79 | 1.0 | [Sic] |
| Clns | 80 | single RV |  |
| FI | 81 | ? single RV | "Multiple re-entry vehicles can be carried". |
| JWS | 81 |  | First Soviet SLBM with PBV, suggesting MRV or MIRV, but only single RV payloads have been tested. |
| SMP | 81 | single RV | '81, '83, '84, '85. |
| US17 | 81 | multiple |  |
| IISS | 82 | single RV |  |
|  |  | 7 MIRVs | Tested [final report] |
| JFS | 82 | MRV or MIRV | Has PBV. |
| SSF | 82 | single RV | PBV. |
| GSN | 83 | single RV | Has PBV; first Soviet SLBM to employ PBV to aim single RV. |
| vanD | 83 | single RV |  |
| CRW+ | 84 | single RV |  |

## NUCLEAR YIELD

|  |  | kT | MT | kT conv | |
|---|---|---|---|---|---|
| CB01 | 78 |  | 0.2 | 200.00 | Estimated, with SS-N-18, 1985. |
| Rock | 78 |  |  |  | MIRV warheads probably in kT range. |
| Nitz | 79 |  | .75 | 750.00 | |
| Clns | 80 | 500 |  | 500.00 | |
| IISS | 82 |  |  |  | "MT", per warhead. |
|  |  | 200 |  | 200.00 | Tested, 7 MIRV version [final report] |
| GSN | 83 |  | c. 1 | 1,000 | |
| vanD | 83 |  | .75 | 750.00 | |
| CRW+ | 84 |  | 1 | 1,000 | |

## THROWWEIGHT

|  |  | lb | t | kg | kg conv |
|---|---|---|---|---|---|
| Nitz | 79 | 2,000 |  |  | 907.20 |
| IISS | 82 | 2,500 |  |  | 1,134 |

## WEIGHT

|  |  | lb | t | kg | kg conv | |
|---|---|---|---|---|---|---|
| Rock | 78 |  | ? 20 |  | 20,000 | Tonnes. |

## LENGTH

|  |  | in | ft | cm | m | m conv | |
|---|---|---|---|---|---|---|---|
| Rock | 78 |  |  |  | ? 17 | 17.00 | |
| FI | 81 |  |  |  | c. 11 | 11.00 | |
| JWS | 81 |  |  |  | c. 11.06 | 11.06 | |
| SMP | 81 |  |  |  | 10.5 | 10.50 | [From graph], '81, '83, '84, '85. |
| JFS | 82 |  | 36.3 |  |  | 11.06 | |
| FI+ | 83 |  |  |  | 11.06 | 11.06 | |
| GSN | 83 |  | c. 35 |  | c. 10.6 | 10.67 | |
| CRW+ | 84 |  |  |  | 10.6 | 10.60 | |

## BALLISTIC MISSILES   SLBMs   SS-N-17

### DIAMETER

|  |  | in | ft | cm | m | cm conv |
|---|---|---|---|---|---|---|
| Rock | 78 |  |  |  | ? 1.8 | 180.00 |
| FI | 81 |  |  | ? 165 |  | 165.00 |
| JWS | 81 |  |  |  | c. 1.65 | 165.00 |

### NUMBER OF STAGES

| Rock | 78 | 1 |
| FI | 81 | 2 |
| AWST | 82 | 1 |
| JFS | 82 | 2 |
| JFS+ | 83 | 1 |

### TYPE OF PROPULSION

| FIar | 76 | solid | First solid-fuel SLBM of USSR. |
| JCS | 77 | solid | First Soviet SLBM using a solid propellant, FY78-79. |
| SecD | 77 | solid | FY78-81. |
| Rock | 78 | solid | First Soviet solid fuel SLBM yet tested for operational deployment by Soviet Navy. |
| Clns | 80 | solid | |
| FI | 81 | solid | |
| JWS | 81 | solid | |
| AWST | 82 | solid | First SLBM solid booster. |
| CFW | 82 | solid | First ballistic missile with solid-fuel propulsion. |
| IISS | 82 | solid | |
| JFS | 82 | solid | |
| SSF | 82 | solid | |
| GSN | 83 | solid | First Soviet solid-fuel SLBM. |

### MIDCOURSE (OR MAIN) GUIDANCE

| JWS | 81 | inertial |
| JFS | 82 | inertial |
| GSN | 83 | inertial |
| CFW+ | 84 | inertial |

### DEVELOPMENT BEGAN

| SSF | 82 | 1969 | Year design began. |
| SMP | 85 | 1969 | Technological development began. [From graph.] |

### DEVELOPMENT UNDERWAY

| SecD | 81 | "Experimental". |
| US17 | 81 | "Under development". |

### PROTOTYPE TESTS

| SecC | 75 | | Some evidence that USSR preparing to test new smaller SLBM in the SS-N-6 class. |
| FIar | 76 | | Tests expected to last at least another year. |
| SecC | 76 | | "In recent months, there have been test launches of a small and a large new SLBM which may be intended as the eventual replacements for the SS-N-6 and SS-N-8". |
| JCS | 77 | 1975 | Flight testing began, FY78-79. May soon be ready for at-sea testing phase. |
| SecD | 77 | | Has been tested. |
| USO6 | 78 | 1975 | Testing began. |
| Clns | 80 | 1977 | Sea trials began. |
| JWS | 81 | 1975 | Testing began. |
| SSF | 82 | 1976 | First flight test. |
| GSN | 83 | 1975 | Flight testing began. |
| JWS+ | 83 | | Believed to have had periodic sea trials since 1977 in one converted Yankee. |
| SMP | 85 | 1976-76 | Engineering and testing began. [From graph.] |

### PRODUCTION UNDERWAY

| SMP | 84 | "The Soviets continue to produce four SLBMs—the SS-N-6, SS-N-8, SS-N-17 and SS-N-18—at the rate of 200 a year. One new system, the SS-N-20/Typhoon, is in series production". |

### IOC

| JCS | 77 | | [Blank in chart, ed], FY78-79. |
| GSN | 83 | c. 1977 | |
| vanD | 83 | 1978 | |

### DEPLOYMENT BEGAN

| Clns | 80 | 1977 | First deployed. |
| IISS | 82 | 1977 | First year deployed. |
| JCS | 83 | c. 1981 | Year of introduction [from graph]. |
| CFW+ | 84 | 1977 | |
| JCS | 84 | 1980-81 | Major strategic system deployments [graph]. |
| SecD | 84 | 1979 | Introduction of strategic systems [graph]. |
| JCS | 85 | 1981-82 | Major strategic system deployments [graph]. |
| SMP | 85 | 1980-81 | [Limited deployment, from graph.] |

## DEPLOYMENT HISTORY

| | | |
|---|---|---|
| USO6 | 78 | Probably destined to replace SS-N-6. |
| FI | 81 | Being developed to replace SS-N-6s in some Yankees. |
| IISS | 82 | "May be solid-fuel successor to SS-N-6" [final report] |
| GSN | 83 | No deployment beyond one Yankee II. |
| SMP | 85 | Will be in limited deployment at least through 1990. [From graph.] |

## NUMBERS IN SERVICE

### NUMBER IN 1970

| | | | |
|---|---|---|---|
| Clns | 80 | none | 1970-1976. |

### NUMBER IN 1977

| | | | |
|---|---|---|---|
| Nitz | 79 | 12 | Launchers. |
| Clns | 80 | 12 | 1977-1979; "Sea trials began in 1977, but this table carries those missiles as active on that date". |

### NUMBER IN 1978

| | | | |
|---|---|---|---|
| IISS | 78 | 16 | July, total deployed, [table]. |
| | | 12 | 1978-83, July, total deployed, [text in 78, table since]. |

### NUMBER IN 1979

| | | | |
|---|---|---|---|
| Nitz | 79 | 12 | 18 June, launchers. |

### NUMBER IN 1980

| | | | |
|---|---|---|---|
| Clns+ | 83 | 12 | In 1980-1982. |

### NUMBER IN 1984

| | | | |
|---|---|---|---|
| JCS | 84 | 12 | As of 1 Jan. |
| SMP | 84 | 12 | |

### NUMBER IN 1985

| | | | |
|---|---|---|---|
| JCS | 85 | 12 | As of 1 Jan. |
| SMP | 85 | 12 | |

## CARRIERS

### CARRIER — TUBES — RELOADS

| | | CARRIER | TUBES | RELOADS |
|---|---|---|---|---|
| JCS | 77 | converted Yankee | | Expected to be platform; "a Yankee". |
| JCS | 78 | Yankee | | One carries SS-NX-17s. |
| SecD | 78 | Yankee | | Fitted in only one to date; believe it may be backfitted into some or all [other] Yankees. |
| SecD | 79 | Yankee | | Was backfitted to only one, FY80-82. |
| Clns | 80 | Yankee II | 12 | |
| FI | 81 | Yankee | 12 | N=1, trials. |
| JWS | 81 | Yankee | | Test platform. |
| SMP | 81 | Yankee II | 12 | '81, '83, '84, '85. |
| USND | 81 | Yankee II | | tubes: '81, '83, '84, '85. |
| CFW | 82 | Yankee II | | N=1. |
| JFS | 82 | Yankee | | "Deployed in one 'Yankee' class from 1977 for test programme". |
| SSF | 82 | Yankee II | 12 | 1 in operation in 1980, for testing. |
| GSN | 83 | Yankee II | | N=1. |
| JWS+ | 83 | Yankee Yankee III | 12 | N=1, trials; sole platform, since 1977, [text]. [Sic, figure caption]. |
| AWST+ | 84 | Yankee 2 | 12 | |
| JCS | 84 | Yankee | | Deployed on one modified Yankee. |
| JCS | 85 | Yankee II | | Carried "on a Yankee-II-class SSBN". |

## DESIGNER

| | | |
|---|---|---|
| SSF | 82 | Chelomei derivative, Design Bureau. |

## DESIGN AND ENGINEERING

| | | |
|---|---|---|
| JWS+ | 83 | Lack of urgency and/or progress "indicative of an essentially R & D oriented effort that may have failed to live up to expectations thus far". |

## USE AND CONFIGURATION

| | | |
|---|---|---|
| vanD | 83 | Reliability: .75; Penetration 1.0. |

## STRATEGY

| | | |
|---|---|---|
| Rock | 78 | "Probably tied in with the strategic requirement of a localized conflict and lacking the range of the existing SS-N-8 SLBM, probably fills a defence niche hitherto carried by surface ships". |

## RELATION TO OTHER MISSILES

| | | |
|---|---|---|
| FIar | 76 | Similar in size to SSN-8. |
| JWS | 81 | Perhaps SSN-X-17 is forerunner of SSN-X-20 for Typhoon subs. |

# SS-N-18

```
Sources are:   CBO1  78    US07  79    FI    81    USND  81    CBO   83    JWS+  83    SecD  84
SecC  75       JCS   78    US08  79    IISS  81    AWST  82    Clns+ 83    SMP   83    SMP   84
FIar  76       Rock  78    AWar  80    JCS   81    CFW   82    FI+   83    vanD  83    JCS   85
SecC  76       SecD  78    Clns  80    JWS   81    IISS  82    GSN   83    ArkS  84    SMP   85
A140  77       US06  78    IISS  80    SecD  81    JCS   82    IISS+ 83    AWST+ 84    BW    86
JCS   77       IISS  79    JCS   80    SMP   81    JFS   82    JCS   83    CFW+  84
SecD  77       SecD  79    SecD  80    US17  81    SSF   82    JFS+  83    JCS   84
```

---------- **US DESIGNATION** ----------

| | | | |
|---|---|---|---|
| FIar | 76 | SS-XN-18 | [Sic] |
| JCS | 77 | SS-NX-18 | FY78-79. |
| SecD | 77 | SS-NX-18 | FY78-79. |
| Rock | 78 | SS-NX-18 | |
| SecD | 79 | SS-N-18 | FY80-82,85. |
| AWar | 80 | SS-N-18 | |
| Clns | 80 | SS-N-18 | |
| JCS | 80 | SS-N-18 | FY81-85. |
| FI | 81 | SS-N-18 | |
| JWS | 81 | SSN-18 | |
| SMP | 81 | SS-N-18 | '81, '83, '84, '85. |
| USND | 81 | SS-N-18 | |
| AWST | 82 | SS-N-18 | |
| CFW | 82 | SS-N-18 | |
| IISS | 82 | SS-N-18 | |
| JFS | 82 | SS-N-18 | |
| SSF | 82 | SS-N-18 | |
| CBO | 83 | SS-N-18 | |
| GSN | 83 | SS-N-18 | |
| vanD | 83 | SS-N-18 | |

---------- **SOVIET DESIGNATION** ----------

| | | | |
|---|---|---|---|
| FI | 81 | RSM-50 | |
| JWS | 81 | ? RSM-50 | |
| GSN | 83 | RSM-50 | Soviet designation. |

---------- **DESCRIPTION** ----------

| | | | |
|---|---|---|---|
| Rock | 78 | | 4th generation SLBM. |
| JCS | 81 | SLBM | FY82,84. |
| JWS | 81 | SLBM | |
| USND | 81 | SLBM | |
| SSF | 82 | | Fifth generation. |

---------- **EDITOR'S NOTES** ----------

| | | |
|---|---|---|
| SecC | 75 | Referred to, not by name, FY76-77. |
| JCS | 77 | Listed FY78-79,81-86; in FY81, appears only in comparison to SS-N-20. |
| SecD | 77 | Listed FY78-82,85. |
| JCS | 78 | Space in chart for 2 mods, but no second one mentioned, FY79. |
| SMP | 81 | Listed '81, '83, '84, '85. |

---------- **RANGE** ----------

| | | mi | nm | km | km conv | |
|---|---|---|---|---|---|---|
| FIar | 76 | | | > 8,000 | 8,000 | Design range. |
| Rock | 78 | | | 8,000+ | 8,000 | [Table] |
| | | | | > 11,000 | 11,000 | [Text] |
| SecD | 78 | | | | | Very long range. |
| US06 | 78 | | 4,600 | | 8,519 | |
| SecD | 79 | | | c. 7,500 | 7,500 | |
| US07 | 79 | | 5,000 | | 9,260 | |
| US08 | 79 | | 5,600 | | 10,371 | |
| Clns | 80 | | 4,500 | | 8,334 | |
| SecD | 80 | | | 6,500-7,700 | 6,500-7,700 | |
| FI | 81 | | | 8,000 | 8,000 | |
| JWS | 81 | | | c. 7,500 | 7,500 | Max. |
| SMP | 81 | | | 7,500 | 7,500 | Range. |
| | | | | c. 6,500-8,000 | 6,500-8,000 | "Maximum operational booster range is about 6,500 to 8,000" km depending on payload; greater range possible if PBV used. |
| US17 | 81 | | 5,200 | | 9,630 | |
| USND | 81 | | c. 5,000 | | 9,260 | |
| AWST | 82 | | 4,000 | | 7,408 | |
| CFW | 82 | | 4,000 | | 7,408 | |
| JFS | 82 | | 5,200 | | 9,630 | Max. |
| JWS+ | 83 | | | c. 7,000 | 7,000 | [Analysis table] |

## CEP

| | | ft | mi | nm | m | km | m conv | |
|---|---|---|---|---|---|---|---|---|
| CB01 | 78 | 3000 | | | | | 914.40 | Estimated, with SS-N-17, 1985. |
| Rock | 78 | | | | | | | "Extremely high accuracy". |
| Clns | 80 | | | 0.75 | | | 1,389 | |
| AWST | 82 | | | 0.76 | | | 1,407 | [Final report] |
| CFW | 82 | | | | 800 | | 800.00 | |
| CBO | 83 | | | 0.3 | | | 555.60 | In 1983; lack of data on future trends. |
| GSN | 83 | | | 0.76 | | | 1,407 | Estimated for MIRV version in Western press. |
| JFS+ | 83 | | | 0.76 | | | 1,407 | [Says "miles", ed] |
| vanD | 83 | | | .76 | | | 1,407 | |
| CFW+ | 84 | | | | | c. 1,100 | 1,100 | |

## WARHEAD TYPE

| | | nuclear | conven | chem |
|---|---|---|---|---|
| CFW | 82 | yes | | |
| JFS | 82 | yes | | |
| GSN | 83 | yes | | |
| JWS+ | 83 | yes | | |

## WARHEAD CHARACTERISTICS

| | | | |
|---|---|---|---|
| SecC | 75 | | "Some evidence" that this new missile [SS-N-8 or follow on] may carry MIRVs. |
| SecC | 76 | MIRV | In recent testing. |
| JCS | 77 | "1-3 (MIRV)" | [Also] with more sophisticated guidance, could deliver 3 MIRVs (probably). Has PBV, FY78-79. |
| SecD | 77 | | It has flown [to date] with 2 MIRVs. |
| Rock | 78 | "? x MIRV" | |
| SecD | 78 | $\leq$ 3 MIRVs | FY79-80. |
| | | | Has PBV, FY79-81. |
| USD8 | 79 | 3 MIRVs | |
| | | 7 MIRVs | According to Zumwalt-Bagley. |
| Clns | 80 | 3 MIRVs | |
| SecD | 80 | 3 or 7 MIRVs | |
| JCS | 81 | MIRVed | FY82,86. Only Soviet MIRVed SLBM. |
| JWS | 81 | 3 MIRVs | Separate versions. |
| | | 7 MIRVs | Separate versions. |
| | | | Large PBV. |
| SecD | 81 | MIRVed | |
| SMP | 81 | MIRV | Has PBV. First Soviet SLBM to demonstrate a MIRV capability. |
| AWST | 82 | 3 MIRVs | "First MIRVed SLBM". |
| CFW | 82 | 3 MRVs | "Triple nuclear MRV-type warhead". |
| JFS | 82 | MRV | [Sic] |
| CBO | 83 | 3 RVs | |
| SMP | 83 | MIRVed | |
| vanD | 83 | 3 RVs | |

## NUCLEAR YIELD

| | | kT | MT | kT conv | |
|---|---|---|---|---|---|
| CB01 | 78 | | 0.2 | 200.00 | Estimated, with SS-N-17, 1985. |
| Rock | 78 | | | | MIRV warheads in kiloton range. |
| USD8 | 79 | | 1+ | 1,000 | |
| Clns | 80 | | 1 | 1,000 | |
| AWST | 82 | | 0.20 | 200.00 | [Final report] |
| CBO | 83 | 500 | | 500.00 | Each. |
| GSN | 83 | c. 200 | | 200.00 | Each multiple warhead. |
| vanD | 83 | | .20 | 200.00 | |

## THROWWEIGHT

| | | lb | t | kg | kg conv |
|---|---|---|---|---|---|
| CBO | 83 | 2,500 | | | 1,134 |

## WEIGHT

| | | lb | t | kg | kg conv | |
|---|---|---|---|---|---|---|
| JFS | 82 | 44,000 | | | 19,958 | Launch weight. |
| GSN | 83 | c. 44,500 | | c. 20,250 | 20,185 | |
| CFW+ | 84 | | | 20,250 | 20,250 | |

BALLISTIC MISSILES  SLBMs  SS-N-18  289

```
------------- LENGTH -------------
              in        ft        cm        m         m conv
FI    81                                    c. 14     14.00              [From graph], '81, '83, '84, '85.
JWS   81                                    c. 14.1   14.10
SMP   81                                    14        14.00
JFS   82                46.3                          14.11
FI+   83                                    14.1      14.10
GSN   83                c. 45               c. 13.6   13.72
CRW+  84                                    13.6      13.60

------------- DIAMETER -------------
              in        ft        cm        m         cm conv
FI    81                          ? 180                180.00
JWS   81                                    c. 1.8     180.00

------------- NUMBER OF STAGES -------------
Rock  78      2
FI    81      ? 2
SMP   81      2
AWST  82      2
CRW   82      2
JFS   82      2
GSN   83      2

------------- TYPE OF PROPULSION -------------
FIar  76      liquid
SecD  77      liquid                FY78-81.
Rock  78      storable liquid       Stages 1 and 2.
Clns  80      liquid
JWS   81      liquid
SMP   81      liquid
AWST  82      liquid storable
CRW   82      liquid
JFS   82      liquid
SSF   82      liquid
GSN   83      liquid
JFS+  83      solid

------------- MIDCOURSE (OR MAIN) GUIDANCE -------------
JCS   77                            Believed more sophisticated than SS-N-8.
JCS   78                            More sophisticated than SS-N-8.
USD6  78                            Believed more refined than SS-N-8, "capability for two celestial
                                    observations".
JFS   82      inertial
SSF   82      stellar inertial
GSN   83      inertial
CRW+  84      inertial

------------- DEVELOPMENT BEGAN -------------
SSF   82      1969                  Year design began.
SMP   85      1964                  Technological development began. [From graph.]

------------- PROTOTYPE TESTS -------------
SecC  75                            This year "rather substantial evidence" that USSR preparing to test new version of
                                    SS-N-8 or a follow on missile.
FIar  76                            First underwater launch conducted recently; trials expected to continue over the
                                    next 12 months.
SecC  76                            "In recent months, there have been test launches of a small and a large new SLBM
                                    which may be intended as the eventual replacements for the SS-N-6 and SS-N-8".
JCS   77      1975                  Testing began "later in 1975" than SS-NX-17.
JCS   78      Nov 1976              First firing from sea platform.
Rock  78      mid-1970s             Tests began, still underway in 1977.
SecD  78                            Soviets are testing it.
Clns  80      1977                  Sea trials began.
FI    81      spring 1977           First test-fired.
JWS   81                            Flight tests began 1975. First sub-launched test in Nov 1976, land-based tests
                                    earlier.
JFS   82      1976                  Sea trials Nov 1976.
GSN   83      1975                  Flight testing began.
SMP   85      1970-71               Engineering and testing began. [From graph.]

------------- PRODUCTION UNDERWAY -------------
SMP   81      late 1970s            "By the late 1970s, the Soviets were producing the MIRVed SS-N-18".
SMP   84                            "The Soviets continue to produce four SLBMs--the SS-N-6, SS-N-8, SS-N-17 and
                                    SS-N-18--at the rate of 200 a year. One new system, the SS-N-20/Typhoon, is in
                                    series production".
```

290  SOVIET MISSILES  IDDS

```
----------- IOC -----------
JCS    77                       [Blank in chart], FY78-79.
FI     81       1977-78         Became operational winter 1977-78.
GSN    83       1978
vanD   83       1978

----------- DEPLOYMENT BEGAN -----------
Rock   78                       May be entering front line service by 1980.
USD6   78       1977
SecD   79                       Soviets have begun deployment.
Clns   80       1977            First deployed.
JWS    81       1979
SMP    81                       "By the late 1970s", Soviets deploying SS-N-18 on modified Deltas.
US17   81       1979            Year of introduction.
IISS   82       1978            First year deployed.
JCS    83       c. 1978         Year of introduction [from graph].
ArkS   84       1976            "Because DoD revised its estimate of the initial deployment date for the Delta
                                III...from 1978 to 1976, full MIRVing [to Mod 3] is likewise now assumed to have
                                taken place earlier than previously estimated".
CFW+   84       1978
JCS    84       1978-79         Major strategic system deployments [graph], FY85-86.
SecD   84       1977            Introduction of strategic systems [graph].
SMP    85       1975-76         [From graph.]

----------- DEPLOYMENT HISTORY -----------
AWar   80       Deployment underway.
AWST   82       Follow-on to SS-N-8 SLBM.
SMP    83       As Delta IIIs deployed, Yankees and Hotels dismantled in compensation [since Sept-81 report].
SMP    84       SS-NX-23 likely to be deployed to replace SS-N-18.
SMP    85       SS-NX-23 likely to be deployed on Delta IIIs to replace SS-N-18, as well as on new Delta IVs.
                Will be deployed at least through 1990. [From graph.]

----------- NUMBERS IN SERVICE -----------

----------- NUMBER IN 1960 -----------
SSF    82       none            In 1960, 1965, 1970, 1975.

----------- NUMBER IN 1970 -----------
Clns   80       none            1970-1976.

----------- NUMBER IN 1977 -----------
Clns   80       80              1977, "Sea trials began in 1977, but this table carries those missiles as active on that
                                date".
Clns+  83       64

----------- NUMBER IN 1978 -----------
Clns   80       128

----------- NUMBER IN 1979 -----------
IISS   79       144             July, total deployed.
Clns   80       160
Clns+  83       144

----------- NUMBER IN 1980 -----------
IISS   80       160             July, total deployed.
SSF    82       176
Clns+  83       160

----------- NUMBER IN 1981 -----------
IISS   81       176             July, total deployed.
Clns+  83       208

----------- NUMBER IN 1982 -----------
IISS   82       208             July, total deployed.
Clns+  83       224

----------- NUMBER IN 1983 -----------
CBO    83       176
Clns+  83                       Warhead count assumes half have 3 warheads and half have 7 warheads.
IISS+  83       224             July, total deployed.

----------- NUMBER IN 1984 -----------
JCS    84       224             As of 1 Jan.
SMP    84       224

----------- NUMBER IN 1985 -----------
JCS    85       224             As of 1 Jan.
SMP    85       224
```

BALLISTIC MISSILES  SLBMs  SS-N-18  291

---------- NUMBER IN 1990 ----------
CBO    83              240 in 1990, 1996.

---------- CARRIERS ----------

---------- CARRIER ---------- TUBES ---- RELOADS ----------
A140   77    new SSBN           24              Probable carrier.
Rock   78    new sub            <_ 20           New post-Delta sub; possibly to take SS-NX-18.

---------- CARRIER ---------- TUBES ---- RELOADS ----------
Rock   78    Delta II           16              Probably designed for employment on Delta IIs.
FI     81                                       May be retrofitted to earlier Deltas [final report].

---------- CARRIER ---------- TUBES ---- RELOADS ----------
SecD   79    Delta III                          FY80-82.
AWar   80    Delta 3            16              N=10 now, 11 by end of year.
Clns   80    Delta III          16
FI     81    Delta III          16              N=11.
JCS    81    Delta III                          FY82-83, 86.
JWS    81    Delta III          16              Up to 12 subs by late 1981.
SMP    81    Delta III          16              '81, '83, '84, '85.
                                                tubes: '81, '83, '84, '85.
USND   81    Delta III
AWST   82    Delta 3
CPW    82    Delta III
JFS    82    Delta III
SSF    82    Delta III          16              11 in operation in 1980.
FI+    83    Delta III                          N=13.
GSN    83    Delta III
JWS+   83    Delta III                          14 estimated by 1983. 14 launched; production believed
                                                nearing an end.
SMP    83    Delta III          16              Each carries 16 MIRVed SLBMs. Production program nearing
                                                completion; 14 launched to date; a few more will probably
                                                be built. Some constructed [since Sept-81].
AWST+  84    Delta 3            16
BW     86    Delta III          16       no     N=14.

---------- DESIGNER ----------
SSF    82              Chelomei derivative, Design Bureau.

---------- DESIGN AND ENGINEERING ----------
FIar   76    Back-up to [SS-XN]-17.
JCS    84    SS-NX-23 is follow on to SS-N-18.

---------- USE AND CONFIGURATION ----------
CBO    83    System availability: 0.40.
vanD   83    Reliability: .75; Penetration 1.0.

---------- STRATEGY ----------
Rock   78    "Very long range and extremely high accuracy"; will be formidable component of Soviet strategic
             missile force.
SecD   78    Permits Soviets to cover US targets from Barents Sea and waters of the North Pacific (as does
             SS-N-8).
SecD   79    Permits Soviets to cover US targets from Barents Sea and waters of the western Pacific (as does
             SS-N-8).
SecD   80    Permits Soviets to cover targets in CONUS from Barents Sea and Sea of Okhotsk.
JCS    81    Can strike US from Soviet waters.
SMP    81    Can hit targets in the US from their home ports. Today Delta III subs "can cover most US targets
             from the relative security of their home waters".
JCS    82    Can strike US from adjacent waters or even home ports.

---------- RELATION TO OTHER MISSILES ----------
JCS    77    Similar to SS-N-8 in some respects. Larger than SS-N-8, FY78-79.
JWS    81    Some similarity to SSN-8. Greatest volume of Soviet SLBMs, comparable to US "Trident II (D-5)".

# SS-N-18 Mod 1

```
Sources are:   AWar  80      IISS  82      FI+   83      JFS+  83      ArkS  84      BW    86
JCS   78       FI    81      SSF   82      GSN   83      JWS   83      CFW+  84
Nitz  79       SMP   81      Clns+ 83      IISS+ 83      Lu01  83      SMP   85
```

## US DESIGNATION

| | | |
|---|---|---|
| Nitz | 79 | SS-N-18 |
| SMP | 81 | SS-N-18 Mod 1 |
| IISS | 82 | SS-N-18 Mod 1 |
| SSF | 82 | SS-N-18 Mod 1 |
| FI+ | 83 | SS-N-18 Mod 1 |
| GSN | 83 | SS-N-18 Mod 1 |
| CFW+ | 84 | SS-N-18 Mod 1 |

## EDITOR'S NOTES

AWar 80 — Source provides no designations for any Mods.

## RANGE

| | | mi | nm | km | km conv | |
|---|---|---|---|---|---|---|
| AWar | 80 | | | 16,600 | 16,600 | [Sic] |
| SMP | 81 | | | 6,500 | 6,500 | '81, '83, '84, '85. |
| IISS | 82 | | | 7,400 | 7,400 | Max. |
| SSF | 82 | | 3,500 | | 6,482 | Mods 1 and 3. |
| FI+ | 83 | | | 6,500 | 6,500 | |
| GSN | 83 | | 3,530 | 6,500 | 6,537 | Mods 1 and 3. |
| IISS+ | 83 | | | 6,500 | 6,500 | |
| JFS+ | 83 | | 3,500 | | 6,482 | Max. |
| JWS | 83 | | | 6,500 | 6,500 | |
| CFW+ | 84 | | 3,530 | | 6,537 | |
| BW | 86 | | | 6,500 | 6,500 | |

## CEP

| | | ft | mi | nm | m | km | m conv | |
|---|---|---|---|---|---|---|---|---|
| Nitz | 79 | | | .3 | | | 555.60 | |
| AWar | 80 | | | 0.76 | | | 1,407 | [Mods 1 and 2]. |
| IISS | 82 | | | | 1,400 | | 1,400 | |
| SSF | 82 | | | 0.76 | | | 1,407 | Mods 1 and 2. |
| Clns+ | 83 | | | 0.5 | | | 926.00 | |
| Lu01 | 83 | | | | 1350 | | 1,350 | |
| BW | 86 | | | | 1,400 | | 1,400 | |

## WARHEAD CHARACTERISTICS

| | | | |
|---|---|---|---|
| JCS | 78 | 3 MIRVs | [Also] "estimated" it can deliver 3 MIRVs. |
| Nitz | 79 | 3 RVs | |
| AWar | 80 | 3 MIRVs | |
| FI | 81 | 3 MIRVs | |
| SMP | 81 | 3 MIRVs | '81, '83, '84, '85. |
| IISS | 82 | 3 MIRVs | |
| SSF | 82 | 3 MIRVs | |
| GSN | 83 | 3 MIRVs | One mod of SS-N-18 is first Soviet SLBM with MIRV warhead; initially 2 RVs. |
| JWS | 83 | 3 MIRVs | |
| CFW+ | 84 | 3 RVs | |
| BW | 86 | 3 MIRVs | |

## NUCLEAR YIELD

| | | kT | MT | kT conv | |
|---|---|---|---|---|---|
| Nitz | 79 | | .5 | 500.00 | |
| AWar | 80 | | 0.20 | 200.00 | |
| FI | 81 | 200 | | 200.00 | Per warhead. |
| IISS | 82 | | | | "KT", per warhead. |
| SSF | 82 | | 0.2 | 200.00 | Per warhead. |
| Clns+ | 83 | 500 | | 500.00 | |
| IISS+ | 83 | ? 200 | | 200.00 | |
| CFW+ | 84 | 200 | | 200.00 | |
| BW | 86 | 200 | | 200.00 | |

## THROWEIGHT

| | | lb | t | kg | kg conv |
|---|---|---|---|---|---|
| Nitz | 79 | 2,500 | | | 1,134 |
| IISS | 82 | 5,000 | | | 2,268 |

## TYPE OF PROPULSION

| | | | |
|---|---|---|---|
| IISS | 82 | solid | [Sic] |
| BW | 86 | storable liquid | |

```
----------  PROTOTYPE TESTS ----------
SSF   82   1976          First flight test.
           ---------- DEPLOYMENT BEGAN ----------
SSF   82   1978          Year operation began.
SMP   85   1975-76       [Assumed Mod 1, from graph.]
BW    86   1978
           ----------  DEPLOYED  ----------
ArkS  84                 While SMP 84 assumes full MIRVing for SLBMs (all SS-N-18s assumed to be Mod 3 with
                         7 MIRVs, "DoD admits that all three versions of the SS-N-18 are fielded".
BW    86                 In service.
           ----------  DEPLOYMENT HISTORY  ----------
IISS  82   "Solid-fuel SS-N-8 successor".
           ----------  NUMBERS IN SERVICE  ----------
           ----------  NUMBER IN 1977  ----------
Nitz  79   16            Launchers.
           ----------  NUMBER IN 1979  ----------
Nitz  79   144           18 June, launchers.
```

# SS-N-18 Mod 2

```
Sources are:   SMP   81      Clns  83      IISS+ 83      LuO1  83      CFW+  84
AWar  80       IISS  82      FI+   83      JFS+  83      SMP   83      BW    86
FI    81       SSF   82      GSN   83      JWS   83      ArkS  84
```

---------- **US DESIGNATION** ----------

| | | |
|---|---|---|
| SMP | 81 | SS-N-18 Mod 2 |
| IISS | 82 | SS-N-18 Mod 2 |
| SSF | 82 | SS-N-18 Mod 2 |
| Clns | 83 | SS-N-18 Mod 2 |
| FI+ | 83 | SS-N-18 Mod 2 |
| GSN | 83 | SS-N-18 Mod 2 |
| CFW+ | 84 | SS-N-18 Mod 2 |

---------- **EDITOR'S NOTES** ----------

| | | |
|---|---|---|
| AWar | 80 | Source provides no designations for any Mods. |

---------- **RANGE** ----------

| | | mi | nm | km | km conv | |
|---|---|---|---|---|---|---|
| AWar | 80 | | | 8,000 | 8,000 | |
| FI | 81 | | | 16,000 | 16,000 | Report of range for single-warhead version. '81, '83, '84, '85. |
| SMP | 81 | | | 8,000 | 8,000 | |
| IISS | 82 | | | 8,300 | 8,300 | Max. |
| SSF | 82 | | 4,300 | | 7,963 | |
| Clns | 83 | | 5,000 | | 9,260 | |
| FI+ | 83 | | | 8,000 | 8,000 | |
| GSN | 83 | | 4,350 | 8,000 | 8,056 | |
| IISS+ | 83 | | | 8,000 | 8,000 | |
| JFS+ | 83 | | 4,300 | | 7,963 | Max. |
| JWS | 83 | | | 8,000 | 8,000 | |
| CFW+ | 84 | | 4,350 | | 8,056 | |
| BW | 86 | | | 8,000 | 8,000 | |

---------- **CEP** ----------

| | | ft | mi | nm | m | km | m conv | |
|---|---|---|---|---|---|---|---|---|
| AWar | 80 | | | 0.76 | | | 1,407 | [Mods 1 and 2]. |
| IISS | 82 | | | | 600 | | 600.00 | Mods 2 and 3. |
| SSF | 82 | | | 0.76 | | | 1,407 | Mods 1 and 2. |
| Clns | 83 | | | 0.5 | | | 926.00 | |
| LuO1 | 83 | | | | 1350 | | 1,350 | |
| BW | 86 | | | | 1,400 | | 1,400 | |

---------- **WARHEAD CHARACTERISTICS** ----------

| | | | |
|---|---|---|---|
| AWar | 80 | single RV | |
| FI | 81 | single RV | |
| SMP | 81 | single RV | '81, '83, '84, '85. |
| IISS | 82 | single RV | |
| SSF | 82 | single RV | |
| Clns | 83 | single RV | |
| GSN | 83 | single RV | |
| JWS | 83 | single RV | |
| CFW+ | 84 | single RV | |
| BW | 86 | single RV | |

---------- **NUCLEAR YIELD** ----------

| | | kT | MT | kT conv | |
|---|---|---|---|---|---|
| AWar | 80 | | 0.45 | 450.00 | |
| FI | 81 | 450 | | 450.00 | |
| IISS | 82 | 450 | | 450.00 | |
| SSF | 82 | | 0.45 | 450.00 | |
| Clns | 83 | 500–1,000 | | 500.00–1,000 | [Literally 500 kT–1 MT, ed] |
| GSN | 83 | c. 450 | | 450.00 | |
| CFW+ | 84 | 450 | | 450.00 | |
| BW | 86 | 450 | | 450.00 | |

---------- **TYPE OF PROPULSION** ----------

| | | |
|---|---|---|
| FI | 81 | liquid |
| Clns | 83 | liquid |
| BW | 86 | storable liquid |

```
------- DEPLOYMENT BEGAN -------
Clns  83   1977         First deployed.
IISS+ 83   1978         First year deployed.
SMP   83                Each Delta III carries 16 MIRVed SLBMs [might imply single RV version is not
                        deployed, ed], '83, '84.
BW    86   1978

------- DEPLOYED -------
ArkS  84                While SMP 84 assumes full MIRVing for SLBMs (all SS-N-18s assumed to be Mod 3 with
                        7 MIRVs, "DoD admits that all three versions of the SS-N-18 are fielded".
BW    86                In service.
```

# SS-N-18 Mod 3

```
Sources are:  FI    81      IISS  82      Clns  83      GSN   83      JWS   83      CPW+  84      BW    86
              Nitz  79      SMP   81      SSF   82      FI+   83      JFS+  83      ArkS  84      SMP   85
```

## US DESIGNATION

|  | | |
|---|---|---|
| Nitz 79 | SS-N-18 | |
| SMP 81 | SS-N-18 Mod 3 | |
| IISS 82 | SS-N-18 Mod 3 | |
| SSF 82 | SS-N-18 Mod 3 | |
| Clns 83 | SS-N-18 Mod 3 | |
| FI+ 83 | SS-N-18 Mod 3 | |
| GSN 83 | SS-N-18 Mod 3 | |
| CPW+ 84 | SS-N-18 Mod 3 | |

## RANGE

| | mi | nm | km | km conv | |
|---|---|---|---|---|---|
| SMP 81 | | | 6,500 | 6,500 | '81, '83, '84, '85. |
| IISS 82 | | | 6,500 | 6,500 | Max. |
| SSF 82 | | 3,500 | | 6,482 | Mods 1 and 3. |
| Clns 83 | | 4,500 | | 8,334 | |
| FI+ 83 | | | 6,500 | 6,500 | |
| GSN 83 | | 3,530 | 6,500 | 6,537 | Mods 1 and 3. |
| JFS+ 83 | | 3,500 | | 6,482 | Max. |
| JWS 83 | | | 6,500 | 6,500 | Max. |
| CPW+ 84 | | 3,530 | | 6,537 | |
| BW 86 | | | 6,500 | 6,500 | |

## CEP

| | ft | mi | nm | m | km | m conv | |
|---|---|---|---|---|---|---|---|
| Nitz 79 | | | .3 | | | 555.60 | |
| IISS 82 | | | | 600 | | 600.00 | Mods 2 and 3. |
| Clns 83 | | | 0.5 | | | 926.00 | |
| BW 86 | | | | ? 1,400 | | 1,400 | |

## WARHEAD CHARACTERISTICS

|  | | |
|---|---|---|
| Nitz 79 | 7 RVs | |
| FI 81 | 7 MIRVs | |
| SMP 81 | 7 MIRVs | '81, '83, '84, '85. |
| IISS 82 | 7 MIRVs | |
| SSF 82 | 7 MIRVs | |
| Clns 83 | 7 MIRVs | |
| GSN 83 | 7 MIRVs | |
| JWS 83 | 7 MIRVs | |
| CPW+ 84 | 7 MIRVs | |
| BW 86 | 7 MIRVs | |

## NUCLEAR YIELD

| | kT | MT | kT conv | |
|---|---|---|---|---|
| Nitz 79 | | .2 | 200.00 | |
| IISS 82 | 200 | | 200.00 | Per warhead. |
| Clns 83 | 500 | | 500.00 | |
| BW 86 | ? 100 | | 100.00 | |

## THROWWEIGHT

| | lb | t | kg | kg conv |
|---|---|---|---|---|
| Nitz 79 | 2,500 | | | 1,134 |

## TYPE OF PROPULSION

|  | |
|---|---|
| Clns 83 | liquid |
| BW 86 | storable liquid |

## DEPLOYMENT BEGAN

|  | | |
|---|---|---|
| Clns 83 | 1978 | First deployed. |
| BW 86 | ? 1978 | |

## DEPLOYED

|  | |
|---|---|
| ArkS 84 | While SMP 84 assumes full MIRVing for SLBMs (all SS-N-18s assumed to be Mod 3 with 7 MIRVs, "DoD admits that all three versions of the SS-N-18 are fielded". |
| BW 86 | In service. |

## DEPLOYMENT HISTORY

|  | |
|---|---|
| SMP 85 | Will be deployed at least through 1990. [Assumed Mod 3, from graph.] |

# SS-N-20

| Sources are: | FI | 81 | US17 | 81 | JCS | 82 | FI+ | 83 | JFS+ | 83 | AWST+ | 84 | SMP | 84 |
|---|---|---|---|---|---|---|---|---|---|---|---|---|---|---|
| Nitz | 79 | JWS | 81 | AWST | 82 | JFS | 82 | GSN | 83 | JWS+ | 83 | CFW+ | 84 | JCS | 85 |
| AWar | 80 | SecD | 81 | CFW | 82 | SSF | 82 | IISS+ | 83 | SMP | 83 | JCS | 84 | SMP | 85 |
| JCS | 80 | SMP | 81 | IISS | 82 | CBO | 83 | JCS | 83 | ArkS | 84 | SecD | 84 | BW | 86 |

## ———— US DESIGNATION ————

| | | |
|---|---|---|
| Nitz | 79 | New MIRV |
| AWar | 80 | SS-NX-20 |
| FI | 81 | SS-NX-20 |
| JWS | 81 | SSN-X-20 |
| SecD | 81 | SS-NX-20 |
| SMP | 81 | SS-NX-20    '81, '83. |
| AWST | 82 | SS-NX-20 |
| CFW | 82 | SS-NX-20 |
| IISS | 82 | SS-NX-20 |
| JCS | 82 | SS-NX-20    FY83-85. |
| JFS | 82 | SS-N-20 |
| SSF | 82 | SS-NX-20 |
| CBO | 83 | SS-NX-20 |
| GSN | 83 | SS-N-20 |
| CFW+ | 84 | SS-N-20 |
| SecD | 84 | SSN20 |
| SMP | 84 | SS-N-20 |
| JCS | 85 | SS-N-20 |

## ———— DESCRIPTION ————

| | | |
|---|---|---|
| JWS | 81 | SLBM |
| SSF | 82 | Sixth generation. |
| JCS | 83 | SLBM |

## ———— EDITOR'S NOTES ————

| | | |
|---|---|---|
| JCS | 80 | Listed FY81,83-86, referred to not by name in FY81. |
| SecD | 81 | Listed FY82,85. |
| SMP | 81 | Listed '81, '83, '84, '85. |

## ———— RANGE ————

| | | mi | nm | km | km conv | |
|---|---|---|---|---|---|---|
| JWS | 81 | | | < 8,000 | 8,000 | "Almost" 8,000 km reputed. |
| SMP | 81 | | | 8,300 | 8,300 | '81, '83, '84, '85. |
| US17 | 81 | | 5,000 | | 9,260 | |
| AWST | 82 | | 4,000+ | | 7,408 | |
| CFW | 82 | | > 4,000 | | 7,408 | |
| IISS | 82 | | | 8,300 | 8,300 | Max. |
| JFS | 82 | | 4,200 | | 7,778 | Max. |
| SSF | 82 | | 4,500 | | 8,334 | |
| FI+ | 83 | | | <_ 8,000 | 8,000 | |
| GSN | 83 | | 4,500 | 8,300 | 8,334 | |
| JFS+ | 83 | | 5,000 | | 9,260 | Max. |
| JWS+ | 83 | | | c. 8,000 | 8,000 | Reputed range. |
| CFW+ | 84 | | 4,300 | | 7,963 | |
| JCS | 84 | | | 8,300 | 8,300 | Booster range. |
| JCS | 85 | | c. 5,000 | | 9,260 | |
| BW | 86 | | | 8,300 | 8,300 | |

## ———— CEP ————

| | | ft | mi | nm | m | km | m conv | |
|---|---|---|---|---|---|---|---|---|
| Nitz | 79 | | | .25 | | | 463.00 | |
| CFW | 82 | | | | [< 800] | | 800.00 | Presumably higher accuracy than SS-N-18, [final report] In 1983; lack of data on future trends. Presumed more accurate than any previous SLBM. |
| CBO | 83 | | | 0.3 | | | 555.60 | |
| GSN | 83 | | | | | | | |
| CFW+ | 84 | | | | 600 | | 600.00 | |
| BW | 86 | | | | ? 1,000 | | 1,000 | |

## ———— WARHEAD TYPE ————

| | | nuclear | conven | chem |
|---|---|---|---|---|
| JFS | 82 | yes | | |
| GSN | 83 | yes | | |
| JWS+ | 83 | yes | | |
| SMP | 83 | yes | | |

## WARHEAD CHARACTERISTICS

| Source | Year | Value | Notes |
|---|---|---|---|
| Nitz | 79 | 14 RVs | |
| JCS | 80 | ? MIRV | Expected to be MIRV-capable. |
| FI | 81 | MIRV | Probably. |
| JWS | 81 | MIRV | Expected. |
| SecD | 81 | MIRV | Almost certainly will be MIRVed. |
| SMP | 81 | 12 RVs | |
| CFW | 82 | MIRV | |
| IISS | 82 | 12 MIRVs | |
| JFS | 82 | MIRV | |
| SSF | 82 | 12 RVs | |
| CBO | 83 | 9 RVs | |
| FI+ | 83 | $\leq$ 12 MIRVs | |
| GSN | 83 | 6-9 MIRVs | |
| IISS+ | 83 | 6-9 MIRVs | |
| JFS+ | 83 | 12 MIRVs | |
| JWS+ | 83 | $\leq$ 12 MIRVs | Conjectured. |
| SMP | 83 | 6-9 MIRVs | '83, '84, '85. [Also:] each Typhoon adds "some 200" RVs to the force. |
| ArkS | 84 | | While SMP 84 assumes 9 MIRVs, "DoD admits that...the SS-N-20 has from six to nine MIRVs". |
| AWST+ | 84 | 6-9 MIRVs | |
| CFW+ | 84 | 6-9 MIRVs | |
| JCS | 84 | 6-9 warheads | |
| JCS | 85 | 6-9 warheads | |
| BW | 86 | 6-8 MIRVs | |

## NUCLEAR YIELD

| Source | Year | kT | MT | kT conv | Notes |
|---|---|---|---|---|---|
| Nitz | 79 | | .2 | 200.00 | |
| CBO | 83 | 100 | | 100.00 | Each. |
| BW | 86 | ? 100 | | 100.00 | |

## THROWWEIGHT

| Source | Year | lb | t | kg | kg conv | Notes |
|---|---|---|---|---|---|---|
| Nitz | 79 | 5,000 | | | 2,268 | |
| FI | 81 | | | | | Greater than SS-N-18. |
| CBO | 83 | 7,500 | | | 3,402 | |

## PAYLOAD

| Source | Year | lb | t | kg | kg conv | Notes |
|---|---|---|---|---|---|---|
| CFW | 82 | | | | | Presumably greater than SS-N-18, [final report] |
| GSN | 83 | | | | | Presumed greater than any previous SLBM. |

## LENGTH

| Source | Year | in | ft | cm | m | m conv | Notes |
|---|---|---|---|---|---|---|---|
| SMP | 81 | | | | 15 | 15.00 | [From graph], '81, '83, '84, '85. |

## NUMBER OF STAGES

| Source | Year | Value |
|---|---|---|
| CFW | 82 | 3 |
| GSN | 83 | 3 |
| JFS+ | 83 | 3 |

## TYPE OF PROPULSION

| Source | Year | Value | Notes |
|---|---|---|---|
| JCS | 80 | ? solid | Expected. |
| FI | 81 | solid | |
| JWS | 81 | solid | |
| CFW | 82 | [liquid] | [All sub-launched ballistic missiles said to be liquid except SS-N-17, ed]. |
| IISS | 82 | solid | |
| SSF | 82 | solid | |
| GSN | 83 | solid | First solid-fuel SLBM to be produced in quantity. |
| JFS+ | 83 | solid | |
| SMP | 83 | solid | '84, '85. |
| BW | 86 | solid | |

## MIDCOURSE (OR MAIN) GUIDANCE

| Source | Year | Value | Notes |
|---|---|---|---|
| GSN | 83 | inertial | |
| AWST+ | 84 | | "Uses steller and inertial navigation for terminal and mid-course updates". |
| CFW+ | 84 | inertial | |

## FIRST OBSERVED

| Source | Year | Value | Notes |
|---|---|---|---|
| US17 | 81 | 1980 | First reported [designation "SS-N-?", ed]. |
| JFS | 82 | 1980 | First reported. |

## DEVELOPMENT BEGAN

| Source | Year | Value | Notes |
|---|---|---|---|
| SSF | 82 | 1973 | Year design began. |
| SMP | 85 | 1972 | Technological development began. [From graph.] |

BALLISTIC MISSILES  SLBMs  SS-N-20  299

```
------------ DEVELOPMENT UNDERWAY ------------
IISS   82                    Under development.

------------ PROTOTYPE TESTS ------------
AWar   80                    First two launches failed this year.
FI     81    Jan 1980        First flight observed. This and next 3 flights reported to be failures.
JWS    81    Jan 1980        All 4 flight tests were failures.
             1981            2 test flights reported to be successful.
SMP    81    1979            First tested.
JCS    82                    Has been test fired.
JFS    82                    Trials not successful in 1980-81.
SSF    82    1980            First flight test.
FI+    83    1981            First successful flights.
GSN    83    Dec 1981        Successful test series, reported in Western press.
JWS+   83    Oct 1982        Simultaneous launch of 4 SS-NX-20s from dived Typhoon.
SMP    83                    Testing continued [since Sept-81 edition]. Typhoon has test fired its new missiles
                             [since late 1981].
SMP    84    1983            Test firings from Typhoon.
                             "Based on past Soviet practice, before the end of the 1980s, they may initiate
                             testing of modified versions of the SS-NX-23 and the SS-N-20"; these "are likely to
                             be more accurate than their predecessors".
SMP    85    1983            Test firings from Typhoon.
                             "The Soviets will probably begin flight testing a modified version of the SS-N-20."
             1978-79         Engineering and testing began. [From graph.]

------------ PRODUCTION UNDERWAY ------------
SMP    84                    "The Soviets continue to produce four SLBMs—the SS-N-6, SS-N-8, SS-N-17 and
                             SS-N-18—at the rate of 200 a year. One new system, the SS-N-20/Typhoon, is in
                             series production".

------------ IOC ------------
GSN    83    1982-1983
JWS+   83    1983            Anticipated, official US sources.

------------ DEPLOYMENT BEGAN ------------
FI     81                    Not expected to be operational until mid-1980s.
JWS    81                    US DoD says unlikely to enter service before mid-1980s.
SecD   81                    Will not be deployed before mid-1980s at the earliest.
SMP    81                    Will probably reach operational status by the mid-1980s; will be operational in the
                             mid 1980s.
IISS   82    ? 1981          First year deployed.
JCS    82                    Complete [SS-NX-20/Typhoon] weapon system probably not operational until mid-1980s.
IISS+  83                    ? Now operational.
JCS    83    c. 1984-85      Year of introduction [from graph].
SMP    83    end 1983        First Typhoon "will be fully operational by the end of 1983".
CRW+   84    1983
JCS    84    1982-83         Major strategic system deployments [graph], FY85-86.
SecD   84    1983            Introduction of strategic systems [graph].
SMP    84                    Now fully operational.
SMP    85    1983-84         [From graph.]
BW     86    1983

------------ DEPLOYED ------------
BW     86                    In service.

------------ DEPLOYMENT HISTORY ------------
SMP    85                    Will be deployed at least through 1990. [From graph.]

------------ NUMBERS PRODUCED ------------
GSN    83                    First solid-fuel SLBM to be produced in quantity.

------------ NUMBERS IN SERVICE ------------

------------ NUMBER IN 1982 ------------
IISS   82    20              1982-83, July, total deployed.

------------ NUMBER IN 1983 ------------
CBO    83    20

------------ NUMBER IN 1984 ------------
JCS    84    40              As of 1 Jan; "includes SLBMs potentially carried on...Typhoon SSBNs on sea trial".
SMP    84    40

------------ NUMBER IN 1985 ------------
JCS    85    60              As of 1 Jan; "includes SLBMs potentially carried on...Typhoon SSBNs on sea trial".
SMP    85    60
```

SOVIET MISSILES IDDS

---------- **NUMBER IN 1990** ----------
CBO  83              200 in 1990.

---------- **NUMBER IN 1996** ----------
CBO  83              440 in 1996.

---------- **CARRIERS** ----------

---------- **CARRIER** ---------- TUBES --- RELOADS ----------
JCS  80                                      New SSBN, expected to be larger than Delta III.
FI   81   Typhoon           20
JWS  81   Typhoon           20               Expected to arm Typhoon; one launched 1980, at least 3
                                             under construction 1981.
SecD 81   Typhoon
SMP  81   Typhoon           20               '81, '83, '84, '85. Launched in 1980.
                                             tubes: '81, '83, '84, '85.
AWST 82   Typhoon           20
CFW  82   Typhoon                            Expected.
JCS  82   Typhoon                            To be carried by Typhoon, FY83-84.
JFS  82   Typhoon
SSF  82   Typhoon           20               None in operation in 1980.
FI+  83   Sierra                             Often referred to as Typhoon.
GSN  83   Typhoon
JWS+ 83   Typhoon                            First launched Sept 1980; second Sept 1982.
SMP  83   Typhoon                            First has completed sea trials and now based on Kola
                                             Peninsula; second has been launched. SSBN force continues
                                             to be modernized with production "of additional units" of
                                             Typhoon class.
CFW+ 84   Typhoon
JCS  84   Typhoon           20               2 launched, 1 now operational.
                                             tubes: FY85-86.
SMP  84   Typhoon                            2 built; 1 is operational, the other soon will be. 3-4
                                             others probably now under construction, '84, '85. Could be
                                             up to 8 operational by early 1990s, '84, '85.
JCS  85   Typhoon                            3 launched, 2 now operational.
SMP  85   Typhoon                            3 built.
BW   86   Typhoon           20      no       N=3.

---------- **DESIGNER** ----------
SSF  82              Chelomei derivative, Design Bureau.

---------- **DESIGN AND ENGINEERING** ----------
CFW  82   Expected to enter full operational service aboard Typhoon, "but has reportedly experienced
          developmental difficulties".
GSN  83   Reported to have experienced "significant development problems", but Western press reports of
          succesful test series in December 1981.

---------- **USE AND CONFIGURATION** ----------
CBO  83   System availability: 0.40.
SMP  84   Typhoon can operate under Arctic ice pack, '84, '85.

---------- **STRATEGY** ----------
SMP  81   Assumed Typhoon's missile will be more capable than SS-N-18, "possibly having greater range,
          better accuracy, higher payload and more warheads".
JCS  83   Like Delta, Typhoon will provide ability to launch against targets in the US from "Soviet home
          waters".
JWS+ 83   Number of Typhoons built may depend in part on progress with SS-NX-21.
SMP  83   All of NATO Europe, North American and Asia in range of Typhoon even when operating in home
          waters.

---------- **GEOGRAPHICAL DEPLOYMENTS** ----------
JWS+ 83              First Typhoon has completed sea trials and is based on Kola Peninsula.

---------- **RELATION TO OTHER MISSILES** ----------
JCS  80   Expected to be larger than SS-N-18.
FI   81   Larger than SS-N-18.
GSN  83   Largest Soviet SLBM yet produced.

# SS-N-23

Sources are: JCS 84　　SMP 84　　JCS 85　　SMP 85　　BW 86

## ――――― US DESIGNATION ―――――

| | | | |
|---|---|---|---|
| JCS | 84 | SS-NX-23 | FY85-86. |
| SMP | 84 | SS-NX-23 | '84, '85. |
| JCS | 85 | SS-NX-23 | |

## ――――― DESCRIPTION ―――――

| | | |
|---|---|---|
| SMP | 84 | New SLBM, '84, '85. |

## ――――― EDITOR'S NOTES ―――――

| | | |
|---|---|---|
| SMP | 84 | Listed '84, '85. |

## ――――― RANGE ―――――

| | | mi | nm | km | km conv | |
|---|---|---|---|---|---|---|
| SMP | 84 | | | | | Long-range, '84, '85. |
| BW | 86 | | ? 9,300 | | 9,300 | |

## ――――― CEP ―――――

| | | ft | mi | nm | m | km | m conv | |
|---|---|---|---|---|---|---|---|---|
| SMP | 84 | | | | | | | More accurate than SS-N-18, '84, '85. |
| SMP | 85 | | | | | | | USSR's "most accurate" SLBM. |
| BW | 86 | | | ? 1,000 | | | 1,000 | |

## ――――― WARHEAD CHARACTERISTICS ―――――

| | | |
|---|---|---|
| SMP | 84 | More than SS-N-18, '84, '85. |
| SMP | 85 | Has MIRVs. |
| BW | 86 | MIRVs |

## ――――― NUCLEAR YIELD ―――――

| | | kT | MT | kT conv |
|---|---|---|---|---|
| BW | 86 | ? 100 | | 100.00 |

## ――――― THROWWEIGHT ―――――

| | | lb | t | kg | kg conv | |
|---|---|---|---|---|---|---|
| SMP | 84 | | | | | Greater than SS-N-18, '84, '85. |

## ――――― LENGTH ―――――

| | | in | ft | cm | m | m conv | |
|---|---|---|---|---|---|---|---|
| SMP | 84 | | | | 14 | 14.00 | [From graph], '84, '85. |

## ――――― TYPE OF PROPULSION ―――――

| | | | |
|---|---|---|---|
| SMP | 84 | liquid | '84, '85. |
| BW | 86 | storable liquid | |

## ――――― DEVELOPMENT BEGAN ―――――

| | | | |
|---|---|---|---|
| SMP | 85 | 1976 | Technological development began. [From graph.] |

## ――――― PROTOTYPE TESTS ―――――

| | | | |
|---|---|---|---|
| SMP | 84 | | Being tested; flight tests have begun. "Based on past Soviet practice, before the end of the 1980s, they may initiate testing of modified versions of the SS-NX-23 and the SS-N-20"; these "are likely to be more accurate than their predecessors". |
| JCS | 85 | | Now being tested. |
| SMP | 85 | | Now being flight-tested. "Based on past Soviet practice, they [the USSR] may initiate testing of a modified version of the SS-NX-23 before the end of the 1980s." |
| | | 1983-84 | Engineering and testing began. [From graph.] |

## ――――― IOC ―――――

| | | | |
|---|---|---|---|
| JCS | 85 | ? 1986 | On Delta IV. |

## ――――― DEPLOYMENT BEGAN ―――――

| | | | |
|---|---|---|---|
| SMP | 85 | 1987-88 | [Possible deployment, from graph.] |
| BW | 86 | 1987 | |

## ――――― DEPLOYMENT HISTORY ―――――

| | | |
|---|---|---|
| SMP | 84 | "Likely to be deployed as a replacement for the SS-N-18" carried by Delta III, '84, '85. |
| SMP | 85 | Will be deployed at least through 1990. [From graph.] |

## ――――― NUMBERS IN SERVICE ―――――

## ――――― NUMBER IN 1985 ―――――

| | | |
|---|---|---|
| SMP | 85 | 16 |

------------- **CARRIERS** -------------

------------- **CARRIER** ------------- TUBES ---- RELOADS -----------

| | | | | | |
|---|---|---|---|---|---|
| JCS | 85 | Delta IV | | | Soviets have launched first Delta-IV class, "which should become operational in 1986", intended to carry SS-N-23. |
| SMP | 85 | Delta IV | 16 | | "A new class of SSBN", "will be fitted with the SS-NX-23 SLBM." The "likely platform" for SS-N-23, 2 units have been launched. |
| BW | 86 | Delta IV | 16 | no | N=2. |

------------- **CARRIER** ------------- TUBES ---- RELOADS -----------

| | | | | | |
|---|---|---|---|---|---|
| BW | 86 | ? Delta III | 16 | no | |

------------- **DESIGN AND ENGINEERING** -------------

JCS 84  "SLBM projections include...development of the SS-NX-23, follow-on missile to the SS-N-18".

------------- **STRATEGY** -------------

SMP 85  "Most capable long-range" MIRVd SLBM.

# SS-1

Sources are:   AWST* 71     HRST 75     FIar 76     Rock 78

---------- **US DESIGNATION** ----------
| | |
|---|---|
| AWST* 71 | SS-1a |
| FIar 76 | SS-1 |
| Rock 78 | SS-1 |

---------- **NATO CODENAME** ----------
| | |
|---|---|
| AWST* 71 | Scunner |
| HRST 75 | Scunner |
| FIar 76 | Scunner |
| Rock 78 | Scunner |

---------- **DESCRIPTION** ----------
| | |
|---|---|
| AWST* 71 | [Listed with strategic missiles] |
| Rock 78 | strategic |

---------- **RANGE** ----------
| | mi | nm | km | km conv |
|---|---|---|---|---|
| AWST* 71 | 150 | | | 241.35 |
| HRST 75 | 150 | | | 241.35 |
| Rock 78 | | | 350 | 350.00 |

---------- **WARHEAD TYPE** ----------
| | nuclear | conven | chem | |
|---|---|---|---|---|
| Rock 78 | | yes | | HE. |

---------- **WARHEAD CHARACTERISTICS** ----------
| | |
|---|---|
| Rock 78 | single |

---------- **WEIGHT** ----------
| | lb | t | kg | kg conv | |
|---|---|---|---|---|---|
| Rock 78 | | 13 | | 13,000 | Tonnes. |

---------- **LENGTH** ----------
| | in | ft | cm | m | m conv |
|---|---|---|---|---|---|
| HRST 75 | | 46 | | | 14.02 |
| Rock 78 | | | | 14.5 | 14.50 |

---------- **DIAMETER** ----------
| | in | ft | cm | m | cm conv |
|---|---|---|---|---|---|
| HRST 75 | 65 | | | | 165.10 |
| Rock 78 | | | | 1.6 | 160.00 |

---------- **NUMBER OF STAGES** ----------
| | |
|---|---|
| AWST* 71 | 1 |

---------- **TYPE OF PROPULSION** ----------
| | | |
|---|---|---|
| AWST* 71 | liquid | |
| HRST 75 | liquid | |
| Rock 78 | liquid | [Unstorable]. Nose payload could be "very small solid propellant boost motor". Thrust, tonnes: 28. |

---------- **DEPLOYMENT BEGAN** ----------
| | | |
|---|---|---|
| FIar 76 | 1947 | First long-range rocket weapon to enter service with Russian Army. Small number deployed in 1947. |
| Rock 78 | 1947 | Service intro. |

---------- **RETIRED** ----------
| | |
|---|---|
| Rock 78 | 1955 |

---------- **CARRIERS** ----------

---------- **CARRIER** ---------- **TUBES** ---- **RELOADS** ----------
| | | | |
|---|---|---|---|
| FIar 76 | | | Mobile weapon. |

---------- **DESIGN AND ENGINEERING** ----------
| | |
|---|---|
| AWST* 71 | "Soviet built German V-2". |
| HRST 75 | "Essentially a V-2". |
| FIar 76 | Believed to have been copy of German A-4 (V-2) rocket. |
| Rock 78 | Generally regarded as development of German A-4 (V-2); approximately same dimensions, but with additional payload in nose. "Matured from technical dissection of the German V-2". |

---------- **USE AND CONFIGURATION** ----------
| | |
|---|---|
| FIar 76 | Mobile weapon launched from a simple firing table [from SS-3 entry, ed]. |

---------- **STRATEGY** ----------

FIar 76  "The small number deployed in 1947, two years before the explosion of Russia's first nuclear bomb, were probably intended more for training purposes than for a serious operational role".

Rock 78  Extra nose payload [over V-2] could have been extra solid booster or "instrumental package without propulsion, for the purpose of obtaining atmospheric data". Since 1947 appearance 2 years before nuclear charge available, more likely SS-1 was "a precursor tactical ballistic surface-to-surface weapon, if indeed it did play a military role". More likely it was "a high-altitude sounding rocket used for gathering precise air temperature and pressure data as a preliminary step to acquiring the information for later technical developments".

---------- **OTHER INFORMATION** ----------

Rock 78  "Considerable speculation surrounds the very existence of this rocket".

# SS-2

Sources are:   AWST* 71      GA01 75      HRST 75      FIar 76      JCS 76      Rock 78

------------ **US DESIGNATION** ------------
|  | |
|---|---|
| AWST* 71 | SS-2 |
| GA01 75 | SS-2 |
| FIar 76 | SS-2 |
| JCS 76 | SS-2 |
| Rock 78 | SS-2 |

------------ **NATO CODENAME** ------------
|  | |
|---|---|
| AWST* 71 | Sibling |
| GA01 75 | Sibling |
| HRST 75 | Sibling |
| FIar 76 | Sibling |
| Rock 78 | Sibling |

------------ **DESCRIPTION** ------------
|  | |
|---|---|
| AWST* 71 | [Listed with Strategic missiles, ed] |
| Rock 78 | strategic |

------------ **EDITOR'S NOTES** ------------
|  | |
|---|---|
| JCS 76 | Listed FY77. |

------------ **RANGE** ------------
| Source | mi | nm | km | km conv | |
|---|---|---|---|---|---|
| AWST* 71 | 300 | | | 482.70 | |
| HRST 75 | 200-300 | | | 321.80-482.70 | |
| FIar 76 | | | | | SS-2 embodied Russian modifications to increase range and reliability [over SS-1]. |
| Rock 78 | | | ?450 | 450.00 | |

------------ **WARHEAD TYPE** ------------
| Source | nuclear | conven | chem | |
|---|---|---|---|---|
| Rock 78 | | yes | | HE. |

------------ **WARHEAD CHARACTERISTICS** ------------
|  | |
|---|---|
| Rock 78 | single RV |

------------ **LENGTH** ------------
| Source | in | ft | cm | m | m conv | |
|---|---|---|---|---|---|---|
| GA01 75 | | 53 | | | 16.15 | Version A. |
| | | 58.5 | | | 17.83 | Version A-2. |

------------ **TYPE OF PROPULSION** ------------
|  | | |
|---|---|---|
| HRST 75 | liquid | |
| Rock 78 | liquid | [Unstorable] |

------------ **OBSERVED** ------------
|  | | |
|---|---|---|
| FIar 76 | late 1950s | "Photographs released in the late 1950s showing the launch of 'geophysical rockets' included pictures of a V-2-type rocket with extended tanks; this may have been Sibling". |
| Rock 78 | late 1950s | "A somewhat obscure development, the Sibling was evident in Soviet photographs released in the late 1950's". |

------------ **DEPLOYMENT BEGAN** ------------
|  | | |
|---|---|---|
| Rock 78 | 1948 | Service intro. |

------------ **DEPLOYMENT HISTORY** ------------
|  | |
|---|---|
| Rock 78 | Used by military forces. |

------------ **UNDATED NUMBER** ------------
|  | |
|---|---|
| FIar 76 | Saw only limited service. |
| Rock 78 | "Only a few were developed". |

------------ **CARRIERS** ------------

------------ **CARRIER** ------------ TUBES --- RELOADS ------------
| Source | Carrier | | |
|---|---|---|---|
| FIar 76 | mobile | | |

---------- **DESIGN AND ENGINEERING** ----------
AWST* 71   "Development of V-2".
HRST  75   Derived from V-2.
FIar  76   Based on SS-1. SS-2 embodied Russian modifications to increase range and reliability [over SS-1].
Rock  78   Took form of modified SS-1. Longer propellant tanks than V-2; may have been V-2 successor; similar design and construction. "Matured from technical dissection of the German V-2".

---------- **USE AND CONFIGURATION** ----------
FIar  76   Mobile weapon launched from a simple firing table [from SS-3 entry, ed].

---------- **STRATEGY** ----------
GAO1  75   [Notes:] "Missile development; geophysical, IGY".
Rock  78   Had recoverable pods for safe return of atmospheric data. Probably responsible for gathering upper atmosphere data, more and higher than SS-1. When Soviets displayed photos, they called it 'geophysical rocket'.

---------- **RELATION TO OTHER MISSILES** ----------
JCS   76   PRC MRBM force is a 1-stage, liquid, transportable system developed from the Soviet SS-2.

---------- **OTHER INFORMATION** ----------
GAO1  75   2 versions, A, A-2.

# SS-3

```
Sources are:  Zaeh  61      AWST* 71    FIar  76    Rock  78    SSF   82    SMP   85
COTS  58      IA01  65      HRST  75    MOW   76    USO9  79    Su01  83    BW    86
IMSG  60      Ley   68      JWS*  75    SWM   77    SMP   81    Meyr  84
```

### US DESIGNATION

|  | |  |
|---|---|---|
| AWST* 71 | SS-3 | |
| JWS* 75 | SS-3 | |
| FIar 76 | SS-3 | |
| MOW 76 | SS-3 | |
| SWM 77 | SS-3 | |
| Rock 78 | SS-3 | |
| SMP 81 | SS-3 | '81, '85. |
| SSF 82 | SS-3 | |
| Meyr 84 | SS-3 | |

### NATO CODENAME

|  | |  |
|---|---|---|
| Ley 68 | Shyster | [Shyster entry] |
| AWST* 71 | Shyster | |
| HRST 75 | Shyster | |
| JWS* 75 | Shyster | |
| FIar 76 | Shyster | |
| MOW 76 | Shyster | |
| SWM 77 | Shyster | |
| Rock 78 | Shyster | |

### OTHER DESIGNATION

|  | |  |
|---|---|---|
| COTS 58 | T-1 | |
| IMSG 60 | T-1 | |
| | M-101 | |
| Zaeh 61 | T-1 | |
| Ley 68 | T-1 | [T-1 entry] |
| | M-101 | [T-1 entry] |

### DESCRIPTION

|  | |  |
|---|---|---|
| COTS 58 | | [Listed under:] "Medium Range: 200 to 700 Miles". |
| AWST* 71 | | "First generation IRBM". [Listed with Strategic missiles] |
| JWS* 75 | | 1st-generation "Low medium-range ballistic missile". |
| Rock 78 | MRBM | Strategic ballistic missile. |
| SMP 81 | MRBM | |
| SMP 85 | LRIRBM | Longer Range Intermediate-Range Ballistic Missile. |

### EDITOR'S NOTES

|  |  |
|---|---|
| Ley 68 | The source distinguishes Shyster and T-1 as different missiles. Data is common to both unless indicated. |
| MOW 76 | Data are contained in the Sandal entry. |
| SMP 81 | Listed '81, '85. |
| Meyr 84 | Note to chart: "Values pertain to initial service period". |

### RANGE

| Source | mi | nm | km | km conv | Notes |
|---|---|---|---|---|---|
| COTS 58 | 400 | | | 643.60 | |
| IMSG 60 | 500 | | | 804.50 | |
| Zaeh 61 | 250 | | | 402.25 | |
| Ley 68 | 500-600 | | | 804.50-965.40 | [Shyster entry] |
| | 600-750 | | | 965.40-1,206 | [T-1 entry] |
| AWST* 71 | 750 | | | 1,206 | |
| HRST 75 | 700-800 | | | 1,126-1,287 | |
| JWS* 75 | | | 800-1,200 | 800.00-1,200 | Originally using "LOX/alcohol" fuel, later "LOX/kerosene". Range estimates vary widely, perhaps due to propellant change. |
| FIar 76 | | | 900 | 900.00 | "An earlier, shorter-range version using alcohol instead of kerosene has been reported. This would explain the large variation in the published range figures for this weapon". |
| | | | ? 700-900 | 700.00-900.00 | |
| Rock 78 | | | 900 | 900.00 | [Table] |
| | | | <_ 900 | 900.00 | [Text] |
| USO9 79 | | | 800-1,200 | 800.00-1,200 | |
| Meyr 84 | | | 1,200 | 1,200 | |
| BW 86 | | | ? 1,200 | 1,200 | |

308 SOVIET MISSILES IDDS

```
------------ SPEED ------------
              mph       kmph       Mach    kmph conv
IMSG   60   c. 5,000                        8,045

------------ CEP ------------
              ft        mi         nm       m          km         m conv
Meyr   84                                              2.5-5.0    2,500-5,000
BW     86                                 ? 3,700                 3,700

------------ WARHEAD TYPE ------------
              nuclear   conven     chem
Flar   76     yes       yes                HE.
Rock   78     yes       yes                HE.
US09   79     yes       yes                HE.
Meyr   84     yes

------------ WARHEAD CHARACTERISTICS ------------
Rock   78    single RV
Meyr   84    single RV
BW     86    single RV

------------ NUCLEAR YIELD ------------
              kT                   MT       kT conv
Meyr   84    100-500                        100.00-500.00
BW     86   ? 300                           300.00

------------ PAYLOAD ------------
              lb                   t        kg         kg conv
IMSG   60    2,650                                     1,202

------------ WEIGHT ------------
              lb                   t        kg         kg conv
IMSG   60    37,850                                    17,168
Zaeh   61    38,000                                    17,236
Ley    68                          26                  26,000     Tons, [Shyster entry]
                                   20                  20,000     Tons, [T-1 entry]
JWS*   75                                   c. 26,000  26,000
US09   79                                   c. 26,000  26,000

------------ LENGTH ------------
              in        ft         cm       m          m conv
IMSG   60               62                             18.90
Zaeh   61               50-52                          15.24-15.85
Ley    68               69                             21.03      [Shyster entry]
                        50                             15.24      [T-1 entry]
AWST*  71               70                             21.34
HRST   75               75                             22.86
JWS*   75                                   21         21.00
Flar   76                                   21         21.00
Rock   78                                   21         21.00
US09   79                                   21         21.00

------------ DIAMETER ------------
              in        ft         cm       m          cm conv
IMSG   60     66                                       167.64
Zaeh   61               5.15                           156.97
Ley    68               5 1/4                          160.02     [Shyster entry]
                        5-6                            152.40-182.88  [T-1 entry]
AWST*  71               5                              152.40
HRST   75     60                                       152.40
JWS*   75                                   1.6        160.00
Flar   76                                   1.7        170.00
Rock   78                                   1.7        170.00
US09   79                                   1.6        160.00

------------ WINGSPAN ------------
              in        ft         cm       m          cm conv
IMSG   60               12                             365.76

------------ NUMBER OF STAGES ------------
Zaeh   61    1
Ley    68    1         [T-1 entry]
AWST*  71    1
JWS*   75    1
Flar   76    1
Rock   78    1
```

## TYPE OF PROPULSION

| | | | |
|---|---|---|---|
| IMSG | 60 | liquid | Liquid oxygen and alcohol. Thrust: 77,000 lb. |
| Zaeh | 61 | | Thrust: 78,000 lb. |
| Ley | 68 | liquid | [T-1 entry] |
| AWST* | 71 | liquid | |
| HRST | 75 | liquid | |
| JWS* | 75 | liquid | Originally using "LOX/alcohol" fuel, later "LOX/kerosene". |
| FIar | 76 | liquid | RD-103 rocket, burning liquid oxygen and kerosene. "An earlier, shorter-range version using alcohol instead of kerosene has been reported". |
| SWM | 77 | liquid | "A single-chamber RD-103 developed between 1952-53", propellants were probably liquid oxygen and hydrocarbon (kerosene?). |
| Rock | 78 | liquid | [Unstorable]. Motor is single-chamber development of RD-103, first tested in 1952 using oxygen and kerosene propellants. |
| BW | 86 | unstorable liquid | |

## MIDCOURSE (OR MAIN) GUIDANCE

| | | |
|---|---|---|
| JWS* | 75 | radio command |
| FIar | 76 | radio command |
| Rock | 78 | ? radio |
| USD9 | 79 | radio command |

## CONTROL SYSTEM

| | | |
|---|---|---|
| IMSG | 60 | 4 control vanes submerged in vehicle's exhaust. |
| JWS* | 75 | Control by vanes in efflux. |
| FIar | 76 | Aerodynamic surfaces on fins, and vanes in motor exhaust. |
| Rock | 78 | Aerodynamic on rear fins, and vanes in efflux; this control system reminiscent of SS-1, SS-2, and SS-1B/-1C. |

## OBSERVED IN PARADE

| | | | |
|---|---|---|---|
| IMSG | 60 | | [Pictures show it on trailer, ed] |
| FIar | 76 | 1957 | First displayed in parade 7 Nov 1957. |
| MOW | 76 | Nov 1957 | |
| Rock | 78 | 7 Nov 1957 | First publicly revealed. |

## DEVELOPMENT BEGAN

| | | | |
|---|---|---|---|
| SMP | 85 | 1946 | Technological development began. [From graph.] |

## PROTOTYPE TESTS

| | | | |
|---|---|---|---|
| SMP | 85 | 1952-53 | Engineering and testing began. [From graph.] |

## IOC

| | | |
|---|---|---|
| USD9 | 79 | before 1960 |

## DEPLOYMENT BEGAN

| | | | |
|---|---|---|---|
| FIar | 76 | 1955 | First entered service. |
| Rock | 78 | 1955 | Service intro. |
| Meyr | 84 | 1955 | Service. |
| SMP | 85 | 1956-57 | [From graph.] |
| BW | 86 | 1955 | |

## DEPLOYED

| | | |
|---|---|---|
| COTS | 58 | Operational. |
| IMSG | 60 | Operational. |

## RETIRED

| | | | |
|---|---|---|---|
| JWS* | 75 | | "Almost certainly no longer in operational service" in USSR. "Probably obsolete". |
| FIar | 76 | | Now withdrawn from service. May be held in reserve, as Russians tend not to scrap a weapon as long as it remains of potential use. |
| SMP | 81 | | [No implication it is still deployed, ed.] |
| SMP | 85 | 1963 | [End of "deployment", from graph, ed.] |
| BW | 86 | ? 1967 | |

## NUMBERS IN SERVICE

### NUMBER IN 1955

| | | |
|---|---|---|
| SSF | 82 | 24 |

### NUMBER IN 1960

| | | |
|---|---|---|
| SSF | 82 | 48 |

### NUMBER IN 1965

| | | | |
|---|---|---|---|
| SMP | 81 | | More than 700 fixed launchers for SS-3,4,5 [together] were operational at peak deployment in the mid-1960s". |
| SSF | 82 | 28 | |

### NUMBER IN 1970

| | | | |
|---|---|---|---|
| SSF | 82 | none | In 1970, 1975, 1980. |

310  SOVIET MISSILES  IDDS

---------- UNDATED NUMBER ----------
SuO1  83    709    From late 1950s to mid-1960s "up to" 709 SS-3,4,5 missiles deployed around Soviet periphery, "all but about 100 arrayed against the European theatre".
BW    86    ? 48   Peak.

---------- CARRIERS ----------

---------- CARRIER ---------- TUBES ---- RELOADS ----------
FIar  76    mobile
Rock  78    mobile launcher                    Mobile launcher pulled by tracked vehicle carrying 20 men (in 1957 parade).

---------- DESIGN AND ENGINEERING ----------
COTS  58    "A version of the German V-2".
IMSG  60    Way to T-1 paved by captured German V-2 rockets. Many T-1 components tested in German V-2; some V-2 components incorporated without change. Principal improvements over V-2 found in more powerful propulsion unit and T-1's increased propellant space. "These factors, coupled with structural-design advances, have resulted in a considerably better mass ratio being attained". Being used as component test vehicle for more advanced rockets. T-1's motor has been used as upper-stage unit for several missiles, including T-3 ICBM.
Zaeh  61    Improved V-2.
JWS*  75    Development from German V-2 through Scunner, Sibling. Marked departure from squat V-2 configuration.
FIar  76    First Russian-designed long-range rocket.
MOW   76    An improved version of captured German V-2s having "a longer, cylindrical body, permitting increased propellant tankage".
SWM   77    Incorporated V-2 technology.
Rock  78    Shyster, with 900 km range, "owes its success to an earlier model of unknown designation".

---------- USE AND CONFIGURATION ----------
IA01  65    "Various special vehicles are required to service these missiles [Shyster and Sandal], such as cranes for erecting and positioning the rocket on its launcher, ancillary vehicles for systems proving and fire control, tankers, etc.".
FIar  76    Mobile weapon launched from a simple firing table.
Meyr  84    Penetration: 1.0; reliability: 0.2-0.5, "includes system reliability multiplied by operational readiness of deployed system"; [literally 0.2-5.0, ed].

---------- RELOADS IN GENERAL ----------
BW    86    Yes.

---------- STRATEGY ----------
HRST  75    First of IRBM arsenal.
Rock  78    The first operational Soviet strategic missile. "Capable of reaching targets in London and Paris from E. Germany".
Meyr  84    [Ed: chart is labeled] Systems "assigned primary operational and strategic missions within the European theatre(s) of military operations".

---------- OTHER INFORMATION ----------
IMSG  60    Equipped with 4 triangular fins.

# SS-4

| Sources are: | SecC | 70 | JCS | 76 | Rock | 78 | SecD | 80 | AFM | 82 | Clns+ | 83 | AFM+ | 84 |
|---|---|---|---|---|---|---|---|---|---|---|---|---|---|---|
| Zaeh | 61 | SecC | 71 | MOW | 76 | JCS | 79 | A168 | 81 | AWST | 82 | IISS+ | 83 | JCS | 84 |
| SecC | 64 | IISS | 73 | NA03 | 76 | USD9 | 79 | IISS | 81 | IISS | 82 | JCS | 83 | Meyr | 84 |
| IA01 | 65 | FI* | 74 | SecD | 76 | BA04 | 80 | JCS | 81 | JCS | 82 | JWS+ | 83 | SecD | 84 |
| SecC | 65 | HRST | 75 | JCS | 77 | Clns | 80 | JWS | 81 | Pr01 | 82 | NWG | 83 | SMP | 84 |
| Ley | 68 | JCS | 75 | SWM | 77 | IISS | 80 | SecD | 81 | SecD | 82 | SMP | 83 | SMP | 85 |
| SecC | 69 | FIar | 76 | JCS | 78 | JCS | 80 | SMP | 81 | SSF | 82 | Su01 | 83 | BW | 86 |

---------- **US DESIGNATION** ----------

| SecC | 69 | SS-4 | |
| FI* | 74 | SS-4 | FY70-72. |
| JCS | 75 | SS-4 | |
| FIar | 76 | SS-4 | FY76-79,80-86. |
| MOW | 76 | SS-4 | |
| SecD | 76 | SS-4 | FY77-78,80,82-85. |
| SWM | 77 | SS-4 | |
| JCS | 78 | SS-4 | |
| Rock | 78 | SS-4 | |
| Clns | 80 | SS-4 | |
| JWS | 81 | SS-4 | |
| SMP | 81 | SS-4 | '81, '83, '84, '85. |
| AFM | 82 | SS-4 | |
| AWST | 82 | SS-4 | |
| IISS | 82 | SS-4 | |
| SSF | 82 | SS-4 | |
| Meyr | 84 | SS-4 | |

---------- **NATO CODENAME** ----------

| Ley | 68 | Sandal |
| FI* | 74 | Sandal |
| HRST | 75 | Sandal |
| FIar | 76 | Sandal |
| MOW | 76 | Sandal |
| SWM | 77 | Sandal |
| Rock | 78 | Sandal |
| Clns | 80 | Sandal |
| JWS | 81 | Sandal |
| AFM | 82 | Sandal |
| AWST | 82 | Sandal |
| IISS | 82 | Sandal |
| SSF | 82 | Sandal |

---------- **OTHER DESIGNATION** ----------

| Zaeh | 61 | T-1A |

---------- **DESCRIPTION** ----------

| SecC | 64 | MRBM | FY65-71. |
| MOW | 76 | | "Medium-range ballistic missile". |
| JCS | 77 | MRBM | FY78-81. |
| Rock | 78 | IRBM | |
| Clns | 80 | MRBM | |
| JWS | 81 | MRBM | |
| SMP | 81 | MRBM | Long-Range Theater Nuclear Weapon. |
| AFM | 82 | MRBM | |
| SSF | 82 | | First generation. Medium range. |
| SMP | 83 | | Longer-Range Intermediate-Range Nuclear Forces, '83, '84, '85. |
| SecD | 84 | | Listed as included in INF totals. |

---------- **EDITOR'S NOTES** ----------

| SecC | 64 | Listed FY65-72, not by name FY65-69. |
| JCS | 75 | Listed FY76-86; referred to only as MRBM in FY79. |
| SecD | 76 | Listed FY76-78,80,82-85; referred to only as "older missile launchers" in FY81. |
| SMP | 81 | Listed '81, '83, '84, '85. |
| Pr01 | 82 | Referred to as MRBM of Cuban crisis, not by name. |
| Meyr | 84 | Note to chart: "Values pertain to initial service period". |

## 312 SOVIET MISSILES IDDS

### RANGE

| | | mi | nm | km | km conv | |
|---|---|---|---|---|---|---|
| Zaeh | 61 | 350 | | | 563.15 | |
| IAD1 | 65 | 900-1,100 | | | 1,448-1,769 | |
| SecC | 65 | | 1,020 | | 1,889 | FY66-71. |
| Ley | 68 | 900-1,000 | | | 1,448-1,609 | |
| SecC | 71 | | 1,050 | | 1,944 | |
| FI* | 74 | | | c. 1,800 | 1,800 | |
| HRST | 75 | 1,000 | | | 1,609 | |
| FIar | 76 | | | 1,800 | 1,800 | |
| MOW | 76 | c. 1,100 | | c. 1,750 | 1,769 | Max. |
| NAO3 | 76 | 1,200 | | | 1,930 | |
| JCS | 77 | | 1,050 | | 1,944 | FY78-79. |
| SWM | 77 | 1,100 | | 1,770 | 1,769 | |
| Rock | 78 | | | 1,800 | 1,800 | |
| Clns | 80 | | 1,200 | | 2,222 | |
| SecD | 80 | | | 1900-4100 | 1,900-4,100 | [SS-4,5]. |
| JWS | 81 | | | 1500-1800 | 1,500-1,800 | Depends on weight of warhead. |
| SMP | 81 | | | 2,000 | 2,000 | '81, '83, '84, '85. |
| AFM | 82 | 1,200 | | | 1,930 | Max. |
| AWST | 82 | 1,100 | | | 1,769 | |
| IISS | 82 | | | 2,000 | 2,000 | Max [overall listing] |
| | | | | 1,900 | 1,900 | Range [European theater] |
| JCS | 82 | | | 2,000 | 2,000 | FY83-86. |
| SSF | 82 | | 1,100 | | 2,037 | |
| Clns+ | 83 | | 1,400 | | 2,592 | |
| JWS+ | 83 | | | c. 1,800 | 1,800 | |
| Meyr | 84 | | | 1,900 | 1,900 | |
| BW | 86 | | | 2,000 | 2,000 | |

### SPEED

| | | mph | kmph | Mach | kmph conv | |
|---|---|---|---|---|---|---|
| MOW | 76 | 4,300 | 6,900 | | 6,918 | Max. |
| JWS | 81 | | | 6-7 | | |
| AFM | 82 | | | 6.7 | | Max. |
| JCS | 82 | | | | | Operational flight time: minutes, FY83-86. |

### CEP

| | | ft | mi | nm | m | km | m conv | |
|---|---|---|---|---|---|---|---|---|
| BAO4 | 80 | 6-12,000 | | | | | 1,83-3,657 | |
| Clns | 80 | | | 1.25 | | | 2,315 | [Unit assumed; not included, ed] |
| IISS | 82 | | | | 2,300 | | 2,300 | |
| SSF | 82 | | | 1.5 | | | 2,778 | |
| JWS+ | 83 | | | | 2,400 | | 2,400 | Typically. |
| Meyr | 84 | | | | | 1.5-3.0 | 1,500-3,000 | |
| BW | 86 | | | | 2,300 | | 2,300 | |

### WARHEAD TYPE

| | | nuclear | conven | chem | |
|---|---|---|---|---|---|
| FI* | 74 | yes | yes | | HE. |
| JCS | 75 | yes | | | All MR/IRBMs (including spares and refires) assumed to have nuclear warheads available, FY76-77. |
| FIar | 76 | yes | yes | | HE. |
| MOW | 76 | yes | yes | | HE. |
| SWM | 77 | yes | yes | | HE. |
| Rock | 78 | yes | yes | | HE. |
| JWS | 81 | yes | yes | | HE. |
| AFM | 82 | yes | yes | | HE. |
| Meyr | 84 | yes | | | |

### WARHEAD CHARACTERISTICS

| | | | |
|---|---|---|---|
| Rock | 78 | single RV | |
| Clns | 80 | single RV | |
| SecD | 80 | single RV | FY81-82. |
| JCS | 81 | single RV | FY82-86. |
| SMP | 81 | single RV | '81, '83, '84, '85. |
| IISS | 82 | single RV | |
| SSF | 82 | single RV | |
| Meyr | 84 | single RV | |
| BW | 86 | single RV | |

BALLISTIC MISSILES  IRBMs  SS-4  313

## ---------- NUCLEAR YIELD ----------

|  |  | kT | MT | kT conv |  |
|---|---|---|---|---|---|
| SecC | 65 |  |  |  | "The kiloton to the six MT range" [MR/IRBMs generally]. |
| FI* | 74 |  | 1 | 1,000 |  |
| FIar | 76 |  | 1 | 1,000 |  |
| MOW | 76 |  | 1 | 1,000 |  |
| SWM | 77 |  | 1 | 1,000 |  |
| Rock | 78 |  | 1.0 | 1,000 | [Table] |
|  |  |  | $\leq$ 1.0 | 1,000 | [Text] |
| Clns | 80 |  | 1 | 1,000 |  |
| JWS | 81 |  | 1 | 1,000 |  |
| AFM | 82 |  | 1 | 1,000 |  |
| IISS | 82 |  | 1 | 1,000 |  |
| SSF | 82 |  | 2.0 | 2,000 | Per warhead. |
| SuO1 | 83 |  | 2 | 2,000 | Up to. |
| Meyr | 84 |  | 1-3 | 1,000-3,000 |  |
| BW | 86 | 1,000 |  | 1,000 |  |

## ---------- THROWWEIGHT ----------

|  |  | lb | t | kg | kg conv |
|---|---|---|---|---|---|
| Clns | 80 | 3,000 |  |  | 1,360 |
| IISS | 82 | 3,000 |  |  | 1,360 |

## ---------- WEIGHT ----------

|  |  | lb | t | kg | kg conv |  |
|---|---|---|---|---|---|---|
| Zaeh | 61 | 44,000 |  |  | 19,958 |  |
| IAO1 | 65 |  | c. 28 |  | 28,000 |  |
| Ley | 68 |  | 28 |  | 28,000 | Tons. |
| FI* | 74 |  |  | c. 27,000 | 27,000 |  |
| MOW | 76 | c. 60,000 |  | c. 27,200 | 27,216 |  |
| JWS | 81 |  |  | c. 27,000 | 27,000 |  |
| AFM | 82 | 60,000 |  |  | 27,216 |  |

## ---------- LENGTH ----------

|  |  | in | ft | cm | m | m conv |  |
|---|---|---|---|---|---|---|---|
| Zaeh | 61 |  | 50-52 |  |  | 15.24-15.85 |  |
| IAO1 | 65 |  | 72 |  |  | 21.95 |  |
| Ley | 68 |  | 72 |  |  | 21.95 |  |
| SecC | 70 |  | 75 |  |  | 22.86 |  |
| FI* | 74 |  |  |  | c. 21 | 21.00 |  |
| HRST | 75 |  | 85 |  |  | 25.91 |  |
| FIar | 76 |  |  |  | 23.5 | 23.50 |  |
| MOW | 76 |  | 68'0" |  | 20.8 | 20.73 |  |
| SWM | 77 |  | 73.5 |  | 22.4 | 22.40 |  |
| Rock | 78 |  |  |  | 23.5 | 23.50 |  |
| JWS | 81 |  |  |  | c. 21 | 21.00 |  |
| SMP | 81 |  |  |  | 20 | 20.00 | [From graph], '81, '83, '84, '85. |
| AFM | 82 |  | 77'0" |  |  | 23.47 |  |
| AWST | 82 |  | 73.3 |  |  | 22.34 |  |
| AFM+ | 84 |  | 68'0" |  |  | 20.73 |  |

## ---------- DIAMETER ----------

|  |  | in | ft | cm | m | cm conv |  |
|---|---|---|---|---|---|---|---|
| Zaeh | 61 |  | 5.15 |  |  | 156.97 |  |
| IAO1 | 65 |  | 5.25 |  |  | 160.02 |  |
| Ley | 68 |  | 5 1/4 |  |  | 160.02 |  |
| HRST | 75 | 65 |  |  |  | 165.10 |  |
| FIar | 76 |  |  |  | 1.7 | 170.00 |  |
| MOW | 76 |  | 5'3" |  | 1.60 | 160.02 | Max. |
| SWM | 77 |  | 5.41 |  | 1.65 | 164.90 |  |
| Rock | 78 |  |  |  | 1.7 | 170.00 |  |
| JWS | 81 |  |  | c. 160 |  | 160.00 |  |
| AFM | 82 |  | 5'7" |  |  | 170.18 |  |
| AWST | 82 |  | 5.4 |  |  | 164.59 |  |
| AFM+ | 84 |  | 5'3" |  |  | 160.02 |  |

## ---------- WINGSPAN ----------

|  |  | in | ft | cm | m | cm conv |
|---|---|---|---|---|---|---|
| FI* | 74 |  |  | c. 160 |  | 160.00 |

```
                 ---------- NUMBER OF STAGES ----------
Zaeh    61       1
FI*     74       1
MOW     76                          No booster.
SWM     77       1
Rock    78       1
JWS     81       1
AWST    82       1

                 ---------- TYPE OF PROPULSION ----------
Zaeh    61                              Thrust: 99,000 lb.
SecC    64       liquid                 FY65,72.
FI*     74       liquid
HRST    75       liquid
Flar    76       cryogenic              Lox/kerosene in original version.
                 storable liquid        Nitric acid/kerosene, on the later version. Flared skirt houses 4-chamber
                                        RD-214 engine, thrust 72,000 kg.
MOW     76       liquid
SWM     77       storable liquid        Nitric acid/kerosene. 4 chamber RD-214 engine of 74 tonnes vacuum thrust,
                                        chamber pressure of 45 atmospheres and specific impulse of 264 sec.
Rock    78       storable liquid        Nitric acid and kerosene. Thrust, tonnes: 72 (at liftoff). Variant of RD-214
                                        motor; 4 fixed combustion chambers. Specific impulse 264 sec, chamber
                                        pressures at 46.51 kg/cm2. Motor developed by GDL-OKB between 1952 and 1958.
BA04    80       Nonstorable liquid.    Must be fed in shortly before launch.
Clns    80       liquid
JWS     81       liquid
SMP     81       liquid
AFM     82       liquid                 One 4-chamber liquid sustainer (nitric acid/kerosene), 163,142 lb thrust in
                                        vacuo.
AWST    82       liquid
SSF     82       liquid
BW      86       storable liquid

                 ---------- MIDCOURSE (OR MAIN) GUIDANCE ----------
FI*     74       inertial
Flar    76       command                Similar to Shyster's, on "early rounds".
                 inertial               On the later version.
MOW     76       inertial
SWM     77       radio command          Formerly.
                 inertial               [Now]
Rock    78       radio command          Earlier.
                 inertial               Currently; replaced radio "in recent years".
JWS     81       radio command          Originally.
                 inertial               By time of Cuba crisis.
AFM     82       inertial
SSF     82       radio command          [Originally]
                 inertial               Later.

                 ---------- CONTROL SYSTEM ----------
FI*     74       Elevators and vanes in eflux.
SWM     77       Steerable exhaust vanes and aerodynamic rudders.
Rock    78       Four triangular fins with aerodynamic control surfaces and vanes which project into the exhaust
                 efflux.
JWS     81       Elevators and vanes in efflux.

                 ---------- FIRST OBSERVED ----------
MOW     76       1961
SWM     77       1961
JWS     81       1961

                 ---------- DEVELOPMENT BEGAN ----------
SSF     82       1949-50       Year design began.
SMP     85       1948          Technological development began. [From graph.]

                 ---------- PROTOTYPE TESTS ----------
SSF     82       1957          First flight test.
SMP     85       1954-55       Engineering and testing began. [From graph.]

                 ---------- IOC ----------
US09    79       before 1961
JCS     82       1960          Year operational.
JCS     83       1959          Year operational.
JCS     84       1958          Year operational, FY85-86.
```

BALLISTIC MISSILES  IRBMs  SS-4  315

```
                    ------- DEPLOYMENT BEGAN -------
FIar   76    1959              Deployed from 1959 onwards.
MOW    76    1959
SWM    77    1959              Service entry.
Rock   78    1959              Service intro; "operational in 1959 as the standard Soviet [IRBM]".
Clns   80    1959              First deployed.
JWS    81    1958
AFM    82    1959
IISS   82    1959              First year deployed.
SSF    82    1958              Year operation began.
Meyr   84    1959              Service.
SMP    85    1959-60           [From graph.]
BW     86    1959

                    ------- DEPLOYED -------
FIar   76                      Remains in service.
MOW    76    in service.
SecD   82                      [Listed as dismantlable, therefore deployed, ed], FY83-84.
BW     86                      In service.

                    ------- DEPLOYMENT HISTORY -------
SecC   71    "Deactivation of SS-4 MRBM soft launchers sites" may indicate SS-11s being deployed to replace
             aging and more vulnerable SS-4s and 5s.
FIar   76    Expected to be replaced by SS-X-20s in 1980s.
MOW    76    Sandal replaced Shyster as USSR's MRBM.
SWM    77    ? Being replaced by SS-20.
Rock   78    Being replaced by SS-20.
JCS    79    "Some of the older SS-4 and SS-5 launchers" phased out in past year. SS-20s expected to replace
             entire SS-4 and SS-5 force, FY80-81.
JCS    80    Some of older SS-4s and SS-5s phased out in past year. SS-20s [generically] replace SS-4s and 5s.
A168   81    About a quarter of SS-4s and SS-5s transferred east, mostly in 1968.
JCS    81    Phase-out of SS-4s and SS-5s is ongoing. "By mid-1980s, probable phase out of some of the older
             single-RV SS-4 and SS-5 missiles". SS-20 "is augmenting" older SS-4s and 5s.
JWS    81    Was standard Soviet missile. Replacement by SS-20 expected.
SecD   81    Some being retired, but substantial number remain [SS-4,5].
SMP    81    USSR "withdrawing older SS-4s and SS-5s from the forces as the SS-20s are deployed".
AFM    82    "Replacement with SS-20s is being maintained at the rate of one every five days" [final report].
IISS   82    Being withdrawn.
JCS    82    "Total number of [SS-4,5] launchers has remained constant" with SS-20s increasing.
JWS+   83    Numbers of SS-4,5s declining since 1977, according to US official source.
SuO1   83    Some 180 SS-11 VRBM deployed at two former SS-4,5 fields in European USSR. 60 silos begun each
             year in 1968, 1970 and 1971. These "initially supplemented, and then replaced, SS-4 and SS-5
             deployments in those fields." First "100 or so" SS-4,5 deactivated from 1968 to 1972, as replaced
             by SS-11 VRBM. In 1969-70 all 70 SS-4,5 missiles in Far East deactivated, replaced in early 1970s
             by 120 SS-11s.

                    ------- NUMBERS IN SERVICE -------

                    ------- NUMBER IN 1955 -------
SSF    82    none

                    ------- NUMBER IN 1960 -------
SSF    82    200

                    ------- NUMBER IN 1965 -------
SMP    81                      "More than 700 fixed launchers" for SS-3,4,5 "were operational at peak deployment in the
                                mid-1960s"; drawdown began in late 60s.
SSF    82    608

                    ------- NUMBER IN 1969 -------
SecC   69                      "Four new, soft SS-4 launchers have been constructed at an existing MRBM site in the
                                Soviet Far East".

                    ------- NUMBER IN 1970 -------
SecC   71    424               Soft, 30 Dec 1970 and 1 Mar 1971.
             84                Hard, 30 Dec 1970 and 1 Mar 1971.
Clns   80    550
SSF    82    508               In 1970, 1975.
SuO1   83    649               January, "actual" figure of SS-4 and SS-5 together, along with "official estimate" of
                                650.

                    ------- NUMBER IN 1971 -------
SecC   71                      Soft, 380-380 [sic], mid-1971.
                               72-60 hard, mid-1971.
Clns   80    500               1971-1976.
```

### 316  SOVIET MISSILES   IDDS

```
---------- NUMBER IN 1972 ----------
SecC  71              Soft, 260-300, mid-1972.
                      44-56 hard, mid-1972.

---------- NUMBER IN 1973 ----------
IISS  73      500     1973-79, July, total deployed.

---------- NUMBER IN 1975 ----------
JCS   75              Believed total number of MR/IRBMs on operational launchers will be about 600 by
                      mid-1975.
                      "Allowing for refires and spares", estimated inventory totals over 1000 MR/IRBMs, all
                      assumed to have nuclear warheads available, FY76-77.
                      Number of SS-4 and SS-5 launchers remained essentially unchanged over the last year,
                      FY76-77.
SWM   77      500     Mid-1975.

---------- NUMBER IN 1976 ----------
FIar  76      500     Deployed at present.
JCS   76              Believed total number of MR/IRBMs on operational launchers will be about 600 by
                      mid-1976.
MOW   76      500     10% in Far East.

---------- NUMBER IN 1977 ----------
JCS   77              SS-4 and SS-5 force slightly reduced this year with deactivation of some SS-4s.
Clns  80      480

---------- NUMBER IN 1978 ----------
Rock  78      c. 500  In USSR and along Chinese border.
Clns  80      420
SMP   83      490     [Read from graph], '83, '84, '85.

---------- NUMBER IN 1979 ----------
Clns  80      350
SMP   83      430     [Read from graph], '83, '84, '85.
SuO1  83      433     January, "actual" figure of SS-4 and SS-5 together, along with "official estimate" of
                      450.

---------- NUMBER IN 1980 ----------
BA04  80      400     500 deployed, 100 now retired.
IISS  80      380     July, total deployed.
SecD  80              450 total [of SS-4,5].
SSF   82      360
Clns+ 83      350
SMP   83      390     [Read from graph], '83, '84, '85.
SuO1  83      391     January, "actual" figure of SS-4 and SS-5 together, along with "official estimate" of
                      400.

---------- NUMBER IN 1981 ----------
IISS  81      340     July, total deployed.
JCS   81      [450]   "There are 450 older SS-4 MRBMs, older SS-5 IRBMs, and 180 new" SS-20s in Strategic
                      Rocket Force.
JWS   81      450     Remaining, official US estimate.
                      400-500 launchers [from Skean entry, ed], SS-4s + SS-5s, early 1981.
SecD  81              C. 400 "older MRBMs and IRBMs".
SMP   81      320
JCS   82      300     Deployed (globally) as of Dec 1981.
Clns+ 83      280
SMP   83      365     [Read from graph], '83, '84, '85.
SuO1  83      355     January, "actual" figure of SS-4 and SS-5 together, along with "official estimate" of
                      over 350.

---------- NUMBER IN 1982 ----------
AFM   82      320     Remaining operational, mostly near western borders.
IISS  82      275     July, total deployed.
JCS   82              Since 1977, "over 200 SS-4 and SS-5 launchers" deactivated.
Clns+ 83      260
JWS+  83      275     Independent observer, late 1982.
SMP   83      300     [Read from graph], '83, '84, '85.
SuO1  83      ? 300   January, "actual" figure of SS-4 and SS-5 together, along with "official estimate" of
                      335.
              232     End of year.
```

BALLISTIC MISSILES  IRBMs  SS-4  317

```
------------   NUMBER IN 1983 ------------
IISS+  83   223      July, total deployed, Europe (including European USSR) only.
JCS    83   232      Deployed (globally) as of 1 Jan.
JWS+   83   232      US estimates, early 1983.
                     C. 320 for SS-4,5.
SMP    83   232
            245      [Read from graph], '83, '84, '85.

------------   NUMBER IN 1984 ------------
AFM+   84   c. 200   Remain operational.
JCS    84   224      Deployed (globally) as of 1 Jan.
SMP    84   220      1984 [read from graph], '84, '85.
            224      All are deployed in western USSR opposite NATO.
SMP    85   120      All are deployed in western USSR opposite NATO.

------------   NUMBER IN 1985 ------------
SecD   80            50 (est.) for "mid-1980's" based on current trends [SS-4,5].

------------   UNDATED NUMBER ------------
SSF    82   600      [Literally 100, but almost certainly transposed with SS-5], "Number of missiles
                     deployed".
SuO1   83   709      From late 1950s to mid-1960s "up to" 709 SS-3,4,5 missiles deployed around Soviet
                     periphery, "all but about 100 arrayed against the European theatre".
BW     86   ? 600    Peak.

------------   LAUNCHER/SILO ------------
SecC   65            Deployed in 4 launcher per site soft configuration (plus a re-fire capability) [MRBM/IRBM
                     generally]. "4 launcher per site configuration for the hardened MRBMs".
SecC   71            Soft and hard basing.
FIar   76            Some deployed in silos.
Rock   78            Mobile and fixed-base rocket.
JWS    81            Based in both soft and hardened sites.
SMP    81            "Based at fixed, vulnerable sites" [SS-3,4,5].
JCS    82            Operational mode: fixed, FY83-86.
SSF    82            Basing mode: "Soft site, with 4 launchers and ability to refire; hard site, with 4 launchers".
ClnS+  83            Launch pad/silo.
SuO1   83            Two-thirds of launchers were unprotected open pads deployed in close groups of four.  Remaining
                     135 missiles in underground silos deployed in close groups of three. [Described with SS-5, ed.]

------------   CARRIERS ------------

------------   CARRIER ------------ TUBES ---- RELOADS ------------
Rock   78                                  Mobile and fixed-base rocket.

------------   DESIGNER ------------
SSF    82   Yangel       Design Bureau.

------------   DESIGN AND ENGINEERING ------------
Zaeh   61   Improved T-1.
FI *   74   MRBM origins can be traced to German V2. Operated by WTO.
HRST   75   "Somewhat advanced version of Shyster".
FIar   76   Development of Shyster. SS-4 does not have tapered nose section of SS-3.
MOW    76   Longer with more propellant and greater range than SS-3 Shyster.
SWM    77   MRBM follow-on to SS-3 Shyster.
Rock   78   Derived from SS-3.
JWS    81   Developed version of SS-3 Shyster.
AFM    82   Development via SS-3 drew heavily on German V-2 technology.
AWST   82   Development of Shyster.
```

318   SOVIET MISSILES   IDDS

---------- **USE AND CONFIGURATION** ----------

| | | |
|---|---|---|
| IAD1 | 65 | "Various special vehicles are required to service these missiles [Shyster and Sandal], such as cranes for erecting and positioning the rocket on its launcher, ancillary vehicles for systems proving and fire control, tankers, etc". |
| FIar | 76 | "In the mobile role, the complete weapon system consists of a convoy of about a dozen vehicles carrying ground-support equipment and propellants. A 20-man firing crew is required to erect and fuel the missile". |
| MOW | 76 | Crew carried in truck that hauls missile on its trailer. Other vehicles carry ground support equipment and propellants. |
| BA04 | 80 | "Very poor" reaction time, resulting from fuel which must be fed in shortly before launch. |
| A168 | 81 | A day or more is required to prepare for firing. |
| JWS | 81 | Launch system needs 12 tractors and trailers, and 20 men. |
| AFM | 82 | 12 tractors with special trailers, and 20 men, needed for operation. |
| PrO1 | 82 | According to Special National Intelligence Estimate of 11-19-62, reconnaisance of Cuban sites showed four sites, each with four missile launchers and one reload. Reloading might take four to six hours. Could be launched within eight hours of a decision to do so. |
| NWG | 83 | Operational echelon: Theater. |
| SuO1 | 83 | Requires several hours to prepare for firing. |
| Meyr | 84 | Penetration: 1.0; reliability: 0.3-0.7, "includes system reliability multiplied by operational readiness of deployed system". |

---------- **RELOADS IN GENERAL** ----------

| | | |
|---|---|---|
| SecC | 65 | Soft configuration has re-fire capability. |
| JCS | 81 | "Soft SS-4 and SS-5 launchers" have a refire capability. |
| JWS | 81 | Reloading possible for soft sites, not for hard. |
| SMP | 81 | "Based at fixed, vulnerable sites", "although provisions for force reconstitution and refire were made", [SS-3,4,5]. |
| AWST | 82 | Has refire capability. |
| PrO1 | 82 | Reloading might take four to six hours. [See USE AND CONFIGURATION, ed.] |
| SSF | 82 | Ability to refire from soft sites. Reload capability of about 200 SS-4s in 1960; reload capability (SS-4s and SS-5s): 514 in 1965, 466 in 1970 and 1975, 265 in 1980. |
| BW | 86 | Some. |

---------- **STRATEGY** ----------

| | | |
|---|---|---|
| SecD | 76 | "Could be used for nuclear attacks on targets in Europe or Asia". |
| JCS | 79 | SS-4s, 5s, 20s "constitute the main nuclear threat to NATO", FY80-81. |
| SuO1 | 83 | Of value only for a first strike, due to low survivability. |
| Meyr | 84 | [Ed: chart is labeled] Systems "assigned primary operational and strategic missions within the European theatre(s) of military operations". |

---------- **COMBAT REPORTS/EFFECTIVENESS** ----------

| | | |
|---|---|---|
| FIar | 76 | During 1962, a number installed in Cuba but withdrawn again as a result of US pressure. |
| MOW | 76 | Sandals were behind [Cuban] missile crisis. |
| SWM | 77 | Weapon of the Cuban crisis of 1962. |
| Rock | 78 | Deployed in Cuba in 1962. "If it had remained operational" could have reached line from Houston to Washington. |
| JWS | 81 | Missile behind the 1962 Cuba crisis. |
| AFM | 82 | Precipitated Cuban missile crisis. |
| IISS | 82 | Serviceability: 0.7; Survivability: 0.5; Reliability: 0.65. |

---------- **GEOGRAPHICAL DEPLOYMENTS** ----------

| | | |
|---|---|---|
| FIar | 76 | Those presently deployed are in Western USSR and along the Sino-Soviet border in Central Asia, some in silos. |
| MOW | 76 | [50] deployed in Far East [10% of 500 total]. |
| JCS | 78 | MRBM/IRBM "Located mainly in the western USSR". |
| JCS | 79 | Bulk of SS-4,5,20 force located in the western USSR. |
| JCS | 80 | Majority of SS-4,5,20 force "apparently are targeted against Western Europe", FY81-82. |
| SecD | 80 | 450 (of 450 total) in European theater [SS-4,5]. |
| JCS | 81 | The rest [not against Europe] can strike Middle East, China, and Japan. |
| JWS | 81 | Reported extensively deployed in Central Asia and facing China. |
| JWS+ | 83 | Reported extensively deployed in Central Asia region and facing China. |
| SMP | 83 | [All shown in Western Theater on map labeled "Deployment ... 1983"; a few were in southern regions on map labeled "Deployment 1978"], '83, '84. |

---------- **EXPORTED TO** ----------

| | | | |
|---|---|---|---|
| FI* | 74 | | Operated by WTO. |
| FIar | 76 | Cuba | During 1962, a number installed but withdrawn again as a result of US pressure. |
| Rock | 78 | Cuba | Deployed in 1962. "If it had remained operational" could have reached line from Houston to Washington. |

---------- **OTHER INFORMATION** ----------

| | | |
|---|---|---|
| FIar | 76 | With upper stage added, SS-4 has been used as launch vehicle for Cosmos satellites weighing up to 400 kg. This combination known in US as B-1, and has flown from Plesetsk and Kapustin Yar. |
| Rock | 78 | Beginning in 1958, GDL-OKB developed rocket motor (RD-119) for a stage to be placed above basic Sandal, enabling 400 kg payloads to reach low earth orbit. Together known as B-1; to date 150 satellites launched with it; still used for orbiting small payloads. |

# SS-5

```
Sources are:   FI*    74    NAO3  76    JCS   79    IISS  81    IISS  82    JCS   83    Meyr  84
SecC  64       GAO1   75    SecD  76    BAO4  80    JCS   81    JCS   82    JWS+  83    OrO1  84
IAO1  65       HRST   75    JCS   77    Clns  80    JWS   81    PrO1  82    NWG   83    SecD  84
SecC  65       JCS    75    SWM   77    IISS  80    SecD  81    SecD  82    SMP   83    SMP   84
IDO1  68       FIar   76    IISS  78    JCS   80    SMP   81    SSF   82    SuO1  83    SMP   85
SecC  71       JCS    76    JCS   78    SecD  80    AFM   82    Clns+ 83    AFM+  84    BW    86
IISS  73       MOW    76    Rock  78    A168  81    AWST  82    IISS+ 83    JCS   84
```

---------- **US DESIGNATION** ----------

| | | | |
|---|---|---|---|
| SecC | 71 | SS-5 | |
| FI* | 74 | SS-5 | |
| JCS | 75 | SS-5 | FY76-78,80-85. |
| FIar | 76 | SS-5 | |
| MOW | 76 | SS-5 | |
| SecD | 76 | SS-5 | FY77-78,80,82-85. |
| SWM | 77 | SS-5 | |
| JCS | 78 | SS-5 | |
| Rock | 78 | SS-5 | |
| Clns | 80 | SS-5 | |
| JWS | 81 | SS-5 | |
| SMP | 81 | SS-5 | '81, '83, '84, '85. |
| AFM | 82 | SS-5 | |
| AWST | 82 | SS-5 | |
| IISS | 82 | SS-5 | |
| SSF | 82 | SS-5 | |
| Meyr | 84 | SS-5 | |

---------- **NATO CODENAME** ----------

| | | |
|---|---|---|
| FI* | 74 | Skean |
| HRST | 75 | Skean |
| FIar | 76 | Skean |
| MOW | 76 | Skean |
| SWM | 77 | Skean |
| Rock | 78 | Skean |
| Clns | 80 | Skean |
| JWS | 81 | Skean |
| AFM | 82 | Skean |
| AWST | 82 | Skean |
| IISS | 82 | Skean |
| SSF | 82 | Skean |

---------- **DESCRIPTION** ----------

| | | | |
|---|---|---|---|
| SecC | 64 | IRBM | FY65-71. |
| GAO1 | 75 | MRBM | |
| MOW | 76 | IRBM | |
| JCS | 77 | IRBM | FY78-81. |
| Rock | 78 | LRBM | "Long Range Ballistic Missile" [table]. |
| | | | "Long range IRBM" [text]. |
| Clns | 80 | IRBM | |
| JWS | 81 | IRBM | |
| SMP | 81 | IRBM | Long-Range Theater Nuclear Weapon. |
| SSF | 82 | | First generation. Intermediate range. |
| SMP | 83 | | Longer-Range Intermediate-Range Nuclear Forces, '83, '84, '85. |
| SecD | 84 | | Listed as included in INF totals. |

---------- **EDITOR'S NOTES** ----------

| | | |
|---|---|---|
| SecC | 64 | Listed FY65-72, not by name FY65-71. |
| IAO1 | 65 | Data not clearly identified, and may not refer to SS-5. |
| JCS | 75 | Listed FY76-85; referred to only as IRBM in FY79. |
| SecD | 76 | Listed FY77-78,80-85; referred to only as "older missile launchers" in FY81. |
| SMP | 81 | Listed '81, '83, '84, '85. |
| PrO1 | 82 | Referred to only as IRBM of Cuban crisis. |
| Meyr | 84 | Note to chart: "Values pertain to initial service period". |

## SOVIET MISSILES

### RANGE

| | | mi | nm | km | km conv | |
|---|---|---|---|---|---|---|
| IA01 | 65 | 6,000+ | | | 9,654 | |
| SecC | 65 | | 2,200 | | 4,074 | FY66-72. |
| ID01 | 68 | 2,200+ | | | 3,539 | |
| FI* | 74 | | | c. 3,000 | 3,000 | |
| GA01 | 75 | 2,200 | | | 3,539 | |
| HRST | 75 | 1,500-2,000 | | | 2,413-3,218 | |
| FIer | 76 | | | 3,500 | 3,500 | |
| MOW | 76 | 2,300 | | 3,700 | 3,700 | Max. |
| NA03 | 76 | 2,300 | | | 3,700 | |
| JCS | 77 | | 2,200 | | 4,074 | FY78-79. |
| SWM | 77 | 2,000 | | 3,220 | 3,218 | |
| Rock | 78 | | | 3,500 | 3,500 | |
| Clns | 80 | | 2,300 | | 4,259 | |
| SecD | 80 | | | 1900-4100 | 1,900-4,100 | [SS-4,5]. |
| JWS | 81 | | | 3,500 | 3,500 | Est. |
| SMP | 81 | | | 4,100 | 4,100 | '81, '83, '84. |
| AFM | 82 | 2,500 | | | 4,022 | Max. |
| AWST | 82 | 2,000 | | | 3,218 | "2,000 mi. class". |
| IISS | 82 | | | 4,100 | 4,100 | Max. |
| JCS | 82 | | | 4,100 | 4,100 | FY83-85. |
| SSF | 82 | | 2,200 | | 4,074 | |
| Clns+ | 83 | | 2,650 | | 4,907 | |
| Meyr | 84 | | | 3,900 | 3,900 | |
| BW | 86 | | | 4,100 | 4,100 | |

### SPEED

| | | mph | kmph | Mach | kmph conv | |
|---|---|---|---|---|---|---|
| JCS | 82 | | | | | Operational flight time: minutes, FY83-85. |

### CEP

| | | ft | mi | nm | m | km | m conv | |
|---|---|---|---|---|---|---|---|---|
| BA04 | 80 | 6,000 | | | | | 1,828 | |
| Clns | 80 | | | 0.6 | | | 1,111 | [Unit assumed, not included, ed] |
| IISS | 82 | | | | 1,100 | | 1,100 | |
| SSF | 82 | | | 1.0 | | | 1,852 | |
| Meyr | 84 | | | | | 0.9-1.9 | 900.00-1,900 | |
| BW | 86 | | | | 1,900 | | 1,900 | |

### WARHEAD TYPE

| | | nuclear | conven | chem | |
|---|---|---|---|---|---|
| FI* | 74 | yes | | | |
| JCS | 75 | yes | | | All MR/IRBMs (including spares and refires) assumed to have nuclear warheads available, FY76-77. |
| MOW | 76 | yes | | | |
| JWS | 81 | yes | | | |
| AFM | 82 | yes | | | |
| Meyr | 84 | yes | | | |

### WARHEAD CHARACTERISTICS

| | | | |
|---|---|---|---|
| Rock | 78 | single RV | Warhead has "distinctly rounded-nose section, signifying a more developed re-entry vehicle to those used with the SS-3 and SS-4". |
| Clns | 80 | single RV | |
| SecD | 80 | single RV | FY81-82. |
| JCS | 81 | single RV | FY82-85. |
| SMP | 81 | single RV | '81, '83, '84. |
| IISS | 82 | single RV | |
| SSF | 82 | single RV | |
| Meyr | 84 | single RV | |
| BW | 86 | single RV | |

## NUCLEAR YIELD

| | | kT | MT | kT conv | |
|---|---|---|---|---|---|
| SecC | 65 | | | | "The kiloton to the six MT range" [MR/IRBMs in general]. |
| FIar | 76 | | 1 | 1,000 | |
| MOW | 76 | | 1 | 1,000 | |
| SWM | 77 | | 1 | 1,000 | |
| Rock | 78 | | 1.0-5.0 | 1,000-5,000 | "May also have been used to test a variety of multi-megaton warhead configurations", but never deployed operationally. |
| Clns | 80 | | 1 | 1,000 | |
| JWS | 81 | | ? 1 | 1,000 | |
| AFM | 82 | | 1 | 1,000 | |
| IISS | 82 | | 1 | 1,000 | |
| SSF | 82 | | 4.0-6.0 | 4,000-6,000 | Per warhead. |
| SuO1 | 83 | | 5 | 5,000 | "Up to." |
| Meyr | 84 | 500-2,000 | | 500.00-2,000 | 500 kT-2 MT. |
| OrO1 | 84 | | 4.0-6.0 | 4,000-6,000 | |
| BW | 86 | 1,000 | | 1,000 | |

## THROWEIGHT

| | | lb | t | kg | kg conv |
|---|---|---|---|---|---|
| Clns | 80 | 3,500 | | | 1,587 |
| IISS | 82 | 3,500 | | | 1,587 |

## WEIGHT

| | | lb | t | kg | kg conv |
|---|---|---|---|---|---|
| IAO1 | 65 | | > 100 | | 100,000 |
| IDO1 | 68 | | c. 80 | | 80,000 |

## LENGTH

| | | in | ft | cm | m | m conv | |
|---|---|---|---|---|---|---|---|
| IAO1 | 65 | | 88.5 | | | 26.97 | |
| IDO1 | 68 | | 77 | | | 23.47 | |
| FI* | 74 | | | c. 25 | | 25.00 | |
| HRST | 75 | | 90 | | | 27.43 | |
| FIar | 76 | | | | 24.5 | 24.50 | |
| MOW | 76 | | 75'0" | | 23.0 | 22.86 | |
| SWM | 77 | | 80+ | | 24.4+ | 24.38 | |
| Rock | 78 | | | | 24.5 | 24.50 | "There are intelligence reports which seem to confirm a basic length of 24.5 m". |
| | | | | | 24.4-26 | 24.40-26.00 | Various sources indicate. |
| JWS | 81 | | | c. 25 | | 25.00 | |
| SMP | 81 | | | | 22 | 22.00 | [From graph], '81, '83, '84. |
| AFM | 82 | | 80'0" | | | 24.38 | |
| AWST | 82 | | 80 | | | 24.38 | |
| AFM+ | 84 | | 72'0" | | | 21.95 | |

## DIAMETER

| | | in | ft | cm | m | cm conv | |
|---|---|---|---|---|---|---|---|
| IAO1 | 65 | | 8.9 | | | 271.27 | |
| IDO1 | 68 | | 7.7 | | | 234.70 | Without tail cone. |
| FI* | 74 | | | c. 240 | | 240.00 | |
| HRST | 75 | 96 | | | | 243.84 | |
| FIar | 76 | | | | 2.6 | 260.00 | |
| MOW | 76 | | 8'0" | | 2.40 | 243.84 | Max. |
| SWM | 77 | | 8.0 | | 2.44 | 243.84 | |
| Rock | 78 | | | | 2.6 | 260.00 | |
| JWS | 81 | | | | c. 2.4 | 240.00 | |
| AFM | 82 | | 8'6" | | | 259.08 | |
| AWST | 82 | | 8 | | | 243.84 | |
| AFM+ | 84 | | 8'0" | | | 243.84 | |

## NUMBER OF STAGES

| | | |
|---|---|---|
| FI* | 74 | 1 |
| FIar | 76 | 1 |
| SWM | 77 | 1 |
| JWS | 81 | 1 |
| AFM | 82 | 1 |
| AWST | 82 | 1 |

---------- TYPE OF PROPULSION ----------
SecC   64    liquid              FY65,72.
FI*    74    liquid
HRST   75    liquid
FIar   76    storable liquid     4-chamber engine.
MOW    76    liquid
SWM    77    storable liquid     Probably 2 x RD-216 twin chamber GDL-OKB rocket engines—possibly nitric acid
                                 and kerosene.
Rock   78    liquid              [Unstorable]. Probably two GDL-OKB RD-216 motors. Thrust, tonnes: 178 [first
                                 stage].
BA04   80    Storable liquid.
Clns   80    liquid
JWS    81    liquid
SMP    81    liquid
AFM    82    liquid              4 chambers.
AWST   82    storable
SSF    82    liquid
BW     86    storable liquid

---------- MIDCOURSE (OR MAIN) GUIDANCE ----------
FI*    74    ? inertial
FIar   76    inertial            All subsequent Russian long-range ballistic missiles also have inertial
                                 guidance.
MOW    76    inertial            Probably.
SWM    77    inertial
Rock   78    inertial
JWS    81    inertial            Probably.
AFM    82    inertial
SSF    82    radio command       [Originally]
             inertial            Later.

---------- CONTROL SYSTEM ----------
FIar   76    No external control fins, relies on vanes in motor exhaust.
Rock   78    Vanes in exhaust, no aerodynamic fins.
AFM    82    Vanes in motor exhaust.

---------- OBSERVED IN PARADE ----------
MOW    76    Nov 1964
Rock   78    Nov 1964            First public display.
JWS    81    1964

---------- DEVELOPMENT BEGAN ----------
SSF    82    1952-53             Year design began.
SMP    85    1950                Technological development began. [From graph.]

---------- DEVELOPMENT UNDERWAY ----------
Rock   78                        Developed in late 1950s.

---------- PROTOTYPE TESTS ----------
SSF    82    1959-60             First flight test.
SMP    85    1957-58             Engineering and testing began. [From graph.]

---------- IOC ----------
JWS    81    1961                Operational.
JCS    82    1961                Year operational, FY83-85.

---------- DEPLOYMENT BEGAN ----------
FIar   76    1961                Entered service.
MOW    76    1961
SWM    77    ? 1961              Service entry.
             ? 1964              First deployed.
Rock   78    1961                Service intro.
             1961                Entered operational service.
Clns   80    1961                First deployed.
IISS   82    1961                First year deployed.
SSF    82    1961                Year operation began.
Meyr   84    1961                Service.
SMP    85    1961-62             [From graph.]
BW     86    1961

---------- DEPLOYED ----------
MOW    76                        In service.
SecD   82                        [Listed as dismantlable, therefore deployed, ed], FY83-84.

---------- RETIRED ----------
SMP    84                        "No longer operational"; "by the end of 1983, all SS-5 LRINF missiles were being
                                 retired".
BW     86    1984

## DEPLOYMENT HISTORY

| | | |
|---|---|---|
| FI* | 74 | A successor to SS-4. |
| SWM | 77 | ? Being replaced by SS-20. |
| Rock | 78 | Indications it is in line for replacement by SS-20. "Has not, as expected at that time [1964?, ed], totally superseded Sandal"; plays complementary role. |
| JCS | 79 | "Some of the older SS-4 and SS-5 launchers" phased out in past year. SS-20s expected to replace entire SS-4 and SS-5 force, FY80-81. |
| JCS | 80 | Some of older SS-4s and SS-5s phased out in past year. SS-20s [generically] replace SS-4s and 5s. |
| JCS | 81 | Phase-out of SS-4s and SS-5s is ongoing. "By mid-1980s, probable phase out of some of the older single-RV SS-4 and SS-5 missiles". SS-20 "is augmenting" older SS-4s and 5s. |
| JWS | 81 | Successor to Shyster and Sandal. Being replaced by SS-20. |
| SecD | 81 | Some being retired, but substantial number remain [SS-4,5]. |
| SMP | 81 | USSR "withdrawing older SS-4s and SS-5s from the forces as the SS-20s are deployed". |
| IISS | 82 | Being withdrawn. |
| JCS | 82 | "Total number of [SS-4,5,20] launchers has remained constant" with SS-20s increasing. |
| Su01 | 83 | Some 180 SS-11 VRBM deployed at two former SS-4,5 fields in European USSR. 60 silos begun each year in 1968, 1970 and 1971. These "initially supplemented, and then replaced, SS-4 and SS-5 deployments in those fields." First "100 or so" SS-4,5 deactivated from 1968 to 1972, as replaced by SS-11 VRBM. In 1969-70 all 70 SS-4,5 missiles in Far East deactivated, replaced in early 1970s by 120 SS-11s. |

## NUMBERS IN SERVICE

### NUMBER IN 1955

| | | | |
|---|---|---|---|
| SSF | 82 | none | In 1955, 1960. |

### NUMBER IN 1965

| | | | |
|---|---|---|---|
| SMP | 81 | | "More than 700 fixed launchers" for SS-3,4,5 "were operational at peak deployment in the mid-1960s"; drawdown began in late 60s. |
| SSF | 82 | 97 | |

### NUMBER IN 1970

| | | | |
|---|---|---|---|
| SecC | 71 | 42 | Soft, for 30 Dec 1970, 1 Mar 1971, Mid 1971, and Mid 1972. |
| | | 48 | Hard, 30 Dec 1970 and 1 Mar 1971. |
| Clns | 80 | 100 | 1970-1973. |
| SSF | 82 | 90 | In 1970, 1975. |
| Su01 | 83 | 649 | January, "actual" figure of SS-4 and SS-5 together, along with "official estimate" of 650. |

### NUMBER IN 1971

| | | | |
|---|---|---|---|
| SecC | 71 | | "48-39" hard, Mid 1971. |

### NUMBER IN 1972

| | | | |
|---|---|---|---|
| SecC | 71 | | "39-48" hard, Mid 1972. |

### NUMBER IN 1973

| | | | |
|---|---|---|---|
| IISS | 73 | 100 | 1973-77, July, total deployed. |

### NUMBER IN 1974

| | | | |
|---|---|---|---|
| Clns | 80 | 90 | 1974-1977. |

### NUMBER IN 1975

| | | | |
|---|---|---|---|
| JCS | 75 | | Believed total number of MR/IRBMs on operational launchers will be about 600 by mid-1975. |
| | | | "Allowing for refires and spares", estimated inventory totals over 1000 MR/IRBMs, all assumed to have nuclear warheads available, FY76-77. |
| | | | Number of SS-4 and SS-5 launchers remained essentially unchanged over the last year, FY76-77. |
| SWM | 77 | 100 | Mid-1975. |

### NUMBER IN 1976

| | | | |
|---|---|---|---|
| FIar | 76 | 100 | In Western USSR. |
| JCS | 76 | | Believed total number of MR/IRBMs on operational launchers will be about 600 by mid-1976. |
| MOW | 76 | $\leq$ 100 | |

### NUMBER IN 1977

| | | | |
|---|---|---|---|
| JCS | 77 | | SS-4 and SS-5 force slightly reduced this year with deactivation of some SS-4s. |

### NUMBER IN 1978

| | | | |
|---|---|---|---|
| IISS | 78 | 90 | 1978-79, July, total deployed. |
| Rock | 78 | 100 | In Western USSR. |
| Clns | 80 | 80 | |
| SMP | 83 | 85 | [Read from graph], '83, '84, '85. |

324  SOVIET MISSILES  IDDS

```
----------  NUMBER IN 1979 ------------
Clns    80    60
SMP     83    75        [Read from graph], '83, '84, '85.
SuO1    83    433       January, "actual" figure of SS-4 and SS-5 together, along with "official estimate" of
                        450.

----------  NUMBER IN 1980 ------------
BA04    80    100       Number deployed.
IISS    80    60        July, total deployed.
SecD    80              450 total [of SS-4,5].
SSF     82    40
Clns+   83    50
SMP     83    60        [Read from graph], '83, '84, '85.
SuO1    83    391       January, "actual" figure of SS-4 and SS-5 together, along with "official estimate" of
                        400.

----------  NUMBER IN 1981 ------------
IISS    81    40        July, total deployed.
JCS     81              "There are 450 older SS-4 MRBMs, older SS-5 IRBMs, and 180 new" SS-20s in Strategic
                        Rocket Force.
JWS     81    c. 100    In IRBM fields.
SecD    81              C. 400 "older MRBMs and IRBMs".
SMP     81    35
JCS     82    35        35 deployed (globally) as of Dec 1981.
Clns+   83    35
SMP     83    35        [Read from graph], '83, '84, '85.
SuO1    83    355       January, "actual" figure of SS-4 and SS-5 together, along with "official estimate" of
                        over 350.

----------  NUMBER IN 1982 ------------
AFM     82    35        All are in West USSR.
IISS    82    16        July, total deployed.
JCS     82              Since 1977, "over 200 SS-4 and SS-5 launchers" deactivated.
SSF     82    [100]     [Literally 600, but almost certainly transposed with SS-4]. "Number of missiles
                        deployed".
Clns+   83    15
SMP     83    35        [Read from graph], '83, '84, '85.
SuO1    83    ? 300     January, "actual" figure of SS-4 and SS-5 together, along with "official estimate" of
                        335.
              16        End of year.

----------  NUMBER IN 1983 ------------
IISS+   83    16        July, total deployed, Europe (including European USSR) only.
JCS     83    16        Deployed (globally) as of 1 Jan 1983.
JWS+    83    16        US and independent estimates, early 1983 and late 1982.
                        C. 320 for SS-4,5.
SMP     83    16
              12        [Read from graph], '83, '84, '85.

----------  NUMBER IN 1984 ------------
AFM+    84    c. 16
                        570 total for IRBM/MRBMs.
JCS     84    13        Deployed (globally) as of 1 Jan.

----------  NUMBER IN 1985 ------------
SecD    80              50 (est.) for "mid-1980's" based on current trends [SS-4,5].

----------  UNDATED NUMBER
SuO1    83    709       From late 1950s to mid-1960s "up to" 709 SS-3,4,5 missiles deployed around Soviet
                        periphery, "all but about 100 arrayed against the European theatre".
BW      86    ? 97      Peak.

----------  LAUNCHER/SILO ------------
SecC    65    4 launcher per site soft configuration (plus a re-fire capability). 3 launcher per site
              configuration for hardened IRBMs.
FIar    76    Some are in silos, but majority in soft sites.
MOW     76    Soviet films show it deployed in silos.
SWM     77    Silo-launched.
Rock    78    Most deployed in western USSR at surface sites; "a number" fitted to hot launch silos like Sandal.
              Large mass prevents it from being mobile.
JWS     81    Based in both hard and soft sites. Shown in silos in official Soviet films.
SMP     81    "Based at fixed, vulnerable sites" [SS-3,4,5].
AFM     82    Some deployed in silos.
JCS     82    Operational mode: fixed, FY83-85.
SSF     82    Basing mode: Soft site, with 4 launchers and ability to refire; hard site with 3 launchers.
Clns+   83    Launch pad/silo.
SuO1    83    Two-thirds of launchers were unprotected open pads deployed in close groups of four. Remaining
              135 missiles in underground silos deployed in close groups of three. [Described with SS-4, ed.]
```

BALLISTIC MISSILES  IRBMs  SS-5  325

---------- CARRIERS ----------

---------- CARRIER ---------- TUBES ---- RELOADS ----------
Rock 78                                                    Large mass prevents it from being mobile.

---------- DESIGNER ----------
SSF  82   Yangel       Design Bureau.

---------- DESIGN AND ENGINEERING ----------
HRST 75   "Stretch-range IRBM derived from Shyster".
MOW  76   "Infinitely more formidable" than German V-2, but the ultimate development of that concept.
          Scale-up of Shyster-Sandal family.
SWM  77   Basically an enlarged Sandal.
Rock 78   Rear section reminscent of SS-4; Skean is probably "advanced first generation derivative" of it.
AFM  82   SS-5 is further development of SS-3/SS-4 concept.
AWST 82   Shyster-Sandal follow-on.

---------- USE AND CONFIGURATION ----------
BA04 80   "Moderate" reaction time.
A168 81   A day or more is required to prepare for firing.
JWS  81   Successor to Shyster and Sandal, transported differently.
Pr01 82   According to Special National Intelligence Estimate of 11-19-62, reconnaisance of Cuban sites
          showed two sites, each with four missile launchers and one reload. Reloading might take six to
          eight hours.
NWG  83   Operational echelon: Theater.
Su01 83   Requires several hours to prepare for firing.
Meyr 84   Penetration: 1.0; reliability: 0.3-0.7, "includes system reliability multiplied by operational
          readiness of deployed system".

---------- RELOADS IN GENERAL ----------
SecC 65   Soft configuration has re-fire capability.
JCS  81   "Soft SS-4 and SS-5 launchers" have a refire capability.
JWS  81   Soft sites have reload capability.
SMP  81   "Based at fixed, vulnerable sites", "although provisions for force reconstitution and refire were
          made", [SS-3,4,5].
AWST 82   Has refire capability.
Pr01 82   Reloading might take six to eight hours. [See USE AND CONFIGURATION, ed.]
SSF  82   Ability to refire from soft sites: Reload capability (SS-4s and SS-5s): 514 in 1965, 466 in 1970
          and 1975, 265 in 1980.
BW   86   Some.

---------- STRATEGY ----------
SecD 76   "Could be used for nuclear attacks on targets in Europe or Asia".
JCS  79   SS-4s, 5s, 20s "constitute the main nuclear threat to NATO", FY80-81.
Su01 83   Of value only for a first strike, due to low survivability.
Meyr 84   [Ed: chart is labeled] Systems "assigned primary operational and strategic missions within the
          European theatre(s) of military operations".

---------- COMBAT REPORTS/EFFECTIVENESS ----------
FIar 76   "Although launch sites for the SS-5 were under construction in Cuba during the 1962 crisis, there
          is no evidence that the missiles had arrived before the US blockade took effect".
Rock 78   In 1962 Russians prepared "what appeared to be SS-5 launch sites" in Cuba, but never actually set
          up there.
IISS 82   Serviceability: 0.75; Survivability: 0.6; Reliability: 0.7.

---------- GEOGRAPHICAL DEPLOYMENTS ----------
MOW  76             Deployment divided between West and East fronts.
SWM  77             Was deployed with SS-4s in West USSR.
JCS  78             MRBM/IRBM "Located mainly in the western USSR".
Rock 78             Most deployed in western USSR at surface sites.
JCS  79             Bulk of SS-4,5,20 force located in the western USSR.
JCS  80             Majority of SS-4,5,20 force "apparently are targeted against Western Europe",
                    FY81-82.
SecD 80             450 (of 450 total) are in European theater [SS-4,5].
JCS  81             The rest (not against Europe) can strike Middle East, China, and Japan.
SMP  83             [2 remaining sites on Kola Peninsula and Baltic state on map labeled "Deployment
                    ... 1983"; 7 additional sites were in southern and western regions on map labeled
                    "Deployment 1978"].

---------- EXPORTED TO ----------
FI*  74             Operated by WTO.

---------- OTHER INFORMATION ----------
FIar 76   Has been modified for use as satellite launch vehicle, orbiting payloads up to 1,500 kg. Launch
          vehicle has restartable 3rd stage, and may be usued to orbit military reconn. satellites from
          Plesetsk; in this form it is known as C-1.
Rock 78   C-1 space launcher is SS-5 with upper liquid stage; can place 1,100 kg in low earth orbit, but
          usually puts payload(s) in "high inclination orbits".

# SS-14

```
Sources are:   SecC  69      SecD  71      IA06  75      SWM   77      FI    81      SSF   82      Meyr  84
IA04  66       SecC  70      GA01  75      FIar  76      Rock  78      JWS   81      JWS+  83      SMP   84
ID01  68       SecC  71      HRST  75      MOW   76      AFM*  80      AWST  82      Su01  83      SMP   85
```

---------- **US DESIGNATION** ----------

| | | |
|---|---|---|
| SecC | 69 | SS-14 |
| SecD | 71 | SS-14 |
| FIar | 76 | SS-14 |
| MOW | 76 | SS-14 |
| SWM | 77 | SS-14 |
| Rock | 78 | SS-14 |
| AFM* | 80 | SS-14 |
| FI | 81 | SS-14 |
| JWS | 81 | SS-14 |
| AWST | 82 | SS-14 |
| SSF | 82 | SS-14 |
| Meyr | 84 | SS-14 |
| SMP | 84 | SS-X-14 | '84, '85 |

---------- **NATO CODENAME** ----------

| | | | |
|---|---|---|---|
| SecD | 71 | Scamp | |
| HRST | 75 | Scamp | |
| FIar | 76 | Scapegoat | |
| MOW | 76 | Scamp | |
| | | Scapegoat | Weapon system first sighted 9 May 1965 and called Scamp. Missile was later sighted separately and called Scapegoat. Only later did Soviet films reveal that Scamp carried Scapegoat. |
| SWM | 77 | Scapegoat | |
| Rock | 78 | Scapegoat | Missile itself. |
| | | Scamp | In transporter/container configuration. |
| AFM* | 80 | Scapegoat | Is IRBM. |
| | | Scamp | Refers to entire system. |
| FI | 81 | Scamp | |
| | | Scapegoat | |
| JWS | 81 | Scapegoat | |
| | | Scamp | System (container and vehicle). |
| AWST | 82 | Scamp | |
| SSF | 82 | Scapegoat | |

---------- **DESCRIPTION** ----------

| | | | |
|---|---|---|---|
| SecC | 69 | MRBM | FY70-72. |
| MOW | 76 | | Ballistic missile. |
| Rock | 78 | LRBM | "Long Range Ballistic Missile", [table] |
| | | IRBM | [Text] |
| JWS | 81 | mobile IRBM | |
| SSF | 82 | | Second generation. Medium range. |
| SMP | 84 | | "LRINF missile". |

---------- **EDITOR'S NOTES** ----------

| | | |
|---|---|---|
| SecC | 69 | Listed FY70-72. |
| SecD | 71 | Listed FY72. |
| FI | 81 | Final report. |
| Meyr | 84 | Note to chart: "Values pertain to initial service period". |
| SMP | 84 | Listed '84, '85. |

---------- **RANGE** ----------

| | | mi | nm | km | km conv | |
|---|---|---|---|---|---|---|
| IA04 | 66 | 600+ | | | 965.40 | |
| SecC | 69 | | 1,050 | | 1,944 | Being tested to this range. |
| | | | c. 1,500 | | 2,778 | Operational range estimate, FY70-72. |
| SecC | 70 | | c. 1,000 | | 1,852 | Being tested to this range. |
| GA01 | 75 | 4,500 | | | 7,240 | |
| HRST | 75 | 1,500-2,000 | | | 2,413-3,218 | |
| FIar | 76 | | | ? 2,000 | 2,000 | |
| MOW | 76 | 2,500 | | 4,000 | 4,022 | Max. |
| SWM | 77 | 2,200 | | 3,540 | 3,539 | |
| Rock | 78 | | | 4,000 | 4,000 | |
| AFM* | 80 | c. 2,500 | | | 4,022 | |
| FI | 81 | | | 4,000 | 4,000 | |
| JWS | 81 | | | c. 4,000 | 4,000 | |
| AWST | 82 | | | | | IRBM range. |
| SSF | 82 | | 1,500 | | 2,778 | |
| Meyr | 84 | | | 4,000 | 4,000 | |

BALLISTIC MISSILES  IRBMs  SS-14  327

## CEP

|       |    | ft | mi | nm | m | km | m conv |  |
|-------|----|----|----|----|----|----|--------|---|
| FIar  | 76 |    |    |    |    | [2] | 2,000 | Same poor accuracy as SS-13. |
| SSF   | 82 |    |    | 0.8-1.0 |    |    | 1,481-1,852 |  |
| Meyr  | 84 |    |    |    |    | 0.6-1.2 | 600.00-1,200 |  |

## WARHEAD TYPE

|       |    | nuclear | conven | chem |
|-------|----|---------|--------|------|
| MOW   | 76 | yes     |        |      |
| SWM   | 77 | yes     |        |      |
| FI    | 81 | yes     |        | Thermonuclear. |
| JWS   | 81 | yes     |        |      |
| Meyr  | 84 | yes     |        |      |

## WARHEAD CHARACTERISTICS

| Rock | 78 | single RV |
| JWS  | 81 |           | Warhead differs from SS-13. |
| SSF  | 82 | single RV |
| Meyr | 84 | single RV |

## NUCLEAR YIELD

|       |    | kT | MT | kT conv |  |
|-------|----|----|----|---------|---|
| ID01  | 68 |    | 2-3 | 2,000-3,000 |  |
| IA06  | 75 |    | ? 1 | 1,000 |  |
| FIar  | 76 |    | 1   | 1,000 |  |
| MOW   | 76 |    | c. 1 | 1,000 |  |
| Rock  | 78 |    | 1.0 | 1,000 |  |
| JWS   | 81 |    | [< 1] | 1,000 | ? < SS-13. |
| SSF   | 82 |    | 0.6 | 600.00 |  |
| JWS+  | 83 |    | 1   | 1,000 | [Analysis table] |
| Meyr  | 84 | 500-1,000 |    | 500.00-1,000 | 500 kT-1 MT. |

## PAYLOAD

|       |    | lb | t | kg | kg conv |  |
|-------|----|----|---|----|---------|---|
| JWS   | 81 |    |   |    |         | ? < SS-13. |

## WEIGHT

|       |    | lb | t | kg | kg conv |  |
|-------|----|----|---|----|---------|---|
| ID01  | 68 |    | >_ 15 |    | 15,000 | "15- ". |
| FI    | 81 |    |   | c. 12,000 | 12,000 |  |

## LENGTH

|       |    | in | ft | cm | m | m conv |  |
|-------|----|----|----|----|----|--------|---|
| IA04  | 66 |    | c. 39 |    |    | 11.89 |  |
| ID01  | 68 |    | 40 |    |    | 12.19 | Container. |
| HRST  | 75 |    | 34 |    |    | 10.36 |  |
| FIar  | 76 |    |    |    | 11 | 11.00 |  |
| MOW   | 76 |    | 35'0" |    | 10.6 | 10.67 |  |
| SWM   | 77 |    | 35 |    | 10.7 | 10.67 |  |
| Rock  | 78 |    |    |    | 11 | 11.00 | [Table] |
|       |    |    |    |    | < 11 | 11.00 | "Nearly" 11 m [text] |
| AFM*  | 80 |    | c. 35 |    |    | 10.67 |  |
| FI    | 81 |    |    |    | 11 | 11.00 |  |
| JWS   | 81 |    |    |    | 10.6 | 10.60 | Stage 1, 4 m; stage 2, 3.5 m; nosecone and RV, 2.3 m. |
| AWST  | 82 |    | 34 |    |    | 10.36 |  |

## DIAMETER

|       |    | in | ft | cm | m | cm conv |  |
|-------|----|----|----|----|----|---------|---|
| IA04  | 66 |    | c. 6 |    |    | 182.88 | First stage. |
| ID01  | 68 |    | 6.25 |    |    | 190.50 | Container. |
| FIar  | 76 |    |    |    | 1.4 | 140.00 |  |
| MOW   | 76 |    | 4'7.5" |    | 1.40 | 140.97 | First stage. |
|       |    |    | 3'2.5" |    | 0.98 | 97.79 | Second stage. |
| SWM   | 77 |    | 4.6 |    | 1.4 | 140.21 |  |
| Rock  | 78 |    |    |    | 1.4 | 140.00 |  |
| FI    | 81 |    |    | 140 |    | 140.00 |  |
| JWS   | 81 |    |    |    | 1.9 | 190.00 | Stage 1 base. |
|       |    |    |    |    | 1.4 | 140.00 | Stage 1 and stage 2 base. |
|       |    |    |    |    | 1   | 100.00 | Stage 2, nosecone, and RV. |

## NUMBER OF STAGES

| MOW   | 76 | 2 |
| SWM   | 77 | 2 |
| FI    | 81 | 2 |
| JWS   | 81 | 2 |
| AWST  | 82 | 2 |

328  SOVIET MISSILES  IDDS

```
----------   TYPE OF PROPULSION   ----------
SecC    69      solid                       FY70-72.
HRST    75      solid
FIar    76      solid
MOW     76      solid                       4 nozzles per stage.
SWM     77      solid
Rock    78      solid                       Stages 1 and 2.
FI      81      solid
JWS     81      solid
AWST    82      solid
SSF     82      solid

----------   MIDCOURSE (OR MAIN) GUIDANCE   ----------
MOW     76      inertial                    Probably.
SWM     77      inertial
FI      81      inertial
JWS     81      inertial                    Presumed.
SSF     82      inertial

----------   CONTROL SYSTEM   ----------
FI      81      ? Moveable nozzles.

----------   OBSERVED   ----------
SMP     84      [Shown in picture]

----------   OBSERVED IN PARADE   ----------
ID01    68      Nov 1967        First shown, missile.
                May 1965        First shown, mobile system.
MOW     76      9 May 1965      Weapon system first sighted 9 May 1965 and called Scamp. Missile was later sighted
                                separately and called Scapegoat. Only later did Soviet films reveal that Scamp
                                carried Scapegoat.
Rock    78      Nov 1967        First public display, showing missile.
                May 1965        First appeared shrouded in transporter/container.

----------   DEVELOPMENT BEGAN   ----------
SSF     82      1958            Year design began.
SMP     85      1958            Technological development began. [From graph.]

----------   PROTOTYPE TESTS   ----------
SecC    71      1965            3 tests in 1965, 2 undetermined and then 1 success; none in 1966.
                1967-1970       3 successes in 1967; 7 successes and 1 failure in 1968; 3 successes in 1969; 2
                                successes in 1970.
AWST    82                      "Tested not deployed".
SSF     82      1965            First flight test.
SMP     84      late 1960s      In test.
SMP     85      1964-65         Engineering and testing began. [From graph.]

----------   DEPLOYMENT BEGAN   ----------
SecC    69      mid-69          Possible for mobile mode.
                mid-70          Possible for fixed sites.
SecC    70                      No evidence to date of deployment.
                mid-70          Last 2 firings appear to have been for troop training; if so could be operational
                                by mid-70 as mobile system.
SecC    71                      Crew training occurred during the past year but no deployment identified to date.
SecD    71                      No deployment has been identified to date.
FIar    76                      [1968 or 1971], entered service same year as SS-13.
Rock    78      1968            Service intro.
AWST    82                      "Tested not deployed".
Meyr    84      1970            Service.
SMP     84                      Never deployed.

----------   DEPLOYED   ----------
MOW     76      in service

----------   DEPLOYMENT HISTORY   ----------
SecC    70      Might be used to replace SS-4.
Rock    78      "Its most prolific deployment seems to have been along the Sino-Soviet border, where it
                complements the SS-15, a missile which appeared later than Scapegoat and may have replaced it in
                increasing numbers".
Su01    83      Limited deployment in Asia, was not placed into series production and deployment, [with SS-15].
SMP     85      Cancelled in 1970 without deployment. [From graph.]

----------   NUMBERS IN SERVICE   ----------

----------   NUMBER IN 1955   ----------
SSF     82      none        In 1955, 1960, 1965.
```

BALLISTIC MISSILES  IRBMs  SS-14

```
----------  NUMBER IN 1970 ----------
SecC   71    none      30 Dec 1970 and 1 Mar 1971.
SSF    82    29

----------  NUMBER IN 1971 ----------
SecC   71              0-18, mid-71 projection.

----------  NUMBER IN 1972 ----------
SecC   71              18-27, mid-72 projection.

----------  NUMBER IN 1975 ----------
SSF    82    none      In 1975, 1980.

----------  NUMBER IN 1978 ----------
Rock   78              Unknown number deployed.

----------  LAUNCHER/SILO ----------
SecC   69    May be deployed in both fixed and mobile mode, FY70-71.

----------  CARRIERS ----------

----------  CARRIER ---------- TUBES ---- RELOADS ----------
SecC   69                                            May be deployed in both fixed and mobile mode, FY70-71.
HRST   75    tracked vehicle                         Mobile system.
MOW    76    tracked vehicle                         Older and newer types. Called "Iron Maiden" because
                                                     enclosed in split container.
FI     81                                            Enclosed in split container.
AWST   82                                            Mobile, in container on tracked transporter.
SSF    82                                            Basing mode: Mobile.
SMP    84    mobile

----------  CARRIER ---------- TUBES ---- RELOADS ----------
FIar   76    mod JSIII chassis                       Missile enclosed in split container.
SWM    77    Scamp (mod JS-III)                      Mobile 'Iron Maiden' transported/erector.
Rock   78    JS 111 chassis                          In parade.
AFM*   80    JS III tank chassis
FI     81    JS III chassis                          (Mod).

----------  CARRIER ---------- TUBES ---- RELOADS ----------
IAD4   66    T-10 chassis                            Chassis of T-10 heavy tank.

----------  DESIGNER ----------
SSF    82    Nadiradize      Design Bureau.

----------  DESIGN AND ENGINEERING ----------
FIar   76    Consists of the 2 upper stages of SS-13.
SWM    77    Consists of 2 upper stages of SS-13.
Rock   78    Consists of 2nd and 3rd stages of SS-13.
AFM*   80    Appears to be top 2 stages of SS-13.
FI     81    Consists of upper 2 stages of SS-13.
JWS    81    Appears to be top 2 stages of SS-13.

----------  USE AND CONFIGURATION ----------
IAD4   66    Reaction time after vehicle positioning could be very short.
FIar   76    Method of operation similar to SS-12.
SWM    77    After erection, covers are opened and moved aside.
AFM*   80    Container moved away from missile after erection.
JWS    81    Missile is erected, then container removed before firing.
Meyr   84    Penetration: 1.0; reliability: 0.3-0.7, "includes system reliability multiplied by operational
             readiness of deployed system".

----------  STRATEGY ----------
SWM    77    Soviet commanders enthusiastic about this mobile type of IRBM.
Meyr   84    [Ed: chart is labeled] Systems "assigned primary operational and strategic missions within the
             European theatre(s) of military operations".

----------  GEOGRAPHICAL DEPLOYMENTS ----------
MOW    76              Reported deployed in Mongolia.
Rock   78              "Its most prolific deployment seems to have been along the Sino-Soviet border,
                       where it complements the SS-15, a missile which appeared later than Scapegoat and
                       may have replaced it in increasing numbers".
AFM*   80              Deployment includes Outer Mongolia.
FI     81              Deployed on Mongolian/Chinese border.
JWS    81              Reported deployed Eastern USSR.

----------  RELATION TO OTHER MISSILES ----------
SecC   69    Has upper two stages of SS-13, FY70-71.
```

---------- **OTHER INFORMATION** ----------
SuO1  83     "Not considered satisfactory", [with SS-13 and SS-15].

# SS-15

```
Sources are:   SecC  71      HRST  75      MOW   76      Rock  78      FI*   79      SuO1  83      SMP   84
               ID01  68      GA01  75      FIar  76      SWM   77      AFM*  79      SSF   82      Meyr  84      SMP   85
```

### ———— US DESIGNATION ————

| | | |
|---|---|---|
| SecC | 71 | PL-1 |
| FIar | 76 | SS-15 |
| MOW | 76 | SS-XZ |
| SWM | 77 | SS-15 |
| Rock | 78 | SS-15 |
| AFM* | 79 | SS-15 |
| FI* | 79 | SS-XZ [SS-15] |
| SSF | 82 | SS-15 |
| Meyr | 84 | SS-15 |
| SMP | 84 | SS-X-15     '84, '85 |

### ———— NATO CODENAME ————

| | | |
|---|---|---|
| HRST | 75 | Scrooge |
| FIar | 76 | Scrooge |
| MOW | 76 | Scrooge |
| Rock | 78 | Scrooge |
| AFM* | 79 | Scrooge |
| FI* | 79 | Scrooge |
| SSF | 82 | Scrooge |

### ———— DESCRIPTION ————

| | | | |
|---|---|---|---|
| SecC | 71 | | Mobile ICBM candidate. |
| MOW | 76 | | "Mobile strategic weapon system". |
| Rock | 78 | LRBM | "Long Range Ballistic Missile", [table] |
| | | ICBM | [Text] |
| SSF | 82 | | Second generation. Intermediate range. |
| SMP | 84 | | "Shorter-range ICBM". |
| SMP | 85 | ICBM | |

### ———— EDITOR'S NOTES ————

| | | |
|---|---|---|
| SecC | 71 | Listed FY72. |
| Meyr | 84 | Note to chart: "Values pertain to initial service period". |
| SMP | 84 | Listed '84, '85. |

### ———— RANGE ————

| | | mi | nm | km | km conv | |
|---|---|---|---|---|---|---|
| SecC | 71 | | | | | "It...[cut off in DoD xeroxing, ed] range is not yet firmly established, but it appears to be a... With this range it would not be suitable as an ICBM unless... deployed in the far north, which would present serious oper...". |
| GA01 | 75 | 3,000 | | | 4,827 | |
| HRST | 75 | 2,000 | | | 3,218 | |
| FIar | 76 | | | 5,600 | 5,600 | |
| MOW | 76 | 3,500 | | 5,600 | 5,631 | Max. |
| SWM | 77 | 3,500 | | 5,630 | 5,631 | [Table] |
| Rock | 78 | | | 5,600 | 5,600 | [Table] |
| | | | | 5,500 | 5,500 | [Text] |
| AFM* | 79 | ≤ 3,500 | | | 5,631 | Derived from size of tube. |
| SSF | 82 | | 3,000-4,000 | | 5,556-7,408 | |
| Meyr | 84 | | | 5,000 | 5,000 | |

### ———— CEP ————

| | | ft | mi | nm | m | km | m conv |
|---|---|---|---|---|---|---|---|
| SSF | 82 | | | 0.8-1.0 | | | 1,481-1,852 |
| Meyr | 84 | | | | | 0.6-1.2 | 600.00-1,200 |

### ———— WARHEAD TYPE ————

| | | nuclear | conven | chem | |
|---|---|---|---|---|---|
| MOW | 76 | yes | | | "Thermonuclear or nuclear". |
| Meyr | 84 | yes | | | |

### ———— WARHEAD CHARACTERISTICS ————

| | | |
|---|---|---|
| Rock | 78 | single RV |
| SSF | 82 | single RV |
| Meyr | 84 | single RV |

```
---------- NUCLEAR YIELD ----------
              kT              MT           kT conv
FIar  76                    ? 1            1,000
Rock  78                      1.0          1,000
SSF   82                      0.6            600.00
Meyr  84     500-1,000                       500.00-1,000      500 kT-1 MT.

---------- WEIGHT ----------
              lb              t            kg           kg conv
FI*   79                                   60,000       60,000       "All-up weight", would create
                                                                     high track pressures and poor
                                                                     off-road mobility.

---------- LENGTH ----------
              in              ft           cm           m            m conv
ID01  68                      c. 65                                  19.81
FIar  76                                                ? 18.5       18.50
                                                        19           19.00        Launch tube.
MOW   76                      62'0"                     18.9         18.90        Launch tube.
SWM   77                      60                        18.3         18.29        [Table]
                                                        19           18.90        [Text]
Rock  78                                                18.5         18.50
                                                        19           19.00        Container.
AFM*  79                      62                                     18.90        Dimensions of launch tube.
FI*   79                                                19           19.00

---------- DIAMETER ----------
              in              ft           cm           m            cm conv
ID01  68                      c. 6.9                                 210.31
FIar  76                                                ? 2.0        200.00
                                                        2.2          220.00       Launch tube.
MOW   76                      6'6"                      2.00         198.12       Launch tube.
SWM   77                      5.5-6.0                   1.7-1.8      167.64-182.88
Rock  78                                                2.0          200.00
                                                        2.2          220.00       Container.
AFM*  79                      6'6"                                   198.12       Dimensions of launch tube.

---------- NUMBER OF STAGES ----------
Rock  78     3

---------- TYPE OF PROPULSION ----------
FIar  76     ? solid
SWM   77     solid
Rock  78     solid              Stages 1, 2 and 3.
SSF   82     solid

---------- MIDCOURSE (OR MAIN) GUIDANCE ----------
SSF   82     inertial

---------- OBSERVED ----------
SMP   84                        [Shown in picture]

---------- OBSERVED IN PARADE ----------
ID01  68     Nov 1965           First shown.
FIar  76     1965               First displayed.
Rock  78     Nov 1965           First public appearance.

---------- DEVELOPMENT BEGAN ----------
SSF   82     1958-61            Year design began.
SMP   85     1962               Technological development began. [From graph.]

---------- PROTOTYPE TESTS ----------
SecC  71     1968-69            No tests in 1965, 1966, or 1967; 3 failures and 2 successes in 1968; 3 successes in
                                1969; no tests in 1970. "There have been no...[cut off in DoD xeroxing, ed] and the
                                program may have been abandoned". "Another year of testing will probably be
                                required before th... Could be operational in a mobile role".
SSF   82     1968               First flight test.
SMP   84     late 1960s         In test.
SMP   85     1968-69            Engineering and testing began. [From graph.]

---------- DEPLOYMENT BEGAN ----------
FIar  76     <_ 1970            "May have entered service as late as 1970".
Rock  78     1969               Service intro, [table].
             late 1960s         First appeared in service, [text].
Meyr  84     1970               Service.
SMP   84                        Never deployed.
SMP   85     1973-74            [From graph.]
```

BALLISTIC MISSILES  IRBMs  SS-15  333

---------- DEPLOYED ----------
FIar 76                     Units believed deployed in Outer Mongolia.
MOW  76    in service

---------- DEPLOYMENT HISTORY ----------
Rock 78    Due for replacement with SS-20. "[SS-14's] most prolific deployment seems to have been along the
           Sino-Soviet border, where it complements the SS-15, a missile which appeared later than Scapegoat
           and may have replaced it in increasing numbers", [from SS-14 entry]. "Several SS-14 units may have
           been re-equipped with the Scrooge".
SuO1 83    Limited deployment in Asia, was not placed into series production and deployment, [with SS-14].
SMP  85    Cancelled in 1976. [From graph.]

---------- CARRIERS ----------

---------- CARRIER ---------- TUBES ---- RELOADS ----------
HRST 75    tracked vehicle                    Mobile system.
FIar 76    tracked vehicle
MOW  76                                       Mobile.
SSF  82                                       Basing mode: Mobile.
SMP  84    mobile

---------- CARRIER ---------- TUBES ---- RELOADS ----------
Rock 78    JS 111 chassis                     Used for display.
AFM* 79    JS III                             Same basic vehicle as SS-14.
FI*  79    JS-III chassis                     60,000 kg "all-up weight", would create high track
                                              pressures and poor off-road mobility.

---------- CARRIER ---------- TUBES ---- RELOADS ----------
SWM  77    Scrooge                            Mobile transporter/erector.

---------- DESIGNER ----------
SSF  82    Nadiradize     Design Bureau.

---------- USE AND CONFIGURATION ----------
FIar 76    Launched directly from tube.
MOW  76    Container serves as launch tube.
SWM  77    Assumed that missile is launched from its tube.
Rock 78    Hydraulic jacks on JS 111 lift to vertical, missile fired from container, allowing great mobility.
AFM* 79    Missile is fired from launch tube, which is raised to vertical for firing.
Meyr 84    Penetration: 1.0; reliability: 0.3-0.7, "includes system reliability multiplied by operational
           readiness of deployed system".

---------- STRATEGY ----------
Rock 78    Scrooge would probably be used against western Europe and the UK in the event of a strategic
           European conflict"; also many are deployed in outer Mongolia.
Meyr 84    [Ed: chart is labeled] Systems "assigned primary operational and strategic missions within the
           European theatre(s) of military operations".

---------- GEOGRAPHICAL DEPLOYMENTS ----------
FIar 76                Units believed deployed in Outer Mongolia.
MOW  76                Observed in service in Mongolia.
SWM  77                Scrooge launchers identified in Outer Mongolia.
Rock 78                "[SS-14's] most prolific deployment seems to have been along the Sino-Soviet
                       border, where it complements the SS-15, a missile which appeared later than
                       Scapegoat and may have replaced it in increasing numbers", [from SS-14 entry].
                       "Several SS-14 units may have been re-equipped with the Scrooge".
                       Many are deployed in outer Mongolia.

---------- RELATION TO OTHER MISSILES ----------
ID01 68    [For range and payload, entry says "see ICBM Savage", implying they know that is what it contains,
           ed].
FIar 76    Missile may be a modified SS-13.
MOW  76    Similar in concept to Scamp/Scapegoat.
SWM  77    Possibly related to SS-13 Savage. Missile has better performance than SS-14.
Rock 78    Thought to be derivative of SS-13; but 20 m length and 1.7 m diameter of SS-13 is greater than
           SS-15 container, so substantial modification is indicated.
FI*  79    Probably this system is transporter/loader for SS-13 Savage and not a weapon. SS-13 without
           warhead would fit neatly in container.
SuO1 83    Based on two stages of SS-13.

---------- OTHER INFORMATION ----------
AFM* 79    Little known of missile.
SuO1 83    "Not considered satisfactory", [with SS-13 and SS-14].

# SS-20

```
Sources are:    SecD 77     JCS  79     SecD 80     SMP  81     SecD 82     NWG  83     Meyr 84
Flar 76         SWM  77     SecD 79     A168 81     US15 81     SSF  82     SecD 83     Or01 84
JCS  76         IISS 78     BA03 80     FI   81     AFM  82     Clns+ 83    SMP  83     SecD 84
MOW  76         JCS  78     BA04 80     IISS 81     AWST 82     FI+  83     Su01 83     SMP  84
SecD 76         Rock 78     Clns 80     JCS  81     IISS 82     IISS+ 83    AFM+ 84     SMP  85
IISS 77         SecD 78     IISS 80     JWS  81     JCS  82     JCS  83     ArkS 84
JCS  77         IISS 79     JCS  80     SecD 81     JP01 82     JWS+ 83     JCS  84
```

---------- **US DESIGNATION** ----------

| | | | |
|---|---|---|---|
| Flar | 76 | SS-X-20 | [Text] |
| | | SS-20 | [Table] |
| JCS | 76 | SS-X-20 | FY77-78. |
| MOW | 76 | SS-X-20 | |
| SecD | 76 | SS-X-20 | FY77-78. |
| SWM | 77 | SS-20 | |
| JCS | 78 | SS-20 | FY79-86. |
| Rock | 78 | SS-20 | |
| SecD | 78 | SS-20 | FY79-85. |
| Clns | 80 | SS-20 | |
| FI | 81 | SS-20 | |
| JWS | 81 | SS-20 | |
| SMP | 81 | SS-20 | '81, '83, '84, '85. |
| AFM | 82 | SS-20 | |
| AWST | 82 | SS-20 | |
| IISS | 82 | SS-20 | |
| SSF | 82 | SS-20 | |
| Meyr | 84 | SS-20 | |

---------- **DESCRIPTION** ----------

| | | | |
|---|---|---|---|
| MOW | 76 | | Mobile IRBM. |
| SecD | 76 | | "Development of a mobile IRBM". |
| JCS | 77 | | "Currently assessed as an IRBM". |
| SWM | 77 | | "Mini-ICBM in a land-mobile role". |
| JCS | 78 | IRBM | FY79-82. |
| Rock | 78 | LRBM | "Long Range Ballistic Missile"; "Land mobile weapons system of true ICBM proportions". |
| SecD | 78 | IRBM | FY79-82. |
| Clns | 80 | IRBM | |
| SecD | 80 | | Part of Soviet LRTNF, FY81-82. |
| FI | 81 | mobile IRBM. | |
| JWS | 81 | | Mobile IRBM. |
| SMP | 81 | | Intermediate Range Ballistic Missiles. Long-Range Theater Nuclear Weapon. |
| AFM | 82 | mobile IRBM | |
| AWST | 82 | | "Intermediate range missile". |
| SSF | 82 | | Third generation Intermediate range. |
| SMP | 83 | | Longer-Range Intermediate-Range Nuclear Forces, '83, '85. |

---------- **EDITOR'S NOTES** ----------

| | | |
|---|---|---|
| Flar | 76 | Source distinguishes 3 unnamed Mods, all "employ the same basic missile". |
| JCS | 76 | Listed FY77-86. |
| SecD | 76 | Listed FY77-85. |
| SMP | 81 | Listed '81, '83, '84, '85. |
| Meyr | 84 | Note to chart: "Values pertain to initial service period". |

BALLISTIC MISSILES   IRBMs   SS-20   335

```
          ———— RANGE ————
              mi          nm           km           km conv
SecD  77                  >_ 2,000                  3,704          Demonstrated range.
                                       5,500        5,500          Range could equal SALT limit of
                                                                   5,500 km with 3rd stage or MIRV
                                                                   offloading.
Rock  78                               5,700        5,700          [Table]
SecD  78                               >_ 3,000     3,000          Estimated at least 3,000 km.
SecD  79                               > 3,000      3,000          Well over 3,000 km.
BA04  80                  2,000-2,400               3,704-4,444
Clns  80                  2,700                     5,000
SecD  80                               >_ 4,400     4,400          FY81-82.
                                                                   Exceeds SS-4,5 range, FY81-82.
FI    81                               4,000        4,000
JCS   81                               4,400        4,400
JWS   81                               3,700-5,500  3,700-5,500    Lower figure is certain, upper
                                                                   figure for offloading MIRV
                                                                   payload or adding a stage.
SMP   81                               5,000        5,000          '81, '83, '84, '85.
US15  81      3,107                                 4,999          According to IISS.
AWST  82                  2,500                     4,630
JCS   82                               5,000        5,000          FY83-86.
JP01  82                               5,000        5,000
SSF   82                  2,700                     5,000
Clns+ 83                  3,100                     5,741
JWS+  83                               4000-5000    4,000-5,000    Generally accepted range.
                                       4,000-4,500  4,000-4,500    Soviet figure.
                                       5,000        5,000          US estimate.
                                       3,700        3,700          "Has been recorded".
AFM+  84      3,100                                 4,987
Meyr  84                               5,000        5,000

          ———— SPEED ————
              mph         kmph         Mach         kmph conv
JCS   82                                                           Operational flight time: minutes, FY83-86.
JP01  82                                                           6 minutes for SS-20 to reach "least distant"
                                                                   targets.

          ———— CEP ————
              ft    mi    nm           m            km    m conv
JCS   79                                                           "Improved".
BA04  80      1,500                                       457.20
Clns  80                  0.2                             370.40   [Unit assumed, ed]
JCS   80                                                           [Accuracy] greater than SS-4
                                                                   and SS-5, FY81-84.
SecD  80                                                           "Substantially more
                                                                   accurate" than SS-4,5 (each
                                                                   warhead), FY81-82.
A168  81      440                                         134.11   Over 2500 mile range.
FI    81                               750                750.00   If launched from
                                                                   pre-surveyed sites.
SMP   81                                                           "Very accurate".
US15  81      c. 300                                      91.44
AFM   82      c. 2,500                                    762.00   When launched from
                                                                   pre-surveyed site.
JP01  82                               c. 150             150.00   At max range, better at
                                                                   3,000 km range.
SSF   82                  0.16                            296.32
Clns+ 83                  0.23                            425.96
JWS+  83                               400                400.00
SMP   83                                                           "Accuracy and reaction time"
                                                                   are "marked improvement over
                                                                   the older systems".
Meyr  84                                            0.1-0.4  100.00-400.00
SecD  84                                                           Highly accurate.
SMP   84                                                           Accuracy very significantly
                                                                   increased over older SS-4s
                                                                   and SS-5s.
SMP   85                                                           "Accuracy and reaction time"
                                                                   are "significant"
                                                                   improvement the older
                                                                   systems.
```

## WARHEAD TYPE

|  |  | nuclear | conven | chem |  |
|---|---|---|---|---|---|
| JWS+ | 83 | yes | | | |
| SMP | 83 | yes | | | |
| ArkS | 84 | yes | | | "The SS-20 is not conventionally-capable". |
| Meyr | 84 | yes | | | |

## WARHEAD CHARACTERISTICS

| | | | |
|---|---|---|---|
| FIar | 76 | | Has PBV; could carry MIRVs. |
| MOW | 76 | MIRVed | |
| SecD | 76 | MIRVed | FY77,84. |
| SecD | 77 | 3 MIRVs | FY78-82. |
| SWM | 77 | | Probably single RV or MRVs. |
| JCS | 78 | MIRVs | FY79-81. |
| Rock | 78 | single or 3 MRV | |
| JCS | 79 | 3 MIRVs | FY80,82-86. |
| Clns | 80 | 3 warheads | |
| FI | 81 | 3 MIRVs | |
| SecD | 81 | | [Estimates of total warheads reflect exactly 3 per launcher, ed]. |
| SMP | 81 | 3 MIRVs | '81, '83, '84, '85. |
| US15 | 81 | 3 MIRVs | |
| AFM | 82 | 3 MIRVs | |
| AWST | 82 | 3 MIRVs | |
| SSF | 82 | 3 RVs | |
| JWS+ | 83 | 3 MIRVs | |
| | | single RV | Alternative. |
| Meyr | 84 | 3 MIRVs | |
| SecD | 84 | 3 warheads | |

## NUCLEAR YIELD

|  |  | kT | MT | kT conv |  |
|---|---|---|---|---|---|
| Rock | 78 | | 1.5 | 1,500 | |
| BA03 | 80 | 150-500 | | 150.00-500.00 | Estimates of yield vary over range. |
| BA04 | 80 | 150 | | 150.00 | [In text, 1 kT in table.] |
| Clns | 80 | 150 | | 150.00 | |
| FI | 81 | 151 | | 151.00 | [Sic] |
| US15 | 81 | 150 | | 150.00 | |
| AFM | 82 | 150 | | 150.00 | |
| SSF | 82 | | 0.15-0.50 | 150.00-500.00 | Per warhead. |
| JWS+ | 83 | 150 | | 150.00 | |
| Meyr | 84 | 100-300 | | 100.00-300.00 | |
| Or01 | 84 | | .15-.5 | 150.00-500.00 | |

## THROWEIGHT

|  |  | lb | t | kg | kg conv |
|---|---|---|---|---|---|
| Clns | 80 | 3,500 | | | 1,587 |
| US15 | 81 | 12,000 | | | 5,443 |

## WEIGHT

|  |  | lb | t | kg | kg conv |
|---|---|---|---|---|---|
| FI | 81 | | | c. 25,000 | 25,000 |

## LENGTH

|  |  | in | ft | cm | m | m conv |  |
|---|---|---|---|---|---|---|---|
| FIar | 76 | | | | ? 11 | 11.00 | |
| SWM | 77 | | 55.0 | | 16.8 | 16.76 | |
| Rock | 78 | | | | 11.5 | 11.50 | |
| FI | 81 | | | | ? 16 | 16.00 | |
| SMP | 81 | | | | 16 | 16.00 | [From graph], '81, '83, '84, '85. |
| AWST | 82 | | 55 | | | 16.76 | |
| AFM+ | 84 | | 55 | | | 16.76 | |

## DIAMETER

|  |  | in | ft | cm | m | cm conv |
|---|---|---|---|---|---|---|
| FIar | 76 | | | | ? 1.7 | 170.00 |
| SWM | 77 | | 6.5 | | 2.0 | 198.12 |
| Rock | 78 | | | | 1.8 | 180.00 |
| FI | 81 | | | c. 170 | | 170.00 |
| AWST | 82 | | 6.5 | | | 198.12 |

## NUMBER OF STAGES

|  |  |  |  |
|---|---|---|---|
| JCS | 77 | 2 | FY78,80. |
| SWM | 77 | 2 | |
| FI | 81 | 2 | |
| AWST | 82 | 2 | |
| JWS+ | 83 | 2 | [Analysis table] |

```
---------- TYPE OF PROPULSION ----------
FIar    76    solid
MOW     76    solid
JCS     77    solid          FY78,80.
SWM     77    solid
Rock    78    solid          Stages 1 and 2.
Clns    80    solid
FI      81    solid
SMP     81    solid
AFM     82    solid
AWST    82    solid
SSF     82    solid
JWS+    83    solid          [Analysis table]

---------- MIDCOURSE (OR MAIN) GUIDANCE ----------
FI      81    inertial
SMP     81    inertial
SSF     82    inertial

---------- CONTROL SYSTEM ----------
FI      81    ? Fluid injection.

---------- DEVELOPMENT BEGAN ----------
SSF     82    1965-68        Year design began.
SMP     85    1966           Technological development began. [From graph.]

---------- DEVELOPMENT UNDERWAY ----------
MOW     76                   Under development.

---------- PROTOTYPE TESTS ----------
FIar    76                   Trials program on Kamchatka Peninsula is almost complete.
SecD    76                   Recent testing emphasis on SS-X-20 [over SS-X-16].
JCS     77    Fall, 1974     R&D flight testing began fall 1974, now essentially complete. Program has proceeded
                             at a rapid pace; highly successful. Has featured all milestones of a typical R&D
                             program.
Rock    78                   "Tests have been carried out in the Kamchatka Peninsula".
SSF     82    1974-75        First flight test.
SMP     85    1972-73        Engineering and testing began. [From graph.]

---------- IOC ----------
JCS     82    1977           Year operational, FY83-86.

---------- DEPLOYMENT BEGAN ----------
FIar    76                   Not yet in service.
JCS     77                   Deployment could begin early in 1977. Initial deployment is expected "in the near
                             future". "Evidence that SS-X-20s will soon be deployed".
SecD    77                   "Prospective deployment" [implies not deployed].
SWM     77    1976           Said to be deployed Kamchatka late spring 1976.
JCS     78                   "Initial deployment of the SS-20 has begun".
Rock    78    1976           Service intro.
                             "Operational deployment is still awaited".
SecD    78                   "In our judgment" "already being deployed". Assessed to be operational.
JCS     79                   Deployment "has been underway for the last year".
Clns    80    1977           First deployed.
JWS     81    1977
SMP     81    1977           Introduced, "First reached operational status".
IISS    82    1977           First year deployed.
SSF     82    1977           Year operation began.
SecD    83    1977
Meyr    84    1977           Service.
SMP     84    1977           Initial deployment, '84, '85.
SMP     85    1976-77        [From graph.]

---------- DEPLOYED ----------
SecD    79                   In the field.

---------- DEPLOYMENT COMPLETE ----------
SuO1    83                   Moratorium on further European deployment announced March 1982 "probably reflected
                             conclusion of the intended deployment". No new deployment sites started since
                             moratorium announced [as of May/June 1983, ed.]
```

## DEPLOYMENT HISTORY

| | | |
|---|---|---|
| SecD | 77 | Expected to be deployed as replacement for SS-4s and SS-5s. |
| SWM | 77 | Replacing SS-4 and SS-5. |
| Rock | 78 | Expected to replace SS-4, maybe also SS-5. |
| SecD | 78 | MRBM and IRBM force being modernized with SS-20. Expected to "replace or augment" MRBM and IRBM force, FY79-80. |
| JCS | 79 | More SS-20s expected to replace entire SS-4 and -5 force. |
| JCS | 80 | No sign of abatement in deployment; deployment continues. |
| SecD | 80 | Gradually augmenting older missiles. Already substantial base structure for SS-20 and it is expanding. |
| JCS | 81 | Additional deployments of SS-20 in past year, FY82-83. Deployment continues. "Widely deployed". |
| JWS | 81 | Probable role is to replace SS-4 and SS-5. Now undergoing operational deployment. |
| SecD | 81 | Clear impression that SS-20 is augmenting (not replacing) SS-4,5 at least for forseeable future. Substantial base structure expanding "at a rapid pace". |
| SMP | 81 | "There is no sign that the deployment is slackening"; since Jan 1981, "pace of SS-20 base construction has increased, particularly opposite the NATO nations". |
| JCS | 82 | In 1981, "Soviets initiated construction of additional new SS-20 bases". |
| SecD | 82 | USSR continues to deploy SS-20s. |
| JWS+ | 83 | Pace of deployment increased after Jan 1981. In May 1983, stated 2 or 3 new based under construction in Central Asia, within range of NATO targets. |
| SecD | 83 | USSR engaged since 1977 in a significant buildup of its LRINF [longer-range INF] with SS-20 deployment. |
| SMP | 83 | New launchers added during 1982. "The Soviets continue to produce and deploy" SS-20s "in both the Western and Far Eastern Theaters and are likely to construct additional complexes beyond those presently under construction". |
| SuO1 | 83 | From publicly released figures possible to extrapolate "construction of about 9 new deployment bases (each having one regiment of 9 launchers) per year in 1978 and 1979". Number of new bases started began to decline in 1980, no new bases begun after early 1982 [as of May/June 1983, ed.] "Deployment group starts (nine launch positions to each group)" peaked in 1978-79, completions in 1980-81. Completions in 1980-81 based on deployment decisions made in 1972-75. Initial and major part of deployment programme established in "framework of the tenth Five Year Plan in 1975-76". Original military proposals in early and mid-1970s may have been for 450-500 launchers, with decision to level off made in eleventh Five-Year Plan, probably in 1980. |
| SecD | 84 | Soviets continuing to construct new sites. |
| SMP | 84 | Construction of new SS-20 facilities in western USSR has resumed. |
| SMP | 85 | Soviets have "pressed ahead" with construction of new SS-20 bases in western and eastern USSR. |

## NUMBERS PRODUCED

| | | | |
|---|---|---|---|
| BA03 | 80 | 60 | New missiles being produced at the rate of "about 60 per year". |
| JCS | 83 | | At least as many refire warheads again as on launchers (1200, SS-4,5,20 total). |

## NUMBERS IN SERVICE

### NUMBER IN 1955

| | | | |
|---|---|---|---|
| SSF | 82 | none | In 1955, 1960, 1965, 1970, 1975. |

### NUMBER IN 1970

| | | | |
|---|---|---|---|
| Clns | 80 | none | 1970-1977. |

### NUMBER IN 1977

| | | | |
|---|---|---|---|
| IISS | 77 | c. 20 | July, total deployed. |
| SecD | 77 | | Deployment of SS-20 could give USSR 3 times previous warheads (in MRBMs and IRBMs). |
| Or01 | 84 | 10 | December. |

### NUMBER IN 1978

| | | | |
|---|---|---|---|
| IISS | 78 | 100 | July, total deployed. |
| SecD | 78 | | "With a successful multiple refire capability", deployment of SS-20 could give USSR 3 times previous warheads (in MRBMs and IRBMs). |
| Clns | 80 | 60 | |
| SMP | 83 | 12 | [Read from graph], '83, '84, '85. |
| Or01 | 84 | 70 | December. |

### NUMBER IN 1979

| | | | |
|---|---|---|---|
| IISS | 79 | 120 | July, total deployed. |
| JCS | 79 | < 100 | "Somewhat less than 100" operational SS-20 launchers deployed in Western USSR and Soviet Far East. |
| Clns | 80 | 120 | |
| SMP | 83 | 75 | [Read from graph], '83, '84, '85. |
| Or01 | 84 | 140 | December. |
| SecD | 84 | 140 | At time of NATO deployment decision [1979]; 230 added over past 4 years. |

BALLISTIC MISSILES  IRBMs  SS-20  339

---------- **NUMBER IN 1980** ----------
| | | | |
|---|---|---|---|
| BA03 | 80 | 150 | Launchers deployed, two-thirds targeted on Western Europe, rest on China. |
| BA04 | 80 | 100 | "More planned replacements." |
| IISS | 80 | 160 | July, total deployed. |
| SecD | 80 | > 100 | More than 100 launchers. |
| | | 100 | 100 launchers (1980). |
| SSF | 82 | 180 | |
| Clns+ | 83 | 160 | |
| SMP | 83 | 130 | [Read from graph], '83, '84, '85. |
| Or01 | 84 | 200 | December. |

---------- **NUMBER IN 1981** ----------
| | | | |
|---|---|---|---|
| FI | 81 | 180 | Operational in West USSR. |
| IISS | 81 | 230 | July, total deployed. |
| JCS | 81 | 180 | In Strategic Rocket Force. |
| JWS | 81 | 180 | By early 1981. |
| SecD | 81 | c. 180 | About 180 launchers. 180 launchers worldwide, 1 Jan 1981. |
| | | | About 80 operational launchers added since FY 1981 report. |
| SMP | 81 | c. 250 | With 750 nuclear warheads deployed as of July 1981. |
| | | 250+ | "At bases known to be under construction, another 65 launchers" (195 warheads) will be deployed; "perhaps as many as 100-to-150 additional launchers—300-to-450 warheads—could be fielded before the deployment program reaches its conclusion". |
| AFM | 82 | 250 | By July 1981. |
| JCS | 82 | 260+ | Deployed, global, Dec 1981. |
| Clns+ | 83 | 240 | |
| JWS+ | 83 | c. 300 | 1981, [analysis table] |
| SMP | 83 | 205 | [Read from graph], '83, '84, '85. |
| Or01 | 84 | 270 | December. |

---------- **NUMBER IN 1982** ----------
| | | | |
|---|---|---|---|
| AWST | 82 | 180 | Deployed. |
| IISS | 82 | 315 | July, total deployed. |
| JP01 | 82 | >_ 260 | |
| | | | One installed every five days. |
| Clns+ | 83 | 325 | |
| SMP | 83 | 280 | [Read from graph], '83, '84, '85. |
| Or01 | 84 | 300 | March. |

---------- **NUMBER IN 1983** ----------
| | | | |
|---|---|---|---|
| FI+ | 83 | > 330 | Now operational. |
| | | | Up to 465 launchers could be deployed by the time production is completed. |
| IISS+ | 83 | c. 360 | July, total deployed, Europe (including European USSR) only. |
| JCS | 83 | 333 | Deployed globally, 1 Jan. |
| | | > 300 | Launchers deployed, carrying nearly 1,000 warheads. |
| JWS+ | 83 | 351 | Operational as of May 1983, official US estimates. |
| SecD | 83 | > 300 | Deployed. |
| SMP | 83 | 333 | |
| | | > 330 | |
| | | 340 | [Read from graph], '83, '84, '85. |
| Su01 | 83 | 351 | Launchers, May/June. |
| Or01 | 84 | 351 | March. |
| | | 369 | December. |
| | | 378 | With additional base discovered East of Urals after article went to press, December 1984. |

---------- **NUMBER IN 1984** ----------
| | | | |
|---|---|---|---|
| AFM+ | 84 | c. 360 | |
| JCS | 84 | 378 | Deployed, global, 1 Jan. |
| SecD | 84 | 378 | Already deployed. |
| SMP | 84 | 378 | |

---------- **NUMBER IN 1985** ----------
| | | | |
|---|---|---|---|
| BA04 | 80 | 200 | "By the mid-1980s, according to Pentagon sources, barring any successful effort to negotiate new limits on theater weapons, the current Soviet force of some 450 medium-range missiles will be replaced by a maximum of some 200 SS-20s." |
| SecD | 80 | | 250-300 (mid-1980's, estimated). |
| JCS | 81 | | > 300 expected by mid-80s. |
| SecD | 81 | | 300+ (mid-1980's, estimated); may be larger because of continuing construction program. |
| SMP | 85 | c. 400 | |

---------- **NUMBER IN 1988** ----------
| | | | |
|---|---|---|---|
| SMP | 84 | | Deployed launchers could increase by at least 50% by the late 1980s. |

## UNDATED NUMBER

| | | | |
|---|---|---|---|
| JWS | 81 | | 300 missiles expected eventually, perhaps 2/3 against Europe. |
| AFM | 82 | | Eventual force total expected to be "300/400 plus reloads" [final report]. |
| JCS | 82 | | Number of launchers likely "will be less than SS-4 and SS-5 levels". |
| SSF | 82 | 180 | "Number of missiles deployed". |

## CARRIERS

### CARRIER —— TUBES —— RELOADS

| | | |
|---|---|---|
| FIar | 76 | Will be deployed on mobile launchers as was SS-14. |
| JCS | 76 | "Developed to be capable of deployment in mobile variants". |
| MOW | 76 | Mobile. |
| SecD | 76 | Believed it "will be deployed in a mobile or road-transportable mode". |
| JCS | 77 | "Developed with mobile deployment capabilities". |
| SWM | 77 | Mobile launch system. |
| JCS | 78 | Deployed as a mobile system. |
| | | Mobile system, FY79-80,82-86. |
| SecD | 78 | Mobile, FY79-82,85. |
| JCS | 80 | Mobility and survivability greater than SS-4 and SS-5, FY81-83. |
| SecD | 80 | Mobile and difficult to target, FY81-82. |
| FI | 81 | Tracked vehicle. |
| JWS | 81 | Mobile. |
| SMP | 81 | Mobile launcher, '81, '83, '85. "Highly mobile". |
| AFM | 82 | Tracked vehicle. |
| AWST | 82 | Mobile. |
| SSF | 82 | Mobile. |
| CLns+ | 83 | "Mobile (no survey)". |
| FI+ | 83 | reloads: Each operational launcher has a reload round in reserve. |
| JCS | 84 | reloads: "SS-20 refires" [exist] FY85-86. [Chart implies c. 1 reload per launcher, based on "warheads on refire missiles", ed]. |
| SecD | 84 | reloads: Refire missiles available. |
| SMP | 85 | Designed for road-mobile deployment. |

### CARRIER —— TUBES —— RELOADS

| | | |
|---|---|---|
| JWS+ | 83 | Wheeled transporter/erector/launcher. |
| | | reloads: Said to be capable of refire, but according to some, considerable heat is involved in each launch, so it may not be possible to prepare for second launch in less than a few hours. |

## DESIGNER

| | | |
|---|---|---|
| SSF | 82 | Nadiradize Design Bureau. |

## DESIGN AND ENGINEERING

| | | |
|---|---|---|
| FIar | 76 | Uses upper 2 stages of SS-16. |
| MOW | 76 | Derived from SS-X-16, using its first two stages. |
| SecD | 76 | Uses "first two booster stages" of SS-X-16. Derivative of SS-X-16, comprises first two stages, FY77-78. |
| JCS | 77 | Comprised of first two stages of SS-X-16. |
| SWM | 77 | Employs first two stages of SS-16. |
| Rock | 78 | Developed from first 2 stages of SS-16. |
| SecD | 78 | Consists of first 2 stages of SS-16, FY79-80. |
| SecD | 80 | Derivative of SS-16. |
| JWS | 81 | SS-20 is top 2 stages of SS-16. |
| AFM | 82 | Consists of first 2 stages of SS-X-16. |
| AWST | 82 | Uses first two stages of SS-16. |
| SMP | 84 | Development of SS-20 missiles founded on technologies represented by SS-X-14 and SS-X-15. "The Soviets are likely to improve the SS-20". |

## USE AND CONFIGURATION

| | | |
|---|---|---|
| BA04 | 80 | "Good" reaction time. |
| SMP | 81 | "Reaction time [max.]" "1 hour+". |
| JCS | 82 | Responsiveness improved over SS-4 and SS-5. |
| JP01 | 82 | "Every mobile SS-20 base" made up of one launcher and two vehicles each with a reload. Can be launched roughly 1/2 hour after the first one. Mobile bases are permanently deployed—almost impossible to spot them on the ground. |
| JWS+ | 83 | CEP must vary with care of initial site survey and with range. Some reports claim there are significant number of pre-surveyed sites in USSR, but not absolutely essential for SS-20 operation. Earlier report, said to be based on secret CIA report, said re-targeting and re-siting could probably be effected in seconds if new target within 1 or 2 degrees of original launch azimuth. If targets widely separated, might require 20-30 minutes for retargeting. US sources report complexes of (up to) 9 missiles. Unit also has command, support, and reload vehicles (one per launcher). May be meteorological unit among support vehicles. |
| NWG | 83 | Operational echelon: Theater. |
| SMP | 83 | "Accuracy and reaction time" are "marked improvement over the older systems". |
| Meyr | 84 | Penetration: 1.0; reliability: 0.6-0.9, "includes system reliability multiplied by operational readiness of deployed system". |
| SMP | 84 | Mobility enables both on- and off-road operation, '84, '85. Reaction time very significantly increased over older SS-4s and SS-5s. Condor transport will be able to carry "major weapons systems such as the SS-20 missile launcher". |
| SMP | 85 | "Accuracy and reaction time" are "significant" improvement the older systems. |

## RELOADS IN GENERAL

| | | |
|---|---|---|
| JCS | 79 | Has refire capability, FY80,82. |
| BA03 | 80 | "Each launcher may be reloadable and refireable and eventually provided with several missiles." |
| SecD | 80 | "Estimated reload capability". |
| JWS | 81 | Has refire capability. |
| SecD | 81 | Launchers have a refire capability. |
| SMP | 81 | 1 per launcher, '81, '83, '84, '85. |
| AFM | 82 | Multiple reloads. |
| IISS | 82 | ? 1 reload per system. |
| SMP | 83 | "Reloads for each launcher". |
| SMP | 84 | SS-20 launcher has capability of being reloaded and refired, Soviets stockpile refire missiles. |
| SMP | 85 | "Each [launcher] with a MIRVed three-warhead missile and reload". |

## STRATEGY

| | | |
|---|---|---|
| FIer | 76 | Ability to substitute lighter warhead and get greater range could allow USSR to upgrade during a crisis, changing strategic balance with US. |
| JCS | 77 | Upgrading to SS-X-16 could be relatively quick, FY78-79. |
| SecD | 77 | Both SS-X-20 and Backfire "indefinite as to range capabilities and missions". Does not fit either central or non-central arms control category. |
| JCS | 79 | SS-20 could be upgraded to SS-16, FY80-81. |
| BA03 | 80 | Because of yield and accuracy of warheads and relative invulnerability, SS-20 is nuclear war fighting rather than war deterring weapon. |
| BA04 | 80 | Large collateral damage from 150-kiloton warhead "would appear to limit preemptive 'first use' of the SS-20 in support of a conventional military attack upon NATO". More likely used to retaliate for nuclear attacks on Soviet Union or its allies. |
| JCS | 81 | Could be upgraded to ICBM configuration "rather rapidly". |
| SecD | 81 | LRTNF "greatly augmented" by SS-20 and Backfire. |
| SMP | 81 | "Three warheads per missile greatly increases Soviet firepower; mobility increases survivability". |
| AFM | 82 | Formidable threat to Europe, but not covered by SALT II, since range < 5,500 km. |
| SecD | 82 | Much of [TNF as a whole] "could be quickly shifted or retargeted to be concentrated against any potential theater of conflict". |
| JCS | 83 | Can strike targets anywhere in Europe or Asia from Soviet territory. |
| SecD | 83 | In past 5 years, Soviet deployments of SS-20s targeted on Europe have posed grave threat to credibility of NATO's deterrent posture. |
| SMP | 83 | SS-20 mobility enhances survivability, '83, '84, '85. |
| AFM+ | 84 | Was subject of a round of strategic arms talks in 1983. |
| JCS | 84 | Significantly greater threat than SS-4s, 5s. Can cover entire European theater, all of East Asia, and other vital areas. Can reach significant targets in Europe, even when deployed beyond NATO LRINF range FY85-86. |
| Meyr | 84 | [Ed: chart is labeled] Systems "assigned primary operational and strategic missions within the European theatre[s] of military operations". |
| SecD | 84 | Can reach Europe, but not the US. Deployments "over the past six years" have eroded credibility of NATO's deterrent posture. [Some or all] states in Asia, Middle East, North Africa and Europe are potential targets. |
| SMP | 85 | "Mobility enables on- and off-road operation." |

## COMBAT REPORTS/EFFECTIVENESS

| | | |
|---|---|---|
| IISS | 82 | Serviceability: 0.9; Survivability: 0.9; Reliability: 0.8. |

## GEOGRAPHICAL DEPLOYMENTS

| | | |
|---|---|---|
| SecD | 78 | Could hit targets in Europe, Asia, and the Middle East. |
| JCS | 79 | Preponderance of SS-4,5,20 force located in western USSR. "About one-half" of SS-20s located in western USSR.<br>"Somewhat less than 100" operational SS-20 launchers deployed in Western USSR and Soviet Far East. |
| SecD | 80 | 60 (of 100, 1980) in European theater (normally based in Europe or within striking range of Europe). 2/3 of total "could be deployed against NATO" in mid-1980s, FY81-82. |
| FI | 81 | Also deployed against Asian targets. 180 operational in West USSR. |
| JCS | 81 | Some SS-20s based in eastern and central USSR [growing?]. Majority of SS-4,5,20 force believed targeted on Western Europe; rest can reach Middle East, China, and Japan.<br>[Map shows areas of coverage of world, ed], FY82-86. |
| JWS | 81 | 300 missiles expected eventually, perhaps 2/3 against Europe. |
| SecD | 81 | 110 (of 180, 1 Jan 1981) in European theater [caveat as in FY81]. Some deployed to Asia. |
| SMP | 81 | Increase in SS-20 sites in western USSR intensifies tactical nuclear strike capability targeted against Western Europe, [map gives coverage of Europe]. Of 250 launchers, "175 with 525 warheads are deployed opposite the NATO countries". Since Jan 1981, "pace of SS-20 base construction has increased, particularly opposite the NATO nations" SS-20 force will be concentrated primarily against European theater. |
| IISS | 82 | 33% assumed in West, and 33% in central "swing-zone" with range to target Europe. |
| JCS | 82 | "Wider deployment pattern" than SS-4,5 (plus range increase) enable extension of LRTNF to Mideast, Southwest Asia, and East Asia. In 1981, "Soviets initiated construction of additional new SS-20 bases", from which missiles can hit NATO Europe; remaining SS-20 deployments likely will be located in western USSR. |
| SecD | 82 | USSR continues to deploy SS-20s both on "Western Front" and "in East Asia and the Pacific". |
| FI+ | 83 | Some two-thirds located in western USSR; remainder for use against Asian targets. |
| JWS+ | 83 | 243 (of 351 of May 1983) based west of Urals. Remainder presumed targeted on oChina, Japan and South Korea. |
| SecD | 83 | Majority are deployed at bases from which they could "attack targets throughout Western Europe". |
| SMP | 83 | [Map shows deployment and target coverage, ed], '83, '84, '85. 230+ listed for Western Theater; 95+ for Far East. First deployment in 1978 was east of Lake Baikal [read from map]. "Starting in 1977, the mobile SS-20 was deployed to the Far East"; only LRINF missile in region. "The Soviets continue to produce and deploy" SS-20s "in both the Western and Far Eastern Theaters and are likely to construct additional complexes beyond those presently under construction". |
| SuO1 | 83 | Deployed in thirds [of 351] in three areas: facing Europe-Mediterranean, facing East-South Asia, and east of Urals in 'swing' position capable of supporting either theater. |
| SMP | 84 | 243 SS-20s with 729 warheads "and an equal number for refire" in place opposite NATO. 135 deployed in Far Eastern Theater. |
| SMP | 85 | Soviets have "pressed ahead" with construction of new SS-20 bases in western and eastern USSR. Two-thirds of all SS-20s [c. 400] "are opposite European NATO". "Some shifting of the SS-20 force has recently been observed as the Soviets prepare for deployment of the SS-X-25 ICBM; however no reduction in the SS-20 force is expected from this activity." |

## RELATION TO OTHER MISSILES

| | | |
|---|---|---|
| Rock | 78 | Considerable improvement over existing mobile IRBMs. |
| SecD | 80 | Substantially more capable than its predecessors, FY81-82. |
| SMP | 84 | Closely related to SS-16. |

# SS-20 Mod 1

```
Sources are:   Flar  76      SWM  77      Rock  78      IISS  82      JWS  83
```

------------ **US DESIGNATION** ------------

IISS 82   SS-20 Mod 1

------------ **EDITOR'S NOTES** ------------

JWS 83   Source calls this Mod 2.

------------ **RANGE** ------------

|  | mi | nm | km | km conv | |
|---|---|---|---|---|---|
| Flar 76 |  |  | 5,700 | 5,700 | |
| SWM 77 | 1,710 |  | 2,750 | 2,751 | For 1.5 MT, 1000 lb warhead. |
| Rock 78 |  |  | 2,800 | 2,800 | With single 1.5 MT warhead. |
| IISS 82 |  |  | 5,000 | 5,000 | Max. |

------------ **WARHEAD CHARACTERISTICS** ------------

IISS 82   single RV
JWS  83   single RV

------------ **NUCLEAR YIELD** ------------

|  | kT | MT | kT conv |
|---|---|---|---|
| Flar 76 |  | 1.5 | 1,500 |
| SWM 77 |  | 1.5 | 1,500 |
| IISS 82 |  | 1.5 | 1,500 |
| JWS 83 |  | 1.5 | 1,500 |

------------ **WARHEAD WEIGHT** ------------

|  | lb | t | kg | kg conv |
|---|---|---|---|---|
| SWM 77 | 1,000 |  |  | 453.60 |

------------ **THROWWEIGHT** ------------

|  | lb | t | kg | kg conv |
|---|---|---|---|---|
| Flar 76 |  |  | 500 | 500.00 |

# SS-20 Mod 2

Sources are:  FIar 76    SecD 78    SecD 79    IISS 82    IISS+ 83    JWS 83    BW 86

### US DESIGNATION
| | |
|---|---|
| IISS 82 | SS-20 Mod 2 |

### EDITOR'S NOTES
| | |
|---|---|
| JWS 83 | Source calls this Mod 1. |

### RANGE

| | mi | nm | km | km conv | |
|---|---|---|---|---|---|
| FIar 76 | | | 5,700 | 5,700 | Estimate it can carry 3 MIRVs to 3,000 km. |
| SecD 78 | | | 3,000 | 3,000 | Well over 3,000 km with 3 MIRV payload. |
| SecD 79 | | | > 3,000 | 3,000 | Max. |
| IISS 82 | | | 5,000 | 5,000 | |
| BW 86 | | | 5,000 | 5,000 | |

### CEP

| | ft | mi | nm | m | km | m conv |
|---|---|---|---|---|---|---|
| IISS 82 | | | | 400 | | 400.00 |
| BW 86 | | | | 430 | | 430.00 |

### WARHEAD CHARACTERISTICS
| | | |
|---|---|---|
| FIar 76 | 3 MRVs | [Sic] |
| IISS 82 | 3 MIRVs | |
| JWS 83 | 3 MIRVs | |
| BW 86 | 3 MIRVs | |

### NUCLEAR YIELD

| | kT | MT | kT conv | |
|---|---|---|---|---|
| IISS 82 | 150 | | 150.00 | Per warhead. |
| JWS 83 | 150 | | 150.00 | |
| BW 86 | 150 | | 150.00 | |

### THROWWEIGHT

| | lb | t | kg | kg conv |
|---|---|---|---|---|
| FIar 76 | | | 500 | 500.00 |

### TYPE OF PROPULSION
| | |
|---|---|
| BW 86 | solid |

### DEPLOYMENT BEGAN
| | | |
|---|---|---|
| IISS+ 83 | 1977 | First year deployed. |
| BW 86 | 1977 | |

### DEPLOYED
| | |
|---|---|
| BW 86 | In service. |

### UNDATED NUMBER
| | | |
|---|---|---|
| BW 86 | ? 400 | Peak. |

### CARRIERS

### CARRIER — TUBES — RELOADS
| | CARRIER | TUBES | RELOADS | |
|---|---|---|---|---|
| BW 86 | | yes | | Mobile. |

# SS-20 Mod 3

```
Sources are:   FIar 76    SWM 77    Rock 78    Clns 80    IISS 82    JWS 83
```

## US DESIGNATION
|  |  |
|---|---|
| Rock 78 | SS-20 Mod |
| IISS 82 | SS-20 Mod 3 |

## DESCRIPTION
|  | |
|---|---|
| Rock 78 | ICBM |

## EDITOR'S NOTES
|  | |
|---|---|
| IISS 82 | Final report. |

## RANGE

| Source | mi | nm | km | km conv | Notes |
|---|---|---|---|---|---|
| FIar 76 | | | 7,500 | 7,500 | |
| SWM 77 | 4,600 | | 7,400 | 7,401 | For 50-75 kT, 300 lb warhead. |
| Rock 78 | | | 7,500 | 7,500 | |
| Clns 80 | | 4,000 | | 7,408 | With one warhead, range is 4,000 nm, sufficient to hit 60% of the US; no evidence any have been deployed in this mode. |
| IISS 82 | | | 7,400 | 7,400 | Max. |
| JWS 83 | | | 7,400 | 7,400 | Max; this must be debatable. |

## WARHEAD CHARACTERISTICS
|  | |
|---|---|
| Rock 78 | single RV |
| IISS 82 | single RV |
| JWS 83 | single RV |

## NUCLEAR YIELD

| Source | kT | MT | kT conv |
|---|---|---|---|
| FIar 76 | 50 | | 50.00 |
| SWM 77 | 50-75 | | 50.00-75.00 |
| Rock 78 | | 0.05 | 50.00 |
| IISS 82 | 50 | | 50.00 |
| JWS 83 | 50 | | 50.00 |

## WARHEAD WEIGHT

| Source | lb | t | kg | kg conv |
|---|---|---|---|---|
| SWM 77 | 300 | | | 136.08 |

## THROWWEIGHT

| Source | lb | t | kg | kg conv |
|---|---|---|---|---|
| FIar 76 | | | 150 | 150.00 |

## LENGTH

| Source | in | ft | cm | m | m conv |
|---|---|---|---|---|---|
| Rock 78 | | | | 11.5 | 11.50 |

## DIAMETER

| Source | in | ft | cm | m | cm conv |
|---|---|---|---|---|---|
| Rock 78 | | | | 1.8 | 180.00 |

## TYPE OF PROPULSION
|  | | |
|---|---|---|
| Rock 78 | solid | Stages 1 and 2. |

## DEPLOYMENT BEGAN
|  | | |
|---|---|---|
| Rock 78 | 1976 | Service intro. |

# FROG

```
Sources are:   JCS  75      Rock  78      JCS   80     SecD  81     NWG   83     Or01  84
SecD  71       JCS  76      SecD  78      JWS*  80     SMP   81     SA01  83     SMP   84
IISS  73       SecD 76      JCS   79      SecD  80     JCS   82     SMP   83     SMP   85
SecD  73       Bras 77      Clns  80      JCS   81     Clns+ 83     JCS   84
```

### US DESIGNATION
|  | | |
|---|---|---|
| SecD | 71 | FROG |
| JCS | 75 | FROG |
| Bras | 77 | FROG |
| Rock | 78 | Frog series |
| Clns | 80 | Frog |
| SMP | 81 | FROG |

### SOVIET DESIGNATION
| | | |
|---|---|---|
| Bras | 77 | ? LUNA |

### DESCRIPTION
| | | |
|---|---|---|
| SecD | 71 | Listed as tactical rocket or missile assigned to ground forces. |
| SecD | 76 | Tactical rocket of TNF. |
| Rock | 78 | Short range battlefield support missiles. |
| SecD | 81 | Medium-range [FROG, Scud and Scaleboard] launchers. |
| SMP | 81 | Tactical Nuclear Surface-to-Surface Missile. Short-range ballistic missile. |
| JCS | 84 | Listed as "short-range nuclear forces", Europe, both as Soviet and NSWP system. |

### EDITOR'S NOTES
| | | |
|---|---|---|
| SecD | 71 | Listed FY72,74,77,79-82. Pertains to FROGs in general, no specific model. |
| JCS | 75 | Listed FY76-83,85. Pertains to FROGs in general, no specific model. |
| Clns | 80 | Pertains to FROGs in general, no specific model. |
| SMP | 81 | Listed '81, '83, '84, '85. Pertains to FROGs in general, no specific model. |

### RANGE
| Source | | mi | nm | km | km conv | |
|---|---|---|---|---|---|---|
| Clns | 80 | 37 | | | 59.53 | |
| JCS | 80 | | | | | Improved in range [since 1960]. |
| SMP | 81 | 40 | | | 64.36 | |
| JCS | 82 | | | 70 | 70.00 | |
| NWG | 83 | | | 14-70 | 14.00-70.00 | |
| SA01 | 83 | | | 75 | 75.00 | |
| SMP | 85 | | | | | Has shorter range than SS-21. |

### CEP
| Source | | ft | mi | nm | m | km | m conv | |
|---|---|---|---|---|---|---|---|---|
| JCS | 80 | | | | | | | Improved in accuracy [since 1960]. |
| Clns+ | 83 | | | 0.25 | | | 463.00 | |
| SMP | 85 | | | | | | | Less accurate than SS-21. |

### WARHEAD TYPE
| Source | | nuclear | conven | chem | |
|---|---|---|---|---|---|
| JCS | 75 | yes | | | Nuclear-capable, FY76-81. |
| Bras | 77 | yes | | | |
| Clns | 80 | yes | yes | | HE. |
| JWS* | 80 | yes | yes | maybe | All six missiles. |
| SecD | 80 | yes | yes | yes | Nuclear, chemical, or conventional munitions, FY81-82. |
| SMP | 81 | yes | yes | yes | [FROG, Scud, Scaleboard and replacements together]. |

### WARHEAD CHARACTERISTICS
| | | |
|---|---|---|
| Bras | 77 | "Size and payload could allow conventional multiple warheads to be fitted". |

### NUCLEAR YIELD
| Source | | kT | MT | kT conv | |
|---|---|---|---|---|---|
| Clns | 80 | 1 | | 1.00 | |
| Clns+ | 83 | 200 | | 200.00 | |
| SA01 | 83 | 200 | | 200.00 | Maximum. |

### PAYLOAD
| Source | | lb | t | kg | kg conv | |
|---|---|---|---|---|---|---|
| JCS | 80 | | | | | Improved in payload [since 1960]. |

### NUMBER OF STAGES
| | | |
|---|---|---|
| Rock | 78 | 1 or 2 |

## TYPE OF PROPULSION

| | | |
|---|---|---|
| Bras 77 | solid | |
| Rock 78 | solid | "Solid propellant boost with supplementary thrust from annular nozzles". |
| Clns 80 | solid | |
| JWS* 80 | solid | |

## MIDCOURSE (OR MAIN) GUIDANCE

| | | |
|---|---|---|
| SecD 73 | unguided | |
| JCS 75 | unguided | FY76-79. |
| Bras 77 | unguided | "Stabilisation is by spin which is sustained in flight by fins". |
| Rock 78 | unguided | Spin-stabilized. |
| JWS* 80 | unguided | |
| SMP 85 | unguided | |

## OBSERVED

Rock 78   1957   "First emerged".

## DEPLOYMENT BEGAN

| | | |
|---|---|---|
| Clns 80 | 1965 | First deployed. |
| JWS* 80 | | Introduced since mid-1950s. |
| SA01 83 | 1950s | First deployed. |

## DEPLOYED

| | |
|---|---|
| SecD 73 | Deployed with ground forces. |
| JCS 75 | Deployed by USSR. |
| Bras 77 | "FROG 1-7" in service. |

## DEPLOYMENT HISTORY

| | |
|---|---|
| JCS 76 | "New models of free rockets" continue to be added to forces. |
| JCS 80 | Deployed in ever larger numbers [since 1960]. |
| JWS* 80 | Widely deployed throughout WTO, but nuclear warheads almost certainly in Soviet custody. |
| JCS 82 | Currently no indication USSR will draw down these older systems to counterbalance introduction of new ones. |
| SMP 84 | Being replaced. |
| SMP 85 | Soviets have begun to replace FROGS with "more accurate" SS-21. |

## NUMBERS IN SERVICE

### NUMBER IN 1970
Clns 80   500

### NUMBER IN 1971
Clns 80   550

### NUMBER IN 1972
Clns 80   600

### NUMBER IN 1973

| | | |
|---|---|---|
| IISS 73 | c. 600 | 1973-76, July, total deployed. |
| Clns 80 | 650 | In 1973-1977. |

### NUMBER IN 1978

| | | |
|---|---|---|
| Clns 80 | 660 | |
| Clns+ 83 | 670 | In 1978-1979. |

### NUMBER IN 1979
Clns 80   670

### NUMBER IN 1980
Clns+ 83   700   In 1980-1982.

### NUMBER IN 1981
SMP 83   650 "FROG/SS-21", at end of 1981, Warsaw Pact forces "facing NATO Europe".

### NUMBER IN 1982
JCS 82   Growth in numbers has continued.

### NUMBER IN 1983

| | | |
|---|---|---|
| NWG 83 | | 4 FROG launchers are assigned to every Soviet and East German division. |
| SA01 83 | 375 | With SS-21, launchers in Central Europe. Includes all systems stationed outside the USSR and in the western military districts of the USSR. |
| Or01 84 | 250 | "In Eastern satellites", [from British Foreign Office report.] |
| SMP 84 | | 700 "FROG/SS-21", at end of 1983, Warsaw Pact forces facing NATO Europe. |

## NUMBER IN 1985

| | | | |
|---|---|---|---|
| SMP | 85 | 750 | Total. "Currently there are some 375 FROG and SS-21 launchers opposite NATO". "Two hundred FROG launchers are opposite the Sino-Soviet border and in the Far East." ["Over 200" later, ed.] "About 100 [FROG launchers] are opposite southwest Asia and eastern Turkey." "About 75 [FROG launchers] are in the Strategic Reserve MDs." |

## CARRIERS

| | | CARRIER | TUBES | RELOADS | |
|---|---|---|---|---|---|
| Bras | 77 | | | | Launcher/transporter. |
| Clns | 80 | wheeled vehicle | | | |
| JWS* | 80 | | | | All but latest variant seen on "track-laying vehicle". |

## USE AND CONFIGURATION

| | | |
|---|---|---|
| Bras | 77 | In Soviet army FROG is deployed with tank and motorized rifle divisions in battalions of 4 launchers. |
| SecD | 78 | Tactical nuclear capability organic to ground forces (at divisional and higher levels) [FROG, Scud and Scaleboard], FY79-80. |
| JCS | 80 | At disposal of Front commander (normally). |
| JCS | 81 | "A Front normally has" FROGs. |
| SMP | 81 | The Front has FROGs. |
| NWG | 83 | "The basic division nuclear fire support system within the Warsaw Pact". 4 FROG launchers are assigned to every Soviet and East German division. |
| SMP | 85 | Found in a battalion of four launchers. |

## STRATEGY

| | | |
|---|---|---|
| Rock | 78 | Operate in highly mobile mode, "to bombard front line troops or battle tank concentrations", complementing Scud and Scaleboard. |

## COMBAT REPORTS/EFFECTIVENESS

| | | |
|---|---|---|
| Bras | 77 | Ineffective in 1973 war. |
| Rock | 78 | Frogs fired "in anger" against Israeli forces during 1973 war. |

## GEOGRAPHICAL DEPLOYMENTS

| | | | |
|---|---|---|---|
| JCS | 79 | | Launchers are deployed in Eastern Europe; more are based in USSR, probably earmarked for NATO. |
| NWG | 83 | East Germany | 104 deployed. |
| SMP | 85 | | Opposite NATO, opposite southwest Asia and eastern Turkey, opposite Sino-Soviet border, in Far East. |

## EXPORTED TO

| | | | |
|---|---|---|---|
| Rock | 78 | | 300 launchers exported (all countries outside USSR); all have conventional warheads. |
| | | Bulgaria | |
| | | Czech | |
| | | Egypt | Has largest number of export Frogs, at least 50 launchers by 1977. |
| | | E. Germany | |
| | | Hungary | |
| | | Iraq | |
| | | N Korea | |
| | | Poland | |
| | | Romania | |
| | | Syria | Has about 40 launchers; expecting to exceed Egypt's total by 1980. |
| SecD | 78 | | FROG and Scud launchers possessed by other WTO nations; warheads "remain under Soviet control", FY79-80. |
| JWS* | 80 | Warsaw Pact | |
| | | Middle East | |
| | | ? Africa | |

## RELATION TO OTHER MISSILES

| | | |
|---|---|---|
| Bras | 77 | Similar to Honest John, the nearest equivalent Western weapon. |
| NWG | 83 | Similar to US Honest John. |

## OTHER INFORMATION

| | | |
|---|---|---|
| JWS* | 80 | FROG stands for "Free (or Free Flight) Range Over Ground". |
| SMP | 85 | FROG stands for "Free Rocket Over Ground". |

# FROG 1

Sources are:
| | | | | | | | | | | | | | |
|---|---|---|---|---|---|---|---|---|---|---|---|---|---|
| | Zaeh | 61 | Ley | 68 | MOW | 76 | SWM | 77 | JWS* | 80 | NWG | 83 | BW 86 |
| IMSG | 60 | IAD1 | 65 | HRST | 75 | Bras | 77 | Rock | 78 | AWST | 82 | Meyr | 84 |

---------- US DESIGNATION ----------
| | | |
|---|---|---|
| JWS* | 80 | FROG 1 |
| Meyr | 84 | FROG-1 |

---------- NATO CODENAME ----------
| | | |
|---|---|---|
| Ley | 68 | Frog 1 |
| HRST | 75 | Frog 1 |
| MOW | 76 | Frog-1 |
| Bras | 77 | FROG 1 |
| SWM | 77 | Frog 1 |
| Rock | 78 | Frog 1 |
| AWST | 82 | Frog 1 |

---------- OTHER DESIGNATION ----------
| | | |
|---|---|---|
| IMSG | 60 | T-5B |
| Zaeh | 61 | T-5B |
| Ley | 68 | T-5B |

---------- DESCRIPTION ----------
| | | |
|---|---|---|
| Rock | 78 | tactical |

---------- EDITOR'S NOTES ----------
| | | |
|---|---|---|
| Ley | 68 | While we believe Frog 1 and T-5B to be the same missile, the source clearly didn't, with separate entries that we have merged here. |
| Meyr | 84 | Note to chart: "Values pertain to initial service period". |

---------- RANGE ----------
| | | mi | nm | km | km conv | |
|---|---|---|---|---|---|---|
| IMSG | 60 | c. 20 | | | 32.18 | |
| Zaeh | 61 | 10-15 | | | 16.09-24.14 | |
| IAD1 | 65 | <_ 40 | | | 64.36 | |
| Ley | 68 | 25 | | | 40.23 | [T-5B entry] |
| | | 40 | | | 64.36 | [Frog 1 entry] |
| HRST | 75 | 15 | | | 24.14 | |
| MOW | 76 | c. 15 | | c. 24 | 24.14 | Max. |
| Bras | 77 | | | 25-65 | 25.00-65.00 | |
| SWM | 77 | 20 | | 32 | 32.18 | |
| Rock | 78 | | | 32 | 32.00 | |
| AWST | 82 | 20 | | | 32.18 | |
| Meyr | 84 | | | 35 | 35.00 | |
| BW | 86 | | | 32 | 32.00 | |

---------- CEP ----------
| | | ft | mi | nm | m | km | m conv |
|---|---|---|---|---|---|---|---|
| Meyr | 84 | | | | | 0.6-1.0 | 600.00-1,000 |
| BW | 86 | | | | 800 | | 800.00 |

---------- WARHEAD TYPE ----------
| | | nuclear | conven | chem | |
|---|---|---|---|---|---|
| Zaeh | 61 | yes | | | "Atomic warhead". |
| MOW | 76 | yes | yes | | HE. |
| Bras | 77 | yes | yes | yes | HE. |
| SWM | 77 | yes | yes | yes | |
| Rock | 78 | yes | yes | | HE. |
| AWST | 82 | yes | | | |
| Meyr | 84 | yes | | | |

---------- WARHEAD CHARACTERISTICS ----------
| | | |
|---|---|---|
| Rock | 78 | single |
| Meyr | 84 | single RV |
| BW | 86 | single RV |

---------- NUCLEAR YIELD ----------
| | | kT | MT | kT conv |
|---|---|---|---|---|
| Rock | 78 | <_ 25 | | 25.00 |
| AWST | 82 | 25 | | 25.00 |
| Meyr | 84 | 20-100 | | 20.00-100.00 |
| BW | 86 | 25 | | 25.00 |

```
------------- WARHEAD WEIGHT -------------
              lb              t         kg        kg conv
SWM   77    2,600                       1,180     1,179
Rock  78                                1,200     1,200

------------- WEIGHT -------------
              lb              t         kg        kg conv
IMSG  60    6,000                                 2,721
IA01  65                    c. 4                  4,000        Tons.
Ley   68                      4                   4,000        Tons, [Frog 1 entry]
HRST  75    6,000                                 2,721
MOW   76    6,000                       2,700     2,721
Bras  77                                3,000     3,000
SWM   77    6600-7000                   3000-3175 2,993-3,175
Rock  78                     3.1                  3,100        Tonnes.

------------- LENGTH -------------
              in              ft        cm        m         m conv
IMSG  60                     31                             9.45
Zaeh  61                     25-30                          7.62-9.14
IA01  65                     31.5                           9.60
Ley   68                     30                             9.14      [T-5B entry]
                             31.5                           9.60      [Frog 1 entry]
HRST  75                     31                             9.45
MOW   76                     31'0"                9.50      9.45
Bras  77                                          10.0      10.00
SWM   77                     33.4                 10.2      10.18
Rock  78                                          10.2      10.20
AWST  82                     32                             9.75

------------- DIAMETER -------------
              in              ft        cm        m         cm conv
Zaeh  61                     c. 2.5                         76.20
IA01  65                     1.8                            54.86     Excluding warhead.
Ley   68                     c. 2                           60.96     [T-5B entry]
                             1.8                            54.86     [Frog 1 entry]
HRST  75    24                                              60.96
Bras  77                                85                  85.00     Calibre, [850 mm in source]
SWM   77    24                          61                  60.96
                                        84                  84.00     Warhead.
Rock  78                                          0.6       60.00
AWST  82                     2.0                            60.96

------------- WINGSPAN -------------
              in              ft        cm        m         cm conv
IMSG  60                     3.4                            103.63
MOW   76                     3'3"                 1.0       99.06

------------- NUMBER OF STAGES -------------
IMSG  60       1
Zaeh  61     ? 1
Ley   68     ? 2    [T-5B entry]
Bras  77       1
SWM   77       1
AWST  82       1

------------- TYPE OF PROPULSION -------------
IMSG  60    solid
Zaeh  61              "6 canted nozzles. Double-base propellant".
IA01  65    solid     7 nozzles, burn time 5-6 secs.
HRST  75    solid
MOW   76    solid     7 nozzles.
Bras  77    solid
SWM   77    solid
Rock  78    solid     Stage 1.
JWS*  80    solid     Propelled by solid fuel, but there has been change of propellant, since
                      earliest missile [of FROG series] had heating jacket on missile to prime
                      propellant.
BW    86    solid

------------- MIDCOURSE (OR MAIN) GUIDANCE -------------
IMSG  60    unguided    Spin-stabilized.
MOW   76    unguided    Spin-stabilized.
Bras  77    Nil
SWM   77    unguided    Spin-stabilized.

------------- OBSERVED -------------
IMSG  60              [Shown in 3 pictures].
```

## DEPLOYMENT BEGAN

| | | | |
|---|---|---|---|
| MOW | 76 | | Has been in large-scale service in USSR and WTO armies since 1957. |
| Bras | 77 | 1957 | First introduced. |
| SWM | 77 | 1957 | |
| Rock | 78 | 1957 | Service intro. |
| NWG | 83 | late 1950s | Initially deployed. |
| Meyr | 84 | 1958 | Service. |
| BW | 86 | 1957 | |

## DEPLOYED

MOW 76     In service.

## RETIRED

| | | | |
|---|---|---|---|
| JWS* | 80 | | ? Withdrawn from service, ? still used for training |
| BW | 86 | ?? 1970 | |

## NUMBERS IN SERVICE

## NUMBER IN 1957

MOW 76     Has been in large-scale service in USSR and WTO armies since 1957.

## CARRIERS

### CARRIER — TUBES — RELOADS

| | | | |
|---|---|---|---|
| IMSG | 60 | tracked carrier | |
| Ley | 68 | tank-launched | [T-5B entry] |
| HRST | 75 | tracked vehicle | |
| SWM | 77 | tracked vehicle | |

### CARRIER — TUBES — RELOADS

| | | | | | |
|---|---|---|---|---|---|
| MOW | 76 | JS-3 | | | |
| Bras | 77 | JS 3 | 1 | | Vehicle based on JS 3 chassis; has 6 road wheels, 3 return rollers; engine is V 12 diesel of 520 bhp; All-up weight c. 36,000 kg. |
| JWS* | 80 | JS-III | 1 | | |
| BW | 86 | JS-3 | 1 | yes | |

## USE AND CONFIGURATION

IMSG 60     Transported and fired from fully tracked tank carrier, on which is mounted T-5B's launching ramp. Firing crew is transported in the tank carrier; probably they can elevate and fire missile without leaving the transporter. Heated casing is wrapped around weapon to ensure solid-propellant motor stays at appropriate firing temperature.

MOW 76     Ribbed casing on JS-3 elevates to form launch tube.

Bras 77     Launched from ribbed box container, which could be the delivery container for the motor section of the missile. Container forms heater jacket to maintain constant temperature for solid fuel. Tubular framework protects warhead (which overhangs vehicle) when traveling. Built-up superstructure of vehicle has enclosed cabin; carrying cradle is also launch rail, elevated by hydraulic ram; "There are steadying jacks at the rear so that the suspension does not lock for firing" as newer Western systems do; traverse is probably very limited.

Meyr 84     Penetration: 1.0; reliability: 0.4–0.8, "includes system reliability multiplied by operational readiness of deployed system".

## STRATEGY

IMSG 60     Intended for use against tactical targets within 20-mile range.

Meyr 84     [Ed: chart is labeled] Systems "assigned primary operational and strategic missions within the European theatre(s) of military operations".

## RELATION TO OTHER MISSILES

Zaeh 61     Similar to Honest John.

MOW 76     USSR counterpart to Honest John.

## OTHER INFORMATION

IMSG 60     Has 4 tail fins, warhead contained in swollen nose piece, which protrudes several inches in diameter beyond main body.

MOW 76     FROG = Free Rocket Over Ground.

SWM 77     Bulbous warhead. Fixed cruciform tail fins.

# FROG 2

```
Sources are:  Zaeh  61    Ley   68    MOW   76    SWM   77    JWS*  80    BW    86
IMSG  60      IA01  65    HRST  75    Bras  77    Rock  78    AWST  82
```

---------- **US DESIGNATION** ----------

| | | |
|---|---|---|
| Ley   | 68 | Frog 2    [Frog 2 entry] |
| HRST  | 75 | Frog 2 |
| MOW   | 76 | Frog-2 |
| Bras  | 77 | FROG 2 |
| SWM   | 77 | Frog 2 |
| Rock  | 78 | Frog 2 |
| JWS*  | 80 | FROG 2 |
| AWST  | 82 | Frog 2 |

---------- **OTHER DESIGNATION** ----------

| | | |
|---|---|---|
| IMSG | 60 | T-5C |
| Zaeh | 61 | T-5C |
| Ley  | 68 | T-5C    [T-5C entry] |

---------- **DESCRIPTION** ----------

Rock 78   tactical

---------- **EDITOR'S NOTES** ----------

Ley 68   The source has separate entries for Frog 2 and T-5C, which we judge to be the same missile; data below is labeled with the source's designation.

---------- **RANGE** ----------

| | | mi | nm | km | km conv | |
|---|---|---|---|---|---|---|
| IMSG | 60 | 15    |   |    | 24.14       | |
| Zaeh | 61 | 5-10  |   |    | 8.05-16.09  | |
| IA01 | 65 | 15-17 |   |    | 24.14-27.35 | |
| Ley  | 68 | 15-17 |   |    | 24.14-27.35 | [Frog 2 entry] |
|      |    | 15-20 |   |    | 24.14-32.18 | [T-5C entry] |
| HRST | 75 | 30    |   |    | 48.27       | |
| Bras | 77 |       |   | 25 | 25.00       | |
| SWM  | 77 | 12    |   | 19 | 19.31       | [Table] |
|      |    | 7.5   |   | 12 | 12.07       | [Text] |
| Rock | 78 |       |   | 20 | 20.00       | |
| AWST | 82 | 12    |   |    | 19.31       | |
| BW   | 86 |       |   | 25 | 25.00       | |

---------- **CEP** ----------

| | | ft | mi | nm | m | km | m conv |
|---|---|---|---|---|---|---|---|
| BW | 86 | | | | 800 | | 800.00 |

---------- **WARHEAD TYPE** ----------

| | | nuclear | conven | chem | |
|---|---|---|---|---|---|
| IMSG | 60 | yes | yes |     | HE. |
| Zaeh | 61 | yes |     |     | "Atomic warhead". |
| Bras | 77 | yes | yes | yes | HE. |
| SWM  | 77 | yes | yes |     | |
| Rock | 78 |     | yes | yes | HE, [n.b.: only this Frog listed with chemical warhead and not nuclear; reiterated in both table and text, ed]. |

---------- **WARHEAD CHARACTERISTICS** ----------

| | | |
|---|---|---|
| Rock | 78 | single |
| BW   | 86 | single RV |

---------- **NUCLEAR YIELD** ----------

| | | kT | MT | kT conv |
|---|---|---|---|---|
| BW | 86 | ? 10 | | 10.00 |

---------- **WARHEAD WEIGHT** ----------

| | | lb | t | kg | kg conv |
|---|---|---|---|---|---|
| SWM  | 77 | 1,200 | | 545 | 544.32 |
| Rock | 78 |       | | 550 | 550.00 |

---------- **PAYLOAD** ----------

| | | lb | t | kg | kg conv |
|---|---|---|---|---|---|
| IMSG | 60 | 1,100 | | | 498.96 |

BALLISTIC MISSILES  SRBMs  FROG 2    353

```
─────────── WEIGHT ───────────
              lb           t          kg        kg conv
IMSG  60    4,400                                1,995
IAD1  65                 2.35                    2,350        Tons.
Ley   68                 2.3                     2,300        Tons, [Frog 2 entry].
HRST  75    4,400                                1,995
Bras  77                            2,400        2,400
SWM   77    5,400                   2,450        2,449
Rock  78                 2.5                     2,500        Tonnes.

─────────── LENGTH ───────────
              in          ft         cm     m    m conv
IMSG  60                 25                      7.62
Zaeh  61                 25-35                   7.62-10.67
IAD1  65                 29.5                    8.99
Ley   68                 29.5                    8.99         [Frog 2 entry]
                         25                      7.62         [T-5C entry]
HRST  75                 33.5                    10.21
MOW   76                 29'6"          9.0      8.99
Bras  77                                9.0      9.00
SWM   77                 31.2           9.5      9.51
Rock  78                                9.5      9.50
AWST  82                 32                      9.75

─────────── DIAMETER ───────────
              in          ft         cm     m    cm conv
IMSG  60    13.1                                 33.27
Zaeh  61               c. 2.0                    60.96
IAD1  65                 1                       30.48        Excluding warhead.
Ley   68                 1.0                     30.48        [Frog 2 entry]
                       c. 2                      60.96        [T-5C entry]
HRST  75    14.5                                 36.83
Bras  77                            60           60.00        Calibre, [600 mm]
SWM   77    12                      30.5         30.48
Rock  78                                  0.3    30.00
AWST  82                 1.2                     36.58

─────────── WINGSPAN ───────────
              in          ft         cm     m    cm conv
IMSG  60                 3                       91.44
SWM   77                 3.1              1.05   94.49

─────────── NUMBER OF STAGES ───────────
Zaeh  61    ? 1
Ley   68    1
HRST  75    2
MOW   76    1
Bras  77    1
AWST  82    2

─────────── TYPE OF PROPULSION ───────────
IMSG  60    solid             "Motor...can be fired efficiently over a wide temperature range".
HRST  75    solid
Bras  77    solid
SWM   77    solid
Rock  78    solid             Stages 0 and 1.
BW    86    solid

─────────── MIDCOURSE (OR MAIN) GUIDANCE ───────────
IMSG  60    unguided          "Unguided but rotates during flight".
MOW   76    unguided
Bras  77    Nil
SWM   77    unguided          Spin-stabilized.

─────────── OBSERVED ───────────
IMSG  60                      [Shown in picture].

─────────── OBSERVED IN PARADE ───────────
MOW   76    Nov 1957          8 paraded in Moscow.

─────────── DEPLOYMENT BEGAN ───────────
SWM   77    1957
Rock  78    1957              Service intro.
BW    86    1957

─────────── DEPLOYED ───────────
IMSG  60                      Fully operational.
MOW   76                      In service.
```

354  SOVIET MISSILES  IDDS

```
----------  RETIRED  ----------
JWS*  80                    ? Withdrawn from service, ? still used for training.
BW    86    ?? 1970

----------  NUMBERS PRODUCED  ----------
MOW   76              Delivered in large numbers to USSR and WTO.

----------  CARRIERS  ----------

----------  CARRIER  ----------  TUBES ---- RELOADS ----------
IMSG  60   tank                                 Tracked amphibious tank.
Ley   68   tank-launched                        [T-5C entry]
HRST  75   tracked vehicle

----------  CARRIER  ----------  TUBES ---- RELOADS ----------
MOW   76   PT-76
Bras  77   PT 76              1                 PT 76 chassis, 6 road wheels, no return rollers; 6
                                                cylinder diesel engine developing 240 bhp. PT 76 is "an
                                                inherent swimmer", FROG 2 mounting may be too; road speed
                                                35 kmph, endurance of 250 km. Entire system weight 15,000
                                                kg.
SWM   77   PT-76 chassis                        Air-transportable by An-22.
JWS*  80   PT-76              1
BW    86   PT-76              1         yes

----------  USE AND CONFIGURATION  ----------
IMSG  60   Transporter-Launcher provides full crew protection when the missile is fired.

----------  EXPORTED TO  ----------
MOW   76   WTO           Delivered in large numbers.

----------  RELATION TO OTHER MISSILES  ----------
IMSG  60   Smaller than T-5B, but similar in many other respects.
Zaeh  61   Smaller version of T-5B.
Bras  77   Lighter, smaller than FROG 1.

----------  OTHER INFORMATION  ----------
IMSG  60   4 fixed tail fins stabilize missile. Has long cylindrical configuration, enlarged warhead
           compartment in the nose.
MOW   76   Bulbous warhead.
SWM   77   Fixed cruciform tail fins.
```

# FROG 3

```
Sources are:  IA02  65     ID01  68     HRST  75    Bras  77    Rock  78    JCS   81    Meyr  84
              IA01  65     IA05  66     Ley   68    MOW   76    SWM   77    JWS*  80    AWST  82    BW    86
```

## US DESIGNATION

|  | |  |
|---|---|---|
| Ley | 68 | Frog 3 |
| HRST | 75 | Frog 3 |
| MOW | 76 | Frog-3 |
| Bras | 77 | FROG 3 |
| SWM | 77 | Frog 3 |
| Rock | 78 | Frog 3 |
| JWS* | 80 | FROG 3 |
| JCS | 81 | FROG 3 |
| AWST | 82 | Frog 3 |
| Meyr | 84 | FROG-3 |

## DESCRIPTION

| | | |
|---|---|---|
| Rock | 78 | tactical |

## EDITOR'S NOTES

| | | |
|---|---|---|
| JCS | 81 | Listed FY82. |
| Meyr | 84 | Note to chart: "Values pertain to initial service period". |

## RANGE

| Source | | mi | nm | km | km conv | |
|---|---|---|---|---|---|---|
| IA01 | 65 | 22-28 | | | 35.40-45.05 | |
| IA02 | 65 | 20-30 | | | 32.18-48.27 | |
| IA05 | 66 | c. 25 | | | 40.23 | |
| Ley | 68 | 22-28 | | | 35.40-45.05 | |
| HRST | 75 | 50 | | | 80.45 | |
| Bras | 77 | | | 40 | 40.00 | |
| SWM | 77 | 22-28 | | 36-45 | 35.40-45.05 | |
| Rock | 78 | | | 80 | 80.00 | [Table] |
| | | | | 45 | 45.00 | "Two separate stages increasing the range to 45 km", [text]. |
| JWS* | 80 | | | c. 40 | 40.00 | |
| AWST | 82 | 45-55 | | | 72.41-88.50 | |
| Meyr | 84 | | | 40 | 40.00 | |
| BW | 86 | | | 40 | 40.00 | |

## CEP

| Source | | ft | mi | nm | m | km | m conv |
|---|---|---|---|---|---|---|---|
| Meyr | 84 | | | | | 0.5-0.8 | 500.00-800.00 |
| BW | 86 | | | | 650 | | 650.00 |

## WARHEAD TYPE

| Source | | nuclear | conven | chem | |
|---|---|---|---|---|---|
| Bras | 77 | yes | yes | yes | HE. |
| SWM | 77 | yes | yes | yes | |
| Rock | 78 | yes | yes | | HE; "similar payload capability to Frog 2...also able to carry a nuclear warhead" [text, no mention of chemical capability]. |
| JWS* | 80 | yes | yes | | HE. |
| Meyr | 84 | yes | | | |

## WARHEAD CHARACTERISTICS

| | | |
|---|---|---|
| Rock | 78 | single |
| Meyr | 84 | single RV |
| BW | 86 | single RV |

## NUCLEAR YIELD

| Source | | kT | MT | kT conv |
|---|---|---|---|---|
| Meyr | 84 | 20-100 | | 20.00-100.00 |
| BW | 86 | ? 20 | | 20.00 |

## WARHEAD WEIGHT

| Source | | lb | t | kg | kg conv | |
|---|---|---|---|---|---|---|
| Bras | 77 | | | 250 | 250.00 | Weight of nuclear warhead. |
| SWM | 77 | 1,000 | | 454 | 453.60 | |
| JWS* | 80 | | | 450 | 450.00 | |

SOVIET MISSILES

```
------------ WEIGHT ------------
              lb           t            kg       kg conv
IA01   65                  2.1                    2,100      Tons.
Ley    68                  2.1                    2,100      Tons.
Bras   77                               2,280     2,280
SWM    77     5,000                     2,266     2,268
Rock   78                  2.3                    2,300      Tonnes.
JWS*   80                            c. 2,250     2,250

------------ LENGTH ------------
              in           ft           cm        m        m conv
IA02   65                  34.5                            10.52
Ley    68                  34.5                            10.52
HRST   75                  33.5                            10.21
Bras   77                                         10.5     10.50
SWM    77                  34.4                   10.5     10.49
Rock   78                                         10.5     10.50
JWS*   80                                         10.5     10.50
AWST   82                  34                              10.36

------------ DIAMETER ------------
              in           ft           cm        m        cm conv
IA01   65                  1.3                             39.62     Excluding warhead.
Ley    68                  1.3                             39.62
HRST   75     14.5                                         36.83
Bras   77                               55                 55.00     Calibre, [550 mm in source]
SWM    77     15.8                      40                 40.13
Rock   78                                         0.4      40.00
JWS*   80                               40                 40.00
AWST   82                  1.1       c. 55                 55.00     Warhead.
                                                           33.53

------------ NUMBER OF STAGES ------------
ID01   68     2       Tandem.
Bras   77     2       First rocket of its type with 2 stages.
SWM    77     2       Tandem motors.
JWS*   80     2
AWST   82     2

------------ TYPE OF PROPULSION ------------
ID01   68     solid      12 single jets around periphery of booster nozzle.
HRST   75     solid
Bras   77     solid      First stage motor made of 12 jets built around central nozzle.
SWM    77     solid
Rock   78     solid      Stages 0 and 1.
JWS*   80     solid
BW     86     solid

------------ MIDCOURSE (OR MAIN) GUIDANCE ------------
MOW    76     unguided
Bras   77     Nil
SWM    77     unguided   Spin-stabilized.
JWS*   80     unguided   Spin-stabilized.

------------ DEPLOYMENT BEGAN ------------
Bras   77     1960         Introduced.
Rock   78     1959         Service intro.
JWS*   80     c. 1960      First introduced.
Meyr   84     1960         Service.
BW     86     1960

------------ DEPLOYED ------------
MOW    76                  In service.
BW     86                  In service.

------------ DEPLOYMENT HISTORY ------------
JWS*   80     Almost certainly obsolescent.

------------ NUMBERS PRODUCED ------------
MOW    76             Delivered in large numbers to USSR and WTO.

------------ CARRIERS ------------

------------ CARRIER ------------ TUBES ---- RELOADS ------------
HRST   75     amphib. vehicle
```

|          |    | CARRIER       | TUBES | RELOADS |                                                                                              |
|----------|----|---------------|-------|---------|----------------------------------------------------------------------------------------------|
| Bras     | 77 | [PT 76]       | 1     |         | Exactly as for FROG 2, except 2 track return rollers fitted above road wheels.               |
| SWM      | 77 | PT-76 chassis |       |         | Air-transportable by An-22.                                                                  |
| JWS*     | 80 | PT-76         |       |         | 35 kmph on road, road range 240 km.                                                          |
| AWST     | 82 | PT76 chassis  |       |         |                                                                                              |
| BW       | 86 | PT-76         | 1     | yes     |                                                                                              |

## USE AND CONFIGURATION

| IAD5 | 66 | Crew of 5 and commander can ready the missile for firing "within minutes". |
| JWS* | 80 | Believed to remain in service at divisional level. |
| Meyr | 84 | Penetration: 1.0; reliability: 0.4-0.8, "includes system reliability multiplied by operational readiness of deployed system". |

## STRATEGY

| Meyr | 84 | [Ed: chart is labeled] Systems "assigned primary operational and strategic missions within the European theatre(s) of military operations". |

## EXPORTED TO

| MOW  | 76 | WTO   | Delivered in large numbers. |
| Bras | 77 | Egypt |                             |
| JCS  | 81 |       | NSWP countries have "older FROG-3s and 5s". Being replaced gradually in NSWP forces by FROG 7. |

## RELATION TO OTHER MISSILES

| MOW | 76 | A more cylindrical bulbous warhead [than Frog 2]. |

## OTHER INFORMATION

| SWM | 77 | Fixed cruciform tail fins. Cylindrical, bulbous warhead. |

# FROG 4

```
Sources are:  Ley   68      MOW   76      SWM   77      JWS*  80      BW    86
IAD1  65      HRST  75      Bras  77      Rock  78      AWST  82
```

------------ **US DESIGNATION** ------------

|  | |  |
|---|---|---|
| Ley | 68 | Frog 4 |
| HRST | 75 | Frog 4 |
| MOW | 76 | Frog-4 |
| Bras | 77 | FROG 4 |
| SWM | 77 | Frog 4 |
| Rock | 78 | Frog 4 |
| JWS* | 80 | FROG 4 |
| AWST | 82 | Frog 4 |

------------ **SOVIET DESIGNATION** ------------

SWM  77    T-5E          "Soviet designation is T-5E".

------------ **DESCRIPTION** ------------

Rock  78   tactical

------------ **EDITOR'S NOTES** ------------

MOW  76    Data from entry labeled "Frog-2,3,4 and 5". Above list of characteristics is "Data apply to Frog-4"; all that data is in this record.

------------ **RANGE** ------------

| Source | | mi | nm | km | km_conv | |
|---|---|---|---|---|---|---|
| IAD1 | 65 | 30+ | | | 48.27 | |
| Ley | 68 | 30+ | | | 48.27 | |
| HRST | 75 | 70 | | | 112.63 | |
| MOW | 76 | 30 | | 50 | 48.27 | Max. |
| Bras | 77 | | | 45 | 45.00 | |
| SWM | 77 | 30+ | | | 48.27 | [Table] |
| | | 31 | | 50 | 49.88 | [Text] |
| Rock | 78 | | | 100 | 100.00 | [Table] |
| | | | | 50 | 50.00 | [Text] |
| JWS* | 80 | | | c. 45 | 45.00 | |
| AWST | 82 | 70 | | | 112.63 | |
| BW | 86 | | | 50 | 50.00 | |

------------ **CEP** ------------

| Source | | ft | mi | nm | m | km | m_conv |
|---|---|---|---|---|---|---|---|
| BW | 86 | | | | 650 | | 650.00 |

------------ **WARHEAD TYPE** ------------

| Source | | nuclear | conven | chem | |
|---|---|---|---|---|---|
| MOW | 76 | yes | yes | | HE. |
| Bras | 77 | yes | yes | yes | HE. |
| SWM | 77 | yes | yes | yes | |
| Rock | 78 | yes | yes | | HE. |
| JWS* | 80 | yes | yes | | |

------------ **WARHEAD CHARACTERISTICS** ------------

|  | | |
|---|---|---|
| Rock | 78 | single |
| BW | 86 | single RV |

------------ **NUCLEAR YIELD** ------------

| Source | | kT | MT | kT_conv |
|---|---|---|---|---|
| BW | 86 | ? 20 | | 20.00 |

------------ **WARHEAD WEIGHT** ------------

| Source | | lb | t | kg | kg_conv | |
|---|---|---|---|---|---|---|
| Bras | 77 | | | 300 | 300.00 | Weight of nuclear warhead. |

------------ **WEIGHT** ------------

| Source | | lb | t | kg | kg_conv | |
|---|---|---|---|---|---|---|
| IAD1 | 65 | | 2.04 | | 2,040 | Tons. |
| Ley | 68 | | 2.0 | | 2,000 | Tons. |
| MOW | 76 | 4,400 | | 2,000 | 1,995 | |
| Bras | 77 | | | 2,040 | 2,040 | |
| Rock | 78 | | 2.0 | | 2,000 | Tonnes. |
| JWS* | 80 | | | c. 2,000 | 2,000 | |

## BALLISTIC MISSILES  SRBMs  FROG 4

### LENGTH

|      |    | in   | ft      | cm | m     | m conv |
|------|----|------|---------|----|-------|--------|
| IAD1 | 65 |      | 33.46   |    |       | 10.20  |
| Ley  | 68 |      | 33.5    |    |       | 10.21  |
| HRST | 75 |      | 32      |    |       | 9.75   |
| MOW  | 76 |      | 33'6"   |    | 10.20 | 10.21  |
| Bras | 77 |      |         |    | 10.2  | 10.20  |
| SWM  | 77 |      | 33.5    |    | 10.2  | 10.21  |
| Rock | 78 |      |         |    | 10.2  | 10.20  |
| JWS* | 80 |      |         |    | 10.2  | 10.20  |
| AWST | 82 |      | 32      |    |       | 9.75   |

### DIAMETER

|      |    | in   | ft     | cm   | m    | cm conv |                   |
|------|----|------|--------|------|------|---------|-------------------|
| IAD1 | 65 |      | 1.3    |      |      | 39.62   |                   |
| Ley  | 68 |      | 1.3    |      |      | 39.62   |                   |
| HRST | 75 | 14.5 |        |      |      | 36.83   |                   |
| MOW  | 76 |      | 1'3.7" |      | 0.40 | 39.88   |                   |
| Bras | 77 |      |        | 40   |      | 40.00   | Calibre, [400 mm] |
| SWM  | 77 | 15.8 |        | 40   |      | 40.13   |                   |
| Rock | 78 |      |        |      | 0.4  | 40.00   |                   |
| JWS* | 80 |      |        | c. 40|      | 40.00   | Warhead.          |
| AWST | 82 |      | 1.2    |      |      | 36.58   |                   |

### WINGSPAN

|     |    | in | ft    | cm | m    | cm conv |
|-----|----|----|-------|----|------|---------|
| MOW | 76 |    | 3'6"  |    | 1.05 | 106.68  |

### NUMBER OF STAGES

| HRST | 75 | 2 |                              |
| MOW  | 76 | 2 |                              |
| Bras | 77 | 2 |                              |
| SWM  | 77 | 2 | Tandem motors.               |
| JWS* | 80 | 2 | Booster and sustainer in tandem. |
| AWST | 82 | 2 |                              |

### TYPE OF PROPULSION

| HRST | 75 | solid |                                               |
| MOW  | 76 | solid | 12 small nozzles around main nozzle.          |
| Bras | 77 | solid |                                               |
| SWM  | 77 | solid |                                               |
| Rock | 78 | solid | Stages 0 and 1. Propulsion similar to Frog 3 [text]. |
| JWS* | 80 | solid | Booster and sustainer.                        |
| BW   | 86 | solid |                                               |

### MIDCOURSE (OR MAIN) GUIDANCE

| MOW  | 76 | unguided | Spin-stabilised. |
| Bras | 77 | Nil      |                  |
| SWM  | 77 | unguided | Spin-stabilized. |
| JWS* | 80 | unguided | Spin-stabilized. |

### DEPLOYMENT BEGAN

| Rock | 78 | 1959   | Service intro. |
| BW   | 86 | ? 1959 |                |

### DEPLOYED

| MOW  | 76 | In service. |
| JWS* | 80 | In service. |

### RETIRED

| BW | 86 | ?? 1970 |

### NUMBERS PRODUCED

| MOW | 76 | Delivered in large numbers to USSR and WTO. |

### CARRIERS

#### CARRIER — TUBES — RELOADS

| HRST | 75 | tracked vehicle |   |
| Bras | 77 |                 | 1 |

#### CARRIER — TUBES — RELOADS

| MOW  | 76 | PT-76        | 1 |     |                                                         |
| SWM  | 77 | PT-76 chassis|   |     | Air-transportable by An-22.                             |
| JWS* | 80 | PT-76        |   |     | 35 kmph on road, road range 240 km.                     |
| AWST | 82 | PT76 chassis |   |     |                                                         |
| BW   | 86 | PT-76        | 1 | yes |                                                         |

---------- USE AND CONFIGURATION ----------
Bras 77    FROGs 4 and 5 may have "built-in north-seeking gyro to assist in laying".
JWS* 80    Believed to remain in service at divisional level.

---------- EXPORTED TO ----------
MOW  76    WTO            Delivered in large numbers.
           Cuba           Serves.
Bras 77    Algeria
           Cuba
JWS* 80    Warsaw Pact    In service.
           Cuba           In service.

---------- RELATION TO OTHER MISSILES ----------
Bras 77    FROGs 4 and 5 appear identical to 3 except in warhead shape.
JWS* 80    Like FROG 3, but warhead has same diameter as body of missile.

---------- OTHER INFORMATION ----------
MOW  76    Warhead same diameter as 2nd stage casing.
SWM  77    Fixed cruciform tail fins.

# FROG 5

```
Sources are:  Bras  77     Rock  78     JCS   81     Meyr  84
MOW   76      SWM   77     JWS*  80     AWST  82     BW    86
```

---------- **US DESIGNATION** ----------

|  | |  |
|---|---|---|
| MOW | 76 | Frog-5 |
| Bras | 77 | FROG 5 |
| SWM | 77 | Frog 5 |
| Rock | 78 | Frog 5 |
| JWS* | 80 | FROG 5 |
| JCS | 81 | FROG 5 |
| AWST | 82 | Frog 5 |
| Meyr | 84 | FROG-5 |

---------- **DESCRIPTION** ----------

| Rock | 78 | tactical |
|---|---|---|

---------- **EDITOR'S NOTES** ----------

| JCS | 81 | Listed FY82. |
|---|---|---|
| Meyr | 84 | Note to chart: "Values pertain to initial service period". |

---------- **RANGE** ----------

| Source | | mi | nm | km | km conv |
|---|---|---|---|---|---|
| Bras | 77 | | | 45 | 45.00 |
| SWM | 77 | 21.7 | | 35 | 34.92 |
| Rock | 78 | | | 50 | 50.00 |
| JWS* | 80 | | | c. 55 | 55.00 |
| AWST | 82 | 30 | | | 48.27 |
| Meyr | 84 | | | 60 | 60.00 |
| BW | 86 | | | 50 | 50.00 |

---------- **CEP** ----------

| Source | | ft | mi | nm | m | km | m conv |
|---|---|---|---|---|---|---|---|
| Meyr | 84 | | | | | 0.4-0.7 | 400.00-700.00 |
| BW | 86 | | | | 550 | | 550.00 |

---------- **WARHEAD TYPE** ----------

| Source | | nuclear | conven | chem | |
|---|---|---|---|---|---|
| Bras | 77 | yes | yes | yes | HE. |
| SWM | 77 | yes | yes | yes | |
| Rock | 78 | yes | yes | | HE. |
| JWS* | 80 | yes | yes | | |
| Meyr | 84 | yes | | | |

---------- **WARHEAD CHARACTERISTICS** ----------

| Rock | 78 | single |
|---|---|---|
| Meyr | 84 | single RV |
| BW | 86 | single RV |

---------- **NUCLEAR YIELD** ----------

| Source | | kT | MT | kT conv |
|---|---|---|---|---|
| Meyr | 84 | 20-100 | | 20.00-100.00 |
| BW | 86 | ? 20 | | 20.00 |

---------- **WARHEAD WEIGHT** ----------

| Source | | lb | t | kg | kg conv | |
|---|---|---|---|---|---|---|
| Bras | 77 | | | 300 | 300.00 | Weight of nuclear warhead. |
| SWM | 77 | 1,000 | | 454 | 453.60 | |

---------- **WEIGHT** ----------

| Source | | lb | t | kg | kg conv |
|---|---|---|---|---|---|
| Bras | 77 | | | 2,040 | 2,040 |
| JWS* | 80 | | | c. 3,000 | 3,000 |

---------- **LENGTH** ----------

| Source | | in | ft | cm | m | m conv |
|---|---|---|---|---|---|---|
| Bras | 77 | | | | 10.2 | 10.20 |
| SWM | 77 | | 31.2 | | 9.5 | 9.51 |
| Rock | 78 | | | | 9.5 | 9.50 |
| JWS* | 80 | | | | 9.1 | 9.10 |

## DIAMETER

|  |  | in | ft | cm | m | cm conv |  |
|---|---|---|---|---|---|---|---|
| Bras | 77 |  |  | 40 |  | 40.00 | Calibre, [400 mm in source] |
| SWM | 77 | 15.8 |  | 40 |  | 40.13 |  |
| Rock | 78 |  |  |  | 0.4 | 40.00 |  |
| JWS* | 80 |  |  | c. 55 |  | 55.00 | Warhead. |

## NUMBER OF STAGES

| Bras | 77 | 2 |  |
|---|---|---|---|
| JWS* | 80 | 2 | Booster and sustainer in tandem. |

## TYPE OF PROPULSION

| Bras | 77 | solid |  |
|---|---|---|---|
| SWM | 77 | solid |  |
| Rock | 78 | solid | Stages 0 and 1. |
| JWS* | 80 | solid | Booster and sustainer. |
| BW | 86 | solid |  |

## MIDCOURSE (OR MAIN) GUIDANCE

| MOW | 76 | unguided |  |
|---|---|---|---|
| Bras | 77 | Nil |  |
| SWM | 77 | unguided | Spin-stabilized. |
| JWS* | 80 | unguided | Spin-stabilized. |

## DEPLOYMENT BEGAN

| Rock | 78 | 1961 | Service intrc. |
|---|---|---|---|
| Meyr | 84 | 1964 | Service. |
| BW | 86 | ? 1964 |  |

## DEPLOYED

| MOW | 76 | In service. |
|---|---|---|
| JWS* | 80 | Operational. |
| BW | 86 | In service. |

## NUMBERS PRODUCED

| MOW | 76 | Delivered in large numbers to USSR and WTO. |
|---|---|---|

## CARRIERS

### CARRIER — TUBES — RELOADS

| Bras | 77 |  | 1 |  |
|---|---|---|---|---|

### CARRIER — TUBES — RELOADS

| JWS* | 80 | PT-76 |  |  | 35 kmph on road, road range 240 km. |
|---|---|---|---|---|---|
| AWST | 82 | PT76 |  |  |  |
| BW | 86 | PT-76 | 1 | yes |  |

## USE AND CONFIGURATION

| Bras | 77 | FROGs 4 and 5 may have "built-in north-seeking gyro to assist in laying". |
|---|---|---|
| JWS* | 80 | Believed to remain in service at divisional level. |
| Meyr | 84 | Penetration: 1.0; reliability: 0.4-0.8, "includes system reliability multiplied by operational readiness of deployed system". |

## STRATEGY

| Meyr | 84 | [Ed: chart is labeled] Systems "assigned primary operational and strategic missions within the European theatre(s) of military operations". |
|---|---|---|

## EXPORTED TO

| MOW | 76 | WTO | Delivered in large numbers. |
|---|---|---|---|
|  |  | North Korea | Also serves. |
| Bras | 77 | North Korea |  |
| JWS* | 80 | Warsaw Pact |  |
| JCS | 81 |  | NSWP countries have "older FROG-3s and 5s". Being replaced gradually in NSWP forces by FROG 7. |

## RELATION TO OTHER MISSILES

| MOW | 76 | "Yet another variation in warhead" [over Frog-2,3,4] |
|---|---|---|
| Bras | 77 | FROGs 4 and 5 appear identical to 3 except in warhead shape. |
| SWM | 77 | Similar to Frog's 3 and 4. |
| JWS* | 80 | Shorter, thicker, and heavier than FROG 4, and nose cone of warhead is conical. |

# FROG 6

Sources are:   Bras 77     JWS* 77     SWM 77     Rock 78     AWST 82

---------- **US DESIGNATION** ----------
| | |
|---|---|
| Bras 77 | FROG 6 |
| JWS* 77 | FROG 6 — FROG 8 could well be training missile, as was FROG 6. |
| SWM 77 | Frog 6 |
| Rock 78 | Frog 6 |
| AWST 82 | Frog 6 |

---------- **DESCRIPTION** ----------
Rock 78    Tactical.

---------- **EDITOR'S NOTES** ----------
JWS* 77    This information follows description of FROG 9, see SS-21.

---------- **OBSERVED** ----------
Bras 77    Rarely seen in public.

---------- **DEPLOYMENT BEGAN** ----------
Rock 78    1965    Service intro.

---------- **STRATEGY** ----------
Bras 77    Training vehicle with dummy rocket. Production of FROG 6 may have been necessary because of increasing complexity of sighting systems attributed to FROGs 4 and 5.
JWS* 77    Training missile. FROG 8 could well be training missile, as was FROG 6.
SWM 77    Dummy missile for training purposes.
AWST 82    Training missile.

# FROG 7

```
Sources are:  Bras  77    Rock  78    IISS  81    IISS  82    NWG   83    SMP   84    BW    86
HRST  75      IISS  77    FI*   79    JCS   81    JWS   82    SMP   83    JCS   85
MOW   76      SWM   77    FI    81    AWST  82    IISS+ 83    Meyr  84    SMP   85
```

------- **US DESIGNATION** -------

| | | | |
|---|---|---|---|
| HRST | 75 | Frog 7 | |
| MOW | 76 | Frog-7 | |
| Bras | 77 | FROG 7 | |
| SWM | 77 | Frog 7 | |
| Rock | 78 | Frog 7 | |
| FI* | 79 | Frog 7 | |
| FI | 81 | Frog 7 | |
| JCS | 81 | FROG 7 | FY82,86. |
| AWST | 82 | Frog 7 | |
| IISS | 82 | FROG-7 | [USSR listing] |
| | | FROG-3/-7 | [NSWP listing] |
| JWS | 82 | FROG-7 | |
| IISS+ | 83 | FROG-3/-5/-7 | [NSWP listing] |
| SMP | 83 | FROG 7 | '83, '84. |
| Meyr | 84 | FROG-7 | |

------- **SOVIET DESIGNATION** -------

SWM  77    Luna

------- **DESCRIPTION** -------

| | | | |
|---|---|---|---|
| MOW | 76 | | "Bombardment missile". |
| Rock | 78 | tactical | |
| JWS | 82 | | "Surface-to-surface, spin-stabilized unguided tactical missile". |

------- **EDITOR'S NOTES** -------

| | | |
|---|---|---|
| FI | 81 | Final report. |
| JCS | 81 | Listed FY82,86. |
| IISS | 82 | FROG 7 is an entry under USSR weapons; FROG-3/-7 is an entry under Non-Soviet Warsaw Pact (NSWP) weapons. |
| SMP | 83 | Listed '83, '84. See also generic FROG entry. |
| Meyr | 84 | Note to chart: "Values pertain to initial service period". |

------- **RANGE** -------

| | | mi | nm | km | km conv | |
|---|---|---|---|---|---|---|
| MOW | 76 | c. 45 | | c. 72 | 72.41 | |
| Bras | 77 | | | 60 | 60.00 | |
| SWM | 77 | 37 | | 60 | 59.53 | [Table] |
| | | 35 | | 56 | 56.32 | [Text] |
| Rock | 78 | | | 60 | 60.00 | |
| FI* | 79 | | | 60 | 60.00 | [Table] |
| | | | | c. 55 | 55.00 | [Text] |
| FI | 81 | | | 60 | 60.00 | |
| AWST | 82 | 36 | | | 57.92 | |
| IISS | 82 | | | 70 | 70.00 | Max [USSR listing] |
| | | | | 40-60 | 40.00-60.00 | Max [NSWP listing, FROG-3/-7] |
| JWS | 82 | | | c. 60 | 60.00 | |
| SMP | 83 | | | [70] | 70.00 | SS-21 range of about 120 km is 50 km more than FROG-7. |
| Meyr | 84 | | | 60 | 60.00 | |
| SMP | 84 | | | c. 70 | 70.00 | '84, '85. |
| BW | 86 | | | 70 | 70.00 | |

------- **CEP** -------

| | | ft | mi | nm | m | km | m conv | |
|---|---|---|---|---|---|---|---|---|
| IISS | 82 | | | | 400 | | 400.00 | [USSR listing] |
| | | | | | 380 | | 380.00 | [NSWP listing, FROG-3/-7] |
| Meyr | 84 | | | | | 0.4-0.7 | 400.00-700.00 | |
| BW | 86 | | | | ? 400 | | 400.00 | |

------- **WARHEAD TYPE** -------

| | | nuclear | conven | chem | |
|---|---|---|---|---|---|
| MOW | 76 | yes | yes | | HE. |
| Bras | 77 | yes | yes | yes | HE. |
| SWM | 77 | yes | yes | yes | |
| Rock | 78 | yes | yes | yes | HE. |
| FI* | 79 | yes | yes | | HE. |
| FI | 81 | yes | yes | | "450 kg nuclear or HE" [?] |
| JWS | 82 | yes | yes | | HE. |
| Meyr | 84 | yes | | | |

BALLISTIC MISSILES  SRBMs  FROG 7  365

## WARHEAD CHARACTERISTICS

| Source | Year | Characteristic | Notes |
|---|---|---|---|
| Rock | 78 | single | |
| IISS | 82 | single | [Both entries] |
| Meyr | 84 | single RV | |
| BW | 86 | single RV | |

## NUCLEAR YIELD

| Source | Year | kT | MT | kT conv | Notes |
|---|---|---|---|---|---|
| Bras | 77 | c. 25 | | 25.00 | Warhead of 25 kT class. |
| IISS | 82 | 200 | | 200.00 | [Both entries] |
| IISS+ | 83 | c. 200 | | 200.00 | |
| NWG | 83 | 10-200 | | 10.00-200.00 | "The FROG-7 can deliver three separate nuclear warheads, in the one, 100 and 200 kiloton range". |
| Meyr | 84 | 50-300 | | 50.00-300.00 | |
| BW | 86 | ? 200 | | 200.00 | |

## WARHEAD WEIGHT

| Source | Year | lb | t | kg | kg conv | Notes |
|---|---|---|---|---|---|---|
| SWM | 77 | 990 | | | 449.06 | [Table] |
|  |  | 1000 | | 450 | 453.60 | [Text] |
| Rock | 78 | | | 500 | 500.00 | Weight of both conventional and nuclear warhead. |
| FI* | 79 | 1,000 | | 450 | 453.60 | |

## WEIGHT

| Source | Year | lb | t | kg | kg conv | Notes |
|---|---|---|---|---|---|---|
| Bras | 77 | | | 6,300 | 6,300 | |
| SWM | 77 | 5,070 | | 2,300 | 2,299 | |
| Rock | 78 | | 2.5 | | 2,500 | Tonnes. |
| FI* | 79 | | | c. 2,500 | 2,500 | |
| FI | 81 | | | c. 2,500 | 2,500 | |
| JWS | 82 | | c. 2 | c. 2,000 | 2,000 | |

## LENGTH

| Source | Year | in | ft | cm | m | m conv |
|---|---|---|---|---|---|---|
| HRST | 75 | | 31 | | | 9.45 |
| MOW | 76 | | 29'6" | | 9 | 8.99 |
| Bras | 77 | | | | 9.0 | 9.00 |
| SWM | 77 | | 29.8 | | 9.1 | 9.08 |
| Rock | 78 | | | | 9.0 | 9.00 |
| FI* | 79 | | | | 9.0 | 9.00 |
| FI | 81 | | | | 9.0 | 9.00 |
| AWST | 82 | | 31 | | | 9.45 |
| JWS | 82 | | | c. 900 | | 9.00 |

## DIAMETER

| Source | Year | in | ft | cm | m | cm conv | Notes |
|---|---|---|---|---|---|---|---|
| HRST | 75 | 18 | | | | 45.72 | |
| MOW | 76 | | 1'11.6" | | 0.60 | 59.94 | |
| Bras | 77 | | | 60 | | 60.00 | Calibre, [600 mm in source] |
| SWM | 77 | 21.6 | | 55 | | 54.86 | Warhead same diameter as missile body. |
| Rock | 78 | | | | 0.6 | 60.00 | |
| FI* | 79 | | | 60 | | 60.00 | |
| FI | 81 | | | 60 | | 60.00 | |
| JWS | 82 | | | c. 55 | | 55.00 | |

## WINGSPAN

| Source | Year | in | ft | cm | m | cm conv |
|---|---|---|---|---|---|---|
| FI* | 79 | | | ? 200 | | 200.00 |
| FI | 81 | | | ? 200 | | 200.00 |
| AWST | 82 | | 1.5 | | | 45.72 |

## NUMBER OF STAGES

| Source | Year | Stages | Notes |
|---|---|---|---|
| HRST | 75 | 1 | |
| MOW | 76 | 1 | |
| Bras | 77 | 1 | |
| SWM | 77 | 1 | |
| Rock | 78 | 1 | |
| FI* | 79 | 1 | [Table] |
|  |  | 2 | [Text] |
| FI | 81 | 1 | |
| AWST | 82 | 1 | |
| JWS | 82 | 1 | |

```
------------ TYPE OF PROPULSION ------------
MOW    76    solid         18 small nozzles around main nozzle.
Bras   77    solid         Main motor venturi surrounded by ring of 20 nozzles.
SWM    77    solid         ? Improved propellants.
Rock   78    solid
FI*    79    solid
FI     81    solid
JWS    82    solid         Main nozzle surrounded by ring of much smaller nozzles.
BW     86    solid

------------ MIDCOURSE (OR MAIN) GUIDANCE ------------
MOW    76    unguided      Spin stabilised.
Bras   77    Nil
SWM    77    unguided      Spin-stabilized. Range adjustment by speed brakes. Guided version has been
                           mentioned.
FI*    79    unguided      Spin-stabilised, range adjusted by speed brakes.
FI     81    unguided      Spin stabilized.
JWS    82    unguided      Spin-stabilized.

------------ FIRST OBSERVED ------------
MOW    76    7 Nov 1965
Bras   77    1967          First shown to public.
JWS    82    1967          First shown to public.

------------ DEPLOYMENT BEGAN ------------
Bras   77    ? 1965        Probably in service since 1965.
SWM    77    1965
Rock   78    1965          Service intro.
IISS   82    1965          First year deployed [USSR listing]
              1957-65      First year deployed [NSWP listing, FROG-3/-7]
Meyr   84    1965          Service.
BW     86    1965

------------ DEPLOYED ------------
MOW    76                  In service.
SWM    77                  Deployed.
JWS    82                  Deployed.
BW     86                  In service.

------------ DEPLOYMENT HISTORY ------------
MOW    76    Frog-7s were deployed rapidly in and out of WTO.
IISS   82    Being replaced by SS-21 [USSR listing]. FROG-3 obsolescent [NSWP listing]
JWS    82    Other FROGs in service are mostly FROGs 3 and 5; being replaced gradually by FROG-7.
SMP    83    Being replaced by SS-21 in Western Theater.
SMP    84    Being replaced by SS-21.
JCS    85    Being replaced by SS-21, [a new missile].

------------ NUMBERS IN SERVICE ------------

------------ NUMBER IN 1977 ------------
IISS   77    c. 450        July, total deployed.

------------ NUMBER IN 1981 ------------
IISS   81    482           1981-82, July, total deployed, Europe (including European USSR) only.

------------ NUMBER IN 1982 ------------
IISS   82    205           July, Total deployed, [NSWP listing, FROG-3/-7]

------------ NUMBER IN 1983 ------------
IISS+  83    620           July, total deployed, 440 of these in Europe.

------------ UNDATED NUMBER ------------
BW     86    ? 480         Peak.

------------ CARRIERS ------------

------------ CARRIER ------------ TUBES ---- RELOADS ------------
MOW    76    wheeled vehicle    1         2
                                                    reloads: On latest pictures.
AWST   82                                           Modernized wheeled transporter.

             ------------ CARRIER ------------ TUBES ---- RELOADS ------------
Bras   77    ZIL 135            1                   First FROG with wheeled vehicle, 8-wheeled.
SWM    77    ZIL-135                                Onboard crane for swift reloading.
JWS    82    ZIL-135                                Modern wheeled erector launcher; has crane onboard for use
                                                    in reloading.
BW     86    ZIL-135            1         yes
```

## USE AND CONFIGURATION

| | | |
|---|---|---|
| Bras | 77 | Steadying jacks fold down to provide stable firing platform. For first time, missile sits on launch rail, not cradle. Tender based on ZIL 135 carries 3 missiles "ready and prepared for firing"; crane vehicle required to transfer to launch rail. |
| SWM | 77 | Warheads carried on separate trailers. |
| Rock | 78 | Unlike earlier Frogs, 7 uses transporter-erector "capable of rapidly changing rounds after each launch. Support vehicles carry batteries of additional missiles and a selection of warheads to suit the preference of battle commanders". |
| NWG | 83 | Operational echelon: Division. Warheads are not stored aat division level, but reported at Army level, "where they are centrally controlled". |
| Meyr | 84 | Penetration: 1.0; reliability: 0.6-0.9, "includes system reliability multiplied by operational readiness of deployed system". |

## STRATEGY

| | | |
|---|---|---|
| Meyr | 84 | [Ed: chart is labeled] Systems "assigned primary operational and strategic missions within the European theatre(s) of military operations". |

## COMBAT REPORTS/EFFECTIVENESS

| | | |
|---|---|---|
| JWS | 82 | Some Syrian missiles were fired against Israelis in 1973; had HE warheads and not very effective. "At one time suggested that these particular missiles may have been a guided version"; not confirmed or repeated recently. |

## GEOGRAPHICAL DEPLOYMENTS

| | | |
|---|---|---|
| NWG | 83 | East Germany    104 deployed. |

## EXPORTED TO

| | | | |
|---|---|---|---|
| MOW | 76 | WTO | |
| | | Egypt | |
| | | Syria | |
| | | Iraq | |
| | | North Korea | |
| Bras | 77 | Bulgaria | |
| | | Czech | |
| | | Egypt | |
| | | East Germany | |
| | | Hungary | |
| | | Iraq | |
| | | North Korea | |
| | | Romania | |
| | | Poland | |
| | | Syria | |
| SWM | 77 | WTO | Deployed. |
| | | Egypt | Deployed. |
| | | Syria | Deployed. |
| FI* | 79 | WTO | |
| | | North Korea | |
| | | Mideast | [Detail] |
| JCS | 81 | | NSWP countries have "older FROG-3s and 5s". Being replaced gradually in NSWP forces by FROG 7. |
| IISS | 82 | all NSWP | [NSWP listing, FROG-3/-7] |
| JWS | 82 | WTO | Operational in armies of several WTO nations. Supplied "in quantity" to those [below, ed]. |
| | | Egypt | |
| | | Iraq | |
| | | Libya | |
| | | North Korea | |
| | | Syria | |
| SMP | 85 | Kuwait | Soviets have signed contracts with Kuwait "that include weapons such as" FROG 7. |

## RELATION TO OTHER MISSILES

| | | |
|---|---|---|
| SWM | 77 | Shorter and fatter than Frog 4. A "much-improved" FROG. |
| JWS | 82 | Wheeled vehicle new departure [in FROG series], as was missile. Much cleaner appearance than predecessors. Probably last in FROG series. |

## OTHER INFORMATION

| | | |
|---|---|---|
| SWM | 77 | Fixed cruciform fins. Warhead same diameter as missile body. |

# SS-1 Scud

### US DESIGNATION
| | | |
|---|---|---|
| JCS | 84 | SS-1 |

### NATO CODENAME
| | | | |
|---|---|---|---|
| SecD | 71 | Scud | FY72,74-75,77. |
| HRST | 75 | Scuds A,B,C | |
| JCS | 75 | Scud | FY76-82. |
| Bras | 77 | SCUD A and B | |
| SMP | 81 | | Scud and Scud B, '81, '83, '84, '85. |
| IISS | 82 | Scud | |
| SMP | 85 | Scud-1 B | [In photo caption, ed.] |

### DESCRIPTION
| | | |
|---|---|---|
| SecD | 71 | Listed as tactical missile assigned to ground forces. |
| SecD | 73 | Short range missile. |
| HRST | 75 | Medium-range series. |
| JCS | 75 | Short-range tactical ballistic missile, FY76-79. |
| SecD | 76 | SSM of TNF. |
| JCS | 79 | Tactical nuclear missile. |
| JCS | 80 | Operational-tactical missiles, FY81-82. |
| SMP | 83 | Tactical Nuclear Missile (Scud B). |
| JCS | 84 | Listed as "short-range nuclear forces", Europe, both as Soviet and NSWP system. |

### EDITOR'S NOTES
| | | |
|---|---|---|
| SecD | 71 | Listed FY72,74-75,77. For later years, see Scud B. |
| JCS | 75 | Listed FY76-82,85. For other years, see Scud B. |
| SMP | 81 | Listed '81, '83, '84, '85. |

### RANGE
| | | mi | nm | km | km conv | |
|---|---|---|---|---|---|---|
| HRST | 75 | 200-400 | | | 321.80-643.60 | |
| JCS | 75 | | 85-160 | | 157.42-296.32 | FY76-79. |
| JCS | 81 | | | | | Improved in range [since 1960]. |
| SMP | 83 | | | 300 | 300.00 | '83, '84, '85. |
| SMP | 85 | | | | | Shorter range than SS-23. |

### CEP
| | | ft | mi | nm | m | km | m conv | |
|---|---|---|---|---|---|---|---|---|
| JCS | 81 | | | | | | | Improved in accuracy [since 1960]. |

### WARHEAD TYPE
| | | nuclear | conven | chem | |
|---|---|---|---|---|---|
| SecD | 74 | yes | | | Nuclear-capable. |
| JCS | 75 | yes | | | Nuclear-capable, FY76-82. |
| JCS | 79 | yes | | | |
| SMP | 81 | yes | yes | yes | [FROG, Scud, Scaleboard and replacements together]. |

### WARHEAD CHARACTERISTICS
| | | |
|---|---|---|
| IISS | 82 | single RV |

### NUCLEAR YIELD
| | | kT | MT | kT conv |
|---|---|---|---|---|
| IISS | 82 | | < 1 | 1,000 |

### PAYLOAD
| | | lb | t | kg | kg conv | |
|---|---|---|---|---|---|---|
| JCS | 81 | | | | | Improved in payload [since 1960]. |

### LENGTH
| | | in | ft | cm | m | m conv |
|---|---|---|---|---|---|---|
| HRST | 75 | | 31-38 | | | 9.45-11.58 |

### DIAMETER
| | | in | ft | cm | m | cm conv |
|---|---|---|---|---|---|---|
| HRST | 75 | 30-36 | | | | 76.20-91.44 |

### TYPE OF PROPULSION
| | | |
|---|---|---|
| HRST | 75 | liquid |

BALLISTIC MISSILES  SRBMs  SS-1 Scud  369

――――――――――― **MIDCOURSE (OR MAIN) GUIDANCE** ―――――――
JWS 82    inertial           A form of inertial guidance, "possibly allied to radio command for the early
                             stages of flight" is generally assumed method of aiming Scud. Post-burnout
                             control is open question, but trouble taken for extensive meteorological data
                             collection suggests there is none. Possible but unlikely that radio command
                             could extend or curtail rocket burn.

――――――――――― **CONTROL SYSTEM** ―――――――
JWS 82    Almost certainly by vanes or spoilers in efflux.

――――――――――― **RADAR** ―――――――
JWS 82    End Tray           Weather radar.

――――――――――― **DEPLOYED** ―――――――
SecD 74                      Deployed.
JCS 75                       Deployed.

――――――――――― **DEPLOYMENT HISTORY** ―――――――
JCS 76    Continue to be added to forces, FY77-78.
JCS 81    "Deployed in ever larger numbers" [since 1960].
IISS 82   Being replaced by SS-23.
JWS 82    "Clearly a successful weapon from its length of service and steady improvement programmes", Scud
          now has successor SS-X-23 in development.
SMP 84    Expected to be replaced by SS-23, '84, '85.
SMP 85    Deployed with armies in all theaters.

――――――――――― **NUMBERS IN SERVICE** ―――――――

――――――――――― **NUMBER IN 1970** ―――――――
Clns 80   300                In 1970-1972.
                             About one-fourth were SCUD-As in 1970; SCUD-B replaced them all by 1978.

――――――――――― **NUMBER IN 1973** ―――――――
IISS 73   c. 300             1973-76, July, total deployed.
Clns 80   350

――――――――――― **NUMBER IN 1974** ―――――――
Clns 80   400                In 1974-1975.

――――――――――― **NUMBER IN 1976** ―――――――
Clns 80   450

――――――――――― **NUMBER IN 1977** ―――――――
IISS 77   c. 750             July, total deployed.
Clns 80   500

――――――――――― **NUMBER IN 1978** ―――――――
Clns 80   530                In 1978-1979.

――――――――――― **NUMBER IN 1980** ―――――――
JCS 80                       During past year, USSR "completed a program to increase the number of SCUD launchers
                             available to the Front commander of the" GSFG.

――――――――――― **NUMBER IN 1981** ―――――――
IISS 81   410                July, total deployed, Europe (including European USSR) only.

――――――――――― **NUMBER IN 1982** ―――――――
IISS 82   450                July, total deployed, Europe (including European USSR) only.

――――――――――― **NUMBER IN 1983** ―――――――
OrO1 84   250                "In Eastern satellites", [British Foreign Office report, ed.]

――――――――――― **NUMBER IN 1985** ―――――――
SMP 85    575                400 opposite European NATO. 100 opposite Sino-Soviet border and in Far East. 75 opposite
                             southwest Asia and eastern Turkey. "One brigade [12-18 missiles, ed.] is in strategic
                             reserve."

――――――――――― **CARRIERS** ―――――――

――――――――― **CARRIER** ――――― **TUBES** ―― **RELOADS** ―――――
JWS 82    land-mobile

――――――――― **CARRIER** ――――― **TUBES** ―― **RELOADS** ―――――
SMP 84    TEL                [1]                '84, '85
                                                tubes: [From picture, '84, '85.]

---------- USE AND CONFIGURATION ----------
JCS  80  During past year, USSR "completed a program to increase the number of SCUD launchers available to the Front commander of the" GSFG. "At the army and front level", Scud A and B provide "nuclear delivery means that is organic to ground forces".
JCS  81  Scud A and B "are nuclear delivery means organic to Warsaw Pact ground froces".
SMP  81  A Front has Scuds.
JWS  82  Land-mobile; vehicle is transporter and erector. Differences within JS-III and MAZ-543 configurations noted. "The non-Soviet Union Scud launcher/erector vehicles of this type [MAZ-543] are more complex than those supplied to other users of Scud missiles" [sic]. Launcher vehicles can be reloaded in the field. For re-supply purposes Scud is towed tail-first by Type ZIL-157V tractor with special semi-trailer. Loading onto launcher is by Type 8T210 (6.3 t) mobile crane on Ural-375 vehicle. Preparation time quoted by Soviet sources as c. 1 hour. Scud units also have metereological mobile unit with End Tray weather radar, ZIL-157V tanker with trailerised pump unit, and command and control vehicles. End tray assumed to be for "radio-sonde tracking and data collection for use in obtaining upper-air information for ballistics computation prior to launch". Fairly length survey procedure also entailed on arrival at firing position; tripod-mounted theodolites and optical devices attached to launcher vehicle used for this.
SMP  83  "Normally deployed in brigades at army and front level", '83, '84.
SMP  84  [Picture caption:] "chemical/biological decontamination of a SCUD-B tactical missile during a Soviet field-training exercise".
SMP  85  12-18 missiles in a brigade. [Picture caption:] "Scud-1B nuclear brigades are deployed with armies in all theaters. In field-training exercises, chemical troops practice chemical/biological decontamination of SCUD launchers."

---------- STRATEGY ----------
SecD  74  Deployed with ground forces.

---------- COMBAT REPORTS/EFFECTIVENESS ----------
SecD  74  In 1973 war, "deployment of nuclear-capable SCUD missile launchers".
IISS  82  Serviceability: 0.8; Survivability: 0.7; Reliability: 0.75; for "Scud A/B" and "Scud B/C", [same values, ed].

---------- GEOGRAPHICAL DEPLOYMENTS ----------
JCS  79  Launchers deployed in Eastern Europe; more are based in USSR and probably earmarked for NATO.
SMP  85  In USSR opposite NATO, Sino-Soviet border, and southwest Asia and eastern Turkey. [See NUMBER IN 1985, ed.]

---------- EXPORTED TO ----------
Bras 77  Bulgaria
         Czech
         East Germany
         Egypt
         Hungary
         Iraq
         Poland
         Rumania
         Syria
JWS  82                                Only Soviet missiles likely to have nuclear warheads.
         Bulgaria
         Czech
         East Germany
         Egypt
         Hungary
         Iraq
         Libya
         Poland
         Romania
         Syria

# SS-1b Scud A

```
Sources are:  IAD1  65    FI*   74    Bres  77    Clns  80    JWS   82    SAD1  83
IMSG  60      IAD5  66    HRST  75    SWM   77    AWST  82    Clns+ 83    Meyr  84
Zaeh  61      Ley   68    MOW   76    Rock  78    IISS  82    IISS+ 83    BW    86
```

### US DESIGNATION

| | | |
|---|---|---|
| FI* | 74 | SS-1 |
| MOW | 76 | SS-1b |
| Bras | 77 | SS-1B |
| SWM | 77 | SS-1B |
| Rock | 78 | SS-1B |
| AWST | 82 | SS-1B |
| IISS | 82 | SS-1b |
| JWS | 82 | SS-1B |
| Meyr | 84 | SS-1b |

### NATO CODENAME

| | | |
|---|---|---|
| Ley | 68 | Scud A |
| FI* | 74 | Scud A |
| MOW | 76 | Scud-A |
| Bras | 77 | SCUD A |
| SWM | 77 | Scud A |
| Rock | 78 | Scud A |
| Clns | 80 | Scud-A |
| AWST | 82 | Scud A |
| IISS | 82 | Scud A |
| JWS | 82 | Scud A |

### OTHER DESIGNATION

| | | |
|---|---|---|
| IMSG | 60 | T-7A |
| Zaeh | 61 | T-7A |

### DESCRIPTION

| | | |
|---|---|---|
| IMSG | 60 | | Short-range ballistic missile. |
| MOW | 76 | | Heavy artillery rocket. |
| Rock | 78 | tactical | [Also] medium range tactical ballistic missile. |

### EDITOR'S NOTES

| | | |
|---|---|---|
| Meyr | 84 | Note to chart: "Values pertain to initial service period". |

### RANGE

| | | mi | nm | km | km conv | |
|---|---|---|---|---|---|---|
| IMSG | 60 | 50-100 | | | 80.45-160.90 | |
| Zaeh | 61 | c. 50 | | | 80.45 | |
| IAD1 | 65 | > 60 | | | 96.54 | |
| IAD5 | 66 | c. 100 | | | 160.90 | |
| Ley | 68 | 50 | | | 80.45 | Min. |
| | | 102 | | | 164.12 | Max. |
| FI* | 74 | | | 80-150 | 80.00-150.00 | |
| MOW | 76 | 50-93 | | 80-150 | 80.45-149.64 | Max. |
| Bras | 77 | | | 150 | 150.00 | |
| SWM | 77 | 50-110 | | 80-180 | 80.45-176.99 | [Table] |
| | | 112 | | 180 | 180.21 | [Text] |
| Rock | 78 | | | 130 | 130.00 | |
| Clns | 80 | 50 | | | 80.45 | Scud A. |
| AWST | 82 | 110 | | | 176.99 | |
| IISS | 82 | | | 150 | 150.00 | Max. |
| JWS | 82 | | | 80-150 | 80.00-150.00 | [Analysis table] |
| | | | | c. 130 | 130.00 | [Text] |
| Meyr | 84 | | | 150 | 150.00 | |
| BW | 86 | | | 150 | 150.00 | |

### SPEED

| | | mph | kmph | Mach | kmph conv | |
|---|---|---|---|---|---|---|
| IMSG | 60 | 3,200 | | | 5,148 | Max. |
| MOW | 76 | | | 5 | | [1,100 mps] cutoff velocity, Scud B's 1,500 mps |
| Rock | 78 | | | | | is 400 mps more than Scud A. |

### CEP

| | | ft | mi | nm | m | km | m conv |
|---|---|---|---|---|---|---|---|
| Clns+ | 83 | | | 0.5 | | | 926.00 |
| Meyr | 84 | | | | | 0.8-1.5 | 800.00-1,500 |
| BW | 86 | | | | 930 | | 930.00 |

## WARHEAD TYPE

|  |  | nuclear | conven | chem |  |
|---|---|---|---|---|---|
| Zaeh | 61 | yes | | | |
| FI* | 74 | yes | yes | | HE. |
| MOW | 76 | yes | maybe | | HE. |
| Bras | 77 | yes | | | |
| SWM | 77 | yes | yes | yes | |
| Rock | 78 | yes | yes | | HE. |
| JWS | 82 | yes | yes | | HE. |
| Clns+ | 83 | yes | | | |
| Meyr | 84 | yes | | | |

## WARHEAD CHARACTERISTICS

| Rock | 78 | single |
|---|---|---|
| Meyr | 84 | single RV |
| BW | 86 | single RV |

## NUCLEAR YIELD

|  |  | kT | MT | kT conv |  |
|---|---|---|---|---|---|
| Bras | 77 | c. 100 | | 100.00 | 100 kT class. |
| Clns+ | 83 | 40 | | 40.00 | |
| SA01 | 83 | 1 | | 1.00 | Maximum. |
| Meyr | 84 | 20-100 | | 20.00-100.00 | |
| BW | 86 | ? 40 | | 40.00 | |

## WARHEAD WEIGHT

|  |  | lb | t | kg | kg conv |  |
|---|---|---|---|---|---|---|
| SWM | 77 | 1,500 | | 680 | 680.40 | Includes nose cone. |
| Rock | 78 | | | 680 | 680.00 | |

## PAYLOAD

|  |  | lb | t | kg | kg conv |
|---|---|---|---|---|---|
| IMSG | 60 | 175 | | | 79.38 |

## WEIGHT

|  |  | lb | t | kg | kg conv |  |
|---|---|---|---|---|---|---|
| IMSG | 60 | 4,510 | | | 2,045 | |
| Zaeh | 61 | | c. 7 1/2 | | 7,500 | Tons. |
| IA01 | 65 | | 4.5-5 | | 4,500-5,000 | Tons [one place in article] |
| | | | c. 5 | | 5,000 | Tons [another place in article] |
| Ley | 68 | | c. 5 | | 5,000 | Tons. |
| FI* | 74 | | | c. 4,500 | 4,500 | |
| MOW | 76 | 10,000 | | 4,500 | 4,536 | |
| Bras | 77 | | | 4,500 | 4,500 | |
| SWM | 77 | 9,920 | | 4,500 | 4,499 | |
| Rock | 78 | | 4.4 | | 4,400 | Tonnes. |
| JWS | 82 | | | c. 4,500 | 4,500 | |

## LENGTH

|  |  | in | ft | cm | m | m conv |  |
|---|---|---|---|---|---|---|---|
| IMSG | 60 | | c. 30 | | | 9.14 | |
| Zaeh | 61 | | 25 | | | 7.62 | |
| IA01 | 65 | | 34.8 | | | 10.61 | |
| Ley | 68 | | 34.8 | | | 10.61 | |
| FI* | 74 | | | | c. 10.5 | 10.50 | |
| MOW | 76 | | 35'0" | | 10.66 | 10.67 | |
| Bras | 77 | | | | 10.5 | 10.50 | |
| SWM | 77 | | 35.1 | | 10.7 | 10.70 | |
| Rock | 78 | | | | 10.7 | 10.70 | |
| AWST | 82 | | 31 | | | 9.45 | |
| JWS | 82 | | | c. 1,050 | | 10.50 | [Analysis table] |
| | | | | | [10.75] | 10.75 | [Text: 0.5 m shorter than 11.25 m of Scud B] |

## DIAMETER

|  |  | in | ft | cm | m | cm conv |  |
|---|---|---|---|---|---|---|---|
| IMSG | 60 | 32 | | | | 81.28 | |
| Zaeh | 61 | | 3.0 | | | 91.44 | |
| IA01 | 65 | | 2.8 | | | 85.34 | |
| Ley | 68 | | 2.8 | | | 85.34 | |
| FI* | 74 | | | c. 85 | | 85.00 | |
| MOW | 76 | | 2'6" | | 0.75 | 76.20 | |
| Bras | 77 | | | 85 | | 85.00 | |
| SWM | 77 | 33 | | 84 | | 83.82 | Calibre, [850 mm in source] |
| Rock | 78 | | | | 0.85 | 85.00 | |
| AWST | 82 | | 2.5 | | | 76.20 | |
| JWS | 82 | | | 85 | | 85.00 | |

## BALLISTIC MISSILES  SRBMs  SS-1b Scud A

```
------------- WINGSPAN -------------
              in        ft       cm       m     cm conv
IMSG  60                9.6                     292.61

------------- NUMBER OF STAGES -------------
IMSG  60      1
Zaeh  61      1
FI*   74      1
MOW   76      1
Bras  77      1
SWM   77      1
Rock  78      1
AWST  82      1

------------- TYPE OF PROPULSION -------------
IMSG  60      liquid              Liquid oxygen and hydrocarbon.
Zaeh  61                          "17,000 lb thrust for 30-60 seconds". "Composite propellant".
FI*   74      liquid
MOW   76      storable liquid
Bras  77      liquid
SWM   77      storable liquid
Rock  78      storable liquid     "Unsymmetrical dimethyl hydrazine/inhibited red fuming nitric acid
                                  combination"; oxidizer tank at front end of missile. Thrust, tonnes: 18.
AWST  82      storable
JWS   82      storable liquid
Clns+ 83      liquid
BW    86      storable liquid

------------- MIDCOURSE (OR MAIN) GUIDANCE -------------
IMSG  60      radio inertial
FI*   74      radio command
MOW   76      command             ? Superceded by simplified inertial.
Bras  77      radio command
SWM   77      radio command       Simplified inertial; accelerometer senses that pre-set velocity reached and
                                  turns off motor.
Rock  78      radio command       "Probably controlled by a tracking system which permits the operators to shut
                                  down the motor by single radio command".
JWS   82      radio command

------------- CONTROL SYSTEM -------------
FI*   74      Rear fins.
MOW   76      Tail surfaces.
Bras  77      Control surfaces at rear of motor, also act as stabilizing fins.
SWM   77      Rear fins.
Rock  78      Movable vanes in efflux. Fins don't look movable.

------------- FIRST OBSERVED -------------
MOW   76      Nov 1957
Bras  77      1957

------------- OBSERVED -------------
IMSG  60                          [Shown in picture].

------------- DEVELOPMENT UNDERWAY -------------
Rock  78                          Developed in early 1950s.

------------- DEPLOYMENT BEGAN -------------
SWM   77      1957
Rock  78      1957                Service intro.
IISS  82      1957                First year deployed.
Clns+ 83      1957                First deployed.
SA01  83      1957                First deployed.
Meyr  84      1957                Service.
BW    86      1957

------------- DEPLOYED -------------
IMSG  60                          Fully operational.
MOW   76                          In service.
Bras  77                          In service.

------------- RETIRED -------------
Rock  78      1976
Clns  80      1978                Scud-Bs replaced all Scud-As by 1978.
JWS   82                          Believed withdrawn from first-line service.
BW    86      ? 1973
```

## 374 SOVIET MISSILES IDDS

### DEPLOYMENT HISTORY
| | | |
|---|---|---|
| Rock | 78 | Replaced by Scud B. |
| JWS | 82 | Possibly obsolescent. Believed withdrawn from first-line service. |

### NUMBERS IN SERVICE

#### NUMBER IN 1970
| | | | |
|---|---|---|---|
| Clns | 80 | [c. 75] | About one-fourth of 300 Scuds in 1970 were Scud As. |
| Clns+ | 83 | 50 | |

#### NUMBER IN 1971
| | | |
|---|---|---|
| Clns+ | 83 | 40 |

#### NUMBER IN 1972
| | | |
|---|---|---|
| Clns+ | 83 | 20 |

#### NUMBER IN 1973
| | | |
|---|---|---|
| Clns+ | 83 | 10 |

#### NUMBER IN 1974
| | | | |
|---|---|---|---|
| Clns+ | 83 | none | In 1974-1982. |

#### NUMBER IN 1983
| | | | |
|---|---|---|---|
| IISS+ | 83 | 440 | July, total deployed, Europe (including European USSR) only, [might be European total for both Scud A and B, ed]. |
| SA01 | 83 | 250 | Launchers in Central Europe. Includes all systems stationed outside the USSR and in the western military districts of the USSR. |

#### UNDATED NUMBER
| | | | |
|---|---|---|---|
| BW | 86 | ? 100 | Peak. |

### CARRIERS

#### CARRIER — TUBES — RELOADS
| | | | | | |
|---|---|---|---|---|---|
| IMSG | 60 | tracked tank | | | |
| Zaeh | 61 | tank-transported | | | |
| HRST | 75 | tracked carrier | | | A and B versions. |
| Rock | 78 | tracked vehicle | | | "Tracked vehicle modified from World War II tank chassis". |
| AWST | 82 | | | | Tracked transporter-erector. |
| Clns+ | 83 | | | | "Wheeled, Tracked". |

#### CARRIER — TUBES — RELOADS
| | | | | | |
|---|---|---|---|---|---|
| FI* | 74 | JS III | | | Tracked vehicle. |
| MOW | 76 | JS-3 | | | Same as Frog-1. |
| Bras | 77 | JS 3 | 1 | | Vehicle "basically that of" FROG 1. |
| SWM | 77 | JS-III | | | |
| JWS | 82 | JS-III | 1 | | |
| BW | 86 | JS-3 | 1 | yes | |

### DESIGN AND ENGINEERING
| | | |
|---|---|---|
| Rock | 78 | "Matured from technical dissection of the German V-2". |

### USE AND CONFIGURATION
| | | |
|---|---|---|
| IMSG | 60 | Part of highly mobile missile system. Tracked tank transporter also carries weapon's crew. |
| MOW | 76 | Preparation for launching said to take more than one hour after reaching site. |
| Bras | 77 | Launched vertically. Protective framework protects rocket, raises rocket to vertical, lowered for firing. Although details of launch procedure not released, "the fuelling and guidance check-out must involve a considerable number of men and vehicles at the launch site". Weapon is "most unlikely to travel after fuelling"; "the radar and radio vehicles needed by the type of guidance used increases the size of the firing unit". "Reaction time is claimed as one hour". |
| SWM | 77 | Raised to verticial by hydraulic jacks for launching. Operation to include concealment, movement after firing. |
| Rock | 78 | Transported in horizontal position, erected for launch by raising it to vertical; supporting framework lowered preparatory to ignition. |
| Meyr | 84 | Penetration: 1.0; reliability: 0.4-0.8, "includes system reliability multiplied by operational readiness of deployed system". |

### STRATEGY
| | | |
|---|---|---|
| Rock | 78 | "Designed to saturate rear echelons and/or supply dumps". |
| Meyr | 84 | [Ed: chart is labeled] Systems "assigned primary operational and strategic missions within the European theatre(s) of military operations". |

```
------------    EXPORTED TO ------------
FI*    74    WTO
             Egypt
             Syria
MOW    76    WTO          Deployed.
             Egypt        Formerly deployed.
             Syria        Formerly deployed.
JWS    82    Warsaw Pact

------------    RELATION TO OTHER MISSILES ------------
Zaeh   61    Similar to U.S. Sergeant.

------------    OTHER INFORMATION ------------
IMSG   60    Constant diameter, capped with ogival nose, has 4 movable tail fins.
```

# SS-1c Scud B

| Sources are: | | | | | | | | | | |
|---|---|---|---|---|---|---|---|---|---|---|
| | | Ley | 68 | SWM | 77 | Clns | 80 | AWST | 82 | Clns+ 83 | JCS 85 |
| IA04 | 66 | HRST | 75 | Rock | 78 | SecD | 80 | IISS | 82 | IISS+ 83 | SMP 85 |
| IA05 | 66 | MOW | 76 | SecD | 78 | FI | 81 | JCS | 82 | NWG 83 | BW 86 |
| ID01 | 68 | Bras | 77 | FI* | 79 | SecD | 81 | JWS | 82 | Meyr 84 | |

---------- US DESIGNATION ----------

| MOW | 76 | SS-1c |
| Bras | 77 | SS-1C |
| SWM | 77 | SS-1C |
| Rock | 78 | SS-1C |
| FI* | 79 | SS-1C |
| FI | 81 | SS-1C |
| AWST | 82 | SS-1C |
| IISS | 82 | SS-1c |
| JWS | 82 | SS-1C |
| Meyr | 84 | SS-1c |
| JCS | 85 | SS-1 Scud B |

---------- NATO CODENAME ----------

| Ley | 68 | Scud B | |
| MOW | 76 | Scud-B | |
| Bras | 77 | SCUD B | |
| SWM | 77 | Scud B | |
| Rock | 78 | Scud B | |
| SecD | 78 | Scud B | FY79-82. |
| FI* | 79 | Scud B | |
| Clns | 80 | Scud-B | |
| FI | 81 | Scud B | |
| AWST | 82 | Scud B | |
| IISS | 82 | Scud B | Also KY-3 Scud C [in NSWP listing, final report] |
| JCS | 82 | Scud B | |
| JWS | 82 | Scud B | |

---------- DESCRIPTION ----------

| Rock | 78 | tactical |
| SecD | 81 | Medium-range [FROG, Scud, Scaleboard] launchers. |
| JWS | 82 | "Surface-to-surface, artillery missile". |
| JCS | 85 | Listed in "short-range nuclear forces", Europe, both as Soviet and NSWP system. |

---------- EDITOR'S NOTES ----------

| SecD | 78 | Listed FY79-82. For earlier years, see Scud. |
| FI | 81 | Final report. |
| IISS | 82 | Scud B is a model of SS-1 Scud in List of Soviet weapons, but "SS-1c Scud B, KY-3 Scud C" is a separate entry in Non-Soviet Warsaw Pact weapon listing. Data is accordingly labeled "USSR Listing" and "NSWP Listing". |
| JCS | 82 | Listed FY83. For earlier years, see Scud. |
| Meyr | 84 | Note to chart: "Values pertain to initial service period". |

---------- RANGE ----------

| | | mi | nm | km | km conv | |
|---|---|---|---|---|---|---|
| IA04 | 66 | 60+ | | | 96.54 | |
| IA05 | 66 | c. 125 | | | 201.13 | |
| Ley | 68 | 60 | | | 96.54 | Min. |
| | | 136 | | | 218.82 | Max. |
| MOW | 76 | 185 | | 300 | 297.67 | Enlarged (over Scud A) for more propellant and greater range. |
| Bras | 77 | | | 165-280 | 165.00-280.00 | |
| SWM | 77 | 100-170 | | 160-280 | 160.90-273.53 | [Table] |
| | | 174 | | 280 | 279.97 | [Text] |
| Rock | 78 | | | 280 | 280.00 | |
| FI* | 79 | | | 280 | 280.00 | Max. |
| Clns | 80 | 185 | | | 297.67 | Scud B. |
| FI | 81 | | | 280 | 280.00 | Max. |
| AWST | 82 | 180 | | | 289.62 | |
| IISS | 82 | | | 160-300 | 160.00-300.00 | Max, [USSR listing]. |
| | | | | 160-450 | 160.00-450.00 | Max, [NSWP listing]. |
| JCS | 82 | | | 300 | 300.00 | |
| JWS | 82 | | | 160-270 | 160.00-270.00 | |
| IISS+ | 83 | | | 300 | 300.00 | |
| Meyr | 84 | | | 280 | 280.00 | |
| BW | 86 | | | 300 | 300.00 | |

## SPEED

| | | mph | kmph | Mach | kmph conv | |
|---|---|---|---|---|---|---|
| Rock | 78 | | | | | 1,500 mps Cutoff velocity. |

## CEP

| | | ft | mi | nm | m | km | m conv | |
|---|---|---|---|---|---|---|---|---|
| Clns+ | 83 | | | 0.5 | | | 926.00 | |
| Meyr | 84 | | | | | 0.5-1.0 | 500.00-1,000 | |
| BW | 86 | | | | 930 | | 930.00 | |

## WARHEAD TYPE

| | | nuclear | conven | chem | |
|---|---|---|---|---|---|
| MOW | 76 | maybe | maybe | | HE. |
| Bras | 77 | yes | | | |
| SWM | 77 | yes | yes | | |
| Rock | 78 | yes | yes | | HE. |
| FI* | 79 | yes | yes | | HE. |
| Clns | 80 | yes | yes | | HE. |
| SecD | 80 | yes | yes | yes | Nuclear, chemical, or conventional munitions, FY81-82. |
| FI | 81 | yes | yes | | |
| JWS | 82 | yes | yes | | HE; "nuclear, optional non-nuclear warhead". |
| Meyr | 84 | yes | | | |

## WARHEAD CHARACTERISTICS

| | | | |
|---|---|---|---|
| Rock | 78 | single | |
| IISS | 82 | single | [NSWP listing] |
| Meyr | 84 | single RV | |
| BW | 86 | single RV | |

## NUCLEAR YIELD

| | | kT | MT | kT conv | |
|---|---|---|---|---|---|
| Clns | 80 | 1 | | 1.00 | |
| IISS | 82 | | < 1 | 1,000 | [NSWP listing] |
| NWG | 83 | 1-10 | | 1.00-10.00 | |
| Meyr | 84 | 100-500 | | 100.00-500.00 | |
| BW | 86 | ? 10 | | 10.00 | |

## WARHEAD WEIGHT

| | | lb | t | kg | kg conv |
|---|---|---|---|---|---|
| SWM | 77 | 1697-1896 | | 770-860 | 769.76-860.03 |
| Rock | 78 | | | <_ 860 | 860.00 |

## WEIGHT

| | | lb | t | kg | kg conv | |
|---|---|---|---|---|---|---|
| IA04 | 66 | | c. 6 | | 6,000 | Tons. |
| Ley | 68 | | c. 6 | | 6,000 | Tons. |
| MOW | 76 | c. 13,890 | | c. 6,300 | 6,300 | |
| Bras | 77 | | | 6,300 | 6,300 | |
| SWM | 77 | 14,046 | | 6,370 | 6,371 | |
| Rock | 78 | | 6.3 | | 6,300 | Tonnes. |
| FI* | 79 | | | 6,300 | 6,300 | |
| FI | 81 | | | 6,300 | 6,300 | |
| JWS | 82 | | | c. 6,300 | 6,300 | Estimated. |

## LENGTH

| | | in | ft | cm | m | m conv |
|---|---|---|---|---|---|---|
| IA04 | 66 | | c. 37.8 | | | 11.52 |
| ID01 | 68 | | c. 38 | | | 11.58 |
| Ley | 68 | | 37.8 | | | 11.52 |
| MOW | 76 | | 37'0" | | 11.25 | 11.28 |
| Bras | 77 | | | | 11.0 | 11.00 |
| SWM | 77 | | 37.4 | | 11.4 | 11.40 |
| Rock | 78 | | | | 11.2 | 11.20 |
| FI* | 79 | | | | 11.2 | 11.20 |
| FI | 81 | | | | 11.2 | 11.20 |
| AWST | 82 | | 33 | | | 10.06 |
| JWS | 82 | | | c. 1,125 | | 11.25 |

## SOVIET MISSILES   IDDS

```
------------ DIAMETER ------------
              in      ft      cm      m      cm conv
IA04   66             2.79                    85.04
Ley    68             3.0                     91.44
Bras   77                     85              85.00       Calibre, [850 mm in source]
SWM    77     33              84              83.82
Rock   78                            0.85     85.00
FI*    79                     85              85.00
FI     81                     85              85.00
AWST   82             2.5                     76.20
JWS    82                     85              85.00

------------ WINGSPAN ------------
              in      ft      cm      m      cm conv
FI*    79                     180             180.00
FI     81                     180             180.00

------------ NUMBER OF STAGES ------------
Bras   77     1
SWM    77     1
FI*    79     1
AWST   82     1

------------ TYPE OF PROPULSION ------------
IA05   66     solid
MOW    76     storable liquid
Bras   77     storable liquid
SWM    77     storable liquid
Rock   78     storable liquid        Thrust, tonnes: 25.
FI*    79     storable liquid        UDMH/IRFNA.
Clns   80     liquid
FI     81     storable liquid        ? UDMH/IRFNA.
AWST   82     storable
JWS    82     storable liquid
BW     86     storable liquid

------------ MIDCOURSE (OR MAIN) GUIDANCE ------------
MOW    76     simplified inertial  Probably.
Bras   77     inertial
SWM    77     simplified inertial
FI*    79     simplified inertial
FI     81     simplified inertial
JWS    82     simplified inertial

------------ CONTROL SYSTEM ------------
MOW    76     Tail surfaces.
Bras   77     Tail fins (as in SCUD A).
SWM    77     Rear fins control.
FI*    79     Vanes in efflux.
FI     81     Vanes in efflux.

------------ OBSERVED IN PARADE ------------
ID01   68     Nov 1965      First shown.

------------ DEPLOYMENT BEGAN ------------
SWM    77     1965
Rock   78     1965          Service intro.
Clns   80     1965          First deployed.
IISS   82     1965          First year deployed [both USSR and NSWP listings]
NWG    83     1968          Introduced, replacing Scud A.
Meyr   84     1965          Service.
BW     86     1965

------------ DEPLOYED ------------
MOW    76                   In service.
Bras   77                   In service.
SWM    77                   Deployed.
BW     86                   In service.

------------ DEPLOYMENT HISTORY ------------
Clns   80     Scud-Bs replaced all Scud-As by 1978.
JWS    82     The standard model is now widely deployed.

------------ NUMBERS IN SERVICE ------------

------------ NUMBER IN 1970 ------------
Clns   80     [c. 225]  About one fourth of 300 Scuds in 1970 were Scud As.
Clns+  83     250
```

```
---------- NUMBER IN 1971 ----------
Clns+ 83    260

---------- NUMBER IN 1972 ----------
Clns+ 83    280

---------- NUMBER IN 1973 ----------
Clns+ 83    340

---------- NUMBER IN 1974 ----------
Clns+ 83    400       In 1974-1975.

---------- NUMBER IN 1976 ----------
Clns+ 83    450

---------- NUMBER IN 1977 ----------
Clns+ 83    500

---------- NUMBER IN 1978 ----------
Clns  80    530       1978-79.
Clns+ 83    530       In 1978-1979.

---------- NUMBER IN 1979 ----------
FI *  79              18 launchers per Front (USSR).

---------- NUMBER IN 1980 ----------
Clns+ 83    550       In 1980-1982.

---------- NUMBER IN 1982 ----------
IISS  82    143       July, total deployed, [in NSWP listing].

---------- NUMBER IN 1983 ----------
IISS+ 83    570       July, total deployed, national total, [might be national total for both Scud A and B,
                      ed].
            137       Total deployed, 7/83 [in NSWP listing].

---------- UNDATED NUMBER ----------
BW    86    ? 575     Peak.

---------- CARRIERS ----------

---------- CARRIER ---------- TUBES --- RELOADS ----------
HRST  75    tracked carrier                    A and B versions.
Clns  80    wheeled vehicle
JCS   85    wheeled vehicle    1
                                               tubes: Photo with caption "Soviet SS-1 Scud B" shows
                                                      wheeled vehicle with one missile.

---------- CARRIER ---------- TUBES --- RELOADS ----------
MOW   76    Scud-A vehicle
Bras  77    SCUD A vehicle     1               First seen on SCUD A's tracked vehicle.
SWM   77    JS-3
JWS   82    JS-III             1
BW    86    JS-3               1       yes

---------- CARRIER ---------- TUBES --- RELOADS ----------
MOW   76    MAZ-543                            This 8-wh vehicle is now the standard.
Bras  77    MAZ 543            1               8-wh, first seen in 1965.
SWM   77    MAZ-543
Rock  78    MAZ-543                            8-wh twin-cab vehicle.
FI *  79    MAZ-543                            8-wh vehicle.
AWST  82    MAZ-543
JWS   82    MAZ-543            1               8-wh vehicle; since 1965.
BW    86    MAZ-543            1       yes

---------- DESIGN AND ENGINEERING ----------
Rock  78    Very similar to Scud A. 0.5 m length increase facilitates increase in propellant capacity and
            reverse positioning of tanks. "Matured from technical dissection of the German V-2".
```

## USE AND CONFIGURATION

| | | |
|---|---|---|
| Bras | 77 | Launch pad integral to erector frame; rear of vehicle supported on jacks while missile is elevated to vertical for launching. "SCUD is deployed in brigades of nine launch sections and attached to fronts". |
| SWM | 77 | Enclosed in ribbed, split casing, erected to vertical for launch. |
| Rock | 78 | Protective casing removed prior to erection. |
| SecD | 78 | Tactical nuclear capability organic to ground forces (at divisional and higher levels) [FROG, Scud, and Scaleboard], FY79-80. |
| FI* | 79 | 18 launchers per Front [USSR]. |
| NWG | 83 | Operational echelon: Army. "The only Army nuclear armed missile system above the division level which is assigned to East Germany". "Nine launchers are assigned to each of the five Soviet armies and two East German armies" [in East Germany]. Warheads thought stored in central Army depots. |
| Meyr | 84 | Penetration: 1.0; reliability: 0.6-0.9, "includes system reliability multiplied by operational readiness of deployed system". |
| SMP | 85 | "Armies and fronts have missile brigades equipped with 12-18 SS-1C Scud SRBMs." |

## STRATEGY

| | | |
|---|---|---|
| Rock | 78 | "Designed to saturate rear echelons and/or supply dumps". |
| Meyr | 84 | [Ed: chart is labeled] Systems "assigned primary operational and strategic missions within the European theatre(s) of military operations". |

## COMBAT REPORTS/EFFECTIVENESS

| | | |
|---|---|---|
| SWM | 77 | All 3 Scuds fired in 1973 missed. |
| Rock | 78 | Scud rockets fired during 1973 war. "Scud is the only Soviet long range battlefield ballistic missile to have been fired in anger". |

## GEOGRAPHICAL DEPLOYMENTS

| | | | |
|---|---|---|---|
| NWG | 83 | East Germany | 54 deployed, with 150 warheads; [Ed: but conflicts with] "Nine launchers are assigned to each of the five Soviet armies and two East German armies" [in East Germany]. |

## EXPORTED TO

| | | | |
|---|---|---|---|
| Bras | 77 | East Germany | |
| SWM | 77 | WTO | Deployed in all WTO armies, though nuclear warheads only in USSR. |
| Rock | 78 | Bulgaria | |
| | | Czech | |
| | | E Germany | |
| | | Egypt | |
| | | Poland | |
| | | Romania | |
| | | Syria | |
| SecD | 78 | | FROG and Scud launchers possessed by other WTO nations; warheads "remain under Soviet control", FY79-80. |
| FI* | 79 | WTO | |
| | | Mideast | |
| IISS | 82 | All NSWP | [NSWP listing] |
| JWS | 82 | | Nuclear warheads probably only for Soviet formations. |
| | | Warsaw Pact | |
| | | Egypt | |
| | | Iraq | |
| | | Syria | |

## RELATION TO OTHER MISSILES

| | | |
|---|---|---|
| Bras | 77 | Direct development of SCUD A. |
| NWG | 83 | "The latest (third) modification, the SCUD-b..." [sic; other two?, ed]. |

## OTHER INFORMATION

| | | |
|---|---|---|
| Bras | 77 | Rumours of SCUD C believed to arise from confusion over 2 launch vehicles used for SCUD B. |
| Rock | 78 | "Some specialists have identified a third, interim, variant displaying the structural characteristics of the Scud A, but with the increased length of the Scud B". |
| JWS | 82 | Has been noted in 2 versions, "one of which is thought to represent an interim stage of development leading to the standard model now widely deployed" in WTO and beyond. |

# SS-1c Scud C

Sources are:  HRST 75    AWST 82    JWS 82

------ US DESIGNATION ------
AWST 82    SS-1C

------ NATO CODENAME ------
AWST 82    Scud C
JWS  82    Scud C

------ OTHER DESIGNATION ------
JWS  82    KY-3 Scud

------ RANGE ------
|  | mi | nm | km | km conv |
|---|---|---|---|---|
| AWST 82 | 450 | | | 724.05 |
| JWS 82 | | | c. 450 | 450.00 |

------ CEP ------
|  | ft | mi | nm | m | km | m conv |
|---|---|---|---|---|---|---|
| JWS 82 | | | | | | |

"Lower" CEP than Scud B.

------ WARHEAD TYPE ------
|  | nuclear | conven | chem |
|---|---|---|---|
| AWST 82 | yes | | |

------ LENGTH ------
|  | in | ft | cm | m | m conv |
|---|---|---|---|---|---|
| AWST 82 | | 38 | | | 11.58 |

------ DIAMETER ------
|  | in | ft | cm | m | cm conv |
|---|---|---|---|---|---|
| AWST 82 | | 3.0 | | | 91.44 |

------ NUMBER OF STAGES ------
AWST 82    1

------ TYPE OF PROPULSION ------
AWST 82    storable

------ DEPLOYMENT BEGAN ------
JWS  82    1965    First deployed.

------ CARRIERS ------

------ CARRIER ------ TUBES --- RELOADS ------
HRST 75    wheeled carrier
AWST 82    Wheeled transporter-erector.

------ RELATION TO OTHER MISSILES ------
JWS  82    "The existence of a longer range Scud C was confirmed in a US Armed Services Committee reference in hearings of April 1978 to the KY-3 Scud, when it was stated that this version was first deployed in 1965. It was added that the later version has a longer range than Scud B, but a lower CEP accuracy".

# SS-12

| Sources are: | NA01 | 72 | JCS | 76 | Rock | 78 | SecD | 80 | AWST | 82 | Clns+ | 83 | Meyr | 84 |
|---|---|---|---|---|---|---|---|---|---|---|---|---|---|---|
| SecC | 69 | SecD | 72 | MOW | 76 | SecD | 78 | IISS | 81 | IISS | 82 | IISS+ | 83 | SMP | 84 |
| SecC | 70 | SecC | 73 | SecD | 76 | FI* | 79 | JCS | 81 | JCS | 82 | JWS+ | 83 | SMP | 85 |
| SecC | 71 | SecD | 73 | Bras | 77 | JCS | 79 | SecD | 81 | JWS | 82 | NWG | 83 | BW | 86 |
| SecD | 71 | FIar | 76 | SWM | 77 | Clns | 80 | SMP | 81 | SSF | 82 | SMP | 83 | | |

---------- US DESIGNATION ----------

| SecC | 70 | SS-12 | FY71-72,74. |
|---|---|---|---|
| SecD | 71 | SS-12 | FY72,80-82. |
| FIar | 76 | SS-12 | |
| MOW | 76 | SS-12 | |
| Bras | 77 | SS-12 | |
| SWM | 77 | SS-12 | |
| Rock | 78 | SS-12 | |
| FI* | 79 | SS-12 | |
| Clns | 80 | SS-12 | |
| JCS | 81 | SS-12 | FY82-83. |
| SMP | 81 | SS-12 | '81, '83, '84, '85. |
| AWST | 82 | SS-12 | |
| IISS | 82 | SS-12 | |
| JWS | 82 | SS-12 | |
| SSF | 82 | SS-12 | |
| Meyr | 84 | SS-12 | |

---------- NATO CODENAME ----------

| SecD | 72 | Scaleboard | FY73-74,77,79-82. |
|---|---|---|---|
| FIar | 76 | Scaleboard | |
| JCS | 76 | Scaleboard | FY77-80,82. |
| MOW | 76 | Scaleboard | |
| Bras | 77 | Scaleboard | |
| SWM | 77 | Scaleboard | |
| Rock | 78 | Scaleboard | |
| FI* | 79 | Scaleboard | |
| Clns | 80 | Scaleboard | |
| SMP | 81 | Scaleboard | '81, '83, '84. |
| AWST | 82 | Scaleboard | |
| IISS | 82 | Scaleboard | |
| JWS | 82 | Scaleboard | |

---------- DESCRIPTION ----------

| SecC | 71 | SRBM | Short Range Ballistic Missile. |
|---|---|---|---|
| SecD | 71 | | Listed as tactical missile assigned to ground forces. |
| SecD | 72 | | Short range missile, FY73-74. Listed as part of TNF. |
| JCS | 76 | | Short-range missiles, FY77-78. |
| MOW | 76 | | Tactical missile. |
| SecD | 76 | | SSM of TNF. |
| Rock | 78 | MRBM | Also "medium range tactical missile". |
| SecD | 81 | | Medium-range [FROG, Scud, Scaleboard] launchers. |
| SMP | 81 | | Short range ballistic missile, '81, '85. |
| AWST | 82 | | Ballistic missile. |
| JWS | 82 | | "Short medium-range ballistic missile". "Surface-to-surface tactical". |

---------- EDITOR'S NOTES ----------

| SecC | 69 | Listed FY70-72,74, not by name FY70. |
|---|---|---|
| SecD | 71 | Listed FY72-74,77,79-82. |
| JCS | 76 | Listed FY77-80,82-83. |
| SMP | 81 | See also SS-22/Scaleboard. Listed '81, '83, '85. |
| JWS+ | 83 | From SS-22 listing. |
| Meyr | 84 | Note to chart: "Values pertain to initial service period". |

BALLISTIC MISSILES  SRBMs  SS-12  383

## RANGE

| | | mi | nm | km | km conv | |
|---|---|---|---|---|---|---|
| SecC | 70 | | c. 500 | | 926.00 | FY71-722. |
| SecD | 71 | | 500 | | 926.00 | The missile system is a mobile launcher "estimated to carry a 500 nm missile". |
| NA01 | 72 | | | 1,000 | 1,000 | Past FEBA. |
| FIar | 76 | | | ? 800 | 800.00 | [Table] |
| | | | | ? 700-800 | 700.00-800.00 | [Text] |
| JCS | 76 | | | | | Longer range SSM [than FROG and Scud], FY77-79. |
| MOW | 76 | 500 | | 805 | 804.50 | Max. |
| | | | | | | Larger than Scuds, implying longer range. |
| Bras | 77 | | | 700-800 | 700.00-800.00 | |
| SWM | 77 | 435-497 | | 700-800 | 699.92-799.67 | |
| Rock | 78 | | | 750 | 750.00 | [Table] |
| | | | | > 700 | 700.00 | [Text] |
| FI* | 79 | | | 700-800 | 700.00-800.00 | |
| Clns | 80 | 500 | | | 804.50 | |
| JCS | 81 | | | | | Exceeds Scud and SS-X-23. |
| SMP | 81 | | | 900 | 900.00 | |
| | | 500 | | | 804.50 | |
| AWST | 82 | 510 | | | 820.59 | |
| IISS | 82 | | | 490-900 | 490.00-900.00 | Max, [overall listing] |
| | | | | 900 | 900.00 | Range [European theater] |
| JCS | 82 | | | 900 | 900.00 | |
| JWS | 82 | | | c. 800 | 800.00 | |
| JWS+ | 83 | | | c. 900 | 900.00 | According to 1983 SMP. |
| NWG | 83 | 500 | | | 804.50 | |
| SMP | 83 | | | c. 900 | 900.00 | '83, '84. |
| Meyr | 84 | | | 800 | 800.00 | |
| BW | 86 | | | 900 | 900.00 | |

## CEP

| | | ft | mi | nm | m | km | m conv | |
|---|---|---|---|---|---|---|---|---|
| Rock | 78 | | | | | | | "Extremely good accuracy". |
| IISS | 82 | | | | 900 | | 900.00 | |
| Clns+ | 83 | | | 0.4 | | | 740.80 | |
| Meyr | 84 | | | | | 0.4-1.0 | 400.00-1,000 | |
| BW | 86 | | | | 740 | | 740.00 | |

## WARHEAD TYPE

| | | nuclear | conven | chem | |
|---|---|---|---|---|---|
| JCS | 76 | yes | | | Nuclear-capable, FY77-80. |
| MOW | 76 | yes | | | Probably. |
| Bras | 77 | yes | | | |
| SWM | 77 | yes | yes | | |
| Clns | 80 | yes | | | |
| SecD | 80 | yes | yes | yes | Nuclear, chemical, or conventional munitions, FY81-82. |
| SMP | 81 | yes | yes | yes | [FROG, Scud, Scaleboard and replacements together, ed]. |
| AWST | 82 | yes | | | |
| JWS | 82 | yes | | | "Presumed nuclear". |
| Meyr | 84 | yes | | | |

## WARHEAD CHARACTERISTICS

| | | | |
|---|---|---|---|
| Rock | 78 | single RV | |
| IISS | 82 | single RV | |
| Meyr | 84 | single RV | |
| BW | 86 | single RV | |

## NUCLEAR YIELD

| | | kT | MT | kT conv | |
|---|---|---|---|---|---|
| FIar | 76 | | ? 1 | 1,000 | [Table] |
| | | | | | Megaton range, [text] |
| Bras | 77 | | | | Megaton class. |
| Rock | 78 | | 1.0 | 1,000 | [Also,] can send warhead "of up to 1 MT" to its target. |
| Clns | 80 | | 1 | 1,000 | |
| AWST | 82 | 500 | | 500.00 | |
| IISS | 82 | 200 | | 200.00 | |
| JWS | 82 | | | | MT range, yield rather greater than comparable Pershing missile. |
| Meyr | 84 | 200-1,000 | | 200.00-1,000 | 200 kT-1 MT. |
| BW | 86 | ? 500 | | 500.00 | |

```
              ————— WEIGHT —————
                       lb            t            kg         kg conv
JWS    82                                    c.  6,800       6,800         [Analysis table]
                                                 6,800+      6,800         [Text]

              ————— LENGTH —————
                       in            ft           cm    m    m conv
FIar   76                                              ? 11   11.00
                                                    c.  11   11.00         Container.
MOW    76                            37'0"        11.25      11.28
Bras   77                                         11.0+      11.00
SWM    77                            36           11         10.97
Rock   78                                         10.5       10.50
FI*    79                                      c.  11        11.00
JWS    82                       c. 1,125                     11.25

              ————— DIAMETER —————
                       in            ft           cm    m    cm conv
FIar   76                                              c. 1  100.00        Container.
Bras   77                                         85+        85.00         Calibre, [850+ mm in source]
Rock   78                                               1.4  140.00
FI*    79                                      c. 190        190.00
JWS    82                                         100        100.00

              ————— NUMBER OF STAGES —————
Bras   77      1
SWM    77      1
Rock   78      1

              ————— TYPE OF PROPULSION —————
SecC   71      liquid
FIar   76      ? solid
MOW    76      storable liquid
Bras   77      liquid
SWM    77      storable liquid
Rock   78      storable solid
FI*    79      probably solid
Clns   80      solid
JWS    82      storable liquid      Presumed, probably.
BW     86      storable liquid

              ————— MIDCOURSE (OR MAIN) GUIDANCE —————
FIar   76      ? inertial
MOW    76      inertial             Probably, set up by equipment on launcher.
Bras   77      inertial
SWM    77      inertial
Rock   78      inertial
FI*    79      inertial
JWS    82      ? inertial           Probably. Nothing known of guidance system but "presumed to be some form of
                                    inertial guidance suitable for use with a missile that is elevated from the
                                    horizontal to the vertical shortly before firing".

              ————— OBSERVED —————
FIar   76                           SS-12 has never been shown to Western observers.
JWS    82                           Missile has not been seen outside of casing.

              ————— OBSERVED IN PARADE —————
MOW    76      Nov 1967
Bras   77      Nov 1967             First introduced to public.
SWM    77      Nov 1967
Rock   78      Nov 1967             First publicly displayed.
JWS    82      1967                 First reported in the West; but since stated it has been operational since 1965.

              ————— IOC —————
JWS    82      1965                 Operational since 1965.
```

BALLISTIC MISSILES  SRBMs  SS-12  385

---------- DEPLOYMENT BEGAN ----------
SecC  69            "The Soviets are deploying what is believed to be a new short-range land-mobile
                    missile system [deleted] near the Chinese border" [not said to be SS-12, ed].
SecC  70            There has been deployment in the past year; deployments have been principally
                    concentrated in the Far East.
SWM   77   1969
Rock  78   1968     Service intro.
Clns  80   1969     First deployed.
IISS  82   1969     First year deployed.
Clns+ 83   1965     First deployed, [listing on p36].
           1969     First deployed, [listing on p58].
Meyr  84   1969     Service.
BW    86   1969

---------- DEPLOYED ----------
SecD  73            Deployed with ground forces.
Bras  77            In service.
Rock  78   1969     In extensive service by 1969.
JWS   82            Operational status uncertain. Believed fully operational in USSR.
BW    86            In service.

---------- DEPLOYMENT HISTORY ----------
JCS   76   Continue to be added to the forces, FY77-78.
SMP   81   Being replaced by SS-22, '81, '85.
IISS  82   Being replaced by SS-22.
JWS   82   In 1977, American assessment reported Scaleboards still being added to Soviet inventory.
SMP   83   Expected to be replaced by SS-22, '83, '84.
SMP   84   In modernization, older missiles to be replaced with "improved SS-12/22s".

---------- NUMBERS IN SERVICE ----------

---------- NUMBER IN 1955 ----------
SSF   82   none     In 1955, 1960, 1965.

---------- NUMBER IN 1970 ----------
SecC  70   [c 50]   Deactivation of MRBM/IRBM sites has brought launchers down to 650, but compensated by
                    deployment of SS-12, keeping total force level at about 700.
Clns  80   60
SSF   82   54

---------- NUMBER IN 1971 ----------
SecC  71   54       Operational for 30 Dec 1970 and 1 Mar 1971.
           54-72    Operational for mid-71.
SecD  71            Estimated that USSR has "a number of SS-12 launchers operational".
Clns  80   80

---------- NUMBER IN 1972 ----------
SecC  71   72-108   Operational for mid-72.
SecD  72            "Small" numbers.
Clns  80   100      1972-1973.

---------- NUMBER IN 1974 ----------
Clns  80   120      1974-1979.

---------- NUMBER IN 1975 ----------
SSF   82   72       In 1975, 1980.

---------- NUMBER IN 1976 ----------
MOW   76            "An increasing proportion of the total of 300 short-range ballistic missiles in service
                    with Soviet land forces is likely to consists of 'Scaleboards'".

---------- NUMBER IN 1979 ----------
Clns+ 83   110      In 1979-1980.

---------- NUMBER IN 1981 ----------
IISS  81   65       July, total deployed, Europe (including European USSR) only.
Clns+ 83   60       In 1981-1982.

---------- NUMBER IN 1982 ----------
IISS  82   70       July, total deployed, Europe (including European USSR) only.

---------- NUMBER IN 1983 ----------
IISS+ 83   120      July, total deployed, 70 of these in Europe.

---------- NUMBER IN 1985 ----------
SMP   85   > 100    "Over 60 launchers are opposite European NATO." "40 are opposite the Sino-Soviet border
                    in the Far East." "One brigade [12-18 missiles, ed.] is in the Strategic Reserve
                    military district." [With SS-22, ed.]

## 386 SOVIET MISSILES IDDS

---------- UNDATED NUMBER ----------
BW    86    ? 120    Peak.

---------- CARRIERS ----------

---------- CARRIER ---------- TUBES ---- RELOADS ----------
SecC  70                                          Mobile, FY71-72.
Clns  80    wheeled vehicle

---------- CARRIER ---------- TUBES ---- RELOADS ----------
FIar  76    MAZ-543                               8-wh vehicle, enclosed in ribbed container.
MOW   76    MAZ-543
Bras  77    MAZ 543         1                     Wheeled erector/launcher similar to that used for SCUD A
                                                  and FROG 7.
SWM   77    MAZ-543
Rock  78    MAZ-543
FI *  79    MAZ-543                               8-wh vehicle.
AWST  82    MAZ 53 truck                          [Sic]
JWS   82    MAZ-543                               8-wh vehicle.
BW    86    MAZ-543         1         yes

---------- DESIGN AND ENGINEERING ----------
FIar  76    May be based on SS-14.
SWM   77    Derivative of Scud family.

---------- USE AND CONFIGURATION ----------
SecC  71    Listed as tactical missile for immediate support of ground forces.
SecC  73    Short range missile deployed with ground forces.
FIar  76    Container is lifted to vertical, opened, and moved aside before firing.
Bras  77    Transported in heavily ribbed container; removed only after erection to vertical.
SWM   77    Enclosed in ribbed split casing, elevates with missile for launch.
Rock  78    Normally housed in a ribbed container. Transported on MAZ-543 "and erected for launch at will".
SecD  78    Tactical nuclear capability organic to ground forces (at divisional and higher levels) [FROG, Scud
            and Scaleboard], FY79-80.
FI *  79    Missile is elevated and casing removed before firing.
JCS   81    Designed to support Front operations.
JWS   82    Enclosed in ribbed split casing, elevated with missile to firing position.
NWG   83    Operational echelon: Front.
Meyr  84    Penetration: 1.0; reliability: 0.4-0.8, "includes system reliability multiplied by operational
            readiness of deployed system".

---------- STRATEGY ----------
SecD  73    Deployed with ground forces.
Rock  78    "Designed to saturate rear echelons and/or supply dumps". "The missile would be rapidly deployed
            to destroy support and supply stores at the rear of front line or battle troop positions".
Meyr  84    [Ed: chart is labeled] Systems "assigned primary operational and strategic missions within the
            European theatre(s) of military operations".

---------- COMBAT REPORTS/EFFECTIVENESS ----------
IISS  82    Serviceability: 0.8; Survivability: 0.7; Reliability: 0.75.

---------- GEOGRAPHICAL DEPLOYMENTS ----------
SecC  69                "The Soviets are deploying what is believed to be a new short-range land-mobile
                        missile system [deleted] near the Chinese border" [not said to be SS-12, ed].
SecC  70                There has been deployment in the past year; deployments have been principally
                        concentrated in the Far East.
JCS   79                Deployed in Eastern Europe; more based in USSR are probably earmarked for NATO
                        area.
NWG   83                Has never been deployed outside USSR.
SMP   85                Opposite NATO, Sino-Soviet border in Far East, and southwest Asia/eastern Turkey.
                        [See NUMBER IN 1985, ed.]

# SS-21

| Sources are: | | | | | | | |
|---|---|---|---|---|---|---|---|
| | SecD 79 | FI 81 | AWST 82 | SecD 82 | JWS+ 83 | JCS 84 | JCS 85 |
| JWS* 77 | Clns 80 | JCS 81 | IISS 82 | Clns+ 83 | NWG 83 | Meyr 84 | SMP 85 |
| Rock 78 | JCS 80 | SecD 81 | JCS 82 | IISS+ 83 | SA01 83 | SecD 84 | BW 86 |
| JCS 79 | SecD 80 | SMP 81 | JWS 82 | JCS 83 | SMP 83 | SMP 84 | |

## US DESIGNATION

| | | | |
|---|---|---|---|
| JWS* | 77 | Frog-9 | |
| Rock | 78 | Frog 9 | |
| JCS | 79 | SS-21 | FY80-86. |
| SecD | 79 | SS-21 | FY80-83,85. |
| Clns | 80 | SS-21 | |
| FI | 81 | SS-21 | |
| SMP | 81 | SS-21 | '81, '83, '84, '85. |
| AWST | 82 | SS-21 | |
| IISS | 82 | SS-21 | |
| JWS | 82 | SS-21 | |
| JWS+ | 83 | SS-21 | |
| Meyr | 84 | SS-21 | |

## DESCRIPTION

| | | |
|---|---|---|
| JCS | 79 | Tactical nuclear missile, FY80-81. |
| SecD | 81 | One of medium-range launchers. |
| SMP | 81 | Short range ballistic missile, '81, '83, '85. |
| JCS | 84 | Listed as "short-range nuclear forces", Europe, both as Soviet and NSWP system, FY85-86. |

## EDITOR'S NOTES

| | | |
|---|---|---|
| JWS* | 77 | Evidence that Frog-9 referred to SS-21 is tenuous. |
| Rock | 78 | Evidence that Frog-9 referred to SS-21 is tenuous. |
| JCS | 79 | Listed FY80-86. |
| SecD | 79 | Listed FY80-83,85. |
| SMP | 81 | Listed '81, '83, '84, '85. |
| AWST | 82 | AWST has 2 separate SS-N-21 entries in its chart; data from the two is merged here; final year for duplicate entries. |
| Meyr | 84 | Note to chart: "Values pertain to initial service period". |

## RANGE

| | | mi | nm | km | km conv | |
|---|---|---|---|---|---|---|
| Clns | 80 | 65 | | | 104.59 | |
| FI | 81 | | | c. 100 | 100.00 | |
| SMP | 81 | | | | | Longer-range [than FROG], '81, '85. |
| AWST | 82 | | 65 | | 120.38 | [Final report] "70% range increase" [over Frog] |
| IISS | 82 | | | 120 | 120.00 | Max. |
| JCS | 82 | | | 120 | 120.00 | Improved over FROG. |
| JWS | 82 | | | c. 60 | 60.00 | |
| Clns+ | 83 | 75 | | | 120.68 | |
| JWS+ | 83 | | | ≤ 120 | 120.00 | Official US reports. |
| NWG | 83 | | | 14-120 | 14.00-120.00 | |
| SA01 | 83 | | | 155 | 155.00 | |
| SMP | 83 | | | c. 120 | 120.00 | '83, '84, '85. |
| Meyr | 84 | | | 120 | 120.00 | |
| BW | 86 | | | 120 | 120.00 | |

## CEP

| | | ft | mi | nm | m | km | m conv | |
|---|---|---|---|---|---|---|---|---|
| FI | 81 | | | | | | | Accuracy improved over Frog. |
| JCS | 81 | | | | | | | Accuracy improved over FROG. |
| SMP | 81 | | | | | | | More accurate [than FROG], '81, '85. |
| AWST | 82 | | | | | | | "33% accuracy increase" [over Frog]. |
| IISS | 82 | | | | 300 | | 300.00 | |
| Clns+ | 83 | | | 0.15 | | | 277.80 | |
| SMP | 83 | | | | | | | More accurate than FROG-7, '83, '84. |
| Meyr | 84 | | | | | 0.1-0.3 | 100.00-300.00 | |
| BW | 86 | | | | 300 | | 300.00 | |

---------- **WARHEAD TYPE** ----------
|        |    | nuclear | conven | chem |                                                    |
|--------|----|---------|--------|------|----------------------------------------------------|
| JCS    | 79 | yes     |        |      | Tactical nuclear missile, FY80-81.                 |
| SecD   | 79 |         |        |      | [Implies nuclear only, ed].                        |
| Clns   | 80 | yes     |        |      |                                                    |
| SecD   | 80 | yes     | yes    | yes  | FY81-83.                                           |
| SMP    | 81 | yes     | yes    | yes  | [FROG, Scud, Scaleboard and replacements together, ed]. |
| IISS   | 82 | yes     | yes    |      | "Dual capable".                                    |
| JWS    | 82 | yes     | yes    |      | HE.                                                |
| JWS+   | 83 | yes     | yes    | yes  | HE; submunitions possible.                         |
| SMP    | 83 | yes     |        |      | Nuclear capable.                                   |
| Meyr   | 84 | yes     |        |      |                                                    |

---------- **WARHEAD CHARACTERISTICS** ----------
| Meyr | 84 | single RV |
| BW   | 86 | single RV |

---------- **NUCLEAR YIELD** ----------
|       |    | kT     | MT | kT conv      |          |
|-------|----|--------|----|--------------|----------|
| Clns  | 80 | 1      |    | 1.00         |          |
| Clns+ | 83 | 100    |    | 100.00       |          |
| NWG   | 83 | 10-100 |    | 10.00-100.00 |          |
| SA01  | 83 | 200    |    | 200.00       | Maximum. |
| Meyr  | 84 | 20-100 |    | 20.00-100.00 |          |
| BW    | 86 | ? 100  |    | 100.00       |          |

---------- **PAYLOAD** ----------
|    |    | lb | t | kg | kg conv |                    |
|----|----|----|---|----|---------|--------------------|
| FI | 81 |    |   |    |         | Improved over Frog. |

---------- **LENGTH** ----------
|     |    | in | ft | cm  | m | m conv |
|-----|----|----|----|-----|---|--------|
| JWS | 82 |    |    | 944 |   | 9.44   |

---------- **DIAMETER** ----------
|     |    | in | ft | cm    | m | cm conv |
|-----|----|----|----|-------|---|---------|
| JWS | 82 |    |    | c. 46 |   | 46.00   |

---------- **NUMBER OF STAGES** ----------
| JWS | 82 | 1 |

---------- **TYPE OF PROPULSION** ----------
| Clns | 80 | solid |
| JWS  | 82 | solid |
| BW   | 86 | solid |

---------- **MIDCOURSE (OR MAIN) GUIDANCE** ----------
| AWST | 82 |        | Guided [final report]         |
| JWS  | 82 |        | Said to be guided; not confirmed. |
| JWS+ | 83 | guided | Specifics unknown.            |

---------- **DEVELOPMENT BEGAN** ----------
| JWS+ | 83 | early 70s | Believed under development from early 70s. |

---------- **DEVELOPMENT UNDERWAY** ----------
| JCS | 79 | Now being developed.  |
| JWS | 82 | Development [status]. |

---------- **PROTOTYPE TESTS** ----------
| Rock | 78 | Has been tested. |

---------- **IOC** ----------
| JCS  | 81 | 1976 |                              |
| SMP  | 81 | 1976 |                              |
| JWS  | 82 | 1976 | "Ready for operation".       |
| JWS+ | 83 | 1976 | Potentially ready for operation. |
| NWG  | 83 | 1976 | First declared operational.  |

BALLISTIC MISSILES  SRBMs  SS-21  389

```
----------- DEPLOYMENT BEGAN -----------
JWS*  77                    "Early in 1977 it was reported that a new Russian" weapon "designated the Frog 9
                            was entering service with the Russian Army".
Rock  78                    "Expected to replace the Frog 7 by 1980".
Clns  80      1978          First deployed.
JCS   80                    "Introduction of...SS-21" [indicates...], FY81-82.
FI    81      1976          First deployed.
SMP   81                    Introduced.
AWST  82                    Operational [final report]
IISS  82      1978          First year deployed.
JWS   82                    Said to be operational; not confirmed.
Clns+ 83      1976          First deployed, [listing on p36].
              1978          First deployed, [listing on p58].
JCS   83                    Has been introduced.
SA01  83      1977          First deployed.
SMP   83                    Operational in Eastern Europe.
Meyr  84      1978          Service.
SecD  84                    Soviets developing, and beginning to deploy [SS-21,22,X-23].
SMP   85                    Soviets have begun replacing FROG in some divisions.
BW    86      ? 1983

----------- DEPLOYED -----------
SecD  82                    Deployed.
BW    86                    In service.

----------- DEPLOYMENT HISTORY -----------
SecD  79                    Replacement for FROG (follow-on), FY80-82.
FI    81                    Frog replacement.
JCS   81                    Follow-on to FROG, FY82-83.
SMP   81                    Replacement for FROG.
JWS   82                    Replacement for FROG-7.
Clns+ 83                    Replacing FROG.
JWS+  83                    Replacement for FROG-7; now replacing FROG-7 in USSR's western theater according to official US
                            sources.
SMP   83                    SS-21 is replacing FROG-7 in the Western Theater.
SMP   84                    Replacing older FROG-7.
JCS   85                    Is replacing FROG 7.
SMP   85                    "...new SS-21 short-range ballistic missiles are now deployed with Soviet divisions in East
                            Germany..."

----------- NUMBERS PRODUCED -----------
SMP   84                    "New production capability in the USSR is available to support large-scale deployment of
                            the SS-21 and SS-23 into the 1990s at an annual production rate of over 200 missiles".

----------- NUMBERS IN SERVICE -----------

----------- NUMBER IN 1970 -----------
Clns  80      none          1970-1977.

----------- NUMBER IN 1977 -----------
Clns+ 83      4

----------- NUMBER IN 1978 -----------
Clns  80      10            1978-1979.

----------- NUMBER IN 1980 -----------
Clns+ 83      10

----------- NUMBER IN 1981 -----------
FI    81                    "First deployed in 1976, but has entered service in only small numbers".
JCS   81                    Only a few have been deployed.
SMP   81                    Only a few have been deployed.
JWS   82                    Only a few deployed according to US sources, early 1981.
Clns+ 83      20
JWS+  83                    "Only a few", official US sources in early 1981; delay could be funding problems or
                            serviceability of deployed FROG-7s.
SMP   83                    650 "FROG/SS-21", at end of 1981, Warsaw Pact forces "facing NATO Europe".

----------- NUMBER IN 1982 -----------
IISS  82      c. 10         July, total deployed, Europe (including European USSR) only.
Clns+ 83      25

----------- NUMBER IN 1983 -----------
IISS+ 83      c. 62         July, total deployed.
SA01  83      375           With FROG, launchers in Central Europe. Includes all systems stationed outside the USSR
                            and in the western military districts of the USSR.
SMP   84                    700 FROG/SS-21 launchers, at end of 1983, systems facing NATO Europe.
SMP   85                    375 FROG/SS-21 launchers currently opposite NATO Europe.
```

```
------------ UNDATED NUMBER ------------
BW    86    ? 130    Peak.

------------ CARRIERS ------------

------------ CARRIER ------------ TUBES --- RELOADS ------------
Clns  80    wheeled vehicle
FI    81    wheeled vehicle                               Same as that first used for SA-8 Gecko system.
JWS+  83                                                  Picture shows transporter/erector/launcher.
SMP   83    mobile system
            [wheeled vehicle]                             [Picture of 6-wh Transporter-Erector-Launcher, ed].
JCS   84                                                  [Picture of "SS-21 TEL", ed].
BW    86    6-wh vehicle          1         yes

------------ DESIGN AND ENGINEERING ------------
JCS   84    Designed to replace FROG 7.

------------ USE AND CONFIGURATION ------------
SecD  79    Launchers at divisional and higher levels.
SMP   81    More mobile [than FROG].
NWG   83    Operational echelon: Division.
Meyr  84    Penetration: 1.0; reliability: 0.6-0.99, "includes system reliability multiplied by operational
            readiness of deployed system".
SMP   84    Division-level system, '84, '85.

------------ STRATEGY ------------
Rock  78    "Very much improved weapon" [over earlier Frogs].
SMP   83    More accurate than FROG-7, "enabling greater targeting flexibility and deeper strikes", '83, '84.
Meyr  84    [Ed: chart is labeled] Systems "assigned primary operational and strategic missions within the
            European theatre[s] of military operations".

------------ GEOGRAPHICAL DEPLOYMENTS ------------
JWS+  83                    Now replacing FROG-7 in USSR's western theater according to official US sources.
NWG   83    East Germany    Has been reported there.
SMP   83                    Operational in Eastern Europe. Nuclear forces strengthened during 1982 by forward
                            deployment of additional SS-21s. SS-21 is replacing FROG-7 in the Western Theater.
SMP   85    East Germany    Now deployed with Soviet divisions.

------------ EXPORTED TO ------------
SMP   84    Syria           The only deployment outside Warsaw Pact.

------------ RELATION TO OTHER MISSILES ------------
JWS*  77    Frog 8 could well be training missile, as was Frog 6.
FI    81    Similar configuration to US Lance.
AWST  82    Frog follow-on; replacement for Frog.

------------ OTHER INFORMATION ------------
JWS   82    "The closeness of these [range, length, diameter] figures to FROG-7 prompt the question as to
            their origin; are they derived, on photographic or visual evidence, or are they extrapolations
            based on what is thought reasonable for a FROG-7 successor?".
```

# SS-22

| Sources are: | JCS | 80 | SecD | 81 | IISS | 82 | SecD | 82 | JCS | 83 | SMP | 83 | SMP | 84 |
|---|---|---|---|---|---|---|---|---|---|---|---|---|---|---|
| JCS | 79 | | SecD | 80 | SMP | 81 | JCS | 82 | Clns | 83 | JWS+ | 83 | Meyr | 84 | SMP | 85 |
| SecD | 79 | | FI | 81 | AWST | 82 | JWS | 82 | IISS+ | 83 | NWG | 83 | SecD | 84 | BW | 86 |

---------- **US DESIGNATION** ----------

| | | | |
|---|---|---|---|
| JCS | 79 | SS-22 | FY80-84. |
| SecD | 79 | SS-22 | FY80-83,85. |
| FI | 81 | SS-22 | |
| SMP | 81 | SS-22 | '81, '83, '84, '85. |
| AWST | 82 | SS-22 | |
| IISS | 82 | SS-22 | |
| JWS | 82 | SS-22 | |
| Clns | 83 | SS-22 | |
| JWS+ | 83 | SS-22 | |
| Meyr | 84 | SS-22 | |

---------- **NATO CODENAME** ----------

| | | | |
|---|---|---|---|
| SMP | 85 | Scaleboard | "SS-22/Scaleboard missiles." |

---------- **DESCRIPTION** ----------

| | | |
|---|---|---|
| JCS | 79 | Tactical nuclear missile, FY80-81. |
| SecD | 81 | One of medium-range launchers. |
| SMP | 81 | Short range ballistic missile, '84, '85. |
| AWST | 82 | "Tactical surface-to-surface". |

---------- **EDITOR'S NOTES** ----------

| | | |
|---|---|---|
| JCS | 79 | Listed FY80-84. |
| SecD | 79 | Listed FY80-83,85. |
| FI | 81 | Final report. |
| SMP | 81 | Listed '81, '83, '84, '85. |
| Meyr | 84 | Note to chart: "Values pertain to initial service period". |

---------- **RANGE** ----------

| | | mi | nm | km | km conv | |
|---|---|---|---|---|---|---|
| FI | 81 | | | < 800 | 800.00 | [Also:] > SS-12's. |
| SMP | 81 | | | c. 900 | 900.00 | '81, '83, '85. |
| IISS | 82 | | | 1,000 | 1,000 | Max. |
| JCS | 82 | | | 900 | 900.00 | |
| JWS | 82 | | | 700-800 | 700.00-800.00 | |
| Clns | 83 | 550 | | | 884.95 | |
| IISS+ | 83 | | | 900 | 900.00 | |
| SMP | 83 | | | [c. 900] | 900.00 | Range similar to "about 900" km of SS-12, '83, '84. |
| Meyr | 84 | | | 900 | 900.00 | |
| BW | 86 | | | 900 | 900.00 | |

---------- **CEP** ----------

| | | ft | mi | nm | m | km | m conv | |
|---|---|---|---|---|---|---|---|---|
| FI | 81 | | | | | | | Better than that of SS-12. |
| SMP | 81 | | | | | | | More accurate than SS-12, '81, '83, '84, '85. |
| Clns | 83 | | | 0.2 | | | 370.40 | |
| IISS+ | 83 | | | | 300 | | 300.00 | |
| JWS+ | 83 | | | | | | | Accuracy increased over SS-12, according to 1983 SMP. |
| Meyr | 84 | | | | | 0.2-0.4 | 200.00-400.00 | |
| BW | 86 | | | | 370 | | 370.00 | |

---------- **WARHEAD TYPE** ----------

| | | nuclear | conven | chem | |
|---|---|---|---|---|---|
| JCS | 79 | yes | | | Tactical nuclear missile, FY80-81. |
| SecD | 79 | yes | | | [Implies nuclear only, ed]. |
| SecD | 80 | yes | yes | yes | FY81-83. |
| SMP | 81 | yes | yes | yes | [FROG, Scud, Scaleboard and replacements together, ed]. |
| JWS | 82 | yes | | | HE. |
| Clns | 83 | yes | yes | | HE. |
| Meyr | 84 | yes | | | |
| SMP | 85 | | | | "The new generation of shorter range missiles [SS-22, SS-23, ed.] can be employed effectively with conventional and improved conventional munitions warheads in light of their greatly increased accuracy." |

```
------------ WARHEAD CHARACTERISTICS ------------
Meyr   84      single RV
BW     86      single RV

------------ NUCLEAR YIELD ------------
               kT              MT         kT conv
IISS   82      500                        500.00
Clns   83      500                        500.00
Meyr   84      100-1,000                  100.00-1,000      100 kT-1 MT.
BW     86      500                        500.00

------------ PAYLOAD ------------
               lb              t          kg         kg conv
FI     81                                                        > SS-12's.

------------ NUMBER OF STAGES ------------
JWS    82              Not known.

------------ TYPE OF PROPULSION ------------
FI     81      solid
JWS    82                      Not known.
Clns   83      solid
BW     86      solid

------------ OBSERVED ------------
JWS    82              US DoD reported it in 1978.

------------ DEVELOPMENT UNDERWAY ------------
JCS    79              Being developed.

------------ DEPLOYMENT BEGAN ------------
JCS    80              "Introduction of...SS-22" indicates Soviet seeking expanded options, FY81-82.
SMP    81              Introduced.
IISS   82      1979    First year deployed [overall listing]
               1978    First deployment [European theater]
JWS    82              US DoD reported it in 1978. "Despite the fact that no additional information had
                       been made public by early 1982, the SS-22 programme is still believed to be
                       active".
Clns   83      1977    First deployed.
JCS    83              Has been introduced [since mid-70s].
SMP    83              Expected to replace SS-12, '83, '84.
Meyr   84      1980    Service.
SecD   84              Soviets developing, and beginning to deploy [SS-21,22,X-23].
BW     86      ? 1984

------------ DEPLOYED ------------
AWST   82              Operational.
JWS    82              Operational.
SecD   82              Deployed.
BW     86              In service.

------------ DEPLOYMENT HISTORY ------------
SecD   79      Replacement and follow-on to SS-12 Scaleboard, FY80-82.
FI     81      Intended as replacement for SS-12 Scaleboard.
SMP    81      Replacement for SS-12.
JCS    82      "Continuing Soviet deployment of the modern SS-22". SS-12 follow on.
JWS    82      Probably replacement for SS-12 Scaleboard.
Clns   83      Rapidly replacing SS-12.
SMP    83      Expected to replace SS-12, '83, '84.
SMP    84      In modernization, older missiles to be replaced with "improved SS-12/22s".
SMP    85      SS-22 is replacing SS-12. 1984 marked first forward deployment of Scaleboard SRBMs to Eastern
               Europe.

------------ NUMBERS IN SERVICE ------------

------------ NUMBER IN 1970 ------------
Clns   83      none    In 1970-1976.

------------ NUMBER IN 1977 ------------
Clns   83      12

------------ NUMBER IN 1978 ------------
Clns   83      24

------------ NUMBER IN 1979 ------------
Clns   83      36
```

---------- **NUMBER IN 1980** ----------
Clns 83    48

---------- **NUMBER IN 1981** ----------
Clns 83    60         In 1981-1982.

---------- **NUMBER IN 1982** ----------
IISS 82    c. 100     July, total deployed, Europe (including European USSR) only.

---------- **NUMBER IN 1983** ----------
IISS+ 83   c. 100     July, total deployed.

---------- **NUMBER IN 1985** ----------
SMP 85     > 100      "Over 60 launchers are opposite European NATO." "40 are opposite the Sino-Soviet border in the Far East." "One brigade [12-18 missiles, ed.] is in the Strategic Reserve military district." [With SS-12, ed.]

---------- **UNDATED NUMBER** ----------
BW 86      ? 20       Peak.

---------- **CARRIERS** ----------

---------- **CARRIER** ---------- **TUBES** ---- **RELOADS** ----------
JWS 82                                              To replace SS-12 on Scaleboard launchers.
Clns 83    wheeled vehicle

---------- **CARRIER** ---------- **TUBES** ---- **RELOADS** ----------
BW 86      ? MAZ-543              1             yes

---------- **USE AND CONFIGURATION** ----------
SecD 79    Launchers at divisional and higher levels.
NWG 83     Operational echelon: Front.
Meyr 84    Penetration: 1.0; reliability: 0.6-0.9, "includes system reliability multiplied by operational readiness of deployed system".
SMP 85     12-18 missiles in a brigade.

---------- **STRATEGY** ----------
Meyr 84    [Ed: chart is labeled] Systems "assigned primary operational and strategic missions within the European theatre(s) of military operations".
SMP 85     Forward deployment in Eastern Europe in position to "strike deep into NATO's rear area from their new launch sites without having first to deploy forward."

---------- **COMBAT REPORTS/EFFECTIVENESS** ----------
IISS 82    Serviceability: 0.8; Survivability: 0.8; Reliability: 0.8.

---------- **GEOGRAPHICAL DEPLOYMENTS** ----------
NWG 83     Most likely will not be deployed outside USSR, as with SS-12.
SMP 85     Opposite NATO, Sino-Soviet border in Far East, and southwest Asia/eastern Turkey. [See NUMBER IN 1985, ed.] "More accurate 900-kilometer-range SS-22/Scaleboard missiles have been forward-deployed to East Germany and Czechoslovakia". "These missile units redeployed from the Western USSR."

---------- **RELATION TO OTHER MISSILES** ----------
JWS+ 83    Replacement for SS-12.

# SS-23

| Sources are: | JCS | 81 | AWST | 82 | JWS | 82 | IISS+ | 83 | NWG | 83 | JCS | 84 | SMP | 84 |
|---|---|---|---|---|---|---|---|---|---|---|---|---|---|---|
| | SecD | 80 | SecD | 81 | IISS | 82 | SecD | 82 | JCS | 83 | SMP | 83 | Meyr | 84 | SMP | 85 |
| | FI | 81 | SMP | 81 | JCS | 82 | Clns | 83 | JWS+ | 83 | AWST+ | 84 | SecD | 84 | BW | 86 |

### US DESIGNATION

| | | | |
|---|---|---|---|
| SecD | 80 | SS-X-23 | FY81-82,85. |
| FI | 81 | SS-X-23 | |
| JCS | 81 | SS-X-23 | FY82-83. |
| SMP | 81 | SS-X-23 | '81, '83. |
| AWST | 82 | SS-X-23 | |
| IISS | 82 | SS-23 | |
| JWS | 82 | SS-X-23 | |
| SecD | 82 | SS-23 | |
| Clns | 83 | SS-23 | |
| JCS | 83 | SS-23 | FY84-86. |
| JWS+ | 83 | SS-23 | |
| AWST+ | 84 | SS-23 | |
| Meyr | 84 | SS-23 | |
| SMP | 84 | SS-23 | '84, '85. |

### DESCRIPTION

| | | |
|---|---|---|
| SecD | 81 | One of medium-range launchers. |
| SMP | 83 | Tactical surface-to-surface missile, '83, '84, '85. |

### EDITOR'S NOTES

| | | |
|---|---|---|
| SecD | 80 | Listed FY81-83,85. |
| JCS | 81 | Listed FY82-86. |
| SMP | 81 | Listed '81, '83, '84, '85. |
| Meyr | 84 | Note to chart: "Values pertain to initial service period". |

### RANGE

| | | mi | nm | km | km conv | |
|---|---|---|---|---|---|---|
| FI | 81 | | | $\leq$ 500 | 500.00 | |
| JCS | 81 | | | | | Expected improved over Scud. |
| SMP | 81 | | | | | Longer-range [than Scud], '81, '85. |
| AWST | 82 | | | | | "100% range increase" [over Scud] |
| IISS | 82 | | | 350 | 350.00 | Max. |
| JCS | 82 | | | 500 | 500.00 | |
| Clns | 83 | 300 | | | 482.70 | |
| IISS+ | 83 | | | 500 | 500.00 | |
| JWS+ | 83 | | | $\leq$ 500 | 500.00 | |
| SMP | 83 | | | 500 | 500.00 | '83, '84, '85. |
| Meyr | 84 | | | 350 | 350.00 | |
| BW | 86 | | | 500 | 500.00 | |

### CEP

| | | ft | mi | nm | m | km | m conv | |
|---|---|---|---|---|---|---|---|---|
| FI | 81 | | | | | | | Accuracy increased over Scud B. |
| JCS | 81 | | | | | | | Accuracy expected improved over Scud. |
| SMP | 81 | | | | | | | More accurate [than Scud]. |
| AWST | 82 | | | | | | | "50% accuracy increase" [over Scud] |
| Clns | 83 | | | 0.2 | | | 370.40 | |
| JWS+ | 83 | | | | | | | Accuracy increased over Scud. |
| SMP | 83 | | | | | | | Accuracy improved over Scud, '83, '84, '85. |
| JCS | 84 | | | | | | | "Improved", FY85-86. |
| Meyr | 84 | | | | | 0.2-0.6 | 200.00-600.00 | |
| BW | 86 | | | | ? 370 | | 370.00 | |

### WARHEAD TYPE

| | | nuclear | conven | chem | |
|---|---|---|---|---|---|
| SecD | 80 | yes | yes | yes | FY81-83. |
| SMP | 81 | yes | yes | yes | [FROG, Scud, Scaleboard and replacements together, ed]. |
| IISS | 82 | yes | yes | | "Dual-capable". |
| Clns | 83 | yes | yes | | HE. |
| JWS+ | 83 | yes | yes | | HE, [analysis table] |
| Meyr | 84 | yes | | | |

## BALLISTIC MISSILES  SRBMs  SS-23

```
------------ WARHEAD CHARACTERISTICS ------------
Meyr    84      single RV
BW      86      single RV

------------ NUCLEAR YIELD ------------
                kT              MT              kT conv
Clns    83      100                             100.00
Meyr    84      100-500                         100.00-500.00
BW      86      ? 100                           100.00

------------ PAYLOAD ------------
                lb              t               kg              kg conv
FI      81                                                                      Increased over Scud B.

------------ TYPE OF PROPULSION ------------
Clns    83      solid
JWS+    83      solid           [Analysis table]
BW      86      solid

------------ DEVELOPMENT UNDERWAY ------------
JCS     82              Development continues.
JWS     82              Now in development.

------------ IOC ------------
NWG     83      1980    First declared operational.

------------ DEPLOYMENT BEGAN ------------
SMP     81              Introduced.
IISS    82      1979-80 First year deployed, [overall listing]
                1980    First deployment, [European theater]
Clns    83      1981    First deployed.
JCS     83              Has been introduced.
Meyr    84      1982    Service.
SecD    84              Soviets developing, and beginning to deploy [SS-21,22,X-23].
BW      86      ? 1986

------------ DEPLOYED ------------
SecD    82              Deployed.
BW      86              In service.

------------ DEPLOYMENT HISTORY ------------
SecD    80      Follow-on for Scud launchers.
FI      81      Intended as replacement for Scud B.
JCS     81      Thought to be a replacement for Scud. Follow-on to Scud, FY82-83.
SecD    81      Replacement for Scud B.
SMP     81      Replacement for Scud B.
AW ST   82      Scud follow-on.
JWS     82      Successor to Scud.
Clns    83      Replacing Scud B.
JWS+    83      Thought to be replacing Scud in Soviet forces.
SMP     83      Scud "is being replaced by" SS-23.
SMP     84      Scud "is expected to be replaced by" SS-23, '84, '85.
SMP     85      "Initial deployment is anticipated opposite NATO and China." [As] "Soviet SS-23s replace the Scuds
                in Europe, the Soviet Union will at least maintain its substantial numerical superiority in
                shorter range non-strategic nuclear missiles while improving the qualitative characteristics of
                its forces."

------------ NUMBERS PRODUCED ------------
SMP     84              "New production capability in the USSR is available to support large-scale deployment of
                        the SS-21 and SS-23 into the 1990s at an annual production rate of over 200 missiles".

------------ NUMBERS IN SERVICE ------------

------------ NUMBER IN 1970 ------------
Clns    83      none    In 1970-1980.

------------ NUMBER IN 1981 ------------
Clns    83      18

------------ NUMBER IN 1982 ------------
IISS    82      c. 10   July, total deployed, Europe (including European USSR) only.
Clns    83      48

------------ NUMBER IN 1983 ------------
IISS+   83      c. 10   July, total deployed.

------------ CARRIERS ------------
```

---------- CARRIER ---------- TUBES ---- RELOADS ------------
Clns 83    wheeled vehicle
BW   86    wheeled vehicle      1         yes

---------- USE AND CONFIGURATION ------------
FI   81    "Time into action and refire time have been shortened", great improvement over Scud.
JCS  81    Expected "reduced reaction and refire times" over Scud.
SMP  81    More mobile [than Scud].
JWS  82    "Expected to have improved accuracy and range performance as well as shorter reaction and refire times" [compared to Scud].
NWG  83    Operational echelon: Army.
Meyr 84    Penetration: 1.0; reliability: 0.6-0.9, "includes system reliability multiplied by operational readiness of deployed system".

---------- STRATEGY ------------
JCS  84    "Deployment...will significantly increase the number and capability of SRINF systems facing NATO", FY85-86.
Meyr 84    [Ed: chart is labeled] Systems "assigned primary operational and strategic missions within the European theatre(s) of military operations".

---------- COMBAT REPORTS/EFFECTIVENESS ------------
IISS 82    Serviceability: 0.8; Survivability: 0.8; Reliability: 0.8.
SMP  85    "The new generation of shorter range missiles [SS-22, SS-23, ed.] can be employed effectively with conventional and improved conventional munitions warheads in light of their greatly increased accuracy."

---------- GEOGRAPHICAL DEPLOYMENTS ------------
NWG  83    East Germany    Now or future.
                           None observed outside USSR as of early 1983.
SMP  85                    Initial deployment anticipated opposite NATO and China.

# SSC-1

```
Sources are:   SecD  72       IISS  73      SWM   77      FI    81      IISS  82      GSN   83      Meyr  84
SecD  71       USO2  72       SecD  73      Rock  78      AWST  82      JWS   82      IISS+ 83      BW    86
```

### US DESIGNATION

|  | |  |
|---|---|---|
| SWM | 77 | SSC-1 |
| Rock | 78 | SSC-1 |
| FI | 81 | SSC-1B |
| AWST | 82 | SSC-1 |
| IISS | 82 | SS-C-1b |
| JWS | 82 | SSC-1B |
| GSN | 83 | SS-C-1b |
| Meyr | 84 | SSC-1b |

### NATO CODENAME

| Source | | Codename | |
|---|---|---|---|
| SecD | 71 | Shaddock | FY72-74. |
| SWM | 77 | Shaddock | |
| Rock | 78 | Shaddock | |
| FI | 81 | Sepal | |
| AWST | 82 | Shaddock | |
| IISS | 82 | Sepal | |
| JWS | 82 | Sepal | |
| GSN | 83 | Sepal | |

### DESCRIPTION

| Source | | Description |
|---|---|---|
| SWM | 77 | Coastal defense missile. |
| IISS | 82 | "GLCM". |

### EDITOR'S NOTES

|  | |  |
|---|---|---|
| SecD | 71 | Listed FY72-74. |
| FI | 81 | Final report for specifications in table; text entry remains. |
| Meyr | 84 | Note to chart: "Values pertain to initial service period". |

### RANGE

| Source | | mi | nm | km | km conv | |
|---|---|---|---|---|---|---|
| SWM | 77 | 280 | | 450 | 450.52 | |
| Rock | 78 | | | 450 | 450.00 | |
| FI | 81 | | | $\leq 850$ | 850.00 | Max. |
| | | | | 200 | 200.00 | Optimum. |
| AWST | 82 | 200 | | | 321.80 | |
| IISS | 82 | | | 450 | 450.00 | Max. |
| JWS | 82 | | | c. 450 | 450.00 | Midcourse guidance by aircraft would be necessary for such a long mission. |
| GSN | 83 | | $\geq 250$ | $\geq 460$ | 463.00 | |
| Meyr | 84 | | | 300 | 300.00 | |
| BW | 86 | | | 450 | 450.00 | |

### SPEED

| Source | | mph | kmph | Mach | kmph conv |
|---|---|---|---|---|---|
| SWM | 77 | | | 1.5 | |
| Rock | 78 | | | 2.5 | |
| FI | 81 | | | 1.4 | |
| BW | 86 | | | 1.0 | |

### CEP

| Source | | ft | mi | nm | m | km | m conv |
|---|---|---|---|---|---|---|---|
| Meyr | 84 | | | | | 0.5-0.8 | 500.00-800.00 |
| BW | 86 | | | | ? 650 | | 650.00 |

### WARHEAD TYPE

| Source | | nuclear | conven | chem |
|---|---|---|---|---|
| SWM | 77 | yes | yes | |
| Rock | 78 | yes | yes | HE. |
| FI | 81 | yes | yes | HE. |
| JWS | 82 | yes | yes | HE. |
| Meyr | 84 | yes | | |

### WARHEAD CHARACTERISTICS

| Source | | Characteristics |
|---|---|---|
| IISS | 82 | single warhead |
| Meyr | 84 | single RV |
| BW | 86 | single RV |

## NUCLEAR YIELD

|  |  | kT | MT | kT conv |  |
|---|---|---|---|---|---|
| SWM | 77 | ? 1 |  | 1.00 |  |
| FI | 81 |  |  |  | KT range |
| IISS | 82 |  | < 1 | 1,000 |  |
| Meyr | 84 | 50-200 |  | 50.00-200.00 |  |
| BW | 86 | ? 100 |  | 100.00 |  |

## WARHEAD WEIGHT

|  |  | lb | t | kg | kg conv |
|---|---|---|---|---|---|
| SWM | 77 | 2,205 |  | 1,000 | 1,000 |
| FI | 81 |  |  | c. 1,000 | 1,000 |

## WEIGHT

|  |  | lb | t | kg | kg conv |
|---|---|---|---|---|---|
| SWM | 77 | 26,000 |  | 11,790 | 11,793 |
| Rock | 78 |  |  | 11,790 | 11,790 |
| FI | 81 |  |  | c. 4,500 | 4,500 |

## LENGTH

|  |  | in | ft | cm | m | m conv |  |
|---|---|---|---|---|---|---|---|
| SWM | 77 |  | 42 |  | 13.8 | 12.80 | [Table] |
|  |  |  | > 40 |  | > 12 | 12.19 | [Text] |
|  |  |  |  |  | 15.7 | 15.70 | Carrying 'tank'. |
| Rock | 78 |  |  |  | 13.8 | 13.80 |  |
| FI | 81 |  |  | c. 10 |  | 10.00 |  |
| AWST | 82 |  | 42 |  |  | 12.80 |  |
| JWS | 82 |  |  | c. 1,000 |  | 10.00 |  |

## DIAMETER

|  |  | in | ft | cm | m | cm conv |  |
|---|---|---|---|---|---|---|---|
| SWM | 77 | 39.4 |  | 100 |  | 100.08 |  |
| Rock | 78 |  |  | 100 |  | 100.00 |  |
| FI | 81 |  |  | c. 100 |  | 100.00 |  |
| JWS | 82 |  |  |  | c. 1 | 100.00 | Probably. |

## WINGSPAN

|  |  | in | ft | cm | m | cm conv |
|---|---|---|---|---|---|---|
| SWM | 77 | 82.7 |  | 210 |  | 210.06 |
| Rock | 78 |  |  | 210 |  | 210.00 |
| FI | 81 |  |  | c. 210 |  | 210.00 |

## NUMBER OF STAGES

| Rock | 78 | "2 + 1" |  |
|---|---|---|---|
| AWST | 82 | 3 | 2 booster, 1 sustainer. |

## TYPE OF PROPULSION

| SWM | 77 | 2 JATOs |  |
|---|---|---|---|
|  |  | turbojet | [Table] |
|  |  | turbojet or ramjet | [Text] |
| Rock | 78 | Jato |  |
|  |  | turbojet |  |
| FI | 81 | 2 JATOs | Boost. |
|  |  | turbojet or ramjet | Cruise. |
| AWST | 82 | solid |  |
|  |  | air breathing |  |
| JWS | 82 |  | Initial boost by 2 rocket units under rear of fuselage. |
|  |  | ramjet or turbojet | Air breathing. |
| BW | 86 | TJ |  |

## MIDCOURSE (OR MAIN) GUIDANCE

| SWM | 77 | radio command |  |
|---|---|---|---|
| Rock | 78 | radio |  |
| FI | 81 | autopilot | And midcourse update. |
| JWS | 82 | radio command |  |

## TERMINAL GUIDANCE

| SWM | 77 | IR or active radar |  |
|---|---|---|---|
| Rock | 78 | radar |  |
| FI | 81 | active radar |  |
| JWS | 82 | IR homing | [Analysis table] |
|  |  | active radar | [Text] |

## CONTROL SYSTEM

| Rock | 78 | Aerodynamic. |
|---|---|---|
| FI | 81 | Aeroplane configuration. |

CRUISE MISSILES  Ground-Launched SSMs  SSC-1  399

```
---------- DEPLOYMENT BEGAN ----------
Rock  78    1962          Service intro.
IISS  82    1962          First year deployed.
Meyr  84    1962          Service.
BW    86    1962

---------- DEPLOYED ----------
SecD  73                  Deployed.
JWS   82                  Deployed.
BW    86                  In service.

---------- NUMBERS IN SERVICE ----------

---------- NUMBER IN 1971 ----------
GSN   83          Each Sepal battalion has 15-18 missiles. In early 1970s [and before], missile battalions
                  per fleet: 3, Northern; 6, Baltic; 5, Black Sea; 5, Pacific. Missile force has declined
                  from about 18,000 men in early 1970s to about 10,000 now; not known how many battalions
                  there are now.

---------- NUMBER IN 1972 ----------
SecD  72          "Small" numbers.

---------- NUMBER IN 1973 ----------
IISS  73   c. 100   1973-81,83, July, total deployed.

---------- NUMBER IN 1982 ----------
IISS  82   c. 100   July, total deployed, Europe (including European USSR) only.

---------- NUMBER IN 1983 ----------
IISS+ 83   c. 100   Total deployed, 7/83.

---------- UNDATED NUMBER ----------
BW    86   ? 100    Peak.

---------- CARRIERS ----------

---------- CARRIER ---------- TUBES --- RELOADS ----------
SWM   77    8-wh vehicle                          Crosscountry vehicle; container on vehicle elevates.
Rock  78    wheeled transporter
FI    81    8-wh vehicle                          Crosscountry vehicle.
AWST  82    transporter-erector
JWS   82    wheeled vehicle                       [Picture shows wheeled vehicle with elevating canister on
                                                  rear to fire missile out over cab at low angle].
GSN   83    8-wh transporter                      Road speed up to 50 mph.
BW    86    8-wh vehicle         1        yes

---------- USE AND CONFIGURATION ----------
SecD  73    Deployed with ground forces.
SWM   77    Wings appear to be hinged flip-out type.
FI    81    "15-18 Shaddocks/reloads" per battalion.
JWS   82    Hinged wings open out when missile leaves launcher.
            Appears it is used by land forces of USSR. Each coastal defence battalion reported equipped with
            15-18 missiles "(inclusive of reloads)".
GSN   83    Each Sepal battalion has 15-18 missiles.
Meyr  84    Penetration: 1.0; reliability: 0.2-0.5, "includes system reliability multiplied by operational
            readiness of deployed system".

---------- TARGET TYPE ----------
SWM   77    [ship]
FI    81    ship

---------- STRATEGY ----------
SecD  71    Listed as tactical missile assigned to ground forces.
SecD  72    Short range missile, FY73-74. Listed as part of TNF.
USD2  72    This land-based "ground-to-ground" role preceded naval use.
Rock  78    "Coastal defence from wheeled transporter".
FI    81    Coast-defense variant of SS-N-3.
JWS   82    Though could be used in ground-to-ground role, "rather more likely" used as coastal defence
            weapon, because of ability "to cruise at speed at low altitude over water".
Meyr  84    [Ed: chart is labeled] Systems "assigned primary operational and strategic missions within the
            European theatre(s) of military operations".

---------- GEOGRAPHICAL DEPLOYMENTS ----------
GSN   83          Each Sepal battalion has 15-18 missiles. In early 1970s [and before], missile
                  battalions per fleet: 3, Northern; 6, Baltic; 5, Black Sea; 5, Pacific. Missile
                  force has declined from about 18,000 men in early 1970s to about 10,000 now; not
                  known how many battalions there are now.
```

------------ **RELATION TO OTHER MISSILES** ------------
```
FI      81    Coast-defense variant of SS-N-3.
AW ST   82    "Naval version for cruisers, submarines".
IISS    82    Similar to SS-N-3.
JWS     82    Land-based version of shipboard Shaddock.
GSN     83    Land-launched version of SS-N-3/12 Shaddock.
```

# SSC-2α

Sources are:  MOW 76   SWM 77   AWST 82   JWS 82

## US DESIGNATION
| | | |
|---|---|---|
| SWM | 77 | SSC-2A |
| AWST | 82 | SSC-2A |
| JWS | 82 | SSC-2A |

## NATO CODENAME
| | | |
|---|---|---|
| MOW | 76 | Salish |
| SWM | 77 | Salish |
| AWST | 82 | Salish |
| JWS | 82 | Salish |

## DESCRIPTION
MOW 76 — Cruise missile.

## RANGE

| | | mi | nm | km | km conv | |
|---|---|---|---|---|---|---|
| MOW | 76 | c. 62 | | c. 100 | 99.76 | |
| JWS | 82 | | | 100-200 | 100.00-200.00 | Range "has been estimated as anything from 100 to 200 km". Since Samlet more modern, "wise to assume not more than 100 km for Salish". |
| | | | | ≤ 100 | 100.00 | |

## SPEED

| | | mph | kmph | Mach | kmph conv | |
|---|---|---|---|---|---|---|
| JWS | 82 | | | c 0.8-0.9 | | Cruising speed. |

## WEIGHT

| | | lb | t | kg | kg conv |
|---|---|---|---|---|---|
| JWS | 82 | | c. 3 | | 3,000 |

## LENGTH

| | | in | ft | cm | m | m conv |
|---|---|---|---|---|---|---|
| JWS | 82 | | | | c. 7 | 7.00 |

## WINGSPAN

| | | in | ft | cm | m | cm conv |
|---|---|---|---|---|---|---|
| JWS | 82 | | | | c. 5 | 500.00 |

## TYPE OF PROPULSION
| | | | |
|---|---|---|---|
| MOW | 76 | turbojet | |
| | | JATO | "Also seen with JATO booster". |
| SWM | 77 | turbojet | |
| | | solid | Boost. |
| JWS | 82 | turbojet | |

## TERMINAL GUIDANCE
| | | | |
|---|---|---|---|
| MOW | 76 | SAR | Probably semi-active homing. Radome above air intake probably contains SAR. |
| SWM | 77 | ? active radar | |
| JWS | 82 | SAR | Has radome above air intake in nose. Believed to cover semi-active homing equipment. No other obvious indications of guidance equipment. |

## DEPLOYED
MOW 76 — In service.

## LAUNCHER/SILO
| | | |
|---|---|---|
| MOW | 76 | Ramp launched. |
| JWS | 82 | When shown publicly, on short ramp on low trailer. Mounted for launching, has been seen on larger ramp with JATO bottle attached. |

## STRATEGY
| | | |
|---|---|---|
| SWM | 77 | Tactical use by Soviet Army. |
| AWST | 82 | Tactical use by Soviet army. |

## RELATION TO OTHER MISSILES
| | | |
|---|---|---|
| MOW | 76 | Appears older than Samlet. |
| SWM | 77 | Land-based version of Samlet. |
| JWS | 82 | Similar to and probably derived from Kennel ASM. Appears to differ from Samlet only in guidance equipment. |

# SSC-2b

Sources are: AJ01 70    MOW 76    SWM 77    Rock 78    AWST 82    JWS 82

## US DESIGNATION

|  | |  |
|---|---|---|
| SWM | 77 | SSC-2B |
| Rock | 78 | SSC-2B |
| AWST | 82 | SSC-2B |
| JWS | 82 | SSC-2B |

## NATO CODENAME

| Source | | Codename |
|---|---|---|
| MOW | 76 | Samlet |
| SWM | 77 | Samlet |
| Rock | 78 | Samlet |
| AWST | 82 | Samlet |
| JWS | 82 | Samlet |

## DESCRIPTION

|  | | |
|---|---|---|
| MOW | 76 | Cruise missile. |

## RANGE

| Source | | mi | nm | km | km conv | |
|---|---|---|---|---|---|---|
| AJ01 | 70 | 400 | | | 643.60 | Max. |
| | | 250 | | | 402.25 | Against ship targets [sic, may refer to Sepal?, ed] |
| MOW | 76 | 63 | | 100 | 101.37 | Max. |
| SWM | 77 | 124 | | 200 | 199.52 | |
| Rock | 78 | | | 180 | 180.00 | |
| AWST | 82 | 50 | | | 80.45 | |
| JWS | 82 | | | c. 200 | 200.00 | "Possibly as much as 200 km with mid-course guidance". |

## SPEED

| Source | | mph | kmph | Mach | kmph conv | |
|---|---|---|---|---|---|---|
| MOW | 76 | | | | | subsonic |
| SWM | 77 | | | 0.8-0.9 | | |
| Rock | 78 | | | 0.9 | | |
| JWS | 82 | | | c 0.8-0.9 | | Cruising speed. |

## WARHEAD TYPE

| Source | | nuclear | conven | chem | |
|---|---|---|---|---|---|
| MOW | 76 | | yes | | HE. |
| SWM | 77 | | yes | | |
| Rock | 78 | | yes | | HE. |
| JWS | 82 | | yes | | HE. |

## WEIGHT

| Source | | lb | t | kg | kg conv |
|---|---|---|---|---|---|
| MOW | 76 | 6,615 | | 3,000 | 3,000 |
| SWM | 77 | 6,614 | | 3,000 | 3,000 |
| Rock | 78 | | | 3,100 | 3,100 |
| JWS | 82 | | | c. 3,000 | 3,000 |
| | | | c. 3 | | 3,000 |

## LENGTH

| Source | | in | ft | cm | m | m conv |
|---|---|---|---|---|---|---|
| MOW | 76 | | 27'0" | | 8.2 | 8.23 |
| SWM | 77 | | 23 | | 7 | 7.01 |
| Rock | 78 | | | | 7 | 7.00 |
| AWST | 82 | | 25 | | | 7.62 |
| JWS | 82 | | | c. 700 | 7 | 7.00 |
| | | | | | c. 7 | 7.00 |

## DIAMETER

| Source | | in | ft | cm | m | cm conv |
|---|---|---|---|---|---|---|
| Rock | 78 | | | 120 | | 120.00 |

## WINGSPAN

| Source | | in | ft | cm | m | cm conv |
|---|---|---|---|---|---|---|
| MOW | 76 | 16'0" | | | 4.9 | 40.64 |
| SWM | 77 | 197 | | 500 | | 500.38 |
| Rock | 78 | | | 500 | | 500.00 |
| AWST | 82 | | 14 | | | 426.72 |
| JWS | 82 | | | 500 | | 500.00 |
| | | | | | c. 5 | 500.00 |

## CRUISE MISSILES Ground-Launched SSMs SSC-2b

```
------------ NUMBER OF STAGES ------------
Rock  78    "1 + 1"

------------ TYPE OF PROPULSION ------------
MOW   76    turbojet
            JATO
SWM   77    turbojet
            solid              Boost, [table]
                               Launch assisted by undertail JATO [text].
Rock  78    solid
            turbojet
AWST  82    air breathing
JWS   82    turbojet
            boosters           JATO bottle necessary for launching.

------------ MIDCOURSE (OR MAIN) GUIDANCE ------------
MOW   76    command            Probably beam-riding/command guidance.
SWM   77    radio              Autopilot.
Rock  78    radio
JWS   82    command            Electronics pod mounted on tail fin; may contain apparatus for command
                               guidance link.

------------ TERMINAL GUIDANCE ------------
MOW   76    radar
SWM   77    active radar       Compared to Salish, larger radome and "electronic pod 3 on fin".
Rock  78    radar
JWS   82    ? active radar     Probably. Has radome over jet air intake; larger than on Salish; assumed to
                               cover radar homing seeker.

------------ CONTROL SYSTEM ------------
MOW   76    Aerodynamic control surfaces.
Rock  78    Aerodynamic.

------------ RADAR ------------
SWM   77    Sheet Bend (on van)

------------ DEPLOYED ------------
MOW   76              In service.
SWM   77              Deployed.
JWS   82              Deployed.

------------ DEPLOYMENT HISTORY ------------
SWM   77    Widely deployed by USSR. Has replaced the 130 mm coastal gun.
Rock  78    Replacing 130-mm gun.

------------ CARRIERS ------------

------------ CARRIER ------------ TUBES ---- RELOADS ------------
JWS   82                                          When shown in publicly, appears "on what appears to be its
                                                  launching ramp constructed as a trailer". But other photos
                                                  indicate more permanent launch emplacement also exists.

------------ CARRIER ------------ TUBES ---- RELOADS ------------
SWM   77    Zil-157V                              Carried on single-axle semi-trailer towed by Zil-157V;
                                                  rail launcher.

------------ TARGET TYPE ------------
SWM   77    [ship]
AWST  82    ship

------------ STRATEGY ------------
MOW   76    Used primarily for coastal defense.
SWM   77    Coastal defense missile.
Rock  78    Coastal defence.
AWST  82    Coast defense.

------------ EXPORTED TO ------------
MOW   76    Poland
            Egypt
            Cuba
SWM   77    Cuba
            Egypt
            WTO
JWS   82    Egypt
            Poland
            Cuba
```

---------- **RELATION TO OTHER MISSILES** ----------
MOW   76   Differs little from Kennel ASM.
SWM   77   Developed from AS-1 Kennel.
AWST  82   Coast defense version of Kennel.
JWS   82   Similar to and probably derived from Kennel ASM. Appears to differ from Samlet only in guidance equipment.

# SSC-4

Sources are: FI+ 83   ABFS 84   AWST+ 84   SMP 84   SMP 85   BW 86

---------- **US DESIGNATION** ----------
| | | |
|---|---|---|
| AWST+ | 84 | SSC-X-4 |
| SMP | 84 | SSC-X-4   '84, '85. |

---------- **OTHER DESIGNATION** ----------
| | | |
|---|---|---|
| FI+ | 83 | Long-range cruise missile [only designation provided, ed]. |

---------- **DESCRIPTION** ----------
| | | |
|---|---|---|
| AWST+ | 84 | Ground launched cruise missile. |
| SMP | 84 | Cruise missile, '84, '85. |

---------- **EDITOR'S NOTES** ----------
| | | |
|---|---|---|
| SMP | 84 | Listed '84, '85. |

---------- **RANGE** ----------

| | | mi | nm | km | km conv | |
|---|---|---|---|---|---|---|
| SMP | 84 | | | 3,000 | 3,000 | '84, '85. |
| BW | 86 | | | 3,000 | 3,000 | |

---------- **CEP** ----------

| | | ft | mi | nm | m | km | m conv | |
|---|---|---|---|---|---|---|---|---|
| SMP | 84 | | | | | | | "Capable of threatening hardened targets"; "could eventually be accurate enough to permit the use of conventional warheads", '84, '85. |

---------- **WARHEAD TYPE** ----------

| | | nuclear | conven | chem | |
|---|---|---|---|---|---|
| SMP | 84 | yes | maybe | | Initially to be fitted with nuclear warheads; could eventually be accurate enough to permit use of conventional warheads, depending on guidance and future munitions developments, '84, '85. ["[P]robably will be fitted..." in '85 report, ed.] |

---------- **WARHEAD CHARACTERISTICS** ----------
| | | |
|---|---|---|
| SMP | 84 | 1 warhead   '84, '85. |
| BW | 86 | single RV |

---------- **LENGTH** ----------

| | | in | ft | cm | m | m conv | |
|---|---|---|---|---|---|---|---|
| SMP | 84 | | | | 7 | 7.00 | [From graph], '84, '85. |

---------- **TYPE OF PROPULSION** ----------
| | | |
|---|---|---|
| BW | 86 | TJ |

---------- **DEVELOPMENT BEGAN** ----------
| | | | |
|---|---|---|---|
| SMP | 85 | 1976 | Technological development began. [From graph.] |

---------- **DEVELOPMENT UNDERWAY** ----------
| | | |
|---|---|---|
| SMP | 84 | In development, '84, '85. |

---------- **PROTOTYPE TESTS** ----------
| | | |
|---|---|---|
| FI+ | 83 | USSR is testing it, according to Andropov. Likely to be in the very early stages of development. |

---------- **DEPLOYMENT BEGAN** ----------
| | | | |
|---|---|---|---|
| FI+ | 83 | | Unlikely to appear before late 1980s. |
| SMP | 84 | | "Nearly deployed"; may not be ready for operational deployment until about 1985. |
| SMP | 85 | | May not be ready for operational deployment until "late this year or next". |
| | | 1987-88 | [Possible deployment, from graph.] |
| BW | 86 | ? 1985 | |

---------- **DEPLOYMENT HISTORY** ----------
| | | |
|---|---|---|
| SMP | 85 | Possible deployment may last through 1990. [From graph.] |

---------- **CARRIERS** ----------

---------- **CARRIER** ---------- TUBES --- RELOADS ----------
| | | |
|---|---|---|
| SMP | 84 | Mobile, '84, '85. |

---------- **DESIGN AND ENGINEERING** ----------
| | | |
|---|---|---|
| SMP | 85 | 1982-83 Engineering and testing began. [From graph.] |

---------- **USE AND CONFIGURATION** ----------
SMP  84  Ground launched; will "probably follow operational procedures like those of the SS-20 LRINF missile", '84, '85.

---------- **TARGET TYPE** ----------
SMP  84  "Land-attack".

---------- **STRATEGY** ----------
SMP  84  Its range and likelihood it will not be deployed outside USSR indicate its mission will be support of theater operations, '84, '85.

---------- **RELATION TO OTHER MISSILES** ----------
FI+  83  Similar in concept to the US cruise missile.
ABFS 84  "A ground-launched version of the SS-NX-21".

CRUISE MISSILES  Ground-Launched SSMs  GLCM  407

# GLCM

Sources are: SMP 84    SMP 85

---------- **US DESIGNATION** ----------

SMP 84        "Larger cruise missile" has not yet been designated. [Ground-based version labeled "GLCM" in chart, ed.]

---------- **DESCRIPTION** ----------

SMP 84        Cruise missile, '84, '85.

---------- **EDITOR'S NOTES** ----------

SMP 84    Described as version of "Larger cruise missile".
SMP 85    Described as variant of "Larger system".

---------- **RANGE** ----------

| | mi | nm | km | km conv | |
|---|---|---|---|---|---|
| SMP 84 | | | | | Long-range, '84, '85. |
| SMP 85 | | | | | Variant of larger system which is "probably designed for long-range operations." |

---------- **CEP** ----------

| | ft | mi | nm | m | km | m conv | |
|---|---|---|---|---|---|---|---|
| SMP 84 | | | | | | | "Capable of threatening hardened targets"; "could eventually be accurate enough to permit the use of conventional warheads", '84, '85. |

---------- **WARHEAD TYPE** ----------

| | nuclear | conven | chem | |
|---|---|---|---|---|
| SMP 84 | yes | maybe | | Initially to be fitted with nuclear warheads; could eventually be accurate enough to permit use of conventional warheads, depending on guidance and future munitions developments, '84, '85. ["[P]robably will be fitted..." in '85 report, ed.] |

---------- **LENGTH** ----------

| | in | ft | cm | m | m conv | |
|---|---|---|---|---|---|---|
| SMP 84 | | | | 12.5 | 12.50 | [From graph], '84, '85. |

---------- **DEVELOPMENT UNDERWAY** ----------

SMP 84        In development, '84, '85.

---------- **IOC** ----------

SMP 84    Could be operational within next 2 years.
SMP 85    "The SS-NX-24 could be operational within the next two years, and the ground-based version sometime after that."

---------- **RELATION TO OTHER MISSILES** ----------

SMP 84    Ground-launched version of "Larger cruise missile" which has sea-based version [SS-N-24, labeled "SLCM" in chart, ed].
SMP 85    Ground-launched version of "Larger system" which has sea-based version ["SS-NX-24" in chart, ed]. [Elsewhere:] "A ground-based variant of this missile [SS-N-24] may be in development." Has no US counterpart.

# AT-1

```
Sources are:  IA03  65      JCS  73     NA02  74     JCS  75     JCS  77     AWST* 78    Rock  78
IA01  65      Ley   68      FI*  74     HRST 75     MOW  76     SWM  77     JCS   78    JWS   82
```

**US DESIGNATION**

|  | |  |
|---|---|---|
| FI* | 74 | AT-1 |
| JCS | 77 | AT-1 |
| SWM | 77 | AT-1 |
| AWST* | 78 | AT-1 |
| Rock | 78 | AT-1 |
| JWS | 82 | AT-1 |

**NATO CODENAME**

| Source | | Codename | |
|---|---|---|---|
| Ley | 68 | Snapper | |
| JCS | 73 | Snapper | FY74-79. |
| FI* | 74 | Snapper | |
| HRST | 75 | Snapper | |
| MOW | 76 | Snapper | |
| SWM | 77 | Snapper | |
| AWST* | 78 | Snapper | |
| Rock | 78 | Snapper | |
| JWS | 82 | Snapper | |

**SOVIET DESIGNATION**

|  | | |
|---|---|---|
| AWST* | 78 | Shmell |
| JWS | 82 | Shmell or 3M6 (Bumblebee). |

**DESCRIPTION**

|  | | |
|---|---|---|
| FI* | 74 | "First-generation" weapon. |
| MOW | 76 | Light anti-tank missile. |
| AWST* | 78 | First generation antitank missile. |
| JWS | 82 | "Surface-to-surface guided anti-tank missile". |

**EDITOR'S NOTES**

|  | | |
|---|---|---|
| JCS | 73 | Listed FY74-79. |

**RANGE**

| Source | | mi | ft | km | m | m conv | |
|---|---|---|---|---|---|---|---|
| IA01 | 65 | | ≤ 6,500 | | | 1,981 | |
| Ley | 68 | | 6,000+ | | | 1,828 | |
| FI* | 74 | | | | 500 | 500.00 | Min. |
| | | | | | 2,300 | 2,300 | Max. |
| NA02 | 74 | | 1,650-7,200 | | 500-2,300 | 502.92-2,194 | [550-2,400 yd in source] |
| HRST | 75 | 2 | | | | 3,218 | |
| MOW | 76 | | 1,640-6,560 | | 500-2,000 | 499.87-1,999 | |
| SWM | 77 | | 1,640-7,550 | | 500-2,300 | 499.87-2,301 | [Table] |
| | | | 6,560 | | 2,000 | 1,999 | Max, [text] |
| AWST* | 78 | | 7,650 | | | 2,331 | [2,550 yd in source] |
| Rock | 78 | | | | 500-2,300 | 500.00-2,300 | |
| JWS | 82 | | | | 500-2,300 | 500.00-2,300 | |

**SPEED**

| Source | | fps | mph | mps | kmph | mps conv | |
|---|---|---|---|---|---|---|---|
| NA02 | 74 | | 200 | | 320 | 89.40 | |
| MOW | 76 | | 120 | | 323 | 53.64 | [Sic] |
| SWM | 77 | 164 | | 50 | | 49.99 | |
| Rock | 78 | | | 90 | | 90.00 | |
| JWS | 82 | | | | 320 | 88.90 | |

**WARHEAD TYPE**

| Source | | nuclear | conven | chem | |
|---|---|---|---|---|---|
| FI* | 74 | | yes | | Hollow charge. |
| MOW | 76 | | yes | | Hollow charge, contact fuse. |
| SWM | 77 | | | | Hollow charge. |
| JWS | 82 | | yes | | Hollow charge. |

**PENETRATION**

| Source | | in | cm | mm | cm conv |
|---|---|---|---|---|---|
| MOW | 76 | 13.7 | 35 | | 34.80 |
| SWM | 77 | | 35 | | 35.00 |
| AWST* | 78 | | | 300 | 30.00 |
| JWS | 82 | | | 356-380 | 35.60-38.00 |

CRUISE MISSILES ATMs AT-1

```
---------- WARHEAD WEIGHT ----------
             lb           t        kg        kg conv
FI*   74                            5.2         5.20
MOW   76     11.5                   5.25        5.22
SWM   77     11.6                   5.25        5.26
Rock  78                            5.2         5.20
JWS   82                            5.2         5.20

---------- WEIGHT ----------
             lb           t        kg        kg conv
IA03  65    > 44                               19.96
Ley   68     44                                19.96
FI*   74                           22.25       22.25
MOW   76     49                    22.25       22.23
SWM   77     49                    22.3        22.23
Rock  78                           22.3        22.30
JWS   82                           22.25       22.25
                                   22          22.00

---------- LENGTH ----------
             in      ft      cm      m       m conv
IA03  65            c. 4                       1.22
Ley   68     48                                1.22
FI*   74                    113                1.13
HRST  75             4                         1.22
MOW   76            3'8.5"          1.13       1.13
SWM   77     44.5           113                1.13
AWST* 78             3.8                       1.16
Rock  78                    113                1.13
JWS   82                    113                1.13

---------- DIAMETER ----------
             in      ft      cm      m       cm conv
IA03  65      5                                12.70
Ley   68      5                                12.70
FI*   74                     14                14.00
HRST  75      5.5                              13.97
MOW   76      5.5                   0.14       13.97
SWM   77      5.5            14                13.97
AWST* 78      5.5                              13.97
Rock  78                     14                14.00
JWS   82                     14                14.00

---------- WINGSPAN ----------
             in      ft      cm      m       cm conv
IA03  65            2'9"                       83.82
Ley   68     33                                83.82
MOW   76    2'5.5"                  0.75        6.24
SWM   77     29              74                73.66
Rock  78                     78                78.00
JWS   82                     75                75.00

---------- NUMBER OF STAGES ----------
FI*   74     ? 1
AWST* 78       1
Rock  78       1

---------- TYPE OF PROPULSION ----------
FI*   74     ? solid
HRST  75     solid
MOW   76     solid
SWM   77     solid
AWST* 78     solid
Rock  78     solid
JWS   82     solid

---------- MIDCOURSE (OR MAIN) GUIDANCE ----------
FI*   74     CLOS
MOW   76
SWM   77
Rock  78     command
JWS   82     CLOS
```

[Text]
[Analysis table]

Control by joystick, aided by tracking flares. Spin-stabilized.
Command line-of-sight joystick control assisted by tracking flares.

Optical tracking. Operator has "periscope binoculars embodying an illuminated variable-brightness reticle with which to sight the target", and uses joystick to control missile, keeping on line of sight with aid of tracking flares on two of the wings.

```
------------ TRANSMISSION METHOD ------------
HRST    75      wire-guided
MOW     76      wire-guided
SWM     77      wire-guided
AWST*   78      wire-guided
JWS     82      wire-guided

------------ CONTROL SYSTEM ------------
FI*     74      Spoilers.
MOW     76      Vibrating spoilers on trailing edge of wings.
SWM     77      Trailing-edge vibrating spoilers.
Rock    78      Aerodynamic.
JWS     82      Vibrating trailing-edge spoilers.

------------ PRODUCTION ENDED ------------
SWM     77                      No longer in production.

------------ DEPLOYED ------------
JCS     75      deployed        FY76-79.
MOW     76                      In service.

------------ RETIRED ------------
JWS     82                      Previously thought withdrawn from service in WTO, reported as deployed by
                                [countries exported to].

------------ DEPLOYMENT HISTORY ------------
SWM     77      Being replaced by AT-3 Sagger.
JWS     82      Considered obsolete, being replaced by AT-3 Sagger.

------------ CARRIERS ------------

------------ CARRIER ------------ TUBES --- RELOADS ------------
FI*     74      vehicle-mounted
HRST    75      armored vehicle
JCS     77      vehicle             3                   [Shown in picture].
                                                        tubes: [Inferred from picture].
Rock    78      vehicle-mounted

------------ CARRIER ------------ TUBES --- RELOADS ------------
MOW     76      GAZ-69              4                   Original.
SWM     77      GAZ-69              4
JWS     82      GAZ-69                                  Formerly.

------------ CARRIER ------------ TUBES --- RELOADS ------------
MOW     76      BRDM                3                   Current.
SWM     77      BTR-40P [BRDM]      3                   Stored under hinged armour plates.
JWS     82      BRDM                3                   Usually.
                                                        tubes: Standard in BRDG [sic, misprint?, ed]; triple mount
                                                              is retractable.

------------ CARRIER ------------ TUBES --- RELOADS ------------
JCS     78                                              Generally ground or vehicle mounted.
JWS     82      ground launcher     4                   For infantry.

------------ USE AND CONFIGURATION ------------
IAD3    65      On GAZ-69, launcher normally stowed with missiles vertically. Rotates through 60 degrees toward
                the rear for firing; vehicle must point away from target.
FI*     74      Operator can fire from 50 m away.
SWM     77      Firings from up to 50 m from launcher.
JWS     82      Operator up to 50 m from launcher can fire and guide missile.

------------ COMBAT REPORTS/EFFECTIVENESS ------------
MOW     76      Examples were captured by Israel in 1967.
JWS     82      Western analyses of Snapper report missile is stable in flight but slow to respond to control
                commands, indicating significant aerodynamic or electronic damping of the control loop. High
                number of simulated or live practice tracking exercises claimed necessary to maintain operator
                competence.
```

```
------------ EXPORTED TO ------------
FI *    74    WTO
              Egypt
              Cuba
MOW     76    WTO
              Afghanistan
              Cuba
              Egypt
              Mongolia
              Syria
              Yugoslavia
JWS     82    Egypt
              Syria
              Yugoslavia
              Afghanistan
              Bulgaria
              Czech
              E Germany
              Hungary
              Mongolia
              Poland
              Romania

------------ RELATION TO OTHER MISSILES ------------
MOW     76    Similar in configuration to Nord SS.10 and MBB Cobra.
JWS     82    Similar in general configuration to MBB Cobra and Contraves-Oerlikon Mosquito.
```

# AT-2

```
Sources are:  AJ02  74     Bras  77     Rock  78     SMP   81     JWS   82     AFM+  84
IA03  65      HRST  75     JCS   77     Clns  80     AFM   82     Clns+ 83
Ley   68      JCS   75     SWM   77     GU01  80     AWST  82     FI+   83
JCS   73      MOW   76     JCS   78     FI    81     JAWA  82     JWS+  83
```

## US DESIGNATION

| | | |
|---|---|---|
| JCS | 77 | AT-2 |
| SWM | 77 | AT-2 |
| Rock | 78 | AT-2 |
| FI | 81 | AT-2 |
| SMP | 81 | AT-2 |
| AFM | 82 | AT-2 |
| AWST | 82 | AT-2 |
| JAWA | 82 | AT-2 |
| JWS | 82 | AT-2 |

## NATO CODENAME

| | | | |
|---|---|---|---|
| Ley | 68 | Swatter | |
| JCS | 73 | Swatter | FY74-79. |
| HRST | 75 | Swatter | |
| MOW | 76 | Swatter | |
| Bras | 77 | Swatter | |
| SWM | 77 | Swatter | |
| Rock | 78 | Swatter | |
| Clns | 80 | Swatter | |
| FI | 81 | Swatter | |
| SMP | 81 | Swatter | |
| AFM | 82 | Swatter | |
| AWST | 82 | Swatter | |
| JAWA | 82 | Swatter | |
| JWS | 82 | Swatter | |

## DESCRIPTION

| | | |
|---|---|---|
| MOW | 76 | Second-generation missile. |
| Clns | 80 | Type: Heavy. |
| JWS | 82 | "Anti-tank, surface-to-surface and air-to-surface". |

## EDITOR'S NOTES

| | | |
|---|---|---|
| JCS | 73 | Listed FY74-79. |
| SWM | 77 | Most data is for Swatter A, at least from table entry? |
| SMP | 81 | Listed '81. |

## RANGE

| | | mi | ft | km | m | m conv | |
|---|---|---|---|---|---|---|---|
| Ley | 68 | | 6,000+ | | | 1,828 | |
| Bras | 77 | | | | 3,500 | 3,500 | Max. |
| | | | | | 600 | 600.00 | Min. |
| SWM | 77 | | 1,640-7,550 | | 500-2,300 | 499.87-2,301 | [Table] |
| | | | 8,200 | | 2,500 | 2,499 | [Text] |
| Rock | 78 | | | | 300-2,200 | 300.00-2,200 | |
| Clns | 80 | | | | 2,500 | 2,500 | Effective Range. |
| GU01 | 80 | | | | 2,500 | 2,500 | Possibly more in air-launched role. |
| FI | 81 | | | | 600 | 600.00 | Min. |
| | | | | | 2,500 | 2,500 | Max. |
| AFM | 82 | | 7,220 | | | 2,200 | |
| | | | 985 | | | 300.23 | Min. |
| AWST | 82 | | 10,500 | | | 3,200 | [3,500 yd in source] |
| JWS | 82 | | | c. 2,500 | 2,500 | 2,500 | [Analysis table] |
| | | | | | 500-3,500 | 500.00-3,500 | [Text] |
| AFM+ | 84 | | 1640-11,500 | | | 499.87-3,505 | |

## SPEED

| | | fps | mph | mps | kmph | mps conv | |
|---|---|---|---|---|---|---|---|
| Bras | 77 | | | 160 | | 160.00 | 20 sec time of flight to max range. |
| Rock | 78 | | | 150 | | 150.00 | |
| Clns | 80 | | | | | | 27 Secs flight time to Max Range. |
| FI | 81 | | | 750 | | 750.00 | [Sic] |
| AFM | 82 | | 335 | | | 149.75 | |
| JWS | 82 | | | 150 | | 150.00 | Average velocity. |
| Clns+ | 83 | | | | | | 25 sec flight to max range. |

```
------------- WARHEAD TYPE -------------
            nuclear   conven    chem
MOW   76              yes                  Hollow charge.
Bras  77              yes                  HEAT, delayed action fuze.
GUD1  80              yes                  Hollow charge.
JWS   82              yes                  HEAT [analysis table]. HEAP [text].

------------- PENETRATION -------------
            in        cm              mm        cm conv
Bras  77                               500       50.00     "Said to achieve about 500 mm
                                                           penetration of armour plate".
SWM   77              40                         40.00     [Table]
                    > 40                         40.00     [Text]
GUD1  80              30                         30.00
FI    81              50                         50.00
AWST  82                               400       40.00
JWS   82                               500       50.00

------------- WEIGHT -------------
            lb        t               kg        kg conv
IA03  65   > 55                                  24.95
Ley   68     55                                  24.95
MOW   76     33                        15        14.97
Bras  77                               26.5      26.50     Before launch.
Rock  78                               25        25.00
Clns  80     45                                  20.41
GUD1  80     44                        20        19.96
FI    81                               26.5      26.50
AFM   82     55                                  24.95
JWS   82                               27        27.00     [Analysis table]
                                       29.4      29.40     [Text]
AFM+  84     65                                  29.48

------------- LENGTH -------------
            in        ft       cm     m         m conv
IA03  65              c. 4                       1.22
Ley   68     48                                  1.22
AJ02  74                        90                0.90
HRST  75              3                           0.91
MOW   76              3'8"              1.12      1.12
Bras  77                                1.13      1.13
SWM   77    35.4                90                0.90
Rock  78                        90                0.90
GUD1  80              3'9"              1.14      1.14
FI    81                       112                1.12
AFM   82              2'11 1/2"                   0.90
AWST  82              3                           0.91
JWS   82                       112                1.12     [Analysis table]
                                        1.16     1.16     [Text]
AFM+  84              3'9 3/4"                   1.16

------------- DIAMETER -------------
            in        ft       cm     m         cm conv
IA03  65    5 1/2                                13.97
Ley   68    5 1/2                                13.97
HRST  75    6                                    15.24
MOW   76    5.9"                       .15       14.99
Bras  77                       13                13.00     [130 mm in source]
SWM   77    5.9                15                14.99
Rock  78                       15                15.00
GUD1  80    5.2                13.2              13.21
FI    81                       15                15.00
AWST  82    6                                    15.24
JWS   82                       15                15.00     [Analysis table]
                               13.2              13.20     [Text]
Clns+ 83    5.5                                  13.97     Caliber.

------------- WINGSPAN -------------
            in        ft       cm     m         cm conv
Ley   68    26                                   66.04
MOW   76    2'2"                       .65        5.50
Bras  77                       65                65.00     [650 mm in source]
SWM   77    26                 66                66.04
Rock  78                       66                66.00
FI    81                       66                66.00
AFM   82              2'2"                       66.04
JWS   82                       66                66.00     [Max].
```

414  SOVIET MISSILES   IDDS

```
------------ NUMBER OF STAGES ------------
Bras  77      2
Rock  78      1
FI    81    ? 1
AWST  82      1

------------ TYPE OF PROPULSION ------------
HRST  75    solid
MOW   76    solid
Bras  77    solid
SWM   77    solid
Rock  78    solid
FI    81    ? solid
AWST  82    solid
JWS   82    solid

------------ MIDCOURSE (OR MAIN) GUIDANCE ------------
IAD3  65              Tracer material (stored in a pair of fairings) to assist tracking.
Bras  77    CLOS      Manual command to line of sight.
SWM   77    CLOS
Rock  78    command
GUD1  80              Semi-automatic version reported as well.
FI    81    CLOS
JWS   82    CLOS

------------ TERMINAL GUIDANCE ------------
HRST  75    ? IR      Probably.
MOW   76    ? IR      Possibly.
Rock  78    IR
FI    81    IR
AFM   82              Has terminal homing.
JAWA  82              "Embodies terminal homing".
JWS   82    ? IR      Possible.

------------ TRANSMISSION METHOD ------------
MOW   76    command guidance    Does not use wire guidance.
Bras  77    radio               Radio command link; only anti-tank guided weapon with this system.
                                Radio guidance avoids problems of wire breaks, but makes vulnerable to ECM.
SWM   77    wire-guided
Clns  80    Radio
AWST  82    wire-guided
JWS   82    wire                [Analysis table]
            radio               Command link has 3 frequencies as ECM protection.

------------ CONTROL SYSTEM ------------
SWM   77    Elevons on rear-mounted wings and also foreplanes.
Rock  78    Aerodynamic.
FI    81    Elevons + canard fins.
AFM   82    Elevons on trailing-edges of rear-mounted cruciform wings.
JWS   82    Elevons on trailing edges of rear-mounted cruciform wings.

------------ FIRST OBSERVED ------------
MOW   76              "First observed some years after 'Snapper' entered service".

------------ DEPLOYED ------------
JCS   75              Deployed, FY76-79.
MOW   76              In service.
Bras  77              In service.
SWM   77              Still in service in WTO.

------------ DEPLOYMENT HISTORY ------------
GUD1  80    To be replaced [on helicopters] with AT-6.
FI    81    Deployed USSR.

------------ NUMBERS IN SERVICE ------------

------------ NUMBER IN 1982 ------------
JAWA  82              Standard Soviet anti-tank weapon.

------------ CARRIERS ------------

------------ CARRIER ------- TUBES ---- RELOADS ------------
JCS   77    vehicle           4                     [Shown in picture].
JCS   78                                            Generally ground or vehicle mounted.
Rock  78    vehicle mounted
```

|  |  | CARRIER | TUBES | RELOADS |  |
|---|---|---|---|---|---|
| MOW | 76 | BRDM | 4 | | |
| Bras | 77 | BRDM | 4 | | |
| SWM | 77 | BTR-40P (BRDM) | 4 | | |
| Clns | 80 | BRDM | 4 | | |
| FI | 81 | BRDM-1 | 3 | | |
| JAWA | 82 | BRDM | | | [Shown in picture] |
| JWS | 82 | BRDM | 4 | | Traverse of 45 degrees. |

|  |  | CARRIER | TUBES | RELOADS |
|---|---|---|---|---|
| JWS | 82 | BRDM-2 | 4 | 4 |

|  |  | CARRIER | TUBES | RELOADS |  |
|---|---|---|---|---|---|
| MOW | 76 | Mil Mi-24 Hind-A | 2 x 2 | | |
| FI | 81 | Mi-24 Hind | | | Interim armament. |
| SMP | 81 | Hind A | 4 | | |
| AFM | 82 | Mi-24 Hind-A and -D | | | |
| JAWA | 82 | Mi-24 Hind-A/D | | | |
| JWS | 82 | Mi-24 Hind-A | | | "Suitable for" this role. Terminal homing and radio guidance suit AT-2 to airborne role. |
| FI+ | 83 | Mi-24 Hind | | | Arms most versions. |
| JWS+ | 83 | Hind A and D | | | "Operational in manually-guided B version" [A, ed?] |

|  |  | CARRIER | TUBES | RELOADS |
|---|---|---|---|---|
| SMP | 81 | Hip E | 4 | |
| AFM | 82 | Mi-8 Hip-E | | |
| JAWA | 82 | Mi-8 Hip-E | | |

---------- USE AND CONFIGURATION ----------
SWM 77  Missiles fully retractable.
JWS 82  In ground deployment, does not arm until 500 m from launch site.

---------- EXPORTED TO ----------
MOW 76  WTO              Except Bulgaria.
        Egypt
        Syria
Bras 77 Czech
        Egypt
        East Germany
        Hungary
        Poland
        Rumania
        Syria
SWM 77  WTO              Still in service.
FI 81   Hungary
        Romania
JWS 82                   In service with Warsaw Pact, Syria and Egypt only.
        Warsaw Pact
        Egypt
        Syria
FI+ 83  Afghanistan
        Egypt
        Iraq
        Libya
        Syria
        ? Czech
        ? South Yemen
JWS+ 83 Hungary
        Iraq
        Libya

---------- RELATION TO OTHER MISSILES ----------
MOW 76  'Swatter' is more advanced than 'Snapper'.
JWS 82  In class of French SS11.

# AT-2 Swatter A

Sources are: ID09 78

|  | RANGE | | | | | |
|---|---|---|---|---|---|---|
|  | mi | ft | km | m | m conv | |
| ID09 78 | | | | c. 2,500 | 2,500 | Older Swatter A missile. |

# AT-2 Swatter B

Sources are: SWM 77  ID09 78  JWS 82  FI+ 83  JWS+ 83

## NATO CODENAME

| Source | | Codename |
|---|---|---|
| SWM | 77 | Swatter B |
| JWS | 82 | Swatter 2 |
| FI+ | 83 | Swatter 2 |

## RANGE

| Source | | mi | ft | km | m | m conv | |
|---|---|---|---|---|---|---|---|
| ID09 | 78 | | | | c. 3,500 | 3,500 | Swatter B of Swatter 2 system. |

## SPEED

| Source | | fps | mph | mps | kmph | mps conv | |
|---|---|---|---|---|---|---|---|
| ID09 | 78 | | | 150 | | 150.00 | Swatter B of Swatter 2 system. |

## WEIGHT

| Source | | lb | t | kg | | kg conv | |
|---|---|---|---|---|---|---|---|
| ID09 | 78 | | | 29 | | 29.00 | Swatter B of Swatter 2 system. |

## LENGTH

| Source | | in | ft | cm | m | m conv | |
|---|---|---|---|---|---|---|---|
| ID09 | 78 | | | | 1.16 | 1.16 | Swatter B of Swatter 2 system. |
| JWS | 82 | | | | 1.56 | 1.56 | |
| FI+ | 83 | | | | 1.56 | 1.56 | Swatter 2 [text] |

## DIAMETER

| Source | | in | ft | cm | m | cm conv | |
|---|---|---|---|---|---|---|---|
| ID09 | 78 | | | 16 | | 16.00 | Swatter B of Swatter 2 system. |
| JWS | 82 | | | 16 | | 16.00 | Swatter 2 [text] |
| FI+ | 83 | | | 16 | | 16.00 | |

## MIDCOURSE (OR MAIN) GUIDANCE

|  | | | |
|---|---|---|---|
| SWM | 77 | | " 'B' version of Swatter has second-generation auto-guidance system" [over wire CLOS]. |
| JWS+ | 83 | SACLOS | |

## TERMINAL GUIDANCE

|  | | | |
|---|---|---|---|
| ID09 | 78 | semi-active IR | Reported to have been converted to "semi-active IR/radio command guidance" (Swatter 2), those on BMP, BRDM, and Hind-D. |

## TRANSMISSION METHOD

|  | | | |
|---|---|---|---|
| ID09 | 78 | radio command | Reported to have been converted to "semi-active IR/radio command guidance" (Swatter 2), those on BMP, BRDM, and Hind-D. |

## DEPLOYMENT HISTORY

|  | | |
|---|---|---|
| JWS+ | 83 | Replacing manually-guided Swatter [A]. |

## CARRIERS

| Source | | CARRIER | TUBES | RELOADS | |
|---|---|---|---|---|---|
| ID09 | 78 | BRDM 2 | 4 | | Swatter 2. |
| JWS | 82 | BRDM-2 | | | Later Swatter 2 on BRDM-2 has different launcher and wider traverse. |

| Source | | CARRIER | TUBES | RELOADS | |
|---|---|---|---|---|---|
| ID09 | 78 | Hind-D | 4 | | Swatter B. |

## RELATION TO OTHER MISSILES

|  | | |
|---|---|---|
| FI+ | 83 | Improved version of Swatter. |

## AT-3

```
Sources are:   HRST  75    Bras  77    Rock  78    SMP   81    JWS   82    AFM+  84
JCS   73       JCS   75    JCS   77    JCS   79    AFM   82    Clns+ 83
ID04  74       MOW   76    SWM   77    Clns  80    AWST  82    FI+   83
NA02  74       AW01  77    ID09  78    FI    81    JAWA  82    JWS+  83
```

---------- **US DESIGNATION** ----------

|  | |  |
|---|---|---|
| JCS | 77 | AT-3 |
| SWM | 77 | AT-3 |
| Rock | 78 | AT-3 |
| FI | 81 | AT-3 |
| SMP | 81 | AT-3 |
| AFM | 82 | AT-3 |
| AWST | 82 | AT-3 |
| JAWA | 82 | AT-3 |
| JWS | 82 | AT-3 |

---------- **NATO CODENAME** ----------

| Source | | Codename | |
|---|---|---|---|
| JCS | 73 | Sagger | FY74-80. |
| HRST | 75 | Sagger | |
| MOW | 76 | Sagger | |
| Bras | 77 | Sagger | |
| SWM | 77 | Sagger | |
| Rock | 78 | Sagger | |
| Clns | 80 | Sagger | |
| FI | 81 | Sagger | |
| SMP | 81 | Sagger | |
| AFM | 82 | Sagger | |
| AWST | 82 | Sagger | |
| JAWA | 82 | Sagger | |
| JWS | 82 | Sagger | |

---------- **SOVIET DESIGNATION** ----------

| FI | 81 | Milinka |
|---|---|---|

---------- **DESCRIPTION** ----------

| Clns | 80 | Type: medium. |
|---|---|---|
| JWS | 82 | "Surface-to-surface guided anti-tank missile". |

---------- **EDITOR'S NOTES** ----------

| JCS | 73 | Listed FY74-80. |
|---|---|---|
| SMP | 81 | Listed '81. |

---------- **RANGE** ----------

| Source | | mi | ft | km | m | m conv | |
|---|---|---|---|---|---|---|---|
| NA02 | 74 | | 9,510 | | 2,900 | 2,898 | [3,170 yd in source] |
| HRST | 75 | 2 | | | | 3,218 | |
| MOW | 76 | | 1,640-9,840 | | 500-3,000 | 499.87-2,999 | |
| AW01 | 77 | | 8,202 | | 2,500 | 2,499 | |
| Bras | 77 | | | | 3,300 | 3,300 | Max. |
| | | | | | 550 | 550.00 | Min. |
| SWM | 77 | | 1,640-7,550 | | 500-2,500 | 499.87-2,301 | |
| | | | | | 1,000 | 1,000 | Hits possible with naked eye. |
| | | | | | 3,000 | 3,000 | Hits possible with telescopic sight. |
| ID09 | 78 | | | | 2,000 | 2,000 | Before guidance conversion, firings went out only this far. |
| | | | | | 3,000 | 3,000 | Now full range should be possible. |
| Rock | 78 | | | | 500-3,000 | 500.00-3,000 | |
| Clns | 80 | | | | 3,000 | 3,000 | Effective range. |
| FI | 81 | | | | 500 | 500.00 | Min. |
| | | | | | 3,000 | 3,000 | Max. |
| AFM | 82 | | 9,850 | | | 3,002 | |
| | | | 1,650 | | | 502.92 | Min range. |
| AWST | 82 | | 7,500 | | | 2,286 | [2,500 yd in source] |
| JWS | 82 | | | | 500-3,000 | 500.00-3,000 | |
| | | | | | 1,000-3,000 | 1,000-3,000 | Effective engagement range for infantry. |

## SPEED

|  |  | fps | mph | mps | kmph | mps conv |  |
|---|---|---|---|---|---|---|---|
| Bras | 77 |  |  | 150 |  | 150.00 | 22 secs flight to max range. |
| SWM | 77 | 394 |  | 120 |  | 120.09 | Takes 12.5 sec to travel 1,500 m. |
| Rock | 78 |  |  | 120 |  | 120.00 |  |
| Clns | 80 |  |  |  |  |  | 27 secs flight time to max range. |
| FI | 81 |  |  | 120 |  | 120.00 | 25 secs to max range. |
| AFM | 82 |  | 270 |  |  | 120.69 |  |
| JWS | 82 |  |  | 120 |  | 120.00 | Average velocity. |
| Clns+ | 83 |  |  |  |  |  | 25 sec flight to max range. |

## WARHEAD TYPE

|  |  | nuclear | conven | chem |  |
|---|---|---|---|---|---|
| MOW | 76 |  | yes |  | Hollow charge. |
| Bras | 77 |  | yes |  | HEAT. |
| JAWA | 82 |  | yes |  | Hollow charge. |
| JWS | 82 |  | yes |  | HEAT [analysis table], hollow charge [text]. As powerful as AT-1 or AT-2. |

## PENETRATION

|  |  | in | cm | mm | cm conv |  |
|---|---|---|---|---|---|---|
| Bras | 77 |  |  | > 400 | 40.00 | Of armour plate. |
| SWM | 77 |  | 40 |  | 40.00 | [Table] |
|  |  |  | > 40 |  | 40.00 | [Text] |
| FI | 81 |  | > 40 |  | 40.00 |  |
| JWS | 82 |  |  | 400 | 40.00 |  |

## WARHEAD WEIGHT

|  |  | lb | t | kg | kg conv |
|---|---|---|---|---|---|
| NAO2 | 74 | 6 |  | 2.7 | 2.72 |
| SWM | 77 | 5.95 |  | 2.7 | 2.70 |
| Rock | 78 |  |  | 2.7 | 2.70 |
| FI | 81 |  |  | 2.7 | 2.70 |

## WEIGHT

|  |  | lb | t | kg | kg conv |  |
|---|---|---|---|---|---|---|
| MOW | 76 | 24.25 |  | 11 | 11.00 |  |
| Bras | 77 |  |  | 11.3 | 11.30 | Before launch. |
| SWM | 77 | 24.2 |  | 11 | 10.98 |  |
| Rock | 78 |  |  | 11 | 11.00 |  |
| Clns | 80 | 25 |  |  | 11.34 |  |
| FI | 81 |  |  | 11 | 11.00 |  |
| AFM | 82 | 25 |  |  | 11.34 |  |
| JWS | 82 |  |  | 11 | 11.00 | [Analysis table] |
|  |  |  |  | 11.3 | 11.30 | [Text] |

## LENGTH

|  |  | in | ft | cm | m | m conv |  |
|---|---|---|---|---|---|---|---|
| HRST | 75 |  | 2 |  |  | 0.61 |  |
| MOW | 76 |  | 2'6" |  | .76 | 0.76 |  |
| Bras | 77 |  |  | 88 |  | 0.88 | [Literally 880 m, probably intended as mm, ed] |
| SWM | 77 | 34.6 |  | 88 |  | 0.88 |  |
| Rock | 78 |  |  | 87 |  | 0.87 |  |
| FI | 81 |  |  | 87 |  | 0.87 |  |
| AFM | 82 |  | 2'10 1/4" |  |  | 0.87 |  |
| AWST | 82 |  | 2 |  |  | 0.61 |  |
| JWS | 82 |  |  | 86 |  | 0.86 |  |

## DIAMETER

|  |  | in | ft | cm | m | cm conv |  |
|---|---|---|---|---|---|---|---|
| HRST | 75 | 6 |  |  |  | 15.24 |  |
| Bras | 77 |  |  | 12 |  | 12.00 | [Literally 120 m, probably intended as mm, ed] |
| SWM | 77 | 4.7 |  | 12 |  | 11.94 |  |
| Rock | 78 |  |  | 12 |  | 12.00 |  |
| Clns | 80 | 5.5 |  |  |  | 13.97 | "Caliber". |
| FI | 81 |  |  | 12 |  | 12.00 |  |
| AWST | 82 | 6 |  |  |  | 15.24 |  |
| JWS | 82 |  |  | 12 |  | 12.00 | [Analysis table] |
|  |  |  |  | 11.9 |  | 11.90 | [Text] |
| Clns+ | 83 | 4.7 |  |  |  | 11.94 | Caliber. |

## WINGSPAN

|  |  | in | ft | cm | m | cm conv |
|---|---|---|---|---|---|---|
| SWM | 77 | 18 |  | 46 |  | 45.72 |
| Rock | 78 |  |  | 46 |  | 46.00 |
| FI | 81 |  |  | 46 |  | 46.00 |
| AFM | 82 |  | 1'6" |  |  | 45.72 |
| JWS | 82 |  |  | 47 |  | 47.00 |

## NUMBER OF STAGES

| Bras | 77 | 2 |
| Rock | 78 | 1 |
| AWST | 82 | 1 |
| JWS | 82 | 2 |

## TYPE OF PROPULSION

| HRST | 75 | solid |  |
| MOW | 76 | solid |  |
| Bras | 77 | solid |  |
| SWM | 77 | solid | Booster and sustainer. |
| Rock | 78 | solid |  |
| FI | 81 | solid | Booster and sustainer. |
| AWST | 82 | solid |  |
| JWS | 82 | solid |  |

## MIDCOURSE (OR MAIN) GUIDANCE

| Bras | 77 |  | Manual command to line of sight. |
| SWM | 77 |  | Command line-of-sight. |
| Rock | 78 | command |  |
| FI | 81 | CLOS | "A Saclos version has been reported". |
| JWS | 82 | CLOS | Optical tracking. "Official US reports speak of later versions of Sagger having semi-automatic guidance but corroboration or details have not been obtained". |

## TERMINAL GUIDANCE

| ID09 | 78 | IR | Several reports confirm converted to "semi-automatic IR/wire guidance", but only those on BRDMs and BMPs. |
| Clns | 80 | wire |  |

## TRANSMISSION METHOD

| HRST | 75 | wire-guided |  |
| MOW | 76 | wire-guided |  |
| SWM | 77 | wire-guided |  |
| ID09 | 78 | wire | Several reports confirm converted to "semi-automatic IR/wire guidance", but only those on BRDMs and BMPs. |
| AWST | 82 | wire-guided |  |
| JAWA | 82 | wire-guided |  |
| JWS | 82 | wire |  |

## CONTROL SYSTEM

| Rock | 78 | "Jetevator". |
| FI | 81 | Jetevator nozzles. |

## FIRST OBSERVED

| MOW | 76 | May 1965 |

## OBSERVED

| SWM | 77 | May 1965 |

## DEPLOYMENT BEGAN

| Rock | 78 | 1966 | Service intro. |

## DEPLOYED

| JCS | 75 | deployed | FY78-79. |
| MOW | 76 |  | In service. |
| Bras | 77 |  | In service. |
| SWM | 77 |  | In service. |

## DEPLOYMENT HISTORY

| Bras | 77 | "Sagger has now replaced Snapper in Soviet equipped Armies". |
| JWS | 82 | Extensively deployed by USSR. |

## NUMBERS PRODUCED

| SWM | 77 | Produced in large numbers. |

## NUMBERS IN SERVICE

```
------------ NUMBER IN 1981 ------------
FI     81              Each USSR armoured division: 12 infantry launchers, 132 BMPs, and 9 BRDMs. Each USSR
                       mechanized division: 36 infantry launchers, 102 BMPs, and 27 BRDMs.
       ------------ CARRIERS ------------

       ------------ CARRIER ------------ TUBES ---- RELOADS ------------
HRST   75     armored vehicle
JCS    77     vehicle                                           [Picture shows mounted on vehicle], FY78-79.
Rock   78     vehicle mounted
AWST   82     vehicle                   6                       Missiles rail-mounted.

       ------------ CARRIER ------------ TUBES ---- RELOADS ------------
MOW    76     BRDM                      6                       Armored plate protects launcher during transport.
Bras   77     BRDM                      6                       Can be elevated from inside vehicle.
SWM    77     BTR-40P                                           (BRDM) M1967.
              BRDM-1                    6
Clns   80     BRDM                      6
JWS    82     BRDM-1                    2 x 3    14 total       Mountings are retractable; shielded in firing position.

       ------------ CARRIER ------------ TUBES ---- RELOADS ------------
FI     81     "BRDM-2/BTR-40P"          6
JWS    82     BRDM-2/BTR40PB

       ------------ CARRIER ------------ TUBES ---- RELOADS ------------
SWM    77     BMD-1 APC                 1
FI     81     BMD-1                     1
JWS    82     BMD                                               Light tank.

       ------------ CARRIER ------------ TUBES ---- RELOADS ------------
MOW    76     BMP                       1                       First seen 1967 and becoming WTO standard.
Bras   77     BMP                       1                       Can be fired from within vehicle.
SWM    77     BMP-1                     1        4 total
JCS    79     BMP                                               BMP has "a Sagger" launcher.
Clns   80     BMP                       1
JWS    82     BMP-1                     1        4 or 5
                                                                reloads: Total.

       ------------ CARRIER ------------ TUBES ---- RELOADS ------------
FI     81     BMP-76PB                  1        4 total

       ------------ CARRIER ------------ TUBES ---- RELOADS ------------
JWS+   83     BOV-1                     2 x 3    >_ 6           Yugoslav 4 x 4 vehicle; reloading is manual.

       ------------ CARRIER ------------ TUBES ---- RELOADS ------------
SWM    77     Polish Skot APC           2
FI     81     Skot 8-wh APC             2                       Polish.
JWS    82     Czech vehicle                                     SKOT OT-64 Model 5 APC.

       ------------ CARRIER ------------ TUBES ---- RELOADS ------------
ID04   74     Mi-24 Hind A              2 x 2
JWS    82     Hind-A                                            "Can also be deployed on Hind-A".

       ------------ CARRIER ------------ TUBES ---- RELOADS ------------
SMP    81     Hip F                     6                       Export version of Hip E.
AFM    82     export Mi-8 Hip-F
JAWA   82     Mi-8 Hip-F

       ------------ CARRIER ------------ TUBES ---- RELOADS ------------
AFM    82     Polish Mi-2
JAWA   82     Mi-2

       ------------ CARRIER ------------ TUBES ---- RELOADS ------------
JAWA   82     Gazelle                                           Yugoslavia.
JWS+   83     Gazelle                   4                       Yugoslav helicopter; gunner/copilot has roof sight to
                                                                assist guidance.
AFM+   84     Gazelle                                           Yugoslav.

       ------------ CARRIER ------------ TUBES ---- RELOADS ------------
Bras   77     manpack
SWM    77     2-3 man teams
FI     81     2-man teams
JWS    82     manpack                                           Single mounting.

       ------------ CARRIER ------------ TUBES ---- RELOADS ------------
JWS+   83     Hoplite                                           Soviet helicopter.
```

---------- **USE AND CONFIGURATION** ----------
Bras 77   Team of 3 can set out up to 4 missiles, connected to single control sight, lid of each missile's
          carrying case acts as launching stand.
SWM  77   2-3 man infantry teams, remote firing up to 100 m from launcher. More compact than previous Soviet
          AT missiles.
FI   81   Portable rail launcher up to 100 m from vehicles. One man of infantry team positions missile,
          tensions the wire, and attaches fuzes; other operates missile from "a few metres" away.
          Each USSR armoured division: 12 infantry launchers, 132 BMPs, and 9 BRDMs. Each USSR mechanized
          division: 36 infantry launchers, 102 BMPs, and 27 BRDMs.
JWS  82   More compact than AT-1 or AT-2. Manpack version has one 'suitcase' for missile and launch rail;
          another unit has launch equipment. Operator can be up to 15 m from missile launcher. Periscopic
          sight has 10x magnification. 3-man firing team normally employed for 4 missiles. 2 operate
          missiles; other deploys ahead with RPG-7 for closer-range protection. Team can be deployed and
          ready to fire in c. 12-15 mins. In BRDM setup, gunner can be within or up to 80m from vehicle. 6
          missiles can be fired rapidly; sight has 8x magnification.

---------- **COMBAT REPORTS/EFFECTIVENESS** ----------
Bras 77   Used in 1973 war, fame as "suitcase missile".
SWM  77   Used by Egypt and Syria in 1973.
JWS  82   Used in "some thousands" by Egyptian Army in 1973 with considerable effect against Israeli tanks.

---------- **EXPORTED TO** ----------
MOW  76   WTO
          Egypt
          Syria
          North Vietnam
          Yugoslavia
Bras 77   Algeria
          Bulgaria
          Czech
          Egypt
          East Germany
          Hungary
          Israel
          Poland
          Rumania
          Syria
          Vietnam
          Yugoslavia
SWM  77   WTO              In service.
FI   81   Bulgaria
          Czech
          Egypt
          East Germany
          Hungary
          Jugoslavi        License manufacture reported.
          Poland           Skot APCs.
          Romania
          Syria            C. 6,000 rounds.
          Vietnam          BTR-40P and BMP.
          Libya
          Uganda
          Algeria
          Iraq
          Angola
          Mozambique
          Afghanistan
          Ethiopia
                           Iran reported to have ordered, but status uncertain [final report].
          Israel           May be using captured stocks.
JWS  82   Warsaw Pact
          others
          China            Existence of Chinese-built Sagger reported in Sept-79. Believed "essentially
                           identical" to original Soviet model.
                           Extensively deployed by WTO forces and other nations, particularly in Mideast.
FI+  83   Cuba
          India
          North Korea
          Zambia

---------- **OTHER INFORMATION** ----------
MOW  76   Latest of 3 Soviet anti-tank missiles.

# AT-4

```
Sources are:   ID09  78      FI  81      JWS  82     Clns  83      FI+  83
```

### ──────── US DESIGNATION ────────
| | | |
|---|---|---|
| FI | 81 | AT-4 |
| JWS | 82 | AT-4 |

### ──────── NATO CODENAME ────────
| | | |
|---|---|---|
| FI | 81 | Spigot |
| JWS | 82 | Spigot |
| Clns | 83 | Spigot |

### ──────── DESCRIPTION ────────
| | | |
|---|---|---|
| JWS | 82 | "Man-portable surface-to-surface wire-guided anti-tank missile". |
| Clns | 83 | Type: Medium. |

### ──────── RANGE ────────

| | | mi | ft | km | m | m conv | |
|---|---|---|---|---|---|---|---|
| JWS | 82 | | | | c. 2,000 | 2,000 | |
| Clns | 83 | | | | 2,000 | 2,000 | Effective range. |
| FI+ | 83 | | | | 4,000 | 4,000 | Max, suggested by some sources, but seems optimistic given launch tube dimensions. |

### ──────── SPEED ────────

| | | fps | mph | mps | kmph | mps conv | |
|---|---|---|---|---|---|---|---|
| JWS | 82 | | | c 150-250 | | 150.00-250.00 | |
| Clns | 83 | | | | | | 15 sec flight to max range. |

### ──────── WARHEAD TYPE ────────

| | | nuclear | conven | chem | |
|---|---|---|---|---|---|
| JWS | 82 | | yes | | HEAT. |

### ──────── PENETRATION ────────

| | | in | cm | mm | cm conv |
|---|---|---|---|---|---|
| JWS | 82 | | | 500 | 50.00 |
| FI+ | 83 | | | 500 | 50.00 |

### ──────── WEIGHT ────────

| | | lb | t | kg | kg conv | |
|---|---|---|---|---|---|---|
| ID09 | 78 | | | 7 | 7.00 | Missile. |
| | | | | 5 | 5.00 | Launcher. |
| JWS | 82 | | | 10-12 | 10.00-12.00 | Est. ["launcher weight" in text; unmarked in analysis table, ed] |
| Clns | 83 | 30 | | | 13.61 | |
| FI+ | 83 | | | 10-12 | 10.00-12.00 | |

### ──────── LENGTH ────────

| | | in | ft | cm | m | m conv | |
|---|---|---|---|---|---|---|---|
| JWS | 82 | | | c. 98 | | 0.98 | [Analysis table] |
| | | | | | 1.2 | 1.20 | Launcher length, [text] |
| FI+ | 83 | | | | 1.2 | 1.20 | Launch tube. |

### ──────── DIAMETER ────────

| | | in | ft | cm | m | cm conv | |
|---|---|---|---|---|---|---|---|
| JWS | 82 | | | 12 | | 12.00 | [Analysis table] |
| | | | | 13.5 | | 13.50 | Launcher diameter [text] |
| Clns | 83 | 4.7 | | | | 11.94 | Caliber. |
| FI+ | 83 | | | 13.5 | | 13.50 | Launch tube. |

### ──────── TYPE OF PROPULSION ────────
| | | |
|---|---|---|
| JWS | 82 | solid |

### ──────── MIDCOURSE (OR MAIN) GUIDANCE ────────
| | | | |
|---|---|---|---|
| JWS | 82 | SACLOS | Semi-automatic CLOS; optical tracking. |

### ──────── TRANSMISSION METHOD ────────
| | | |
|---|---|---|
| JWS | 82 | wire-guided |
| Clns | 83 | wire |

### ──────── OBSERVED ────────
| | | | |
|---|---|---|---|
| ID09 | 78 | May 1975 | Yugoslav report of new AT weapon serving with WTO forces. |
| JWS | 82 | | In service and observed in USSR, Czech, E Germany and Czech. |

424  SOVIET MISSILES  IDDS

---------- DEPLOYED ----------
FI   81            Deployed.
JWS  82            Deployed.

---------- DEPLOYMENT HISTORY ----------
JWS  82   Expected to become replacement for AT-3, in course of next few years.
FI+  83   Replacing AT-3 in anti-tank platoons of Soviet Army on 3 for 2 basis.

---------- CARRIERS ----------

---------- CARRIER ---------- TUBES --- RELOADS ----------
JWS  82                                          [Shown in back of jeep, ed]

---------- CARRIER ---------- TUBES --- RELOADS ----------
FI   81   infantry
JWS  82   infantry
Clns 83   manpack           1

---------- USE AND CONFIGURATION ----------
FI   81   Consists of tripod-mounted launch tube, below which is guidance electronics pack. Aimer uses
          periscopic sight which seems to contain 2 optical systems -- perhaps day and night sights.
JWS  82   3 men normally constitute section; carry total of 4 rounds plus sight and mount; missile is
          tripod-mounted. Sight appears to use separate optical paths for tracking target and missile;
          trajectory corrections automatic in response to gunner following target.
FI+  83   Replacing AT-3 in anti-tank platoons of Soviet Army on 3 for 2 basis. Launch tube suggests ejected
          by gas generator, like Hot.

---------- EXPORTED TO ----------
FI   81   Czech           Deployed.
JWS  82   Czech           In service.
          Germany (DR)    In service.
          Poland          In service.
FI+  83   East Germany
          Poland
          Syria
          ? Hungary
          ? South Yemen

---------- RELATION TO OTHER MISSILES ----------
FI   81   Similar in concept to US Tow.
JWS  82   AT-4 [designation] "provisionally assigned" to man-portable version of AT-5.

# AT-5

```
Sources are:   ID09  78      FI   81      JWS  82      Clns  83      FI+  83
```

```
------------ US DESIGNATION ------------
FI      81      AT-5
JWS     82      AT-5

------------ NATO CODENAME ------------
FI      81      Spandrel
JWS     82      Spandrel
Clns    83      Spandrel

------------ OTHER DESIGNATION ------------
FI      81              Was referred to as 'Fagot' and 'AT-4?', before real designations known.

------------ DESCRIPTION ------------
JWS     82              "Surface-to-surface guided anti-tank missile".
Clns    83              Type: Heavy.
```

### RANGE

|  |  | mi | ft | km | m | m conv |  |
|---|---|---|---|---|---|---|---|
| ID09 | 78 |  |  |  | [1950-2600] | 1,950-2,600 | [Derived from speed and flight time, ed]. Slightly shorter minimum range [than earlier USSR missiles] |
| FI | 81 |  |  |  | 3000-4000 | 3,000-4,000 | "Seems a little optimistic". |
| JWS | 82 |  |  | c. 4,000 | | 4,000 | Estimated. |
| Clns | 83 |  |  |  | 4,000 | 4,000 | Effective range. |

### SPEED

|  |  | fps | mph | mps | kmph | mps conv |  |
|---|---|---|---|---|---|---|---|
| ID09 | 78 |  |  | ? 150-200 |  | 150.00-200.00 | Max flight time "thus" c. 13 sec. |
| JWS | 82 |  |  | c 150-250 |  | 150.00-250.00 |  |
| Clns | 83 |  |  |  |  |  | 15 sec flight to max range. |

### WARHEAD TYPE

|  |  | nuclear | conven | chem |  |  |
|---|---|---|---|---|---|---|
| JWS | 82 |  | yes |  | HEAT. |  |

### PENETRATION

|  |  | in | cm | mm | cm conv |  |
|---|---|---|---|---|---|---|
| ID09 | 78 |  | 50-60 |  | 50.00-60.00 |  |
| JWS | 82 |  |  | 500 | 50.00 |  |

### WEIGHT

|  |  | lb | t | kg | kg conv |  |
|---|---|---|---|---|---|---|
| FI | 81 |  |  |  |  | Launch tube dimensions suggest comparable in size and weight to AT-3 Sagger. |
| JWS | 82 |  |  | 10-12 | 10.00-12.00 | [Estimated]. |
| Clns | 83 | 50 |  |  | 22.68 |  |

### LENGTH

|  |  | in | ft | cm | m | m conv |  |
|---|---|---|---|---|---|---|---|
| ID09 | 78 |  |  |  | 1.2 | 1.20 | Launcher. |
| FI | 81 |  |  |  |  |  | Launch tube dimensions suggest comparable in size and weight to AT-3 Sagger. |
| JWS | 82 |  |  | c. 100 |  | 1.00 | [Analysis table] |
|  |  |  |  |  | 1.2 | 1.20 | Launcher length [text] |

### DIAMETER

|  |  | in | ft | cm | m | cm conv |  |
|---|---|---|---|---|---|---|---|
| ID09 | 78 |  |  | 13.4 |  | 13.40 | Launcher. |
| FI | 81 |  |  |  |  |  | Launch tube dimensions suggest comparable in size and weight to AT-3 Sagger. |
| JWS | 82 |  |  | 13 |  | 13.00 | [Analysis table] |
|  |  |  |  | 13.5 |  | 13.50 | Launcher diameter [text] |
| Clns | 83 | 5.5 |  |  |  | 13.97 | Caliber. |

|       |    | WINGSPAN | in | ft | cm | m | cm conv | |
|-------|----|----------|----|----|----|----|---------|---|
| FI    | 81 |          |    |    |    |    |         | Launch tube dimensions suggest comparable in size and weight to AT-3 Sagger. |

|     |    | TYPE OF PROPULSION |
|-----|----|---------------------|
| JWS | 82 | solid               |

|      |    | MIDCOURSE (OR MAIN) GUIDANCE | |
|------|----|------------------------------|---|
| ID09 | 78 |                              | Tracking of missile "Likely to be semi-automatic using IR". |
| FI   | 81 | Saclos                       | "Like Milan, Hot, Dragon, and Tow, Spandrel uses semi-automatic command-to-line-of-sight (Saclos) guidance". |
| JWS  | 82 | CLOS                         | Optical tracking. Rotatable optical sighting/tracking head in BRDM-2 in vehicle roof. |

|      |    | TRANSMISSION METHOD |
|------|----|---------------------|
| Clns | 83 | wire                |

|    |    | FIRST OBSERVED |
|----|----|----------------|
| FI | 81 | Nov 1977       |

|     |    | DEPLOYED |
|-----|----|----------|
| JWS | 82 | In service with Soviet forces. |

|     |    | DEPLOYMENT HISTORY |
|-----|----|--------------------|
| FI  | 81 | AT-5 replaces Swatter and Sagger installations. |
| FI+ | 83 | Is replacing AT-2 Swatter. |

CARRIERS

|      |    | CARRIER        | TUBES | RELOADS | |
|------|----|----------------|-------|---------|---|
| ID09 | 78 | BRDM-2 (BTR-40PB) | 5  | >_ 15   | reloads: "At least 15 spare rounds" likely. |
| FI   | 81 | BRDM           | 5     |         | |
| JWS  | 82 | BRDM-2 AFVs    | 5     | 10      | Only the launchers on these vehicles seen by Western observers. Launchers on presumably trainable turret; hatch behind them for reloading. reloads: Total, according to official US estimates. |
| Clns | 83 | BRDM           | 5     |         | |
| FI+  | 83 | BRDM           | 5     |         | tubes: Deployed. |

|     |    | CARRIER | TUBES | RELOADS | |
|-----|----|---------|-------|---------|---|
| FI+ | 83 | BMP     | 1     |         | Some carry AT-5s. |

|      |    | USE AND CONFIGURATION |
|------|----|-----------------------|
| ID09 | 78 | Fired without recoil, probably. Can be reloaded with armor protection. |
| JWS  | 82 | Shape of launch tubes has led to conjecture that gas generator used to eject missile. Light color of tubes led to conjecture of need to minimize temperature in tube, possibly to protect propellant. |

|    |    | GEOGRAPHICAL DEPLOYMENTS |
|----|----|--------------------------|
| FI | 81 | Operational with USSR forces in E. Germany. |

|      |    | RELATION TO OTHER MISSILES |
|------|----|----------------------------|
| ID09 | 78 | Designed along the lines of Milan, has about same performance. |
| JWS  | 82 | Spigot and Spandrel differ in details of launch tubes, specifically front end caps, reason unknown. |
| FI+  | 83 | Probably uses same missile as AT-4; launch tubes mounted on BRDM differ only slightly from those for AT-4. |

# AT-6

```
Sources are:  JCS   80    FI    81    SMP   81    AWST  82    JWS   82    FI+   83
              GUD1  80    AJ03  81    JCS   81    AFM   82    JAWA  82    Clns  83    JWS+  83
```

```
─────────── US DESIGNATION ───────────
JCS   80    AT-6        FY81-82.
FI    81    AT-6
SMP   81    AT-6
AFM   82    AT-6
AWST  82    AT-6
JAWA  82    AT-6
JWS   82    AT-6

─────────── NATO CODENAME ───────────
JCS   80    Spiral      FY81-82.
FI    81    Spiral
SMP   81    Spiral
AFM   82    Spiral
AWST  82    Spiral
JAWA  82    Spiral
JWS   82    Spiral
Clns  83    Spiral

─────────── DESCRIPTION ───────────
Clns  83                Type: Heavy.

─────────── EDITOR'S NOTES ───────────
JCS   80    Listed FY81-82.
SMP   81    Listed '81.

─────────── RANGE ───────────
            mi         ft              km          m           m conv
GUD1  80    4.4-6.3                    7-10                    7,079-10,136
FI    81                              ? 7-10                   7,000-10,000
AFM   82    4.2-6.2                                            6,757-9,975
JAWA  82    4.3-6.2                    7-10                    6,918-9,975      3.75-5.3 nm
Clns  83                                           5,000       5,000            Effective range.
FI+   83                                7-10                   7,000-10,000     Has not been confirmed.

─────────── SPEED ───────────
            fps        mph             mps         kmph        mps conv
Clns  83                                                                        20 sec flight to max range.

─────────── WARHEAD TYPE ───────────
            nuclear    conven          chem
GUD1  80               yes                         ? Hollow charge.
AWST  82                                           Shaped charge.
JWS   82               yes                         HEAT.

─────────── WEIGHT ───────────
            lb         t               kg          kg conv
Clns  83    50                                     22.68

─────────── DIAMETER ───────────
            in         ft              cm          m           cm conv
Clns  83    5.5                                                13.97            Caliber.

─────────── NUMBER OF STAGES ───────────
AWST  82    1

─────────── TYPE OF PROPULSION ───────────
GUD1  80    solid
AWST  82    solid
JWS   82    solid

─────────── MIDCOURSE (OR MAIN) GUIDANCE ───────────
GUD1  80                ? Semi-active laser, designators from ground vehicles or infantry.
FI    81                Semi-active seeker.
AFM   82                Target illuminated by laser.
AWST  82    line-of-sight
JAWA  82                Homes on targets illuminated by a laser designator.
JWS   82    CLOS        Thought semi-active homing; either IR or laser target marking.

─────────── TRANSMISSION METHOD ───────────
Clns  83    radio
```

428  SOVIET MISSILES  IDDS

---------- **DEPLOYMENT BEGAN** ----------
JWS   82            Status: "Early operational deployment".
FI+   83            Technical problems seem to have delayed large-scale deployment, and may have been
                    temporarily withdrawn from front-line use.

---------- **DEPLOYED** ----------
JCS   80            Deployed on Hind D.
JWS   82            Deployed.

---------- **CARRIERS** ----------

          **CARRIER** ---------- TUBES --- RELOADS ----------
AFM   82            Unlike other AT missiles, this has no surface-launched
                    application.

          **CARRIER** ---------- TUBES --- RELOADS ----------
JCS   80   Hind D
FI    81   [Mi-24 Hind D]                    "Replacement for the airborne AT-2 Swatter variant carried
                                             by Mi-24 Hind D helicopters".
JWS   82   Mi-24 Hind D

          **CARRIER** ---------- TUBES --- RELOADS ----------
JCS   81   Hind E
SMP   81   Hind E           4
                                             tubes: [Inferred from picture, ed].
AFM   82   Mi-24 Hind-E
JAWA  82   Mi-24 Hind-E
Clns  83   Hind E           4
JWS+  83   Hind E
                                             tubes: 2 carried on stub wings.

---------- **USE AND CONFIGURATION** ----------
AFM   82   Tube launched.
JAWA  82   Tube launched.
FI+   83   Carried in a slim launch tube.

---------- **STRATEGY** ----------
JCS   80   "The AT-6/SPIRAL-configured HIND D...has greater standoff range as well as launch-and-leave
           capability".
AJ03  81   "Launch and leave capability" [according to Carter's Defense R&E chief William Perry, disputed by
           others quoted in article].
JCS   81   "The AT-6/SPIRAL-configured HIND E...has greater standoff range as well as launch-and-leave
           capability".

---------- **EXPORTED TO** ----------
FI    81   ? Czech        [Final report]

---------- **RELATION TO OTHER MISSILES** ----------
JWS   82   May be missile "to which some sources have applied the designation AS-8".

---------- **OTHER INFORMATION** ----------
JWS   82   Little known of missile.

# AS-1

```
Sources are:  MOW  76     ID10 78     CH01 79     JWS  82
FI *  74      SWM  77     ID11 78     AWST 82     GSN  83
```

### US DESIGNATION
| | | |
|---|---|---|
| FI *  | 74 | AS-1 |
| MOW   | 76 | AS-1 |
| SWM   | 77 | AS-1 |
| AWST  | 82 | AS-1 |
| JWS   | 82 | AS-1 |
| GSN   | 83 | AS-1 |

### NATO CODENAME
| | | |
|---|---|---|
| FI *  | 74 | Kennel |
| MOW   | 76 | Kennel |
| SWM   | 77 | Kennel |
| AWST  | 82 | Kennel |
| JWS   | 82 | Kennel |
| GSN   | 83 | Kennel |

### SOVIET DESIGNATION
| | | | |
|---|---|---|---|
| GSN | 83 | Komet | In Soviet service. |

### RANGE
| | | mi | nm | km | km conv | |
|---|---|---|---|---|---|---|
| MOW  | 76 | 63  |      | 100  | 101.37 | Max. |
| SWM  | 77 | 62+ |      | 100+ | 99.76  | Range [table] |
|      |    | 93  |      | 150  | 149.64 | Max range [text] |
| ID10 | 78 |     |      | 150  | 150.00 | High profile. |
|      |    |     |      | 90   | 90.00  | Low profile. |
| CH01 | 79 | 93  |      | 150  | 149.64 | High profile. |
|      |    | 56  |      | 90   | 90.10  | Low profile. |
| AWST | 82 | 63  |      |      | 101.37 | |
| JWS  | 82 |     | <_ 50 | <_ 90 | 92.60 | |
| GSN  | 83 |     | >  50 | > 92.5 | 92.60 | |

### SPEED
| | | mph | kmph | Mach | kmph conv | |
|---|---|---|---|---|---|---|
| MOW  | 76 | | | 0.9 | | |
| SWM  | 77 | | | 0.9 | | |
| ID10 | 78 | | | 0.9 | | Over aircraft speed. |
| AWST | 82 | | | 0.9 | | |

### WARHEAD TYPE
| | | nuclear | conven | chem | |
|---|---|---|---|---|---|
| MOW  | 76 | | yes | | HE. |
| ID10 | 78 | | | | HE, cannot carry nuclear warhead. |
| CH01 | 79 | | | | HE. |

### WARHEAD WEIGHT
| | | lb | t | kg | kg conv |
|---|---|---|---|---|---|
| ID10 | 78 |       | | 900 | 900.00 |
| CH01 | 79 | 1,984 | | 900 | 899.94 |

### WEIGHT
| | | lb | t | kg | kg conv | |
|---|---|---|---|---|---|---|
| SWM  | 77 | 6,614 | | 3,000 | 3,000 | |
| ID10 | 78 |       | | 2,850 | 2,850 | All-up, 500 kg is fuel. |
| CH01 | 79 | 6,614 | | 3,000 | 3,000 | At launch. |
|      |    | 5,511 | | 2,500 | 2,499 | Burnout. |

### LENGTH
| | | in | ft | cm | m | m conv |
|---|---|---|---|---|---|---|
| MOW  | 76 | | 27'0"     |       | 8.2  | 8.23 |
| SWM  | 77 | | 27        |       | 8.2  | 8.23 |
| ID10 | 78 | |           |       | 8.25 | 8.23 |
| CH01 | 79 | | 27'8 1/4" |       | 8.44 | 8.44 |
| AWST | 82 | | 26        |       |      | 7.92 |
| JWS  | 82 | |           | c. 844 |     | 8.44 |

### DIAMETER
| | | in | ft | cm | m | cm conv |
|---|---|---|---|---|---|---|
| ID10 | 78 | | | | 1.20 | 120.00 |

430 SOVIET MISSILES  IDDS

```
------------ WINGSPAN ------------
                 in      ft       cm       m      cm conv
MOW   76                 16'0"            4.9      487.68
SWM   77                 16               4.9      487.68
CH01  79                 15'9"            4.8      480.06
AWST  82                 14                        426.72
JWS   82                          480              480.00

------------ NUMBER OF STAGES ------------
AWST  82    1

------------ TYPE OF PROPULSION ------------
MOW   76    turbojet          Unidentified turbojet.
SWM   77    turbojet
AWST  82    turbojet
JWS   82    turbojet

------------ MIDCOURSE (OR MAIN) GUIDANCE ------------
MOW   76                      Possibly beam-riding or radio command.
SWM   77    autopilot         Beam riding or radio command (2 radomes under Badger).
ID10  78    beam riding
JWS   82                      Beam riding or radio command link.

------------ TERMINAL GUIDANCE ------------
MOW   76                      Passive or active terminal homing.
SWM   77    IR                Active or passive IR homing.
ID10  78    radar             Passive or active homing head.
JWS   82    radar             Either passive or active; parent Badger has at least 2 radars on underside of
                              fuselage "which could serve such a mode of operation", [SAR?, ed].

------------ CONTROL SYSTEM ------------
MOW   76    Aerodynamic surfaces.

------------ FIRST OBSERVED ------------
CH01  79    1961

------------ DEVELOPMENT BEGAN ------------
ID10  78    end of 1950s      Probably.
GSN   83    1946              And possibly earlier.

------------ PRODUCTION BEGAN ------------
ID11  78    1959              First produced.

------------ IOC ------------
SWM   77    early 1960s       Operational.
GSN   83    1958-1959         "Was operational from 1958-1959".

------------ DEPLOYMENT BEGAN ------------
SWM   77                      "Oldest of the Soviet family" of cruise missiles.
CH01  79    1963              Entered service.
GSN   83                      First fully operational Soviet ASM.

------------ DEPLOYED ------------
FI *  74                      Operated by USSR.

------------ RETIRED ------------
ID10  78    early 1970s       Not in service, USSR. Still in service in Indonesia and Egypt.
GSN   83                      No longer used by Soviet forces.

------------ DEPLOYMENT HISTORY ------------
MOW   76    Is thought to have been superceded by Kelt.
SWM   77    Replaced by AS-5 Kelt.

------------ CARRIERS ------------

------------ CARRIER ------------ TUBES ---- RELOADS ------------
GSN   83                                     "Gorshkov has implied that the missile was developed for
                                             use aboard surface ships, but limitations of the engine
                                             dictated that it be used only from aircraft".

------------ CARRIER ------------ TUBES ---- RELOADS ------------
GSN   83    Tu-4 Bull                        Developed for Bull.

------------ CARRIER ------------ TUBES ---- RELOADS ------------
FI *  74    Tu-16 Badger
SWM   77    Tu-16 Badger B      2            Naval Air Force.
JWS   82    Tu-16 Badger        2
```

---------- TARGET TYPE ----------
FI*    74   ship              Mainly.
MOW    76   ship
SWM    77   ship, "shore"
ID10   78   ship              Exclusively.
AWST   82   ship
JWS    82   ship              Probably.

---------- STRATEGY ----------
JWS    82   Now obsolete.

---------- EXPORTED TO ----------
FI*    74   Egypt
MOW    76   Indonesia         No longer operational there.
            Egypt             No longer operational there.
SWM    77   Indonesia
            Egypt
ID10   78   Indonesia         Still in service.
            Egypt             Still in service.
JWS    82   Egypt             And other client nations; probably remains in service with some client nations.

---------- RELATION TO OTHER MISSILES ----------
FI*    74   Very similar to Samlet.
SWM    77   Surface-to-surface version is Samlet.
JWS    82   Appearance almost identical to Samlet coastal defense missile; may be versions of same missile.
GSN    83   Land-launched SS-C-2b Samlet was a variant of AS-1.

# AS-2

```
Sources are:   SWM   77      USND  81     CRW   82    JFS   82    IISS+ 83    CRW+  84
FI*   74       ID10  78      AFM   82     IISS  82    JWS   82    JFS+  83    Meyr  84
MOW   76       CH01  79      AWST  82     JAWA  82    GSN   83    AFM+  84    BW    86
```

## US DESIGNATION

|  | |  |
|---|---|---|
| FI* | 74 | AS-2 |
| MOW | 76 | AS-2 |
| SWM | 77 | AS-2 |
| USND | 81 | AS-2 |
| AFM | 82 | AS-2 |
| AWST | 82 | AS-2 |
| CRW | 82 | AS 2 |
| IISS | 82 | AS-2 |
| JAWA | 82 | AS-2 |
| JFS | 82 | AS-2 |
| JWS | 82 | AS-2 |
| GSN | 83 | AS-2 |
| Meyr | 84 | AS-2 |

## NATO CODENAME

|  | |  |
|---|---|---|
| FI* | 74 | Kipper |
| MOW | 76 | Kipper |
| SWM | 77 | Kipper |
| USND | 81 | Kipper |
| AFM | 82 | Kipper |
| AWST | 82 | Kipper |
| CRW | 82 | Kipper |
| IISS | 82 | Kipper |
| JAWA | 82 | Kipper |
| JFS | 82 | Kipper |
| JWS | 82 | Kipper |
| GSN | 83 | Kipper |

## DESCRIPTION

IISS 82   ALCM

## EDITOR'S NOTES

Meyr 84   Note to chart: "Values pertain to initial service period". Footnote to air-launched missiles: "Includes delivery vehicle performance".

## RANGE

| Source | | mi | nm | km | km conv | |
|---|---|---|---|---|---|---|
| MOW | 76 | 132 | | 213 | 212.39 | Max. |
| SWM | 77 | 112-130 | | 180-210 | 180.21-209.17 | [Table] |
| | | > 100 | | > 160 | 160.90 | [Text:] "stand-off range may exceed 160 km (100 miles)". |
| ID10 | 78 | | | c. 100 | 100.00 | "Low or very low level flight" [text] |
| | | | | 210 | 210.00 | High altitude [table] |
| USND | 81 | | c. 100 | | 185.20 | |
| AFM | 82 | 130 | | | 209.17 | |
| AWST | 82 | 31 | | | 49.88 | |
| CRW | 82 | | 100 | | 185.20 | |
| IISS | 82 | | | 200 | 200.00 | Max. |
| JFS | 82 | | 115 | | 212.98 | Max. |
| JWS | 82 | | | c. 200 | 200.00 | [Analysis table] |
| | | | | 180-210 | 180.00-210.00 | Most Western estimates [text] |
| GSN | 83 | | 100+ | 185+ | 185.20 | |
| AFM+ | 84 | 132 | | | 212.39 | |
| Meyr | 84 | | | 200 | 200.00 | |
| BW | 86 | | | 210 | 210.00 | 100 km at low altitude. |

## SPEED

| Source | | mph | kmph | Mach | kmph conv | |
|---|---|---|---|---|---|---|
| MOW | 76 | | | 1.2 | | |
| SWM | 77 | | | 1.2 | | |
| ID10 | 78 | | | 1.2 | | Over aircraft speed. |
| AFM | 82 | | | 1.2 | | |
| JFS | 82 | | | 1.4 | | |
| BW | 86 | | | 1.2 | | |

## CEP

| Source | | ft | mi | nm | m | km | m conv |
|---|---|---|---|---|---|---|---|
| Meyr | 84 | | | | | 0.5-1.5 | 500.00-1,500 |
| BW | 86 | | | | ? 1,000 | | 1,000 |

```
---------- WARHEAD TYPE ----------
              nuclear    conven    chem
MOW    76     yes
ID10   78     yes        yes                HE. Nuclear warhead is presumably the standard 1,000 kg nuclear bomb
                                            of Tactical Air Forces.
CH01   79     yes        yes                HE.
AFM    82                yes
CFW    82     yes        yes
IISS   82     yes        yes
JAWA   82     yes                           Nuclear warhead can be fitted [final report].
JFS    82                yes                HE.
JWS    82                yes                HE.
GSN    83     yes        yes
JFS+   83     yes        yes
Meyr   84     yes
```

```
---------- WARHEAD CHARACTERISTICS ----------
IISS   82     single warhead
Meyr   84     single RV
BW     86     single RV
```

```
---------- NUCLEAR YIELD ----------
              kT              MT          kT conv
IISS   82                     < 1         1,000
Meyr   84     200-600                     200.00-600.00
BW     86     ? 400                       400.00
```

```
---------- WARHEAD WEIGHT ----------
              lb              t           kg            kg conv
ID10   78                                 1,000         1,000         Nuclear warhead.
AFM    82     2,200                                     997.92
JFS    82     2,200                                     997.92
GSN    83     2,200                       1,000         997.92
CFW+   84                                 1,000         1,000
```

```
---------- THROWWEIGHT ----------
              lb              t           kg            kg conv
IISS   82     2,200                                     997.92
```

```
---------- WEIGHT ----------
              lb              t           kg            kg conv
SWM    77     13,228+                     6,000+        6,000
CH01   79     13,228+                     6,000+        6,000         Launch.
              10,362                      4,700         4,700         Burnout.
AFM    82     9,260                                     4,200
CFW    82                                 900-1,000     900.00-1,000  [Sic]
JFS    82     7,700                                     3,492         Launch.
GSN    83     9,260                       4,200         4,200
JFS+   83     9,100                                     4,127         Launch.
CFW+   84                                 4,200         4,200
```

```
---------- LENGTH ----------
              in              ft          cm            m       m conv
MOW    76                     31'0"                     9.5     9.45
SWM    77                     31                        9.4     9.45     [Table]
                                                        9.5     9.50     [Text]
CH01   79                     30'10"                    9.4     9.40
AFM    82                     31'0"                             9.45
AWST   82                     31                                9.45
JFS    82                     33                                10.06
JWS    82                                 c. 1,000              10.00
GSN    83                     31                        9.5     9.45
JFS+   83                     31                                9.45
```

```
---------- DIAMETER ----------
              in              ft          cm            m       cm conv
ID10   78                                               .90     90.00
CH01   79     35 2/5                      90                    89.92
JWS    82                                 100                   100.00
```

```
------------ WINGSPAN ------------
              in      ft        cm       m        cm conv
MOW    76             16                 4.88     487.68
SWM    77             15-16              4.6-4.9  457.20-487.68
CH01   79             15'1"              4.6      459.74
AFM    82             16'0"                       487.68
AWST   82             16                          487.68
JWS    82                       460               460.00
GSN    83             16                 4.9      487.68

------------ NUMBER OF STAGES ------------
AWST   82     1
JFS    82     1

------------ TYPE OF PROPULSION ------------
MOW    76     turbojet          Unidentified turbojet.
SWM    77     turbojet          Underslung.
ID10   78     turbojet          With afterburner.
CH01   79     turbojet          Probably.
AFM    82     turbojet          Underslung.
AWST   82     turbojet
CFW    82     solid             [Sic; final report]
JFS    82     turbojet
JWS    82     turbojet
GSN    83     turbojet
CFW+   84     turbojet
BW     86     TJ

------------ MIDCOURSE (OR MAIN) GUIDANCE ------------
SWM    77     autopilot
              ? radio command
ID10   78                       Beam riding only in initial phases, then launch aircraft can turn away.
                                Autopilot or pre-programmed flight path.
CH01   79     radio command
AFM    82                       "Initial beam-riding, subsequent pre-programmed flight under autopilot
                                control, and active radar terminal homing".
AWST   82                       "Autopilot-command override, midcourse".
CFW    82                       Inertial guidance or automatic pilot.
JAWA   82                       Initial beam-riding, subsequent pre-programmed flight under autopilot control.
JWS    82                       Programmed autopilot (optional command override).
GSN    83                       Inertial or autopilot.
JFS+   83     autopilot         Command override.

------------ TERMINAL GUIDANCE ------------
SWM    77     radar.
ID10   78     IR
CH01   79     IR
AFM    82     active radar
AWST   82     active
CFW    82     radar
JAWA   82     active radar
JFS    82     active radar
JWS    82     active radar     Possibly alternative of passive radar in some models.
GSN    83     radar homing

------------ FIRST OBSERVED ------------
MOW    76     1961
ID10   78     1961
AFM    82     1961
JWS    82     1961

------------ IOC ------------
JFS    82     1960     Operational.
GSN    83     1961

------------ DEPLOYMENT BEGAN ------------
ID10   78     1965     Entered service.
CH01   79     1965     Entered service.
IISS   82     1961     First year deployed.
CFW+   84     1961
Meyr   84     1960     Service.
BW     86     1961

------------ DEPLOYED ------------
FI *   74              Operated by USSR.
MOW    76              In service.
JWS    82              Deployment: USSR.
BW     86              In service.
```

---------- NUMBERS IN SERVICE ----------

---------- NUMBER IN 1983 ----------
IISS+ 83    90         July, total deployed.

---------- CARRIERS ----------

---------- CARRIER ---------- TUBES ---- RELOADS ----------
FI*   74   Tu-16 Badger C
MOW   76   Tu-16 Badger-C                  Has radar in wide nose radome, presumably for detecting
                                           and locating target before missile is launched.
SWM   77   Tu-16 Badger C       [1]
ID10  78   Tu-16 Badger C       1
USND  81   Badger C
AWST  82   Badger
CFW   82   Badger-C
JAWA  82   Tu-16 Badger-C
JFS   82   Badger C
JWS   82   Tu-16 Badger-C
GSN   83   Badger-C
IISS+ 83   Badger C             1
CFW+  84   Badger C and G
BW    86   Badger C, G          1           no

---------- CARRIER ---------- TUBES ---- RELOADS ----------
USND  81   Badger G
GSN   83   Badger-G

---------- USE AND CONFIGURATION ----------
Meyr  84   Penetration: 0.7-0.9; reliability: 0.2-0.5, "includes system reliability multiplied by operational
           readiness of deployed system".

---------- TARGET TYPE ----------
FI*   74   ship             Mainly.
MOW   76   ship or land
SWM   77   ship             And coastal targets.
ID10  78   carrier          Perhaps IR guidance indicates carrier as target--good heat source but small radar
                            cross-section.
AFM   82   ship
JAWA  82   shipping
JWS   82   shipping         Considered by Western observers essentially an anti-shipping missile; can't
                            completely discount its use for strategic bombing of land targets.

---------- STRATEGY ----------
Meyr  84   [Ed: chart is labeled] Systems "assigned primary operational and strategic missions within the
           European theatre[s] of military operations".

---------- COMBAT REPORTS/EFFECTIVENESS ----------
SWM   77   "None of these early winged missiles would be particularly effective against the defence systems
           of modern warships".

---------- RELATION TO OTHER MISSILES ----------
FI*   74   Similar in configuration to Hound Dog.
MOW   76   Similar to US Hound Dog but less refined in form.
SWM   77   Resembles US Hound Dog, though smaller and less advanced.
JWS   82   Resembles US Hound Dog in general appearance.

# AS-3

| Sources are: | MOW | 76 | JCS | 79 | IISS | 81 | IISS | 82 | GSN | 83 | Meyr | 84 |
|---|---|---|---|---|---|---|---|---|---|---|---|---|
| SecC | 68 | JCS | 77 | Clns | 80 | SMP | 81 | JAWA | 82 | IISS+ | 83 | SMP | 84 |
| SecC | 70 | SWM | 77 | GUO1 | 80 | USND | 81 | JFS | 82 | JFS+ | 83 | BW | 86 |
| FI* | 74 | ID10 | 78 | JCS | 80 | AFM | 82 | JWS | 82 | venD | 83 |
| JCS | 75 | CHO1 | 79 | SecD | 80 | AWST | 82 | Clns+ | 83 | AFM+ | 84 |

---------- **US DESIGNATION** ----------

| SecC | 70 | AS-3 | FY71-72. |
|---|---|---|---|
| FI* | 74 | AS-3 | |
| JCS | 75 | AS-3 | FY76-82. |
| MOW | 76 | AS-3 | |
| SWM | 77 | AS-3 | |
| Clns | 80 | AS-3 | |
| SecD | 80 | AS-3 | |
| SMP | 81 | AS-3 | '81, '84, '85. |
| USND | 81 | AS-3 | |
| AFM | 82 | AS-3 | |
| AWST | 82 | AS-3 | |
| IISS | 82 | AS-3 | |
| JAWA | 82 | AS-3 | |
| JFS | 82 | AS-3 | |
| JWS | 82 | AS-3 | |
| GSN | 83 | AS-3 | |
| venD | 83 | AS-3 | |
| Meyr | 84 | AS-3 | |

---------- **NATO CODENAME** ----------

| SecC | 70 | Kangaroo | FY71-72. |
|---|---|---|---|
| FI* | 74 | Kangaroo | |
| JCS | 75 | Kangaroo | FY76-78. |
| MOW | 76 | Kangaroo | |
| SWM | 77 | Kangaroo | |
| Clns | 80 | Kangeroo | [Sic] |
| SMP | 81 | Kangaroo | |
| USND | 81 | Kangaroo | |
| AFM | 82 | Kangaroo | |
| AWST | 82 | Kangaroo | |
| IISS | 82 | Kangaroo | |
| JAWA | 82 | Kangaroo | |
| JFS | 82 | Kangaroo | |
| JWS | 82 | Kangaroo | |
| GSN | 83 | Kangaroo | |

---------- **DESCRIPTION** ----------

| FI* | 74 | | Long-range standoff weapon. |
|---|---|---|---|
| JCS | 75 | | Large air-to-surface guided missile, FY76-77. |
| MOW | 76 | | Strategic missile. |
| SecD | 80 | ASM | |
| IISS | 82 | ALCM | |
| GSN | 83 | | Strategic attack missile. |

---------- **EDITOR'S NOTES** ----------

| SecC | 68 | Listed FY69-72, not by name FY69-70. |
|---|---|---|
| JCS | 75 | Listed FY76-82. |
| SecD | 80 | Listed FY81. |
| SMP | 81 | Listed '81, '84, '85. |
| Meyr | 84 | Footnote to air-launched missiles: "Includes delivery vehicle performance" Note to chart: "Values pertain to initial service period". |

CRUISE MISSILES  ASMs  AS-3

## RANGE

|       |    | mi      | nm      | km      | km conv         |                                                                                                                                      |
|-------|----|---------|---------|---------|-----------------|--------------------------------------------------------------------------------------------------------------------------------------|
| SecC  | 70 |         | c. 275  |         | 509.30          | FY71-72.                                                                                                                             |
| FI *  | 74 |         |         | 500+    | 500.00          |                                                                                                                                      |
| MOW   | 76 | 400     |         | 650     | 643.60          | Max.                                                                                                                                 |
| JCS   | 77 |         | 350     |         | 648.20          | Max, FY78-79.                                                                                                                        |
| SWM   | 77 | 115-403 |         | 185-650 | 185.04-648.43   | [Table]                                                                                                                              |
|       |    | 110-404 |         |         | 176.99-650.04   | [Text]                                                                                                                               |
| ID10  | 78 |         |         | 650     | 650.00          | With warhead of 2,300 kg or more, at high altitude.                                                                                  |
| CH01  | 79 | 115-404 |         | 185-650 | 185.04-650.04   |                                                                                                                                      |
| CLns  | 80 | 350     |         |         | 563.15          | [Fighter/attack listing, ed]                                                                                                         |
|       |    | 400     |         |         | 643.60          | [Strategic listing, ed]                                                                                                              |
| SMP   | 81 |         |         | 650     | 650.00          |                                                                                                                                      |
| USND  | 81 |         | 200-300 |         | 370.40-555.60   |                                                                                                                                      |
| AFM   | 82 | 400     |         |         | 643.60          |                                                                                                                                      |
| AWST  | 82 | 350     |         |         | 563.15          |                                                                                                                                      |
| IISS  | 82 |         |         | 650     | 650.00          | Max.                                                                                                                                 |
| JFS   | 82 |         | 350     |         | 648.20          | Max.                                                                                                                                 |
| JWS   | 82 |         |         | 185-650 | 185.00-650.00   | Range of estimates in specialist sources; all seem possible.                                                                         |
|       |    |         | 350     | 648     | 648.20          | Recent US Senate report stated nuclear warhead could be carried to this range.                                                       |
| GSN   | 83 |         | c. >200 | c. >370 | 370.40          |                                                                                                                                      |
| Meyr  | 84 |         |         | 600     | 600.00          |                                                                                                                                      |
| BW    | 86 |         |         | 650     | 650.00          |                                                                                                                                      |

## SPEED

|       |    | mph | kmph | Mach    | kmph conv |                                                             |
|-------|----|-----|------|---------|-----------|-------------------------------------------------------------|
| MOW   | 76 |     |      | 2       |           |                                                             |
| ID10  | 78 |     |      | 1.6-1.8 |           | Inlet cone is not adjustable; speed depends on flight profile. |
| CLns  | 80 |     |      | 2.0     |           |                                                             |
| GU01  | 80 |     |      | 1.6-1.8 |           |                                                             |
| AFM   | 82 |     |      | 1.8     |           | Max.                                                        |
| AWST  | 82 |     |      | 2       |           |                                                             |
| JFS   | 82 |     |      | 1.8     |           |                                                             |
| JWS   | 82 |     |      | c. 2    |           |                                                             |
| CLns+ | 83 |     |      | 2.0     |           |                                                             |
| SMP   | 84 |     |      |         |           | Subsonic [sic], '84, '85.                                   |
| BW    | 86 |     |      | 0.8     |           |                                                             |

## CEP

|      |    | ft | mi | nm  | m      | km      | m conv        |
|------|----|----|----|-----|--------|---------|---------------|
| vanD | 83 |    |    | 1.0 |        |         | 1,852         |
| Meyr | 84 |    |    |     |        | 0.5-1.5 | 500.00-1,500  |
| BW   | 86 |    |    |     | ? 1,000 |        | 1,000         |

## WARHEAD TYPE

|       |    | nuclear | conven | chem |                        |
|-------|----|---------|--------|------|------------------------|
| FI *  | 74 | yes     |        |      |                        |
| MOW   | 76 | yes     |        |      |                        |
| JCS   | 77 | yes     |        |      | FY78-79.               |
| SWM   | 77 | maybe   |        |      |                        |
| ID10  | 78 | yes     | yes    |      | HE.                    |
| CLns  | 80 | yes     |        |      |                        |
| GU01  | 80 | yes     |        |      | Thermonuclear.         |
| AFM   | 82 | yes     | yes    |      | HE.                    |
| JAWA  | 82 | yes     | yes    |      | HE.                    |
| JFS   | 82 | yes     |        |      |                        |
| JWS   | 82 | yes     | yes    |      | HE.                    |
| CLns+ | 83 | yes     |        |      |                        |
| GSN   | 83 | yes     |        |      | Nuclear warhead only.  |
| IISS+ | 83 | yes     | yes    |      |                        |
| Meyr  | 84 | yes     |        |      |                        |

## WARHEAD CHARACTERISTICS

| IISS | 82 | single warhead |
| vanD | 83 | single RV      |
| Meyr | 84 | single RV      |
| BW   | 86 | single RV      |

## NUCLEAR YIELD

| | | kT | MT | kT conv | |
|---|---|---|---|---|---|
| SecC | 70 | | 1-3 | 1,000-3,000 | FY71-72. |
| Clns | 80 | | 1 | 1,000 | |
| AFM | 82 | 500 | | 500.00 | |
| IISS | 82 | | >1 | 1,000 | |
| JAWA | 82 | 800 | | 800.00 | |
| JFS | 82 | 500 | | 500.00 | |
| JWS | 82 | c. 800 | | 800.00 | |
| vanD | 83 | | 2.0 | 2,000 | |
| AFM+ | 84 | 800 | | 800.00 | |
| Meyr | 84 | | 1-3 | 1,000-3,000 | |
| BW | 86 | 2,000 | | 2,000 | |

## WARHEAD WEIGHT

| | | lb | t | kg | kg conv | |
|---|---|---|---|---|---|---|
| ID10 | 78 | | | >_ 2,300 | 2,300 | For a 650 km range. |
| Clns | 80 | 5,000 | | | 2,268 | |
| GU01 | 80 | 5,070 | | 2,300 | 2,299 | |
| AFM | 82 | 5,070 | | | 2,299 | |
| JAWA | 82 | 5,070 | | 2,300 | 2,299 | |

## WEIGHT

| | | lb | t | kg | kg conv | |
|---|---|---|---|---|---|---|
| ID10 | 78 | | 11 | | 11,000 | All-up ["t" for tons]; 3,700 kg is fuel. |
| CH01 | 79 | c. 24,250 | | c. 11,000 | 10,999 | Launch. |
| | | c. 16,094 | | c. 7,300 | 7,300 | Burnout. |
| GU01 | 80 | c. 24,250 | | c. 11,000 | 10,999 | |
| AFM | 82 | 17,600 | | | 7,983 | |
| JFS | 82 | 17,600 | | | 7,983 | Launch. |
| JFS+ | 83 | 24,450 | | | 11,090 | Launch. |

## LENGTH

| | | in | ft | cm | m | m conv | |
|---|---|---|---|---|---|---|---|
| FI* | 74 | | | | 14.9 | 14.90 | |
| MOW | 76 | | 48'11" | | 14.9 | 14.91 | |
| SWM | 77 | | 49.2 | | 15 | 15.00 | [Table] |
| | | | 49 | | 15 | 14.94 | [Text] |
| CH01 | 79 | | 49 | | 14.96 | 14.94 | |
| GU01 | 80 | | 49 | | 14.9 | 14.94 | |
| AFM | 82 | | 48'11" | | | 14.91 | |
| AWST | 82 | | 50 | | | 15.24 | |
| JFS | 82 | | 49.2 | | | 15.00 | |
| JWS | 82 | | | 1,500 | | 15.00 | [Analysis table and text] |
| | | | | | 14.96 | 14.96 | US official sources quote 14.96 m. |
| JFS+ | 83 | | 46.6 | | | 14.20 | |
| AFM+ | 84 | | 49'1" | | | 14.96 | |

## DIAMETER

| | | in | ft | cm | m | cm conv |
|---|---|---|---|---|---|---|
| CH01 | 79 | | 6'1" | | 1.85 | 185.42 |
| GU01 | 80 | | 6 | | 1.85 | 182.88 |
| JWS | 82 | | | 130 | | 130.00 |

## WINGSPAN

| | | in | ft | cm | m | cm conv |
|---|---|---|---|---|---|---|
| FI* | 74 | | | 900 | | 900.00 |
| MOW | 76 | | 30 | | 9.15 | 914.40 |
| SWM | 77 | | 29.5 | | 9.0 | 899.16 |
| GU01 | 80 | | 32'10" | | 9.15 | 1,000 |
| AFM | 82 | | 30 | | | 914.40 |
| AWST | 82 | | 30 | | | 914.40 |
| JWS | 82 | | | 900 | | 900.00 |
| AFM+ | 84 | | 29'6" | | | 899.16 |

## NUMBER OF STAGES

| | | |
|---|---|---|
| AWST | 82 | 1 |
| JFS | 82 | 1 |

CRUISE MISSILES  ASMs  AS-3

```
----------- TYPE OF PROPULSION -----------
FI*    74    ? turbojet
MOW    76    turbojet        Unidentified turbojet; compressed air from bomber may start ramjet before
                             release.
SWM    77    turbojet
ID10   78    turbojet        With afterburner.
GUD1   80    turbojet        After-burning.
AWST   82    turbojet
JFS    82    turbojet
JWS    82    turbojet        Probably.
BW     86    TJ

----------- MIDCOURSE (OR MAIN) GUIDANCE -----------
FI*    74    command
MOW    76                    Radar scanner in Tu-95 nose may give initial bearing.
SWM    77                    Autopilot, mid-course by radio command.
ID10   78                    Beam-riding in initial phase, then autopilot for pre-programmed path.
Clns   80                    Preprogrammed autopilot.
AFM    82                    Beam-riding, followed by autopilot.
JAWA   82                    Assumed to be initial beam-riding and subsequent pre-programmed flight under
                             autopilot control.
JWS    82                    "Autopilot with command over-ride facilities for navigational corrections".
Clns+  83    programmed

----------- TERMINAL GUIDANCE -----------
GUD1   80    none
JAWA   82    none
JFS    82    non-homing
JWS    82    none
GSN    83                    "Not believed to have effective anti-ship homing system".

----------- CONTROL SYSTEM -----------
MOW    76    Aerodynamic control surfaces.

----------- FIRST OBSERVED -----------
MOW    76    1961
JWS    82    1961

----------- OBSERVED IN PARADE -----------
AFM    82    1961            Observed on Tu-95 on Aviation Day.

----------- IOC -----------
JFS    82    1961            Operational.
JFS+   83    1960            Operational.
vanD   83    1961

----------- DEPLOYMENT BEGAN -----------
CHO1   79    1963            Entered service.
Clns   80    1960            First deployed [fighter/attack listing, ed]
             1961            First deployed [strategic listing, ed]
IISS   82    1961            First year deployed.
Meyr   84    1960            Service.
BW     86    1960

----------- DEPLOYED -----------
FI*    74                    Operated by USSR.
MOW    76                    In service.
JWS    82                    Deployment: USSR. Assumed to have retained its operational status.
BW     86                    In service.

----------- RETIRED -----------
SWM    77                    "Must now be on the retirement list".

----------- NUMBERS IN SERVICE -----------

----------- NUMBER IN 1981 -----------
IISS   81    c. 70           1981-83, July, total deployed.

----------- NUMBER IN 1982 -----------
IISS   82    ? 70            July, total deployed.

----------- CARRIERS -----------

----------- CARRIER ----------- TUBES --- RELOADS -----------
JCS    79                                    [Unclear wording might mean Bison carries AS-3 too,
                                             unlikely, ed], FY80,82.
```

|  |  | CARRIER | TUBES | RELOADS |  |
|---|---|---|---|---|---|
| SecC | 68 | Bear |  |  | Most of old Bear bombers already modified to include ASM capability, FY69-70. |
| SecC | 70 | Bear |  |  | About 2/3 of Bears carry AS-3, FY71-72. |
| MOW | 76 | Tu-95 |  |  |  |
| Clns | 80 | TU-95 Bear |  |  |  |
| JCS | 80 | Bear |  |  | More than half the Bears are equipped with AS-3. |
| SecD | 80 | Bear | 1 |  | "About two-thirds of the BEAR aircraft are configured to carry one AS-3" [of intercontinental (LRA) BEARs?, ed]. |
| SMP | 81 | Bear |  |  | 2 of the 3 strike versions are configured to carry AS-3. |
| AFM | 82 | Tu-95 Bear B/C |  |  | Arms about 75 [final report of number]. |
| AWST | 82 | Bear |  |  |  |
| JFS | 82 | Bear |  |  |  |
| JWS | 82 | Tu-95 Bear |  |  |  |
| SMP | 84 | Bear |  |  | AS-4 is replacing AS-3 on older ASM-equipped Bears, [appears in Strategic Aviation section, ed]. "Several of these reconfigurations (Bear G) have been completed", '84, '85. |

|  |  | CARRIER | TUBES | RELOADS |  |
|---|---|---|---|---|---|
| JCS | 75 | Bear B |  |  | FY76-77. |
| SWM | 77 | Tu-95 Bear B | [1] |  |  |
| ID10 | 78 | Tu-95 Bear B | 1 |  |  |
| USND | 81 | Bear B |  |  |  |
| JAWA | 82 | Tu-95 Bear-B |  |  |  |
| GSN | 83 | Bear-B |  |  | Of Soviet strategic aviation. |
| IISS+ | 83 | Bear B | 1 |  |  |
| JFS+ | 83 | Bear B |  |  |  |
| BW | 86 | Bear B | 1 | no |  |

|  |  | CARRIER | TUBES | RELOADS |  |
|---|---|---|---|---|---|
| JCS | 75 | Bear C |  |  | FY76-77. |
| USND | 81 | Bear C |  |  |  |
| JAWA | 82 | Tu-95 Bear-C |  |  |  |
| GSN | 83 | Bear-C |  |  | Of Soviet strategic aviation. |
| IISS+ | 83 | Bear C | 1 |  |  |
| JFS+ | 83 | Bear C |  |  |  |
| BW | 86 | Bear C | 1 | no |  |

|  |  | CARRIER | TUBES | RELOADS |
|---|---|---|---|---|
| FI* | 74 | Tu-22 Blinder |  |  |

**USE AND CONFIGURATION**

| GUD1 | 80 | Thought that AS-3 launched 11,000-12,000 m, climbs to 18,000 m at M1.6 using beam riding. Descent at M1.8 begins 100 mi from target. |
|---|---|---|
| vanD | 83 | Reliability: .90; Penetration .9. |
| Meyr | 84 | Penetration: 0.7-0.9; reliability: 0.3-0.7, "includes system reliability multiplied by operational readiness of deployed system". |

**TARGET TYPE**

| FI* | 74 | Probably for use against soft targets. |
|---|---|---|

**STRATEGY**

| SWM | 77 | Chief targets would be ports and industrial centers. |
|---|---|---|
| ID10 | 78 | Targets are fixed-coordinate objectives. Low accuracy means nuclear warhead required for satisfactory results. |
| Clns | 80 | [Listed with fighter/attack and strategic aircraft, ed] |
| GUD1 | 80 | Arms both Long-Range and Naval Aviation. |
| JWS | 82 | Most likely role is "medium/long-range stand-off strategic nuclear weapon for use against area targets, such as major cities or ports". |
| Meyr | 84 | "Initially developed for intercontinental strikes, but likely to be assigned to missions in the European theatre of military operations". [Ed: chart is labeled] Systems "assigned primary operational and strategic missions within the European theatre(s) of military operations". |

**RELATION TO OTHER MISSILES**

| SWM | 77 | Largest Soviet AS weapon. Resembles Fitter interceptor. |
|---|---|---|
| ID10 | 78 | A pilotless combat aircraft, closely based on Su-7 Fitter A. |
| AFM | 82 | Largest current Soviet ASM. |
| JWS | 82 | Largest Soviet ASM yet revealed. |

# AS-4

```
Sources are:  JCS  77    Clns 80    JCS  81    CFW  82    Clns+ 83   SMP  83    JCS  85
SecC 68       SWM  77    GU01 80    SMP  81    IISS 82    GSN  83    ABFS 84    SecD 85
IA06 75       ID10 78    SecD 80    USND 81    JAWA 82    IISS+ 83   CFW+ 84    SMP  85
IISS 76       Rock 78    FI   81    AFM  82    JFS  82    JCS  83    Meyr 84    BW   86
MOW  76       CH01 79    IISS 81    AWST 82    JWS  82    JFS+  83   SMP  84
```

```
---------- US DESIGNATION ----------
MOW  76    AS-4
JCS  77    AS-4       FY78-79,81-82,84,86.
SWM  77    AS-4
Rock 78    AS-4
Clns 80    AS-4
SecD 80    AS-4       FY81,86.
FI   81    AS-4
SMP  81    AS-4       '81, '83, '84, '85.
USND 81    AS-4
AFM  82    AS-4
AWST 82    AS-4
CFW  82    AS 4
IISS 82    AS-4
JAWA 82    AS-4
JFS  82    AS-4
JWS  82    AS-4
GSN  83    AS-4
Meyr 84    AS-4

---------- NATO CODENAME ----------
MOW  76    Kitchen
JCS  77    Kitchen    FY78-79,81-82.
SWM  77    Kitchen
Rock 78    Kitchen
Clns 80    Kitchen
FI   81    Kitchen
USND 81    Kitchen
AFM  82    Kitchen
AWST 82    Kitchen
CFW  82    Kitchen
IISS 82    Kitchen
JAWA 82    Kitchen
JFS  82    Kitchen
JWS  82    Kitchen
GSN  83    Kitchen
SMP  83    Kitchen

---------- DESCRIPTION ----------
MOW  76               "Strategic missile". 1961 Tushino commentator referred to Tu-22 as spearhead
                      of strategic rocket force, and most of the 22 Tu-22s in 1967 air show carried
                      Kitchen. [Is this the evidence for "strategic" designation?, ed]
SecD 80    ASM        FY81,86.
JCS  81               Air-to-surface cruise missile.
IISS 82    ALCM
JAWA 82               Stand-off missile.
JCS  83               "Supersonic anti-ship cruise missiles".
JCS  85               air-to-surface missile.

---------- EDITOR'S NOTES ----------
SecC 68    Listed FY69, but not by name, only as Blinder B ASM.
JCS  77    Listed FY78-79,81-82,84,86.
SecD 80    Listed FY81,86.
SMP  81    Listed '81, '83, '84, '85.
Meyr 84    Footnote to air-launched missiles: "Includes delivery vehicle performance" Note to chart: "Values
           pertain to initial service period".
```

## SOVIET MISSILES  IDDS

### RANGE

| | | mi | nm | km | km conv | |
|---|---|---|---|---|---|---|
| SecC | 68 | | 300 | | 555.60 | |
| IAD6 | 75 | | | 300-800 | 300.00-800.00 | |
| MOW | 76 | 460 | | 740 | 740.14 | Almost certainly too great. |
| SWM | 77 | 186+ | | 300+ | 299.27 | |
| ID10 | 78 | | | c. 300 | 300.00 | Low-level flight, 1,000 kg payload. |
| | | | | c. 720 | 720.00 | High flight, 1,000 kg payload. |
| Rock | 78 | | | 300 | 300.00 | |
| CH01 | 79 | c. 185 | | 298 | 297.67 | Low flight profile. |
| | | 447 | | 720 | 719.22 | High flight profile. |
| Clns | 80 | 250 | | | 402.25 | [Fighter/attack listing] |
| | | 300 | | | 482.70 | [Strategic listing] |
| GUD1 | 80 | 450 | | 720 | 724.05 | Max. |
| FI | 81 | | | $\leq$ 300 | 300.00 | |
| USND | 81 | | 150-250 | | 277.80-463.00 | |
| AFM | 82 | 185 | | | 297.67 | At low altitude. |
| AWST | 82 | 250 | | | 402.25 | |
| CFW | 82 | | 170 | | 314.84 | |
| IISS | 82 | | | 300 | 300.00 | Max. |
| JFS | 82 | | 300 | | 555.60 | Max. |
| JWS | 82 | | | 300-800 | 300.00-800.00 | Range of different estimates. |
| | | | | 298 | 298.00 | UK MoD, early 1976; thought to have been an underestimate. |
| GSN | 83 | | 150-250 | 278-463 | 277.80-463.00 | |
| JFS+ | 83 | | 250 | | 463.00 | Max. |
| Meyr | 84 | | | 500 | 500.00 | |
| SecD | 85 | | | | | Long-range. |
| BW | 86 | | | ? 720 | 720.00 | 300 km at low altitude. |

### SPEED

| | | mph | kmph | Mach | kmph conv | |
|---|---|---|---|---|---|---|
| ID10 | 78 | | | 2.5 | | Over launch aircraft. |
| CH01 | 79 | | | 2.5 | | |
| Clns | 80 | | | 4.0 | | |
| GUD1 | 80 | | | $\leq$ 2.5 | | |
| FI | 81 | | | 2+ | | |
| AFM | 82 | | | > 2 | | |
| CFW | 82 | | | 3.5 | | |
| JFS | 82 | | | 3+ | | |
| JWS | 82 | | | | | Aerodynamic features suggest high cruising speed. |
| Clns+ | 83 | | | 3.3 | | |
| JCS | 83 | | | | | Supersonic, FY84,86. |
| SMP | 84 | | | | | Supersonic, '84, '85. |
| JCS | 85 | | | | | Supersonic. |
| BW | 86 | | | ? 2.5 | | |

### CEP

| | | ft | mi | nm | m | km | m conv |
|---|---|---|---|---|---|---|---|
| Meyr | 84 | | | | | 0.5-1.5 | 500.00-1,500 |
| BW | 86 | | | | ? 1,000 | | 1,000 |

### WARHEAD TYPE

| | | nuclear | conven | chem | |
|---|---|---|---|---|---|
| MOW | 76 | yes | | | |
| JCS | 77 | yes | | | Can carry nuclear warhead, FY78-79. |
| ID10 | 78 | yes | yes | | HE. |
| Clns | 80 | yes | yes | | HE. |
| GUD1 | 80 | yes | yes | | HE. |
| FI | 81 | yes | | | |
| AFM | 82 | yes | yes | | |
| CFW | 82 | yes | | | |
| JAWA | 82 | yes | yes | | HE. |
| JFS | 82 | yes | yes | | |
| JWS | 82 | yes | yes | | HE. |
| Clns+ | 83 | yes | yes | | HE. |
| GSN | 83 | yes | yes | | |
| IISS+ | 83 | yes | yes | | |
| Meyr | 84 | yes | | | |

### WARHEAD CHARACTERISTICS

| | | |
|---|---|---|
| IISS | 82 | single warhead |
| Meyr | 84 | single RV |
| BW | 86 | single RV |

```
------------- NUCLEAR YIELD -------------
              kT            MT        kT conv
Clns   80                                              "? KT".
AFM    82     350                     350.00
IISS   82                    < 1      1,000
JAWA   82     350                     350.00
JFS    82     350                     350.00
JWS    82     c. 200                  200.00
IISS+  83     c. 200                  200.00
Meyr   84     200-600                 200.00-600.00
BW     86     350                     350.00

------------- WARHEAD WEIGHT -------------
              lb            t         kg        kg conv
ID10   78                             1,000     1,000
Clns   80     2,200                             997.92
GUD1   80     2,205                   1,000     1,000
AFM    82     2,200                             997.92
JAWA   82     2,200                   1,000     997.92
JFS    82     2,200                             997.92
GSN    83     2,200                   1,000     997.92
CFW+   84                             1,000     1,000      [Unit inferred]

------------- WEIGHT -------------
              lb            t         kg        kg conv
SWM    77     13,228+                 6,000+    6,000
ID10   78                             6,000     6,000       All-up; 2,600 kg fuel.
Rock   78                             6,000+    6,000
CH01   79     13,228+                 6,000+    6,000       Launch.
              c. 7,496                c. 3,400  3,400       Burnout.
GUD1   80     13,230                  6,000     6,001
FI     81                             6,000+    6,000
AFM    82     13,225                            5,998
JFS    82     ? 13,000                          5,896       Launch.
GSN    83     14,330                  6,500     6,500
JFS+   83                                                   12,875 ft Launch.
CFW+   84                             6,500     6,500

------------- LENGTH -------------
              in     ft         cm    m         m conv
MOW    76            37'0"            11.3      11.28
SWM    77            37               11.3      11.28
Rock   78                             11.3      11.30
FI     81                             11-11.3   11.00-11.30
AFM    82            37'0"                      11.28
AWST   82            36                         10.97
JFS    82            37                         11.28
JWS    82                       1,130           11.30
GSN    83            37               11.3      11.28
JFS+   83            37.5                       11.43

------------- DIAMETER -------------
              in     ft         cm    m         cm conv
Rock   78                       50              50.00
CH01   79     35 2/5            90              89.92
GUD1   80            3          90              91.44
FI     81                    c. 50              50.00
JWS    82                       95              95.00

------------- WINGSPAN -------------
              in     ft         cm    m         cm conv
SWM    77            8                2.4       243.84
ID10   78                             3.00      300.00
Rock   78                       248             248.00
CH01   79            7'10 1/2"        2.4       240.03
GUD1   80            10               3         304.80
FI     81                    c. 245             245.00
AFM    82            9'10"                      299.72
AWST   82            8                          243.84
GSN    83            11               3.35      335.28

------------- NUMBER OF STAGES -------------
Rock   78     1
AWST   82     1
JFS    82     1
```

444  SOVIET MISSILES  IDDS

```
-------------- TYPE OF PROPULSION --------------
MOW   76   rocket
SWM   77   liquid
Rock  78   liquid
FI    81   liquid rocket
AFM   82   liquid rocket
AWST  82   liquid rocket
JAWA  82   liquid
JFS   82   liquid rocket
JWS   82   solid              [Analysis table]
           liquid rocket      According to US sources.
GSN   83   turbojet
BW    86   storable liquid

-------------- MIDCOURSE (OR MAIN) GUIDANCE --------------
SWM   77                      ? Inertial plus mid-course.
ID10  78   inertial
Rock  78   inertial
CH01  79                      Probably inertial, and mid-course radio command correction.
Clns  80                      Preprogrammed auto & radar.
GUD1  80   inertial
FI    81   ? inertial
AFM   82   inertial
CFW   82   inertial
JAWA  82   inertial
JFS   82                      Pre-programmed guidance.
JWS   82   inertial
Clns+ 83   programmed
GSN   83                      Inertial or autopilot.

-------------- TERMINAL GUIDANCE --------------
ID10  78   IR
CH01  79   IR
GUD1  80   IR or SAR
AFM   82   radar
CFW   82   radar
JAWA  82   radar
JFS   82   passive radar
           active radar
JWS   82   active radar
           passive radar      Version reported in 1981.
Clns+ 83   homing
GSN   83   radar or IR        J-band radar.

-------------- CONTROL SYSTEM --------------
ID10  78   Differential tail surfaces.
FI    81   Aeroplane configuration.

-------------- FIRST OBSERVED --------------
MOW   76   1961
AFM   82   1961
GSN   83   1961               On Blinder-B of strategic air arm.

-------------- OBSERVED IN PARADE --------------
MOW   76   1967               Most of the 22 Tu-22s in 1967 air show carried Kitchen.
AFM   82   1967               Most of 22 Tu-22s in 1967 display carried an AS-4.

-------------- IOC --------------
JFS   82   1965               Operational.
GSN   83   1967

-------------- DEPLOYMENT BEGAN --------------
Rock  78   1966               Service intro.
CH01  79   1967               Entered service.
Clns  80   1967               First deployed.
GUD1  80   mid-1960s          Entered service; various sources say 1962-1967.
IISS  82   1962               First year deployed.
CFW+  84   1967
Meyr  84   1962               Service.
BW    86   ? 1967

-------------- DEPLOYED --------------
MOW   76                      In service.
GUD1  80                      Operated by USSR.
JWS   82                      Deployment: USSR.
BW    86                      In service.
```

## DEPLOYMENT HISTORY

| | | |
|---|---|---|
| SecD | 80 | AS-4 equipped BACKFIRE replacing BADGER [for anti-ship missions]. |

## NUMBERS PRODUCED

| | | |
|---|---|---|
| AFM | 82 | 1,000 UK MoD 76. |
| JWS | 82 | [c 1,000] Total production "somewhat less" than "well in excess of 1000". UK MoD Mason, early 1976. |

## NUMBERS IN SERVICE

### NUMBER IN 1976

| | | | |
|---|---|---|---|
| IISS | 76 | c. 800 | 1976-80, July, total deployed. |

### NUMBER IN 1981

| | | | |
|---|---|---|---|
| IISS | 81 | c. 135 | July, total deployed. |

### NUMBER IN 1982

| | | | |
|---|---|---|---|
| IISS | 82 | c. 180 | July, total deployed. |

### NUMBER IN 1983

| | | | |
|---|---|---|---|
| IISS+ | 83 | <_ 645 | July, total deployed. |

## CARRIERS

| Source | Year | CARRIER | TUBES | RELOADS | |
|---|---|---|---|---|---|
| SecC | 68 | Blinder B | | | Will carry ASM. |
| MOW | 76 | Tu-22 | | | |
| SWM | 77 | Tu-22 Blinder B | [1] | | |
| ID10 | 78 | Tu-22 Blinder C | 1 | | [Text] |
| | | Tu-22 Blinder B | 1 | | [Table] |
| Rock | 78 | Tu-22 Blinder | | | |
| Clns | 80 | TU-22 Blinder | 1 | | |
| FI | 81 | Tu-22 Blinder B | 1 | | Since 1967. Missile semi-recessed in weapons bay. |
| SMP | 81 | Tu-22 Blinder | 1 | | "Missile-carrying variant" of Blinder. |
| USND | 81 | Blinder B | | | |
| AFM | 82 | Tu-22 Blinder B | 1 | | |
| AWST | 82 | Blinder B | | | |
| CFW | 82 | Blinder-B | | | |
| JAWA | 82 | Tu-22 | | | |
| JFS | 82 | Blinder | | | |
| JWS | 82 | Tu-22 Blinder | | | According to UK MoD Mason, early 1976. [Also:] First seen on Blinder. |
| GSN | 83 | Blinder | | | |
| IISS+ | 83 | Blinder B | 1 | | |
| JFS+ | 83 | Blinder B | | | |
| ABFS | 84 | | | | Blinder only carries bombs [in SNA]. |
| BW | 86 | Blinder B | 1 | no | |

| Source | Year | CARRIER | TUBES | RELOADS | |
|---|---|---|---|---|---|
| MOW | 76 | Backfire | | | |
| JCS | 77 | Backfire | | | "Has been identified with Backfire", FY78-79. |
| ID10 | 78 | Tu-26 Backfire | 2 | | |
| Clns | 80 | TU-26 Backfire | 2 | | |
| SecD | 80 | Backfire | 2 | | |
| FI | 81 | Tu-26 Backfire | | | As an interim weapon. |
| JCS | 81 | Backfire | 1 or 2 | | |
| USND | 81 | Backfire | | | |
| AFM | 82 | Backfire | | | |
| CFW | 82 | Backfire-B | | | |
| JAWA | 82 | Tu-22M | | | Has been seen in more than one form on Backfire-B. |
| JFS | 82 | Backfire | | | |
| JWS | 82 | Tu-26 Backfire B | | | |
| GSN | 83 | Backfire-B | [1] | | |
| IISS+ | 83 | Backfire B | 1 or 2 | | |
| JFS+ | 83 | Backfire B | | | |
| SMP | 83 | Backfire | 1 | | Carries either bombs or AS-4s.<br>tubes: [From picture]. |
| SMP | 84 | Backfire | 1 or 2 | | tubes: 1 mounted "partially in its fuselage", or 2 on wing pylons. |
| SecD | 85 | Backfire | | | n = 100 assigned to SNA, dedicated to antiship mission. |
| SMP | 85 | Backfire | | | tubes: Can carry "AS-4 missile attached to its wings". |
| BW | 86 | Backfire B | 1-2 | no | |

---------- CARRIER ---------- TUBES --- RELOADS ----------

| | | | | | |
|---|---|---|---|---|---|
| AFM | 82 | Tu-95 | | | |
| JAWA | 82 | Tu-95 | | | |
| JWS | 82 | Tu-95 Bear | | | |
| GSN | 83 | Bear-B | | | Of strategic aviation. |
| | | Bear-C | | | Of strategic aviation. |
| IISS+ | 83 | Bear B | 2 | | tubes: Carries 2 with no AS-3. |
| SMP | 84 | Bear | | | AS-4 is replacing AS-3 on older ASM-equipped Bears, [appears in Strategic Aviation section, ed]. "Several of these reconfigurations (Bear G) have been completed", '84, '85. |
| JCS | 85 | Bear | | | "Some older, long-range Bear aircraft" being reconfigured to carry AS-4. |
| SecD | 85 | Bear | | | "The Bear bomber... is being reconfigured to carry the AS-4 missile." |
| BW | 86 | Bear G | 2 | no | |

---------- DESIGN AND ENGINEERING ----------

JWS  82  One of most technically advanced of Soviet ASMs.

---------- USE AND CONFIGURATION ----------

SWM  77  "Tu-95 Bear D may be mid-course spotter and control platform for this and other long-range missiles".

GUD1 80  Normally launched at 11,000 m, dive or climb depends on mission. Climb (for long-range target) at M1.8 to 27,000 m at 150 mi from target, then descend in shallow dive reaching M2.5. For target less than 200 mi, descends for low-level cruise at M1.2.

GSN  83  Ventral fin folds to starboard while carried on launch a/c. After launch from medium altitude (c. 20,000 ft), missile climbs steeply to achieve cruise altitude and speed then dives steeply at its target.

Meyr 84  Penetration: 0.8-0.9; reliability: 0.3-0.7, "includes system reliability multiplied by operational readiness of deployed system".

---------- TARGET TYPE ----------

| | | | |
|---|---|---|---|
| GSN | 83 | ship | |
| JCS | 83 | ship | |
| SecD | 85 | ship | "Backfire bombers armed with long-range, supersonic AS-4 air-to-surface missiles are the greatest menace" to US naval forces. |

---------- STRATEGY ----------

Clns 80  [Listed with fighter/attack and strategic aircraft].

JWS  82  US DoD sources say a number of versions in use, two have been identified for strategic and naval use, nuclear and conventional warheads respectively.

GSN  83  Developed as stand-off, anti-ship missile for Bear and Blinder.

Meyr 84  [Ed: chart is labeled] Systems "assigned primary operational and strategic missions within the European theatre(s) of military operations".

JCS  85  Bear aircraft with AS-4 are "capable of threatening the continental United States."

---------- EXPORTED TO ----------

GUD1 80  Libya

---------- RELATION TO OTHER MISSILES ----------

SMP  84  Newer than AS-3, '84, '85.

---------- OTHER INFORMATION ----------

AFM  82  Missile has been seen in more than one form.

JAWA 82  Has been seen in more than one form on Backfire-B.

# AS-5

```
Sources are:   MOW   76      Rock  78      FI    81      AWST  82      JWS   82      AFM+  84
SecC   68      SWM   77      CH01  79      SMP   81      CFW   82      Clns  83      CFW+  84
SecC   69      ID10  78      Clns  80      USND  81      JAWA  82      GSN   83      BW    86
IA06   75      ID11  78      GU01  80      AFM   82      JFS   82      JFS+  83
```

---------- US DESIGNATION ------------

|  | |  |
|---|---|---|
| MOW | 76 | AS-5 |
| SWM | 77 | AS-5 |
| Rock | 78 | AS-5 |
| Clns | 80 | AS-5 |
| FI | 81 | AS-5 |
| SMP | 81 | AS-5 |
| USND | 81 | AS-5 |
| AFM | 82 | AS-5 |
| AWST | 82 | AS-5 |
| CFW | 82 | AS 5 |
| JAWA | 82 | AS-5 |
| JFS | 82 | AS-5 |
| JWS | 82 | AS-5 |
| Clns | 83 | AS-5 |
| GSN | 83 | AS-5 |

---------- NATO CODENAME ------------

| Source | | Codename |
|---|---|---|
| MOW | 76 | Kelt |
| SWM | 77 | Kelt |
| Rock | 78 | Kelt |
| Clns | 80 | Kelt |
| FI | 81 | Kelt |
| SMP | 81 | Kelt |
| USND | 81 | Kelt |
| AFM | 82 | Kelt |
| AWST | 82 | Kelt |
| CFW | 82 | Kelt |
| JAWA | 82 | Kelt |
| JFS | 82 | Kelt |
| JWS | 82 | Kelt |
| Clns | 83 | Kelt |
| GSN | 83 | Kelt |

---------- DESCRIPTION ------------

| SWM | 77 | Stand-off missile. |
|---|---|---|

---------- EDITOR'S NOTES ------------

| SecC | 68 | Listed FY69-70, but only as Badger ASM. |
|---|---|---|
| SMP | 81 | Listed '81. |

---------- RANGE ------------

| Source | | mi | nm | km | km conv | Notes |
|---|---|---|---|---|---|---|
| SecC | 69 | | 120 | | 222.24 | |
| MOW | 76 | ?> 200 | | ?> 320 | 321.80 | |
| SWM | 77 | 112+ | | 180+ | 180.21 | |
| ID10 | 78 | | | 320 | 320.00 | High profile. |
| | | | | 180 | 180.00 | Low profile. |
| Rock | 78 | | | 320 | 320.00 | |
| CH01 | 79 | 199 | | 320 | 320.19 | High profile. |
| | | 100 | | 160 | 160.90 | Low profile. |
| Clns | 80 | 150 | | | 241.35 | |
| GU01 | 80 | ? 125 | | ? 200 | 201.13 | Effective operation limit. |
| FI | 81 | | | <_ 320 | 320.00 | |
| USND | 81 | | c. 100 | | 185.20 | |
| AFM | 82 | 200 | | | 321.80 | |
| AWST | 82 | 100+ | | | 160.90 | |
| CFW | 82 | | 100 | | 185.20 | |
| JFS | 82 | | 125 | | 231.50 | Max. |
| JWS | 82 | | | <_ 180 | 180.00 | [Analysis table] |
| | | | | 160-320 | 160.00-320.00 | And above; range of Western estimates [text]. |
| | | | | 160 | 160.00 | British MoD of April-76. |
| Clns | 83 | 100 | | | 160.90 | |
| GSN | 83 | | 100+ | 185+ | 185.20 | |
| AFM+ | 84 | 100 | | | 160.90 | At low altitude. |
| | | 200 | | | 321.80 | "At height". |
| BW | 86 | | | 320 | 320.00 | 160 km, M0.9 at low altitude. |

```
------------ SPEED ------------
                mph       kmph      Mach     kmph conv
ID10   78                           0.95                   Over aircraft speed.
CH01   79                           0.95
Clns   80                           1.0
FI     81                           0.9-1.2
AFM    82                           1.2                    30,000 ft.
                                    0.9                    Low level.
JFS    82                           1.2
Clns   83                           1.3
AFM+   84                           0.9                    At low altitude.
BW     86                           1.2                    160 km, M0.9 at low altitude.

------------ WARHEAD TYPE ------------
                nuclear   conven    chem
IA06   75       yes       yes                HE.
ID10   78                 yes                HE.
Rock   78                 yes                HE; [n.b., this is the only ASM with warhead type specified; implies
                                             nonnuclear?, ed].
Clns   80       yes       yes                HE.
GUD1   80                 yes                HE.
FI     81                 yes                HE.
AFM    82                 yes
CFW    82       yes       yes
JAWA   82                 yes                HE.
JFS    82                 yes                HE.
GSN    83       yes       yes

------------ WARHEAD CHARACTERISTICS ------------
BW     86       single RV

------------ NUCLEAR YIELD ------------
                kT        MT        kT conv
Clns   83       500                 500.00
BW     86       ? 500               500.00

------------ WARHEAD WEIGHT ------------
                lb        t         kg        kg conv
ID10   78                           1,000     1,000
Clns   80       2,200                         997.92
GUD1   80       2,205               1,000     1,000
AFM    82       2,200                         997.92
JFS    82       2,200                         997.92
JWS    82                           c. 160    160.00       [Sic]
GSN    83       2,200               1,000     997.92

------------ WEIGHT ------------
                lb        t         kg        kg conv
MOW    76                                                  Underwing pylon on Tu-16 is more
                                                           massive than for Kennel,
                                                           suggesting Kelt is a heavier
                                                           weapon.
SWM    77       10,580              4,800     4,799
ID10   78                           3,500     3,500        All-up; 1,100 kg fuel.
CH01   79       10,580              4,800     4,799        Launch.
                8,157               3,700     3,700        Burnout.
GUD1   80       7,715               3,500     3,499
GSN    83       c. 10,500           c. 4,725  4,782
JFS+   83       8,000                         3,628        Launch.
AFM+   84       7,715                         3,499
CFW+   84                           4,700     4,700
```

CRUISE MISSILES ASMs AS-5

```
------------- LENGTH -------------
              in      ft          cm       m         m conv
MOW   76              c. 31'0"             9.45      9.45
SWM   77              30.8                 9.4       9.39
ID10  78                                   9.5       9.50
Rock  78                                   10        10.00
CH01  79              31                   9.45      9.45        Length.
                      28'2 2/5"            8.59      8.60        "Some reports suggest".
GUD1  80              31                   9.5       9.45
FI    81                                   9.4-10    9.40-10.00
AFM   82              31'0"                          9.45
AW ST 82              31                             9.45
JFS   82              28                             8.53
JWS   82                          859                8.59
GSN   83              31                   9.45      9.45
JFS+  83              28.2                           8.60
AFM+  84              28'2"                          8.59

------------- DIAMETER -------------
              in      ft          cm       m         cm conv
Rock  78                          100                100.00
CH01  79      c. 35               c. 90.0            88.90
GUD1  80              3           90                 91.44
FI    81                          100                100.00

------------- WINGSPAN -------------
              in      ft          cm       m         cm conv
MOW   76              c. 15'0"             c. 4.57   457.20
SWM   77              15                   4.6       457.20
ID10  78                          455                455.00
Rock  78                          490                490.00
CH01  79              c. 15                4.57      457.20      Span.
                      14'1 1/4"            4.3       429.90      "Some reports suggest".
FI    81                          455-495            455.00-495.00
AFM   82              15'0"                          457.20
AW ST 82              15                             457.20
JWS   82                          430                430.00
GSN   83              15                   4.6       457.20
AFM+  84              14'1 1/4"                      429.90

------------- NUMBER OF STAGES -------------
Rock  78      1
AW ST 82      1
JFS   82      1
JWS   82      1

------------- TYPE OF PROPULSION -------------
MOW   76      rocket            Unidentified rocket.
SWM   77      liquid
Rock  78      liquid
FI    81      liquid
AFM   82      liquid
AW ST 82      liquid
CFW   82      solid             [Final report]
JAWA  82      liquid
JFS   82      liquid
JWS   82      liquid
GSN   83      liquid
BW    86      storable liquid

------------- MIDCOURSE (OR MAIN) GUIDANCE -------------
SWM   77      autopilot
Clns  80      autopilot         Preprogrammed autopilot.
GUD1  80      autopilot
AFM   82                        Autopilot or pre-programmed path.
CFW   82                        Inertial or autopilot.
JAWA  82      autopilot         On pre-programmed flight path.
JWS   82      autopilot         "[Inertial] autopilot" for mid-course phase.
Clns  83      programmed
GSN   83                        Inertial or autopilot.
```

```
                ------ TERMINAL GUIDANCE ------
MOW    76    radar                Possibly.
SWM    77    radar                Active or passive.
Rock   78    radar
GUD1   80    radar                Active or passive.
FI     81    active radar
             passive radar        Also fitted.
AFM    82    radar                Switch between active and passive possible.
CFW    82    radar
JAWA   82    radar                Can be switched from active to passive as required.
JFS    82    active radar
             passive radar
JWS    82    active radar
             passive radar        Passive version reported.
Clns   83    homing
GSN    83    radar                J-band radar.
CFW+   84                         J-band radar.

                ------ CONTROL SYSTEM ------
MOW    76    Aerodynamic.
Rock   78    Aerodynamic.
FI     81    Aeroplane configuration.

                ------ RADAR ------
JAWA   82                         Hemispherical nose fairing probably houses larger radar [than Kennel].

                ------ FIRST OBSERVED ------
ID10   78    July 1966

                ------ IOC ------
JFS    82    1966                 Operational.
GSN    83    1965-1966

                ------ DEPLOYMENT BEGAN ------
ID10   78    1966                 Entered service.
ID11   78    1968                 Appeared on Badger.
Rock   78    1969                 Service intro.
CH01   79    1966                 Entered service.
Clns   80    1965                 First deployed.
GUD1   80    1966-67              Deployed.
Clns   83    1965                 First deployed.
CFW+   84    1965
BW     86    1966

                ------ DEPLOYED ------
MOW    76                         In service. "Published reports imply" that Kelt is a standard anti-shipping
                                  missile.
JWS    82                         Deployment: USSR.
BW     86                         In service.

                ------ DEPLOYMENT HISTORY ------
SWM    77    Supercedes Kennel.
AFM    82    Supercedes AS-1 Kennel.
AWST   82    Replaces AS-1.

                ------ NUMBERS PRODUCED ------
SWM    77    > 1,000   Well over 1,000.
AFM    82    > 1,000   Well over 1,000 by Spring 1976, UK MoD.
JWS    82    > 1,000   Total production "well in excess of 1000", UK MoD Mason, early 1976.

                ------ NUMBERS IN SERVICE ------

                ------ NUMBER IN 1976 ------
JWS    82                         Most widely deployed [of AS-4,5,6] according to UK MoD Roy Mason in early 1976.

                ------ NUMBER IN 1977 ------
SWM    77                         "Most widely deployed of all Soviet stand off missiles".

                ------ CARRIERS ------
```

|  |  | CARRIER | TUBES | RELOADS | |
|---|---|---|---|---|---|
| SecC | 68 | Badger | | | Significant portion of force being provided with ASM capability, FY69-70. |
| MOW | 76 | Tu-16 | | | Underwing pylon on Tu-16 is more massive than for Kennel, suggesting Kelt is a heavier weapon. |
| Rock | 78 | Tu-16 Badger | | | |
| Clns | 80 | TU-16 Badger | ? | 2 | |
| AFM | 82 | Tu-16 | | | tubes: "2 AS-5, AS-6". |
| AWST | 82 | Badger | | | |
| JWS | 82 | Tu-16 Badger | | | |

|  |  | CARRIER | TUBES | RELOADS | |
|---|---|---|---|---|---|
| JFS | 82 | Badger C | | | [Final report] |
| GSN | 83 | Badger-C | | | |
| BW | 86 | ? Badger C | 2 | no | |

|  |  | CARRIER | TUBES | RELOADS | |
|---|---|---|---|---|---|
| SWM | 77 | Tu-16 Badger G | 2 | | |
| FI | 81 | Tu-16 Badger G | 2 | | Badger G with AS-5 superceded Badger B with AS-1. |
| SMP | 81 | Badger G | 2 | | |
| USND | 81 | Badger G | | | |
| CFW | 82 | Badger-G | | | |
| JAWA | 82 | Tu-16 Badger-G | | | |
| JFS | 82 | Badger G | | | |
| GSN | 83 | Badger-G | | | |
| CFW+ | 84 | Badger C and G | | | |
| BW | 86 | Badger G | 2 | no | |

---------- **USE AND CONFIGURATION** ----------
GUD1  80  Seeker used for last 30-40 km of attack. Normally launched at 9,000 m; climbs for long-range mission, dives for short-range. Climb: M0.9 to 18,000 m, begins descent 100 mi from target.

---------- **TARGET TYPE** ----------
MOW   76  ship
SWM   77  ship
AWST  82  ship
JWS   82  ship            And others.
GSN   83  ship

---------- **STRATEGY** ----------
Clns  80  [Listed with fighter/attack aircraft].
JWS   82  Role thought confined to service with SNA in anti-ship and other tactical roles.

---------- **COMBAT REPORTS/EFFECTIVENESS** ----------
MOW   76  25 Kelt launched from Egyptian Tu-16s in 1973, 20 destroyed by defenses.
SWM   77  25 used in 1973 war, most destroyed by defenses.
ID10  78  In 1973 war, 5 hit targets (2 radar stations and supply depot). Successful radar attacks indicate passive head works.
AFM   82  25 used in Oct 1973, only 5 hit targets.

---------- **EXPORTED TO** ----------
JAWA  82  Egypt           [Picture shows one under Egyptian Tu-16 wing, ed]
JWS   82  Egypt           [Picture shows AS-5 on Egyptian plane, ed]

---------- **RELATION TO OTHER MISSILES** ----------
SWM   77  Resembles Styx.
AFM   82  Configuration similar to AS-1 Kennel.
JAWA  82  Externally similar to Kennel, resembles scaled-down MiG-15.
JWS   82  Superficial but pronounced similarities with Kennel and with Styx; fuselage centre-body and wings same or very similar to Kennel; nose section (and guidance) of Styx is employed.

# AS-6

| Sources are: | ID10 | 78 | Clns | 80 | USND | 81 | IISS | 82 | Clns+ | 83 | JFS+ | 83 | Meyr | 84 |
|---|---|---|---|---|---|---|---|---|---|---|---|---|---|---|
| IA06 | 75 | Rock | 78 | GU01 | 80 | AFM | 82 | JAWA | 82 | FI+ | 83 | JWS+ | 83 | BW | 86 |
| MOW | 76 | AW05 | 79 | FI | 81 | AWST | 82 | JFS | 82 | GSN | 83 | AFM+ | 84 | | |
| SWM | 77 | CH01 | 79 | IISS | 81 | CFW | 82 | JWS | 82 | IISS+ | 83 | CFW+ | 84 | | |

---------- **US DESIGNATION** ----------

| IA06 | 75 | AS-6 |
| MOW | 76 | AS-6 |
| SWM | 77 | AS-6 |
| Rock | 78 | AS-6 |
| Clns | 80 | AS-6 |
| FI | 81 | AS-6 |
| USND | 81 | AS-6 |
| AFM | 82 | AS-6 |
| AWST | 82 | AS-6 |
| CFW | 82 | AS 6 |
| IISS | 82 | AS-6 |
| JAWA | 82 | AS-6 |
| JFS | 82 | AS-6 |
| JWS | 82 | AS-6 |
| GSN | 83 | AS-6 |
| Meyr | 84 | AS-6 |

---------- **NATO CODENAME** ----------

| IA06 | 75 | Kitchen | [Sic] |
| Rock | 78 | Kingfish | |
| Clns | 80 | Kingfish | |
| FI | 81 | Kingfish | |
| USND | 81 | Kingfish | |
| AFM | 82 | Kingfish | |
| AWST | 82 | Kingfish | |
| CFW | 82 | Kingfish | |
| IISS | 82 | Kingfish | |
| JAWA | 82 | Kingfish | |
| JFS | 82 | Kingfish | |
| JWS | 82 | Kingfish | |
| GSN | 83 | Kingfish | |

---------- **DESCRIPTION** ----------

| Clns | 80 | | Listed with fighter/attack and strategic aircraft. |
| IISS | 82 | ALCM | |
| GSN | 83 | | Advanced anti-ship missile. |

---------- **EDITOR'S NOTES** ----------

Meyr 84  Footnote to air-launched missiles: "Includes delivery vehicle performance" Note to chart: "Values pertain to initial service period".

CRUISE MISSILES  ASMs  AS-6  453

## RANGE

| | | mi | nm | km | km conv | |
|---|---|---|---|---|---|---|
| MOW | 76 | 155 | | 250 | 249.40 | Low altitude. |
| | | 435-500 | | 700-800 | 699.92-804.50 | High altitude. |
| SWM | 77 | 435-497 | | 700-800 | 699.92-799.67 | |
| ID10 | 78 | | | 700 | 700.00 | High altitude. |
| | | | | 250 | 250.00 | Low altitude. |
| Rock | 78 | | | 200 | 200.00 | |
| CH01 | 79 | 497 | | 800 | 799.67 | High altitude. |
| | | 135+ | | 220+ | 217.22 | Low altitude. |
| Clns | 80 | 200 | | | 321.80 | |
| GUD1 | 80 | <_ 135 | | | 217.22 | Probably 135 or less; up to 500 mi quoted. Range estimates vary wildly, solid fuel questionable if long range. |
| FI | 81 | | | 220 | 220.00 | |
| USND | 81 | | 150-250 | | 277.80-463.00 | |
| AFM | 82 | 135 | | | 217.22 | At low altitude. |
| AW ST | 82 | 150 | | | 241.35 | |
| CFW | 82 | | > 100 | | 185.20 | |
| IISS | 82 | | | 250 | 250.00 | Max. |
| JFS | 82 | | 300 | | 555.60 | Max. |
| JWS | 82 | | | c. 200 | 200.00 | [Unmarked] |
| | | 135+ | | 220+ | 217.22 | UK MoD Mason, early 1976. |
| GSN | 83 | | 150-250 | 278-463 | 277.80-463.00 | |
| JFS+ | 83 | | 150 | | 277.80 | Max. |
| CFW+ | 84 | | 150-250 | | 277.80-463.00 | |
| Meyr | 84 | | | 300 | 300.00 | |
| BW | 86 | | | ? 700 | 700.00 | 280 km, (M1.2) at low altitude. |

## SPEED

| | | mph | kmph | Mach | kmph conv | |
|---|---|---|---|---|---|---|
| SWM | 77 | | | 2.5 | | |
| Clns | 80 | | | 3.0 | | |
| GUD1 | 80 | | | 2.5 | | Cruise. |
| | | | | <_ 3.5 | | Terminal. |
| | | | | 1.2 | | Low level mission, cruise. |
| FI | 81 | | | 3 | | |
| AFM | 82 | | | 3 | | |
| AW ST | 82 | | | 3 | | |
| CFW | 82 | | | 2.5 | | |
| JFS | 82 | | | 3 | | |
| JWS | 82 | | | 3 | | Max; US sources confirmed it is capable of supersonic flight. |
| GSN | 83 | | | 2.5-3.5 | | Estimated max speed. |
| CFW+ | 84 | | | 2.5-3.5 | | |
| BW | 86 | | | 3.0 | | 280 km, (M1.2) at low altitude. |

## CEP

| | | ft | mi | nm | m | km | m conv |
|---|---|---|---|---|---|---|---|
| Meyr | 84 | | | | | 0.2-0.5 | 200.00-500.00 |
| BW | 86 | | | ? 350 | | | 350.00 |

## WARHEAD TYPE

| | | nuclear | conven | chem | |
|---|---|---|---|---|---|
| MOW | 76 | yes | yes | | HE. |
| SWM | 77 | yes | yes | | Nuclear capability for shore-target strikes. |
| ID10 | 78 | yes | yes | | HE. |
| CH01 | 79 | yes | yes | | Thermonuclear; HE. |
| Clns | 80 | yes | yes | | HE. |
| FI | 81 | yes | | | |
| AFM | 82 | yes | yes | | |
| CFW | 82 | yes | yes | | |
| JAWA | 82 | yes | yes | | HE. |
| JFS | 82 | yes | yes | | HE. |
| JWS | 82 | yes | yes | | HE. |
| Clns+ | 83 | yes | yes | | HE. |
| GSN | 83 | yes | yes | | |
| IISS+ | 83 | yes | yes | | |
| Meyr | 84 | yes | | | |

## WARHEAD CHARACTERISTICS

| | | |
|---|---|---|
| Meyr | 84 | single RV |
| BW | 86 | single RV |

454  SOVIET MISSILES  IDDS

## NUCLEAR YIELD

| | | kT | MT | kT conv |
|---|---|---|---|---|
| Clns | 80 | | 1 | 1,000 |
| FI | 81 | 200 | | 200.00 |
| AFM | 82 | 200 | | 200.00 |
| IISS | 82 | 200 | | 200.00 |
| JAWA | 82 | 200 | | 200.00 |
| JFS | 82 | 200 | | 200.00 |
| JWS | 82 | c. 350 | | 350.00 |
| Meyr | 84 | 100–500 | | 100.00–500.00 |
| BW | 86 | ? 200 | | 200.00 |

## WARHEAD WEIGHT

| | | lb | t | kg | kg conv | |
|---|---|---|---|---|---|---|
| MOW | 76 | 1,000 | | 450 | 453.60 | "Alternative 1,000 lb (450 kg) nuclear or high-explosive". |
| ID10 | 78 | | | 1,000 | 1,000 | |
| CH01 | 79 | c. 992 | | c. 450 | 449.97 | |
| Clns | 80 | 1,100 | | | 498.96 | |
| AFM | 82 | 2,200 | | | 997.92 | |
| JAWA | 82 | 2,200 | | 1,000 | 997.92 | |
| JFS | 82 | 2,200 | | | 997.92 | |
| GSN | 83 | 1,100 | | 500 | 498.96 | |
| CFW+ | 84 | | | 500 | 500.00 | |

## WEIGHT

| | | lb | t | kg | kg conv | |
|---|---|---|---|---|---|---|
| IA06 | 75 | | | 4,500 | 4,500 | |
| MOW | 76 | c. 10,000 | | c. 4,500 | 4,536 | |
| SWM | 77 | 10,580 | | 4,800 | 4,799 | |
| ID10 | 78 | | | 4,800 | 4,800 | All-up; 1,900 kg fuel. |
| Rock | 78 | | | ? 5,000 | 5,000 | |
| CH01 | 79 | 10,580 | | 4,800 | 4,799 | Launch. |
| | | 6,393 | | 2,900 | 2,899 | Burnout. |
| GUD1 | 80 | 10,500 | | 4,800 | 4,762 | |
| FI | 81 | | | c. 5,000 | 5,000 | |
| AFM | 82 | 11,000 | | | 4,989 | |
| JFS | 82 | 11,000 | | | 4,989 | Launch weight. |
| GSN | 83 | c. 11,000 | | c. 4,950 | 4,989 | |
| JFS+ | 83 | 8,400 | | | 3,810 | Launch. |
| CFW+ | 84 | | | 4,900 | 4,900 | |

## LENGTH

| | | in | ft | cm | m | m conv |
|---|---|---|---|---|---|---|
| SWM | 77 | | 29.5 | | 9.0 | 8.99 |
| ID10 | 78 | | | | 10.0 | 10.00 |
| Rock | 78 | | | | 9.0 | 9.00 |
| CH01 | 79 | | 29'6 1/3" | | 9.0 | 9.00 |
| GUD1 | 80 | | 33 | | 10 | 10.06 |
| AFM | 82 | | 34'6" | | | 10.52 |
| JFS | 82 | | 33 | | | 10.06 |
| JWS | 82 | | | 1,050 | | 10.50 |
| FI+ | 83 | | | | c. 10.5 | 10.50 |
| GSN | 83 | | 34'7" | | 10.5 | 10.54 |
| JFS+ | 83 | | 23.5 | | | 7.16 |

## DIAMETER

| | | in | ft | cm | m | cm conv |
|---|---|---|---|---|---|---|
| SWM | 77 | 35.4 | | 90 | | 89.92 |
| Rock | 78 | | | 90 | | 90.00 |
| GUD1 | 80 | | 3 | 90 | | 91.44 |

## WINGSPAN

| | | in | ft | cm | m | cm conv | |
|---|---|---|---|---|---|---|---|
| SWM | 77 | | 10.5 | | 3.2 | 320.04 | |
| ID10 | 78 | | | | 2.90 | 290.00 | |
| Rock | 78 | | | 3.2 | | 3.20 | [Sic] |
| GUD1 | 80 | | 9'6" | | 2.9 | 289.56 | |
| AFM | 82 | | 8'2 1/2" | | | 250.19 | |
| JWS | 82 | | | 250 | | 250.00 | Max span. |
| FI+ | 83 | | | | c. 2.5 | 250.00 | [Literally cm] |
| GSN | 83 | | 8'2" | | 2.5 | 248.92 | |

## NUMBER OF STAGES

| | | |
|---|---|---|
| Rock | 78 | 1 |
| AWST | 82 | 1 |
| JFS | 82 | 1 |

CRUISE MISSILES ASMs AS-6

```
------------- TYPE OF PROPULSION -------------
MOW    76    liquid rocket
SWM    77    ? solid
ID10   78    solid
Rock   78    liquid
GUD1   80    solid
AFM    82    liquid rocket
JAWA   82    liquid
JFS    82    solid rocket
JWS    82    liquid
GSN    83    turbojet
BW     86    ? L

------------- MIDCOURSE (OR MAIN) GUIDANCE -------------
MOW    76    inertial
SWM    77    inertial         ? Mid-course, homing.
ID10   78    inertial
Rock   78    inertial
Clns   80    autopilot        Preprogrammed autopilot.
FI     81    inertial
AFM    82    inertial         Inertial midcourse guidance, plus active radar homing, giving "exceptional
                              degree of accuracy".
AWST   82    inertial
JAWA   82    inertial
JFS    82    inertial
JWS    82    inertial
Clns+  83    programmed
GSN    83    inertial

------------- TERMINAL GUIDANCE -------------
MOW    76    active radar
ID10   78                     Active or passive homing.
GUD1   80    active radar     In last 50 km.
FI     81    active radar     Or passive radiation seeker.
AFM    82    active radar
JAWA   82    active radar
JFS    82    active radar
             passive radar
JWS    82    active radar
             passive radar    Possibly this variant exists.
Clns+  83    homing
GSN    83    active radar     J-band.

------------- CONTROL SYSTEM -------------
FI+    83    Aeroplane configuration.

------------- FIRST OBSERVED -------------
ID10   78    early 1975
CHO1   79    1975             First seen.
AFM    82    Dec 1977         First seen by Japanese pilot.

------------- IOC -------------
GUD1   80    1970             Reported operational as early as 1970.
JFS    82    1970-71          Operational.
GSN    83    1970
JFS+   83    1970             Operational.

------------- DEPLOYMENT BEGAN -------------
ID10   78    1976             Entered service.
Rock   78    1970             Service intro.
CHO1   79    1976             Entered service.
Clns   80    1970             First deployed.
GUD1   80    1975-76          "Definitive version" entered service.
IISS   82    1977             First year deployed.
CPW+   84    1970
Meyr   84    1977             Service.
BW     86    ? 1977

------------- DEPLOYED -------------
JWS    82                     Deployment: USSR.
BW     86                     In service.

------------- NUMBERS IN SERVICE -------------

------------- NUMBER IN 1981 -------------
IISS   81    c. 65            1981-82, July, total deployed.
```

456  SOVIET MISSILES  IDDS

---------- **NUMBER IN 1982** ----------
IISS  82    ? 65        July, total deployed.

---------- **NUMBER IN 1983** ----------
IISS+ 83    <_ 880      July, total deployed.

---------- **CARRIERS** ----------

---------- **CARRIER** ---------- TUBES ---- RELOADS ----------
Rock  78    Tu-16 Badger
Clns  80    TU-16 Badger    [2]
                                                    tubes: "2 AS-5, AS-6".
FI    81    Tu-16
AFM   82    Tu-16            2                      Expected carrier was Backfire, but "there has been no
                                                    evidence of this".
AWST  82    modified Tu-16
JWS   82    Tu-16 Badger

---------- **CARRIER** ---------- TUBES ---- RELOADS ----------
AW05  79    Tu-16 Badger C                          Can carry 2 AS-5s under wings or 1 AS-2 centerline or 1
                                                    AS-6 on wing pylon.
USND  81    Badger C
CFW   82    Badger-C         2
JFS   82    Badger C
GSN   83    Badger-C
IISS+ 83    Badger C         2
BW    86    Badger C         2          no

---------- **CARRIER** ---------- TUBES ---- RELOADS ----------
JFS+  83    Badger D

---------- **CARRIER** ---------- TUBES ---- RELOADS ----------
ID10  78    Tu-16 Badger G   1
USND  81    Badger G
CFW   82    Badger-G         2
JAWA  82    Tu-16 Badger-G mod
GSN   83    Badger-G
IISS+ 83    Badger G         2
JFS+  83    Badger G
BW    86    Badger G         2          no

---------- **CARRIER** ---------- TUBES ---- RELOADS ----------
MOW   76    Backfire
SWM   77    Tupelov Backfire B  2                   New weapons for Backfire that fit in weapons bay are
                                                    expected.
ID10  78    Tu-26 Backfire B  2
Rock  78    Backfire
FI    81    Backfire
JWS   82    Tu-26 Backfire                          Was thought to arm Backfire; more likely it was mistaken
                                                    for AS-4.
GSN   83                                            Believed developed for use with Backfire, but up to press
                                                    time, AS-6 observed only on Badger.
AFM+  84    Backfire                                Little evidence, but SMP refers to Backfire carrying 3 300
                                                    km, Mach 3 anti-shipping ASMs; AS-6 must have this range
                                                    capability if not low profile for entire mission.

---------- **DESIGN AND ENGINEERING** ----------
ID10  78    Developed from AS-4 Kitchen.
JWS   82    Configuration somewhat like AS-4.

---------- **USE AND CONFIGURATION** ----------
GU01  80    Max range mission: launched at 11,000 m from M0.8 bomber, accelerates to M1.8 climbing to 18,000
            m. Reaches M2.5, up to M3.5 before impact.
FI    81    Usually launched at some 11,000 m altitude, then climbs to >18,000 m for mid-course cruise.
JWS   82    US report revealed one role as anti-shipping missions where "high-altitude supersonic cruise
            approach trajectory is followed by a diving attack onto the target".
GSN   83    Flight profile similar to AS-4 estimated.
JWS+  83    Reliable source has quoted release altitude of 11,000 m, followed by climb to cruise height of up
            to 18,000 m.
Meyr  84    Penetration: 0.8-0.9; reliability: 0.6-0.9, "includes system reliability multiplied by operational
            readiness of deployed system".

---------- **TARGET TYPE** ----------
SWM   77    land, ?ship
GU01  80                    Shipping, land, ? seaboard US.
JWS   82    ship            US report revealed one role as anti-shipping missions.
GSN   83    ship

---------- **STRATEGY** ----------
SWM 77 "For stand-off strike against land targets for high and low-level". "Low-altitude capability is uncertain". Soviets deny capability to hit US proper, but is possible with refueling.
JWS 82 "High performance". Deployed with SNA.
Meyr 84 [Ed: chart is labeled] Systems "assigned primary operational and strategic missions within the European theatre(s) of military operations".

---------- **OTHER INFORMATION** ----------
MOW 76 All info is provisional.

# AS-7

```
Sources are:  Rock  78      GUO1  80      AWST 82      JFS  82      JFS+  83
MOW   76      CHO1  79      FI    81      CFW  82      JWS  82      JWS+  83
SWM   77      Clns  80      AFM   82      JAWA 82      JAWA+ 83      CFW+  84
```

---------- **US DESIGNATION** ----------

| | | |
|---|---|---|
| MOW | 76 | AS-7 |
| SWM | 77 | AS-7 |
| Rock | 78 | AS-7 |
| Clns | 80 | AS-7 |
| FI | 81 | AS-7 |
| AFM | 82 | AS-7 |
| AWST | 82 | AS-7 |
| CFW | 82 | AS 7 |
| JAWA | 82 | AS-7 |
| JFS | 82 | AS-7 |
| JWS | 82 | AS-7 |

---------- **NATO CODENAME** ----------

| | | |
|---|---|---|
| MOW | 76 | Kerry |
| SWM | 77 | Kerry |
| Rock | 78 | Kerry |
| Clns | 80 | Kerry |
| FI | 81 | Kerry |
| AFM | 82 | Kerry |
| AWST | 82 | Kerry |
| JAWA | 82 | Kerry |
| JFS | 82 | Kerry |
| JWS | 82 | Kerry |

---------- **RANGE** ----------

| | | mi | nm | km | km conv | |
|---|---|---|---|---|---|---|
| SWM | 77 | 11 | | 18 | 17.70 | |
| Rock | 78 | | | 10 | 10.00 | |
| Clns | 80 | 6 | | | 9.65 | |
| GUO1 | 80 | 6 1/4 | | | 10.06 | |
| FI | 81 | | | 10 | 10.00 | |
| AFM | 82 | 7 | | | 11.26 | |
| AWST | 82 | 6 | | | 9.65 | |
| CFW | 82 | | 6 | | 11.11 | |
| JFS | 82 | | 6 | | 11.11 | Max. |
| JWS | 82 | | | c. 10 | 10.00 | |

---------- **SPEED** ----------

| | | mph | kmph | Mach | kmph conv |
|---|---|---|---|---|---|
| Clns | 80 | | | 1.0 | |
| FI | 81 | | | 0.6 | |
| AFM | 82 | | | 0.6 | |
| AWST | 82 | | | 0.6 | |
| CFW | 82 | | | 1 | |
| JFS | 82 | | | 0.6 | |

---------- **WARHEAD TYPE** ----------

| | | nuclear | conven | chem | | |
|---|---|---|---|---|---|---|
| Clns | 80 | | yes | | HE. | |
| FI | 81 | | yes | | HE. | |
| AFM | 82 | | yes | | | |
| CFW | 82 | | yes | | | |
| JAWA | 82 | | yes | | HE. | |
| JFS | 82 | | yes | | HE. | |
| JWS | 82 | | yes | | HE. | |

---------- **WARHEAD WEIGHT** ----------

| | | lb | t | kg | kg conv | |
|---|---|---|---|---|---|---|
| Clns | 80 | 200 | | | 90.72 | |
| AFM | 82 | 220 | | | 99.79 | |
| CFW | 82 | | | 100 | 100.00 | [Weight not identified, ed] |
| JAWA | 82 | 220 | | 100 | 99.79 | |
| JFS | 82 | 220 | | | 99.79 | |
| JWS | 82 | | | c. 100 | 100.00 | |
| JWS+ | 83 | | | 300-400 | 300.00-400.00 | Recent reports, more likely than older 1,200 kg value. |

---------- WEIGHT ----------
|         |    | lb       | t | kg        | kg conv |                    |
|---------|----|----------|---|-----------|---------|--------------------|
| Rock    | 78 |          |   | 1,200     | 1,200   |                    |
| CHO1    | 79 | c. 2,650 |   | c. 1,200  | 1,202   |                    |
| GUD1    | 80 | <_ 2,650 |   | <_ 1,200  | 1,202   | Probably less.     |
| FI      | 81 |          |   | c. 1,200  | 1,200   |                    |
| AFM     | 82 | 2,640    |   |           | 1,197   |                    |
| JFS     | 82 | 2,640    |   |           | 1,197   | Launch weight.     |
| JWS     | 82 |          |   | <_ 1,200  | 1,200   | Early reports.     |
|         |    |          |   | 300-400   | 300.00-400.00 | Recent reports. |

---------- LENGTH ----------
|       |    | in | ft    | cm | m | m conv |
|-------|----|----|-------|----|---|--------|
| AFM   | 82 |    | 11'6" |    |   | 3.51   |
| AWST  | 82 |    | 11.5  |    |   | 3.51   |
| JFS   | 82 |    | 11.5  |    |   | 3.51   |

---------- DIAMETER ----------
|       |    | in | ft | cm | m | cm conv |
|-------|----|----|----|----|---|---------|
| AWST  | 82 |    | 1  |    |   | 30.48   |

---------- NUMBER OF STAGES ----------
| Rock  | 78 | 1 |
| AFM   | 82 | 1 |
| AWST  | 82 | 1 |
| JAWA  | 82 | 1 |
| JFS   | 82 | 1 |

---------- TYPE OF PROPULSION ----------
| FI    | 81 | solid |
| AFM   | 82 | solid |
| AWST  | 82 | solid |
| CFW   | 82 | solid |
| JAWA  | 82 | solid |
| JFS   | 82 | solid |
| JWS   | 82 | solid |

---------- MIDCOURSE (OR MAIN) GUIDANCE ----------
| SWM   | 77 | TV command    | "Anti-radiation or TV command (?)". |
| CHO1  | 79 | radio command | Probable. |
| Clns  | 80 | command       | |
| GUD1  | 80 | command       | Visually command-guided, joystick. Semi-automatic command guidance or semi-active laser homing may appear later. |
| FI    | 81 | radio command | |
| AFM   | 82 | radio command | |
| AWST  | 82 | beam rider    | |
| JAWA  | 82 | radio command | |
| JFS   | 82 | radio command | |
| JWS   | 82 | radio command | Not confirmed. |
| JFS+  | 83 |               | Beam rider. |

---------- TERMINAL GUIDANCE ----------
| SWM   | 77 | anti-radiation | "Anti-radiation or TV command (?)". |
| CHO1  | 79 | anti-radiation | "Possibly an anti-radiation homing head". |
| GUD1  | 80 | laser          | Command guided now; semi-automatic command guidance or semi-active laser homing may appear later. |
| CFW   | 82 | pencil radar   | "Pencil-beam radar terminal homing". |

---------- DEPLOYMENT BEGAN ----------
| Clns  | 80 | 1971       | First deployed. |
| GUD1  | 80 | mid-1970s  | Deployed. |
| CFW+  | 84 | late 1970s | |

---------- DEPLOYED ----------
| JWS   | 82 | Deployment: USSR. |

---------- CARRIERS ----------

---------- CARRIER ---------- TUBES ---- RELOADS ----------
| FI    | 81 | other attack a/c |

---------- CARRIER ---------- TUBES ---- RELOADS ----------
| MOW | 76 | Su-17 |
| | | Su-20 |
| SWM | 77 | Su-7B Fitter A |
| | | Su-17 Fitter C |
| Rock | 78 | Su-7B Fitter |
| GUD1 | 80 | Su-7B Fitter A |
| | | Su-17 Fitter C,D |
| AFM | 82 | Su-17 Fitter |
| JAWA | 82 | Su-17 |
| JFS | 82 | Fitter |
| JWS | 82 | ? Fitter |

---------- CARRIER ---------- TUBES ---- RELOADS ----------
| SWM | 77 | ? Su-19 Fencer |
| GUD1 | 80 | Su-19 Fencer |
| FI | 81 | Su-24 Fencer | Possibly only an interim weapon.
| JWS | 82 | Su-24 Fencer |

---------- CARRIER ---------- TUBES ---- RELOADS ----------
| GUD1 | 80 | Yak-36 Forger |
| AFM | 82 | Yak-36 Forger |
| CFW | 82 | Forger |
| JAWA | 82 | Yak-36 |
| JFS | 82 | Forger |
| JWS | 82 | ? Forger |
| JAWA+ | 83 | Yak-36MP |

---------- CARRIER ---------- TUBES ---- RELOADS ----------
| MOW | 76 | MiG-23 |
| Clns | 80 | MIG-23 Flogger | 4 |
| GUD1 | 80 | MiG-27 Flogger D |

---------- DESIGN AND ENGINEERING ----------
GUD1  80   Most surprising aspect of AS-7 is that it wasn't developed "10 years or more" earlier.

---------- USE AND CONFIGURATION ----------
FI    81   Normally launched in 300-3,000 height band.

---------- STRATEGY ----------
MOW   76   Reported to be tactical weapon for ground-attack aircraft.
SWM   77   Tactical missile.
Clns  80   [Listed with fighter/attack aircraft].
CFW   82   Tactical weapon.
JAWA  82   Tactical ASM.

---------- RELATION TO OTHER MISSILES ----------
MOW   76   Much smaller than other Soviet ASMs.

# AS-8

```
Sources are:   SWM   77     Rock  78     CH01  79     AFM*  80     GU01  80     FI  81     JWS  82
```

## US DESIGNATION

|  | |  |
|---|---|---|
| SWM | 77 | AS-? |
| Rock | 78 | AS-8 |
| AFM* | 80 | AS-8 |
| FI | 81 | AS-8 |
| JWS | 82 | AS-8 |

## RANGE

| Source | | mi | nm | km | km conv |
|---|---|---|---|---|---|
| Rock | 78 | | | 8 | 8.00 |
| CH01 | 79 | c. 5 | | c. 8 | 8.05 |
| GU01 | 80 | 5-6 | | | 8.05-9.65 |

## SPEED

| Source | | mph | kmph | Mach | kmph conv |
|---|---|---|---|---|---|
| GU01 | 80 | | | 0.5-0.8 | |

## WARHEAD TYPE

| Source | | nuclear | conven | chem | |
|---|---|---|---|---|---|
| GU01 | 80 | | yes | | HE. |

## TYPE OF PROPULSION

|  | | | |
|---|---|---|---|
| CH01 | 79 | solid | Probably. |
| GU01 | 80 | solid | |

## MIDCOURSE (OR MAIN) GUIDANCE

|  | | |
|---|---|---|
| Rock | 78 | inertial |

## TERMINAL GUIDANCE

|  | | | |
|---|---|---|---|
| CH01 | 79 | laser-seeking | Possible. |
| GU01 | 80 | passive radiation | |

## OBSERVED

|  | | |
|---|---|---|
| JWS | 82 | 1977 |

## DEVELOPMENT UNDERWAY

|  | | |
|---|---|---|
| FI | 81 | Under development. |

## CARRIERS

### CARRIER — TUBES — RELOADS

| Source | | Carrier | | | |
|---|---|---|---|---|---|
| Rock | 78 | helicopter | | | |
| FI | 81 | helicopter | | | "So-called A-10" [final report] |

### CARRIER — TUBES — RELOADS

| Source | | Carrier | Tubes | Reloads | |
|---|---|---|---|---|---|
| SWM | 77 | Mi-24 Hind A | 4 | | |
| CH01 | 79 | Mi-24 Hind A, C | | | Primary armament. |
| GU01 | 80 | Mi-24 Hind | | | |
| FI | 81 | Mi-24 Hind | | | |
| JWS | 82 | Mi-24 | | | "Provisional designation of new missile on Mi-24 helicopters", noted in 1977. |

### CARRIER — TUBES — RELOADS

| Source | | Carrier | | | |
|---|---|---|---|---|---|
| GU01 | 80 | MiG-27 Flogger D | | | Could arm it. |

## TARGET TYPE

|  | | |
|---|---|---|
| SWM | 77 | ? tank |

## RELATION TO OTHER MISSILES

|  | | |
|---|---|---|
| SWM | 77 | ? Swatter derivative. |
| AFM* | 80 | "There is reason to believe that" this designation was "applied, mistakenly, to the missile now designated At-6 'Spiral'". |
| FI | 81 | "Fire-and-forget weapon, similar in concept to Hellfire". "Reports that the weapon entered service in 1977 suggest that 'AS-8' is a garbled report of the helicopter-mounted AT-6 Spiral". |

# AS-9

```
Sources are:  Clns   80      FI    81      AWST  82      JAWA  82      Clns+ 83      JWS+  83
              CH01   79      GUD1  80      AFM   82      CFW   82      JWS   82      JFS   83      CFW+  84
```

## US DESIGNATION

|  |  |
|---|---|
| Clns 80 | AS-9 |
| FI 81 | AS-X-9 |
| AFM 82 | AS-X-9 |
| AWST 82 | AS-9 |
| CFW 82 | AS 9 |
| JAWA 82 | AS-X-9 |
| JWS 82 | AS-9 |
| Clns+ 83 | AS-9 |
| JFS 83 | AS-9 |

## DESCRIPTION

Clns 80 — [Listed with fighter/attack aircraft].

## RANGE

| Source | mi | nm | km | km conv | |
|---|---|---|---|---|---|
| CH01 79 | <_ 48 | | <_ 90 | 77.23 | |
| Clns 80 | 50 | | | 80.45 | |
| GUD1 80 | 53-62 | | 85-100 | 85.28-99.76 | |
| FI 81 | | | 85-100 | 85.00-100.00 | [Table] |
| | | | 80-90 | 80.00-90.00 | [Text] |
| AFM 82 | 50-56 | | | 80.45-90.10 | |
| AWST 82 | 55 | | | 88.50 | |
| CFW 82 | | 60 | | 111.12 | |
| JAWA 82 | 50-56 | 43-48 | 80-90 | 80.45-90.10 | |
| JWS 82 | | | ? 80-90 | 80.00-90.00 | |
| Clns+ 83 | | 40 | | 74.08 | |
| JFS 83 | | 60 | | 111.12 | Max. |

## SPEED

| Source | mph | kmph | Mach | kmph conv |
|---|---|---|---|---|
| Clns 80 | | | | Subsonic. |
| FI 81 | | | | subsonic |
| AWST 82 | | | 0.8 | |
| CFW 82 | | | 3.0 | |
| Clns+ 83 | | | 3.0 | |
| JFS 83 | | | 3 | |

## WARHEAD TYPE

| Source | nuclear | conven | chem | |
|---|---|---|---|---|
| Clns 80 | | yes | | HE. |
| FI 81 | | yes | | HE. |
| CFW 82 | | yes | | |
| Clns+ 83 | yes | yes | | HE. |
| JFS 83 | | yes | | HE. |

## NUCLEAR YIELD

| Source | kT | MT | kT conv |
|---|---|---|---|
| Clns+ 83 | 1 | | 1.00 |

## WARHEAD WEIGHT

| Source | lb | t | kg | kg conv | |
|---|---|---|---|---|---|
| Clns 80 | 330 | | | 149.69 | |
| CFW 82 | | | 150 | 150.00 | [Weight not identified, ed] |
| JFS 83 | 330 | | | 149.69 | |

## LENGTH

| Source | in | ft | cm | m | m conv |
|---|---|---|---|---|---|
| AWST 82 | | 19.8 | | | 6.04 |

## WINGSPAN

| Source | in | ft | cm | m | cm conv |
|---|---|---|---|---|---|
| AWST 82 | | 1.6 | | | 48.77 |

## NUMBER OF STAGES

|  | |
|---|---|
| AWST 82 | 1 |

```
----------- TYPE OF PROPULSION -----------
GUO1    80    solid
FI      81    solid
AWST    82    liquid
CFW     82    turbojet
JFS     83    turbojet

----------- MIDCOURSE (OR MAIN) GUIDANCE -----------
Clns+   83    programmed

----------- TERMINAL GUIDANCE -----------
Clns    80    yes
FI      81                          Passive radiation seeker.
AFM     82    anti-radiation
AWST    82    anti-radiation
CFW     82    anti-radar            Passive homing on electromagnetic radiation.
JAWA    82    anti-radiation
JWS     82    anti-radiation        Possibly.
Clns+   83    homing
JFS     83                          Passive homing.

----------- DEVELOPMENT UNDERWAY -----------
JWS     82                          US officials claim at development stage.

----------- DEPLOYMENT BEGAN -----------
Clns    80    1975        First deployed.
CFW+    84    late 1970s

----------- CARRIERS -----------

----------- CARRIER ----------- TUBES ---- RELOADS -----------
CHO1    79    Su-19
AFM     82    Su-24 Fencer
JAWA    82    Su-24 Fencer
JWS     82    Su-19 Fencer

----------- CARRIER ----------- TUBES ---- RELOADS -----------
CFW     82    Badger
JFS     83    Badger                                In service.

----------- CARRIER ----------- TUBES ---- RELOADS -----------
CFW     82    Backfire
JFS     83    Backfire                              In service.

----------- CARRIER ----------- TUBES ---- RELOADS -----------
CFW     82    Fitter-C
              Fitter-D
JFS     83    Fitter C,D                            In service.

----------- TARGET TYPE -----------
FI      81    ? ship
JWS+    83    ? ship       Anti-radiation, though other sources attribute anti-ship role.
```

# AS-10

```
Sources are:  GUD1  80       AWST  82       JWS   82       JAWA  83       JWS+  83            CFW+  84
AFM*  80      FI    81       CFW   82       Clns+ 83       JFS   83       AFM   84
```

```
------------ US DESIGNATION ------------
AFM*  80      AS-X-10
FI    81      AS-X-10
AWST  82      AS-X-10
CFW   82      AS 10
JWS   82      AS-X-10
Clns+ 83      AS-10
JFS   83      AS-10
AFM   84      AS-10

------------ RANGE ------------
              mi            nm             km              km conv
FI    81                                   <_ 10           10.00
AWST  82      6                                             9.65
CFW   82                    6                              11.11
JWS   82                                   c. 10           10.00
Clns+ 83      4                                             6.44
JAWA  83      6.2           5.4            10               9.98      Max.
JFS   83                    65                            120.38      Max.
JWS+  83                                   <_ 40           40.00
AFM   84      6.2                                           9.98      Max.

------------ SPEED ------------
              mph           kmph           Mach            kmph conv
FI    81                                   0.8
AWST  82                                   0.8
CFW   82                                   1.0
JWS   82                                   0.8
Clns+ 83                                                   Subsonic.
JAWA  83                                   0.8             Cruising.
JFS   83                                   1+
AFM   84                                   0.8             Max.

------------ WARHEAD TYPE ------------
              nuclear       conven         chem
FI    81                    yes                            HE.
CFW   82                    yes
Clns+ 83                    yes                            HE.
JFS   83                    yes                            HE.

------------ WARHEAD WEIGHT ------------
              lb                           t               kg              kg conv
CFW   82                                                   100            100.00
Clns+ 83      100                                                          45.36
JFS   83      220                                                          99.79

------------ WEIGHT ------------
              lb                           t               kg              kg conv
JFS   83      660                                                         299.38      Launch.

------------ LENGTH ------------
              in            ft             cm              m               m conv
FI    81                                                   c. 3            3.00
AWST  82                    10                                             3.05
JWS   82                                                   c. 3            3.00
JAWA  83                    9'10"                          3.0             3.00
AFM   84                    9'10"                                          3.00

------------ DIAMETER ------------
              in            ft             cm              m               cm conv
AWST  82                    1                                             30.48

------------ NUMBER OF STAGES ------------
AWST  82      1
JFS   83      1
```

## TYPE OF PROPULSION

| | | |
|---|---|---|
| FI | 81 | solid |
| AWST | 82 | solid |
| CPW | 82 | solid |
| JWS | 82 | solid |
| JAWA | 83 | solid |
| JFS | 83 | solid |
| AFM | 84 | solid |

## MIDCOURSE (OR MAIN) GUIDANCE

| | | |
|---|---|---|
| JFS | 83 | Electro-optical, semi-active laser. |

## TERMINAL GUIDANCE

| | | | |
|---|---|---|---|
| GUD1 | 80 | television | Probably locked on aided by powerful electro-optical sensors of parent aircraft. |
| FI | 81 | semi-active laser | [Text] "Electro-optical [laser seeker]", [table] |
| AWST | 82 | semi-active laser | |
| CPW | 82 | electro-optical | |
| JWS | 82 | laser | |
| CLns+ | 83 | semi-active | |
| JAWA | 83 | | Semi-active laser homing. |
| AFM | 84 | | Semi-active laser homing. |

## DEPLOYMENT BEGAN

| | | | |
|---|---|---|---|
| CLns+ | 83 | 1976 | First deployed. |
| CPW+ | 84 | 1980 | |

## CARRIERS

### CARRIER — TUBES — RELOADS

| | | | |
|---|---|---|---|
| GUD1 | 80 | MiG-27 Flogger D | Suitable also for MiG-27 Flogger D. |
| FI | 81 | MiG-27 | |
| AWST | 82 | MiG-27 | |
| JWS | 82 | MiG-27 | |
| JAWA | 83 | MiG-27 | |
| AFM | 84 | MiG-27 | Said to be operational. |

### CARRIER — TUBES — RELOADS

| | | | |
|---|---|---|---|
| FI | 81 | Su-17 | |
| AWST | 82 | Su-17 | |
| CPW | 82 | Fitter-D | |
| JWS | 82 | Su-17 | |
| JAWA | 83 | Su-17 | |
| JFS | 83 | Fitter D | In service. |
| AFM | 84 | Su-17 | Said to be operational. |

### CARRIER — TUBES — RELOADS

| | | | |
|---|---|---|---|
| FI | 81 | Su-24 | |
| AWST | 82 | Su-19 | |
| JWS | 82 | Su-19 | |
| JAWA | 83 | Su-24 | |
| JWS+ | 83 | ? Su-24 Fencer | |
| AFM | 84 | Su-24 | Said to be operational. |

### CARRIER — TUBES — RELOADS

| | | |
|---|---|---|
| GUD1 | 80 | Mi-24 |

### CARRIER — TUBES — RELOADS

| | | | |
|---|---|---|---|
| JFS | 83 | Forger | In service. |

## STRATEGY

| | | |
|---|---|---|
| JWS+ | 83 | Different sources suggest anti-radiation role. |

## RELATION TO OTHER MISSILES

| | | |
|---|---|---|
| AFM* | 80 | "There is reason to believe that" this designation was "applied, mistakenly, to the missile now designated AT-6 'Spiral'". |
| GUD1 | 80 | Complement to AS-8. |

# AS-11

Sources are:  FI 81    AWST 82    JWS 82    JFS 83    JWS+ 83

### US DESIGNATION
| | | |
|---|---|---|
| FI | 81 | AS-X-? |
| AWST | 82 | AS-11 |
| JFS | 83 | AS-11 |

### RANGE
| | | mi | nm | km | km conv | |
|---|---|---|---|---|---|---|
| FI | 81 | | | c. 800 | 800.00 | |
| AWST | 82 | 500 | | | 804.50 | |
| JWS | 82 | | | <_ 800 | 800.00 | |
| JFS | 83 | | 500 | | 926.00 | Max. |

### SPEED
| | | mph | kmph | Mach | kmph conv |
|---|---|---|---|---|---|
| FI | 81 | | | 3.5 | |
| AWST | 82 | | | 3.5 | |
| JWS | 82 | | | 3.5 | |
| JFS | 83 | | | 3.5 | |

### LENGTH
| | | in | ft | cm | m | m conv |
|---|---|---|---|---|---|---|
| AWST | 82 | | 16.5 | | | 5.03 |
| JFS | 83 | | 16.5 | | | 5.03 |

### DIAMETER
| | | in | ft | cm | m | cm conv |
|---|---|---|---|---|---|---|
| AWST | 82 | | 1 | | | 30.48 |

### TYPE OF PROPULSION
| | | |
|---|---|---|
| AWST | 82 | solid |
| JFS | 83 | solid |

### DEPLOYMENT HISTORY
| | | |
|---|---|---|
| FI | 81 | Being developed to replace AS-6 Kingfish. |
| AWST | 82 | Replaces AS-4. |
| JWS | 82 | Possible replacement for AS-4 Kitchen. |
| JFS | 83 | Replacement for AS-4 Kitchen. |

### CARRIERS

**CARRIER** --- **TUBES** --- **RELOADS**

| | | | | |
|---|---|---|---|---|
| JWS+ | 83 | Su-24 Fencer | | West German sources. |

### STRATEGY
| | | |
|---|---|---|
| JWS+ | 83 | Anti-radar, West German sources. |

# AS-12

Sources are: JWS 83

---------- **US DESIGNATION** ----------
JWS 83   AS-12

---------- **CARRIERS** ----------

---------- **CARRIER** ---------- **TUBES** --- **RELOADS** ----------
JWS 83   Su-24 Fencer                              West German sources.

---------- **STRATEGY** ----------
JWS 83   Anti-radar, West German sources.

# AS-15

```
Sources are:  AWST  82    JWS   82    ABFS  84    AWST+ 84    JCS   85    SMP   85
FI    81      JAWA  82    JAWA+ 83    AFM   84    SMP   84    SecD  85    BW    86
```

---------- **US DESIGNATION** ----------

| | | |
|---|---|---|
| FI | 81 | AS-X-? |
| AWST | 82 | ALCM |
| AFM | 84 | AS-X-15 |
| AWST+ | 84 | AS-X-15 |
| SMP | 84 | AS-X-15 |
| JCS | 85 | AS-15 |
| SecD | 85 | AS-15 |
| SMP | 85 | AS-15 |

---------- **DESCRIPTION** ----------

| | | | |
|---|---|---|---|
| FI | 81 | | A new air-launched cruise missile. |
| AFM | 84 | | Air-launched cruise missile. |
| SMP | 84 | ALCM | [Also] cruise missile, '84, '85. |

---------- **EDITOR'S NOTES** ----------

| | | |
|---|---|---|
| SMP | 84 | Listed '84, '85. |
| JCS | 85 | Listed FY86. |
| SecD | 85 | Listed FY86. |

---------- **MINIMUM ALTITUDE** ----------

| | | mi | ft | km | m | m_conv | |
|---|---|---|---|---|---|---|---|
| SMP | 85 | | | | | | "Low-altitude." |

---------- **RANGE** ----------

| | | mi | nm | km | km_conv | |
|---|---|---|---|---|---|---|
| FI | 81 | | | 1,200 | 1,200 | |
| AWST | 82 | | 646 | | 1,196 | |
| JAWA | 82 | 750 | 650 | 1,200 | 1,206 | Ranges of tests. |
| | | 1,500 | 1,300 | 2,400 | 2,413 | Max range could be this high. |
| JWS | 82 | | | ≤ 1,200 | 1,200 | Max. |
| ABFS | 84 | | > 1,500 | | 2,778 | |
| AFM | 84 | c. 1,850 | | | 2,976 | |
| AWST+ | 84 | 1,500 | | | 2,413 | |
| SMP | 84 | | | 3,000 | 3,000 | '84, '85. |
| JCS | 85 | | | | | Long-range. |
| BW | 86 | | | 3,000 | 3,000 | |

---------- **CEP** ----------

| | | ft | mi | nm | m | km | m_conv | |
|---|---|---|---|---|---|---|---|---|
| AFM | 84 | 150 | | | | | 45.72 | Possible with TERCOM-type guidance system. |
| SMP | 84 | | | | | | | "Capable of threatening hardened targets"; "could eventually be accurate enough to permit the use of conventional warheads", '84, '85. |

---------- **WARHEAD TYPE** ----------

| | | nuclear | conven | chem | |
|---|---|---|---|---|---|
| AWST | 82 | yes | | | [Final report] |
| SMP | 84 | yes | maybe | | Initially to be fitted with nuclear warheads; could eventually be accurate enough to permit use of conventional warheads, depending on guidance and future munitions developments, '84, '85. |

---------- **WARHEAD CHARACTERISTICS** ----------

| | | | |
|---|---|---|---|
| SMP | 84 | 1 warhead | '84, '85. |
| BW | 86 | single RV | |

---------- **LENGTH** ----------

| | | in | ft | cm | m | m_conv | |
|---|---|---|---|---|---|---|---|
| SMP | 84 | | | | 7 | 7.00 | [From graph], '84, '85. |

---------- **TYPE OF PROPULSION** ----------

| | | |
|---|---|---|
| FI | 81 | air-breathing |
| AWST | 82 | turbojet |
| JWS | 82 | turbojet |
| BW | 86 | TJ |

## DEVELOPMENT UNDERWAY

| | | | |
|---|---|---|---|
| FI | 81 | | Under development. |
| SMP | 84 | | In development. |

## PROTOTYPE TESTS

| | | | |
|---|---|---|---|
| JAWA | 82 | | "US administration officials are reported to have said, on 1 February 1979" that USSR had begun to test missile "in the category of the USAF's ALCM". Officials claimed at least 8 test-launched from Backfires in preceding months, over ranges of c. 650 nm. Said to be first suggestion USSR testing long-range cruise missiles from bombers. |
| JAWA+ | 83 | | Soviet testing of weapon in category of US ALCM confirmed by Premier Andropov. |
| JCS | 85 | | Currently in flight testing and production phases, [described with SS-NX-21.] |

## PRODUCTION UNDERWAY

| | | | |
|---|---|---|---|
| JCS | 85 | | Currently in flight testing and production phases, [described with SS-NX-21.] |

## IOC

| | | | |
|---|---|---|---|
| AFFS | 84 | late 1980s | Expected operational. |
| SMP | 84 | | "Could reach initial operational status this year on...Bear H". |
| JCS | 85 | | Now operational. |
| SecD | 85 | 1984 | Introduction, [from graph]. |
| SMP | 85 | 1984 | With Bear H. |

## DEPLOYMENT BEGAN

| | | | |
|---|---|---|---|
| AFM | 84 | | Predicted by mid-1980s. |
| SMP | 84 | | "Nearly deployed". |
| BW | 86 | 1984 | |

## CARRIERS

### CARRIER — TUBES — RELOADS

| | | | | | |
|---|---|---|---|---|---|
| FI | 81 | bombers | | | |
| AWST | 82 | bombers | | | |
| AWST+ | 84 | | | | "New supercruise missile carrying aircraft". |

### CARRIER — TUBES — RELOADS

| | | | | | |
|---|---|---|---|---|---|
| AFM | 84 | Backfire | | | |
| AWST+ | 84 | Backfire | | | |

### CARRIER — TUBES — RELOADS

| | | | | | |
|---|---|---|---|---|---|
| AFM | 84 | Blackjack | | | |
| AWST+ | 84 | Blackjack | | | |
| SMP | 84 | Blackjack | | | When operational, could carry AS-X-15. |
| JCS | 85 | Blackjack | | | "Both bombers [Bear H and Blackjack] will carry AS-15." |
| SMP | 85 | Blackjack | | | Could be deployed on Blackjack when operational, [possibly by 1988]. |
| BW | 86 | ? Blackjack | | | |

### CARRIER — TUBES — RELOADS

| | | | | | |
|---|---|---|---|---|---|
| AFM | 84 | ? Bear | | | New version. |
| SMP | 84 | Bear H | | | "Assessed to be the initial carrier". |
| JCS | 85 | Bear H | | | "Both bombers [Bear H and Blackjack] will carry AS-15." |
| SecD | 85 | Bear H | | | "modified Bear H bombers" |
| SMP | 85 | Bear H | 4 | | "Newly built Bear H bombers have become the launch platform". tubes: From illustration, four on under-wing pylons. |
| BW | 86 | Bear H | 4 | | |

## USE AND CONFIGURATION

| | | | |
|---|---|---|---|
| FI | 81 | | To arm bombers of Soviet Long-Range Aviation. "Not yet clear if it is in the performance class of the US ALCM". |
| JAWA | 82 | | US officials said unlikely to have high accuracy of US designs or ability to approach targets at tree-top heights. |
| JWS | 82 | | "Roughly equivalent" to US ALCM. |

## TARGET TYPE

| | | | |
|---|---|---|---|
| SMP | 84 | | "Land-attack". |
| JCS | 85 | | Land-attack. |

## STRATEGY

| | | | |
|---|---|---|---|
| AFM | 84 | | Will provide Soviets with greatly improved capabilities for low-level and stand-off attack in both theatre and intercontinental operations. Later generation ALCMs with longer range expected to follow to arm Blackjack. All weapons believed to use guidance system similar to US TERCOM. |

## RELATION TO OTHER MISSILES

| | | | |
|---|---|---|---|
| SMP | 85 | | Similar in design to US Tomahawk. Two variants being developed, SS-NX-21 and SSC-X-4. |

# TASM

Sources are: GUO1 80   FI 81   AWST 82   JWS 82

### US DESIGNATION
| | | |
|---|---|---|
| GUO1 | 80 | AS-11 |
| FI | 81 | AS-X-? |
| JWS | 82 | AS-X-11 |

### OTHER DESIGNATION
| | | |
|---|---|---|
| GUO1 | 80 | Also known as "Advanced TASM". |
| FI | 81 | "Advanced TASM". |
| AWST | 82 | "Advanced ASM". |

### EDITOR'S NOTES
| | | |
|---|---|---|
| AWST | 82 | Final report. |

### RANGE

| | | mi | nm | km | km conv | |
|---|---|---|---|---|---|---|
| GUO1 | 80 | 25 | | 40 | 40.23 | |
| FI | 81 | | | 40 | 40.00 | |
| AWST | 82 | 25 | | | 40.23 | |
| JWS | 82 | | | c. 40 | 40.00 | Max. |

### SPEED

| | | mph | kmph | Mach | kmph conv | |
|---|---|---|---|---|---|---|
| GUO1 | 80 | | | | | Subsonic. |
| FI | 81 | | | | | high subsonic |

### WARHEAD TYPE

| | | nuclear | conven | chem | |
|---|---|---|---|---|---|
| GUO1 | 80 | | yes | | HE. |
| FI | 81 | | yes | | HE. |

### TYPE OF PROPULSION
| | | |
|---|---|---|
| FI | 81 | solid |

### MIDCOURSE (OR MAIN) GUIDANCE
| | | |
|---|---|---|
| FI | 81 | Command or inertial. |

### TERMINAL GUIDANCE
| | | | |
|---|---|---|---|
| GUO1 | 80 | | Television guidance with data link. |
| FI | 81 | electro-optical | |
| AWST | 82 | | EO command link homing. |
| JWS | 82 | | EO guidance with command link. |

### CARRIERS

### CARRIER — TUBES — RELOADS
| | | | | |
|---|---|---|---|---|
| GUO1 | 80 | Su-19 Fencer | | Assumed. |

### RELATION TO OTHER MISSILES
| | | |
|---|---|---|
| JWS | 82 | "Possible Soviet equivalent to US Maverick". |

# SS-N-1

```
Sources are:  AJ01  70      FI*   74      SWM   77      Clns  80      AWST  82      ABFS  84
MR04  66      US01  70      MOW   76      ID11  78      SMP   81      CFW   82      BW    86
MR05  66      SecC  71      JWS*  77      Rock  78      USND  81      GSN   83
```

### US DESIGNATION

| | | |
|---|---|---|
| SecC | 71 | SS-N-1 |
| FI* | 74 | SS-N-1 |
| MOW | 76 | SS-N-1 |
| JWS* | 77 | SS-N-1 |
| SWM | 77 | SS-N-1 |
| Rock | 78 | SS-N-1 |
| Clns | 80 | SS-N-1 |
| SMP | 81 | SS-N-1 |
| USND | 81 | SS-N-1 |
| AWST | 82 | SS-N-1 |
| CFW | 82 | SS-N-1 |
| GSN | 83 | SS-N-1 |

### NATO CODENAME

| | | |
|---|---|---|
| FI* | 74 | Scrubber |
| MOW | 76 | Scrubber |
| JWS* | 77 | Scrubber |
| SWM | 77 | Scrubber |
| Rock | 78 | Scrubber |
| USND | 81 | Scrubber |
| AWST | 82 | Scrubber |
| CFW | 82 | Scrubber |
| GSN | 83 | Scrubber |

### OTHER DESIGNATION

| | | | |
|---|---|---|---|
| FI* | 74 | Strela | |
| JWS* | 77 | Strela | Also associated with SS-N-1, but not NATO name. |
| SWM | 77 | Strela | This is not NATO name. |
| GSN | 83 | Strela | "American intelligence agencies tended to use the name Strela". |

### DESCRIPTION

| | | |
|---|---|---|
| MOW | 76 | Cruise missile. |

### EDITOR'S NOTES

| | | |
|---|---|---|
| SecC | 71 | Listed FY72. |
| SMP | 81 | Mentioned '81. |
| CFW | 82 | Final report of SS-N-1. |

### RANGE

| | | mi | nm | km | km conv | |
|---|---|---|---|---|---|---|
| MR05 | 66 | | | 100-180 | 100.00-180.00 | |
| AJ01 | 70 | 30 | | | 48.27 | [Normal] |
| | | 130 | | | 209.17 | With forward-observing aircraft. |
| US01 | 70 | 60-110 | | | 96.54-176.99 | |
| SecC | 71 | | 45 | | 83.34 | |
| FI* | 74 | | | 110-210 | 110.00-210.00 | |
| MOW | 76 | c. 115 | | c. 185 | 185.04 | Max. |
| JWS* | 77 | | 100 | 185 | 185.20 | Max; practical operating range is probably much less. |
| SWM | 77 | 149 | | 240 | 239.74 | [Text] |
| | | 68-115 | | 110-185 | 109.41-185.04 | [Table] |
| ID11 | 78 | | | 230 | 230.00 | |
| Rock | 78 | | | 240 | 240.00 | |
| Clns | 80 | | 130 | | 240.76 | |
| USND | 81 | | c. 100 | | 185.20 | |
| AWST | 82 | 13.8-150 | | | 22.20-241.35 | |
| CFW | 82 | | 25 | | 46.30 | Surface targets. |
| | | | 120-130 | | 222.24-240.76 | Land targets. |
| GSN | 83 | | c. 100 | c. 185 | 185.20 | Max. |
| BW | 86 | | | 85 | 85.00 | |

### SPEED

| | | mph | kmph | Mach | kmph conv | |
|---|---|---|---|---|---|---|
| MR05 | 66 | | | 1.2 | | |
| FI* | 74 | | | ? 0.95 | | |
| JWS* | 77 | | | | | Subsonic. |
| SWM | 77 | | | 0.9 | | |
| Rock | 78 | | | 0.9 | | |
| CFW | 82 | | | | | Subsonic. |
| BW | 86 | | | 0.9 | | |

```
------------- WARHEAD TYPE -------------
              nuclear   conven   chem
MR05  66      yes                         Capable of carrying nuclear warhead.
FI*   74                yes               HE.
MOW   76                yes               HE.
JWS*  77                yes               HE.
SWM   77                yes
Rock  78                yes               HE.
Clns  80      yes       yes               HE.
CFW   82      yes       yes

------------- WARHEAD CHARACTERISTICS -------------
BW    86      single RV

------------- NUCLEAR YIELD -------------
              kT               MT         kT conv
Clns  80                                                 KT range, "KT".
ABFS  84                                                 Low kT.
BW    86      ? 10                        10.00

------------- WARHEAD WEIGHT -------------
              lb               t          kg        kg conv
Clns  80      < 1,000                               453.60

------------- WEIGHT -------------
              lb               t          kg        kg conv
MR05  66                       c. 6                 6,000              Tons.
FI*   74                       c. 6,500             6,500
SWM   77      9,000                       4,080     4,082
Rock  78                                  ? 6,000   6,000

------------- LENGTH -------------
              in               ft         cm        m         m conv
MR05  66                                            9         9.00
FI*   74                                            6.8-7.6   6.80-7.60
MOW   76                       c. 25                c. 7.6    7.62
JWS*  77                                            7.6       7.60
SWM   77                       23.3-24.9            6.8-7.6   7.10-7.59
Rock  78                                            ? 7       7.00

------------- DIAMETER -------------
              in               ft         cm        m         cm conv
FI*   74                                  100-150             100.00-150.00
JWS*  77                                  100                 100.00
SWM   77      39.4-59                     100-150             100.08-149.86
Rock  78                                  ? 130               130.00

------------- WINGSPAN -------------
              in               ft         cm        m         cm conv
FI*   74                                  350-460             350.00-460.00
JWS*  77                                            4.6       460.00
SWM   77      138-181                     350-460             350.52-459.74
Rock  78                                  ? 400               400.00

------------- NUMBER OF STAGES -------------
Rock  78      "1 + 1"

------------- TYPE OF PROPULSION -------------
FI*   74      solid                       Booster.
              ? turbojet
MOW   76      solid                       Booster.
              ? ramjet                    Sustainer.
JWS*  77      solid                       For launch and acceleration.
              ? ramjet
SWM   77      solid                       Booster. Launched with help of undertail JATO.
              ? turbojet
Rock  78      solid
              turbojet
CFW   82      liquid, turbojet
BW    86      TJ

------------- MIDCOURSE (OR MAIN) GUIDANCE -------------
FI*   74      radio command
MOW   76      radio command
SWM   77      autopilot/radio
Rock  78      radio
CFW   82      radio                       Radio-directed for initial trajectory.
```

---------- **TERMINAL GUIDANCE** ----------

| | | | |
|---|---|---|---|
| FI* | 74 | IR | |
| MOW | 76 | IR | |
| JWS* | 77 | ? IR | Others disagree. alternative homing heads over history likely. |
| SWM | 77 | radar or IR | [Table] |
| | | IR | [Text] |
| Rock | 78 | IR | |
| CFW | 82 | active radar | "Active automatic radar". |

---------- **CONTROL SYSTEM** ----------

| | | |
|---|---|---|
| Rock | 78 | Aerodynamic. |

---------- **PRODUCTION BEGAN** ----------

| | | | |
|---|---|---|---|
| ID11 | 78 | 1959 | Soviets produced their first missiles [SS-N-1 specifically?, ed]. |

---------- **PRODUCTION ENDED** ----------

| | | |
|---|---|---|
| CFW | 82 | No longer produced. |

---------- **IOC** ----------

| | | | |
|---|---|---|---|
| MR05 | 66 | 1957-58 | Operational. |

---------- **DEPLOYMENT BEGAN** ----------

| | | | |
|---|---|---|---|
| USD1 | 70 | 1961 | Krupny [n.b.] introduced. |
| MOW | 76 | c. 1958-59 | Entered service. |
| JWS* | 77 | | Earliest known Soviet weapon of its kind. Service entry 1958-59. |
| SWM | 77 | 1958-59 | |
| Rock | 78 | 1958 | Service intro. |
| Clns | 80 | 1958 | First deployed. |
| SMP | 81 | late 1950s | Installed on Kildin and Krupnyy classes. |
| GSN | 83 | 1959 | SS-N-1 was the first SSM deployed by Soviet navy. |
| BW | 86 | 1959 | |

---------- **DEPLOYED** ----------

| | | |
|---|---|---|
| FI* | 74 | Operated by USSR. |
| MOW | 76 | In service. |
| BW | 86 | In service. |

---------- **RETIRED** ----------

| | | |
|---|---|---|
| JWS* | 77 | By 1975 only a single Krupny remained - presumed withdrawn by now. |
| GSN | 83 | By late 1960s, Scrubber-armed ships being discarded or converted. |

---------- **DEPLOYMENT HISTORY** ----------

| | | |
|---|---|---|
| SWM | 77 | Being superceded by SS-N-11. |

---------- **CARRIERS** ----------

---------- **CARRIER** ---------- TUBES ---- RELOADS ----------

| | | | | | |
|---|---|---|---|---|---|
| SecC | 71 | | | | Part of armament of 20 "Guided Missile Combatants". |
| Rock | 78 | cruisers | | | |
| AWST | 82 | | | | Destroyer weapon. |

---------- **CARRIER** ---------- TUBES ---- RELOADS ----------

| | | | | | |
|---|---|---|---|---|---|
| MR04 | 66 | Krupnyi | 2 | 14 total | |
| FI* | 74 | Krupny | 2 | | N=4. |
| MOW | 76 | Krupny | 2 | | |
| JWS* | 77 | Krupny | | | |
| SWM | 77 | Krupny | 2 | | |
| SMP | 81 | Krupnyy | | | |
| USND | 81 | Krupnyy | | | |
| GSN | 83 | Krupnyy | 2 | | 8 ships, starting in 1960. |
| BW | 86 | Krupnyy | 2 | yes | N=? 8. |

---------- **CARRIER** ---------- TUBES ---- RELOADS ----------

| | | | | | |
|---|---|---|---|---|---|
| MR04 | 66 | Kildin | 1 | 7 total | Kildin is modified Kotlin with SS-N-1; perhaps testing missile for Krupnyi class. |
| FI* | 74 | Kildin | 1 | | N=4. |
| MOW | 76 | Kildin | 1 | | N=4. |
| JWS* | 77 | Kildin | | | |
| SWM | 77 | Kildin | 1 | | |
| SMP | 81 | Kildin | | | |
| USND | 81 | Kilden | | | [Sic] |
| CFW | 82 | Neulovimyy | | | May still be fitted in the remaining Kildin-class Neulovimyy. |
| GSN | 83 | Kildin | 1 | | 4 ships, starting in 1959. |
| BW | 86 | Kildin | 1 | yes | N=4. |

---------- **USE AND CONFIGURATION** ----------
FI*   74   Stored in hangar with 17m attached launch rail. Assembly can be trained through 200 degrees and elevated.
MOW   76   Against long-range targets, Krupny's helicopter controls the missile, as ship's radar cannot be used.
JWS*  77   Launch ramp is about 17m long. Mount can be trained over 200 degrees, and elevated; mount has hangar at one end. Deckhouse to rear of launcher presumably for reloads.

---------- **TARGET TYPE** ----------
SecC  71   ship

---------- **COMBAT REPORTS/EFFECTIVENESS** ----------
GSN   83   "Had several limitations and was awkward to handle aboard ship".

---------- **RELATION TO OTHER MISSILES** ----------
FI*   74   Appears to be longer-range, larger version of Styx.
MOW   76   US Naval Institute suggests it is based on German V-1.

---------- **OTHER INFORMATION** ----------
MOW   76   Little known about it.

# SS-N-2

```
Sources are:  MR05  66     US01  70     Clns  80     CFW   82     Clns+ 83     JFS+  83     SMP   84
              IA01  65     Ley   68     US02  72     FI    81     JFS   82     FI+   83     SMP   83
              MR04  66     AF01  70     MOW   76     USND  81     JWS   82     GSN   83     CFW+  84
```

---------- **US DESIGNATION** ----------

| | | |
|---|---|---|
| MOW | 76 | SS-N-2 |
| Clns | 80 | SS-N-2 |
| FI | 81 | SS-N-2 |
| USND | 81 | SS-N-2 |
| CFW | 82 | SS-N-2A and B |
| JFS | 82 | SS-N-2A and B |
| JWS | 82 | SS-N-2 |
| GSN | 83 | SS-N-2 |
| SMP | 83 | SS-N-2         '83, '84. |

---------- **NATO CODENAME** ----------

| | | |
|---|---|---|
| MOW | 76 | Styx |
| FI | 81 | Styx |
| USND | 81 | Styx |
| CFW | 82 | Styx |
| JFS | 82 | Styx |
| JWS | 82 | Styx |
| GSN | 83 | Styx |
| SMP | 83 | Styx |

---------- **OTHER DESIGNATION** ----------

Ley   68        "Patrol-boat launched", [source provides no other designation, ed].

---------- **DESCRIPTION** ----------

| | | |
|---|---|---|
| MOW | 76 | Cruise missile. |
| USND | 81 | SSM. |
| JWS | 82 | "Short/medium range weapon". |
| SMP | 83 | Naval cruise missile. |

---------- **EDITOR'S NOTES** ----------

| | | |
|---|---|---|
| FI | 81 | Final report for specifications in table; text entry remains. |
| SMP | 83 | Mentioned '83, '84. |

---------- **RANGE** ----------

| | | mi | nm | km | km conv | |
|---|---|---|---|---|---|---|
| IA01 | 65 | 40 | | | 64.36 | |
| MR04 | 66 | | | 20-30 | 20.00-30.00 | |
| Ley | 68 | 40 | | | 64.36 | |
| AF01 | 70 | 22 | | | 35.40 | |
| US01 | 70 | c. 12 | | | 19.31 | |
| US02 | 72 | | 12-24 | | 22.22-44.45 | Varies for different versions. |
| MOW | 76 | 26 | | 42 | 41.83 | Max. |
| Clns | 80 | | 30 | | 55.56 | |
| FI | 81 | | | $\leq$ 42 | 42.00 | |
| USND | 81 | | c. 25 | | 46.30 | |
| CFW | 82 | | 25 | | 46.30 | Max. |
| | | | 16 | | 29.63 | Practical range. |
| JFS | 82 | | 25 | | 46.30 | Max. |
| JWS | 82 | | 20 | 40 | 37.04 | Max, for early models. |

---------- **SPEED** ----------

| | | mph | kmph | Mach | kmph conv | |
|---|---|---|---|---|---|---|
| US02 | 72 | | | | | Subsonic; "some versions may be supersonic";. |
| | | | | 1 | | Later versions capable of reaching Mach 1. |
| FI | 81 | | | 0.9 | | |
| JFS | 82 | | | 0.9 | | |
| JWS | 82 | | | c. 0.9 | | |
| GSN | 83 | | | 0.9 | | |

---------- **WARHEAD TYPE** ----------

| | | nuclear | conven | chem | |
|---|---|---|---|---|---|
| MOW | 76 | | yes | | HE. |
| Clns | 80 | | yes | | HE. |
| FI | 81 | | yes | | HE, "Linear or polygon charge". |
| CFW | 82 | | yes | | |
| JFS | 82 | | yes | | HE. |
| JWS | 82 | | yes | | HE. |

## SOVIET MISSILES  IDDS

| | | ─── WARHEAD WEIGHT ─── | | | | |
|---|---|---|---|---|---|---|
| | | lb | t | kg | kg conv | |
| CLns | 80 | 1,000 | | | 453.60 | |
| FI | 81 | | | 360-400 | 360.00-400.00 | |
| CFW | 82 | | | 400-450 | 400.00-450.00 | |
| JFS | 82 | 900 | | | 408.24 | |
| GSN | 83 | 1,100 | | 500 | 498.96 | |
| CFW+ | 84 | | | 500 | 500.00 | |

| | | ─── WEIGHT ─── | | | | |
|---|---|---|---|---|---|---|
| | | lb | t | kg | kg conv | |
| IA01 | 65 | | 3.9 | | 3,900 | Tons. |
| MR05 | 66 | | 1.2 | | 1,200 | Tons. |
| Ley | 68 | | 3.9 | | 3,900 | Tons. |
| USD2 | 72 | 3,000 | | | 1,360 | |
| FI | 81 | | | 2500-3000 | 2,500-3,000 | |
| JFS | 82 | 5,100 | | | 2,313 | Launch weight. |
| JWS | 82 | | | 2,300 | 2,300 | |

| | | ─── LENGTH ─── | | | | | |
|---|---|---|---|---|---|---|---|
| | | in | ft | cm | m | m conv | |
| IA01 | 65 | | 24.6 | | | 7.50 | |
| MR05 | 66 | | | | c. 8 | 8.00 | |
| Ley | 68 | | 24.6 | | | 7.50 | |
| MOW | 76 | | 21'4" | | 6.5 | 6.50 | |
| FI | 81 | | | | 6.25-6.5 | 6.25-6.50 | |
| JFS | 82 | | 21 | | | 6.40 | |
| JWS | 82 | | | 625 | | 6.25 | Est. |
| JFS+ | 83 | | 15 | | | 4.57 | |

| | | ─── DIAMETER ─── | | | | | |
|---|---|---|---|---|---|---|---|
| | | in | ft | cm | m | cm conv | |
| MR05 | 66 | | | | c. 1 | 100.00 | |
| Ley | 68 | | 2.7 | | | 82.30 | |
| MOW | 76 | 2'3" | | 0.70 | | 5.72 | Max. |
| FI | 81 | | | 75 | | 75.00 | |
| JWS | 82 | | | 75 | | 75.00 | |

| | | ─── WINGSPAN ─── | | | | | |
|---|---|---|---|---|---|---|---|
| | | in | ft | cm | m | cm conv | |
| IA01 | 65 | | 8.2 | | | 249.94 | |
| MOW | 76 | 8'10" | | | 2.70 | 22.44 | |
| FI | 81 | | | 275 | | 275.00 | |
| JWS | 82 | | | 275 | | 275.00 | Max. |
| GSN | 83 | | 9'2" | | 2.8 | 279.40 | |

| | | ─── NUMBER OF STAGES ─── |
|---|---|---|
| JFS | 82 | 2 |

| | | ─── TYPE OF PROPULSION ─── | |
|---|---|---|---|
| MOW | 76 | | Nozzle for sustainer in tail. |
| | | solid | Booster, jettisonable. |
| FI | 81 | solid | Boost. |
| | | turbojet | Cruise. |
| CFW | 82 | liquid | |
| JFS | 82 | solid | Booster. |
| | | ? liquid | Sustainer, "probably liquid bi-propellant sustainer". |
| JWS | 82 | | Internal cruise motor. |
| | | solid | Jettisonable booster rocket used for launch and acceleration. |
| GSN | 83 | turbojet | |
| | | solid | Booster. |
| CFW+ | 84 | solid | Booster. |

| | | ─── MIDCOURSE (OR MAIN) GUIDANCE ─── | |
|---|---|---|---|
| USD2 | 72 | inertial | "Inertial nav.". |
| MOW | 76 | | Probably autopilot/radio command. |
| FI | 81 | | Autopilot or radio command. |
| JWS | 82 | ? radio command | Probably. Autopilot or radio command; probably different guidance combinations used over missile's life. |

| | | ─── TERMINAL GUIDANCE ─── | |
|---|---|---|---|
| MOW | 76 | radar | Reported. |
| CFW | 82 | active radar | "I-band active radar guidance in targeting". |
| JFS | 82 | active radar | |
| | | IR | |
| JWS | 82 | IR/radar | IR or active radar; continuation of command guidance also possible. |

|     |    | ────── CONTROL SYSTEM ────── |
|-----|----|------|
| MOW | 76 | Ailerons and tail control surfaces. |
| FI  | 81 | Ailerons and rudder. |

────── RADAR ──────

| GSN | 83 | Square Tie | Used to detect targets. |
|-----|----|------------|--------------------------|

────── PRODUCTION UNDERWAY ──────

| SMP | 84 | In production. |
|-----|----|----------------|

────── IOC ──────

| MR05 | 66 | 1959-60 | Became operational. |
|------|----|---------|---------------------|

────── DEPLOYMENT BEGAN ──────

| Clns | 80 | 1960 | First deployed. |
|------|----|------|-----------------|
| CFW+ | 84 | "1958/1964" | |

────── DEPLOYED ──────

| MOW | 76 | in service | |
|-----|----|------------|---|
| JWS | 82 | | Deployment: USSR. |

────── DEPLOYMENT HISTORY ──────

JWS 82  Despite age of system, new Styx installations are apparently still being made. Most widespread fittings on Osa and Komar boats.

────── NUMBERS PRODUCED ──────

JWS 82  Total of more than 150 Osa boats transferred to foreign navies.

────── UNDATED NUMBER ──────

GSN 83  Including 3rd world sales, most widely used anti-ship missile prior to Exocet becoming operational in 1973.

────── CARRIERS ──────

────── CARRIER ────── TUBES ──── RELOADS ──────

JWS 82  [Salish entry] shows picture of missile "apparently an SS-N-2" in unlikely-looking land-based launch configuration. Raises question of whether Styx used in land-based role.

────── CARRIER ────── TUBES ──── RELOADS ──────

| MR04 | 66 | Komar | 2 x 1 | 2 total | |
|------|----|-------|-------|---------|---|
| MOW  | 76 | Komar | 2 | | 130 Osa and Komar ships, total. |
| Clns | 80 | Komar | 2 | | |
| FI   | 81 | Komar (Mosquito) | 2 | | |
| USND | 81 | Komar | | | |
| JWS  | 82 | Komar | | | |

────── CARRIER ────── TUBES ──── RELOADS ──────

| MR04 | 66 | Osa | 4 x 1 | 4 total | |
|------|----|-----|-------|---------|---|
| MOW  | 76 | Osa | 4 | | 130 Osa and Komar ships, total. |
| Clns | 80 | Osa | 4 | | |
| FI   | 81 | Osa I (Wasp) | 4 | | N=120. |
| USND | 81 | Osa I | | | [Listed twice, ed] |
| CFW  | 82 | Osa-I | | | |
| JWS  | 82 | Osa | | | |

────── CARRIER ────── TUBES ──── RELOADS ──────

| CFW | 82 | Osa-II |
|-----|----|--------|

────── CARRIER ────── TUBES ──── RELOADS ──────

| Clns | 80 | Kashin | 4 |
|------|----|--------|---|
| GSN  | 83 | Kashin | |

────── CARRIER ────── TUBES ──── RELOADS ──────

| GSN | 83 | Kildin |
|-----|----|--------|

────── CARRIER ────── TUBES ──── RELOADS ──────

| Clns+ | 83 | Matka | 2 |
|-------|----|-------|---|
| GSN   | 83 | Matka | |

────── CARRIER ────── TUBES ──── RELOADS ──────

| GSN | 83 | Terantul |
|-----|----|----------|

────── CARRIER ────── TUBES ──── RELOADS ──────

| USND | 81 | Nanuchka II | | |
|------|----|-------------|---|---|
| GSN  | 83 | Nanuchka | | Those for foreign transfer. |

|      |    | ——— CARRIER ——— TUBES —— RELOADS ——— |
|------|----|---|
| JWS  | 82 | Luda — Chinese; first installation on larger boat. |

——— USE AND CONFIGURATION ———
| USD2 | 72 | Not a wave skimmer. |
|------|----|---|
| CFW  | 82 | "Altitude can be preset at 100, 150, 200, 250, or 300 m". |
| JFS  | 82 | "Altitude can be preset up to 300 m". |

——— STRATEGY ———
| GSN | 83 | Developed to provide anti-ship capability for small boats in coastal defense role. |
|-----|----|---|

——— COMBAT REPORTS/EFFECTIVENESS ———
| MOW | 76 | Proven effective in sinking Israeli Eilat, at range of 15 miles. |
|-----|----|---|
| JWS | 82 | Styx first weapon of its type used in combat, in 1967 and later by Indian ships against Pakistan in December 1971. |
| GSN | 83 | Egyptian Komar boats sank Israeli Elath steaming 12 miles offshore on 25 Oct 1967. Subsequently used by Indian Navy to sink a number Pakistani ships in 1971 conflict. Used by Egypt and Syria in 1973 without effect. |

## CRUISE MISSILES  Sea-Launched SSMs  SS-N-2

|  |  | EXPORTED TO |  |
|---|---|---|---|
| USO2 | 72 | Poland | Probably. |
|  |  | Indonesia | Probably. |
|  |  | Egypt | Probably. |
| MOW | 76 | Algeria |  |
|  |  | Bulgaria |  |
|  |  | China |  |
|  |  | Cuba |  |
|  |  | Egypt |  |
|  |  | East Germany |  |
|  |  | India |  |
|  |  | Indonesia |  |
|  |  | Iraq |  |
|  |  | North Korea |  |
|  |  | Poland |  |
|  |  | Romania |  |
|  |  | Syria |  |
|  |  | Yugoslavia |  |
| FI | 81 | Algeria | 14 Osa. |
|  |  | Bulgaria | 3 Osa. |
|  |  | China | 17 Osa, 6 Komar, and unlicensed copies [final report] |
|  |  | Cuba | 1 Osa, 18 Komar. |
|  |  | E Germany | 12 Osa. |
|  |  | Egypt | 12 Osa, 1 Komar, war losses not replaced. |
|  |  | Finland | 4 Osa, plus Isku. |
|  |  | India | "70 + frigates Talwar and Trishul", Osa launchers replace forward gun turret. |
|  |  | Indonesia | 12 Komar, probably not operational [final report] |
|  |  | Iraq | 10 Osa. |
|  |  | Jugoslavia | 10 Osa, 10 locally designed Type 211 missile boats, under construction. |
|  |  | Libya | 24 Osa. |
|  |  | ? Morocco | ? 6 Osa [final report] |
|  |  | North Korea | 8 Osa, 10 Komar, and new missile boats. |
|  |  | Poland | 12 Osa. |
|  |  | Romania | 5 Osa. |
|  |  | Somalia | 3 [Osa?, ed] |
|  |  | South Yemen | ? 2 Osa. |
|  |  | Sri Lanka | Up to 5 Osa. |
|  |  | Syria | 6 Osa, 6 Komar. |
|  |  | Vietnam | 2 Komar. |
| JWS | 82 | Algeria |  |
|  |  | Bulgaria |  |
|  |  | China | China has own production lines; also has a number of Styx missiles deployed as coastal defence weapons. |
|  |  | Cuba |  |
|  |  | Egypt |  |
|  |  | Germany (DR) |  |
|  |  | Ethiopia |  |
|  |  | Finland |  |
|  |  | India |  |
|  |  | Iraq |  |
|  |  | Indonesia |  |
|  |  | N Korea |  |
|  |  | Libya |  |
|  |  | Poland |  |
|  |  | Romania |  |
|  |  | Somalia |  |
|  |  | Syria |  |
|  |  | South Yemen |  |
|  |  | Vietnam |  |
|  |  | Yugoslavia |  |
| FI+ | 83 | Algeria | 3 Osa I, 9 Osa II, 6 Komar. |
|  |  | Bulgaria | 3 Osa I, 1 Osa II. |
|  |  | Cuba | 5 Osa I, 13 Osa II, 10 Komar. |
|  |  | East Germany | 15 Osa I. |
|  |  | Egypt | 8 Osa I, 4 Komar, 1 Skory. |
|  |  | Ethiopia | 3 Osa II. |
|  |  | India | 8 Osa I, 8 Osa II, 2 Kashin, 2 Leader, 2 Type 12, and (on order) 6 Godavari. |
|  |  | Iraq | 4 Osa I, 8 Osa II. |
|  |  | Libya | 12 Osa II. |
|  |  | North Korea | 8 Osa I, 10 Komar. |
|  |  | Poland | 13 Osa I. |
|  |  | Somalia | 2 Osa II. |
|  |  | Vietnam | 8 Osa II; 3 Komar (? no longer operational). |
|  |  | South Yemen | 6 Osa II. |
|  |  | Syria | 6 Osa I, 6 Osa II, 6 Komar. |
|  |  | Jugoslavia | 10 Osa I, 6 Koncar. |
| SMP | 83 | Cuba |  |

---------- **OTHER INFORMATION** ----------
JFS 82  2A has fixed wings, 2B folding wings.
JWS 82  3 versions, A, B, C. Main difference between A and B is that A has fixed wings; B's wings fold; reflected in housing. SS-N-2B [sic] was known as SS-N-11.

# SS-N-2α

```
Sources are:  SWM  77     Rock 78      FI  81      AWST 82      JFS  82      JWS  82      GSN  83
```

## US DESIGNATION

| | | |
|---|---|---|
| SWM | 77 | SS-N-2A |
| Rock | 78 | SS-N-2A |
| AWST | 82 | SS-N-2A |
| JFS | 82 | SS-N-2A |
| JWS | 82 | SS-N-2a |
| GSN | 83 | SS-N-2a |

## NATO CODENAME

| | | |
|---|---|---|
| SWM | 77 | Styx |
| Rock | 78 | Styx |
| AWST | 82 | Styx |

## RANGE

| | | mi | nm | km | km conv |
|---|---|---|---|---|---|
| SWM | 77 | 5.7–25 | | 9–40 | 9.17–40.23 |
| Rock | 78 | | | 42 | 42.00 |
| AWST | 82 | 5.7–26 | | | 9.17–41.83 |
| GSN | 83 | | c. 25 | c. 46 | 46.30 |

## SPEED

| | | mph | kmph | Mach | kmph conv |
|---|---|---|---|---|---|
| SWM | 77 | | | 0.9 | |
| Rock | 78 | | | 0.9 | |

## WARHEAD TYPE

| | | nuclear | conven | chem | |
|---|---|---|---|---|---|
| Rock | 78 | | yes | | HE. |
| GSN | 83 | | yes | | |

## WARHEAD WEIGHT

| | | lb | t | kg | kg conv | |
|---|---|---|---|---|---|---|
| SWM | 77 | 794 | | 360 | 360.16 | |
| Rock | 78 | | | 400 | 400.00 | |
| JWS | 82 | | | 400 | 400.00 | SS-N-2A and B. |

## WEIGHT

| | | lb | t | kg | kg conv |
|---|---|---|---|---|---|
| SWM | 77 | 5,500 | | 2,945 | 2,494 |
| Rock | 78 | | | ? 3,000 | 3,000 |
| GSN | 83 | c. 5,500 | | c. 2,500 | 2,494 |

## LENGTH

| | | in | ft | cm | m | m conv |
|---|---|---|---|---|---|---|
| SWM | 77 | | 20.5–21.3 | | 6.25–6.50 | 6.25–6.49 |
| Rock | 78 | | | | ? 6.5 | 6.50 |
| AWST | 82 | | 15 | | | 4.57 |
| GSN | 83 | | 19 | | 5.8 | 5.79 |

## DIAMETER

| | | in | ft | cm | m | cm conv | |
|---|---|---|---|---|---|---|---|
| SWM | 77 | 29.5 | | 75 | | 74.93 | |
| Rock | 78 | | | 75 | | 75.00 | |
| AWST | 82 | | 2 | | | 60.96 | |
| GSN | 83 | 29.5 | | 75 | | 74.93 | [750 mm in source] |

## WINGSPAN

| | | in | ft | cm | m | cm conv |
|---|---|---|---|---|---|---|
| SWM | 77 | 108 | | 27 | | 274.32 |
| Rock | 78 | | | 275 | | 275.00 |

## NUMBER OF STAGES

| | | | |
|---|---|---|---|
| Rock | 78 | "1 + 1" | |
| AWST | 82 | 2 | 1 booster, 1 sustainer. |

## TYPE OF PROPULSION

| | | | |
|---|---|---|---|
| SWM | 77 | rocket | |
| | | solid | Boost [table]; launch assisted by undertail JATO, [text]. |
| Rock | 78 | solid | |
| | | turbojet | |
| AWST | 82 | solid | |

## SOVIET MISSILES

---------- **MIDCOURSE (OR MAIN) GUIDANCE** ----------
SWM 77 "Autopilot, radio command".
Rock 78 radio

---------- **TERMINAL GUIDANCE** ----------
SWM 77 active radar
                                    Capable of homing on target's defensive radar.
Rock 78 IR
FI 81 active radar   Some sources distinguish SS-N-2A and -2B, latter having additional IR terminal homing.
GSN 83 active radar   Terminal radar seeker automatically switched on some 5 mi from estimated target position in SS-N-2a/b. Missile will home on largest target.

---------- **CONTROL SYSTEM** ----------
SWM 77 Ailerons and tail control surfaces.
Rock 78 Aerodynamic.

---------- **IOC** ----------
GSN 83 1958

---------- **DEPLOYMENT BEGAN** ----------
SWM 77 1960
Rock 78 1960       Service intro.
JFS 82 1959       In service.
JWS 82 1959 or 1960   Service entry.

---------- **CARRIERS** ----------

---------- **CARRIER** ---------- **TUBES** ---- **RELOADS** ----------
Rock 78 patrol boats                     "Mounted on fast patrol boats", SS-N-2A; "on patrol boats", SS-N-2B.

---------- **CARRIER** ---------- **TUBES** ---- **RELOADS** ----------
SWM 77 Komar           2
AWST 82 Komar
GSN 83 Komar                      First deployment of SS-N-2a.

---------- **CARRIER** ---------- **TUBES** ---- **RELOADS** ----------
SWM 77 Osa             4
AWST 82 Osa
GSN 83 Osa I

---------- **USE AND CONFIGURATION** ----------
GSN 83 Soviet doctrine appears to call for launching SS-N-2a/b at range of 10-13 nm (18-24 km), or about half of max range. No data link after launch in SS-N-2a/b.

---------- **COMBAT REPORTS/EFFECTIVENESS** ----------
SWM 77 Sank Israeli Eilat on 21 Oct 1967.

# SS-N-2b

```
Sources are:   Rock  78      AWST  82      JFS  82      GSN  83
SWM    77      FI    81      CFW   82      JWS  82
```

### US DESIGNATION

|  | |  |
|---|---|---|
| SWM | 77 | SS-N-2B |
| Rock | 78 | SS-N-2B |
| AWST | 82 | SS-N-2B |
| CFW | 82 | SS-N-2B |
| JFS | 82 | SS-N-2B |
| JWS | 82 | SS-N-2b |
| GSN | 83 | SS-N-2b |

### NATO CODENAME

|  | |  |
|---|---|---|
| SWM | 77 | Styx |
| Rock | 78 | Styx |
| AWST | 82 | Styx |

### RANGE

| Source | | mi | nm | km | km conv |
|---|---|---|---|---|---|
| SWM | 77 | 5.2-25 | | 8.5-40 | 8.37-40.23 |
| Rock | 78 | | | 8-40 | 8.00-40.00 |
| AWST | 82 | 5.2-25 | | | 8.37-40.23 |
| GSN | 83 | | c. 27 | c. 50 | 50.00 |

### WARHEAD TYPE

| Source | | nuclear | conven | chem | |
|---|---|---|---|---|---|
| Rock | 78 | | yes | | HE. |
| GSN | 83 | | yes | | |

### WARHEAD WEIGHT

| Source | | lb | t | kg | kg conv | |
|---|---|---|---|---|---|---|
| Rock | 78 | | | 400 | 400.00 | |
| JWS | 82 | | | 400 | 400.00 | SS-N-2A and B. |

### WEIGHT

| Source | | lb | t | kg | kg conv |
|---|---|---|---|---|---|
| GSN | 83 | c. 5,500 | | c. 2,500 | 2,494 |

### LENGTH

| Source | | in | ft | cm | m | m conv |
|---|---|---|---|---|---|---|
| AWST | 82 | | 15 | | | 4.57 |
| GSN | 83 | | 19 | | 5.8 | 5.79 |

### DIAMETER

| Source | | in | ft | cm | m | cm conv | |
|---|---|---|---|---|---|---|---|
| AWST | 82 | | 2 | | | 60.96 | |
| GSN | 83 | 29.5 | | 75 | | 74.93 | [750 mm in source] |

### NUMBER OF STAGES

|  | | | |
|---|---|---|---|
| AWST | 82 | 2 | 1 booster, 1 sustainer. |

### TYPE OF PROPULSION

|  | | |
|---|---|---|
| AWST | 82 | solid |

### MIDCOURSE (OR MAIN) GUIDANCE

|  | | | |
|---|---|---|---|
| SWM | 77 | | Autopilot/radio command. |
| Rock | 78 | radio | |
| AWST | 82 | autopilot | |

### TERMINAL GUIDANCE

|  | | | |
|---|---|---|---|
| SWM | 77 | IR | Has additional IR homing [over SS-N-2a] |
| Rock | 78 | IR | |
| FI | 81 | active radar | |
| | | IR | Some sources distinguish SS-N-2A and -2B, latter having additional IR terminal homing. |
| AWST | 82 | active radar | |
| CFW | 82 | IR or radar | In most recent version, SS-N-2B. |
| GSN | 83 | active radar | Terminal radar seeker automatically switched on some 5 mi from estimated target position in SS-N-2a/b. Missile will home on largest target. |

### CONTROL SYSTEM

|  | | |
|---|---|---|
| Rock | 78 | Aerodynamic. |

### IOC

|  | | |
|---|---|---|
| GSN | 83 | 1964 |

------- **DEPLOYMENT BEGAN** -------
| | | | |
|---|---|---|---|
| Rock | 78 | 1961 | Service intro. |
| JFS | 82 | 1965 | In service. |
| JWS | 82 | 1965 | Appeared. |

------- **CARRIERS** -------

------- **CARRIER** ------- TUBES ---- RELOADS -------
| | | | |
|---|---|---|---|
| Rock | 78 | patrol boats | "Mounted on fast patrol boats", SS-N-2A; "on patrol boats", SS-N-2B. |

------- **CARRIER** ------- TUBES ---- RELOADS -------
| | | |
|---|---|---|
| SWM | 77 | Komar |
| AWST | 82 | Komar |

------- **CARRIER** ------- TUBES ---- RELOADS -------
| | | |
|---|---|---|
| SWM | 77 | Osa |
| AWST | 82 | Osa |

------- **CARRIER** ------- TUBES ---- RELOADS -------
| | | | |
|---|---|---|---|
| JWS | 82 | Osa II | |
| GSN | 83 | Osa II | SS-N-2b, later SS-N-2c. |

------- **USE AND CONFIGURATION** -------
GSN 83  Soviet doctrine appears to call for launching SS-N-2a/b at range of 10-13 nm (18-24 km), or about half of max range. No data link after launch in SS-N-2a/b.

------- **OTHER INFORMATION** -------
CFW 82  SS-N-2B has folding wings.

# SS-N-2c

```
Sources are:   JWS*  77      Rock  78     USND  81     JFS   82     JFS+  83
SecC  71       SWM   77      JWS*  80     AWST  82     JWS   82     CFW+  84
MOW   76       USD3  77      FI    81     CFW   82     GSN   83
```

## US DESIGNATION

|  | |  |  |
|---|---|---|---|
| SecC | 71 | SS-N-11 | |
| MOW | 76 | SS-N-11 | |
| JWS* | 77 | SS-N-11 | |
| SWM | 77 | SS-N-11 | |
| Rock | 78 | SS-N-11 | |
| JWS* | 80 | SS-N-11 | Should be termed SS-N-2 (Mod), to reflect relationship to that missile. |
| FI | 81 | SS-N-11 | |
| USND | 81 | SS-N-2 (Imp) | |
| | | SS-N-11 | "Cancelled designator" [no other information under this heading, ed] |
| AWST | 82 | SS-N-11 | |
| CFW | 82 | SS-N-2C | |
| | | SS-N-11 | Formerly. |
| JFS | 82 | SS-N-2C | |
| JWS | 82 | SS-N-2c | |
| GSN | 83 | SS-N-2c | |
| | | SS-N-11 | Used briefly to indicate SS-N-2c. |
| JFS+ | 83 | SS-N-11 | Originally. |

## NATO CODENAME

|  | |  |  |
|---|---|---|---|
| USND | 81 | Styx | "STYX (Improved)". |

## DESCRIPTION

|  | | |
|---|---|---|
| USND | 81 | SSM |

## EDITOR'S NOTES

|  | |  |
|---|---|---|
| SecC | 71 | Listed FY72. |
| FI | 81 | Final report under name SS-N-11 for text section. |

## RANGE

| Source | | mi | nm | km | km conv | Notes |
|---|---|---|---|---|---|---|
| MOW | 76 | c. 43.5 | | c. 70 | 69.99 | |
| JWS* | 77 | | | | | ? Similar to Styx. |
| SWM | 77 | 31 | | 50 | 49.88 | |
| Rock | 78 | | | 50 | 50.00 | |
| FI | 81 | | | c. 50 | 50.00 | |
| USND | 81 | | c. 40 | | 74.08 | |
| AWST | 82 | 30 | | | 48.27 | |
| CFW | 82 | | 45 | | 83.34 | Max. In order to employ fully the over-the-horizon max range, necessary to have forward observer. |
| JFS | 82 | | 45 | | 83.34 | Max Third party control is required to utilize full range. |
| JWS | 82 | | | c. 80 | 80.00 | |
| GSN | 83 | | c. 40-45 | c. 74-83 | 74.08-83.34 | |

## SPEED

| Source | | mph | kmph | Mach | kmph conv | Notes |
|---|---|---|---|---|---|---|
| MOW | 76 | | | | | High subsonic. |
| JWS* | 77 | | | | | ? Similar to Styx. |
| SWM | 77 | | | 0.9 | | |
| Rock | 78 | | | 0.9 | | |
| FI | 81 | | | 0.9 | | |
| AWST | 82 | | | 0.9 | | |
| JFS | 82 | | | 0.9 | | |

## WARHEAD TYPE

| Source | | nuclear | conven | chem | | |
|---|---|---|---|---|---|---|
| Rock | 78 | | yes | | HE. | |
| FI | 81 | | yes | | HE. | |
| JFS | 82 | | yes | | HE. | |

## WARHEAD WEIGHT

| Source | | lb | t | kg | kg conv |
|---|---|---|---|---|---|
| FI | 81 | | | c. 500 | 500.00 |
| JFS | 82 | 1,000 | | | 453.60 |
| JWS | 82 | | | 450 | 450.00 |

## WEIGHT

| | | lb | t | kg | kg conv | |
|---|---|---|---|---|---|---|
| JFS | 82 | 5,100 | | | 2,313 | Launch weight. |

## LENGTH

| | | in | ft | cm | m | m conv | |
|---|---|---|---|---|---|---|---|
| MOW | 76 | | c. 22 | | c. 6.7 | 6.71 | |
| SWM | 77 | | 21 | | 6.4 | 6.40 | |
| Rock | 78 | | | | 6.4 | 6.40 | |
| FI | 81 | | | | c. 6.4 | 6.40 | |
| JFS | 82 | | 21 | | | 6.40 | |
| GSN | 83 | | 21'6" | | 6.55 | 6.55 | |
| CFW+ | 84 | | | | 6.5 | 6.50 | |

## DIAMETER

| | | in | ft | cm | m | cm conv | |
|---|---|---|---|---|---|---|---|
| GSN | 83 | 31 | | 78.8 | | 78.74 | [788 mm in source] |

## NUMBER OF STAGES

| Rock | 78 | "1 + 1" |
| JFS | 82 | 2 |

## TYPE OF PROPULSION

| SWM | 77 | rocket | |
| | | solid | Boost. |
| Rock | 78 | solid | |
| | | turbojet | |
| FI | 81 | | ? As SS-N-2, [solid boost, turbojet cruise]. |
| AWST | 82 | solid | |
| CFW | 82 | liquid | |
| JFS | 82 | solid | Booster. |
| | | ? liquid | Sustainer, "probably liquid bi-propellant sustainer". |

## MIDCOURSE (OR MAIN) GUIDANCE

| Rock | 78 | radio | |
| FI | 81 | | ? As SS-N-2, [autopilot or radio command]. |
| GSN | 83 | inertial | In addition to active radar [SS-N-2 generally], IR homing and "additional inertial" in SS-N-2c. |

## TERMINAL GUIDANCE

| MOW | 76 | radar + ? IR | Osa 2 believed to carry more sophisticated Styx, with IR homing in addition to radar. This may be SS-N-11 [from SS-N-2 entry, ed]. |
| JWS* | 77 | | ? More advanced than Styx. |
| SWM | 77 | ?active radar | |
| Rock | 78 | IR | |
| FI | 81 | | ? As SS-N-2, [active radar or IR]. |
| CFW | 82 | radar and IR | "Combined radar and infrared terminal homing" [final report] |
| JFS | 82 | active radar | |
| | | IR | |
| GSN | 83 | active radar | |
| | | IR | |
| CFW+ | 84 | radar or IR | Separate versions. |

## CONTROL SYSTEM

| Rock | 78 | Aerodynamic. |
| FI | 81 | ? As SS-N-2 [ailerons and rudder]. |

## IOC

| GSN | 83 | 1967 |

## DEPLOYMENT BEGAN

| SWM | 77 | 1968 | |
| Rock | 78 | 1968 | Service intro. |
| CFW+ | 84 | 1967 | |

## DEPLOYED

| MOW | 76 | In service. |

## DEPLOYMENT HISTORY

| Rock | 78 | "Replacement for SS-N-2 on fast patrol boats". |
| FI | 81 | Developed from and replaces SS-N-2 Styx. Succeeds SS-N-1 on destroyers. |

## NUMBERS IN SERVICE

## NUMBER IN 1977

| JWS* | 77 | >_ 244 | Total of at least 244 launchers in service [4 too low based on carrier information, ed]. |

## CARRIERS

|  |  | CARRIER | TUBES | RELOADS |  |
|---|---|---|---|---|---|
| SecC | 71 |  |  |  | Part of armament of 130 "Small Cruise Missile Patrol Boats". |
| MOW | 76 | other ships |  |  |  |
| Rock | 78 | patrol boats |  |  |  |

|  |  | CARRIER | TUBES | RELOADS |  |
|---|---|---|---|---|---|
| MOW | 76 | Osa 2 |  |  | Osa 2 believed to carry more sophisticated Styx, with IR homing in addition to radar. This may be SS-N-11 [from SS-N-2 entry, ed]. |
|  |  | Osa-2 |  |  | SS-N-11 is designation given by one source to missile in new launcher on Osa-2. |
| JWS* | 77 | Osa-II | 4 |  | N=55. |
| SWM | 77 | Osa II |  |  | Replacing Styx. |
| JWS* | 80 | Osa II | 4 |  |  |
| FI | 81 | Osa II | 4 x 1 |  | N=55. |
| USND | 81 | Osa II |  |  |  |
| GSN | 83 | Osa II |  |  | SS-N-2b, later SS-N-2c. |

|  |  | CARRIER | TUBES | RELOADS |  |
|---|---|---|---|---|---|
| AWST | 82 | Osa-3 |  |  |  |

|  |  | CARRIER | TUBES | RELOADS |  |
|---|---|---|---|---|---|
| JWS* | 77 | mod Kashin | 4 |  | N>=2. |
| SWM | 77 | Kashin |  |  | Replacing Scrubber. |
| USO3 | 77 | mod Kashin | 4 |  | Missiles rearward facing. |
| FI | 81 | Mod Kashin | 4 x 1 |  | N>=2. |
| USND | 81 | Mod Kashin |  |  |  |
| CFW | 82 | Modified Kashin |  |  |  |

|  |  | CARRIER | TUBES | RELOADS |  |
|---|---|---|---|---|---|
| JWS* | 77 | mod Kildin | 4 |  | N>=5, replaces Scrubber. |
| SWM | 77 | Kildin |  |  | Replacing Scrubber. |
| USO3 | 77 | mod Kildin | 4 |  | Missiles rearward facing. |
| USND | 81 | Mod Kilden |  |  | [Sic] |
| CFW | 82 | Modified Kildin |  |  |  |

|  |  | CARRIER | TUBES | RELOADS |  |
|---|---|---|---|---|---|
| FI | 81 | Matka | 2 x 1 |  |  |
| USND | 81 | Matka |  |  |  |

|  |  | CARRIER | TUBES | RELOADS |  |
|---|---|---|---|---|---|
| USND | 81 | Tarantul |  |  |  |
| CFW | 82 | Tarantul |  |  |  |

|  |  | CARRIER | TUBES | RELOADS |  |
|---|---|---|---|---|---|
| JWS* | 77 | Nanuchka (India) | 4 |  |  |
| CFW | 82 | Nanuchka II |  |  | Export model. |

|  |  | USE AND CONFIGURATION |
|---|---|---|
| CFW | 82 | Altitude can be preset at 100, 150, 200, 250, or 300 m, [final report]. During final approach, missile descends to an altitude of 2 to 5 m, [final report]. In order to employ fully the over-the-horizon max range, necessary to have forward observer. |
| JFS | 82 | Much updated version of Styx which completes its run as a sea-skimmer. Altitude can be preset up to 300 m. Third party control is required to utilize full range. |
| JWS | 82 | SS-N-2C has sea-skimming terminal trajectory; "can be preset to a limit of 300 m"; aircraft may be employed for guidance at extended ranges. |

|  |  | TARGET TYPE |
|---|---|---|
| SecC | 71 | ship |

|  |  | EXPORTED TO |  |
|---|---|---|---|
| FI | 81 | Iraq | Deployed, 4 Osa II. |
|  |  | India | Deployed, 6-8 Vijadurg (Nanuchka); also for 3 locally designed frigates. |
|  |  | Finland | 5 Osa II on order. |

|  |  | RELATION TO OTHER MISSILES |
|---|---|---|
| MOW | 76 | Uncertain whether it is new missile or version of Styx. |
| JWS* | 77 | New container/launchers. Missiles believed to be an advanced version of Styx. |

# SS-N-3

| Sources are: | SecC | 69 | IISS | 73 | SecD | 76 | Clns | 80 | CFW | 82 | GSN | 83 | SMP | 85 |
|---|---|---|---|---|---|---|---|---|---|---|---|---|---|---|
| MR04 | 66 | AJ01 | 70 | IISS | 74 | IISS | 77 | IISS | 80 | IISS | 82 | IISS+ | 83 | | |
| MR05 | 66 | US01 | 70 | IISS | 75 | SWM | 77 | FI | 81 | JFS | 82 | JFS+ | 83 | | |
| SecC | 66 | SecC | 71 | JCS | 76 | ID11 | 78 | IISS | 81 | JWS | 82 | CFW+ | 84 | | |
| SecC | 67 | JCS | 72 | MOW | 76 | Rock | 78 | SMP | 81 | US20 | 82 | Meyr | 84 | | |
| SecC | 68 | US02 | 72 | NA03 | 76 | US05 | 78 | USND | 81 | Clns+ | 83 | SMP | 84 | | |

---------- **US DESIGNATION** ----------

| SecC | 68 | SS-N-3 | FY69-70,72. |
|---|---|---|---|
| JCS | 72 | SS-N-3 | FY73-77. |
| MOW | 76 | SS-N-3 | |
| SecD | 76 | SS-N-3 | |
| SWM | 77 | SS-N-3 | [Text] |
| | | SS-N-3A | [Table, but does not distinguish model A as other sources do, ed]. |
| Rock | 78 | SSC-1/SS-N-3 | [SSC-1 also has separate chart entry; why that designation repeated here is unclear, ed]. |
| Clns | 80 | SS-N-3 | |
| FI | 81 | SS-N-3 | |
| SMP | 81 | SS-N-3 | '81, '84, '85. |
| USND | 81 | SS-N-3b | [N.b. Another listing (same information) is for SS-N-3c, NATO Code: Shaddock]. |
| CFW | 82 | SS-N-3 | |
| IISS | 82 | SS-N-3 | |
| JFS | 82 | SS-N-3 | |
| JWS | 82 | SS-N-3 | |
| GSN | 83 | SS-N-3 | |
| Meyr | 84 | SS-N-3 | |

---------- **NATO CODENAME** ----------

| MOW | 76 | Shaddock | |
|---|---|---|---|
| SWM | 77 | Shaddock | |
| Rock | 78 | Shaddock | |
| FI | 81 | Shaddock | |
| USND | 81 | Sepal | |
| CFW | 82 | Shaddock | |
| IISS | 82 | Shaddock | |
| JFS | 82 | Sepal | |
| JWS | 82 | Shaddock | |
| GSN | 83 | Shaddock | |
| | | Sepal | "SS-N-3b was also assigned the NATO code name Sepal, but all ship-launched versions are generally referred to as Shaddock". |
| JFS+ | 83 | Shaddock | |
| | | Sepal | |

---------- **DESCRIPTION** ----------

| JCS | 72 | | Listed as 1st generation strategic missile. |
|---|---|---|---|
| MOW | 76 | | Cruise missile. |
| USND | 81 | SSM | |
| IISS | 82 | SLCM | |

---------- **EDITOR'S NOTES** ----------

| SecC | 66 | Listed FY67-70,72, not by name FY67-68. |
|---|---|---|
| JCS | 72 | Listed FY73-77. |
| SecD | 76 | Listed FY77. |
| FI | 81 | Final report for specifications in table; text entry remains. |
| SMP | 81 | Mentioned '81, '84, '85. |
| Meyr | 84 | Note to chart: "Values pertain to initial service period". |

## RANGE

|       |    | mi  | nm       | km      | km conv         |                                                                                                     |
|-------|----|-----|----------|---------|-----------------|-----------------------------------------------------------------------------------------------------|
| MR05  | 66 |     |          | 400-650 | 400.00-650.00   |                                                                                                     |
| SecC  | 66 |     | 300      |         | 555.60          |                                                                                                     |
| SecC  | 67 |     | 350-450  |         | 648.20-833.40   |                                                                                                     |
| SecC  | 68 |     | 450      |         | 833.40          | Max, FY69-70.                                                                                       |
|       |    |     | 250      |         | 463.00          | Normal operating range against ships, FY69-70.                                                      |
| AJ01  | 70 | 400 |          |         | 643.60          | [Unmarked]                                                                                          |
|       |    | 250 |          |         | 402.25          | Without mid-course correction.                                                                      |
| US01  | 70 | 400 |          |         | 643.60          |                                                                                                     |
| SecC  | 71 |     | 220      |         | 407.44          | Arming subs.                                                                                        |
|       |    |     | 150      |         | 277.80          | Arming surface combatants.                                                                          |
| US02  | 72 |     | 350-400  |         | 648.20-740.80   |                                                                                                     |
|       |    | 25  |          |         | 40.23           | Has been mentioned for subs.                                                                        |
| MOW   | 76 | 280 |          | 450     | 450.52          | Max.                                                                                                |
| NA03  | 76 | 300 |          |         | 482.70          |                                                                                                     |
| SWM   | 77 | 280 |          | 450     | 450.52          |                                                                                                     |
| ID11  | 78 |     |          | 300-450 | 300.00-450.00   |                                                                                                     |
| Rock  | 78 |     |          | 800     | 800.00          |                                                                                                     |
| US05  | 78 |     | 250      |         | 463.00          | Range.                                                                                              |
|       |    |     | 400      |         | 740.80          | "Has been flown to".                                                                                |
| Clns  | 80 |     | 150-250  |         | 277.80-463.00   |                                                                                                     |
| FI    | 81 |     |          | ≤ 850   | 850.00          | Max.                                                                                                |
|       |    |     |          | 200     | 200.00          | Optimum.                                                                                            |
| USND  | 81 |     | c. 250   |         | 463.00          |                                                                                                     |
| CPW   | 82 |     | 220      |         | 407.44          | Max.                                                                                                |
| IISS  | 82 |     |          | 450     | 450.00          | Max.                                                                                                |
| JFS   | 82 |     | 250+     |         | 463.00          | Max.                                                                                                |
| JWS   | 82 |     |          | 840     | 840.00          | Max (est) [analysis table].                                                                         |
|       |    |     |          | ≥ 450   | 450.00          | With midcourse guidance.                                                                            |
|       |    |     |          | 180     | 180.00          | Practical range nearer 180 km for cruisers; less for subs. Range limited mainly by radio/radar horizons. |
| US20  | 82 |     | 25-30    |         | 46.30-55.56     | Effective autonomous range.                                                                         |
| JFS+  | 83 |     | 250      |         | 463.00          | Max.                                                                                                |
| Meyr  | 84 |     |          | 300     | 300.00          |                                                                                                     |

## SPEED

|       |    | mph | kmph | Mach | kmph conv |                                                         |
|-------|----|-----|------|------|-----------|---------------------------------------------------------|
| MR05  | 66 |     |      | 2    |           |                                                         |
| MOW   | 76 |     |      | 0.95 |           | Max.                                                    |
| Rock  | 78 |     |      | 2.5  |           |                                                         |
| FI    | 81 |     |      | 1.4  |           |                                                         |
| JFS   | 82 |     |      | 1.5  |           |                                                         |
| JWS   | 82 |     |      |      |           | transonic                                               |
| GSN   | 83 |     |      |      |           | [See Midcourse Guidance for indirect evidence on speed, ed]. |
| JFS+  | 83 |     |      | 1.4  |           |                                                         |

## CEP

|       |    | ft | mi | nm | m | km      | m conv         |
|-------|----|----|----|----|---|---------|----------------|
| Meyr  | 84 |    |    |    |   | 0.5-1.0 | 500.00-1,000   |

## WARHEAD TYPE

|       |    | nuclear | conven | chem |                                               |
|-------|----|---------|--------|------|-----------------------------------------------|
| MOW   | 76 | yes     |        |      | Probably nuclear.                             |
| Rock  | 78 | yes     | yes    |      | HE.                                           |
| US05  | 78 | yes     | yes    |      |                                               |
| Clns  | 80 | yes     | yes    |      | HE.                                           |
| FI    | 81 | yes     | yes    |      | HE.                                           |
| CPW   | 82 | yes     | yes    |      |                                               |
| IISS  | 82 | yes     | yes    |      |                                               |
| JFS   | 82 | yes     | yes    |      | HE.                                           |
| JWS   | 82 | yes     | yes    |      | HE; standard warhead is 350 kT nuclear.       |
| US20  | 82 | yes     | yes    |      | HE.                                           |
| GSN   | 83 | yes     | yes    |      |                                               |
| Meyr  | 84 | yes     |        |      |                                               |

## WARHEAD CHARACTERISTICS

| IISS | 82 | single warhead |
| Meyr | 84 | single RV      |

## SOVIET MISSILES IDDS

```
------------ NUCLEAR YIELD ------------
              kT            MT          kT conv
Clns  80                                              KT range, "KT".
FI    81                                              KT range.
IISS  82     350                        350.00
JFS   82     350                        350.00        [Unmarked]
             800                        800.00        Strategic.
JWS   82  c. 350                        350.00        Standard.
             800                        800.00        Strategic.
US20  82     350                        350.00
Clns+ 83      10                         10.00
Meyr  84     50-200                      50.00-200.00

------------ WARHEAD WEIGHT ------------
              lb            t           kg            kg conv
US05  78    2,000                                     907.20
Clns  80    2,000                                     907.20
FI    81                             c. 1,000         1,000
JFS   82    2,200                                     997.92
JWS   82                             c. 1,000         1,000
GSN   83    2,200                                     997.92
CPW+  84                                1,000         1,000

------------ THROWWEIGHT ------------
              lb            t           kg            kg conv
IISS  82    2,000                                     907.20

------------ WEIGHT ------------
              lb            t           kg            kg conv
MR05  66                 c. 12                        12,000       Tons.
Rock  78                                11,790        11,790
FI    81                             c. 4,500         4,500
JFS   82    9,900                                     4,490        Launch weight.
JWS   82                                4,700         4,700
GSN   83 c. 12,000                   c. 5,400         5,443
CPW+  84                                5,400         5,400

------------ LENGTH ------------
              in           ft          cm            m            m conv
MOW   76                  35'9"                      10.0         10.90
SWM   77                  42                         13.8         12.80
Rock  78                                             13.8         13.80
FI    81                                          c. 10           10.00
JFS   82                  36                                      10.97
JWS   82                              1,100                       11.00
JFS+  83                  42                                      12.80
CPW+  84                                             10.2         10.20

------------ DIAMETER ------------
              in           ft          cm            m            cm conv
MR05  66                                             1.8          180.00
SWM   77     29.4                     100                         74.68
Rock  78                              100                         100.00
FI    81                           c. 100                         100.00
JWS   82                              86                          86.00    [Est].
GSN   83  c. 39                    c. 97.5                        99.06    [975 mm in source]

------------ WINGSPAN ------------
              in           ft          cm            m            cm conv
SWM   77     82.7                     210                         210.06
Rock  78                              210                         210.00
FI    81                           c. 210                         210.00
GSN   83               c. 6'10"                   c. 2.1          208.28
CPW+  84                                             5            500.00

------------ NUMBER OF STAGES ------------
Rock  78    "2 + 1"
```

## TYPE OF PROPULSION

| | | | |
|---|---|---|---|
| MOW | 76 | ramjet or turbojet solid | Plus 2 large solid JATO boosters. |
| SWM | 77 | turbojet 2 JATOs | |
| Rock | 78 | turbojet Jato | |
| FI | 81 | turbojet or ramjet 2 JATOs | Cruise. Boosters. |
| CFW | 82 | liquid, turbojet | |
| JFS | 82 | 1 liquid turbojet 2 solid | Sustainer. Boosters. |
| JWS | 82 | solid | Internal cruise motor. Boosters. |
| GSN | 83 | turbojet 2 solid | Boosters. |

## MIDCOURSE (OR MAIN) GUIDANCE

| | | | |
|---|---|---|---|
| USO2 | 72 | midcourse command | "Inertial nav.". |
| MOW | 76 | command | |
| SWM | 77 | autopilot radio command | ? Gathered automatically to beam-riding for flights beyond horizon, mid-course via cooperative aircraft or helicopter. |
| Rock | 78 | radio | |
| FI | 81 | autopilot | Plus midcourse update; "errors detected early in flight are corrected via the command data link". |
| CFW | 82 | inertial | With midcourse correction by radio. |
| JFS | 82 | command | Probably. |
| JWS | 82 | radio command | Probably; course corrections transmitted by radio. "Quite probable that a measure of terrain following capability is provided" for overland role. Relies upon radio altimeter. |
| GSN | 83 | inertial | With mid-course correction. "Requires mid-course guidance for over-the-horizon use. This is sent as a radar picture via Video Data Link (VDL) from the targeting ship or aircraft to the launching ship and then relayed—with target indicated—to the missile in flight. The launching submarine is thus required to remain on the surface after launch, for as long as 25 minutes when firing against targets at a range of some 250 n.miles". |

## TERMINAL GUIDANCE

| | | | |
|---|---|---|---|
| USO2 | 72 | active radar | |
| MOW | 76 | active radar | Possibly. |
| SWM | 77 | yes | |
| Rock | 78 | IR | |
| FI | 81 | active radar | |
| CFW | 82 | active radar | |
| JFS | 82 | active radar | |
| JWS | 82 | IR active radar | A likely improvement over the years. |
| US20 | 82 | | Some sources suggest terminal phase link to sub platform to sort out target. |
| GSN | 83 | active radar | |

## CONTROL SYSTEM

| | | |
|---|---|---|
| Rock | 78 | Aerodynamic. |
| FI | 81 | Aeroplane configuration. |

## RADAR

| | | | |
|---|---|---|---|
| SWM | 77 | Scoop Pair | On ships. |
| FI | 81 | Scoop Pair | On ships. |
| JWS | 82 | Scoop Pair | Track missile, for surface installations. |
| GSN | 83 | Scoop Pair Front Door Front Pierce | Fire control on ships. Fire control on subs. Fire control on subs. |

## OBSERVED

| | | |
|---|---|---|
| MOW | 76 | Never has been revealed in public. |

## PRODUCTION UNDERWAY

| | | |
|---|---|---|
| SMP | 84 | In production. |

## IOC

| | | | |
|---|---|---|---|
| JFS | 82 | 1960 | Operational. |

492  SOVIET MISSILES  IDDS

## DEPLOYMENT BEGAN

| | | | |
|---|---|---|---|
| US01 | 70 | 1962 | Appeared on Kynda. |
| SWM | 77 | 1958-62 | [Entry states "entered service 1962", but first on Whisky subs in 1958, ed] |
| ID11 | 78 | 1961 | Appeared. |
| Rock | 78 | 1962 | Service intro. |
| Clns | 80 | 1962 | First deployed. |
| SMP | 81 | | "The Soviets began their submarine cruise missile programs in the 1950s converting existing submarines to fire the long-range SS-N-3 missile". |
| IISS | 82 | 1962 | First year deployed. |
| Meyr | 84 | 1958 | Service. |

## DEPLOYED

MOW  76  in service

## DEPLOYMENT HISTORY

FI  81  All but Twin Cylinder and 4 Echo I's thought to remain in service. Being replaced by SS-N-12.
GSN  83  SS-N-12 has replaced SS-N-3 on some Echo II subs.

## NUMBERS IN SERVICE

### NUMBER IN 1968
SecC  68  301-329  On [subs] as of 1 Oct.

### NUMBER IN 1973
IISS  73  48  1973-76, July, total deployed, only those on surface ships.
         338  July, total deployed, only those on subs.

### NUMBER IN 1974
IISS  74  314  July, total deployed, only those on subs.

### NUMBER IN 1975
IISS  75  264  1975-76, July, total deployed, only those on subs.
         312  1975-76, July, total deployed.
SWM  77  264  Subs.
         48  Surface ships.

### NUMBER IN 1976
MOW  76  100  Land-based.
              Carried by at least 55 subs.
SecD  76       "A large inventory".

### NUMBER IN 1977
IISS  77  324  1977-79, July, total deployed.

### NUMBER IN 1980
IISS  80  342  July, total deployed.

### NUMBER IN 1981
IISS  81  324  July, total deployed.

### NUMBER IN 1982
IISS  82  356  July, total deployed.

### NUMBER IN 1983
IISS+ 83  316  July, total deployed.

## CARRIERS

### CARRIER — TUBES — RELOADS

| | | | | |
|---|---|---|---|---|
| SecC | 66 | | | Carried on 39-43 boats, of which 16-18 nuclear-powered. |
| SecC | 67 | | | Carried on 43-48 boats, of which 21-23 nuclear-powered. |
| SecC | 68 | | | Carried on 52-57 boats, of which 29-31 nuclear-powered, as of 1 October. Construction of boats apparently coming to an end; last expected to be delivered by close of 1969, FY69-70. |
| SecC | 71 | | | Part of armament of 64 "cruise missile subs", . Part of armament of 20 "guided missile combatants". |
| SecD | 76 | | | Serve to support SSGNs/SSGs/guided-missile cruisers. |

### CARRIER — TUBES — RELOADS
Rock  78  subs

### CARRIER — TUBES — RELOADS

| | | | | |
|---|---|---|---|---|
| SWM | 77 | E1 | 3 x 2 | |
| FI | 81 | Echo I | 6 | N<_5, from 1960. |
| SMP | 81 | Echo I | | Designed to carry SS-N-3. |
| USND | 81 | Echo I | | |
| GSN | 83 | Echo I | | N=5, all since converted to SSNs. |

CRUISE MISSILES  Sea-Launched SSMs  SS-N-3  493

```
            ------- CARRIER -------  TUBES ---- RELOADS -----------
MR05   66   E2
JCS    76   Echo II                                    "Carries versions of the SS-N-3 surface launched cruise
                                                       missile".
MOW    76   E II                     8
SWM    77   E2                       4 x 2
Clns   80   Echo II                  8                 "SS-N-3/12".
FI     81   Echo II                  8                 N=27, from 1963.
SMP    81   Echo II                                    Designed to carry SS-N-3.
USND   81   Echo II
CFW    82   Echo-II                                    Unmodified.
SMP    85   Echo II                                    USSR proceeding with conversion of "older 1960 vintage
                                                       SS-N-3-equipped ECHO II SSGNs" to SS-N-12.

            ------- CARRIER -------  TUBES ---- RELOADS -----------
MR05   66   J                                          May be assumed J class subs developed to test weapon
                                                       system for E2 subs; J subs are probably converted W subs;
                                                       missile fitting same as E2.
MOW    76   J                        4
Clns   80   Juliett                  4                 "SS-N-3/12".
FI     81   Juliet                   4                 N=16, from 1962.
SMP    81   Juliett                                    Designed to carry SS-N-3.
USND   81   Juliett
CFW    82   Juliett

            ------- CARRIER -------  TUBES ---- RELOADS -----------
MOW    76   W                        2 or 4
USND   81   Whiskey Conversion

            ------- CARRIER -------  TUBES ---- RELOADS -----------
SWM    77   Whisky Twin Cyl.         2
FI     81   Whisky Twin Cyl.         2                 N=5, doubtful if ever operational; first deployed 1958.

            ------- CARRIER -------  TUBES ---- RELOADS -----------
FI     81   Whisky Long Bin          4                 N=7.
CFW    82   Whiskey Long Bin

            ------- CARRIER -------  TUBES ---- RELOADS -----------
Rock   78   cruisers

            ------- CARRIER -------  TUBES ---- RELOADS -----------
MR04   66   Kynda                    2 x 4    16 total
MOW    76   Kynda                    2 x 4    8        N=4.
SWM    77   Kynda                    2 x 4
Clns   80   Kynda                    8
FI     81   Kynda                    2 x 4             N=4, from 1962; separate guidance for each quad launcher.
USND   81   Kynda
CFW    82   Kynda                    1 x 4

            ------- CARRIER -------  TUBES ---- RELOADS -----------
MOW    76   Kresta I                 2 x 2    no       N=4.
SWM    77   Kresta                   2 x 2
Clns   80   Kresta I                 4
FI     81   Kresta I                 2 x 2             N=4.
USND   81   Kresta I
CFW    82   Kresta-I                 1 x 2

            ------- USE AND CONFIGURATION -----------
JCS    76   Surface launched.
SWM    77   Approaches target in terminal dive; high or low trajectories to target.
ID11   78   ? Could descend steeply onto target.
FI     81   Cruise height 3,000-6,000 m.
CFW    82   Subs launch from surface.
JWS    82   Launchers on Kresta, Kynda, and all subs can be elevated. Only Kynda launchers can be trained.
US20   82   Altitude (cruising) 900-12,000 ft, depending on range.
GSN    83   Subs launch from surface.
Meyr   84   Penetration: 0.7-0.9; reliability: 0.2-0.5, "includes system reliability multiplied by operational
            readiness of deployed system".

            ------- TARGET TYPE -----------
SecC   71   ship
US02   72              Land-based role preceded sea-based.
FI     81   ship
```

---------- **STRATEGY** ----------
SecC 66  Soviets do not appear to consider cruise missile subs as primarily a strategic attack system, FY67-69.
SecC 69  Soviets do not appear to consider cruise missile subs as a strategic attack system.
Meyr 84  [Ed: chart is labeled] Systems "assigned primary operational and strategic missions within the European theatre(s) of military operations".

---------- **COMBAT REPORTS/EFFECTIVENESS** ----------
MOW 76  Most formidable of Soviet subsonic "flying bombs".

---------- **EXPORTED TO** ----------
JWS 82  USSR only [not exported]

---------- **RELATION TO OTHER MISSILES** ----------
SWM 77  Sea-launched version of SSC-1.
SMP 81  SS-N-12 are improvement over older SS-N-3 antiship missiles.
JWS 82  SS-N-12 is related and newer missile.
GSN 83  SS-C-1 is similar. SS-N-12 is improved version.
CFW+ 84  Variant of the SS-C-1.

---------- **OTHER INFORMATION** ----------
MOW 76  May be one different land- and several different sea-based varieties.
IISS 82  Numerous versions.
JWS 82  Largest Soviet cruise missiles.

# SS-N-3α

Sources are:   AWST 82    JFS 82    JWS 82    US20 82    GSN 83    CFW 84    BW 86

## US DESIGNATION
|  | |  |
|---|---|---|
| AWST | 82 | SS-N-3A |
| JFS | 82 | SS-N-3A |
| GSN | 83 | SS-N-3a |
| CFW | 84 | SS-N-3A |

## NATO CODENAME
|  | | |
|---|---|---|
| AWST | 82 | "Shaddock type". |

## RANGE
| Source | | mi | nm | km | km conv | |
|---|---|---|---|---|---|---|
| AWST | 82 | 250 | | | 402.25 | |
| US20 | 82 | | 250 | | 463.00 | Version A with cooperating aircraft. |
| GSN | 83 | | 250 | 463 | 463.00 | SS-N-3a/b. |
| CFW | 84 | | 250 | | 463.00 | |
| BW | 86 | | | 410 | 410.00 | 50 km from Lone subs. |

## SPEED
| Source | | mph | kmph | Mach | kmph conv |
|---|---|---|---|---|---|
| AWST | 82 | | | 0.9-1.4 | |
| BW | 86 | | | 1.0 | |

## WARHEAD CHARACTERISTICS
|  | | |
|---|---|---|
| BW | 86 | single RV |

## NUCLEAR YIELD
| Source | | kT | MT | kT conv |
|---|---|---|---|---|
| BW | 86 | 350 | | 350.00 |

## WARHEAD WEIGHT
| Source | | lb | t | kg | kg conv | |
|---|---|---|---|---|---|---|
| US20 | 82 | 1,000-2,000 | | | 453.60-907.20 | Version A. |

## LENGTH
| Source | | in | ft | cm | m | m conv | |
|---|---|---|---|---|---|---|---|
| AWST | 82 | | 42 | | | 12.80 | |
| GSN | 83 | | 33'6" | | 10.2 | 10.21 | SS-N-3a/b. |

## TYPE OF PROPULSION
|  | | |
|---|---|---|
| BW | 86 | TJ |

## MIDCOURSE (OR MAIN) GUIDANCE
|  | | | |
|---|---|---|---|
| CFW | 84 | inertial | With mid-course correction. |

## TERMINAL GUIDANCE
|  | | |
|---|---|---|
| CFW | 84 | active radar |

## IOC
|  | | | |
|---|---|---|---|
| GSN | 83 | 1962 | SS-N-3a and 3b. |

## DEPLOYMENT BEGAN
|  | | |
|---|---|---|
| CFW | 84 | 1962 |
| BW | 86 | 1958 |

## DEPLOYED
|  | | |
|---|---|---|
| BW | 86 | In service. |

## CARRIERS

### CARRIER — TUBES — RELOADS
| Source | | Carrier | Tubes | Reloads | |
|---|---|---|---|---|---|
| JFS | 82 | subs | | | |

### CARRIER — TUBES — RELOADS
| Source | | Carrier | Tubes | Reloads | |
|---|---|---|---|---|---|
| BW | 86 | Echo I | 6 | no | N=5. |

### CARRIER — TUBES — RELOADS
| Source | | Carrier | Tubes | Reloads | |
|---|---|---|---|---|---|
| AWST | 82 | Echo 2 | | | |
| JWS | 82 | E2 | 4 x 2 | | tubes: "Eight missiles in pair". |
| GSN | 83 | Echo II | | | SS-N-3a or c. |
| CFW | 84 | Echo II | | | |
| BW | 86 | Echo II | 8 | no | N=27. |

| | | CARRIER | TUBES | RELOADS | |
|---|---|---|---|---|---|
| AWST | 82 | Julliett | | | |
| JWS | 82 | J | 2 x 2 | | |
| GSN | 83 | Juliett | | | SS-N-3a or c. |
| CFW | 84 | Juliett | | | |
| BW | 86 | Juliett | 4 | no | N=16. |

| | | CARRIER | TUBES | RELOADS | |
|---|---|---|---|---|---|
| JWS | 82 | W Twin Cylinder | 2 | | Still a current configuration. |
| BW | 86 | W Twin Cyl | 2 | no | N=5. |

| | | CARRIER | TUBES | RELOADS | |
|---|---|---|---|---|---|
| JWS | 82 | W Long Bin | 4 | | 6.5m section inserted into hull with launchers mounted on it. Still a current configuration. |
| BW | 86 | W Long Bin | 4 | no | N=7. |

| | | USE AND CONFIGURATION | |
|---|---|---|---|
| JFS | 82 | Surface launched from submarines. | |

# SS-N-3b

```
Sources are:  AWST 82    JFS 82    JWS 82    GSN 83    CFW 84    BW 86
```

## US DESIGNATION

|  |  |
|---|---|
| AWST 82 | SS-N-3B |
| JFS 82 | SS-N-3B |
| GSN 83 | SS-N-3b |
| CFW 84 | SS-N-3B |

## NATO CODENAME

|  |  |  |
|---|---|---|
| GSN 83 | Sepal | "SS-N-3b was also assigned the NATO code name Sepal, but all ship-launched versions are generally referred to as Shaddock". |

## RANGE

| Source | mi | nm | km | km conv | |
|---|---|---|---|---|---|
| AWST 82 | 250 | | | 402.25 | |
| GSN 83 | | 250 | 463 | 463.00 | SS-N-3a/b. |
| CFW 84 | | 250 | | 463.00 | |
| BW 86 | | | 280 | 280.00 | |

## SPEED

| Source | mph | kmph | Mach | kmph conv |
|---|---|---|---|---|
| BW 86 | | | 1.0 | |

## WARHEAD CHARACTERISTICS

|  |  |
|---|---|
| BW 86 | single RV |

## NUCLEAR YIELD

| Source | kT | MT | kT conv |
|---|---|---|---|
| BW 86 | 350 | | 350.00 |

## LENGTH

| Source | in | ft | cm | m | m conv | |
|---|---|---|---|---|---|---|
| GSN 83 | | 33'6" | | 10.2 | 10.21 | SS-N-3a/b. |

## TYPE OF PROPULSION

|  |  |
|---|---|
| BW 86 | TJ |

## MIDCOURSE (OR MAIN) GUIDANCE

|  |  |  |
|---|---|---|
| AWST 82 | | Requires aircraft for midcourse update. |
| CFW 84 | [inertial] | [With mid-course correction], similar guidance [to SS-N-3a]. |

## IOC

|  |  |  |
|---|---|---|
| GSN 83 | 1962 | SS-N-3a and 3b. |

## DEPLOYMENT BEGAN

|  |  |
|---|---|
| CFW 84 | 1962 |
| BW 86 | 1962 |

## DEPLOYED

|  |  |
|---|---|
| BW 86 | In service. |

## CARRIERS

### CARRIER — TUBES — RELOADS

| Source | Carrier | Tubes | Reloads | Note |
|---|---|---|---|---|
| AWST 82 | Kynda | | | |
| JFS 82 | Kynda | | | |
| JWS 82 | Kynda | 2 x 4 | yes | Launchers can be trained. |
| GSN 83 | Kynda | | | |
| CFW 84 | Kynda | | | |
| BW 86 | Kynda | 8 | 8 | N=4. |

### CARRIER — TUBES — RELOADS

| Source | Carrier | Tubes | Reloads | Note |
|---|---|---|---|---|
| AWST 82 | Kresta | | | |
| JFS 82 | Kresta I | | | |
| JWS 82 | Kresta I | | | |
| GSN 83 | Kresta I | | | |
| CFW 84 | Kresta I | | | |
| BW 86 | Kresta I | 4 | no | N=4. |

## SS-N-3c

Sources are: JFS 82  JWS 82  US20 82  GSN 83  CFW 84

------- **US DESIGNATION** -------
| | | |
|---|---|---|
| JFS | 82 | SS-N-3c |
| JWS | 82 | SS-N-3C   Thought to be strategic variant. |
| GSN | 83 | SS-N-3c |
| CFW | 84 | SS-N-3C |

------- **RANGE** -------

| | | mi | nm | km | km conv | |
|---|---|---|---|---|---|---|
| US20 | 82 | | 170 | | 314.84 | C version with cooperating aircraft. |
| GSN | 83 | | 400+ | 740+ | 740.80 | |
| CFW | 84 | | 400 | | 740.80 | |

------- **NUCLEAR YIELD** -------

| | | kT | MT | kT conv | |
|---|---|---|---|---|---|
| JFS | 82 | 800 | | 800.00 | "Strat." [version] |

------- **WARHEAD WEIGHT** -------

| | | lb | t | kg | kg conv | |
|---|---|---|---|---|---|---|
| US20 | 82 | 2,000 | | | 907.20 | Version C. |

------- **LENGTH** -------

| | | in | ft | cm | m | m conv | |
|---|---|---|---|---|---|---|---|
| GSN | 83 | | 38'6" | | 11.75 | 11.73 | |
| CFW | 84 | | | | 11.8 | 11.80 | [Literally 111.8 m] |

------- **MIDCOURSE (OR MAIN) GUIDANCE** -------
CFW  84  inertial       Only.

------- **IOC** -------
GSN  83  1960

------- **DEPLOYMENT BEGAN** -------
CFW  84  1960

------- **RETIRED** -------
CFW  84  "Possibly still in use from submarines".

------- **CARRIERS** -------

------- **CARRIER** ------- TUBES ---- RELOADS -------
GSN  83  Echo II                              SS-N-3a or c.

------- **CARRIER** ------- TUBES ---- RELOADS -------
GSN  83  Juliett                              SS-N-3a or c.

------- **CARRIER** ------- TUBES ---- RELOADS -------
GSN  83  Whiskey (mod.)                  Most have been discarded.

------- **STRATEGY** -------
JFS  82  Strategic version is 3C.
GSN  83  Originally developed for strategic attack role in SS-N-3c variant.

------- **RELATION TO OTHER MISSILES** -------
GSN  83  SS-N-3c is contemporary of US Regulus missile.

# SS-N-7

Sources are:
| | | | | | | | | | | | | |
|---|---|---|---|---|---|---|---|---|---|---|---|---|
| | | NA03 | 76 | IISS | 80 | USND | 81 | JFS | 82 | FI+ | 83 | JFS+ | 83 |
| SecC | 71 | SWM | 77 | US11 | 80 | AWST | 82 | JWS | 82 | FI+ | 83 | CFW+ | 84 |
| ID04 | 74 | Rock | 78 | FI | 81 | CFW | 82 | US20 | 82 | GSN | 83 | SMP | 84 |
| MOW | 76 | Clns | 80 | IISS | 81 | IISS | 82 | Clns+ | 83 | IISS+ | 83 | BW | 86 |

---------- **US DESIGNATION** ----------

| | | |
|---|---|---|
| SecC | 71 | SS-N-7 |
| MOW | 76 | SS-N-7 |
| SWM | 77 | SS-N-7 |
| Rock | 78 | SS-N-7 |
| Clns | 80 | SS-N-7 |
| FI | 81 | SS-N-7 |
| USND | 81 | SS-N-7 |
| AWST | 82 | SS-N-7 |
| CFW | 82 | SS-N-7 |
| IISS | 82 | SS-N-7 |
| JFS | 82 | SS-N-7 |
| JWS | 82 | SS-N-7 |
| GSN | 83 | SS-N-7 |
| SMP | 84 | SS-N-7 |

---------- **NATO CODENAME** ----------

| | | |
|---|---|---|
| IISS | 82 | Siren |
| GSN | 83 | Siren |
| CFW+ | 84 | Siren |

---------- **DESCRIPTION** ----------

| | | |
|---|---|---|
| USND | 81 | SSM |
| IISS | 82 | SLCM |
| JFS | 82 | SLCM |

---------- **EDITOR'S NOTES** ----------

| | | |
|---|---|---|
| SecC | 71 | Listed FY72. |
| SMP | 84 | Listed '84. |

---------- **RANGE** ----------

| | | mi | nm | km | km conv | |
|---|---|---|---|---|---|---|
| ID04 | 74 | | c. 26 | | 48.15 | |
| MOW | 76 | 35 | | 56 | 56.32 | Est. |
| NA03 | 76 | 30 | | | 48.27 | |
| SWM | 77 | 28-34 | | 45-55 | 45.05-54.71 | |
| Rock | 78 | | | 55 | 55.00 | |
| Clns | 80 | | 30 | | 55.56 | |
| US11 | 80 | | 30 | | 55.56 | |
| FI | 81 | | | c. 50 | 50.00 | [Text] |
| | | | | 55-60 | 55.00-60.00 | [Table] Subs detect targets with passive sonar at about 50 km, identify by acoustic signature. Under favorable conditions works out to 100 km. |
| USND | 81 | | c. 30 | | 55.56 | |
| AWST | 82 | 35 | | | 56.32 | |
| CFW | 82 | | 35 | | 64.82 | Max. |
| IISS | 82 | | | 45 | 45.00 | Max. |
| JFS | 82 | | 30+ | | 55.56 | Max. |
| JWS | 82 | | | 45-55 | 45.00-55.00 | [Analysis table] |
| | | | | 45-53 | 45.00-53.00 | Variously reported in this range. |
| US20 | 82 | | 35 | | 64.82 | In theory. |
| | | | 25-30 | | 46.30-55.56 | In practice (sonar acquisition range). |
| GSN | 83 | | 30-35 | 55-64 | 55.56-64.82 | |
| JFS+ | 83 | | 35 | | 64.82 | Max. |
| BW | 86 | | | 50 | 50.00 | |

---------- **SPEED** ----------

| | | mph | kmph | Mach | kmph conv | |
|---|---|---|---|---|---|---|
| SWM | 77 | | | 1.5 | | |
| Rock | 78 | | | 1.5 | | |
| FI | 81 | | | 1.5 | | |
| JFS | 82 | | | 0.9 | | |
| JWS | 82 | | | | | High subsonic. |
| US20 | 82 | | | | | Subsonic. |
| GSN | 83 | | | 0.9 | | Max. |
| BW | 86 | | | 0.9 | | |

## WARHEAD TYPE

|      |    | nuclear | conven | chem |      |
|------|----|---------|--------|------|------|
| Clns | 80 | yes     | yes    |      | HE.  |
| FI   | 81 | yes     | yes    |      | HE.  |
| CFW  | 82 |         | yes    |      |      |
| IISS | 82 | yes     | yes    |      |      |
| JFS  | 82 | yes     | yes    |      | HE.  |
| JWS  | 82 | yes     | yes    |      | HE.  |
| US20 | 82 | yes     | yes    |      | HE.  |
| GSN  | 83 | yes     | yes    |      |      |
| CFW+ | 84 | yes     | yes    |      |      |

## WARHEAD CHARACTERISTICS

| IISS | 82 | single warhead |
|------|----|----------------|
| BW   | 86 | single RV      |

## NUCLEAR YIELD

|       |    | kT  | MT | kT conv |                |
|-------|----|-----|----|---------|----------------|
| Clns  | 80 |     |    |         | KT range, "KT". |
| IISS  | 82 | 200 |    | 200.00  |                |
| JFS   | 82 | 200 |    | 200.00  |                |
| JWS   | 82 | 200 |    | 200.00  |                |
| US20  | 82 | 200 |    | 200.00  |                |
| Clns+ | 83 | 10  |    | 10.00   |                |
| FI+   | 83 | 200 |    | 200.00  |                |
| BW    | 86 | 200 |    | 200.00  |                |

## WARHEAD WEIGHT

|      |    | lb    | t | kg     | kg conv |
|------|----|-------|---|--------|---------|
| Clns | 80 | 1,200 |   |        | 544.32  |
| FI   | 81 |       |   | c. 500 | 500.00  |
| JFS  | 82 | 1,100 |   |        | 498.96  |
| JWS  | 82 |       |   | 500    | 500.00  |
| US20 | 82 | 1,100 |   |        | 498.96  |
| GSN  | 83 | 1,100 |   | 500    | 498.96  |
| CFW+ | 84 |       |   | 500    | 500.00  |

## THROWWEIGHT

|      |    | lb    | t | kg | kg conv |
|------|----|-------|---|----|---------|
| IISS | 82 | 1,200 |   |    | 544.32  |

## WEIGHT

|      |    | lb       | t | kg       | kg conv |         |
|------|----|----------|---|----------|---------|---------|
| FI   | 81 |          |   | c. 3,500 | 3,500   |         |
| GSN  | 83 | c. 7,500 |   | c. 3,375 | 3,402   |         |
| JFS+ | 83 | 7,400    |   |          | 3,356   | Launch. |
| CFW+ | 84 |          |   | 2,900    | 2,900   |         |

## LENGTH

|      |    | in | ft         | cm | m    | m conv |
|------|----|----|------------|----|------|--------|
| MOW  | 76 |    | 24'11.25"  |    | 7.6  | 7.60   |
| SWM  | 77 |    | 22         |    | 6.7  | 6.71   |
| Rock | 78 |    |            |    | 6.7  | 6.70   |
| FI   | 81 |    |            |    | c. 7 | 7.00   |
| JFS  | 82 |    | 22         |    |      | 6.71   |
| GSN  | 83 |    | 23         |    | 7    | 7.01   |
| CFW+ | 84 |    |            |    | 7    | 7.00   |

## DIAMETER

|    |    | in | ft | cm    | m | cm conv      |
|----|----|----|----|-------|---|--------------|
| FI | 81 |    |    | 50-55 |   | 50.00-55.00 |

## TYPE OF PROPULSION

| FI   | 81 | 1 solid rocket |                                          |
|------|----|----------------|------------------------------------------|
|      |    |                | Some sources say turbofan [final report] |
| AWST | 82 | solid          |                                          |
| CFW  | 82 | solid          |                                          |
| JFS  | 82 | solid          |                                          |
| JWS  | 82 | solid          | Probably.                                |
| GSN  | 83 | solid          |                                          |
| BW   | 86 | solid          |                                          |

## CRUISE MISSILES  Sea-Launched SSMs  SS-N-7

### ──── MIDCOURSE (OR MAIN) GUIDANCE ────
| | | | |
|---|---|---|---|
| SWM | 77 | autopilot | |
| Rock | 78 | radio | |
| FI | 81 | autopilot | |
| JFS | 82 | autopilot | |
| JWS | 82 | autopilot | |
| US20 | 82 | autopilot | With midcourse correction. |

### ──── TERMINAL GUIDANCE ────
| | | | |
|---|---|---|---|
| SWM | 77 | yes | |
| FI | 81 | active radar | |
| AWST | 82 | radar | |
| JFS | 82 | active radar | |
| JWS | 82 | radar | |
| US20 | 82 | active radar | |
| GSN | 83 | radar homing | J-band. |

### ──── CONTROL SYSTEM ────
| | | |
|---|---|---|
| Rock | 78 | Aerodynamic. |

### ──── PRODUCTION UNDERWAY ────
| | | |
|---|---|---|
| SMP | 84 | In production. |

### ──── IOC ────
| | | | |
|---|---|---|---|
| JFS | 82 | 1969-70 | Operational. |
| GSN | 83 | 1971 | |
| JFS+ | 83 | 1968 | Operational. |

### ──── DEPLOYMENT BEGAN ────
| | | | |
|---|---|---|---|
| SWM | 77 | 1969-70 | |
| Rock | 78 | 1969 | Service intro. |
| Clns | 80 | 1968 | First deployed. |
| FI | 81 | 1967-68 | |
| IISS | 82 | 1968 | First year deployed. |
| CPW+ | 84 | 1970 | |
| BW | 86 | 1968 | |

### ──── DEPLOYED ────
| | | |
|---|---|---|
| MOW | 76 | In service. |
| BW | 86 | In service. |

### ──── DEPLOYMENT HISTORY ────
| | | |
|---|---|---|
| GSN | 83 | Succeeded in later Charlie SSGNs by SS-N-9. |

### ──── NUMBERS IN SERVICE ────

### ──── NUMBER IN 1980 ────
| | | | |
|---|---|---|---|
| IISS | 80 | 120 | July, total deployed. |

### ──── NUMBER IN 1981 ────
| | | | |
|---|---|---|---|
| IISS | 81 | ≤ 138 | July, total deployed. |

### ──── NUMBER IN 1982 ────
| | | | |
|---|---|---|---|
| IISS | 82 | 154 | July, total deployed. |

### ──── NUMBER IN 1983 ────
| | | | |
|---|---|---|---|
| IISS+ | 83 | c. 144 | July, total deployed. |

### ──── CARRIERS ────

### ──── CARRIER ──── TUBES ──── RELOADS ────
| | | | | | |
|---|---|---|---|---|---|
| SecC | 71 | | | | Part of armament of 64 "Cruise Missile Submarines". |
| Rock | 78 | sub | | | |

### ──── CARRIER ──── TUBES ──── RELOADS ────
| | | | | | |
|---|---|---|---|---|---|
| ID04 | 74 | Victor | | | "An IDR source has stated". |

### ──── CARRIER ──── TUBES ──── RELOADS ────
| | | | | | |
|---|---|---|---|---|---|
| Clns | 80 | Papa | 8 | | |
| FI | 81 | Papa | | | |
| FI+ | 83 | ? Papa | 10 | | N=1, may carry SS-N-9 instead. |

|      |    | CARRIER      | TUBES | RELOADS |                                    |
|------|----|--------------|-------|---------|------------------------------------|
| MOW  | 76 | C            | 8     |         | N=11.                              |
| SWM  | 77 | C            | 8     |         |                                    |
| Clns | 80 | Charlie      | 8     |         |                                    |
| FI   | 81 | Charlie I    | 8     |         | N=12.                              |
| USND | 81 | Charlie I    |       |         |                                    |
| AWST | 82 | C            |       |         |                                    |
| CFW  | 82 | Charlie I    | 8     |         |                                    |
| JFS  | 82 | Charlie      |       |         |                                    |
| JWS  | 82 | C            | 8     |         |                                    |
| FI+  | 83 | Charlie I    | 10    |         | N=12.                              |
| GSN  | 83 | Charlie I    |       |         |                                    |
| BW   | 86 | Charlie I    | 8     | no      | N=12.                              |

|     |    | CARRIER       | TUBES | RELOADS |                                    |
|-----|----|---------------|-------|---------|------------------------------------|
| FI  | 81 | Charlie II    | 8     |         | N$\geq$2.                          |
| FI+ | 83 | ? Charlie II  | 8     |         | N=6, may carry SS-N-9 instead.     |

---------- DESIGN AND ENGINEERING ----------
FI+   83   Thought to use SS-N-9 technology.

---------- USE AND CONFIGURATION ----------
MOW   76   Possibly underwater launched.
SWM   77   Underwater launch. "Possible surface skimmer in terminal approach to target".
Rock  78   Underwater launch.
FI    81   Can be fired from submerged subs. Subs detect targets with passive sonar at about 50 km, identify
           by acoustic signature. Under favorable conditions works out to 100 km. [Final report]
AWST  82   Underwater-launched cruise missile.
CFW   82   Can be launched while submerged.
JFS   82   Submarine launched from dived Charlies.
JWS   82   Can be launched while submerged. Has been suggested it operates as surface-skimmer over large part
           of its range. C sub designed from outset to carry both missiles and torpedoes.
US20  82   90 ft altitude, not true surface skimmer.
FI+   83   Probably flies in sea-skimming mode for most of flight.
GSN   83   Soviet Navy's first underwater-launched, antiship cruise missile.

---------- TARGET TYPE ----------
SecC  71   ship
FI    81   ship
GSN   83   ship

---------- RELATION TO OTHER MISSILES ----------
FI    81   "May be a development of the SS-N-2/SS-N-9 missiles".

# SS-N-9

```
Sources are:  SWM   77      Clns  80      SMP   81      CFW   82      US20  82      IISS+ 83      SMP   84
SecC  71      USD3  77      IISS  80      US18  81      IISS  82      Clns+ 83      JFS+  83      BW    86
MOW   76      ID11  78      FI    81      USND  81      JFS   82      FI+   83      ABFS  84
NAD3  76      Rock  78      IISS  81      AWST  82      JWS   82      GSN   83      CFW+  84
```

---------- US DESIGNATION ----------

| | | |
|---|---|---|
| SecC | 71 | SS-N-9 |
| MOW | 76 | SS-N-9 |
| SWM | 77 | SS-N-9 |
| Rock | 78 | SS-N-9 |
| Clns | 80 | SS-N-9 |
| FI | 81 | SS-N-9 |
| USND | 81 | SS-N-9 |
| AWST | 82 | SS-N-9 |
| CFW | 82 | SS-N-9 |
| IISS | 82 | SS-N-9 |
| JFS | 82 | SS-N-9 |
| JWS | 82 | SS-N-9 |
| GSN | 83 | SS-N-9 |

---------- NATO CODENAME ----------

| | | | |
|---|---|---|---|
| FI | 81 | Siren | |
| USND | 81 | Siren | |
| AWST | 82 | Siren | |
| JFS | 82 | Siren | [Nanuchka listing only, ed] |
| JWS | 82 | Siren | |

---------- DESCRIPTION ----------

| | | | |
|---|---|---|---|
| SWM | 77 | | Medium-range naval cruise missile intermediate between Styx and Shaddock. |
| USND | 81 | SSM | |
| IISS | 82 | SLCM | |
| JFS | 82 | SLCM | [Papa listing only, ed] |
| | | SSM | [Nanuchka listing only, ed] |
| SMP | 84 | Cruise missile. | |

---------- EDITOR'S NOTES ----------

| | | |
|---|---|---|
| SecC | 71 | Listed FY72. |
| SMP | 81 | Mentioned but not by name (p46) '81. |
| JFS | 82 | Source has 2 separate table entries, main distinction being the carrier: Papa (sub) vs Nanuchka (corvette). Data is common to both unless marked. |

## RANGE

| | | mi | nm | km | km conv | |
|---|---|---|---|---|---|---|
| MOW | 76 | ≤ 170 | | ≤ 275 | 273.53 | |
| NA03 | 76 | 45 | | | 72.41 | |
| SWM | 77 | 47–171 | | 75–275 | 75.62–275.14 | |
| USO3 | 77 | | 40–60 | | 74.08–111.12 | "Normal range". |
| Rock | 78 | | | 275 | 275.00 | |
| Clns | 80 | | 150 | | 277.80 | |
| FI | 81 | | | 110 | 110.00 | |
| SMP | 81 | | | 100 | 100.00 | Missiles can be fired "at a range of up to 100 kilometers from the intended target". |
| USND | 81 | | c. 60 | | 111.12 | |
| AWST | 82 | 70 | | | 112.63 | |
| CFW | 82 | | 30 | | 55.56 | Max [final report] |
| | | | 60 | | 111.12 | "Can reach 60 miles with an aerial relay (aircraft fitted with a Video Data Link system)", [final report]. |
| IISS | 82 | | | 280 | 280.00 | Max. |
| JFS | 82 | | 60 | | 111.12 | Max. |
| JWS | 82 | | ≤ 150 | | 277.80 | Original estimate, with external midcourse guidance by a/c; subsequently revised downward. |
| | | | 60 | | 111.12 | Quoted by American and British sources as maximum range. |
| | | | 40 | | 74.08 | Likely normal operating range. |
| US20 | 82 | | 60 | | 111.12 | In theory. |
| | | | 25–30 | | 46.30–55.56 | In practice due to acquisition range [in subs] |
| Clns+ | 83 | | 150 | | 277.80 | [Ship entry] |
| | | | 60 | | 111.12 | [Sub entry] |
| GSN | 83 | | 60 | 111 | 111.12 | |
| JFS+ | 83 | | 70 | | 129.64 | Max, requires third party to reach maximum range. |
| BW | 86 | | | 110 | 110.00 | 50 km from lone subs. |

## SPEED

| | | mph | kmph | Mach | kmph conv | |
|---|---|---|---|---|---|---|
| SWM | 77 | | | 1.4 | | |
| FI | 81 | | | 0.8 | | |
| AWST | 82 | | | 1.4 | | |
| JFS | 82 | | | 0.9 | | |
| GSN | 83 | | | 0.9 | | [Earlier in text, called "supersonic", sic] |
| JFS+ | 83 | | | 1.4 | | |
| BW | 86 | | | 1.4 | | |

## WARHEAD TYPE

| | | nuclear | conven | chem |
|---|---|---|---|---|
| SWM | 77 | | yes | |
| Rock | 78 | | yes | HE. |
| Clns | 80 | yes | yes | HE. |
| FI | 81 | yes | yes | HE. |
| CFW | 82 | yes | yes | |
| IISS | 82 | yes | yes | |
| JFS | 82 | yes | yes | HE. |
| US20 | 82 | | yes | HE. |
| GSN | 83 | yes | yes | |

## WARHEAD CHARACTERISTICS

| | | |
|---|---|---|
| IISS | 82 | single warhead |
| BW | 86 | single RV |

## NUCLEAR YIELD

| | | kT | MT | kT conv | |
|---|---|---|---|---|---|
| Clns | 80 | | | | KT range, "KT". |
| IISS | 82 | 200 | | 200.00 | [Nanuchka listing only, ed] |
| JFS | 82 | 200 | | 200.00 | |
| US20 | 82 | 200 | | 200.00 | |
| BW | 86 | 200 | | 200.00 | |

CRUISE MISSILES  Sea-Launched SSMs  SS-N-9

```
---------- WARHEAD WEIGHT ----------
              lb          t           kg         kg conv
Clns  80    1,000                                453.60
FI    81                           c. 500        500.00
JFS   82    1,100                                498.96      [Nanuchka listing only, ed]
US20  82    1,100                                498.96
GSN   83    1,100                     500        498.96
CFW+  84                              500        500.00

---------- WEIGHT ----------
              lb          t           kg         kg conv
FI    81                           c. 3,000      3,000
GSN   83   c. 6,500                c. 2,950      2,948
CFW+  84                              3,300      3,300

---------- LENGTH ----------
              in         ft          cm     m        m conv
SWM   77                 30                 9.1      9.14
Rock  78                                    9.1      9.10
FI    81                                 c. 9        9.00
JFS   82                 30                 9.14
GSN   83                 29                 8.84     8.84
CFW+  84                                    8.8      8.80

---------- NUMBER OF STAGES ----------
AWST  82    1
JFS   82    1

---------- TYPE OF PROPULSION ----------
US03  77    solid       But difficult to reconcile with 150 nm range with helicopter guidance.
FI    81    liquid      ? Air-breathng.
AWST  82    solid
CFW   82    solid
JFS   82    solid
JWS   82    solid
GSN   83    solid
BW    86    solid

---------- MIDCOURSE (OR MAIN) GUIDANCE ----------
SWM   77    autopilot   With radio command link guidance, mid-course by aircraft or helicopter.
FI    81    autopilot   Plus midcourse update.
CFW   82    inertial
JWS   82    autopilot   With or without radio command link [text]; probably radio command [analysis
                        table].
US20  82    autopilot   With command override.
GSN   83                Mid-course guidance required for maximum range.
JFS+  83    inertial

---------- TERMINAL GUIDANCE ----------
SWM   77    ? active radar
Rock  78    radar
FI    81    active radar
            IR
AWST  82    IR
            active radar
CFW   82    active radar
JFS   82    active radar
JWS   82    ? active radar
GSN   83    IR
            active radar
JFS+  83    ? IR        Possibly.

---------- CONTROL SYSTEM ----------
Rock  78    Aerodynamic.

---------- RADAR ----------
FI    81    Band Stand
JWS   82    Band Stand  Search and fire control radar on Nanuchka.
GSN   83    Band Stand  On surface ships, search and fire control.

---------- OBSERVED ----------
JWS   82                To date, no pictures or details of missiles themselves available.

---------- PRODUCTION UNDERWAY ----------
SMP   84                In production.
```

```
------------ IOC ------------
JFS   82   1968-69          Operational [Nanuchka listing only, ed]
GSN   83   1968-1969

------------ DEPLOYMENT BEGAN ------------
SWM   77   1968-69
ID11  78   1968             Nanuchka appeared.
Rock  78   1969             Service intro.
Clns  80   1971             First deployed.
FI    81   1969
IISS  82   "1968/9"         First year deployed.
JWS   82   1969             Nanuchkas "made their appearance".
CFW+  84   1969
BW    86   1969

------------ DEPLOYED ------------
MOW   76                    In service.
BW    86                    In service.

------------ DEPLOYMENT HISTORY ------------
GSN   83   Launched initially from surface ships, "reported to have been subsequently fitted" in later
           Charlies for underwater launch.

------------ NUMBERS IN SERVICE ------------

------------ NUMBER IN 1980 ------------
IISS  80   118       July, total deployed.

------------ NUMBER IN 1981 ------------
IISS  81   >_ 130    July, total deployed.

------------ NUMBER IN 1982 ------------
IISS  82   >_ 136    July, total deployed.

------------ NUMBER IN 1983 ------------
IISS+ 83   c. 154    July, total deployed.

------------ CARRIERS ------------

------------ CARRIER ------------ TUBES ---- RELOADS ------------
SecC  71                                      Part of armament of 130 "Small Cruise Missile Patrol
                                              Boats".

------------ CARRIER ------------ TUBES ---- RELOADS ------------
Rock  78   corvettes

------------ CARRIER ------------ TUBES ---- RELOADS ------------
MOW   76   Nanuchka           6
SWM   77   Nanuchka           x 3
Clns  80   Nanuchka           6                [Literally, "SA-N-9" arms Nanuchka, ed]
FI    81   Nanuchka           2 x 3            N>_14.
USND  81   Nanuchka
AWST  82   Nanuchka
CFW   82   Nanuchka-I
JFS   82   Nanuchka                            [Nanuchka listing only, ed]
JWS   82   Nanuchka           2 x 3            Only vessel definitely associated with SS-N-9. 4 SS-N-2Bs
                                               replace SS-N-9 on those exported to India.
Clns+ 83   Nanuchka           6
FI+   83   Nanuchka                            N>_16.
GSN   83   Nanuchka                            SS-N-2c replaces SS-N-9 on those exported to Algeria,
                                               India, Libya.
BW    86   Nanuchka I         6       no

------------ CARRIER ------------ TUBES ---- RELOADS ------------
SMP   81   Charlie I          8                [Link to SS-N-9 uncertain, ed]; First Charlie I completed
                                               in 1968.
CFW   82   Nanuchka-III
FI+   83   Nanuchka III                        N=5, heavier gun armament [than Nanuchka I]
BW    86   Nanuchka III       6       no

------------ CARRIER ------------ TUBES ---- RELOADS ------------
SMP   81   Charlie II         8                [Link to SS-N-9 uncertain, ed]
```

|  |  | CARRIER | TUBES | RELOADS |  |
|---|---|---|---|---|---|
| USND | 81 | Sarancha | | | |
| CFW | 82 | Sarancha | | | |
| JWS | 82 | Sarancha | | | Hydrofoils. |
| CLns+ | 83 | Sarancha | 4 | | |
| GSN | 83 | Sarancha | | | |
| JFS+ | 83 | Sarancha | | | |
| BW | 86 | Sarancha | ? 4 | no | |

|  |  | CARRIER | TUBES | RELOADS |  |
|---|---|---|---|---|---|
| ABFS | 84 | Sovremennyy | | | N=2. |

|  |  | CARRIER | TUBES | RELOADS |  |
|---|---|---|---|---|---|
| MOW | 76 | possibly some subs | | | |

|  |  | CARRIER | TUBES | RELOADS |  |
|---|---|---|---|---|---|
| CFW | 82 | Charlie I | | | |

|  |  | CARRIER | TUBES | RELOADS |  |
|---|---|---|---|---|---|
| USND | 81 | Charlie II | | | |
| CFW | 82 | Charlie II | | | |
| US20 | 82 | Charlie II | | | Some of them. |
| CLns+ | 83 | Charlie II | 8 | | |
| FI+ | 83 | ? Charlie II | 8 | | N=6, may carry SS-N-7 instead. |
| GSN | 83 | Charlie II | | | |
| JFS+ | 83 | Charlie II | | | |
| SMP | 84 | Charlie II | | | |
| BW | 86 | Charlie II | 8 | no | |

|  |  | CARRIER | TUBES | RELOADS |  |
|---|---|---|---|---|---|
| GSN | 83 | Charlie III | | | |

|  |  | CARRIER | TUBES | RELOADS |  |
|---|---|---|---|---|---|
| US18 | 81 | improved Charlie | | | |

|  |  | CARRIER | TUBES | RELOADS |  |
|---|---|---|---|---|---|
| USND | 81 | Papa | | | |
| CFW | 82 | Papa | | | |
| JFS | 82 | ? Papa | | | [Papa listing only, ed] |
| JWS | 82 | Papa | | | Conjectured Papa may have SS-N-9 or a derivative, but existing SS-N-7 may be retained. |
| US20 | 82 | Papa | | | |
| FI+ | 83 | ? Papa | 10 | | N=1, may carry SS-N-7 instead. |
| GSN | 83 | Papa | | | |
| JFS+ | 83 | Papa | | | |
| BW | 86 | Papa | ? 10 | no | |

|  |  | USE AND CONFIGURATION |
|---|---|---|
| US03 | 77 | Never exceeds c. 400 ft altitude. |
| SMP | 81 | Can be fired submerged. |
| US18 | 81 | Underwater launch [for Charlie]. |
| CFW | 82 | Submerged launch from subs. |
| US20 | 82 | Altitude, 225 ft. |

|  |  | TARGET TYPE |
|---|---|---|
| SecC | 71 | ship |
| FI | 81 | ship |
| SMP | 81 | ship |
| SMP | 84 | ship |

# SS-N-10

```
Sources are:  FI*   74    MOW   76    JWS*  77    USO3  77    USND  81    GSN   83
              SecC  71    ID04  74    NA03  76    SWM   77    USO4  77    AWST  82
```

### US DESIGNATION

| | | |
|---|---|---|
| SecC | 71 | SS-N-10 |
| FI* | 74 | SS-N-10 |
| MOW | 76 | SS-N-10 |
| JWS* | 77 | SS-N-10 |
| SWM | 77 | SS-N-10 |
| USND | 81 | SS-N-10   "Cancelled designator" [no other information, ed] |
| AWST | 82 | SS-N-10 |
| GSN | 83 | SS-N-10 |

### EDITOR'S NOTES

| | | |
|---|---|---|
| SecC | 71 | Listed FY72. |

### RANGE

| | | mi | nm | km | km conv | |
|---|---|---|---|---|---|---|
| SecC | 71 | | 15-25 | | 27.78-46.30 | |
| FI* | 74 | | | ? c. 50 | 50.00 | |
| ID04 | 74 | | 29 | | 53.71 | |
| MOW | 76 | c. 43.5 | | c. 70 | 69.99 | |
| NA03 | 76 | 35 | | | 56.32 | |
| JWS* | 77 | | c. 30 | c. 55 | 55.56 | Appears modest compared to Shaddock range. |
| SWM | 77 | 31 | | 50 | 49.88 | |
| AWST | 82 | 17-29 | | | 27.35-46.66 | |

### SPEED

| | | mph | kmph | Mach | kmph conv | |
|---|---|---|---|---|---|---|
| FI* | 74 | | | | | ? Supersonic. |
| MOW | 76 | | | | | Supersonic. |
| JWS* | 77 | | | > 1 | | |
| SWM | 77 | | | 1.2 | | |
| USO3 | 77 | | | 1.2-2.0 | | |
| AWST | 82 | | | 1.9 | | |

### WARHEAD TYPE

| | | nuclear | conven | chem | |
|---|---|---|---|---|---|
| JWS* | 77 | maybe | | | In the absence of precise long-range target location, doubtful if conventional warhead for ASW role would suffice. |

### WEIGHT

| | | lb | t | kg | kg conv |
|---|---|---|---|---|---|
| SWM | 77 | 6,000 | | 2,720 | 2,721 |

### LENGTH

| | | in | ft | cm | m | m conv |
|---|---|---|---|---|---|---|
| MOW | 76 | | 27'11" | | 8.5 | 8.51 |
| SWM | 77 | | 25 | | 7.6 | 7.62 |

### MIDCOURSE (OR MAIN) GUIDANCE

| | | |
|---|---|---|
| AWST | 82 | command |

### TERMINAL GUIDANCE

| | | | |
|---|---|---|---|
| JWS* | 77 | ? active radar | |
| | | passive radar | Possible variant. |
| SWM | 77 | radar | Radar guidance/anti-radiation passive homing. |
| AWST | 82 | IR/radar | Possibly. |

### RADAR

| | | | |
|---|---|---|---|
| USO3 | 77 | Eye Bowl | Derived from Head Light, sometimes called Head Light B. |

### IOC

| | | | |
|---|---|---|---|
| JWS* | 77 | 1968 | Became operational. |

### DEPLOYMENT BEGAN

| | | |
|---|---|---|
| SWM | 77 | 1968 |

### DEPLOYED

| | | |
|---|---|---|
| FI* | 74 | Operated by Soviet Navy. |
| MOW | 76 | In service. |

### CARRIERS

|       |    | CARRIER   | TUBES | RELOADS |                                              |
|-------|----|-----------|-------|---------|----------------------------------------------|
| SecC  | 71 |           |       |         | Part of armament of 20 "Guided Missile Combatants". |

|       |    | CARRIER | TUBES | RELOADS |      |
|-------|----|---------|-------|---------|------|
| IDO4  | 74 | Kara    | 2 x 4 |         |      |
| MOW   | 76 | Kara    | 8     |         | N=3. |
| JWS*  | 77 | Kara    | 8     |         | N=4. |

|       |    | CARRIER   | TUBES | RELOADS |      |
|-------|----|-----------|-------|---------|------|
| FI*   | 74 | Kresta II | 2 x 4 |         | N=3. |
| MOW   | 76 | Kresta II | 8     |         | N=8. |
| JWS*  | 77 | Kresta II | 8     |         | N=9. |
| SWM   | 77 | Kresta II | 2 x 4 |         |      |
| AWST  | 82 | Kresta 2  |       |         |      |

|       |    | CARRIER | TUBES | RELOADS |       |
|-------|----|---------|-------|---------|-------|
| FI*   | 74 | Krivak  | 1 x 4 |         |       |
| MOW   | 76 | Krivak  | 4     |         | N=9.  |
| JWS*  | 77 | Krivak  | 4     |         | N=11. |
| SWM   | 77 | Krivak  | 1 x 4 |         |       |
| AWST  | 82 | Krivak  |       |         |       |

|       |    | TARGET TYPE |                                                                                                                                                                                                                                                                        |
|-------|----|-------------|------------------------------------------------------------------------------------------------------------------------------------------------------------------------------------------------------------------------------------------------------------------------|
| SecC  | 71 | ship        |                                                                                                                                                                                                                                                                        |
| MOW   | 76 | surface     |                                                                                                                                                                                                                                                                        |
| JWS*  | 77 | ship, sub   | Recent suggestion that SS-N-10 has dual role, both anti-ship and ASW. Near press deadline, "authoritatively" asserted that launchers do not carry anti-ship missiles. Therefore SS-N-10 designation must be spurious. Perhaps SS-N-14 missile is "alternative round" for SS-N-10 launchers. |
| USO4  | 77 |             | "Apparently the SS-N-10 simply does not exist".                                                                                                                                                                                                                        |
| AWST  | 82 | ship        |                                                                                                                                                                                                                                                                        |
| GSN   | 83 |             | The SS-N-14 ASW missile originally assigned this designation when Western intelligence evaluated it as an anti-ship missile.                                                                                                                                           |

# SS-N-12

| Sources are: | Clns | 80 | SMP | 81 | IISS | 82 | Clns+ | 83 | JFS+ | 83 | SMP | 84 |
|---|---|---|---|---|---|---|---|---|---|---|---|---|
| SWM | 77 | | IISS | 80 | USND | 81 | JFS | 82 | FI+ | 83 | SMP | 83 | SMP | 85 |
| USO3 | 77 | | FI | 81 | AWST | 82 | JWS | 82 | GSN | 83 | CFW+ | 84 | BW | 86 |
| ID11 | 78 | | IISS | 81 | CFW | 82 | US20 | 82 | IISS+ | 83 | Meyr | 84 | | |

---------- **US DESIGNATION** ----------

| SWM | 77 | SS-NX-12 | |
|---|---|---|---|
| Clns | 80 | SS-N-12 | |
| FI | 81 | SS-N-12 | [Table] |
| | | SS-NX-12 | [Text] |
| SMP | 81 | SS-N-12 | '81, '83, '84, '85. |
| USND | 81 | SS-N-12 | |
| AWST | 82 | SS-NX-12 | |
| CFW | 82 | SS-N-12 | |
| IISS | 82 | SS-N-12 | |
| JFS | 82 | SS-N-12 | |
| JWS | 82 | SS-N-12 | |
| GSN | 83 | SS-N-12 | |
| Meyr | 84 | SS-N-12 | |

---------- **NATO CODENAME** ----------

| SMP | 81 | Sandbox |
|---|---|---|
| IISS | 82 | Sandbox |
| JWS | 82 | Sandbox |
| GSN | 83 | Sandbox |
| CFW+ | 84 | Sandbox |

---------- **DESCRIPTION** ----------

| USND | 81 | SSM |
|---|---|---|
| IISS | 82 | SLCM |

---------- **EDITOR'S NOTES** ----------

| SMP | 81 | Listed '81, '83, '84, '85. |
|---|---|---|
| IISS | 82 | Source has two unlabeled versions for SS-N-12 on separate lines. Data is common to both unless noted as "1,000 km version" or "550 km version". |
| Meyr | 84 | Note to chart: "Values pertain to initial service period". |

---------- **RANGE** ----------

| | | mi | nm | km | km conv | |
|---|---|---|---|---|---|---|
| SWM | 77 | 280-1,553 | | 450-2,500 | 450.52-2,498 | |
| USO3 | 77 | | 300 | | 555.60 | Cruising M2.5 at 35,000 ft. |
| | | | <_ 2,000 | | 3,704 | If flying higher at transonic speed. |
| ID11 | 78 | | | 450 | 450.00 | |
| Clns | 80 | | 345 | | 638.94 | |
| FI | 81 | | | c. 500 | 500.00 | |
| | | | | 3,000 | 3,000 | Similar to Shaddock, but could be adapted for 3,000 km at transonic speed. |
| SMP | 81 | | | c. 550 | 550.00 | '81, '85. Max. |
| USND | 81 | | c. 300 | | 555.60 | |
| AWST | 82 | | 300 | | 555.60 | |
| CFW | 82 | | 300 | | 555.60 | Max. |
| IISS | 82 | | | 1,000 | 1,000 | Max, [1,000 km version, ed] |
| | | | | 550 | 550.00 | Max, [550 km version, ed] |
| JFS | 82 | | 300 | | 555.60 | Max. |
| JWS | 82 | | | 550 | 550.00 | Estimated max. Payload/range inferred from size of launcher. |
| US20 | 82 | | 300 | | 555.60 | Altitude 30,000 ft for 300 nm range. |
| Clns+ | 83 | | 345 | | 638.94 | [Ship entry] |
| | | | 350 | | 648.20 | [Sub entry] |
| GSN | 83 | | 300 | 555 | 555.60 | |
| Meyr | 84 | | | 600 | 600.00 | |
| SMP | 85 | | | 550 | 550.00 | |
| BW | 86 | | | 550 | 550.00 | |

## SPEED

|  |  | mph | kmph | Mach | kmph conv |  |
|---|---|---|---|---|---|---|
| SWM | 77 |  |  | 2.5 |  |  |
| FI | 81 |  |  | 2.5 |  |  |
| AWST | 82 |  |  | 2.5 |  |  |
| JFS | 82 |  |  | 2.5 |  |  |
| GSN | 83 |  |  |  |  | Approximately twice as fast as SS-N-3. |
| SMP | 85 |  |  |  |  | Supersonic. |
| BW | 86 |  |  | 2.5 |  |  |

## CEP

|  |  | ft | mi | nm | m | km | m conv |
|---|---|---|---|---|---|---|---|
| Meyr | 84 |  |  |  |  | 0.3-0.7 | 300.00-700.00 |

## WARHEAD TYPE

|  |  | nuclear | conven | chem |
|---|---|---|---|---|
| US03 | 77 | yes |  |  |
| Clns | 80 | yes | yes | HE. |
| FI | 81 | yes | yes |  |
| CFW | 82 | yes | yes |  |
| IISS | 82 | yes | yes |  |
| JFS | 82 | yes | yes | HE. |
| US20 | 82 | yes | yes | HE. |
| GSN | 83 | yes | yes |  |
| Meyr | 84 | yes |  |  |

## WARHEAD CHARACTERISTICS

| IISS | 82 | single warhead |
|---|---|---|
| Meyr | 84 | single RV |
| BW | 86 | single RV |

## NUCLEAR YIELD

|  |  | kT | MT | kT conv |  |
|---|---|---|---|---|---|
| US03 | 77 |  |  |  | Large nuclear warhead. |
| Clns | 80 |  |  |  | KT range, "KT". |
| IISS | 82 | 350 |  | 350.00 |  |
| JFS | 82 | 350 |  | 350.00 |  |
| US20 | 82 | 350 |  | 350.00 |  |
| Clns+ | 83 | 10 |  | 10.00 | [Sub entry] |
| Meyr | 84 | 100-200 |  | 100.00-200.00 |  |
| BW | 86 | 350 |  | 350.00 |  |

## WARHEAD WEIGHT

|  |  | lb | t | kg | kg conv |
|---|---|---|---|---|---|
| Clns | 80 | 2,000 |  |  | 907.20 |
| FI | 81 |  |  | c. 1,000 | 1,000 |
| JFS | 82 | 2,200 |  |  | 997.92 |
| US20 | 82 | 2,200 |  |  | 997.92 |
| GSN | 83 | 2,200 |  | 1,000 | 997.92 |
| CFW+ | 84 |  |  | 1,000 | 1,000 |

## THROWWEIGHT

|  |  | lb | t | kg | kg conv |  |
|---|---|---|---|---|---|---|
| IISS | 82 | 2,200 |  |  | 997.92 | [1,000 km version only, ed] |

## WEIGHT

|  |  | lb | t | kg | kg conv |
|---|---|---|---|---|---|
| FI | 81 |  |  | c. 5,000 | 5,000 |

## LENGTH

|  |  | in | ft | cm | m | m conv |  |
|---|---|---|---|---|---|---|---|
| US03 | 77 |  | 50 |  |  | 15.24 | Launch tube on Kiev. |
| FI | 81 |  |  |  | c. 10 | 10.00 |  |
| GSN | 83 |  | 38'6" |  | 11.7 | 11.73 |  |

## DIAMETER

|  |  | in | ft | cm | m | cm conv |  |
|---|---|---|---|---|---|---|---|
| US03 | 77 |  | 8 |  |  | 243.84 | Launch tube on Kiev. |

## WINGSPAN

|  |  | in | ft | cm | m | cm conv |
|---|---|---|---|---|---|---|
| FI | 81 |  |  | c. 250 |  | 250.00 |

512  SOVIET MISSILES  IDDS

### TYPE OF PROPULSION
| | | | |
|---|---|---|---|
| SWM | 77 | turbojet | |
| FI | 81 | turbojet or ramjet | [Table] |
| | | turbojet | [Text] |
| | | JATO units | Boost. |
| CFW | 82 | liquid | |
| GSN | 83 | turbojet | |
| BW | 86 | TJ | |

### MIDCOURSE (OR MAIN) GUIDANCE
| | | | |
|---|---|---|---|
| FI | 81 | [autopilot] | [Plus midcourse update], ? as SS-N-3. |
| SMP | 81 | | Hormone B helicopter (seen on Kiev) is "capable of providing over-the-horizon targeting information for" SS-N-12, '81, '85. |
| JWS | 82 | autopilot/inertial | Most likely; probably aided by command updating at intervals. Long range implies mid-course guidance by a/c would be useful, and especially so from subs. |
| US20 | 82 | autopilot | With midcourse correction. |
| GSN | 83 | radio command | |
| JFS+ | 83 | inertial | |

### TERMINAL GUIDANCE
| | | | |
|---|---|---|---|
| FI | 81 | [active radar] | ? As SS-N-3. |
| JFS | 82 | active radar | |
| JWS | 82 | active radar | |
| US20 | 82 | active radar | |
| GSN | 83 | active radar | |

### CONTROL SYSTEM
| | | |
|---|---|---|
| FI | 81 | Aeroplane configuration. |

### RADAR
| | | | |
|---|---|---|---|
| FI | 81 | | On retractable mount, E or F band (C or D according to one source). |
| GSN | 83 | Trap Door | On Kiev. |

### OBSERVED
| | | |
|---|---|---|
| JWS | 82 | No publicly released photos of actual missile. Payload/range inferred from size of launcher. |

### PRODUCTION UNDERWAY
| | | |
|---|---|---|
| SMP | 84 | In production. |

### IOC
| | | | |
|---|---|---|---|
| JFS | 82 | 1973 | Operational. |
| JWS | 82 | 1973 | Believed operational. |
| GSN | 83 | 1973 | |

### DEPLOYMENT BEGAN
| | | | |
|---|---|---|---|
| Clns | 80 | 1976 | First deployed. |
| IISS | 82 | 1976 | First year deployed, [1,000 km version only, ed] |
| CFW+ | 84 | 1973 | |
| Meyr | 84 | 1975 | Service. |
| BW | 86 | 1973 | |

### DEPLOYED
| | | |
|---|---|---|
| BW | 86 | In service. |

### DEPLOYMENT HISTORY
| | | |
|---|---|---|
| SWM | 77 | Shaddock successor. |
| FI | 81 | Replacement for SS-N-3 Shaddock. Able to use Shaddock launchers. |
| SMP | 81 | Improvement over SS-N-3, '81, '85. |
| AWST | 82 | Follow-on for SS-N-3. |
| IISS | 82 | SS-N-3 replacement. |
| JFS | 82 | Development of SS-N-3. |
| JWS | 82 | Generally presumed to be improved SS-N-3. |
| GSN | 83 | Advanced version of SS-N-3 Shaddock. |

### NUMBERS IN SERVICE

### NUMBER IN 1980
| | | | |
|---|---|---|---|
| IISS | 80 | 48 | July, total deployed. |

### NUMBER IN 1981
| | | | |
|---|---|---|---|
| IISS | 81 | 56 | July, total deployed. |

### NUMBER IN 1982
| | | | |
|---|---|---|---|
| IISS | 82 | 32 | July, total deployed. |

CRUISE MISSILES Sea-Launched SSMs SS-N-12

```
------------ NUMBER IN 1983 ------------
IISS+ 83    80          July, total deployed.
           ------------ CARRIERS ------------

           ------------ CARRIER ------------ TUBES ---- RELOADS ------------
SWM  77    others                                              [Than Echo II], possibly ships.

           ------------ CARRIER ------------ TUBES ---- RELOADS ------------
JFS  82    subs

           ------------ CARRIER ------------ TUBES ---- RELOADS ------------
SWM  77    Echo II
Clns 80    Echo II          8                    "SS-N-3/12".
USND 81    Echo II
AWST 82    Echo 2
CFW  82    Echo II                               Replacing SS-N-3.
JWS  82    ? Echo II                             Possibly.
FI+  83    ? Echo II                             N=29, may be deployed on several of these.
GSN  83    Echo II
JFS+ 83    Echo 2
SMP  85    Echo II                               USSR proceeding with conversion of "older 1960 vintage
                                                 SS-N-3-equipped ECHO II SSGNs" to SS-N-12.
BW   86    Echo II          8      no

           ------------ CARRIER ------------ TUBES ---- RELOADS ------------
Clns 80    Juliett          4                    "SS-N-3/12".

           ------------ CARRIER ------------ TUBES ---- RELOADS ------------
JFS  82    surface ships

           ------------ CARRIER ------------ TUBES ---- RELOADS ------------
FI   81    Kiev, Minsk      8      maybe
SMP  81    Kiev             8      yes           '81, '85.
USND 81    Kiev
CFW  82    Kiev
JFS  82    Kiev
JWS  82    Kiev           [4 x 2]                New style launcher.
                                                 tubes: 8 on foredeck [appear in picture as 4 x 2]
GSN  83    Kiev
SMP  83    Minsk          4 x 2                  On foredeck, '83, '84.
                                                 tubes: '83, '84.
BW   86    Kiev             8      yes

           ------------ CARRIER ------------ TUBES ---- RELOADS ------------
FI   81                                          ? New cruiser class [final report]
CFW  82    BLK-COM-1
CFW+ 84    Slava
SMP  85    Slava           16
                                                 reloads: Total.
BW   86    Slava                  yes            N=1.

           ------------ USE AND CONFIGURATION ------------
SWM  77    Altitude 10,670 m.
ID11 78    Could descend steeply onto target.
AWST 82    Altitude 35,000 ft.
US20 82    Altitude 30,000 ft for 300 nm range.
JFS+ 83    Altitude 35,000 ft.
Meyr 84    Penetration: 0.7-0.9; reliability: 0.5-0.9, "includes system reliability multiplied by operational
           readiness of deployed system".

           ------------ TARGET TYPE ------------
FI   81    ship
SMP  81    ship            '84, '85.
AWST 82    ship
JWS  82    ship

           ------------ STRATEGY ------------
JWS  82    Long-range anti-ship engagement thought to be role, as for SS-N-3.
Meyr 84    [Ed: chart is labeled] Systems "assigned primary operational and strategic missions within the
           European theatre(s) of military operations".
```

## SS-N-13

```
Sources are:  JWS*  75    JCS   77    Rock  78    US06  78    US11  80    USND  81    CFW   82
              JCS   75    MOW   76    ID11  78    US05  78    FI*   79    US12  80    AWST  82    GSN   83
```

#### US DESIGNATION

| | | |
|---|---|---|
| JCS  | 75 | SS-NX-13   FY76,78. |
| JWS* | 75 | SSN(X)-13 |
| MOW  | 76 | SS-N-13 |
| Rock | 78 | SS-NX-13 |
| FI*  | 79 | SS-NX-13 |
| USND | 81 | SS-NX-13 |
| AWST | 82 | SS-NX-13 |
| CFW  | 82 | SS-N-13 |
| GSN  | 83 | SS-NX-13 |

#### DESCRIPTION

| | | |
|---|---|---|
| MOW  | 76 | Missile could be ballistic. |
| JCS  | 77 | Tactical ballistic missile. |
| FI*  | 79 | SLBM |
| USND | 81 | SLBM |
| AWST | 82 | Ballistic missile. |
| CFW  | 82 | Tactical ballistic missile. |
| GSN  | 83 | Tactical ballistic missile. |

#### EDITOR'S NOTES

| | | |
|---|---|---|
| JCS | 75 | Listed FY76,78. |
| CFW | 82 | Final report of SS-N-13. |

#### RANGE

| | | mi | nm | km | km conv | |
|---|---|---|---|---|---|---|
| JWS* | 75 |  |  | c. 750 | 750.00 | |
| MOW  | 76 | 400 |  | 645 | 643.60 | |
| ID11 | 78 |  |  | ? 700 | 700.00 | |
| Rock | 78 |  |  | 750 | 750.00 | |
| US06 | 78 | 350-400 |  |  | 563.15-643.60 | |
| FI*  | 79 |  |  | 750 | 750.00 | |
| US11 | 80 |  | 300 |  | 555.60 | |
| US12 | 80 |  | 400 |  | 740.80 | [One place in article] |
|      |    |  | 100-600 |  | 185.20-1,111 | [Other place in article] |
| AWST | 82 |  | 100-600 |  | 185.20-1,111 | |
| CFW  | 82 |  | 370 |  | 685.24 | |
| GSN  | 83 |  | c. 370 | c. 685 | 685.24 | |

#### SPEED

| | | mph | kmph | Mach | kmph conv | |
|---|---|---|---|---|---|---|
| JWS* | 75 |  |  | 4.0 |  | Terminal. |
| Rock | 78 |  |  | 4 |  | |
| FI*  | 79 |  |  | 4 |  | |

#### WARHEAD TYPE

| | | nuclear | conven | chem |
|---|---|---|---|---|
| JWS* | 75 | yes | | |
| Rock | 78 | yes | | |
| AWST | 82 | yes | | |
| CFW  | 82 | yes | | |
| GSN  | 83 | yes | | Nuclear warhead would have to be used with operational missile. |

#### WARHEAD CHARACTERISTICS

| | | |
|---|---|---|
| US12 | 80 | single RV |

#### NUMBER OF STAGES

| | | |
|---|---|---|
| US12 | 80 | 2 |
| AWST | 82 | 2 |

#### TYPE OF PROPULSION

| | | |
|---|---|---|
| US12 | 80 | liquid |
| CFW  | 82 | liquid |

#### MIDCOURSE (OR MAIN) GUIDANCE

| | | |
|---|---|---|
| JWS* | 75 | Terminal guidance in conjunction with satellite targeting. |
| Rock | 78 | Satellite updates. |
| US06 | 78 | Initial reports suggest either aircraft or satellite. In-flight guidance could change course by up to 30 nm. |
| US12 | 80 | inertial |
| GSN  | 83 | Possibly with satellite targeting at launch. |

## CRUISE MISSILES Sea-Launched SSMs SS-N-13

### TERMINAL GUIDANCE
| | | | |
|---|---|---|---|
| JWS* | 75 | yes | |
| US12 | 80 | radar | "A sensor (undefined) which 'locks on' to the target near apogee"; course change by up to 50 km (35 nm). |
| AWST | 82 | radar | "Radar homing maneuvers". |
| GSN | 83 | radar | Terminal maneuvering capability of some 30 nm (55 km). |

### OBSERVED
| | | |
|---|---|---|
| US06 | 78 | "The press" began describing SS-NX-13 in 1973. |

### PROTOTYPE TESTS
| | | | |
|---|---|---|---|
| JCS | 75 | | Currently under development. |
| MOW | 76 | | Has been tested. |
| JCS | 77 | | Advanced technology of it is significant; project could be resurrected. Not tested since Nov 1973. |
| FI* | 79 | | No tests reported since Nov 1973, so presumably cancelled. |
| US12 | 80 | | Program "dormant since 1973". Cancelled. Tests completed Nov 1973. |
| USND | 81 | | Experimental, inactive. |
| CRW | 82 | | "Program suspended in 1973 due to inadequate technology, but the concept may yet be revived in a new program". |
| GSN | 83 | late 1960s | Flight-tested in USSR from late 1960s until Nov 1973. |

### IOC
| | | |
|---|---|---|
| US12 | 80 | 1969 |

### DEPLOYMENT BEGAN
| | | | |
|---|---|---|---|
| MOW | 76 | 1975 | May have been put in service. |
| JCS | 77 | | Not operational. |
| AWST | 82 | | "Not yet operational". |
| GSN | 83 | | Has not been deployed. |

### CARRIERS

#### CARRIER --- TUBES --- RELOADS
| | | |
|---|---|---|
| MOW | 76 | sub |
| AWST | 82 | sub |

#### CARRIER --- TUBES --- RELOADS
| | | |
|---|---|---|
| JWS* | 75 | G |

#### CARRIER --- TUBES --- RELOADS
| | | | |
|---|---|---|---|
| JCS | 77 | Yankee | May have been intended for Yankee. |
| US05 | 78 | Yankee | Perhaps. |
| FI* | 79 | Yankee | Intended to be compatible with SS-N-6 tubes. |
| US12 | 80 | ? Yankee | |
| GSN | 83 | Yankee | Developed for Yankee. |

### USE AND CONFIGURATION
| | | |
|---|---|---|
| Rock | 78 | Underwater launch. |
| US12 | 80 | Apogee is 150 nm. |
| AWST | 82 | Subsurface launched. Apogee is 150 nm. |

### TARGET TYPE
| | | |
|---|---|---|
| JCS | 75 | ship |
| JCS | 77 | ship |
| AWST | 82 | ship |

### STRATEGY
| | | |
|---|---|---|
| JWS* | 75 | US DoD sees it as potential threat to USN fleet. Standard SM-2 missile with nuclear warhead (with Aegis) being considered as a defence. Perhaps depressed-trajectory. Unofficial reports are inconsistent on range-speed. [Source provides much more information, ed]. |
| Rock | 78 | Anti-aircraft carrier or FBMS. |
| US06 | 78 | Initially target thought to be surface ships, later SSBNs. |
| FI* | 79 | Intended for use against CVs and missile subs. |
| GSN | 83 | Intended for anti-carrier role. |

# SS-N-14

```
Sources are:   Rock 78    US14 80    CFW  82    US19 82    IISS+ 83   SMP  85
SWM  77        AW03 79    FI   81    IISS 82    Clns+ 83   JFS+  83   BW   86
US03 77        US10 79    USND 81    JFS  82    FI+   83   JWS+  83
US04 77        Clns 80    AWST 82    JWS  82    GSN   83   CFW+  84
```

## ————— US DESIGNATION —————

|  | |  | |
|---|---|---|---|
| SWM | 77 | SS-N-14 | |
| Rock | 78 | SS-N-14 | |
| Clns | 80 | SS-N-14 | |
| FI | 81 | SS-N-14 | |
| USND | 81 | SS-N-14 | |
| AWST | 82 | SS-N-14 | |
| CFW | 82 | SS-N-14 | |
| IISS | 82 | SS-N-14 | |
| JFS | 82 | SSN-14 | [Missile entry, ed] |
|  |  | SN-N-14 | [Sic, ASW entry, ed] |
| JWS | 82 | SS-N-14 | |
| GSN | 83 | SS-N-14 | |
| SMP | 85 | SS-N-4 | |

## ————— NATO CODENAME —————

| Source | | Codename |
|---|---|---|
| USND | 81 | Silex |
| IISS | 82 | Silex |
| JWS | 82 | Silex |
| FI+ | 83 | Silex |
| GSN | 83 | Silex |
| CFW+ | 84 | Silex |

## ————— DESCRIPTION —————

| Source | | Description |
|---|---|---|
| IISS | 82 | SLCM |
| JFS | 82 | Rocket-assisted torpedo launcher [ASW entry, ed] |

## ————— EDITOR'S NOTES —————

|  | |  |
|---|---|---|
| JFS | 82 | Source has separate listings under ASW Weapons and under Missiles. Data is from the Missiles entry unless otherwise marked. |
| SMP | 85 | Listed '85. |

## ————— RANGE —————

| Source | | mi | nm | km | km conv | |
|---|---|---|---|---|---|---|
| US03 | 77 | c. 20 | | | 32.18 | |
| Rock | 78 | | | 55 | 55.00 | |
| US10 | 79 | | 25 | | 46.30 | |
| Clns | 80 | | 30 | | 55.56 | |
| FI | 81 | | | 40 | 40.00 | [Table] |
|  |  | | | 30 | 30.00 | [Text] |
| USND | 81 | | c. 25 | | 46.30 | |
| AWST | 82 | 30 | | | 48.27 | |
| CFW | 82 | | 30 | | 55.56 | Max. |
|  |  | | 4 | | 7.41 | Min. |
| IISS | 82 | | | 55 | 55.00 | Max. |
| JFS | 82 | | 30 | | 55.56 | Max. |
| JWS | 82 | | | 55 | 55.00 | Max. |
| GSN | 83 | | 30 | 55 | 55.56 | |
|  |  | | 4 | 7.4 | 7.41 | Minimum effective range, estimated. |
| JFS+ | 83 | | 4 | | 7.41 | Min [ASW entry, ed] |
|  |  | | 30 | | 55.56 | Max [ASW and Missile entries, ed] |
| BW | 86 | | | 55 | 55.00 | |

## ————— SPEED —————

| Source | | mph | kmph | Mach | kmph conv | |
|---|---|---|---|---|---|---|
| FI | 81 | | | | Subsonic. | |
| JFS | 82 | | | 0.9 | | |
| JWS | 82 | | | 0.95 | | At 750 m above sea. |
| BW | 86 | | | 0.9 | | |

## ————— WARHEAD TYPE —————

| Source | | nuclear | conven | chem | |
|---|---|---|---|---|---|
| US04 | 77 | yes | yes | | |
| Rock | 78 | yes | | | |
| Clns | 80 | yes | yes | | HE. |
| JFS | 82 | yes | | | "Homing torpedo or nuclear". |
| JWS | 82 | yes | yes | | "Nuclear warhead or homing torpedo". |
| GSN | 83 | | yes | | |

CRUISE MISSILES  Sea-Launched SSMs  SS-N-14

## WARHEAD CHARACTERISTICS

| | | | |
|---|---|---|---|
| Rock | 78 | torpedo | |
| FI | 81 | torpedo | Acoustic-homing. |
| AWST | 82 | torpedo | Acoustic homing. |
| JFS | 82 | | "Homing torpedo or nuclear". |
| JWS | 82 | | "Believed that a nuclear warhead can be carried instead of the torpedo". |
| GSN | 83 | torpedo | Acoustic ASW homing torpedo. |
| BW | 86 | single RV | |

## NUCLEAR YIELD

| | | kT | MT | kT conv | |
|---|---|---|---|---|---|
| Clns | 80 | | | | KT range, "KT". |
| IISS | 82 | | < 1 | 1,000 | |
| JWS | 82 | | | | Low kT range. |
| BW | 86 | ? 10 | | 10.00 | |

## WARHEAD WEIGHT

| | | lb | t | kg | kg conv | |
|---|---|---|---|---|---|---|
| JWS | 82 | | | 100 | 100.00 | Weight of nuclear warhead. |

## LENGTH

| | | in | ft | cm | m | m conv |
|---|---|---|---|---|---|---|
| FI | 81 | | | | 7-8 | 7.00-8.00 |
| JFS | 82 | | 24.9 | | | 7.59 |
| JWS | 82 | | | | 7.6 | 7.60 |
| GSN | 83 | | c. 25 | | c. 7.6 | 7.62 |

## TYPE OF PROPULSION

| | | | |
|---|---|---|---|
| Rock | 78 | solid | |
| FI | 81 | solid | |
| CFW | 82 | solid | |
| JFS | 82 | solid | |
| GSN | 83 | solid | Rocket-propelled. |
| JFS+ | 83 | | Cruise-missile assisted [ASW entry, ed] |
| BW | 86 | solid | |

## MIDCOURSE (OR MAIN) GUIDANCE

| | | | |
|---|---|---|---|
| AWST | 82 | autopilot | Command override. |
| JFS | 82 | radio command | |
| JWS | 82 | autopilot | "Programmed/radio command flight to target area under autopilot control with command override capability". |
| GSN | 83 | inertial | |

## CONTROL SYSTEM

| | | |
|---|---|---|
| FI | 81 | Aeroplane configuration. |

## RADAR

| | | | |
|---|---|---|---|
| US14 | 80 | | [On Kirov] SS-N-14 is controlled by 2 FCS. |
| JWS | 82 | Eye Bowl | Normally associated with system, but not on Kresta II. 2 on Kirov and Krivak. |
| | | Bass Tilt | On Kresta II may perform control function normally assigned to Eye Bowl. |
| US19 | 82 | | 2 "Directors" [FCS]. |
| CFW+ | 84 | Head Lights | Radar director. |
| | | Eye Bowl | Radar director. |

## OBSERVED

| | | |
|---|---|---|
| JWS | 82 | No confirmed photos of SS-N-14 missile released. |

## IOC

| | | | |
|---|---|---|---|
| JFS | 82 | 1968 | Operational [Missile entry, ed] |
| | | 1970 | [ASW entry, ed] |
| JWS | 82 | 1968 | Operational. |
| GSN | 83 | 1968 | |

## DEPLOYMENT BEGAN

| | | | |
|---|---|---|---|
| SWM | 77 | | Entering service. |
| Rock | 78 | ? 1975 | Service intro. |
| Clns | 80 | 1974 | First deployed. |
| IISS | 82 | 1974 | First year deployed. |
| CFW+ | 84 | 1974 | |
| BW | 86 | 1968 | |

## DEPLOYED

| | | |
|---|---|---|
| BW | 86 | In service. |

## NUMBERS IN SERVICE

```
------- NUMBER IN 1982 -------
IISS  82    292       July, total deployed.

------- NUMBER IN 1983 -------
IISS+ 83    c. 288    July, total deployed.

------- CARRIERS -------

------- CARRIER ------- TUBES --- RELOADS -------
Rock  78    cruisers
            destroyers
AWST  82    cruiser

------- CARRIER ------- TUBES --- RELOADS -------
SWM   77    Kara
US10  79    Kara                      8 total
Clns  80    Kara          8
FI    81    Kara          2 x 4                 N=5, launched from container.
USND  81    Kara
CFW   82    Kara
JFS   82    Kara
JWS   82    Kara
GSN   83    Kara                      no
BW    86    Kara          8           no

------- CARRIER ------- TUBES --- RELOADS -------
SWM   77    Krivak
AW03  79    Krivak 2      4
Clns  80    Krivak        4
FI    81    Krivak        1 x 4                 N=10, launched from container.
USND  81    Krivak I/II
CFW   82    Krivak-I, II
JFS   82    Krivak
JWS   82    Krivak        1 x 4
GSN   83    Krivak                    no
BW    86    Krivak I,II   4           no

------- CARRIER ------- TUBES --- RELOADS -------
SWM   77    Moskva
USO3  77    Moskva        1 x 2
FI    81    Moskva                              N=2, launched from multi-purpose SUW-N-1 launcher.

------- CARRIER ------- TUBES --- RELOADS -------
USO3  77    Kiev          1 x 2
FI    81    Kiev                                Launched from multi-purpose SUW-N-1 launcher.

------- CARRIER ------- TUBES --- RELOADS -------
SWM   77    Kresta II
US10  79    Kresta II                 8 total
Clns  80    Kresta II     8
FI    81    Kresta II     2 x 4                 N=10.
USND  81    Kresta II
CFW   82    Kresta-II
JFS   82    Kresta II
JWS   82    Kresta II     2 x 4
GSN   83    Kresta II     x 4         no
                                                tubes: [Quadruple launcher shows in picture, ed]
BW    86    Kresta II     8           no

------- CARRIER ------- TUBES --- RELOADS -------
US14  80    Kirov         1 x 2       yes
FI    81    Kirov
USND  81    Kirov
CFW   82    Kirov
JFS   82    Kirov
JWS   82    Kirov         1 x 2
Clns+ 83    Kirov         2
GSN   83    Kirov                     yes
                                                reloads: Onboard reload capability.
SMP   85    Kirov                               Principal armament of 20 SS-N-19s "complemented by
                                                launchers for the SS-N-14 antisubmarine missile in the
                                                first ship of the class [Kirov, ed.] only."
BW    86    Kirov         2           yes       N=1.
```

# CRUISE MISSILES Sea-Launched SSMs SS-N-14

|  |  | **CARRIER** | **TUBES** | **RELOADS** |
|---|---|---|---|---|
| US19 | 82 | Udaloy | 8 | [no] |
| GSN | 83 | Udaloy |  | no |
| JWS+ | 83 | Udaloy | 2 x 4 |  |
| BW | 86 | Udaloy | 8 | no |

|  |  | **CARRIER** | **TUBES** | **RELOADS** |
|---|---|---|---|---|
| JWS+ | 83 | ? Krasina |  |  |

|  |  | **CARRIER** | **TUBES** | **RELOADS** |
|---|---|---|---|---|
| AWST | 82 | submarine |  |  |

### USE AND CONFIGURATION

| | | |
|---|---|---|
| USD4 | 77 | Underwater speed of 30 knots possible with a high re-entry speed and modern propellants. |
| FI | 81 | A small winged missile drops a homing torpedo or nuclear depth charge near target sub. |
| CFW | 82 | Aerodynamic cruise missile drops a parachute-retarded hominng torpedo. |
| JFS | 82 | 4 cells, elevatable [ASW entry, ed] |
| JWS | 82 | Believed possible SS-N-14 has anti-ship capability; may be able to drop torpedo outside "ship's normal defensive cover". No confirmation of this possibility seen. Flies 750 m above sea. |
| GSN | 83 | Carries homing torpedo "out to the first sonar convergence zone". Water entry is slowed by parachute. |

### TARGET TYPE

| | | | |
|---|---|---|---|
| FI | 81 | sub | |
| USND | 81 | sub | |
| CFW | 82 | [sub] | Also surface ships. |
| IISS | 82 | sub | |
| JFS | 82 | sub | ASW torpedo, also reported to have surface-to-surface capability [ASW entry, ed] |
|  |  | sub, ? ship | "Probably has an anti-surface ship capability (originally SS-N-10)". |
| JWS | 82 | sub, ? ship | For many years was thought anti-ship rather than ASW system; evidence causing change of assessment has not been released but is believed to be conclusive. Also believed possible SS-N-14 has anti-ship capability. |
| GSN | 83 | sub, ?ship | ASW weapon; probably could be employed against surface ships too. Initially evaluated by Western intelligence as antiship, and labeled SS-N-10. |
| SMP | 85 | sub | |

### RELATION TO OTHER MISSILES

| | | |
|---|---|---|
| FI | 81 | Similar in concept to Ikara and Malafon. |
| CFW | 82 | Conceptually resembling French Malafon. |
| JWS | 82 | Somewhat similar in concept to French Malafon or Australian Ikara, in which subsonic winged vehicle carries homing torpedo to position of target submarine, with course correction possible in flight. |
| CFW+ | 84 | Conceptually resembles Australian Ikara. |

# SS-N-15

```
Sources are:   FI    81    JFS   82    GSN   83    ABFS  84    SMP   85
Rock  78       USND  81    JWS   82    JFS+  83    CFW+  84    BW    86
JCS   80       CFW   82    Clns+ 83    JWS+  83    SMP   84
```

------------ **US DESIGNATION** ------------

|  | |  |
|---|---|---|
| Rock | 78 | SS-N-15 |
| JCS | 80 | SS-N-15 |
| FI | 81 | SS-N-15 |
| USND | 81 | SS-N-15 |
| CFW | 82 | SS-N-15 |
| JFS | 82 | SSN-15  [Missile entry, ed] |
|  |  | SS-N-15 [ASW entry, ed] |
| JWS | 82 | SS-N-15 |
| Clns+ | 83 | SS-N-15 |
| GSN | 83 | SS-N-15 |
| SMP | 84 | SS-N-15 |
| SMP | 85 | SS-N-15 |

------------ **DESCRIPTION** ------------

| | | |
|---|---|---|
| JCS | 80 | Existing Soviet stand-off weapon [ASW, sub-launched]. |
| JFS | 82 | Rocket-assisted torpedo launcher [ASW entry, final report, ed] |
| SMP | 85 | nuclear depth-bomb |

------------ **EDITOR'S NOTES** ------------

| | | |
|---|---|---|
| JCS | 80 | Listed FY81. |
| JFS | 82 | Source has separate listings under ASW Weapons and under Missiles. Data is from the Missiles entry unless otherwise marked. |
| Clns+ | 83 | Missile is listed both with sub-based and surface ship-based weapons. Data is common to both unless marked. |
| SMP | 84 | Listed '84, '85. |

------------ **RANGE** ------------

| Source | | mi | nm | km | km conv | |
|---|---|---|---|---|---|---|
| Rock | 78 | | | 40 | 40.00 | |
| JCS | 80 | | | | | If US subs don't get new nuclear-capable weapon after SUBROC phased out, US will have to rely on Mk 48 torpedo, whose use will force US subs to close within Soviet detection envelope "and within range" of SS-N-15 and SS-NX-16. |
| FI | 81 | | | 40 | 40.00 | |
| USND | 81 | | | | | Short range. |
| CFW | 82 | | 20 | | 37.04 | Max. |
| JFS | 82 | | 25 | | 46.30 | Max. |
| JWS | 82 | | | c. 35 | 35.00 | Max; unlikely all can be used because of difficulty of targeting sub in range as well as bearing [Final report of qualification on range, ed]. |
| Clns+ | 83 | | 25 | | 46.30 | [Ship entry] |
|  |  | | 20 | | 37.04 | [Sub entry] |
| GSN | 83 | | 20 | 37 | 37.04 | |
| JFS+ | 83 | | 20 | | 37.04 | [ASW entry, ed] |
| JWS+ | 83 | | | 45-50 | 45.00-50.00 | Official US figures, superceding 35 km. |
| BW | 86 | | | 45 | 45.00 | |

------------ **WARHEAD TYPE** ------------

| Source | | nuclear | conven | chem | |
|---|---|---|---|---|---|
| Rock | 78 | yes | | | |
| FI | 81 | yes | | | |
| CFW | 82 | yes | | | |
| JFS | 82 | yes | | | [Both ASW and Missile entries, ed] |
| JWS | 82 | yes | | | Depth bomb. |
| Clns+ | 83 | yes | | | [Ship entry] |
|  |  | yes | yes | | HE, [sub entry] |
| GSN | 83 | yes | | | |

------------ **WARHEAD CHARACTERISTICS** ------------

| | | |
|---|---|---|
| CFW | 82 | nuclear depth bomb [From SS-N-16 entry, ed] |
| JFS+ | 83 | Depth bomb [ASW entry, ed] |
| BW | 86 | single RV |

# CRUISE MISSILES Sea-Launched SSMs SS-N-15

```
------------ NUCLEAR YIELD ------------
              kT            MT         kT conv
Clns+  83     10                        10.00
BW     86   ? 10                        10.00

------------ DIAMETER ------------
              in     ft    cm     m    cm conv
GSN    83     21          53.3         53.34        Max. [533 mm in source]

------------ TYPE OF PROPULSION ------------
Rock   78    solid
GSN    83    solid
BW     86    solid

------------ MIDCOURSE (OR MAIN) GUIDANCE ------------
GSN    83    inertial

------------ IOC ------------
JFS    82    1974        Operational [Missile entry, ed]
             1972        [ASW entry, ed]
GSN    83    1972

------------ DEPLOYMENT BEGAN ------------
Rock   78  ? 1975        Service intro.
Clns+  83    1973        First deployed.
CFW+   84    1972
BW     86    1972

------------ DEPLOYED ------------
FI     81                Deployed on Victors.
BW     86                In service.

------------ CARRIERS ------------

------------ CARRIER ------------ TUBES ---- RELOADS ------------
USND   81    subs                                "Newer classes of submarines".

------------ CARRIER ------------ TUBES ---- RELOADS ------------
FI     81    Victor
GSN    83    Victor
ABFS   84    Victor                              N=38.
BW     86    Victor 1,2,3           yes          N=38.

------------ CARRIER ------------ TUBES ---- RELOADS ------------
CFW    82    Victor I
JWS+   83  ? Victor I

------------ CARRIER ------------ TUBES ---- RELOADS ------------
CFW    82    Victor II
JFS    82    Victor II
JWS+   83  ? Victor II

------------ CARRIER ------------ TUBES ---- RELOADS ------------
CFW    82    Victor III
JWS    82    Victor III

------------ CARRIER ------------ TUBES ---- RELOADS ------------
JFS    82    Charlie II
JWS+   83  ? Charlie II                          Maybe some equipped.
ABFS   84    Charlie                             N=18.

------------ CARRIER ------------ TUBES ---- RELOADS ------------
ABFS   84    Echo                                N=5.

------------ CARRIER ------------ TUBES ---- RELOADS ------------
CFW    82    Alfa
JWS    82    Alfa
GSN    83    Alfa
ABFS   84    Alfa                                N=6.
SMP    84    Alfa                                '84, '85.
BW     86    Alfa                   yes          N=6.

------------ CARRIER ------------ TUBES ---- RELOADS ------------
JFS    82    Tango
JWS    82    Tango
```

|  |  | CARRIER | TUBES | RELOADS |
|---|---|---|---|---|
| JWS | 82 | Papa | | |
| AB FS | 84 | Papa | | N=1. |

|  |  | CARRIER | TUBES | RELOADS |
|---|---|---|---|---|
| SMP | 85 | Mike | | |

|  |  | USE AND CONFIGURATION |
|---|---|---|
| Rock | 78 | Submarine-launch, atmospheric flight. |
| CFW | 82 | Submerged launch from sub torpedo tubes. |
| JWS | 82 | [Subroc description; unclear whether applies to SS-N-15, ed:] sub-launched missile follows short underwater path, broaches surface, flies airborne for major part of distance to target; depth charge bomb released to continue on ballistic trajectory, sinking to optimum depth before detonation. "This type of weapon" relies on accurate localization of target, rapid launch and flight to target before target has time to travel far. |
| GSN | 83 | Fired from standard submarine torpedo tubes. Weapon fired with range and bearing derived from launching submarine's sonar. |

|  |  | TARGET TYPE | |
|---|---|---|---|
| JCS | 80 | sub | |
| FI | 81 | sub | |
| USND | 81 | sub | |
| CFW | 82 | sub | |
| JFS | 82 | sub | [Both ASW and Missile entries, ed] |
| CFW+ | 84 | | Also useable against surface targets. |
| SMP | 84 | sub | |
| SMP | 85 | sub | [Implied, ed.] |

|  |  | RELATION TO OTHER MISSILES |
|---|---|---|
| FI | 81 | Similar to Subroc. |
| CFW | 82 | Similar to US SUBROC. |
| JWS | 82 | Same general type as American Subroc. |
| GSN | 83 | Similar to US SUBROC. |

# SS-N-16

```
Sources are:   FI    81      CFW   82      JWS   82      JFS+  83      CFW+  84      SMP   85
               JCS   80      USND  81      JFS   82      GSN   83      JWS+  83      SMP   84
```

──────────── **US DESIGNATION** ────────────

|  | |  |
|---|---|---|
| JCS | 80 | SS-NX-16 |
| FI | 81 | SS-N-16 |
| USND | 81 | SS-NX-16 |
| CFW | 82 | SS-N-16 |
| JFS | 82 | SS-N-16 |
| JWS | 82 | SS-N-16 |
| GSN | 83 | SS-N-16 |
| SMP | 84 | SS-N-16 |
| SMP | 85 | SS-N-16 |

──────────── **DESCRIPTION** ────────────

| | | |
|---|---|---|
| JCS | 80 | Existing Soviet stand-off weapon [ASW, sub-launched]. |
| JFS | 82 | Rocket-assisted torpedo launcher. |
| SMP | 85 | ASW missile. |

──────────── **EDITOR'S NOTES** ────────────

| | | |
|---|---|---|
| JCS | 80 | Listed FY81. |
| SMP | 84 | Listed '84, '85. |

──────────── **RANGE** ────────────

| Source | | mi | nm | km | km conv | |
|---|---|---|---|---|---|---|
| JCS | 80 | | | | | If US subs don't get new nuclear-capable weapon after SUBROC phased out, US will have to rely on Mk 48 torpedo, whose use will force US subs to close within Soviet detection envelope "and within range" of SS-N-15 and SS-NX-16. |
| USND | 81 | | | | | Short range. |
| CFW | 82 | | 50 | | 92.60 | Max. |
| GSN | 83 | | 30-50 | 55-92 | 55.56-92.60 | |
| JFS+ | 83 | | 50 | | 92.60 | |

──────────── **WARHEAD TYPE** ────────────

| Source | | nuclear | conven | chem |
|---|---|---|---|---|
| JFS | 82 | | yes | |
| GSN | 83 | | yes | |

──────────── **WARHEAD CHARACTERISTICS** ────────────

| | | | |
|---|---|---|---|
| FI | 81 | torpedo | |
| CFW | 82 | homing torpedo | |
| JWS | 82 | torpedo | Like SS-N-15 but acoustic homing torpedo replaces nuclear depth bomb. |
| GSN | 83 | | ASW homing torpedo. |

──────────── **TYPE OF PROPULSION** ────────────

| | | |
|---|---|---|
| GSN | 83 | solid |

──────────── **MIDCOURSE (OR MAIN) GUIDANCE** ────────────

| | | | |
|---|---|---|---|
| GSN | 83 | inertial | To torpedo release point. |

──────────── **IOC** ────────────

| | | |
|---|---|---|
| GSN | 83 | early 1970s |

──────────── **DEPLOYMENT BEGAN** ────────────

| | | |
|---|---|---|
| CFW+ | 84 | c. 1980 |

──────────── **CARRIERS** ────────────

──────────── **CARRIER** ──── **TUBES** ──── **RELOADS** ────────────

| | | | | | |
|---|---|---|---|---|---|
| USND | 81 | | | | "Newer classes of submarines". |

──────────── **CARRIER** ──── **TUBES** ──── **RELOADS** ────────────

| | | | | | |
|---|---|---|---|---|---|
| GSN | 83 | Victor II | | | |

──────────── **CARRIER** ──── **TUBES** ──── **RELOADS** ────────────

| | | | | | |
|---|---|---|---|---|---|
| GSN | 83 | Victor III | | | |
| SMP | 84 | Victor III | | | '84, '85. |

──────────── **CARRIER** ──── **TUBES** ──── **RELOADS** ────────────

| | | | | | |
|---|---|---|---|---|---|
| GSN | 83 | ? Alfa | | | |

---------- CARRIER ------------ TUBES ---- RELOADS ------------
SMP   85    Mike

---------- USE AND CONFIGURATION ------------
JFS   82    ASW torpedoes use homing device.
GSN   83    Carries ASW homing torpedo in lieu of nuclear warhead. Parachute lowers torpedo into water, and "a protective nosecap separates upon water entry, and when the torpedo reaches a prescribed depth, a programmed search maneuver is begun, with the torpedo homing on any target detected during the search". Weapon is launched from tubes larger than standard 533-mm torpedo tubes.

---------- TARGET TYPE ------------
JCS   80    sub
FI    81    sub
USND  81    sub
CFW   82    [sub]       Also useful against surface targets.
JFS   82    sub         ASW torpedo.
SMP   84    sub         '84, '85.

---------- STRATEGY ------------
JWS+  83    Official US source has confirmed SS-N-16's existence.

---------- COMBAT REPORTS/EFFECTIVENESS ------------
JWS   82    Difficulties of targeting even more acute in this case [than SS-N-15]; "Operational performance of this missile must be viewed with considerable scepticism".

---------- RELATION TO OTHER MISSILES ------------
FI    81    May be a variant of SS-N-15.
CFW   82    Derived from SS-N-15.
JWS   82    Like SS-N-15 but acoustic homing torpedo replaces nuclear depth bomb.
GSN   83    Further development of SS-N-15.

# SS-N-19

| Sources are: | SMP | 81 | IISS | 82 | US20 | 82 | GSN | 83 | JWS+ | 83 | CFW+ | 84 | SMP | 85 |
|---|---|---|---|---|---|---|---|---|---|---|---|---|---|---|
| FI | 81 | US18 | 81 | JFS | 82 | Clns | 83 | JCS | 83 | SMP | 83 | JCS | 84 | BW | 86 |
| IISS | 81 | CFW | 82 | JWS | 82 | FI+ | 83 | JFS+ | 83 | ABFS | 84 | SMP | 84 | | |

------- **US DESIGNATION** -------

| FI | 81 | SS-N-? | [Oscar entry only, ed] |
|---|---|---|---|
| | | SS-NX-19 | [Kirov entry only, ed] |
| SMP | 81 | SS-N-19 | '83, '84, '85. |
| CFW | 82 | SS-N-19 | |
| IISS | 82 | SS-N-19 | |
| JFS | 82 | SS-N-19 | |
| JWS | 82 | SS-NX-19 | |
| Clns | 83 | SS-N-19 | |
| GSN | 83 | SS-N-19 | |
| JWS+ | 83 | SS-N-19 | |
| JCS | 84 | SS-N-19 | |

------- **DESCRIPTION** -------

| IISS | 82 | SLCM |
|---|---|---|
| GSN | 83 | Improved long-range anti-ship cruise missile. |
| JCS | 83 | Long-range antiship cruise missiles. |
| SMP | 84 | Cruise missile, '84, '85. |

------- **EDITOR'S NOTES** -------

| FI | 81 | Source has 2 separate records for Oscar and Kirov-carried missile. Data applies to both unless otherwise marked. Final report of separate entries. |
|---|---|---|
| SMP | 81 | Listed '83, '84, '85; mentioned (not by name, p46) in '81. |
| Clns | 83 | Missile is listed both with sub-based and surface ship-based weapons. Data is common to both unless marked. |
| JCS | 83 | Listed FY84-85; not by name in FY84. |

------- **RANGE** -------

| | | mi | nm | km | km conv | |
|---|---|---|---|---|---|---|
| FI | 81 | | | 300-400 | 300.00-400.00 | [Kirov entry only, ed] |
| SMP | 81 | | | >450 | 450.00 | Estimated. |
| US18 | 81 | | >250 | | 463.00 | |
| CFW | 82 | | 300 | | 555.60 | Max. |
| IISS | 82 | | | 460 | 460.00 | Max. |
| JFS | 82 | | 250+ | | 463.00 | Max. |
| JWS | 82 | | | <_500 | 500.00 | At supersonic speed (at least Mach 2.5). |
| US20 | 82 | 25-30 | | | 46.30-55.56 | In practice [by sub], due to acquisition range (unassisted). |
| Clns | 83 | | 300 | | 555.60 | [Ship entry] |
| | | | 250 | | 463.00 | [Sub entry] |
| FI+ | 83 | | | 500 | 500.00 | |
| GSN | 83 | | 240+ | 445+ | 444.48 | |
| JFS+ | 83 | | 270+ | | 500.04 | Max. |
| SMP | 83 | | | 500 | 500.00 | |
| ABFS | 84 | | | c. 460 | 460.00 | |
| JCS | 84 | | 300 | | 555.60 | Max. |
| SMP | 85 | | | 550 | 550.00 | |
| BW | 86 | | | 500 | 500.00 | |

------- **SPEED** -------

| | | mph | kmph | Mach | kmph conv | |
|---|---|---|---|---|---|---|
| JWS | 82 | | | 2.5 | | |
| US20 | 82 | | | | | [Above subsonic] |
| FI+ | 83 | | | <_2.5 | | |
| GSN | 83 | | | 1+ | | |
| ABFS | 84 | | | 2.5 | | |
| BW | 86 | | | 2.5 | | |

------- **WARHEAD TYPE** -------

| | | nuclear | conven | chem | |
|---|---|---|---|---|---|
| CFW | 82 | yes | yes | | |
| JWS | 82 | yes | yes | | HE. |
| Clns | 83 | yes | yes | | HE. |
| FI+ | 83 | yes | yes | | HE. |
| GSN | 83 | yes | yes | | |
| JFS+ | 83 | yes | yes | | HE. |

------- **WARHEAD CHARACTERISTICS** -------

| BW | 86 | single RV |
|---|---|---|

526  SOVIET MISSILES  IDDS

---------- **NUCLEAR YIELD** ----------
|       |    | kT    | MT | kT conv |           |
|-------|----|-------|----|---------|-----------|
| CLns  | 83 | 500   |    | 500.00  | [Sub entry] |
| BW    | 86 | ? 350 |    | 350.00  |           |

---------- **WARHEAD WEIGHT** ----------
|       |    | lb    | t | kg | kg conv |              |
|-------|----|-------|---|----|---------|--------------|
| CLns  | 83 | 2,000 |   |    | 907.20  | [Ship entry] |

---------- **TYPE OF PROPULSION** ----------
| CFW | 82 | liquid |
| FI+ | 83 |        | US DoD drawing suggests a winged missile "with under-fuselage powerplant and twin rocket boosters, but such a configuration is incompatible with the size and shape of the hatches covering the Kirov's launch tubes, or with the launch tubes of the Oscar class". |
| GSN | 83 | turbojet |
| BW  | 86 | TJ |

---------- **MIDCOURSE (OR MAIN) GUIDANCE** ----------
| JWS  | 82 | ? inertial | Lack of radar, and long range, suggests that "inertial guidance, allied to an external detection and designation system (satellite or aircraft, for instance) with self-contained homing is the likely guidance technique". |
| US20 | 82 | autopilot  | No midcourse guidance; relies on ocean surveillance satellite. |
| GSN  | 83 | inertial   | High speed alleviates need for mid-course guidance because of limited distance target ship can travel during flight time. |

---------- **TERMINAL GUIDANCE** ----------
| JWS  | 82 | self-contained |
| US20 | 82 | active radar   |
| GSN  | 83 | radar          |

---------- **RADAR** ----------
| FI  | 81 | Not yet identified [Kirov entry only, ed] |
| JWS | 82 | No radar associated with SS-N-19 identified on Oscar or Kirov, but probably some of many "sensors and antennas" on Kirov could provide data or command for outgoing SS-N-19. |

---------- **PRODUCTION UNDERWAY** ----------
| SMP | 84 | In production. |

---------- **IOC** ----------
| GSN | 83 | 1971 | [Sic] |

---------- **DEPLOYMENT BEGAN** ----------
| FI    | 81 |      | About to enter service [Oscar entry only, ed]. |
| IISS  | 82 | 1980 | First year deployed. |
| CLns  | 83 | 1980 | First deployed, [ship entry]. |
|       |    | 1981 | First deployed, [sub entry]. |
| CFW+  | 84 | 1971 | [Sic] |
| BW    | 86 | 1980 |      |

---------- **DEPLOYED** ----------
| BW | 86 | In service. |

---------- **NUMBERS IN SERVICE** ----------

---------- **NUMBER IN 1981** ----------
| IISS | 81 | 40 | July, total deployed. |

---------- **NUMBER IN 1982** ----------
| IISS | 82 | 44 | 1982-83, July, total deployed. |

---------- **CARRIERS** ----------

CRUISE MISSILES  Sea-Launched SSMs  SS-N-19  527

|  |  | CARRIER | TUBES | RELOADS |  |
|---|---|---|---|---|---|
| FI | 81 | Oscar | 24 |  | [Oscar entry only, ed] |
| SMP | 81 | Oscar | 24 |  | '81, '83, '84, '85. Oscar class introduced in 1980. Oscar's missile is probably submarine variant of new Kirov missile.<br>tubes: '81, '83. |
| US18 | 81 | Oscar | 24 |  |  |
| CFW | 82 | Oscar |  |  |  |
| IISS | 82 | O |  |  |  |
| JFS | 82 | Oscar |  |  |  |
| JWS | 82 | Oscar | 20 |  | tubes: [Sic] |
| Clns | 83 | Oscar | 24 |  |  |
| FI+ | 83 | Oscar |  |  | N >_ 1. |
| GSN | 83 | Oscar |  |  |  |
| JCS | 83 | Oscar | 24 |  | FY84-85.<br>tubes: FY84-85. |
| BW | 86 | Oscar | 24 | no |  |

|  |  | CARRIER | TUBES | RELOADS |  |
|---|---|---|---|---|---|
| ABFS | 84 | Y |  |  | N=8, SSNs. |

|  |  | CARRIER | TUBES | RELOADS |  |
|---|---|---|---|---|---|
| FI | 81 | Kirov | 20 |  | [Kirov entry only, ed] |
| US18 | 81 | Kirov |  |  |  |
| CFW | 82 | Kirov |  |  |  |
| IISS | 82 | Kirov |  |  |  |
| JFS | 82 | Kirov |  |  |  |
| JWS | 82 | Kirov | 20 |  |  |
| Clns | 83 | Kirov | 20 |  |  |
| FI+ | 83 | Kirov |  |  | N=2. |
| GSN | 83 | Kirov |  |  |  |
| SMP | 85 | Kirov |  |  | Principal armament is battery of 20 SS-N-19s. [Photos of Frunze, second unit of Kirov-class, equipped with 20 SS-N-19s.] |
| BW | 86 | Kirov | 20 |  |  |

|  |  | CARRIER | TUBES | RELOADS |  |
|---|---|---|---|---|---|
| FI+ | 83 | Bal-Com 1 |  |  | N=3 under construction. |

|  |  | USE AND CONFIGURATION |
|---|---|---|
| FI | 81 | Vertical launch [Kirov entry only, ed]. |
| SMP | 81 | Missiles can be launched submerged. |
| CFW | 82 | Submerged launch from Oscar. |
| JWS | 82 | Vertically-launched. |
| US20 | 82 | Flies at c. 225 ft altitude [not a true surface skimmer] |
| GSN | 83 | Submerged launched from Oscar. |
| JCS | 83 | Submerged launch from Oscar, FY84-85. |

|  |  | TARGET TYPE |  |
|---|---|---|---|
| FI | 81 | ship |  |
| JWS | 82 | ship | Main role is anti-ship weapon. |
| GSN | 83 | ship |  |
| JCS | 83 | ship |  |
| SMP | 83 | ship | '83, '84, '85. |

|  |  | STRATEGY |
|---|---|---|
| JWS | 82 | Oscar and Kirov could be intended to operate together as nucleus of "'ocean superiority' task force able to challenge western fleets over wide areas". |

|  |  | RELATION TO OTHER MISSILES |
|---|---|---|
| CFW | 82 | Evidently has improved characteristics over SS-N-12. |
| GSN | 83 | Evolved from SS-N-3/12. |

|  |  | OTHER INFORMATION |
|---|---|---|
| SMP | 83 | 24 missiles on Oscar is more than 3 times as many cruise missiles as fitted "on previous classes of series-produced Soviet" subs. |

# SS-N-21

```
Sources are:  GSN   83      JWS   83      ABFS  84      CFW   84      JCS   85      BW    86
              JFS   82      JFS+  83      SMP   83      AWST+ 84      SMP   84      SMP   85
```

---------- US DESIGNATION ----------

|  | |  | |
|---|---|---|---|
| JFS | 82 | SS-N-21 | |
| GSN | 83 | SS-N-21 | |
| JWS | 83 | SS-NX-21 | |
| SMP | 83 | SS-NX-21 | '83, '84, '85. |
| AWST+ | 84 | SS-NX-21 | |
| CFW | 84 | SS-NX-21 | |
| JCS | 85 | SS-NX-21 | |

---------- DESCRIPTION ----------

| Source | | Description |
|---|---|---|
| JFS | 82 | Cruise missile. |
| GSN | 83 | Anti-ship cruise missile. |
| JWS | 83 | Submarine-launched cruise missile. |
| AWST+ | 84 | Sub launched cruise missile. |
| SMP | 84 | Cruise missile. '84, '85. |
| JCS | 85 | Cruise missile. |

---------- EDITOR'S NOTES ----------

|  | |  |
|---|---|---|
| JWS | 83 | From SS-N-20 entry. |
| SMP | 83 | Mentioned '83, '84, '85. |

---------- RANGE ----------

| Source | | mi | nm | km | km conv | |
|---|---|---|---|---|---|---|
| GSN | 83 | | c. 1,620 | c. 3,000 | 3,000 | Estimated. |
| JWS | 83 | | | c. 3,000 | 3,000 | |
| SMP | 83 | | | 3,000 | 3,000 | "Estimated maximum range on the order of 3,000" km. |
| CFW | 84 | | 900-1,200 | | 1,666-2,222 | |
| SMP | 84 | | | 3,000 | 3,000 | '84, '85. |
| JCS | 85 | | | 3,000 | 3,000 | Long-range. |
| BW | 86 | | | 3,000 | 3,000 | |

---------- SPEED ----------

| Source | | mph | kmph | Mach | kmph conv |
|---|---|---|---|---|---|
| CFW | 84 | | | 0.7 | |
| BW | 86 | | | 0.7 | |

---------- CEP ----------

| Source | | ft | mi | nm | m | km | m conv | |
|---|---|---|---|---|---|---|---|---|
| SMP | 84 | | | | | | | "Capable of threatening hardened targets", '84, '85, ["probably" in '85 report, ed.]; "could eventually be accurate enough to permit the use of conventional warheads", '84, '85. |

---------- WARHEAD TYPE ----------

| Source | | nuclear | conven | chem | |
|---|---|---|---|---|---|
| JFS+ | 83 | yes | | | |
| JWS | 83 | yes | | | |
| SMP | 83 | yes | | | '83, '84. |
| SMP | 84 | yes | maybe | | Initially to be fitted with nuclear warheads; could eventually be accurate enough to permit use of conventional warheads, depending on guidance and future munitions developments, '84, '85. |
| SMP | 85 | maybe | | | "When first deployed, these cruise missiles [SS-NX-21, SSC-X-4, SS-NX-24 and its ground-based variant, ed.] probably will be fitted with nuclear warheads." |

---------- WARHEAD CHARACTERISTICS ----------

|  | | | |
|---|---|---|---|
| SMP | 84 | 1 warhead | '84, '85. |
| BW | 86 | single RV | |

---------- NUCLEAR YIELD ----------

| Source | | kT | MT | kT conv |
|---|---|---|---|---|
| ABFS | 84 | 200 | | 200.00 |
| BW | 86 | ? 200 | | 200.00 |

---------- LENGTH ----------

| Source | | in | ft | cm | m | m conv | |
|---|---|---|---|---|---|---|---|
| SMP | 84 | | | | 7 | 7.00 | [From graph], '84, '85. |

## CRUISE MISSILES  Sea-Launched SSMs  SS-N-21

### ———————— DEVELOPMENT UNDERWAY ————————
| | | |
|---|---|---|
| GSN | 83 | Development has been reported. |
| SMP | 83 | Currently under development. |
| SMP | 84 | In development. |

### ———————— PROTOTYPE TESTS ————————
| | | |
|---|---|---|
| JCS | 85 | In the flight testing and production phases, [described with AS-15.] |

### ———————— PRODUCTION UNDERWAY ————————
| | | |
|---|---|---|
| JCS | 85 | In the flight testing and production phases, [described with AS-15.] |

### ———————— IOC ————————
| | | |
|---|---|---|
| CFW | 84 | Will probably become operational by 1985. |
| SMP | 85 | Expected to become operational this year. |

### ———————— DEPLOYMENT BEGAN ————————
| | | | |
|---|---|---|---|
| GSN | 83 | mid-1980s | Expected to enter service. |
| SMP | 83 | | "Soviet Navy will soon be the recipient of" SS-N-21. |
| SMP | 84 | | "Nearly deployed"; probably will become operational this year. |
| JCS | 85 | | "Deployments are imminent." |
| BW | 86 | ? 1986 | |

### ———————— CARRIERS ————————

#### ———— CARRIER ———— TUBES ———— RELOADS ————
| | | | | | |
|---|---|---|---|---|---|
| SMP | 85 | | | | Four classes of SSNs in production capable of carrying SS-NX-21. |

#### ———— CARRIER ———— TUBES ———— RELOADS ————
| | | | | | |
|---|---|---|---|---|---|
| CFW | 84 | | | | Probably intended for subs, but may be developed for surface and air launch too. |
| BW | 86 | attack subs | | yes | |

#### ———— CARRIER ———— TUBES ———— RELOADS ————
| | | | | | |
|---|---|---|---|---|---|
| SMP | 84 | sub | | '84, '85. | |

#### ———— CARRIER ———— TUBES ———— RELOADS ————
| | | | | | |
|---|---|---|---|---|---|
| AWST+ | 84 | Victor 3 | | | |
| SMP | 84 | Victor III | | Possible, '84, '85. | |

#### ———— CARRIER ———— TUBES ———— RELOADS ————
| | | | | | |
|---|---|---|---|---|---|
| AWST+ | 84 | Yankee | | | |
| SMP | 84 | new Yankee | | Possible, '84, '85. | |

#### ———— CARRIER ———— TUBES ———— RELOADS ————
| | | | | | |
|---|---|---|---|---|---|
| SMP | 84 | Mike | | Possible, '84, '85. | |

#### ———— CARRIER ———— TUBES ———— RELOADS ————
| | | | | | |
|---|---|---|---|---|---|
| SMP | 84 | Sierra | | Possible, '84, '85. | |

#### ———— CARRIER ———— TUBES ———— RELOADS ————
| | | | | | |
|---|---|---|---|---|---|
| SMP | 85 | Akula | | | Joined Soviet Navy in 1984, [from photo caption]. |

### ———————— USE AND CONFIGURATION ————————
| | | |
|---|---|---|
| GSN | 83 | Launched from standard submarine torpedo tubes. May also be suitable for launching from surface ship torpedo tubes. Launch similar to US Harpoon and Tomahawk. |
| JFS+ | 83 | For use from submarine torpedo tubes. |
| SMP | 83 | Size is compatible with submarine torpedo tubes. |
| CFW | 84 | Torpedo-tube launched weapon. |
| SMP | 84 | Small enough to be fired from standard torpedo tubes, '84, '85. Could be deployed near US coasts, '84, '85. |

### ———————— TARGET TYPE ————————
| | | |
|---|---|---|
| GSN | 83 | ship |
| SMP | 84 | "Land-attack". '84, '85. |
| JCS | 85 | Land-attack. |

### ———————— STRATEGY ————————
| | | |
|---|---|---|
| JWS | 83 | Number of Typhoons built may depend in part on progress with SS-NX-21. |
| SMP | 83 | Mission is primarily nuclear strike. |

### ———————— RELATION TO OTHER MISSILES ————————
| | | |
|---|---|---|
| CFW | 84 | Similar in concept to US Tomahawk. |

# SS-N-22

```
Sources are:   FI    83     IISS   83     ABFS   84     SMP    84     BW    86
JFS    82      GSN   83     JFS+   83     CFW    84     SMP    85
```

---------- **US DESIGNATION** ----------
| | | |
|---|---|---|
| JFS | 82 | SS-N-22 |
| GSN | 83 | SS-N-22 |
| IISS | 83 | SS-NX-22 |
| CFW | 84 | SS-N-22 |
| SMP | 84 | SS-N-22 |

---------- **DESCRIPTION** ----------
SMP  85    Antiship missile.

---------- **EDITOR'S NOTES** ----------
SMP  84    Listed '84, '85.

---------- **RANGE** ----------
|      |     | mi | nm     | km     | km conv       |      |
|------|-----|----|--------|--------|---------------|------|
| GSN  | 83  |    | c. 60  | c. 111 | 111.12        |      |
| JFS+ | 83  |    | 120    |        | 222.24        | Max. |
| CFW  | 84  |    | 55-68  |        | 101.86-125.94 |      |
| BW   | 86  |    |        | 110    | 110.00        |      |

---------- **SPEED** ----------
|     |    | mph | kmph | Mach | kmph conv |
|-----|----|-----|------|------|-----------|
| GSN | 83 |     |      |      |           |
| CFW | 84 |     |      | 2.5  |           |
| SMP | 85 |     |      |      |           |
| BW  | 86 |     |      | 2.5  |           |

Reported to be improved and much higher speed version of SS-N-9.

Supersonic.

---------- **WARHEAD TYPE** ----------
|      |    | nuclear | conven | chem |               |
|------|----|---------|--------|------|---------------|
| GSN  | 83 | yes     | yes    |      |               |
| IISS | 83 |         |        |      | ? Dual-capable. |

---------- **WARHEAD CHARACTERISTICS** ----------
BW  86    single RV

---------- **NUCLEAR YIELD** ----------
|    |    | kT    | MT | kT conv |
|----|----|-------|----|---------|
| BW | 86 | ? 200 |    | 200.00  |

---------- **TYPE OF PROPULSION** ----------
| GSN | 83 | solid |
| BW  | 86 | solid |

---------- **MIDCOURSE (OR MAIN) GUIDANCE** ----------
GSN  83    Mid-course guidance.

---------- **TERMINAL GUIDANCE** ----------
GSN  83    radar homing

---------- **PRODUCTION UNDERWAY** ----------
SMP  84    In production.

---------- **IOC** ----------
GSN  83    1981

---------- **DEPLOYMENT BEGAN** ----------
| IISS | 83 | ? 1982 | First year deployed. |
| BW   | 86 | 1982   |                      |

---------- **DEPLOYED** ----------
BW  86    In service.

---------- **NUMBERS IN SERVICE** ----------

---------- **NUMBER IN 1983** ----------
IISS  83    c. 20    July, total deployed.

---------- **CARRIERS** ----------

---------- **CARRIER** ---------- **TUBES** --- **RELOADS** ----------
CFW  84    Not used by submarines to date.

|       |    | **CARRIER**   | **TUBES** | **RELOADS** |                                              |
|-------|----|---------------|-----------|-------------|----------------------------------------------|
| JFS   | 82 | Sovremenny    |           |             |                                              |
| FI    | 83 | Sovremenny    | 4 x 4     |             | N=2 in service, at least 3 under construction. |
| GSN   | 83 | Sovremennyy   | [? 1] x 3 |             |                                              |
| CFW   | 84 | Sovremennyy   |           |             |                                              |
| SMP   | 85 | Sovremennyy   |           | 8           |                                              |
|       |    |               |           |             | reloads: Estimated.                          |
| BW    | 86 | Sovremennyy   |           |             |                                              |

|       |    | **CARRIER**   | **TUBES** | **RELOADS** |                                              |
|-------|----|---------------|-----------|-------------|----------------------------------------------|
| JFS   | 82 | Tarantul      |           |             | "Fitted in 'Sovremenny' and later 'Tarantul' classes". |
| FI    | 83 | Tarantul II   | 2 x 2     |             | Only a handful thought operational.          |
| CFW   | 84 | Tarantul II   |           |             |                                              |
| BW    | 86 | Tarantul II   | ? 4       |             |                                              |

|       |    | **CARRIER** | **TUBES** | **RELOADS** |       |
|-------|----|-------------|-----------|-------------|-------|
| ABFS  | 84 | Krasina     |           |             | N=1.  |

**DESIGN AND ENGINEERING**

FI   83   Probably developed from SS-N-9.

**USE AND CONFIGURATION**

CFW  84   Flies at "sea-skimming" altitudes to range of 55-68 nm.

**TARGET TYPE**

SMP  85   Ship.

**RELATION TO OTHER MISSILES**

JFS  82   Improvement on SS-N-9.
IISS 83   ? Improved SS-N-9.
CFW  84   Successor to SS-N-9.

## SS-N-24

Sources are: SMP 84   SMP 85

---------- **US DESIGNATION** ----------
SMP 84      "Larger cruise missile" has not yet been designated. [Sea-based version labeled "SLCM" in chart, ed.]
SMP 85      SS-NX-24

---------- **DESCRIPTION** ----------
SMP 84      Cruise missile, '84, '85.

---------- **EDITOR'S NOTES** ----------
SMP 84      Described as version of "Larger cruise missile".
SMP 85      Described as variant of "Larger system".

---------- **RANGE** ----------
            mi        nm        km        km conv
SMP 85                                              Variant of larger system which is "probably designed for long-range operations."

---------- **CEP** ----------
            ft        mi        nm        m         km        m conv
SMP 84                                                                "Capable of threatening hardened targets"; "could eventually be accurate enough to permit the use of conventional warheads", '84, '85.

---------- **WARHEAD TYPE** ----------
            nuclear   conven    chem
SMP 84      yes       maybe              Initially to be fitted with nuclear warheads; could eventually be accurate enough to permit use of conventional warheads, depending on guidance and future munitions developments, '84, '85. ["[P]robably will be fitted..." in '85 report, ed.]

---------- **LENGTH** ----------
            in        ft        cm        m         m conv
SMP 84                                    12.5      12.50        [From graph], '84, '85.

---------- **DEVELOPMENT UNDERWAY** ----------
SMP 84      In development, '84, '85.

---------- **IOC** ----------
SMP 84      Could be operational within next 2 years, '84, '85.

---------- **CARRIERS** ----------

---------- **CARRIER** ---------- TUBES ---- RELOADS ----------
SMP 85      Yankee                              "A newly converted YANKEE-Class nuclear-powered cruise missile attack submarine (SSGN) will be the test platform for the SS-NX-24." "In this case, the ballistic missile tubes were removed from a YANKEE SSBN in a process that converted the unit to an attack submarine."

---------- **RELATION TO OTHER MISSILES** ----------
SMP 84      Sea-launched version of "Larger cruise missile" which has ground-based version [labeled "GLCM" in chart, ed].
SMP 85      Sea-launched version of "Larger system" which has ground-based version [labeled "GLCM" in chart, ed]. [Elsewhere:] "A ground-based variant of this missile may be in development." Has no US counterpart.

# ABM-1

```
Sources are:  ID01  68    SecC  71    JCS   74    SWM   77    JCS   80    JCS   82    JCS   85
SecC  63      SecC  68    SecD  71    JCS   75    JCS   78    FI    81    JWS   82    SecD  85
SecC  64      SecC  69    JCS   72    JWS*  75    Rock  78    JCS   81    Clns+ 83    BW    86
IA01  65      SecC  70    SecC  72    SecC  75    SecD  78    SMP   81    FI+   83
SecC  65      SecD  70    SecD  72    JCS   76    CH01  79    SMP   81    SMP   83
SecC  66      JCS   71    JCS   73    MOW   76    SecD  79    AFM   82    AFM+  84
SecC  67                  SecC  73    JCS   77    Clns  80    AWST  82    SMP   84
```

---------- US DESIGNATION ----------

| | | | |
|---|---|---|---|
| SecC | 70 | ABM-1 | FY71-73,76. |
| SecD | 70 | ABM-1 | FY71-72. |
| JCS | 72 | ABM-1b | FY76-83. |
| | | ABM-1 | |
| SWM | 77 | ABM-1 | |
| Clns | 80 | ABM-1 | |
| FI | 81 | ABM-1B | |
| SMP | 81 | ABM-1B | Moscow defense includes the ABM-1B/GALOSH interceptor missiles, '81, '83, '84, '85. |
| | | ABM-1 | '81, '83. |
| AFM | 82 | ABM-1B | |
| AWST | 82 | ABM-1 | |
| JWS | 82 | ABM-1B | |

---------- NATO CODENAME ----------

| | | | |
|---|---|---|---|
| SecC | 66 | Galosh | FY67-71,76. |
| SecD | 69 | Galosh | FY70-73,79-81. |
| JCS | 71 | Galosh | FY72-83,86. |
| JWS* | 75 | Galosh | "Improved Galosh". |
| MOW | 76 | Galosh | |
| SWM | 77 | Galosh | |
| Rock | 78 | Galosh | |
| Clns | 80 | Galosh | |
| FI | 81 | Galosh | |
| SMP | 81 | Galosh | '81, '83, '84, '85. |
| AFM | 82 | Galosh | |
| AWST | 82 | Galosh | |
| JWS | 82 | Galosh | |

---------- DESCRIPTION ----------

| | | | |
|---|---|---|---|
| MOW | 76 | ABM | |
| Rock | 78 | | Listed under ABMs. |
| JWS | 82 | | "Surface-to-air anti-ballistic missile missile". |
| | | ABM | |

---------- EDITOR'S NOTES ----------

| | | |
|---|---|---|
| SecC | 63 | Listed FY64-74,76-77; not by name FY64-66,74,77. One page missing from FY70 report. |
| SecD | 69 | Listed FY70-73,79-81,86. |
| JCS | 71 | Listed FY72-83,86. |
| SMP | 81 | Listed '81, '83, '84, '85. |
| JWS | 82 | This entry contains information on ABM systems more broadly than ABM-1 itself. |

---------- MAXIMUM ALTITUDE ----------

| | | mi | ft | km | m | m conv | |
|---|---|---|---|---|---|---|---|
| Clns+ | 83 | 200 | | | | 321,800 | Combat ceiling, "Stat. miles". |

---------- RANGE ----------

| | | mi | nm | km | km conv | |
|---|---|---|---|---|---|---|
| JWS* | 75 | | 200-400 | 350-700 | 370.40-740.80 | |
| MOW | 76 | > 200 | | > 320 | 321.80 | |
| SWM | 77 | 186 | | 300 | 299.27 | |
| Rock | 78 | | | 300 | 300.00 | Slant range. |
| CH01 | 79 | 186+ | | 300+ | 299.27 | |
| Clns | 80 | 200 | | | 321.80 | Slant range. |
| AFM | 82 | > 200 | | | 321.80 | |
| AWST | 82 | | | | | Several hundred miles. |
| JWS | 82 | | | >_ 300 | 300.00 | |
| Clns+ | 83 | 350 | | | 563.15 | Slant range. |
| BW | 86 | | | ? 300 | 300.00 | |

534  SOVIET MISSILES  IDDS

```
———————— WARHEAD TYPE ————————
             nuclear    conven    chem
MOW   76     yes
JCS   77     yes                          Believed nuclear.
CH01  79     yes
Clns  80     yes
AFM   82     yes
JWS   82     yes

———————— WARHEAD CHARACTERISTICS ————————
BW    86     single RV

———————— NUCLEAR YIELD ————————
             kT         MT         kT conv
Rock  78                                       MT size.
CH01  79                2-3        2,000-3,000  Suitable for use outside atmosphere.
JWS   82                                       Multi-megaton.
BW    86     ? 3,000               3,000

———————— WEIGHT ————————
             lb         t          kg         kg conv
CH01  79     72,000                32,660     32,659

———————— LENGTH ————————
             in         ft         cm         m          m conv
IA01  65                c. 74                            22.56    Container [n.b. missile called ICBM,
                                                                  ed]
ID01  68                66                               20.12    Container without nosecap.
MOW   76                67                    20.4       20.42    Container.
SWM   77                59-62                 18-19      17.98-18.90
                                              20.4       20.40    Display container.
Rock  78                                      18.5       18.50
                                              19         19.00    Display container.
JWS   82                           c. 2,000   20.00               Container.
AFM+  84                c. 65                 19.81               Container.

———————— DIAMETER ————————
             in         ft         cm         m          cm conv
IA01  65                c. 10                            304.80   Container [n.b. missile called ICBM,
                                                                  ed]
ID01  68                9.2                              280.42   Container.
MOW   76                9                     2.75       274.32   Container.
SWM   77     94.5-106              240-270               240.03-269.24
Rock  78                                      2.4        240.00
                                              2.7        270.00   Display container.
CH01  79                c 8'10.3"             c. 2.75    270.00   [8'10 3/10" in source]
JWS   82                           275                   275.00   Internal diameter of container.

———————— NUMBER OF STAGES ————————
ID01  68     2
SWM   77     Multi.
Rock  78     ? 2        Probably.
CH01  79     2-3
AWST  82     Multi.
JWS   82     3

———————— TYPE OF PROPULSION ————————
ID01  68     solid
MOW   76                Unknown; 4 first-stage nozzles.
SWM   77                4 first stage nozzles.
Rock  78                Unknown for stages 1 and 2. 4 exhaust nozzles showed through rear of container
                        in Nov 1964 display.
CH01  79     solid      Booster and sustainers.
AFM   82                4 combustion chambers on first stage.
JWS   82                4 first stage nozzles.
BW    86     ? S

———————— MIDCOURSE (OR MAIN) GUIDANCE ————————
MOW   76     radar command
JWS   82     radar command
```

CRUISE MISSILES  ABMs  ABM-1  535

| | | RADAR | |
|---|---|---|---|
| SecC | 65 | | Large Moscow radar, apparently phase-array, appears "associated with satellite tracking efforts", but data provisional. |
| SecC | 66 | Triad | At several of the SA-1 complexes (one large and two small radars operating together), FY67-68. |
| | | | "A large phased array radar southwest of Moscow oriented toward our ICBM threat corridor", FY67-69. |
| | | | "Dual purpose early warning tracking radars sited at two locations to the northwest" also have capability to track satellites. |
| SecC | 67 | Dog House | FY68-70. |
| | | Hen House | FY68-71. |
| | | Triad | Each complex has two, FY68-72. |
| | | Hen House | Large phased-array radars sited at two locations to the northwest, FY68-69. [Dog House and 2 Hen House] "may be intended as forward acquisition radars for the Moscow system, while the Triad radars handle the target and interceptor missile tracking functions", FY68-69. |
| | | Triad | Question whether single Triad with mechanically steered radars could handle more than 8 launchers; US abandoned this type of radar for ABM defense because of grave limitations. |
| SecC | 69 | Triad | 3 of the 8 Triads appear to have reached operational status; the other 5 will probably by 1970. |
| | | Hen House | A new dual Hen House "large array radar" being added to the 2 already in place north of Moscow. New radar could improve coverage of likely ballistic missile corridors to Moscow area. Probably an early warning radar. |
| | | Dog House | 2nd one may be under construction south of Moscow. Could provide some additional coverage against POLARIS/POSEIDON threat, and a ballistic missile threat from China. Dog House is "a forward acquisition radar for use in controlling the assignment of targets to the GALOSH launch complexes". |
| SecC | 70 | Hen House | Probably at least 7 either operating or under construction, expanding surveillance coverage to include most areas of concern to them. For ballistic missile early warning, and initial tracking. Located on Kola Peninsula, Baltic coast, and in southern USSR. Most recent addition is near Sevastopol, covering part of Mediterranean. Provide warning against ICBMs from US and SLBMs launched from Atlantic, and portions of Mediterranean and Pacific, FY71-72. |
| | | | "TAT" provided by Dog House and Try Add in Moscow area; new TAT radar is being built south of Moscow. |
| SecC | 71 | | Numbers of surveillance radars: 5-6, 1 Oct 1970 and Jan 1971; 6-8, mid-71; 10-11, mid-72. |
| JCS | 72 | Hen House | FY73,75-83. Early warning, phased array, located peripherally in USSR, FY73,75. |
| | | Try Add | Engagement radar, FY73-75,78-83. |
| | | Dog House | FY73-83. Acquisition and tracking radar. FY73-75. |
| SecC | 72 | | Numbers of surveillance radars: 8, 1 Nov 1971; 10-11, mid-72; 11-12, mid-73. |
| JCS | 73 | Dog House | Second Dog House under construction. |
| JCS | 74 | Try Add | Each complex has 2 Try Add radar sites, "one large target tracking radar, and two smaller interceptor tracking and guidance radars, per site". Total of 8 Try Add sites. |
| | | Dog House | 2 Dog House radars. |
| JCS | 75 | Cat House | [Distribution suggests Chekhov and Cat House probably same radar; maybe same as 2nd Dog House, ed]. Chekhov radar listed, FY76-78. |
| JCS | 76 | | Early warning capabilities upgraded over past few years; [radar not stated]. |
| MOW | 76 | Try Add | Engagement. |
| | | Hen House | Early warning. |
| | | Dog House | Acquisition and tracking. |
| SWM | 77 | Try Add | |
| | | Dog House | |
| | | Hen House | |
| JCS | 78 | Hen House | Hen House BMEW network has been expanding [in past few years], FY79-80. |
| | | Dog House | Battle management radar, FY79-82. |
| | | Cat House | Battle management radar, FY79-82. |
| Rock | 78 | Try-Add | "Try-Add engagement radars comprising Chekhov target-tracking radars and two guidance and interception radars". [Implies Chekhov part of Try-Add]. |
| | | Dog House | Phased array unit in the vicinity of Moscow, "acquire and track the target" on information received from Hen House. |
| | | Hen House | Early warning radar. |
| JCS | 80 | large phased array | Work is continuing on large phased array radars, to provide better target-handling and tracking data and more accurate impact predictions, FY81-82. |
| JCS | 81 | large phased array | Large phased array radars under construction at various locations throughout USSR. Could perform some battle management functions as well as redundant BMEW coverage. First of them expected operational in early 1980s. |
| SMP | 81 | Hen House | Peripherally located BMEW radar. |
| | | | System also includes battle management radars and missile engagement radars. |
| | | | Soviets continue to improve BMEW capability "by constructing large phased-array radars to supplement the old HEN HOUSE network and to close existing gaps in coverage". |
| AWST | 82 | Hen House | Early warning. |
| | | Dog/Cat House | Battle management. |
| | | Try Add | Engagement. |

| | | | |
|---|---|---|---|
| JCS | 82 | large phased array | Not yet fully operational; superior to Hen House; probably designed to close existing gaps in coverage. A new large phased-array radar is being constructed near Moscow; probably will serve in a battle management role for the upgraded Moscow system, augmenting or possibly replacing Dog House and Cat House systems,. |
| JWS | 82 | Dog House | Battle management radar. |
| | | Cat House | Battle management radar. |
| | | Triad | Engagement radar. |
| | | Hen House | Early warning radars (remotely sited; the other radar types are in Moscow complexes). |
| | | | 2 operational OTH radars facing the US reported to be operational. Phased array radars under construction in 1981 at various sites throughout USSR, service entry expected in early 1980s [relation to ABM-1 system of OTH and phased array radars is not explicit, ed]. |
| FI+ | 83 | | Remaining 32 sites upgraded by addition of phased array radars to supplement existing Hen House. |
| SMP | 83 | | System includes battle management radars and (at launch complexes) engagement radars. Soviet work [? since Sept-81] includes additional construction of large phased-array radars at USSR periphery. [Map shows global radar coverage, ed], '83, '84, '85. |
| AFM+ | 84 | | New ABM phased array radars being built. |
| SMP | 84 | Try Add | 6, guidance and engagement radars, '84 '85. |
| | | Dog House | Target-tracking radars, south of Moscow, '84 '85. |
| | | Cat House | Target-tracking radars, south of Moscow, '84 '85. |
| | | | New system includes new large radar at Pushkino "designed to control ABM engagements", '84 '85. |

---------- OBSERVED ----------

| | | | |
|---|---|---|---|
| JWS | 82 | | Since observed only in enveloping container, little known about it. Soviet pictures of ABM launch that could be ABM-1 show only that it is "more or less conical in shape". |

---------- OBSERVED IN PARADE ----------

| | | | |
|---|---|---|---|
| SecC | 66 | 1964 | Displayed. |
| MOW | 76 | 7 Nov 1964 | |
| SWM | 77 | 1964 | |
| Rock | 78 | Nov 1964 | |
| AFM | 82 | 1964 | |
| JWS | 82 | 1964 | First shown in parade. |

---------- DEVELOPMENT UNDERWAY ----------

| | | | |
|---|---|---|---|
| MOW | 76 | | Under development at least as long as US Safeguard. |
| SMP | 83 | | Soviet work [? since Sept-81] includes "continued development of" "a modified version of the older GALOSH ABM interceptor". |

---------- PROTOTYPE TESTS ----------

| | | | |
|---|---|---|---|
| SecC | 70 | | During past 2 years, testing of improved GALOSH noted. Could be available as early as this year. No firm estimate of possible capabilities available. |
| SecD | 72 | | During past 3 years, tests of improved Galosh. Has controlled coast capability and restartable engine, providing more flexibility. |
| JCS | 73 | | Some decline in testing last year from 1971. |
| JCS | 74 | | Testing continued during 1973 [refers to testing of ABM; whether ABM-1 or not is unclear, ed]. |

---------- IOC ----------

| | | | |
|---|---|---|---|
| SecC | 65 | | Was previously stated Soviets possibly constructing ABM at Moscow to be operational about mid-1967. |
| SecC | 66 | | IOC against primary ICBM threat corridor might be late 1967, but more likely 1968. Capability against Polaris would require at least another year or two. At one time thought might be operational at Moscow around mid-1967. |
| SecC | 67 | | Could have IOC in 1967 or early 1968, "and a full operational capability with six complexes (96 launchers) by 1970-71 (By that time the Soviets could also construct two more complexes to fill out the southern part of the ring, for a total of 128 launchers)". |
| SecC | 68 | | Expect Moscow IOC sometime in 1968 and a full operational capability with its 6 complexes and 96 reliable launchers sometime in 1971. Somewhat later than previous estimate of IOC by 1967 or early 1968 and full operation in 1970-71. |
| SecC | 70 | | During past year Soviets appear to have brought 3 complexes to operational status. Remaining complex expected operational this year. |
| SMP | 81 | 1968 | Year operational. |

CRUISE MISSILES  ABMs  ABM-1  537

---------- **DEPLOYMENT BEGAN** ----------
SecC 63                ABM system designed "against all types of strategic ballistic missiles" not yet
                       ready for deployment; US analysts have assumed limited deployment by mid-1968.
SecC 64                Soviets may be starting to deploy ABM system around Moscow.
SecC 66                No ABM facilities completed.
SecD 71                Four ABM complexes near Moscow are now operational.
Rock 78   1970         Service intro.
Clns 80   1964         First deployed.
SMP  81   1987         First launchers, "new defenses could be fully operational by 1987".
SMP  84   late 1980s   New system likely to reach fully operational status.
BW   86   1968

---------- **DEPLOYED** ----------
JCS  71                Deployed around Moscow, FY72-75.
MOW  76                In service.
Rock 78                Still deployed.
JWS  82                Deployed.
BW   86                In service.

---------- **DEPLOYMENT HISTORY** ----------
SecC 67   US reasonably certain deployment of GALOSH around Moscow will be completed.
SecC 68   Construction around Moscow proceeding at a moderate pace. No effort during last year to expand
          GALOSH system or extend it to other cities. Work on 7th complex south of the city stoped 2 years
          ago, has not resumed.
SecC 69   During past year construction stopped at 2 of 6 Moscow complexes; work on 2 others [of 8 total]
          had been abandoned earlier.
SecD 69   Construction apparently curtailed during the past year on some of the Galosh sites around Moscow.
SecD 70   Deployment at Moscow is nearly complete. A number of Moscow ABM complexes brought to operational
          status in the past year.
JCS  72   Construction has begun at previously uncompleted sites.
JCS  73   Work has progressed markedly; "reasonable to assume" USSR will deploy to 100 launcher limit (and 6
          radars).
SecC 73   Continued construction near Moscow system could be for additional launchers, or for command and
          control and communications.
JCS  74   Early in 1971, following 3-year lapse, new construction began at 3 previously abandoned ABM
          complexes near Moscow. 36 launchers and 4 [new, additional?] radars expected by end of decade.
JCS  75   8 complexes originally under construction, FY76-77.
JWS* 75   USSR is deploying 36 more [over 64] ABM missiles; perhaps they plan a mix of 2 Galosh versions.
JCS  76   No evidence of increase to 100 launchers, FY77-79.
FI   81   "Deployment of a replacement is expected, probably the SH-4/SH-8 combination".
JWS  82   Deployment slowed in 1968; thought Soviets abandoning idea of ABM defenses; now appears slowdown
          associated with radar reconfiguration to take account of threat from China, and improvements in
          ABM system itself. 8 complexes originally under construction; only 4 completed.
AFM+ 84   Major updating of entire Galosh system now underway.
SMP  84   Soviets expanding and upgrading the system since 1980 within limits of ABM treaty, '84, '85.
JCS  85   "Deployment at Moscow within ABM Treaty." [From summary chart, ed.]

---------- **NUMBERS IN SERVICE** ----------

---------- **NUMBER IN 1967** ----------
SecC 67                6 complexes under active construction, a 7th is now dormant.

---------- **NUMBER IN 1968** ----------
SecC 68                Still consists of 6 complexes at outer SA-1 ring sites.

---------- **NUMBER IN 1969** ----------
SecC 69                4 complexes where work is continuing will provide only 64 missiles on launchers,
                       FY70-71.

---------- **NUMBER IN 1970** ----------
SecC 70                During past year Soviets appear to have brought 3 complexes to operational status.
                       Remaining complex expected operational this year.
                       Construction of 4 complexes appears complete, may not have received full complement of
                       GALOSH missiles. All 4, with over 60 launchers could be operational this year.
SecD 70   Some 60      Launchers.
SecC 71   64           Launchers (1 Oct 1970 and Jan 1971, mid-71, and mid-72).
Clns 80   64           1970-1979.

---------- **NUMBER IN 1971** ----------
SecD 71   64           Missiles on launchers.
SecC 72   64           Launchers (1 Nov 1971, mid-72, and mid-73).

---------- **NUMBER IN 1972** ----------
JCS  72   [64]         4 operational complexes of 16 launchers, FY73-81.

---------- **NUMBER IN 1973** ----------
SecC 73   64           Launchers (mid-73, mid-74, and mid-75).

538  SOVIET MISSILES  IDDS

---------- **NUMBER IN 1976** ----------
MOW    76   64         4 sites around Moscow.

---------- **NUMBER IN 1977** ----------
SWM    77   64         4 sites near Moscow.

---------- **NUMBER IN 1978** ----------
Rock   78   [64]       4 sites, 16 missiles per site, around Moscow.
SecD   78   64         ABM system still consists of 64 Galosh launchers, FY79-81. "Does not include test and
                       training launchers, but does include launchers at test sites that are thought to be part
                       of the operational force".

---------- **NUMBER IN 1979** ----------
JCS    82   [32]       In late 1979, 32 launchers (of original 64) were dismantled.
JWS    82              Half of 64 ABM-1B Moscow launchers removed in 1979/80.

---------- **NUMBER IN 1980** ----------
AFM    82   32         Half of the 64 launchers around Moscow deactivated in 1980.
AFM+   84   32         Reduced from 64 in 1980.

---------- **NUMBER IN 1981** ----------
FI     81   32         "Two of the four 16-missile sites around Moscow have been dismantled".
SMP    81   32         Operational launchers.
                       Deployed in 4 operational ABM complexes near Moscow, '81, '83.

---------- **NUMBER IN 1982** ----------
AWST   82   64         At 4 sites near Moscow.
JWS    82   64         Launchers in 4 sites around Moscow.

---------- **NUMBER IN 1984** ----------
SMP    84              System being upgraded to 100 launchers permitted by treaty, '84, '85.

---------- **UNDATED NUMBER** ----------
SMP    84   64         Launchers, originally, '84, '85.
BW     86   64         Peak.

---------- **LAUNCHER/SILO** ----------
Clns   80              Single launchers. Launch site: Fixed.
AFM+   84              Silo launchers being built.
SMP    84              Originally 64 reloadable above-ground launchers at 4 complexes. New system will have silo-based
                       Galosh launchers, '84, '85.

---------- **DESIGN AND ENGINEERING** ----------
SecC   66              Probably Soviet difficulties with ABM program, shown by "a series of changes and modifications to
                       the ABM facilities and equipment at their test ranges", and highly sporadic activity at deployment
                       sites.
JCS    85              Continued modernization of ABM system around Moscow. Upgrade projected for 1985.
SecD   85              Soviet Union currently upgrading the capability of its Moscow ABM defense system.

---------- **USE AND CONFIGURATION** ----------
SecC   66              Believed intended for exo-atmospheric intercepts, FY67-68.
SecC   67              "The GALOSH itself is a large, relatively slow acceleration missile", FY68-69. "Each complex
                       has...16 launch positions apparently designed for the 'GALOSH' missile", FY68-71.
SecC   68              Not very suitable for terminal defense and cannot take advantage of atmospheric discrimination of
                       decoys. 8 complexes required, in present pattern of deployment, to complete the ring around
                       Moscow, FY69-70.
JCS    71              Exo-atmospheric intercept, FY72-73.
JWS*   75              An important feature is 'loiter' capability. Motor can stop and start 4 or 5 times at upper
                       altitudes while decoys are sorted from warheads by ground radars. Interception in RV's terminal
                       phase.
SecC   75              Deployed in 4 complexes, 16 per complex. "Still uses dish radars for tracking incoming RVs and for
                       commanding the ABM interceptors". GALOSH interceptor has relatively slow rate of acceleration and
                       thus must be launched while target RV is still well above the atmosphere.
Rock   78              Designed to intercept and destroy incoming missiles "before they re-enter through the dense layers
                       of the atmosphere". Most probably launched at high inclination "from the container in which it was
                       first displayed".
JWS    82              Long-range operation implying exo-atmospheric intercept reported. Each of 4 Moscow complexes
                       consists of 2 large tracking radars, 4 small interceptor guidance and tracking radars, and 16
                       missile launchers.

---------- **RELOADS IN GENERAL** ----------
SecC   67              "There is a real question whether the reloading speed of the GALOSH [Now estimated at 10-30
                       minutes after arrival of the missile at the launcher] would be fast enough to be of any use in a
                       single engagement".
SecC   69              Estimated 30 minutes to load a Galosh launcher.
SecC   70              Launcher reload apparently possible; estimated to take 15-30 minutes to reload launcher.
SecD   71              Launcher reload apparently is possible.
SMP    84              Silo-based launchers may be reloadable, '84, '85.

## STRATEGY

| | | |
|---|---|---|
| SecC | 66 | Probable ABM system. |
| SecC | 67 | Now identified as ABM system. "If used for both area and terminal defense, the GALOSH system would be very expensive, at least $15 million per missile on launcher [dividing the total investment cost by the number of missiles on launchers] where only 16 missiles are provided per complex. Even if two reload missiles were provided for each launcher, the cost per missile would still amount to about 46 million". |
| SecC | 69 | Significance of latest ABM cutback not yet clear. Could be Soviets are reconsidering entire ABM policy; or could be GALOSH is too expensive or technically inadequate. Activity at main ABM test site indicates R&D is continuing on this problem. |
| SecD | 69 | As presently deployed, could provide only limited defense of Moscow area, and could be seriously degraded by programmed US systems. |
| JCS | 71 | Long-range, FY72-73. |
| JCS | 75 | Little defense against large attack; defense credible against small, accidental, or unauthorized attack, FY76-78. |
| SWM | 77 | System modified to counter Chinese ICBM. Exo-atmospheric. ? Launched from container at high angle. Reports of improved 'loitering' Galosh. |
| JCS | 78 | Could provide adequate protection against small attacks using unsophisticated missiles without penetration aids. |
| Rock | 78 | All 64 ABMs have been modified to counter Chinese ICBMs. |
| SecD | 79 | More an area than a point defense system. |
| SMP | 81 | Intercept range: long. |
| JWS | 82 | Moscow system capable of 'thin' defense of the capital. Appears 1968 deployment slowdown associated with radar reconfiguration to take account of threat from China, and improvements in ABM system itself. |
| SMP | 83 | The system cannot presently cope with a massive attack. |
| SMP | 84 | "This system is intended to afford a layer of defense for Soviet civil and military command authorities in the Moscow environs during a nuclear war rather than blanket protection for the city itself". When completed, new system will be "a two-layer defense composed of silo-based long-range modified Galosh interceptors designed to engage targets outside the atmosphere; silo-based high-acceleration interceptors designed to engage targets within the atmosphere", and new radars, '84, '85. |

## COMBAT REPORTS/EFFECTIVENESS

| | | |
|---|---|---|
| SecC | 64 | Moscow system probably effective against single missiles, but not against those with "even elementary" penetration aids. |
| SecC | 67 | 17 Nov 1966 intelligence estimate concludes Moscow ABM system will have "a good capability against a numerically limited attack on the Moscow area by currently operational missiles, but that its capabilities could be degraded by advanced penetration systems and it could not cope with a very heavy attack. Moreover, the present deployment will not cover all of the multi-directional POLARIS threat to Moscow". |
| SecC | 68 | Consensus of intelligence community that GALOSH "could provide a limited defense of the Moscow area but that it could be seriously degraded by sophisticated penetration aids, precursor bursts and the vulnerability of the radars to nuclear detonations". Further, present Dog House deployment will not cover all of the multi-directional POLARIS [or POSEIDON] threats to Moscow. |
| JCS | 85 | "Limited capability." |

## GEOGRAPHICAL DEPLOYMENTS

| | | |
|---|---|---|
| SecC | 67 | Moscow ABM system appears to consist of complexes deployed at "some of the outer ring SA-1 sites, about 45 n.mi. from the center of the city", FY68-69. |

## RELATION TO OTHER MISSILES

| | | |
|---|---|---|
| SecD | 69 | Galosh resembles in certain important respects NIKE-ZEUS system US abandoned years ago due to poor effectiveness. |
| CHO1 | 79 | Thought inferior to US Spartan. |
| JWS | 82 | In official US defense circles, said to more closely resemble Spartan than Sprint. |

## OTHER INFORMATION

| | | |
|---|---|---|
| SMP | 81 | World's only deployed ABM system. |
| SMP | 83 | World's only operational ABM system, '83, '84, '85. |
| SecD | 85 | World's only operational ABM sytem. |

# ABM-3

Sources are:  SWM 77    AWST 82    JWS 82    AWST+ 84

---------- US DESIGNATION ----------
SWM 77    ABM-?
AWST 82   ABM-X-3
JWS 82    ABM-X-3

---------- EDITOR'S NOTES ----------
AWST 82   Use of phased array radars is only element in entry carrying over to 1984.

---------- WARHEAD TYPE ----------
          nuclear    conven    chem
AWST 82   yes

---------- NUMBER OF STAGES ----------
SWM 77              Multi.

---------- RADAR ----------
AWST 82   phased array    "Works with movable phased array radar and ten 400x600-ft. phased array radars ringing USSR".
JWS 82    phased array    And missile tracking.
AWST+ 84  Flat Twin       Phased array.
          Pawn Shop       Phased array.

---------- PROTOTYPE TESTS ----------
SWM 77              Being tested at Sary Sagan.

---------- IOC ----------
AWST 82   1980    Operational.

---------- DEPLOYMENT BEGAN ----------
JWS 82              Described as 'near-term' project.

---------- CARRIERS ----------

---------- CARRIER ---------- TUBES --- RELOADS ----------
AWST+ 84                                Movable ABM system.

---------- USE AND CONFIGURATION ----------
JWS 82    Reported to be rapidly deployable system with phased array radar, missile tracking radar and a new missile.

---------- STRATEGY ----------
SWM 77    "High acceleration ABM similar to US Sprint for defence of point targets within the atmosphere". Believed intended for defence against tactical missiles.

---------- RELATION TO OTHER MISSILES ----------
SWM 77    Similar to US Sprint.
AWST 82   Sprint-like interceptor.

---------- OTHER INFORMATION ----------
JWS 82    No clues about missing ABM-2 designation obtained.

# SH-4

Sources are:   SWM  77     Rock  78     FI  81     AWST  82

---------- **US DESIGNATION** ----------
| | | |
|---|---|---|
| SWM | 77 | SH-4 |
| Rock | 78 | SH-4 |
| FI | 81 | SH-4 |
| AWST | 82 | SH-4   [Possibly SH-X] |

---------- **EDITOR'S NOTES** ----------
Rock  78    Information under Galosh text entry.

---------- **PROTOTYPE TESTS** ----------
SWM  77    1974
                   Testing at Sary Sagan.
Rock  78    1974   Tested from Sary Shagan.

---------- **DEPLOYMENT BEGAN** ----------
Rock  78           May already be deployed.

---------- **USE AND CONFIGURATION** ----------
Rock  78    "Capability of stopping and re-starting a terminal manoeuvering platform to give ground-based radars sufficient time to discriminate between warheads and decoys".
FI  81      Is thought able to stop and start propulsion at very high altitude 4 or 5 times, while ground radar and/or atmospheric re-entry sort warheads from decoys.

---------- **STRATEGY** ----------
SWM  77     "Long-range". Exo-atmospheric.
Rock  78    Capable of exo-atmospheric interception.
AWST  82    "New ABM designed for exo-atmosphere intercepts".

# SH-8

Sources are:  Rock 78    FI 81    SMP 81    AWST 82    AWST+ 84

### ———— US DESIGNATION ————
| | | |
|---|---|---|
| FI | 81 | SH-8 |
| SMP | 81 | SH-X |
| AWST | 82 | SH-X |
| AWST+ | 84 | SH-8 |

### ———— EDITOR'S NOTES ————
| | | |
|---|---|---|
| Rock | 78 | Information under Galosh text entry. |
| SMP | 81 | Listed '81. |

### ———— SPEED ————
|  |  | mph | kmph | Mach | kmph conv | |
|---|---|---|---|---|---|---|
| FI | 81 | | | | | Hypersonic. |

### ———— RADAR ————
| | | | |
|---|---|---|---|
| FI | 81 | phased array | C-band. Known in US as X-3 [final report of US equivalent] |

### ———— DEVELOPMENT UNDERWAY ————
| | | |
|---|---|---|
| SMP | 81 | "Developmental". |

### ———— STRATEGY ————
| | | |
|---|---|---|
| Rock | 78 | "A new high acceleration, mobile missile is thought to be under development capable of destroying incoming short-range attack missiles within the atmosphere". |
| FI | 81 | "A hypersonic surface-to-air missile capable of intercepting incoming US missiles". Could be second component of new ABM system for warheads getting past SH-4. |
| SMP | 81 | Intercept range: short. |
| AWST | 82 | "High acceleration missile for endo-atmospheric intercepts of SRAM" [Final report for "of SRAM"]. |

# SA-1

```
Sources are:   ID01  68      SecC  71      JCS   75     Rock  78     AFM   82     AFM+  84
MR03  66       Ley   68      AWST* 73      MOW   76     Clns  80     JWS   82     SMP   85
SecC  66       SecC  68      FI*   74      NA03  76     JCS   81     Clns+ 83
SecC  67       SecC  70      HRST  75      SWM   77     SMP   81     SMP   83
```

---------- US DESIGNATION ----------

|  | |  |  |
|---|---|---|---|
| SecC | 66 | SA-1 | |
| AWST* | 73 | SA-1 | |
| FI* | 74 | SA-1 | |
| JCS | 75 | SA-1 | FY76-83. |
| MOW | 76 | SA-1 | |
| SWM | 77 | SA-1 | |
| Rock | 78 | SA-1 | |
| Clns | 80 | SA-1 | |
| SMP | 81 | SA-1 | '81, '83, '84, '85. |
| AFM | 82 | SA-1 | |
| JWS | 82 | SA-1 | |

---------- NATO CODENAME ----------

| Source | | Codename | Notes |
|---|---|---|---|
| AWST* | 73 | Guild | |
| FI* | 74 | Guild | |
| HRST | 75 | Guild | |
| JCS | 75 | Guild | FY76-80. |
| MOW | 76 | Guild | |
| SWM | 77 | Guild | |
| Rock | 78 | Guild | |
| Clns | 80 | Guild | |
| SMP | 81 | Guild | '81, '83. |
| AFM | 82 | Guild | |
| JWS | 82 | Guild | |

---------- OTHER DESIGNATION ----------

Ley  68    "Medium [range] (new type)" [only designation provided by source, ed].

---------- DESCRIPTION ----------

SMP  81    Strategic SAM, '81, '85.
JWS  82    Seen on transporters, but classified as part of Soviet strategic air defence force.
           "Anti-aircraft guided missile system".

---------- EDITOR'S NOTES ----------

SecC  66    Listed FY67-73.
JCS   75    Listed FY76-83.
SMP   81    Listed '81, '83, '84, '85.

---------- MAXIMUM ALTITUDE ----------

| Source | | mi | ft | km | m | m conv | Notes |
|---|---|---|---|---|---|---|---|
| Rock | 78 | | | 20 | | 20,000 | "Altitude". |
| Clns+ | 83 | | 60,000 | | | 18,288 | Combat ceiling. |

---------- RANGE ----------

| Source | | mi | nm | km | km conv | Notes |
|---|---|---|---|---|---|---|
| MR03 | 66 | 22 | | | 35.40 | Slant range. |
| ID01 | 68 | c. 25 | | | 40.23 | |
| Ley | 68 | 25 | | | 40.23 | |
| HRST | 75 | 10-15 | | | 16.09-24.14 | |
| NA03 | 76 | 20 | | | 32.18 | Max. |
| Rock | 78 | | | 32 | 32.00 | |
| SMP | 81 | | | 50 | 50.00 | '81, '83, '84, '85. |
| AFM | 82 | 31 | | | 49.88 | |
| JWS | 82 | | | c. 32 | 32.00 | British MoD report, April 1976. |
| Clns+ | 83 | 25 | | | 40.23 | Slant range. |

---------- SPEED ----------

| Source | | mph | kmph | Mach | kmph conv |
|---|---|---|---|---|---|
| MR03 | 66 | | | 2.5 | |

---------- WARHEAD TYPE ----------

| Source | | nuclear | conven | chem | |
|---|---|---|---|---|---|
| MOW | 76 | | yes | | HE. |
| Clns | 80 | yes | yes | | HE. |

## 544 SOVIET MISSILES  IDDS

```
---------- WEIGHT ----------
             lb           t              kg          kg conv
MR03  66                  3                          3,000        Tons.
ID01  68                  3.5-4                      3,500-4,000
Ley   68                  c. 3                       3,000        Tons.
Rock  78                                 3,000       3,000

---------- LENGTH ----------
             in      ft           cm          m       m conv
MR03  66             38                              11.58
Ley   68             23                               7.01
AWST* 73             40                              12.19
FI*   74                               c. 12         12.00
HRST  75             40                              12.19
MOW   76             39'0"                  12.0     11.89
SWM   77             39                     12       11.89
Rock  78                                    11.5     11.50
SMP   81                                    12       12.00      [From graph], '81, '83, '84, '85.
AFM   82             39'0"                           11.89
JWS   82                        c. 1,200             12.00

---------- DIAMETER ----------
             in      ft           cm          m      cm conv
MR03  66             2                                60.96
Ley   68             1.8                              54.86
AWST* 73             2.3                              70.10
FI*   74                              c. 70           70.00
HRST  75     28                                       71.12
MOW   76             2'3.5"                 0.70      69.85
SWM   77     27.6                     70              70.10
Rock  78                              60              60.00
AFM   82             2'3 1/2"                         69.85
JWS   82                              71              71.00     [Analysis table]
                                      c. 70           70.00     [Text]

---------- WINGSPAN ----------
             in      ft           cm          m      cm conv
Ley   68             8 1/4                           251.46
SWM   77     110                      280            279.40
Rock  78                              270            270.00

---------- NUMBER OF STAGES ----------
MR03  66     1
MOW   76              No separate booster.
Rock  78     1
JWS   82              Has no separate booster stage.

---------- TYPE OF PROPULSION ----------
MR03  66     liquid
ID01  68     solid
FI*   74     "dual-thrust"
HRST  75     solid
JCS   75     liquid             FY76-77.
MOW   76     liquid
SWM   77     dual-thrust solid
Rock  78     solid
JWS   82     liquid

---------- MIDCOURSE (OR MAIN) GUIDANCE ----------
HRST  75     radio command
SWM   77     radio command
Rock  78     radio

---------- CONTROL SYSTEM ----------
FI*   74     Cruciform canards.
MOW   76     Foreplanes and wing trailing-edge surfaces.
SWM   77     "All-moving" foreplanes and small ailerons on wings.
Rock  78     Aerodynamic.
JWS   82     Movable foreplanes [distinguishes Guild from Guideline].

---------- RADAR ----------
SWM   77     Yo-yo
```

CRUISE MISSILES  Ground-Launched SAMs  SA-1

---------- OBSERVED IN PARADE ----------
MOW   76   7 Nov 1960
                       "'Guild' was the second type of Soviet surface-to-air missile displayed in a Moscow parade".
SWM   77   Nov 1960
Rock  78   1960           First public display.
AFM   82   7 Nov 1960
           1968           Next appeared in parade [after 1960].
JWS   82   1960           First shown in Moscow.
           May 1968       Shown in Moscow parade.

---------- PRODUCTION ENDED ----------
SWM   77                  No longer produced.

---------- IOC ----------
SecC  71   1954           Year operational.

---------- DEPLOYMENT BEGAN ----------
JCS   75   1954           First deployed, FY76-77.
MOW   76                  Has been a standard weapon since 1954.
SWM   77   1954
Rock  78   1954           Service intro.
Clns  80   1956           First deployed.
JCS   81                  In place for almost 30 years.
AFM   82                  After 1960, reported to be standard anti-aircraft weapon.
SMP   83                  SA-10 is replacing "30-year-old SA-1s around Moscow".

---------- DEPLOYED ----------
FI *  74                  Operated by USSR.
MOW   76                  In service.
JCS   81                  In place for almost 30 years.
JWS   82                  Deployed.

---------- RETIRED ----------
JWS   82                  Doubtful if it remains operational; possible a small number might be retained on reserve basis in Soviet Union.

---------- DEPLOYMENT HISTORY ----------
SecC  67                  No significant changes in deployment in the last few years. Soviets will perhaps begin to thin out the SA-1 system.
SecC  68                  No significant changes in deployment, FY69-70.
MOW   76                  Has been a standard weapon since 1954.
AFM   82                  After 1960, reported to be standard anti-aircraft weapon.
SMP   83                  SA-10 is replacing "30-year-old SA-1s around Moscow".
AFM+  84                  Replacement by SA-10 has started.
SMP   85                  Operational.

---------- NUMBERS PRODUCED ----------
SWM   77                  Large numbers were built for test and training purposes.

---------- NUMBERS IN SERVICE ----------

---------- NUMBER IN 1962 ----------
SecC  70   3,200          Launchers, 1962 through 1971 [read from graph, shows a level line].
SecC  71   3,200          Launchers, 1962 through 1972 [read from hand-drawn graph].

---------- NUMBER IN 1970 ----------
Clns  80   3200           1970-1979.

---------- NUMBER IN 1974 ----------
FI *  74                  Not nearly so widely deployed as Guideline.

---------- NUMBER IN 1975 ----------
JCS   75                  Mid-1975 level of sites expected to decrease gradually, FY76-77.

---------- NUMBER IN 1976 ----------
MOW   76                  Number expected to decline gradually.

---------- NUMBER IN 1977 ----------
SWM   77                  Not widely deployed.

---------- NUMBER IN 1980 ----------
Clns+ 83   3200           In 1980-1982.

---------- NUMBER IN 1982 ----------
AFM   82                  Phase-out in USSR has probably started.
JWS   82   3,200          Estimated still available for use around Moscow.

---------- **LAUNCHER/SILO** ----------
Clns 80        Single launchers. Launch site: Fixed.

---------- **CARRIERS** ----------

---------- **CARRIER** ---------- TUBES --- RELOADS ----------
JWS  82                                         Seen on transporters.

---------- **USE AND CONFIGURATION** ----------
MRO3 66  System includes launcher fueling vehicle, radar equipment and computers (for guidance), and "other special vehicles".

---------- **RELOADS IN GENERAL** ----------
SecC 71  "2 missiles per launcher" [appears without explanation on chart of launcher numbers, ed].

---------- **STRATEGY** ----------
SecC 67  Believe Soviets will perhaps begin to thin out the SA-1 system "which was designed to defend against mass bomber attacks".
SMP  81  Effective altitude: medium, '81, '83, '84, '85.
SMP  85  Incapable of defending against targets with "small radar-cross-section[s] such as cruise missiles".

---------- **GEOGRAPHICAL DEPLOYMENTS** ----------
SecC 67  Deployed in two rings around Moscow only, FY68-70.
         Moscow ABM system is at "some of the outer ring SA-1 sites, about 45 n.mi from the center of the city".
SecC 70  Emplaced only around Moscow, FY71-73.
JCS  81  Continues to provide primary air defense for Moscow from sites deployed in two rings around the city, FY82-83.
SMP  83  SA-10 is replacing "30-year-old SA-1s around Moscow".

---------- **EXPORTED TO** ----------
MOW  76        No evidence it was ever exported.
AFM  82        Thought not supplied outside USSR.

---------- **RELATION TO OTHER MISSILES** ----------
FI*  74  Superficially similar to Guideline.

# SA-2

| Sources are: | MR03 | 66 | SecC | 69 | HRST | 75 | Bras | 77 | Clns | 80 | AWST | 82 | JWS+ | 83 |
|---|---|---|---|---|---|---|---|---|---|---|---|---|---|---|
| SecC | 63 | | SecC | 66 | SecC | 70 | JCS | 75 | JCS | 77 | FI | 81 | JCS | 82 | AFM+ | 84 |
| SecC | 64 | | SecC | 67 | SecC | 71 | SecD | 75 | SWM | 77 | ID12 | 81 | JWS | 82 | SMP | 85 |
| IA01 | 65 | | ID01 | 68 | JCS | 72 | MOW | 76 | AF01 | 78 | SMP | 81 | Clns+ | 83 | SMP | 85 |
| SecC | 65 | | Ley | 68 | JCS | 74 | NA03 | 76 | AWO2 | 78 | US16 | 81 | FI+ | 83 | BW | 86 |
| IA05 | 66 | | SecC | 68 | SecD | 74 | SecC | 76 | Rock | 78 | AFM | 82 | FI+ | 83 | | |

---------- US DESIGNATION ----------

| SecC | 64 | SA-2 | FY65-73,75-77. |
|---|---|---|---|
| JCS | 72 | SA-2 | FY73,75-81,83. |
| SecD | 74 | SA-2 | FY75-77. |
| HRST | 75 | SA-2 | |
| MOW | 76 | SA-2 | |
| Bras | 77 | SA-2 | |
| SWM | 77 | SA-2 | |
| Rock | 78 | SA-2 | |
| Clns | 80 | SA-2 | |
| FI | 81 | SA-2 | |
| SMP | 81 | SA-2 | '81, '83, '84, '85. |
| AFM | 82 | SA-2 | |
| AWST | 82 | SA-2 | SA-2B,C,D,E,F. |
| JWS | 82 | SA-2 | |

---------- NATO CODENAME ----------

| Ley | 68 | Guideline | |
|---|---|---|---|
| JCS | 72 | Guideline | FY76-80. |
| HRST | 75 | Guideline | |
| MOW | 76 | Guideline | |
| Bras | 77 | Guideline | |
| SWM | 77 | Guideline | |
| Rock | 78 | Guideline | |
| Clns | 80 | Guideline | |
| FI | 81 | Guideline | |
| SMP | 81 | Guideline | '81, '83. |
| AFM | 82 | Guideline | |
| AWST | 82 | Guideline | |
| JWS | 82 | Guideline | |

---------- SOVIET DESIGNATION ----------

| MOW | 76 | V750VK | |
|---|---|---|---|
| | | V75SM | Applies to entire weapon system. |
| SWM | 77 | V750VK | Missile. |
| | | V75SM | System. |
| FI | 81 | V750VK | |
| | | V75SM | Complete system. |
| JWS | 82 | V750VK | One version of missile at least. |
| | | V75SM | Complete system including radar and power supplies. |

---------- DESCRIPTION ----------

| SecC | 63 | 2nd generation SAM, FY64-65. |
|---|---|---|
| Bras | 77 | Medium/high level SAM. |
| SMP | 81 | Strategic SAM, '84, '85. |
| JWS | 82 | "Medium-range, anti-aircraft guided weapon system". |
| | | "Surface-to-air tactical guided missile". |

---------- EDITOR'S NOTES ----------

| SecC | 63 | Listed FY64-73,75-77, not by name FY64. |
|---|---|---|
| JCS | 72 | Listed FY73,75-81,83. |
| SecD | 74 | Listed FY75-77. |
| SMP | 81 | Listed '81, '83, '84, '85. |
| JWS | 82 | Specifications apply to a model supplied to Egypt; source indicates other and later versions exist. |

## MINIMUM ALTITUDE

| | | mi | ft | km | m | m conv | |
|---|---|---|---|---|---|---|---|
| SecC | 66 | | c. 1,500 | | | 457.20 | Some sites improved down to this level. |
| SecC | 67 | | c. 1,500 | | | 457.20 | Through successive modifications of missiles and radars at some sites, has low altitude capability down to this level, FY68-69. |
| | | | c. 3,000 | | | 914.40 | Unmodified sites. |
| SecC | 69 | | | | | | Some versions of SA-2 "may, under optimum conditions" have effective minimum altitude down to [deleted] feet. |
| JCS | 75 | | | | | | Like virtually all Soviet strategic SAMs, SA-2 modernized during lifetime, new versions increasing range and improving low-altitude performance, FY76-77. |
| AWST | 82 | | 300 | | | 91.44 | SA-2F. |

## MAXIMUM ALTITUDE

| | | mi | ft | km | m | m conv | |
|---|---|---|---|---|---|---|---|
| MOW | 76 | 11.2 | | 18 | | 18,020 | |
| Bras | 77 | | | | 20,000 | 20,000 | Max effective altitude. |
| SWM | 77 | | 82,020 | | 25,000 | 24,999 | |
| AF01 | 78 | | | 24.4 | | 24,400 | |
| Rock | 78 | | | 28 | | 28,000 | "Altitude". |
| Clns | 80 | | 80,000 | | | 24,384 | Combat Ceiling "Med-80,000 ft". |
| FI | 81 | | | | ≤ 28000 | 28,000 | |
| AFM | 82 | | 82,000 | | | 24,993 | |
| AWST | 82 | | 90,000 | | | 27,432 | |
| JWS | 82 | | | | 18,000 | 18,000 | Ceiling. |
| FI+ | 83 | | | | ≤ 18,000 | 18,000 | |

## RANGE

| | | mi | nm | km | km conv | |
|---|---|---|---|---|---|---|
| IA01 | 65 | 20-25 | | | 32.18-40.23 | |
| IA05 | 66 | c. 20-25 | | | 32.18-40.23 | |
| MR03 | 66 | 19 | | | 30.57 | 1957 version, slant range. |
| | | 25 | | | 40.23 | 1960 version, slant range. |
| ID01 | 68 | 18-22 | | | 28.96-35.40 | |
| Ley | 68 | 20-25 | | | 32.18-40.23 | |
| HRST | 75 | 30 | | | 48.27 | |
| JCS | 75 | | | | | Like virtually all Soviet strategic SAMs, SA-2 modernized during lifetime, new versions increasing range and improving low-altitude performance, FY76-77. |
| MOW | 76 | c. 30 | | c. 45 | 48.27 | Max. |
| NA03 | 76 | 25-30 | | | 40.23-48.27 | Max. |
| Bras | 77 | | | 35 | 35.00 | Max effective slant range, [35,000 m in source]. |
| SWM | 77 | 24+ | | 40+ | 38.62 | |
| AF01 | 78 | | | 45 | 45.00 | Slant range. |
| AW02 | 78 | | c. 27 | | 50.00 | |
| Rock | 78 | | | 50 | 50.00 | |
| Clns | 80 | 25 | | | 40.23 | Slant range. |
| FI | 81 | | | 40-50 | 40.00-50.00 | Max. |
| SMP | 81 | | | 50 | 50.00 | '81, '83, '84, '85. |
| AFM | 82 | 28 | | | 45.05 | |
| AWST | 82 | 19-27 | | | 30.57-43.44 | |
| JWS | 82 | | | 40-50 | 40.00-50.00 | Slant range. |
| Clns+ | 83 | 30 | | | 48.27 | Slant range. |
| AFM+ | 84 | 31 | | | 49.88 | Slant range. |
| BW | 86 | | | ? 50 | 50.00 | |

CRUISE MISSILES  Ground-Launched SAMs  SA-2  549

### SPEED

| | | mph | kmph | Mach | kmph conv | |
|---|---|---|---|---|---|---|
| MR03 | 66 | | | 2.5 | | 1957 version. |
| | | | | 3 | | 1960 version. |
| MOW | 76 | | | 3.5 | | |
| SWM | 77 | | | 3.5 | | |
| FI | 81 | | | 3.5 | | |
| AFM | 82 | | | 3.5 | | |
| JWS+ | 83 | | | c. 3.5 | | |
| BW | 86 | | | 3.5 | | |

### WARHEAD TYPE

| | | nuclear | conven | chem | |
|---|---|---|---|---|---|
| MOW | 76 | | yes | | HE, contact or proximity fuze. Detonated by command. Mk 4: Large white warhead may be nuclear. |
| Bras | 77 | | yes | | "HE blast". |
| SWM | 77 | | yes | | |
| AW02 | 78 | | yes | | HE. |
| Rock | 78 | yes | yes | | HE. |
| Clns | 80 | yes | yes | | HE. |
| FI | 81 | yes | yes | | A version (first seen 1967) reported to have nuclear warhead. High-explosive (fragmentation) with internally-grooved casing. Various warhead/fuze combinations exist. |
| AFM | 82 | maybe | yes | | Version displayed Nov 1967 had enlarged white-painted warhead, claimed to be more effective; may be nuclear. |
| AWST | 82 | | yes | | HE. |
| JWS | 82 | yes | yes | | HE; most have HE warheads. Version seen in 1967 believed to have nuclear warhead. Contact, proximity, and command fuzes have been reported. |
| JWS+ | 83 | | | | Proximity fuze. |

### WARHEAD CHARACTERISTICS

| BW | 86 | single RV |
|---|---|---|

### WARHEAD WEIGHT

| | | lb | t | kg | kg conv | |
|---|---|---|---|---|---|---|
| MOW | 76 | 288 | | 131 | 130.64 | |
| Bras | 77 | | | 130 | 130.00 | |
| SWM | 77 | 288 | | 154 | 130.64 | [Sic] |
| AW02 | 78 | 287 | | | 130.18 | |
| FI | 81 | | | 130 | 130.00 | |
| AFM | 82 | 288 | | | 130.64 | |
| JWS | 82 | | | 130 | 130.00 | |

### WEIGHT

| | | lb | t | kg | kg conv | |
|---|---|---|---|---|---|---|
| IA01 | 65 | | c. 1.4 | | 1,400 | Tons. |
| MR03 | 66 | | 2.5 | | 2,500 | Tons. |
| ID01 | 68 | | c. 1.5 | | 1,500 | |
| Ley | 68 | | 1.4 | | 1,400 | Tons. |
| HRST | 75 | 3,000 | | | 1,360 | |
| MOW | 76 | c. 5,070 | | c. 2,300 | 2,299 | |
| Bras | 77 | | | 2,300 | 2,300 | |
| SWM | 77 | 5,000 | | 2,300 | 2,268 | |
| Rock | 78 | | | 2,300 | 2,300 | |
| FI | 81 | | | 2,300 | 2,300 | |
| AFM | 82 | 5,070 | | | 2,299 | |
| JWS | 82 | | | c. 2,300 | 2,300 | |

### LENGTH

| | | in | ft | cm | m | m conv | |
|---|---|---|---|---|---|---|---|
| IA01 | 65 | | 23 | | | 7.01 | |
| IA05 | 66 | | 32 1/3 | | | 9.86 | |
| MR03 | 66 | | 34 | | | 10.36 | |
| Ley | 68 | | 23 | | | 7.01 | |
| HRST | 75 | | 35.5 | | | 10.82 | |
| MOW | 76 | | 35'1.5" | | 10.70 | 10.71 | |
| | | | [36'4.5"] | | | 11.09 | Mk 4 is 15 in longer than Mk 2. |
| Bras | 77 | | | | 10.6 | 10.60 | |
| SWM | 77 | | 35 | | 10.7 | 10.67 | |
| Rock | 78 | | | | 10.7 | 10.70 | |
| FI | 81 | | | | 10.7 | 10.70 | |
| SMP | 81 | | | | 10 | 10.00 | [From graph], '81, '83, '84, '85. |
| AFM | 82 | | 34'9" | | | 10.59 | |
| AWST | 82 | | 35.5 | | | 10.82 | |
| JWS | 82 | | | c. 1,070 | | 10.70 | |

```
------------ DIAMETER ------------
              in        ft         cm       m      cm conv
IA01   65               1.5                         45.72          Excluding booster.
MR03   66               1 1/2                       45.72
Ley    68               1.5                         45.72
HRST   75     26                                    66.04
MOW    76               2'2"              .66       66.04          Booster.
                        1'8"              .51       50.80          Second stage.
Bras   77                          50                50.00         [500 mm in source]
SWM    77     27.6                 70                70.10
Rock   78                          70                70.00
FI     81                          50                50.00         Missile.
                                   70                70.00         Booster.
AFM    82               1'8"                         50.80
AWST   82     26                                     66.04
JWS    82                          70                70.00         Booster.
                                   50                50.00         Second stage.

------------ WINGSPAN ------------
              in        ft         cm       m      cm conv
IA01   65               6.25                        190.50
Ley    68               6 1/4                       190.50
MOW    76               5'7"              1.70      170.18
SWM    77     67                   170               170.18
Rock   78                          220               220.00
FI     81                          170               170.00        Missile.
                                   220               220.00        Booster.
AFM    82               5'7"                         170.18

------------ NUMBER OF STAGES ------------
ID01   68     2
Rock   78     "1 + 1"
AWST   82     2

------------ TYPE OF PROPULSION ------------
ID01   68     solid
HRST   75     solid
MOW    76     solid              Booster.
              liquid             Sustainer.
Bras   77     solid              Boosters.
              liquid             Propellant.
SWM    77     solid              Boost. Booster burns 4-5 sec.
              liquid             Sustainer (nitric acid and ? kerosene). Sustainer burns 22 sec.
Rock   78     solid
              ramjet
FI     81     solid              Booster, c. 5 sec burn.
              liquid             Sustainer, nitric acid/hydrocarbons, c. 22 sec burn.
AFM    82     solid              Boost.
              liquid             Sustainer.
AWST   82     solid              Booster.
              liquid             Sustainer.
JWS    82     solid
              liquid             Sustainer.
BW     86     storable liquid

------------ MIDCOURSE (OR MAIN) GUIDANCE ------------
HRST   75     radio command
JCS    75     command            FY76-77.
                                 Some guidance systems modified with optics, providing improved ECCM and
                                 low-altitude capabilities.
MOW    76     radio command      Automatic.
Bras   77     radio command
SWM    77     radio command      (UHF) automatic, with radar tracking of target.
Rock   78     radio
FI     81     radio command
AFM    82     radio command      Auto, radar tracks target.
AWST   82     command            "Radar tracking G-band, acquisition E/F band".
JWS    82     radio command      Targets tracked by [ground] radar, feeds data to computer; produces signals
                                 which modulate output of command transmitter. 2 sets of 4 strip antennas
                                 mounted fore and aft of missile wings to receive these signals. Missile must
                                 be gathered to radar beam in first 6 secs of flight, or never.
                                 Recent reports refer to optical guidance for improved ECCM and low-altitude
                                 performance.

------------ TERMINAL GUIDANCE ------------
Bras   77                        Several versions (possibly with terminal guidance) in Soviet Army.
AFM    82                        Some late versions have terminal homing.
```

## CONTROL SYSTEM

| | | |
|---|---|---|
| MR03 | 66 | 1960 version has guidance fins on noses of casing—presumably supplements earlier missile against low-flying aircraft. |
| MOW | 76 | Movable tail surfaces and control surfaces on booster fins. |
| Rock | 78 | Aerodynamic. |
| FI | 81 | Cruciform rear fins. |
| JWS | 82 | Movable tail surfaces. |

## RADAR

| | | | |
|---|---|---|---|
| SecC | 76 | | "Although the number of older SA-2s is declining, the Soviets have continued to introduce into the remaining active sites a new radar that" [substantial section deleted]. |
| Bras | 77 | Fan Song | Fire control. |
| SWM | 77 | Fan Song | Mods A, B are E/F band; D, E are G-band. |
| AWO2 | 78 | Fan Song | Deployed with B,C,D,E,F models of Fan Song, provides frequency diversity. |
| JWS | 82 | Fan Song | |

## OBSERVED IN PARADE

| | | | |
|---|---|---|---|
| ID01 | 68 | Nov 1957 | First shown. |
| MOW | 76 | 1957 | |
| SWM | 77 | 1957 | |
| | | 1967 | Mk 4 (with large white warhead) first displayed. |
| Rock | 78 | 1957 | First public display. |
| FI | 81 | 1967 | Version reported to have nuclear warhead. |
| AFM | 82 | Nov 1967 | Version displayed had enlarged white-painted warhead, claimed to be more effective; may be nuclear. |
| JWS | 82 | 1967 | A version observed in Moscow, somewhat longer with larger warhead. |

## IOC

| | | | |
|---|---|---|---|
| SecC | 71 | 1958 | Year operational. |
| SMP | 81 | 1959 | Initially operational. |
| AFM | 82 | 1959 | Operational. |

## DEPLOYMENT BEGAN

| | | | |
|---|---|---|---|
| JCS | 75 | 1958 | First introduced, FY76-77. |
| Bras | 77 | 1950s | Has been in service since the 1950s. |
| Rock | 78 | 1957 | Service intro. |
| Clns | 80 | 1958 | First deployed. |
| JWS | 82 | 1958 | [1058 in source] |
| JWS+ | 83 | 1958 | Soviet forces. |
| BW | 86 | 1958 | |

## DEPLOYED

| | | |
|---|---|---|
| MOW | 76 | In service. |
| Bras | 77 | In service. |
| JWS | 82 | Deployed. |
| SMP | 85 | Operational. |
| SMP | 85 | Operational. |
| BW | 86 | In service. |

## DEPLOYMENT HISTORY

| | | |
|---|---|---|
| SecC | 63 | Estimated deployment will continue in large numbers. |
| SecC | 64 | Now estimate USSR will deploy considerably more SA-2s (but fewer SA-3s). |
| SecC | 65 | Buildup under way for some years, now leveling off. |
| MR03 | 66 | 1957 model available in large numbers in USSR and WTO. |
| SecC | 67 | No significant change in deployment during last few years. Believe Soviets will tend to retain SA-2 systems. |
| SecC | 68 | No significant change in deployment, FY69-70. |
| JCS | 74 | "Continued deactivation" of SA-2 exceeded new SA-3s and SA-5s. |
| SecD | 74 | The number of active SA-2 sites is declining, FY75-77. |
| JCS | 75 | Expected to continue to decline, FY76-77. |
| JCS | 77 | SA-2 sites have declined in number, FY78-81. |
| FI | 81 | Being replaced by SA-3/SA-5. |
| SMP | 81 | "Has been the backbone of Soviet SAM defenses". |
| JCS | 82 | "Throughout the Soviet Union, the long-term drawdown of the SA-2 may resume due to replacement with the more capable SA-10". |
| JWS | 82 | Standard equipment in Soviet forces. Likely to remain in service for some years, but number deployed expected to decline. In Jan 1976 US SecDef reported further decline in active SA-2 sites. |

## NUMBERS IN SERVICE

### NUMBER IN 1961

| | | | |
|---|---|---|---|
| SecC | 70 | 2600 | Operational launchers [perhaps end-year, ed], [from graph]. |
| SecC | 71 | 2600 | Operational launchers [perhaps end-year, ed], [from hand-drawn graph]. |

### NUMBER IN 1962

| | | | |
|---|---|---|---|
| SecC | 70 | 2900 | Operational launchers [perhaps end-year, ed], [from graph]. |
| SecC | 71 | 2900 | Operational launchers [perhaps end-year, ed], [from hand-drawn graph]. |

## SOVIET MISSILES IDDS

```
------------ NUMBER IN 1963 ------------
SecC  70   3200      Operational launchers [perhaps end-year, ed], [from graph].
SecC  71   3200      Operational launchers [perhaps end-year, ed], [from hand-drawn graph].

------------ NUMBER IN 1964 ------------
SecC  70   3500      Operational launchers [perhaps end-year, ed], [from graph].
SecC  71   3500      Operational launchers [perhaps end-year, ed], [from hand-drawn graph].

------------ NUMBER IN 1965 ------------
SecC  65             Leveling off at about 1100-1200 sites.
SecC  70   3800      Operational launchers [perhaps end-year, ed], [from graph].
SecC  71   3800      Operational launchers [perhaps end-year, ed], [from hand-drawn graph].

------------ NUMBER IN 1966 ------------
SecC  66             SA-2 deployment within the USSR appears to have leveled out at some 800-900 operational
                     sites, fewer than earlier estimate.
SecC  70   4100      Operational launchers [perhaps end-year, ed], [from graph].
SecC  71   4100      Operational launchers [perhaps end-year, ed], [from hand-drawn graph].

------------ NUMBER IN 1967 ------------
SecC  67             Deployed at about 900 primary sites around the rest of the country [other than SA-1
                     rings around Moscow].
SecC  70   4300      Operational launchers [perhaps end-year, ed], [from graph].
SecC  71   4000      Operational launchers [perhaps end-year, ed], [from hand-drawn graph].

------------ NUMBER IN 1968 ------------
SecC  68             Deployed at about 870 primary sites around the rest of the country [other than SA-1
                     rings around Moscow].
SecC  70   4600      Operational launchers [perhaps end-year, ed], [from graph].
SecC  71   3600      Operational launchers [perhaps end-year, ed], [from hand-drawn graph].

------------ NUMBER IN 1969 ------------
SecC  69             Deployed at about 840 primary operational sites throughout the country.
SecC  70   4800      Operational launchers [perhaps end-year, ed], [from graph].
SecC  71   3100      Operational launchers [perhaps end-year, ed], [from hand-drawn graph].

------------ NUMBER IN 1970 ------------
SecC  70             4900 in 1970, operational launchers [perhaps end-year, ed], [from graph].
SecC  71   2300      Operational launchers [perhaps end-year, ed], [from hand-drawn graph].
                     7842, 1 Oct, "SA-1/2".
Clns  80   4600
JWS   82   4,600     10 years ago [1970].

------------ NUMBER IN 1971 ------------
SecC  70             5000 in 1971, operational launchers [perhaps end-year, ed], [from graph].
SecC  71             1800 in 1971, operational launchers [perhaps end-year, ed], [from hand-drawn graph].
                     7536-8376, mid-71, "SA-1/2".
Clns  80   4500

------------ NUMBER IN 1972 ------------
SecC  71             1800 in 1972, operational launchers [perhaps end-year, ed], [from hand-drawn graph].
                     7356-8376 [sic], mid-72, "SA-1/2".
Clns  80   4300

------------ NUMBER IN 1973 ------------
Clns  80   4100

------------ NUMBER IN 1974 ------------
Clns  80   3700

------------ NUMBER IN 1975 ------------
Clns  80   3500

------------ NUMBER IN 1976 ------------
MOW   76   c. 4,500  In USSR. Number deployed is expected to decline gradually.
Clns  80   3400

------------ NUMBER IN 1977 ------------
Clns  80   3300

------------ NUMBER IN 1978 ------------
Clns  80   3000

------------ NUMBER IN 1979 ------------
Clns  80   2800
```

## CRUISE MISSILES  Ground-Launched SAMs  SA-2

---------- **NUMBER IN 1980** ----------
| | | | |
|---|---|---|---|
| JWS | 82 | 2,800 | Today [1980]. |
| Clns+ | 83 | 2800 | In 1980-1982. |

---------- **NUMBER IN 1981** ----------
| | | | |
|---|---|---|---|
| FI | 81 | c. 3,500 | Declining 100-200/year. |

---------- **NUMBER IN 1982** ----------
| | | | |
|---|---|---|---|
| AFM | 82 | 3,500 | Number declines yearly. |

---------- **NUMBER IN 1984** ----------
| | | | |
|---|---|---|---|
| AFM+ | 84 | c. 3,000 | Thought to remain operational. |

---------- **UNDATED NUMBER** ----------
| | | | |
|---|---|---|---|
| SWM | 77 | 4,500 | In USSR at one time. |

---------- **CARRIERS** ----------

---------- **CARRIER** ---------- **TUBES** --- **RELOADS** ----------
| | | | |
|---|---|---|---|
| AW02 | 78 | | Transported on truck-towed launcher. |

---------- **CARRIER** ---------- **TUBES** --- **RELOADS** ----------
| | | | | | |
|---|---|---|---|---|---|
| MOW | 76 | ZiL 157 | | | Also needs radar van and generators. |
| Bras | 77 | ZIL 157 | | | Mounted on articulated trailer, towed by ZIL 157 truck. |
| JWS | 82 | Zil 157 | | | Cross-country semi-trailer transporter-erector. |

---------- **USE AND CONFIGURATION** ----------
| | | |
|---|---|---|
| IA05 | 66 | Transported on trailer, can be set up anywhere in a few hours; "considerably longer" to set up essential components of radar, computer, and power supply. |
| Bras | 77 | Poor cross-country ability, would not be expected to be deployed in forward areas. |
| AF01 | 78 | "Field mobile". Has ECCM in USSR. 3 batteries totalling 18 launchers [in Army]. |
| Clns | 80 | Single launchers. Launch site: Fixed. |
| ID12 | 81 | Normally deployed in rear areas [not with troops]. |
| US16 | 81 | 3 batteries, 18 launchers in typical GSFG Army Group. |
| JWS | 82 | Land-mobile system. |

---------- **RELOADS IN GENERAL** ----------
| | | |
|---|---|---|
| SecC | 71 | "4 missiles per launcher", [appears without explanation on chart of launcher numbers, ed]. |

---------- **STRATEGY** ----------
| | | |
|---|---|---|
| JCS | 75 | High altitude, FY76-77. |
| SMP | 81 | Effective altitude: medium, '81, '83, '84, '85. |
| SMP | 85 | Incapable of defending against targets with "small radar-cross-section[s] such as cruise missiles". |
| SMP | 85 | Incapable of defending against targets with "small radar-cross-section[s] such as cruise missiles". |

---------- **COMBAT REPORTS/EFFECTIVENESS** ----------
| | | |
|---|---|---|
| SecC | 64 | Moderately effective against bombers at high altitudes, but "of very limited effectiveness against low altitude attacks". |
| SecC | 65 | Moderately effective against bombers at medium and high altitudes, but "of very limited effectiveness against low altitude attacks". |
| SecD | 74 | Deployed in 1973 war in a heavy barrier defense with SA-3 and SA-6. |
| JCS | 75 | Employed by North Vietnam with reasonable effectiveness during Linebacker II, FY76-77. |
| SWM | 77 | Early version downed Gary Powers' U-2. |
| FI | 81 | Has seen extensive action in SE Asia and Mideast. |
| ID12 | 81 | In 1967, Egypt had 18 battalions of 6 launchers each. Israelis attacked below SA-2 effective altitude. |
| AFM | 82 | Used extensively in North Vietnam and the Mideast. |
| JWS | 82 | Initially successfully against B-52s over North Vietnam; subsequently B-52 ECM equipment was more than a match for North Vietnamese ECCM. Israeli ECM devices at first adequate to deflect "the beam-riding missiles"; subsequently, improved missiles were introduced having terminal guidance with wider range of frequencies than Israeli ECM could handle; Then Israeli obtained US ECM pods which successfully jammed "missile acquisition, tracking, and guidance systems"; [years unclear, ed]. Technologically obsolescent; unlikely to be very effective against enemy able to deploy "reasonably sophisticated" ECM. |

---------- **GEOGRAPHICAL DEPLOYMENTS** ----------
| | | |
|---|---|---|
| IA05 | 66 | Deployed in Cuba during missile crisis. |
| SecC | 70 | Deployed throughout USSR; [map shows it along southern fringe of country from Pacific to Caspian; and in broad bands throughout European USSR]. |
| SecC | 71 | Deployed throughout USSR, FY72-73. |
| JCS | 75 | GSFG has 50-75 "battalions of SA-2s, 3s, and 4s". |
| SMP | 81 | Deployed throughout USSR. |

|  |  | EXPORTED TO |  |
|---|---|---|---|
| JCS | 74 | Arabs | Provided to Arabs. |
|  |  | PRC | PRC's only SAM is copy of SA-2. |
| JCS | 75 | PRC | PRC variant of SA-2 is basic PRC operational SAM, FY76-79. |
| MOW | 76 | WTO |  |
|  |  | Afghanistan |  |
|  |  | China |  |
|  |  | Cuba |  |
|  |  | Egypt |  |
|  |  | India |  |
|  |  | Indonesia |  |
|  |  | Iraq |  |
|  |  | North Vietnam |  |
|  |  | North Korea |  |
|  |  | Syria |  |
|  |  | Yugoslavia |  |
| Bras | 77 | Afghanistan |  |
|  |  | Albania |  |
|  |  | Bulgaria |  |
|  |  | Cuba |  |
|  |  | Czech |  |
|  |  | Egypt |  |
|  |  | East Germany |  |
|  |  | Hungary |  |
|  |  | India |  |
|  |  | Iraq |  |
|  |  | Libya |  |
|  |  | Mongolia |  |
|  |  | North Korea |  |
|  |  | Poland |  |
|  |  | Rumania |  |
|  |  | Syria |  |
|  |  | Vietnam |  |
|  |  | Yugoslavia |  |
| SWM | 77 |  | Widely deployed in WTO, also SE Asia and Mideast. |
| FI | 81 | Afghanistan |  |
|  |  | Albania | [Final report] |
|  |  | Algeria | [Final report] |
|  |  | Bulgaria | 2 battalions. |
|  |  | China | [Final report] |
|  |  | Cuba | 24 battalions, 144 launchers. |
|  |  | Czech | C. 10 sites, [final report] |
|  |  | Egypt | 24-30 batteries. |
|  |  | E Germany | 2 battalions. |
|  |  | Hungary | 2 battalions. |
|  |  | India | C. 20 sites. |
|  |  | Iraq |  |
|  |  | Jugoslavia | 8 batteries. |
|  |  | North Korea | 20 battalions. |
|  |  | Libya | 8 SA-2,3,6 batteries protect Okba Ben Nafi air base. |
|  |  | Mongolia | 1 battery. |
|  |  | Poland | 30 sites. |
|  |  | Romania |  |
|  |  | Syria |  |
|  |  | Vietnam | C. 300 launchers. |
| SMP | 81 |  | "Used by non-Soviet Warsaw Pact and other communist and Third World nations as well". |
| AFM | 82 |  | Standard SAM in about 20 countries. |
| JWS | 82 | Afghanistan |  |
|  |  | Albania |  |
|  |  | Algeria |  |
|  |  | Bulgaria |  |
|  |  | China | Thought to have established its own production line at one time. |
|  |  | Cuba |  |
|  |  | Czech |  |
|  |  | Egypt |  |
|  |  | East Germany |  |
|  |  | Hungary |  |
|  |  | India |  |
|  |  | Indonesia |  |
|  |  | Iraq |  |
|  |  | North Korea |  |
|  |  | Libya |  |
|  |  | Poland |  |
|  |  | Romania |  |
|  |  | Vietnam |  |
|  |  | Yugoslavia |  |

"As Soviet use of this weapon declines (and support facilities become increasingly difficult) it is probable that its use among many of the client states equipped

|     |    |              | with it will diminish also". |
|-----|----|--------------|------------------------------|
| FI+ | 83 | Ethiopia     |                              |
|     |    | Jordan       |                              |
|     |    | Peru         |                              |
|     |    | Somalia      |                              |
|     |    | Sudan        |                              |
|     |    | North Yemen  |                              |
|     |    | South Yemen  |                              |
| SMP | 85 | Angola       | "[In 1984] Angola received large quantities of Soviet equipment, including... deliveries of the SA-2 SAM system." |
|     |    | Algeria      | Soviet military advisors assigned to "air defense units with Soviet SA-2, SA-3, and SA-6 surface-to-air missiles." |
| SMP | 85 | Angola       | "[In 1984] Angola received large quantities of Soviet equipment, including... deliveries of the SA-2 SAM system." |
|     |    | Algeria      | Soviet military advisors assigned to "air defense units with Soviet SA-2, SA-3, and SA-6 surface-to-air missiles." |

---------- **RELATION TO OTHER MISSILES** ----------

| SecC | 63 | Similar to US Nike-Hercules. |
| SecD | 75 | PRC's SA-1 is PRC version of SA-2; production has declined from earlier levels [PRC]. |
| MOW  | 76 | Mk 1, original version, similar to first US Nike Ajax. |

---------- **OTHER INFORMATION** ----------

| MOW  | 76 | Data apply to 'Mk 2' version. |
| AFM  | 82 | There are several versions, incorporating improvements. Data above for standard export version. |
| AWST | 82 | Variety of missiles used with the system. |

# SA-3

| Sources are: | SecC | 67 | SecD | 71 | JCS | 74 | Bras | 77 | Clns | 80 | AWST | 82 | JWS+ | 83 |
|---|---|---|---|---|---|---|---|---|---|---|---|---|---|---|
| SecC | 63 | Ley | 68 | JCS | 72 | SecD | 74 | JCS | 77 | JCS | 80 | JCS | 82 | SMP | 83 |
| SecC | 64 | SecC | 68 | SecC | 72 | HRST | 75 | SWM | 77 | FI | 81 | JWS | 82 | AFM+ | 84 |
| SecC | 65 | SecC | 69 | ID03 | 73 | JCS | 75 | AWO2 | 78 | ID12 | 81 | Clns+ | 83 | SMP | 85 |
| MR03 | 66 | SecC | 70 | JCS | 73 | MOW | 76 | Rock | 78 | SMP | 81 | FI+ | 83 | | |
| SecC | 66 | SecC | 71 | SecD | 73 | NA03 | 76 | JCS | 79 | AFM | 82 | FI+ | 83 | | |

---------- **US DESIGNATION** ----------

| SecC | 64 | SA-3 | FY65-73,75-77. |
|---|---|---|---|
| SecD | 71 | SA-3 | FY72,74-77. |
| JCS | 72 | SA-3 | FY73-83. |
| MOW | 76 | SA-3 | |
| Bras | 77 | SA-3 | |
| SWM | 77 | SA-3 | |
| Rock | 78 | SA-3 | |
| Clns | 80 | SA-3 | |
| FI | 81 | SA-3 | |
| SMP | 81 | SA-3 | '81, '83, '84, '85. |
| AFM | 82 | SA-3 | |
| AWST | 82 | SA-3 | |
| JWS | 82 | SA-3 | |

---------- **NATO CODENAME** ----------

| HRST | 75 | Goa | |
| JCS | 75 | Goa | FY76-78. |
| MOW | 76 | Goa | |
| Bras | 77 | Goa | |
| SWM | 77 | Goa | |
| Rock | 78 | Goa | |
| Clns | 80 | Goa | |
| FI | 81 | Goa | |
| SMP | 81 | Goa | '81, '83, '85. |
| AFM | 82 | Goa | |
| AWST | 82 | Goa | |
| JWS | 82 | Goa | |

---------- **OTHER DESIGNATION** ----------

Ley  68     "Short-range", [no other designation provided, ed].

---------- **DESCRIPTION** ----------

SMP  81     Strategic SAM, '84, '85.
JWS  82     "Shipborne or land-based surface-to-air guided missile".

---------- **EDITOR'S NOTES** ----------

SecC  63    Listed FY64-73,75-77, not by name in FY64.
SecD  71    Listed FY72,74-77.
JCS   72    Listed FY73-83.
SMP   81    Listed '81, '83, '84, '85.

---------- **MINIMUM ALTITUDE** ----------

| | | mi | ft | km | m | m conv | |
|---|---|---|---|---|---|---|---|
| SecC | 69 | | | | | | "Further analysis of the SA-3, including radar simulations, has led us to estimate a somewhat greater effectiveness at lower altitudes for this system than heretofore, down to about [deleted] feet under favorable conditions". |
| SecC | 70 | | | | | | "Under optimum conditions it could intercept aircraft at altitudes as low as [deleted]". |
| Rock | 78 | | | 0.2 | | 200.00 | |
| FI | 81 | | | | 100-300 | 100.00-300.00 | |
| AWST | 82 | | 150 | | | 45.72 | |

CRUISE MISSILES  Ground-Launched SAMs  SA-3

### MAXIMUM ALTITUDE

| | | mi | ft | km | m | m conv | |
|---|---|---|---|---|---|---|---|
| MOW | 76 | | 40,000 | | 12,200 | 12,192 | |
| Bras | 77 | | | | 20,000 | 20,000 | Max effective altitude, est. |
| SWM | 77 | | 39,370+ | | 12,000+ | 11,999 | |
| AWO2 | 78 | | > 40,000 | | | 12,192 | |
| Rock | 78 | | | 15 | | 15,000 | |
| Clns | 80 | | 40,000 | | | 12,192 | Combat Ceiling, "Lo-40,000" ft. |
| FI | 81 | | | | 10000-15000 | 10,000-15,000 | |
| AFM | 82 | | 49,200 | | | 14,996 | |
| AWST | 82 | | 60,000 | | | 18,288 | |
| FI+ | 83 | | | 25-30 | | 25,000-30,000 | Max. |
| JWS+ | 83 | | | | > 13,000 | 13,000 | Ceiling. |

### RANGE

| | | mi | nm | km | km conv | |
|---|---|---|---|---|---|---|
| MR03 | 66 | 12 | | | 19.31 | Max slant range. |
| Ley | 68 | 18 | | | 28.96 | |
| IDO3 | 73 | | | 5-30 | 5.00-30.00 | "Coverage" range. |
| MOW | 76 | 18.5 | | 30 | 29.77 | Max. |
| NA03 | 76 | 16-19 | | | 25.74-30.57 | Max. |
| Bras | 77 | | | 25 | 25.00 | Max effective slant range, [25,000 m in source], est. |
| SWM | 77 | 18 | | 29 | 28.96 | |
| AWO2 | 78 | | 3-12 | | 5.56-22.22 | Effective range. |
| Rock | 78 | | | 35 | 35.00 | |
| Clns | 80 | 18 | | | 28.96 | Slant range. |
| FI | 81 | | | 30-35 | 30.00-35.00 | Max. |
| SMP | 81 | | | 20 | 20.00 | '81, '83, '84, '85. |
| AFM | 82 | 21.75 | | | 35.00 | |
| AWST | 82 | 3.5-12 | | | 5.63-19.31 | |
| JWS | 82 | | | 25-30 | 25.00-30.00 | |
| Clns+ | 83 | 15 | | | 24.14 | Slant range. |
| AFM+ | 84 | 12.5 | | | 20.11 | Slant range. |

### SPEED

| | | mph | kmph | Mach | kmph conv |
|---|---|---|---|---|---|
| MR03 | 66 | | | 2-2.5 | |
| SWM | 77 | | | 2.0 | |
| FI | 81 | | | 2 | |
| AFM | 82 | | | 2 | |
| JWS+ | 83 | | | c. 2+ | |

### WARHEAD TYPE

| | | nuclear | conven | chem | |
|---|---|---|---|---|---|
| MOW | 76 | | yes | | HE. |
| Bras | 77 | | yes | | HE blast. |
| SWM | 77 | | yes | | |
| AWO2 | 78 | | yes | | HE, proximity-fuzed. |
| Rock | 78 | | yes | | HE. |
| Clns | 80 | yes | yes | | HE; "An improved version of SA-3 [sic], first displayed in 1967, may have a nuclear warhead". |
| FI | 81 | | yes | | HE. |
| AFM | 82 | | yes | | |
| JWS+ | 83 | | | | HE, proximity fuze. |

### WARHEAD WEIGHT

| | | lb | t | kg | kg conv |
|---|---|---|---|---|---|
| Bras | 77 | | | 70 | 70.00 |
| FI | 81 | | | 60 | 60.00 |
| AFM | 82 | 132 | | | 59.88 |

### WEIGHT

| | | lb | t | kg | kg conv | |
|---|---|---|---|---|---|---|
| MR03 | 66 | | 1/2 | | 500.00 | Ton. |
| Ley | 68 | | 0.7 | | 700.00 | Tons. |
| Bras | 77 | | | 950 | 950.00 | |
| SWM | 77 | 882 | | 400 | 400.08 | |
| Rock | 78 | | | 600 | 600.00 | |
| FI | 81 | | | 600 | 600.00 | |
| AFM | 82 | 1,323 | | | 600.11 | |
| JWS+ | 83 | | | 636 | 636.00 | Launch weight. |

558  SOVIET MISSILES  IDDS

```
------------ LENGTH ------------
              in    ft      cm     m    m conv
MR03  66            20                  6.10
Ley   68            17 1/4               5.26
HRST  75            22                  6.71
MOW   76            22            6.7   6.71
Bras  77                          6.7   6.70
SWM   77            22            6.7   6.71
Rock  78                          6.7   6.70
FI    81                          6.7   6.70
SMP   81                          6     6.00     [From graph], '81, '83, '84, '85.
AFM   82            22'0"               6.71
AWST  82            22                  6.71
JWS   82                    c. 670      6.70

------------ DIAMETER ------------
              in    ft      cm     m    cm conv
MR03  66            1                   30.48
Ley   68            1.0                 30.48
HRST  75      20                        50.80
MOW   76            2'3"          0.70  68.58    Booster.
                    1'6"          0.45  45.72    2nd stage.
Bras  77                    45          45.00    [450 mm in source]
SWM   77      18            46          45.72
Rock  78                    70          70.00
FI    81                    46          46.00    Missile.
                            70          70.00    Booster.
AFM   82            1'6"                45.72
AWST  82            1.9                 57.91
JWS   82                    60          60.00    Booster.
                            25-45       25.00-45.00  2nd stage.
FI+   83                    60          60.00    Booster.

------------ WINGSPAN ------------
              in    ft      cm     m    cm conv
MOW   76            4'0"          1.22  121.92
SWM   77      47            120         119.38
Rock  78                    150         150.00
FI    81                    122         122.00  Missile.
                            150         150.00  Booster.
AFM   82            4'0"                121.92

------------ NUMBER OF STAGES ------------
MOW   76     2
Bras  77     2
AWO2  78     3      2-stage solid fuel booster and solid sustainer.
Rock  78    "1 + 1"
AFM   82     2
AWST  82     2
JWS   82     2

------------ TYPE OF PROPULSION ------------
HRST  75    solid
JCS   75    solid         FY76-77.
MOW   76    solid
Bras  77    solid
SWM   77    solid         Booster and sustainer.
AWO2  78    solid
Rock  78    solid+solid
FI    81    solid         Booster and sustainer.
AFM   82    solid
AWST  82    solid         Booster.
JWS   82    solid

------------ MIDCOURSE (OR MAIN) GUIDANCE ------------
JCS   75    command              FY76-77.
MOW   76    radio command
Bras  77    radar command
SWM   77    radio command        Automatic.
Rock  78    radio
FI    81    radio command
AFM   82    radio
AWST  82    command
JWS   82    ? radio command      Probably.
```

## CRUISE MISSILES  Ground-Launched SAMs  SA-3

### ---------- TERMINAL GUIDANCE ----------
| | | | |
|---|---|---|---|
| MOW | 76 | radar | |
| SWM | 77 | radar | |
| AFM | 82 | radar | |
| JWS | 82 | | Role of missile would suggest homing, but fairly definite it is command guided. [elsewhere, ed:] "a homing system may be incorporated". |

### ---------- CONTROL SYSTEM ----------
| | | |
|---|---|---|
| MOW | 76 | Movable foreplanes. |
| SWM | 77 | Steerable foreplanes. |
| Rock | 78 | Aerodynamic. |
| FI | 81 | Cruciform fins. |
| JWS | 82 | Movable foreplane surfaces. |

### ---------- RADAR ----------
| | | | |
|---|---|---|---|
| Bras | 77 | Low Blow | Fire control; a surveillance radar is also associated. |
| SWM | 77 | Flat Face | Truck-mounted UHF (810-850 MHz and 880-950 MHz) P-15 radar with dual parabolic antennae, for target acquisition. |
| | | Low Blow | I-band (9,000-9,400 MHz) fire control, operating on principle of SA-2's Fan Song. |
| AWO2 | 78 | Low Blow | I-band, high clutter-resistance; up to 6 a/c tracked simultaneously, 1 or 2 missiles fired at same target. |
| AWST | 82 | Low Blow | Fire control radar, I-band. |
| JWS | 82 | Low Blow | I/J-band fire control radar. |
| | | Flat Face | Goa "commonly associated with" Flat Face acquisition radar. |

### ---------- OBSERVED IN PARADE ----------
| | | | |
|---|---|---|---|
| Rock | 78 | 1964 | First public display. |

### ---------- IOC ----------
| | | | |
|---|---|---|---|
| SecC | 71 | 1961 | Year operational. |

### ---------- DEPLOYMENT BEGAN ----------
| | | | |
|---|---|---|---|
| SecC | 63 | | USSR estimated to have deployed "HAWK-type systems by the 1966-1968 period". |
| JCS | 75 | early 1961 | First introduced, FY76-77. |
| Bras | 77 | c. 1960 | In service. |
| Clns | 80 | 1961 | First deployed. |
| JWS | 82 | 1961 | Introduced into service. |

### ---------- DEPLOYED ----------
| | | |
|---|---|---|
| MOW | 76 | In service. |
| Bras | 77 | In service. |

### ---------- DEPLOYMENT HISTORY ----------
| | | |
|---|---|---|
| SecC | 65 | Deployment is still continuing on a modest scale. |
| SecC | 66 | Deployment continues at a slow pace. Force not expected to grow much. |
| SecC | 67 | No significant change in deployment in last few years. |
| SecC | 68 | No significant changes in deployment, FY69-70. |
| SecC | 70 | In the past year, deployment of SA-3 was accelerated, "and has now expanded to about [deleted] sites, of which some [deleted] are now operational". |
| JCS | 72 | Being deployed [inferred or stated, ed], FY73-83. |
| JCS | 73 | Long-term buildup may be nearing completion. |
| SecD | 73 | Deployment continuing at a slow pace. |
| SecD | 74 | Additional SA-3s being deployed, FY75-77. |
| JCS | 75 | Estimate mid-1975 level of sites will increase. |
| JCS | 77 | SA-2 decline largely offset by additional SA-3 and SA-5 deployment, FY78-79. |
| JCS | 79 | 2-rail SA-3 launchers being replaced by 4-rail launchers, FY80-83. |
| JCS | 80 | SA-3 launcher replacement offsets SA-2 decline. |
| SMP | 81 | "Over half the sites use newer four-rail launchers, rather than the two-rail launchers", doubling missiles ready to launch. |
| AFM | 82 | Numbers increasing with USSR, allies and friends. |
| JCS | 82 | Over half the SA-3 sites now have 4-rail launchers. |
| JWS | 82 | Numbers of SA-3 increasing steadily, confirmed in early 1981 and early 1982 by US reports. |
| SMP | 85 | Operational. |

### ---------- NUMBERS IN SERVICE ----------

### ---------- NUMBER IN 1961 ----------
| | | | |
|---|---|---|---|
| SecC | 71 | 40 | "Operational launchers" [perhaps end-year, ed], [from hand-drawn graph]. |

### ---------- NUMBER IN 1962 ----------
| | | | |
|---|---|---|---|
| SecC | 71 | 120 | "Operational launchers" [perhaps end-year, ed], [from hand-drawn graph]. |

### ---------- NUMBER IN 1963 ----------
| | | | |
|---|---|---|---|
| SecC | 71 | 200 | "Operational launchers" [perhaps end-year, ed], [from hand-drawn graph]. |

------------ **NUMBER IN 1964** ------------
SecC 64              Estimate USSR will deploy "considerably more SA-2's but fewer SA-3's".
SecC 71     280      "Operational launchers" [perhaps end-year, ed], [from hand-drawn graph].

------------ **NUMBER IN 1965** ------------
SecC 65              Present deployments suggest most likely to be deployed in comparatively limited numbers
                     [relative to SA-2] as SA-2 supplement.
SecC 71     320      "Operational launchers" [perhaps end-year, ed], [from hand-drawn graph].

------------ **NUMBER IN 1966** ------------
SecC 66              Current size is about 110 sites, force not expected to grow much, FY67-68.
SecC 71     335      "Operational launchers" [perhaps end-year, ed], [from hand-drawn graph].

------------ **NUMBER IN 1967** ------------
SecC 71     350      "Operational launchers" [perhaps end-year, ed], [from hand-drawn graph].

------------ **NUMBER IN 1968** ------------
SecC 68              Deployed at about 115 sites in selected areas.
SecC 71     365      "Operational launchers" [perhaps end-year, ed], [from hand-drawn graph].

------------ **NUMBER IN 1969** ------------
SecC 69              Deployed at about 120 primary sites [does not count those deployed by Soviet forces in
                     Eastern Europe].
SecC 71     430      "Operational launchers" [perhaps end-year, ed], [from hand-drawn graph].
JWS  82     1,800    Tubes in 900 positions [1969].

------------ **NUMBER IN 1970** ------------
SecC 71     820      1 Oct, "Estimated Deployment".
            600      "Operational launchers" [perhaps end-year, ed], [from hand-drawn graph].
Clns 80     1800     Rails on 900 launchers.

------------ **NUMBER IN 1971** ------------
SecC 71              920-1160 in mid-71, "Estimated Deployment".
                     780 in 1971, "Operational launchers" [perhaps end-year, ed], [from hand-drawn graph].
Clns 80     2000     Rails on 1000 launchers.

------------ **NUMBER IN 1972** ------------
SecC 71              960-1280 in mid-72, "Estimated Deployment".
                     950 in 1972, "Operational launchers" [perhaps end-year, ed], [from hand-drawn graph].
Clns 80     2200     Rails on 1100 launchers.

------------ **NUMBER IN 1973** ------------
Clns 80     2600     Rails on 1100 launchers.

------------ **NUMBER IN 1974** ------------
Clns 80     3100     Rails on 1150 launchers.

------------ **NUMBER IN 1975** ------------
Clns 80     3500     Rails on 1200 launchers.

------------ **NUMBER IN 1976** ------------
Clns 80     3700     Rails on 1300 launchers.

------------ **NUMBER IN 1977** ------------
Clns 80     4000     Rails on 1300 launchers.

------------ **NUMBER IN 1978** ------------
Clns 80     4300     Rails on 1300 launchers.

------------ **NUMBER IN 1979** ------------
Clns  80    4500     Rails on 1400 launchers.
JWS   82    4,500    Tubes in 1,400 positions.
Clns+ 83    4200     Rails on 1400 launchers.

------------ **NUMBER IN 1980** ------------
Clns+ 83    4500     Rails on 1400 launchers.

------------ **NUMBER IN 1981** ------------
SMP   81             "Now deployed throughout the USSR and Warsaw Pact at over 400 sites".
Clns+ 83    4600     Rails on 1400 launchers.

------------ **NUMBER IN 1982** ------------
Clns+ 83    4500     Rails on 1300 launchers.

## CRUISE MISSILES  Ground-Launched SAMs  SA-3

### LAUNCHER/SILO

| | | |
|---|---|---|
| Bras | 77 | [Picture shows 2 on launcher, not truck, ed]. |
| JCS | 77 | [Picture shows 4-rail launcher, ed]. |
| Clns | 80 | "A 4-rail version began replacing the standard 2-rail system beginning in 1973". Each arm/rail holds one missile. |
| FI | 81 | Ground-mounted trainable launcher. 4-rail launcher deployed since 1973. |
| SMP | 81 | Newer 4-rail launcher and older 2-rail launcher. |
| AWST | 82 | Launcher with 4 rails. |
| JWS | 82 | Originally twin launcher; now triple missile launcher, first noted at Yugoslavia air defence batteries (not known if Soviet or Yugoslav initiative); still more recently 4-round launcher replacing twin launchers in USSR. |
| FI+ | 83 | Quad launcher now equips more than half the Soviet sites. |
| AFM+ | 84 | Deployed on 2, 3, and 4-round launchers. |

### CARRIERS

#### CARRIER

| | | CARRIER | TUBES | RELOADS |
|---|---|---|---|---|
| SWM | 77 | tracked vehicles | | |
| JCS | 79 | | | [Picture with 2 rails, on truck, ed], FY80-81. |
| Clns | 80 | | | Launch site: Mobile. |
| FI | 81 | tracked vehicles | | |
| JWS | 82 | | | Usually transported on specially converted military trucks. |

#### CARRIER

| | | CARRIER | TUBES | RELOADS |
|---|---|---|---|---|
| MRO3 | 66 | Zil-157 chassis | 2 | Launcher. |
| Bras | 77 | ZIL 157 truck | 2 | 6 wheels. |
| SWM | 77 | ZIL-157 tractor | | |
| FI | 81 | Zil-157 tractor | 2 | Carrier [not launcher]. |

### USE AND CONFIGURATION

| | | |
|---|---|---|
| SecC | 70 | "Located in specific areas where the low altitude threat is considered most severe", FY71-72. |
| SecC | 72 | "Located in specific areas around military targets and military command centers". |
| JCS | 75 | Road-transportable, FY76-77. Employed in both point defense and barrier role, FY76-77. Capable of intercepting at low altitudes within a limited range of the launch site, FY76-77. |
| Bras | 77 | Provides low with medium level coverage at divisional level; is the link between SA-2,4 and SA-6. |
| SWM | 77 | Mobile. 6 aircraft can be tracked simultaneously, 2 missiles can be fired at same time. |
| ID12 | 81 | Normally deployed in rear areas. |
| SMP | 81 | "Provides low-altitude coverage and point defense to selected strategic areas". |

### STRATEGY

| | | |
|---|---|---|
| SecC | 64 | Earlier thought capable of "intercepting low-altitude penetrators (including high-speed, low-altitude air-to-surface missiles)"; "may not have this capability". |
| SecC | 65 | Apparently designed to engage low altitude penetrators. |
| SecC | 66 | Specifically designed against low altitude threat, FY67-69,71-73. |
| SecC | 67 | Limited deployment suggests not much better than SA-2 against low altitude threat. Logically, seems Soviets would eventually deploy improved low-altitude SAM system, but as yet no evidence of such a deployment, FY68-69. |
| SecC | 70 | Accelerated SA-3 deployment apparently reflects increased confidence in SA-3 by Soviets. But still no evidence of development of "a more effective, long-range Soviet low-altitude SAM". |
| JCS | 73 | Low altitude, FY74-75,78-80. |
| JCS | 75 | Medium and low altitude, FY76-77. |
| MOW | 76 | "Close-range missile". |
| SWM | 77 | Low-altitude. |
| SMP | 81 | Effective altitude: low-to-medium, '81, '83, '84, '85. |
| AFM | 82 | "Mobile low-altitude system to complement SA-2". |
| JWS | 82 | Intended for short-range defence against low-flying targets. |

### COMBAT REPORTS/EFFECTIVENESS

| | | |
|---|---|---|
| SecD | 74 | Deployed in 1973 war in a heavy barrier defense with SA-2 and SA-6. |
| FI | 81 | Extensive use in SE Asia and Mideast. |
| ID12 | 81 | First combat use was in 1973 war. |
| SMP | 85 | Incapable of defending against targets with "small radar-cross-section[s] such as cruise missiles". |

### GEOGRAPHICAL DEPLOYMENTS

| | | |
|---|---|---|
| SecC | 69 | SA-3 is widely deployed with Soviet forces in Eastern Europe, primarily for the defense of airfields. |
| JCS | 75 | GSFG has 50-75 "battalions of SA-2s, 3s, and 4s". |

---------- **EXPORTED TO** ----------

| | | | |
|---|---|---|---|
| SecD | 71 | UAR | Reportedly a 1970 USSR-UAR agreement covers introduction of SA-3 to UAR. |
| JCS | 74 | Arabs | Provided to Arabs. |
| Bras | 77 | Bulgaria | |
| | | Czech | |
| | | Egypt | |
| | | East Germany | |
| | | Hungary | |
| | | Iraq | |
| | | Libya | |
| | | Poland | |
| | | Rumania | |
| | | Syria | |
| | | Vietnam | |
| SWM | 77 | | Widely deployed in WTO, SE Asia and Mideast. |
| FI | 81 | Poland | |
| | | East Germany | |
| | | Jugoslavia | |
| | | Czech | [Final report] |
| | | India | Pichora is local name. |
| | | Egypt | C. 60 twin ramps. |
| | | Ethiopia | |
| | | Iraq | |
| | | Libya | 8 SA-2,3,6 batteries protect Okba Ben Nafi air base. |
| | | Peru | |
| | | Syria | |
| | | Vietnam | |
| | | Uganda | [Final report] |
| | | Finland | |
| AFM | 82 | | Numbers increasing with USSR, allies and friends. |
| JWS | 82 | Egypt | [List of other than WTO states] |
| | | Finland | |
| | | Iraq | |
| | | Libya | |
| | | Syria | |
| | | Vietnam | |
| | | Uganda | |
| | | Yugoslavia | |
| | | | Now widely deployed in USSR and WTO states; use expected to increase both within and beyond WTO. |
| FI+ | 83 | Afghanistan | |
| | | Algeria | |
| | | Angola | |
| | | Bulgaria | |
| | | Cuba | |
| | | Hungary | |
| | | Mali | |
| | | Mozambique | |
| | | Somalia | |
| | | Tanzania | |
| | | Zambia | |
| SMP | 83 | Cuba | USSR "has continued to add to Cuba's" strength with...SA-3. |
| SMP | 85 | Algeria | [Soviet military advisors in Algeria are] "assigned to equipment repair installations and individual combat units. These include... air defense units with Soviet SA-2, SA-3, and SA-6 surface-to-air missiles." |
| | | Peru | "Approximately 150 Soviet military advisers and technicians provide maintenance and instruction on Soviet-made military equipment in Peru, including instruction on the Soviet SA-3/GOA missile in the Peruvian air defense school." |

---------- **RELATION TO OTHER MISSILES** ----------

| | | |
|---|---|---|
| MOW | 76 | USSR 1961 counterpart to US Hawk. |
| Bras | 77 | A naval version exists. |
| SWM | 77 | Naval version is SA-N-1. US equivalent is MIM-23A Hawk. |
| AFM | 82 | Counterpart of US Hawk. |
| JWS | 82 | Missile also deployed on ships; see separate entry. Same role as US Hawk. |

# SA-4

```
Sources are:  ID01  68      JCS   75      SecD  76      AWO2  78      SMP   81      JWS   82      SMP   83
IA01  65      Ley   68      SecD  75      Bras  77      Rock  78      US16  81      Clns+ 83      JCS   84
IA02  65      SecD  71      JCS   76      JCS   77      Clns  80      AFM   82      FI+   83      SMP   84
IA05  66      SecD  74      MOW   76      SWM   77      FI    81      AWST  82      FI+   83      SMP   85
MR03  66      HRST  75      NA03  76      AF01  78      ID12  81      JCS   82      JWS+  83
```

---------- **US DESIGNATION** ----------

| | | |
|---|---|---|
| SecD | 71 | SA-4 | FY72,75-77. |
| JCS  | 75 | SA-4 | FY76-79,83-85. |
| MOW  | 76 | SA-4 | |
| Bras | 77 | SA-4 | |
| SWM  | 77 | SA-4 | |
| Rock | 78 | SA-4 | |
| Clns | 80 | SA-4 | |
| FI   | 81 | SA-4 | |
| SMP  | 81 | SA-4 | '81, '83, '84, '85. |
| AFM  | 82 | SA-4 | |
| AWST | 82 | SA-4 | |
| JWS  | 82 | SA-4 | |

---------- **NATO CODENAME** ----------

| | | |
|---|---|---|
| SecD | 71 | Ganef | |
| HRST | 75 | Ganef | |
| MOW  | 76 | Ganef | |
| Bras | 77 | Ganef | |
| SWM  | 77 | Ganef | |
| Rock | 78 | Ganef | |
| Clns | 80 | Ganef | |
| FI   | 81 | Ganef | |
| SMP  | 81 | Ganef | '81, '83, '84. |
| AFM  | 82 | Ganef | |
| AWST | 82 | Ganef | |
| JWS  | 82 | Ganef | |

---------- **OTHER DESIGNATION** ----------

Ley  68        "Tank-launched", [no other designation provided, ed].

---------- **DESCRIPTION** ----------

SMP  81        Tactical surface-to-air missile, '81, '83.
JWS  82        Land-mobile surface-to-air tactical guided missile.

---------- **EDITOR'S NOTES** ----------

SecD 71        Listed FY72,75-77.
JCS  75        Listed FY76-79,83-85.
SMP  81        Listed '81, '83, '84, '85.
SMP  84        Source recognizes the Mods SA-4a and SA-4b.

---------- **MINIMUM ALTITUDE** ----------

| | | mi | ft | km | m | m_conv | |
|---|---|---|---|---|---|---|---|
| AWST | 82 | | 1,000 | | | 304.80 | |

---------- **MAXIMUM ALTITUDE** ----------

| | | mi | ft | km | m | m_conv | |
|---|---|---|---|---|---|---|---|
| MOW  | 76 | | 80,000  |       | 24,400  | 24,384        | |
| Bras | 77 | |         |       | 30,000  | 30,000        | Max effective altitude, est. |
| SWM  | 77 | | 59,000+ |       | 18,000+ | 17,983        | |
| AF01 | 78 | |         | 24-28 |         | 24,000-28,000 | |
| Rock | 78 | |         | 25    |         | 25,000        | "Altitude". |
| Clns | 80 | | 80,000  |       |         | 24,384        | Ceiling. |
| FI   | 81 | |         |       | 25,000  | 25,000        | |
| AFM  | 82 | | 80,000  |       |         | 24,384        | |
| AWST | 82 | | 80,000  |       |         | 24,384        | |
| JWS+ | 83 | |         |       | 24,000  | 24,000        | Ceiling. |

## 564  SOVIET MISSILES  IDDS

```
---------- RANGE ----------
          mi            nm          km          km conv
IA01  65  > 45                                  72.41
IA02  65  > 45                                  72.41
IA05  66  > 30                                  48.27
MR03  66    62                                  99.76      Max slant range.
Ley   68    45                                  72.41
MOW   76  c. 43.5                   c. 70       69.99
NA03  76    40                                  64.36      Max.
Bras  77                            40          40.00      Max effective slant range,
                                                           [40,000 m in source], est.
SWM   77    43                      70          69.19
Rock  78                            70          70.00
Clns  80    50                                  80.45
FI    81                            c. 70       70.00      Max.
ID12  81                            > 70        70.00
SMP   81                            70          70.00      '81, '83, '84, '85.
AFM   82    43                                  69.19
AWST  82    5-45                                8.05-72.41
JWS   82                            c. 70       70.00

---------- SPEED ----------
          mph          kmph         Mach        kmph conv
MR03  66                            2.5
Clns  80                            4
FI    81                            2.5
JWS+  83                            2.5

---------- WARHEAD TYPE ----------
          nuclear      conven       chem
MOW   76               yes                      HE.
Bras  77               yes                      "HE blast".
SWM   77               yes
Rock  78               yes                      HE.
Clns  80               yes                      HE.
FI    81               yes                      HE.
AFM   82               yes
JWS+  83                                        HE, proximity fuze.

---------- WARHEAD WEIGHT ----------
          lb           t            kg          kg conv
Clns  80   300                                  136.08

---------- WEIGHT ----------
          lb           t            kg          kg conv
IA01  65               c. 2.2                   2,200          Tons.
IA05  66               2                        2,000          Tons; probably over 2 tons at
                                                               launch.
MR03  66               1 1/2-2                  1,500-2,000    Tons.
ID01  68               c. 2.5                   2,500          Tons.
Ley   68               2.2                      2,200          Tons.
MOW   76  c. 2,200                  c. 1,000    997.92
Bras  77                            2,000       2,000
SWM   77   2,205+                   1,000+      1,000
Rock  78                            1,000       1,000
FI    81                            1,000       1,000
AFM   82   3,975                                1,803
JWS   82                            c. 1,800    1,800
FI+   83                            1,800       1,800

---------- LENGTH ----------
          in           ft           cm          m         m conv
IA01  65               24.6                               7.50
IA02  65               c. 24.5                            7.47
MR03  66               28                                 8.53
Ley   68               24.6                               7.50
HRST  75               26                                 7.92
MOW   76               30'0"                    9.15      9.14
Bras  77                                        9         9.00
SWM   77               30.2                     9.2       9.20
Rock  78                                        9.0       9.00
FI    81                                        9.0       9.00
SMP   81                                        8.5       8.50      [From graph], '81, '83, '84, '85.
AFM   82               28'10.5"                           8.80      [28'10 1/2" in source].
AWST  82               26                                 7.92
JWS   82                            c. 880               8.80
FI+   83                                        8.8       8.80
```

# CRUISE MISSILES  Ground-Launched SAMs  SA-4

```
------------ DIAMETER ------------
              in      ft     cm     m     cm conv
IA01   65             2.8                  85.34
MR03   66             3                    91.44
Ley    68             2.8                  85.34
HRST   75   32.5                           82.55
MOW    76             2'8"          0.80   81.28
Bras   77                    80             80.00    [800 mm in source]
SWM    77   31.5             80             80.01
Rock   78                    80             80.00
FI     81                    80             80.00
AFM    82             2'8"                 81.28
AWST   82             2.7                  82.30
JWS    82                    90             90.00
FI+    83                    90             90.00

------------ WINGSPAN ------------
              in      ft     cm     m     cm conv
IA01   65             7.9                 240.79
Ley    68             7.9                 240.79
MOW    76             7'6"         2.30   228.60
SWM    77   90              230            228.60
Rock   78                   260            260.00
FI     81                   260            260.00
AFM    82             7'6"                 228.60
JWS    82                   260            260.00    Tail.
                                    2.3    230.00    Wings.

----------- NUMBER OF STAGES -----------
Rock   78   "4 + R"

----------- TYPE OF PROPULSION -----------
HRST   75   solid
MOW    76   solid              4 wrap-around boosters jettison after burnout.
            ramjet             Sustainer.
Bras   77   solid              4 boosters wrap-around.
            ramjet
SWM    77   integral ramjet    4 wrap-round rocket boosters, ramjet sustainer.
Rock   78   solid+ramjet
FI     81   solid              4 boosters.
            ramjet             Sustainer.
AFM    82   solid              4 wrap-around boosters.
            ramjet
AWST   82   solid              4 strap-ons.
            ramjet             Sustainer.
JWS    82   solid              4 boosters.
            ramjet

----------- MIDCOURSE (OR MAIN) GUIDANCE -----------
MOW    76   command
Bras   77   radar command
SWM    77   radio command     Auto.
Rock   78   radio
Clns   80   command guidance
FI     81   radio command
AFM    82   radio command
AWST   82                     "Salvo and guide two missiles per target; missiles do not have to remain in
                              radar beam".
JWS    82   radio command

----------- TERMINAL GUIDANCE -----------
AFM    82   SAR
JWS    82   SAR

----------- CONTROL SYSTEM -----------
IA05   66   Nose cruciform fins and trailing-edge spoilers.
MOW    76   Pivoted wings.
SWM    77   All-moving wings.
Rock   78   Aerodynamic.
FI     81   Cruciform canard wings.
JWS    82   Moving wings on forepart.
```

566  SOVIET MISSILES  IDDS

```
----------  RADAR -----------
Bras   77    Pat Hand            C-band Fire control.
                                 Other surveillance and height-finding radars also associated.
SWM    77    Long Track          E-band scanning.
             Pat Hand            H-band acquisition.
AF01   78                        "Extended range radar".
AW02   78    Pat Hand            Operates in G/H Band.
             Long Track          On tracked vehicle, also used by SA-6.
Clns   80                        Target acquisition: Surveillance and acquisition radar.
FI     81    Long Track          E-band long-range surveillance; beam is c. 7.5 deg in elevation and 3.5 deg in
                                 azimuth, completes scan every 4 sec.
             H-band Pat Hand     Target acquisition and fire control.
ID12   81    Pat Hand            "Pat Hand target acquisition and fire-control radar".
AWST   82                        E-band surveillance.
                                 H-band acquisition.
JWS    82    Pat Hand            Separately mounted G/H-band target acquisition and fire control radars.
             Long Track          Longer range surveillance; also used with other mobile missiles.

----------  OBSERVED -----------
SMP    84                        Picture shows reloading procedure.

----------  OBSERVED IN PARADE -----------
MOW    76    May 1964
SWM    77    1964
Rock   78    1964                First public display.
AFM    82    1964
JWS    82    1964                First seen in public in Moscow.

----------  PRODUCTION UNDERWAY -----------
JCS    76    early 1970s         Continued to be produced in the early 1970s.

----------  DEPLOYMENT BEGAN -----------
SecD   71    1967                "Has been in service with Soviet forces in the USSR and Eastern Europe since 1967".
JCS    76    late 1960s          First introduced in substantial numbers.
Clns   80    1967                First deployed.
SMP    81    c. 1967             Introduced.

----------  DEPLOYED -----------
MOW    76                        In service.
SecD   76                        Deployed.
Bras   77                        In service.
JCS    77                        Deployed, FY78-79.
FI     81                        Deployed.
AFM    82                        Operational.
JWS    82                        Deployed.

----------  DEPLOYMENT HISTORY -----------
MOW    76    Widely deployed USSR.
AFM    82    Standard weapon for defence of combat areas.
JWS    82    Widely deployed in USSR.
JCS    84    SA-X-12 will replace SA-4 as Army and Front level air-defense system.
SMP    84    To be replaced or augmented by SA-X-12.
SMP    85    "SA-4... should shortly begin being replaced by the SA-X-12."

----------  NUMBERS IN SERVICE -----------

----------  NUMBER IN 1970 -----------
Clns   80    200

----------  NUMBER IN 1971 -----------
Clns   80    300

----------  NUMBER IN 1972 -----------
Clns   80    400

----------  NUMBER IN 1973 -----------
Clns   80    500

----------  NUMBER IN 1974 -----------
Clns   80    600

----------  NUMBER IN 1975 -----------
Clns   80    700

----------  NUMBER IN 1976 -----------
Clns   80    800
```

## CRUISE MISSILES  Ground-Launched SAMs  SA-4

```
---------- NUMBER IN 1977 ----------
Clns   80   1000
---------- NUMBER IN 1978 ----------
Clns   80   1100
---------- NUMBER IN 1979 ----------
Clns   80   1200
---------- NUMBER IN 1980 ----------
Clns+  83   1200
---------- NUMBER IN 1981 ----------
Clns+  83   1250
---------- NUMBER IN 1982 ----------
Clns+  83   1400
---------- CARRIERS ----------
```

| | | CARRIER | TUBES | RELOADS | |
|---|---|---|---|---|---|
| Ley | 68 | tank | | | |
| JCS | 76 | Tracked vehicle | | | |
| MOW | 76 | tracked launcher | 2 | | |
| Bras | 77 | | 2 | | Heavy tracked vehicle; all radars similarly mounted. |
| SWM | 77 | tracked vehicle | 2 | | |
| Clns | 80 | | 1 x 2 | | |
| FI | 81 | | 2 | | Tracked transporter/launcher. |
| AFM | 82 | tracked vehicle | 2 | | |
| AWST | 82 | mobile carrier | | | |
| JWS | 82 | tracked vehicle | | | |

| | | CARRIER | TUBES | RELOADS |
|---|---|---|---|---|
| MR03 | 66 | mod PT76 chassis | | |

---------- USE AND CONFIGURATION ----------

| | | |
|---|---|---|
| IA05 | 66 | Launcher moves up and down and side-to-side; but locked in horizontal position while vehicle is moving. |
| MOW | 76 | Air-transportable by An-22. |
| Bras | 77 | Appears to have good cross-country mobility. Could on occasion be deployed at brigade/regimental level. |
| SWM | 77 | 2 vehicles can be airlifted by An-22. SA-4 battery has 3 twin-launchers, one loading vehicle, and one Pat Hand radar. |
| AF01 | 78 | 9 SA-4 batteries [in Army] |
| ID12 | 81 | Furthest from FEBA of tactical SAMS, medium to high coverage. Coverage would extend 45 km beyond FEBA, and overlap with SA-4s of adjoining armies. Can be deployed within 15 km of FEBA. Normally deployed in batteries of 3 twin launchers, one Pat Hand radar, and one loader vehicle. |
| SMP | 81 | First truly mobile tactical SAM. |
| US16 | 81 | 9 batteries, 27 launchers in typical GSFG Army Group. |
| AFM | 82 | Mobile, mounted on twin-round tracked vehicle air transportable in An-22. |
| JCS | 82 | Employed at Front and Army levels, FY83-84. |
| JWS | 82 | Has 360 degree trainable mount with elevation mechanism. Clamps support missile during transit released by manual operation. Reloading is by means of mobile crane; special 2-point attachment yoke for lifting. Wheeled transporter vehicles used for resupply purposes; each carries one missile. Launcher with 2 missiles can be airlifted by An-22. |
| SMP | 83 | "Mobile and a functional part of ground force units". |
| SMP | 84 | Deployed in SAM brigades at the front level. |
| | | Echelon assignment [for SA-4a and SA-4b]: Front/Army. |
| SMP | 85 | Standard SAM at army and front levels. |

---------- STRATEGY ----------

| | | |
|---|---|---|
| SecD | 74 | With SA-6, the principal elements of Soviet [low to medium altitude?] mobile air defense. |
| SecD | 75 | Part of theater air defense. |
| MOW | 76 | Probably evolved to provide highly mobile air cover up to a/c service ceiling. May have surface-to-surface capability. |
| Bras | 77 | Medium/high level SAM. |
| FI | 81 | Possible secondary surface-to-surface capability. |
| ID12 | 81 | Medium to high coverage. |
| SMP | 81 | Effective altitude: medium-to-high, '81, '83, '84, '85. |
| JCS | 82 | Medium to high altitude system, FY83-84. |
| JWS | 82 | Probably for use in forward areas. Believed can be used in ground-to-ground role too. Likely operational role is "long- to medium-range interception of high flying targets". |

---------- GEOGRAPHICAL DEPLOYMENTS ----------

| | | |
|---|---|---|
| JCS | 75 | GSFG has 50-75 "battalions of SA-2s, 3s, and 4s". |

---------- **EXPORTED TO** ----------
| | | | |
|---|---|---|---|
| SecD | 71 | | "Now entering service with several of the subordinate armies", [meaning WTO?, ed]. |
| MOW | 76 | ? Egypt | |
| FI | 81 | Czech | |
| | | E. Germany | |
| AFM | 82 | E. Germany | Operational. |
| | | Czech | Operational. |
| JWS | 82 | Egypt | Was at one time deployed, but since not used in 1973 war, likely missile were withdrawn, possibly in exchange for SA-6 Gainful. |
| | | Germany (DR) | Only recently seen. |
| | | Czech | Only recently seen. |
| FI+ | 83 | Poland | |

---------- **RELATION TO OTHER MISSILES** ----------
| | | |
|---|---|---|
| Bras | 77 | Equivalent to US Nike Hercules. |

CRUISE MISSILES  Ground-Launched SAMs  SA-5  569

# SA-5

```
Sources are:   SecC  67    SecC  70   JCS   73   NAO3  76   Clns  80   JWS   82   SMP   83
SecC  64       IDO1  68    SecD  70   SecD  73   JCS   77   FI    81   PrO1  82   AFM+  84
IAO1  65       Ley   68    SecC  71   SecD  74   SWM   77   JCS   81   Clns+ 83   SMP   84
SecC  65       SecC  68    SecD  71   HRST  75   AWO2  78   SMP   81   FI+   83   JCS   85
MRO3  66       SecC  69    JCS   72   JCS   75   Rock  78   AFM   82   FI+   83   SMP   85
SecC  66       SecD  69    SecD  72   MOW   76   CHO1  79   AWST  82   JWS+  83   BW    86
```

```
---------- US DESIGNATION ----------
SecC  68    SA-5         FY69-77.
SecD  70    SA-5         FY71-77.
JCS   72    SA-5         FY73-83,86.
MOW   76    SA-5
SWM   77    SA-5
Rock  78    SA-5
Clns  80    SA-5
FI    81    SA-5
SMP   81    SA-5         '81, '83, '84, '85.
AFM   82    SA-5
AWST  82    SA-5
JWS   82    SA-5

---------- NATO CODENAME ----------
JCS   75    Gammon       FY76-78.
MOW   76    Gammon
SWM   77    Gammon
Rock  78    Gammon
Clns  80    Gammon
FI    81    Gammon
SMP   81    Gammon       '81, '83, '85.
AFM   82    Gammon
AWST  82    Gammon
JWS   82    Gammon
PrO1  82    Griffon
            Gammon       Changed from Griffon due to controversy over ABM capabilities.

---------- OTHER DESIGNATION ----------
SecC  67    Tallinn      FY68-70.
IDO1  68    Griffon
Ley   68                 "Long-range two-stage", [no other designation provided, ed].
SecD  69    Tallinn      "So called 'Tallinn' system".
HRST  75    Griffon
MOW   76    Griffon      Known originally as Griffon.
SWM   77    Griffon      In some reports, but this is not NATO code.
JWS   82    Griffon      Widely accepted as NATO codename until recently.
PrO1  82    Tallinn      Associated with air-defense sites near Tallinn; "Tallinn-type missiles such as
                         Griffon."

---------- DESCRIPTION ----------
SecD  72                 Strategic SAM.
SMP   81                 Strategic SAM, '84, '85.
JWS   82                 "Variously described as an unmanned long-range interceptor, an anti-aircraft
                         missile, and an anti-missile missile".
                         "Long-range surface-to-air guided missile".

---------- EDITOR'S NOTES ----------
SecC  64    Listed FY65-77, not by name FY65-68.
SecD  69    Listed FY70-77.
JCS   72    Listed FY73-83,86.
SMP   81    Listed '81, '83, '84, '85.

---------- MAXIMUM ALTITUDE ----------
            mi        ft          km          m          m conv
MOW   76    c. 18                 c. 29                  28,962      [Unmarked].
SWM   77              98,425                  30,000     29,999      May now exceed this.
                      100,000                 30,500     30,480      "Altitude".
Rock  78                          30                     30,000
CHO1  79              c. 95,143               c. 29,000  28,999
Clns  80              95,000                              28,956     Combat Ceiling.
FI    81                          c. 30,000              30,000
AFM   82              95,000                             28,956
AWST  82              100,000                            30,480
JWS   82                          c. 29                  29,000
```

```
------------ RANGE ------------
              mi           nm         km       km conv
MR03   66    112                               180.21      Max slant range.
ID01   68    c. 160                            257.44
Ley    68    150                               241.35
SecC   69                 c. 100                185.20      Max effective range against high
                                                           flying aircraft probably "on the
                                                           order of 100 n.mi.".
SecC   70                 50-100                92.60-185.20
MOW    76    c. 155                  c. 250    249.40
NA03   76    100                               160.90      Max.
SWM    77    100+                    160+      160.90
AW02   78                 >_ 100                185.20
Rock   78                            80-250    80.00-250.00
CH01   79    c. 155                  c. 250    249.40      Range.
             100+                    160+      160.90      Slant range.
Clns   80    50-150                            80.45-241.35 Slant range.
FI     81                            80        80.00       Min.
                                     250       250.00      Max.
SMP    81                            300       300.00      '81, '83, '84, '85.
AFM    82    185                               297.67
AWST   82    150                               241.35
JWS    82                            c. 250    250.00      [Analysis table]
                                     c. 300    300.00      [Text]
Clns+  83    175                               281.58      Slant range.
JWS+   83                            c. 300    300.00      [Analysis table]
BW     86                            300       300.00
```

```
----------- SPEED -----------
              mph          kmph       Mach     kmph conv
MR03   66                            3-5
FI     81                            3.5+
AFM    82                            3.5
BW     86                            3.5
```

```
---------- WARHEAD TYPE ----------
              nuclear      conven     chem
MOW    76    maybe         maybe
SWM    77    yes           yes
Rock   78    yes           yes        HE.
Clns   80    yes           yes        HE.
FI     81    yes           yes        HE; [final report of nuclear]
```

```
---------- WARHEAD CHARACTERISTICS ----------
BW     86    single RV
```

```
---------- WARHEAD WEIGHT ----------
              lb           t          kg       kg conv
FI+    83                            60        60.00
```

```
----------- WEIGHT -----------
              lb           t          kg       kg conv
MR03   66                 6-9                  6,000-9,000    Tons.
ID01   68                 c. 10                10,000
Ley    68                 10                   10,000         Tons.
MOW    76    c. 22,050              c. 10,000  10,001
Rock   78                            9,000     9,000
FI     81                            c. 9,000  9,000
AFM    82    44,090                            19,999         [Final report]
JWS    82                            c. 10,000 10,000
FI+    83                            c. 10,000 10,000
```

```
----------- LENGTH -----------
              in           ft         cm       m        m conv
IA01   65                 48.55                         14.80
MR03   66                 52                            15.85
Ley    68                 48.5                          14.78
HRST   75                 49                            14.94
MOW    76                 54'0"                16.50    16.46
SWM    77                 54                            16.4    16.46
Rock   78                                      16.5     16.50
FI     81                                      16.5     16.50
SMP    81                                      10       10.00   [From graph], '81, '83, '84, '85.
AFM    82                 54'0"                         16.46
AWST   82                 49                            14.94
JWS    82                            1,650              16.50
AFM+   84                 34'9"                         10.59
```

## CRUISE MISSILES Ground-Launched SAMs SA-5

```
------------ DIAMETER ------------
              in       ft         cm        m      cm conv
IA01   65              2.46                        74.98      Excluding booster.
MR03   66              3                           91.44
Ley    68              2.5                         76.20
HRST   75     35                                   88.90
MOW    76              3'6"                 1.07   106.68     Booster.
                       3'10"                 .85   116.84     2nd stage.
SWM    77     34                  87.5             86.36
Rock   78                         100              100.00
CH01   79              3'3 2/5"              1.0   100.08     Booster.
                       2'7 1/2"              0.8   80.01      2nd stage.
FI     81                         100              100.00     Booster.
                                  80               80.00      Missile.
AFM    82              2'10"                       86.36
AWST   82              2.9                         88.39
JWS    82                         100              100.00     Booster.
                                             0.8   80.00      2nd stage.

------------ WINGSPAN ------------
              in       ft         cm        m      cm conv
IA01   65              13                          396.24
Ley    68              13                          396.24
MOW    76              12'0"                3.65   365.76
SWM    77     144                 365              365.76
Rock   78                         396              396.00
CH01   79              11'11.75"            3.65   365.13     [11'11 3/4" in source].
FI     81                         396              396.00
AFM    82              12'0"                       365.76
AFM+   84              9'6"                        289.56

------------ NUMBER OF STAGES ------------
Ley    68     2
MOW    76              Multistage.
Rock   78     "1 + 1"
AFM    82     2
AWST   82     2
JWS    82     2 or 3  2 stages evident; has been suggested it has 3rd in warhead section used during final
                      stages of interception.

------------ TYPE OF PROPULSION ------------
MR03   66                         Both solid and liquid.
SecC   68                         Strap-on boosters.
MOW    76     solid               1st stage.
                                  Warhead may contain "in-built" rocket motor.
SWM    77     solid               Booster and sustainer.
Rock   78     solid+solid
FI     81     solid               Booster and sustainer.
AFM    82     solid               Possibly with terminal propulsion for warhead.
AWST   82     solid               Booster.
JWS    82     solid
AFM+   84                         4 wrap-around jettisonable boosters.
BW     86     solid

------------ MIDCOURSE (OR MAIN) GUIDANCE ------------
SWM    77     radio command       Automatic.
FI+    83     radio command

------------ TERMINAL GUIDANCE ------------
MOW    76     radar
SWM    77     SAR
Rock   78     radar
FI     81     SAR
AFM    82     SAR
AWST   82     radar
JWS    82     radar               Large nose radar reflector "can be combined with an active radar target
                                  seeking system".

------------ CONTROL SYSTEM ------------
MOW    76     Wing trailing-edge surfaces and 2nd-stage tail-fins.
SWM    77     Steerable rear fins and ailerons.
Rock   78     Aerodynamic.
JWS    82     Moving tail surfaces on wings and tail.
FI+    83     Cruciform wings and rear fins.
```

## RADAR

| | | | |
|---|---|---|---|
| SecC | 67 | | Local radars of limited capability, would need support of Hen House if system expected to perform "with a reasonable degree of effectiveness" as ABM. 17 Nov 1966 intelligence estimate concludes that some Tallinn complexes located so that Hen House and Dog House radars could not furnish useful target tracking data to them. |
| SWM | 77 | Square Pair | Improved radar was tested at Sary Sagan in early 1970s. Reports suggest operation with a new mobile ABM radar. |
| AWO2 | 78 | Square Pair | Traget tracking. |
| | | Back Net | Acquisition. |
| | | Side Net | Height-finding. |
| FI | 81 | Square Pair | |
| JWS | 82 | | Nose houses radar reflector of at least 60 cm diameter. |

## OBSERVED

| | | |
|---|---|---|
| SecC | 67 | Very little known of the interceptor missile itself. |

## OBSERVED IN PARADE

| | | | |
|---|---|---|---|
| MOW | 76 | Nov 1963 | |
| SWM | 77 | 1963 | |
| Rock | 78 | 1963 | First public display. |
| JWS | 82 | 1963 | First publicly displayed. |

## PROTOTYPE TESTS

| | | |
|---|---|---|
| AWST | 82 | Tested as ABM. |

## IOC

| | | | |
|---|---|---|---|
| SecC | 65 | | Previously stated Leningrad system operational as early as mid-1965, FY66-67. |
| SecC | 67 | | Could be operational by 1967-68. |
| SecC | 68 | | A few may now be operational. |
| SecC | 71 | 1967 | Year operational. |
| JCS | 75 | 1967 | Initially operational. |
| JWS | 82 | 1967 | Operational. |

## DEPLOYMENT BEGAN

| | | | |
|---|---|---|---|
| SecC | 64 | | Appears Soviets are deploying ABM around Leningrad. |
| CHO1 | 79 | 1967 | Introduced. |
| Clns | 80 | 1963 | First deployed. |
| SMP | 81 | 1963 | First deployed. |
| Clns+ | 83 | 1967 | First deployed. |
| BW | 86 | 1967 | |

## DEPLOYED

| | | |
|---|---|---|
| MOW | 76 | In service. |
| FI | 81 | Deployed. |
| JWS | 82 | Deployed. |
| BW | 86 | In service. |

## DEPLOYMENT HISTORY

| | | |
|---|---|---|
| SecC | 66 | Construction proceeding. |
| SecC | 67 | More may be under construction than identified complexes, FY68-69. Being deployed. |
| SecD | 69 | "As expected", deployment is continuing. |
| SecC | 70 | Being deployed, FY71-72. |
| SecD | 70 | Being installed in various locations to supplement existing SAMs. |
| JCS | 72 | Being deployed, [said or implied], FY73-83. |
| SecD | 72 | Has been deployed throughout the Soviet Union". |
| SecD | 73 | Deployments continue at a slow pace. |
| SecD | 74 | Additional SA-5s being deployed, FY75-77. |
| JCS | 75 | USSR deploying "a product-improved SA-5", FY76-77. |
| JCS | 77 | SA-2 decline largely offset by additional [SA-3 and] SA-5 deployment, FY78-81. |
| SWM | 77 | Has replaced many SA-2 sites. |
| FI | 81 | SA-3 and SA-5 are building up as SA-2 declines. |
| JCS | 81 | Deployment has increased slightly, FY82-83. |
| SMP | 81 | "Deployment continues today". |
| JWS | 82 | Numbers of SA-5 increasing steadily, confirmed in early 1981 and early 1982 by US reports. |
| SMP | 84 | Deployment continues at a very slow pace within USSR. |
| SMP | 85 | Operational. |

## NUMBERS IN SERVICE

### NUMBER IN 1961

| | | | |
|---|---|---|---|
| SecC | 71 | none | 1961 through 1965 "Operational launchers" [perhaps end-year, ed], [from hand-drawn graph]. |

### NUMBER IN 1966

| | | | |
|---|---|---|---|
| SecC | 71 | 335 | "Operational launchers" [perhaps end-year, ed], [from hand-drawn graph]. |

------- NUMBER IN 1967 -------
SecC 67            At least 22 complexes definitely identified, most with 3 launch sites, each with six launch positions and one radar.
SecC 71   350      "Operational launchers" [perhaps end-year, ed], [from hand-drawn graph].

------- NUMBER IN 1968 -------
SecC 68            More than 40 complexes identified (double last year's estimate), most consisting of 3 launch sites, each with 6 launch positions and one radar. A few may now be operational.
SecC 71   365      "Operational launchers" [perhaps end-year, ed], [from hand-drawn graph].

------- NUMBER IN 1969 -------
SecC 69   [c. 450] 60 complexes now identified of which some 25 appear operational. Typical complex contains 3 sites of 6 launchers each.
SecC 71   430      "Operational launchers" [perhaps end-year, ed], [from hand-drawn graph].

------- NUMBER IN 1970 -------
SecC 71   1086     Estimated deployment, 1 Oct.
          600      "Operational launchers" [perhaps end-year, ed], [from hand-drawn graph].
Clns 80   1100
JWS 82    1,100

------- NUMBER IN 1971 -------
SecC 71            1230-1500 estimated deployment, mid-71.
                   780, 1971 "Operational launchers" [perhaps end-year, ed], [from hand-drawn graph].
Clns 80   1200

------- NUMBER IN 1972 -------
SecC 71            1380-1800 estimated deployment, mid-72.
                   950, 1972 "Operational launchers" [perhaps end-year, ed], [from hand-drawn graph].
Clns 80   1300

------- NUMBER IN 1973 -------
Clns 80   1400

------- NUMBER IN 1974 -------
Clns 80   1500

------- NUMBER IN 1975 -------
SWM 77    1,100    Mid-1975.
Clns 80   1600

------- NUMBER IN 1976 -------
Clns 80   1800

------- NUMBER IN 1977 -------
Clns 80   1800

------- NUMBER IN 1978 -------
Clns 80   1900

------- NUMBER IN 1979 -------
Clns 80   1900

------- NUMBER IN 1980 -------
JWS 82    1,900    Today [10 years later than 1970].
Clns+ 83  1900

------- NUMBER IN 1981 -------
FI 81     1,100
SMP 81             "Over 100 complexes operational throughout" USSR.
Clns+ 83  2000     In 1981-1982.

------- NUMBER IN 1982 -------
AFM 82    1,200

------- NUMBER IN 1983 -------
FI+ 83    c. 1,100 In service at more than 100 Soviet sites.

------- NUMBER IN 1984 -------
AFM+ 84   c. 1,200 Deployed at more than 100 sites.

------- LAUNCHER/SILO -------
AWO2 78   Used only in fixed hardened sites in the USSR.
Clns 80   Single launchers. Launch site: Fixed.

------- CARRIERS -------

|           |    | CARRIER ———————— TUBES —— RELOADS ———————                                                                                                                                                                                                                                                                                                                                                                                      |
|-----------|----|---------------------|
| Clnst     | 83 | Mobile.             |

## USE AND CONFIGURATION
| JCS | 75 | USSR deploying "a product-improved SA-5", FY76-77. |

## STRATEGY
| SecC | 65 | Considerable uncertainty, but evidence primarily against aerodynamic vehicles rather than ballistic missiles. |
| SecC | 66 | Evidence does not permit confident judgment of specific mission of these sites: could be for ABM defense or aerodynamic vehicles, or both. |
| SecC | 67 | One view is that it is primarily advanced SAM system designed against high altitude, high speed aircraft. Pattern of deployment, configuration of the sites and their equipment, and the apparent characteristics of the radars lend credence to this view. As SAM, most effective against high-altitude aircraft such as B-70 or SR-71. "It now appears that it would be ineffective against low-altitude penetrating bombers" such as B-52 or FB-111. Incongruity of designing SAM against B-70 or SR-71, plus its being cheaper than GALOSH supports view it is ABM, with anti-air capability in addition. Weight of evidence at the moment tends to support conclusion that primary mission is air defense. |
| ID01 | 68 | Can also be used against ASMs. |
| SecC | 68 | Almost complete agreement of intelligence community that system "designed primarily for defense against high speed aerodynamic vehicles flying at high and medium altitudes". Last year little known of missile, this year it is known to have "strap-on boosters and delta wings -- clearly a missile designed for use within the atmosphere, most likely against an aerodynamic rather than a ballistic missile threat". Majority of intelligence community no longer believes it has "any significant ABM capability". |
| SecC | 69 | Considered unlikely Soviets will modify 'Tallinn' system for an ABM role; "modifications to give it such a capability would be very costly and would involve very difficult technical problems. For example, as presently configured these complexes are highly vulnerable to nuclear attack; it would require deployment of a totally new engagement radar and a higher performance, high-acceleration missile to give the system an effective self-defense or terminal defense capability against missile attack". |
| SecD | 69 | It is "designed against fast, high flying aerodynamic vehicles, rather than ballistic missiles" according to majority of intelligence community; ABM role can't be excluded, though. |
| SecC | 70 | Possible, though unlikely, that SA-5 has limited ABM capability. |
| SecD | 70 | Long-range all weather SAM. Data from past year reinforces U.S. Intelligence Board view that SA-5 unlikely to have ABM capability at this time. |
| SecD | 71 | "Some of my technical experts" think SA-5 "might be capable of adaptation for certain ABM roles", FY72-73. |
| SecD | 72 | General agreement it is not intended as ABM. |
| JCS | 73 | High altitude system, FY74-75,78-80. Long range system, FY74-75. |
| JCS | 75 | "Provides point defense for certain vital areas of strategic importance", FY76-77. |
| MOW | 76 | USSR described in 1963 as ABM, but "does not appear" to be able to intercept ICBM. Limited effectiveness against tactical ballistic missiles and ASMs. |
| SWM | 77 | Limited ABM capability. Reports suggest operation with a new mobile ABM radar. |
| FI | 81 | May have some ABM capability. |
| SMP | 81 | Effective altitude: medium-to-high, '81, '83, '84, '85. "Long-range interceptor designed to counter the threat of high-performance aircraft". |
| AFM | 82 | Suggestion of ABM capability denied during SALT II. Long-range, high-altitude defence. |
| JWS | 82 | May have some anti-missile capability, but primarily suited for long-range anti-aircraft operations. Thought to have a smaller anti-missile capability than US Nike Zeus. Aerodynamic maneuver indicates homing ability on targets traveling at missile speed must be very limited. Provides point defense for "a number of major areas of strategic importance". Soviets appear to plan to use SA-3 for low- and SA-5 for high-altitude strategic air defense. |
| JCS | 85 | May have potential to intercept some types of US strategic ballistic missiles, with SA-10 and SA-X-12. |

## COMBAT REPORTS/EFFECTIVENESS
| SecC | 70 | Represents considerable improvement over older systems in terms of range, velocity and firepower. |
| SMP | 85 | Incapable of defending against targets with "small radar-cross-section[s] such as cruise missiles". |

## GEOGRAPHICAL DEPLOYMENTS
| SecC | 64 | Appears Soviets are deploying ABM around Leningrad. |
| SecC | 66 | Construction is proceeding at Leningrad and "three additional places on the northwest periphery of the Soviet Union". |
| SecC | 67 | Being deployed "across the northwestern approaches to the Soviet Union and in a few other places". Several 'farms' of launchers located in a barrier line across northwestern part of European Russia and around Leningrad and Moscow, and some parts of the southern approaches, FY68-69. |
| SecC | 69 | Being deployed as barrier defense around European USSR and for point defense of selected targets. |
| SecC | 70 | Being deployed throughout Soviet Union, FY71-72. |
| SMP | 84 | The most significant deployments have occurred outside USSR in Eastern Europe, Mongolia, and Syria. |

## CRUISE MISSILES  Ground-Launched SAMs  SA-5

---------- **EXPORTED TO** ----------

| | | | |
|---|---|---|---|
| FI+ | 83 | Syria | 2 sites under construction. |
| JWS+ | 83 | Syria | In late 1982/early 1983, USSR supplied (and reportedly manned) missile installations in Syria, following fighting in Lebanon. Two sites identified, near Damascus and city of Homs. Later supplemented by early warning radars and SA-2,3,6, according to May 1983 reports. Doubtful these exported missiles the same as SA-5 Gammon reported in 1983 SMP; probably Syria's are the original SA-5 known as Griffon. |
| SMP | 83 | Syria | Soviets have given "advanced air defense equipment, most notably SA-5 [SAMs], with Soviet technicians in the country"; is in addition to replacement of Syria's losses of June-82. |
| SMP | 84 | Syria | The most significant deployments have occurred outside USSR in Eastern Europe, Mongolia, and Syria. Currently manned by Soviets in Syria. |
| SMP | 85 | Syria | "The SA-5 surface-to-air missile equipment in Syria, in addition to enhancing Syria's air defense, provides a dramatic symbol of Soviet support." "Highlighting the Soviet upgrading of the Syrian air defense are two operational SA-5/GAMMON missile complexes located at Dumayr and Homs -- the first operational SA-5s outside the Soviet Union." |

---------- **RELATION TO OTHER MISSILES** ----------

| | | |
|---|---|---|
| CH01 | 79 | Performance generally thought inferior to US Nike Zeus. |
| JWS | 82 | Similar in size and weight to US Nike Zeus, but thought somewhat inferior in performance. |

# SA-6

```
Sources are:  JCS   74    JCS   76    Bras  77    Rock  78    SMP   81    JWS   82    SMP   83
SecD  71      SecD  74    MOW   76    HI01  77    Clns  80    US16  81    Clns+ 83    AFM+  84
ID03  73      HRST  75    NA03  76    SWM   77    FI    81    AFM   82    FI+   83    SMP   84
SecD  73      JCS   75    SecD  76    AF01  78    ID12  81    AWST  82    FI+   83    SMP   85
ID05  74      SecD  75    AW01  77    AW02  78    JCS   81    JCS   82    JWS+  83
```

―――――――― **US DESIGNATION** ――――――――

```
SecD  71    SA-6        FY72,74-77.
JCS   74    SA-6        FY75-79,82-84.
MOW   76    SA-6
Bras  77    SA-6
SWM   77    SA-6
Rock  78    SA-6
Clns  80    SA-6
FI    81    SA-6
SMP   81    SA-6        '81, '83, '84, '85.
AFM   82    SA-6
AWST  82    SA-6
JWS   82    SA-6
SMP   84    SA-6a
            SA-6b
```

―――――――― **NATO CODENAME** ――――――――

```
SecD  71    Gainful
HRST  75    Gainful
MOW   76    Gainful
Bras  77    Gainful
SWM   77    Gainful
Rock  78    Gainful
Clns  80    Gainful
FI    81    Gainful
SMP   81    Gainful     '81, '83.
AFM   82    Gainful
AWST  82    Gainful
JWS   82    Gainful
```

―――――――― **DESCRIPTION** ――――――――

```
SMP   81                Tactical surface-to-air missile, '81, '83.
JWS   82                "Surface-to-air tactical guided missile".
```

―――――――― **EDITOR'S NOTES** ――――――――

```
SecD  71    Listed FY72,74-77.
JCS   74    Listed FY75-79,82-84.
SMP   81    Listed '81, '83, '84, '85.
```

―――――――― **MINIMUM ALTITUDE** ――――――――

|  | mi | ft | km | m | m conv | |
|---|---|---|---|---|---|---|
| ID03 73 | | | | ≥ 100 | 100.00 | "Appears feasible" with aid of optical target and missile tracking. |
| | | | | 50 | 50.00 | |
| AW01 77 | | 300 | | 91 | 91.44 | |
| HI01 77 | | | | | | "Treetop level". |
| SWM 77 | | 328 | | 100 | 99.97 | |
| AW02 78 | | 300 | | | 91.44 | |
| Rock 78 | | | 0.1 | | 100.00 | |
| FI 81 | | | | 100 | 100.00 | |
| AWST 82 | | 50 | | | 15.24 | |

―――――――― **MAXIMUM ALTITUDE** ――――――――

|  | mi | ft | km | m | m conv | |
|---|---|---|---|---|---|---|
| ID03 73 | | | 15-18 | | 15,000-18,000 | |
| AW01 77 | | 33,000 | | 10,058 | 10,058 | |
| Bras 77 | | | | 15,000 | 15,000 | Max effective altitude, est. |
| HI01 77 | | c. 70,000 | | | 21,336 | |
| SWM 77 | | 59,055 | | 18,000 | 17,999 | |
| AW02 78 | | 33,000 | | | 10,058 | |
| Rock 78 | | | 15 | | 15,000 | |
| Clns 80 | | 40,000 | | | 12,192 | Ceiling. |
| FI 81 | | | | 13,000 | 13,000 | |
| AFM 82 | | 59,000 | | | 17,983 | |
| AWST 82 | | 30,000 | | | 9,144 | |
| JWS 82 | | | c. 18 | | 18,000 | |
| JWS+ 83 | | | | 18,000 | 18,000 | Ceiling. |

CRUISE MISSILES  Ground-Launched SAMs  SA-6  577

|  |  | RANGE | | | | |
|---|---|---|---|---|---|---|
|  |  | mi | nm | km | km conv |  |
| ID03 | 73 |  |  | 30-35 | 30.00-35.00 | At Mach 2.5 at low altitude. |
|  |  |  |  | 60 | 60.00 | Theoretically at medium and high altitude. |
|  |  |  |  | 3-4 | 3.00-4.00 | Assumed for "inner 'dead' zone". |
| MOW | 76 | c. 37 |  | c. 60 | 59.53 |  |
| NA03 | 76 | 20 |  |  | 32.18 | Max. |
| AW01 | 77 | 18.6-21.7 |  | 30-35 | 29.93-34.92 | Slant range. |
| Bras | 77 |  |  | 25 | 25.00 | Max effective slant range, est, [25,000 m in source]. |
| SWM | 77 | 37 |  | 59.5 | 59.53 |  |
| AF01 | 78 |  |  | 35 | 35.00 | Slant range. |
| AW02 | 78 |  | 12 |  | 22.22 | Max. |
|  |  |  | 2 |  | 3.70 | Min. |
| Rock | 78 |  |  | 35 | 35.00 |  |
| Clns | 80 | 15 |  |  | 24.14 |  |
| FI | 81 |  |  | 35 | 35.00 | Max. |
| ID12 | 81 |  |  | c. 30 | 30.00 | Max effective range. |
| SMP | 81 |  |  | 30 | 30.00 | '81, '83, '84, '85. |
| AFM | 82 | 18.5 |  |  | 29.77 |  |
| AWST | 82 | 2-12 |  |  | 3.22-19.31 |  |
| JWS | 82 |  |  | <_ 60 | 60.00 | Max, possible, high-altitude. |
|  |  |  |  | 30 | 30.00 | Probable low-altitude. |
|  |  |  |  | 4 | 4.00 | Minimum engagement. |
| FI+ | 83 |  |  | 4 | 4.00 | Min. |

|  |  | SPEED | | | |
|---|---|---|---|---|---|
|  |  | mph | kmph | Mach | kmph conv |
| ID03 | 73 |  |  | c. 2.5 |  |
| MOW | 76 |  |  | 2.8 |  |
| SWM | 77 |  |  | 2.8 |  |
| Clns | 80 |  |  | 3 |  |
| FI | 81 |  |  | 2.8-3 |  |
| AFM | 82 |  |  | 2.8 |  |
| JWS | 82 |  |  | 2.8 | Max. |
|  |  |  |  | 1.5 | After solid boost. |
| JWS+ | 83 |  |  | 2.8 |  |

|  |  | WARHEAD TYPE | | | |
|---|---|---|---|---|---|
|  |  | nuclear | conven | chem |  |
| ID03 | 73 |  | yes |  | Ratio of fragmentation to HE uncertain, but "based on conventional design", 40 kg (50%) explosive is expected. Radio proximity fuze, on command link; passive proximity fuze on semi-active homing; impact fuze probably incorporated. |
| MOW | 76 |  | yes |  | HE. |
| AW01 | 77 |  | yes |  | HE fragmentation. |
| Bras | 77 |  | yes |  | "HE blast" with proximity fuze. |
| SWM | 77 |  |  |  | Proximity/impact fuse. |
| Clns | 80 |  | yes |  | HE. |
| FI | 81 |  | yes |  |  |
| AFM | 82 |  | yes |  |  |
| JWS | 82 |  | yes |  | HE; Fuze is "proximity and impact; possibly also command". |

|  |  | WARHEAD WEIGHT | | | |
|---|---|---|---|---|---|
|  |  | lb | t | kg | kg conv |
| MOW | 76 | 176 |  | 80 | 79.83 |
| Bras | 77 |  |  | 80 | 80.00 |
| SWM | 77 | 176 |  | 80 | 79.83 |
| Rock | 78 |  |  | 80 | 80.00 |
| Clns | 80 | 100 |  |  | 45.36 |
| FI | 81 |  |  | 80 | 80.00 |
| AFM | 82 | 176 |  |  | 79.83 |
| JWS | 82 |  |  | c. 40 | 40.00 | [Analysis table]
|  |  |  |  | c. 80 | 80.00 | [Text], "80 kg total with 40 kg HE".

|  |  | WEIGHT | | | |
|---|---|---|---|---|---|
|  |  | lb | t | kg | kg conv |
| MOW | 76 | 1,212 |  | 550 | 549.76 |
| Bras | 77 |  |  | 550 | 550.00 |
| SWM | 77 | 1,212 |  | 550 | 549.76 |
| Rock | 78 |  |  | 550 | 550.00 |
| FI | 81 |  |  | 550 | 550.00 |
| AFM | 82 | 1,212 |  |  | 549.76 |
| JWS | 82 |  |  | c. 550 | 550.00 |

```
------------ LENGTH ------------
              in       ft        cm      m    m conv
HRST   75              10                     3.05
MOW    76              20'4"            6.2   6.20
Bras   77                               6     6.00
SWM    77              20.3             6.2   6.19
Rock   78                               6.2   6.20
FI     81                               6.2   6.20
SMP    81                               6.3   6.30    [From graph], '81, '83, '84, '85.
AFM    82              20'4"                  6.20
AWST   82              19                     5.79
JWS    82                       620           6.20    Including tail cone.
                                        6     6.00    Without tail cone.

------------ DIAMETER ------------
              in       ft        cm      m    cm conv
MOW    76              1'1.25"          0.335 33.66
Bras   77                       33.5          33.50   [335 mm in source]
SWM    77     13.2              33.5          33.53
Rock   78                       33.5          33.50
FI     81                       33.5          33.50
AFM    82              1'1.2"                 33.53
AWST   82     6                               15.24
JWS    82                       33            33.00   [Analysis table]
                                33.5          33.50   [Text]

------------ WINGSPAN ------------
              in       ft        cm      m    cm conv
SWM    77     48.8              124           123.95
Rock   78                       152           152.00
FI     81                       152           152.00
JWS    82                       124           124.00  Tail.

------------ NUMBER OF STAGES ------------
Rock   78     1
AWST   82     1
JWS    82     1        Single-stage body.

------------ TYPE OF PROPULSION ------------
ID03   73     liquid         Integral rocket/ramjet, ramjet is liquid fueled. A rough estimate of the drag
                             of Gainful shows c. 1.2=1.5 kN thrust to maintain Mach 2.5 at low altitude.
                             Booster must deliver 320 kN. Ramjet engine must have "fixed diffuser",
                             implying fixed Mach # and max 40 secs flight at low altitude.
HRST   75     solid
MOW    76     solid          Integral rocket-ramjet, both use solid propellant.
Bras   77     ramjet
              solid          Booster.
SWM    77     solid          Integral ramjet.
Rock   78     ramjet         "R/R", [R means ramjet, ed].
FI     81     solid          All-solid-propellant integral rocket/ramjet.
SMP    81                    World's first integral rocket ramjet.
AFM    82     solid          Integral all-solid rocket/ramjet propulsion. After burnout of solid booster,
                             its casing becomes ramjet combustion chamber for ram air mixed with the
                             exhaust from a solid-propellant gas generator.
AWST   82     solid          Integral.
JWS    82                    "Dual-thrust integral rocket-ramjet"; solid boost accelerates at c. 20 g to c.
                             Mach 1.5; than tail cone jettisoned and rocket propellant chamber becomes
                             combustion chamber for ramjet; air comes in through 4 intakes symmetrically
                             around center section. Ramjet takes missile to almost Mach 2.8.

------------ MIDCOURSE (OR MAIN) GUIDANCE ------------
ID05   74                    Earlier assumption that command guidance could be used has proved incorrect.
MOW    76     radio command
SWM    77     radio command  I-band.
Rock   78     radio
Clns   80     command
FI     81     radio command
AFM    82     radio command
AWST   82     command
JWS    82     command
```

## CRUISE MISSILES  Ground-Launched SAMs  SA-6

```
----------- TERMINAL GUIDANCE -----------
MOW    76    SAR
Bras   77    SAR
SWM    77    SAR              CW radar.
Rock   78    radar
Clns   80    SAR
FI     81    CW SAR
AFM    82    SAR
AWST   82    semi-active      CW terminal.
JWS    82    SAR

----------- CONTROL SYSTEM -----------
Rock   78    Aerodynamic.
FI     81    Cruciform centrebody wings.
JWS    82    Aerodynamic by centre and tail fins.

----------- RADAR -----------
ID05   74    Straight Flush   Fire control. [See source for detail on radar operations, ed].
             Flat Face        Surveillance [in Egypt].
Bras   77    Straight Flush   Fire control.
                              Also a separate surveillance radar and height finder.
SWM    77    Long Track       Acquisition.
             Straight Flush   [H-band] fire control.
AW02   78    Straight Flush   Tracking.
             Long Track       Acquisition.
Clns   80                     Search and acquisition radar.
                              Tracking radar.
AWST   82    E-band           Acquisition.
JWS    82    Straight Flush   Fire control system has "primary search and acquisition radar, a target
                              tracking and illuminating radar, a command link with secondary radar response
                              for missile tracking and a missile-borne semi-active homing system". In ECM
                              conditions some tracking functions can be performed optically; sometimes
                              Straight Flush includes long-range (up to 30 km) EO tracker. Radar limitation
                              would appear to be "restricted search capability when operating without the
                              support of other types of surveillance radar"; but normal practice is to
                              include such longer range units in combat formations. [See source for more
                              radar detail, ed].
             Flat Face        Acquisition.
             Long Track       Acquisition.
             Thin Skin        Height-finder.

----------- OBSERVED IN PARADE -----------
MOW    76    7 Nov 1967
Rock   78    1967             First public display.
SMP    81    1967             Unveiled.
AFM    82    Nov 1967
JWS    82    1967             First publicly shown in Moscow.

----------- DEPLOYMENT BEGAN -----------
JCS    76    early 1970s      Appeared.
Clns   80    1970             First deployed.
SMP    81    1970s            Introduced "in the early and mid-1970s".

----------- DEPLOYED -----------
JCS    75                     Deployed, FY76,78-79.
MOW    76                     In service. Standard USSR weapon.
SecD   76                     Deployed.
Bras   77                     In service.
JWS    82                     Deployed.

----------- DEPLOYMENT HISTORY -----------
SecD   71    Currently being deployed.
SecD   73    Being deployed [implied, ed].
JCS    81    Soviets and NSWP countries continue to replace older S-60, 57mm AAA with SA-6 in tank and
             motorized rifle divisions.

----------- NUMBERS IN SERVICE -----------

----------- NUMBER IN 1970 -----------
Clns   80    30

----------- NUMBER IN 1971 -----------
Clns   80    100

----------- NUMBER IN 1972 -----------
Clns   80    200
```

---------- NUMBER IN 1973 ----------
Clns  80   300

---------- NUMBER IN 1974 ----------
Clns  80   400        In 1974-1975.

---------- NUMBER IN 1976 ----------
Clns  80   500

---------- NUMBER IN 1977 ----------
Clns  80   600

---------- NUMBER IN 1978 ----------
Clns  80   700        In 1978-1979.

---------- NUMBER IN 1980 ----------
Clns+ 83   800

---------- NUMBER IN 1981 ----------
Clns+ 83   750

---------- NUMBER IN 1982 ----------
Clns+ 83   800

---------- CARRIERS ----------

---------- CARRIER ---------- TUBES --- RELOADS ----------
JCS   75   fully tracked
JCS   76   tracked vehicle
MOW   76   tracked vehicle       3
Bras  77   tracked vehicle
Clns  80                         1 x 3
AFM   82   tracked vehicle       3
JWS   82   tracked vehicle       3

---------- CARRIER ---------- TUBES --- RELOADS ----------
SWM   77   mod PT-76 tank        3
FI    81   mod PT-76 chassis     3

---------- USE AND CONFIGURATION ----------
ID03  73   Launch acceleration averages 23g [table, c. 20g]. Lateral acceleration at low altitude scarcely
           over 15g.
ID05  74   In Egypt, SA-6 battery consists of 8 triple launchers, Straight Flush fire control, and Flat Face
           long range surveillance radar.
JCS   75   Highly mobile, responsive air defense to moving armored column, FY76-77.
Bras  77   Used with SA-7,9 and S-60 and ZSU-23-4 to provide comprehensive air defence of forward areas.
SWM   77   Missile group consists of 3 triple launchers, one loading vehicle, and one Straight Flush radar.
AF01  78   Radar and launchers linked by radar rather than cable. 10 batteries defend Army.
AW02  78   Designed for rapid movement.
FI    81   Rocket accelerates at 20g to M1.5; tail cone [incl. rocket nozzle] jettisoned at burnout. Ramjet
           takes missile to M2.8.
ID12  81   A division's air defense regiment has 20 SA-6 or SA-8 launchers. Firing battery normally consists
           of 3 launchers, replenishment vehicle, and Straight Flush radar.
JCS   81   Soviets and NSWP countries continue to replace older S-60, 57mm AAA with SA-6 in tank and
           motorized rifle divisions.
SMP   81   Introduced at maneuver division level. [SA-6 and SA-8] can keep pace with "rapidly advancing
           maneuver forces".
US16  81   5 batteries, 15 launchers, in typical GSFG Army Group.
JCS   82   Employed at division level, FY83-84.
JWS   82   Basic SA-6 unit is regiment, with 5 missile-firing batteries. Battery has Straight Flush radar and
           4 TELs, each with 3 missiles; 6 reload missiles held in reserve on two transporter-loader vehicles
           [unclear whether for each TEL or each battery, ed]. Fully mobile; launchers and fire control radar
           systems mounted on separate tracked vehicles.
SMP   83   "Mobile and a functional part of ground force units".
SMP   84   Echelon assignment [of SA-6a and SA-6b]: Division.
SMP   85   Standard SAM at division level, "though some divisions still have an AAA-equipped regiment".

########## STRATEGY ##########

| | | |
|---|---|---|
| SecD | 71 | Purpose is to upgrade theater air defense. |
| SecD | 73 | "Air defense capabilities are improving with the deployment of the SA-6 mobile SAM system". |
| SecD | 74 | With SA-4, principal element of Soviet [low to medium altitude?] mobile air defense. |
| JCS | 75 | Not a strategic system, FY76-77. |
| SecD | 75 | Part of theater air defense. |
| MOW | 76 | Rapid-reaction defense against low and med altitude attack. |
| Bras | 77 | Low with medium SAM. |
| AWO2 | 78 | Used with SA-4 for coverage beyond FEBA. |
| ID12 | 81 | For use at low-medium altitudes. |
| SMP | 81 | Effective altitude: low-to-medium, '81, '83, '84, '85. |
| AFM | 82 | Mobile, low-altitude system. |
| JCS | 82 | Low-to-medium altitude system, FY83-84. |

########## COMBAT REPORTS/EFFECTIVENESS ##########

| | | |
|---|---|---|
| SecD | 74 | Deployed in 1973 war in a heavy barrier defense with SA-2 and SA-3. |
| JCS | 75 | Employed with notable effectiveness in recent Middle East hostilities, FY76-77. Impressive capabilities, especially against low-level attacks, FY76-77. |
| MOW | 76 | Very effective in 1973 war. |
| Bras | 77 | Most successful air defence missile of 1973 war. |
| HI01 | 77 | According to Israel's air chief Peled, main effect of SAMs in 1973 was to decrease pilots' loiter time over targets for "eyeball intelligence". |
| SWM | 77 | Effective in 1973 war against Israeli strike aircraft, in the face of ECM. |
| ID12 | 81 | First combat use was 1973 war. |
| AFM | 82 | Unexpectedly effective in 1973 war. US ECM was ineffective. |
| JWS | 82 | Little known until 1973 war; used extensively in that war (so far as known, first operational use); scored a number of successes in early phases of hostilities. |

########## GEOGRAPHICAL DEPLOYMENTS ##########

| | | |
|---|---|---|
| JCS | 75 | GSFG has "several regiments of SA-6s". |

---------- EXPORTED TO ----------
| | | |
|---|---|---|
| JCS | 74 | Arabs — Provided to Arabs. |
| MOW | 76 | Egypt |
| | | Iraq |
| | | Syria |
| Bras | 77 | Bulgaria |
| | | Czech |
| | | East Germany |
| | | Egypt |
| | | Hungary |
| | | Iraq |
| | | Libya |
| | | Poland |
| | | Rumania |
| | | Syria |
| SWM | 77 | Egypt |
| | | Syria |
| | | Libya |
| | | North Vietnam |
| FI | 81 | Egypt — Several dozen vehicles. |
| | | Iraq — 25 launchers. |
| | | Libya — 8 SA-2,3,6 batteries protect Okba Ben Nafi air base. |
| | | Syria — 60 vehicles, maybe got improved version in 1978. |
| | | Vietnam |
| | | Bulgaria |
| | | Czech |
| | | Hungary |
| | | Poland |
| | | Mozambique — 24 vehicles. |
| | | Finland reported negotiating [final report] |
| SMP | 81 | "A recent export". |
| AFM | 82 | Supplied within and beyond Warsaw Pact. |
| JWS | 82 | Deployed "USSR & others". |
| FI+ | 83 | Angola |
| | | Algeria |
| | | Cuba |
| | | Ethiopia |
| | | East Germany |
| | | India |
| | | Jugoslavia |
| | | Kuwait |
| | | Mali |
| | | Peru |
| | | Romania |
| | | Somalia |
| | | Tanzania |
| | | North Yemen |
| | | Zambia |
| AFM+ | 84 | Algeria |
| | | Angola |
| | | East Germany |
| | | India |
| | | Romania |
| | | South Yemen |
| | | Yugoslavia |
| SMP | 85 | Algeria — Soviet military advisors "air defense units with Soviet SA-2, SA-3, and SA-6 surface-to-air missiles." |

---------- RELATION TO OTHER MISSILES ----------
| | | |
|---|---|---|
| MOW | 76 | Somewhat larger than Hawk, with similar role. |
| Bras | 77 | Equivalent to the US HAWK. |

---------- OTHER INFORMATION ----------
| | | |
|---|---|---|
| ID03 | 73 | At least 6 SA-6 vehicles with missiles at USAF Hanscom Field base. |

… CRUISE MISSILES  Ground-Launched SAMs  SA-7

# SA-7

```
Sources are:   SecD  75    SecD  76    SWM   77    US13  80    AFM   82    FI+   83
SecD  73       JCS   76    Bras  77    AF01  78    FI    81    AWST  82    JAWA  83
ID04  74       MOW   76    HI01  77    AW02  78    ID12  81    JWS   82    AFM+  84
JCS   74       NA03  76    JCS   77    Rock  78    SMP   81    FI+   83    SMP   84
```

---------- US DESIGNATION ----------

|  | |  |  |
|---|---|---|---|
| SecD | 73 | SA-7 | FY74,76-77. |
| JCS | 74 | SA-7 | FY75,77-79. |
| MOW | 76 | SA-7 | |
| Bras | 77 | SA-7 | |
| SWM | 77 | SA-7 | |
| Rock | 78 | SA-7 | |
| FI | 81 | SA-7 | |
| | | SA-N-7 | [Sic, this designation also, no distinct information, but see SA-N-5]. |
| SMP | 81 | SA-7 | '81, '83. |
| AFM | 82 | SA-7 | |
| AWST | 82 | SA-7 | |
| JWS | 82 | SA-7 | |
| JAWA | 83 | SA-7 | |
| SMP | 84 | SA-7a | |

---------- NATO CODENAME ----------

| Source | | Codename | Notes |
|---|---|---|---|
| SecD | 73 | Grail | |
| MOW | 76 | Grail | |
| Bras | 77 | GRAIL | |
| SWM | 77 | Grail | |
| Rock | 78 | Grail | |
| FI | 81 | Grail | |
| SMP | 81 | Grail | '81, '83. |
| AFM | 82 | Grail | |
| AWST | 82 | Grail | |
| JWS | 82 | Grail | |
| JAWA | 83 | Grail | |

---------- OTHER DESIGNATION ----------

|  | |  |  |
|---|---|---|---|
| SWM | 77 | Strela | (Arrow) is not NATO name. |
| FI | 81 | Strella | |
| JWS | 82 | Strela | (Arrow); once widely known by this name. |

---------- DESCRIPTION ----------

| Source | | Description |
|---|---|---|
| SMP | 81 | Tactical surface-to-air missile. |

---------- EDITOR'S NOTES ----------

|  | |  |
|---|---|---|
| SecD | 73 | Listed FY74,76-77. |
| JCS | 74 | Listed FY75,77-79. |
| SMP | 81 | Listed '81. |

---------- MINIMUM ALTITUDE ----------

| Source | | mi | ft | km | m | m conv |
|---|---|---|---|---|---|---|
| MOW | 76 | | 165 | | 50 | 50.29 |
| SWM | 77 | | 148 | | 45 | 45.11 |
| FI | 81 | | | | 45 | 45.00 |
| AWST | 82 | | 50 | | | 15.24 |

---------- MAXIMUM ALTITUDE ----------

| Source | | mi | ft | km | m | m conv | Notes |
|---|---|---|---|---|---|---|---|
| ID04 | 74 | | | | 2,700 | 2,700 | |
| MOW | 76 | | 5,000 | | 1,500 | 1,524 | |
| Bras | 77 | | | | 900 | 900.00 | Max effective altitude. |
| SWM | 77 | | 4,920 | | 1,500 | 1,499 | |
| AF01 | 78 | | | | 3,050 | 3,050 | |
| AW02 | 78 | | 15,000 | | | 4,572 | |
| Rock | 78 | | | 1.5 | | 1,500 | "Altitude". |
| FI | 81 | | | | 1,500 | 1,500 | |
| AFM | 82 | | 5000 | | | 1,524 | Basic version. |
| AWST | 82 | | 10,000 | | | 3,048 | |

## RANGE

| | | mi | nm | km | km conv | |
|---|---|---|---|---|---|---|
| MOW | 76 | | | 3.5 | 3.50 | [11,500 ft (3,500 m) in source] |
| NA03 | 76 | 6 | | | 9.65 | Max. |
| Bras | 77 | | | 3 | 3.00 | Max effective slant range, [3,000 m in source]. |
| SWM | 77 | 1.8\|2.5 | | 2.9\|4.0 | 2.90 | [Slash appears in source without explanation, ed] |
| AF01 | 78 | | | 3.5 | 3.50 | Slant range. |
| AW02 | 78 | | .5-2 | | 0.93-3.70 | |
| Rock | 78 | | | 3.6 | 3.60 | |
| FI | 81 | | | 3.6 | 3.60 | Max. |
| AFM | 82 | 2.15 | | | 3.46 | |
| AWST | 82 | .5-3 | | | 0.80-4.83 | |
| JWS | 82 | | | 9-10 | 9.00-10.00 | UK MoD report of April, 1976. |

## SPEED

| | | mph | kmph | Mach | kmph conv |
|---|---|---|---|---|---|
| SWM | 77 | | | 1.5 | |
| FI | 81 | | | 1.5 | |
| AFM | 82 | | | 1.5 | |

## WARHEAD TYPE

| | | nuclear | conven | chem | |
|---|---|---|---|---|---|
| ID04 | 74 | | yes | | HE. |
| Bras | 77 | | yes | | "HE blast". |
| FI | 81 | | yes | | Fragmentation with smooth fragmentation casing and contact and graze fuzes. |
| AFM | 82 | | yes | | |

## WARHEAD WEIGHT

| | | lb | t | kg | kg conv |
|---|---|---|---|---|---|
| ID04 | 74 | | | 1.8 | 1.80 |
| SWM | 77 | 5.5 | | 2.5 | 2.49 |
| Rock | 78 | | | 2.5 | 2.50 |
| FI | 81 | | | 2.5 | 2.50 |
| AFM | 82 | 5.5 | | | 2.49 |

## WEIGHT

| | | lb | t | kg | kg conv | |
|---|---|---|---|---|---|---|
| ID04 | 74 | | | 10.6 | 10.60 | Max weight of system [includes launcher] |
| Bras | 77 | | | 9.2 | 9.20 | |
| SWM | 77 | 22+ | | 10+ | 9.98 | |
| Rock | 78 | | | 9.2 | 9.20 | |
| FI | 81 | | | 9.2 | 9.20 | |
| AFM | 82 | 20 | | | 9.07 | |
| JWS | 82 | | | c. 9.2 | 9.20 | |

## LENGTH

| | | in | ft | cm | m | m conv |
|---|---|---|---|---|---|---|
| MOW | 76 | | 4'5" | 135 | | 1.35 |
| Bras | 77 | | | | 1.3 | 1.30 |
| SWM | 77 | | 4.4 | | 1.35 | 1.34 |
| Rock | 78 | | | | 1.35 | 1.35 |
| FI | 81 | | | | 1.35 | 1.35 |
| AFM | 82 | | 4'5" | | | 1.35 |
| AWST | 82 | | 4.5 | | | 1.37 |
| JWS | 82 | | | 129 | | 1.29 |
| FI+ | 83 | | | | 1.29 | 1.29 |
| AFM+ | 84 | | 4'3" | | | 1.30 |

## DIAMETER

| | | in | ft | cm | m | cm conv | |
|---|---|---|---|---|---|---|---|
| MOW | 76 | c. 2.75 | | 7 | | 6.99 | |
| Bras | 77 | | | 7 | | 7.00 | [70 mm in source] |
| SWM | 77 | 2.75 | | 7.0 | | 6.99 | |
| Rock | 78 | | | 7 | | 7.00 | |
| FI | 81 | | | 7 | | 7.00 | [Final report] |
| AFM | 82 | 2.75 | | | | 6.99 | |
| AWST | 82 | 2.75 | | | | 6.99 | |

## NUMBER OF STAGES

| | | |
|---|---|---|
| Bras | 77 | 3 |
| Rock | 78 | "1 + 1" |
| AWST | 82 | 1 |

CRUISE MISSILES  Ground-Launched SAMs  SA-7

```
--------------- TYPE OF PROPULSION ---------------
Bras   77   solid
SWM    77   solid                    Booster and sustainer.
Rock   78   solid+solid
FI     81   solid                    Booster and sustainer.
AFM    82   solid                    Booster and sustainer.
AWST   82   solid
JWS    82   solid

--------------- TERMINAL GUIDANCE ---------------
ID04   74                            New version has additional manual guidance mode.
MOW    76   IR                       Initial optical aiming.
Bras   77   IR                       Fire control: none.
SWM    77   IR                       Radar-aiming system has been introduced some models have filters for flares.
Rock   78   IR
FI     81   IR                       Has filter to screen out decoy flares.
AFM    82   IR                       Has filter to screen out decoy flares.
AWST   82   IR
JWS    82   IR

--------------- CONTROL SYSTEM ---------------
SWM    77   "Small cruciform (steerable?) wings and tail fins flick out as missile leaves launch tube".
Rock   78   Aerodynamic.
FI     81   Cruciform canard fins.

--------------- PRODUCTION UNDERWAY ---------------
JCS    76   early 1970s     Continued to be produced.

--------------- DEPLOYMENT BEGAN ---------------
SecD   73                   "Tenuous evidence of the introduction of the SA-7 GRAIL, a man-portable SAM, into
                            Soviet units".
JCS    76   late 1960s      First introduced in substantial numbers.
SMP    81   1968            Introduced.
AFM    82   1968            Standard weapon in WTO since 1968.
JWS    82                   In service well over 10 years.

--------------- DEPLOYED ---------------
MOW    76               In service.
SecD   76               Deployed.
Bras   77               In service.
JCS    77               Deployed, FY78-79.
JWS    82               Operational.

--------------- CARRIERS ---------------

--------------- CARRIER --------------- TUBES --- RELOADS ---------------
SWM    77   vehicles              4,6,8
AFM    82   vehicles and ships    4,6,8              Vehicle-launched version has radar aiming.
AWST   82   vehicle                                  Radar-aimed launcher.
JWS    82   jeep                  1         4        Mounting observed in Oct 1975 Egyptian parade.

--------------- CARRIER --------------- TUBES --- RELOADS ---------------
JCS    74   ? BRDMs                                  Arabs.

--------------- CARRIER --------------- TUBES --- RELOADS ---------------
JCS    74   hand-held                                Arabs.
MOW    76                                            Shoulder-fired.
Bras   77   hand held
SWM    77   man-portable
SMP    81   shoulder-fired
AFM    82   infantry
AWST   82   shoulder-launched
JWS    82   man-portable                             Original and principal form of deployment.

--------------- CARRIER --------------- TUBES --- RELOADS ---------------
JAWA   83   Gazelle                                  Yugoslav, first known helicopter fitting.
AFM+   84   Gazelle

--------------- CARRIER --------------- TUBES --- RELOADS ---------------
JAWA   83   Mi-24 Hind                               Some, French report.
AFM+   84   Mi-24                                    Reported on some.

--------------- CARRIER --------------- TUBES --- RELOADS ---------------
JWS    82   ship fittings                            Under name SA-N-5.

--------------- CARRIER --------------- TUBES --- RELOADS ---------------
US13   80   Osa II                                   "A number" of USSR navy Osas equipped with SA-7.
```

## USE AND CONFIGURATION

| | | |
|---|---|---|
| ID04 | 74 | Has no IFF. |
| SecD | 75 | Has tail-chase-only engagement capability. |
| Bras | 77 | Robust and simple to operate. Homes automatically on jet exhaust. |
| ID12 | 81 | In Soviet doctrine, deployed at company level and nearest FEBA. Usually deployed as a section, with company commander retaining control. Other SA-7 teams en route to battle area would ride in BMPs or other armored carriers. SA-7 gunners would dismount after infantry troops and follow just behind, keeping within 20-30 m of company commander and 15-20 m of each other. |
| JWS | 82 | Simple optical sighting and tracking are employed. IR seeker is activated when operator has acquired target. An indicator light denotes seeker acquisition and operator is then free to fire. Relies on tail pursuit interception. [Entry includes diagram with labeled parts, ed]. |
| AFM+ | 84 | Carried by some helicopters for use against other helicopters. |
| SMP | 84 | Echelon assignment [of SA-7a and SA-7b]: Company/Battalion. |

## TARGET TYPE

| | | |
|---|---|---|
| JAWA | 83 | Probably both anti-armor and air-to-air roles. |

## STRATEGY

| | | |
|---|---|---|
| Bras | 77 | Very low level hand held SAM. |

## COMBAT REPORTS/EFFECTIVENESS

| | | |
|---|---|---|
| JCS | 74 | Massive numbers deployed by Arabs in 1973 war. |
| MOW | 76 | Used in Vietnam and 1973 war. |
| Bras | 77 | Disadvantages are that as tail-chaser, normally only effective. After attack delivered, and has only small warhead "whose effect can be negated by armour plate over the aircraft's jet pipe". |
| HI01 | 77 | First used in North Vietnam in 1972. After effective ECM introduced in Vietnam, SA-7 was mounted on a vehicle and fired in salvos, complicating ECM. In 1973 war, able to avoid decoy flares. |
| SWM | 77 | Used in Vietnam and 1973 wars. Often effective despite countermeasures. Few kills against jets, just damaged jetpipes. |
| AF01 | 78 | C. 5,000 fired in 1973 war, 3 certain and 7 possible planes downed. |
| AW02 | 78 | "Not considered effective against high-speed aircraft pulling high g forces". |
| Rock | 78 | Infantry weapon used in Vietnam. |
| FI | 81 | Widely used in SE Asia and Mideast. Effective against targets flying at less than "50 kt". |
| ID12 | 81 | First used in Vietnam in 1972. |
| AFM | 82 | Effective against helicopters and slow, low-flying planes, despite countermeasures. |
| JWS | 82 | Especially effective against helicopters. Flares were for a time an effective countermeasure; then later models were equipped with filters to combat the tactic. |

## EXPORTED TO

| | | | |
|---|---|---|---|
| SecD | 73 | Vietnam | Has been deployed. |
| JCS | 74 | Arabs | Provided to Arabs. |
| MOW | 76 | Angola | |
| | | Bulgaria | |
| | | Egypt | |
| | | Poland | |
| | | Syria | |
| | | North Vietnam | |
| Bras | 77 | Bulgaria | |
| | | Czech | |
| | | Egypt | |
| | | East Germany | |
| | | Hungary | |
| | | Iraq | |
| | | Poland | |
| | | Rumania | |
| | | Syria | |
| | | Vietnam | |
| | | South Yemen | |
| SWM | 77 | | Possessed by terrorists and guerillas. |
| FI | 81 | WTO | |
| | | Angola | |
| | | Cuba | |
| | | Egypt | Infantry and jeep-mounted. |
| | | Kuwait | |
| | | Syria | |
| | | Vietnam | |
| | | Yemen (PDRY) | |
| | | Jugoslavia | |
| | | North Korea | |
| | | Libya | |
| | | Iraq | |
| | | Ethiopia | |
| | | Finland | |
| | | Mozambique | |
| | | Peru | |
| | | ? Kuwait | |
| | | | Various rebel/terrorist organizations [final report] |
| AFM | 82 | | Supplied to 20 other nations. |
| JWS | 82 | Algeria | |

|     |    |             |                                           |
|-----|----|-------------|-------------------------------------------|
|     |    | Angola      |                                           |
|     |    | Bulgaria    |                                           |
|     |    | Cuba        | Possibly [analysis table]; probably [text]. |
|     |    | China       | Probably.                                 |
|     |    | Czech       |                                           |
|     |    | Egypt       |                                           |
|     |    | GDR         |                                           |
|     |    | Hungary     |                                           |
|     |    | India       |                                           |
|     |    | Kuwait      |                                           |
|     |    | Libya       |                                           |
|     |    | Morocco     | Probably.                                 |
|     |    | Mozambique  |                                           |
|     |    | Poland      |                                           |
|     |    | Romania     |                                           |
|     |    | Syria       |                                           |
|     |    | Vietnam     |                                           |
|     |    | South Yemen |                                           |
|     |    | Yugoslavia  | Probably.                                 |
| FI+ | 83 | Afghanistan |                                           |
|     |    | Algeria     |                                           |
|     |    | Argentina   |                                           |
|     |    | Botswana    |                                           |
|     |    | Bulgaria    |                                           |
|     |    | East Germany |                                          |
|     |    | Guinea-Bissau |                                         |
|     |    | Guyana      |                                           |
|     |    | Hungary     |                                           |
|     |    | Iran        |                                           |
|     |    | Mauritania  |                                           |
|     |    | Morocco     |                                           |
|     |    | Nicaragua   |                                           |
|     |    | Poland      |                                           |
|     |    | Seychelles  |                                           |
|     |    | Sudan       |                                           |
|     |    | Tanzania    |                                           |
|     |    | Uganda      |                                           |
|     |    | North Yemen |                                           |
|     |    | South Yemen |                                           |
|     |    | Zambia      |                                           |
|     |    | Zimbabwe    |                                           |

------------- **RELATION TO OTHER MISSILES** -------------

| MOW | 76 | Similar in principle to US Redeye. |
| SWM | 77 | Equivalent of US Redeye. |
| FI  | 81 | Russian equivalent of Redeye. |
| AFM | 82 | Counterpart of US Redeye. |
| JWS | 82 | Similar in concept to US Redeye. Same missiles used in SA-9 system [but see SA-9 entry for contradiction, ed]; others say SA-9 missiles larger and heavier; MoD report of April-76 gave lower range for SA-9 than SA-7. |

# SA-7b

```
Sources are:   AW02  78      FI   81      AFM   82      JWS   82      SMP   84
```

**———————— US DESIGNATION ————————**

| | | | |
|---|---|---|---|
| FI | 81 | SA-7 Mk 2 | May be known as SA-7B. |
| SMP | 84 | SA-7b | |

**———————— MAXIMUM ALTITUDE ————————**

|  |  | mi | ft | km | m | m conv | |
|---|---|---|---|---|---|---|---|
| FI | 81 | | | | 14,000 | 14,000 | |
| AFM | 82 | | 14000 | | | 4,267 | Uprated version. |

**———————— RANGE ————————**

|  |  | mi | nm | km | km conv | |
|---|---|---|---|---|---|---|
| JWS | 82 | | | | | Mk 2 has been confirmed, principal difference being boosted propellant charge for increased range and speed. |

**———————— TYPE OF PROPULSION ————————**

| FI | 81 | Has uprated rocket motor, enabling greater speed, but burn time is not increased. |
|---|---|---|
| AFM | 82 | Uprated version has more powerful motor, giving higher speed. |

**———————— PRODUCTION UNDERWAY ————————**

| AW02 | 78 | Being produced in improved version in record numbers. |
|---|---|---|

**———————— DEPLOYMENT BEGAN ————————**

| AW02 | 78 | 1974 | Mod version (with improvements) in inventory since 1974. |
|---|---|---|---|

**———————— USE AND CONFIGURATION ————————**

| SMP | 84 | Echelon assignment: Company/Battalion. |
|---|---|---|

# SA-8

| Sources are: | JCS | 76 | Bras | 77 | AFO1 | 78 | FI | 81 | AFM | 82 | Clns+ | 83 | SMP | 83 |
|---|---|---|---|---|---|---|---|---|---|---|---|---|---|---|
| ID06 | 75 | | MOW | 76 | HI01 | 77 | AW02 | 78 | ID12 | 81 | AWST | 82 | FI+ | 83 | AFM+ | 84 |
| JCS | 75 | | SecD | 76 | JCS | 77 | Rock | 78 | JCS | 81 | JCS | 82 | FI+ | 83 | SMP | 84 |
| SecD | 75 | | AW01 | 77 | SWM | 77 | Clns | 80 | SMP | 81 | JWS | 82 | JWS+ | 83 | SMP | 85 |

---------- US DESIGNATION ----------

| JCS | 75 | SA-8 | FY76-79,82-84. |
|---|---|---|---|
| SecD | 75 | SA-8 | FY76-77. |
| MOW | 76 | SA-8 | |
| Bras | 77 | SA-8 | |
| SWM | 77 | SA-8 | |
| Rock | 78 | SA-8 | |
| Clns | 80 | SA-8 | |
| FI | 81 | SA-8 | |
| SMP | 81 | SA-8 | '81, '83, '84, '85. |
| AFM | 82 | SA-8 | |
| AWST | 82 | SA-8 | |
| JWS | 82 | SA-8 | |
| SMP | 84 | SA-8a | |

---------- NATO CODENAME ----------

| MOW | 76 | Gecko | |
|---|---|---|---|
| SWM | 77 | Gecko | |
| Rock | 78 | Gecko | |
| Clns | 80 | Gecko | |
| FI | 81 | Gecko | |
| SMP | 81 | Gecko | '81, '83, '85. |
| AFM | 82 | Gecko | |
| AWST | 82 | Gech | [Sic] |
| JWS | 82 | Gecko | |

---------- DESCRIPTION ----------

| SMP | 81 | Tactical surface-to-air missile, '81, '83, '84, '85. |
|---|---|---|
| JWS | 82 | "Low-altitude surface-to-air missile". |
| JWS+ | 83 | Mobile autonomous tactical air defence missile. |

---------- EDITOR'S NOTES ----------

| JCS | 75 | Listed FY76-79,82-84. |
|---|---|---|
| SecD | 75 | Listed FY76-77. |
| SMP | 81 | Listed '81, '83, '84, '85. |

---------- MINIMUM ALTITUDE ----------

| | | mi | ft | km | m | m conv | |
|---|---|---|---|---|---|---|---|
| AW01 | 77 | | 150 | | 46 | 45.72 | |
| SWM | 77 | | 164 | | 50 | 49.99 | |
| FI | 81 | | | | 50 | 50.00 | |
| AWST | 82 | | 150 | | | 45.72 | |

---------- MAXIMUM ALTITUDE ----------

| | | mi | ft | km | m | m conv | |
|---|---|---|---|---|---|---|---|
| AW01 | 77 | | 20,000 | | 6,100 | 6,096 | |
| Bras | 77 | | | | 3,500 | 3,500 | Max effective altitude, est. |
| SWM | 77 | | 19,685 | | 6,000 | 5,999 | |
| AFO1 | 78 | | | | 6,100 | 6,100 | |
| AW02 | 78 | | c. 20,000 | | | 6,096 | |
| Rock | 78 | | | 6 | | 6,000 | "Altitude". |
| Clns | 80 | | 40,000 | | | 12,192 | Ceiling. |
| FI | 81 | | | | 6,000 | 6,000 | |
| AFM | 82 | | 32,800 | | | 9,997 | |
| AWST | 82 | | 30,000 | | | 9,144 | |
| JWS+ | 83 | | | | 6,000 | 6,000 | Ceiling. |

590  SOVIET MISSILES  IDDS

## RANGE

| | | mi | nm | km | km conv | |
|---|---|---|---|---|---|---|
| ID06 | 75 | | | <_ 10-12 | 10.00-12.00 | |
| MOW | 76 | 7.5 | | 12 | 12.07 | |
| AW01 | 77 | 6-9 | | 10-15 | 9.65-14.48 | |
| Bras | 77 | | | 6.5 | 6.50 | Max effective slant range, est, [6,500 m in source]. |
| SWM | 77 | 7.5 | | 12.0 | 12.07 | |
| AF01 | 78 | | | 10-15 | 10.00-15.00 | Slant range. |
| | | | | 10-12 | 10.00-12.00 | Defend reserve and logistics centers out to 10-12 km. |
| AW02 | 78 | | c. 8 | | 14.82 | |
| Rock | 78 | | | 8.0 | 8.00 | |
| Clns | 80 | 7.5 | | | 12.07 | |
| FI | 81 | | | 8 | 8.00 | [Text] |
| | | | | 12 | 12.00 | Max, [table] |
| ID12 | 81 | | | 10-15 | 10.00-15.00 | Max effective range. |
| SMP | 81 | | | 10-15 | 10.00-15.00 | '81, '83. |
| AFM | 82 | 1.8-7.5 | | | 2.90-12.07 | |
| AWST | 82 | 1-6 | | | 1.61-9.65 | [Final report]; [AWST says SA-9 has greater range than SA-8, but values it gives for SA-8 (1-6 mi) are greater than for SA-9 (.4-4 mi), ed]. |
| JWS | 82 | | | 8-16 | 8.00-16.00 | [Analysis table] |
| | | | | 12 | 12.00 | "Officially estimated", [text] |
| AFM+ | 84 | 6-9 | | | 9.65-14.48 | |
| SMP | 84 | | | 12 | 12.00 | '84, '85. |

## SPEED

| | | mph | kmph | Mach | kmph conv |
|---|---|---|---|---|---|
| Clns | 80 | | | 2 | |
| FI | 81 | | | ? 1.5 | |
| JWS | 82 | | | c. 2 | |
| FI+ | 83 | | | 2.0 | |

## WARHEAD TYPE

| | | nuclear | conven | chem | |
|---|---|---|---|---|---|
| MOW | 76 | | yes | | HE. |
| Bras | 77 | | yes | | HE with proximity fuze. |
| Rock | 78 | | yes | | HE. |
| Clns | 80 | | yes | | HE. |
| FI | 81 | | yes | | HE; and proximity fuze. |
| AFM | 82 | | yes | | |
| JWS+ | 83 | | | | HE, proximity fuze. |

## WARHEAD WEIGHT

| | | lb | t | kg | kg conv | |
|---|---|---|---|---|---|---|
| MOW | 76 | 90-110 | | 40-50 | 40.82-49.90 | |
| Rock | 78 | | | 50 | 50.00 | |
| Clns | 80 | 50 | | | 22.68 | |
| FI | 81 | | | 40-50 | 40.00-50.00 | |
| AFM | 82 | 90-110 | | | 40.82-49.90 | |
| AWST | 82 | | | | | Lighter than SA-9. |

## WEIGHT

| | | lb | t | kg | kg conv |
|---|---|---|---|---|---|
| ID06 | 75 | | | 180-200 | 180.00-200.00 |
| Bras | 77 | | | c. 200 | 200.00 |
| Rock | 78 | | | 195 | 195.00 |
| FI | 81 | | | 180-200 | 180.00-200.00 |
| AFM | 82 | 440 | | | 199.58 |
| JWS | 82 | | | c. 190 | 190.00 |
| FI+ | 83 | | | c. 190 | 190.00 |

## LENGTH

| | | in | ft | cm | m | m conv | |
|---|---|---|---|---|---|---|---|
| MOW | 76 | | 10'6" | | 3.2 | 3.20 | |
| AW01 | 77 | | 10 | | 3 | 3.05 | |
| Bras | 77 | | | | 3.20 | 3.20 | |
| SWM | 77 | | 10.5 | | 3.2 | 3.20 | |
| Rock | 78 | | | | 3.2 | 3.20 | |
| FI | 81 | | | | 3.2 | 3.20 | |
| SMP | 81 | | | | 3 | 3.00 | [From graph], '81, '83, '84, '85. |
| AFM | 82 | | 10'6" | | | 3.20 | |
| JWS | 82 | | | c. 320 | | 3.20 | |

CRUISE MISSILES  Ground-Launched SAMs  SA-8

```
------------ DIAMETER ------------
              in     ft      cm      m      cm conv
MOW   76    8.25                    .21     20.96
Bras  77                    21              21.00       [210 mm in source]
SWM   77    8.25            20.2            20.96
Rock  78                    21              21.00
FI    81                    21              21.00
AFM   82    8.25"                           20.96
JWS   82                    21              21.00

------------ WINGSPAN ------------
              in     ft      cm      m      cm conv
ID06  75                            0.38    38.00       Forward fins.
                                    0.64    64.00       Rear fins.
SWM   77    25              64              63.50
Rock  78                    64              64.00
FI    81                    64              64.00
JWS   82                    60              60.00       Tail.
FI+   83                    60              60.00

------------ NUMBER OF STAGES ------------
Bras  77    ? 2
Rock  78    1

------------ TYPE OF PROPULSION ------------
MOW   76    ? solid
Bras  77    ? solid
SWM   77    ? solid          Dual-thrust.
Rock  78    solid
FI    81    solid            Dual-thrust.
AFM   82    solid            Dual-thrust.
JWS   82    solid

------------ MIDCOURSE (OR MAIN) GUIDANCE ------------
MOW   76    command          By proportional navigation.
AW01  77    EO tracker       "Most likely television".
Bras  77    ? SACLOS         Probably; (inferred from positioning of radar dishes and antennae).
SWM   77    command
Rock  78    command
Clns  80    command
FI    81    command
AFM   82    command          By proportional navigation.
JWS   82    command
JWS+  83    radar command

------------ TERMINAL GUIDANCE ------------
Rock  78    IR
FI    81    ? IR
AFM   82    SAR
JWS   82    SAR or IR        Both have been postulated; either is feasible; which is not known.

------------ CONTROL SYSTEM ------------
SWM   77    Cruciform canard fins. Steerable foreplanes.
Rock  78    Aerodynamic.
FI    81    Cruciform canard fins.
AFM   82    Canard foreplane control surfaces.
JWS   82    4 small canard surfaces.
```

---------- RADAR ----------
| | | | |
|---|---|---|---|
| ID06 | 75 | | Search radar sweeps 360 degrees over 30 km range; probably 4-8 GHz. Tracking to 20-25 km, must use 13-15 GHz [derived from its function]. 2 small antennas to pick up missile's beacon. Extra antenna possibly for guidance in ECM conditions [See source for more information, ed]. |
| MOW | 76 | | Surveillance radar, with range c. 18 mi, folds behind launcher. Tracking radar is of pulsed type, est. range 12-15 mi. |
| Bras | 77 | | Fire control radar. |
| SWM | 77 | | Surveillance, acquisition and tracking, and two guidance radars operating with low-light electro-optical tracker. 2 guidance radars can guide 2 missiles against target on different frequencies, frustrating ECM. |
| Clns | 80 | | Target acquisition by search and tracking radar. |
| FI | 81 | | Folding surveillance radar "surmounts" 4-round launcher, believed to operate in 4-8 GHz band, effective range some 30 km "against a typical target". Forward-mounted target-tracking radar flanked by 2 command dishes, 13-15 GHz, range of 20-25 km. Two missiles can be launched at same target, controlled by twin I-band command link. Low-light TV camera used for optical tracking and probably automatic missile gathering. |
| AFM | 82 | | Pulsed type; estimated range of radar: 12-15 miles. |
| JWS | 82 | Land Roll | Has antennas for search and tracking, plus 2 separate beacon tracking antennas and 2 transmitting horns for command guidance signals. [See source for more radar information, ed]. |

---------- OBSERVED IN PARADE ----------
| | | | |
|---|---|---|---|
| ID06 | 75 | 1975 | The 12 systems displayed appear to be prototypes. |
| MOW | 76 | 7 Nov 1975 | |
| Bras | 77 | Nov 1975 | Parade in Moscow. |
| SWM | 77 | Nov 1975 | |
| Rock | 78 | 1975 | First public display. |
| AFM | 82 | 7 Nov 1975 | |

---------- DEVELOPMENT UNDERWAY ----------
| | | |
|---|---|---|
| MOW | 76 | Under development. |

---------- PRODUCTION UNDERWAY ----------
| | | |
|---|---|---|
| JCS | 75 | Still in production, FY76-78. |

---------- DEPLOYMENT BEGAN ----------
| | | | |
|---|---|---|---|
| JCS | 76 | | Appeared in early 70s (or more recently). |
| MOW | 76 | early 1976 | Service evaluation stage. |
| Bras | 77 | | Now entering service with Soviet Army. |
| SWM | 77 | | Believed about to enter USSR Divisional service. |
| Rock | 78 | 1977 | Service intro. |
| Clns | 80 | 1975 | First deployed. |
| FI | 81 | | Now entering service. |
| SMP | 81 | 1970s | Introduced "in the early and mid-1970s". |

---------- DEPLOYED ----------
| | | |
|---|---|---|
| JCS | 75 | Deployed, FY76-79. |
| SecD | 76 | Developed and deployed. |
| JWS | 82 | In service. |

---------- DEPLOYMENT HISTORY ----------
| | | |
|---|---|---|
| JCS | 75 | Currently being deployed, FY76-78. |
| JCS | 81 | Soviets and NSWP countries continue to replace older, towed S-60, 57mm antiaircraft artillery with SA-8 in tank and motorized rifle divisions. |
| JWS | 82 | Gecko is initially replacing S60 57 mm AA guns. |

---------- NUMBERS IN SERVICE ----------

---------- NUMBER IN 1970 ----------
| | | | |
|---|---|---|---|
| Clns | 80 | none | In 1970-1973. |

---------- NUMBER IN 1974 ----------
| | | |
|---|---|---|
| Clns | 80 | 20 |

---------- NUMBER IN 1975 ----------
| | | |
|---|---|---|
| Clns | 80 | 50 |

---------- NUMBER IN 1976 ----------
| | | |
|---|---|---|
| Clns | 80 | 100 |

---------- NUMBER IN 1977 ----------
| | | |
|---|---|---|
| Clns | 80 | 200 |

---------- NUMBER IN 1978 ----------
| | | |
|---|---|---|
| Clns | 80 | 250 |

CRUISE MISSILES  Ground-Launched SAMs  SA-8  593

---------- **NUMBER IN 1979** ----------
Clns 80  300

---------- **NUMBER IN 1980** ----------
Clns+ 83  400    In 1980-1981.

---------- **NUMBER IN 1982** ----------
Clns+ 83  550

---------- **CARRIERS** ----------

---------- **CARRIER** ---------- TUBES ---- RELOADS ----------
| | | | | | |
|---|---|---|---|---|---|
| JCS | 75 | vehicle | | | Self-contained system, FY76-77. |
| JCS | 76 | wh vehicle | | | Lightly armored, amphibious. |
| Clns | 80 | | 2 x 2 | | |
| AWST | 82 | | | | Fully mobile, self-contained. |
| SMP | 85 | 6-wh vehicle | | | From photo with caption reading "The SA-8 tactical air defense system..." |

---------- **CARRIER** ---------- TUBES ---- RELOADS ----------
| | | | | | |
|---|---|---|---|---|---|
| ID06 | 75 | 6-wh vehicle | 4 | ≤ 8 | reloads: 12 rounds total, probably reloaded from inside vehicle. |
| MOW | 76 | 6-wh vehicle | 4 | | Rotating turret on the amphib. vehicle carries missiles and fire control equipment. |
| Bras | 77 | 6-wh vehicle | 4 | | Same vehicle mounts search and tracking radars and fire control unit. |
| SWM | 77 | 6-wh vehicle | [4 or 8] | | Radars on same vehicle as missile. |
| FI | 81 | 6-wh vehicle | 4 | < 10 totl | Amphibious, with rotating launcher with various radars. reloads: [Final report] |
| AFM | 82 | 6-wh vehicle | 4 or 6 | ≤ 6 | Missiles on rotating turret of new 3-axle amphibious vehicle. |
| JWS | 82 | 6-wh vehicle | 2 x 2 | 4 to 8 | Two versions noted, both c. 9 m long, but differing slightly in length. reloads: Various estimates. |
| FI+ | 83 | | 6 | | Latest version carries ready-to-fire rounds in storage/launch boxes. |

---------- **USE AND CONFIGURATION** ----------
| | | |
|---|---|---|
| ID06 | 75 | Vehicle can be airlifted by An-22 and Mi-12, but not An-12 and Il-76. 1973 prototype of vehicle had waterjet propulsion unit. Launcher and fire control unit rotate somewhat separately [assuming proportional navigation]. Perhaps can be reloaded from inside vehicle. Lateral acceleration should be 20g. [Possible operating procedure outlined, ed] |
| JCS | 75 | Provide highly mobile, responsive air defense to moving armored column, FY76-77. |
| MOW | 76 | Can be airlifted. |
| Bras | 77 | Appears to have all-weather 24 hour capability. |
| HI01 | 77 | Can be airlifted by training [sic] aircraft. |
| SWM | 77 | Can engage one target with two missiles simultaneously on different control frequencies. Air transportable. |
| AF01 | 78 | Believed able to fire 2 missiles under independent control. 5 batteries deployed from Army, 20-30 km deep. Expected to become division-level weapon, replacing S-60 or SA-6 regiments. |
| ID12 | 81 | Even the newest Soviet SAMs cannot fire on the move, but SA-8 can fire very shortly after halting. A Soviet division's air defense regiment has 20 SA-8 or SA-6 launchers. |
| JCS | 81 | Soviets and NSWP countries continue to replace older, towed S-60, 57mm antiaircraft artillery with SA-8 in tank and motorized rifle divisions. |
| SMP | 81 | Introduced at maneuver division level. "Unique among Soviet tactical air defense systems in that all the components needed to conduct a target engagement are on a single vehicle". [SA-6 and SA-8] can keep pace with "rapidly advancing maneuver forces". |
| AFM | 82 | "Unique among Soviet tactical air defence weapons" in that all components needed for engagement are on one vehicle. System can be airlifted by transport aircraft. |
| JCS | 82 | Employed at division level, FY83-84. |
| JWS | 82 | Radar arrangement indicates possible simultaneous firing of 2 missiles with separate guidance. Seems certain that a multiple target capability exists. Missile launcher arms and the search radar antenna fold down for reduced transit height. Design probably optimized for high acceleration, max speed, and maneuver rather than range. Operates with armored and motorized rifle divisions. |
| JWS+ | 83 | Usually deployed in batteries of 4 vehicles, each motor rifle and tank division having 5 batteries forming an air defence regiment. |
| SMP | 83 | "Mobile and a functional part of ground force units". |
| SMP | 84 | Echelon assignment [of SA-8a and SA-8b]: Division. |
| SMP | 85 | Standard SAM at division level, "though some divisions still have an AAA-equipped regiment". Part of USSR's integrated system of surface-to-air missiles and antiaircraft artillery. |

|        |    | **STRATEGY** |
|--------|----|------|
| IDO6   | 75 | All-terrain, all-weather defense against low level attack. |
| JCS    | 75 | [Listed as significant initiative in general purpose forces], FY76-78. |
|        |    | Provide excellent defense against air threats at all altitudes [could mean SA-8 and SA-9 together, ed], FY76-77. |
| SecD   | 75 | Part of theater air defense. |
| Bras   | 77 | Low level air defence system designed to bridge gap between SA-7,9 and SA-6. |
| SMP    | 81 | Effective altitude: low, '81, '83, '84, '85. |
| JCS    | 82 | Low to medium altitude system, FY83-84. |

|     |    | **COMBAT REPORTS/EFFECTIVENESS** |
|-----|----|------|
| FI+ | 83 | Saw combat last year in Bekaa Valley; several Syrian systems were destroyed by Israeli air attacks. |

|      |    | **EXPORTED TO** | |
|------|----|------|------|
| FI   | 81 | Syria | On order. |
| JWS  | 82 | Warsaw Pact | |
| FI+  | 83 | Jordan | |
| JWS+ | 83 | Jordan | Confirmed in 1983 that these units operational. |
|      |    | Syria | Along with others such as SA-2,3,5,6. |
| SMP  | 85 | Kuwait | The Soviets have signed arms contracts with Kuwait "that include weapons such as the SA-8/Gecko". |

|      |    | **RELATION TO OTHER MISSILES** |
|------|----|------|
| IDO6 | 75 | Resembles US Sea Sparrow in volume and weight. |
| MOW  | 76 | Soviet counterpart to Roland. May have same missile as SA-N-4. |
| Bras | 77 | Guidance probably similar to Roland and Rapier and performance "likely to be much the same". Could be same missile as naval SA-N-4. |
| SWM  | 77 | Comparable to European Roland. |
| FI   | 81 | Roughly equivalent to Crotale. |
| AFM  | 82 | Missile believed same as for SA-N-4 system. |

# SA-8b

Sources are:   JWS  82      Clns  83      JWS  83      SMP  84

### US DESIGNATION
| | |
|---|---|
| Clns 83 | SA-8B |
| JWS 83 | SA-8B |
| SMP 84 | SA-8b |

### CARRIERS

| | CARRIER | TUBES | RELOADS | |
|---|---|---|---|---|
| Clns 83 | | 2 x 3 | | |

| | CARRIER | TUBES | RELOADS | |
|---|---|---|---|---|
| JWS 82 | 6-wh vehicle | 2 x 3 | | New version with 6-tube vehicle appeared in 1980 parade. |

### USE AND CONFIGURATION
| | |
|---|---|
| JWS 83 | Launch containers on vehicle look slimmer than original missiles, may have folding fins. A special vehicle carrying reload missiles accompanies SA-8B fire unit vehicles. |
| SMP 84 | Echelon assignment: Division. |

### RELATION TO OTHER MISSILES
| | |
|---|---|
| Clns 83 | Same characteristics as SA-8 except number of launchers per vehicle. |

# SA-9

```
Sources are:  SecD  75    SecD  76    AFO1  78    FI    81    AWST  82    FI+   83    AFM+  84
ID07  75      JCS   76    AW01  77    AW02  78    ID12  81    JCS   82    FI+   83    SMP   84
JCS   75      MOW   76    Bras  77    Rock  78    SMP   81    JWS   82    JWS+  83    SMP   85
JWS*  75      NA03  76    SWM   77    Clns  80    AFM   82    Clns+ 83    SMP   83
```

---------- **US DESIGNATION** ----------

| | | |
|---|---|---|
| JCS | 75 | SA-9 | FY76-79,83-84. |
| JWS* | 75 | SA-10 or -8 | [Actually closest to SA-9, ed]. |
| SecD | 75 | SA-9 | FY76-77. |
| MOW | 76 | SA-9 |
| Bras | 77 | SA-9 |
| SWM | 77 | SA-9 |
| Rock | 78 | SA-9 |
| Clns | 80 | SA-9 |
| FI | 81 | SA-9 |
| SMP | 81 | SA-9 | '81, '83, '84, '85. |
| AFM | 82 | SA-9 |
| AWST | 82 | SA-9 |
| JWS | 82 | SA-9 |

---------- **NATO CODENAME** ----------

| | | |
|---|---|---|
| MOW | 76 | Gaskin |
| Bras | 77 | Gaskin |
| SWM | 77 | Gaskin |
| Rock | 78 | Gaskin |
| Clns | 80 | Gaskin |
| FI | 81 | Gaskin |
| SMP | 81 | Gaskin | '81, '83. |
| AFM | 82 | Gaskin |
| AWST | 82 | Gaskin |
| JWS | 82 | Gaskin |

---------- **DESCRIPTION** ----------

| | | |
|---|---|---|
| SMP | 81 | Tactical surface-to-air missile, '81, '83. |
| JWS+ | 83 | Mobile short-range tactical air defence missile. |
| SMP | 85 | Standard air defense system. |

---------- **EDITOR'S NOTES** ----------

| | | |
|---|---|---|
| JCS | 75 | Listed FY76-79,83-84. |
| SecD | 75 | Listed FY76-77. |
| SMP | 81 | Listed '81, '83, '84, '85. |

---------- **MINIMUM ALTITUDE** ----------

| | | mi | ft | km | m | m conv |
|---|---|---|---|---|---|---|
| AWST | 82 | | 50 | | | 15.24 |

---------- **MAXIMUM ALTITUDE** ----------

| | | mi | ft | km | m | m conv | |
|---|---|---|---|---|---|---|---|
| Bras | 77 | | | | 900 | 900.00 | Max effective altitude. |
| Clns | 80 | | 15,000 | | | 4,572 | Ceiling. |
| FI | 81 | | | | 5,000 | 5,000 | |
| AFM | 82 | | 16,400 | | | 4,998 | |
| AWST | 82 | | 15,000 | | | 4,572 | |
| JWS | 82 | | | | 4,000 | 4,000 | |
| FI+ | 83 | | | | 4,000 | 4,000 | |
| JWS+ | 83 | | | | 4,000-5,000 | 4,000-5,000 | Ceiling. |

CRUISE MISSILES  Ground-Launched SAMs  SA-9  **597**

### RANGE

| | | mi | nm | km | km conv | |
|---|---|---|---|---|---|---|
| MOW | 76 | | | | | "Larger, with a heavier warhead and longer range" than Grail. |
| NA03 | 76 | 5 | | | 8.05 | Max. |
| AW01 | 77 | c. 4 | | c. 7 | 6.44 | |
| Bras | 77 | | | 3 | 3.00 | Max effective slant range, [3,000 m in source]. |
| SWM | 77 | 3.1 | | 5 | 4.99 | |
| AF01 | 78 | | | 7 | 7.00 | Slant range. |
| AW02 | 78 | | [4] | | 7.41 | Double SA-7. |
| Rock | 78 | | | 4 | 4.00 | |
| Clns | 80 | 5 | | | 8.05 | |
| FI | 81 | | | 6 | 6.00 | Max. |
| SMP | 81 | | | 8 | 8.00 | '81, '83, '84, '85. |
| AFM | 82 | 4.35 | | | 7.00 | |
| AWST | 82 | .4-4 | | | 0.64-6.44 | [AWST says SA-9 has greater range than SA-8, but values it gives for SA-8 (1-6 mi) are greater than for SA-9 (.4-4 mi), ed]. |
| JWS | 82 | | | c. 8 | 8.00 | From official US statements, and UK MoD report of April-76. |
| FI+ | 83 | | | 8 | 8.00 | Max. |
| AFM+ | 84 | 5 | | | 8.05 | |

### SPEED

| | | mph | kmph | Mach | kmph conv |
|---|---|---|---|---|---|
| Clns | 80 | | | 2 | |
| FI | 81 | | | ? 1.5+ | |
| JWS+ | 83 | | | 1.5+ | |

### WARHEAD TYPE

| | | nuclear | conven | chem | |
|---|---|---|---|---|---|
| Bras | 77 | | yes | | HE blast. |
| Rock | 78 | | yes | | HE. |
| Clns | 80 | | yes | | HE. |
| FI | 81 | | yes | | HE. |
| JWS | 82 | | yes | | HE fragmentation. |

### WARHEAD WEIGHT

| | | lb | t | kg | kg conv | |
|---|---|---|---|---|---|---|
| MOW | 76 | | | | | "Larger, with a heavier warhead and longer range" than Grail. |
| SWM | 77 | | | | | Larger than SA-7's. |
| Clns | 80 | 15 | | | 6.80 | |
| AWST | 82 | | | | | Heavier than SA-8. |

### WEIGHT

| | | lb | t | kg | kg conv | |
|---|---|---|---|---|---|---|
| Bras | 77 | | | 9.2 | 9.20 | |
| Rock | 78 | | | 30 | 30.00 | |
| FI | 81 | | | ? 30 | 30.00 | |
| AFM | 82 | 66 | | | 29.94 | |
| JWS+ | 83 | | | 30+ | 30.00 | Launch weight. |

### LENGTH

| | | in | ft | cm | m | m conv | |
|---|---|---|---|---|---|---|---|
| Bras | 77 | | | | 1.3 | 1.30 | |
| Rock | 78 | | | | 1.8 | 1.80 | |
| FI | 81 | | | | 1.8 | 1.80 | |
| SMP | 81 | | | | 2 | 2.00 | [From graph], '81, '83, '84, '85. |
| AFM | 82 | | 5'9" | | | 1.75 | |
| JWS+ | 83 | | | | c. 2 | 2.00 | |

### DIAMETER

| | | in | ft | cm | m | cm conv | |
|---|---|---|---|---|---|---|---|
| Bras | 77 | | | 7 | | 7.00 | [70 mm in source] |
| Rock | 78 | | | 11.5 | | 11.50 | |
| FI | 81 | | | 11 | | 11.00 | |
| AFM | 82 | 4.33 | | | | 11.00 | |
| JWS+ | 83 | | | 12 | | 12.00 | |

### WINGSPAN

| | | in | ft | cm | m | cm conv |
|---|---|---|---|---|---|---|
| Rock | 78 | | | 32 | | 32.00 |
| FI | 81 | | | 30 | | 30.00 |

```
                  ----------- NUMBER OF STAGES -----------
Bras   77          3
Rock   78          "1 + 1"
AWST   82          1

                  ----------- TYPE OF PROPULSION -----------
Bras   77          solid
SWM    77          solid
Rock   78          solid+solid
FI     81          solid                    Booster and sustainer.
JWS    82          solid
JWS+   83          solid

                  ----------- TERMINAL GUIDANCE -----------
ID07   75          IR                       Same IR head as SA-7 grail.
Bras   77          IR
SWM    77          IR                       Targets optically acquired by operator, who is alerted by "surveillance data
                                            link" perhaps same as for SA-7. Radar direction on BRDM.
AW02   78          IR
Rock   78          IR
Clns   80          IR
FI     81          IR
SMP    81          IR
AFM    82          IR
JWS    82          IR

                  ----------- CONTROL SYSTEM -----------
Rock   78          Aerodynamic.
FI     81          Cruciform canard fins.

                  ----------- RADAR -----------
Bras   77          Gundish                  Fire control radar on ZSU-23-4.
ID12   81                                   Launch vehicle is not equipped with radar.
JWS    82          Gun Dish                 Associated with some units; mounted in front of SA-9 turret.
JWS+   83          Gun Dish                 Earlier reports probably referred to SA-13, which resembles SA-9.

                  ----------- OBSERVED IN PARADE -----------
SWM    77          Nov 1975
Rock   78          1975                     First public display.

                  ----------- DEPLOYMENT BEGAN -----------
JCS    76          early 70s                Appeared.
Rock   78          1976                     Service intro.
Clns   80          1968                     First deployed.
SMP    81          1968                     Was deployed in 1968.
AFM    82          1968
JWS    82          1968                     First deployed, according to US sources.

                  ----------- DEPLOYED -----------
JCS    75                                   Deployed, FY76-79.
SecD   76                                   Deployed.
Bras   77                                   In service.
FI     81                                   Deployed.

                  ----------- DEPLOYMENT HISTORY -----------
SMP    85          Being selectively replaced and augmented by SA-13.

                  ----------- NUMBERS IN SERVICE -----------

                  ----------- NUMBER IN 1970 -----------
Clns   80          none       In 1970-1973.

                  ----------- NUMBER IN 1974 -----------
Clns   80          500

                  ----------- NUMBER IN 1975 -----------
Clns   80          600

                  ----------- NUMBER IN 1976 -----------
Clns   80          700

                  ----------- NUMBER IN 1977 -----------
Clns   80          900

                  ----------- NUMBER IN 1978 -----------
Clns   80          1000
```

## CRUISE MISSILES Ground-Launched SAMs SA-9

| | | ——— NUMBER IN 1979 ——— |
|---|---|---|
| Clns | 80 | 1100 |

| | | ——— NUMBER IN 1980 ——— |
|---|---|---|
| Clns+ | 83 | 1100 |

| | | ——— NUMBER IN 1981 ——— |
|---|---|---|
| Clns+ | 83 | 1050 |

| | | ——— NUMBER IN 1982 ——— |
|---|---|---|
| Clns+ | 83 | 850 |

### CARRIERS

| | | CARRIER | TUBES | RELOADS | |
|---|---|---|---|---|---|
| JCS | 75 | | | | Self-contained [vehicle], FY76-77. |
| JWS* | 75 | | | | Self-contained. |
| JCS | 76 | | | | Lightly armored wheeled amphibious vehicle. |
| Clns | 80 | | 4 | | |
| SMP | 81 | a scout car | | | |
| AWST | 82 | | | | "Fully mobile, self-contained battlefield air-defense system". |
| SMP | 85 | wheeled vehicle | | | "The SA-9 system [is] mounted on a wheeled transporter-erector-launcher [TEL]." |

| | | CARRIER | TUBES | RELOADS | |
|---|---|---|---|---|---|
| MOW | 76 | BRDM | 2 x 4 | | |
| SWM | 77 | BRDM [mod] | 2 x 4 | | |
| FI | 81 | modified BRDM | 2 x 4 | | |
| AFM+ | 84 | BRDM-2 | | 4 | |
| | | | | | reloads: Stowed in vehicle. |

| | | CARRIER | TUBES | RELOADS | |
|---|---|---|---|---|---|
| AW01 | 77 | BRDM-2 | x 4 | | Vehicle is 18' (5.5 m) long, manned by crew of 4. tubes: "Grouped in fours" in canister launchers. |
| Bras | 77 | BRDM-2 [mod] | 2 x 2 | | |
| AFM | 82 | BRDM-2 | 2 x 2 | | tubes: Box launcher for 2 pairs of missiles. |
| JWS | 82 | BRDM-2 | 4 | 8 total | Normal configuration. |
| | | BRDM-2 | 2 | | A small number observed. |

### USE AND CONFIGURATION

| | | |
|---|---|---|
| ID07 | 75 | If halted, SA-9s may operate off a ZSU-23-4 Gun Dish radar, otherwise threat warning probably passed by radio. |
| JCS | 75 | Provide highly mobile, responsive air defense to moving armored column, FY76-77. |
| MOW | 76 | Mobile system. |
| AW01 | 77 | Used with ZSU-23-4 gun. |
| Bras | 77 | Mobile very low/with low level system, complements ZSU-23-4 gun for forward area defence; when in position uses ZSU-23-4's radar for local warning and fire control. Turret has 360 degree traverse. |
| AF01 | 78 | 4 missiles mounted on powered turret. 64 "troops [batteries]" of SA-9 support a typical combined arms or tank Army. |
| AW02 | 78 | Must have target in field of view for IR seeker to be effective. Targets optically acquired by operator, who is alerted by "surveillance data link". |
| Clns | 80 | Optical target acquisition and tracking. |
| ID12 | 81 | Even the newest Soviet SAMs do not fire while moving, but SA-9 can fire very shortly after halting. Operates in 4-vehicle platoon. |
| JCS | 82 | Employed at regiment level, FY83-84. |
| JWS | 82 | Launch containers are individually loaded manually; they fold down and lie flat during transit. When operating in autonomous mode, acquisition assumed to be visual, subsequent aiming by optical means. Reported that as part of air defense column, SA-9 vehicles can be linked to search radars to assist in acquisition; could also use "radio communication links to 'tell off' targets to individual SA-9 firing units". Gun Dish radar on some units would provide night and adverse weather capability. |
| SMP | 83 | "Mobile and a functional part of ground force units". |
| SMP | 84 | Echelon assignment: Regiment. |
| SMP | 85 | Battery of SA-9/SA-13 SAMs is standard air defense system for tank or motorized rifle regiment along with ZSU-23/4 self-propelled AAA pieces. |

### STRATEGY

| | | |
|---|---|---|
| JCS | 75 | Provide excellent defense against air threats at all altitudes [could mean SA-8 and SA-9 together], FY76-77. |
| SecD | 75 | Part of theater air defense. |
| Bras | 77 | Very low/ith low level system. |
| SWM | 77 | Battlefield "Low-level defense". |
| SMP | 81 | Effective altitude: Low, '81, '83, '84, '85. |
| JCS | 82 | Low altitude system, FY83-84. |

```
                ------ EXPORTED TO ------
Bras    77      Egypt
                Syria
FI      81      WTO
                Egypt
                ? Syria
SMP     81                          "A recent export".
JWS     82      Algeria
                Hungary
                Poland
                Syria
                Vietnam
                Yugoslavia
FI+     83      Algeria
                East Germany
                Hungary
                India
                Iraq
                Jugoslavia
                Libya
                Poland
                Syria
                Vietnam
                North Yemen
```

------ RELATION TO OTHER MISSILES ------

| | | |
|---|---|---|
| JWS* | 75 | [Either SA-8 or SA-10] may relate to a vehicle-mounted multiple launcher version of man-portable Grail. |
| MOW | 76 | Tube-launched missile similar to SA-7 Grail. |
| Bras | 77 | Appears to be improved version of SA-7 and probably uses the same IR homing head. |
| SWM | 77 | Development of SA-7 with larger warhead and greater range. |
| FI | 81 | "Vehicle-mounted development of SA-8" [sic], SA-9 heavier than it. Has fixed fins, and "may be derived from AA-2 Atoll rather than SA-7 Grail". |
| JWS | 82 | SA-9 missile originally thought to be SA-7 Grail; now this seems incorrect. Family descent (if any) more likely to be through AAMs, such as AA-2 Atoll. |

------ OTHER INFORMATION ------

JWS*    75      Little known, but source stating existence is authentic.

# SA-10

```
Sources are:  SecD  79      FI    81      SMP   81      JCS   82      FI+   83      SMP   83      SMP   84
AWO2  78      JCS   80      JCS   81      AFM   82      JWS   82      FI+   83      AFM+  84      JCS   85
JCS   79      SecD  80      SecD  81      AWST  82      Clns  83      JWS+  83      JCS   84      SMP   85
```

### US DESIGNATION

| | | |
|---|---|---|
| JCS  79 | SA-X-10 | FY80-82. |
|         | SA-10   | FY83, 85-86. |
| SecD 79 | SA-X-10 | FY80-81. |
| FI   81 | SA-10 | |
| SecD 81 | SA-10 | |
| SMP  81 | SA-10 | '81, '83, '84, '85. |
| AFM  82 | SA-10 | |
| AWST 82 | SA-10 | |
| JWS  82 | SA-10 | |
| Clns 83 | SA-10 | |

### DESCRIPTION

| | |
|---|---|
| JCS 79 | New strategic SAM, FY80-82. |
| JCS 81 | "All purpose" SAM. |
| SMP 81 | Strategic SAM, '84, '85. |
| JWS 82 | "New air-defence missile". |

### EDITOR'S NOTES

| | |
|---|---|
| JCS 79  | Listed FY80-83, 85-86. |
| SecD 79 | Listed FY80-82. |
| SMP 81  | Listed '81, '83, '84, '85. |

### MINIMUM ALTITUDE

| | mi | ft | km | m | m conv | |
|---|---|---|---|---|---|---|
| FI   81 |   |       |   | 300 | 300.00 | |
| AFM  82 |   | 1,000 |   |     | 304.80 | [Final report] |
| AWST 82 |   | 1,000 |   |     | 304.80 | |
| JWS  82 |   |       |   | 300 | 300.00 | |
| AFM+ 84 |   |       |   |     |        | All-altitude capability. |

### MAXIMUM ALTITUDE

| | mi | ft | km | m | m conv | |
|---|---|---|---|---|---|---|
| FI   81 |   |        |   | 5,000 | 5,000  | |
| AFM  82 |   | 16,500 |   |       | 5,029  | [Final report] |
| AWST 82 |   | 15,000 |   |       | 4,572  | |
| JWS  82 |   |        |   | 4,500 | 4,500  | |
| Clns 83 |   | 80,000 |   |       | 24,384 | Combat ceiling. |
| FI+  83 |   |        |   | 4,500 | 4,500  | |
| AFM+ 84 |   |        |   |       |        | All-altitude capability. |

### RANGE

| | mi | nm | km | km conv | |
|---|---|---|---|---|---|
| FI   81 |      |    | 50  | 50.00  | Max. |
| SMP  81 |      |    | 100 | 100.00 | '81, '83, '84, '85. |
| AFM  82 | ≤ 60 |    |     | 96.54  | |
| AWST 82 |      | 27 |     | 50.00  | |
| JWS  82 |      |    | 50  | 50.00  | |
| Clns 83 | 60   |    |     | 96.54  | Slant range. |
| JWS+ 83 |      |    | 100 | 100.00 | Max, "has been postulated". |

### SPEED

| | mph | kmph | Mach | kmph conv |
|---|---|---|---|---|
| FI   81 |   |   | 5.6 | |
| AFM  82 |   |   | ? 6 | |
| AWST 82 |   |   | 6   | |

### WARHEAD TYPE

| | nuclear | conven | chem | | |
|---|---|---|---|---|---|
| FI   81 |     | maybe |   | ? HE. | |
| Clns 83 | yes | yes   |   | HE.   | |

### WEIGHT

| | lb | t | kg | kg conv |
|---|---|---|---|---|
| FI 81 |   |   | c. 1,500 | 1,500 |

```
              ---------- LENGTH ----------
                        in      ft      cm      m       m conv
FI      81                                      c. 7    7.00
SMP     81                                      7       7.00        [From graph], '81, '83, '84, '85.
AFM     82              23                              7.01
AWST    82              23                              7.01
JWS     82                                      c. 7    7.00

              ---------- DIAMETER ----------
                        in      ft      cm      m       cm conv
FI      81                              c. 45                   45.00
AFM     82              17.7"                                   44.96
AWST    82                      1.5                             45.72
JWS     82                              c. 45                   45.00

              ---------- NUMBER OF STAGES ----------
FI      81              1
AFM     82              1
AWST    82              1

              ---------- TYPE OF PROPULSION ----------
FI      81              solid
AWST    82              solid

              ---------- TERMINAL GUIDANCE ----------
FI      81              SAR
                        ? active radar
AFM     82              active radar
AWST    82              active radar
JWS     82              active radar

              ---------- RADAR ----------
FI      81              3 CW radars.
AWST    82              Uses 3 radars.
JWS     82              Suggested feature is "a target detection/designation/guidance system relying
                        upon three types of radar". Radars more advanced than previous systems.
FI+     83              Tower-mounted surveillance radar to detect low-flying targets.
                        A lower and more easily transported radar tower has been developed "to make
                        the system mobile rather than transportable".

              ---------- OBSERVED ----------
SMP     84              Drawing shows artist's conception of mobile SA-10.

              ---------- DEVELOPMENT UNDERWAY ----------
JCS     79              Development continues, FY80-81.
SecD    79              They continue to develop it.
FI      81              Development "seems to have been unduly protracted", suggesting it is being adopted
                        for use against cruise missiles.
JWS     82              Development [stage].

              ---------- PRODUCTION UNDERWAY ----------
SMP     84              In series production.

              ---------- IOC ----------
AFM     82              IOC prediction varied from 'about now' to mid-1980s.
SMP     84              Mobile version could be operational by 1985, '84, '85.

              ---------- DEPLOYMENT BEGAN ----------
SecD    79                      Expected to be deployed soon.
JCS     80                      "Initial deployment is expected in the near future".
FI      81                      "Initial development [sic] could begin some time this year".
JCS     81                      Modified version of SA-X-10 also could become available.
                                Deployment has begun.
SMP     81                      "Is now becoming operational".
AWST    82      1979-80          Deployment [final report]
JWS     82      1981            Entered operational service with PVO Strany.
                                Omission from 'X' in designation could indicate no longer regarded as experimental.
Clns    83      1980            First deployed.
FI+     83                      Some now deployed.
SMP     84      1980            First deployed.
SMP     85      1980            First site reached operational status.
```

CRUISE MISSILES  Ground-Launched SAMs  SA-10  603

---------- DEPLOYMENT HISTORY ----------
FI      81   "Not expected to be in widespread service until the mid-1980s".
JCS     81   Expected to be widely deployed in the years ahead.
AFM     82   Full deployment likely to be protracted, as DoD thinks effective anti-ALCM system would required
             500-1000 sites with 10 launchers each.
JCS     82   Long-term SA-2 drawdown may resume due to replacement with more capable SA-10.
JWS     82   Probable SA-10 will replace SA-2 Guideline SAMs.
FI+     83   Small numbers now deployed near high-value targets in the Soviet Union.
SMP     83   Deployment has steadily increased. "In addition to deployment around the USSR, the system is
             replacing" SA-1s around Moscow.
AFM+    84   Deployment continuing.
SMP     84   Being deployed.
SMP     85   "Nearly 60 sites are now operational and work is underway on at least another 30."

---------- NUMBERS IN SERVICE ----------

---------- NUMBER IN 1970 ----------
Clns    83   none    In 1970-1979.

---------- NUMBER IN 1980 ----------
Clns    83   30

---------- NUMBER IN 1981 ----------
Clns    83   600

---------- NUMBER IN 1982 ----------
Clns    83   1200

---------- NUMBER IN 1983 ----------
FI+     83           Small numbers deployed.

---------- NUMBER IN 1984 ----------
SMP     84   [< 1,400] "Now operational at some 40 sites with nearly 350 launchers and four SA-10s per
             launcher".

---------- NUMBER IN 1985 ----------
SMP     85           Numbers of Soviet strategic and tactical air defense forces increasing with "continuing
             deployment of new systems like the SA-10".

---------- CARRIERS ----------

---------- CARRIER ---------- TUBES ---- RELOADS ----------
JWS+    83                                       Mobile version being developed, according to 1983 report
                                                 from US.

---------- CARRIER ---------- TUBES ---- RELOADS ----------
Clns    83                     6                 Mobile.
AFM+    84                                       A mobile version is being tested.

---------- USE AND CONFIGURATION ----------
AW02    78   Highly maneuverable high supersonic speed weapon.
FI      81   Rapid-acceleration SAM. May be launched vertically.
AFM     82   ? Acceleration at 100g.
SMP     83   "Development of a mobile SA-10 is underway". New program since 1981: "testing of a mobile version
             of the SA-10 SAM".
AFM+    84   Multiple target engagement capabilities.
SMP     84   Soviets developing mobile version, '84, '85.
SMP     85   Further deployment and upgrading of SA-5 to enhance its capability to work in conjunction with
             low-altitude systems like SA-10 likely in future.

## STRATEGY

| | | |
|---|---|---|
| SecD | 79 | A new SAM for low-altitude intercepts. |
| SecD | 80 | "Will be able to engage aircraft-sized targets at any altitude". |
| FI | 81 | Anti-cruise missile system would need 500-1,000 10-launcher sites. Development "seems to have been unduly protracted", suggesting it is being adopted for use against cruise missiles. |
| JCS | 81 | "Will have good low-altitude capabilities against aircraft"; low altitude capability of SA-X-10 "will enhance the already substantial Soviet air defense network with a multi-target handling capability". |
| SecD | 81 | Significant improvements in air defense include the new SA-10 SAM. |
| SMP | 81 | Effective altitude: low-to-high, '81, '83, '84, '85. Latest strategic SAM, "designed for increased low-altitude capability. With radars which are more advanced than previous systems, the SA-10 was designed to counter low-altitude manned aircraft, although it may have some capability against cruise missiles". |
| AFM | 82 | According to US press, SA-10 threatens viability of cruise missile. Full deployment likely to be protracted, as DoD thinks effective anti-ALCM system would required 500-1000 sites with 10 launchers each. Would cost $50 billion in US. |
| JWS | 82 | Attracting interest because of suspected potential as possible defense against US cruise missiles. Official US circles regard SA-10 as within 'Strategic' category. CJCS Jones said certain SA-10 features indicate "designed specifically to engage low-altitude targets", and indicated good range performance. Was designed to counter low-altitude manned aircraft; may be capable of use against cruise missiles. |
| FI+ | 83 | Last autumn was tested against Soviet RVs, suggesting it may have limited ABM role. |
| SMP | 83 | "Can engage multiple aircraft and possibly cruise missiles at any altitude". |
| JCS | 84 | "Estimated to be effective against small, low-altitude targets", FY85-86. |
| SMP | 84 | "May have the potential to intercept some types of US strategic ballistic missiles". "Could, if properly supported, add significant point-target coverage to a wide-spread ABM deployment". Mobile version could be used to support theater forces; but perhaps more importantly if deployed with territorial defense forces, would allow Soviets to change location of those SA-10s, '84, '85. |
| JCS | 85 | May have potential to intercept some types of US strategic ballistic missiles, with SA-5 and SA-X-12. |
| SMP | 85 | Capability of Soviet strategic and tactical air defense forces increasing with "continuing deployment of new systems like the SA-10". Of operational strategic SAMs [SA-1, SA-2, SA-3, SA-5, SA-10], only SA-10 capable of defending against targets with small radar-cross-sections, such as cruise missiles. "This emphasis on Moscow and the patterns noted for the other SA-10 sites suggest a first priority on terminal defense of wartime command and control, military, and key industrial complexes." [See GEOGRAPHIC DEPLOYMENTS, ed.] |

## GEOGRAPHICAL DEPLOYMENTS

| | | |
|---|---|---|
| FI+ | 83 | Deployment expected on territory of other WTO nations. |
| AFM+ | 84 | Deployment continuing around Moscow and throughout USSR. |
| SMP | 85 | "More than half of these sites [60 operational and 30 under construction] are located near Moscow. [See STRATEGY, ed.] |

## OTHER INFORMATION

| | | |
|---|---|---|
| JWS | 82 | Very little known of missile. Only evidence available in Western defense circles is "gleaned from heavily censored US DoD statements and observations", based on undisclosed intelligence efforts. |

# SA-11

```
Sources are:  ID12  81       AFM   82      JCS  82      Clns  83      JWS+  83      AFM+  84      SMP  85
FI   81       SMP   81       AWST  82      JWS  82      FI+   83      SMP   83      SMP   84
```

## US DESIGNATION
|  | |  | |
|---|---|---|---|
| FI | 81 | SA-11 | |
| SMP | 81 | SA-11 | '81, '83, '84, '85. |
| AFM | 82 | SA-11 | |
| AWST | 82 | SA-11 | |
| JCS | 82 | SA-11 | |
| JWS | 82 | SA-11 | |
| Clns | 83 | SA-11 | |

## NATO CODENAME
| | | |
|---|---|---|
| JWS | 82 | No NATO name revealed yet. |

## DESCRIPTION
| | | |
|---|---|---|
| SMP | 81 | Tactical surface-to-air missile, '81, '83. |
| JCS | 82 | Warsaw Pact SAM system. |
| JWS | 82 | "New land-mobile short-range surface-to-air missile system". |
| SMP | 85 | New division level SAM. |

## EDITOR'S NOTES
| | | |
|---|---|---|
| SMP | 81 | Listed '81, '83, '84, '85. |
| JCS | 82 | Listed FY83. |

## MINIMUM ALTITUDE
| Source | | mi | ft | km | m | m conv | |
|---|---|---|---|---|---|---|---|
| FI | 81 | | | | 30 | 30.00 | |
| AFM | 82 | | 80 | | | 24.38 | |
| AWST | 82 | | 100 | | | 30.48 | |
| JWS | 82 | | | | 30 | 30.00 | |
| AFM+ | 84 | | 100 | | | 30.48 | |

## MAXIMUM ALTITUDE
| Source | | mi | ft | km | m | m conv | |
|---|---|---|---|---|---|---|---|
| FI | 81 | | | | 14,000 | 14,000 | |
| AFM | 82 | | 49,000 | | | 14,935 | |
| AWST | 82 | | 45,000 | | | 13,716 | |
| JWS | 82 | | | | 14,000 | 14,000 | |
| Clns | 83 | | 50,000 | | | 15,240 | Ceiling. |
| AFM+ | 84 | | 46,000 | | | 14,020 | |

## RANGE
| Source | | mi | nm | km | km conv | |
|---|---|---|---|---|---|---|
| FI | 81 | | | 25-30 | 25.00-30.00 | Max. |
| AFM | 82 | ≤ 12 | | | 19.31 | |
| AWST | 82 | | 1.6-15 | | 2.96-27.78 | |
| JWS | 82 | | | 3 | 3.00 | Min. |
| | | | | 28 | 28.00 | Max. |
| Clns | 83 | 25 | | | 40.23 | |
| SMP | 83 | | | 30 | 30.00 | '83, '84, '85. |
| AFM+ | 84 | 18.5 | | | 29.77 | |

## SPEED
| Source | | mph | kmph | Mach | kmph conv | |
|---|---|---|---|---|---|---|
| FI | 81 | | | 3 | | |
| AFM | 82 | | | 3 | | |
| JWS | 82 | | | 3 | | Probably. |
| Clns | 83 | | | 2 | | |

## WARHEAD TYPE
| Source | | nuclear | conven | chem | | |
|---|---|---|---|---|---|---|
| Clns | 83 | | yes | | HE. | |

## WARHEAD WEIGHT
| Source | | lb | t | kg | kg conv |
|---|---|---|---|---|---|
| Clns | 83 | 50 | | | 22.68 |

## LENGTH
| Source | | in | ft | cm | m | m conv | |
|---|---|---|---|---|---|---|---|
| AWST | 82 | | 18.4 | | | 5.61 | |
| SMP | 83 | | | | 5.5 | 5.50 | [From graph], '83, '84, '85. |

## 606 SOVIET MISSILES IDDS

### DIAMETER

|  |  | in | ft | cm | m | cm conv |
|---|---|---|---|---|---|---|
| AWST | 82 |  | 1.32 |  |  | 40.23 |

### NUMBER OF STAGES

| AWST | 82 | 2 |
|---|---|---|

### TYPE OF PROPULSION

| AWST | 82 | solid |
|---|---|---|

### TERMINAL GUIDANCE

| FI | 81 | radar | |
|---|---|---|---|
| AFM | 82 | radar | |
| JWS | 82 | radar | Has been assumed. |
| Clns | 83 | semi-active | |
| FI+ | 83 | SAR | Perhaps with electro-optical back-up mode. |

### RADAR

| FI | 81 | Clam Shell | Acquisition radar. |
|---|---|---|---|
|  |  | Flap Lid | Tracking radar, may be mounted on missile launching vehicle. |
| AWST | 82 | Straight Flush | "Used with SA-6, Straight Flush radar". |
| JWS | 82 | Clamshell | 3-D acquisition radar, mounted on accompanying ZSU-23-4 vehicle. |
|  |  | Flap Lid | Tracking radar (on Clamshell vehicle). |
| FI+ | 83 | Clam Shell | 3-D acquisition radar. |
|  |  | Flap Lid | Tracking radar. |
| SMP | 85 |  | Onboard radar that increases mobility and target-handling capability. |

### DEVELOPMENT UNDERWAY

| JWS | 82 | Thought still under development. |
|---|---|---|

### PRODUCTION UNDERWAY

| SMP | 84 | In series production. |
|---|---|---|

### DEPLOYMENT BEGAN

| FI | 81 |  | US sources say still under development, but absence of 'X' in designation suggests already operational [final report]. |
|---|---|---|---|
| ID12 | 81 |  | Now entering service. |
| AFM | 82 |  | Deployed alongside SA-6s. |
| JWS | 82 | 1978-79 | Initial troop trials thought to have begun. |
| Clns | 83 | 1982 | First deployed. |
| FI+ | 83 |  | Field trials started in late 1970s, but probably not yet in operational service. |
| JWS+ | 83 |  | Probably early deployment [analysis table]. |
| SMP | 85 |  | Beginning to enter inventory. |

### NUMBERS IN SERVICE

### NUMBER IN 1970

| Clns | 83 | none | In 1970-1981. |
|---|---|---|---|

### NUMBER IN 1982

| Clns | 83 | 50 |
|---|---|---|

### CARRIERS

### CARRIER — TUBES — RELOADS

| FI | 81 | launch vehicle | 3 |  | Observed operating in conjunction with SA-6. |
|---|---|---|---|---|---|
|  |  | launch vehicle | 4 |  | Observed operating in conjunction with SA-6. |
| AFM | 82 |  | 3 or 4 |  | tubes: [Final report of 3 rails] |
| Clns | 83 |  | 4 |  | tubes: Launcher containers. |

### CARRIER — TUBES — RELOADS

| JWS | 82 | ZSU-24-4 [sic] | 4 |  | Vehicle adapted for SA-11. |
|---|---|---|---|---|---|
| FI+ | 83 | ZSU-23-4 chassis | 4 |  | Tracked, trainable launcher; second chassis carries radars. |

### USE AND CONFIGURATION

| AFM | 82 | Deployed alongside SA-6s. |
|---|---|---|
| JCS | 82 | Employed at division level. |
| JWS | 82 | Missiles mounted on turntable rotatable through 360 degrees. |
| Clns | 83 | Target acquisition by radar. |
| SMP | 83 | "Mobile and a functional part of ground force units". |
| SMP | 84 | Echelon assignment: Division. |
| SMP | 85 | Division level SAM. |

--------- **STRATEGY** ---------
ID12 81  For use against low- and middle-altitude targets.
JCS  82  Low-to-medium altitude system.
SMP  83  Effective altitude: low-to-medium, '83, '84, '85.

--------- **RELATION TO OTHER MISSILES** ---------
AFM  82  SA-11 may be improved version of SA-6.

# SA-12

```
Sources are:   AFM   82      AFM+  84      JCS   84      JCS   85
FI    81       AWST  82      AWST+ 84      SMP   84      SMP   85
```

---------- **US DESIGNATION** ----------

| | | | |
|---|---|---|---|
| FI | 81 | SA-X-12 | |
| AFM | 82 | SA-12 | |
| AWST | 82 | SA-12 | |
| JCS | 84 | SA-X-12 | |
| SMP | 84 | SA-X-12 | '84, '85. |

---------- **DESCRIPTION** ----------

| | | |
|---|---|---|
| JCS | 84 | Tactical SAM system. |
| JCS | 85 | Mobile tactical SAM system. |

---------- **EDITOR'S NOTES** ----------

| | | |
|---|---|---|
| JCS | 84 | Listed FY85-86. |
| SMP | 84 | Listed '84, '85. |

---------- **MINIMUM ALTITUDE** ----------

| | | mi | ft | km | m | m_conv | |
|---|---|---|---|---|---|---|---|
| FI | 81 | | | | 30 | 30.00 | |
| AFM | 82 | | 100 | | | 30.48 | [Final report] |
| AWST | 82 | | 300 | | | 91.44 | |

---------- **MAXIMUM ALTITUDE** ----------

| | | mi | ft | km | m | m_conv | |
|---|---|---|---|---|---|---|---|
| FI | 81 | | | | 30,000 | 30,000 | |
| AFM | 82 | | 100,000 | | | 30,480 | [Final report] |
| AWST | 82 | | 100,000 | | | 30,480 | |

---------- **RANGE** ----------

| | | mi | nm | km | km_conv | |
|---|---|---|---|---|---|---|
| FI | 81 | | | 100 | 100.00 | Max. |
| AFM | 82 | 60 | | | 96.54 | [Final report] |
| AWST | 82 | | 55 | | 101.86 | |
| SMP | 84 | | | 100 | 100.00 | '84, '85. |

---------- **LENGTH** ----------

| | | in | ft | cm | m | m_conv | |
|---|---|---|---|---|---|---|---|
| SMP | 84 | | | | 7 | 7.00 | [From graph], '84, '85. |

---------- **TYPE OF PROPULSION** ----------

| | | |
|---|---|---|
| AWST | 82 | solid |

---------- **TERMINAL GUIDANCE** ----------

| | | | |
|---|---|---|---|
| AFM | 82 | active radar | [Final report] |

---------- **RADAR** ----------

| | | | |
|---|---|---|---|
| FI | 81 | phased array | Can handle several targets simultaneously. |
| AFM | 82 | phased array | [Final report] |
| AWST | 82 | phased array | For multiple target handling. |
| AFM+ | 84 | | Truck-mounted, estimated range up to 150 miles. |
| SMP | 84 | | Target-tracking. |
| | | | Fire control. |

---------- **PROTOTYPE TESTS** ----------

| | | |
|---|---|---|
| SMP | 84 | In flight testing. |

---------- **PRODUCTION UNDERWAY** ----------

| | | |
|---|---|---|
| AFM+ | 84 | In production. |

---------- **IOC** ----------

| | | | |
|---|---|---|---|
| SMP | 85 | ? 1986 | Nearing operational status. |

---------- **DEPLOYMENT BEGAN** ----------

| | | |
|---|---|---|
| JCS | 84 | "Could be deployed in the mid-1980s", FY85-86. |

---------- **DEPLOYMENT HISTORY** ----------

| | | |
|---|---|---|
| JCS | 84 | Will replace SA-4. |
| SMP | 84 | To augment or replace SA-4. |

---------- **NUMBERS IN SERVICE** ----------

## NUMBER IN 1985

| | | |
|---|---|---|
| SMP | 85 | Numbers of Soviet strategic and tactical air defense forces increasing with "impending deployment of the SA-X-12". |

## CARRIERS

### CARRIER ———— TUBES ——— RELOADS

| | | CARRIER | TUBES | |
|---|---|---|---|---|
| AWST+ | 84 | | | Track mounted. |
| SMP | 85 | tracked vehicle | 2 | Illustration with caption "SA-X-12 air defense system" shows three different types of tracked vehicles with two launchers each, along with three accompanying vehicles. |

## USE AND CONFIGURATION

| | | |
|---|---|---|
| AFM+ | 84 | Radar said "to offer capability of engaging intermediate-range and submarine-launched missiles, but to offer only limited defence against ICBMs". |
| SMP | 84 | "Fired from mobile launchers, accompanied by reload vehicles and supported by target-tracking and fire control radars". To augment or replace SA-4 in SAM brigades at the front level. |
| SMP | 85 | Soon to replace SA-4; SA-4 "the standard weapon [SAM] at army and front levels". |

## STRATEGY

| | | |
|---|---|---|
| FI | 81 | Probably intended to replace SA-2 and SA-5. |
| AFM | 82 | Intended to replace SA-2 and SA-5 [final report]. |
| AWST | 82 | "Has ABM capability". |
| AFM+ | 84 | Considered capable against aircraft and ballistic missiles. |
| JCS | 84 | Will replace SA-4 as Army and Front level air defense system; could also serve as gap-filler for strategic defense. |
| SMP | 84 | Effective altitude: low-to-high, '84, '85. Both a tactical SAM and "anti-tactical ballistic missile". "May have the potential to intercept some types of US strategic ballistic missiles". "Could, if properly supported, add significant point-target coverage to a wide-spread ABM deployment". "To counter high-performance aircraft and theater warfare missiles...". Capable of engaging high-performance aircraft and SRBMs like US Lance; "It may also be used to attempt to intercept longer-range INF missiles". |
| JCS | 85 | May have potential to intercept some types of US strategic ballistic missiles, with SA-5 and SA-X-10. |
| SMP | 85 | "SA-4... should shortly begin being replaced by the SA-X-12." "Incorporates ballistic missile defense capabilities." Capability of Soviet strategic and tactical air defense forces increasing with "impending deployment of the SA-X-12". Has "good" low-altitude air defense capabilities. "Will probably also have a capability against tactical ballistic missiles." "The surface-to-air missiles of the SA-X-12 air defense system are designed to counter high performance aircraft and will also have a capability against tactical ballistic missiles." |

# SA-13

```
Sources are:  SMP   81      AWST  82      JWS   82      FI+   83      AFM+  84      SMP   85
FI    81      AFM   82      JCS   82      Clns  83      SMP   83      SMP   84
```

```
------------- US DESIGNATION -------------
FI    81      SA-X-13
SMP   81      SA-13           '81, '83, '84, '85.
AFM   82      SA-13
AWST  82      SA-13
JCS   82      SA-13           FY83-84.
JWS   82      SA-13
Clns  83      SA-13
FI+   83      SA-13

------------- NATO CODENAME -------------
JWS   82                      No NATO name yet released.

------------- OTHER DESIGNATION -------------
AWST  82      Strella 10
JWS   82      Strella 10      Used by some US observers.

------------- DESCRIPTION -------------
SMP   81                      Tactical surface-to-air missile, '81, '83.
JCS   82                      Warsaw Pact SAM, FY83-84.
SMP   84                      Current tactical air defense system.
SMP   85                      Standard air defense system.

------------- EDITOR'S NOTES -------------
SMP   81      Listed '81, '83, '84, '85.
JCS   82      Listed FY83-84.

------------- MINIMUM ALTITUDE -------------
              mi      ft         km      m       m conv
AWST  82              30                         9.14    "Can intercept at 2 naut.
                                                         mi. at 30-ft. altitude".
JWS   82                                 50      50.00   In clear weather.
FI+   83                                 50      50.00
AFM+  84              165                        50.29

------------- MAXIMUM ALTITUDE -------------
              mi      ft         km      m       m conv
FI    81                                 10,000  10,000
JWS   82                                 10,000  10,000  In clear weather.
Clns  83              20,000                     6,096   Ceiling.
AFM+  84              32,000                     9,753

------------- RANGE -------------
              mi      nm         km              km conv
FI    81                         5-7             5.00-7.00   Max.
SMP   81                         8               8.00        '81, '83, '84, '85.
AFM   82      c. 5                               8.05
AWST  82              3-4                        5.56-7.41   "Can intercept at 2 naut. mi. at
                                                             30-ft. altitude".
JWS   82                         7.5             7.50        Max slant range.
Clns  83      7.5                                12.07
FI+   83                         7.5             7.50        Max, in clear weather.

------------- SPEED -------------
              mph     kmph       Mach    kmph conv
Clns  83                         2

------------- WARHEAD TYPE -------------
              nuclear conven     chem
Clns  83              yes                HE.

------------- WARHEAD WEIGHT -------------
              lb      t          kg      kg conv
Clns  83      10                         4.54

------------- LENGTH -------------
              in      ft         cm      m       m conv
SMP   81                                 2       2.00        [From graph], '81, '83, '84, '85.

------------- NUMBER OF STAGES -------------
AWST  82      1
```

## CRUISE MISSILES Ground-Launched SAMs SA-13

### TYPE OF PROPULSION
| | | |
|---|---|---|
| AWST | 82 | solid |
| JWS | 82 | solid |

### TERMINAL GUIDANCE
| | | | |
|---|---|---|---|
| FI | 81 | passive IR | Seeker thought to be cooled, operating in 2 frequency bands. For discrimination against flares and other IR countermeasures. |
| AWST | 82 | IR | "Passive RF detectors; cooled-IR or dual-band IR seekers". |
| JWS | 82 | passive IR | Assumed more advanced than SA-9; probably incorporating dual-wavelength sensor to provide protection from IR countermeasures. |
| Clns | 83 | IR | |

### RADAR
| | | | |
|---|---|---|---|
| FI | 81 | range-only radar | |
| AWST | 82 | range-only radar | |
| JWS | 82 | range-only radar | With passive radar emission sensors also mounted on SA-13 vehicle for target detection and aid in acquisition. |
| FI+ | 83 | | Claims that radar has surveillance capability seem unlikely. |

### DEPLOYMENT BEGAN
| | | | |
|---|---|---|---|
| SMP | 81 | late 70s | Fielded in the late 70s. |
| JWS | 82 | late 1970s | Introduced into service. |
| Clns | 83 | 1981 | First deployed. |
| FI+ | 83 | late 70s | |

### DEPLOYED
| | | | |
|---|---|---|---|
| AFM | 82 | late 70s | Deployed in late 70s. |
| JWS | 82 | | Deployed. |

### DEPLOYMENT HISTORY
| | | |
|---|---|---|
| FI | 81 | Being developed to replace current SA-9. |
| SMP | 81 | Probably a replacement for SA-9. |
| AFM | 82 | Replacement for SA-9. |
| JWS | 82 | Probably replacement for SA-9 Gaskin. |
| SMP | 85 | Selectively replacing and augmenting SA-9 system. |

### NUMBERS IN SERVICE

### NUMBER IN 1970
| | | | |
|---|---|---|---|
| Clns | 83 | none | In 1970-1980. |

### NUMBER IN 1981
| | | |
|---|---|---|
| Clns | 83 | 100 |

### NUMBER IN 1982
| | | |
|---|---|---|
| Clns | 83 | 200 |

### CARRIERS

#### CARRIER — TUBES — RELOADS
| | | | | |
|---|---|---|---|---|
| FI | 81 | tracked vehicle | 4 | |
| SMP | 81 | tracked vehicle | | |
| AFM | 82 | tracked vehicle | | |
| AWST | 82 | tracked vehicle | 4 | |
| Clns | 83 | | 4 | |
| | | | | tubes: Launcher containers. |
| SMP | 85 | tracked vehicle | | Tracked TEL, [Transporter-Erector-Launcher, ed.] |

#### CARRIER — TUBES — RELOADS
| | | | | |
|---|---|---|---|---|
| FI+ | 83 | AT-P chassis | 2 x 2 | Tracked. |
| | | | | tubes: Some sources claim 6 ready-to-fire rounds. |
| AFM+ | 84 | MT-LB | | Tracked vehicle. |

### USE AND CONFIGURATION
| | | |
|---|---|---|
| SMP | 81 | "Deployed along with the ZSU-23-4 in the antiaircraft battery of motorized rifle and tank regiments". |
| AFM | 82 | Together with ZSU-23-4, SA-13 equips AA batteries of motorised rifle an tank regiments. |
| JCS | 82 | Fielded at regiment level, FY83-84. |
| JWS | 82 | Introduced in late 1970s into AA batteries of motorized infantry and tank regiments. |
| Clns | 83 | Target acquisition by radar. |
| SMP | 83 | "Mobile and a functional part of ground force units". |
| AFM+ | 84 | Provides improved capability in rough terrain and increased storage for re-load missiles over SA-9. |
| SMP | 84 | Echelon assignment: Regiment. |
| SMP | 85 | Battery of SA-9/SA-13 SAMs is standard air defense system for tank or motorized rifle regiment along with ZSU-23/4 "self-propelled AAA pieces". |

————————— **STRATEGY** —————————
FI    81   For point defense of ground forces.
SMP   81   Effective altitude: low, '81, '83, '84, '85.
JCS   82   Low altitude system, FY83-84.

————————— **RELATION TO OTHER MISSILES** —————————
JWS   82   Many similarities to SA-9.

# SA-14

Sources are: FI 81  AFM 82  SMP 84

## US DESIGNATION
- FI 81 — SA-X-?
- SMP 84 — SA-14

## OTHER DESIGNATION
- AFM 82 — "New Infantry SAM".

## DESCRIPTION
- SMP 84 — Current tactical air defense system.

## EDITOR'S NOTES
- FI 81 — Final report.
- SMP 84 — Listed '84.

## TERMINAL GUIDANCE
- FI 81 — laser beam
- AFM 82 — laser beam  For beam-riding.

## DEVELOPMENT UNDERWAY
- FI 81 — In mid-70s USSR was reported to be working on laser-beam riding missile which could be the same weapon.

## DEPLOYMENT BEGAN
- FI 81 — May be about to enter service.
- AFM 82 — Deployment about to start.

## CARRIERS

| Source | CARRIER | TUBES | RELOADS |
|---|---|---|---|
| FI 81 | man-portable | | |

## USE AND CONFIGURATION
- SMP 84 — Echelon assignment: Company/Battalion.

## STRATEGY
- FI 81 — New man-portable SAM to replace SA-7 Grail.

# SA-N-1

```
Sources are:  MRO5  66    MOW   76    FI    81    AWST  82    JWS   82    JFS+  83    CRW+  84
IA01  65      ID01  68    NA03  76    USND  81    CRW   82    FI+   83    JWS+  83
MR04  66      US01  70    SWM   77    AFM   82    JFS   82    GSN   83    AFM+  84
```

## US DESIGNATION

|  | |  |
|---|---|---|
| MOW | 76 | SA-N-1 |
| SWM | 77 | SA-N-1 |
| FI | 81 | SA-N-1 |
| USND | 81 | SA-N-1 |
| AFM | 82 | SA-N-1 |
| AWST | 82 | SA-N-1 |
| CRW | 82 | SA-N-1 |
| JFS | 82 | SA-N-1 |
| JWS | 82 | SA-N-1 |
| GSN | 83 | SA-N-1 |

## NATO CODENAME

| Source | | Codename |
|---|---|---|
| MOW | 76 | Goa |
| FI | 81 | Goa |
| AFM | 82 | Goa |
| AWST | 82 | Goa |
| CRW | 82 | Goa |
| JFS | 82 | Goa |
| JWS | 82 | Goa |
| GSN | 83 | Goa |

## MINIMUM ALTITUDE

| Source | | mi | ft | km | m | m_conv | |
|---|---|---|---|---|---|---|---|
| FI | 81 | | | | 100-300 | 100.00-300.00 | |
| CRW | 82 | | 300 | | | 91.44 | |
| JFS+ | 83 | | 300 | | | 91.44 | |

## MAXIMUM ALTITUDE

| Source | | mi | ft | km | m | m_conv | |
|---|---|---|---|---|---|---|---|
| FI | 81 | | | | 10000-15000 | 10,000-15,000 | |
| AFM | 82 | | 49,200 | | | 14,996 | |
| CRW | 82 | | 50,000 | | | 15,240 | |
| JWS | 82 | | | c. 12,000 | | 12,000 | Ceiling. |
| FI+ | 83 | | | 25-30 | | 25,000-30,000 | Max. |
| JFS+ | 83 | | 50,000 | | | 15,240 | |

## RANGE

| Source | | mi | nm | km | km_conv | |
|---|---|---|---|---|---|---|
| IA01 | 65 | c. 18 | | | 28.96 | |
| MRO5 | 66 | | | 15-27 | 15.00-27.00 | |
| ID01 | 68 | c. 20 | | | 32.18 | |
| MOW | 76 | 18.5 | | 30 | 29.77 | Max. |
| NA03 | 76 | 9 | | | 14.48 | |
| FI | 81 | | | 30-35 | 30.00-35.00 | Max. |
| USND | 81 | | c. 10 | | 18.52 | |
| AFM | 82 | 21.75 | | | 35.00 | |
| CRW | 82 | | | 20 | 20.00 | [20,000 m in source] |
| JFS | 82 | | 17 | | 31.48 | Max slant range. |
| JWS | 82 | | | c. 15 | 15.00 | Max slant range. |
| GSN | 83 | | c. 17 | c. 31.5 | 31.48 | |

## SPEED

| Source | | mph | kmph | Mach | kmph_conv | |
|---|---|---|---|---|---|---|
| FI | 81 | | | 2 | | |
| AFM | 82 | | | 2 | | |
| JFS | 82 | | | 1+ | | |
| GSN | 83 | | | 2.0 | | Approximate max speed. |
| JFS+ | 83 | | | 3+ | | |

## WARHEAD TYPE

| Source | | nuclear | conven | chem | | |
|---|---|---|---|---|---|---|
| MOW | 76 | | yes | | HE. | |
| FI | 81 | | yes | | HE. | |
| AFM | 82 | | yes | | | |
| CRW | 82 | | yes | | | |
| JFS | 82 | | yes | | HE. | |
| JWS | 82 | | yes | | HE. | |
| GSN | 83 | | yes | | | |
| JWS+ | 83 | | | | HE. | |

```
------------ WARHEAD WEIGHT ------------
                lb              t           kg          kg conv
FI     81                                   60          60.00
AFM    82       132                                     59.88
CFW    82                                   60          60.00
GSN    83       132                         60          59.88
JFS+   83       160                                     72.58
JWS+   83                                   70          70.00

------------ WEIGHT ------------
                lb              t           kg          kg conv
IA01   65                       .7                      700.00      Tons.
MR05   66                       .8                      800.00      Tons.
FI     81                                   600         600.00
AFM    82       1,323                                   600.11
GSN    83       882                         400         400.08
JFS+   83       2,100                                   952.56      Launch.
CFW+   84                                   400         400.00

------------ LENGTH ------------
                in              ft          cm          m           m conv
IA01   65                       17.25                               5.26
MR05   66                                   c. 4                    4.00
ID01   68                       c. 20                               6.10        Includes booster.
MOW    76                       22                      6.7         6.71
FI     81                                               6.7         6.70
AFM    82                       22'0"                               6.71
JFS    82                       22                                  6.71        Booster.
JWS    82                                   c. 590                  5.90
GSN    83                       22                      6.6         6.71

------------ DIAMETER ------------
                in              ft          cm          m           cm conv
IA01   65                       1                                   30.48       Excluding booster.
MR05   66                                   c. .3                   30.00
MOW    76                       2'3"                    0.70        68.58       Booster.
                                1'6"                    0.45        45.72       2nd stage.
FI     81                                   46                      46.00       Missile.
                                            70                      70.00       Booster.
AFM    82                       1'6"                                45.72
FI+    83                                   60                      60.00       Booster.
GSN    83       18.1                        46                      45.97       Missile, [460 mm in source].
                27.6                        70.1                    70.10       Booster, [701 mm in source].

------------ WINGSPAN ------------
                in              ft          cm          m           cm conv
MOW    76                       4'0"                    1.22        121.92
FI     81                                   122                     122.00      Missile.
                                            150                     150.00      Booster.
AFM    82                       4'0"                                121.92
JWS    82                                   120                     120.00
GSN    83                       4'11"                   1.5         149.86

------------ NUMBER OF STAGES ------------
MOW    76       2
AFM    82       2
AW ST  82       2
JFS    82       2

------------ TYPE OF PROPULSION ------------
MOW    76       solid
FI     81       solid           Booster and sustainer.
AFM    82       solid
AW ST  82       solid           Booster.
JFS    82       solid
JWS    82       solid
GSN    83       solid           With tandem solid boosters.

------------ MIDCOURSE (OR MAIN) GUIDANCE ------------
MOW    76       radio command
FI     81       radio command
AFM    82       radio
CFW    82       command
JFS    82       radio command
JWS    82       beam rider
GSN    83       radio command
```

---------- TERMINAL GUIDANCE ----------
MOW    76    radar
AFM    82    radar
CFW    82    radar
JWS    82    semi-active

---------- CONTROL SYSTEM ----------
MOW    76    Movable foreplanes.
FI     81    Cruciform fins.

---------- RADAR ----------
SWM    77    Reel              [Table], E/I band.
              Peel Group        [Text]
FI     81    Peel Group        E/I-band.
CFW    82    Peel Group        Radar directors.
JWS    82    Peel Group
GSN    83    Peel Group        Fire control radar.

---------- OBSERVED ----------
ID01   68    1964              First shown.
MOW    76    1971              Tass picture showed Goa missiles with additional tail fins and control surfaces on wings.

---------- IOC ----------
JFS    82    1961              Operational.
GSN    83    1961              Also, derived from land-based SA-3, "which became operational in 1964".
JFS+   83    1960              Operational.

---------- DEPLOYMENT BEGAN ----------
SWM    77    1962
FI     81    1962              In service since 1962.
JWS    82    1961-62           Entry into service.
CFW+   84    1961

---------- DEPLOYED ----------
MOW    76                      In service.
FI     81                      Deployed.

---------- DEPLOYMENT HISTORY ----------
AFM    82    Most widely used naval SAM of USSR.
JWS    82    Principal SAM of Soviet Navy.
GSN    83    First SAM widely fitted in Soviet warships.

---------- NUMBERS IN SERVICE ----------

---------- NUMBER IN 1982 ----------
JWS    82    66                Twin launchers in Soviet fleet, latest figures.

---------- CARRIERS ----------

---------- CARRIER ---------- TUBES ---- RELOADS ----------
AFM+   84                      x 2               43 ships of Soviet Navy.

---------- CARRIER ---------- TUBES ---- RELOADS ----------
MOW    76    Kresta            2 x 2             N=4.
SWM    77    Kresta I
FI     81    Kresta I                            N=4.
USND   81    Kresta I
AWST   82    Kresta
CFW    82    Kresta-I
JWS    82    Kresta            2 x 2
GSN    83    Kresta I

---------- CARRIER ---------- TUBES ---- RELOADS ----------
MR04   66    Kashin            2 x 2    30 total
US01   70    Kashin                               Appeared 1964.
MOW    76    Kashin            2 x 2              N=19.
SWM    77    Kashin
FI     81    Kashin                               And mod Kashin, n=19.
USND   81    Kashin
AWST   82    Kashin
CFW    82    Kashin
JWS    82    Kashin            2 x 2
GSN    83    Kashin            1 x 2

# CRUISE MISSILES  Sea-Launched SAMs  SA-N-1

|        |    | CARRIER    | TUBES | RELOADS  |       |
|--------|----|------------|-------|----------|-------|
| MRO4   | 66 | Kynda      | 1 x 2 | 40 total |       |
| MOW    | 76 | Kynda      | 1 x 2 |          | N=4.  |
| SWM    | 77 | Kynda      |       |          |       |
| FI     | 81 | Kynda      |       |          | N=4.  |
| USND   | 81 | Kynda      |       |          |       |
| AWST   | 82 | Kynda      |       |          |       |
| CFW    | 82 | Kynda      |       |          |       |
| JWS    | 82 | Kynda      | 1 x 2 |          |       |
| GSN    | 83 | Kynda      |       |          |       |

|        |    | CARRIER    | TUBES | RELOADS  |       |
|--------|----|------------|-------|----------|-------|
| MOW    | 76 | Kanin      | 1 x 2 |          | N=6.  |
| SWM    | 77 | Kanin      |       |          |       |
| FI     | 81 | Kanin      |       |          | N=7.  |
| USND   | 81 | Kanin      |       |          |       |
| CFW    | 82 | Kanin      |       |          |       |
| JWS    | 82 | Kanin      | 1 x 2 |          |       |
| GSN    | 83 | Kanin      | 1 x 2 |          |       |

|        |    | CARRIER    | TUBES | RELOADS  |       |
|--------|----|------------|-------|----------|-------|
| MRO4   | 66 | Kotlin     | 1 x 2 | 40 total |       |
| MOW    | 76 | Kotlin     | 1 x 2 |          | N=8.  |
| SWM    | 77 | Kotlin     |       |          |       |
| FI     | 81 | SAM Kotlin |       |          | N=8.  |
| USND   | 81 | Kotlin     |       |          |       |
| CFW    | 82 | Kotlin     |       |          |       |
| JWS    | 82 | Kotlin     | 1 x 2 |          |       |
| GSN    | 83 | SAM Kotlin | 1 x 2 |          |       |

## USE AND CONFIGURATION

| MRO4 | 66 | Task group of 2 Kashins and one Kynda has approx 100 air defense missiles at their disposal, [sic, text, cf "carriers" section]. |
|------|----|---|
| USO1 | 70 | "It is assumed that 30 to 40 missiles are stored in the magazines of each launcher". |
| FI   | 81 | Fired from roll-stabilised twin launcher, mounted atop magazine. |
| AFM  | 82 | Fired from roll-stabilised twin-round launcher. |
| CFW  | 82 | Twin launcher. |
| JWS  | 82 | Launcher is clearly roll-stabilized. Launcher mounted on magazine and reloaded vertically through small hatches. |
| GSN  | 83 | Fired from twin-armed launcher, which is loaded at 90 degrees elevation from below deck magazine. Four tail fins are folded until missile leaves launcher. |
| AFM+ | 84 | Roll-stabilized twin launchers. |

## TARGET TYPE

| CFW | 82 |         | Also has surface-to-surface capability. |
|-----|----|---------|-----------------------------------------|
| GSN | 83 | surface | Also.                                   |

## STRATEGY

| MOW | 76 | Close-range missile. |
|-----|----|----------------------|
| GSN | 83 | "Considered effective at low to medium altitudes and in the surface-to-surface mode". |

## EXPORTED TO

| FI  | 81 | Poland   | Deployed on 1 SAM Kotlin. |
|-----|----|----------|---------------------------|
|     |    | ? India  | ? Deployed on several Kashin. |
| JWS | 82 | Poland   | "One Polish, ex-Soviet 'Kotlin' class, ship is fitted with Goa". |

## RELATION TO OTHER MISSILES

| MOW | 76 | Twin-round launcher "very like" that of US Tartar. |
| SWM | 77 | Naval version of SA-3. |
| FI  | 81 | Naval variant of SA-3. |
| AFM | 82 | Counterpart of US Hawk. |
| JWS | 82 | Missile assumed identical to ground-launched missile, but equipment very different. |

# SA-N-2

```
Sources are:  US01  70      SWM  77      USND 81      CFW  82      JWS  82      GSN  83
              MR04  66      MOW  76      FI   81      AFM  82      JFS  82      FI+  83      CFW+ 84
```

## US DESIGNATION

|  | |  |
|---|---|---|
| MOW | 76 | SA-N-2 |
| SWM | 77 | SA-N-2 |
| FI | 81 | SA-N-2 |
| USND | 81 | SA-N-2 |
| AFM | 82 | SA-N-2 |
| CFW | 82 | SA-N-2 |
| JFS | 82 | SA-N-2 |
| JWS | 82 | SA-N-2 |
| GSN | 83 | SA-N-2 |

## NATO CODENAME

| Source | | Codename |
|---|---|---|
| MOW | 76 | Guideline |
| FI | 81 | Guideline |
| AFM | 82 | Guideline |
| CFW | 82 | Guideline |
| JFS | 82 | Guideline |
| JWS | 82 | Guideline |
| GSN | 83 | Guideline |

## MINIMUM ALTITUDE

| Source | | mi | ft | km | m | m conv | |
|---|---|---|---|---|---|---|---|
| CFW | 82 | | 300 | | | 91.44 | |

## MAXIMUM ALTITUDE

| Source | | mi | ft | km | m | m conv | |
|---|---|---|---|---|---|---|---|
| FI | 81 | | | | <_ 28000 | 28,000 | |
| CFW | 82 | | 90,000 | | | 27,432 | |
| FI+ | 83 | | | | <_ 18,000 | 18,000 | |
| GSN | 83 | | | | | | Considered medium-altitude weapon. |

## RANGE

| Source | | mi | nm | km | km conv | |
|---|---|---|---|---|---|---|
| FI | 81 | | | 40-50 | 40.00-50.00 | Max. |
| USND | 81 | | c. 30 | | 55.56 | |
| CFW | 82 | | | 50 | 50.00 | [50,000 m in source] |
| JFS | 82 | | 25 | | 46.30 | Max. |
| JWS | 82 | | | c. 45 | 45.00 | |
| GSN | 83 | | c. 25 | c. 44 | 46.30 | |

## SPEED

| Source | | mph | kmph | Mach | kmph conv | |
|---|---|---|---|---|---|---|
| FI | 81 | | | 3.5 | | |
| JFS | 82 | | | 1+ | | |
| GSN | 83 | | | c. 2.5 | | Max speed. |
| CFW+ | 84 | | | 2.5 | | |

## WARHEAD TYPE

| Source | | nuclear | conven | chem | |
|---|---|---|---|---|---|
| CFW | 82 | | yes | | |
| JFS | 82 | | yes | | HE. |
| JWS | 82 | | yes | | HE. |
| GSN | 83 | | yes | | |

## WARHEAD WEIGHT

| Source | | lb | t | kg | kg conv |
|---|---|---|---|---|---|
| CFW | 82 | | | 150 | 150.00 |
| JFS | 82 | 290 | | | 131.54 |
| GSN | 83 | 287 | | 130 | 130.18 |

## WEIGHT

| Source | | lb | t | kg | kg conv | |
|---|---|---|---|---|---|---|
| FI | 81 | | | 2,300 | 2,300 | |
| JFS | 82 | 5,000 | | | 2,268 | Launch weight. |
| JWS | 82 | | | 2,300 | 2,300 | |
| GSN | 83 | 5,070 | | 2,300 | 2,299 | |
| CFW+ | 84 | | | 2,300 | 2,300 | |

## LENGTH

|  |  | in | ft | cm | m | m conv |  |
|---|---|---|---|---|---|---|---|
| FI | 81 |  |  |  | 10.7 | 10.70 |  |
| JFS | 82 |  | 34.7 |  |  | 10.58 |  |
| JWS | 82 |  |  | 1,070 |  | 10.70 |  |
| GSN | 83 |  | 35'2" |  | 10.7 | 10.72 |  |

## DIAMETER

|  |  | in | ft | cm | m | cm conv |  |
|---|---|---|---|---|---|---|---|
| FI | 81 |  |  | 50 |  | 50.00 | Missile. |
|  |  |  |  | 70 |  | 70.00 | Booster. |
| JWS | 82 |  |  | 50 |  | 50.00 |  |

## WINGSPAN

|  |  | in | ft | cm | m | cm conv |  |
|---|---|---|---|---|---|---|---|
| FI | 81 |  |  | 170 |  | 170.00 | Missile. |
|  |  |  |  | 220 |  | 220.00 | Booster. |
| JWS | 82 |  |  | 170 |  | 170.00 |  |
| GSN | 83 |  | 7'2.2" |  | 2.2 | 218.95 |  |

## TYPE OF PROPULSION

| JFS | 82 | solid | Booster. |
|---|---|---|---|
|  |  | liquid | Sustainer. |
| JWS | 82 | solid | Booster. |
| GSN | 83 | solid | Rocket. |
|  |  | liquid | Booster. |

## MIDCOURSE (OR MAIN) GUIDANCE

| CFW | 82 | "radar/command" |
|---|---|---|
| JFS | 82 | radio command |
| JWS | 82 | radio command |
| GSN | 83 | radio command |

## CONTROL SYSTEM

| FI | 81 | Cruciform rear fins. |
|---|---|---|

## RADAR

| SWM | 77 | Fan Song | C-band. |
|---|---|---|---|
| CFW | 82 | Fan Song | Radar director. |
| JWS | 82 | Fan Song |  |
| GSN | 83 | Fan Song-E | Fire control. |
|  |  | High Lark | Height-finding radar. |
| CFW+ | 84 | Fan Song E |  |

## IOC

| USO1 | 70 | 1962 | Dzerzhinski converted. |
|---|---|---|---|
| GSN | 83 | 1961 | Limited basis. |
|  |  | 1957 | Land-based system became operational. |

## DEPLOYMENT BEGAN

| CFW+ | 84 | 1961 |
|---|---|---|

## DEPLOYED

| CFW | 82 | Obsolescent. |
|---|---|---|
| JFS | 82 | Non-operational. |
| GSN | 83 | Dzerzhinskiy system not operational. |

## CARRIERS

### CARRIER — TUBES — RELOADS

| MRO4 | 66 | Dzerzhinski | 1 x 2 |  |
|---|---|---|---|---|
| MOW | 76 | Dzerzhinski |  |  |
| SWM | 77 | Dzerzhinski | 1 x 2 | 1 ship. |
| FI | 81 | Dzerzhinski |  | One ship only, Sverdlov class. |
| USND | 81 | Dzerzhinskiy |  | Sverdlov class. |
| AFM | 82 | Dzerzhinski |  | 1 ship. |
| CFW | 82 | Dzerhinskiy | x 2 | Only. |
| JFS | 82 | Dzerzhinsky |  | Only. |
| JWS | 82 | Dzerjinski |  | Only 1 ship fitted. |
| GSN | 83 | Dzerzhinskiy |  | 1 Sverdlov class ship. |

## STRATEGY

| CFW | 82 | Some surface-to-surface capability. |
|---|---|---|

---------- **COMBAT REPORTS/EFFECTIVENESS** ----------
SWM 77 Not widely deployed, probably due to problems with guidance on ships.
FI  81 Apparently unsuccessful, since only on one ship.
JWS 82 Reason not deployed more could be difficulty of providing suitable stable platforms for flapping Fan Song radars; or relative difficulty of gathering missiles to the flight path, even more of a problem in shipborne installations than on the ground. Performance as shipborne missile substantially the same as ground version.
GSN 83 Lack of success indicated by fitting in just one ship.

---------- **RELATION TO OTHER MISSILES** ----------
GSN 83 Adopted from land-based SA-2.

# SA-N-3

```
Sources are:   NA03  76     US10  79     AFM   82     JFS   82     GSN   83     SMP   83
JCS   76       JCS   77     FI    81     AWST  82     JWS   82     JFS+  83     AFM+  84
MOW   76       SWM   77     USND  81     CFW   82     FI+   83     JWS+  83     CFW+  84
```

## US DESIGNATION

|  | |  |  |
|---|---|---|---|
| JCS | 76 | SA-N-3 | FY77-79. |
| MOW | 76 | SA-N-3 | |
| SWM | 77 | SA-N-3 | |
| FI | 81 | SA-N-3 | |
| USND | 81 | SA-N-3 | |
| AFM | 82 | SA-N-3 | |
| AWST | 82 | SA-N-3 | |
| CFW | 82 | SA-N-3 | |
| JFS | 82 | SA-N-3 | |
| JWS | 82 | SA-N-3 | |
| GSN | 83 | SA-N-3 | |
| SMP | 83 | SA-N-3 | '83, '84. |

## NATO CODENAME

| Source | | Codename |
|---|---|---|
| MOW | 76 | Goblet |
| SWM | 77 | Goblet |
| FI | 81 | Goblet |
| AFM | 82 | Goblet |
| AWST | 82 | Goblet |
| CFW | 82 | Goblet |
| JFS | 82 | Goblet |
| JWS | 82 | Goblet |
| GSN | 83 | Goblet |

## EDITOR'S NOTES

|  | |  |
|---|---|---|
| JCS | 76 | Listed FY77-79. |
| SMP | 83 | Listed '83, '84. |

## MINIMUM ALTITUDE

| Source | | mi | ft | km | m | m conv | Notes |
|---|---|---|---|---|---|---|---|
| CFW | 82 | | 300 | | | 91.44 | |
| JWS | 82 | | | | 150 | 150.00 | Early version aboard Moskva, Kresta III [sic], and Kara. |
| GSN | 83 | | | | | | Low-to-medium altitude system. |
| JFS+ | 83 | | 300 | | | 91.44 | |

## MAXIMUM ALTITUDE

| Source | | mi | ft | km | m | m conv | Notes |
|---|---|---|---|---|---|---|---|
| AFM | 82 | | 82,000 | | | 24,993 | Earlier version. |
| CFW | 82 | | 80,000 | | | 24,384 | |
| JWS | 82 | | | | 25,000 | 25,000 | Early version aboard Moskva, Kresta III [sic], and Kara. |
| JFS+ | 83 | | 80,000 | | | 24,384 | |

## RANGE

| Source | | mi | nm | km | km conv | Notes |
|---|---|---|---|---|---|---|
| MOW | 76 | c. 37 | | c. 60 | 59.53 | |
| USND | 81 | | c. 30 | | 55.56 | |
| AFM | 82 | 18.6 | | | 29.93 | Earlier version. |
| | | 34 | | | 54.71 | Later version. |
| CFW | 82 | | | 30 | 30.00 | [Basic version], [30,000 m in source]. |
| | | | | 55 | 55.00 | Improved version on Kiev, [55,000 m]. |
| JFS | 82 | | 20-30 | | 37.04-55.56 | Max. |
| JWS | 82 | | | c. 30 | 30.00 | Early version aboard Moskva, Kresta III [sic], and Kara. |
| GSN | 83 | | c. 30 | c. 55.5 | 55.56 | |

## SPEED

| Source | | mph | kmph | Mach | kmph conv | Notes |
|---|---|---|---|---|---|---|
| JFS | 82 | | | 1+ | | |
| GSN | 83 | | | 2.5 | | Approximate max speed. |
| JFS+ | 83 | | | 3+ | | |
| JWS+ | 83 | | | > 3 | | Max. |
| CFW+ | 84 | | | 2.5 | | |

## SOVIET MISSILES

```
------------ WARHEAD TYPE ------------
             nuclear    conven    chem
AFM   82     yes        yes                [Final report of nuclear warhead]
CRW   82                yes
JFS   82                yes                HE.
JWS   82                yes                HE.
GSN   83                yes

------------ WARHEAD WEIGHT ------------
             lb              t         kg         kg conv
AFM   82     132                                  59.88
CRW   82                                60        60.00
JFS   82     90                                   40.82
JWS   82                                c. 40     40.00
GSN   83     176                        80        79.83
JWS+  83                                > 200     200.00     "Other estimates" [supplements 40
                                                             kg estimate, ed]
AFM+  84     88                                   39.92

------------ WEIGHT ------------
             lb              t         kg         kg conv
JFS   82     1,200                                544.32     Launch weight.
JWS   82                                c. 540    540.00
GSN   83     1,213                      550       550.22
AFM+  84     1,200                                544.32
CRW+  84                                550       550.00

------------ LENGTH ------------
             in         ft         cm         m         m conv
JFS   82                20                              6.10
JWS   82                                      c. 6     6.00
GSN   83                20'4"                 6.2      6.20
AFM+  84                19'8"                          5.99

------------ DIAMETER ------------
             in         ft         cm         m         cm conv
GSN   83     13.2                  33.5                 33.53      [335 mm in source]

------------ WINGSPAN ------------
             in         ft         cm         m         cm conv
GSN   83                5                     1.5       152.40

------------ NUMBER OF STAGES ------------
JFS   82     2
JWS   82     2

------------ TYPE OF PROPULSION ------------
JFS   82     solid
JWS   82     solid
GSN   83     ramjet
             solid                Booster.

------------ MIDCOURSE (OR MAIN) GUIDANCE ------------
CRW   82     "radar/command"
JFS   82     radio command
JWS   82     command

------------ TERMINAL GUIDANCE ------------
GSN   83     semiactive

------------ RADAR ------------
MOW   76     Head Light          Fire control.
FI    81     Headlight           G,H,I band radars [final report]
CRW   82     Head Lights         Head Lights-series radar director.
JWS   82     Head Light
GSN   83     Head Light          Fire control radar.

------------ IOC ------------
JFS   82     1967                Operational.
GSN   83     1967
             1967                Land-based SA-6 Gainful became operational.

------------ DEPLOYMENT BEGAN ------------
JWS   82     1967                Entered service.
CRW+  84     1967

------------ DEPLOYED ------------
MOW   76                         In service.
```

CRUISE MISSILES  Sea-Launched SAMs  SA-N-3  623

---------- CARRIERS ----------

|  |  | CARRIER | TUBES | RELOADS |  |
|---|---|---|---|---|---|
| JCS | 76 |  |  |  | Fitted to major surface combatants entering force since 1968, FY77-79. |
| SWM | 77 | helicopter carriers |  |  |  |
| US10 | 79 | cruisers |  | 22/lnchr |  |
|  |  |  |  |  | reloads: "22 missiles/launcher" magazine capability. |
| AWST | 82 | helicopter carriers |  |  |  |

|  |  | CARRIER | TUBES | RELOADS |  |
|---|---|---|---|---|---|
| JCS | 76 | Moskva |  |  | FY77,79. |
| MOW | 76 | Moskva |  |  |  |
| FI | 81 | Moskva | 2 x 2 | 180 | N=2. |
| USND | 81 | Moskva |  |  |  |
| AFM | 82 | Moskva |  |  |  |
|  |  | Leningrad |  |  |  |
| CFW | 82 | Moskva |  |  |  |
| JWS | 82 | Moskva |  |  | Moskva and Kresta II were first ships fitted. |
| GSN | 83 | Moskva |  |  |  |

|  |  | CARRIER | TUBES | RELOADS |  |
|---|---|---|---|---|---|
| JCS | 76 | Kiev |  |  | FY77-79. Will be on Kiev. |
| MOW | 76 | Kiev |  |  |  |
| JCS | 77 | Kiev |  |  | Utilized on Kiev. |
| FI | 81 | Kiev | 2 x 2 |  | N=2. |
| USND | 81 | Kiev |  |  |  |
| AFM | 82 | Kiev |  |  |  |
|  |  | Minsk |  |  |  |
| CFW | 82 | Kiev |  |  | Improved version of Goblet. |
| JWS | 82 | Kiev |  |  |  |
| GSN | 83 | Kiev |  |  |  |
| SMP | 83 | Minsk |  |  | [Launcher shown in picture on foredeck and identified in caption], '83, '84. |

|  |  | CARRIER | TUBES | RELOADS |  |
|---|---|---|---|---|---|
| JCS | 76 | Kresta II |  |  | FY77-79. |
| MOW | 76 | Kresta II |  |  |  |
| SWM | 77 | Kresta II |  |  |  |
| FI | 81 | Kresta II | 2 x 2 |  | N=10. |
| USND | 81 | Kresta II |  |  |  |
| AFM | 82 | Kresta II |  |  |  |
| CFW | 82 | Kresta-II |  |  |  |
| JWS | 82 | Kresta II |  |  | Moskva and Kresta II were first ships fitted. |
| GSN | 83 | Kresta II |  |  |  |

|  |  | CARRIER | TUBES | RELOADS |  |
|---|---|---|---|---|---|
| JCS | 76 | Kara |  |  | FY77-79. |
| NA03 | 76 | Kara | [3 x] |  |  |
|  |  |  |  |  | tubes: "3 x SA-N-3". |
| FI | 81 | Kara | 2 x 2 |  | N=5. |
| USND | 81 | Kara |  |  |  |
| AFM | 82 | Kara |  |  |  |
| CFW | 82 | Kara |  |  |  |
| JWS | 82 | Kara |  |  |  |
| GSN | 83 | Kara | x 2 |  |  |
|  |  |  |  |  | tubes: [1 twin launcher shows in picture] |

---------- USE AND CONFIGURATION ----------
| CFW | 82 | Twin launcher. |
| JWS | 82 | Launcher does not appear to be roll-stabilized, suggesting missile gathering is efficient. 4 reload hatches generally indicate launcher is dual-purpose and magazine contains both SA-N-3 and possibly SS-N-14. |

---------- TARGET TYPE ----------
| JCS | 77 | Also has an antisurface capability, FY78-79. |
| AFM | 82 | Has antiship capability. |
| CFW | 82 | Has an anti-surface target capability. |
| GSN | 83 | Reported to have effective surface-to-surface capability. |

---------- **RELATION TO OTHER MISSILES** ----------

| | | |
|---|---|---|
| MOW | 76 | Has been suggested that Goblet related to Gainful. |
| SWM | 77 | Naval version of Army SA-6. |
| FI | 81 | A drawing of a 6 m long, 60 cm diameter weapon appeared; now thought to be configuration of SS-N-14 [previously thought to be SA-N-3?, ed]. Most sources think SA-6 missile is used in the Goblet system. |
| JWS | 82 | Similar size to Goa, but different launcher and radar. |
| FI+ | 83 | Was thought for a long time to be shipboard version of SA-6. Many sources now think missile is 6.0 m weapon originally believed to be SS-N-14 round. If so, it is only naval SAM with custom-designed missile. |
| GSN | 83 | Derived from land-based SA-6 Gainful. Improved over SA-N-1, which it succeeds. |

# SA-N-4

```
Sources are:  MOW   76     SWM   77     FI    81     AWST  82     JWS   82     GSN   83     SMP   85
              ID06  75     NA03  76     AW03  79     USND  81     CFW   82     Clns+ 83     JFS+  83
              JCS   76     JCS   77     US14  80     AFM   82     JFS   82     FI+   83     CFW+  84
```

|       |    | US DESIGNATION |          |
|-------|----|----------------|----------|
| JCS   | 76 | SA-N-4         | FY77-79. |
| MOW   | 76 | SA-N-4         |          |
| SWM   | 77 | SA-N-4         |          |
| FI    | 81 | SA-N-4         |          |
| USND  | 81 | SA-N-4         |          |
| AFM   | 82 | SA-N-4         |          |
| AWST  | 82 | SA-N-4         |          |
| CFW   | 82 | SA-N-4         |          |
| JFS   | 82 | SA-N-4         |          |
| JWS   | 82 | SA-N-4         |          |
| GSN   | 83 | SA-N-4         |          |
| SMP   | 85 | SA-N-4         |          |

|     |    | NATO CODENAME |
|-----|----|---------------|
| CFW | 82 | Gecko         |

|     |    | EDITOR'S NOTES |
|-----|----|----------------|
| JCS | 76 | Listed FY77-79. |
| SMP | 85 | Listed '85. |

|     |    | MINIMUM ALTITUDE |    |    |   |        |
|-----|----|------|----|----|---|--------|
|     |    | mi   | ft | km | m | m conv |
| CFW  | 82 |      | 30 |    |   | 9.14   |
| JFS+ | 83 |      | 30 |    |   | 9.14   |

|     |    | MAXIMUM ALTITUDE |    |    |   |        |
|-----|----|------|--------|----|---|--------|
|     |    | mi   | ft     | km | m | m conv |
| CFW  | 82 |      | 10,000 |    |   | 3,048  |
| JFS+ | 83 |      | 10,000 |    |   | 3,048  |

|      |    | RANGE |      |      |         |                         |
|------|----|-------|------|------|---------|-------------------------|
|      |    | mi    | nm   | km   | km conv |                         |
| USND | 81 |       | c. 8 |      | 14.82   |                         |
| CFW  | 82 |       |      | 9    | 9.00    | [9,000 m in source]     |
| JFS  | 82 |       | 8    |      | 14.82   | Max.                    |
| GSN  | 83 |       | c. 8 | 14.8 | 14.82   | [Literally 14.8 m, typo]|
| JFS+ | 83 |       | 6    |      | 11.11   | Max.                    |

|     |    | SPEED |      |      |          |
|-----|----|-------|------|------|----------|
|     |    | mph   | kmph | Mach | kmph conv |
| JFS | 82 |       |      | ? 2  |          |

|     |    | WARHEAD TYPE |        |      |              |
|-----|----|---------|--------|------|--------------|
|     |    | nuclear | conven | chem |              |
| CFW | 82 |         | yes    |      |              |
| JFS | 82 |         | yes    |      | HE.          |
| JWS | 82 |         | yes    |      | Presumed HE. |
| GSN | 83 |         | yes    |      |              |

|      |    | WARHEAD WEIGHT |   |    |         |
|------|----|-----|---|----|---------|
|      |    | lb  | t | kg | kg conv |
| GSN  | 83 | 110 |   | 50 | 49.90   |
| JFS+ | 83 | 27  |   |    | 12.25   |

|      |    | WEIGHT |   |        |         |
|------|----|--------|---|--------|---------|
|      |    | lb     | t | kg     | kg conv |
| GSN  | 83 | c. 420 |   | c. 190 | 190.51  |
| CFW+ | 84 |        |   | 190    | 190.00  |

|     |    | LENGTH |       |     |     |        |
|-----|----|----|-------|----|-----|--------|
|     |    | in | ft    | cm | m   | m conv |
| JFS | 82 |    | 10.5  |    |     | 3.20   |
| GSN | 83 |    | 10'6" |    | 3.2 | 3.20   |

|     |    | DIAMETER |    |    |   |         |                    |
|-----|----|------|----|----|---|---------|--------------------|
|     |    | in   | ft | cm | m | cm conv |                    |
| GSN | 83 | 8.25 |    | 21 |   | 20.96   | [210 mm in source] |

```
------------ WINGSPAN ------------
                    in      ft        cm        m      cm conv
GSN    83                   2'1.2"              0.64    64.01

------------ TYPE OF PROPULSION ------------
JWS    82    solid
GSN    83    solid
JFS+   83    solid                     Booster and sustainer.

------------ MIDCOURSE (OR MAIN) GUIDANCE ------------
CFW    82    "radar/command"
JFS    82    radio command

------------ TERMINAL GUIDANCE ------------
GSN    83    semiactive

------------ RADAR ------------
SWM    77    Pop Group
FI     81    Pop Group       Almost identical to radar for Gecko, but only single command-link antenna.
CFW    82    Pop Group       Radar director.
JWS    82    Pop Group       At first, Soviets took much effort to hide radar from West. "Apparent
                             similarities of some consequence" to SA-8 radar.
GSN    83    Pop Group       Missile control radar, similar to SA-8 Grechko [sic] radar.

------------ IOC ------------
JCS    76    1971            Became operational.
JFS    82    after 1970      Operational.
GSN    83    early 1970s

------------ DEPLOYMENT BEGAN ------------
CFW+   84    1969

------------ DEPLOYED ------------
MOW    76                    In service.
FI     81                    Deployed.

------------ NUMBERS IN SERVICE ------------

------------ NUMBER IN 1976 ------------
MOW    76            At least 31 installations operational on 5 ship classes.

------------ CARRIERS ------------

------------ CARRIER ------------ TUBES --- RELOADS ------------
JCS    76                                      Placed on virtually all new construction combatants over
                                               900 tons built since 1971.
MOW    76                                      At least 31 installations operational on 5 ship classes.
SWM    77                                      Others [besides Krivak]
AFM    82    8 ship classes
JWS    82                                      Mostly post-1970 Soviet ships.
CFW+   84                            20
                                               reloads: Cylindrical magazine for twin launcher holds 20
                                                        missiles.

------------ CARRIER ------------ TUBES --- RELOADS ------------
NAO3   76    Kiev              3-4 x
FI     81    Kiev                              N=2.
USND   81    Kiev
AW ST  82    Kiev
GSN    83    Kiev

------------ CARRIER ------------ TUBES --- RELOADS ------------
NAO3   76    Kara              4 x
JCS    77    Kara                              FY78-79.
AWO3   79    Kara              2 x 2
FI     81    Kara                              N=5.
USND   81    Kara
AW ST  82    Kara
CFW    82    Kara
CLns+  83    Kara
GSN    83    Kara
```

CRUISE MISSILES  Sea-Launched SAMs  SA-N-4

|  |  | CARRIER | TUBES | RELOADS |  |
|---|---|---|---|---|---|
| US14 | 80 | Kirov | 2 x |  |  |
|  |  |  |  |  | tubes: Separate FCS for each launcher. |
| FI | 81 | Kirov |  |  |  |
| USND | 81 | Kirov |  |  |  |
| CFW | 82 | Kirov |  |  |  |
| CInst | 83 | Kirov | [2 x 2] |  |  |
|  |  |  |  |  | tubes: "2 SA-N-4 (Twin)". |
| GSN | 83 | Kirov |  |  |  |

|  |  | CARRIER | TUBES | RELOADS |  |
|---|---|---|---|---|---|
| FI | 81 | Sverdlov |  |  | N=2, converted to command ships. |
| USND | 81 | Sverdlov |  |  |  |
| CFW | 82 | Sverdlov |  |  | N=2. |
| GSN | 83 | Sverdlov |  |  | Command ships. |

|  |  | CARRIER | TUBES | RELOADS |  |
|---|---|---|---|---|---|
| FI | 81 | Koni |  |  |  |
| USND | 81 | Koni |  |  |  |
| GSN | 83 | Koni |  |  |  |

|  |  | CARRIER | TUBES | RELOADS |  |
|---|---|---|---|---|---|
| FI | 81 | Grisha I |  |  |  |
| USND | 81 | Grisha I/III |  |  |  |
| CFW | 82 | Grisha |  |  |  |
| GSN | 83 | Grisha |  |  |  |

|  |  | CARRIER | TUBES | RELOADS |  |
|---|---|---|---|---|---|
| ID06 | 75 | Krivak |  |  | Since 1971. |
| NA03 | 76 | Krivak | 2 x 2 |  |  |
| SWM | 77 | Krivak |  |  |  |
| AW03 | 79 | Krivak 2 | 2 x 2 |  |  |
| FI | 81 | Krivak |  |  | N>=5. |
| USND | 81 | Krivak I/II |  |  |  |
| AWST | 82 | Krivac |  |  |  |
| CFW | 82 | Krivak |  |  |  |
| JWS | 82 | Krivak |  |  | First fitted on Nanuchka and Krivak. |
| GSN | 83 | Krivak |  |  |  |

|  |  | CARRIER | TUBES | RELOADS |  |
|---|---|---|---|---|---|
| ID06 | 75 | Nanuchka |  |  | Since 1970. |
| NA03 | 76 | Nanuchka | 1 x 2 |  |  |
| FI | 81 | Nanuchka |  |  |  |
| USND | 81 | Nanuchka |  |  |  |
| CFW | 82 | Nanuchka |  |  |  |
| JWS | 82 | Nanuchka |  |  | First fitted on Nanuchka and Krivak. |
| GSN | 83 | Nanuchka |  |  |  |

|  |  | CARRIER | TUBES | RELOADS |  |
|---|---|---|---|---|---|
| USND | 81 | Sarancha |  |  |  |
| CFW | 82 | Sarancha |  |  |  |
| GSN | 83 | Sarancha |  |  |  |

|  |  | CARRIER | TUBES | RELOADS |  |
|---|---|---|---|---|---|
| FI | 81 | Ivan Rogov |  |  |  |
| USND | 81 | Ivan Rogov |  |  |  |
| CFW | 82 | Ivan Rogov |  |  |  |
| GSN | 83 | Ivan Rogov |  |  |  |

|  |  | CARRIER | TUBES | RELOADS |  |
|---|---|---|---|---|---|
| FI | 81 | Berezina |  |  |  |
| USND | 81 | Berezina |  |  |  |
| CFW | 82 | Berezina |  |  |  |
| GSN | 83 | Berezina |  |  |  |

|  |  | CARRIER | TUBES | RELOADS |  |
|---|---|---|---|---|---|
| FI | 81 | Rapuchka |  |  |  |

|  |  | CARRIER | TUBES | RELOADS |  |
|---|---|---|---|---|---|
| SMP | 85 | Slava |  | 40 | First of class "entered inventory" in 1982. |
|  |  |  |  |  | reloads: Total. |

------------ **USE AND CONFIGURATION** ------------
| | | |
|---|---|---|
| MOW | 76 | Retractable twin-round launcher is housed in a bin on deck. |
| SWM | 77 | Fast-reacting missile. Retractable twin-round pop-up launcher in weather-protective bin. |
| FI | 81 | Fired from twin launcher housed in circular bin when not in use. |
| AFM | 82 | Retractable twin 'pop-up' launcher is housed inside a bin on deck. |
| AWST | 82 | High acceleration. |
| CFW | 82 | Twin launcher, retracting into vertical drum. |
| JWS | 82 | Bin-type launcher has retracting mechanism so that 'pop-up' missile launching mode is possible. Twin launcher. |
| GSN | 83 | Has fully retractable, twin-arm launcher. |
| CFW+ | 84 | Launcher retracts into cylindrical magazine. |

------------ **TARGET TYPE** ------------
| | | |
|---|---|---|
| CFW | 82 | Can be used against surface targets. |
| JFS+ | 83 | Has surface capability. |

------------ **STRATEGY** ------------
| | | |
|---|---|---|
| JCS | 76 | "An integrated radar, fire control, and missile system for close-in point defense against aircraft and possibly cruise missiles", FY77-78. |
| FI | 81 | Short-range system. |
| AFM | 82 | Close-range system. |
| JWS | 82 | Close-in air defence role, some observers stress anti-helicopter aspect. |
| GSN | 83 | For point defense. |

------------ **EXPORTED TO** ------------
| | | | |
|---|---|---|---|
| FI | 81 | India | Deployed. |
| FI+ | 83 | Algeria | 2 Koni. |
| | | East Germany | 2 Koni. |
| | | India | 6 Godavari, 3 Nanuchka II. |
| | | Libya | Nanuchka II, ? N=4. |
| | | Jugoslavia | 1 Koni. |

------------ **RELATION TO OTHER MISSILES** ------------
| | | |
|---|---|---|
| MOW | 76 | Missiles may be same as for SA-8 system. |
| SWM | 77 | Possibly related to SA-8. |
| US14 | 80 | Generally similar to NATO Sea Sparrow. |
| FI | 81 | Uses same missile as SA-8 Gecko. |
| AFM | 82 | Missiles similar to those used with SA-8 system. |
| JWS | 82 | Compatibility of radar, mission, and launcher size suggest "broadly common system" with SA-8. |

------------ **OTHER INFORMATION** ------------
| | | |
|---|---|---|
| JFS | 82 | "Probably PDMS". |

# SA-N-5

```
Sources are:  USND  81      CFW   82      JWS   82      GSN   83      CFW+  84
FI     81     AFM   82      JFS   82      FI+   83      JFS+  83
```

## US DESIGNATION

|  |    |  |
|---|---|---|
| FI   | 81 | SA-N-5 |
| USND | 81 | SA-N-5 |
| AFM  | 82 | SA-N-5 |
| CFW  | 82 | SA-N-5 |
| JFS  | 82 | SA-N-5 |
| JWS  | 82 | SA-N-5 |
| GSN  | 83 | SA-N-5 |

## RANGE

| Source |    | mi | nm | km | km conv | |
|---|---|---|---|---|---|---|
| USND | 81 |   | c. 3 |       | 5.56  |      |
| JFS  | 82 |   | 5.6  |       | 10.37 | Max. |
| GSN  | 83 |   | 5.6  | 10.36 | 10.37 |      |

## SPEED

| Source |    | mph | kmph | Mach | kmph conv |
|---|---|---|---|---|---|
| JFS | 82 |   |   | ? 1.5 |   |

## WARHEAD TYPE

| Source |    | nuclear | conven | chem |     |
|---|---|---|---|---|---|
| JFS | 82 |   | yes |   | HE. |
| GSN | 83 |   | yes |   |     |

## WARHEAD WEIGHT

| Source |    | lb  | t | kg  | kg conv |
|---|---|---|---|---|---|
| GSN | 83 | 5.5 |   | 2.5 | 2.49 |

## WEIGHT

| Source |    | lb   | t | kg  | kg conv |               |
|---|---|---|---|---|---|---|
| JFS | 82 | 32   |   |     | 14.52 | Launch weight. |
| GSN | 83 | 20.3 |   | 9.2 | 9.21  |                |

## LENGTH

| Source |    | in | ft | cm | m | m conv |
|---|---|---|---|---|---|---|
| JFS | 82 |        | 4.75 |   |      | 1.45 |
| GSN | 83 | 2'5.25"|      |   | 0.76 | 0.74 |

## DIAMETER

| Source |    | in | ft | cm | m | cm conv | |
|---|---|---|---|---|---|---|---|
| GSN | 83 | 2.75 |   | 6.99 |   | 6.99 | [69.9 mm in source] |

## NUMBER OF STAGES

|  |    |   |
|---|---|---|
| JFS+ | 83 | 1 |

## TYPE OF PROPULSION

|  |    |        |
|---|---|---|
| JFS  | 82 | rocket |
| GSN  | 83 | solid  |
| JFS+ | 83 | solid  |

## MIDCOURSE (OR MAIN) GUIDANCE

|  |    |               |
|---|---|---|
| CFW | 82 | visually aimed |
| JFS | 82 | manual aiming  |

## TERMINAL GUIDANCE

|  |    |    |
|---|---|---|
| CFW | 82 | IR |
| JFS | 82 | IR |
| GSN | 83 | IR |

## IOC

|  |    |      |                                       |
|---|---|---|---|
| GSN | 83 | 1966 | [SA-7] operational in Soviet Ground Forces. |

## DEPLOYMENT BEGAN

|  |    |      |
|---|---|---|
| CFW+ | 84 | 1974 |

## DEPLOYED

|  |    |           |
|---|---|---|
| FI | 81 | Deployed. |

## CARRIERS

## CARRIER — TUBES — RELOADS

| Source |    | CARRIER | TUBES | RELOADS |
|---|---|---|---|---|
| AFM | 82 | many small ships | x 4 |   |

|       |    | CARRIER        | TUBES | RELOADS |                                                          |
|-------|----|----------------|-------|---------|----------------------------------------------------------|
| FI    | 81 | Osa I          |       |         |                                                          |
|       |    | Osa II         |       |         |                                                          |
| CFW   | 82 | some Osas      |       |         |                                                          |
| GSN   | 83 | Osa            |       |         |                                                          |

|       |    | CARRIER        | TUBES | RELOADS |   |
|-------|----|----------------|-------|---------|---|
| CFW   | 82 | Pauk           |       |         |   |
| GSN   | 83 | Pauk           |       |         |   |

|       |    | CARRIER        | TUBES | RELOADS |   |
|-------|----|----------------|-------|---------|---|
| CFW   | 82 | Tarantul       |       |         |   |
| GSN   | 83 | Tarantul       |       |         |   |

|       |    | CARRIER        | TUBES | RELOADS |                             |
|-------|----|----------------|-------|---------|-----------------------------|
| USND  | 81 | Polnocny       |       |         |                             |
| JWS   | 82 | Polnochniy     | x 4   |         | And other similar vessels.  |
| GSN   | 83 | Polnocny       |       |         |                             |

|       |    | CARRIER        | TUBES | RELOADS |   |
|-------|----|----------------|-------|---------|---|
| USND  | 81 | Ropucha        |       |         |   |
| GSN   | 83 | Ropocha        |       |         |   |

|       |    | CARRIER        | TUBES | RELOADS |   |
|-------|----|----------------|-------|---------|---|
| JWS   | 82 | Balzam         |       |         |   |

|       |    | CARRIER        | TUBES | RELOADS |                                                          |
|-------|----|----------------|-------|---------|----------------------------------------------------------|
| USND  | 81 | some AGIs      |       |         |                                                          |
| JWS   | 82 | AGIs           |       |         | About 50% of all Soviet AGIs, according to Jane's Ships. |
| GSN   | 83 | various AGIs   |       |         |                                                          |

|       |    | CARRIER        | TUBES | RELOADS |   |
|-------|----|----------------|-------|---------|---|
| CFW   | 82 | landing ships  |       |         |   |

|       |    | CARRIER           | TUBES | RELOADS |   |
|-------|----|-------------------|-------|---------|---|
| CFW   | 82 | some minesweepers |       |         |   |

|       |    | CARRIER          | TUBES | RELOADS |   |
|-------|----|------------------|-------|---------|---|
| CFW   | 82 | many auxiliaries |       |         |   |

---------- USE AND CONFIGURATION ----------
| FI  | 81 | Uses simple pivoted mount. |
| AFM | 82 | Simple air-defense system. Carries 4 SA-7 'Grail' launch tubes in "a framework that can be slewed for aiming". |
| CFW | 82 | Employs either 4-missile launch rack with operator, or shoulder-fired, singly. |
| JFS | 82 | In light and amphibious forces. |
| JWS | 82 | Consists of framework on which 4 Grail launcher tubes can be fixed side-by-side, and supported on central pedestal which permits gunner to manually slew and train missiles. |
| GSN | 83 | Fired from 4-missile launch rack or single, shoulder-held launch tube. |

---------- COMBAT REPORTS/EFFECTIVENESS ----------
| JWS | 82 | Performance likely to be substantially same as in land-based use, although task of manual acquisition and aiming "cannot have been made any easier for the operator by being mounted on a platform subject to sea motion as well as the ship's own course changes during any engagement". Nevertheless, deterrect to close investigations by hostile air forces, especially helicopters, is probably well worth cost. |

---------- EXPORTED TO ----------
| FI  | 81 | Syria         | Deployed. |
|     |    | others        |           |
| FI+ | 83 | Egypt         | Osa.      |
|     |    | East Germany  | 5 Parchim.|

---------- RELATION TO OTHER MISSILES ----------
| FI  | 81 | Navalised version of SA-7 Grail. |
| CFW | 82 | Naval version of SA-7 Grail. |
| JFS | 82 | Seaborne form of SA-7 Grail. |
| JWS | 82 | Naval version of SA-7 Grail. |
| GSN | 83 | Small, shipborne form of SA-7 Grail. |

# SA-N-6

```
Sources are:  FI     81    CFW   82    JFS  82    Clns+ 83    JFS+  83    ABFS  84    CFW+  85
              US14   80    AWST  82    JCS  82    JWS   82    GSN   83    JWS+  83    AFM   84    SMP   85
```

```
─────────── US DESIGNATION ───────────
FI     81    SA-NX-6
AWST   82    SA-NX-6
CFW    82    SA-N-6
JCS    82    naval SA-10
JFS    82    SA-N-6
JWS    82    SA-N-6
GSN    83    SA-N-6
AFM    84    SA-N-6
SMP    85    SA-N-6

─────────── EDITOR'S NOTES ───────────
SMP    85    Listed '85.
```

### MAXIMUM ALTITUDE

|         | mi | ft       | km | m        | m conv |
|---------|----|----------|----|----------|--------|
| JWS  82 |    |          |    | >_ 30,000 | 30,000 |
| AFM  84 |    | >_ 100,000 |    |          | 30,480 |
| CFW+ 84 |    | 90,000   |    |          | 27,432 |

### RANGE

|          | mi | nm  | km     | km conv |                           |
|----------|----|-----|--------|---------|---------------------------|
| CFW  82  |    |     | >_ 80  | 80.00   | [80,000 m in source]      |
| JCS  82  |    |     |        |         | Long-range, naval version.|
| JFS  82  |    | 31+ |        | 57.41   | Max.                      |
| JWS  82  |    |     | <_ 60  | 60.00   | On the low side.          |
| GSN  83  |    | 30+ | 55.5+  | 55.56   |                           |
| JFS+ 83  |    | 42  |        | 77.78   | Max.                      |
| AFM  84  | 37 |     |        | 59.53   | At Mach 6.                |
| SMP  85  |    |     |        |         | Long-range.               |

### SPEED

|          | mph | kmph | Mach | kmph conv |
|----------|-----|------|------|-----------|
| JFS  82  |     |      | 6    |           |
| JWS  82  |     |      | 6    |           |
| JFS+ 83  |     |      | 3    |           |
| JWS+ 83  |     |      | 3    |           |
| AFM  84  |     |      | 6    |           |

Max, revised downward to this by "at least one respected authority".

### WARHEAD TYPE

|          | nuclear | conven | chem |
|----------|---------|--------|------|
| GSN  83  |         | yes    |      |
| JFS+ 83  |         | yes    |      |
| ABFS 84  |         |        |      |

HE.
"It is known that there are nuclear-capable surface-to-air naval missiles either deployed or under development and near deployment. The SA-N-6 is thought to be adapted from the dual-capable SA-10 missile, though there is no indication that it itself is dual-capable. If the SA-N-6 is not nuclear-capable, then a nuclear-capable naval surface-to-air missile is near deployment stage".

### WARHEAD WEIGHT

|          | lb  | t | kg | kg conv |
|----------|-----|---|----|---------|
| JWS  82  |     |   | 90 | 90.00   |
| AFM  84  | 200 |   |    | 90.72   |

### LENGTH

|          | in | ft    | cm | m    | m conv |
|----------|----|-------|----|------|--------|
| JFS  82  |    | 23    |    |      | 7.01   |
| JWS  82  |    |       |    | c. 7 | 7.00   |
| GSN  83  |    | 23    |    | 7    | 7.01   |
| AFM  84  |    | c. 23 |    |      | 7.01   |

### NUMBER OF STAGES

JFS  82    1

### TYPE OF PROPULSION

JFS  82    rocket
GSN  83    solid

## MIDCOURSE (OR MAIN) GUIDANCE

| | | |
|---|---|---|
| JWS  | 82 | Expect mid-course guidance and terminal homing (as in Aegis). |
| JFS+ | 83 | Track-via-missile. |
| JWS+ | 83 | Track-via-missile. |
| AFM  | 84 | Likely. |

## TERMINAL GUIDANCE

| | | | |
|---|---|---|---|
| CFW | 82 | | "Reportedly uses track-via-missile guidance via the Top Dome radar system". |
| JFS | 82 | Homing | |
| JWS | 82 | | Expect mid-course guidance and terminal homing (as in Aegis). |
| GSN | 83 | | "Guidance provides for track-via-missile, with the missile in flight providing radar data to the launching ship". |
| AFM | 84 | | Likely. |

## RADAR

| | | | |
|---|---|---|---|
| US14 | 80 | | 2 radars. |
| FI   | 81 | Top Dome | 2 on Kirov. |
| CFW  | 82 | Top Dome | |
| JWS  | 82 | Top Dome | 2 on Kirov. [Source has much speculation on other Kirov radars: Top Pair, Top Steer, Round House, Rum Tub, ed]. |
| GSN  | 83 | Top Dome | |

## OBSERVED

| | | |
|---|---|---|
| JWS | 82 | Very little information available; no photos of missile available. |

## IOC

| | | | |
|---|---|---|---|
| JFS  | 82 | 1979      | Operational. |
| GSN  | 83 | 1977-1978 | |
| JWS+ | 83 | 1979      | Operational. |

## DEPLOYMENT BEGAN

| | | |
|---|---|---|
| CFW+ | 84 | 1981 |

## CARRIERS

### CARRIER ----- TUBES --- RELOADS

| | | |
|---|---|---|
| FI | 81 | Expected to arm a new class of guided missile cruiser. |

### CARRIER ----- TUBES --- RELOADS

| | | CARRIER | TUBES | RELOADS | |
|---|---|---|---|---|---|
| US14 | 80 | Kirov | 12 | | |
| FI   | 81 | Kirov | 12 | | |
| AWST | 82 | Kirov |    | | |
| CFW  | 82 | Kirov |    | | |
| JCS  | 82 | Kirov |    | | Carries a naval version of SA-10. |
| JFS  | 82 | Kirov |    | | |
| JWS  | 82 | Kirov | 12 | | |
| CLns+| 83 | Kirov | 12 | | |
| GSN  | 83 | Kirov | 12 | | |
| JWS+ | 83 | Kirov |    | | |
| AFM  | 84 | Kirov | 12 | | Vertical launch. |
| SMP  | 85 | Kirov |    | 96 | reloads: Total. "The Kirov is outfitted with an array of air defense weapons, including 96 long-range SA-N-6 missiles." |
|      |    | Frunze |   | 96 | [Second of Kirov class, ed.] reloads: Total. |

### CARRIER ----- TUBES --- RELOADS

| | | CARRIER | TUBES | RELOADS | |
|---|---|---|---|---|---|
| CFW  | 82 | BLK-COM-1 | | | |
| GSN  | 83 | Krasina   | | | |
| JFS+ | 83 | Krasina   | | | |
| JWS+ | 83 | Krasina   | | | |
| CFW+ | 84 | Slava     | | | |
| SMP  | 85 | Slava     | | 64 | First of class "entered inventory" in 1982. reloads: Total. |

### CARRIER ----- TUBES --- RELOADS

| | | CARRIER | |
|---|---|---|---|
| FI  | 81 | Kara-class Azov | Trials installation only. |
| GSN | 83 | Kara | 1 ship, the initial platform. |

## USE AND CONFIGURATION

| | | |
|---|---|---|
| FI   | 81 | Vertically launched weapon. |
| CFW  | 82 | "Vertical launch from 8-missile rotating magazines". |
| JWS  | 82 | Vertically launched. |
| GSN  | 83 | Vertically launched from below deck rotary magazine, 8 missiles per launcher. |
| JFS+ | 83 | Vertical launch from 8-missile magazines. |
| SMP  | 85 | Vertically-launched, [on Frunze, ed.] |

---------- **STRATEGY** ------------
| | | |
|---|---|---|
| CFW | 82 | Probably has an anti-ship capability. |
| JCS | 82 | SA-10 is first Soviet naval SAM to offer "a simultaneous, multiple target tracking and engagement capability". |
| JFS | 82 | Anti-missile capability. |
| JWS | 82 | Expected role: all-round hemispherical cover, to long ranges, multiple target detection and tracking, high resistance to ECM and jamming. |
| GSN | 83 | Anti-cruise capabilities. |
| JWS+ | 83 | Conjectured it has "anti-missile capability for defence against anti-ship missiles". |
| AFM | 84 | Assumed to deal with same multiple threats as US Aegis. Likely has multiple target detection and tracking features, and high resistance to ECM and jamming. |

---------- **RELATION TO OTHER MISSILES** ------------
| | | |
|---|---|---|
| FI | 81 | Thought to be derivative of SA-10. |
| CFW | 82 | Navalized version of SA-10. |
| JFS | 82 | Probably based on SA-10. |
| JWS | 82 | Thought to be naval version of SA-X-10. While not necessarily inspired by Aegis, reasonable guess that Soviet requirement was for system to meet very similar threat. |
| GSN | 83 | Appears to be adopted from SA-10. |
| JWS+ | 83 | Suggested it is naval version of SA-10. |

# SA-N-7

```
Sources are:  JWS  82    GSN  83    JWS  83    CFW+ 84
CFW 82        FI   83    JFS  83    AFM  84    SMP  85
```

---------- **US DESIGNATION** ----------

| | | | |
|---|---|---|---|
| CFW | 82 | SA-N-7 | |
| JWS | 82 | SA-NX-7 | [Analysis table] |
|     |    | SA-N-7  | [Text] |
| GSN | 83 | SA-N-7 | |
| JFS | 83 | SA-N-7 | |
| JWS | 83 | SA-N-7 | [Analysis table] |
| AFM | 84 | SA-N-7 | |
| SMP | 85 | SA-N-7 | |

---------- **DESCRIPTION** ----------

SMP 85   SAM

---------- **EDITOR'S NOTES** ----------

SMP 85   Listed '85.

---------- **MINIMUM ALTITUDE** ----------

| | | mi | ft | km | m | m conv |
|---|---|---|---|---|---|---|
| CFW | 82 | | 100 | | | 30.48 |

---------- **MAXIMUM ALTITUDE** ----------

| | | mi | ft | km | m | m conv | |
|---|---|---|---|---|---|---|---|
| CFW | 82 | | 46,000 | | | 14,020 | |
| JFS | 83 | | 45,000 | | | 13,716 | Altitude. |

---------- **RANGE** ----------

| | | mi | nm | km | km conv | |
|---|---|---|---|---|---|---|
| CFW | 82 | | | 28 | 28.00 | [28,000 m in source] |
|     |    | | | 3  | 3.00  | Minimum, [3,000 m in source]. |
| GSN | 83 | | c. 15 | c. 28 | 27.78 | [28 m in source, typo] |
| JFS | 83 | | ? 15  |       | 27.78 | Max. |
| SMP | 85 | | | | | Short-range. |

---------- **SPEED** ----------

| | | mph | kmph | Mach | kmph conv |
|---|---|---|---|---|---|
| CFW | 82 | | | 3 | |
| JFS | 83 | | | 3 | |

---------- **WARHEAD TYPE** ----------

| | | nuclear | conven | chem | |
|---|---|---|---|---|---|
| GSN | 83 | | yes | | |
| JFS | 83 | | yes | | HE. |

---------- **NUMBER OF STAGES** ----------

JFS 83   1

---------- **TYPE OF PROPULSION** ----------

GSN 83   solid

---------- **TERMINAL GUIDANCE** ----------

| | | | |
|---|---|---|---|
| JWS | 82 | SAR | Also likely missiles home by SAR, "operating virtually autonomously after launch". |
| FI  | 83 | SAR | |
| JFS | 83 | radar | |

---------- **RADAR** ----------

| | | | |
|---|---|---|---|
| CFW | 82 | Front Dome | Radar tracker/illuminators. |
| JWS | 82 | Front Dome | 6 on Sovremennyi; 8 on Provorny. |
| FI  | 83 | Top Dome   | 6 on Sovremenny, fire-control/target-illumination radars. |
| GSN | 83 | Front Dome | 6 on Sovremenny-class. |
| AFM | 84 |            | 6 fire control/target illuminating radars. |

---------- **DEVELOPMENT UNDERWAY** ----------

JWS 82   Development [stage].

---------- **IOC** ----------

GSN 83   1981

---------- **DEPLOYMENT BEGAN** ----------

CFW+ 84   1981

---------- **CARRIERS** ----------

|        | CARRIER      | TUBES    | RELOADS |                           |
|--------|--------------|----------|---------|---------------------------|
| CFW 82 | Sovremennyy  |          |         |                           |
| JWS 82 | Sovremennyi  | 2 x 1    |         |                           |
| FI  83 | Sovremenny   | 2 x 1    |         |                           |
| GSN 83 | Sovremennyy  | [2 lchrs]|         |                           |
| JFS 83 | Sovremennyy  |          |         |                           |
| AFM 84 | Sovremennyy  | 2 x 1    |         | Each ship of class.       |
| SMP 85 | Sovremennyy  |          | 44      | reloads: Total, estimated.|

|        | CARRIER    | TUBES | RELOADS |                              |
|--------|------------|-------|---------|------------------------------|
| CFW 82 | Provornyy  |       |         |                              |
| JWS 82 | Provorny   | 2 x 1 |         | A modified Kashin used for trials. |
| FI  83 | Provorny   |       |         | First tested.                |
| GSN 83 | Provornyy  | 1 x 1 |         | Of Kashin class, 1 unit [trials]. |
| JFS 83 | Provorny   |       |         |                              |

---------- USE AND CONFIGURATION ----------

CFW 82  Single-armed launchers.

JWS 82  Many radars and single rail launcher imply SA-N-7 designed to cope with multiple targets; implying rapid firing and reloading; also probable the overall system highly sophisticated.

FI  83  Top Dome radars assigned to individual engagements, taking command of rounds from rapid-fire launchers.

JFS 83  Single-arm launcher.

AFM 84  Sophistication and rapid-fire potential indicated by requirement for 6 radars.

---------- TARGET TYPE ----------

CFW 82  ship          Probably has anti-ship capability.

---------- RELATION TO OTHER MISSILES ----------

CFW 82  Navalized version of SA-11.
JWS 82  US observers have said SA-N-7 is naval equivalent to SA-11.
FI  83  May be naval adaptation of SA-11.
GSN 83  Shipboard version of SA-11.
AFM 84  Thought to be naval equivalent of SA-11.

---------- OTHER INFORMATION ----------

JWS 82  Little has been disclosed for publication.

# SA-N-8

Sources are:  CFW 82   GSN 83   JFS 83   JWS 83   AFM 84

### ────────── US DESIGNATION ──────────
| | | |
|---|---|---|
| CFW | 82 | SA-N-? |
| JFS | 83 | SA-N-8 |
| JWS | 83 | SA-N-8 |
| AFM | 84 | SA-N-8 |

### ────────── RANGE ──────────
|  |  | mi | nm | km | km conv |  |
|---|---|---|---|---|---|---|
| JFS | 83 |  |  |  |  | Short-range. |

### ────────── WARHEAD TYPE ──────────
|  |  | nuclear | conven | chem |  |
|---|---|---|---|---|---|
| JFS | 83 |  | yes |  | HE. |

### ────────── DIAMETER ──────────
|  |  | in | ft | cm | m | cm conv |  |
|---|---|---|---|---|---|---|---|
| JWS | 83 |  |  |  |  |  | 2 m for covers of forward silos on Udaloy. |

### ────────── RADAR ──────────
CFW 82   Not yet operational, due to lack of radar directors.

### ────────── OBSERVED ──────────
JWS 83   Missile has not been seen; virtually nothing known of it.

### ────────── DEVELOPMENT UNDERWAY ──────────
JWS 83   Development [stage, analysis table]

### ────────── DEPLOYMENT BEGAN ──────────
CFW 82   Not yet operational, due to lack of radar directors.

### ────────── CARRIERS ──────────

### ────────── CARRIER ────────── TUBES ─── RELOADS ──────────
| | | CARRIER | TUBES | RELOADS | |
|---|---|---|---|---|---|
| CFW | 82 | Udaloy | | | |
| GSN | 83 | Udaloy | [2 x 4] | | Will be fitted. |
| JFS | 83 | Udaloy | | | |
| JWS | 83 | Udaloy | x 4 | | Vertical launch silos. |
| | | | | | tubes: Forward launcher; for aft launchers not specified. |
| AFM | 84 | Udaloy | | | Vertically launched. |

### ────────── USE AND CONFIGURATION ──────────
CFW 82   New vertically launched, short-ranged system. To be carried in 2-m diameter launch cylinders aboard Udaloy.
GSN 83   Advanced SAM employing vertical launch. Cover plates are 2 m in diameter.
JFS 83   Vertically launched.

### ────────── STRATEGY ──────────
JWS 83   Missile characteristics are inferred from Udaloy's other armament, from likelihood of only 2 available radar stands, and vertical launch technique: Short to medium range performance, high speed missiles using sophisticated semi-active or active radar plus IR homing.

### ────────── RELATION TO OTHER MISSILES ──────────
CFW 82   Probably intended as successor to SA-N-4.
GSN 83   Probably successor to SA-N-4.

### ────────── OTHER INFORMATION ──────────
AFM 84   Nothing positive known of missile.

# SA-N-9

Sources are: SMP 85

---------- **US DESIGNATION** ----------
SMP 85   SA-NX-9

---------- **EDITOR'S NOTES** ----------
SMP 85   Listed '85.

---------- **RANGE** ----------
          mi          nm          km         <u>km conv</u>
SMP 85                                                       "shorter range" [than SA-N-6, ed.]

---------- **CARRIERS** ----------

---------- **CARRIER** ---------- **TUBES** ---- **RELOADS** ----------
SMP 85   Frunze                              128        "The Kirov is outfitted with an array of air defense weapons, including 96 long-range SA-N-6 missiles and, on the second unit, provisions for 128 SA-N-X-9 shorter range SAMs." "In 1984, Frunze, the second unit of the Kirov-Class, became operational."
reloads: Provisions for, total.

# AA-1

```
Sources are:  SecC  70     IDO8  76     SWM   77     Rock  78     Clns  80     AFM   82
SecC  69      FI*   74     MOW   76     AWST* 78     CH01  79     GU01  80     JWS   82
```

---------- **US DESIGNATION** ----------
| | |
|---|---|
| SWM 77 | AA-1 |
| AWST* 78 | AA-1 |
| Rock 78 | AA-1 |
| Clns 80 | AA-1 |
| AFM 82 | AA-1 |
| JWS 82 | AA-1 |

---------- **NATO CODENAME** ----------
| | |
|---|---|
| FI* 74 | Alkali |
| MOW 76 | Alkali |
| SWM 77 | Alkali |
| AWST* 78 | Alkali |
| Rock 78 | Alkali |
| Clns 80 | Alkali |
| AFM 82 | Alkali |
| JWS 82 | Alkali |

---------- **EDITOR'S NOTES** ----------
| | |
|---|---|
| SecC 69 | Listed FY70-71, but not by name, only as Su-9 missile. Conceivably could be the AA-3 instead that is intended. |
| Clns 80 | Listed separately with interceptor aircraft and fighter/attack aircraft. |

---------- **RANGE** ----------
| | mi | nm | km | km conv | |
|---|---|---|---|---|---|
| SecC 69 | | 3-6 | | 5.56-11.11 | |
| SecC 70 | | 2-4 | | 3.70-7.41 | |
| FI* 74 | | | 6-8 | 6.00-8.00 | |
| MOW 76 | c. 5 | | c. 8 | 8.05 | |
| SWM 77 | 3.7-5 | | 5.9-8 | 5.95-8.05 | |
| AWST* 78 | 3-4 | | | 4.83-6.44 | |
| Rock 78 | | | 8 | 8.00 | |
| CH01 79 | 3 3/4-5 | | 6-8 | 6.03-8.05 | |
| Clns 80 | 5 | | | 8.05 | [Interceptor listing] |
| | 3 | | | 4.83 | [Fighter listing] |
| AFM 82 | 3.7-5 | | | 5.95-8.05 | |
| JWS 82 | | | 6-8 | 6.00-8.00 | |

---------- **SPEED** ----------
| | mph | kmph | Mach | kmph conv | |
|---|---|---|---|---|---|
| FI* 74 | | | 1-2 | | |
| SWM 77 | | | 1-2 | | |
| Rock 78 | | | 2 | | |
| Clns 80 | | | 1-2 | | [Interceptor listing] |
| | | | 2.0 | | [Fighter listing] |

---------- **WARHEAD TYPE** ----------
| | nuclear | conven | chem | |
|---|---|---|---|---|
| Clns 80 | | yes | | HE. |
| GU01 80 | | | | HE. |

---------- **WARHEAD WEIGHT** ----------
| | lb | t | kg | kg conv |
|---|---|---|---|---|
| Clns 80 | 30 | | | 13.61 |

---------- **WEIGHT** ----------
| | lb | t | kg | kg conv |
|---|---|---|---|---|
| SWM 77 | 198 | | 90 | 89.81 |
| Rock 78 | | | 90 | 90.00 |
| AFM 82 | 200 | | | 90.72 |

---------- **LENGTH** ----------
| | in | ft | cm | m | m conv | |
|---|---|---|---|---|---|---|
| FI* 74 | | | c. 188 | | 1.88 | |
| MOW 76 | | c. 6'2" | | c. 1.88 | 1.88 | |
| SWM 77 | | 7.9 | | 2.4 | 2.41 | [Earlier versions] |
| | | 6.1 | | 1.86 | 1.86 | Latest version. |
| AWST* 78 | | 8 | | | 2.44 | |
| Rock 78 | | | | 1.9 | 1.90 | |
| GU01 80 | | 6'6" | | 2 | 1.98 | |
| AFM 82 | | 6'2" | | | 1.88 | |
| JWS 82 | | | 188 | | 1.88 | |

## CRUISE MISSILES AAMs AA-1

```
---------- DIAMETER ----------
              in       ft       cm       m       cm conv
FI*    74                       c. 18            18.00
MOW    76    c. 7                       c. 0.18  17.78
SWM    77    7                  18               17.78
AWST*  78    7                                   17.78
Rock   78                       18               18.00
CH01   79    7                  17.8             17.78
AFM    82    7                                   17.78
JWS    82                       18               18.00

---------- WINGSPAN ----------
              in       ft       cm       m       cm conv
FI*    74                       c. 58            58.00
MOW    76    c. 12.6                    c. 0.32  32.00
SWM    77    32                 58               81.28     [Earlier versions]
             12.6               22.8             32.00     Latest version.
Rock   78                       58               58.00
GU01   80             2         61               60.96
AFM    82             1'10 3/4"                  57.79
JWS    82                       58               58.00

---------- NUMBER OF STAGES ----------
AWST*  78    1
Rock   78    1

---------- TYPE OF PROPULSION ----------
FI*    74    solid
MOW    76    solid
SWM    77    solid
AWST*  78    solid
Rock   78    solid
AFM    82    solid
JWS    82    solid

---------- MIDCOURSE (OR MAIN) GUIDANCE ----------
SWM    77              Some reports suggest beam riding from launch aircraft fire control radar.

---------- TERMINAL GUIDANCE ----------
FI*    74    SAR
ID08   76    radio              Originally.
             IR                 Later.
MOW    76    radar
SWM    77    semi-active homing
Rock   78    radar
Clns   80    radar
             IR
AFM    82    SAR                I/J band.
JWS    82    radar              Probably.

---------- CONTROL SYSTEM ----------
FI*    74    Cruciform canard fins.
MOW    76    Control surfaces in wing trailing-edges.
SWM    77    Steerable foreplanes and wing-mounted surfaces.
Rock   78    Aerodynamic.

---------- RADAR ----------
GU01   80    Spin Scan          On Su-11 Fishpot-B.
             Scan Fix           On MiG-17PFU Frescoe E (Izumrud (Emerald)).
             Scan Odd           On MiG-19PM Farmer D.

---------- DEPLOYMENT BEGAN ----------
ID08   76              "Equipped as long ago as 1956".
Clns   80    1960      First deployed [interceptor listing]
             1959      First deployed [fighter listing]

---------- DEPLOYED ----------
MOW    76              In service.
Rock   78              [Not in service; "Armed MiG-17", other missiles described in present tense, ed].
AFM    82              Expected to disappear from service soon.

---------- CARRIERS ----------

---------- CARRIER ---------- TUBES --- RELOADS ----------
AFM    82    others                              [Than Su-9 and MiG-19]
```

```
----------  CARRIER  ----------  TUBES --- RELOADS ----------
FI*   74    MiG-17
SWM   77    all-weather MiG-17                              [Text:] MiG-17 and all-weather MiG-19.
AWST* 78    MiG-17
Rock  78    MiG-17
Clns  80    MIG-17 Fresco            4
GUO1  80    MiG-17PFU Frescoe-E 4

----------  CARRIER  ----------  TUBES --- RELOADS ----------
FI*   74    MiG-19
MDW   76    MiG-19                                          All-weather interceptor versions; first-generation
                                                            armament of this plane.
SWM   77    MiG-19                                          [Text:] MiG-17 and all-weather MiG-19.
AWST* 78    MiG-19
Clns  80    MIG-19 Farmer            4
GUO1  80    MiG-19PM Farmer-D        4
AFM   82    MiG-19

----------  CARRIER  ----------  TUBES --- RELOADS ----------
FI*   74    Su-9
MDW   76    Su-9                                            First-generation armament of this plane.
SWM   77    Su-9 Fishpot B
AWST* 78    Su-7 Fishpot B
Clns  80    Su-9 Fishpot             4
GUO1  80    Su-11 Fishpot B          4
AFM   82    Su-9

----------  USE AND CONFIGURATION ----------
SecC  69    Capable of tail attack only, FY70-71.

----------  EXPORTED TO ----------
FI*   74    WTO
            others

----------  OTHER INFORMATION ----------
AFM   82    First operational Soviet AAM.
```

# AA-2

```
Sources are:  MOW  76      Rock  78     Clns  80     FI   81     AWST 82    JWS  82    SMP  83
              ID08 76      SWM   77     CH01  79     GUD1 80     AFM  82    JAWA 82    FI+  83
```

## US DESIGNATION

|  | |  |
|---|---|---|
| SWM | 77 | AA-2 |
| Rock | 78 | AA-2 |
| Clns | 80 | AA-2 |
| FI | 81 | AA-2 |
| AFM | 82 | AA-2 |
| AWST | 82 | AA-2 |
| JAWA | 82 | AA-2 |
| JWS | 82 | AA-2 |

## NATO CODENAME

| Source | | Codename |
|---|---|---|
| MOW | 76 | Atoll |
| SWM | 77 | Atoll |
| Rock | 78 | Atoll |
| Clns | 80 | Atoll |
| FI | 81 | Atoll |
| AFM | 82 | Atoll |
| AWST | 82 | Atoll |
| JAWA | 82 | Atoll |
| JWS | 82 | Atoll |
| SMP | 83 | Atoll |

## SOVIET DESIGNATION

|  | |  | |
|---|---|---|---|
| SWM | 77 | K13A | |
| FI | 81 | K-13A | |
| AFM | 82 | K-13A | |
| JAWA | 82 | K-13A | |
| JWS | 82 | SB06 | And/or K13A. |

## EDITOR'S NOTES

|  | |  |
|---|---|---|
| SMP | 83 | Listed '83. |

## RANGE

| Source | | mi | nm | km | km conv | |
|---|---|---|---|---|---|---|
| MOW | 76 | c. 3-4 | | c. 5-6.5 | 4.83-6.44 | |
| SWM | 77 | 3-4 | | 4.8-6.4 | 4.83-6.44 | |
| Rock | 78 | 3-4 | | 6 | 6.00 | |
| Clns | 80 | 3 | | | 4.83 | |
| GUD1 | 80 | 4 | | 7 | 6.44 | Max. |
| FI | 81 | | | 5.7 | 5.70 | |
| AFM | 82 | 3-4 | | | 4.83-6.44 | |
| AWST | 82 | 3-4 | | | 4.83-6.44 | |
| JWS | 82 | | | 15 | 15.00 | |

## SPEED

| Source | | mph | kmph | Mach | kmph conv |
|---|---|---|---|---|---|
| CH01 | 79 | | | 2+ | |
| Clns | 80 | | | 2.5 | |
| FI | 81 | | | 2.5 | |
| AFM | 82 | | | 2.5 | |

## WARHEAD TYPE

| Source | | nuclear | conven | chem | |
|---|---|---|---|---|---|
| MOW | 76 | | yes | | HE. |
| Clns | 80 | | yes | | HE. |
| FI | 81 | | yes | | Fragmentation with smooth casing. |
| JWS | 82 | | yes | | HE. |

## WARHEAD WEIGHT

| Source | | lb | t | kg | kg conv |
|---|---|---|---|---|---|
| Clns | 80 | 25 | | | 11.34 |
| FI | 81 | | | 6 | 6.00 |

## WEIGHT

| Source | | lb | t | kg | kg conv |
|---|---|---|---|---|---|
| SWM | 77 | 154 | | 70 | 69.85 |
| Rock | 78 | | | 70 | 70.00 |
| GUD1 | 80 | 155 | | 70 | 70.31 |
| FI | 81 | | | 70 | 70.00 |
| AFM | 82 | 154 | | | 69.85 |

## LENGTH

|        |    | in | ft | cm | m | m conv |
|--------|----|----|----|----|---|--------|
| MOW    | 76 |    | 9'2" |  | 2.80 | 2.79 |
| SWM    | 77 |    | 9.2 |   | 2.8 | 2.80 |
| Rock   | 78 |    |    |    | 2.8 | 2.80 |
| FI     | 81 |    |    |    | 2.8 | 2.80 |
| AFM    | 82 |    | 9'2" |  |   | 2.79 |
| AWST   | 82 |    | 9.2 |   |   | 2.80 |
| JWS    | 82 |    |    | 280 |  | 2.80 |

## DIAMETER

|        |    | in | ft | cm | m | cm conv |
|--------|----|----|----|----|---|---------|
| MOW    | 76 | 4.72 |  |   | 0.12 | 11.99 |
| SWM    | 77 | 4.72 |  | 12 |  | 11.99 |
| Rock   | 78 |    |    | 12 |  | 12.00 |
| CH01   | 79 | 4 3/4 |  | 12 |  | 12.07 |
| GU01   | 80 | 4.7 |  | 12 |  | 11.94 |
| FI     | 81 |    |    | 12 |  | 12.00 |
| AFM    | 82 | 4.72 |  |   |  | 11.99 |
| AWST   | 82 | 4.75 |  |   |  | 12.07 |
| JWS    | 82 |    |    | 12 |  | 12.00 |

## WINGSPAN

|        |    | in | ft | cm | m | cm conv |  |
|--------|----|----|----|----|---|---------|---|
| MOW    | 76 |    | 1'8.75" |  | 0.53 | 52.71 |  |
| SWM    | 77 | 20.8 |  | 53 |  | 52.83 | Fins. |
|        |    | 17.7 |  | 45 |  | 44.96 | Wings. |
| Rock   | 78 |    |    | 53 |  | 53.00 |  |
| GU01   | 80 |    | 1'9" | 53 |  | 53.34 |  |
| FI     | 81 |    |    | 53 |  | 53.00 |  |
| AFM    | 82 |    | 1' 8 3/4" |  |  | 52.71 |  |
| JWS    | 82 |    |    | 53 |  | 53.00 | Tail. |
|        |    |    |    | 45 |  | 45.00 | Forward surfaces. |

## NUMBER OF STAGES

| Rock | 78 | 1 |
| AWST | 82 | 1 |

## TYPE OF PROPULSION

| MOW  | 76 | solid |
| SWM  | 77 | solid |
| Rock | 78 | solid |
| FI   | 81 | solid |
| AFM  | 82 | solid |
| AWST | 82 | solid |
| JAWA | 82 | solid |
| JWS  | 82 | solid |

## TERMINAL GUIDANCE

| MOW  | 76 | IR |
| SWM  | 77 | IR |
| Rock | 78 | IR |
| Clns | 80 | Radar |
|      |    | IR |
| GU01 | 80 | All aspect seeker may be under development. |
| FI   | 81 | IR |
| AFM  | 82 | IR |
| AWST | 82 | IR |
| JAWA | 82 | IR |
| JWS  | 82 | IR |

## CONTROL SYSTEM

| MOW | 76 | Cruciform foreplanes and gyroscopically-controlled tab at tip of trailing edge of tail-fins. Span of control surfaces: 1'5.75" (0.45m). |
| SWM | 77 | Foreplanes. |
| Rock | 78 | Aerodynamic. |
| FI | 81 | Cruciform canard fins. |
| JWS | 82 | Control surfaces. |

## DEPLOYMENT BEGAN

| ID08 | 76 | c. 1960 | Introduced. |
| Clns | 80 | 1960 | First deployed. |

## DEPLOYED

| MOW | 76 | In service. |

## DEPLOYMENT HISTORY

| JWS | 82 | Now thought obsolescent with Soviet forces; probably to be replaced by AA-8. |

## CRUISE MISSILES AAMs AA-2

```
---------- CARRIERS ----------

---------- CARRIER ---------- TUBES ---- RELOADS ----------
MOW    76    Yak-28P

---------- CARRIER ---------- TUBES ---- RELOADS ----------
SWM    77    MiG-17 Fresco
Rock   78    MiG-17

---------- CARRIER ---------- TUBES ---- RELOADS ----------
MOW    76    MiG-21 (old)          2
SWM    77    MiG-21 Fishbed
Rock   78    MiG-21
Clns   80    MIG-21 Fishbed        4
GUD1   80    MiG-21F Fishbed-C     2
             MiG-21                            Currently on export MiG-21s.
FI     81    MiG-21 Fishbed C      2           AA-2.
             MiG-21PF Fishbed D    2           AA-2.
AFM    82    MiG-21
AWST   82    MiG-21 Fishbed
JAWA   82    MiG-21
JWS    82    MiG-21 Fishbed        2 + 2
                                               tubes: Advanced Atoll + Atoll.

---------- CARRIER ---------- TUBES ---- RELOADS ----------
MOW    76    MiG-21 (new)          2 + 2
                                               tubes: Atoll + Advanced Atoll.
GUD1   80    MiG-21PFMA            4           Fishbed J.
FI+    83    Fishbed J/K/L

---------- CARRIER ---------- TUBES ---- RELOADS ----------
SWM    77    MiG-23
Rock   78    MiG-23
GUD1   80    MiG-23 Flogger B                  Currently on export models.
FI     81    Flogger E                         [Does not say whether it carries AA-2 or AA-2-2, ed]
AFM    82    export MiG-23
JAWA   82    MiG-23
JWS    82    MiG-23 Flogger E      4           [Picture shows Libyan jet with it, ed]

---------- CARRIER ---------- TUBES ---- RELOADS ----------
Clns   80    SU-7 Fitter           2
             SU-17 Fitter-C        2

---------- CARRIER ---------- TUBES ---- RELOADS ----------
GUD1   80    Su-22                             Peru's version.
FI     81    Su-22                             [Does not say whether it carries AA-2 or AA-2-2, ed]
AFM    82    export Su-22
JAWA   82    Su-22
FI+    83    Su-22 Fitter
```

---------- DESIGN AND ENGINEERING ----------
SMP    83    "Reflects near mirror-imaging of deployed Western systems and their technologies" [as do other Soviet weapon systems].

---------- USE AND CONFIGURATION ----------
GUD1   80    Present versions thought capable of pursuit course interception only.

---------- COMBAT REPORTS/EFFECTIVENESS ----------
SWM    77    Saw action in SE Asia, India-Pakistan, and 1973 wars. Problem of missile not locking on; deficiency assumed remedied.
FI     81    Has seen widespread use in Mideast, Indo-Pakistan wars, and SE Asia. "It has poor performance, even in the advanced version [AA-2-2], and the seeker does not always lock on the target" even if aircraft optimally positioned.

```
----------- EXPORTED TO -----------
MOW   76    India             Produced in India too.
SWM   77                      Widely exported.
FI    81    WTO
            Afghanistan
            Algeria
            Bangladesh
            China
            Cuba
            Iraq
            Jugoslavia
            Laos
            Libya
            Mozambique
            Nigeria
            North Korea
            Peru
            Somalia
            Syria
            Uganda
            Vietnam
            Yemen (PDRY)
            ? Albania
JWS   82    WTO countries
            Egypt
            India             Has Atoll production facilities under license.
            Afghanistan
            Algeria
            China             Also an indigenous Chinese model.
            Cuba
            Finland
            Iraq
            North Korea
            Syria
            Vietnam
FI+   83    Angola
            Finlend
            India
            Sudan
            North Yemen
            South Yemen

----------- RELATION TO OTHER MISSILES -----------
MOW   76    Similar to US Sidewinder.
SWM   77    Resembles US Sidewinder.
FI    81    Russian equivalent of Sidewinder.
AFM   82    Counterpart to US Sidewinder 1A (AIM-9B).
JAWA  82    Almost identical to Sidewinder (AIM-9B) in size and configuration, and has similar IR guidance.
JWS   82    Closely resembles US AIM-9B, IR Sidewinder.
```

# AA-2-2

```
Sources are:  SWM  77     GU01  80     AFM   82     JAWA  82     AFM+  84
MOW   76      Clns  80    FI    81     ID13  82     JWS   82
```

**————— US DESIGNATION —————**

|  | |  |
|---|---|---|
| SWM | 77 | AA-2-2 |
| FI | 81 | AA-2-2 |
| AFM | 82 | AA-2-2 |
| JAWA | 82 | AA-2-2 |

**————— NATO CODENAME —————**

| Source | | Codename |
|---|---|---|
| MOW | 76 | Advanced Atoll |
| SWM | 77 | Advanced Atoll |
| Clns | 80 | Advanced Atoll |
| FI | 81 | Advanced Atoll |
| AFM | 82 | Advanced Atoll |
| JAWA | 82 | Advanced Atoll |
| JWS | 82 | Advanced Atoll |

**————— RANGE —————**

| Source | | mi | nm | km | km conv |
|---|---|---|---|---|---|
| Clns | 80 | 5 | | | 8.05 |

**————— SPEED —————**

| Source | | mph | kmph | Mach | kmph conv |
|---|---|---|---|---|---|
| Clns | 80 | | | 2.5 | |

**————— WARHEAD TYPE —————**

| Source | | nuclear | conven | chem | |
|---|---|---|---|---|---|
| Clns | 80 | | yes | | HE. |

**————— WARHEAD WEIGHT —————**

| Source | | lb | t | kg | kg conv |
|---|---|---|---|---|---|
| Clns | 80 | 25 | | | 11.34 |

**————— LENGTH —————**

| Source | | in | ft | cm | m | m conv | |
|---|---|---|---|---|---|---|---|
| ID13 | 82 | | | | > 3 | 3.00 | |
| JAWA | 82 | | > 9'10" | | > 3.0 | 3.00 | |
| JWS | 82 | | | [c. 310] | | 3.10 | Advanced Atoll is est. 30 cm longer than Atoll. |
| AFM+ | 84 | | >_ 9'10" | | | 3.00 | |

**————— TERMINAL GUIDANCE —————**

| Source | | Guidance | Notes |
|---|---|---|---|
| MOW | 76 | radar | |
| SWM | 77 | SAR | |
| Clns | 80 | IR | |
| GU01 | 80 | ? IR and SAR | May have interchangeable IR and SAR, not confirmed. |
| FI | 81 | IR and SAR | Some sources suggest Advanced Atoll has both IR and SAR versions. An all-aspect IR seeker could be operational on Atoll by mid-1980s. |
| AFM | 82 | radar | Atoll with radar homing. |
| JAWA | 82 | radar | |
| JWS | 82 | ? SAR | Advanced Atoll may exist with SAR; nose presumed to house radar receiver and antenna package. US DIA refers to versions a-d, lending support to claims of SAR model. |

**————— DEPLOYMENT BEGAN —————**

| Source | | Year | |
|---|---|---|---|
| Clns | 80 | 1973 | First deployed. |

**————— CARRIERS —————**

| Source | | CARRIER | TUBES | RELOADS | |
|---|---|---|---|---|---|
| Clns | 80 | YAK-28 Firebar | 2 + 2 | | tubes: Atoll + Advanced Atoll. |
| FI | 81 | MiG-17 Frescoe E | 2 | | AA-2-2. |
| FI | 81 | MiG-21PFMA | 2 or 4 | | Fishbed F, AA-2-2. |

|  |  | CARRIER | TUBES | RELOADS |
|---|---|---|---|---|
| FI | 81 | MiG-21M Fishbed J | 4 | AA-2-2. |
|  |  | MiG-21SMT Fishbed K | 4 | AA-2-2. |
| AFM | 82 | MiG-21 | 2 + 2 | Fishbed J,K,L,N. tubes: Advanced Atoll + Atoll. |
| JAWA | 82 | Fishbed-J,K,L,N |  | Carry in mix with standard Atolls. |

|  |  | CARRIER | TUBES | RELOADS |
|---|---|---|---|---|
| Clns | 80 | SU-15 Flagon | 2 |  |

--------- COMBAT REPORTS/EFFECTIVENESS ------------
FI   81   Poor performance.

# AA-2

```
Sources are:   NA03  76      Rock  78      Clns  80      FI    81      AWST  82      JWS   82
MOW   76       SWM   77      CH01  79      GU01  80      AFM   82      JAWA  82      JWS+  83
```

```
------------ US DESIGNATION ------------
SWM   77    AA-3
Rock  78    AA-3
Clns  80    AA-3
FI    81    AA-3    Or AA-3-2 [Source implies AA-3-2 does not refer to IR vs SAR models, ed].
AFM   82    AA-3
AWST  82    AA-3
JAWA  82    AA-3
JWS   82    AA-3
```

```
------------ NATO CODENAME ------------
MOW   76    Anab
SWM   77    Anab
Rock  78    Anab
Clns  80    Anab
FI    81    Anab
AFM   82    Anab
AWST  82    Anab
JAWA  82    Anab
JWS   82    Anab
```

```
------------ RANGE ------------
            mi          nm          km          km conv
MOW   76    c. 6.2                  c. 10       9.98
NA03  76    10+                                 16.09
SWM   77    5-6                     8-9.7       8.05-9.65
Rock  78                            16          16.00
CH01  79    5-6                     8-10        8.05-9.65
Clns  80    15                                  24.14
FI    81                            16+         16.00
AFM   82    > 10                                16.09
AWST  82    12                                  19.31
JWS   82                            16+         16.00       UK MoD.
```

```
------------ SPEED ------------
            mph         kmph        Mach        kmph conv
Clns  80                            2.5
GU01  80                            c. 2                    Probably.
```

```
------------ WARHEAD TYPE ------------
            nuclear     conven      chem
MOW   76                yes                     HE.
Clns  80                yes                     HE.
FI    81                yes                     HE.
```

```
------------ WARHEAD WEIGHT ------------
            lb          t           kg          kg conv
Clns  80    80                                  36.29
```

```
------------ WEIGHT ------------
            lb          t           kg          kg conv
SWM   77    606                     275         274.88
Rock  78                            275         275.00
FI    81                            c. 275      275.00
```

```
------------ LENGTH ------------
            in          ft          cm          m           m conv
Rock  78                                        3.8         3.80
GU01  80                11'10"                  3.6-4.0     3.61
FI    81                                        3.6-4.0     3.60-4.00
AWST  82                12                                  3.66
```

```
------------ DIAMETER ------------
            in          ft          cm          m           cm conv
SWM   77    11                      28                      27.94
Rock  78                            30                      30.00
FI    81                            c. 28                   28.00
AFM   82    11                                              27.94
AWST  82    6                                               15.24
JWS   82                            28                      28.00       Both versions.
```

```
--------- WINGSPAN ---------
              in     ft      cm     m      cm conv
SWM   77      51             130           129.54
Rock  78                     135           135.00
FI    81                 c.  130           130.00
AFM   82           4'3"                    129.54
JWS   82                     130           130.00           Both versions.

--------- NUMBER OF STAGES ---------
Rock  78      1
AWST  82      1

--------- TYPE OF PROPULSION ---------
SWM   77      solid
Rock  78      solid
FI    81      solid
AWST  82      solid
JWS   82      solid

--------- TERMINAL GUIDANCE ---------
Rock  78      IR or radar
Clns  80      Radar, IR
AWST  82      IR

--------- CONTROL SYSTEM ---------
MOW   76      Movable foreplanes.
SWM   77      All-moving foreplanes.
Rock  78      Aerodynamic.
FI    81      Cruciform canard fins.

--------- RADAR ---------
CH01  79      Skip Spin
FI    81      Skip Spin          On Su-11 Fishpot C, Su-15 Flagon, and Yak-28.

--------- FIRST OBSERVED ---------
AFM   82      1961
JWS   82      1961          On Firebar.

--------- DEPLOYMENT BEGAN ---------
Clns  80      1961          First deployed.

--------- DEPLOYED ---------
MOW   76                    In service.

--------- DEPLOYMENT HISTORY ---------
GU01  80      Both AA-3 and AA-3-2 obsolescent.
JWS   82      Adopted as standard weapon by Soviet forces. Believed being withdrawn from service; could take
              considerable time.

--------- NUMBERS PRODUCED ---------
FI    81              Several thousand.
AFM   82      1000s   UK MoD.
JWS   82      1000s   Total production "certainly in the thousands", UK MoD.

--------- CARRIERS ---------

--------- CARRIER --------- TUBES --- RELOADS ---------
MOW   76      Yak-28                  1 + 1
SWM   77      Yak-28P Firebar
Clns  80      YAK-28 Firebar          2 + 2
                                                     tubes: Anab + Atoll.
FI    81      Yak-28                  2
                                                     tubes: IR + SAR (normally), either AA-3 or AA-3-2.
AFM   82      Yak-28P                 1 + 1          First observed on Yak-28P.
AWST  82      Firebar
JAWA  82      Yak-28P
JWS   82      Yak-28 Firebar                         First seen [picture labeled Yak-28P, ed]
JWS+  83      Yak-28 Firebar

--------- CARRIER --------- TUBES --- RELOADS ---------
SWM   77      MiG-23 Flogger
Rock  78      MiG-23

--------- CARRIER --------- TUBES --- RELOADS ---------
SWM   77      MiG-25 Foxbat
```

|  |  | CARRIER | TUBES | RELOADS |
|---|---|---|---|---|
| MOW | 76 | Su-11 | 1 + 1 | |
| SWM | 77 | Su-11 Fishpot C | | |
| Rock | 78 | Fishpot C | | |
|  |  | Su-11 | | |
| FI | 81 | Su-11 Fishpot C | 1 + 1 | |
| AFM | 82 | Su-11 | | |
| AWST | 82 | Fishpot C | | |
| JAWA | 82 | Su-11 | | |
| JWS | 82 | Su-9 Fishpot | | |
| JWS+ | 83 | Su-11 Fishpot | | |

tubes: IR + SAR (normally), either AA-3 or AA-3-2. Has become standard on Su-11 [final report of Su-11].

Later seen.

|  |  | CARRIER | TUBES | RELOADS |
|---|---|---|---|---|
| MOW | 76 | Su-15 | 1 + 1 | |
| SWM | 77 | Su-15 Flagon | | |
| Rock | 78 | Su-15 | | |
| Clns | 80 | SU-15 Flagon | 2 | |
| FI | 81 | Su-15 Flagon | 2 | |
| AFM | 82 | Su-15 | | |
| JAWA | 82 | Su-15 | | |
| JWS+ | 83 | Su-15 Flagon | | |

tubes: IR + SAR (normally), AA-3-2. Has become standard on Su-15.

---------- USE AND CONFIGURATION ----------
| MOW | 76 | Like US, USSR has learned by experience the value of carrying both IR and radar missiles. |
| SWM | 77 | Most aircraft carry a mix of IR and SAR. |
| JWS | 82 | In service some time; reasonable to assume has undergone periodic improvements and updates. |

---------- EXPORTED TO ----------
| JWS | 82 | client states |
|  |  | Bulgaria |
|  |  | Czech |
|  |  | East Germany |
|  |  | Hungary |
|  |  | Poland |
|  |  | Romania |

# AA-3IR

Sources are:   MOW 76    SWM 77    FI 81    AFM 82    JAWA 82    JWS 82

```
------- LENGTH -------
              in      ft        cm     m      m conv
MOW   76              13'5"            4.1    4.09
SWM   77              13.4             4.1    4.08
AFM   82              13'5"                   4.09
JWS   82                        360           3.60

------- TERMINAL GUIDANCE -------
MOW   76    IR
SWM   77    IR
FI    81    IR
AFM   82    IR
JAWA  82    IR
JWS   82    IR
```

# AA-3RADAR

Sources are:  MOW  76   SWM  77   FI  81   AFM  82   JAWA  82   JWS  82   JWS+  83

### LENGTH

|  |  | in | ft | cm | m | m_conv |
|---|---|---|---|---|---|---|
| MOW | 76 | | 13'1" | | 4.0 | 3.99 |
| SWM | 77 | | 13.1 | | 4.0 | 3.99 |
| AFM | 82 | | 13'1" | | | 3.99 |
| JWS | 82 | | | 360 | | 3.60 |

### TERMINAL GUIDANCE

| MOW | 76 | SAR | |
|---|---|---|---|
| SWM | 77 | SAR | |
| FI | 81 | SAR | I/J-band. |
| AFM | 82 | SAR | I/J-band. |
| JAWA | 82 | SAR | I/J-band. |
| JWS | 82 | radar | I-band. |

### RADAR

| JWS+ | 83 | Skip Scan | Has been employed with this radar. |
|---|---|---|---|

# AA-4

Sources are:  FI* 74    AWST* 75    JWS* 75    ID08 76

---------- **NATO CODENAME** ----------
| | |
|---|---|
| FI* 74 | Awl |
| AWST* 75 | Awl |
| JWS* 75 | Awl |

---------- **LENGTH** ----------

|  | in | ft | cm | m | m conv |
|---|---|---|---|---|---|
| FI* 74 | | | c. 500 | | 5.00 |
| JWS* 75 | | | | c. 5 | 5.00 |

---------- **TERMINAL GUIDANCE** ----------
JWS* 75    radar and IR    Reasonable to assume Soviets "have examined" both.

---------- **DEPLOYMENT BEGAN** ----------
FI* 74                Operated by USSR.
ID08 76   c. 1960     Introduced. Did not meet requirements, soon withdrawn.

---------- **CARRIERS** ----------

---------- **CARRIER** ---------- TUBES --- RELOADS ----------
| | | | |
|---|---|---|---|
| FI* 74 | MiG-23 | | |
| AWST* 75 | Flipper | | |
| JWS* 75 | MiG-23 Flipper | | Seen on this plane only. |

---------- **CARRIER** ---------- TUBES --- RELOADS ----------
FI* 74    Tu-28

---------- **RELATION TO OTHER MISSILES** ----------
JWS* 75   Similar in configuration to US Sparrow IIIB, but larger and heavier.

# AA-5

```
Sources are:  MOW  76      SWM   77    CH01  79    GU01  80    AFM   82    JAWA  82    AFM+  84
              ID08 76      NA03  76    Rock  78    Clns  80    FI    81    AWST  82
                                                               JWS   82
```

## US DESIGNATION

|  | |  |
|---|---|---|
| SWM | 77 | AA-5 |
| Rock | 78 | AA-5 |
| Clns | 80 | AA-5 |
| FI | 81 | AA-5 |
| AFM | 82 | AA-5 |
| AWST | 82 | AA-5 |
| JAWA | 82 | AA-5 |
| JWS | 82 | AA-5 |

## NATO CODENAME

| Source | | Codename |
|---|---|---|
| MOW | 76 | Ash |
| SWM | 77 | Ash |
| Rock | 78 | Ash |
| Clns | 80 | Ash |
| FI | 81 | Ash |
| AFM | 82 | Ash |
| AWST | 82 | Ash |
| JAWA | 82 | Ash |
| JWS | 82 | Ash |

## RANGE

| Source | | mi | nm | km | km conv | |
|---|---|---|---|---|---|---|
| NA03 | 76 | 18 | | | 28.96 | |
| SWM | 77 | 13 | | 22 | 20.92 | |
| Rock | 78 | | | 30 | 30.00 | |
| CH01 | 79 | 18.5 | | 30 | 29.77 | |
| Clns | 80 | 15 | | | 24.14 | |
| GU01 | 80 | c. 19 | | c. 30 | 30.57 | |
| FI | 81 | | | 30 | 30.00 | |
| AFM | 82 | 18.5 | | | 29.77 | |
| AWST | 82 | | 12 | | 22.22 | |
| JWS | 82 | | | c. 30 | 30.00 | UK MoD, April, 1979 [sic]. |

## SPEED

| Source | | mph | kmph | Mach | kmph conv | |
|---|---|---|---|---|---|---|
| Clns | 80 | | | 2.5 | | |
| GU01 | 80 | 1,980 | 3,168 | c. 3 | 3,185 | At 40,000 ft, probably. |

## WARHEAD TYPE

| Source | | nuclear | conven | chem | |
|---|---|---|---|---|---|
| MOW | 76 | | yes | | HE. |
| Clns | 80 | | yes | | HE. |
| FI | 81 | | yes | | HE. |

## WARHEAD WEIGHT

| Source | | lb | t | kg | kg conv |
|---|---|---|---|---|---|
| Clns | 80 | 150 | | | 68.04 |

## WEIGHT

| Source | | lb | t | kg | kg conv |
|---|---|---|---|---|---|
| SWM | 77 | 441 | | 200 | 200.04 |
| Rock | 78 | | | 200 | 200.00 |
| FI | 81 | | | c. 200 | 200.00 |

## LENGTH

| Source | | in | ft | cm | m | m conv |
|---|---|---|---|---|---|---|
| Rock | 78 | | | | 5.5 | 5.50 |
| AWST | 82 | | 18 | | | 5.49 |
| JWS | 82 | | | 530 | | 5.30 |

## DIAMETER

| Source | | in | ft | cm | m | cm conv |
|---|---|---|---|---|---|---|
| SWM | 77 | 11.8 | | 30 | | 29.97 |
| Rock | 78 | | | 30 | | 30.00 |
| CH01 | 79 | 11 | | 28.0 | | 27.94 |
| FI | 81 | | | c. 30 | | 30.00 |
| AWST | 82 | 8 | | | | 20.32 |
| JWS | 82 | | | 30 | | 30.00 |
| AFM+ | 84 | 12 | | | | 30.48 |

654  SOVIET MISSILES   IDDS

```
------------ WINGSPAN ------------
              in       ft      cm        m      cm_conv
SWM   77      51               130              129.54
Rock  78                       130              130.00
FI    81                    c. 130              130.00
JWS   82                       130              130.00
AFM+  84               4'3"                     129.54

------------ NUMBER OF STAGES ------------
Rock  78      1
AWST  82      1

------------ TYPE OF PROPULSION ------------
SWM   77      solid
Rock  78      solid
FI    81      solid
AWST  82      solid
JWS   82      solid

------------ TERMINAL GUIDANCE ------------
Rock  78      IR
Clns  80      radar
              IR

------------ CONTROL SYSTEM ------------
SWM   77      Movable tail fins.
Rock  78      Aerodynamic.
FI    81      Cruciform rear fins.

------------ RADAR ------------
FI    81      Big Nose            On Tu-28P.

------------ DEPLOYMENT BEGAN ------------
IDDS  76      c. 1960     Introduced.
Rock  78      1962        Service intro.
Clns  80      1965        First deployed.

------------ DEPLOYED ------------
MOW   76                  In service.
JWS   82                  In service.

------------ NUMBERS PRODUCED ------------
FI    81      > 1,000     Certainly exceeds 1,000.
AFM   82                  Several thousand produced.
JWS   82      1000s       "Some thousands produced", UK MoD, April, 1979 [sic].

------------ CARRIERS ------------

------------ CARRIER ------------ TUBES ---- RELOADS ------------
MOW   76      fighters                          Normally carried in a 'mix' (IR and radar) by all-weather
                                                fighters.

------------ CARRIER ------------ TUBES ---- RELOADS ------------
MOW   76      Tu-28P              2 + 2
SWM   77      Tu-28P Fiddler      2 + 2         Still in service in WTO countries.
Rock  78      Tu-28
Clns  80      TU-28 Fiddler       4
GUO1  80      Tu-28P Fiddler      2 + 2
                                                tubes: Originally carried only 2 missiles.
FI    81      Tu-28P              2 + 2
AFM   82      Tu-28P              2 + 2         Of Voyska PVO.
AWST  82      Fiddler             4
JAWA  82      Tu-28P Fiddler
JWS   82      Fiddler             2 + 2

------------ CARRIER ------------ TUBES ---- RELOADS ------------
SWM   77      MiG-25 Foxbat A     2             As of 1972 — armament later shifted to AA-6.
Rock  78      Foxbat                            Initial versions.
JAWA  82      ? MiG-25 Foxbat-A                 May be an alternate weapon [final report]

------------ STRATEGY ------------
SWM   77      High-altitude missile.
GUO1  80      Role of Tu-28 Fiddler was to patrol USSR borders where no SAMs were.
JWS   82      Size of missiles, parent aircraft, and Fiddler's large nose radome suggest intended for long-range
              interception.
```

---------- EXPORTED TO ----------
JWS 82  Bulgaria
        Czech
        East Germany
        Poland
        Romania

# AA-5IR

Sources are:  MOW 76   SWM 77   FI 81   AFM 82   JAWA 82   JWS 82   AFM+ 84

```
------------- LENGTH -------------
              in        ft          cm        m        m conv
MOW   76                18'0"                 5.5       5.49
SWM   77                18                    5.5       5.49
FI    81                            c. 5.5              5.50
AFM   82                18'0"                           5.49
AFM+  84                17'4 1/2"                       5.30

------------- TERMINAL GUIDANCE -------------
MOW   76   IR
FI    81   IR
AFM   82   IR
JAWA  82   IR
JWS   82   IR
```

# AA-5RADAR

```
Sources are:   MOW   76      SWM   77      FI   81      AFM   82      JAWA  82      JWS  82

----------- LENGTH -----------
               in         ft          cm         m        m conv
MOW   76                17'0"                    5.2       5.18
SWM   77                17                       5.2       5.18
FI    81                            c. 5.2                 5.20
AFM   82                17'0"                              5.18

----------- TERMINAL GUIDANCE -----------
MOW   76    radar              "Semi-active or active".
FI    81    SAR                I/J-band.
AFM   82    SAR                I/J-band.
JAWA  82    SAR                I/J-band.
JWS   82    radar              I-band.
```

# AA-6

```
Sources are:    SWM  77      Clns 80     AFM  82     JWS  82     AFM+ 84
ID08 76         Rock 78      GU01 80     AWST 82     FI+  83     SMP  85
MOW  76         CH01 79      FI   81     JAWA 82     JWS+ 83
```

---------- **US DESIGNATION** ----------

| | | |
|---|---|---|
| SWM  | 77 | AA-6 |
| Rock | 78 | AA-6 |
| Clns | 80 | AA-6 |
| FI   | 81 | AA-6 |
| AFM  | 82 | AA-6 |
| AWST | 82 | AA-6 |
| JAWA | 82 | AA-6 |
| JWS  | 82 | AA-6 |
| SMP  | 85 | AA-6 |

---------- **NATO CODENAME** ----------

| | | |
|---|---|---|
| MOW  | 76 | Acrid |
| SWM  | 77 | Acrid |
| Rock | 78 | Acrid |
| Clns | 80 | Acrid |
| FI   | 81 | Acrid |
| AFM  | 82 | Acrid |
| AWST | 82 | Acrid |
| JAWA | 82 | Acrid |
| JWS  | 82 | Acrid |

---------- **RANGE** ----------

|      |    | mi   | nm | km    | km conv |                  |
|------|----|------|----|-------|---------|------------------|
| Rock | 78 |      |    | 45    | 45.00   |                  |
| CH01 | 79 | 23   |    | 37    | 37.01   |                  |
| Clns | 80 | 15   |    |       | 24.14   |                  |
| AFM  | 82 | >_ 23|    |       | 37.01   |                  |
| AWST | 82 |      | 20 |       | 37.04   |                  |
| JWS  | 82 |      |    | c. 37 | 37.00   | [Analysis table] |

---------- **SPEED** ----------

|      |    | mph | kmph | Mach   | kmph conv |                                                      |
|------|----|-----|------|--------|-----------|------------------------------------------------------|
| SWM  | 77 |     |      | 2.2    |           | Must be well above M3.2 of launch aircraft [text]    |
| Rock | 78 |     |      | 2.2    |           |                                                      |
| Clns | 80 |     |      | 2.2    |           |                                                      |
| FI   | 81 |     |      | 4.5    |           | [Sic], speed M2.2 greater than speed of carrier a/c. |
| JWS  | 82 |     |      | c. 2.2 |           |                                                      |
| AFM+ | 84 |     |      | 2.2    |           | Cruising speed.                                      |

---------- **WARHEAD TYPE** ----------

|      |    | nuclear | conven | chem |                    |
|------|----|---------|--------|------|--------------------|
| MOW  | 76 |         | yes    |      | HE.                |
| Rock | 78 |         | yes    |      | HE.                |
| Clns | 80 |         | yes    |      | HE.                |
| FI   | 81 |         | yes    |      | Probably HE.       |
| JWS  | 82 |         | yes    |      | ? HE fragmentation.|

---------- **WARHEAD WEIGHT** ----------

|      |    | lb  | t | kg     | kg conv       |
|------|----|-----|---|--------|---------------|
| SWM  | 77 | 220 |   | 100    | 99.79         |
| Rock | 78 |     |   | 100    | 100.00        |
| Clns | 80 | 90  |   |        | 40.82         |
| FI   | 81 |     |   | 60-100 | 60.00-100.00  |
| JWS  | 82 |     |   | <_ 100 | 100.00        |
| AFM+ | 84 | 220 |   |        | 99.79         |

---------- **WEIGHT** ----------

|      |    | lb        | t | kg       | kg conv       |                                              |
|------|----|-----------|---|----------|---------------|----------------------------------------------|
| ID08 | 76 |           |   | 850      | 850.00        |                                              |
| SWM  | 77 | 1433-1874 |   | 650-850  | 650.01-850.05 |                                              |
| Rock | 78 |           |   | 800      | 800.00        |                                              |
| JWS  | 82 |           |   | c. 750   | 750.00        | Radar version somewhat heavier than IR model.|

CRUISE MISSILES  AAMs  AA-6   659

```
------------  LENGTH ------------
              in      ft      cm          m       m conv
MOW   76              c. 19           c. 5.80      5.79
Rock  78                              6.3          6.30
AWST  82              19                           5.79

------------  DIAMETER ------------
              in      ft      cm          m       cm conv
ID08  76                                 .39      39.00
SWM   77     15.7            40                   39.88
Rock  78                     40                   40.00
CH01  79     15 3/4          40.0                 40.01
GUD1  80           c. 1'2"  c. 36                 35.56
FI    81                     40                   40.00
AWST  82              1                           30.48
JWS   82                     30                   30.00

------------  WINGSPAN ------------
              in      ft      cm          m       cm conv
SWM   77     88.5            225                  224.79
Rock  78                     225                  225.00
CH01  79           7'4 3/5"              2.25     225.04
FI    81                     225                  225.00

------------  NUMBER OF STAGES ------------
Rock  78     1
AWST  82     1

------------  TYPE OF PROPULSION ------------
SWM   77     solid
Rock  78     solid
FI    81     solid
AWST  82     solid
JWS   82     solid

------------  MIDCOURSE (OR MAIN) GUIDANCE ------------
JWS   82              For max efficiency, "would require an inertial- or autopilot-controlled
                      mid-course phase", and would enable radar version to fly to target without
                      requiring target illumination by aircraft's radar, if active homing head is
                      used. Size and weight of AA-6 support this suggestion.

------------  TERMINAL GUIDANCE ------------
Rock  78     radar
Clns  80     radar
             IR
JWS   82     radar & IR      Separate versions.

------------  CONTROL SYSTEM ------------
SWM   77     Steerable foreplanes and wing-mounted control surfaces.
Rock  78     Aerodynamic.
FI    81     ?Ailerons + canard fins.
JWS   82     Wing and canard control surfaces.

------------  RADAR ------------
ID08  76                     Track radar on MiG-25 out to 100 km.
SWM   77     Fox Fire (for SAR)   Has 'look-down' capability.
FI    81     Fox Fire        On MiG-25, derived from Big Nose radar of Tu-28s.

------------  FIRST OBSERVED ------------
SWM   77     1975
AFM   82     1975
JWS   82     late 1976       First seen by West, on MiG-25, picture published in Western magazines.

------------  DEPLOYMENT BEGAN ------------
Rock  78     1975            Service intro.
Clns  80     1970            First deployed.

------------  DEPLOYED ------------
MOW   76                     In service.
FI    81                     Deployed USSR.

------------  CARRIERS ------------
```

## 660 SOVIET MISSILES  IDDS

|  |  | CARRIER | TUBES | RELOADS |  |
|---|---|---|---|---|---|
| MOW | 76 | MiG-25 Foxbat-A | 1 + 1 | | |
| SWM | 77 | MiG-25 Foxbat A | 2 + 2 | | |
| Rock | 78 | Foxbat | 2 | | |
| Clns | 80 | MIG-25 Foxbat | 4 | | |
| FI | 81 | MiG-25 Foxbat | 2 + 2 | | |
| AFM | 82 | MiG-25 Foxbat A | 2 + 2 | | |
| AW ST | 82 | Foxbat | 4 | | |
| JAW A | 82 | MiG-25 Foxbat-A | | | Carries both IR and SAR version. |
| JWS | 82 | MiG-25 Foxbat | 2 + 2 | | |

|  |  | CARRIER | TUBES | RELOADS |  |
|---|---|---|---|---|---|
| SMP | 85 | Foxbat E | | | Photo caption: "...FOXBAT E, seen with the AA-6 missile." |

|  |  | CARRIER | TUBES | RELOADS |
|---|---|---|---|---|
| GUO1 | 80 | Su-15 Flagon D, E | 2 | |
| FI | 81 | Su-15 Flagon D, E | | |
| JWS+ | 83 | Su-15 Flagon D, E | | |

|  |  | CARRIER | TUBES | RELOADS |  |
|---|---|---|---|---|---|
| ID08 | 76 | Su-19 | | | Could be used on Su-19 or Yak-28. |

|  |  | CARRIER | TUBES | RELOADS |  |
|---|---|---|---|---|---|
| ID08 | 76 | Yak-28 | | | Could be used on Su-19 or Yak-28. |

---------- **DESIGN AND ENGINEERING** ------------

GUO1  80  Probably built of titanium.

---------- **USE AND CONFIGURATION** -----------

ID08  76  As with previous generation [of AAMs], doctrine is to launch 1 IR, then 1 SAR less than 1 sec later.

FI  81  "It is likely that the weapons are ripple-fired in pairs, the IR round preceding the radar missile by about 1 sec".

JWS  82  Possible IR version launched very shortly after the initial radar-guided Acrid "to a similar point in space where it is anticipated that the IR seeker will detect and lock-on to its own target". Would be useful against multiple targets approaching together. For max efficiency, "would require an inertial- or autopilot-controlled mid-course phase", and would enable radar version to fly to target without requiring target illumination by aircraft's radar, if active homing head is used. Size and weight of AA-6 support this suggestion. Reports indicate Foxbat may have 'Markham' data link, relaying ground radar information to cockpit, allowing pilot to vector to assigned targets; significant development in Soviet practice over tight-ground control.

---------- **STRATEGY** -------------

SWM  77  "High-performance". Designed for long range and maneuverability at medium and high altitudes.

---------- **EXPORTED TO** -------------

FI  81  Libya
        ? Algeria
        ? Iraq

JWS  82  Libya       Possible, [picture also shows it on Libyan plane]
         Algeria     Possible.
         Iraq        Possible.

FI+  83  ? India
         ? Syria

---------- **RELATION TO OTHER MISSILES** -------------

SWM  77  Resembles enlarged Anab.
FI   81  Similar to AA-3 Anab but larger.
AFM  82  Configuration like Anab, but larger.
JWS  82  Resemblance to AA-3 is clear. Soviet approximation to American Phoenix; doubted it has Phoenix's multiple target capability.

---------- **OTHER INFORMATION** -------------

MOW  76  Largest Soviet AAM in service.
JWS  82  First of new family of Soviet AAMs.

# AA-6IR

```
Sources are:   MOW   76      CH01  79      FI    81      JWS   82
               ID08  76      SWM   77      GUD1  80      JAWA  82      AFM+  84
```

|  | RANGE | | | | |  |
|---|---|---|---|---|---|---|
|  | mi | nm |  | km | km conv |  |
| ID08 76 |  |  |  | 22.5 | 22.50 |  |
| SWM 77 | 12.4 |  |  | 20 | 19.95 |  |
| FI 81 |  |  |  | 20-25 | 20.00-25.00 |  |
| JWS 82 |  |  |  | c. 20 | 20.00 | [Text]; may be conservative. |

|  | WEIGHT | | | | |  |
|---|---|---|---|---|---|---|
|  | lb | t |  | kg | kg conv |  |
| CH01 79 | 1,433 |  |  | 550 | 650.01 |  |
| GUD1 80 | 1,545 |  |  | 700 | 700.81 |  |
| FI 81 |  |  |  | [650-750] | 650.00-750.00 | "700-800 650-750" kg [SAR and IR respectively?] |

|  | LENGTH | | | | |
|---|---|---|---|---|---|
|  | in | ft | cm | m | m conv |
| ID08 76 |  |  |  | 5.8 | 5.80 |
| SWM 77 |  | 19 |  | 5.8 | 5.79 |
| CH01 79 |  | 19 |  | 5.8 | 5.79 |
| GUD1 80 |  | 19'5" |  | 5.9 | 5.92 |
| FI 81 |  |  |  | 5.9 | 5.90 |
| JWS 82 |  |  |  | c. 5.8 | 5.80 |
| AFM+ 84 |  | 19'0" |  |  | 5.79 |

```
---------- TERMINAL GUIDANCE ----------
MOW   76    IR
SWM   77    IR
FI    81    IR       Both IR homing and SAR version exist.
JAWA  82    IR
```

# AA-6 RADAR

```
Sources are:  MOW   76      CH01  79      FI    81      JAWA  82      AFM+  84
              ID08  76      SWM   77      GU01  80      AFM   82      JWS   82
```

### RANGE

| Source |  | mi | nm | km | km conv | |
|---|---|---|---|---|---|---|
| ID08 | 76 | | | 50 | 50.00 | |
| SWM  | 77 | 10 | | 37 | 16.09 | |
| FI   | 81 | | | 45-50 | 45.00-50.00 | |
| JWS  | 82 | | | c. 40-50 | 40.00-50.00 | [Text]; may be conservative. |

### WEIGHT

| Source |  | lb | t | kg | kg conv | |
|---|---|---|---|---|---|---|
| CH01 | 79 | 1,874 | | 750 | 850.05 | |
| GU01 | 80 | 1,655 | | 750 | 750.71 | |
| FI   | 81 | | | [700-800] | 700.00-800.00 | "700-800 650-750" kg [SAR and IR respectively?] |

### LENGTH

| Source |  | in | ft | cm | m | m conv |
|---|---|---|---|---|---|---|
| ID08 | 76 | | | | 6.15 | 6.15 |
| SWM  | 77 | | 20 | | 6.09 | 6.10 |
| CH01 | 79 | | 20 | | 6.1 | 6.10 |
| GU01 | 80 | | 20'8" | | 6.3 | 6.30 |
| FI   | 81 | | | | 6.3 | 6.30 |
| AFM  | 82 | | 20'0" | | | 6.10 |
| JWS  | 82 | | | 629 | | 6.29 |
| AFM+ | 84 | | 20'7 1/2" | | | 6.29 |

### TERMINAL GUIDANCE

|  |  |  |  |
|---|---|---|---|
| MOW  | 76 | SAR | |
| SWM  | 77 | SAR | Aircraft wingtip fairings believed to house continuous wave target illuminating equipment. |
| FI   | 81 | SAR | Both IR homing and SAR version exist. |
| AFM  | 82 | | ? Aircraft wingtip fairings house continuous-wave illuminating equipment. |
| JAWA | 82 | SAR | Wingtip fairings thought to house continuous wave illuminating equipment. |

# AA-7

```
Sources are:  MOW   76      Rock  78      Clns  80      FI    81      AWST  82      JAWA  82      AFM+  84
              ID08  76      SWM   77      CH01  79      GU01  80      AFM   82      ID13  82      JWS   82      SMP   85
```

```
----------- US DESIGNATION -----------
MOW    76      AA-7
SWM    77      AA-7
Rock   78      AA-7
Clns   80      AA-7
FI     81      AA-7
AFM    82      AA-7
AWST   82      AA-7
JAWA   82      AA-7
JWS    82      AA-7      US DIA refers to AA-7a and AA-7b versions.
SMP    85      AA-7

----------- NATO CODENAME -----------
MOW    76      Apex
SWM    77      Apex
Rock   78      Apex
Clns   80      Apex
FI     81      Apex
AFM    82      Apex
AWST   82      Apex
JAWA   82      Apex
JWS    82      Apex

----------- EDITOR'S NOTES -----------
AFM    82      Dimensions, range, weight to be regarded as provisional.

----------- RANGE -----------
               mi            nm            km            km conv
ID08   76                                  21            21.00       For "missile homing" of MiG-23.
MOW    76      17.25                       27.75         27.76
SWM    77      17                          28            27.35
Rock   78                                  30            30.00
Clns   80      15                                        24.14
AFM    82      ? 17                                      27.35
AWST   82                    15                          27.78
                             12                          22.22       Look down, shoot down range.
AFM+   84      20                                        32.18

----------- SPEED -----------
               mph           kmph          Mach          kmph conv
ID08   76                                  2                         Above launch aircraft.
Rock   78                                  2
Clns   80                                  3.0
FI     81                                  3.5

----------- WARHEAD TYPE -----------
               nuclear       conven        chem
Rock   78                    yes                         HE.
Clns   80                    yes                         HE.
FI     81                    yes                         HE.

----------- WARHEAD WEIGHT -----------
               lb            t             kg            kg conv
Rock   78                                  6             6.00
CH01   79      <_ 88                       <_ 40         39.92
Clns   80      100                                       45.36
FI     81                                  40            40.00
JWS    82                                  c. 40         40.00       [Analysis table]
                                           <_ 40         40.00

----------- WEIGHT -----------
               lb            t             kg            kg conv
Rock   78                                  330           330.00
CH01   79      <_ 772                      <_ 350        350.18
FI     81                                  320           320.00
AFM    82      ? 705                                     319.79
JWS    82                                  c. 350        350.00      [Analysis table]
                                           300-350       300.00-350.00  [Text]
```

```
---------- LENGTH ----------
              in        ft           cm        m       m conv
SWM    77               14.1                   4.3     4.30
Rock   78                                      4.3     4.30
CH01   79               14'1 1/4"              4.3     4.30
AFM    82               14'1 1/4"                      4.30
ID13   82                                      4.6     4.60
AFM+   84               15'1 1/4"                      4.60

---------- DIAMETER ----------
              in        ft           cm        m       cm conv
SWM    77     9.45                   24                24.00
Rock   78                            24                24.00
CH01   79     9 2/5                  24.0              23.88
FI     81                            26                26.00
AFM    82     ? 9.4                                    23.88
ID13   82                            22                22.00     Main section.
                                     20                20.00     Nose section.
                                     24-26             24.00-26.00  Previously published estimates which
                                                                    are wrong.
JWS    82                            24                24.00
AFM+   84     8.75                                     22.23

---------- WINGSPAN ----------
              in        ft           cm        m       cm conv
SWM    77     41.3                   105               104.90
Rock   78                                      1.0     100.00   "1.0 cm" [source typo].
CH01   79               3'5 1/2"               1.05    105.41
FI     81                            140               140.00
AFM    82               3'5 1/2"                       105.41
ID13   82                            100               100.00   Wing.
                                     65                65.00    Tail.
                                     40                40.00    Foreplanes.
                                               1.4     140.00   Previously published estimate which is
                                                                wrong.
JWS    82                            105               105.00

---------- NUMBER OF STAGES ----------
Rock   78     1
AWST   82     1
ID13   82     ? 2

---------- TYPE OF PROPULSION ----------
MOW    76     solid
SWM    77     solid
Rock   78     solid
FI     81     solid
AFM    82     solid
AWST   82     solid
ID13   82     solid      Perhaps dual-thrust for boost and cruise.
JAWA   82     solid
JWS    82     solid

---------- TERMINAL GUIDANCE ----------
Rock   78     IR
Clns   80     Radar
              IR
FI     81     SAR
              IR
ID13   82     SAR        J-band. Exemplar of missile shown in Soviet Military Power is clearly
                         radar-guided.
              passive IR

---------- CONTROL SYSTEM ----------
Rock   78     Aerodynamic.
FI     81     Rear fins.
JWS    82     2 sets (rear and forward) of control surfaces.

---------- RADAR ----------
ID08   76     High Lark   On MiG-23, 50 km track range.
FI     81     High Lark   On Flogger B has limited look-down search and tracking capability--radar used
                          for SAR version only.
ID13   82     High Lark   Power at least 150 kW, c. 75 cm antenna diameter.

---------- FIRST OBSERVED ----------
JWS    82     1976        Became known to West.
```

```
---------- DEVELOPMENT BEGAN ----------
ID13   82   ? mid-1960s    "Assuming that development began in the mid-1960s...".

---------- PRODUCTION UNDERWAY ----------
MOW    76                  In production.

---------- DEPLOYMENT BEGAN ----------
Rock   78   1975           Service intro.
Clns   80   1974           First deployed.
AFM+   84                  Both IR and SAR versions deployed.

---------- DEPLOYED ----------
MOW    76                  In service.

---------- CARRIERS ----------

---------- CARRIER ---------- TUBES --- RELOADS ----------
JWS    82   MiG-21

---------- CARRIER ---------- TUBES --- RELOADS ----------
MOW    76   MiG-23                              Interceptors.
SWM    77   MiG-23 Flogger    1 + 1 + 2
                                              tubes: IR + SAR + AA-8s.
Rock   78   MiG-23            2
Clns   80   MIG-23 Flogger    2 + 2
                                              tubes: Apex + Aphid.
FI     81   MiG-23S Flogger   1 + 1
AFM    82   MiG-23                              One of 2 standard missiles for MiG-23.
AWST   82   MiG-23
JAWA   82   MiG-23
JWS    82   MiG-23 Flogger    1 + 1             Flogger B carries AA-8s too.
SMP    85   Flogger G                           Photo caption: "...FLOGGER G, seen at top with AA-7 and
                                                AA-8 air-to-air missiles under wing."

---------- CARRIER ---------- TUBES --- RELOADS ----------
FI     81   Foxbat A
AFM    82   MiG-25                              Alternate missile for MiG-25.
JAWA   82   ? MiG-25
JWS    82   MiG-25 Foxbat

---------- DESIGN AND ENGINEERING ----------
ID13   82   Weapon is large and bulky by modern standards; design team faced some kind of constraint; vacuum
            tube technology may have been used, or volume given to ECCM or even dual-mode guidance system.

---------- USE AND CONFIGURATION ----------
ID08   76   As with earlier generation [of AAMs], doctrine is to launch 1 IR, then 1 SAR less than 1 sec
            later.

---------- STRATEGY ----------
FI     81   Designed for use at low and medium altitudes.
JAWA   82   Long-range AAM.
JWS    82   2 sets of control surfaces indicate high maneuverability.

---------- COMBAT REPORTS/EFFECTIVENESS ----------
ID13   82   Similar in concept to US AIM-7 Sparrow, but lower in performance; Sparrow is 100 kg lighter and
            carries 40 kg warhead weight over much greater ranges.

---------- RELATION TO OTHER MISSILES ----------
ID08   76   Direct successor to AA-3 Anab.
ID13   82   Basic configuration like AA-5 Ash.
JWS    82   Comparable to US Sparrow. Thought successor to AA-3 Anab, resembles it somewhat; performance is
            clearly better.
```

# AA-7IR

```
Sources are:  SWM   77      GUO1  80      AFM   82      JWS   82
              IDO8  76      CHO1  79      FI    81      JAWA  82
```

## RANGE

|       |      | mi    | nm | km | km conv |
|-------|------|-------|----|----|---------|
| IDO8  | 76   |       |    | 15 | 15.00   |
| CHO1  | 79   | 9 1/3 |    | 15 | 15.02   |
| GUO1  | 80   | 10    |    | 16 | 16.09   |
| FI    | 81   |       |    | 15 | 15.00   |
| JWS   | 82   |       |    | 15 | 15.00   |

## LENGTH

|      |    | in | ft | cm     | m    | m conv |
|------|----|----|----|--------|------|--------|
| IDO8 | 76 |    |    |        | 4.22 | 4.22   |
| FI   | 81 |    |    |        | 4.2  | 4.20   |
| JWS  | 82 |    |    | [< 430]|      | 4.30   |

IR slightly shorter than radar version.

## TERMINAL GUIDANCE

| SWM  | 77 | IR   |
|------|----|------|
| AFM  | 82 | ? IR |
| JAWA | 82 | IR   |
| JWS  | 82 | IR   |

# AA-7RADAR

```
Sources are:  SWM   77      GUO1  80      AFM   82      JWS   82
ID08  76      CHO1  79      FI    81      JAWA  82
```

### RANGE

|        |    | mi     | nm | km    | km conv |
|--------|----|--------|----|-------|---------|
| ID08   | 76 |        |    | 32.5  | 32.50   |
| CHO1   | 79 | 18 2/3 |    | 30    | 30.03   |
| GUO1   | 80 | 20     |    | 32    | 32.18   |
| FI     | 81 |        |    | 33    | 33.00   |
| JWS    | 82 |        |    | 35    | 35.00   |

### LENGTH

|       |    | in | ft | cm  | m    | m conv |
|-------|----|----|----|-----|------|--------|
| ID08  | 76 |    |    |     | 4.50 | 4.50   |
| FI    | 81 |    |    |     | 4.5  | 4.50   |
| JWS   | 82 |    |    | 430 |      | 4.30   |

### TERMINAL GUIDANCE

| SWM  | 77 | SAR   |
| AFM  | 82 | radar |
| JAWA | 82 | SAR   |
| JWS  | 82 | radar |

# AA-8

| Sources are: | SWM | 77 | Clns | 80 | AFM | 82 | JAWA | 82 | FI+ | 83 | AFM+ | 84 |
|---|---|---|---|---|---|---|---|---|---|---|---|---|
| | ID08 | 76 | Rock | 78 | GU01 | 80 | AWST | 82 | JWS | 82 | JAWA+ | 83 | 85 |
| | MOW | 76 | CH01 | 79 | FI | 81 | ID13 | 82 | Clns | 83 | JWS+ | 83 | |

### ———— US DESIGNATION ————

| | | |
|---|---|---|
| MOW | 76 | AA-8 |
| SWM | 77 | AA-8 |
| Rock | 78 | AA-8 |
| Clns | 80 | AA-8 |
| FI | 81 | AA-8 |
| AFM | 82 | AA-8 |
| AWST | 82 | AA-8 |
| JAWA | 82 | AA-8 |
| JWS | 82 | AA-8 |
| Clns | 83 | AA-8 |
| | 85 | AA-8 |

### ———— NATO CODENAME ————

| | | |
|---|---|---|
| MOW | 76 | Aphid |
| SWM | 77 | Aphid |
| Rock | 78 | Aphid |
| Clns | 80 | Aphid |
| FI | 81 | Aphid |
| AFM | 82 | Aphid |
| AWST | 82 | Aphid |
| JAWA | 82 | Aphid |
| JWS | 82 | Aphid |
| Clns | 83 | Aphid |

### ———— RANGE ————

| | | mi | nm | km | km conv | |
|---|---|---|---|---|---|---|
| ID08 | 76 | | | 9 | 9.00 | For "missile homing" of MiG-21. |
| MOW | 76 | 3-5 | | 5-6.5 | 4.83-8.05 | |
| SWM | 77 | 3.1-5 | | 5-8 | 4.99-8.05 | |
| Rock | 78 | | | 8 | 8.00 | |
| CH01 | 79 | 3 1/2-5 | | 5.5-8 | 5.63-8.05 | |
| Clns | 80 | 3-4 | | | 4.83-6.44 | |
| GU01 | 80 | 4.4 | | 7 | 7.08 | |
| AFM | 82 | 3.5-5 | | | 5.63-8.05 | |
| AWST | 82 | 3-4 | | | 4.83-6.44 | |
| ID13 | 82 | | | 5-7 | 5.00-7.00 | Max. |
| | | | | 0.5 | 0.50 | [500 m in source], minimum range could be this or less. |
| JAWA | 82 | | | < 0.5 | 0.50 | [500 m (1,640 ft) in source], minimum range. Minimum range. |
| Clns | 83 | 4 | | | 6.44 | |
| AFM+ | 84 | 3-4.3 | | | 4.83-6.92 | Max; under 1,650 ft min. |

### ———— SPEED ————

| | | mph | kmph | Mach | kmph conv | |
|---|---|---|---|---|---|---|
| ID08 | 76 | | | 2 | | Above launch aircraft. |
| Rock | 78 | | | 2 | | |
| Clns | 80 | | | 3.0 | | |
| FI | 81 | | | 3 | | |
| Clns | 83 | | | 2.5 | | |

### ———— WARHEAD TYPE ————

| | | nuclear | conven | chem | |
|---|---|---|---|---|---|
| Rock | 78 | | yes | | HE. |
| CH01 | 79 | | | | HE. |
| Clns | 80 | | yes | | HE. |
| FI | 81 | | yes | | HE. |
| JWS | 82 | | yes | | HE. |
| Clns | 83 | | yes | | HE. |

### ———— WARHEAD WEIGHT ————

| | | lb | t | kg | kg conv |
|---|---|---|---|---|---|
| Rock | 78 | | | 6 | 6.00 |
| CH01 | 79 | 13 1/4 | | 6 | 6.01 |
| Clns | 80 | 20 | | | 9.07 |
| FI | 81 | | | 6 | 6.00 |
| JWS | 82 | | | 7-9 | 7.00-9.00 |
| Clns | 83 | 18 | | | 8.16 |
| AFM+ | 84 | 13.2 | | | 5.99 |

## CRUISE MISSILES AAMs AA-8

### WEIGHT

|         |    | lb  | t | kg    | kg conv |
|---------|----|-----|---|-------|---------|
| Rock    | 78 |     |   | 55    | 55.00   |
| CH01    | 79 | 121 |   | 55    | 54.89   |
| FI      | 81 |     |   | 55    | 55.00   |
| AFM     | 82 | 121 |   |       | 54.89   |
| ID13    | 82 |     |   |       |         |
| JWS     | 82 |     |   | c. 54 | 54.00   |

Probably the lightest AAM fielded since mid-1950s.

### LENGTH

|        |    | in | ft       | cm  | m   | m conv |
|--------|----|----|----------|-----|-----|--------|
| SWM    | 77 |    | 6.9      |     | 2.1 | 2.10   |
| Rock   | 78 |    |          |     | 2.1 | 2.10   |
| CH01   | 79 |    | 6'6 3/4" |     | 2.0 | 2.00   |
| AFM    | 82 |    | 6'6 3/4" |     |     | 2.00   |
| AFM+   | 84 |    | 7'2 1/2" |     |     | 2.20   |

### DIAMETER

|        |    | in       | ft | cm   | m | cm conv |
|--------|----|----------|----|------|---|---------|
| Rock   | 78 |          |    | 13   |   | 13.00   |
| CH01   | 79 | c. 5 1/10|    | 13.0 |   | 12.95   |
| FI     | 81 |          |    | 13   |   | 13.00   |
| AFM    | 82 | 5.12     |    |      |   | 13.00   |
| ID13   | 82 |          |    | 12   |   | 12.00   |
| JWS    | 82 |          |    | 13   |   | 13.00   |
| AFM+   | 84 | 4.75     |    |      |   | 12.07   |

### WINGSPAN

|        |    | in | ft       | cm | m | cm conv |           |
|--------|----|----|----------|----|---|---------|-----------|
| Rock   | 78 |    |          | 52 |   | 52.00   |           |
| FI     | 81 |    |          | 52 |   | 52.00   |           |
| ID13   | 82 |    |          | 21 |   | 21.00   | Foreplane.|
|        |    |    |          | 28 |   | 28.00   | Canard.   |
|        |    |    |          | 40 |   | 40.00   | Tail.     |
| AFM+   | 84 |    | 1'3 3/4" |    |   | 40.01   |           |

### NUMBER OF STAGES

| Rock | 78 | 1 |
| AWST | 82 | 1 |

### TYPE OF PROPULSION

| MOW  | 76 | solid |
| SWM  | 77 | solid |
| Rock | 78 | solid |
| FI   | 81 | solid |
| AFM  | 82 | solid |
| AWST | 82 | solid |
| ID13 | 82 | solid | Probably single-burn solid rocket, burn time 1.5-2.0 secs. |
| JAWA | 82 | solid |
| JWS  | 82 | solid |

### TERMINAL GUIDANCE

| Rock | 78 | IR    |          |
| CH01 | 79 | IR    |          |
|      |    | SAR   | Perhaps. |
| Clns | 80 | radar |          |
|      |    | IR    |          |
| AFM  | 82 | IR    |          |
| JAWA | 82 | IR    |          |
| Clns | 83 | IR    |          |

### CONTROL SYSTEM

| Rock | 78 | Aerodynamic.   |
| FI   | 81 | ? Canard fins. |

### PRODUCTION UNDERWAY

| MOW | 76 | In production. |

### DEPLOYMENT BEGAN

| ID08 | 76 | c. 1975 | Introduced.     |
| Rock | 78 | 1975    | Service intro.  |
| Clns | 80 | 1975    | First deployed. |
| Clns | 83 | 1975    | First deployed. |

### DEPLOYED

| MOW | 76 | In service. |

---------- **DEPLOYMENT HISTORY** ----------
JWS  82  Almost certainly AA-2 replacement.

---------- **CARRIERS** ----------

---------- **CARRIER** ------------ **TUBES** ---- **RELOADS** ----------
JWS  82                                2 + 2           If AA-2 replacement, will be able to operate with more
                                                       aircraft than just MiG-21 and MiG-23.
                                                       tubes: AA-8 + Advanced Atoll.

---------- **CARRIER** ------------ **TUBES** ---- **RELOADS** ----------
SWM   77  MiG Fishbed
AFM   82  MiG-21                                       Late models.
JAWA  82  MiG-21
JWS   82  MiG-21 Fishbed
FI+   83  MiG-21s                      2               Late-series.

---------- **CARRIER** ------------ **TUBES** ---- **RELOADS** ----------
MOW   76  MiG-23                                       Interceptors.
SWM   77  MiG-23S Flogger              2 + 2
                                                       tubes: AA-8 + AA-7.
Rock  78  MiG-23
Clns  80  MiG-23 Flogger               2 + 2
                                                       tubes: Aphid + Apex.
GUD1  80  MiG-23S Flogger B            2
FI    81  MiG-23                       2
AFM   82  MiG-23                                       One of 2 standard missiles for MiG-23.
AWST  82  MiG-23
JAWA  82  MiG-23
JWS   82  MiG-23 Flogger
FI+   83  MiG-23                       2 or 4
      85  Flogger G                                    Photo caption: "...FLOGGER G, seen at top with AA-7 and
                                                       AA-8 air-to-air missiles under wing."

---------- **CARRIER** ------------ **TUBES** ---- **RELOADS** ----------
GUD1  80  Yak-36 Forger                <_ 4
JAWA+ 83  Yak-36MP
JWS+  83  Yak-36 Flogger                               [Sic]
AFM+  84  Yak-36MP

---------- **CARRIER** ------------ **TUBES** ---- **RELOADS** ----------
JAWA+ 83  Su-15
AFM+  84  Su-15

---------- **USE AND CONFIGURATION** ----------
ID08  76  As with earlier missiles [of AAMs], doctrine is to launch 1 IR, then 1 SAR less than 1 sec apart.
AFM+  84  Highly manoeuvrable.

---------- **STRATEGY** ----------
ID08  76  "Pure short-range air combat missile, configured for extreme lateral accelerations and minimum
          radius turns".
SWM   77  Dog-fight weapon.
FI    81  Dogfight weapon.
ID13  82  Designed for use against low altitude targets, so designed to "see" target against terrain.
JAWA  82  Close-range weapon. Configuration should ensure high manoeuvrability.
JWS   82  Close combat missile.

---------- **RELATION TO OTHER MISSILES** ----------
SWM   77  Atoll development.
FI    81  "May be similar in performance to AIM-9L Sidewinder". "Thought to be derived from AA-2 Atoll".
JWS   82  Possibly derived from AA-2 Atoll.

---------- **OTHER INFORMATION** ----------
ID13  82  Dual foreplanes were ignored until US released Soviet Military Power.

# AA-8IR

```
Sources are:   ID08  76      GU01  80      FI    81      ID13  82      JWS   82
```

```
----------- RANGE -----------
              mi              nm              km          km conv
ID08  76                                      7              7.00
FI    81                                      7              7.00
JWS   82                                   c. 8              8.00

----------- LENGTH ----------
              in        ft        cm         m          m conv
ID08  76                                    2.00           2.00
FI    81                                    2.0            2.00
ID13  82                                    2.2            2.20
JWS   82                                [< 210]            2.10        Slightly less than radar version.

----------- TERMINAL GUIDANCE -----------
GU01  80      IR
FI    81      IR
ID13  82      passive IR
JWS   82      ? IR              Probably.
```

# AA-8 RADAR

```
Sources are:   ID08  76      GU01  80      FI   81      ID13  82      JWS  82
```

|  | RANGE | | | | |  |
|---|---|---|---|---|---|---|
|  | mi | nm | km | km conv | |
| ID08 76 | | | 15 | 15.00 | |
| FI 81 | | | 15 | 15.00 | |
| JWS 82 | | | [> 8] | 8.00 | "Rather more" than IR version. |

|  | LENGTH | | | | |
|---|---|---|---|---|---|
|  | in | ft | cm | m | m conv |
| ID08 76 | | | | 2.15 | 2.15 |
| FI 81 | | | | 2.2 | 2.20 |
| JWS 82 | | | c. 210 | | 2.10 |

**TERMINAL GUIDANCE**

| | | |
|---|---|---|
| GU01 80 | SAR | Though IR and SAR exist, small diameter must have caused problems with SAR. [Guidance listed as IR, though SAR and IR versions referred to, ed]. |
| FI 81 | | |
| ID13 82 | radar | Deployment [of IR version] with radar-guided Atoll suggests radar-guided Aphid (J-band) has lagged behind IR version. |
| JWS 82 | | Radar and IR versions reported, but to date photos are of what is assumed to be IR-guided. |

# AA-9

| Sources are: | SMP | 81 | AW ST | 82 | FI+ | 83 | AFM+ | 84 | SMP | 84 |
|---|---|---|---|---|---|---|---|---|---|---|
| | FI | 81 | AFM | 82 | JAWA | 82 | JAWA+ | 83 | AWST+ | 84 |

---------- US DESIGNATION ----------

| FI | 81 | AA-9 |
| SMP | 81 | AA-X-9 |
| AFM | 82 | AA-X-9 |
| AW ST | 82 | AA-X-9 |
| JAWA | 82 | AA-X-9 |
| JAWA+ | 83 | AA-9 |
| AFM+ | 84 | AA-9 |
| AW ST+ | 84 | AA-9 |
| SMP | 84 | AA-9 | '84, '85. |

---------- EDITOR'S NOTES ----------

| SMP | 81 | Listed '81, '84, '85. |

---------- RANGE ----------

| | | mi | nm | km | km conv | |
|---|---|---|---|---|---|---|
| FI | 81 | | | 40-45 | 40.00-45.00 | High altitude. |
| | | | | c. 20 | 20.00 | Low altitude. |
| AW ST | 82 | | 25 | | 46.30 | |
| | | | 10-12.5 | | 18.52-23.15 | Low-altitude head-on. |
| | | | 4.5-5 | | 8.33-9.26 | Rear. |
| | | | 21.5-24.5 | | 39.82-45.37 | High altitude head-on. |
| JAWA+ | 83 | 25-28 | 21-24 | 40-45 | 40.23-45.05 | At height. |
| | | 12.5 | 11 | 20 | 20.11 | At low altitude. |
| AFM+ | 84 | 25-28 | | | 40.23-45.05 | "At height". |
| | | 12.5 | | | 20.11 | At sea level. |
| SMP | 84 | | | | | Long-range missile, '84, '85. |

---------- NUMBER OF STAGES ----------

| AW ST | 82 | 1 |

---------- TYPE OF PROPULSION ----------

| AW ST | 82 | solid |

---------- TERMINAL GUIDANCE ----------

| AW ST | 82 | radar |

---------- DEPLOYMENT BEGAN ----------

| FI | 81 | Expected to enter service this year. |
| FI+ | 83 | Now in service. |

---------- CARRIERS ----------

---------- CARRIER ---------- TUBES ---- RELOADS ----------

| FI | 81 | MiG-25 Foxbat | | | On improved version. |
| SMP | 81 | Modified Foxbat | 4 | | Will be the Soviets' first look-down/shoot-down fighter. |
| AFM | 82 | modified MiG-25 | | | |
| JAWA | 82 | modified MiG-25 | | | |
| FI+ | 83 | MiG-25M Foxhound | | | |
| JAWA+ | 83 | Foxhound | | | Standard armament. |
| AFM+ | 84 | MiG-25M | | | |
| SMP | 84 | Foxhound | | | '84,'85. |

---------- CARRIER ---------- TUBES ---- RELOADS ----------

| AFM+ | 84 | MiG-29 | | | Expected. |

---------- CARRIER ---------- TUBES ---- RELOADS ----------

| AFM+ | 84 | MiG-31 | | | Standard armament. |

---------- CARRIER ---------- TUBES ---- RELOADS ----------

| AFM+ | 84 | Su-27 | | | Expected. |

---------- STRATEGY ----------

| FI+ | 83 | Flight trials against simulated cruise missile targets are continuing. |
| SMP | 84 | "Long-range missile that can be used against low-flying aircraft", '84, '85. |

---------- COMBAT REPORTS/EFFECTIVENESS ----------

| FI | 81 | During snap-down attacks, has worked against drones at 50 m altitude. |
| AFM | 82 | Reported success against cruise missiles after "look-down/snap-down" launch from modified MiG-25. |
| JAWA | 82 | Reported to have achieved successes against simulated cruise missiles, after 'lookdown/snapdown' launch. No details available. |

---------- **RELATION TO OTHER MISSILES** ----------
FI    81    Seems to be Soviet equivalent to the AIM-7M or Sky Flash.

# AA-10

Sources are: SMP 84   SMP 85

###### US DESIGNATION
| | | |
|---|---|---|
| SMP | 84 | AA-X-10 |
| SMP | 85 | AA-10 |

###### EDITOR'S NOTES
SMP 84   Listed '84, '85.

###### RANGE

| | | mi | nm | km | km conv | |
|---|---|---|---|---|---|---|
| SMP | 84 | | | | | Medium-range missile, '84, '85. |
| SMP | 85 | | | | | "Beyond-visual-range", on Su-27 Flanker. |

###### RADAR
SMP 85   On Mig-29 Fulcrum, "improved" AA-10 integrated with "true" look-down/shoot-down radar, "USSR's most modern look-down/shoot-down radar". On Su-27 Flanker, has "true" look-down/shoot-down weapon system.

###### DEPLOYMENT BEGAN
| | | | |
|---|---|---|---|
| SMP | 85 | c. 1984 | Introduced on Fulcrum. |
| | | c. 1985? | On Flanker, which is "nearing deployment". |

###### CARRIERS

| | | CARRIER | TUBES | RELOADS | |
|---|---|---|---|---|---|
| SMP | 84 | Fulcrum | | | '84, '85. |

| | | CARRIER | TUBES | RELOADS | |
|---|---|---|---|---|---|
| SMP | 84 | Flanker | | | '84, '85. |

###### DESIGN AND ENGINEERING
SMP 85   AA-10 and radar technology on Fulcrum "made possible, in part, by thefts from the West".

###### STRATEGY
SMP 84   "Medium-range missile with similar capabilities [to AA-9]"; AA-9 "can be used against low-flying aircraft", '84, '85.

SMP 85   Fulcrum designed as an all-weather counterair fighter-interceptor. Flanker designed as an all-weather air-superiority fighter, similiar to US F-15.

# AA-XP-1

Sources are: FI 81    AWST 82

--------- **US DESIGNATION** ---------
| | | |
|---|---|---|
| FI | 81 | AA-XP-1 |
| AWST | 82 | AA-XP-1 |

--------- **RANGE** ---------

|  |  | mi | nm | km | km conv | |
|---|---|---|---|---|---|---|
| FI | 81 | | | 35 | 35.00 | Max at high altitude. |
| | | | | 20 | 20.00 | Max at low altitude. |
| AWST | 82 | | 11-19 | | 20.37-35.19 | |

--------- **NUMBER OF STAGES** ---------
AWST 82    1

--------- **TYPE OF PROPULSION** ---------
AWST 82    solid

--------- **PROTOTYPE TESTS** ---------
FI    81              Currently on trials.

--------- **IOC** ---------
AWST 82    1984

--------- **DEPLOYMENT BEGAN** ---------
FI    81    1984         Expected to enter service.

--------- **STRATEGY** ---------
FI    81    Medium-range, all-aspect missile.
AWST 82    "All-aspect, look down, shoot down".

# AA-XP-2

Sources are:  FI  81    AWST  82    AWST+ 84

###### US DESIGNATION
FI    81    AA-XP-2
AWST  82    AA-XP-2

###### RANGE
|          | mi | nm | km | km conv | |
|----------|----|----|----|---------|-|
| FI 81    |    |    | 70 | 70.00   | High altitude. |
|          |    |    | 40 | 40.00   | Low altitude. |
| AWST 82  |    | 21.5-38 |  | 39.82-70.38 | |
| AWST+ 84 |    |    |    |         | Medium-range. |

###### NUMBER OF STAGES
AWST 82    1

###### TYPE OF PROPULSION
AWST 82    solid

###### TERMINAL GUIDANCE
AWST+ 84   active radar

###### IOC
AWST 82    1984

###### DEPLOYMENT BEGAN
FI 81      Expected in service by mid-1980s.

###### STRATEGY
AWST 82    "All-aspect, look down, shoot down" [final report]

# AA-X

Sources are: FI 81   FI+ 83

---------- **US DESIGNATION** ----------
FI    81    AA-X-?

---------- **TERMINAL GUIDANCE** ----------
FI    81    IR

---------- **DEVELOPMENT UNDERWAY** ----------
FI    81    Under development.

---------- **DEPLOYMENT BEGAN** ----------
FI+   83    May enter service in near future.

---------- **STRATEGY** ----------
FI    81    "An infrared 'dogfighting' missile is known to be under development".
FI+   83    According to intelligence services, USSR about to field missile comparable in performance to AIM-9L Sidewinder.

# Data Index

|  | History | Best Estimates | Data |
|---|---|---|---|
| AA-1 | 43 |  | 638-640 |
| AA-2 | 43 |  | 641-644 |
| AA-2-2 |  |  | 645-646 |
| AA-3 | 43 |  | 647-649 |
| AA-3IR |  |  | 650 |
| AA-3 RADAR |  |  | 651 |
| AA-4 |  |  | 652 |
| AA-5 | 43 |  | 653-655 |
| AA-5IR |  |  | 656 |
| AA-5 RADAR |  |  | 657 |
| AA-6 | 43 |  | 658-660 |
| AA-6IR |  |  | 661 |
| AA-6 RADAR |  |  | 662 |
| AA-7 | 43 |  | 663-665 |
| AA-7IR |  |  | 666 |
| AA-7 RADAR |  |  | 667 |
| AA-8 | 43 |  | 668-670 |
| AA-8IR |  |  | 671 |
| AA-8 RADAR |  |  | 672 |
| AA-9 | 43 |  | 673-674 |
| AA-10 | 43 |  | 675 |
| AA-XP-1 |  |  | 676 |
| AA-XP-2 |  |  | 677 |
| AA-X |  |  | 678 |
| ABM-1 | 40 | 101 | 533-539 |
| ABM-3 |  |  | 540 |
| AS-1 | 40 |  | 429-431 |
| AS-2 | 40 | 96-97 | 432-435 |
| AS-3 | 38 | 89-90 | 436-440 |
| AS-4 | 38, 40 | 97 | 441-446 |
| AS-5 | 40 | 97-98 | 447-451 |
| AS-6 | 40 | 98-99 | 452-457 |
| AS-7 | 38 |  | 458-460 |
| AS-8 | 38 |  | 461 |
| AS-9 | 38 |  | 462-463 |
| AS-10 | 38 |  | 464-465 |
| AS-11 | 40 |  | 466 |
| AS-12 |  |  | 467 |
| AS-15 | 38 | 90 | 468-469 |
| AT-1 | 39 |  | 408-411 |
| AT-2 | 39 |  | 412-415 |
| AT-2 Swatter A |  |  | 416 |
| AT-2 Swatter B |  |  | 417 |
| AT-3 | 38-39 |  | 418-422 |
| AT-4 | 39 |  | 423-424 |
| AT-5 | 39 |  | 425-426 |
| AT-6 | 38-39 |  | 427-428 |
| FROG | 35 | 83-84 | 346-348 |
| FROG 1 | 35 | 84 | 349-351 |
| FROG 2 | 35 | 84 | 352-354 |
| FROG 3 | 35 | 84 | 355-357 |
| FROG 4 | 35 | 84-85 | 358-360 |
| FROG 5 | 35 | 85 | 361-362 |
| FROG 6 |  |  | 363 |
| FROG 7 | 35 | 85 | 364-367 |
| GLCM |  |  | 407 |
| SA-1 | 41 | 104 | 543-546 |
| SA-2 | 41-43 | 103 | 547-555 |
| SA-3 | 41-43 | 104 | 556-562 |
| SA-4 | 42 |  | 563-568 |
| SA-5 | 42 | 103-104 | 569-575 |
| SA-6 | 43 |  | 576-582 |
| SA-7 | 42-43 |  | 583-587 |
| SA-7b |  |  | 588 |
| SA-8 | 43 |  | 589-594 |
| SA-8b |  |  | 595 |
| SA-9 | 42-43 |  | 596-600 |

|  | History | Best Estimates | Data |
|---|---|---|---|
| SA-10 | 42 | 104 | 601-604 |
| SA-11 | 43 |  | 605-607 |
| SA-12 | 42 |  | 608-609 |
| SA-13 | 43 |  | 610-612 |
| SA-14 | 42 |  | 613 |
| SA-N-1 | 43 |  | 614-617 |
| SA-N-2 | 43 |  | 618-620 |
| SA-N-3 | 43 |  | 621-624 |
| SA-N-4 | 43 |  | 625-628 |
| SA-N-5 | 43 |  | 629-630 |
| SA-N-6 | 43 | 104 | 631-633 |
| SA-N-7 | 43 |  | 634-635 |
| SA-N-8 | 43 |  | 636 |
| SA-N-9 |  |  | 637 |
| SH-4 | 40 |  | 541 |
| SH-8 | 40-41 |  | 542 |
| SS-1 | 31, 35 |  | 303-304 |
| SS-1 Scud |  | 86 | 368-370 |
| SS-1b Scud A | 35 | 86-87 | 371-375 |
| SS-1c Scud B | 35 | 87 | 376-380 |
| SS-1c Scud C |  |  | 381 |
| SS-2 | 31 |  | 305-306 |
| SS-3 | 31, 34 | 79 | 307-310 |
| SS-4 | 31, 34-35 | 79-81 | 311-318 |
| SS-5 | 34-35 | 81 | 319-325 |
| SS-6 | 31, 33 | 57-58 | 107-111 |
| SS-7 | 31-32 | 58-59 | 112-117 |
| SS-7 Mod 1 |  |  | 118 |
| SS-7 Mod 2 |  |  | 119 |
| SS-7 Mod 3 |  |  | 120 |
| SS-8 | 31-32 | 59 | 121-126 |
| SS-9 | 32 | 59-60 | 127-134 |
| SS-9 Mod 1 |  |  | 135-136 |
| SS-9 Mod 2 |  |  | 137-138 |
| SS-9 Mod 3 |  |  | 139-141 |
| SS-9 Mod 4 |  |  | 142-144 |
| SS-9 Mod 5 |  |  | 145 |
| SS-10 | 32 |  | 146-148 |
| SS-11 | 32-35 | 60-61 | 149-157 |
| SS-11 Mod 1 |  | 61 | 158-160 |
| SS-11 Mod 2 |  | 61 | 161-162 |
| SS-11 Mod 3 |  | 61-62 | 163-166 |
| SS-11 Mod 4 |  |  | 167 |
| SS-12 | 35 | 87-88 | 382-386 |
| SS-13 | 32-34 |  | 168-174 |
| SS-13 Mod 1 |  | 62 | 175 |
| SS-13 Mod 2 |  | 62 | 176 |
| SS-14 | 34 |  | 326-330 |
| SS-15 | 34 |  | 331-333 |
| SS-16 | 32, 34 |  | 177-182 |
| SS-17 | 32-33 | 62-63 | 183-189 |
| SS-17 Mod 1 |  | 63 | 190-191 |
| SS-17 Mod 2 |  | 63 | 192-193 |
| SS-17 Mod 3 |  | 63-64 | 194-195 |
| SS-18 | 32-33 | 64 | 196-202 |
| SS-18 Mod 1 |  | 64-65 | 203-205 |
| SS-18 Mod 2 |  | 65 | 206-208 |
| SS-18 Mod 3 |  | 65 | 209-211 |
| SS-18 Mod 4 |  | 66 | 212-214 |
| SS-18 Mod 5 |  |  | 215 |
| SS-19 | 32-33 | 66-67 | 216-223 |
| SS-19 Mod 1 |  | 67 | 224-225 |
| SS-19 Mod 2 |  | 67-68 | 226-227 |
| SS-19 Mod 3 |  | 68 | 228-230 |
| SS-19 VarRange |  |  | 231 |
| SS-20 | 34 | 81-82 | 334-342 |
| SS-20 Mod 1 |  |  | 343 |

|  | History | Best Estimates | Data |
|---|---|---|---|
| SS-20 Mod 2 |  | 81-82 | 344 |
| SS-20 Mod 3 |  |  | 345 |
| SS-21 | 35 | 85-86 | 387-390 |
| SS-22 | 35 | 88 | 391-393 |
| SS-23 | 35 | 87 | 394-396 |
| SS-24 | 33 | 68-69 | 232-233 |
| SS-25 | 33 | 69 | 234-236 |
| SS-26 | 33 |  | 237 |
| SS-27 | 33 |  | 238 |
| SS-N-1 | 39 | 91 | 471-474 |
| SS-N-2 | 39 |  | 475-480 |
| SS-N-2a |  |  | 481-482 |
| SS-N-2b |  |  | 483-484 |
| SS-N-2c |  |  | 485-487 |
| SS-N-3 | 39 | 91-93 | 488-494 |
| SS-N-3a |  |  | 495-496 |
| SS-N-3b |  |  | 497 |
| SS-N-3c |  |  | 498 |
| SS-N-4 | 33 | 71 | 239-245 |
| SS-N-5 | 33 | 71-72 | 246-253 |
| SS-N-6 | 33 | 72 | 254-261 |
| SS-N-6 Mod 1 |  | 72 | 262-264 |
| SS-N-6 Mod 2 |  | 72-73 | 265-267 |
| SS-N-6 Mod 3 |  | 73 | 268-270 |
| SS-N-7 | 39 | 93 | 499-502 |
| SS-N-8 | 33 | 73 | 271-279 |
| SS-N-8 Mod 1 |  | 73-74 | 280 |
| SS-N-8 Mod 2 |  | 74 | 281 |
| SS-N-8 Mod 3 |  |  | 282 |
| SS-N-9 | 39 | 93-94 | 503-507 |
| SS-N-10 |  |  | 508-509 |
| SS-N-12 | 39 | 94 | 510-513 |
| SS-N-13 | 39 |  | 514-515 |
| SS-N-14 | 40 | 99 | 516-519 |
| SS-N-15 | 40 | 99-100 | 520-522 |
| SS-N-16 | 40 |  | 523-524 |
| SS-N-17 | 33-34 | 74-75 | 283-286 |
| SS-N-18 | 33-34 | 75 | 287-291 |
| SS-N-18 Mod 1 | 33 | 75 | 292-293 |
| SS-N-18 Mod 2 | 33 | 75-76 | 294-295 |
| SS-N-18 Mod 3 | 33 | 76 | 296 |
| SS-N-19 | 39 | 95 | 525-527 |
| SS-N-20 | 34 | 76-77 | 297-300 |
| SS-N-21 | 38 | 89 | 528-529 |
| SS-N-22 | 39 | 95 | 530-531 |
| SS-N-23 | 34 | 77 | 301-302 |
| SS-N-24 |  |  | 532 |
| SSC-1 | 39 | 95-96 | 397-400 |
| SSC-2a | 40 |  | 401 |
| SSC-2b | 40 |  | 402-404 |
| SSC-4 | 38 | 89 | 405-406 |
| TASM |  |  | 470 |

# Missile Index

| | SS-6 | SS-7 | SS-7 Mod 1 | SS-7 Mod 2 | SS-7 Mod 3 | SS-8 | SS-9 | SS-9 Mod 1 | SS-9 Mod 2 | SS-9 Mod 3 | SS-9 Mod 4 | SS-9 Mod 5 | SS-10 |
|---|---|---|---|---|---|---|---|---|---|---|---|---|---|
| START OF MISSILE | 107 | 112 | 118 | 119 | 120 | 121 | 127 | 135 | 137 | 139 | 142 | 145 | 146 |
| **DESIGNATIONS** | | | | | | | | | | | | | |
| US DESIGNATION | 107 | 112 | — | — | — | 121 | 127 | 135 | 137 | 139 | 142 | 145 | 146 |
| NATO CODENAME | 107 | 112 | — | — | — | 121 | 127 | — | — | 139 | 142 | — | 146 |
| SOVIET DESIGNATION | — | — | — | — | — | — | — | — | — | — | — | — | — |
| OTHER DESIGNATION | 107 | — | — | — | — | — | — | — | — | 139 | — | 145 | — |
| DESCRIPTION | 107 | 112 | — | — | — | 121 | 127 | — | — | 139 | 142 | — | 146 |
| EDITOR'S NOTES | 107 | 112 | — | — | — | 121 | 127 | 135 | 137 | 139 | 142 | — | 146 |
| **PERFORMANCE** | | | | | | | | | | | | | |
| MINIMUM ALTITUDE | — | — | — | — | — | — | — | — | — | — | — | — | — |
| MAXIMUM ALTITUDE | — | — | — | — | — | — | — | — | — | — | — | — | — |
| RANGE | 107 | 112 | 118 | — | 120 | 121 | 127 | 135 | 137 | 139 | 142 | — | 146 |
| SPEED | 107 | — | — | — | — | — | — | — | — | — | — | — | — |
| CEP | 108 | 112 | 118 | 119 | 120 | 122 | 128 | 135 | 137 | 139 | 142 | — | 146 |
| **WARHEAD** | | | | | | | | | | | | | |
| WARHEAD TYPE | 108 | 113 | — | — | — | 122 | 128 | 135 | 137 | — | — | — | — |
| WARHEAD CHARACTERISTICS | 108 | 113 | 118 | 119 | 120 | 122 | 128 | 135 | 137 | 139 | 142 | — | 146 |
| NUCLEAR YIELD | 108 | 113 | 118 | 119 | 120 | 122 | 128 | 135 | 137 | 139 | 143 | — | 146 |
| WARHEAD WEIGHT | — | 113 | — | — | — | 122 | 129 | — | — | — | — | — | — |
| THROWWEIGHT | 108 | 113 | — | — | — | 122 | 129 | — | — | 139 | 143 | — | 146 |
| PAYLOAD | 108 | 113 | — | — | — | 122 | 129 | — | — | 139 | — | — | — |
| **DIMENSIONS** | | | | | | | | | | | | | |
| WEIGHT | 108 | 113 | — | — | — | 123 | 129 | 136 | — | — | — | — | 146 |
| LENGTH | 108 | 113 | — | — | — | 123 | 129 | 136 | 137 | — | 143 | — | 147 |
| DIAMETER | 109 | 114 | — | — | — | 123 | 129 | 136 | 137 | — | 143 | — | 147 |
| WINGSPAN | — | — | — | — | — | — | — | — | — | — | — | — | — |
| **PROPULSION** | | | | | | | | | | | | | |
| NUMBER OF STAGES | 109 | 114 | — | — | — | 123 | 129 | 136 | 137 | — | — | 145 | 147 |
| TYPE OF PROPULSION | 109 | 114 | 118 | — | 120 | 123 | 129 | 136 | 137 | 139 | 143 | — | 147 |
| **GUIDANCE** | | | | | | | | | | | | | |
| MIDCOURSE (OR MAIN) GUIDANCE | 109 | 114 | — | — | — | 123 | 130 | 136 | 138 | 139 | — | — | 147 |
| TERMINAL GUIDANCE | — | — | — | — | — | — | — | — | — | — | — | — | — |
| CONTROL SYSTEM | — | — | — | — | — | 123 | 130 | — | — | — | — | — | 147 |
| RADAR | — | — | — | — | — | — | — | — | — | — | — | — | — |
| **HISTORY** | | | | | | | | | | | | | |
| FIRST OBSERVED | — | — | — | — | — | — | — | — | — | — | — | — | — |
| OBSERVED | — | 114 | — | — | — | — | 130 | — | — | — | — | — | 147 |
| OBSERVED IN PARADE | — | — | — | — | — | 123 | 130 | 136 | — | — | — | — | 147 |
| DEVELOPMENT BEGAN | 109 | 114 | — | — | — | 124 | 130 | — | — | — | — | — | 147 |
| DEVELOPMENT UNDERWAY | 109 | — | — | — | — | — | — | — | — | 140 | — | — | — |
| PROTOTYPE TESTS | 110 | 114 | — | — | — | 124 | 130 | 136 | 138 | 140 | 143 | 145 | 148 |
| PRODUCTION BEGAN | — | — | — | — | — | — | — | — | — | — | — | — | — |
| PRODUCTION UNDERWAY | 110 | — | — | — | — | — | — | — | — | — | — | — | — |
| PRODUCTION ENDED | — | — | — | — | — | — | — | — | — | — | — | — | — |
| IOC | 110 | 114 | 118 | 119 | 120 | 124 | 130 | 136 | 138 | 140 | 144 | — | 148 |
| DEPLOYMENT BEGAN | 110 | 114 | 118 | 119 | 120 | 124 | 130 | 136 | 138 | 140 | 144 | — | 148 |
| DEPLOYED | — | 114 | — | — | — | 124 | 130 | — | — | — | — | — | — |
| DEPLOYMENT COMPLETE | — | 114 | — | — | — | 124 | 130 | — | — | — | — | — | — |
| RETIRED | 110 | 115 | 118 | — | 120 | 124 | 130 | 136 | — | — | — | — | — |
| DEPLOYMENT HISTORY | 110 | 115 | — | — | — | 124 | 131 | — | — | — | — | — | 148 |
| **NUMBERS; PLATFORM** | | | | | | | | | | | | | |
| NUMBERS PRODUCED | — | 115 | — | — | — | 125 | — | — | — | — | — | — | — |
| NUMBERS IN SERVICE | 110 | 115 | — | — | — | 125 | 131 | 136 | 138 | 140 | 144 | — | — |
| UNDATED NUMBER | 111 | 117 | — | — | — | 126 | 133 | 136 | 138 | — | — | — | — |
| LAUNCHER/SILO | 111 | 117 | — | — | — | 126 | 133 | — | — | 140 | — | — | 148 |
| CARRIERS | — | — | — | — | — | — | — | — | — | — | — | — | — |
| **DESIGN; USE; STRATEGY** | | | | | | | | | | | | | |
| DESIGNER | 111 | 117 | — | — | — | 126 | 133 | — | — | 140 | — | — | 148 |
| DESIGN AND ENGINEERING | 111 | 117 | — | — | — | 126 | — | — | — | 140 | — | 145 | 148 |
| USE AND CONFIGURATION | 111 | 117 | — | — | — | 126 | 133 | — | — | — | — | 145 | — |
| RELOADS IN GENERAL | 111 | 117 | 118 | — | 120 | 126 | 133 | 136 | — | — | 144 | — | — |
| TARGET TYPE | — | — | — | — | — | — | — | — | — | — | — | — | — |
| STRATEGY | 111 | 117 | — | — | — | 126 | 133 | 136 | — | 141 | 144 | 145 | 148 |
| COMBAT REPORTS/EFFECTIVENESS | — | — | — | — | — | — | — | — | — | — | — | — | — |
| GEOGRAPHICAL DEPLOYMENTS | — | 117 | — | — | — | — | 133 | — | — | — | — | — | — |
| EXPORTED TO | — | 117 | — | — | — | 126 | — | — | — | — | — | — | — |
| RELATION TO OTHER MISSILES | — | — | — | — | — | — | 133 | — | 138 | — | — | — | 148 |
| OTHER INFORMATION | 111 | 117 | — | — | — | — | 134 | — | — | — | — | 145 | — |

682 MISSILE INDEX ICBMs

|  | SS-11 | SS-11 Mod 1 | SS-11 Mod 2 | SS-11 Mod 3 | SS-11 Mod 4 | SS-13 | SS-13 Mod 1 | SS-13 Mod 2 | SS-16 | SS-17 | SS-17 Mod 1 | SS-17 Mod 2 | SS-17 Mod 3 |
|---|---|---|---|---|---|---|---|---|---|---|---|---|---|
| START OF MISSILE | 149 | 158 | 161 | 163 | 167 | 168 | 175 | 176 | 177 | 183 | 190 | 192 | 194 |
| **DESIGNATIONS** | | | | | | | | | | | | | |
| US DESIGNATION | 149 | 158 | 161 | 163 | 167 | 168 | 175 | 176 | 177 | 183 | 190 | 192 | 194 |
| NATO CODENAME | 149 | 158 | — | 163 | — | 168 | 175 | — | — | — | — | — | — |
| SOVIET DESIGNATION | — | — | — | — | — | — | — | — | 177 | 183 | — | — | — |
| OTHER DESIGNATION | — | — | — | 163 | — | — | — | — | — | — | — | — | — |
| DESCRIPTION | 149 | 158 | — | 163 | — | 168 | — | — | 177 | 183 | — | — | — |
| EDITOR'S NOTES | 149 | 158 | 161 | — | — | 168 | — | — | 177 | 183 | 190 | — | 194 |
| **PERFORMANCE** | | | | | | | | | | | | | |
| MINIMUM ALTITUDE | — | — | — | — | — | — | — | — | — | — | — | — | — |
| MAXIMUM ALTITUDE | — | — | — | — | — | — | — | — | — | — | — | — | — |
| RANGE | 150 | 158 | 161 | 163 | — | 169 | 175 | 176 | 177 | 183 | 190 | 192 | 194 |
| SPEED | — | — | — | — | — | — | — | — | — | — | — | — | — |
| CEP | 150 | 158 | 161 | 163 | — | 169 | 175 | 176 | 178 | 184 | 190 | 192 | 194 |
| **WARHEAD** | | | | | | | | | | | | | |
| WARHEAD TYPE | 150 | — | — | — | — | 169 | — | — | 178 | 184 | — | — | — |
| WARHEAD CHARACTERISTICS | 150 | 158 | 161 | 164 | 167 | 169 | 175 | 176 | 178 | 184 | 190 | 192 | 194 |
| NUCLEAR YIELD | 151 | 159 | 161 | 164 | — | 170 | 175 | 176 | 178 | 185 | 190 | 192 | 194 |
| WARHEAD WEIGHT | — | — | — | — | — | — | — | — | — | — | — | — | — |
| THROWWEIGHT | 151 | 159 | 161 | 164 | — | 170 | 175 | — | 179 | 185 | 191 | 193 | 194 |
| PAYLOAD | 151 | — | — | — | — | — | — | — | 179 | 185 | — | — | — |
| **DIMENSIONS** | | | | | | | | | | | | | |
| WEIGHT | 151 | — | — | — | — | 170 | — | — | 179 | 185 | — | — | — |
| LENGTH | 151 | 159 | 161 | 164 | — | 170 | — | 176 | 179 | 185 | — | — | 194 |
| DIAMETER | 151 | — | — | 164 | — | 170 | — | — | 179 | 185 | — | — | — |
| WINGSPAN | — | — | — | — | — | — | — | — | — | — | — | — | — |
| **PROPULSION** | | | | | | | | | | | | | |
| NUMBER OF STAGES | 151 | — | — | — | — | 171 | — | — | 179 | 185 | — | — | — |
| TYPE OF PROPULSION | 152 | 159 | 161 | 164 | — | 171 | 175 | 176 | 179 | 186 | 191 | 193 | 194 |
| **GUIDANCE** | | | | | | | | | | | | | |
| MIDCOURSE (OR MAIN) GUIDANCE | 152 | 159 | — | — | — | 171 | — | — | 179 | 186 | — | — | — |
| TERMINAL GUIDANCE | — | — | — | — | — | — | — | — | — | — | — | — | — |
| CONTROL SYSTEM | — | — | — | — | — | 171 | — | — | 179 | 186 | — | — | — |
| RADAR | — | — | — | — | — | — | — | — | — | — | — | — | — |
| **HISTORY** | | | | | | | | | | | | | |
| FIRST OBSERVED | — | — | — | 164 | 167 | — | — | — | 180 | — | — | — | — |
| OBSERVED | 152 | — | — | — | — | — | — | — | — | — | — | — | — |
| OBSERVED IN PARADE | 152 | — | — | — | — | 171 | — | — | — | — | — | — | — |
| DEVELOPMENT BEGAN | 152 | 159 | — | — | — | 171 | — | — | 180 | 186 | — | — | — |
| DEVELOPMENT UNDERWAY | — | — | — | — | — | — | — | — | — | 186 | — | — | — |
| PROTOTYPE TESTS | 152 | 159 | 161 | 164 | 167 | 171 | 175 | 176 | 180 | 186 | 191 | 193 | — |
| PRODUCTION BEGAN | — | — | — | — | — | — | — | — | — | — | — | — | — |
| PRODUCTION UNDERWAY | — | — | — | — | — | — | — | — | — | 186 | — | — | — |
| PRODUCTION ENDED | — | — | — | — | — | — | — | — | — | — | — | — | — |
| IOC | 152 | 159 | 162 | 164 | — | 171 | — | — | 180 | 186 | 191 | — | 194 |
| DEPLOYMENT BEGAN | 153 | 159 | 162 | 165 | — | 172 | 175 | 176 | 180 | 186 | 191 | 193 | 195 |
| DEPLOYED | 153 | 159 | 162 | 165 | — | 172 | — | 176 | — | — | — | 193 | 195 |
| DEPLOYMENT COMPLETE | 153 | — | — | — | — | 172 | — | — | — | 186 | — | — | — |
| RETIRED | — | 159 | — | — | — | — | 175 | — | — | — | 191 | — | — |
| DEPLOYMENT HISTORY | 153 | 159 | 162 | 165 | — | 172 | — | 176 | 181 | 187 | — | — | 195 |
| **NUMBERS; PLATFORM** | | | | | | | | | | | | | |
| NUMBERS PRODUCED | — | — | — | — | — | — | — | — | 181 | 187 | — | — | — |
| NUMBERS IN SERVICE | 153 | 159 | 162 | 165 | — | 172 | 175 | 176 | — | 187 | 191 | 193 | 195 |
| UNDATED NUMBER | — | 160 | 162 | 165 | — | 173 | 175 | 176 | — | 188 | 191 | 193 | 195 |
| LAUNCHER/SILO | 156 | 160 | 162 | 165 | — | 173 | — | 176 | 181 | 188 | 191 | 193 | — |
| CARRIERS | — | — | — | — | — | 173 | — | — | 181 | — | — | — | — |
| **DESIGN; USE; STRATEGY** | | | | | | | | | | | | | |
| DESIGNER | 156 | 160 | — | — | — | 173 | — | — | 181 | 188 | — | — | — |
| DESIGN AND ENGINEERING | — | — | — | — | — | 174 | — | — | 181 | 188 | — | — | — |
| USE AND CONFIGURATION | 156 | 160 | 162 | — | — | 174 | — | — | — | 188 | 191 | — | 195 |
| RELOADS IN GENERAL | 156 | 160 | 162 | 165 | — | 174 | 175 | 176 | 182 | 189 | 191 | 193 | 195 |
| TARGET TYPE | — | — | — | — | — | — | — | — | — | — | — | — | — |
| STRATEGY | 156 | 160 | — | 166 | — | 174 | — | — | 182 | 189 | — | — | — |
| COMBAT REPORTS/EFFECTIVENESS | — | — | — | — | — | 174 | — | — | — | — | — | — | — |
| GEOGRAPHICAL DEPLOYMENTS | 156 | — | — | — | — | 174 | — | — | — | 189 | — | — | — |
| EXPORTED TO | — | — | — | — | — | — | — | — | — | — | — | — | — |
| RELATION TO OTHER MISSILES | 156 | — | — | — | — | 174 | — | — | 182 | 189 | — | — | 195 |
| OTHER INFORMATION | 157 | — | — | — | — | 174 | — | — | 182 | — | — | — | — |

MISSILE INDEX  ICBMs  683

| | SS-18 | SS-18 Mod 1 | SS-18 Mod 2 | SS-18 Mod 3 | SS-18 Mod 4 | SS-18 Mod 5 | SS-19 | SS-19 Mod 1 | SS-19 Mod 2 | SS-19 Mod 3 | SS-19 VarRange | SS-24 | SS-25 | SS-26 | SS-27 |
|---|---|---|---|---|---|---|---|---|---|---|---|---|---|---|---|
| START OF MISSILE | 196 | 203 | 206 | 209 | 212 | 215 | 216 | 224 | 226 | 228 | 231 | 232 | 234 | 237 | 238 |
| **DESIGNATIONS** | | | | | | | | | | | | | | | |
| US DESIGNATION | 196 | 203 | 206 | 209 | 212 | 215 | 216 | 224 | 226 | 228 | 231 | 232 | 234 | 237 | 238 |
| NATO CODENAME | -- | -- | -- | -- | -- | -- | -- | -- | -- | -- | -- | -- | -- | -- | -- |
| SOVIET DESIGNATION | 196 | -- | -- | -- | -- | -- | 216 | -- | -- | -- | -- | 232 | 234 | -- | -- |
| OTHER DESIGNATION | -- | -- | -- | -- | -- | -- | -- | -- | -- | -- | -- | 232 | 234 | -- | -- |
| DESCRIPTION | 196 | -- | 206 | 209 | -- | -- | 216 | -- | -- | -- | 231 | 232 | 234 | -- | 238 |
| EDITOR'S NOTES | 196 | -- | 206 | 209 | 212 | 215 | 216 | 224 | -- | 228 | 231 | 232 | 234 | -- | -- |
| **PERFORMANCE** | | | | | | | | | | | | | | | |
| MINIMUM ALTITUDE | -- | -- | -- | -- | -- | -- | -- | -- | -- | -- | -- | -- | -- | -- | -- |
| MAXIMUM ALTITUDE | -- | -- | -- | -- | -- | -- | -- | -- | -- | -- | -- | -- | -- | -- | -- |
| RANGE | 196 | 203 | 206 | 209 | 212 | 215 | 216 | 224 | 226 | 228 | 231 | 232 | 234 | -- | -- |
| SPEED | -- | -- | -- | -- | -- | -- | -- | -- | -- | -- | -- | -- | -- | -- | -- |
| CEP | 197 | 203 | 206 | 209 | 212 | 215 | 217 | 224 | 226 | 228 | 231 | 232 | 234 | 237 | 238 |
| **WARHEAD** | | | | | | | | | | | | | | | |
| WARHEAD TYPE | 197 | -- | -- | -- | -- | -- | 217 | -- | -- | -- | -- | -- | -- | -- | -- |
| WARHEAD CHARACTERISTICS | 197 | 204 | 207 | 210 | 212 | 215 | 217 | 224 | 226 | 228 | 231 | 232 | 234 | -- | -- |
| NUCLEAR YIELD | 198 | 204 | 207 | 210 | 213 | 215 | 218 | 225 | 226 | 229 | 231 | 232 | 234 | -- | -- |
| WARHEAD WEIGHT | -- | -- | -- | -- | -- | -- | -- | -- | -- | -- | -- | -- | -- | -- | -- |
| THROWEIGHT | 198 | 204 | 207 | 210 | 213 | 215 | 218 | 225 | 226 | 229 | -- | 232 | 234 | 237 | 238 |
| PAYLOAD | 198 | -- | -- | -- | -- | -- | 218 | -- | -- | -- | -- | -- | -- | -- | -- |
| **DIMENSIONS** | | | | | | | | | | | | | | | |
| WEIGHT | 198 | -- | -- | -- | -- | -- | 218 | -- | -- | -- | -- | -- | -- | -- | -- |
| LENGTH | 198 | -- | 207 | -- | 213 | -- | 218 | -- | -- | 229 | -- | 232 | 234 | -- | -- |
| DIAMETER | 198 | -- | 207 | -- | -- | -- | 218 | -- | -- | -- | -- | -- | -- | -- | -- |
| WINGSPAN | -- | -- | -- | -- | -- | -- | -- | -- | -- | -- | -- | -- | -- | -- | -- |
| **PROPULSION** | | | | | | | | | | | | | | | |
| NUMBER OF STAGES | 199 | -- | -- | -- | -- | -- | 219 | -- | -- | -- | -- | -- | -- | -- | -- |
| TYPE OF PROPULSION | 199 | 204 | 207 | 210 | 213 | -- | 219 | 225 | 227 | 229 | 231 | 232 | 234 | 237 | 238 |
| **GUIDANCE** | | | | | | | | | | | | | | | |
| MIDCOURSE (OR MAIN) GUIDANCE | 199 | 204 | -- | 210 | -- | -- | 219 | -- | -- | -- | 231 | -- | -- | -- | -- |
| TERMINAL GUIDANCE | -- | -- | -- | -- | -- | -- | -- | -- | -- | -- | -- | -- | -- | -- | -- |
| CONTROL SYSTEM | 199 | -- | -- | -- | -- | -- | 219 | -- | -- | -- | -- | -- | -- | -- | -- |
| RADAR | -- | -- | -- | -- | -- | -- | -- | -- | -- | -- | -- | -- | -- | -- | -- |
| **HISTORY** | | | | | | | | | | | | | | | |
| FIRST OBSERVED | -- | -- | -- | -- | -- | -- | -- | -- | -- | -- | -- | -- | -- | -- | -- |
| OBSERVED | -- | -- | -- | -- | -- | -- | -- | -- | 227 | -- | -- | -- | -- | -- | -- |
| OBSERVED IN PARADE | -- | -- | -- | -- | -- | -- | -- | -- | -- | -- | -- | -- | -- | -- | -- |
| DEVELOPMENT BEGAN | 199 | -- | -- | -- | 213 | -- | 219 | -- | -- | -- | 231 | 232 | 234 | -- | -- |
| DEVELOPMENT UNDERWAY | 199 | -- | -- | -- | -- | -- | 219 | -- | -- | -- | -- | 232 | 234 | 237 | 238 |
| PROTOTYPE TESTS | 199 | 204 | 207 | 210 | -- | -- | 219 | 225 | 227 | -- | 231 | 233 | 235 | 237 | 238 |
| PRODUCTION BEGAN | -- | -- | -- | -- | -- | -- | -- | -- | -- | -- | -- | -- | -- | -- | -- |
| PRODUCTION UNDERWAY | 199 | -- | -- | -- | -- | -- | -- | -- | -- | -- | -- | 233 | 235 | -- | -- |
| PRODUCTION ENDED | -- | -- | -- | -- | -- | -- | -- | -- | -- | -- | -- | -- | -- | -- | -- |
| IOC | 199 | 204 | 207 | 210 | 213 | -- | 219 | 225 | -- | 229 | -- | 233 | 235 | -- | -- |
| DEPLOYMENT BEGAN | 199 | 204 | 208 | 210 | 213 | 215 | 220 | 225 | 227 | 229 | 231 | 233 | 235 | -- | -- |
| DEPLOYED | -- | -- | -- | -- | 213 | -- | 220 | -- | 227 | 229 | -- | -- | -- | -- | -- |
| DEPLOYMENT COMPLETE | 200 | -- | -- | -- | -- | -- | 220 | -- | -- | -- | -- | -- | -- | -- | -- |
| RETIRED | -- | 205 | 208 | 210 | -- | -- | -- | 225 | -- | -- | -- | -- | -- | -- | -- |
| DEPLOYMENT HISTORY | 200 | 205 | 208 | 210 | 213 | -- | 220 | 225 | -- | 229 | -- | 233 | 235 | -- | 238 |
| **NUMBERS; PLATFORM** | | | | | | | | | | | | | | | |
| NUMBERS PRODUCED | 200 | -- | -- | -- | -- | -- | 220 | -- | -- | -- | -- | -- | -- | -- | -- |
| NUMBERS IN SERVICE | 200 | 205 | 208 | 210 | 213 | -- | 220 | 225 | 227 | 229 | 231 | 233 | 235 | -- | -- |
| UNDATED NUMBER | 201 | 205 | 208 | 211 | 214 | -- | 221 | 225 | 227 | 230 | -- | -- | -- | -- | -- |
| LAUNCHER/SILO | 201 | 205 | -- | 211 | -- | -- | 222 | -- | 227 | 230 | 231 | 233 | 235 | -- | -- |
| CARRIERS | -- | -- | -- | -- | -- | -- | 222 | -- | -- | -- | -- | 233 | 235 | -- | -- |
| **DESIGN; USE; STRATEGY** | | | | | | | | | | | | | | | |
| DESIGNER | 201 | -- | -- | 211 | -- | -- | 222 | -- | -- | -- | 231 | -- | -- | -- | -- |
| DESIGN AND ENGINEERING | 201 | -- | -- | -- | -- | -- | 222 | -- | -- | -- | -- | 233 | 235 | 237 | 238 |
| USE AND CONFIGURATION | 201 | 205 | 208 | 211 | 214 | -- | 222 | 225 | -- | 230 | -- | 233 | 235 | -- | -- |
| RELOADS IN GENERAL | 202 | 205 | 208 | 211 | 214 | -- | 222 | 225 | 227 | 230 | -- | 233 | 235 | -- | -- |
| TARGET TYPE | -- | -- | -- | -- | -- | -- | -- | -- | -- | -- | -- | -- | -- | -- | -- |
| STRATEGY | 202 | 205 | -- | -- | 214 | -- | 223 | -- | -- | 230 | -- | -- | 236 | -- | 238 |
| COMBAT REPORTS/EFFECTIVENESS | -- | -- | -- | -- | -- | -- | -- | -- | -- | -- | -- | -- | -- | -- | -- |
| GEOGRAPHICAL DEPLOYMENTS | 202 | -- | -- | -- | -- | -- | 223 | -- | -- | -- | -- | -- | -- | -- | -- |
| EXPORTED TO | -- | -- | -- | -- | -- | -- | -- | -- | -- | -- | -- | -- | -- | -- | -- |
| RELATION TO OTHER MISSILES | 202 | -- | -- | -- | 214 | -- | 223 | -- | -- | -- | -- | 233 | 236 | 237 | -- |
| OTHER INFORMATION | -- | -- | -- | -- | -- | -- | -- | -- | -- | -- | -- | -- | -- | -- | -- |

MISSILE INDEX  SLBMs

| | SS-N-4 | SS-N-5 | SS-N-6 | SS-N-6 Mod 1 | SS-N-6 Mod 2 | SS-N-6 Mod 3 | SS-N-8 | SS-N-8 Mod 1 | SS-N-8 Mod 2 | SS-N-8 Mod 3 |
|---|---|---|---|---|---|---|---|---|---|---|
| **START OF MISSILE** | 239 | 246 | 254 | 262 | 265 | 268 | 271 | 280 | 281 | 282 |
| **DESIGNATIONS** | | | | | | | | | | |
| US DESIGNATION | 239 | 246 | 254 | 262 | 265 | 268 | 271 | 280 | 281 | 282 |
| NATO CODENAME | 239 | 246 | 254 | -- | -- | -- | 271 | -- | -- | -- |
| SOVIET DESIGNATION | -- | -- | -- | -- | -- | -- | -- | -- | -- | -- |
| OTHER DESIGNATION | -- | -- | 254 | -- | -- | -- | -- | -- | -- | -- |
| DESCRIPTION | 239 | 246 | 254 | 262 | 265 | 268 | 271 | -- | -- | -- |
| EDITOR'S NOTES | 239 | 246 | 255 | 262 | 265 | 268 | 271 | -- | -- | 282 |
| **PERFORMANCE** | | | | | | | | | | |
| MINIMUM ALTITUDE | -- | -- | -- | -- | -- | -- | -- | -- | -- | -- |
| MAXIMUM ALTITUDE | -- | -- | -- | -- | -- | -- | -- | -- | -- | -- |
| RANGE | 240 | 247 | 255 | 262 | 265 | 268 | 272 | 280 | 281 | -- |
| SPEED | -- | -- | -- | -- | -- | -- | -- | -- | -- | -- |
| CEP | 240 | 247 | 255 | 262 | 265 | 268 | 273 | 280 | 281 | 282 |
| **WARHEAD** | | | | | | | | | | |
| WARHEAD TYPE | 240 | 247 | 256 | -- | -- | -- | 273 | -- | -- | -- |
| WARHEAD CHARACTERISTICS | 240 | 248 | 256 | 262 | 266 | 269 | 273 | 280 | 281 | 282 |
| NUCLEAR YIELD | 241 | 248 | 256 | 263 | 266 | 269 | 274 | 280 | 281 | 282 |
| WARHEAD WEIGHT | -- | -- | -- | -- | -- | -- | -- | -- | -- | -- |
| THROWWEIGHT | -- | -- | 256 | 263 | -- | 269 | 274 | 280 | 281 | -- |
| PAYLOAD | -- | -- | -- | -- | -- | -- | -- | -- | -- | -- |
| **DIMENSIONS** | | | | | | | | | | |
| WEIGHT | 241 | 248 | 256 | 263 | 266 | 269 | 274 | -- | -- | -- |
| LENGTH | 241 | 248 | 256 | 263 | 266 | 269 | 274 | -- | -- | -- |
| DIAMETER | 241 | 249 | 257 | 263 | 266 | 269 | 274 | -- | -- | -- |
| WINGSPAN | -- | -- | -- | -- | -- | -- | -- | -- | -- | -- |
| **PROPULSION** | | | | | | | | | | |
| NUMBER OF STAGES | 241 | 249 | 257 | 263 | 266 | 269 | 274 | -- | -- | -- |
| TYPE OF PROPULSION | 242 | 249 | 257 | 263 | 266 | 269 | 275 | 280 | 281 | -- |
| **GUIDANCE** | | | | | | | | | | |
| MIDCOURSE (OR MAIN) GUIDANCE | 242 | 249 | 257 | 263 | 266 | 269 | 275 | -- | -- | -- |
| TERMINAL GUIDANCE | -- | -- | -- | -- | -- | -- | -- | -- | -- | -- |
| CONTROL SYSTEM | -- | -- | -- | -- | -- | -- | -- | -- | -- | -- |
| RADAR | -- | -- | -- | -- | -- | -- | -- | -- | -- | -- |
| **HISTORY** | | | | | | | | | | |
| FIRST OBSERVED | -- | -- | -- | -- | -- | -- | -- | -- | -- | -- |
| OBSERVED | -- | -- | -- | -- | -- | -- | -- | -- | -- | -- |
| OBSERVED IN PARADE | 242 | 249 | 257 | -- | -- | -- | -- | -- | -- | -- |
| DEVELOPMENT BEGAN | 242 | 249 | 257 | -- | -- | -- | 275 | -- | -- | -- |
| DEVELOPMENT UNDERWAY | 242 | 249 | -- | -- | -- | -- | 275 | -- | -- | -- |
| PROTOTYPE TESTS | 242 | 249 | 257 | 263 | 266 | 269 | 275 | 280 | 281 | -- |
| PRODUCTION BEGAN | -- | -- | -- | -- | -- | -- | -- | -- | -- | -- |
| PRODUCTION UNDERWAY | -- | -- | 257 | -- | -- | -- | 275 | -- | -- | -- |
| PRODUCTION ENDED | -- | -- | -- | -- | -- | -- | -- | -- | -- | -- |
| IOC | 242 | 250 | 258 | 263 | 266 | 270 | 275 | 280 | 281 | -- |
| DEPLOYMENT BEGAN | 242 | 250 | 258 | 263 | 266 | 270 | 275 | 280 | 281 | 282 |
| DEPLOYED | 242 | 250 | 258 | 263 | 267 | 270 | 275 | 280 | 281 | -- |
| DEPLOYMENT COMPLETE | 242 | -- | -- | -- | -- | -- | -- | -- | -- | -- |
| RETIRED | 242 | 250 | -- | 263 | 267 | -- | -- | -- | -- | -- |
| DEPLOYMENT HISTORY | 243 | 250 | 258 | 263 | 267 | 270 | 276 | -- | 281 | -- |
| **NUMBERS; PLATFORM** | | | | | | | | | | |
| NUMBERS PRODUCED | -- | -- | -- | -- | -- | -- | 276 | -- | -- | -- |
| NUMBERS IN SERVICE | 243 | 250 | 258 | 263 | 267 | 270 | 276 | -- | -- | -- |
| UNDATED NUMBER | -- | -- | -- | -- | -- | -- | -- | -- | -- | -- |
| LAUNCHER/SILO | -- | -- | 259 | -- | -- | -- | -- | -- | -- | -- |
| CARRIERS | 243 | 251 | 259 | -- | -- | 270 | 277 | -- | -- | -- |
| **DESIGN; USE; STRATEGY** | | | | | | | | | | |
| DESIGNER | 244 | 252 | 260 | -- | -- | -- | 278 | -- | -- | -- |
| DESIGN AND ENGINEERING | -- | 252 | -- | -- | -- | -- | 278 | -- | -- | -- |
| USE AND CONFIGURATION | 244 | 252 | 261 | -- | -- | -- | 278 | -- | -- | -- |
| RELOADS IN GENERAL | -- | -- | -- | -- | -- | -- | -- | -- | -- | -- |
| TARGET TYPE | -- | -- | -- | -- | -- | -- | -- | -- | -- | -- |
| STRATEGY | 244 | 252 | 261 | -- | 267 | -- | 279 | -- | -- | -- |
| COMBAT REPORTS/EFFECTIVENESS | -- | 253 | -- | -- | -- | -- | -- | -- | -- | -- |
| GEOGRAPHICAL DEPLOYMENTS | 244 | 253 | 261 | -- | -- | -- | 279 | -- | -- | -- |
| EXPORTED TO | 245 | -- | -- | -- | -- | -- | -- | -- | -- | -- |
| RELATION TO OTHER MISSILES | 245 | 253 | 261 | -- | -- | -- | 279 | -- | -- | -- |
| OTHER INFORMATION | -- | -- | -- | -- | -- | -- | 279 | -- | -- | -- |

MISSILE INDEX  SLBMs

| | SS-N-17 | SS-N-18 | SS-N-18 Mod 1 | SS-N-18 Mod 2 | SS-N-18 Mod 3 | SS-N-20 | SS-N-23 |
|---|---|---|---|---|---|---|---|
| START OF MISSILE | 283 | 287 | 292 | 294 | 296 | 297 | 301 |
| **DESIGNATIONS** | | | | | | | |
| US DESIGNATION | 283 | 287 | 292 | 294 | 296 | 297 | 301 |
| NATO CODENAME | -- | — | — | — | — | -- | -- |
| SOVIET DESIGNATION | -- | 287 | — | — | — | — | -- |
| OTHER DESIGNATION | -- | — | — | — | — | -- | — |
| DESCRIPTION | 283 | 287 | — | — | -- | 297 | 301 |
| EDITOR'S NOTES | 283 | 287 | 292 | 294 | -- | 297 | 301 |
| **PERFORMANCE** | | | | | | | |
| MINIMUM ALTITUDE | -- | — | — | — | — | -- | -- |
| MAXIMUM ALTITUDE | -- | — | — | — | — | -- | — |
| RANGE | 283 | 287 | 292 | 294 | 296 | 297 | 301 |
| SPEED | -- | — | — | -- | — | — | — |
| CEP | 284 | 288 | 292 | 294 | 296 | 297 | 301 |
| **WARHEAD** | | | | | | | |
| WARHEAD TYPE | 284 | 288 | — | — | — | 297 | — |
| WARHEAD CHARACTERISTICS | 284 | 288 | 292 | 294 | 296 | 298 | 301 |
| NUCLEAR YIELD | 284 | 288 | 292 | 294 | 296 | 298 | 301 |
| WARHEAD WEIGHT | -- | — | — | — | — | -- | — |
| THROWWEIGHT | 284 | 288 | 292 | -- | 296 | 298 | 301 |
| PAYLOAD | -- | — | — | -- | — | 298 | -- |
| **DIMENSIONS** | | | | | | | |
| WEIGHT | 284 | 288 | — | — | — | — | — |
| LENGTH | 284 | 289 | — | — | — | 298 | 301 |
| DIAMETER | 285 | 289 | — | — | — | — | — |
| WINGSPAN | -- | — | — | — | — | -- | — |
| **PROPULSION** | | | | | | | |
| NUMBER OF STAGES | 285 | 289 | — | — | — | 298 | — |
| TYPE OF PROPULSION | 285 | 289 | 292 | 294 | 296 | 298 | 301 |
| **GUIDANCE** | | | | | | | |
| MIDCOURSE (OR MAIN) GUIDANCE | 285 | 289 | -- | — | — | 298 | -- |
| TERMINAL GUIDANCE | -- | — | — | — | — | -- | — |
| CONTROL SYSTEM | -- | — | — | — | — | -- | — |
| RADAR | -- | — | — | — | — | — | — |
| **HISTORY** | | | | | | | |
| FIRST OBSERVED | -- | — | — | — | — | 298 | -- |
| OBSERVED | -- | — | — | — | — | -- | — |
| OBSERVED IN PARADE | -- | — | — | — | — | -- | — |
| DEVELOPMENT BEGAN | 285 | 289 | -- | — | — | 298 | 301 |
| DEVELOPMENT UNDERWAY | 285 | -- | — | — | — | 299 | -- |
| PROTOTYPE TESTS | 285 | 289 | 293 | — | — | 299 | 301 |
| PRODUCTION BEGAN | -- | — | — | — | — | -- | — |
| PRODUCTION UNDERWAY | 285 | 289 | — | — | — | 299 | -- |
| PRODUCTION ENDED | -- | — | — | — | — | -- | — |
| IOC | 285 | 290 | -- | — | -- | 299 | 301 |
| DEPLOYMENT BEGAN | 285 | 290 | 293 | 295 | 296 | 299 | 301 |
| DEPLOYED | -- | — | 293 | 295 | 296 | 299 | -- |
| DEPLOYMENT COMPLETE | -- | — | — | — | — | -- | — |
| RETIRED | -- | — | — | — | — | -- | — |
| DEPLOYMENT HISTORY | 286 | 290 | 293 | — | 296 | 299 | 301 |
| **NUMBERS; PLATFORM** | | | | | | | |
| NUMBERS PRODUCED | -- | — | — | -- | — | 299 | -- |
| NUMBERS IN SERVICE | 286 | 290 | 293 | -- | — | 299 | 301 |
| UNDATED NUMBER | -- | — | — | — | — | -- | — |
| LAUNCHER/SILO | -- | — | — | — | — | -- | — |
| CARRIERS | 286 | 291 | — | — | -- | 300 | 302 |
| **DESIGN; USE; STRATEGY** | | | | | | | |
| DESIGNER | 286 | 291 | -- | — | -- | 300 | — |
| DESIGN AND ENGINEERING | 286 | 291 | — | — | — | 300 | 302 |
| USE AND CONFIGURATION | 286 | 291 | — | — | — | 300 | — |
| RELOADS IN GENERAL | -- | — | — | — | — | -- | — |
| TARGET TYPE | -- | — | — | — | — | — | — |
| STRATEGY | 286 | 291 | -- | — | -- | 300 | 302 |
| COMBAT REPORTS/EFFECTIVENESS | -- | — | — | — | — | -- | — |
| GEOGRAPHICAL DEPLOYMENTS | -- | — | — | — | — | 300 | -- |
| EXPORTED TO | -- | — | — | — | — | -- | — |
| RELATION TO OTHER MISSILES | 286 | 291 | — | — | — | 300 | — |
| OTHER INFORMATION | -- | — | — | — | — | -- | — |

686  MISSILE INDEX  IRBMs

| | SS-1 | SS-2 | SS-3 | SS-4 | SS-5 | SS-14 | SS-15 | SS-20 | SS-20 Mod 1 | SS-20 Mod 2 | SS-20 Mod 3 |
|---|---|---|---|---|---|---|---|---|---|---|---|
| START OF MISSILE | 303 | 305 | 307 | 311 | 319 | 326 | 331 | 334 | 343 | 344 | 345 |
| **DESIGNATIONS** | | | | | | | | | | | |
| US DESIGNATION | 303 | 305 | 307 | 311 | 319 | 326 | 331 | 334 | 343 | 344 | 345 |
| NATO CODENAME | 303 | 305 | 307 | 311 | 319 | 326 | 331 | -- | -- | -- | -- |
| SOVIET DESIGNATION | -- | -- | -- | -- | -- | -- | -- | -- | -- | -- | -- |
| OTHER DESIGNATION | -- | -- | 307 | 311 | -- | -- | -- | -- | -- | -- | -- |
| DESCRIPTION | 303 | 305 | 307 | 311 | 319 | 326 | 331 | 334 | -- | -- | 345 |
| EDITOR'S NOTES | -- | 305 | 307 | 311 | 319 | 326 | 331 | 334 | 343 | 344 | 345 |
| **PERFORMANCE** | | | | | | | | | | | |
| MINIMUM ALTITUDE | -- | -- | -- | -- | -- | -- | -- | -- | -- | -- | -- |
| MAXIMUM ALTITUDE | -- | -- | -- | -- | -- | -- | -- | -- | -- | -- | -- |
| RANGE | 303 | 305 | 307 | 312 | 320 | 326 | 331 | 335 | 343 | 344 | 345 |
| SPEED | -- | -- | 308 | 312 | 320 | -- | -- | 335 | -- | -- | -- |
| CEP | -- | -- | 308 | 312 | 320 | 327 | 331 | 335 | -- | 344 | -- |
| **WARHEAD** | | | | | | | | | | | |
| WARHEAD TYPE | 303 | 305 | 308 | 312 | 320 | 327 | 331 | 336 | -- | -- | -- |
| WARHEAD CHARACTERISTICS | 303 | 305 | 308 | 312 | 320 | 327 | 331 | 336 | 343 | 344 | 345 |
| NUCLEAR YIELD | -- | -- | 308 | 313 | 321 | 327 | 332 | 336 | 343 | 344 | 345 |
| WARHEAD WEIGHT | -- | -- | -- | -- | -- | -- | -- | -- | 343 | -- | 345 |
| THROWWEIGHT | -- | -- | -- | 313 | 321 | -- | -- | 336 | 343 | 344 | 345 |
| PAYLOAD | -- | -- | 308 | -- | -- | 327 | -- | -- | -- | -- | -- |
| **DIMENSIONS** | | | | | | | | | | | |
| WEIGHT | 303 | -- | 308 | 313 | 321 | 327 | 332 | 336 | -- | -- | -- |
| LENGTH | 303 | 305 | 308 | 313 | 321 | 327 | 332 | 336 | -- | -- | 345 |
| DIAMETER | 303 | -- | 308 | 313 | 321 | 327 | 332 | 336 | -- | -- | 345 |
| WINGSPAN | -- | -- | 308 | 313 | -- | -- | -- | -- | -- | -- | -- |
| **PROPULSION** | | | | | | | | | | | |
| NUMBER OF STAGES | 303 | -- | 308 | 314 | 321 | 327 | 332 | 336 | -- | -- | -- |
| TYPE OF PROPULSION | 303 | 305 | 309 | 314 | 322 | 328 | 332 | 337 | -- | 344 | 345 |
| **GUIDANCE** | | | | | | | | | | | |
| MIDCOURSE (OR MAIN) GUIDANCE | -- | -- | 309 | 314 | 322 | 328 | 332 | 337 | -- | -- | -- |
| TERMINAL GUIDANCE | -- | -- | -- | -- | -- | -- | -- | -- | -- | -- | -- |
| CONTROL SYSTEM | -- | -- | 309 | 314 | 322 | 328 | -- | 337 | -- | -- | -- |
| RADAR | -- | -- | -- | -- | -- | -- | -- | -- | -- | -- | -- |
| **HISTORY** | | | | | | | | | | | |
| FIRST OBSERVED | -- | -- | -- | 314 | -- | -- | -- | -- | -- | -- | -- |
| OBSERVED | -- | 305 | -- | -- | -- | 328 | 332 | -- | -- | -- | -- |
| OBSERVED IN PARADE | -- | -- | 309 | -- | 322 | 328 | 332 | -- | -- | -- | -- |
| DEVELOPMENT BEGAN | -- | -- | 309 | 314 | 322 | 328 | 332 | 337 | -- | -- | -- |
| DEVELOPMENT UNDERWAY | -- | -- | -- | -- | 322 | -- | -- | 337 | -- | -- | -- |
| PROTOTYPE TESTS | -- | -- | 309 | 314 | 322 | 328 | 332 | 337 | -- | -- | -- |
| PRODUCTION BEGAN | -- | -- | -- | -- | -- | -- | -- | -- | -- | -- | -- |
| PRODUCTION UNDERWAY | -- | -- | -- | -- | -- | -- | -- | -- | -- | -- | -- |
| PRODUCTION ENDED | -- | -- | -- | -- | -- | -- | -- | -- | -- | -- | -- |
| IOC | -- | -- | 309 | 314 | 322 | -- | -- | 337 | -- | -- | -- |
| DEPLOYMENT BEGAN | 303 | 305 | 309 | 315 | 322 | 328 | 332 | 337 | -- | 344 | 345 |
| DEPLOYED | -- | -- | 309 | 315 | 322 | 328 | 333 | 337 | -- | 344 | -- |
| DEPLOYMENT COMPLETE | -- | -- | -- | -- | -- | -- | -- | 337 | -- | -- | -- |
| RETIRED | 303 | -- | 309 | -- | 322 | -- | -- | -- | -- | -- | -- |
| DEPLOYMENT HISTORY | -- | 305 | -- | 315 | 323 | 328 | 333 | 338 | -- | -- | -- |
| **NUMBERS; PLATFORM** | | | | | | | | | | | |
| NUMBERS PRODUCED | -- | -- | -- | -- | -- | -- | -- | 338 | -- | -- | -- |
| NUMBERS IN SERVICE | -- | -- | 309 | 315 | 323 | 328 | -- | 338 | -- | -- | -- |
| UNDATED NUMBER | -- | 305 | 310 | 317 | 324 | -- | -- | 340 | -- | 344 | -- |
| LAUNCHER/SILO | -- | -- | -- | 317 | 324 | 329 | -- | -- | -- | -- | -- |
| CARRIERS | 303 | 305 | 310 | 317 | 325 | 329 | 333 | 340 | -- | 344 | -- |
| **DESIGN; USE; STRATEGY** | | | | | | | | | | | |
| DESIGNER | -- | -- | -- | 317 | 325 | 329 | 333 | 340 | -- | -- | -- |
| DESIGN AND ENGINEERING | 303 | 306 | 310 | 317 | 325 | 329 | -- | 340 | -- | -- | -- |
| USE AND CONFIGURATION | 303 | 306 | 310 | 318 | 325 | 329 | 333 | 341 | -- | -- | -- |
| RELOADS IN GENERAL | -- | -- | 310 | 318 | 325 | -- | -- | 341 | -- | -- | -- |
| TARGET TYPE | -- | -- | -- | -- | -- | -- | -- | -- | -- | -- | -- |
| STRATEGY | 304 | 306 | 310 | 318 | 325 | 329 | 333 | 341 | -- | -- | -- |
| COMBAT REPORTS/EFFECTIVENESS | -- | -- | -- | 318 | 325 | -- | -- | 341 | -- | -- | -- |
| GEOGRAPHICAL DEPLOYMENTS | -- | -- | -- | 318 | 325 | 329 | 333 | 342 | -- | -- | -- |
| EXPORTED TO | -- | -- | -- | 318 | 325 | -- | -- | -- | -- | -- | -- |
| RELATION TO OTHER MISSILES | -- | 306 | -- | -- | -- | 329 | 333 | 342 | -- | -- | -- |
| OTHER INFORMATION | 304 | 306 | 310 | 318 | 325 | 330 | 333 | -- | -- | -- | -- |

MISSILE INDEX  SRBMs  687

| | FROG | FROG 1 | FROG 2 | FROG 3 | FROG 4 | FROG 5 | FROG 6 | FROG 7 | SS-1 Scud | SS-1b Scud A | SS-1c Scud B | SS-1c Scud C | SS-12 | SS-21 | SS-22 | SS-23 |
|---|---|---|---|---|---|---|---|---|---|---|---|---|---|---|---|---|
| START OF MISSILE | 346 | 349 | 352 | 355 | 358 | 361 | 363 | 364 | 368 | 371 | 376 | 381 | 382 | 387 | 391 | 394 |
| **DESIGNATIONS** | | | | | | | | | | | | | | | | |
| US DESIGNATION | 346 | 349 | 352 | 355 | 358 | 361 | 363 | 364 | 368 | 371 | 376 | 381 | 382 | 387 | 391 | 394 |
| NATO CODENAME | -- | 349 | -- | -- | -- | -- | -- | -- | 368 | 371 | 376 | 381 | 382 | -- | 391 | -- |
| SOVIET DESIGNATION | 346 | -- | -- | -- | 358 | -- | -- | 364 | -- | -- | -- | -- | -- | -- | -- | -- |
| OTHER DESIGNATION | -- | 349 | 352 | -- | -- | -- | -- | -- | -- | 371 | -- | 381 | -- | -- | -- | -- |
| DESCRIPTION | 346 | 349 | 352 | 355 | 358 | 361 | 363 | 364 | 368 | 371 | 376 | -- | 382 | 387 | 391 | 394 |
| EDITOR'S NOTES | 346 | 349 | 352 | 355 | 358 | 361 | 363 | 364 | 368 | 371 | 376 | -- | 382 | 387 | 391 | 394 |
| **PERFORMANCE** | | | | | | | | | | | | | | | | |
| MINIMUM ALTITUDE | -- | -- | -- | -- | -- | -- | -- | -- | -- | -- | -- | -- | -- | -- | -- | -- |
| MAXIMUM ALTITUDE | -- | -- | -- | -- | -- | -- | -- | -- | -- | -- | -- | -- | -- | -- | -- | -- |
| RANGE | 346 | 349 | 352 | 355 | 358 | 361 | -- | 364 | 368 | 371 | 376 | 381 | 383 | 387 | 391 | 394 |
| SPEED | -- | -- | -- | -- | -- | -- | -- | -- | -- | 371 | 377 | -- | -- | -- | -- | -- |
| CEP | 346 | 349 | 352 | 355 | 358 | 361 | -- | 364 | 368 | 371 | 377 | 381 | 383 | 387 | 391 | 394 |
| **WARHEAD** | | | | | | | | | | | | | | | | |
| WARHEAD TYPE | 346 | 349 | 352 | 355 | 358 | 361 | -- | 364 | 368 | 372 | 377 | 381 | 383 | 388 | 391 | 394 |
| WARHEAD CHARACTERISTICS | 346 | 349 | 352 | 355 | 358 | 361 | -- | 365 | 368 | 372 | 377 | -- | 383 | 388 | 392 | 395 |
| NUCLEAR YIELD | 346 | 349 | 352 | 355 | 358 | 361 | -- | 365 | 368 | 372 | 377 | -- | 383 | 388 | 392 | 395 |
| WARHEAD WEIGHT | -- | 350 | 352 | 355 | 358 | 361 | -- | 365 | -- | 372 | 377 | -- | -- | -- | -- | -- |
| THROWWEIGHT | -- | -- | -- | -- | -- | -- | -- | -- | -- | -- | -- | -- | -- | -- | -- | -- |
| PAYLOAD | 346 | -- | 352 | -- | -- | -- | -- | -- | 368 | 372 | -- | -- | -- | 388 | 392 | 395 |
| **DIMENSIONS** | | | | | | | | | | | | | | | | |
| WEIGHT | -- | 350 | 353 | 356 | 358 | 361 | -- | 365 | -- | 372 | 377 | -- | 384 | -- | -- | -- |
| LENGTH | -- | 350 | 353 | 356 | 359 | 361 | -- | 365 | 368 | 372 | 377 | 381 | 384 | 388 | -- | -- |
| DIAMETER | -- | 350 | 353 | 356 | 359 | 362 | -- | 365 | 368 | 372 | 378 | 381 | 384 | 388 | -- | -- |
| WINGSPAN | -- | 350 | 353 | -- | 359 | -- | -- | 365 | -- | 373 | 378 | -- | -- | -- | -- | -- |
| **PROPULSION** | | | | | | | | | | | | | | | | |
| NUMBER OF STAGES | 346 | 350 | 353 | 356 | 359 | 362 | -- | 365 | -- | 373 | 378 | 381 | 384 | 388 | 392 | -- |
| TYPE OF PROPULSION | 347 | 350 | 353 | 356 | 359 | 362 | -- | 366 | 368 | 373 | 378 | 381 | 384 | 388 | 392 | 395 |
| **GUIDANCE** | | | | | | | | | | | | | | | | |
| MIDCOURSE (OR MAIN) GUIDANCE | 347 | 350 | 353 | 356 | 359 | 362 | -- | 366 | 369 | 373 | 378 | -- | 384 | 388 | -- | -- |
| TERMINAL GUIDANCE | -- | -- | -- | -- | -- | -- | -- | -- | -- | -- | -- | -- | -- | -- | -- | -- |
| CONTROL SYSTEM | -- | -- | -- | -- | -- | -- | -- | -- | 369 | 373 | 378 | -- | -- | -- | -- | -- |
| RADAR | -- | -- | -- | -- | -- | -- | -- | -- | 369 | -- | -- | -- | -- | -- | -- | -- |
| **HISTORY** | | | | | | | | | | | | | | | | |
| FIRST OBSERVED | -- | -- | -- | -- | -- | -- | -- | 366 | -- | 373 | -- | -- | -- | -- | -- | -- |
| OBSERVED | 347 | 350 | 353 | -- | -- | -- | 363 | -- | -- | 373 | -- | -- | 384 | -- | 392 | -- |
| OBSERVED IN PARADE | -- | -- | 353 | -- | -- | -- | -- | -- | -- | -- | 378 | -- | 384 | -- | -- | -- |
| DEVELOPMENT BEGAN | -- | -- | -- | -- | -- | -- | -- | -- | -- | -- | -- | -- | -- | 388 | -- | -- |
| DEVELOPMENT UNDERWAY | -- | -- | -- | -- | -- | -- | -- | -- | -- | 373 | -- | -- | -- | 388 | 392 | 395 |
| PROTOTYPE TESTS | -- | -- | -- | -- | -- | -- | -- | -- | -- | -- | -- | -- | -- | 388 | -- | -- |
| PRODUCTION BEGAN | -- | -- | -- | -- | -- | -- | -- | -- | -- | -- | -- | -- | -- | -- | -- | -- |
| PRODUCTION UNDERWAY | -- | -- | -- | -- | -- | -- | -- | -- | -- | -- | -- | -- | -- | -- | -- | -- |
| PRODUCTION ENDED | -- | -- | -- | -- | -- | -- | -- | -- | -- | -- | -- | -- | -- | -- | -- | -- |
| IOC | -- | -- | -- | -- | -- | -- | -- | -- | -- | -- | -- | -- | 384 | 388 | -- | 395 |
| DEPLOYMENT BEGAN | 347 | 351 | 353 | 356 | 359 | 362 | 363 | 366 | -- | 373 | 378 | 381 | 385 | 389 | 392 | 395 |
| DEPLOYED | 347 | 351 | 353 | 356 | 359 | 362 | -- | 366 | 369 | 373 | 378 | -- | 385 | 389 | 392 | 395 |
| DEPLOYMENT COMPLETE | -- | -- | -- | -- | -- | -- | -- | -- | -- | -- | -- | -- | -- | -- | -- | -- |
| RETIRED | -- | 351 | 354 | -- | 359 | -- | -- | -- | -- | 373 | -- | -- | -- | -- | -- | -- |
| DEPLOYMENT HISTORY | 347 | -- | -- | 356 | -- | -- | 366 | 369 | 374 | 378 | -- | 385 | 389 | 392 | 395 |
| **NUMBERS; PLATFORM** | | | | | | | | | | | | | | | | |
| NUMBERS PRODUCED | -- | -- | 354 | 356 | 359 | 362 | -- | -- | -- | -- | -- | -- | -- | 389 | -- | 395 |
| NUMBERS IN SERVICE | 347 | 351 | -- | -- | -- | -- | -- | 366 | 369 | 374 | 378 | -- | 385 | 389 | 392 | 395 |
| UNDATED NUMBER | -- | -- | -- | -- | -- | -- | -- | 366 | -- | 374 | 379 | -- | 386 | 390 | 393 | -- |
| LAUNCHER/SILO | -- | -- | -- | -- | -- | -- | -- | -- | -- | -- | -- | -- | -- | -- | -- | -- |
| CARRIERS | 348 | 351 | 354 | 356 | 359 | 362 | -- | 366 | 369 | 374 | 379 | 381 | 386 | 390 | 393 | 395 |
| **DESIGN; USE; STRATEGY** | | | | | | | | | | | | | | | | |
| DESIGNER | | | | | | | | | | | | | | | | |
| DESIGN AND ENGINEERING | -- | -- | -- | -- | -- | -- | -- | -- | -- | 374 | 379 | -- | 386 | 390 | -- | -- |
| USE AND CONFIGURATION | 348 | 351 | 354 | 357 | 360 | 362 | -- | 367 | 370 | 374 | 380 | -- | 386 | 390 | 393 | 396 |
| RELOADS IN GENERAL | -- | -- | -- | -- | -- | -- | -- | -- | -- | -- | -- | -- | -- | -- | -- | -- |
| TARGET TYPE | -- | -- | -- | -- | -- | -- | -- | -- | -- | -- | -- | -- | -- | -- | -- | -- |
| STRATEGY | 348 | 351 | -- | 357 | -- | 362 | 363 | 367 | 370 | 374 | 380 | -- | 386 | 390 | 393 | 396 |
| COMBAT REPORTS/EFFECTIVENESS | 348 | -- | -- | -- | -- | -- | -- | 367 | 370 | -- | 380 | -- | 386 | -- | 393 | 396 |
| GEOGRAPHICAL DEPLOYMENTS | 348 | -- | -- | -- | -- | -- | -- | 367 | 370 | -- | 380 | -- | 386 | 390 | 393 | 396 |
| EXPORTED TO | 348 | -- | 354 | 357 | 360 | 362 | -- | 367 | 370 | 375 | 380 | -- | -- | 390 | -- | -- |
| RELATION TO OTHER MISSILES | 348 | 351 | 354 | 357 | 360 | 362 | -- | 367 | -- | 375 | 380 | 381 | -- | 390 | 393 | -- |
| OTHER INFORMATION | 348 | 351 | 354 | 357 | 360 | -- | -- | 367 | -- | 375 | 380 | -- | -- | 390 | -- | -- |

## MISSILE INDEX  Ground-Launched SSMs

| | SSC-1 | SSC-2a | SSC-2b | SSC-4 | GLCM |
|---|---|---|---|---|---|
| START OF MISSILE | 397 | 401 | 402 | 405 | 407 |
| **DESIGNATIONS** | | | | | |
| US DESIGNATION | 397 | 401 | 402 | 405 | 407 |
| NATO CODENAME | 397 | 401 | 402 | — | — |
| SOVIET DESIGNATION | — | — | — | — | — |
| OTHER DESIGNATION | — | — | — | 405 | — |
| DESCRIPTION | 397 | 401 | 402 | 405 | 407 |
| EDITOR'S NOTES | 397 | — | — | 405 | 407 |
| **PERFORMANCE** | | | | | |
| MINIMUM ALTITUDE | — | — | — | — | — |
| MAXIMUM ALTITUDE | — | — | — | — | — |
| RANGE | 397 | 401 | 402 | 405 | 407 |
| SPEED | 397 | 401 | 402 | — | — |
| CEP | 397 | — | — | 405 | 407 |
| **WARHEAD** | | | | | |
| WARHEAD TYPE | 397 | — | 402 | 405 | 407 |
| WARHEAD CHARACTERISTICS | 397 | — | — | 405 | — |
| NUCLEAR YIELD | 398 | — | — | — | — |
| WARHEAD WEIGHT | 398 | — | — | — | — |
| THROWWEIGHT | — | — | — | — | — |
| PAYLOAD | — | — | — | — | — |
| **DIMENSIONS** | | | | | |
| WEIGHT | 398 | 401 | 402 | — | — |
| LENGTH | 398 | 401 | 402 | 405 | 407 |
| DIAMETER | 398 | — | 402 | — | — |
| WINGSPAN | 398 | 401 | 402 | — | — |
| **PROPULSION** | | | | | |
| NUMBER OF STAGES | 398 | — | 403 | — | — |
| TYPE OF PROPULSION | 398 | 401 | 403 | 405 | — |
| **GUIDANCE** | | | | | |
| MIDCOURSE (OR MAIN) GUIDANCE | 398 | — | 403 | — | — |
| TERMINAL GUIDANCE | 398 | 401 | 403 | — | — |
| CONTROL SYSTEM | 398 | — | 403 | — | — |
| RADAR | — | — | 403 | — | — |
| **HISTORY** | | | | | |
| FIRST OBSERVED | — | — | — | — | — |
| OBSERVED | — | — | — | — | — |
| OBSERVED IN PARADE | — | — | — | — | — |
| DEVELOPMENT BEGAN | — | — | — | 405 | — |
| DEVELOPMENT UNDERWAY | — | — | — | 405 | 407 |
| PROTOTYPE TESTS | — | — | — | 405 | — |
| PRODUCTION BEGAN | — | — | — | — | — |
| PRODUCTION UNDERWAY | — | — | — | — | — |
| PRODUCTION ENDED | — | — | — | — | — |
| IOC | — | — | — | — | 407 |
| DEPLOYMENT BEGAN | 399 | — | — | 405 | — |
| DEPLOYED | 399 | 401 | 403 | — | — |
| DEPLOYMENT COMPLETE | — | — | — | — | — |
| RETIRED | — | — | — | — | — |
| DEPLOYMENT HISTORY | — | — | 403 | 405 | — |
| **NUMBERS; PLATFORM** | | | | | |
| NUMBERS PRODUCED | — | — | — | — | — |
| NUMBERS IN SERVICE | 399 | — | — | — | — |
| UNDATED NUMBER | 399 | — | — | — | — |
| LAUNCHER/SILO | — | 401 | — | — | — |
| CARRIERS | 399 | — | 403 | 405 | — |
| **DESIGN; USE; STRATEGY** | | | | | |
| DESIGNER | — | — | — | — | — |
| DESIGN AND ENGINEERING | — | — | — | 405 | — |
| USE AND CONFIGURATION | 399 | — | — | 406 | — |
| RELOADS IN GENERAL | — | — | — | — | — |
| TARGET TYPE | 399 | — | 403 | 406 | — |
| STRATEGY | 399 | 401 | 403 | 406 | — |
| COMBAT REPORTS/EFFECTIVENESS | — | — | — | — | — |
| GEOGRAPHICAL DEPLOYMENTS | 399 | — | — | — | — |
| EXPORTED TO | — | — | 403 | — | — |
| RELATION TO OTHER MISSILES | 400 | 401 | 404 | 406 | 407 |
| OTHER INFORMATION | — | — | — | — | — |

| | AT-1 | AT-2 | AT-2 Swatter A | AT-2 Swatter B | AT-3 | AT-4 | AT-5 | AT-6 |
|---|---|---|---|---|---|---|---|---|
| START OF MISSILE | 408 | 412 | 416 | 417 | 418 | 423 | 425 | 427 |
| **DESIGNATIONS** | | | | | | | | |
| US DESIGNATION | 408 | 412 | — | — | 418 | 423 | 425 | 427 |
| NATO CODENAME | 408 | 412 | — | 417 | 418 | 423 | 425 | 427 |
| SOVIET DESIGNATION | 408 | — | — | — | 418 | — | — | — |
| OTHER DESIGNATION | — | — | — | — | — | — | 425 | — |
| DESCRIPTION | 408 | 412 | — | — | 418 | 423 | 425 | 427 |
| EDITOR'S NOTES | 408 | 412 | — | — | 418 | — | — | 427 |
| **PERFORMANCE** | | | | | | | | |
| MINIMUM ALTITUDE | — | — | — | — | — | — | — | — |
| MAXIMUM ALTITUDE | — | — | — | — | — | — | — | — |
| RANGE | 408 | 412 | 416 | 417 | 418 | 423 | 425 | 427 |
| SPEED | 408 | 412 | — | 417 | 419 | 423 | 425 | 427 |
| CEP | — | — | — | — | — | — | — | — |
| **WARHEAD** | | | | | | | | |
| WARHEAD TYPE | 408 | 413 | — | — | 419 | 423 | 425 | 427 |
| PENETRATION | 408 | 413 | — | — | 419 | 423 | 425 | — |
| WARHEAD CHARACTERISTICS | — | — | — | — | — | — | — | — |
| NUCLEAR YIELD | — | — | — | — | — | — | — | — |
| WARHEAD WEIGHT | 409 | — | — | — | 419 | — | — | — |
| THROWWEIGHT | — | — | — | — | — | — | — | — |
| PAYLOAD | — | — | — | — | — | — | — | — |
| **DIMENSIONS** | | | | | | | | |
| WEIGHT | 409 | 413 | — | 417 | 419 | 423 | 425 | 427 |
| LENGTH | 409 | 413 | — | 417 | 419 | 423 | 425 | — |
| DIAMETER | 409 | 413 | — | 417 | 419 | 423 | 425 | 427 |
| WINGSPAN | 409 | 413 | — | — | 420 | — | 426 | — |
| **PROPULSION** | | | | | | | | |
| NUMBER OF STAGES | 409 | 414 | — | — | 420 | — | — | 427 |
| TYPE OF PROPULSION | 409 | 414 | — | — | 420 | 423 | 426 | 427 |
| **GUIDANCE** | | | | | | | | |
| MIDCOURSE (OR MAIN) GUIDANCE | 409 | 414 | — | 417 | 420 | 423 | 426 | 427 |
| TERMINAL GUIDANCE | — | 414 | — | 417 | 420 | — | — | — |
| TRANSMISSION METHOD | 410 | 414 | — | 417 | 420 | 423 | 426 | 427 |
| CONTROL SYSTEM | 410 | 414 | — | — | 420 | — | — | — |
| RADAR | — | — | — | — | — | — | — | — |
| **HISTORY** | | | | | | | | |
| FIRST OBSERVED | — | 414 | — | — | 420 | — | 426 | — |
| OBSERVED | — | — | — | — | 420 | 423 | — | — |
| OBSERVED IN PARADE | — | — | — | — | — | — | — | — |
| DEVELOPMENT BEGAN | — | — | — | — | — | — | — | — |
| DEVELOPMENT UNDERWAY | — | — | — | — | — | — | — | — |
| PROTOTYPE TESTS | — | — | — | — | — | — | — | — |
| PRODUCTION BEGAN | — | — | — | — | — | — | — | — |
| PRODUCTION UNDERWAY | — | — | — | — | — | — | — | — |
| PRODUCTION ENDED | 410 | — | — | — | — | — | — | — |
| IOC | — | — | — | — | — | — | — | — |
| DEPLOYMENT BEGAN | — | — | — | — | 420 | — | — | 428 |
| DEPLOYED | 410 | 414 | — | — | 420 | 424 | 426 | 428 |
| DEPLOYMENT COMPLETE | — | — | — | — | — | — | — | — |
| RETIRED | 410 | — | — | — | — | — | — | — |
| DEPLOYMENT HISTORY | 410 | 414 | — | 417 | 420 | 424 | 426 | — |
| **NUMBERS; PLATFORM** | | | | | | | | |
| NUMBERS PRODUCED | — | — | — | — | 420 | — | — | — |
| NUMBERS IN SERVICE | — | 414 | — | — | 420 | — | — | — |
| LAUNCHER/SILO | — | — | — | — | — | — | — | — |
| CARRIERS | 410 | 414 | — | 417 | 421 | 424 | 426 | 428 |
| **DESIGN; USE; STRATEGY** | | | | | | | | |
| DESIGNER | — | — | — | — | — | — | — | — |
| DESIGN AND ENGINEERING | — | — | — | — | — | — | — | — |
| USE AND CONFIGURATION | 410 | 415 | — | — | 422 | 424 | 426 | 428 |
| RELOADS IN GENERAL | — | — | — | — | — | — | — | — |
| TARGET TYPE | — | — | — | — | — | — | — | — |
| STRATEGY | — | — | — | — | — | — | — | 428 |
| COMBAT REPORTS/EFFECTIVENESS | 410 | — | — | — | 422 | — | — | — |
| GEOGRAPHICAL DEPLOYMENTS | — | — | — | — | — | — | 426 | — |
| EXPORTED TO | 411 | 415 | — | — | 422 | 424 | — | 428 |
| RELATION TO OTHER MISSILES | 411 | 415 | — | 417 | — | 424 | 426 | 428 |
| OTHER INFORMATION | — | — | — | — | 422 | — | — | 428 |

MISSILE INDEX ASMs

| | AS-1 | AS-2 | AS-3 | AS-4 | AS-5 | AS-6 | AS-7 | AS-8 | AS-9 | AS-10 | AS-11 | AS-12 | AS-15 | TASM |
|---|---|---|---|---|---|---|---|---|---|---|---|---|---|---|
| START OF MISSILE | 429 | 432 | 436 | 441 | 447 | 452 | 458 | 461 | 462 | 464 | 466 | 467 | 468 | 470 |
| **DESIGNATIONS** | | | | | | | | | | | | | | |
| US DESIGNATION | 429 | 432 | 436 | 441 | 447 | 452 | 458 | 461 | 462 | 464 | 466 | 467 | 468 | 470 |
| NATO CODENAME | 429 | 432 | 436 | 441 | 447 | 452 | 458 | -- | -- | -- | -- | -- | -- | -- |
| SOVIET DESIGNATION | 429 | -- | -- | -- | -- | -- | -- | -- | -- | -- | -- | -- | -- | -- |
| OTHER DESIGNATION | -- | -- | -- | -- | -- | -- | -- | -- | -- | -- | -- | -- | -- | 470 |
| DESCRIPTION | -- | 432 | 436 | 441 | 447 | 452 | -- | -- | 462 | -- | -- | -- | 468 | -- |
| EDITOR'S NOTES | -- | 432 | 436 | 441 | 447 | 452 | -- | -- | -- | -- | -- | -- | 468 | 470 |
| **PERFORMANCE** | | | | | | | | | | | | | | |
| MINIMUM ALTITUDE | -- | -- | -- | -- | -- | -- | -- | -- | -- | -- | -- | -- | 468 | -- |
| MAXIMUM ALTITUDE | -- | -- | -- | -- | -- | -- | -- | -- | -- | -- | -- | -- | -- | -- |
| RANGE | 429 | 432 | 437 | 442 | 447 | 453 | 458 | 461 | 462 | 464 | 466 | -- | 468 | 470 |
| SPEED | 429 | 432 | 437 | 442 | 448 | 453 | 458 | 461 | 462 | 464 | 466 | -- | -- | 470 |
| CEP | -- | 432 | 437 | 442 | -- | 453 | -- | -- | -- | -- | -- | -- | 468 | -- |
| **WARHEAD** | | | | | | | | | | | | | | |
| WARHEAD TYPE | 429 | 433 | 437 | 442 | 448 | 453 | 458 | 461 | 462 | 464 | -- | -- | 468 | 470 |
| WARHEAD CHARACTERISTICS | -- | 433 | 437 | 442 | 448 | 453 | -- | -- | -- | -- | -- | -- | 468 | -- |
| NUCLEAR YIELD | -- | 433 | 438 | 443 | 448 | 454 | -- | -- | 462 | -- | -- | -- | -- | -- |
| WARHEAD WEIGHT | 429 | 433 | 438 | 443 | 448 | 454 | 458 | -- | 462 | 464 | -- | -- | -- | -- |
| THROWWEIGHT | -- | 433 | -- | -- | -- | -- | -- | -- | -- | -- | -- | -- | -- | -- |
| PAYLOAD | -- | -- | -- | -- | -- | -- | -- | -- | -- | -- | -- | -- | -- | -- |
| **DIMENSIONS** | | | | | | | | | | | | | | |
| WEIGHT | 429 | 433 | 438 | 443 | 448 | 454 | 459 | -- | -- | 464 | -- | -- | -- | -- |
| LENGTH | 429 | 433 | 438 | 443 | 449 | 454 | 459 | -- | 462 | 464 | 466 | -- | 468 | -- |
| DIAMETER | 429 | 433 | 438 | 443 | 449 | 454 | 459 | -- | -- | 464 | 466 | -- | -- | -- |
| WINGSPAN | 430 | 434 | 438 | 443 | 449 | 454 | -- | -- | 462 | -- | -- | -- | -- | -- |
| **PROPULSION** | | | | | | | | | | | | | | |
| NUMBER OF STAGES | 430 | 434 | 438 | 443 | 449 | 454 | 459 | -- | 462 | 464 | -- | -- | -- | -- |
| TYPE OF PROPULSION | 430 | 434 | 439 | 444 | 449 | 455 | 459 | 461 | 463 | 465 | 466 | -- | 468 | 470 |
| **GUIDANCE** | | | | | | | | | | | | | | |
| MIDCOURSE (OR MAIN) GUIDANCE | 430 | 434 | 439 | 444 | 449 | 455 | 459 | 461 | 463 | 465 | -- | -- | -- | 470 |
| TERMINAL GUIDANCE | 430 | 434 | 439 | 444 | 450 | 455 | 459 | 461 | 463 | 465 | -- | -- | -- | 470 |
| CONTROL SYSTEM | 430 | -- | 439 | 444 | 450 | 455 | -- | -- | -- | -- | -- | -- | -- | -- |
| RADAR | -- | -- | -- | -- | 450 | -- | -- | -- | -- | -- | -- | -- | -- | -- |
| **HISTORY** | | | | | | | | | | | | | | |
| FIRST OBSERVED | 430 | 434 | 439 | 444 | 450 | 455 | -- | -- | -- | -- | -- | -- | -- | -- |
| OBSERVED | -- | -- | -- | -- | -- | -- | -- | 461 | -- | -- | -- | -- | -- | -- |
| OBSERVED IN PARADE | -- | -- | 439 | 444 | -- | -- | -- | -- | -- | -- | -- | -- | -- | -- |
| DEVELOPMENT BEGAN | 430 | -- | -- | -- | -- | -- | -- | -- | -- | -- | -- | -- | -- | -- |
| DEVELOPMENT UNDERWAY | -- | -- | -- | -- | -- | -- | -- | 461 | 463 | -- | -- | -- | 469 | -- |
| PROTOTYPE TESTS | -- | -- | -- | -- | -- | -- | -- | -- | -- | -- | -- | -- | 469 | -- |
| PRODUCTION BEGAN | 430 | -- | -- | -- | -- | -- | -- | -- | -- | -- | -- | -- | -- | -- |
| PRODUCTION UNDERWAY | -- | -- | -- | -- | -- | -- | -- | -- | -- | -- | -- | -- | 469 | -- |
| PRODUCTION ENDED | -- | -- | -- | -- | -- | -- | -- | -- | -- | -- | -- | -- | -- | -- |
| IOC | 430 | 434 | 439 | 444 | 450 | 455 | -- | -- | -- | -- | -- | -- | 469 | -- |
| DEPLOYMENT BEGAN | 430 | 434 | 439 | 444 | 450 | 455 | 459 | -- | 463 | 465 | -- | -- | 469 | -- |
| DEPLOYED | 430 | 434 | 439 | 444 | 450 | 455 | 459 | -- | -- | -- | -- | -- | -- | -- |
| DEPLOYMENT COMPLETE | -- | -- | -- | -- | -- | -- | -- | -- | -- | -- | -- | -- | -- | -- |
| RETIRED | 430 | -- | 439 | -- | -- | -- | -- | -- | -- | -- | -- | -- | -- | -- |
| DEPLOYMENT HISTORY | 430 | -- | -- | 445 | 450 | -- | -- | -- | -- | -- | 466 | -- | -- | -- |
| **NUMBERS; PLATFORM** | | | | | | | | | | | | | | |
| NUMBERS PRODUCED | -- | -- | -- | 445 | 450 | -- | -- | -- | -- | -- | -- | -- | -- | -- |
| NUMBERS IN SERVICE | -- | 435 | 439 | 445 | 450 | 455 | -- | -- | -- | -- | -- | -- | -- | -- |
| UNDATED NUMBER | -- | -- | -- | -- | -- | -- | -- | -- | -- | -- | -- | -- | -- | -- |
| LAUNCHER/SILO | -- | -- | -- | -- | -- | -- | -- | -- | -- | -- | -- | -- | -- | -- |
| CARRIERS | 430 | 435 | 439 | 445 | 450 | 456 | 459 | 461 | 463 | 465 | 466 | 467 | 469 | 470 |
| **DESIGN; USE; STRATEGY** | | | | | | | | | | | | | | |
| DESIGNER | -- | -- | -- | -- | -- | -- | -- | -- | -- | -- | -- | -- | -- | -- |
| DESIGN AND ENGINEERING | -- | -- | -- | 446 | -- | 456 | 460 | -- | -- | -- | -- | -- | -- | -- |
| USE AND CONFIGURATION | -- | 435 | 440 | 446 | 451 | 456 | 460 | -- | -- | -- | -- | -- | 469 | -- |
| RELOADS IN GENERAL | -- | -- | -- | -- | -- | -- | -- | -- | -- | -- | -- | -- | -- | -- |
| TARGET TYPE | 431 | 435 | 440 | 446 | 451 | 456 | -- | 461 | 463 | -- | -- | -- | 469 | -- |
| STRATEGY | 431 | 435 | 440 | 446 | 451 | 457 | 460 | -- | -- | 465 | 466 | 467 | 469 | -- |
| COMBAT REPORTS/EFFECTIVENESS | -- | 435 | -- | -- | 451 | -- | -- | -- | -- | -- | -- | -- | -- | -- |
| GEOGRAPHICAL DEPLOYMENTS | -- | -- | -- | -- | -- | -- | -- | -- | -- | -- | -- | -- | -- | -- |
| EXPORTED TO | 431 | -- | -- | 446 | 451 | -- | -- | -- | -- | -- | -- | -- | -- | -- |
| RELATION TO OTHER MISSILES | 431 | 435 | 440 | 446 | 451 | -- | 460 | 461 | -- | 465 | -- | -- | 469 | 470 |
| OTHER INFORMATION | -- | -- | -- | 446 | -- | 457 | -- | -- | -- | -- | -- | -- | -- | -- |

# MISSILE INDEX  Sea-Launched SSMs

| | SS-N-1 | SS-N-2 | SS-N-2a | SS-N-2b | SS-N-2c | SS-N-3 | SS-N-3a | SS-N-3b | SS-N-3c | SS-N-7 | SS-N-9 | SS-N-10 | SS-N-12 | SS-N-13 |
|---|---|---|---|---|---|---|---|---|---|---|---|---|---|---|
| START OF MISSILE | 471 | 475 | 481 | 483 | 485 | 488 | 495 | 497 | 498 | 499 | 503 | 508 | 510 | 514 |
| **DESIGNATIONS** | | | | | | | | | | | | | | |
| US DESIGNATION | 471 | 475 | 481 | 483 | 485 | 488 | 495 | 497 | 498 | 499 | 503 | 508 | 510 | 514 |
| NATO CODENAME | 471 | 475 | 481 | 483 | 485 | 488 | 495 | 497 | — | 499 | 503 | — | 510 | — |
| SOVIET DESIGNATION | — | — | — | — | — | — | — | — | — | — | — | — | — | — |
| OTHER DESIGNATION | 471 | 475 | — | — | — | — | — | — | — | — | — | — | — | — |
| DESCRIPTION | 471 | 475 | — | — | 485 | 488 | — | — | — | 499 | 503 | — | 510 | 514 |
| EDITOR'S NOTES | 471 | 475 | — | — | 485 | 488 | — | — | — | 499 | 503 | 508 | 510 | 514 |
| **PERFORMANCE** | | | | | | | | | | | | | | |
| MINIMUM ALTITUDE | — | — | — | — | — | — | — | — | — | — | — | — | — | — |
| MAXIMUM ALTITUDE | — | — | — | — | — | — | — | — | — | — | — | — | — | — |
| RANGE | 471 | 475 | 481 | 483 | 485 | 489 | 495 | 497 | 498 | 499 | 504 | 508 | 510 | 514 |
| SPEED | 471 | 475 | 481 | — | 485 | 489 | 495 | 497 | — | 499 | 504 | 508 | 511 | 514 |
| CEP | — | — | — | — | 489 | — | — | — | — | — | — | — | 511 | — |
| **WARHEAD** | | | | | | | | | | | | | | |
| WARHEAD TYPE | 472 | 475 | 481 | 483 | 485 | 489 | — | — | — | 500 | 504 | 508 | 511 | 514 |
| WARHEAD CHARACTERISTICS | 472 | — | — | — | — | 489 | 495 | 497 | — | 500 | 504 | — | 511 | 514 |
| NUCLEAR YIELD | 472 | — | — | — | — | 490 | 495 | 497 | 498 | 500 | 504 | — | 511 | — |
| WARHEAD WEIGHT | 472 | 476 | 481 | 483 | 485 | 490 | 495 | — | 498 | 500 | 505 | — | 511 | — |
| THROWWEIGHT | — | — | — | — | — | 490 | — | — | — | 500 | — | — | 511 | — |
| PAYLOAD | — | — | — | — | — | — | — | — | — | — | — | — | — | — |
| **DIMENSIONS** | | | | | | | | | | | | | | |
| WEIGHT | 472 | 476 | 481 | 483 | 486 | 490 | — | — | — | 500 | 505 | 508 | 511 | — |
| LENGTH | 472 | 476 | 481 | 483 | 486 | 490 | 495 | 497 | 498 | 500 | 505 | 508 | 511 | — |
| DIAMETER | 472 | 476 | 481 | 483 | 486 | 490 | — | — | — | 500 | — | — | 511 | — |
| WINGSPAN | 472 | 476 | 481 | — | — | 490 | — | — | — | — | — | — | 511 | — |
| **PROPULSION** | | | | | | | | | | | | | | |
| NUMBER OF STAGES | 472 | 476 | 481 | 483 | 486 | 490 | — | — | — | — | 505 | — | — | 514 |
| TYPE OF PROPULSION | 472 | 476 | 481 | 483 | 486 | 491 | 495 | 497 | — | 500 | 505 | — | 512 | 514 |
| **GUIDANCE** | | | | | | | | | | | | | | |
| MIDCOURSE (OR MAIN) GUIDANCE | 472 | 476 | 482 | 483 | 486 | 491 | 495 | 497 | 498 | 501 | 505 | 508 | 512 | 514 |
| TERMINAL GUIDANCE | 473 | 476 | 482 | 483 | 486 | 491 | 495 | — | — | 501 | 505 | 508 | 512 | 515 |
| CONTROL SYSTEM | 473 | 477 | 482 | 483 | 486 | 491 | — | — | — | 501 | 505 | — | 512 | — |
| RADAR | — | 477 | — | — | — | 491 | — | — | — | — | 505 | 508 | 512 | — |
| **HISTORY** | | | | | | | | | | | | | | |
| FIRST OBSERVED | — | — | — | — | — | — | — | — | — | — | — | — | — | — |
| OBSERVED | — | — | — | — | — | 491 | — | — | — | — | 505 | — | 512 | 515 |
| OBSERVED IN PARADE | — | — | — | — | — | — | — | — | — | — | — | — | — | — |
| DEVELOPMENT BEGAN | — | — | — | — | — | — | — | — | — | — | — | — | — | — |
| DEVELOPMENT UNDERWAY | — | — | — | — | — | — | — | — | — | — | — | — | — | — |
| PROTOTYPE TESTS | — | — | — | — | — | — | — | — | — | — | — | — | — | 515 |
| PRODUCTION BEGAN | 473 | — | — | — | — | — | — | — | — | — | — | — | — | — |
| PRODUCTION UNDERWAY | — | 477 | — | — | — | 491 | — | — | — | 501 | 505 | — | 512 | — |
| PRODUCTION ENDED | 473 | — | — | — | — | — | — | — | — | — | — | — | — | — |
| IOC | 473 | 477 | 482 | 483 | 486 | 491 | 495 | 497 | 498 | 501 | 506 | 508 | 512 | 515 |
| DEPLOYMENT BEGAN | 473 | 477 | 482 | 484 | 486 | 492 | 495 | 497 | 498 | 501 | 506 | 508 | 512 | 515 |
| DEPLOYED | 473 | 477 | — | — | 486 | 492 | 495 | 497 | — | 501 | 506 | 508 | 512 | — |
| DEPLOYMENT COMPLETE | — | — | — | — | — | — | — | — | — | — | — | — | — | — |
| RETIRED | 473 | — | — | — | — | — | — | — | 498 | — | — | — | — | — |
| DEPLOYMENT HISTORY | 473 | 477 | — | — | 486 | 492 | — | — | — | 501 | 506 | — | 512 | — |
| **NUMBERS; PLATFORM** | | | | | | | | | | | | | | |
| NUMBERS PRODUCED | — | 477 | — | — | — | — | — | — | — | — | — | — | — | — |
| NUMBERS IN SERVICE | — | — | — | — | 486 | 492 | — | — | — | 501 | 506 | — | 512 | — |
| UNDATED NUMBER | — | 477 | — | — | — | — | — | — | — | — | — | — | — | — |
| LAUNCHER/SILO | — | — | — | — | — | — | — | — | — | — | — | — | — | — |
| CARRIERS | 473 | 477 | 482 | 484 | 486 | 492 | 495 | 497 | 498 | 501 | 506 | 508 | 513 | 515 |
| **DESIGN; USE; STRATEGY** | | | | | | | | | | | | | | |
| DESIGNER | — | — | — | — | — | — | — | — | — | — | — | — | — | — |
| DESIGN AND ENGINEERING | — | — | — | — | — | — | — | — | — | 502 | — | — | — | — |
| USE AND CONFIGURATION | 474 | 478 | 482 | 484 | 487 | 493 | 496 | — | — | 502 | 507 | — | 513 | 515 |
| RELOADS IN GENERAL | — | — | — | — | — | — | — | — | — | — | — | — | — | — |
| TARGET TYPE | 474 | — | — | — | 487 | 493 | — | — | — | 502 | 507 | 509 | 513 | 515 |
| STRATEGY | — | 478 | — | — | — | 494 | — | — | 498 | — | — | — | 513 | 515 |
| COMBAT REPORTS/EFFECTIVENESS | 474 | 478 | 482 | — | — | 494 | — | — | — | — | — | — | — | — |
| GEOGRAPHICAL DEPLOYMENTS | — | — | — | — | — | — | — | — | — | — | — | — | — | — |
| EXPORTED TO | — | 479 | — | — | 487 | 494 | — | — | — | — | — | — | — | — |
| RELATION TO OTHER MISSILES | 474 | — | — | — | 487 | 494 | — | — | 498 | 502 | — | — | — | — |
| OTHER INFORMATION | 474 | 480 | — | 484 | — | 494 | — | — | — | — | — | — | — | — |

| | SS-N-14 | SS-N-15 | SS-N-16 | SS-N-19 | SS-N-21 | SS-N-22 | SS-N-24 |
|---|---|---|---|---|---|---|---|
| START OF MISSILE | 516 | 520 | 523 | 525 | 528 | 530 | 532 |
| **DESIGNATIONS** | | | | | | | |
| US DESIGNATION | 516 | 520 | 523 | 525 | 528 | 530 | 532 |
| NATO CODENAME | 516 | -- | -- | -- | -- | -- | -- |
| SOVIET DESIGNATION | -- | -- | -- | -- | -- | -- | -- |
| OTHER DESIGNATION | -- | -- | -- | -- | -- | -- | -- |
| DESCRIPTION | 516 | 520 | 523 | 525 | 528 | 530 | 532 |
| EDITOR'S NOTES | 516 | 520 | 523 | 525 | 528 | 530 | 532 |
| **PERFORMANCE** | | | | | | | |
| MINIMUM ALTITUDE | -- | -- | -- | -- | -- | -- | -- |
| MAXIMUM ALTITUDE | -- | -- | -- | -- | -- | -- | -- |
| RANGE | 516 | 520 | 523 | 525 | 528 | 530 | 532 |
| SPEED | 516 | -- | -- | 525 | 528 | 530 | -- |
| CEP | -- | -- | -- | -- | 528 | -- | 532 |
| **WARHEAD** | | | | | | | |
| WARHEAD TYPE | 516 | 520 | 523 | 525 | 528 | 530 | 532 |
| WARHEAD CHARACTERISTICS | 517 | 520 | 523 | 525 | 528 | 530 | -- |
| NUCLEAR YIELD | 517 | 521 | -- | 526 | 528 | 530 | -- |
| WARHEAD WEIGHT | 517 | -- | -- | 526 | -- | -- | -- |
| THROWWEIGHT | -- | -- | -- | -- | -- | -- | -- |
| PAYLOAD | -- | -- | -- | -- | -- | -- | -- |
| **DIMENSIONS** | | | | | | | |
| WEIGHT | -- | -- | -- | -- | -- | -- | -- |
| LENGTH | 517 | -- | -- | -- | 528 | -- | 532 |
| DIAMETER | -- | 521 | -- | -- | -- | -- | -- |
| WINGSPAN | -- | -- | -- | -- | -- | -- | -- |
| **PROPULSION** | | | | | | | |
| NUMBER OF STAGES | -- | -- | -- | -- | -- | -- | -- |
| TYPE OF PROPULSION | 517 | 521 | 523 | 526 | -- | 530 | -- |
| **GUIDANCE** | | | | | | | |
| MIDCOURSE (OR MAIN) GUIDANCE | 517 | 521 | 523 | 526 | -- | 530 | -- |
| TERMINAL GUIDANCE | -- | -- | -- | 526 | -- | 530 | -- |
| CONTROL SYSTEM | 517 | -- | -- | -- | -- | -- | -- |
| RADAR | 517 | -- | -- | 526 | -- | -- | -- |
| **HISTORY** | | | | | | | |
| FIRST OBSERVED | -- | -- | -- | -- | -- | -- | -- |
| OBSERVED | 517 | -- | -- | -- | -- | -- | -- |
| OBSERVED IN PARADE | -- | -- | -- | -- | -- | -- | -- |
| DEVELOPMENT BEGAN | -- | -- | -- | -- | -- | -- | -- |
| DEVELOPMENT UNDERWAY | -- | -- | -- | -- | 529 | -- | 532 |
| PROTOTYPE TESTS | -- | -- | -- | -- | 529 | -- | -- |
| PRODUCTION BEGAN | -- | -- | -- | -- | -- | -- | -- |
| PRODUCTION UNDERWAY | -- | -- | -- | 526 | 529 | 530 | -- |
| PRODUCTION ENDED | -- | -- | -- | -- | -- | -- | -- |
| IOC | 517 | 521 | 523 | 526 | 529 | 530 | 532 |
| DEPLOYMENT BEGAN | 517 | 521 | 523 | 526 | 529 | 530 | -- |
| DEPLOYED | 517 | 521 | -- | 526 | -- | 530 | -- |
| DEPLOYMENT COMPLETE | -- | -- | -- | -- | -- | -- | -- |
| RETIRED | -- | -- | -- | -- | -- | -- | -- |
| DEPLOYMENT HISTORY | -- | -- | -- | -- | -- | -- | -- |
| **NUMBERS; PLATFORM** | | | | | | | |
| NUMBERS PRODUCED | -- | -- | -- | -- | -- | -- | -- |
| NUMBERS IN SERVICE | 517 | -- | -- | 526 | -- | 530 | -- |
| UNDATED NUMBER | -- | -- | -- | -- | -- | -- | -- |
| LAUNCHER/SILO | -- | -- | -- | -- | -- | -- | -- |
| CARRIERS | 518 | 521 | 523 | 526 | 529 | 530 | 532 |
| **DESIGN; USE; STRATEGY** | | | | | | | |
| DESIGNER | -- | -- | -- | -- | -- | -- | -- |
| DESIGN AND ENGINEERING | -- | -- | -- | -- | -- | 531 | -- |
| USE AND CONFIGURATION | 519 | 522 | 524 | 527 | 529 | 531 | -- |
| RELOADS IN GENERAL | -- | -- | -- | -- | -- | -- | -- |
| TARGET TYPE | 519 | 522 | 524 | 527 | 529 | 531 | -- |
| STRATEGY | -- | -- | 524 | 527 | 529 | -- | -- |
| COMBAT REPORTS/EFFECTIVENESS | -- | -- | 524 | -- | -- | -- | -- |
| GEOGRAPHICAL DEPLOYMENTS | -- | -- | -- | -- | -- | -- | -- |
| EXPORTED TO | -- | -- | -- | -- | -- | -- | -- |
| RELATION TO OTHER MISSILES | 519 | 522 | 524 | 527 | 529 | 531 | 532 |
| OTHER INFORMATION | -- | -- | -- | 527 | -- | -- | -- |

|  | ABM-1 | ABM-3 | SH-4 | SH-8 |
|---|---|---|---|---|
| START OF MISSILE | 533 | 540 | 541 | 542 |
| **DESIGNATIONS** | | | | |
| US DESIGNATION | 533 | 540 | 541 | 542 |
| NATO CODENAME | 533 | -- | -- | -- |
| SOVIET DESIGNATION | -- | -- | -- | -- |
| OTHER DESIGNATION | -- | -- | -- | -- |
| DESCRIPTION | 533 | -- | -- | -- |
| EDITOR'S NOTES | 533 | 540 | 541 | 542 |
| **PERFORMANCE** | | | | |
| MINIMUM ALTITUDE | -- | -- | -- | -- |
| MAXIMUM ALTITUDE | 533 | -- | -- | -- |
| RANGE | 533 | -- | -- | -- |
| SPEED | -- | -- | -- | 542 |
| CEP | -- | -- | -- | -- |
| **WARHEAD** | | | | |
| WARHEAD TYPE | 534 | 540 | -- | -- |
| WARHEAD CHARACTERISTICS | 534 | -- | -- | -- |
| NUCLEAR YIELD | 534 | -- | -- | -- |
| WARHEAD WEIGHT | -- | -- | -- | -- |
| THROWWEIGHT | -- | -- | -- | -- |
| PAYLOAD | -- | -- | -- | -- |
| **DIMENSIONS** | | | | |
| WEIGHT | 534 | -- | -- | -- |
| LENGTH | 534 | -- | -- | -- |
| DIAMETER | 534 | -- | -- | -- |
| WINGSPAN | -- | -- | -- | -- |
| **PROPULSION** | | | | |
| NUMBER OF STAGES | 534 | 540 | -- | -- |
| TYPE OF PROPULSION | 534 | -- | -- | -- |
| **GUIDANCE** | | | | |
| MIDCOURSE (OR MAIN) GUIDANCE | 534 | -- | -- | -- |
| TERMINAL GUIDANCE | -- | -- | -- | -- |
| CONTROL SYSTEM | -- | -- | -- | -- |
| RADAR | 535 | 540 | -- | 542 |
| **HISTORY** | | | | |
| FIRST OBSERVED | -- | -- | -- | -- |
| OBSERVED | 536 | -- | -- | -- |
| OBSERVED IN PARADE | 536 | -- | -- | -- |
| DEVELOPMENT BEGAN | -- | -- | -- | -- |
| DEVELOPMENT UNDERWAY | 536 | -- | -- | 542 |
| PROTOTYPE TESTS | 536 | 540 | 541 | -- |
| PRODUCTION BEGAN | -- | -- | -- | -- |
| PRODUCTION UNDERWAY | -- | -- | -- | -- |
| PRODUCTION ENDED | -- | -- | -- | -- |
| IOC | 536 | 540 | -- | -- |
| DEPLOYMENT BEGAN | 537 | 540 | 541 | -- |
| DEPLOYED | 537 | -- | -- | -- |
| DEPLOYMENT COMPLETE | -- | -- | -- | -- |
| RETIRED | -- | -- | -- | -- |
| DEPLOYMENT HISTORY | 537 | -- | -- | -- |
| **NUMBERS; PLATFORM** | | | | |
| NUMBERS PRODUCED | -- | -- | -- | -- |
| NUMBERS IN SERVICE | 537 | -- | -- | -- |
| UNDATED NUMBER | 538 | -- | -- | -- |
| LAUNCHER/SILO | 538 | -- | -- | -- |
| CARRIERS | -- | 540 | -- | -- |
| **DESIGN; USE; STRATEGY** | | | | |
| DESIGNER | -- | -- | -- | -- |
| DESIGN AND ENGINEERING | 538 | -- | -- | -- |
| USE AND CONFIGURATION | 538 | 540 | 541 | -- |
| RELOADS IN GENERAL | 538 | -- | -- | -- |
| TARGET TYPE | -- | -- | -- | -- |
| STRATEGY | 539 | 540 | 541 | 542 |
| COMBAT REPORTS/EFFECTIVENESS | 539 | -- | -- | -- |
| GEOGRAPHICAL DEPLOYMENTS | 539 | -- | -- | -- |
| EXPORTED TO | -- | -- | -- | -- |
| RELATION TO OTHER MISSILES | 539 | 540 | -- | -- |
| OTHER INFORMATION | 539 | 540 | -- | -- |

# MISSILE INDEX  Ground-Launched SAMs

| | SA-1 | SA-2 | SA-3 | SA-4 | SA-5 | SA-6 | SA-7 | SA-7b | SA-8 | SA-8b |
|---|---|---|---|---|---|---|---|---|---|---|
| **START OF MISSILE** | 543 | 547 | 556 | 563 | 569 | 576 | 583 | 588 | 589 | 595 |
| **DESIGNATIONS** | | | | | | | | | | |
| US DESIGNATION | 543 | 547 | 556 | 563 | 569 | 576 | 583 | 588 | 589 | 595 |
| NATO CODENAME | 543 | 547 | 556 | 563 | 569 | 576 | 583 | -- | 589 | -- |
| SOVIET DESIGNATION | -- | 547 | -- | -- | -- | -- | -- | -- | -- | -- |
| OTHER DESIGNATION | 543 | -- | 556 | 563 | 569 | -- | 583 | -- | -- | -- |
| DESCRIPTION | 543 | 547 | 556 | 563 | 569 | 576 | 583 | -- | 589 | -- |
| EDITOR'S NOTES | 543 | 547 | 556 | 563 | 569 | 576 | 583 | -- | 589 | -- |
| **PERFORMANCE** | | | | | | | | | | |
| MINIMUM ALTITUDE | -- | 548 | 556 | 563 | -- | 576 | 583 | -- | 589 | -- |
| MAXIMUM ALTITUDE | 543 | 548 | 557 | 563 | 569 | 576 | 583 | 588 | 589 | -- |
| RANGE | 543 | 548 | 557 | 564 | 570 | 577 | 584 | 588 | 590 | -- |
| SPEED | 543 | 549 | 557 | 564 | 570 | 577 | 584 | -- | 590 | -- |
| CEP | -- | -- | -- | -- | -- | -- | -- | -- | -- | -- |
| **WARHEAD** | | | | | | | | | | |
| WARHEAD TYPE | 543 | 549 | 557 | 564 | 570 | 577 | 584 | -- | 590 | -- |
| WARHEAD CHARACTERISTICS | -- | 549 | -- | -- | 570 | -- | -- | -- | -- | -- |
| NUCLEAR YIELD | -- | -- | -- | -- | -- | -- | -- | -- | -- | -- |
| WARHEAD WEIGHT | -- | 549 | 557 | 564 | 570 | 577 | 584 | -- | 590 | -- |
| THROWWEIGHT | -- | -- | -- | -- | -- | -- | -- | -- | -- | -- |
| PAYLOAD | -- | -- | -- | -- | -- | -- | -- | -- | -- | -- |
| **DIMENSIONS** | | | | | | | | | | |
| WEIGHT | 544 | 549 | 557 | 564 | 570 | 577 | 584 | -- | 590 | -- |
| LENGTH | 544 | 549 | 558 | 564 | 570 | 578 | 584 | -- | 590 | -- |
| DIAMETER | 544 | 550 | 558 | 565 | 571 | 578 | 584 | -- | 591 | -- |
| WINGSPAN | 544 | 550 | 558 | 565 | 571 | 578 | -- | -- | 591 | -- |
| **PROPULSION** | | | | | | | | | | |
| NUMBER OF STAGES | 544 | 550 | 558 | 565 | 571 | 578 | 584 | -- | 591 | -- |
| TYPE OF PROPULSION | 544 | 550 | 558 | 565 | 571 | 578 | 585 | 588 | 591 | -- |
| **GUIDANCE** | | | | | | | | | | |
| MIDCOURSE (OR MAIN) GUIDANCE | 544 | 550 | 558 | 565 | 571 | 578 | -- | -- | 591 | -- |
| TERMINAL GUIDANCE | -- | 550 | 559 | 565 | 571 | 579 | 585 | -- | 591 | -- |
| CONTROL SYSTEM | 544 | 551 | 559 | 565 | 571 | 579 | 585 | -- | 591 | -- |
| RADAR | 544 | 551 | 559 | 566 | 572 | 579 | -- | -- | 592 | -- |
| **HISTORY** | | | | | | | | | | |
| FIRST OBSERVED | -- | -- | -- | -- | -- | -- | -- | -- | -- | -- |
| OBSERVED | -- | -- | -- | 566 | 572 | -- | -- | -- | -- | -- |
| OBSERVED IN PARADE | 545 | 551 | 559 | 566 | 572 | 579 | -- | -- | 592 | -- |
| DEVELOPMENT BEGAN | -- | -- | -- | -- | -- | -- | -- | -- | -- | -- |
| DEVELOPMENT UNDERWAY | -- | -- | -- | -- | -- | -- | -- | -- | 592 | -- |
| PROTOTYPE TESTS | -- | -- | -- | -- | 572 | -- | -- | -- | -- | -- |
| PRODUCTION BEGAN | -- | -- | -- | -- | -- | -- | -- | -- | -- | -- |
| PRODUCTION UNDERWAY | -- | -- | -- | 566 | -- | -- | 585 | 588 | 592 | -- |
| PRODUCTION ENDED | 545 | -- | -- | -- | -- | -- | -- | -- | -- | -- |
| IOC | 545 | 551 | 559 | -- | 572 | -- | -- | -- | -- | -- |
| DEPLOYMENT BEGAN | 545 | 551 | 559 | 566 | 572 | 579 | 585 | 588 | 592 | -- |
| DEPLOYED | 545 | 551 | 559 | 566 | 572 | 579 | 585 | -- | 592 | -- |
| DEPLOYMENT COMPLETE | -- | -- | -- | -- | -- | -- | -- | -- | -- | -- |
| RETIRED | 545 | -- | -- | -- | -- | -- | -- | -- | -- | -- |
| DEPLOYMENT HISTORY | 545 | 551 | 559 | 566 | 572 | 579 | -- | -- | 592 | -- |
| **NUMBERS; PLATFORM** | | | | | | | | | | |
| NUMBERS PRODUCED | 545 | -- | -- | -- | -- | -- | -- | -- | -- | -- |
| NUMBERS IN SERVICE | 545 | 551 | 559 | 566 | 572 | 579 | -- | -- | 592 | -- |
| UNDATED NUMBER | -- | 553 | -- | -- | -- | -- | -- | -- | -- | -- |
| LAUNCHER/SILO | 546 | -- | 561 | -- | 573 | -- | -- | -- | -- | -- |
| CARRIERS | 546 | 553 | 561 | 567 | 573 | 580 | 585 | -- | 593 | 595 |
| **DESIGN; USE; STRATEGY** | | | | | | | | | | |
| DESIGNER | -- | -- | -- | -- | -- | -- | -- | -- | -- | -- |
| DESIGN AND ENGINEERING | -- | -- | -- | -- | -- | -- | -- | -- | -- | -- |
| USE AND CONFIGURATION | 546 | 553 | 561 | 567 | 574 | 580 | 586 | 588 | 593 | 595 |
| RELOADS IN GENERAL | 546 | 553 | -- | -- | -- | -- | -- | -- | -- | -- |
| TARGET TYPE | -- | -- | -- | -- | -- | 586 | -- | -- | -- | -- |
| STRATEGY | 546 | 553 | 561 | 567 | 574 | 581 | 586 | -- | 594 | -- |
| COMBAT REPORTS/EFFECTIVENESS | -- | 553 | 561 | -- | 574 | 581 | 586 | -- | 594 | -- |
| GEOGRAPHICAL DEPLOYMENTS | 546 | 553 | 561 | 567 | 574 | 581 | -- | -- | -- | -- |
| EXPORTED TO | 546 | 554 | 562 | 568 | 575 | 582 | 586 | -- | 594 | -- |
| RELATION TO OTHER MISSILES | 546 | 555 | 562 | 568 | 575 | 582 | 587 | -- | 594 | 595 |
| OTHER INFORMATION | -- | 555 | -- | -- | -- | 582 | -- | -- | -- | -- |

|  | SA-9 | SA-10 | SA-11 | SA-12 | SA-13 | SA-14 |
|---|---|---|---|---|---|---|
| START OF MISSILE | 596 | 601 | 605 | 608 | 610 | 613 |
| **DESIGNATIONS** | | | | | | |
| US DESIGNATION | 596 | 601 | 605 | 608 | 610 | 613 |
| NATO CODENAME | 596 | -- | 605 | -- | 610 | -- |
| SOVIET DESIGNATION | -- | -- | -- | -- | -- | -- |
| OTHER DESIGNATION | -- | -- | -- | -- | 610 | 613 |
| DESCRIPTION | 596 | 601 | 605 | 608 | 610 | 613 |
| EDITOR'S NOTES | 596 | 601 | 605 | 608 | 610 | 613 |
| **PERFORMANCE** | | | | | | |
| MINIMUM ALTITUDE | 596 | 601 | 605 | 608 | 610 | -- |
| MAXIMUM ALTITUDE | 596 | 601 | 605 | 608 | 610 | -- |
| RANGE | 597 | 601 | 605 | 608 | 610 | -- |
| SPEED | 597 | 601 | 605 | -- | 610 | -- |
| CEP | -- | -- | -- | -- | -- | -- |
| **WARHEAD** | | | | | | |
| WARHEAD TYPE | 597 | 601 | 605 | -- | 610 | -- |
| WARHEAD CHARACTERISTICS | -- | -- | -- | -- | -- | -- |
| NUCLEAR YIELD | -- | -- | -- | -- | -- | -- |
| WARHEAD WEIGHT | 597 | -- | 605 | -- | 610 | -- |
| THROWWEIGHT | -- | -- | -- | -- | -- | -- |
| PAYLOAD | -- | -- | -- | -- | -- | -- |
| **DIMENSIONS** | | | | | | |
| WEIGHT | 597 | 601 | -- | -- | -- | -- |
| LENGTH | 597 | 602 | 605 | 608 | 610 | -- |
| DIAMETER | 597 | 602 | 606 | -- | -- | -- |
| WINGSPAN | 597 | -- | -- | -- | -- | -- |
| **PROPULSION** | | | | | | |
| NUMBER OF STAGES | 598 | 602 | 606 | -- | 610 | -- |
| TYPE OF PROPULSION | 598 | 602 | 606 | 608 | 611 | -- |
| **GUIDANCE** | | | | | | |
| MIDCOURSE (OR MAIN) GUIDANCE | -- | -- | -- | -- | -- | -- |
| TERMINAL GUIDANCE | 598 | 602 | 606 | 608 | 611 | 613 |
| CONTROL SYSTEM | 598 | -- | -- | -- | -- | -- |
| RADAR | 598 | 602 | 606 | 608 | 611 | -- |
| **HISTORY** | | | | | | |
| FIRST OBSERVED | -- | -- | -- | -- | -- | -- |
| OBSERVED | -- | 602 | -- | -- | -- | -- |
| OBSERVED IN PARADE | 598 | -- | -- | -- | -- | -- |
| DEVELOPMENT BEGAN | -- | -- | -- | -- | -- | -- |
| DEVELOPMENT UNDERWAY | -- | 602 | 606 | -- | -- | 613 |
| PROTOTYPE TESTS | -- | -- | -- | 608 | -- | -- |
| PRODUCTION BEGAN | -- | -- | -- | -- | -- | -- |
| PRODUCTION UNDERWAY | -- | 602 | 606 | 608 | -- | -- |
| PRODUCTION ENDED | -- | -- | -- | -- | -- | -- |
| IOC | -- | 602 | -- | 608 | -- | -- |
| DEPLOYMENT BEGAN | 598 | 602 | 606 | 608 | 611 | 613 |
| DEPLOYED | 598 | -- | -- | -- | 611 | -- |
| DEPLOYMENT COMPLETE | -- | -- | -- | -- | -- | -- |
| RETIRED | -- | -- | -- | -- | -- | -- |
| DEPLOYMENT HISTORY | 598 | 603 | -- | 608 | 611 | -- |
| **NUMBERS; PLATFORM** | | | | | | |
| NUMBERS PRODUCED | -- | -- | -- | -- | -- | -- |
| NUMBERS IN SERVICE | 598 | 603 | 606 | 608 | 611 | -- |
| UNDATED NUMBER | -- | -- | -- | -- | -- | -- |
| LAUNCHER/SILO | -- | -- | -- | -- | -- | -- |
| CARRIERS | 599 | 603 | 606 | 609 | 611 | 613 |
| **DESIGN; USE; STRATEGY** | | | | | | |
| DESIGNER | -- | -- | -- | -- | -- | -- |
| DESIGN AND ENGINEERING | -- | -- | -- | -- | -- | -- |
| USE AND CONFIGURATION | 599 | 603 | 606 | 609 | 611 | 613 |
| RELOADS IN GENERAL | -- | -- | -- | -- | -- | -- |
| TARGET TYPE | -- | -- | -- | -- | -- | -- |
| STRATEGY | 599 | 604 | 607 | 609 | 612 | 613 |
| COMBAT REPORTS/EFFECTIVENESS | -- | -- | -- | -- | -- | -- |
| GEOGRAPHICAL DEPLOYMENTS | -- | 604 | -- | -- | -- | -- |
| EXPORTED TO | 600 | -- | -- | -- | -- | -- |
| RELATION TO OTHER MISSILES | 600 | -- | 607 | -- | 612 | -- |
| OTHER INFORMATION | 600 | 604 | -- | -- | -- | -- |

## MISSILE INDEX  Sea-Launched SAMs

|  | SA-N-1 | SA-N-2 | SA-N-3 | SA-N-4 | SA-N-5 | SA-N-6 | SA-N-7 | SA-N-8 | SA-N-9 |
|---|---|---|---|---|---|---|---|---|---|
| START OF MISSILE | 614 | 618 | 621 | 625 | 629 | 631 | 634 | 636 | 637 |
| **DESIGNATIONS** | | | | | | | | | |
| US DESIGNATION | 614 | 618 | 621 | 625 | 629 | 631 | 634 | 636 | 637 |
| NATO CODENAME | 614 | 618 | 621 | 625 | -- | -- | -- | -- | -- |
| SOVIET DESIGNATION | -- | -- | -- | -- | -- | -- | -- | -- | -- |
| OTHER DESIGNATION | -- | -- | -- | -- | -- | -- | -- | -- | -- |
| DESCRIPTION | -- | -- | -- | -- | -- | -- | 634 | -- | -- |
| EDITOR'S NOTES | -- | -- | 621 | 625 | -- | 631 | 634 | -- | 637 |
| **PERFORMANCE** | | | | | | | | | |
| MINIMUM ALTITUDE | 614 | 618 | 621 | 625 | -- | -- | 634 | -- | -- |
| MAXIMUM ALTITUDE | 614 | 618 | 621 | 625 | -- | 631 | 634 | -- | -- |
| RANGE | 614 | 618 | 621 | 625 | 629 | 631 | 634 | 636 | 637 |
| SPEED | 614 | 618 | 621 | 625 | 629 | 631 | 634 | -- | -- |
| CEP | -- | -- | -- | -- | -- | -- | -- | -- | -- |
| **WARHEAD** | | | | | | | | | |
| WARHEAD TYPE | 614 | 618 | 622 | 625 | 629 | 631 | 634 | 636 | -- |
| WARHEAD CHARACTERISTICS | -- | -- | -- | -- | -- | -- | -- | -- | -- |
| NUCLEAR YIELD | -- | -- | -- | -- | -- | -- | -- | -- | -- |
| WARHEAD WEIGHT | 615 | 618 | 622 | 625 | 629 | 631 | -- | -- | -- |
| THROWWEIGHT | -- | -- | -- | -- | -- | -- | -- | -- | -- |
| PAYLOAD | -- | -- | -- | -- | -- | -- | -- | -- | -- |
| **DIMENSIONS** | | | | | | | | | |
| WEIGHT | 615 | 618 | 622 | 625 | 629 | -- | -- | -- | -- |
| LENGTH | 615 | 619 | 622 | 625 | 629 | 631 | -- | -- | -- |
| DIAMETER | 615 | 619 | 622 | 625 | 629 | -- | -- | 636 | -- |
| WINGSPAN | 615 | 619 | 622 | 626 | -- | -- | -- | -- | -- |
| **PROPULSION** | | | | | | | | | |
| NUMBER OF STAGES | 615 | -- | 622 | -- | 629 | 631 | 634 | -- | -- |
| TYPE OF PROPULSION | 615 | 619 | 622 | 626 | 629 | 631 | 634 | -- | -- |
| **GUIDANCE** | | | | | | | | | |
| MIDCOURSE (OR MAIN) GUIDANCE | 615 | 619 | 622 | 626 | 629 | 632 | -- | -- | -- |
| TERMINAL GUIDANCE | 616 | -- | 622 | 626 | 629 | 632 | 634 | -- | -- |
| CONTROL SYSTEM | 616 | 619 | -- | -- | -- | -- | -- | -- | -- |
| RADAR | 616 | 619 | 622 | 626 | -- | 632 | 634 | 636 | -- |
| **HISTORY** | | | | | | | | | |
| FIRST OBSERVED | -- | -- | -- | -- | -- | -- | -- | -- | -- |
| OBSERVED | 616 | -- | -- | -- | -- | 632 | -- | 636 | -- |
| OBSERVED IN PARADE | -- | -- | -- | -- | -- | -- | -- | -- | -- |
| DEVELOPMENT BEGAN | -- | -- | -- | -- | -- | -- | -- | -- | -- |
| DEVELOPMENT UNDERWAY | -- | -- | -- | -- | -- | -- | 634 | 636 | -- |
| PROTOTYPE TESTS | -- | -- | -- | -- | -- | -- | -- | -- | -- |
| PRODUCTION BEGAN | -- | -- | -- | -- | -- | -- | -- | -- | -- |
| PRODUCTION UNDERWAY | -- | -- | -- | -- | -- | -- | -- | -- | -- |
| PRODUCTION ENDED | -- | -- | -- | -- | -- | -- | -- | -- | -- |
| IOC | 616 | 619 | 622 | 626 | 629 | 632 | 634 | -- | -- |
| DEPLOYMENT BEGAN | 616 | 619 | 622 | 626 | 629 | 632 | 634 | 636 | -- |
| DEPLOYED | 616 | 619 | 622 | 626 | 629 | -- | -- | -- | -- |
| DEPLOYMENT COMPLETE | -- | -- | -- | -- | -- | -- | -- | -- | -- |
| RETIRED | -- | -- | -- | -- | -- | -- | -- | -- | -- |
| DEPLOYMENT HISTORY | 616 | -- | -- | -- | -- | -- | -- | -- | -- |
| **NUMBERS; PLATFORM** | | | | | | | | | |
| NUMBERS PRODUCED | -- | -- | -- | -- | -- | -- | -- | -- | -- |
| NUMBERS IN SERVICE | 616 | -- | -- | 626 | -- | -- | -- | -- | -- |
| UNDATED NUMBER | -- | -- | -- | -- | -- | -- | -- | -- | -- |
| LAUNCHER/SILO | -- | -- | -- | -- | -- | -- | -- | -- | -- |
| CARRIERS | 616 | 619 | 623 | 626 | 629 | 632 | 634 | 636 | 637 |
| **DESIGN; USE; STRATEGY** | | | | | | | | | |
| DESIGNER | -- | -- | -- | -- | -- | -- | -- | -- | -- |
| DESIGN AND ENGINEERING | -- | -- | -- | -- | -- | -- | -- | -- | -- |
| USE AND CONFIGURATION | 617 | -- | 623 | 628 | 630 | 632 | 635 | 636 | -- |
| RELOADS IN GENERAL | -- | -- | -- | -- | -- | -- | -- | -- | -- |
| TARGET TYPE | 617 | -- | 623 | 628 | -- | -- | 635 | -- | -- |
| STRATEGY | 617 | 619 | -- | 628 | -- | 633 | -- | 636 | -- |
| COMBAT REPORTS/EFFECTIVENESS | -- | 620 | -- | -- | 630 | -- | -- | -- | -- |
| GEOGRAPHICAL DEPLOYMENTS | -- | -- | -- | -- | -- | -- | -- | -- | -- |
| EXPORTED TO | 617 | -- | -- | 628 | 630 | -- | -- | -- | -- |
| RELATION TO OTHER MISSILES | 617 | 620 | 624 | 628 | 630 | 633 | 635 | 636 | -- |
| OTHER INFORMATION | -- | -- | -- | 628 | -- | -- | 635 | 636 | -- |

MISSILE INDEX  AAMs  697

| | AA-1 | AA-2 | AA-2-2 | AA-3 | AA-3 IR | AA-3 Radar | AA-4 | AA-5 | AA-5 IR | AA-5 Radar | AA-6 | AA-6 IR | AA-6 Radar |
|---|---|---|---|---|---|---|---|---|---|---|---|---|---|
| START OF MISSILE | 638 | 641 | 645 | 647 | 650 | 651 | 652 | 653 | 656 | 657 | 658 | 661 | 662 |
| **DESIGNATIONS** | | | | | | | | | | | | | |
| US DESIGNATION | 638 | 641 | 645 | 647 | — | — | — | 653 | — | — | 658 | — | — |
| NATO CODENAME | 638 | 641 | 645 | 647 | — | — | 652 | 653 | — | — | 658 | — | — |
| SOVIET DESIGNATION | — | 641 | — | — | — | — | — | — | — | — | — | — | — |
| OTHER DESIGNATION | — | — | — | — | — | — | — | — | — | — | — | — | — |
| DESCRIPTION | — | — | — | — | — | — | — | — | — | — | — | — | — |
| EDITOR'S NOTES | 638 | 641 | — | — | — | — | — | — | — | — | — | — | — |
| **PERFORMANCE** | | | | | | | | | | | | | |
| MINIMUM ALTITUDE | — | — | — | — | — | — | — | — | — | — | — | — | — |
| MAXIMUM ALTITUDE | — | — | — | — | — | — | — | — | — | — | — | — | — |
| RANGE | 638 | 641 | 645 | 647 | — | — | — | 653 | — | — | 658 | 661 | 662 |
| SPEED | 638 | 641 | 645 | 647 | — | — | — | 653 | — | — | 658 | — | — |
| CEP | — | — | — | — | — | — | — | — | — | — | — | — | — |
| **WARHEAD** | | | | | | | | | | | | | |
| WARHEAD TYPE | 638 | 641 | 645 | 647 | — | — | — | 653 | — | — | 658 | — | — |
| WARHEAD CHARACTERISTICS | — | — | — | — | — | — | — | — | — | — | — | — | — |
| NUCLEAR YIELD | — | — | — | — | — | — | — | — | — | — | — | — | — |
| WARHEAD WEIGHT | 638 | 641 | 645 | 647 | — | — | — | 653 | — | — | 658 | — | — |
| THROWEIGHT | — | — | — | — | — | — | — | — | — | — | — | — | — |
| PAYLOAD | — | — | — | — | — | — | — | — | — | — | — | — | — |
| **DIMENSIONS** | | | | | | | | | | | | | |
| WEIGHT | 638 | 641 | — | 647 | — | — | — | 653 | — | — | 658 | 661 | 662 |
| LENGTH | 638 | 642 | 645 | 647 | 650 | 651 | 652 | 653 | 656 | 657 | 659 | 661 | 662 |
| DIAMETER | 639 | 642 | — | 647 | — | — | — | 653 | — | — | 659 | — | — |
| WINGSPAN | 639 | 642 | — | 648 | — | — | — | 654 | — | — | 659 | — | — |
| **PROPULSION** | | | | | | | | | | | | | |
| NUMBER OF STAGES | 639 | 642 | — | 648 | — | — | — | 654 | — | — | 659 | — | — |
| TYPE OF PROPULSION | 639 | 642 | — | 648 | — | — | — | 654 | — | — | 659 | — | — |
| **GUIDANCE** | | | | | | | | | | | | | |
| MIDCOURSE (OR MAIN) GUIDANCE | 639 | — | — | — | — | — | — | — | — | — | 659 | — | — |
| TERMINAL GUIDANCE | 639 | 642 | 645 | 648 | 650 | 651 | 652 | 654 | 656 | 657 | 659 | 661 | 662 |
| CONTROL SYSTEM | 639 | 642 | — | 648 | — | — | — | 654 | — | — | 659 | — | — |
| RADAR | 639 | — | — | 648 | — | 651 | — | 654 | — | — | 659 | — | — |
| **HISTORY** | | | | | | | | | | | | | |
| FIRST OBSERVED | — | — | — | 648 | — | — | — | — | — | — | 659 | — | — |
| OBSERVED | — | — | — | — | — | — | — | — | — | — | — | — | — |
| OBSERVED IN PARADE | — | — | — | — | — | — | — | — | — | — | — | — | — |
| DEVELOPMENT BEGAN | — | — | — | — | — | — | — | — | — | — | — | — | — |
| DEVELOPMENT UNDERWAY | — | — | — | — | — | — | — | — | — | — | — | — | — |
| PROTOTYPE TESTS | — | — | — | — | — | — | — | — | — | — | — | — | — |
| PRODUCTION BEGAN | — | — | — | — | — | — | — | — | — | — | — | — | — |
| PRODUCTION UNDERWAY | — | — | — | — | — | — | — | — | — | — | — | — | — |
| PRODUCTION ENDED | — | — | — | — | — | — | — | — | — | — | — | — | — |
| IOC | — | — | — | — | — | — | — | — | — | — | — | — | — |
| DEPLOYMENT BEGAN | 639 | 642 | 645 | 648 | — | — | 652 | 654 | — | — | 659 | — | — |
| DEPLOYED | 639 | 642 | — | 648 | — | — | — | 654 | — | — | 659 | — | — |
| DEPLOYMENT COMPLETE | — | — | — | — | — | — | — | — | — | — | — | — | — |
| RETIRED | — | — | — | — | — | — | — | — | — | — | — | — | — |
| DEPLOYMENT HISTORY | — | 642 | — | 648 | — | — | — | — | — | — | — | — | — |
| **NUMBERS; PLATFORM** | | | | | | | | | | | | | |
| NUMBERS PRODUCED | — | — | — | 648 | — | — | — | 654 | — | — | — | — | — |
| NUMBERS IN SERVICE | — | — | — | — | — | — | — | — | — | — | — | — | — |
| UNDATED NUMBER | — | — | — | — | — | — | — | — | — | — | — | — | — |
| LAUNCHER/SILO | — | — | — | — | — | — | — | — | — | — | — | — | — |
| CARRIERS | 639 | 643 | 645 | 648 | — | — | 652 | 654 | — | — | 659 | — | — |
| **DESIGN; USE; STRATEGY** | | | | | | | | | | | | | |
| DESIGNER | — | — | — | — | — | — | — | — | — | — | — | — | — |
| DESIGN AND ENGINEERING | — | 643 | — | — | — | — | — | — | — | — | 660 | — | — |
| USE AND CONFIGURATION | 640 | 643 | — | 649 | — | — | — | — | — | — | 660 | — | — |
| RELOADS IN GENERAL | — | — | — | — | — | — | — | — | — | — | — | — | — |
| TARGET TYPE | — | — | — | — | — | — | — | — | — | — | — | — | — |
| STRATEGY | — | — | — | — | — | — | — | 654 | — | — | 660 | — | — |
| COMBAT REPORTS/EFFECTIVENESS | — | 643 | 646 | — | — | — | — | — | — | — | — | — | — |
| GEOGRAPHICAL DEPLOYMENTS | — | — | — | — | — | — | — | — | — | — | — | — | — |
| EXPORTED TO | 640 | 644 | — | 649 | — | — | — | 655 | — | — | 660 | — | — |
| RELATION TO OTHER MISSILES | — | 644 | — | — | — | — | 652 | — | — | — | 660 | — | — |
| OTHER INFORMATION | 640 | — | — | — | — | — | — | — | — | — | 660 | — | — |

# MISSILE INDEX  AAMs

| | AA-7 | AA-7 IR | AA-7 Radar | AA-8 | AA-8 IR | AA-8 Radar | AA-9 | AA-10 | AA-XP-1 | AA-XP-2 | AA-X |
|---|---|---|---|---|---|---|---|---|---|---|---|
| START OF MISSILE | 663 | 666 | 667 | 668 | 671 | 672 | 673 | 675 | 676 | 677 | 678 |
| **DESIGNATIONS** | | | | | | | | | | | |
| US DESIGNATION | 663 | -- | -- | 668 | -- | -- | 673 | 675 | 676 | 677 | 678 |
| NATO CODENAME | 663 | -- | -- | 668 | -- | -- | -- | -- | -- | -- | -- |
| SOVIET DESIGNATION | -- | -- | -- | -- | -- | -- | -- | -- | -- | -- | -- |
| OTHER DESIGNATION | -- | -- | -- | -- | -- | -- | -- | -- | -- | -- | -- |
| DESCRIPTION | -- | -- | -- | -- | -- | -- | -- | -- | -- | -- | -- |
| EDITOR'S NOTES | 663 | -- | -- | -- | -- | -- | 673 | 675 | -- | -- | -- |
| **PERFORMANCE** | | | | | | | | | | | |
| MINIMUM ALTITUDE | -- | -- | -- | -- | -- | -- | -- | -- | -- | -- | -- |
| MAXIMUM ALTITUDE | -- | -- | -- | -- | -- | -- | -- | -- | -- | -- | -- |
| RANGE | 663 | 666 | 667 | 668 | 671 | 672 | 673 | 675 | 676 | 677 | -- |
| SPEED | 663 | -- | -- | 668 | -- | -- | -- | -- | -- | -- | -- |
| CEP | -- | -- | -- | -- | -- | -- | -- | -- | -- | -- | -- |
| **WARHEAD** | | | | | | | | | | | |
| WARHEAD TYPE | 663 | -- | -- | 668 | -- | -- | -- | -- | -- | -- | -- |
| WARHEAD CHARACTERISTICS | -- | -- | -- | -- | -- | -- | -- | -- | -- | -- | -- |
| NUCLEAR YIELD | -- | -- | -- | -- | -- | -- | -- | -- | -- | -- | -- |
| WARHEAD WEIGHT | 663 | -- | -- | 668 | -- | -- | -- | -- | -- | -- | -- |
| THROWWEIGHT | -- | -- | -- | -- | -- | -- | -- | -- | -- | -- | -- |
| PAYLOAD | -- | -- | -- | -- | -- | -- | -- | -- | -- | -- | -- |
| **DIMENSIONS** | | | | | | | | | | | |
| WEIGHT | 663 | -- | -- | 669 | -- | -- | -- | -- | -- | -- | -- |
| LENGTH | 664 | 666 | 667 | 669 | 671 | 672 | -- | -- | -- | -- | -- |
| DIAMETER | 664 | -- | -- | 669 | -- | -- | -- | -- | -- | -- | -- |
| WINGSPAN | 664 | -- | -- | 669 | -- | -- | -- | -- | -- | -- | -- |
| **PROPULSION** | | | | | | | | | | | |
| NUMBER OF STAGES | 664 | -- | -- | 669 | -- | -- | 673 | -- | 676 | 677 | -- |
| TYPE OF PROPULSION | 664 | -- | -- | 669 | -- | -- | 673 | -- | 676 | 677 | -- |
| **GUIDANCE** | | | | | | | | | | | |
| MIDCOURSE (OR MAIN) GUIDANCE | -- | -- | -- | -- | -- | -- | -- | -- | -- | -- | -- |
| TERMINAL GUIDANCE | 664 | 666 | 667 | 669 | 671 | 672 | 673 | -- | -- | 677 | 678 |
| CONTROL SYSTEM | 664 | -- | -- | 669 | -- | -- | -- | -- | -- | -- | -- |
| RADAR | 664 | -- | -- | -- | -- | -- | -- | 675 | -- | -- | -- |
| **HISTORY** | | | | | | | | | | | |
| FIRST OBSERVED | 664 | -- | -- | -- | -- | -- | -- | -- | -- | -- | -- |
| OBSERVED | -- | -- | -- | -- | -- | -- | -- | -- | -- | -- | -- |
| OBSERVED IN PARADE | -- | -- | -- | -- | -- | -- | -- | -- | -- | -- | -- |
| DEVELOPMENT BEGAN | 665 | -- | -- | -- | -- | -- | -- | -- | -- | -- | -- |
| DEVELOPMENT UNDERWAY | -- | -- | -- | -- | -- | -- | -- | -- | -- | -- | 678 |
| PROTOTYPE TESTS | -- | -- | -- | -- | -- | -- | -- | -- | 676 | -- | -- |
| PRODUCTION BEGAN | -- | -- | -- | -- | -- | -- | -- | -- | -- | -- | -- |
| PRODUCTION UNDERWAY | 665 | -- | -- | 669 | -- | -- | -- | -- | -- | -- | -- |
| PRODUCTION ENDED | -- | -- | -- | -- | -- | -- | -- | -- | -- | -- | -- |
| IOC | -- | -- | -- | -- | -- | -- | -- | -- | 676 | 677 | -- |
| DEPLOYMENT BEGAN | 665 | -- | -- | 669 | -- | -- | 673 | 675 | 676 | 677 | 678 |
| DEPLOYED | 665 | -- | -- | 669 | -- | -- | -- | -- | -- | -- | -- |
| DEPLOYMENT COMPLETE | -- | -- | -- | -- | -- | -- | -- | -- | -- | -- | -- |
| RETIRED | -- | -- | -- | -- | -- | -- | -- | -- | -- | -- | -- |
| DEPLOYMENT HISTORY | -- | -- | -- | 670 | -- | -- | -- | -- | -- | -- | -- |
| **NUMBERS; PLATFORM** | | | | | | | | | | | |
| NUMBERS PRODUCED | -- | -- | -- | -- | -- | -- | -- | -- | -- | -- | -- |
| NUMBERS IN SERVICE | -- | -- | -- | -- | -- | -- | -- | -- | -- | -- | -- |
| UNDATED NUMBER | -- | -- | -- | -- | -- | -- | -- | -- | -- | -- | -- |
| LAUNCHER/SILO | -- | -- | -- | -- | -- | -- | -- | -- | -- | -- | -- |
| CARRIERS | 665 | -- | -- | 670 | -- | -- | 673 | 675 | -- | -- | -- |
| **DESIGN; USE; STRATEGY** | | | | | | | | | | | |
| DESIGNER | -- | -- | -- | -- | -- | -- | -- | -- | -- | -- | -- |
| DESIGN AND ENGINEERING | 665 | -- | -- | -- | -- | -- | -- | 675 | -- | -- | -- |
| USE AND CONFIGURATION | 665 | -- | -- | 670 | -- | -- | -- | -- | -- | -- | -- |
| RELOADS IN GENERAL | -- | -- | -- | -- | -- | -- | -- | -- | -- | -- | -- |
| TARGET TYPE | -- | -- | -- | -- | -- | -- | -- | -- | -- | -- | -- |
| STRATEGY | 665 | -- | -- | 670 | -- | -- | 673 | 675 | 676 | 677 | 678 |
| COMBAT REPORTS/EFFECTIVENESS | 665 | -- | -- | -- | -- | -- | 673 | -- | -- | -- | -- |
| GEOGRAPHICAL DEPLOYMENTS | -- | -- | -- | -- | -- | -- | -- | -- | -- | -- | -- |
| EXPORTED TO | -- | -- | -- | -- | -- | -- | -- | -- | -- | -- | -- |
| RELATION TO OTHER MISSILES | 665 | -- | -- | 670 | -- | -- | 674 | -- | -- | -- | -- |
| OTHER INFORMATION | -- | -- | -- | 670 | -- | -- | -- | -- | -- | -- | -- |

# ABOUT THE AUTHOR AND EDITOR

**Barton Wright** received a B.A. with Distinction from Swarthmore College, an M.A. from the University of Pennsylvania, and a Ph.D. in psycholinguistics from the Massachusetts Institute of Technology. While at the University of Pennsylvania and MIT he was the recipient of a National Science Foundation Graduate Fellowship. Besides compiling and writing *Soviet Missiles* in his three years at IDDS, Dr. Wright developed the data management system for the World Weapon Database. He also wrote and published an index of Chinese and Soviet military affairs in annual US Department of Defense documents. He is currently working as a computer programmer in the Boston area.

**Randall Forsberg**, executive director of the Institute for Defense and Disarmament Studies, has a B.A. in English from Barnard College. From 1968 through 1974 she worked at the Stockholm International Peace Research Institute, where she authored the only published comparative study of postwar worldwide military research and development programs. Her first published article caused the US Department of Defense to withdraw misleading estimates of Soviet expenditure on military research and development. From 1974 to 1982, as a consultant, she compiled the tables of US and Soviet nuclear weapons published in the annual *SIPRI Yearbook* and the estimates of world nuclear weapons published in the United Nations' 1981 *Report on Nuclear Weapons*. Between 1974 and 1979, she did coursework for a Ph.D. in political science at the Massachusetts Institute of Technology; spent a year at the Harvard Program in Science and International Affairs; taught a course at Boston University; and with other members of the Boston Study Group co-authored *The Price of Defense* (New York Times Publishing Co., 1979), recently reissued by W. H. Freeman under the title *Winding Down*.

In 1980 Ms. Forsberg founded the Institute for Defense and Disarmament Studies. The same year she wrote the "Call to Halt the Nuclear Arms Race," the statement that launched the nuclear freeze movement and provided its agreed platform. Over the past five years, she has published many articles on military policy and spending. In July 1983 she received the prestigious MacArthur Award, a five-year, full-support, no-strings-attached grant, "in recognition of her accomplishments in defense studies and arms control."

**John Murphy** has a B.S. in mechanical engineering and an M.S. in civil engineering, both from the University of Notre Dame. He has worked as a researcher at the Institute for Defense and Disarmament Studies since 1984.

# ABOUT THE INSTITUTE

The Institute for Defense and Disarmament Studies (IDDS) approaches the problem of reduction in nuclear and conventional weapons differently from many other organizations. We believe that standard approaches to arms control will *at best* result in a slight decrease in the number of nuclear and conventional weapons and will never lead to a fundamental reduction in armaments or a fundamental change in the present unsuccessful international approach to the problems of war and peace.

The role of most nuclear weapons has always been to deter conventional war—a fact ignored by many analysts and unknown to the public. By posing a threat of escalation from a conventional war to a nuclear war, nuclear weapons make the big powers avoid conventional war with each other. Military planners believe that deep cuts in nuclear arms would increase the risk of conventional war between East and West, which would be terrible in itself and could lead to nuclear war. This view has blocked nuclear reductions for 30 years.

In the past most peace groups have assumed that the superpowers should reduce their nuclear arsenals before they turn to the issues of conventional war. The Institute's work, in contrast, is based on the notion that until *the risk of conventional war* is greatly reduced, it will probably continue to prove impossible to make deep cuts in nuclear arms.

The Institute analyzes long-term trends and projects future developments that show the need to reduce the risk of conventional war and safe ways to do so. We then disseminate our work to peace groups, scholars, journalists, and teachers, whose action is needed to reverse the arms race.

▲ **Program** The Institute conducts three kinds of research not performed elsewhere. First, researchers compile comprehensive, detailed, regularly updated data on world military forces, government arms control talks, and nongovernmental peace efforts. Our World Weapon Database, covering all major weapons produced since 1945, compiles all data published in major sources. Our monthly *Arms Control Reporter* gives a day-by-day chronology of the positions of all countries in all 35 ongoing arms control talks. Our annual *Peace Resource Book* surveys all US national and local peace groups (5,700 groups) and recent literature on war and peace (1,000 publications since 1980).

Second, the weapon and arms control databases make it possible for IDDS staff to provide unique analyses: to correct misleading views of Soviet and Western military strengths; to demonstrate that the role of most nuclear weapons has always been to deter conventional war; and to study the uses of conventional force that create pressure to perpetuate nuclear deterrence of conventional war. The weapon and arms control data also permit Institute scholars to project world nuclear and conventional forces 10 to 20 years in the future; to assess the impact of proposed arms control measures on future forces; and to show that most government proposals, if implemented, would not reverse the arms race.

Finally, the Institute will develop long-term global military policy alternatives that lead to greatly reduced nuclear and conventional arms and military spending. This approach should stimulate many scholars, analysts, and activists to explore their own variations. Ultimately, we expect the debate on alternative futures to create an environment in which gradual steps toward fundamental change are possible.

The results of IDDS work appear not only in forms intended for professional scholars, but also in forms intended for journalists, educators, and the concerned layperson. In 1986, for example, we will publish the first edition of *World Military Forces and Alternatives*, a 100-page illustrated survey. Since March 1985 we have been publishing a popular newsletter, *Defense and Disarmament News*, which combines news and analysis on the peace movement, arms control talks, and the military. A list of other publications follows.

# IDDS PUBLICATIONS

## Books, Periodicals, and Reference Works

**Peace Resource Book 1986.** F. Bernstein *et al.* Cambridge, MA: Ballinger Publishing Co., 1985. 425pp. $14.95 paper. Order from Harper & Row, 1-800/638-3030.

**Soviet Missiles.** B. Wright. Lexington, MA: Lexington Books, 1986. World Weapon Database, Vol. 1.

**The Vulnerability of Strategic Submarines.** T. Stefanick. Forthcoming, 1986.

**Defense & Disarmament News.** Bimonthly coverage of the military, arms control, and the peace movement. M. Goodman, ed. From March 1985. Annual subscription $25 ($15 low income).

**Arms Control Reporter.** Monthly looseleaf chronology of negotiations, proposals, and treaties. G. Hardenbergh, ed. From January 1982. Annual subscription $295 ($310 overseas); sample issue $5 ($10 overseas). Reduced rates for students and local peace groups.

**Index of Soviet and Chinese Military Affairs in Annual US Defense Department Reports 1969-83.** B. Wright. 1983. 74pp. $6.

## Occasional Papers

**Confining the Military to Defense as a Route to Disarmament.** R. Forsberg. No. 1. 1981. Out of print; superseded by revised reprint listed below.

**The Economic and Social Costs of the Nuclear Arms Race.** R. Forsberg. No. 2. 1981. 23pp $2.

**Threats Misperceived: Soviet Military Capabilities and Objectives in the Early Postwar Period 1945-53.** M. Evangelista. No. 3. 1981. Out of print; superseded by M. Evangelista reprint listed below.

**Projected US and Soviet Strategic Forces, 1985-1995, with and without SALT Limits.** P. Braudaway-Bauman, R. Forsberg, and J. Murphy. No. 4. 1985. 50pp. $5.

**The Structure of Military Spending.** R. Forsberg and David S. Meyer. No. 5. 1986. 50pp. $5.

## Reprints

"**Parallel Cuts in Nuclear and Conventional Forces.**" R. Forsberg. *Bulletin of the Atomic Scientists*, August 1985. 4pp $.50.

**A Bilateral Freeze as an Arms Control Objective.** R. Forsberg. Testimony before the Committee on Foreign Affairs, US House of Representatives, 21 June 1984. 3pp. $.50.

"**The Freeze and Beyond: Confining the Military to Defense as a Route to Disarmament.**" R. Forsberg. *World Policy* 1:2, 1984. 36pp. 1-99 copies, $1 each; $35 per hundred.

"**Women and Military Policy-Making.**" R. Forsberg. *Barnard Alumnae Magazine*, Winter 1984. 3pp. $.50.

"**Verification: No Obstacle to Arms Control.**" K. Pieragostini. *Arms Control Today* 13:5, June 1983. 4pp. $.50.

"**What About a Nuclear Freeze?**" R. Forsberg. *Social Education* 47:7, November–December 1983. Based on testimony before the Committee on Foreign Affairs, US House of Representatives, 17 February 1983. 2pp. $.50.

"**Stalin's Postwar Army Reappraised.**" M. Evangelista. *International Security* 7:3, Winter 1982-83. 28pp. $2.

"**RECOVERing Verification.**" K. Pieragostini. *Arms Control Today* 12:11, December 1982. 2pp. $.50.

"**A Bilateral Nuclear-Weapon Freeze.**" R. Forsberg. *Scientific American* 247:5 November 1982. 10pp. $.50.

"**No First Use and the Nuclear Freeze: Complementary or Competing Concepts?**" R. Forsberg. Speech at No First Use Conference, 7 October 1982. 10pp. $1.

"'**START' Proposal Invites a Continued Arms Race.**" R. Forsberg. Hearings before the Subcommittee on International Security and Scientific Affairs, US House of Representatives, 11 May 1982, 12pp. $1.

"**Information on Present Nuclear Arsenals.**" [R. Forsberg] in *Nuclear Weapons: Report of the UN Secretary General*. Brookline, MA: Autumn Press, 1981, 2pp. $.50.

"**Soviet Civil Defense.**" M. Evangelista. *PSR Newsletter* 2:1, Summer 1981. 1p. $.50.

"**Military R&D: A Worldwide Institution?**" R. Forsberg. *Proceedings of the American Philosophical Society* 124:4, August 1980. 5pp. $.50.

**Call to Halt the Nuclear Arms Race.** [R. Forsberg, 1980]. Rev. 1982. 4pp. $.50.

Unless otherwise indicated, order from: Institute for Defense and Disarmament Studies, 2001 Beacon St., Brookline, Massachusetts 02146. 617/734-4216.

## Acronyms and Short References for All Sources

Full Sources, from which *all* information on Soviet missiles was recorded in the Database, are represented by acronyms with letters only (no numerals). The acronyms may refer to either author or title. If they are true acronyms (the first letters of several words), they are all capital letters. If they represent a shortened version of a word, they appear in corresponding upper- and lower-case letters.

Acronyms that end with two or three numerals refer to Partial Sources, from which only unusual data were recorded. Acronyms for articles from journals begin with a standard two-letter designation, such as AW for *Aviation Week and Space Technology* or BA for *Bulletin of the Atomic Scientists*. Books are usually represented by the first two letters of the author's last name, with the second letter lowercase.

The short references listed here begin with the part of the citation the acronym refers to. For citations with complete bibliographic information, see the **Acronyms and Complete References** sections in part I.

| | |
|---|---|
| **A000** | Adelphi papers of the same number. International Institute for Strategic Studies. |
| **ABFS** | W. Arkin, A. Burrows, R. Fieldhouse, and J. Sands. *Nuclearization of the Oceans.* May 1984. |
| **AF00** | *Air Force Magazine* articles. |
| **AFM** | *Air Force Magazine.* "Gallery of Soviet Aerospace Weapons." Annual. |
| **AJ00** | *Armed Forces Journal International* articles. |
| **ArkS** | W. Arkin and J. Sands. "The Soviet Nuclear Stockpile." *Arms Control Today* 14:5, June 1984. |
| **AW00** | *Aviation Week and Space Technology* articles. |
| **Awar** | *Aviation Week* article: "Soviets' Nuclear Arsenal Continues to Proliferate." *Aviation Week and Space Technology*, 16 June 1980. |
| **AWST** | *Aviation Week and Space Technology.* "Soviet Missiles." Annual. |
| **BA00** | *Bulletin of the Atomic Scientists* articles. |
| **Bl01** | Desmond Ball. *Politics and Force Levels.* University of California Press, 1980. |
| **Bras** | *Brassey's Artillery.* Brian Blunt and Talley Taylor. Shelford Bidwell, ed. Bonanza, 1977. |
| **BW** | Barton Wright. Institute for Defense and Disarmament Studies. Best estimates, shown in part III with comments on their derivation from other data in the Database. |
| **CB00** | Congressional Budget Office. US Congress. Various reports. |
| **CBO** | Congressional Budget Office. *Modernizing US Strategic Offensive Forces: The Administration's Programs and Alternatives.* May 1983. |
| **CFW** | *Combat Fleets of the World.* J. L. Couhat, ed. US Naval Institute Press. Irregular. |
| **Ch01** | C. Chant, ed. *The World's Air Forces.* David and Charles, 1979. |
| **Clns** | John M. Collins. *U.S.-Soviet Military Balance.* Congressional Research Service. US Congress. Irregular. |
| **COTS** | *The Challenge of the Sputniks.* R. Witkin, ed. Doubleday, 1958. |
| **FI** | *Flight International.* "World Missile Directory." Annual. |
| **FIar** | *Flight International* article: D. Richardson. "Soviet Strategic Nuclear Rockets Guide," 11 December 1976. |
| **Ga01** | K. Gatland. *Missiles and Rockets.* Macmillan, 1975. |
| **GSN** | *Guide to the Soviet Navy.* N. Polmar. US Naval Institute Press. Irregular. |
| **Gu01** | B. Gunston. *The Encyclopedia of World Air Power.* Crescent Books, 1980. |
| **Hi01** | R. Higham and J. Kipp. *Soviet Aviation and Air Power.* Westview Press, 1977. |
| **HRST** | *History of Rocketry and Space Travel.* W. von Braun and F. I. Ordway. Crowell, 1975. |
| **IA00** | *Interavia* articles. |
| **ID00** | *International Defense Review* articles. |
| **IISS** | International Institute for Strategic Studies. *The Military Balance.* Annual. |
| **IMSG** | *International Missile and Spacecraft Guide.* F. I. Ordway and R. C. Wakeford. McGraw-Hill, 1960. |
| **JAWA** | *Jane's All the World's Aircraft.* J. W. R. Taylor, ed. |
| **JCS** | Joint Chiefs of Staff. *United States Military Posture for FY__.* Organization [earlier, Chairman] of the Joint Chiefs of Staff. US Department of Defense. Annual. (Reviewed for FY 72–86.) |
| **JFS** | *Jane's Fighting Ships.* John Moore, ed. Jane's Publishing Company. Annual. |
| **JP00** | Joint Publications Research Service Publications. |
| **JWS** | *Jane's Weapon Systems.* R. Pretty, ed. Jane's Publishing Company. Annual. |
| **Ley** | Willy Ley. *Rockets, Missiles, and Men in Space.* Viking Press, 1968. |
| **Lu01** | Edward Luttwak. *The Grand Strategy of the Soviet Union.* St. Martin's Press, 1983. |
| **Meyr.** | Stephen M. Meyer. *Soviet Theatre Nuclear Forces: Part II.* Adelphi Paper 188. International Institute for Strategic Studies, 1984. |
| **MOW** | *Missiles of the World.* M. J. H. Taylor and J. W. R. Taylor. Scribners, 1976. |
| **MR00** | *Military Review* articles. |
| **NA00** | *NATO's Fifteen Nations* articles. |